Handbook of Microbiology

HANDBOOK

of

MICROBIOLOGY

Volume III
Microbial Products

EDITORS

Allen I. Laskin, Ph.D.
Esso Research and Engineering Company
Linden, New Jersey

Hubert A. Lechevalier, Ph.D.
Institute of Microbiology
Rutgers University
New Brunswick, New Jersey

Published by

CRC Press, Inc.
18901 Cranwood Parkway · Cleveland, Ohio 44128

HANDBOOK OF MICROBIOLOGY

Volume III: Microbial Products

This book presents data obtained from authentic and highly regarded sources. Reprinted material is quoted with permission, and sources are indicated. A wide variety of references are listed. Every reasonable effort has been made to give reliable data and information, but the editors and the publisher cannot assume responsibility for the validity of all materials or for the consequences of their use.

International Standard Book Number (ISBN)

Complete Set 0-8719-583-1
Volume III   0-87819-583-1

Library of Congress Catalog Card Number 72-88766

PREFACE

The pages of the third volume of the CRC Handbook of Microbiology are dedicated to the tabulation of data on substances known to be formed by microorganisms. Since microbes are hectic and versatile synthesizers, many of the products they form will not be covered. This may be due partly to lack of the most recent information and partly to our failure in finding an author to cover a given group of compounds. However, we feel that the present volume is the most complete compilation of this type, and it is our hope that microbiologists will find it useful. We are urging the users of this Handbook to draw all its shortcomings to our attention. Only with their cooperation can we hope to make subsequent editions of this Handbook increasingly more useful.

We thank all the authors for contributing so generously their time and expertise. Our gratitude also goes to the members of our Advisory Board, who have helped with the planning of this volume. In this case, we are especially indebted to Drs. Nancy N. Gerber and L. C. Vining.

The editorial skill of Mrs. Lisbeth Hammer and the painstaking efforts of the CRC production staff have made our task light, as did the intelligent clerical assistance of Mrs. Verna Lepping.

A. I. Laskin
H. A. Lechevalier
New Jersey 1973

Herman J. Phaff, Ph.D.
Department of Food Technology
University of California
Davis, California

Thomas B. Platt, Ph.D.
Bioanalytical Section
The Squibb Institute of Medical Research
New Brunswick, New Jersey

Otto J. Plescia, Ph.D.
Institute of Microbiology
Rutgers University
New Brunswick, New Jersey

G. Pontecorvo, Ph.D.
Department of Cell Genetics
Imperial Cancer Research Fund
London, England

Chase Van Baalen, Ph.D.
Marine Science Institute
University of Texas
Port Aransas, Texas

Claude Vezina, Ph.D.
Microbiology Department
Ayerst Laboratories
St. Laurent, P. Q., Canada

L. C. Vining, Ph.D. .
National Research Council
Atlantic Regional Laboratory
Halifax, N. S., Canada

E. D. Weinberg, Ph.D.
Department of Microbiology
Indiana University
Bloomington, Indiana

Burton I. Wilner, Ph.D.
Orinda, California

CONTRIBUTORS

Takaaki Aoyagi, Ph.D.
Institute of Microbial Chemistry
Microbial Chemistry Research Foundation
Tokyo, Japan

John P. Arbuthnott, Ph.D.
Department of Bacteriology
University of Glasgow
Bearsden, Glasgow, Scotland

Julius Berger, Ph.D.
Department of Microbiology
Hoffmann-LaRoche Inc.
Nutley, New Jersey

G. S. Bezanson, M.Sc.
Department of Biology
Carleton University
Ottawa, Ontario, Canada

Robert George Brown, Ph.D.
Department of Biology
Dalhousie University
Halifax, Nova Scotia, Canada

Alex Ciegler, Ph.D.
Northern Regional Research Laboratory
U.S. Department of Agriculture
Peoria, Illinois

B. Louise Crandall, M.A.
The Lilly Research Laboratories
Eli Lilly & Company
Indianapolis, Indiana

Donald C. DeLong, Ph.D.
The Lilly Research Laboratories
Eli Lilly & Company
Indianapolis, Indiana

Hans Diekmann, Ph. D.
Institut für Biologie
Universität Tübingen
Tübingen, Germany

Edouard Drouhet, M.D.
Service de Mycologie
Institut Pasteur
Paris, France

Fujio Egami, D.Sc.
Director
Mitsubishi-kasei Institute of Life Sciences
Tokyo, Japan

Edward E. Garcia, Ph.D.
Scientific Department
Hoffmann-LaRoche Inc.
Nutley, New Jersey

Nancy N. Gerber, Ph.D.
Institute of Microbiology
Rutgers University
New Brunswick, New Jersey

T. W. Goodwin, D.Sc.
Department of Biochemistry
University of Liverpool
Liverpool, England

J. F. Grove, D.Sc.
Agricultural Council
Unit of Invertebrate Chemistry and Physiology
The University of Sussex
Falmer, Brighton, Sussex, England

Theodore H. Haskell, Ph.D.
Section of Natural Products
Parke, Davis & Co.
Ann Arbor, Michigan

Derek J. Hook, Ph.D.
Department of Medicinal Chemistry
Purdue University
West Lafayette, Indiana

David M. Isaacson, Ph.D.
Bioanalytical Section
The Squibb Institute for Medical Research
Princeton, New Jersey

Sir Ewart Jones, Ph.D., D.Sc.
Dyson Perrins Laboratory
Oxford University
Oxford, England

Edward Katz, Ph.D.
Department of Microbiology
Georgetown University School of Medicine
Washington, D.C.

Hubert A. Lechevalier, Ph.D.
Institute of Microbiology
Rutgers University
New Brunswick, New Jersey

Willy Leimgruber, Ph.D.
Scientific Department
Hoffmann-LaRoche Inc.
Nutley, New Jersey

James L. Littlejohn, Ph.D.
Department of Microbiology
Clinton Corn Processing Company
Clinton, Iowa

J. C. MacDonald, Ph.D.
Prairie Regional Laboratory
National Research Council of Canada
Saskatoon, Saskatchewan, Canada

L. E. McDaniel, Ph.D.
Institute of Microbiology
Rutgers University
New Brunswick, New Jersey

Witold Mechlinski, Ph.D.
Institute of Microbiology
Rutgers University
New Brunswick, New Jersey

Satoshi Mizutani, Ph.D.
McArdle Laboratories for Cancer Research
University of Wisconsin
Madison, Wisconsin

Saul L. Neidleman, Ph.D.
Cetus Scientific Laboratories Inc.
Berkeley, California

Yoshiro Okami, Ph.D.
Institute of Microbial Chemistry
Microbial Chemistry Research Foundation
Tokyo, Japan

Yasuhide Ota, Ph.D.
Department of Agricultural Chemistry
University of Tokyo
Tokyo, Japan

D. Perlman, Ph.D.
School of Pharmacy
University of Wisconsin
Madison, Wisconsin

T. B. Platt, Ph.D.
Bioanalytical Section
Squibb Institute for Medical Research
New Brunswick, New Jersey

Joseph L. Potter, M.D., Ph.D.
Department of Biochemistry
The Children's Hospital of Akron
Akron, Ohio

Martin H. Rogoff, Ph.D.
Biological Research
Crop Protection Department
Sandoz-Wander, Inc.
Homestead, Florida

Sonia Russell, Ph.D.
Biology Department
Dalhousie University
Halifax, Nova Scotia, Canada

F. J. Simpson, Ph.D.
Atlantic Regional Laboratory
National Research Council of Canada
Halifax, Nova Scotia, Canada

V. R. Srinivasan, Ph.D., Dr. rev. nat. FAIC
Department of Microbiology
Louisiana State University
Baton Rouge, Louisiana

George M. Strunz, Ph.D.
Maritimes Forest Research Centre
Canadian Forestry Service
Fredericton, New Brunswick, Canada

T. Takeuchi, Ph.D.
Institute of Microbial Chemistry
Microbial Chemistry Research Foundation
Tokyo, Japan

Viktor Thaller, Ph.D.
Dyson Perrins Laboratory
Oxford University
Oxford, England

Robert Thomas, Ph.D.
Department of Chemistry
University of Surrey
Guildford, Surrey, England

Daisuke Tsuru, Ph.D.
Faculty of Pharmaceutical Sciences
Nagasaki University
Nagasaki, Japan

W. B. Turner, D.Phil., D.Sc.
Pharmaceuticals Division
Imperial Chemical Industries, Ltd.
Macclesfield, Cheshire, England

Tsuneko Uchida, D.Sc.
Biochemical Preparations Laboratory
Mitsubishi-kasei Institute of Life Sciences
Tokyo, Japan

Hamao Umezawa, M.D.
Institute of Microbial Chemistry
Microbial Chemistry Research Foundation
Tokyo, Japan

Chase Van Baalen, Ph.D.
Marine Science Institute
University of Texas
Port Aransas, Texas

L. C. Vining, Ph.D.
Department of Biology
Dalhousie University
Halifax, Nova Scotia, Canada

Chi-Kit Wat, Ph.D.
Department of Botany
The University of British Columbia
Vancouver, British Columbia, Canada

John W. Westley, Ph.D.
Chemical Research Department
Hoffmann-LaRoche, Inc.
Nutley, New Jersey

J. L. C. Wright, Ph.D.
Department of Biology
Dalhousie University
Halifax, Nova Scotia, Canada

TABLE OF CONTENTS

ANTIBIOTICS, MICROBIAL INHIBITORS, AND FERMENTATION PRODUCTS

MISCELLANEOUS INFORMATION

INDICES

SUBSTANCES RELATED
TO CARBOHYDRATES

SIMPLE ALIPHATIC SUBSTANCES, HYDROCARBONS, ESTERS, ALDEHYDES, KETONES, AND ALCOHOLS

DR. NANCY N. GERBER

The following list has been compiled primarily from References 1 and 2. No claim is made for its completeness. Substances reported from wine, sherry, fusel oil, etc. are not included. However, a list of compounds identified in whisky, wine and beer has been published (see Reference 3).

Name	Producing Organisms	Properties	Reference
Ethylene	*Penicillium digitatum, Blastomyces dermatitidis, B. brasiliensis, Histoplasma capsulatum*	BP: −103°C	1, 2
2-Methyl-2-butene	*Puccinia graminis*	BP: 38°C	1, 2
Octacosane	*Amanita phalloides*	MP: 61°C	1, 2
Methyl formate	*Streptomyces odorifer*	BP: 31°C	4
Methyl acetate	*Streptomyces odorifer*	BP: 59°C	4
Ethyl acetate	*Streptomyces odorifer*	BP: 77°C	4
Propyl acetate	*Ceratocystis fimbriata*	BP: 102°C	2
Isopropyl acetate	*Streptomyces odorifer*	BP: 89°C	4
Isobutyl acetate	*Endoconidiophora coerulescens*	BP: 61°C	1, 2
Methyl heptenyl acetate	*Ceratocystis coerulescens*		5
Actinomycin J$_2$ (dodecyl 5-keto-octadecanoate)	*Actinomyces (Streptomyces) flavus*	MP: 81.5°C	1
Formaldehyde	*Chlamydomonas globosa*	BP: −20°C	6
Acetaldehyde	*Chlamydomonas globosa*	BP: 21°C	6, 8
Propionaldehyde	*Chlorella pyrenoidosa*	BP: 49°C	7
Valeraldehyde	*Cryplomonas ovata, Synura petersenii*	BP: 103°C	8
Heptanal	*Cryplomonas ovata, Synura petersenii*	BP: 153°C	8
Acetone	*Cryplomonas ovata, Synura petersenii*	BP: 56°C	8
2-Butanone (methyl ethyl ketone)	*Chlamydomonas globosa*	BP: 80°C	6
Diacetyl (2,3-butanedione)	Various bacteria	BP: 88°C	9
5-Methyl-3-heptanone	*Streptomyces cinnamoneus*-like		10
2-Methyl-2-heptene-6-one	*Endoconidiophora coerulescens, E. virescens*	BP: 172−174°C	1, 2
3-Octanone	*Aspergillus flavus*		11
Methyl heptenyl ketone	*Ceratocystis coerulescens*		5

Name	Producing Organisms	Properties	Reference
Palmitone (dipentadecyl ketone)	*Corynebacterium diphtheriae*	MP: 82°C	1
cis-Palmitenone (pentadecyl pentadec-7-enyl ketone)	*Corynebacterium diphtheriae*	MP: 40°C	1
Ethanol	Yeasts, fusaria, mucors, penicillia, aspergilli, etc.	BP: 78°C	1
Dihydroxyacetone	*Acetobacter suboxydans*	MP: 75–80°C	1
Glycerol	Yeasts, *Bacillus subtilis, Aspergillus wentii, Clasterosporia, Helminthosporia,* penicillia, etc.	MP: 18°C BP: 290°C (dec.)	1
n-Butanol	*Clostridium acetobutylicum, C. propylbutylicum, C. saccharobutylicum*	BP: 117°C	1
Isobutanol	*Ceratocystis moniliformis, C. coerulescens, C. major, C. fugacearum, Streptomyces odorifer*	BP: 108°C	2, 4, 12
2,3-Butanediol	*Aerobacter aerogenes, Serratia marcescens, Bacillus polymyxa, B. subtilis, B. mesentericus, Pseudomonas hydrophila,* yeasts	BP: 180°C	1
Acetoin (3-hydroxy-2-butanone)	Various bacteria	BP: 148°C	9
3-Methylbutanol	*Aspergillus flavus*		11
1-Octanol	*Aspergillus flavus*		11
3-Octanol	*Aspergillus flavus*	BP: 195°C	11
1-Octen-3-ol	*Aspergillus flavus*		11
cis-2-Octen-1-ol	*Aspergillus flavus*		11
Cetyl alcohol	*Amanita phalloides*	MP: 50°C	1, 2
Stearyl alcohol	*Penicillium notatum*	MP: 59°C	1, 2
d-2-Octadecanol	*Mycobacterium tuberculosis, M. avium, M. phlei*	MP: 56°C	1
d-3-Octadecanol	*Corynebacterium diphtheriae*	MP: 56°C	1
Phthiocerol (a 32 and 34 carbon chain, each with a methoxy, a methyl and 2 hydroxy substituents)	*Mycobacterium tuberculosis*	MP: 73°C	1

REFERENCES

1. Miller, M. W., *The Pfizer Handbook of Microbial Metabolites.* McGraw-Hill, New York (1961).
2. Shibata, S., Natori, S., and Udagawa, S., *List of Fungal Products.* Charles C Thomas, Springfield, Illinois (1964).
3. Kahn, J. H., *J. Assoc. Off. Anal. Chem., 52,* 1166 (1969); *Chem. Abstr., 72,* 30259V (1970).
4. Gaines, H. D., and Collins, R. D., *Lloydia, 26,* 247 (1963).
5. Sprecher, E., and Strachenbrock, K. H., *Z. Naturforsch. Teil B, 18,* 495 (1963).
6. Collins, R. P., and Bean, G. H., *Phycologia, 3,* 55 (1963).
7. Katayama, T., *Kagoshima Daigaku Suisangakubu Kiyo, 15,* 13 (1966); *Chem. Abstr., 67,* 1027j (1967).
8. Collins, R. P., and Kalnins, K., *Lloydia, 28,* 48 (1965); *J. Protozool., 13,* 435 (1966).
9. Henis, Y., Gould, J. R., and Alexander, M., *Appl. Microbiol., 14,* 513 (1966).
10. Henley, D. E., Glaze, W. H., and Silvey, J. K. G., *Environ. Sci. Technol., 3,* 268 (1969).
11. Kaminski, E., Libbey, L. M., Stawicki, S., and Wasowicz, E., *Appl. Microbiol., 24,* 721 (1972).
12. Collins, R. P., and Kalnins, K., *Phyton (Buenos Aires), 22,* 107 (1965).

CYCLITOL ANTIBIOTICS

DR. THEODORE HASKELL

The cyclitol antibiotics are substances of microbial origin which contain a cyclohexane ring substituted in the following manner:

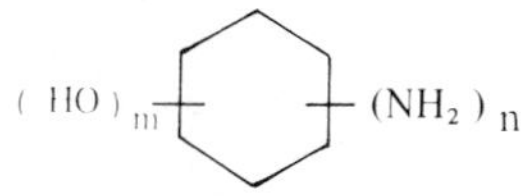

where $6 \geqslant (m + n) \geqslant 3$ and any hydrogen of the hydroxyl and/or the amino groups may be replaced by another substituent. These substances have been classified as "aminoglycoside" antibiotics,[1] since they are for the most part basic molecules containing glycosidic linkages. They are characterized by high solubility in water, poor crystallinity, and high, ill-defined melting points or decomposition ranges. Color reactions are indicative of the type of sugar present. The antibiotics are identified mainly by optical rotation, chromatographic mobilities, elemental analyses, potentiometric titration, chemical degradation, and antimicrobial spectrum. Stereochemistry has been determined by circular dichroism and by proton magnetic-resonance spectra. In chemotherapeutics, the basic antibiotics must be administered as acid addition salts (sulfate, hydrochloride, etc.); they are effective against both Gram-positive and Gram-negative bacteria.

Cyclitol hexols ($\frac{m = 6}{n = 0}$) are termed "inositols" and have been assigned the trivial names *cis, epi, allo, neo, myo, muco, chiro* (D and L), and *scyllo*. The rules of nomenclature have been published[2] and should be consulted in the naming of all cyclitols. The stereochemical feature of cyclitols is exemplified by Formula A, in which the ring is considered as being planar and the vertical lines denote the hydroxyl groups that are designated as above or below the plane of the ring.

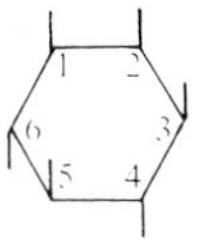

Formula A

The above compound, *myo*-inositol, is thus given the stereochemical prefix 1,2,3,5/4,6, where the hydroxyl groups above the plane are listed in the numerator and those below the plane in the denominator. The larger set of *cis*-hydroxyls is always placed in the numerator; the numbering sequence is then clockwise.

For substituted inositols such as *neo*-inosamine-2 (Formula B), the correct *Chemical Abstracts* nomenclature is 2-amino-2-deoxy-*neo*-inositol.

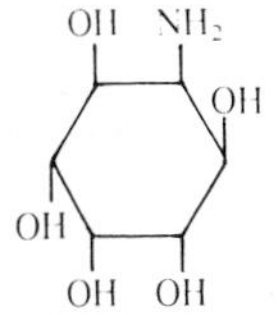

Formula B

Occasionally *Chemical Abstracts* adopts trivial names for substituted inositols that occur most frequently in medically important antibiotics. Such is the case for 2-deoxystreptamine (Formula C), which is 1,3-diamino-1,2,3-trideoxy-*scyllo*-inositol.

For optically active cyclitols, the hydroxyl or amino group at the C_1 position is placed down for the D-form and up for the L. Thus the D-enantiomorph of the tetrol (Formula D) is written as D or (1S)-(1,2,4/5)-1,2,4,5-cyclohexanetetrol.

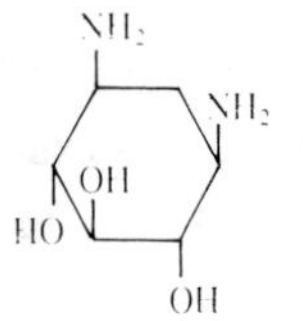

Formula C

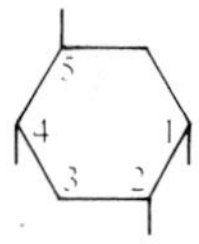

Formula D

Since the D-configuration of carbohydrates is related to the L-form of amino acids, the D- and L-forms of inositols are denoted by the symbols (S) and (R) respectively.[3] Thus validamine (Formula E) is named (1S)-(1,2,4/3,5)-1-amino-5-(hydroxymethyl)-2,3,4-cyclohexanetriol and is stereo-chemically related to D-glucose.

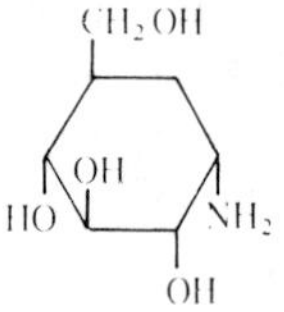

Formula E

No.	Compound (Formula)	Type	Cyclitol	Structure	$[\alpha]_D$ (Solvent)	Reference
1	Kasugamycin ($C_{14}H_{25}N_3O_9$)	m = 6 n = 0	D-chiro-Inositol		+120° (C, 1.6 H_2O)[a]	4
2	Moenomycin, mixture of 7 substances ($C_{70}H_{124}N_6O_{40}P$)	m = 6 n = 0	myo-Inositol	Structure unknown; a phosphorus-containing glycolipid containing D-glucosamine, 6-deoxy-D-glucosamine and D-glucose	None reported	5
3	Antiprotozoin	m = 6 n = 0	myo-Inositol	Structure unknown	+23° (C, 1.3 MeOH)	6
4	Hygromycin A (Homomycin) ($C_{23}H_{29}NO_{12}$)	m = 5 n = 1	2-Amino-2-deoxy-neo-inositol or neo-inosamine-2		−126° (C, 1.0 H_2O)[b]	7
5	Hygromycins C, D, E, F (Not determined)	m = 5 n = 1	2-Amino-2-deoxy-neo-inositol or neo-inosamine-2	Not determined	Not determined	8

No.	Compound (Formula)	Type	Cyclitol	Structure	[a]D (Solvent)	Reference
6	Bluensomycin (Glebomycin) ($C_{21}H_{39}N_5O_{14}$)	m = 5, n = 1	scyllo-Inosamine or 1-Amino-1-deoxy-scyllo-inositol	$R = NH\text{--}C\text{--}NH_2$ (with NH above), $R^1 = O\text{--}C\text{--}NH_2$ (with O below), or vice versa	−89° (C, 1.0 H_2O)[a]	9
7	Streptomycin ($C_{21}H_{39}N_7O_{12}$)	m = 4, n = 2	1,3-Diamino-1,3-dideoxy-scyllo-inositol, or streptamine	A: CHO, B: CH_3, D: H	−87° (C, 1.0 H_2O)[a]	10
8	Dihydrostreptomycin ($C_{21}H_{41}N_7O_{12}$)	m = 4, n = 2	1,3-Diamino-1,3-dideoxy-scyllo-inositol, or streptamine	A: CH_2OH, B: CH_3, D: H	−89.5° (C, 1.0 H_2O)[a]	11

No.	Compound (Formula)	Type	Cyclitol	Structure	$[a]_D$ (Solvent)	Reference
9	Hydroxystreptomycin ($C_{21}H_{39}N_7O_{13}$)	m = 4 n = 2	1,3-Diamino-1,3-dideoxy-*scyllo*-inositol, or streptamine		−91° (H$_2$O)[a]	12
10	Mannosidostreptomycin ($C_{27}H_{49}N_7O_{17}$)	m = 4 n = 2	1,3-Diamino-1,3-dideoxy-*scyllo*-inositol, or streptamine	A: CHO B: CH$_3$ D: CH$_2$OH	−47° (H$_2$O)[a]	13
11	Mannosidohydroxystreptomycin ($C_{27}H_{49}N_7O_{18}$)	m = 4 n = 2	1,3-Diamino-1,3-dideoxy-*scyllo*-inositol, or streptamine	A: CHO B: CH$_2$OH C:	−55° (H$_2$O)[a]	14

No.	Compound (Formula)	Type	Cyclitol	Structure	$[\alpha]_D$ (Solvent)	Reference
12	Hybrimycin A1 ($C_{23}H_{46}N_6O_{14}$)	m = 4 n = 2	1,3-Diamino-1,3-dideoxy-*scyllo*-inositol, or streptamine	$R^1 = H$ $R^2 = OH$ $R^3 = H$ $R^4 = CH_2NH_2$	+44.6° (H₂O)[c]	15
13	Hybrimycin A2 ($C_{23}H_{46}N_6O_{14}$)	m = 4 n = 2	1,3-Diamino-1,3-dideoxy-*scyllo*-inositol, or streptamine	$R^1 = H$ $R^2 = OH$ $R^3 = CH_2NH_2$ $R^4 = H$	+80.5° (H₂O)[c]	15
14	Hybrimycin B1 ($C_{23}H_{46}N_6O_{14}$)	m = 4 n = 2	1,3-Diamino-1,3-dideoxy-*myo*-inositol, or 2-*epi*-streptamine	$R^1 = OH$ $R^2 = H$ $R^3 = H$ $R^4 = CH_2NH_2$	+52.0° (H₂O)[c]	15
15	Hybrimycin B2 ($C_{23}H_{46}N_6O_{14}$)	m = 4 n = 2	1,3-Diamino-1,3-dideoxy-*myo*-inositol, or 2-*epi*-streptamine	$R^1 = OH$ $R^2 = H$ $R^3 = CH_2NH_2$ $R^4 = H$	+94.4° (H₂O)[c]	15
16	Actinospectacin (Spectinomycin) ($C_{14}H_{24}N_2O_7 \cdot 6H_2O$)	m = 4 n = 2	1,3-Diamino-1,3-dideoxy-*myo*-inositol, or 2-*epi*-streptamine		+7.6° (H₂O)[b]	16

No.	Compound (Formula)	Type	Cyclitol	Structure	$[\alpha]_D$ (Solvent)	Reference
17	Hygromycin B ($C_{20}H_{37}N_3O_{13}$)	m = 3 n = 2	1,3-Diamino-1,2,3-trideoxy-*scyllo*-inositol, or hyosamine, 5'-substituted		+20.2° (H_2O)[b]	17
18	Destomycin A ($C_{20}H_{37}N_3O_{13}$)	m = 3 n = 2	1,3-Diamino-1,2,3-trideoxy-*scyllo*-inositol, or hyosamine, 5'-substituted	$R^1 = H$ $R^2 = CH_3$	+7° (H_2O)[b]	18
19	Antibiotic A-396-I ($C_{19}H_{35}N_3O_{13}$)	m = 3 n = 2	1,3-Diamino-1,2,3-trideoxy-*scyllo*-inositol, or hyosamine, 5'-substituted	$R^1 = H$ $R^2 = H$	+11.6° (H_2O)[b]	19
20	Paromomycin ($C_{23}H_{45}N_5O_{14}$)	m = 3 n = 2	1,3-Diamino-1,2,3-trideoxy-*scyllo*-inositol, or 2-deoxystreptamine, 4,5-di-O-substituted		+64° (H_2O)[b]	20

For compound 17: $R^1 = CH_3$ $R^2 = H$

$R^1 = H, R^2 = CH_2NH_2, R^3 = OH$

No.	Compound (Formula)	Type	Cyclitol	Structure	$[\alpha]_D$ (Solvent)	Reference
21	Paromomycin II ($C_{23}H_{45}N_5O_{14}$)	m = 3 n = 2	2-Deoxystreptamine, 4,5-di-O-substituted	 $R^1 = CH_2NH_2, R^2 = H, R^3 = OH$	+104° (H_2O)[a]	21
22	Neomycin B ($C_{23}H_{46}N_6O_{13}$)	m = 3 n = 2	2-Deoxystreptamine, 4,5-di-O-substituted	$R^1 = H, R^2 = CH_2NH_2, R^3 = NH_2$	+83°[b]	21
23	Neomycin C ($C_{23}H_{46}N_6O_{13}$)	m = 3 n = 2	2-Deoxystreptamine, 4,5-di-O-substituted	$R^1 = CH_2NH_2, R^2 = H, R^3 = NH_2$	+121°[b]	21

No.	Compound (Formula)	Type	Cyclitol	Structure	$[a]_D$ (Solvent)	Reference
24	Neomycin LP$_B$ (C$_{25}$H$_{45}$N$_6$O$_{14}$)	m = 3 n = 2	2-Deoxystreptamine, 4,5-di-O-substituted		Not determined	22
				R^1 = H, R^2 = CH$_2$NH$_2$, R^3 = NH$_2$		
				R^1 = CH$_2$NH$_2$, R^2 = H, R^3 = NH$_2$		
25	Neomycin LP$_C$ N (C$_{25}$H$_{48}$N$_6$O$_{14}$)	m = 3 n = 2	2-Deoxystreptamine, 4,5-di-O-substituted		Not determined	22
26	Neomycin A (C$_{12}$H$_{26}$N$_4$O$_6$)	m = 3 n = 2	2-Deoxystreptamine, mono-substituted		+83° (H$_2$O)[a]	23

No.	Compound (Formula)	Type	Cyclitol	Structure	$[a]_D$ (Solvent)	Reference
27	Butirosin A ($C_{21}H_{41}N_5O_{12}$)	m = 3 n = 2	2-Deoxystreptamine, 4,5-di-O-substituted	$R^1 = OH$ $R^2 = H$	+26° ($H_2O)^b$	24
28	Butirosin B ($C_{21}H_{41}N_5O_{12}$)	m = 3 n = 2	2-Deoxystreptamine, 4,5-di-O-substituted	$R^1 = H, R^2 = OH$	+33° ($H_2O)^b$	24
29	Ribostamycin SF733 ($C_{17}H_{34}N_4O_{10}$)	m = 3 n = 2	2-Deoxystreptamine, 4,5-di-O-substituted		+42° ($H_2O)^b$	25

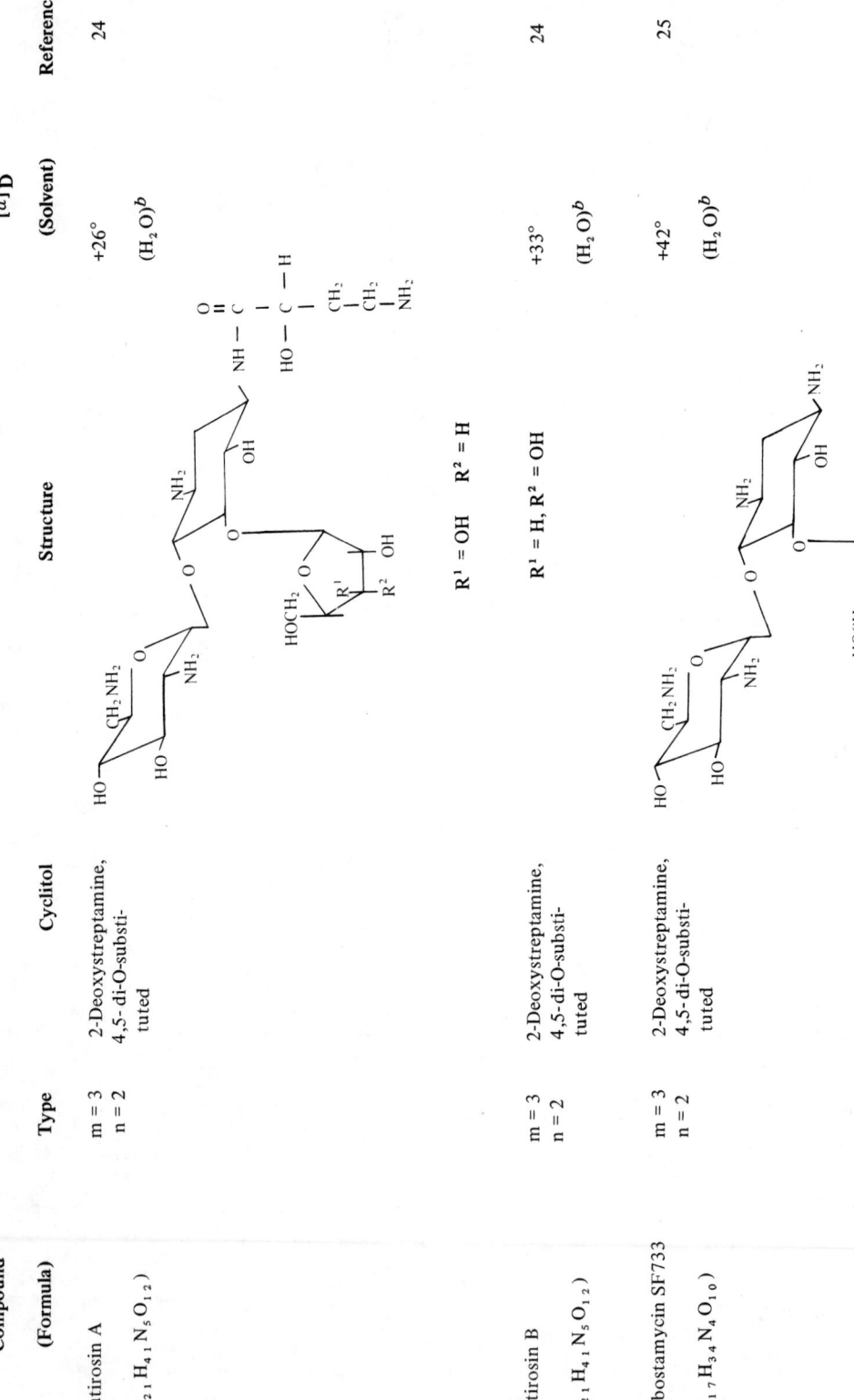

No.	Compound (Formula)	Type	Cyclitol	Structure	$[\alpha]_D$ (Solvent)	Reference
30	Lividomycin ($C_{29}H_{55}N_5O_{18}$)	$m = 3$ $n = 2$	2-Deoxystreptamine, 4,5-di-O-substituted		$+66°$ $(H_2O)^b$	26
31	Kanamycin A ($C_{18}H_{36}N_4O_{11}$)	$m = 3$ $n = 2$	2-Deoxystreptamine, 4,6-di-O-substituted	$R^1 = NH_2, R^2 = OH$	$+146°$ $(0.1N\ H_2SO_4)^a$	27
32	Kanamycin B ($C_{18}H_{37}N_5O_{10}$)	$m = 3$ $n = 2$	2-Deoxystreptamine, 4,6-di-O-substituted	$R^1 = NH_2, R^2 = NH_2$	$+126°$ $(H_2O)^b$	28
33	Kanamycin C ($C_{18}H_{36}N_4O_{11}$)	$m = 3$ $n = 2$	2-Deoxystreptamine, 4,6-di-O-substituted	$R^1 = OH, R^2 = NH_2$	$+126°$ $(H_2O)^b$	27

No.	Compound (Formula)	Type	Cyclitol	Structure	$[a]_D$ (Solvent)	Reference
34	Ebbramycin (Nebramycin) ($C_{18}H_{37}N_5O_9$)	m = 3 n = 2	2-Deoxystreptamine, 4,6-di-O-substituted		+127° (H_2O)[b]	29
35	Gentamicin A ($C_{18}H_{36}N_4O_{10}$)	m = 3 n = 2	2-Deoxystreptamine, 4,6-di-O-substituted		+136° (H_2O)[b]	30
36	Gentamicin C_1 ($C_{21}H_{43}N_5O_7$)	m = 3 n = 2	2-Deoxystreptamine, 4,6-di-O-substituted	$R^1 = R^2 = CH_3$	+158° (H_2O)[b]	31

No.	Compound (Formula)	Type	Cyclitol	Structure	$[\alpha]_D$ (Solvent)	Reference
37	Gentamicin C_{1a} ($C_{19}H_{39}N_5O_7$)	m = 3 n = 2	2-Deoxystreptamine, 4,6-di-O-substituted	$R^1 = R^2 = H$		31
38	Gentamicin C_2 ($C_{20}H_{41}N_5O_7$)	m = 3 n = 2	2-Deoxystreptamine, 4,6-di-O-substituted	$R^1 = CH_3$, $R^2 = H$	+160° (H_2O)[b]	31
39	Sisomicin ($C_{19}H_{37}N_5O_7$)	m = 3 n = 2	2-Deoxystreptamine, 4,6-di-O-substituted		+189° (H_2O)[b]	32
40	Validamycin B ($C_{20}H_{33-37}NO_{14-15}$)	m = 4 n = 1	1-Amino-5-(hydroxymethyl)-2,3,4,6-cyclohexanetetrol	Structure unknown; contains hydroxyvalidamine	+102° (H_2O)[b]	33

No.	Type	Cyclitol	Compound (Formula)	Structure	$[\alpha]_D$ (Solvent)	Reference
41	Validamycin A $(C_{20}H_{33-37}NO_{13-14})$	m = 3 n = 1	(1S)-(1,2,4/3,5)-1-Amino-5-(hydroxymethyl)-2,3,4-cyclohexanetetriol	Structure unknown; contains validamine	+110°	33
		m = 3 n = 0	Validatol, (1S)-(1,2,4/3)-1-(hydroxymethyl)-2,3,4-cyclohexanetriol		$(H_2O)^b$	34

[a]Determined as HCl salt.
[b]Determined as free base.
[c]Determined as N-acetyl derivative.

REFERENCES

1. Hanessian, S., and Haskell, T. H., Antibiotics Containing Sugars, in *The Carbohydrates,* Vol. 2A, p. 159. Academic Press, New York (1970).

2. IUPAC Commission on the Nomelclature of Organic Chemistry, *Biochem. J., 112,* 17 (1969); *J. Biol. Chem., 243,* 5809 (1968).

3. Cahn, R. S., Ingold, C. K., and Prelog, V., *Experientia, 12,* 81 (1956).

4. Suhara, Y., Sasaki, F., Maeda, K., and Umezawa, H., *J. Amer. Chem. Soc., 90,* 6559 (1968).

5. Schacht, U., and Huber, G., *J. Antibiot., 22,* 597 (1969); for related antibiotics, see Slusarchyk, W. A., Osband, J. A., and Weisenborn, F. L., *J. Amer. Chem. Soc., 92,* 4486 (1970).

6. Thirumalachar, M. J., Bringi, N. V., Deshmukh, P. V., and Rahalkar, P. W., *Hindustan Antibiot. Bull., 7,* 25 (1964).

7. Mann, R. L., and Woolf, D. O., *J. Amer. Chem. Soc., 79,* 120 (1957).

8. Holowczak, J., Koffler, H., Garner, H. R., and Elbein, A. D., *J. Biol. Chem., 241,* 3270 (1966).

9. Bannister, B., and Argoudelis, A. D., *J. Amer. Chem. Soc., 85,* 234 (1963).

10. Lemieux, R. U., and Wolfrom, M. L., *Advan. Carbohyd. Chem., 3,* 337 (1948).

11. Peck, R. L., Hoffhine, C. E., and Folkers, K., *J. Amer. Chem. Soc., 68,* 1390 (1946).

12. Stodola, F. H., Shotwell, O. L., Borud, A. M., Benedict, R. G., and Riley, A. C., Jr., *J. Amer. Chem. Soc., 73,* 2290 (1951).

13. Fried, J., and Stavely, H. E., *J. Amer. Chem. Soc., 74,* 5461 (1952).

14. Arcamone, F., Cassinelli, G., D'Amico, G., and Orezzi, P., *Experientia, 24,* 441 (1968).

15. Shier, W. T., Rinehart, K. L., Jr., and Gottlieb, D., *Proc. Nat. Acad. Sci. U.S.A., 63,* 198 (1969).

16. Hoeksema, H., Argoudelis, A. D., Wiley, P., *J. Amer. Chem. Soc., 84,* 3212 (1962).

17. Neuss, N., Koch, K. F., Molloy, B. B., Day, W., Huckstep, L., Dorman, D. E., and Roberts, J. D., *Helv. Chim. Acta, 53,* 2314 (1970).

18. Kondo, S., Akita, E., and Koike, M., *J. Antibiot. Ser. A, 19,* 137 (1966).

19. Shoji, J., and Nakagawa, Y., *J. Antibiot. Ser. A, 23,* 569 (1970).

20. Haskell, T. H., French, J. C., and Bartz, Q. R., *J. Amer. Chem. Soc., 81,* 3481 (1959).

21. Rinehart, K. L., Jr., *The Neomycins and Related Antibiotics.* John Wiley and Sons, New York (1964).

22. Chilton, W. S., *Diss. Abstr., 24,* 4990 (1964); *Chem. Abstr., 61,* 9570 (1964).

23. Peck, R. L., Hoffhine, C. E., Jr., Gale, P., and Folkers, K., *J. Amer. Chem. Soc., 71,* 2590 (1949).

24. Woo, P. W. K., Dion, H. W., and Bartz, Q. R., *Tetrahedron Lett.,* No. 28, p. 2625 (1971).

25. Akita, E., Tsuruoka, T., Ezaki, N., and Niida, T., *J. Antibiot. Ser. A, 23,* 173 (1970).

26. Mori, T., Ichiyanagi, T., Kondo, H., Tokunaga, K., Oda, T., and Munakata, K., *J. Antibiot. Ser. A, 24,* 339 (1971).

27. Umezawa, H., *Recent Advances in the Chemistry and Biochemistry of Antibiotics.* Microbial Chemistry Research Foundation, Tokyo, Japan (1964).

28. Ito, T., Nishio, M., and Ogawa, H., *J. Antibiot. Ser. A, 17,* 189 (1964).

29. Thompson, R. Q., and Presti, E. A., *Antimicrob. Agents Chemother.,* p. 332 (1967).

30. Maehr, H., and Schaffner, C. P., *J. Amer. Chem. Soc., 89,* 6787 (1967).

31. Cooper, D. J., Marigliano, H. M., Yudis, M. D., and Traubel, T., *J. Infect. Dis., 119,* 342 (1969).

32. Cooper, D. J., Jaret, R. S., and Reimann, H., *Chem. Commun.,* p. 285 (1971).

33. Iwasa, T., Kameda, Y., Asai, M., Horii, S., and Mizuno, K., *J. Antibiot. Ser. A, 24,* 119 (1971).

34. Horii, S., Iwasa, T., Mizuta, E., and Kameda, Y., *J. Antibiot. Ser. A, 24,* 59 (1971).

DISACCHARIDES AND TRISACCHARIDES

DR. NANCY N. GERBER

Natural monosaccharides and polysaccharides are listed and discussed in Volume II of the Handbook of Microbiology. A few disaccharides and trisaccharides of microbial origin are listed in the table below. The information was taken from References 1 and 2.

Trivial Name	Systematic Name	Microorganism	Reference
	d-Mannopyranosyl-*l-meso*-erythritol	*Ustilago* sp.	1, 2
Umbilicin	3-β-*d*-Galactopyranosido-*d*-arabitol	*Umbilicaria pustulata*	1
Leucrose	5-O-*a*-*d*-Glucopyranosyl-*d*-fructopyranose	*Leuconostoc mesenteroides*	1
Kojibiose Isomaltose	2-O-*a*-*d*-Glucopyranosyl-*d*-glucopyranose	*Aspergillus niger*	1, 2
Trehalose Mycose	*a*-*d*-Glucosido-*a*-*d*-glucoside	*Amanita muscaria*, other mushrooms and molds, mycobacteria, yeasts, and algae; first isolated from rye ergot (*Claviceps purpurea*).	1, 2
Trehalosamine	*a*-*d*-Glucosido-*a*-*d*-glucosamine	A streptomycete	1
Lactobionic acid	4-β-*d*-Galactopyranosido-*d*-gluconic acid	*Pseudomonas* species, other oxidative bacteria (on lactose)	1
	2-O-*a*-D-Glucopyranosyl-D-glucuronic acid	*Aspergillus niger*	2
Galactosyl-lactose	O-β-D-Galactopyranosyl-(1 → 6)-O-β-D-galactopyranosyl-(1 → 4)-D-glucopyranose	*Penicillium chrysogenum*	2
	O-β-D-Glucopyranosyl-(1 → 2)-O-β-D-fructofuranosyl-(1 → 2)-β-D-fructofuranoside	*Aspergillus niger*	2
	O-β-D-Glucopyranosyl-(1 → 6)-O-β-D-glucopyranosyl-(1 → 4)-D-glucopyranose	*Aspergillus niger*	2
Panose	4-[6-(*a*-D-Glucopyranosyl)-*a*-D-glucopyranosyl]-D-glucose	*Aspergillus niger*	2
6-(*a*-D-Glucosyl)-isomaltose		*Aspergillus oryzae*	2

Trivial Name	Systematic Name	Microorganism	Reference
6-(α-D-Glucosyl)-maltose		*Aspergillus oryzae*	2
	O-α-D-Glucopyranosyl-(1 → 6)-O-α-D-glucopyranosyl-(1 → 2)-β-D-fructofuranoside	*Aspergillus niger*	2
	O-α-D-Glucopyranosyl-(1 → 6)-O-α-D-glucopyranosyl-(1 → 2)-D-glucuronic acid	*Aspergillus niger*	2

REFERENCES

1. Miller, M., *The Pfizer Handbook of Microbial Metabolites.* McGraw-Hill, New York (1961).
2. Shibata, S., Natori, S., and Udagawa, S., *List of Fungal Products.* Charles C Thomas, Springfield, Illinois (1964).

ALIPHATIC AND RELATED COMPOUNDS

CARBOXYLIC ACIDS

DR. J. L. C. WRIGHT and DR. L. C. VINING

The compounds listed in the following table have in common chemical structures possessing one or more carboxyl groups and microbial origin as the free (i.e., non-covalently linked) acid. Carboxylic acids covalently linked in esters, amides, etc., where all carboxyl functions are masked, have not been included. Other carboxylic acids omitted from this table are catalogued elsewhere in this volume under specific titles, e.g., compounds containing an aromatic, furan or pyran ring system or a conjugated polyene and/or polyine system within their structure, steroids, terpenes, and other alicyclic compounds, amino acids and peptides. A few compounds that fall within other specific categories have been included here when their acidic character appeared to warrant it.

Apart from their acidic character, which is usually manifested by an infrared absorption maximum in the region of 1700 cm^{-1}, and an ability to form water-soluble sodium or potassium salts, the substances listed below are quite diverse. Those with a predominantly aliphatic character are normally quite soluble in non-polar organic solvents and can be extracted from acidic aqueous solution. At alkaline pH the ionized form can be back-extracted from the organic to the aqueous phase. However, those substances containing two or more carboxyl groups, and particularly the polyhydroxylated acids, are strongly polar, very soluble in water, and partition into an aqueous phase even at acid pH. Most carboxylic acids are soluble in alcohols, but compounds such as the sugar acids may be poorly soluble in acetone and less polar organic solvents.

Many additional sources of information on microbial carboxylic acids are available. An attempt has been made to give recent or leading references indicating the distribution of the product in nature. More complete descriptions of chemical and physical properties can often be obtained by consulting References 1 to 5.

For many of the compounds listed, chromatographic properties are an essential aid in identification. Whenever possible, this information is given as a reference to the appropriate tables in the *Handbook of Chromatography*, Volume 1;[6] e.g., GC 127; LC 34, 65, 71; PC 77, 79; TLC 29, 134 (in the column listing significant characteristics for formic acid) represent the table numbers of gas, liquid, paper, and thin-layer chromatographic data in the *Handbook of Chromatography*, where the desired information can be found.

Common Name Formula (MW) Structure	Organisms	Significant Characteristics	References

Unsubstituted Monocarboxylic Acids

Common Name Formula (MW) Structure	Organisms	Significant Characteristics	References

Formic acid
CH_2O_2 (46)
HCOOH
	Most Enterobacteriaceae
	Pseudomonas formicans
	Bacillus subtilis
	Serratia marcescens
	Aspergillus oryzae
	Many other organisms
		Boiling point: 100.5°C
		Melting point: 8.4°C
		n_D^{20}: 1.3714
		Chromatographic data:[6]
		  GC 127
		  LC 34, 65, 71
		  PC 77, 79
		  TLC 29, 134
			7

Acetic acid
$C_2H_4O_2$ (60)
CH_3COOH
	Saccharomyces and other yeasts
	Acetobacter spp.
	Clostridium thermoaceticum
		Boiling point: 118°C
		Freezing point: 16.7°C
		n_D^{20}: 1.3718
		Chromatographic data:[6]
		  GC 49, 127
		  LC 4, 34, 36, 57, 58, 61, 65
		  PC 77, 79
		  TLC 29, 132, 134
			7–10

Propionic acid
$C_3H_6O_2$ (74)
CH_3CH_2COOH
	Amanita muscaria
	Certain *Clostridia* and rumen
	  bacteria
	Propionibacteria
		Boiling point: 140.5°C
		Melting point: 21.5°C
		n_D^{25}: 1.3848
		Chromatographic data:[6]
		  GC 127
		  LC 57, 65
		  PC 77, 79
		  TLC 29, 132, 134
			7, 8

Acrylic acid
$C_3H_4O_2$ (72)
$CH_2=CHCOOH$
	Bacillus amaracrylus	Boiling point: 141°C
		  (polymerizes)
		Melting point: 13°C; *p*-
		  toluidide 141°C
		n_D^{20}: 1.4224
			11

n-Butyric acid
$C_4H_8O_2$ (88)
$CH_3(CH_2)_2COOH$
	Clostridium butyricum
	Clostridium and *Bacillus* spp.
		Boiling point: 162°C
		Freezing point: –19°C
		n_D^{20}: 1.3991
		Chromatographic data:[6]
		  LC 7, 57, 65
		  PC 11, 77, 79
		  TLC 29, 132, 134
			7, 8

Isobutyric acid
$C_4H_8O_2$ (88)
$(CH_3)_2CHCOOH$
	Clostridium spp.
	Acetobacter pasteurianum
		Boiling point: 154.3°C
		Melting point: –47°C; S-
		  benzylthiouronium salt
		  143°C
		n_D^{20}: 1.3930
		Chromatographic data:[6]
		  LC 65
		  TLC 134
			8

Common Name Formula (MW) Structure	Organisms	Significant Characteristics	References

Unsubstituted Monocarboxylic Acids (continued)

Common Name Formula (MW) Structure	Organisms	Significant Characteristics	References

n-Valeric acid
$C_5H_{10}O_2$ (102)
$CH_3(CH_2)_3COOH$
	Clostridium sp.	Boiling point: 186–187°C
		Melting point: –34.5°C
		n_D^{20}: 1.4086
		Chromatographic data:[6]
		  LC 57, 65
		  PC 79
		  TLC 132, 134
			12

Isovaleric acid $C_5H_{10}O_2$ (102) $(CH_3)_2CHCH_2COOH$	*Clostridium* spp.	Boiling point: 176.7°C Freezing point: –37.6°C $n_D^{22.4}$: 1.40178 Chromatographic data:[6] PC 77	8

a-Methyl-*n*-butyric acid
$C_5H_{10}O_2$ (102)
$CH_3CH_2CHCOOH$
     |
     CH_3
	Penicillium notatum
	Clostridium sp.
		Melting point: 176°C
		Chromatographic data:[6]
		  TLC 134
			8, 13

n-Caproic acid
$C_6H_{12}O_2$ (116)
$CH_3(CH_2)_4COOH$
	Clostridium kluyveri	Boiling point: 205°C
		Freezing point: –1.5–2°C
		n_D^{20}: 1.4164
		Chromatographic data:[6]
		  LC 7, 61
		  PC 77, 79
		  TLC 29, 132
			7

10-Undecenoic acid $C_{11}H_{20}O_2$ (184) $CH_2=CH(CH_2)_8COOH$	*Rhodotorula glutinis* var. *lusitanica*	Melting point: 24°C n_D^{24}: 1.4464	14
10-Undecynoic acid $C_{11}H_{18}O_2$ (182) $CH\equiv C(CH_2)_8COOH$	*Rhodotorula glutinis* var. *lusitanica*	Melting point: 39°C	14

Stearic acid
$C_{18}H_{36}O_2$ (284)
$CH_3(CH_2)_{16}COOH$
	Lactarius spp.	Melting point: 69–70°C
		Chromatographic data:[6]
		  PC 77
		  TLC 33
			15

Unsubstituted Dicarboxylic and Tricarboxylic Acids

Common Name Formula (MW) Structure	Organisms	Significant Characteristics	References

Oxalic acid
$C_2H_2O_4$ (90)
HOOCCOOH
	Aspergillus niger
	Penicillium oxalicum
	Citromyces spp.
	Many other fungi
		Melting point of dihydrate:
		  101°C with loss of water;
		  sublimes above melting
		  point with decomposition
		Melting point of anhydrous
		  form: 189.5°C (decomposes)
		Chromatographic data:[6]
		  LC 58, 65
		  PC 13, 77
		  TLC 31
			16–18

Common Name Formula (MW) Structure	Organisms	Significant Characteristics	References

Unsubstituted Dicarboxylic and Tricarboxylic Acids (continued)

Common Name Formula (MW) Structure	Organisms	Significant Characteristics	References

Malonic acid
$C_3H_4O_4$ (104)
$HOOCCH_2COOH$
	Penicillium funiculosum
	Penicillium islandicum
	Gibberella fujikuroi
	Hansenula miso
	Other fungi and yeasts
		Melting point: 135°C
		Chromatographic data:[6]
		  LC 65
		  PC 13, 77
		  TLC 31
			19, 20

Succinic acid
$C_4H_6O_4$ (118)
$HOOCCH_2CH_2COOH$
	Aspergillus terreus
	Fusarium oxysporum
	Mucor stolonifer
	Penicillium spp.
	Other fungi
	Acetic acid bacteria
		Melting point: 185—187°C
		Chromatographic data:[6]
		  LC 58, 65
		  PC 13, 77
		  TLC 31
			16, 21

Fumaric acid
$C_4H_4O_4$ (116)
$HOOCCH=CHCOOH$
	Rhizopus spp.
	Mucor spp.
	Aspergillus spp.
	Penicillium spp.
	Fusarium spp.
	Boletus spp.
	Other fungi
		Melting point: 290°C
		Chromatographic data:[6]
		  LC 58, 65
		  PC 113
			9, 10, 17

Ethyl hydrogen fumarate $C_6H_8O_4$ (144) $CH_3CH_2OOCCH=CHCOOH$	*Lasiodiplodia theobromae*	Melting point: 66—67°C	22

Glutaric acid
$C_5H_8O_4$ (132)
$HOOC(CH_2)_3COOH$
	Aspergillus niger
	Acetobacter aceti
	Gluconobacter oxydans
		Melting point: 97°C
		Chromatographic data:[6]
		  LC 65
		  PC 13, 77
		  TLC 31
			21, 23

trans-Glutaconic acid $C_5H_6O_4$ (130) $HOOCCH_2CH=CHCOOH$	*Aspergillus niger* *Peptococcus aerogenes*	Melting point: 138°C	23, 24

Itaconic acid
$C_5H_6O_4$ (130)
$HOOCCH_2CCOOH$
        $\parallel$
        CH_2
	Aspergillus terreus
	Aspergillus itaconicus
	Ustilago zeae
	Helicobasidium mompa
	Penicillium charlesii
	Other fungi
		Melting point: 162—164°C
		Chromatographic data:[6]
		  PC 13
			9, 17, 25

Adipic acid
$C_6H_{10}O_4$ (146)
$HOOC(CH_2)_4COOH$
	Acetobacter aceti
	Gluconobacter oxydans
		Melting point: 152°C
		Chromatographic data:[6]
		  LC 65
		  PC 13, 77
		  TLC 31
			21

Common Name Formula (MW) Structure	Organisms	Significant Characteristics	References

Unsubstituted Dicarboxylic and Tricarboxylic Acids (continued)

Common Name Formula (MW) Structure	Organisms	Significant Characteristics	References

Pimelic acid
$C_7H_{12}O_4$ (160)
$HOOC(CH_2)_5COOH$
	Acetobacter aceti
	Gluconobacter oxydans
		Melting point: 106°C
		Chromatographic data:[6]
		  PC 13, 77
		  TLC 31
			21

2-Decene-1,10-dioic acid $C_{10}H_{16}O_4$ (200) $HOOC(CH_2)_6CH=CHCOOH$	*Penicillium notatum*	Melting point: 172°C	13
Fomentaric acid $C_{41}H_{80}O_4$ (636) $(C_{18}H_{37})_2CCOOH$ $CH_3CHCOOH$	*Fomes fomentarius*	Melting point: 78–80°C	26

cis-Aconitic acid
$C_6H_6O_6$ (174)
$HOOCCH_2C=CHCOOH$
     $\vert$
     COOH
	Aspergillus niger	Melting point: 125°C
		Chromatographic data:[6]
		  PC 13
			17

Hydroxy Acids

Common Name Formula (MW) Structure	Organisms	Significant Characteristics	References

Glycolic acid
$C_2H_4O_3$ (76)
$HOCH_2COOH$
	Acetobacter suboxydans
	Gluconobacter liquefaciens
	Aspergillus niger
		Hygroscopic
		Melting point: 80°C; *p*-bromo-
		  phenacyl ester 138°C
		Chromatographic data:[6]
		  LC 40
		  PC 13
		  TLC 29
			20, 21, 27

Lactic acid
$C_3H_6O_3$ (90)
$CH_3CHCOOH$
   $\vert$
   OH
		Chromatographic data:[6]
		  LC 58, 65
		  PC 13
		  TLC 29, 134
			7, 9, 10

L(+) isomer (Sarcolactic acid)	Many lactobacilli *Rhizopus* sp.	Melting point: 25–26°C $[a]_D^{15}: +3.3°$ (water, c = 5)	
D(−) isomer	*Leuconostoc mesenteroides* Some lactobacilli	Melting point: 26–27°C $[a]_D: -2.3°$ (water, c = 5)	
DL racemate	Many bacteria	Boiling point (15 mm): 122°C Melting point: 18°C	

β-Hydroxypropionic acid
$C_3H_6O_3$ (90)
$HOCH_2CH_2COOH$
	Acetic acid bacteria
	Yeasts
		Syrup
		Melting point of sodium salt:
		  143°C
		Chromatographic data:[6]
		  LC 23
		  TLC 29
			20, 21

Common Name Formula (MW) Structure	Organisms	Significant Characteristics	References

Hydroxy Acids (continued)

Common Name Formula (MW) Structure	Organisms	Significant Characteristics	References
L(–)-Glyceric acid $C_3H_6O_4$ (106) $HOCH_2CHCOOH$ 　　　$\mid$ 　　　OH	Various fungi	Syrup; decomposes on heating Melting point of calcium 　salt: 138°C $[a]_D^{20}$: +14.5° (water, c = 5) Chromatographic data:[6] 　LC 40 　PC 13 　TLC 29	28
2-Phosphoglyceric acid $C_3H_7O_7P$ (186) $HOCH_2CHCOOH$ 　　　$\mid$ 　　　OPO_3H_2	Yeast	Syrup; decomposes on heating Freezing point: –20°C Chromatographic data:[6] 　PC 12, 32	29
DL-β-Hydroxybutyric acid $C_4H_8O_3$ (104) CH_3CHCH_2COOH 　　$\mid$ 　　OH	Acetic acid bacteria Yeasts	Hygroscopic syrup Melting point: sodium salt 　164–165°C; p-phenylphenacyl 　ester 105–106°C Chromatographic data:[6] 　LC 23 　TLC 29, 36	20, 21
a-Hydroxyisovaleric acid $C_5H_{10}O_2$ (118) $(CH_3)_2CHCHCOOH$ 　　　　　$\mid$ 　　　　　OH	*Lactobacillus casei*	Melting point, (±) racemate: 　86°C $[a]_D^{25}$, (+) isomer: –0.5° 　(water)	30
a,β-Dihydroxyisovaleric 　acid $C_5H_{10}O_4$ (134) $(CH_3)_2C - CHCOOH$ 　　　$\mid$　　$\mid$ 　　　OH　OH	*Neurospora crassa* mutant	Syrup Melting point of quinine salt: 　209–210°C (decomposes) $[a]_D^{23}$ of quinine salt: –141° 　(methanol, c = 1)	31
a-Hydroxyisocaproic acid $C_6H_{12}O_3$ (132) $(CH_3)_2CHCH_2CHCOOH$ 　　　　　　　$\mid$ 　　　　　　　OH	*Lactobacillus casei* *Bacillus subtilis* *Proteus vulgaris*	Melting point: (+) isomer 81–82°C; 　(–) isomer 81–82°C; 　(±) racemate 76–77°C $[a]_D^{20}$: (+) isomer +26.3°; 　(–) isomer –27.8°	30
a,β-Dihydroxy-β-methyl- 　valeric acid $C_6H_{12}O_4$ (148) 　　　　CH_3 　　　　$\mid$ $CH_3CH_2 - C - CHCOOH$ 　　　　$\mid$　$\mid$ 　　　　OH　OH	*Neurospora crassa*	Syrup Melting point of quinine salt: 　204°C (decomposes) $[a]_D^{23}$ of quinine salt: –144° 　(methanol, c = 1)	31

Common Name Formula (MW) Structure	Organisms	Significant Characteristics	References

Hydroxy Acids (continued)

Common Name Formula (MW) Structure	Organisms	Significant Characteristics	References
Mevalonic acid (Hiochic acid) $C_6H_{12}O_4$ (148) $\underset{\qquad\quad\ \ OH}{\overset{\qquad CH_3}{HOCH_2CH_2CCH_2COOH}}$	*Saccharomyces* spp.	Oily liquid Melting point: benzhydrylamide 96–97°C; δ-lactone 28°C (hygroscopic) $[\alpha]_D^{20}$ of benzhydrylamide: –2.0° (ethanol, c = 2) ν_{max}: 3390, 1727 cm^{-1} (chloroform)	32
5-Hydroxymethylfuran-2- carboxylic acid $C_6H_6O_4$ (142) $HOCH_2$ __ O __ $COOH$	*Gibberella fujikuroi* *Aspergillus glaucus*	Melting point: 115°C ν_{max}: 1780, 1755, 1663, 1530 cm^{-1} (Nujol)	33
Shikimic acid $C_7H_{10}O_5$ (174) COOH HO, OH, OH	*Penicillium griseofulvum* *Phycomyces blakesleeanus* *Saccharomyces cerevisiae* mutant	Melting point: 190.5°C (sublimes) $[\alpha]_D^{18}$: –183.8° (water, c = 4) Chromatographic data:[6] LC 23 TLC 133	34
Chorismic acid $C_{10}H_{10}O_5$ (226) COOH $\overset{CH_2}{\underset{\ }{O-C-COOH}}$ OH	*Aerobacter aerogenes* mutant	Unstable; isolated as the barium salt, which gives a mixture of 4-hydroxybenzoic acid and prephenic acid on heating at 70°C in water for 1 hour λ_{max}: 272 nm (water); ϵ = 2,700	35
Prephenic acid $C_{10}H_{10}O_6$ (226) HOOC $\quad$ CH$_2$ C COOH $\qquad\qquad\qquad$ O OH	*Escherichia coli* mutant *Neurospora crassa* mutant	Isolated as crystalline barium salt; treatment with acid gives phenylpyruvic acid Melting point of barium salt: 157–158°C	36
a-Hydroxytetracosanoic acid $C_{24}H_{48}O_3$ (384) $\underset{\quad OH}{C_{22}H_{45}CHCOOH}$	*Polyporus umbellatus*		37

Common Name
Formula (MW)
Structure

	Organisms	Significant Characteristics	References

Hydroxy Acids (continued)

Tartronic acid
$C_3H_4O_5$ (120)

HOOCCHCOOH
|
OH

Acetobacter acetosum
Acetobacter melanogenum
Gluconobacter liquefaciens

Melting point: 163°C (decomposes); sublimes at 110–120°C

38

L-Malic acid
$C_4H_6O_5$ (134)
HOOCCH$_2$CHCOOH
|
OH

Aspergilli and many other fungi
Bacillus fluorescens
Escherichia coli

Melting point: 100°C
$[a]_D$: –2.3° (water, c = 8.5)
Chromatographic data:[6]
  LC 58, 65
  PC 13
  TLC 31

18, 39, 40

L(+)-Tartaric acid
$C_4H_6O_6$ (150)
HOOCCH —CHCOOH
| |
OH   OH

Gibberella saubinetti
Fusarium spp.
Acetobacter suboxydans

Melting point: 168–170°C (decomposes)
$[a]_D$: +11.98°
Chromatographic data:[6]
  LC 58, 65
  PC 13, 77
  TLC 31

27

Epoxysuccinic acid
$C_4H_4O_5$ (132)
HOOCCHCHCOOH
 \ /
  O

Aspergillus fumigatus
Penicillium viniferum
Candida formosa
Paecilomyces varioti

Melting point: 185°C (decomposes)
$[a]_D^{18}$: –117° (water, c = 1)

41

Itatartaric acid
$C_5H_8O_6$ (164)
      OH
      |
HOOCCH$_2$CCOOH
      |
      CH$_2$OH

Aspergillus terreus mutant

Isolated as dimethyl ester
Boiling point of dimethyl ester
  (2–3 mm): 129–134°C
Melting point: dibenzylamide
  103–104°C; γ-lactone
  104°C
$[a]_D^{23}$: dimethyl ester –29°;
  dibenzylamide –42° (ethanol,
  c = 3.5)

42

Citric acid
$C_6H_8O_7$ (192)
      OH
      |
HOOCCH$_2$CCH$_2$COOH
      |
      COOH

Aspergillus niger
Many other fungi

Melting point: monohydrate
  ~100°C; anhydrous form
  153°C
Chromatographic data:[6]
  LC 34, 58, 65
  PC 13, 77
  TLC 31

9, 10, 17

L$_s$-*allo*-Isocitric acid
$C_6H_8O_7$ (192)
HOOCCH$_2$CH — CHCOOH
         | |
         COOH OH

Penicillium purpurogenum var.
  rubri-sclerotium
Penicillium duclauxii
Penicillium verruculosum
Penicillium echinulosporum

Isolated as γ-lactone
Melting point of lactone:
  141–142°C
$[a]_D^{19}$ of lactone: +42.3°
  (water, c = 4.83)
ν_{max} of lactone: 1739,
  1701 cm^{-1} (Nujol)

43

Common Name Formula (MW) Structure	Organisms	Significant Characteristics	References

Hydroxy Acids (continued)

2-Phospho-4-hydroxy-4- carboxyadipic acid $C_7H_{11}O_{11}P$ (302)	*Escherichia coli*	Obtained in solution only R_f (PC): 0.24 (methanol– formic acid–water, 16:3:1)	44

$$\text{HOOCCH}_2 - \overset{\overset{\displaystyle OH}{|}}{\underset{\underset{\displaystyle COOH}{|}}{C}} - \text{CH}_2\,\overset{\overset{}{}}{\underset{\underset{\displaystyle OPO_3H_2}{|}}{CH}}\text{COOH}$$

Itaconitin $C_{14}H_{14}O_5$ (262)	*Aspergillus itaconicus*	Melting point: 168°C; 2,4- dinitrophenylhydrazone 169°C λ_{max}: 326 nm Red in concentrated sulfuric acid or ammonia	45

$$\text{CH}_3 - C = C - (CH = CH)_2\,CH = CCH_2COOH$$

Antibiotic 1233 $C_{18}H_{28}O_5$ (324)	*Cephalosporium* sp.	Obtained in two forms, A and B, believed to be C-12 epimers Melting point of A: 76–77°C Melting point of B: 88–94°C λ_{max} of A: 267 nm; $\epsilon = 12,150$ (methanol) ν_{max} of A: 3510, 1830, 1791, 1695, 1680, 1628 cm^{-1} (potassium bromide) ν_{max} of B: 3445, 3225, 1727, 1682, 1614 cm^{-1} (Nujol)	46

$$\text{HOCH}_2\,CH - CH(CH_2)_4\,CHCH_2\,(C = CH)_2\,COOH$$

Decylcitric acids $C_{16}H_{28}O_7$ (332)	*Penicillium spiculisporum*		47

$$\text{CH}_3(CH_2)_9\,CH - \overset{\overset{\displaystyle OH}{|}}{\underset{\underset{\displaystyle COOH}{|}}{C}} - CH_2COOH$$

(+)-Decylcitric acid		Melting point: 133–134°C $[a]_{364\ nm}$: +0.24° (acetone, c = 1.8) ν_{max}: 1680–1700 cm^{-1} Rf (TLC): 0.25 (benzene–dioxane– acetic acid, 90:25:4, on silica gel treated with oxalic acid)	
(–)-Decylcitric acid		$[a]_{364\ nm}$: –0.46° (acetone, c = 1.8) ν_{max}: 1680–1700 cm^{-1} Rf: 0.44	

Common Name Formula (MW) Structure	Organisms	Significant Characteristics	References

Hydroxy Acids (continued)

Spiculisporic acid
$C_{17}H_{28}O_6$ (328)

O —————— C = O
| |
$CH_3(CH_2)_9CH$ — C — CH_2CH_2
| |
COOH   COOH

Penicillium spiculosporum

Melting point: 145°C; hydrate
(opened lactone) 134–135°C
$[a]_{546\ nm}$: –14.8° (ethanol)
Rf (TLC): 0.63 (benzene–dioxane–
acetic acid, 90:25:4, on
silica gel treated with
oxalic acid)

47, 48

Minioluteic acid
$C_{16}H_{26}O_7$ (330)

O —————— C = O
| |
| OH |
| |
$CH_3(CH_2)_9CHC$ — CH
| |
COOH   COOH

Penicillium minioluteum

Melting point: 171°C
$[a]_D^{16}$: +108° (acetone)

49

Norcaperatic acid
$C_{20}H_{36}O_7$ (388)

OH
|
$CH_3(CH_2)_{13}CH$ — C — CH_2COOH
| |
COOH   COOH

Cantharellus floccosus
Polyporus fibrillosus

Melting point: 138–139°C;
potassium salt 173–174°C
ν_{max}: 1720, 1425 cm^{-1}
Gives cherry-red color and
intense green fluorescence
with acetic anhydride and
pyridine

50

Agaricic acid
$C_{22}H_{40}O_7$ (416)

OH
|
$CH_3(CH_2)_{15}CH$ — C — CH_2COOH
| |
COOH   COOH

Polyporus officinalis (Fomes
officinalis, F. laricis)

Melting point: 142°C with
loss of water and carbon
dioxide
$[a]_D^{19}$ of sodium salt: –8.84°
(water, c = 12.4)
Gives pink color and green
fluorescence when warmed
with acetic anhydride, and
deep-red color and green
fluorescence with pyridine–
acetic anhydride

51

Ungulinic acid
$C_{22}H_{38}O_6$ (398)

O —————— C = O
| |
| R″ |
| |
R′ — C — C — CH — R‴
| |
COOH   COOH

R′, R″, or R‴ = $C_{16}H_{33}$

Polyporus benzoinus

Melting point: 78–80°C; anhydride
53–54.5°C
Optically inactive at 546 nm
Gives pink color and green
fluorescence when warmed
with acetic anhydride, and
deep-red color and green
fluorescence with pyridine–
acetic anhydride

52

Common Name Formula (MW) Structure	Organisms	Significant Characteristics	References

Keto Acids

Common Name Formula (MW) Structure	Organisms	Significant Characteristics	References

Glyoxylic acid
$C_2H_2O_3$ (74)
 	
CHCOOH
$\parallel$
O
	Aspergillus niger
	Botrytis cinerea
		Deliquescent crystals
		Melting point: hemihydrate
		  70–75°C; anhydrous form
		  98°C; phenylhydrazone
		  145°C (decomposes)
		Chromatographic data:[6]
		  LC 40
		  PC 81
		  TLC 29, 32
			53

Pyruvic acid
$C_3H_4O_3$ (88)
CH_3CCOOH
$\parallel$
O
	Aspergillus niger
	Pseudomonas saccharophila
	Many other bacteria and fungi
		Boiling point: 165°C (decomposes)
		Melting point: 11.8°C; 2,4-
		  dinitrophenylhydrazone
		  215–216°C
		n_D^{20}:1.4138
		Chromatographic data:[6]
		  PC 77, 81
		  TLC 29, 134
			54–56

Dimethylpyruvic acid
(*a*-Ketoisovaleric acid)
$C_5H_8O_3$ (116)
$(CH_3)_2CHCCOOH$
$\parallel$
O
	Aspergillus spp.
	Piricularia oryzae
	Clostridium spp.
		Boiling point: 76–78°C
		Melting point: ~24°C; 2,4-
		  dinitrophenylhydrazone
		  194°C
		Chromatographic data:[6]
		  PC 81
		  TLC 32
			55, 56

Hydroxypyruvic acid
$C_3H_4O_4$ (104)
$HOCH_2CCOOH$
$\parallel$
O
	Lactobacillus spp.
	Propionibacter spp.
	Clostridium spp.
	Aspergillus niger
		Melting point: monohydrate
		  81–82°C (decomposes);
		  2,4-dinitrophenylhydrazone
		  162°C
		Chromatographic data:[6]
		  TLC 32
			55, 57

a-Ketocaproic acid
$C_6H_{10}O_3$ (130)
$CH_3(CH_2)_3CCOOH$
$\parallel$
O
	Clostridium spp.	Melting point of 2,4-dinitro-
		  phenylhydrazone: 134–136°C
		  (decomposes)
			58

a-Ketoisocaproic acid
$C_6H_{10}O_3$ (130)
$(CH_3)_2CHCH_2CCOOH$
$\parallel$
O
	Clostridium spp.	Oil
		Boiling point (15 mm): 84–85°C
		Melting point of 2,4-dinitro-
		  phenylhydrazone: 155–156°C
		Chromatographic data:[6]
		  PC 81
		  TLC 32
			55

Common Name
Formula (MW)
Structure | Organisms | Significant Characteristics | References

Keto Acids (continued)

α-Keto-β-methyl-*n*-valeric acid
$C_6H_{10}O_3$ (130)

$CH_3CH_2CH \quad CCOOH$
$\quad\quad\quad | \quad\quad ||$
$\quad\quad\quad CH_3 \quad O$

	Clostridium spp.	Melting point of 2,4-dinitro-phenylhydrazone: 170°C Chromatographic data:[6] TLC 32	55

Penicillic acid
$C_8H_{10}O_4$ (170)

$CH_3C - C - C \equiv CHCOOH$
$\quad\quad || \quad || \quad |$
$\quad\quad CH_2 \quad O \quad OCH_3$

$O \text{———} C = O$
$\quad | \quad\quad\quad |$
$CH_3C - C\cdot - C = CH$
$\quad\quad || \quad | \quad |$
$\quad\quad CH_2 \quad OH \quad OCH_3$

	Penicillium spp. *Aspergillus* spp.	Melting point: hydrate 64°C; anhydrous form 82–83°C λ_{max}: 225 nm (ethanol); ϵ = 10,500 Purple color develops on standing in ammonia	59

5-Hydroxy-4-ketohexanoic acid
$C_6H_{10}O_4$ (146)

$CH_3 \, CH - C - (CH_2)_2 \, COOH$
$\quad\quad\quad | \quad\; ||$
$\quad\quad\quad OH \quad O$

	Hansenula miso *Pichia mogii* *Candida albicans* *Candida tropicalis* *Saccharomyces oviformis*	Viscous oil Melting point of 2,4-dinitro-phenylosazone: 278–279°C $[a]_D$: –18° (water, c = 7.5)	60

5-Acetoxy-4-ketohexanoic acid
$C_8H_{12}O_5$ (188)

$\quad\quad OCOCH_3$
$\quad\quad |$
$CH_3CH - C - (CH_2)_2 \, COOH$
$\quad\quad\quad || $
$\quad\quad\quad O$

	Hansenula miso *Candida albicans* *Saccharomyces oviformis*	Oil Melting point of *p*-phenyl-phenacyl ester: 108°C $[a]_D^{20}$: –18° (water, c = 2)	61

Sarkomycin
$C_7H_8O_3$ (140)

	Streptomyces erythrochromogenes	Oil $[a]_D^{15}$: –32.5° (methanol) λ_{max}: 230 nm (water) ν_{max}: 1724, 1639 cm^{-1}	62

Xanthocidin
$C_{11}H_{16}O_4$ (212)

	Streptomyces spp.	Melting point: 185°C (decomposes) $[a]_D^{25}$: +16.7° λ_{max}: 227 nm (methanol); ϵ = 6,420	63

Common Name Formula (MW) Structure	Organisms	Significant Characteristics	References

Keto Acids (continued)

Jasmonic acid

$C_{12}H_{18}O_3$ (210)

Lasiodiplodia theobromae

Oil
Boiling point (0.001 mm):
  125°C
n_D^{19}: 1.4885
$[a]_D$: −83.5° (chloroform,
  c = 0.97)

22

Lactarinic acid

$C_{18}H_{34}O_3$ (298)

$CH_3(CH_2)_{11}\underset{\underset{O}{\|}}{C}(CH_2)_4 COOH$

Lactarius rufus
Other *Lactarius* spp.

Melting point: 87°C; oxime
59−61°C

15

Alternaric acid

$C_{21}H_{30}O_8$ (410)

Alternaria solani

Melting point: monohydrate
  135−136°C; anhydrous form
  138°C
λ_{max}: 210, 274 nm
Orange color develops with
  ferric chloride

64

8,9,13-Triacetoxydocosan-
  oic acid

$C_{28}H_{50}O_8$ (514)

$CH_3(CH_2)_8\underset{\underset{OCOCH_3}{|}}{CH}(CH_2)_3\underset{\underset{OCOCH_3}{|}}{CH}\underset{\underset{OCOCH_3}{|}}{\text{————}CH}(CH_2)_6 COOH$

Yeast

Liquid
$[a]_D^{20}$: −3° (hexane, c = 2.9)
ν_{max}: 1739, 1709, 1370,
  1239, 1020 cm^{-1} (silver
  chloride)

65

Oxalacetic acid

$C_4H_4O_5$ (132)

$HOOCCH_2\underset{\underset{O}{\|}}{C}COOH$

Clostridium spp.

Melting point: *trans*-enol
  form 184°C; *cis*-enol form
  152°C
Chromatographic data:[6]
  PC 81
  TLC 36

55

a-Ketoglutaric acid

$C_5H_6O_5$ (146)

$HOOC(CH_2)_2\underset{\underset{O}{\|}}{C}COOH$

Pseudomonas fluorescens
Pseudomonas saccharophila
Escherichia coli
Proteus vulgaris
Aerobacter aerogenes
Serratia marcescens
Bacillus spp.

Melting point: 115−116°C;
2,4-dinitrophenylhydrazone
  219°C
Chromatographic data:[6]
  PC 13
  TLC 36

9, 56, 66

4-Ketododecan-1,12-dioic
  acid

$C_{12}H_{20}O_5$ (244)

$HOOC(CH_2)_7\underset{\underset{O}{\|}}{C}(CH_2)_2 COOH$

Gibberella fujikuroi

Melting point: 111−112.5°C;
  dimethyl ester 35−36°C
ν_{max}: 3580, 3040, 1710,
  1235, 1080, 940 cm^{-1}

67

Common Name Formula (MW) Structure	Organisms	Significant Characteristics	References

Sugar Acids

Pentonic acids
$C_5H_{10}O_6$ (166)

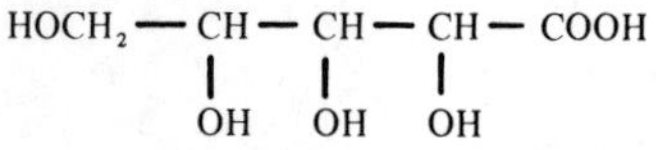

HOCH$_2$ — CH — CH — CH — COOH
           | | |
          OH   OH   OH

34, 68

D-Arabonic acid	*Fusarium lini* *Pseudomonas* spp. *Acetobacter* spp.	Syrup; forms lactone on heating Mealting point: lactone 98–99°C; S-benzylthiouronium salt 144–145°C $[a]_D^{20}$: of lactone: +73.7° (water) on heating	
L-Arabonic acid	*Fusarium lini* *Pseudomonas* spp. *Acetobacter* spp.	Readily forms lactone Melting point: ~118–119°C; lactone 98–100°C $[a]_D$: –10.02° (water) Chromatographic data:[6] PC 72	
D-Ribonic acid	*Pseudomonas* sp.	Forms γ-lactone Melting point: 112–113°C; γ-lactone 80°C $[a]_D^{25}$: –17° (water) Chromatographic data:[6] LC 40	
D-Xylonic acid	*Penicillium corylophilum* *Fusarium lini*	Syrup Melting point: γ-lactone 99–103°C; brucine salt 176°C $[a]_D$: +17.98° (water) Chromatographic data:[6] LC 40	

Hexonic acids
$C_6H_{12}O_7$ (196)

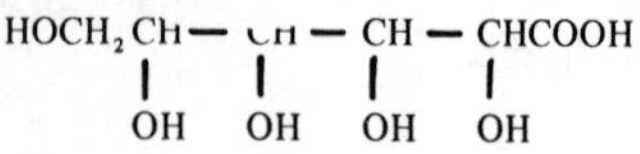

HOCH$_2$Ch — ᴄʜ — CH — CHCOOH
           | | | |
          OH   OH   OH   OH

D-Galactonic acid	*Aspergillus niger* *Aerobacter aerogenes* *Acetobacter* spp.	Lactonizes on heating Melting point: 145–146°C $[a]_D^{20}$: –12.23° Chromatographic data:[6] LC 40	34, 69

Common Name Formula (MW) Structure	Organisms	Significant Characteristics	References

Sugar Acids (continued)

Hexonic acids (*cont.*)

D-Gluconic acid	*Aspergillus niger* *Pseudomonas fragi* Other organisms	Syrup; aqueous solution is an equilibrium mixture with γ- and δ-lactones	9, 10, 17, 34
		Melting point: γ-lactone 134–136°C; δ-lactone 153°C	34, 69
		$[a]_D$: γ-lactone + 67.5°, changing to +6.2°; δ-lactone +63.5°, changing to +6.2°	34, 69
		Chromatographic data:[6] LC 23, 40 (D-gluconic acid) PC 72, 76 (γ- and δ-lactones)	
D-Mannonic acid	*Aspergillus niger* *Penicillium* sp. *Acetobacter xylinum* *Aerobacter aerogenes*	Aqueous solution lactonizes readily Melting point: γ-lactone 151°C; brucine salt 212°C $[a]_D^{20}$ of γ-lactone: +51.8° (water) Chromatographic data:[6] LC 40 (D-mannonic acid) PC 76 (γ-lactone)	69

Bionic acids
$C_{12}H_{22}O_{13}$ (374)

34, 70, 71

HOCH₂CH — CH — CH — CH — CH — O … structural formula

Cellobionic acid	*Pseudomonas quercito-pyrogal- lica* *Pseudomonas calco-acetica* *Pseudomonas aromatica*	Syrup; forms insoluble calcium salt Chromatographic data:[6] LC 40	
Lactobionic acid	*Pseudomonas graveolens* *Pseudomonas aromatica* *Pseudomonas calco-acetica* *Pseudomonas woodsii* *Pseudomonas gluconicum* *Pseudomonas quercito-pyrogal- lica*	Syrup; forms crystalline calcium salt ($5H_2O$) Melting point of δ-lactone: 195–196°C $[a]_D^{20}$ of calcium salt: +23.7° (water) Chromatographic data:[6] LC 40	

Common Name Formula (MW) Structure	Organisms	Significant Characteristics	References

Sugar Acids (continued)

Bionic acids (*cont.*)

    Maltobionic acid

Pseudomonas graveolens
Pseudomonas quercito-pyrogal-lica
Pseudomonas calco-acetica
Pseudomonas aromatica

Syrup
Melting point of brucine salt: 153°C
$[a]_D^{20}$: +98.3° (water)
Chromatographic data:[6]
  LC 40

    Melibionic acid

Pseudomonas calco-acetica
Pseudomonas aromatica
Paracolon bacteria

Syrup; forms amorphous calcium salt
Boiling point of methyl ester (0.06 mm): 173–175°C
n_D^{14} of methyl ester: 1.4640
$[a]_D^{13}$ of methyl ester: +106.4° (water)
Chromatographic data:[6]
  LC 40

D-Lyxuronic acid
$C_5H_8O_6$ (164)

$$HOOC - \underset{\underset{OH}{|}}{\overset{\overset{H}{|}}{C}} - \underset{\underset{H}{|}}{\overset{\overset{OH}{|}}{C}} - \underset{\underset{H}{|}}{\overset{\overset{OH}{|}}{C}} - CHO$$

Acetobacter melanogenum

$[a]_D^{20}$ of calcium salt dihydrate: –23° (water), changing to –53° in 30 minutes
Chromatographic data:[6]
  LC 23

72

D-Glucuronic acid
$C_6H_{10}O_7$ (194)

$$HOOC - \underset{\underset{OH}{|}}{\overset{\overset{H}{|}}{C}} - \underset{\underset{OH}{|}}{\overset{\overset{H}{|}}{C}} - \underset{\underset{H}{|}}{\overset{\overset{OH}{|}}{C}} - \underset{\underset{OH}{|}}{\overset{\overset{H}{|}}{C}} - CHO$$

Penicillium brevi-compactum
Ustulina vulgaris

Melting point: 165°C
$[a]_D^{24}$: +11.7° (water, c = 1), changing to +36.3° in 2 hours
Chromatographic data:[6]
  LC 71
  PC 72, 76

73

Saccharic acid
$C_6H_{10}O_8$ (210)

$$HOOC - \underset{\underset{OH}{|}}{\overset{\overset{H}{|}}{C}} - \underset{\underset{OH}{|}}{\overset{\overset{H}{|}}{C}} - \underset{\underset{H}{|}}{\overset{\overset{OH}{|}}{C}} - \underset{\underset{OH}{|}}{\overset{\overset{H}{|}}{C}} - COOH$$

Aspergillus niger

Melting point: 125°C
$[a]_D^{19}$: +6.86° (water, c = 1), changing to +20.6°
Chromatographic data:[6]
  LC 40

74

Common Name Formula (MW) Structure	Organisms	Significant Characteristics	References

Sugar Acids (continued)

Ascorbic acids
$C_6H_8O_6$ (176)

$$HOCH_2CH - CH - \overset{\overset{OH}{|}}{C} = \overset{\overset{OH}{|}}{C}$$
$$\;\;\;\;\;\;\;\;|\;\;\;\;\;\;|$$
$$\;\;\;\;\;\;\;\;OH\;\;\;\;O \rule{1cm}{0.4pt} C = O$$

Common Name	Organisms	Significant Characteristics	References

L-Xyloascorbic acid
(Ascorbic acid)
	Aspergillus niger
	Aspergillus tamarii
		Melting point: 190–192°C
		(decomposes); 2,4-dinitro-
		phenylhydrazone 282°C
		(decomposes)
		$[a]_D^{23}$: +48° (methanol,
		c = 1)
		λ_{max}: 245 nm (acid); 265 nm
		(neutral)
		Chromatographic data:[6]
		  LC 20, 29
		  PC 13
		  TLC 188
			75

D-Araboascorbic acid (Isoascorbic acid)	*Penicillium* sp.	Melting point: 172°C; 2,4- dinitrophenylhydrazone 238–242°C $[a]_D^{25}$: –23° (water, c = 1) λ_{max}: 265 nm (acid); 245 nm (neutral)	76

4-Keto-D-arabonic acid
$C_5H_8O_6$ (164)

$$HOCH_2\overset{\overset{}{\underset{\underset{O}{\|}}{C}}}{} - \overset{\overset{H}{|}}{\underset{\underset{OH}{|}}{C}} - \overset{\overset{OH}{|}}{\underset{\underset{H}{|}}{C}} - COOH$$

Common Name	Organisms	Significant Characteristics	References
4-Keto-D-arabonic acid	*Acetobacter suboxydans*	Melting point of brucine salt: 154–155°C $[a]_D^{20}$: –10.26°; calcium salt –29°	34, 77

2-Ketohexonic acids
$C_6H_{10}O_7$ (194)

$$HOCH_2\overset{\overset{}{\underset{\underset{OH}{|}}{CH}}}{} - \overset{}{\underset{\underset{OH}{|}}{CH}} - \overset{}{\underset{\underset{OH}{|}}{CH}} - \overset{}{\underset{\underset{O}{\|}}{CCOOH}}$$

Common Name	Organisms	Significant Characteristics	References
2-Keto-D-galactonic acid	*Pseudomonas aeruginosa* *Pseudomonas fluorescens* *Aerobacter aerogenes*	Melting point: 170–171°C; potassium salt 139–140°C	34, 78

2-Keto-D-gluconic acid
(2-Keto-D-mannonic acid)
	Acetobacter suboxydans
	Penicillium brevi-compactum
	Pseudomonas spp.
	Other bacteria
		Melting point: 152°C; phenyl-
		hydrazone 163°C
		$[a]_D$: –99.62°
		Chromatographic data:[6]
		  LC 40
		  PC 72
			9, 27, 34,
			38, 66, 73

Common Name Formula (MW) Structure	Organisms	Significant Characteristics	References

Sugar Acids (continued)

2-Ketohexonic acids (*cont.*)

Common Name Formula (MW) Structure	Organisms	Significant Characteristics	References
2-Keto-L-gulonic acid (2-Keto-L-idonic acid)	*Acetobacter suboxydans* *Aerobacter aerogenes* *Aerobacter* spp. *Pseudomonas* spp. *Serratia* spp. Other bacteria	Melting point: 171°C $[a]_D^{15}$: –48° (water, c = 1)	34, 79

5-Ketohexonic acids
$C_6H_{10}O_7$ (194)

HOCH$_2$C — CH — CH — CHCOOH
‖ | | |
O   OH OH OH

Common Name Formula (MW) Structure	Organisms	Significant Characteristics	References
5-Keto-D-gluconic acid (5-Keto-L-idonic acid)	*Acetobacter* spp. *Pseudomonas* sp. *Aspergillus niger*	Syrup; forms insoluble calcium salt (3H$_2$O) $[a]_D$: –14° (water, c = 2) Chromatographic data:[6] LC 40 PC 72	9, 27, 34, 38, 80
5-Keto-D-mannonic acid (D-Fructuronic acid)	*Gluconobacter cerinus* var. *ammoniacus*		34, 81

2-Keto-3-deoxygluconic acid
$C_6H_{10}O_6$ (178)

     H   H
     | |
HOCH$_2$C — CCH$_2$CCOOH
     | | ‖
    OH OH O

Common Name Formula (MW) Structure	Organisms	Significant Characteristics	References
2-Keto-3-deoxygluconic acid	*Rhodopseudomonas spheroides*	Syrup; forms crystalline calcium salt R_f (PC): 0.49 (acetone–pyridine–water, 2:1:1)	34, 82

2,5-Diketogluconic acid
$C_6H_8O_7$ (192)

      H   OH
      | |
HOCH$_2$C — C — C — CCOOH
    ‖ | | ‖
    O OH H O

Common Name Formula (MW) Structure	Organisms	Significant Characteristics	References
2,5-Diketogluconic acid	*Acetobacter melanogenum* *Gluconobacter liquefaciens* *Pseudomonas albosesamae*	Unstable; reduces Benedict or silver nitrate solution without heating λ_{max} of amorphous calcium salt: 280 nm (acid); 305 nm (neutral) $R_{glucose}$ (PC): 0.59 (*n*-butanol–acetic acid–water, 4:1:5)	34, 83

Glycosylated Acids

Pyolipic acid
$C_{16}H_{30}O_7$ (344)
rhamnose — O
       |
CH$_3$(CH$_2$)$_6$CHCH$_2$COOH

Common Name Formula (MW) Structure	Organisms	Significant Characteristics	References
Pyolipic acid	*Pseudomonas pyocyanea*	Viscous oil; lead salt is soluble in diethyl ether and ethanol, insoluble in water	84

Common Name Formula (MW) Structure	Organisms	Significant Characteristics	References

Glycosylated Acids (continued)

Ustilagic acids

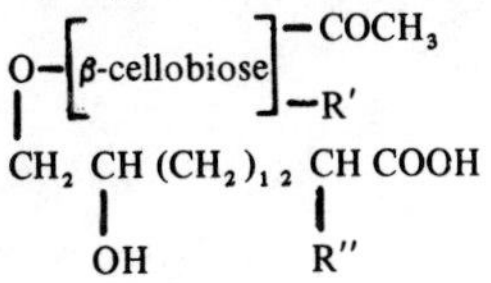

O—[β-cellobiose]—COCH₃ / —R′

$$CH_2\ CH\ (CH_2)_{12}\ CH\ COOH$$
| | OH ... R″

Ustilago zeae

Melting point: $146-147°C$
$[a]_D^{23}$: $+7°$ (pyridine, c = 1)

10, 85

Ustilagic acid I
$C_{36}H_{64}O_{17}$ (768)

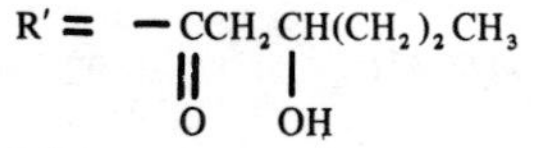

R′ = —CCH₂CH(CH₂)₂CH₃
‖ |
O OH

R″ = H

Ustilagic acid II
$C_{37}H_{66}O_{18}$ (798)

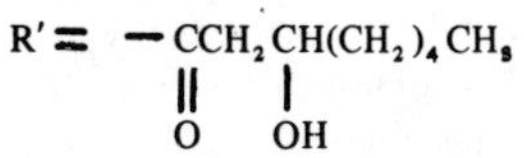

R′ = —CCH₂CH(CH₂)₄CH₃
‖ |
O OH

R″ = CH

Nitrogenous Acids

β-Nitropropionic acid
$C_3H_5O_4N$ (119)
$O_2NCH_2CH_2COOH$

Aspergillus flavus
Aspergillus oryzae
Aspergillus wentii
Aspergillus avenaceus
Penicillium atrovenetum

Melting point: $68-69°C$

86

Antibiotic U-20904
$C_4H_6O_2N_2$ (114)
$H_2NCCH = CHCOOH$
‖
NH

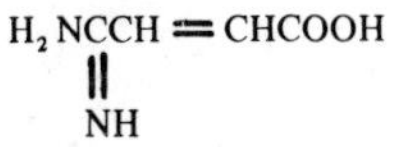

Actinomyces spp.

Decomposes above $223°C$
λ_{max}: 211 nm (water)

87

Antibiotic U-22956
$C_5H_7O_4N$ (145)
$CH_3ONCHCH = CHCOOH$
↓
O

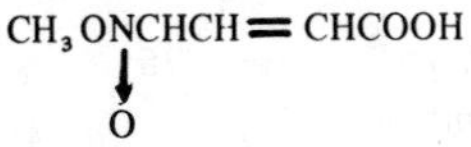

Streptoverticillium fervens var.
    mebrosporus

Melting point: $130-132°C$
λ_{max}: 290 nm; $\epsilon = 152$
    (ethanol)

88

Fumaryl-DL-alanine
$C_7H_9O_5N$ (187)
$CH_3CHNH — CCH = CHCOOH$
| ‖
COOH O

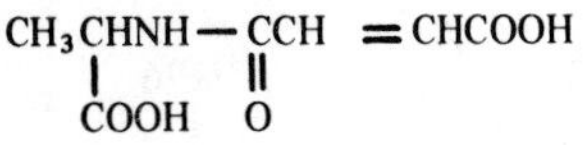

Penicillium resticulosum

Melting point: $229°C$
Gives a strong fluorescein
    reaction when heated with
    resorcinol and concentrated
    sulfuric acid

89

	Common Name Formula (MW) Structure	Organisms	Significant Characteristics	References

Nitrogenous Acids (continued)

Desthiobiotins

Desthiobiotin $C_{10}H_{18}N_2O_3$ (214) R = H	Various microorganisms	Melting point: 156–158°C $[a]_D^{21}$: +10.7° (water, c = 2)	90, 91

a-Methyldesthiobiotin $C_{11}H_{20}N_2O_3$ (228) R = –CH$_3$	*Streptomyces lydicus*	Melting point: 161–162.5°C	92

Jasmonoylisoleucines   *Gibberella fujikuroi*

Isolated as a mixture
Melting point: 147–149°C
ν_{max}: 3439, 1739, 1679, 1673, 1506 cm^{-1} (bromoform)
Hydrogenation produced pure dihydrojasmonoylisoleucine   93

Jasmonoylisoleucine (I) $C_{18}H_{29}O_4N$ (323) R = –CH$_2$ CH=CHCH$_2$ CH$_3$

Dihydrojasmonoylisoleucine (II) $C_{18}H_{31}O_4N$ (325) R = –(CH$_2$)$_4$ CH$_3$		Melting point: 140–141.5°C λ_{max}: 291 nm; ϵ = 33 ν_{max}: 3430, 1733, 1673, 1500 cm^{-1} (chloroform)	

Sulfur-Containing Acids

S-Methylthiopropionic acid $C_4H_8O_2S$ (120) CH$_3$ S(CH$_2$)$_2$COOH	*Neurospora crassa* mutant	R_f (PC): 0.85 (butanol–acetic acid–water, 12:3:5); 0.32 (n-amyl alcohol–5M formic acid, 1:1); 0.68 (ethyl acetate–acetic acid –water, 3:1:1)	94

S-Ethylthiopropionic acid $C_5H_{10}O_2S$ (134) CH$_3$ CH$_2$ S(CH$_2$)$_2$ COOH	*Escherichia coli*	R_f (PC): 0.54 (phenol–water, 4:1); 0.63 (butanol–propionic acid–water, 10:5:7) Mass spectrum: peaks at m/e 134, 117, 73, 61	95

trans-3-(Methylthio)acrylic acid $C_4H_6O_2S$ (118) CH$_3$ SCH=CHCOOH	*Streptomyces lincolnensis*	Melting point: 140°C λ_{max}: 269 nm (methanol); ϵ = 16,780 δ (in acetone-d$_6$): 2.40 (s, 3H), 5.82 (d, 1H, J = 15 Hz), 7.80 (d, 1H, J = 15 Hz), 9.5 (s, 1H)	96

Common Name Formula (MW) Structure	Organisms	Significant Characteristics	References

Sulfur-Containing Acids (continued)

α-Hydroxy-γ-methylthiobutyric
  acid
$C_5H_{10}O_3S$ (134)
$CH_3S(CH_2)_2\underset{\underset{OH}{|}}{CHCOOH}$

			97
D-Isomer	*Oidium lactis*	$[a]_D^{18}$ of zinc salt: +32.8° (water, c = 0.34)	
L-Isomer	*Bacillus subtilis*	$[a]_D^{15}$ of zinc salt: −31.7° (water, c = 1.12)	

α-Keto-γ-methylthiobutyric   *Clostridium* spp.   Unstable, releasing methylmercaptan   55
  acid   Melting point of 2,4-dinitro-
$C_5H_8O_3S$ (148)   phenylhydrazone: 151°C,
$CH_3S(CH_2)_2\underset{\underset{O}{|}}{CCOOH}$

Lipoic acid   Yeast   Melting point: 47°C   98
$C_8H_{14}O_2S_2$ (206)   *Escherichia coli* mutant $[a]_D^{25}$: +96.7° (benzene,
$\underset{\underset{S\text{——}S}{|\quad\quad|}}{CH_2CH_2CH(CH_2)_4COOH}$ c = 2)

Biotins

$$\underset{CH_2SCHR}{\underset{|}{\underset{\underset{\displaystyle CH\text{—}CH}{|\quad\;|}}{\underset{\displaystyle NH\;\;CNH}{|\quad\;|}}}}\;\overset{\displaystyle O}{\overset{\displaystyle \|}{}}$$

Biotin $C_{10}H_{16}O_3N_2S$ (244) R = −(CH$_2$)$_4$COOH	*Phycomyces blakesleeanus*	Melting point: 232−233°C $[a]_D^{21}$: +91° (0.1N sodium hydroxide, c = 1)	91
α-Dehydrobiotin $C_{10}H_{14}O_3N_2S$ (242) R = −(CH$_2$)$_2$CH=CHCOOH	*Streptomyces lydicus*	Melting point: 238−240°C $[a]_D^{25}$: +92° (0.1N sodium hydroxide) ν_{max}: 1645, 985 cm^{-1}	99
α-Methylbiotin $C_{11}H_{18}O_3N_2S$ (258) R = −(CH$_2$)$_3$$\underset{\underset{CH_3}{|}}{CHCOOH}$	*Streptomyces lydicus*	Melting point: 186−188°C (synthetic racemate)	92

Biocytin
$C_{16}H_{28}O_4N_4S$ (376)
R = −(CH$_2$)$_4$CONH(CH$_2$)$_4$$\underset{\underset{NH_2}{|}}{CHCOOH}$
	Phycomyces blakesleeanus	Melting point: 241−243°C
		  (variable)
		$[a]_D^{25}$: +53° (0.1N sodium
		  hydroxide, c = 1.05)
			91

Common Name Formula (MW) Structure	Organisms	Significant Characteristics	References

Sulfur-Containing Acids (continued)

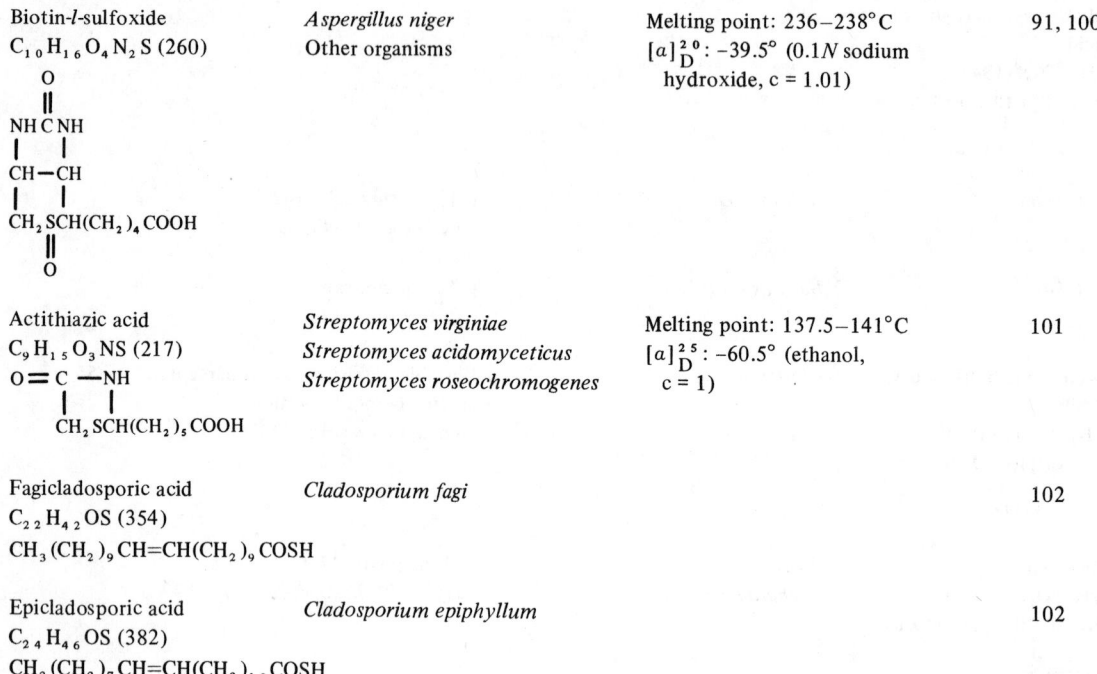

Biotin-*l*-sulfoxide
$C_{10}H_{16}O_4N_2S$ (260)

Aspergillus niger
Other organisms

Melting point: 236–238°C
$[a]_D^{20}$: –39.5° (0.1N sodium hydroxide, c = 1.01)

91, 100

Actithiazic acid
$C_9H_{15}O_3NS$ (217)

Streptomyces virginiae
Streptomyces acidomyceticus
Streptomyces roseochromogenes

Melting point: 137.5–141°C
$[a]_D^{25}$: –60.5° (ethanol, c = 1)

101

Fagicladosporic acid
$C_{22}H_{42}OS$ (354)
$CH_3(CH_2)_9CH=CH(CH_2)_9COSH$

Cladosporium fagi

102

Epicladosporic acid
$C_{24}H_{46}OS$ (382)
$CH_3(CH_2)_7CH=CH(CH_2)_{13}COSH$

Cladosporium epiphyllum

102

REFERENCES

1. Stecher, P. G. (Ed.), *The Merck Index,* 8th ed. Merck and Co., Rahway, New Jersey (1968).
2. Cook, A. H., Bunbury, H. M., and Hey, R. D. (Eds.), *Dictionary of Organic Compounds,* 4th ed. Oxford University Press, Fair Lawn, New Jersey (1965).
3. Miller M. W., *The Pfizer Handbook of Microbial Metabolites.* McGraw-Hill, New York (1961).
4. Rodd, E. H., *Rodd's Chemistry of Carbon Compounds,* 2nd ed., S. Coffey, Ed. American Elsevier, New York (1964).
5. Turner, W. B., *Fungal Metabolites.* Academic Press, New York (1971).
6. Zweig, G., and Sherma, J., *Handbook of Chromatography,* Vol. 1. CRC Press, Cleveland, Ohio (1972).
7. Wood, W. A., in *The Bacteria,* Vol. 2, p. 59, I. C. Gunsalus and R. Y. Stanier, Eds. Academic Press, New York (1961).
8. Barker, H. A., in *The Bacteria,* Vol. 2, p. 151, I. C. Gunsalus and R. Y. Stanier, Eds. Academic Press, New York (1961).
9. Underkofler, L. A., and Hickey, R. J., *Industrial Fermentations,* Vol. 1 and Vol. 2. Chemical Publishing Company, New York (1954).
10. Prescott, S. C., and Dunn, C. G., *Industrial Microbiology,* 2nd ed. McGraw-Hill, New York (1949).
11. Voisenet, E., *Compt. Rend., 151,* 518 (1910).
12. Hardman, J. K., Stadtman, T. C., and Szulmajster, J., *Bacteriol. Proc.,* p. 120 (1958).
13. Cram, D. J., and Tishler, M., *J. Am. Chem. Soc., 70,* 4238 (1948).
14. Prista, N., *An. Fac. Farm. Porto, 14,* 19 (1954).
15. Bougault, J., and Charaux, C., *J. Pharm. Chim., 5,* 65 (1912).
16. Foster, J. W., *Chemical Activities of Fungi.* Academic Press, New York (1949).
17. Martin, S. M., in *Biochemistry of Industrial Microorganisms,* p. 415, C. Rainbow and A. H. Rose, Eds. Academic Press, New York (1963).

18. Takao, S., *Appl. Microbiol., 13,* 732 (1965).
19. Nakamura, Y., Shimomura, T., and Ono, J. *J. Agric. Chem. Soc. Jap., 32,* 800 (1958).
20. Harada, T., and Hirabayashi, T., *Agric. Biol. Chem., 32,* 1175 (1968).
21. Kersters, K., and De Ley, J., *Biochim. Biophys. Acta, 71,* 311 (1963).
22. Aldridge, D. C., Galt, S., Giles, D., and Turner, W. B., *J. Chem. Soc. Sect. C,* p. 1623 (1971).
23. Baba, S., and Sakaguchi, K., *Bull. Agric. Chem. Soc. Jap., 18,* 93 (1942).
24. Horler, D. F., McConnell, W. B., and Westlake, D. W. S., *Can. J. Microbiol., 12,* 1247 (1966).
25. Bentley, R., and Thiessen, C. P., *J. Biol. Chem., 226,* 703 (1957).
26. Singh, P., and Rangaswami, S., *Tetrahedron Lett.,* p. 149 (1967).
27. Yamada, K., Minoda, Y., Kodama, T., and Kotera, U., in *Fermentation Advances,* D. Perlman, Ed. Academic Press, New York (1969).
28. Miller, M. W., *The Pfizer Handbook of Microbial Metabolites,* p. 58. McGraw-Hill, New York (1961).
29. Meyerhof, O., and Kiessling, W., *Biochem. Z., 276,* 239 (1935).
30. Camien, M. N., Fowler, A. V., and Dunn, M. S., *Arch. Biochem. Biophys., 83,* 408 (1959).
31. Sjolander, J. R., Folkers, K., Adelberger, E. A., and Tatum, E. L., *J. Am. Chem. Soc., 76,* 1085 (1954).
32. Wolf, D. E., Hoffman, C. H., Aldrich, P. E., Skeggs, H. R., Wright, L. D., and Folkers, K., *J. Am. Chem. Soc., 79,* 1486 (1957).
33. Kawarada, A., Takahashi, N., Kitamura, H., Seta, Y., Takai, M., and Tamura, S., *Bull. Agric. Chem. Soc. Jap., 19,* 84 (1955).
34. Spencer, J. F. T., and Gorin, P. A. J., *Prog. Ind. Microbiol., 7,* 177 (1968).
35. Gibson, F., and Jackson, L. M., *Nature, 198,* 388 (1963).
36. Plieninger, H., *Angew. Chem. Int. Ed. Engl., 1,* 367 (1962).
37. Yoshioka, L., and Yamamoto, T., *Annu. Meet. Pharm. Soc. Jap.* (1962).
38. Kulka, D., Hall, A. N., and Walker, T. W., *Nature, 167,* 905 (1951).
39. Abe, S., Sito, T., and Takayama, K., *Nippon Nogei Kagaku Kaishi, 34,* 66 (1960); *Chem. Abstr., 58,* 5006 (1961).
40. Virtanen, A. I., and Erkama, J., *Nature, 142,* 954 (1938).
41. Birkinshaw, J. H., Bracken, A., and Raistrick, H., *Biochem. J., 39,* 70 (1945).
42. Stodola, F. H., Friedkin, M., Moyer, A. J., and Coghill, R. D., *J. Biol. Chem., 161,* 739 (1945).
43. Sakaguchi, K., and Beppu, T., *Arch. Biochem. Biophys., 83,* 131 (1959).
44. Umbreit, W. W., *J. Bacteriol., 66,* 74 (1953).
45. Nakajima, S., Kinoshita, K., and Shibata, S., *Chem. Ind.,* p. 805 (1964).
46. Aldridge, D. C., Giles, D., and Turner, W. B., *J. Chem. Soc.,* p. 3888 (1971).
47. Gatenbeck, S., and Mahlen, A., *Acta Chem. Scand., 22,* 2613 (1968).
48. Clutterbuck, P. W., Raistrick, H., and Rintoul, M. L., *Philos. Trans. R. Soc. London Ser. B, 220,* 301 (1931).
49. Birkinshaw, J. H., and Raistrick, H., *Biochem. J., 28,* 828 (1934).
50. Miyata, J. T., Tyler, V. E., Brady, L. R., and Malone, M. H., *Lloydia, 29,* 43 (1966).
51. Thomas, H., and Vogelsang, J., *Ann. Chem. (Justus Liebig's), 357,* 145 (1907).
52. Birkinshaw, J. H., Morgan, E. N., and Findlay, W. P. K., *Biochem. J., 50,* 509 (1952).
53. Challenger, F., Subramaniam, V., and Walker, T. K., *J. Chem. Soc.,* p. 200 (1927).
54. Entner, N., and Doudoroff, M., *J. Biol. Chem., 196,* 853 (1952).
55. Rosenberg, A. J., and Nisman, B., *Biochim. Biophys. Acta, 3,* 348 (1949).
56. Katsuki, H., *J. Am. Chem. Soc., 77,* 4686 (1955).
57. Behal, F. J., *Arch. Biochem. Biophys., 88,* 110 (1960).
58. Nisman, B., and Vinet, G., *Ann. Inst. Pasteur, 77,* 276 (1949).
59. Birkinshaw, J. H., Oxford, A. E., and Raistrick, H., *Biochem. J., 30,* 394 (1936).
60. Hirabayashi, T., and Harada, T., *Agric. Biol. Chem., 33,* 276 (1969).
61. Hirabayashi, T., and Harada, T., *Agric. Biol. Chem., 33,* 1074 (1969).
62. Hooper, I. R., Cheney, L. C., Cron, M. J., Fardig, O. B., Johnson, D. A., Johnson, D. L., Palermiti, F. M., Schmitz, H., and Wheatley, W. B., *Antibiot. Chemother., 5,* 585 (1955).
63. Asahi, K., and Suzuki, S., *Agric. Biol. Chem., 34,* 325 (1970).
64. Grove, J. F., *J. Chem. Soc.,* p. 4056 (1952).
65. Stodola, F. H., Vesonder, R. F., and Wickerman, L. J., *Biochemistry, 4,* 1390 (1965).
66. Koepsell, H. J., Stodola, F. H., and Sharpe, E. S., *J. Am. Chem. Soc. 74,* 5142 (1952).
67. Serebryakov, E. P., Simolin, A. V., Kucherov, V. F., and Rozynov, B. V., *Tetrahedron, 26,* 5215 (1970).
68. Lockwood, L. B., and Nelson, G. E. N., *J. Bacteriol., 52,* 581 (1946).
69. Knoblock, H., and Mayer, H., *Biochem. Z., 307,* 285 (1941).
70. Stodola, F. H., and Lockwood, L. B., *J. Biol. Chem., 171,* 213 (1947).
71. Bentley, R., and Slechta, L., *J. Bacteriol., 79,* 346 (1960).
72. Ameyama, M., and Kondo, K., *Bull. Agric. Chem. Soc. Jap., 22,* 271 (1958).
73. Simonart, P., and Godin, P., *Bull. Soc. Chim. Belg., 60,* 446 (1951).
74. Walker, T. K., Subramaniam, V., and Challenger, F., *J. Chem. Soc.,* p. 3044 (1927).
75. Geiger-Huber, M., and Galli, H., *Helv. Chim. Acta, 28,* 248 (1945).

76. Takahashi, T., Mitsumoto, M., and Kayamori, H., *Nature, 188,* 411 (1960).
77. Liebster, J., Kulhanek, M., and Tadra, M., *Chem. Listy, 47,* 1075 (1953); *Chem. Abstr., 48,* 4044 (1954).
78. Asai, T., Aida, K., and Ueno, Y., *J. Agric. Chem. Soc. Jap., 26,* 625 (1952); *Chem. Abstr., 48,* 12882 (1954).
79. Tengerdy, R. P., *J. Biochem. Microbiol. Technol. Eng., 3,* 241, 255 (1961).
80. De Ley, J., *J. Appl. Bacteriol., 23,* 400 (1960).
81. Terada, O., Suzuki, S., and Kinoshita, S., *Nippon Nogai Kagaku Kaishi, 36,* 212, 217 (1962); *Chem. Abstr., 61,* 7399 (1963).
82. Szymona, M., and Doudoroff, M., *J. Gen. Microbiol., 22,* 167 (1960).
83. Katznelson, H., Tanenbaum, S. W., and Tatum, E. L., *J. Biol. Chem., 204,* 43 (1953).
84. Bergstrom, S., Theorell, H., and Davide, H., *Arch. Biochem., 10,* 165 (1946).
85. Lemieux, R. U., Thorn, J. A., and Bauer, H. F., *Can. J. Chem., 31,* 1054 (1953).
86. Raistrick, A., and Stoessl, A., *Biochem., J., 68,* 647 (1958).
87. Wiley, P. F., and Daniels, E. G., *Antimicrob. Agents Chemother.,* p. 812 (1966).
88. Wiley, P. G., Herr, R. R., Mackellar, F. A., and Argoudelis, A. D., *J. Org. Chem., 30,* 2330 (1965).
89. Birch, A. J., Quereshi, A. A., and Rickards, R. W., *Aust. J. Chem., 21,* 2775 (1968).
90. Ogata, K., Izumi, Y., and Tani, Y., *Agric. Biol. Chem., 37,* 1079 (1973).
91. Eisenberg, M. A., *J. Bacteriol., 86,* 673 (1963).
92. Martin, D. G., Hanka, L. J., and Reineke, L. M., *Tetrahedron Lett.,* p. 3791 (1971).
93. Cross, B. E., and Webster, G. R. B., *J. Chem. Soc. Sect. C,* p. 1839 (1970).
94. Galsworthy, S. B., and Metzenberg, R. L., *Biochemistry, 4,* 1183 (1965).
95. Faith, W. T., and Mallette, M. F., *Arch. Biochem. Biophys., 117,* 75 (1966).
96. Visser, J., and Meyer, H. F., *J. Antibiot., 22,* 510 (1969).
97. Akobe, K., *Z. Physiol. Chem., 244,* 14 (1936).
98. Reed, L. J., Gunsalus, I. C., Schnakenberg, G. H. F., Soper, Q. F., Boaz, H. E., Kern, S. F., and Parke, T. V., *J. Am. Chem. Soc., 75,* 1267 (1953).
99. Hanka, L. J., Bergy, M. E., and Kelly, R. B., *Science, 154,* 1667 (1966).
100. Wright, L. D., Cresson, E. L., Valiant, J., Wolf, D. E., and Folkers, K., *J. Am. Chem. Soc., 76,* 4163 (1954).
101. McLamore, W. M., Celmer, W. E., Bogert, V. V., Pennington, F. C., and Solomons, I. A., *J. Am. Chem. Soc., 74,* 2946 (1952).
102. Olifson, L. E., *Mikotoksikozy Cheloveka Sel'skokhoz Zhivotnykh Sbornik,* p. 47 (1960); *Chem. Abstr., 56,* 5195 (1961).

LACTONES AND LACTAMS

DR. J. L. C. WRIGHT and DR. L. C. VINING

These compounds exhibit characteristic absorption maxima in the carbonyl region of their infrared spectra. γ-Lactones show a peak in the $1740-1800$ cm^{-1} range, which is often definitive; δ-lactones absorb at lower frequencies ($1710-1750$ cm^{-1}), which overlap the region for open-chain esters. β-Lactams and fused-ring γ-lactams also absorb in these regions, but unstrained δ-lactams and larger cyclic amides show peaks near 1650 cm^{-1} in the same region as the amide I absorption of open-chain amides. Lactones are susceptible to hydrolysis with alkali under mild conditions; lactams require more vigorous conditions with an acid catalyst.

The following table does not include large and well-defined groups, such as the macrocyclic lactone (macrolide and polyene macrolide) antibiotics, lactones that contain an aromatic system (coumarins, depsides, etc.), and β-lactam antibiotics of the penicillin and cephalosporin type. Information on lactones that also contain carboxyl functions or that exist in equilibrium with the open-chain hydroxy acids is given in the preceding chapter. Much useful additional information on the properties of compounds listed below can be found in References 1 to 5 for that chapter (see page 48).

In addition to the compounds listed here, five-membered-ring lactones, including tetronic acid derivatives, are discussed on pages 371 to 374, and six-membered-ring (δ)lactones on pages 375 to 381.

Common Name Formula (MW) Structure	Organisms	Significant Characteristics	Ref.

β-Lactones

Antibiotic 1233A — see p. 35

γ-Lactones

Itatartaric acid — see p. 34

| 5-Hydroxyoct-6-en-4-olide $C_8H_{12}O_3$ (156) | *Nigrospora* spp. | Oil Melting point of *p*-nitrobenzoate: $108-109°C$ $[\alpha]_D^{25}$: +49.5° (chloroform) | 1 |

$CH_3CH = CHCH$ ... O ... O

Spiculisporic acid — see p. 36

allo-Isocitric acid — see p. 34

Sugar acids — see pp. 40–44

| Tetrenolin $C_{11}H_{12}O_4$ (208) | Actinomycete strain SS/1018 | Melting point: $126-128°C$ λ_{max}: 340 nm, ϵ = 42,900 | 2 |

$CH = CH CH_2 OH$

$OCH_2 CH = CH \ CH$ ... O ... O

Common Name Formula (MW) Structure	Organisms	Significant Characteristics	Ref.

γ-Lactones (continued)

Patulin
$C_7H_6O_4$ (154)

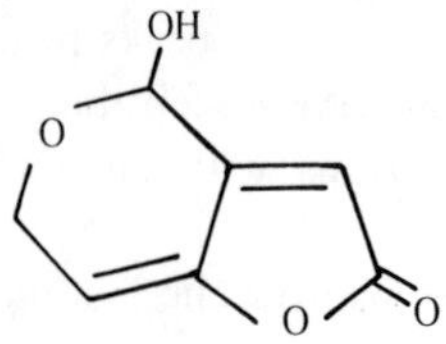

Penicillium spp.
Aspergillus spp.
Other fungi

Melting point: 111°C; phenylhydrazone 152–153°C
λ_{max}: 276.5 nm

3

Dihydrocanadensolide
$C_{11}H_{16}O_4$ (212)

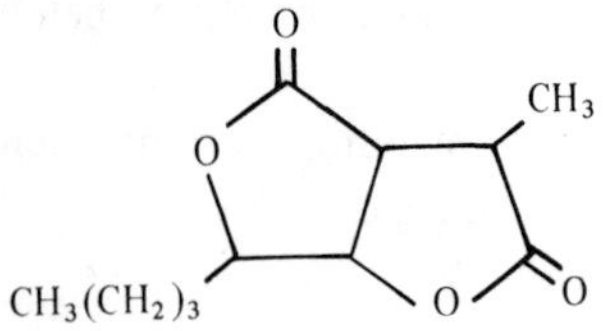

Penicillium canadense
Aspergillus indicus

Melting point: 92°C
$[a]_D^{22}$: −36.1° (chloroform, c = 0.2)

4

Canadensolide
$C_{11}H_{14}O_4$ (210)

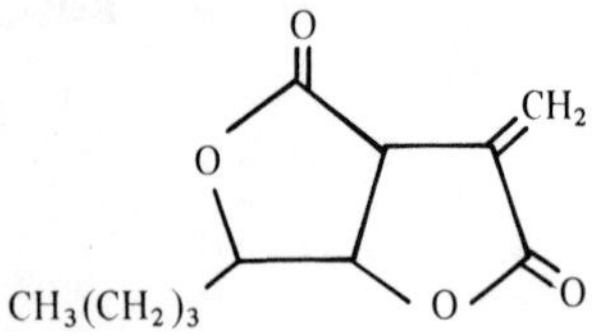

Penicillium canadense

Melting point: 46–47.5°C
$[a]_D$: −141°
λ_{max}: 210 nm, ϵ = 10,000

4

Avenaciolides

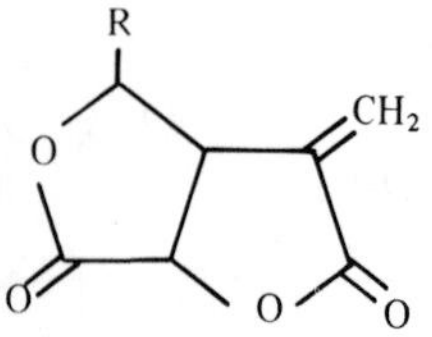

Avenaciolides I and II are epimeric at C-4

Avenaciolide (I)
$C_{15}H_{22}O_4$ (266)
R = −$(CH_2)_7CH_3$

Aspergillus avenaceus

Melting point: 49–50°C and 54–56°C
$[a]_D^{26.5}$: −41.6° (c = 1.2)
λ_{max}: 210 nm, ϵ = 10,000

5

iso-Avenaciolide (II)
$C_{15}H_{22}O_4$ (266)
R = −$(CH_2)_7CH_3$

Aspergillus avenaceus

Melting point: 129–130°C
$[a]_D^{27}$: −154° (ethanol, c = 1.1)
ν_{max}: 1775, 1657 cm^{-1}

6

Ethisolide
$C_9H_{10}O_4$ (182)
R = −CH_2CH_3

Penicillium sp.

Melting point: 122–123°C
$[a]_D^{27}$: −214° (ethanol, c = 1.2)
ν_{max}: 1773, 1658 cm^{-1}

6

Ungulinic acid — see p. 36

Minioluteic acid — see p. 36

Common Name Formula (MW) Structure	Organisms	Significant Characteristics	Ref.

4-Hydroxy-γ-lactones (Tetronic Acids)

γ-Methyltetronic acid
$C_5H_6O_3$ (114)

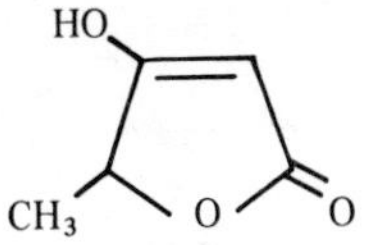

Penicillium charlesii
Penicillium fellutanum (P. cin-erascens)

Melting point: 115°C
$[a]_{546\,nm}$: −21° (water, c = 0.526)

7

Zymonic acid
$C_6H_6O_5$ (158)

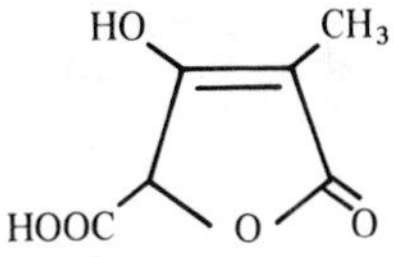

Yeasts

Boiling point (1 mm): 118–123°C
Melting point of methyl ether amide:
  209–210°C
n_D^{28}: 1.4640

8

Carolic acid
$C_9H_{12}O_4$ (184)

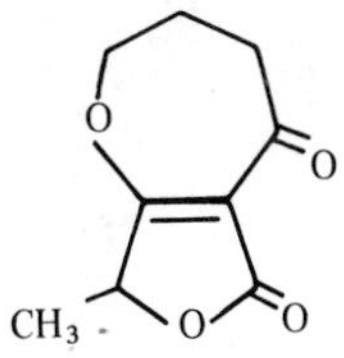

Penicillium charlesii

Melting point: 132°C
$[a]_{546\,nm}$: +84° (water, c = 0.5)

9

Carolinic acid
$C_9H_{10}O_6$ (214)

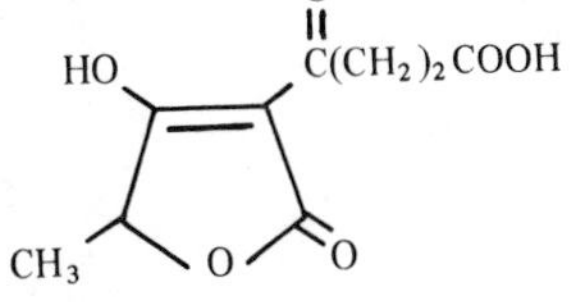

Penicillium charlesii

Melting point: 123°C (decomposes)
$[a]_{546\,nm}$: +60° (water, c = 0.34)

9

Dehydrocarolic acid (hydrate)
$C_9H_{10}O_5$ (198)

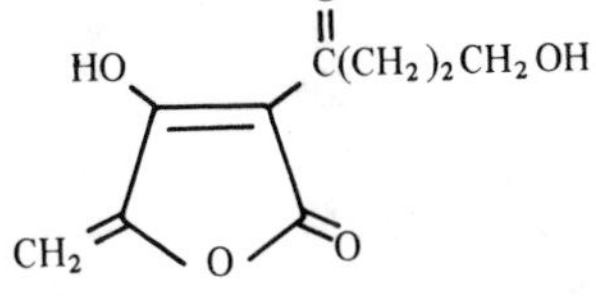

Penicillium fellutanum

Crystallizes from chloroform–
  petroleum ether with loss of
  water; decomposes above
  80°C without melting; bright
  orange with ferric chloride;
  bluish-red on heating with
  concentrated sulfuric acid
Melting point of 2,4-dinitro-
  phenylhydrazone: 157°C

10

Carlosic acid
$C_{10}H_{12}O_6$ (228)

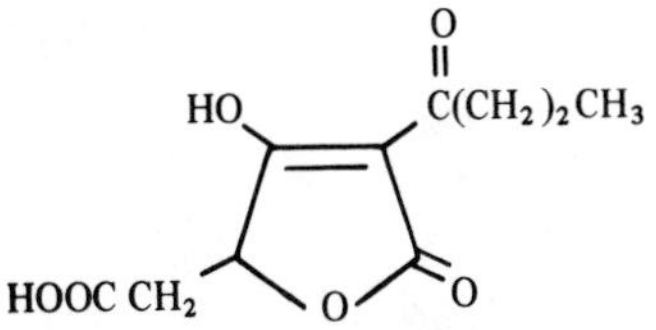

Penicillium charlesii
Penicillium fellutanum

Melting point: 181°C
$[a]_{546\,nm}$: −160° (water, c = 0.21)

9

Common Name
Formula (MW)
   Structure   Organisms   Significant Characteristics   Ref.

4-Hydroxy-γ-lactones (continued)

Carlic acid   *Penicillium charlesii*   Melting point: 176°C   9
$C_{10}H_{12}O_7$ (244) $[a]_{546\ nm}$: −160° (water, c = 0.28)

Terrestric acid   *Penicillium terrestre*   Melting point: 89°C   11
$C_{11}H_{14}O_5$ (226) $[a]_{546\ nm}^{20}$: +61.1° (water, c = 0.53)

Viridicatic acid   *Penicillium viridicatum*   Melting point: 174.5°C   12
$C_{12}H_{16}O_6$ (256) $[a]_{546\ nm}^{20}$: −105° (ethanol, c = 1.0)

Aspertetronins   *Aspergillus rugulosus*   13

Aspertetronin A   Melting point: 72°C
$C_{16}H_{12}O_4$ (276) $[a]_D$: +133° (chloroform, c = 0.3)
R = −CH=CHCH₃ ν_{max}: 1740 (sh), 1705 cm⁻¹

Aspertetronin B   Oil
$C_{16}H_{22}O_5$ (294)   Boiling point (0.4 mm): 100°C
R = CH₂CHCH₃ $[a]_D$: −70.5° (chloroform, c = 5.25)
         | ν_{max}: 1740 (sh), 1705 cm⁻¹
        OH

Ascorbic acids — see p. 43

Common Name Formula (MW) Structure	Organisms	Significant Characteristics	Ref.

γ-Lactols

Acetomycin
$C_{10}H_{14}O_5$ (214)

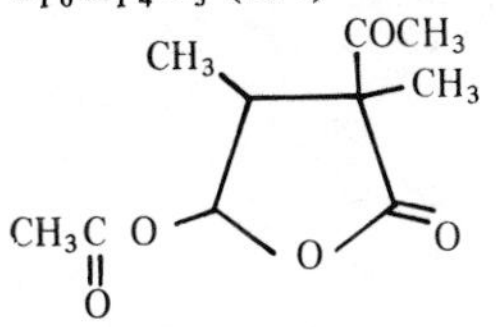

Streptomyces ramulosus — Melting point: 115°C (sublimes at 70°C); $[a]_D$: +167° (ethanol) — 14

3-Carboxy-2,4-pentadienal lactol
$C_6H_6O_3$ (126)

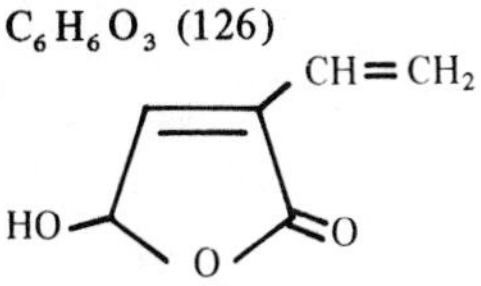

Streptomyces roseochromogenes — Viscous oil; crystalline barium salt; λ_{max}: 245 nm, ϵ = 7,800; barium salt 272 nm, ϵ = 2,200; ν_{max}: 1761, 1600 cm^{-1} — 15

Penicillic acid – see p. 38

(±)-2-Acetamido-2,5-dihydro-5-ketofuran
$C_6H_7O_3N$ (141)

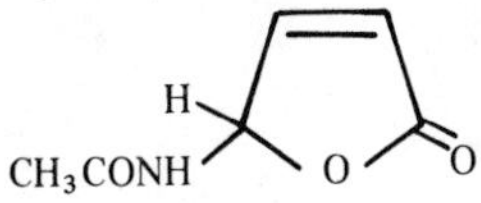

Fusarium equiseti
Fusarium nivale — Melting point: 115°C; ν_{max}: 3297, 3102, 1780, 1755, 1663, 1530 cm^{-1} (Nujol) — 16

δ-Lactones

Ramulosins

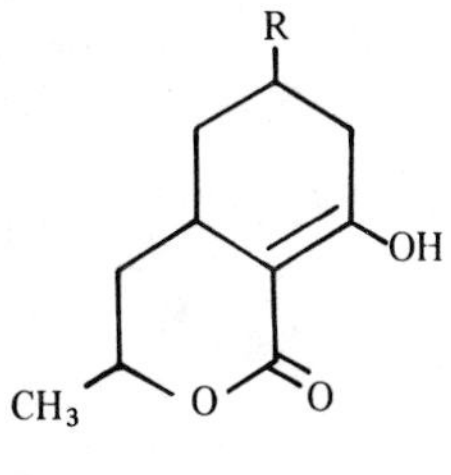

Pestalotia ramulosus

Ramulosin
$C_{10}H_{14}O_3$ (182)
R = –H — Melting point: 120–121°C; $[a]_D^{25}$: +18°; λ_{max}: 264 nm, ϵ = 10,000; Violet color with ferric chloride — 17

Hydroxyramulosin
$C_{10}H_{14}O_4$ (198)
R = –OH — Melting point: 132–133°C; $[a]_D^{25}$: +91.6° (methanol, c = 1); λ_{max}: 262 nm, ϵ = 10,000 (ethanol) — 18

Mevalonic acid lactone –
see p. 33

Common Name Formula (MW) Structure	Organisms	Significant Characteristics	Ref.

δ-Lactones (continued)

Polyketide lactones

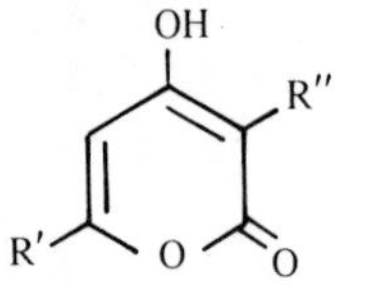

Triacetic acid lactone
$C_6H_6O_3$ (126)
R′ = –CH₃
R″ = –H

Penicillium stipitatum
Penicillium patulum

Melting point: 190°C
λ_{max}: 285 nm, ϵ = 7,500
  (ethanol)

19

Methyltriacetic acid lactone
$C_7H_8O_3$ (140)
R′ = –CH₃
R″ = –CH₃

Penicillium stipitatum

Melting point: 204°C
λ_{max}: 289 nm, ϵ = 8,300
  (ethanol)

20

Tetraacetic acid lactone
$C_8H_8O_4$ (168)
R′ = –CH₂C CH₃
     ‖
     O
R″ = –H.

Penicillium stipitatum

Melting point: 118–119°C
λ_{max}: 284 nm, ϵ = 8,300
  (ethanol)

19

Radicinin
$C_{12}H_{12}O_5$ (236)

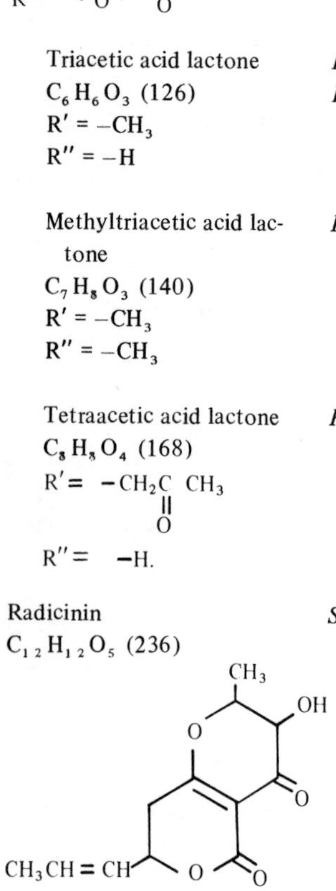

Stemphylium radicinum

Melting point: 238–240°C
  (decomposes)
$[a]_D^{22}$: –187° (chloroform, c = 1.02)
ν_{max}: 1762, 1655, 1605 cm⁻¹
  (chloroform)

21

Citreoviridin
$C_{23}H_{32}O_6$ (404)

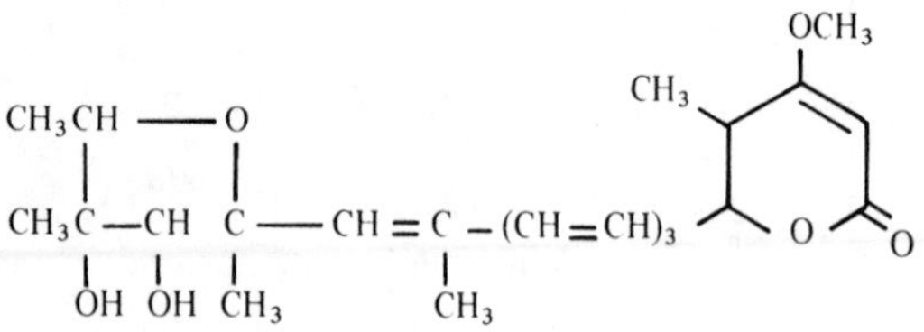

Penicillium citreoviride
Penicillium ochrosalmoneum

Melting point: 107–111°C
λ_{max}: 204, 234, 286 (sh), 294,
  388 nm (ethanol)

22

Alternaric acid — see p. 39

4-Hydroxyocta-2,6-dien-5-olide
$C_8H_{10}O_3$ (154)

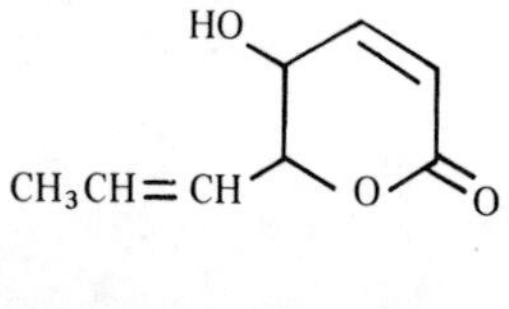

Nigrospora spp.

Melting point: 50–53°C
$[a]_D^{25}$: +175°C
λ_{max}: 204 nm, ϵ = 12,000

23

Common Name Formula (MW) Structure	Organisms	Significant Characteristics	Ref.

δ-Lactones (continued)

Asperline
(Antibiotic U-13,933)
$C_{10}H_{12}O_5$ (212)

CH₃COO

CH₃CH—CH   O   O
O

Aspergillus nidulans

Melting point: 71–73°C
$[a]_D^{25}$: +345° (ethanol, c = 0.9)
λ_{max}: 204 nm, ϵ = 12,000
(ethanol)
ν_{max}: 1735, 1715, 1652 cm⁻¹

24

3(1,2-Epoxypropyl)-5,6-di-
hydro-5-hydroxy-6-methyl-
2H-pyran-2-one
$C_9H_{12}O_3$ (168)

HO

CH—CH CH₃

CH₃   O   O

Aspergillus spp.

Melting point: 110–112°C
$[a]_D^{25}$: –16.5° (chloroform)
λ_{max}: 204 nm, ϵ = 12,000
(methanol)

25

Actinobolin
$C_{13}H_{20}O_6N_2$ (300)

OH

HO

NH₂

CH₃CHCONH   OH

CH₃   O   O

Streptomyces griseoviridus var.
atrofaciens

Isolated as the stable sulfate
salt (½ sulfuric acid, water)
Melting point of N-acetylac-
tinobolamine: 193–194°C
$[a]_D$ of N-acetylactinobola-
mine: +105° (methanol, c =
0.65)

26

Sugar acid lactones – see p. 40–44

Other Lactones

7-Hydroxy-3-keto oct-2-en-
oic acid lactone
$C_8H_{10}O_3$ (154)

O
‖
CH₃CH(CH₂)₂   C CH=CH
|
O ———————— C=O

Stemphylium radicinum

Melting point: 176°C
$[a]_D^{20}$: –47° (acetone, c = 0.36)
λ_{max}: 220 nm, ϵ = 11,200
ν_{max}: 1732, 1703 cm⁻¹ (carbon
tetrachloride)

27

11-Hydroxy-*trans*-8-dodeca-
noic acid lactone
$C_{12}H_{20}O_2$ (196)

CH₃CHCH₂CH = CH(CH₂)₅CH₂
|
O ———————— C=O

Cephalosporium recifei

Oil
$[a]_D^{25}$: +73.2° (chloroform, c = 7.5)
ν_{max}: 1735, 975 cm⁻¹

28

Common Name Formula (MW) Structure	Organisms	Significant Characteristics	Ref.

Other Lactones (continued)

Brefeldin A
$C_{16}H_{24}O_4$ (280)

Penicillium spp.
Curvularia spp.
Nectria radicicola

Melting point: 204–205°C
$[a]_D^{22}$: +96°
λ_{max}: 215 nm, $\epsilon = 11,200$
  (ethanol)
ν_{max}: 1711, 1646 cm^{-1}

29

```
                              CH2
                             /    \
CH3  CH(CH2)3    CH=CH CH    CHOH
     |                  |        |
     O  C  CH=CH  CH—CH —CH2
        ||           |
        O           OH
```

Colletodiol
$C_{14}H_{20}O_6$ (284)

Colletotrichum capsici
Chaetomium funicola

Melting point: 164–167°C
$[a]_D^{22}$: +36° (chloroform, c = 1)
λ_{max}: 212 nm, $\epsilon = 19,000$
  (ethanol)
ν_{max}: 1720, 1660 cm^{-1}

30

```
                        O
                        ||
CH3  CH   CH2  CH=CH  C — O
     |                        \
     O                         CHCH3
     |                        /
O= C    CH=CH   CH  CH  CH2
                |   |
                OH  OH
```

Pyrenophorins

```
                R
                ||
CH3  CH(CH2)2   C  CH=CHC = O
     |                  |
     O                  O
     |                  |
O= C    CH=CH   C  (CH2)2  CH  CH3
                ||
                R
```

Pyrenophorin
$C_{16}H_{20}O_6$ (308)
R = O

Pyrenophora avenae
Stemphylium radicinum

Melting point: 175°C
$[a]_D$: −50°
λ_{max}: 220 nm, $\epsilon = 23,200$

31

Pyrenophorol
$C_{16}H_{24}O_6$ (312)
R = H, OH

Byssochlamys nivea

Melting point: 135°C
$[a]_D^{20}$: −3° (acetone, c = 1)
$\lambda_{inflection}$: 255 nm, $\epsilon = 250$
  (ethanol)
ν_{max}: 1720–1710, 1645 cm^{-1}

32

Lactams

4-Hydroxypyrrolid-2-one
$C_4H_7O_2N$ (101)

Amanita muscaria

Melting point: 153.5–155°C
$[a]_D$: −44.5° (methanol)
ν_{max}: 3257, 3155, 1661 cm^{-1}
  (potassium bromide)

33

```
HO
  \
   [ring]
      N     O
      H
```

Common Name Formula (MW) Structure	Organisms	Significant Characteristics	Ref.

Lactams (continued)

Tetramic acids

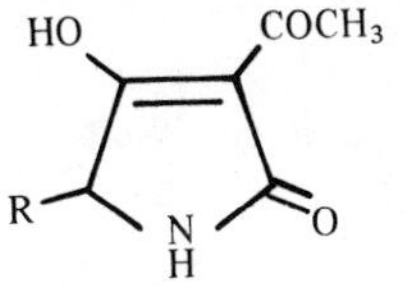

Tenuazonic acid
$C_{10}H_{15}O_3N$ (197)
$R = -CH\ CH_2CH_3$
    $\quad\mid$
    CH_3
	Alternaria tenuis
	Alternaria spp.
	Aspergillus spp.
	Sphaeropsidales
		Boiling point (0.35 mm): 117°C
		$[a]_D$: −136° b 5° (chloroform,
		   c = 0.2)
			34

3-Acetyl-5-isopropyltet- ramic acid $C_9H_{13}O_3N$ (183) $R = -CH(CH_3)_2$	*Alternaria tenuis*	Melting point of 2,4-dinitrophenyl- hydrazone: 222−224°C	35
3-Acetyl-5-isobutyltet- ramic acid $C_{10}H_{15}O_3N$ (197) $R = -CH_2CH(CH_3)_2$	*Alternaria tenuis*	Melting point of semicarbazone: 204−206°C	35
Erythroskyrin $C_{25}H_{31}O_5N$ (425)	*Penicillium islandicum*	Melting point: 130−133°C $[a]_D$: +46.9° (ethanol, c = 0.2) λ_{max}: 260, 409 nm (ethanol) ν_{max}: 1693, 1635, 1569, 1550 cm⁻¹	36

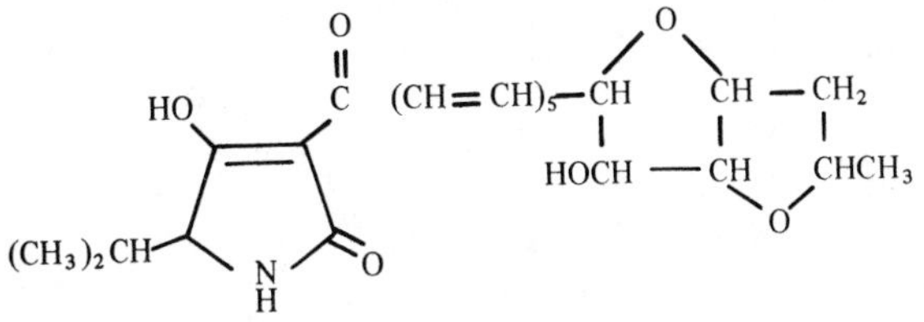

Cytochalasins

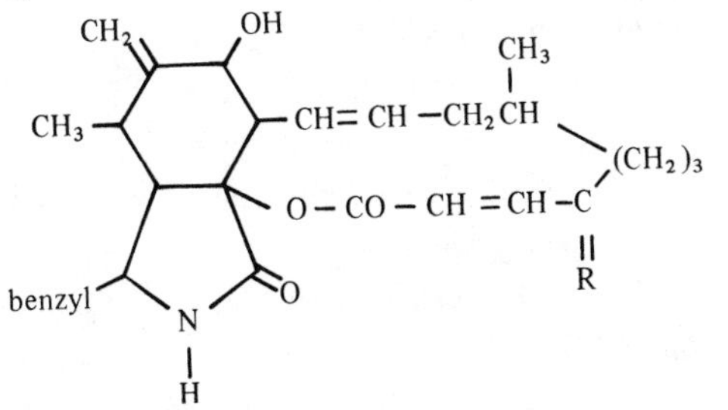

Cytochalasin A
$C_{29}H_{35}O_5N$ (477)
R = O
	Helminthosporium dematioideum	Melting point: 185−187°C
		$[a]_D^{25}$: +92° (ethanol, c = 1.4)
		λ_{max}: 218, 225, 258, 263,
		   268 nm (ethanol)
		ν_{max}: 1814, 1715, 1705,
		   1650, 1638, 1600 cm⁻¹
			37

Common Name Formula (MW) Structure	Organisms	Significant Characteristics	Ref.

Lactams (continued)

Cytochalasins (*cont.*)

Cytochalasin B
(Phomin)
$C_{29}H_{37}O_5N$ (479)
R = H, OH

Helminthosporium dematio-
iodeum
Phoma sp.

Melting point: 218–220°C
$[\alpha]_D^{25}$: +83° (methanol, c = 1.04)
λ_{max}: 213, 219, 258, 264,
267.5 nm (ethanol)
ν_{max}: 1814, 1715, 1690,
1665, 1638, 1600 cm^{-1}

37

Cytochalasin C
$C_{30}H_{37}O_7N$ (507)

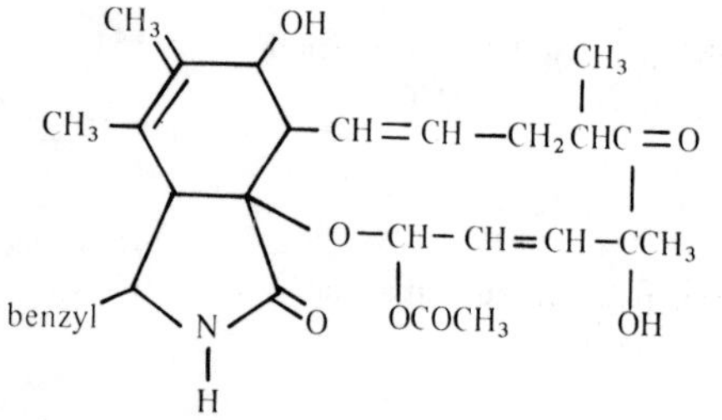

Metarrhizium anisopliae

Melting point: 260–264°C
ν_{max}: 3402, 3186, 3120,
1743, 1708, 1602 cm^{-1}

38

Zygosporins

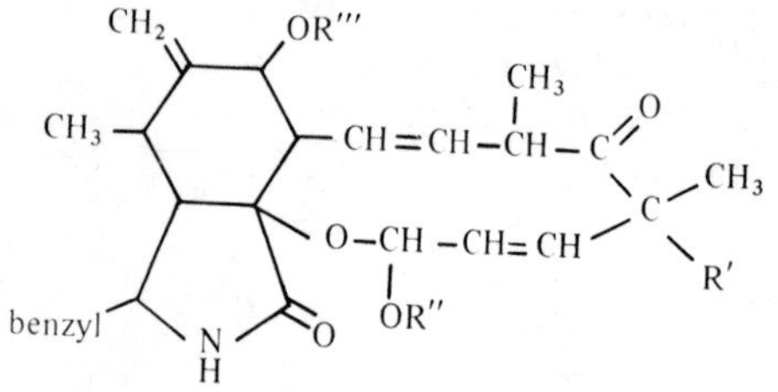

Zygosporin A
(Cytochalasin D)
$C_{29}H_{35}O_7N$ (509)
R' = –OH
R'' = –COCH$_3$
R''' = –H

Metarrhizium anisopliae
Zygosporium masonii

Melting point: 267–271°C
ν_{max}: 3413, 3219, 1741,
1707, 1695, 1644 cm^{-1}

38

Zygosporin D
(Desacetylcytochalasin D)
$C_{27}H_{33}O_6N$ (467)
R' = –OH
R'' = –H
R''' = –H

Zygosporium masonii

Melting point: 180–190°C
$[\alpha]_D^{23}$: –14.9° (dioxane, c = 0.759)
ν_{max}: 3400, 1700 cm^{-1}

39

Zygosporin E
$C_{29}H_{35}O_6N$ (493)
R' = –H
R'' = –H
R''' = –COCH$_3$

Zygosporium masonii

Melting point: 218–223.5°C
$[\alpha]_D^{24}$: +6.2° (dioxane, c = 0.971)
ν_{max}: 3525, 3420, 1743,
1703 cm^{-1}

39

Common Name
Formula (MW)
    Structure   Organisms   Significant Characteristics   Ref.

Lactams (continued)

Zygosporins (*cont.*)

Zygosporin F (Acetylcytochalasin D) $C_{31}H_{37}O_8N$ (551) $R' = -OH$ $R'' = -COCH_3$ $R''' = -COCH_3$	*Zygosporium masonii*	Melting point: $126-129°C$ $[a]_D^{24}: -12.0°$ (dioxane, c = 0.775) ν_{max}: 3420, 1740, 1704 cm^{-1}	39
Zygosporin G $C_{29}H_{35}O_6N$ (493)	*Zygosporium masonii*	Melting point: $115-125°C$ $[a]_D^{24}: -82°$ (dioxane, c = 0.87) ν_{max}: 3410, 1743, 1701 cm^{-1}	39

Bicyclomycin
$C_{12}H_{18}O_7N_2$ (302)
	Streptomyces sapporoensis	Melting point: rhombic form
		  $187-189°C$ (decomposes);
		  monoclinic form $188-191°C$
		  (decomposes)
		$[a]_D^{23}: +63.5°$ (methanol, c = 1)
		ν_{max}: rhombic form 1703,
		  1673 cm^{-1} (Nujol); monoclinic
		  form 1685, 1640 cm^{-1}
		  (Nujol)
			40

REFERENCES

1. Evans, R. H., Ellestad, I. G. A., and Kunstmann, M. P., *Tetrahedron Lett.,* p. 1791 (1969).
2. Pagani, H., Lancini, G., Tamoni, G., and Coronelli, C., *J. Antibiot. (Tokyo), 26,* 1 (1973).
3. Woodward, R. B., and Singh, G., *Experientia, 6,* 238 (1950).
4. McCorkindale, N. J., Wright, J. L. C., Brian, P. W., Clarke, S. M., and Hutchinson, S. A., *Tetrahedron Lett.,* p. 727 (1968).
5. Brookes, D., Tidd, B. K., and Turner, W. B., *J. Chem. Soc.,* p. 5385 (1963).
6. Aldridge, D. C., and Turner, W. B., *J. Chem. Soc. Sect. C,* p. 2431 (1971).
7. Clutterbuck, P. W., Raistrick, H., and Reuter, F., *Biochem. J., 29,* 1300 (1935).
8. Stodola, F. H., Shotwell, O. L., and Lockwood, L. B., *J. Am. Chem. Soc., 74,* 5415 (1952).
9. Clutterbuck, P. W., Haworth, W. N., Raistrick, H., Smith, G., and Stacey, M., *Biochem. J., 28,* 94 (1934).
10. Bracken, A., and Raistrick, H., *Biochem. J., 41,* 569 (1947).
11. Birkinshaw, J. H., and Raistrick, H., *Biochem. J., 30,* 2194 (1936).
12. Birkinshaw, J. H., and Samant, M. S., *Biochem. J., 74,* 369 (1960).
13. Ballantine, J. A., Ferrito, V., Hassall, C. H., and Jones, V. I. P., *J. Chem. Soc. Sect. C,* p. 56 (1969).
14. Ettlinger, L., Gaumann, E., Hutter, R., Keller-Schlierlein, W., Kradolfer, F., Niepp, L., Prelog, V., and Zahner, H., *Helv. Chim. Acta, 41,* 216 (1958).

15.   Els, H., Sobin, B. A., and Celmer, W. D., *J. Am. Chem. Soc., 80,* 878 (1958).
16.   White, E. P., *J. Chem. Soc. Sect. C,* p. 346 (1967).
17.   Stodola, F. H., Cabot, C., and Benjamin, C. R., *Biochem. J., 93,* 92 (1964).
18.   Tanenbaum, S. W., Agarwal, S. G., Williams, T., and Pitcher, R. G., *Tetrahedron Lett.,* p. 2377 (1970).
19.   Bentley, R., and Zwitkowitz, P. M., *J. Am. Chem. Soc., 89,* 676 (1967).
20.   Brenneisen, P. E., Acker, T. E., and Tanenbaum, S. W., *J. Am. Chem. Soc., 88,* 834 (1966).
21.   Grove, J. F., *J. Chem. Soc.,* p. 3234 (1964).
22.   Sakabe, N., Goto, T., and Hirata, Y., *Tetrahedron Lett.,* p. 1825 (1964).
23.   Evans, R. H., Ellestad, I. G. A., and Kunstmann, M. P., *Tetrahedron Lett.,* p. 1791 (1969).
24.   Argoudelis, A. D., and Zieserl, J. F., *Tetrahedron Lett.,* p. 1969 (1966).
25.   Rosenbrook, W., and Carney, R. E., *Tetrahedron Lett.,* p. 1867 (1970).
26.   Munk, M. E., Nelson, D. B., Antosz, F. J., Herald, D. L., and Haskell, T. H., *J. Am. Chem. Soc., 90,* 1087 (1968).
27.   Aldridge, D. C., and Grove, J. F., *J. Chem. Soc.,* p. 3239 (1964).
28.   Vesonder, R. F., Stodola, F. H., Wickerham, L. J., Ellis, J. J., and Rohwedder, W. K., *Can. J. Chem., 49,* 2029 (1971).
29.   Sigg, H. P., *Helv. Chim. Acta, 47,* 1401 (1964).
30.   Grove, J. F., Speake, R. N., and Ward, G., *J. Chem. Soc. Sect. C,* p. 230 (1966).
31.   Nozoe, S., Hirae, K., Tsuda, K., Ishibashi, K., Shirasaka, M., and Grove, J. F., *Tetrahedron Lett.,* p. 4675 (1965).
32.   Zis, K., Furger, P., and Sigg, H. P., *Experientia, 25,* 123 (1969).
33.   Matsumoto, T., Treub, W., Gwinner, R., and Eugster, C. H., *Helv. Chim. Acta, 52,* 716 (1969).
34.   Rosett, T., Sankhala, R. H., Stickings, C. E., Taylor, M. E. V., and Thomas, R., *Biochem. J., 67,* 390 (1957).
35.   Gatenbeck, S., and Sierankiewicz, J., *Antimicrob. Agents Chemother., 3,* 308 (1973).
36.   Shoji, J., and Shibata, S., *Chem. Ind.,* p. 419 (1964).
37.   Rothweiler, W., and Tamm, Ch., *Helv. Chim. Acta, 53,* 696 (1970).
38.   Aldridge, D. C., and Turner, W. B., *J. Chem. Soc., Sect. C,* p. 923 (1969).
39.   Minato, H., and Katayama, T., *J. Chem. Soc., Sect. C,* p. 45 (1970).
40.   Miyoshi, T., Miyairi, N., Aoka, H., Kohsaki, M., Sakai, H., and Imanaka, M., *J. Antibiot. (Tokyo), 25,* 569 (1972).

MICROBIAL POLY-YNES

PROFESSOR SIR EWART R.H. JONES AND DR. V. THALLER

Of the total number (*ca.* 600) of natural poly-ynes known today (for a general account see References 1, 2, and 3) more than 10% have been isolated from higher fungi (Basidiomycetes). All the fungal poly-ynes so far described have been isolated from mycelial cultures, but they have also been reported to occur in sporophores collected in the field.[4]

The poly-ynes can generally be isolated from the culture fluids by continuous ether extraction and subsequent counter-current distribution or/and chromatography. They are relatively stable in dilute solutions, but are heat- and light-sensitive in concentrated solutions and condensed phases. Poly-ynes do crystallize, but many decompose, some very violently, on heating.

The electronic absorption spectra of poly-ynes are their most useful diagnostic feature. Most of these spectra are highly characteristic and possess very sharp vibrational fine structure: when conjugated triple bonds predominate in the chromophore, the band spacing is characteristically about 2200 cm^{-1} (for a table of UV maxima and the associated ϵ-values of a series of standard poly-ynes see Reference 1).

The poly-ynes undergo no chemical reactions of a unique character or of general utility.

Some poly-ynes have been found to have antibiotic activity.

The structural formulas are oriented to indicate actual or probable biogenetic relationship to fatty acids $CH_3 \cdot [CH_2]_n \cdot CO_2H$.

CODE

Organisms

I	*Fomes annosus*	XXVIII	*Clitocybe robusta* Peck
II	*Agrocybe dura*	XXIX	*Clitocybe nebularis* (Fr.) Quel.
III	*Clitocybe diatreta* (Fr.) Quel.	XXX	*Clitocybe rivulosa* (Fr.) Quel.
IV	*Tricholoma nudum* (Bull.) Fr.	XXXI	*Drosophila sarcocephala* (Fr.) Quel.
V	*Tricholoma sordidum*	XXXII	*Tricholoma grammopodium* (Bull.) Fr.
VI	*Lepista diemii*	XXXIII	*Clitocybe obbata*
VII	*Clitocybe cerussata* (Fr.) Quel.	XXXIV	*Clitocybe cyathiformis*
VIII	*Clitocybe gallinacea* (Fr.) Gillet	XXXV	*Clitocybe candida*
IX	*Clitocybe fragrans* (Fr.) Quel.	XXXVI	*Fistulina hepatica* (Huds) Fr.
X	*Clitocybe inversa* (Fr.) Quel.	XXXVII	*Papulospora polyspora*
XI	*Omphalia umbilicata* (Fr.) Quel.	XXXVIII	*B244*
XII	*Polyporus anthracophilus*	XXXIX	*Polyporus guttulatus*
XIII	*Daedalea juniperina* (Murr.)	XL	*Leptoporus kymantodes*
XIV	*Cortinellus berkeleyanus* Ito and Imai	XLI	*Merulius lacrymans*
XV	*Psilocybe sarcocephala*	XLII	*Odontia bicolor*
XVI	*Poria sinuosa* Fr.	XLIII	*Flammula sapinea* Fr.
XVII	*Coprinus quadrifidus*	XLIV	*B285*
XVIII	*Polyporus biformis*	XLV	*Drosophila simivestita*
XIX	*Aleurodiscus roseus* (Per. ex Fr.) v. Höhn *et* Litsch.	XLVI	*Poria subacida* Pk.
		XLVII	*B841*
XX	*Marasmius ramealis*	XLVIII	*Poria corticola*
XXI	*Drosophila (Psathyrella) subatrata*	XLIX	*Poria tenuis*
XXII	*Poria selecta* Karst. ex Rom.	L	*Poria colorea* Overh. et Englerth.
XXIII	*Clitocybe rhizophora* Velen.	LI	*Poria mutans* Pk.
XXIV	*Pleurotus ulmarius* (Fr.) Quel.	LII	*A67*
XXV	*Tricholoma panaeolum* (Fr.) Quel.	LIII	*Nocardia acidophilus* (see References 4 and 40)
XXVI	*Clitocybe dealbata*		
XXVII	*Clitocybe truncicola* (Fr.) Quel.	LIV	*Fistulina pallida* (Berk. and Rav.)

Solvents

Unless stated otherwise, the UV spectra and rotations were measured in ethanol.

Configurations

The symbol *t* above a double bond indicates *trans* configuration; the symbol *c* above a double bond indicates *cis* configuration.

Formula (Common Name)	Organisms	Significant Characteristics	References
C$_6$ Compounds			
H[C≡C]$_3$H	I		5
C$_8$ Compounds			
HOCH$_2$ · [C≡C]$_3$ · CONH$_2$ (Agrocybin)	II	Melting point: 130–140°C (decomposes) UV maxima: 325, 304, 286, 269, 224, 215	6
NC · [C≡C]$_2$ · CH$\overset{t}{=}$CH · CO$_2$H (Diatretyne 2, nudic acid B)	III–XI	Melting point: 179–180°C (decomposes) UV maxima: 322, 302, 284, 269, 253, 241, 231	6, 7, 8, 9
H$_2$NOC · [C≡C]$_2$ · CH$\overset{t}{=}$CH · CO$_2$H (Diatretyne 1)	III, VII, VIII, XI	Decomposes below 195°C UV maxima: 309, 290, 275, 260, 223	6, 8
HO$_2$C · CH=CH · C≡C · CH$\overset{t}{=}$CH · CO$_2$H	XII	Melting point, dimethyl ester: 117–119.5°C UV maxima, dimethyl ester: 307, 292, 278, 240, 214, 205	10
CH$_3$ · C≡C · C=CH · CH=C · CHO ⌐S⌐ (Junipal)	XIII	Melting point: 80°C UV maxima: 320, 286.5, 216.5	11
H[C≡C]$_2$ · CH=C=CH·CH$_2$OH	XIV	[α]$_D$: −380° (CH$_2$Cl$_2$) UV maxima: 277.5, 262.5, 249, 236.5, 226, 207.5	12

Formula (Common Name)	Organisms	Significant Characteristics	References
C₉ Compounds			
$H[C\equiv C]_3 \cdot CH \overset{t}{=} CH \cdot CO_2H$	VI, XV, XVI	Melting point, methyl ester: 20°C (decomposes) UV maxima, methyl ester: 337, 315, 295, 278, 263, 252, 241	9, 13, 14
$H[C\equiv C]_3 \cdot CH \overset{t}{=} CH \cdot CONH \cdot CH \cdot CH(CH_3)_2$ $\qquad\qquad\qquad\qquad\qquad\quad CO_2H$	XVI	UV maxima, methyl ester: 336.5, 314.5, 295, 278, 262.5, 251, 240	14
$H[C\equiv C]_3 \cdot CH \overset{t}{=} CH \cdot CHO$	XVII	UV maxima: 343, 320, 300, 283, 255, 243, 233	15
$H[C\equiv C]_3 \cdot CH \overset{t}{=} CH \cdot CH_2OH$	XVI, XVII	UV maxima: 327, 306, 287, 271, 257, 240, 228, 220 (inflection), 210.5	14, 15
$H[C\equiv C]_3 \cdot CH \overset{t}{=} CH \cdot CH_2OH$ (with epoxide) (Biformyne 1)	XVII, XVIII	UV maxima: 310, 290.5, 274, 258.5, 245, 209	16
$H[C\equiv C]_3 - \underset{HO}{\overset{H}{\underset{|}{C}}} - \underset{OH}{\overset{H}{\underset{|}{C}}} - CH_2OH$ (2D, 3D)	XVII	Melting point: *ca.* 40°C (decomposes) $[a]_D$: +6° UV maxima: 305, 286.5, 269.5, 254, 208	15
$H[C\equiv C]_3 - \underset{HO}{\overset{HO}{\underset{|}{C}}} - \underset{OH}{\overset{H}{\underset{|}{C}}} - CH_2OH$ (2D, 3L)	XIX	$[a]_D$: -8° UV maxima: 305, 284, 267, 253, 208	17
$H[C\equiv C]_2 \cdot CH = C = CH \cdot CH_2 \cdot CO_2H$	XIX	Characterized as the methyl ester (next compound)	17
$H[C\equiv C]_2 \cdot CH = C = CH \cdot CH_2 \cdot CO_2CH_3$	XIX	$[a]_D$: +285° UV maxima: 278, 262, 249, 237, 208	17
$(+)-H[C\equiv C]_2 \cdot CH = C = CH \cdot CH_2 \cdot CH_2OH$ [(+)-Marasin]	XIX	$[a]_D$: +360° UV maxima: 278, 262, 248, 237, 224, 208	17

Formula (Common Name)	Organisms	Significant Characteristics	References
C$_9$ Compounds (continued)			
(−)−H[C≡C]$_2$ · CH=C=CH · CH$_2$ · CH$_2$OH (Marasin)	XIV, XX	[α]$_D$: −325° UV maxima: 278, 263, 249.5, 237, 224	12, 18
H[C≡C]$_2$ · CH=CH · $\overset{\text{H H}}{\underset{\text{HO OH}}{\text{C–C}}}$–CH$_2$OH (2D, 3D)	XVII	[α]$_D$: +2° UV maxima: 280.5, 265, 251, 238.5, 227 (inflection), 208	17
H[C≡C]$_2$ · CH$\overset{t}{=}$CH · $\overset{\text{HO H}}{\underset{\text{H OH}}{\text{C–C}}}$–CH$_2$OH (2D, 3L)	XIX	[α]$_D$: −4° UV maxima: 280, 265, 251, 238.5, 226 (inflection), 206	17
H[C≡C]$_2$ · CH$\overset{c}{=}$CH · CH$_2$ · CH$_2$ · CO$_2$H (Drosophilin E)	XXI	Melting point: 35°C UV maxima: 279.5, 264, 250, 238, 227, 210	19
H[C≡C]$_2$ · CH$\overset{t}{=}$CH · [CH$_2$]$_1$ · CH$_2$OH	XIX	UV maxima: 281, 265, 251, 238, 229	17
CH$_3$ · [C≡C]$_2$ · CH=CH · CH$_2$ · CO$_2$CH$_3$ or CH$_3$ · CH=CH · [C≡C]$_2$ · CH$_2$ · CO$_2$CH$_3$	XIV	UV maxima: 281, 265, 251.5, 237, 227 (inflection), 210	12
HOCH$_2$ · CH$\overset{t}{=}$CH · [C≡C]$_2$ · CH$_2$ · CO$_2$H	XXII	Melting point, methyl ester: 56° (decomposes) UV maxima: 282, 266, 252, 239, 227 (inflection), 213	20
HOCH$_2$ · CH$\overset{t}{=}$CH · [C≡C]$_2$ · CH$_2$ · CH$_2$OH	XVI, XIX	Melting point: 58–59°C (decomposes) UV maxima: 282, 266, 252, 239, 227.5, 212.5	14, 17
CH$_3$ · CH$_2$ · CO · [C≡C]$_2$ · CH(OH) · CH$_2$OH	XXIII	Melting point: 34.5–35°C [α]$_D$: −30° UV maxima: 282.5, 267, 253.5, 241, 227.5	21
CH$_3$ · CH$_2$ · CH(OH) · [C≡C]$_2$ · CH(OH) · CH$_2$OH	XXIII	Melting point: 71.5–72.5°C [α]$_D$: −21° UV maxima: 256, 242, 228, 220	21

Formula (Common Name)	Significant Characteristics	Organisms	References
C_9 Compounds (continued) $CH_3 \cdot CH\overset{t}{=}CH \cdot [C{\equiv}C]_2 \cdot CH_2 \cdot CH_2OH$ (epoxide)	Melting point: 36–37°C $[\alpha]_D^{:}$ −21.5° ($CHCl_3$) UV maxima: 259, 245, 233, 222	XXIII	22
C_{10} Compounds $HO_2C \cdot [C{\equiv}C]_3 \cdot CH\overset{t}{=}CH \cdot CO_2H$	UV maxima, dimethyl ester: 352, 328, 308, 290.5, 256, 244, 236	XVI	14
$HOCH_2 \cdot [C{\equiv}C]_3 \cdot CH\overset{t}{=}CH \cdot CO_2H$ (Diatretyne 3)	The methyl ester decomposes at *ca.* 115°C UV maxima, methyl ester: 343.5, 320.5, 301, 283, 256.5, 245	III, IV, VI, VII–X, XXIV–XXXI	8, 9, 23, 24
$OCH \cdot [C{\equiv}C]_3 \cdot CH\overset{t}{=}CH \cdot CH_2OH$	UV maxima (in ether): 351, 328, 306, 290, 274	XVII	17
$HOCH_2 \cdot [C{\equiv}C]_3 \cdot CH\overset{t}{=}CH \cdot CH_2OH$	Melting point: 138°C (decomposes) UV maxima: 330.5, 309.5, 290.5, 279, 259, 243.5, 231, 212, 205	XVI, XVII, XXXII	14, 15, 25
$CH_3 \cdot [C{\equiv}C]_3 \cdot CH\overset{t}{=}CH \cdot CO_2H$ (Dehydromatricaria acid)	Characterized as the methyl ester (next compound)	XIX, XXIV, XXVI	17, 23, 26
$CH_3 \cdot [C{\equiv}C]_3 \cdot CH\overset{t}{=}CH \cdot CO_2CH_3$ (Dehydromatricaria ester)	Melting point: 105–106°C UV maxima (in hexane): 344, 320, 301, 283, 267.5, 256, 244	XXVII	24
$CH_3 \cdot [C{\equiv}C]_3 \cdot CH\overset{t}{=}CH \cdot CHO$	Melting point: 108–109° (decomposes) UV maxima (in hexane): 350, 326, 306, 288, 272 (inflection), 258, 245.5	XIX, XXIV, XXVII, XXXIII, XXXIV	17, 23, 24, 26, 27
$CH_3 \cdot [C{\equiv}C]_3 \cdot CH\overset{t}{=}CH \cdot CH_2OH$ (Dehydromatricarianol)	Melting point: 128–129°C (decomposes) UV maxima: 328.5, 307, 288.5, 272, 257, 241.5, 230, 221 (inflection), 210	XIII, XVI, XIX, XXIV, XXVII, XXXII, XXXIII, XXXV–XXXVII	12, 14, 17, 23, 24, 26, 28, 29, 30
$CH_3 \cdot [C{\equiv}C]_3 \cdot CH\overset{L}{=}CH \cdot CH_2OCH_3$	Melting point: 40–40.5°C (sinters at 34°C) UV maxima: 328, 307, 288, 271, 257, 240, 229	XXXV	24

Formula (Common Name)	Organisms	Significant Characteristics	References
C₁₀ Compounds (continued)			
$CH_3 \cdot [C{\equiv}C]_3 \cdot CH\overset{c}{=}CH \cdot CH_2OH$ (cis-Dehydromatricarianol)	XXXVI	Cyclized with dilute base to the enol-ether, UV maxima: 320, 270, 255, 240, 218	29
$CH_3O_2C \cdot CH\overset{t}{=}CH \cdot [C{\equiv}C]_2 \cdot CH{=}CH \cdot CO_2CH_3$	XII, XXXVIII	Melting point: 106.5–110°C, UV maxima: 339, 317, 298, 269, 216	10, 31
$HO_2C \cdot CH\overset{t}{=}CH \cdot [C{\equiv}C]_2 \cdot CH\overset{t}{=}CH \cdot CO_2H$	VI, XII, XXI, XXV, XXXIX	Decomposes at *ca.* 200°C, UV maxima: 338, 315, 296, 267, 216	9, 10, 19, 23
$HOCH_2 \cdot CH\overset{t}{=}CH \cdot [C{\equiv}C]_2 \cdot CH\overset{t}{=}CH \cdot CH_2OH$	XII, XXXII, XXXVIII	Melting point: 153–155°C, UV maxima: 311.5, 293, 276, 261, 246.5, 237, 230	25, 31, 32
$CH_3 \cdot CH\overset{t}{=}CH \cdot [C{\equiv}C]_2 \cdot CH\overset{t}{=}CH \cdot CO_2H$ (Matricaria acid)	XII	Melting point: 175–177°C (decomposes), UV maxima: 329, 310, 256, 245	10
$CH_3 \cdot CH\overset{t}{=}CH \cdot [C{\equiv}C]_2 \cdot CH\overset{t}{=}CH \cdot CO_2CH_3$ (Matricaria ester)	XII, XXXVIII	Melting point: 62–63°C, UV maxima: 333, 314, 296 (inflection), 258, 246, 234 (inflection)	10, 31
$CH_3 \cdot CH\overset{t}{=}CH \cdot [C{\equiv}C]_2 \cdot CH\overset{t}{=}CH \cdot CO$ $\overset{\textstyle\mid}{O}$ $CH_3 \cdot CH\overset{t}{=}CH \cdot [C{\equiv}C]_2 \cdot CH\overset{t}{=}CH \cdot \dot{C}H_2$		Melting point: 125–127.5°C, UV maxima: 335, 314, 295, 277.5, 259, 246 (inflection), 238.5, 233 (inflection), 213	10
$CH_3 \cdot CH\overset{t}{=}CH \cdot [C{\equiv}C]_2 \cdot CH\overset{t}{=}CH \cdot CO$ $\overset{\textstyle\mid}{O}$ $CH_3O_2C \cdot CH\overset{t}{=}CH \cdot [C{\equiv}C]_2 \cdot [CH_2]_2 \cdot \dot{C}H_2$	XII	Melting point: 92–94°C, UV maxima: 334, 305, 287, 259, 246.5, 223	10
$CH_3 \cdot CH\overset{t}{=}CH \cdot [C{\equiv}C]_2 \cdot CH\overset{t}{=}CH \cdot CH_2OH$ (Matricarianol)	XII, XXXVIII, XL	Melting point: 105.5–106.5°C, UV maxima: 312, 293, 276, 261, 247, 237, 231.5, 217.5	10, 23, 31
$CH_3 \cdot CH\overset{t}{=}CH \cdot [C{\equiv}C]_2 \cdot CH\overset{c}{=}CH \cdot CH_2OH$ (2-cis-Matricarianol)	XXXIX	Melting point: 15–16°C, UV maxima: 312.5, 293.5, 276.5, 261.5, 246, 237, 229.5, 216.5	23

C_{10} Compounds (continued)

Formula (Common Name)	Organisms	Significant Characteristics	References
$HO_2C \cdot [C{\equiv}C]_3 \cdot [CH_2]_2 \cdot CO_2H$	XLI	Melting point: 170°C (decomposes) UV maxima: 327, 306, 287, 271.5, 257, 244 (inflection), 222.5, 214	23
$HOCH_2 \cdot [C{\equiv}C]_3 \cdot CH_2 \cdot CH(OH) \cdot CO_2H$	XVI	Melting point: 102–103°C (decomposes) $[a]_D$: +44° UV maxima: 283, 267, 253, 211	14
$HOCH_2 \cdot [C{\equiv}C]_2 \cdot CH \overset{t}{=} CH \cdot CH_2 \cdot CH(OH) \cdot CO_2H$	XVI	UV maxima, methyl ester: 280, 264, 251, 238, 230, 208	14
$H[C{\equiv}C]_2 \cdot CH{=}C{=}CH \cdot CH(OH) \cdot CH_2 \cdot CH_2OH$	XLII, XLIII	$[a]_D$: −210° UV maxima: 278.5, 263, 249.5, 237, 225 (inflection), 208	14
$H[C{\equiv}C]_2 \cdot CH{=}C{=}CH \cdot [CH_2]_2 \cdot CH_2OH$	XLII–XLIV	$[a]_D$: −290° UV maxima: 277.5, 262.5, 249, 236.5, 224 (inflection), 207	12
$CH_3 \cdot [C{\equiv}C]_2 \cdot CH{=}C{=}CH \cdot CH_2 \cdot CH_2OH$	XIII	$[a]_D$: +340° UV maxima: 278.5, 263, 249, 237, 225, 208.5	12
$HO_2C \cdot CH \overset{t}{=} CH \cdot [C{\equiv}C]_2 \cdot [CH_2]_2 \cdot CO_2H$	XII, XLI	Characterized as the dimethyl ester (next compound)	10, 23
$CH_3O_2C \cdot CH \overset{t}{=} CH \cdot [C{\equiv}C]_2 \cdot [CH_2]_2 \cdot CO_2CH_3$	XXXVIII	Melting point: 56.5–58°C UV maxima: 303, 285, 270, 255 (inflection), 243 (inflection), 223, 214.5	31
$HO_2C \cdot CH \overset{t}{=} CH \cdot [C{\equiv}C]_2 \cdot [CH_2]_2 \cdot CH_2OH$	XII	Melting point: 154.5–156°C UV maxima: 303, 285, 270, 255 (inflection), 243 (inflection), 222, 215	10
$CH_3O_2C \cdot CH \overset{t}{=} CH \cdot [C{\equiv}C]_2 \cdot [CH_2]_2 \cdot CH_2OH$	XII, XXXVIII	UV maxima: 305, 287, 273, 258 (inflection), 243 (inflection), 223, 215	10, 31
$CH_3 \cdot CH{=}CH \cdot [C{\equiv}C]_2 \cdot [CH_2]_2 \cdot CO_2CH_3$	XXXVIII	UV maxima: 281, 266, 251, 238, 228, 211	31
$HOCH_2 \cdot CH \overset{t}{=} CH \cdot [C{\equiv}C]_2 \cdot [CH_2]_2 \cdot CH_2OH$	XII, XXXVIII	UV maxima: 282, 267, 249, 237, 230, 213	31, 32

Formula (Common Name)	Organisms	Significant Characteristics	References
C$_{10}$ Compounds (continued)			
CH$_3$ · [CH$_2$]$_2$ · [C≡C]$_2$ · CH$\overset{t}{=}$CH · CH$_2$OH	XII, XXXVIII	UV maxima: 282.5, 266.5, 252, 239.5, 228, 213	31, 32
CH$_3$ · CH$_2$ · CH(OH) · [C≡C]$_2$ · CH$\overset{t}{=}$CH · CH$_2$OH	XXXII	Melting point: 55–56°C UV maxima: 283.5, 268, 253, 241	25
CH$_3$ · C≡C · C=CH · CH=CH · CO · CO · CH$_3$ [S]	XIII	UV maxima (in methanol): 347, 295, 250	33
CH$_3$ · C≡C · C=CH · CH=CH · CO · CH(OH) · CH$_3$ [S]	XIII	UV maxima (in methanol): 320, 287 (inflection), 231	33
C$_{11}$ Compounds			
HC≡C · CH$_2$ · [C≡C]$_2$ · CH$\overset{c}{=}$CH · CH$_2$ · CO$_2$H (Drosophilin C)	XXI, XLV	Melting point: 97.5–99°C UV maxima: 280.5, 264.5, 250.5, 238, 226.5, 210.5	19, 34
H$_2$C=C=CH · [C≡C]$_2$ · CH$\overset{c}{=}$CH · CH$_2$ · CO$_2$H (Drosophilin D)	XXI, XLV	Melting point: 22–28°C UV maxima: 308.5, 290.5, 274.5, 259, 217	19, 34
H[C≡C]$_2$ · CH=C=CH · CH(OH) · [CH$_2$]$_2$ · CO$_2$H (Nemotinic acid)	XIX, XLVI–LII	$[\alpha]_D$: +320° UV maxima: 278.5, 263.5, 249.5, 237.5, 209	17, 20, 31, 35
H[C≡C]$_2$ · CH=C=CH · CH(OH) · [CH$_2$]$_2$ · CO$_2$CH$_3$ (Methyl nemotinate)	XLVI	$[\alpha]_D$: +300° UV maxima: 278.5, 263.5, 249.5, 237.5, 209	20
H[C≡C]$_2$ · CH=C=CH · CH · [CH$_2$]$_2$ · CO [O] (Nemotin)	XLVI–LII	$[\alpha]_D$: +350° UV maxima: 278, 262.5, 249, 236.5, 208.5	17, 20, 31, 35
H[C≡C]$_2$ · CH=C=CH · CH · [CH$_2$]$_2$ · CO$_2$H O–(β–D–xylose)	XLVII	$[\alpha]_D$: +237° UV maxima: 279, 264, 250, 237.5, 211	36

Formula (Common Name)	Organisms	Significant Characteristics	References
C₁₁ Compounds (continued)			
$H[C{\equiv}C]_2 \cdot CH{=}C{=}CH \cdot CH(OH) \cdot [CH_2]_2 \cdot CH_2OH$	XLII, XLIV	$[\alpha]_D$: −330° (CH_2Cl_2); UV maxima: 279, 263.5, 250, 237, 225, 208	12, 31
$H[C{\equiv}C]_2 \cdot CH{=}C{=}CH \cdot [CH_2]_3 \cdot CH_2OH$	XLII–XLIV	$[\alpha]_D$: −160°; UV maxima: 278, 263, 249, 237, 224.5, 208	12
$HO_2C \cdot CH\overset{t}{=}CH \cdot [C{\equiv}C]_2 \cdot [CH_2]_3 \cdot CO_2H$	XXI	Melting point: 157−159°C (decomposes); UV maxima: 302, 284, 268, 254 (inflection), 221.5, 215 (inflection)	19
C₁₂ Compounds			
$CH_3 \cdot [C{\equiv}C]_2 \cdot CH{=}C{=}CH \cdot CH(OH) \cdot [CH_2]_2 \cdot CO_2H$ (Odyssic acid)	XLVII	$[\alpha]_D$: +300°; UV maxima: 280.5, 265, 250.5, 238, 211	37
$CH_3 \cdot [C{\equiv}C]_2 \cdot CH{=}C{=}CH \cdot CH \cdot [CH_2]_2 \cdot CO$ — O bridge (Odyssin)	XLVII	$[\alpha]_D$: +360°; UV maxima: 280, 264, 250, 237.5, 210	37
C₁₃ Compounds			
$H[C{\equiv}C]_2 \cdot CH{=}C{=}CH \cdot CH\overset{c}{=}CH \cdot CH\overset{t}{=}CH \cdot CH_2 \cdot CO_2H$ (Mycomycin)	XXI, XLII, LIII	Explodes at 75°C; $[\alpha]_D$: −130°; UV maxima: 281, 267, 256	3, 4, 38
$CH_3 \cdot [C{\equiv}C]_4 \cdot [CH_2]_3 \cdot CH_3$	XXXVI	Melting point: 25−26°C; UV maxima: 357 (inflection), 340.5, 329.5, 318, 308, 298, 289.5, 272.5, 236.5, 224.5, 214	29
$HOCH_2 \cdot CH\overset{t}{=}CH \cdot [C{\equiv}C]_3 \cdot [CH_2]_3 \cdot CH_3$	XXXVI	Melting point: 50−52°C; UV maxima: 330.5, 308.5, 290, 273, 258, 243, 231.5, 223.5 (inflection), 210	29
$CH_3 \cdot [C{\equiv}C]_4 {-}C{-}C{-}C{-}CH_2OH$ (2D, 3L, 4L) with HO HO H / H H OH	XXXVI	Melting point: 155°C (decomposes); $[\alpha]_D$: +1.8°; UV maxima: 358 (inflection), 353, 340 (inflection), 329, 318 (inflection), 308, 289, 272, 239, 227.5, 217, 207.5	29

Formula (Common Name)	Organisms	Significant Characteristics	References
C$_{13}$ Compounds (continued)			
CH$_3 \cdot$ [CH$_2$]$_2 \cdot$ [C≡C]$_3 \cdot$ CH$\overset{t}{=}$CH $\cdot$ CH(OH) $\cdot$ CH$_2$OH	LIV	Melting point: 98–98.5°C (turns red at 68°C) [α]$_D$: +2.1° UV maxima: 331, 309, 290, 273, 259, 244, 232.5, 225 (inflection), 212.5	39
C$_{14}$ Compounds			
HO$_2$C $\cdot$ [C≡C]$_3 \cdot$ CH$_2 \cdot$ CH$\overset{c}{=}$CH $\cdot$ [CH$_2$]$_3 \cdot$ CO$_2$H	XVI	UV maxima of the dimethyl ester: 329, 308, 289.5, 272.5, 257.5, 227, 218, 209.5 (inflection)	14

REFERENCES

1.   Bohlmann, F., Bornowski, H., and Arndt, C., in *Fortschr. Chem. Forsch.,* Vol. 4, p. 138, U. Hofmann, K. Schäfer, and G. Wittig, Eds. Springer-Verlag, Berlin, Germany (1963).
2.   Bohlmann, F., *Fortschr. Chem. Forsch.,* Vol. 6, p. 65, E. Heilbronner, U. Hofmann, K. Schäfer, and G. Wittig, Eds. Springer-Verlag, Berlin, Germany, and New York (1966).
3.   Bu'Lock, J. D., in *Progress in Organic Chemistry,* Vol. 6, p. 86, Sir James Cook and W. Carruthers, Eds. Butterworths, London, England (1964).
4.   Anchel, M., in *Antibiotics,* Vol. 2, p. 190, D. Gottlieb and P. D. Shaw, Eds. Springer-Verlag, Berlin, Germany, and New York (1967).
5.   Glen, A. T., Hutchinson, S. A., and McCorkindale, N. J., *Tetrahedron Lett.,* p. 4223 (1966).
6.   Ashworth, P. J., Jones, E. R. H., Mansfield, G. H., Schlögl, K., Thompson, J. M., and Whiting, M. C., *J. Chem. Soc.,* p. 951 (1958).
7.   Stephenson, J. S., *D. Phil. Thesis,* Oxford, England (1960).
8.   Anchel, M., Silverman, W. B., Valanju, N., and Rogerson, C. T., *Mycologia, 54,* 249 (1962).
9.   Sir Ewart R. H. Jones, Thaller, V., and Turner, J. L., Unpublished Data.
10. Bu'Lock, J. D., Jones, E. R. H., and Turner, W. B., *J. Chem. Soc.,* p. 1607 (1957).
11. Birkinshaw, J. H., and Chaplen, P., *Biochem. J., 60,* 255 (1955).
12. Bew, R. E., Chapman, J. R., Sir Ewart R. H. Jones, Lowe, B. E., and Lowe, G., *J. Chem. Soc. Sect. C,* p. 129 (1966).
13. Jones, E. R. H., *Proc. Chem. Soc.,* p. 199 (1960).
14. Cambie, R. C., Gardner, J. N., Jones, E. R. H., Lowe, G., and Read, G., *J. Chem. Soc.,* p. 2056 (1963).
15. Jones, E. R. H., and Stephenson, J. S., *J. Chem. Soc.,* p. 2197 (1959).
16. Jones, E. R. H., Stephenson, J. S., Turner, W. B., and Whiting, M. C., *J. Chem. Soc.,* p. 2048 (1963).
17. Cambie, R. C., Hirschberg, A., Jones, E. R. H., and Lowe, G., *J. Chem. Soc.,* p. 4120 (1963).
18. Bendz, G., *Ark. Kemi, 14,* 305 (1959).
19. Jones, E. R. H., Leeming, P. R., and Remers, W. A., *J. Chem. Soc.,* p. 2257 (1960).
20. Bew, R. E., Cambie, R. C., Sir Ewart R. H. Jones, and Lowe, G., *J. Chem. Soc. Sect. C,* p. 135 (1966).
21. Sir Ewart R. H. Jones, Lowe, B. E., and Lowe, G., *J. Chem. Soc.,* p. 1476 (1964).
22. Barley, G. C., Day, A. C., Graf, U., Sir Ewart R. H. Jones, O'Neill, I., Tachikawa, R., Thaller, V., and Vere Hodge, R. A., *J. Chem. Soc. Sect. C,* p. 3308 (1971).
23. Gardner, J. N., Jones, E. R. H., Leeming, P. R., and Stephenson, J. S., *J. Chem. Soc.,* p. 691 (1960).
24. McWhorter, E. J., and Anchel, M., *J. Org. Chem., 30,* 2359 (1965).
25. Graf, U., Sir Ewart R. H. Jones, and Thaller, V., Unpublished Data.
26. Cambie, R. C., *D. Phil. Thesis,* Oxford, England (1963).
27. Shannon, P. V. R., *D. Phil. Thesis,* Oxford, England (1964).
28. Bu'Lock, J. D., and Smalley, H. M., *J. Chem. Soc.,* p. 4662 (1962).
29. Sir Ewart R. H. Jones, Lowe, G., and Shannon, P. V. R., *J. Chem. Soc. Sect. C,* p. 139 (1966).
30. Carpenter, J. G., Hodge, P., Jones, Sir E. R. H., and Lowe, G., Unpublished data.
31. Bew, R. E., *D. Phil. Thesis,* Oxford, England (1961).
32. Wright, M. J., *D. Phil. Thesis,* Oxford, England (1970).
33. Curtis, R. F., and Taylor, J. A., *J. Chem. Soc. Sect. C,* p. 1813 (1969).
34. Anchel, M., *Science, 126,* 1229 (1957).
35. Bu'Lock, J. D., Jones, E. R. H., and Leeming, P. R., *J. Chem. Soc.,* p. 4270 (1955).
36. Bu'Lock, J. D., and Gregory, H., *J. Chem. Soc.,* p. 2280 (1960).
37. Bu'Lock, J. D., Jones, E. R. H., and Leeming, P. R., *J. Chem. Soc.,* p. 1097 (1957).
38. Celmer, W. D., and Solomons, I. A., *J. Am. Chem. Soc., 74,* 1870, 2245 (1952); *75,* 1372 (1953).
39. Barley, G. C., Sir Ewart R. H. Jones, and Thaller, V., Unpublished Data.
40. Canhan, S. C., Anchel, M., and Bistis, G. N., *Mycologia, 62,* 599 (1970).

MICROBIAL CAROTENOIDS

DR. T. W. GOODWIN

Carotenoids occur only intracellularly and can be extracted from dried cells or mycelia by lipid solvents, such as ethanol, acetone, or ether. Difficulty can be experienced in extracting carotenoid glucosides. Carotenoids are separated by column or thin-layer chromatography and are usually characterized by their electronic spectra (see Goodwin, T. W. (Ed.), *Chemistry and Biochemistry of Plant Pigments*, Academic Press, London, England, 1965). Further characterization can be made by using infrared, nuclear-magnetic-resonance, and mass spectrometry (see Isler, O. (Ed.), *Carotenoids,* Birkhauser, Basel, Switzerland, 1971).

Carotenoids can be divided into carotenes (hydrocarbons) and xanthophylls (oxygenated carotenoids). Usually they contain forty carbon atoms, with the acyclic lycopene considered as the basic compound from which most, if not all, other carotenoids are derived. According to the new IUPAC-IUB tentative rules on carotenoid nomenclature, the nine carbon end groups are taken as reference points. The acyclic residues (as in lycopene) are given the prefix ψ, so that the systematic name for lycopene is ψ,ψ-carotene. The cyclohexenyl residue with a double bond at position 5,6 is designated β-, and that with a double bond at 4,5 is designated ϵ-. Thus β-carotene becomes β,β-carotene, a-carotene becomes β,ϵ-carotene, and γ-carotene becomes β,ψ-carotene. An aromatic residue is designated either ϕ or χ according to the substituents. Thus leprotene (isorenieratene) is ϕ,ϕ-carotene. The systematic names of the carotenoids tabulated here are given in parentheses after the trivial names.

Lycopene

a-Carotene

β-Carotene

Leprotene

The more saturated derivatives of lycopene, phytoene, phytofluene, ζ-carotene and neurosporene are biosynthetic precursors of lycopene and must be present in trace amounts in all carotenogenic microorganisms. Occasionally they are easily detected, but in other cases biochemical assault, such as treatment with the inhibitor diphenylamine, is necessary to reveal these polyenes.

Table 1 is not intended to be exhaustive in recording the distribution of carotenoids in non-photosynthetic bacteria and fungi, but all the major pigments are indicated with a suitable source. The same carotenes are generally found in both the bacteria and fungi, but divergence is observed in the xanthophylls. In particular, no C_{45} or C_{50} carotenoids have yet been found in fungi.

Table 2 records the distribution of major carotenoids in photosynthetic bacteria. Unlike fungi and non-photosynthetic bacteria where the distribution of carotenoids is sporadic and apparently capricious, all photosynthetic bacteria produce carotenoids. In this they align themselves with higher plants and algae; indeed, no functional photosynthetic tissue is known that does not contain carotenoids as well as chlorophylls. With very few exceptions, e.g., *Rhodomicrobium vannielii,* whose major pigment is β-carotene, all purple photosynthetic bacteria contain acyclic carotenoids. The green photosynthetic bacteria are characterized by aromatic carotenoids.

The numbering of the carotenoids is indicated for lycopene and a-carotene: the plain numerals have priority over prime numerals in the order β,ϵ,ψ. Formulas are given in the tables only when they contain unique features.

CODE DESIGNATIONS OF ORGANISMS

Bacteria

B1	*Flavobacterium dehydrogenans*	F5	*Leucoscypha rutilans*
B2	*Mycobacterium phlei*	F6	*Rhodotorula glutinis*
B3	*Mycobacterium smegmatis*	F7	*Sporidiobolus johnsonii*
B4	*Micrococcus* sp. H5B-2	F8	*Sporobolomyces roseus*
B5	*Micrococcus tetragenus*, pink form	F9	*Verticillium albo-atrum*
B6	*Staphylococcus aureus*	F10	*Caloscypha fulgens*
B7	*Sarcina aurantiaca*	F11	*Lycogola epidendron*
B8	Flexibacteria	F12	*Puccinia graminis*
B9	*Flavobacterium* sp.	F13	*Coleosporium senecionis*
B10	*Streptomyces mediolana*	F14	*Aleuria aurantia*
B11	*Saprospira grandis*	F15	*Plectania coccinea*
B12	*Flexibacter* spp.	F16	*Phillipsia* spp.
B13	*Sarcina lutea*	F17	*Epicoccum nigrum*
B14	*Corynebacterium* spp.		
B15	*Corynebacterium poinsettiae*		
B16	*Halobacterium salinarium*		
B17	Red halophilic bacterium		

Photosynthetic Bacteria

P1	*Rhodopseudomonas spheroides*
P2	*Chlorobium* spp.
P3	*Rhodospirillum rubrum*
P4	*Rhodopseudomonas spheroides,* green mutant
P5	*Chromatium warmingii*
P6	*Rhodomicrobium vannielii*
P7	*Chromatium okenii*

Fungi

F1	*Allomyces javanicus*
F2	*Blakeslea trispora*
F3	*Phycomyces blakesleeanus*
F4	*Neurospora crassa*

TABLE 1
MAJOR CAROTENOIDS IN NON-PHOTOSYNTHETIC BACTERIA AND FUNGI

Common Name (Systematic Name)	Microorganisms		Significant Characteristics[a]	References	
	Bacteria	Fungi		Bacteria	Fungi
Phytoene (7,8,11,12,7′,8′,11′,12′-Octahydro-ψ,ψ-carotene)	B1	F1	Colorless oil UV maxima: 275, 285, 296	1	2
Phytofluene (7,8,11,12,7′,8′-Hexahydro-ψ,ψ-carotene)	B2	F2	Colorless green-fluorescent oil UV maxima: 331, 348, 367	3	1
ζ-Carotene (7,8,7′,8′-Tetrahydro-ψ,ψ-carotene)	B3	F3	Melting point: 42−46°C UV maxima: 378, 400, 425	5	6
Neurosporene (7,8-Dihydro-ψ,ψ-carotene)	B4	F4	Melting point: 117°C UV maxima: 416, 440, 470	7	8
Lycopene (ψ,ψ-Carotene)	B5	F5	Melting point: 172−173°C UV maxima: 446, 472, 505	9	10
β-Zeacarotene (7′,8′-Dihydro-β,ψ-carotene)	−	F6	UV maxima: 406, 428, 454	−	11
γ-Carotene (β,ψ-Carotene)	B6	F7	Melting point: 131−178°C (wide range) UV maxima: 437, 462, 494	12	13
β-Carotene (β,β-Carotene)	B7	F8	Melting point: 178−180°C UV maxima: (425), 451, 482	12	14
Torulene (3′,4′-Dehydro-β,ψ-carotene)	−	F9	Melting point: 183−185°C UV maxima: 460, 484, 518	−	15
(β,γ-Carotene)	−	F10	Melting point: 176−177°C UV maxima: 421, 448, 478	−	16
(3,4-Didehydro-ψ,ψ-carotene)	−	F11	UV maxima: 463, 493, 527	−	17
Isorenieratene; leprotene (φ,φ-Carotene)	B8	−	Melting point: 198−200°C UV maxima: 425, 452, 484	18	−
β-Cryptoxanthin (β,β-Caroten-3-ol)	B9	F12	Melting point: 158−160°C UV maxima: 425, 451, 483	19	20
Rubixanthin (β,ψ-Caroten-3-ol)	B6	F13	UV maxima: 432, 462, 494	12	14
Zeaxanthin (β,β-Carotene-3,3′-diol)	B8	−	Melting point: 205−206°C UV maxima: 423, 451, 483	20	−
Myxobactin (1′-Glucosyloxy-3,4,3′,4′-tetra-dehydro-1′,2′-dihydro-β,ψ-carotene)	B9	−	UV maxima, ethanol: (454), 477, 506	21	−

O—C₆H₁₁O₅

TABLE 1 (Continued)
MAJOR CAROTENOIDS IN NON-PHOTOSYNTHETIC BACTERIA AND FUNGI

Common Name (Systematic Name)	Microorganisms		Significant Characteristics[a]	References	
	Bacteria	Fungi		Bacteria	Fungi
Aleuriaxanthin (1',16'-Didehydro-1',2'-dihydro-β,ψ-caroten-2'-ol)	—	F14	UV maxima: 434, 460, 491	—	24
3-Hydroxyisorenieratene (φ,φ-Caroten-3-ol)	B10	—	UV maxima, diethyl ether: 449, (472)	23	—
Saproxanthin (3',4'-Didehydro-1',2'-dihydro-β,ψ-carotene-3,1'-diol)	B11		Melting point: 178–179°C	24	
Plectaniaxanthin (3',4'-Didehydro-1',2'-dihydro-β,ψ-carotene-1',2'-diol)	—	F15	UV maxima: 445, 470, 502	—	25
Phleixanthophyll [1'-(β-D-Glucopyranosyloxy)-3',4'-didehydro-1',2'-dihydro-β,ψ-caroten-2'-ol]	B2	—	UV maxima, acetone: 454, 478, 509	26	—
(1',2'-Dihydro-β,ψ-caroten-1'-ol)	B2	—	UV maxima:	27	—
3,3'-Dihydroxyisorenieratene (φ,φ-Carotene-3,3'-diol)	B10	—	UV maximum, diethyl ether: 451	28	—
(β,ψ-Caroten-4-one)	B2	—	Melting point: 146°C UV maxima, acetone: 465, (490)	27	—
Deoxyflexixanthin[b] (1'-Hydroxy-3',4'-didehydro-1',2'-dihydro-β,ψ-caroten-4-one)	B8	—	UV maxima: 476.5, 503	21	—
2'-Dehydroplectaniaxanthin (1'-Hydroxy-3',4'-didehydro-1',2'-dihydro-β,ψ-caroten-2'-one)	—	F15	UV maximum: 503	—	10
Flexixanthin (3,1'-Dihydroxy-3',4'-didehydro-1',2'-dihydro-β,ψ-caroten-4-one)	B12	—	UV maxima, acetone: 483, 510	28	—
4-Ketophleixanthophyll (q.v.)	B2	—	UV maxima, acetone: 480, 507	26	—
2'-Dihydrophillipsiaxanthin (q.v.)	—	F16	UV maxima: 482, 510, 544	—	29
Phillipsiaxanthin (1,1'-Dihydroxy-3,4,3',4'-tetrahydro-1,2,1',2'-tetrahydro-ψ,ψ-caroten-2,2'-one)	—	F16	UV maxima: 488, 518, 554	—	29
Rhodoxanthin (4',5'-Didehydro-4,5'-retro-β,β-carotene-3,3'-dione)	B5	F6	Melting point: 219°C UV maxima: 454, 487, 521	9	30

TABLE 1 (Continued)

MAJOR CAROTENOIDS IN NON-PHOTOSYNTHETIC BACTERIA AND FUNGI

Common Name (Systematic Name)	Microorganisms		Significant Characteristics[a]	References	
	Bacteria	Fungi		Bacteria	Fungi
Torularhodin ($3',4'$-Didehydro-β,ψ-caroten-$16'$-oic acid)	—	F17	Melting point: 201–203°C UV maxima: 470, 505, (540)	—	10
Nonaprenoxanthin (C_{45}) [2-(4-Hydroxy-3-methyl-2-butenyl)-$7',8',11',12'$-tetrahydro-ϵ,ψ-carotene]	B1	—	UV maxima: 354, 373, 394	1	—

Sarcinene (C_{50}) [$2,2'$-Bis(3-methyl-2-butenyl)-ϵ,ϵ-carotene]?	B13	—	UV maxima: 415, 440, 469	31	—
Sarcinaxanthin (C_{50})[c] [2-(4-Hydroxy-3-methyl-2-butenyl)-$2'$-(3-methyl-2-butenyl)-ϵ,ϵ-caroten-18-ol]	B13	—	UV maxima: 415, 440, 469	32	—
Decaprenoxanthin (C_{50}) [$2,2'$-Bis(4-hydroxy-3-methyl-2-butenyl)-ϵ,ϵ-carotene]	B1	—	Melting point: 153–155°C UV maxima, acetone: (400), 419, 443, 472	30	—

Corynexanthin (Decaprenoxanthinmonoglucoside)	B14	—	UV maxima, acetone: 418, 440, 468	33, 34	—
Bisanhydrobacterioruberin (C_{50}) [$2,2'$-Bis(3-methyl-2-butenyl-3,4)-$3',4'$-tetradehydro-$1,2,1',2'$-tetrahydro-ψ,ψ-carotene-$1,1'$-diol]	B15	—	Melting point: 170–171°C UV maxima: 368, 387, 465, 493, 527	35	—
Bacterioruberin (C_{50}) [$2,2'$-Bis(3-hydroxy-3-methylbutyl)-$3,4,3',4'$-tetradehydro-$1,2,1',2'$-tetrahydro-ψ,ψ-carotene-$1,1'$-diol]	B16	—	Melting point: 183°C UV maxima, acetone: 372, 388, 468, 498, 532	35	—
2-Isopentenyl-3,4-dehydrorhodopin [2-(3-Methyl-2-butenyl)-3,4-didehydro-1,2-dihydro-ψ,ψ-caroten-1-ol]	B15	—	Melting point: 153°C UV maxima, acetone: 458, 486, 518	35	—

TABLE 1 (Continued)
MAJOR CAROTENOIDS IN NON-PHOTOSYNTHETIC BACTERIA AND FUNGI

Common Name (Systematic Name)	Microorganisms Bacteria	Fungi	Significant Characteristics[a]	References Bacteria	Fungi
Neurosporaxanthin[d] (4'-Apo-β-caroten-4'-oic acid)	—	F4	Melting point: 165–175°C UV maxima: 472.5, (504)	—	36

COOH

| (Methyl-1-mannosyloxy-3,4-didehydro-1,2,-dihydro-8'-apo-ψ-caroten-8'-oate) | B17 | — | UV maxima, acetone: 469, 497 | 7 | — |

$C_6H_{11}O_5O$ COOCH_3

[a] Unless otherwise indicated, the absorption spectra were obtained in light petroleum; the wavelengths in parentheses indicate shoulders in the spectra, not peaks.
[b] Also occurs as the 1'-glucoside myxobactone.
[c] Also occurs as a monoglucoside.
[d] Also occurs as a methyl ester.

TABLE 2
MAJOR CAROTENOIDS IN PHOTOSYNTHETIC BACTERIA

Common Name (Systematic Name)	Organism	Significant Characteristics[a]	Reference
Phytoene (See Table 1)	Widely distributed	Colorless oil UV maxima: 275, 285, 296	
Phytofluene (See Table 1)	Widely distributed	Colorless green-fluorescent oil UV maxima: 331, 348, 367	
Neurosporene (See Table 1)	Widely distributed	Melting point: 117°C UV maxima: 416, 440, 470	
ζ-Carotene, unsymmetrical[b] (7,8,11,12-tetrahydro-ψ,ψ-carotene)	P5	UV maxima: (354), 374, 394, 419.5	37
Chlorobactene (φ,ψ-Carotene)	P2	UV maxima: 434, 460, 491	38
Hydroxychlorobactene[c] (1',2'-Dihydro-φ,ψ-caroten-1'-ol)	P2	UV maxima: 435, 469, 491	38
3,4-Dehydrorhodopin (3,4-Didehydro-1,2-dihydro-ψ,ψ-caroten-1-ol)	P3	UV maxima: 455, 483, 517	39
Rhodopin[c] (1,2-Dihydro-ψ,ψ-caroten-1-ol)	P3	Melting point: 171°C UV maxima: 443, 470, 503.5	39

TABLE 2 (Continued)
MAJOR CAROTENOIDS IN PHOTOSYNTHETIC BACTERIA

Common Name (Systematic Name)	Organism	Significant Characteristics[a]	Reference
Spheroidene, demethylated (3,4-Didehydro-1,2,7′,8′-tetrahydro-ψ,ψ-caroten-1-ol)	P1	UV maxima, benzene: 440.5, 467.5, 501.5	40
Chloroxanthin (1,2,7′,8′-Tetrahydro-ψ,ψ-caroten-1-ol)	P4	UV maxima: 417, 440, 470	41
Anhydrorhodopinol (Anhydrowarmingol) (13-*cis*-ψ,ψ-Caroten-20-ol)	P5	UV maxima, acetone: 362, 450, 470, 501	31
Rhodopinol (Warmingol) (13-*cis*-1,2-Dihydro-ψ,ψ-carotene-1,20-diol)	P5	UV maxima: 445, 471, 503	31

Common Name (Systematic Name)	Organism	Significant Characteristics[a]	Reference
(1′-Methoxy-3′,4′-didehydro-1′,2′-dihydro-β,ψ-carotene)	P6	UV maxima, ethanol: (380), 510	42
Anhydrorhodovibrin (1-Methoxy-3,4-didehydro-1,2-ψ,ψ-carotene)	P3	UV maxima: 455, 482.5, 516	39
Spheroidene (1-Methoxy-3,4-didehydro-1,2,7′,8′-tetrahydro-ψ,ψ-carotene)	P1	Melting point: 141–143°C UV maxima: 429, 455, 486.5	43
Hydroxyspirilloxanthin (1′-Methoxy-3,4,3′4′-tetradehydro-1,2,1′,2′-tetrahydro-ψ,ψ-caroten-1-ol)	P3	UV maxima: 465, 493, 527	39
Rhodovibrin (1′-Methoxy-3′,4′-didehydro-1,2,1′,2′-tetrahydro-ψ,ψ-caroten-1-ol)	P3	Melting point: 168°C UV maxima: 455, 483, 516	39
Hydroxyspheroidene (1′-Methoxy-3′,4′-didehydro-1,2,7,8,1′,2′-hexahydro-ψ,ψ-caroten-1-ol)	P1	UV maxima: 429, 455, 486	44
Spirilloxanthin (1,1′-Dimethoxy-3,4,3′4′-tetradehydro-1,2,1′,2′-tetrahydro-ψ,ψ-carotene)	P3	Melting point: 218°C UV maxima: 468, 499, 534	39

TABLE 2 (Continued)
MAJOR CAROTENOIDS IN PHOTOSYNTHETIC BACTERIA

Common Name (Systematic Name)	Organism	Significant Characteristics[a]	Reference
Rhodopinal (Warmingone)[c] (13-*cis*-1-Hydroxy-1,2-dihydro-ψ,ψ-caroten-20-al)	P5	UV maximum: 498	31
Okenone (1'-Methoxy-1',2'-dihydro-χ,ψ-caroten-4'-one)	P7	UV maxima: 460, 485, 516	45
Spheroidenone (1-Methoxy-3,4-didehydro-1,2,7',8'-tetrahydro-ψ,ψ-caroten-2-one)	P1	UV maxima: 460, 483, 516	43
2,2'-Diketospirilloxanthin (1,1'-Dimethoxy-3,4,3',4'-tetradehydro-1,2,1', 2'-tetrahydro-ψ,ψ-carotene-2,2'-dione)	P1	UV maxima, carbondisulfide: 530, 561, 603	47

[a] Unless otherwise indicated, the absorption spectra were obtained in light petroleum; the wavelengths in parentheses indicate shoulders in the spectra, not peaks.

[b] Replaces ζ-carotene in photosynthetic bacteria.

[c] Also occurs as a glycoside.

REFERENCES

1. Weeks, O. B., and Garner, R. J., *Arch. Biochem. Biophys., 121,* 35 (1967).
2. Turian, G., and Haxo, F. T., *Bot. Gaz., 54,* 254 (1954).
3. Goodwin, T. W., and Jamikorn, M., *Biochem. J., 62,* 269, 275 (1956).
4. Thomas, D. M., and Goodwin, T. W., *Phytochemistry, 6,* 355 (1967).
5. Weeks, O. B., Andrewes, A. G., Brown, B. O., and Weedon, B. C. L., *Nature, 244,* 879 (1969).
6. Goodwin, T. W., *Biochem. J., 50,* 550 (1952).
7. Aasen, A. J., Francis, G. W., and Liaaen-Jensen, S., *Acta Chem. Scand., 23* 2605 (1969).
8. Haxo, F., *Arch. Biochem., 20,* 400 (1949).
9. Reimann, H. A., and Eklund, C. M., *J. Bacteriol., 42,* 605 (1941).
10. Arpin, N., Thesis. University of Lyon, Lyon, France (1968).
11. Simpson, K. L., and Goodwin, T. W., *Phytochemistry, 4,* 193 (1965).
12. Sobin, B., and Stahly, G. L., *J. Bacteriol., 44,* 265 (1942).
13. Fiasson, J. L., *C. R. Acad. Sci. Fr., 264,* 2744 (1967).
14. Lederer, E., *Bull. Soc. Chem. Biol., 20,* 611 (1938).
15. Valadon, L. R. G., and Mummery, R. S., *J. Gen. Microbiol., 45,* 531 (1966).
16. Arpin, N., Fiasson, J. L., Bouchez-Dangye-Caye, M. P., Francis, G. W., and Liaaen-Jensen, S., *Phytochemistry, 10,* 1595 (1971).
17. Liaaen-Jensen, S., *Phytochemistry, 4,* 925 (1965).
18. Liaaen-Jensen, S., *Acta Chem. Scand., 18,* 1562 (1964).
19. Hongen, F., Craig, B., and Ledingham, G., *Can. J. Microbiol., 4,* 521 (1958).
20. McDermott, J. C. D., Britton, G., and Goodwin, T. W., Unpublished Observations (1971).
21. Kleinig, H., Reichenbach, H., and Achenbach, H., *Arch. Mikrobiol., 74,* 223 (1970).
22. Liaaen-Jensen, S., *Phytochemistry, 4,* 925 (1965).

23. Arcamone, F., Camerino, B., Franceschi, G., and Penco, S., *Gazz. Chim. Ital., 100,* 581 (1970).
24. Aasen, A. J., and Liaaen-Jensen, S., *Acta Chem. Scand., 20,* 811 (1966).
25. Arpin, N., and Liaaen-Jensen, S., *Phytochemistry, 6,* 995 (1967).
26. Hertzberg, S., and Liaaen-Jensen, S., *Acta Chem. Scand., 21,* 15 (1967).
27. Hertzberg, S., and Liaaen-Jensen, S., *Acta Chem. Scand., 20,* 1187 (1966).
28. Aasen, A. J., and Liaaen-Jensen, S., *Acta Chem. Scand., 20,* 1970 (1966).
29. Arpin, N., and Liaaen-Jensen, S., *Bull. Soc. Chem. Biol., 49,* 527 (1967).
30. Gribanovski-Sassu, O., and Foppen, F. H., *Phytochemistry, 6,* 907 (1967).
31. Liaaen-Jensen, S., *Pure Appl. Chem., 20,* 421 (1969).
32. Norgård, S., Francis, G. W., Jensen, A., and Liaaen-Jensen, S., *Acta Chem. Scand., 24,* 1460 (1970).
33. Hodgkiss, W., Liston, J., Goodwin, T. W., and Jamikorn, M., *J. Gen. Microbiol., 11,* 438 (1954).
34. Weeks, O. B., and Andrewes, A. G., *Arch. Biochem. Biophys., 137,* 284 (1970).
35. Norgård, S., and Liaaen-Jensen, S., *Acta Chem. Scand., 23,* 1463 (1969).
36. Aasen, A. J., and Liaaen-Jensen, S., *Acta Chem. Scand., 19,* 1843 (1965).
37. Davis, B. H., Holmes, E. A., Loeber, D. E., Toube, T. P., and Weedon, B. C. L., *J. Chem. Soc. Sect. C,* p. 1266 (1969).
38. Liaaen-Jensen, S., Hegge, E., and Jackman, L. M., *Acta Chem. Scand., 18,* 1703 (1964).
39. Liaaen-Jensen, S., *The Constitution of Some Bacterial Carotenoids and Their Bearings on Biosynthetic Problems.* Bruns, Trondheim, Norway (1962).
40. Britton, G., and Goodwin, T. W., *Methods Enzymol., 18C,* 654 (1971).
41. Barber, M. S., Jackman, L. M., Manchand, P. S., and Weedon, B. C. L., *J. Chem. Soc. Sect. C,* p. 2166 (1966).
42. Ben-Aziz, A., Britton, G., and Goodwin, T. W., Unpublished Observations (1970).
43. Liaaen-Jensen, S., *Acta Chem. Scand., 17,* 489 (1963).
44. Liaaen-Jensen, S., *Acta Chem. Scand., 17,* 500 (1963).
45. Schmidt, K., Liaaen-Jensen, S., and Schlegel, H. G., *Arch. Mikrobiol., 46,* 1117 (1963).
46. Land, D. G., quoted by Davies, B. H., in *Chemistry and Biochemistry of Plant Pigments,* p. 489, T. W. Goodwin, Ed. Academic Press, London, England (1965).
47. Jackman, L. M., and Liaaen-Jensen, S., *Acta Chem. Scand., 18,* 1403 (1964).

SIMPLE SULFUR COMPOUNDS

DR. NANCY N. GERBER

Name	Structure	Microorganisms	Reference
Thiourea	H_2NCSNH_2	*Verticillium alboatrum* *Botrytis cinerea*	1
Dimethyl sulfone	$CH_3SO_2CH_2$	*Cladonia deformis*	1
Hydrogen sulfide	H_2S	*Streptomyces odorifer* *Proteus vulgaris* *Escherichia coli*	2
Methyl mercaptan	CH_3SH	Bacterial decomposition of blue-green algae	5
Dimethyl mercaptan	CH_3SCH_3	Bacterial decomposition of blue-green algae	5
Isobutyl mercaptan	$(CH_3)_2CHCH_2SH$	Bacterial decomposition of blue-green algae	5
n-Butyl mercaptan	$CH_3(CH_2)_3SH$	Bacterial decomposition of blue-green algae	5
Isopropyl mercaptan	$(CH_3)_2CHSH$	Actively growing *Microcystis flosaquae*	5
Bis(methylthio) methane	$(CH_3S)_2CH_2$	*Tuber magnatum* (white truffle)	3
N,N-Dimethyl- methioninol	$CH_3SCH_2CH_2CHN(CH_3)_2CH_2OH^a$	*Penicillium camemberti*	4
Fagicladosporic acid	$CH_3(CH_2)_9CH=CH(CH_2)_9COSH$	*Cladosporium fagi*	6
Epicladosporic acid	$CH_3(CH_2)_9CH=CH(CH_2)_{13}COSH$	*Cladosporium fagi*	6
S-Methyl-L- cysteine	$CH_3SCH_2CH(NH_2)COOH$	*Neurospora crassa*	1
S-(2-Amino-ethyl) cysteine, R = H S-(2-Acetamido ethyl) cysteine, R = CH_3CO	$RHNCH_2CH_2SCH_2CH(NH_2)COOH$	*Rozites caperata*	7
β-Methyllanthionine	$HOOCCH(NH_2)CH_2SCH(CH_3)CH(NH_2)COOH$	Yeast	1
Ergothioneine		*Claviceps purpurea* *Coprinus comatus* *Mycobacterium tuberculosis*	1

^{a} Chemically synthesized, not isolated from the microorganism, but the odor is identical to that produced by *P. camemberti* growing on various substrates.

REFERENCES

1. Miller, M. W., *The Pfizer Handbook of Microbial Metabolites.* McGraw–Hill, New York (1961).
2. Collins, R. P., and Gains, H. D., *Appl. Microbiol., 12,* 335 (1964).
3. Frecchi, A., Kienle, M. G., Scala, A., and Caliella, P., *Tetrahedron Lett.,* p. 1681 (1967).
4. Eckert, T., Knieps, A., and Hoffmann, H., *Z. Naturforsch. Teil B, 19,* 1082 (1964).
5. Jenkins, D., Medsker, L. L., and Thomas, J. F., *Environ. Sci. Technol., 1,* 731 (1967).
6. Turner, W. B., *Fungal Metabolites*, Academic Press, New York (1971).
7. Warin, R., Jadot, J., and Casimir, J., *Bull. Soc. R. Sci. Liege, 38,* 5 (1969); *Chem. Abstr., 72,* 39933t (1969).

ALICYCLIC COMPOUNDS

CARBOCYCLIC COMPOUNDS
NOT AROMATIC, NOT TERPENOID OR STEROID, AND NOT INCLUDED ELSEWHERE

DR. NANCY N. GERBER

Name	Structure	Microorganism	Ref.
Dictyopterene A	$BuCH=CH$ ▲ $CH=CH_2$	*Dictyopteris plagiogramma australis*	1
Caldariomycin	[structure: cyclopentane with Cl, Cl, HO, OH]	*Caldariomyces fumago*	2, 3
Sarkomycin	[structure: cyclopentanone with COOH, CH_2, O]	*Streptomyces erythrochromogenes*	2
Terrein	[structure: cyclopentenone with OH, OH]	*Aspergillus terreus* *Penicillium raistrickii*	2, 3, 4
Methyl 2-alkyl-3,5-dichloro-1-hydroxy-4-oxo-cyclopenten-2-enoate	[structure with R^1, R^2, R^3, OH, COOMe] $R^1 = Cl$ $R^2 = O$ $R^3 = Cl$	*Sporormia affinis*	4
Methyl 2-alkyl-5-chloro-1-hydroxy-4-oxocyclopent-2-enoate	$R^1 = Cl$ $R^2 = H, OH$ $R^3 = Cl$	*Sporormia affinis*	4
Methyl 2-alkyl-3,5-dichloro-1,4-dihydroxycyclopent-2-enoate	$R^1 = H$ $R^2 = O$ $R^3 = Cl$	*Periconia macrospinosa*	4
Amidinomycin (Myxoviromycin)	H_2N—[cyclopentane]—$CONHCH_2CH_2C(=NH)NH_2$	*Streptomyces* sp.	5
Vertimycin	$HOCH_2$—[cyclopentanone]—$CHOH$ CH_2OH, O	*Streptomyces* sp.	5

Name	Structure	Microorganism	Ref.
1-Amino-2-nitro-cyclopentanecar-boxylic acid		*Aspergillus wentii*	4
Shikimic acid	R = H, OH	*Escherichia coli*	2
5-Dihydroshikimic acid	R = O	*Escherichia coli* mutants	2
5-Dehydroquinic acid		*Escherichia coli*	2
Prephenic acid		*Escherichia coli* mutants *Neurospora crassa*	2
Chorismic acid		*Aerobacter aerogenes* mutant	6
Frequentin	R = CHO	*Penicillium frequentans*	2, 3
Palitantin	R = CH$_2$OH	*Penicillium frequentans*	2, 3

REFERENCES

1. Moore, R. E., Pettus, J. A., Jr., and Doty, M. S., *Tetrahedron Lett.*, p. 4787 (1968).
2. Miller, M. W., *The Pfizer Handbook of Microbial Metabolites.* McGraw-Hill, New York (1961).
3. Shibata, S., Natori, S., and Udagawa, S., *List of Fungal Products.* Charles C Thomas, Springfield, Illinois (1964).
4. Turner, W. B., *Fungal Metabolites.* Academic Press, New York (1971).
5. Umezawa, H. (Ed.), *Index of Antibiotics from Actinomycetes.* University of Tokyo Press, Tokyo, Japan (1967).
6. Gibson, F., *Biochem. J., 90,* 256 (1964).

THE POLYENE ANTIFUNGAL ANTIBIOTICS

DR. WITOLD MECHLINSKI

The production of polyene antibiotics by streptomycetes is a widespread property. It was found in one study that 130 cultures out of 1,000 produced polyenes.[93]

ISOLATION AND PHYSICAL PROPERTIES

After fermentation, most polyene antibiotics are associated with the mycelium and can be extracted by using wet *n*-butanol, methanol, or other suitable solvents. The extracts are concentrated under reduced pressure, and the crude products are precipitated with acetone, ether or water. Further purificiation depends on the individual properties of the polyene and may involve reprecipitation from different solvents, crystallization, and counter-current distribution. More than sixty different polyenes are reported in the literature; most of them are listed in Table 1 according to their characteristic UV absorption spectra. The UV spectrum of a polyene reflects the presence of a chain of double bonds or chromophore in the molecule and is most useful for classification purposes. In part, the chromophore also determines certain physical properties of a polyene antibiotic, such as solubility in water and organic solvents, the color of the pure substance, stability to light and temperature, and biological activity.

When pure, tetraenes (four conjugated double bonds) are colorless, pentaenes are pale yellow, hexaenes are light yellow, and heptaenes are deep yellow. Solubility in water is limited for all polyenes; it decreases from the tetraenes to the heptaenes in accordance with the increasing hydrophobic character of the chromophore. Stability to light and temperature decreases in the same order. All polyene antibiotics are polyhydroxy compounds with high molecular weight and exhibit absence of a definite melting point; they decompose gradually on heating. Purity of a polyene antibiotic is difficult to define, because most of these substances do not crystallize and, in detailed studies, are found to be mixtures of closely related homologs. The purity of most polyenes can only be estimated on the basis of their biological activity and extinction in the UV region. Additional information concerning the production and some physical properties of the polyenes may be found in References 85 and 60.

CHEMICAL PROPERTIES

The polyene antifungal antibiotics belong to the macrolide (macrocyclic lactone) group of compounds and are chemically related to such antibacterial macrolides as erythromycin, carbomycin (magnamycin), oleandomycin and others, all produced by the streptomycetes. The polyenes differ, however, by the presence of a conjugated chain of double bonds or chromophore in their macrolactone ring, which is larger (26- to 38-membered) than that found in antibacterial macrolides. The high molecular weight (700–1200), the relatively unstable chemical character of the polyenes, and the complexity of their final purification make the structural studies of these antibiotics very complicated. A small number of the isolated polyenes have been investigated chemically in detail and few structures have been proposed. In some instances conflicting results concerning the same antibiotic have been published from different laboratories, and in others the proposed chemical structures were significantly revised by the same authors as more sophisticated analytical and separation methods were made available.

Tables 2 and 3 summarize some characteristics and presently known partial and complete structures of polyenes. The proposed chemical structures from sources indicated by the references were redrawn in order to emphasize relationships within and between groups of polyenes. The structure of amphotericin B was used as the basic pattern in the drawings, because it is the only presently known structure in its absolute configuration. Some additional chemical information dealing with the polyenes may be found in References 11 and 60.

BIOLOGICAL ACTIVITIES

The polyenes are very potent agents that inhibit the growth of fungal species. Their antiprotozoal activity is also very pronounced. None of the polyenes exhibits any appreciable antibacterial activity. The antifungal potency increases with the length of the chromophore from tetraenes to heptaenes. For polyenes with chromophores of the same size, activity is higher if an amino sugar moiety is present in the molecule. This activity is still higher for antibiotics containing, in addition, an aromatic moiety, as in the case of the so-called aromatic heptaenes, viz., candicidin, trichomycin, hamycin, levorin, aureofungin, candimycin, perimycin, and others. However, this very high activity is usually obscured by considerable toxicity.

Tables 4 and 5 show the antifungal MIC's (Minimal Inhibitory Concentrations) and the toxicity in small animals of the better-known polyenes. The data presented here have to be considered with caution in view of the variety of methods employed in the measurements of activity and toxicity and also in view of the uncertain purity and homogeneity of some of the studied polyenes.

TABLE 1
THE POLYENE ANTIFUNGAL ANTIBIOTICS

Chromophore Type	First Three λλ Maxima	Typical Antibiotics of Known Chemical Structure	Antibiotic (Alternate Names)	Producing Microorganism
Tetraenes	291; 304; 318; ± 2 nm	Nystatin Pimaricin Rimocidin Tetrin A and B	Akitamycin	*Streptomyces akitaensis*
			Amphotericin A	*Streptomyces nodosus*
			Chromin	*Streptomyces antibioticus*
			Endomycin A (Helixin A)	*Streptomyces endus*
			Etruscomycin (Lucensomycin)	*Streptomyces lucensis*
			Nystatin (Fungicidin)	*Streptomyces noursei*
			Pimaricin (Tennecetin)	*Streptomyces natalensis*
			Protocidin	*Streptomyces* sp.
			Rimocidin	*Streptomyces rimosus*
			Sistomycosin	*Streptomyces viridosporus*
			Tetraesin	*Streptomyces* sp. No. 5391
			Tetrin A and B	*Streptomyces* sp.
			Unamycin	*Streptomyces fungicidicus*
			Antibiotic PA-166	*Streptomyces* sp.
			Antibiotic 7071 R.P.	*Streptomyces* sp.
			Antibiotic 9971 R.P.	*Streptomyces gascariensis*
Pentaenes	317; 331; 350; ± 2 nm	Eurocidin	Aliomycin	*Streptomyces acidomyceticus*
			Capacidin	*Streptomyces noursei* (variant)
			Distamycine	*Streptomyces distallicus*
			Eurocidin	*Streptomyces albireticuli*
			Fungichromatin	*Streptomyces cellulosae*
			Antibiotic PA-153	*Streptomyces* sp.
			Antibiotic 2814-P	*Streptomyces* sp.

TABLE 1 (Continued)
THE POLYENE ANTIFUNGAL ANTIBIOTICS

Chromophore Type	First Three λλ Maxima	Typical Antibiotics of Known Chemical Structure	Antibiotic (Alternate Names)	Producing Microorganism
Methylpentaenes	323; 340; 357; ± 2 nm	Chainin Filipin Fungichromin Lagosin	Aurenin	*Streptomyces aureorectus*
			Cabicidin	*Streptomyces gougerotii*
			Chainin	*Chainia* sp.
			Durhamycin	*Streptomyces durhamensis*
			Filipin	*Streptomyces filipinensis*
			Fungichromin	*Streptomyces cellulosae*
			Lagosin (Glaxo A-246)	*Streptomyces pentaticus*
			(Pentamycin) (Moldicidin B)	
			Moldicidin A	*Streptomyces* sp.
			Xantholycin	*Streptomyces xantholyticus*
Carbonylpentaenes	364 nm; broad peak	Mycoticin A and B	Flavofungin	*Streptomyces flavofungini*
			Flavomycoin	*Streptomyces roseoflavus*
			Mycoticin	*Streptomyces ruber*
Hexaenes	340; 358; 380; ± 2 nm	None	Cryptocidine	*Streptomyces* sp.
			Endomycin B (Helixin B)	*Streptomyces endus*
			Flavacid	*Streptomyces flavus*
			Hexin	*Streptomyces* sp.
			Mediocidin	*Streptomyces mediocidicus*
Carbonylhexaenes	385 nm; broad peak	None	Dermostatin	*Streptomyces viridogriseus*
Heptaenes	361; 382; 405; ± 2 nm	Amphotericin B Candidin Antibiotic DJ-400	Amphotericin B	*Streptomyces nodosus*
			Antifungin 4915	*Streptomyces* sp.
			Ascosin	*Streptomyces canescus*
			Aureofacin	*Streptomyces aureofaciens*
			Aureofungin	*Streptomyces cinnamoneus*
			Ayfactin A and B	*Streptomyces aureofaciens*
			Candicidin	*Streptomyces griseus*
			Candidin	*Streptomyces viridoflavus*

TABLE 1 (Continued)
THE POLYENE ANTIFUNGAL ANTIBIOTICS

Chromophore Type	First Three λλ Maxima	Typical Antibiotics of Known Chemical Structure	Antibiotic (Alternate Names)	Producing Microorganism
Heptaenes (*cont.*)			Candimycin	*Streptomyces echimensis*
			Eurotin A	*Streptomyces griseus*
			Geobriecin	*Streptomyces jujuy*
			Hamycin	*Streptomyces pimprina*
			Heptamycin	*Streptomyces* sp.
			Levorin	*Streptomyces globisporus*
			Monicamycin	*Streptoverticillum cinnamoneus* var. *monicae*
			Mycoheptin	*Streptomyces netropsis*
			Neoheptaene	*Streptomyces* sp.
			Perimycin (Fungimycin)	*Streptomyces coelicolor* var. *aminophilus*
			Trichomycin	*Streptomyces hachijoensis*
			Antibiotic PA-150	*Streptomyces* sp.
			Antibiotic F-17-C	*Streptomyces cinnamoneus*
			Antibiotic 757	*Streptomyces* sp.
			Antibiotic 2814-H	*Streptomyces* sp.
			Antibiotic DJ-400	*Streptomyces* sp.

TABLE 2
STRUCTURES OF POLYENE ANTIBIOTICS

Antibiotic (Alternate Names)	General Comments	Empirical Formula	M.W.	Structure
Nystatin (Mycostatin®) (Fungicidin)	Nystatin is a mixture of two antibiotics: nystatin A_1 (73—85%) and nystatin A_2 (27—15%).[81] The structure of nystatin A_2, though not fully known, is closely related to that of nystatin A_1.			
	Nystatin A_1 [9]	$C_{47}H_{75}O_{17}N$	926.1	
Pimaricin[20-22] (Mycoprozine®) (Tennecetin)		$C_{33}H_{47}O_{13}N$	665.7	 $R_1 = CH_3$
Etruscomycin[13,14,25] (Lucensomycin)		$C_{36}H_{53}O_{13}N$	707.7	Same as for pimaricin, except $R_1 = CH_2 \cdot CH_2 \cdot CH_2 \cdot CH_3$
Rimocidin[26,27]	The point of attachment of mycosamine to the rimocidin aglycone is not determined.	$C_{38}H_{63}O_{13}N^a$	741.9	
Tetrin A[30]	Tetrin A and B are produced by the same micro-organisms in a ratio of 3:5.[29] Tetrin A is about three times as active against *Penicillium oxalicum* as tetrin B.[30]	$C_{34}H_{51}O_{13}N$	681.8	

TABLE 2 (Continued)
STRUCTURES OF POLYENE ANTIBIOTICS

Antibiotic (Alternate Names)	General Comments	Empirical Formula	M.W.	Structure
Tetrin B[31]	See tetrin A.	$C_{34}H_{51}O_{14}N$	697.8	
Eurocidin[33]	Eurocidin is a mixture of two pentaene antibiotics: eurocidin A and eurocidin B.[33]	$\left. \begin{array}{l} C = 57.97\% \\ H = 8.17\% \\ N = 1.65\% \end{array} \right\} a$		 R = CH₃ R = H
	Eurocidin A			
	Eurocidin B			
Chainin[77]	Crude chainin is a mixture of about 93% chainin, 4% norchainin and 3% homochainin.[77] The latter two are close chemical homologs of chainin.	$C_{33}\overline{H}_{54}O_{10}$	610.8	
Filipin[39-43,77]	Filipin is a mixture of at least four components.[41,42]			 R = H
	Filipin I (4%)	$C_{35}H_{58}O_9$		
	Filipin II (25%)	$C_{35}H_{58}O_{10}$		
	Filipin III (53%)	$C_{35}H_{58}O_{11}$	654.8	
	Filipin IV (18%)	$C_{35}H_{58}O_{11}$	654.8	

TABLE 2 (Continued)
STRUCTURES OF POLYENE ANTIBIOTICS

Antibiotic (Alternate Names)	General Comments	Empirical Formula	M.W.	Structure
Fungichromin[45]		$C_{35}H_{58}O_{12}$	670.8	Same as for filipin, except R = OH
Lagosin[45,49] (Glaxo A-246) (Pentamycin) (Moldicidin B)	Lagosin may be a stereoisomer of fungichromin.[77]	$C_{35}H_{58}O_{12}$	670.8	Same as for filipin, except R = OH
Mycoticin (Flavofungin)	Mycoticin is a mixture of mycoticin A and mycoticin B in a ratio of 1:1.[53] Flavofungin is a mixture of the two components in a ratio of 10:1.[54]			
Mycoticin A[53]		$C_{36}H_{58}O_{10}$	650.8	R = H
Mycoticin B[53]		$C_{37}H_{60}O_{10}$	664.8	R = CH_3
Amphotericin B[62,64] (Fungizone®)	The absolute configuration of all fourteen asymmetric centers of the amphotericin B aglycone was established by X-ray single-crystal analysis[62] as indicated below:	$C_{47}H_{73}O_{17}N$	924.09[62,63]	

Carbon Atom No.	Absolute Configuration
3	R
5	R
8	R
9	R
11	S
13	R
15	S
16	R
17	S
19	R
34	S
35	R
36	R
37	S

TABLE 2 (Continued)
STRUCTURES OF POLYENE ANTIBIOTICS

Antibiotic (Alternate Names)	General Comments	Empirical Formula	M.W.	Structure
Candidin[65]	Crude candidin is a mixture of about 70% candidin, 17% candidoin and 13% candidinin.[65] The latter two are close chemical homologs of candidin.	$C_{47}H_{71}O_{17}N$	922.07	
Antibiotic DJ-400[78]	Antibiotic DJ-400 was separated into two active components: B_1 and B_2.[78]			
	B_1	$C_{66}H_{95}O_{21}N_2$	$1251.0^{a,b}$	R = CH₃
	B_2	$C_{55}H_{88}O_{20}N_2$	$1096.0^{a,b}$	R = CH₃ R = R₁ = H
Perimycin (Nepera)	The structure of perimycin was not proposed. The molecule of the antibiotic contains two basic groups and four C-methyl groups, but no free carboxyl group.[85] It yields p-N-methylaminoacetophenone and a rare amino sugar, 4-amino-4,6-dideoxy-D-mannose (perosamine).	$C_{47}H_{75}O_{14}N_2$	948^a	
	4-Amino-4,6-dideoxy-D-mannose[87]			
	p-N-Methylaminoacetophenone[88]			

a Tentative.
b Calculated.

TABLE 3

SOME PHYSICAL AND CHEMICAL PROPERTIES OF THE POLYENE ANTIBIOTICS

Antibiotic (Alternate Names)	Source (Year of Isolation)	Ionic Character	UV Absorption Characteristics[a]	Specific Rotation,[b] $[\alpha]_D$	Characteristic Structural Features	
					Chromophore	Moiety
Nystatin (Mycostatin®) (Fungicidin)	*Streptomyces noursei*[1] (1950)	Amphoteric[9]	λ_{max}, nm:[80,81] 291, **304**, 318 $E_1^{1\%}$ cm:[80,81] 570, **850**, 775	AcOH:[80] $-10°$ Pyr:[80] $+21°$ DMF:[80] $+12°$	Tetraene	Mycosamine
Pimaricin (Myprozine®) (Tennecetin)[18,19]	*Streptomyces natalensis*[15] (1957)	Amphoteric[20]	λ_{max}, nm:[15] 279, 290, **303**, 318 $E_1^{1\%}$ cm:[15] 360, 710, **1100**, 1020	DMSO:[19] $+180°$	Tetraene	Mycosamine
Etruscomycin (Lucensomycin)	*Streptomyces lucensis*[12] (1957)	Amphoteric[13]	λ_{max}, nm:[12] 290, **303**, 318 $E_1^{1\%}$ cm:[12] 850, **1390**, 1160	Pyr:[12] $+296°$	Tetraene	Mycosamine, *n*-butyl
Rimocidin	*Streptomyces rimosus*[23,24] (1951)	Basic[27]	λ_{max}, nm:[26] 279, 291, **304**, 318 $E_1^{1\%}$ cm:[26] 306, 622, **965**, 890	MeOH:[23] $+75.2°$	Tetraene	Mycosamine, *n*-propyl, α-hydroxylethyl
Tetrin A	*Streptomyces* sp. Ill. No. 155-2[28] (1960)	Amphoteric[30]	λ_{max}, nm:[29] 214, 278, 290, **303**, 318 $E_1^{1\%}$ cm:[29] 194, 442, 812, **1150**, 1109	Pyr:[29] $+27.5°$	Tetraene	Mycosamine
Tetrin B	*Streptomyces* sp. Ill. No. 155-2[28] (1960)	Amphoteric[31]	λ_{max}, nm:[29] 214, 278, 290, **303**, 318 $E_1^{1\%}$ cm:[29] 186, 514, 801, **1128**, 1089	Pyr:[29] $+45°$ EtOH:[29] $+59°$	Tetraene	Mycosamine

TABLE 3 (Continued)
SOME PHYSICAL AND CHEMICAL PROPERTIES OF THE POLYENE ANTIBIOTICS

Antibiotic (Alternate Names)	Source (Year of Isolation)	Ionic Character	UV Absorption Characteristics[a]	Specific Rotation,[b] $[\alpha]_D$	Characteristic Structural Features Chromophore	Moiety
Eurocidin Eurocidin A Eurocidin B	*Streptomyces albireticuli*[32] (1955)	Amphoteric	λ_{max}, nm:[33] 318, **332**, 350	HCl, 0.1N:[32] −200°	Pentaene	Mycosamine *sec*-Butyl Isopropyl
Chainin	*Chainia* sp. No. 3047[76] (1968)	Neutral[77]	λ_{max}, nm:[75] 324, **338**, 357 $E_{1\ cm}^{1\%}$:[75] 840, **1315**, 1276	MeOH:[75] −124.4°	Methylpentaene	*n*-Butyl
Filipin	*Streptomyces filipinensis*[36-38] (1955)	Neutral	λ_{max}, nm:[38] 322, **338**, 355 $E_{1\ cm}^{1\%}$:[38] 1330, **1360**, 910	MeOH:[38] −148.3°	Methylpentaene	*α*-Hydroxyl-*n*-hexyl
Fungichromin	*Streptomyces cellulosae*[44] (1955)	Neutral[45]	λ_{max}, nm:[44] 322, **338**, 356 $E_{1\ cm}^{1\%}$:[44] 960, **1550**, 1460	MeOH:[45] −176° ± 4°	Methylpentaene	*α*-Hydroxyl-*n*-hexyl
Lagosin (Glaxo A-246)[46] (Pentamycin)[47] (Moldicidin B)[48]	*Streptomyces pentaticus*	Neutral[45,49]	λ_{max}, nm:[43] 325, **340**, 357 $E_{1\ cm}^{1\%}$:[43] 1000, **1530**, 1475	MeOH:[45,48] −160° ± 4° Pyr:[45,48] −224°	Methylpentaene	*α*-Hydroxyl-*n*-hexyl
Mycoticin (A + B)	*Streptomyces ruber*[50] (1954)	Neutral[53]	λ_{max}, nm:[53] 262, **364** $E_{1\ cm}^{1\%}$:[53] 79, **948** (broad peak)	Diox:[53] +63.4°	Carbonylpentaene	Isopropyl (mycoticin A), *sec*-butyl (mycoticin B)

TABLE 3 (Continued)
SOME PHYSICAL AND CHEMICAL PROPERTIES OF THE POLYENE ANTIBIOTICS

Antibiotic (Alternate Names)	Source (Year of Isolation)	Ionic Character	UV Absorption Characteristics[a]	Specific Rotation,[b] $[\alpha]_D$	Characteristic Structural Features	
					Chromophore	Moiety
(Flavofungin)	*Streptomyces flavofungini*[51] (1958)			Pyr:[54,82] −95°, MeOH:[54,82] −76°, Diox:[54,82] levorotatory		
Amphotericin B (Fungizone®)	*Streptomyces nodosus*[55,56] (1955)	Amphoteric	λ_{max}, nm:[50,60,61] 345, 363, **382**, 406, $E_{1\,cm}^{1\%}$:[56,60,61] 430, 980, **1670**, 1890	DMF:[61] +238°	Heptaene	Mycosamine
Candidin	*Streptomyces viridoflavus*[66] (1953)	Amphoteric[65]	λ_{max}, nm:[69] 364, **383.5**, 407.5, $E_{1\,cm}^{1\%}$:[69] 985, **1730**, 1910	AcOH: +205° DMF: +365°	Heptaene	Mycosamine
Antibiotic DJ-400	*Streptomyces* sp.[78] (1969)	Amphoteric[78,79]	λ_{max}, nm: 340, 360, **380**, 404	No data	Heptaene, aromatic	
B₁			$E_{1\,cm}^{1\%}$: 416, 760, **1075**, 926			Mycosamine, N-methyl-*p*-aminoacetophenone, 3,5,-heptadienone
B₂			$E_{1\,cm}^{1\%}$: 327, 710, **1036**, 883			Mycosamine, *p*-amino-aceto-phenone
Perimycin (Fungimycin)	*Streptomyces coelicolor* var. *aminophilus*[84,85] (1952) (1968)	Basic[85]	λ_{max}, nm:[85] 361, **383**, 406, $E_{1\,cm}^{1\%}$:[85] **1000**	No data	Heptaene, aromatic	Perosamine, N-methyl-*p*-aminoacetophenone
(Nepera)						

a Values printed in bold-face type are used when only one extinction coefficient is given, and are applied in quantitative analysis.
b Abbreviations used for solvents: AcOH = glacial acetic acid; Diox = dioxane; DMF = dimethylformamide; DMSO = dimethyl sulfoxide; EtOH = ethanol; HCl = hydrochloric acid; MeOH = methanol; Pyr = pyridine.

TABLE 4
ANTIFUNGAL ACTIVITY OF SOME POLYENE ANTIBIOTICS *IN VITRO*

Minimal Inhibitory Concentration, μg/ml

Antibiotic	Candida albicans	Cryptococcus neoformans	Histoplasma capsulatum	Trichophyton rubrum	Saccharomyces cerevisiae	Aspergillus niger	References
Nystatin	1.2–3.1	1.6	1.6	6.3–14.0	0.8–3.1	2.0	2–7
Pimaricin	3.0–6.0	3.0–10.0	1.0–1.6	3.0–15.0	0.9–15.0	1.8	10, 15–17
Etruscomycin	2.5		25.0				12
Rimocidin	3.0–5.0	5.0–10.0	5.0		1.6	1.6	23, 25
Eurocidin	12.5	3.1		25.0		12.5	32
Chainin	1.5	2.5	5.0	10.0	1.5	2.0	75
Filipin	7.7	0.95	7.7	7.7	5.0	3.9	37
Fungichromin	12.5					25.0	44
Lagosin	1.0–2.0	0.6–0.8		1.0–3.0	0.6–0.8	0.6–0.8	46, 47
Mycoticin	4.0–5.0	2.0	10.0	10.0	3.0–20.0	31.0	51, 52
Amphotericin B	0.15–3.7	0.6	0.04–1.0	30.0	0.09–0.5	0.09	2, 10, 55, 57, 58
Candidin	0.6				0.5	0.38–1.0	68
Trichomycin	0.03–0.7	0.17	0.7		0.03–0.25	10.0	10, 58, 71
Perimycin	0.10	0.05			0.07	0.06	85, 86
Candicidin	0.04				0.03	0.83	68
Levorin	0.2	0.025–0.05		5.0	0.05		90, 92
Hamycin	0.01	0.005	0.1–0.5	20.0	0.012–0.015	0.125	89
Aureofungin	0.01–0.4	0.01–0.05		0.4–1.0	0.1–0.4	0.25–0.5	91

TABLE 5
TOXICITY OF SOME POLYENIC ANTIBIOTICS IN SMALL ANIMALS

Antibiotic	LD_{50}, mg/kg	Route	Reference	Antibiotic	LD_{50}, mg/kg	Route	Reference
Nystatin	8000	*per os*	10	Lagosin	1624	*per os*	47
	45	i.p.	10		33.3	i.p.	47
	3	i.v.	10				
Pimaricin	1500	*per os*	15	Mycoticin	> 250	*per os*	51
	250	i.p.	15		25	i.p.	51
	5–10	i.v.	15				
Etruscomycin	1263	*per os*	12	Amphotericin B	> 8000	*per os*	59
	37	i.p.			280–1640	i.p.	10, 59
	45	i.v.			4–6.6	i.v.	59
Rimocidin	50	i.p.	23	Candidin	> 100	*per os*	67
					14	i.p.	67
					1.5	i.v.	
Eurocidin	22–36	i.p.	34	Trichomycin	> 100	*per os*	10
					2.2	i.p.	10
					2.2	i.v.	10
Chainin	> 50	i.p.	75	Perimycin	> 500	*per os*	84
					250	i.v.	84
Fungichromin	> 1000	*per os*	44	Candicidin	98–400	*per os*	3
	16.4	i.p.	44		2.1–7.0	i.p.	3
Levorin	35–40	i.p.	92				

REFERENCES

1. Hazen, E. L., and Brown, R., *Proc. Soc. Exp. Biol. Med., 76,* 93 (1951).
2. Lechevalier, H., in *Antibiotics Annual 1959–1960,* p. 614. Antibiotica, Inc., New York (1960).
3. Hildick-Smith, G., Blank, H., and Sarkany, I., in *Fungus Diseases and Their Treatment,* pp. 372, 388, 476. Little, Brown and Co., Boston, Massachusetts (1964).
4. Winner, H. I., and Hurley, R., in *Candida albicans,* p. 190. Little, Brown and Co., Boston, Massachusetts (1964).
5. Hosoya, S., and Hamamura, N., *J. Antibiot. (Tokyo) Ser. A., 9,* 129 (1956).
6. Kramer, J., *Antibiot. Chemother., 7,* 500 (1957).
7. Baum, C. L., Rubel, H., and Schwarz, J., in *Antibiotics Annual, 1956–1957,* p. 878. Medical Encyclopedia, Inc., New York (1957).
8. Stout, H. A., and Pagano, J. F., *Antibiotics Annual, 1955–1956,* p. 704. Medical Encyclopedia, Inc., New York (1956).
9. Borowski, E., Zielinski, J., Falkowski, L., Ziminski, T., Golik, J., Kolodziejczyk, P., Jereczek, E., Gdulewicz, M., Shenin, Yu., and Kotienko, T., *Tetrahedron Lett.,* p. 685 (1971).
10. Drouhet, E., in *Systemic Mycoses, A CIBA Foundation Symposium,* p. 206, G. E. W. Wolstenholme and R. Rorter, Eds. Little, Brown and Co., Boston, Massachusetts (1967).
11. Dutcher, J. D., in *Kirk-Othmer Encyclopedia of Chemical Technology,* 2nd ed., Vol. 16, pp. 133–143. John Wiley and Sons, New York (1968).
12. Arcamone, F., Bertazzoli, C., Canevazzi, G., DiMarco, A., Ghione, M., and Grein, A., *G. Microbiol., 4,* 119 (1957).
13. Gaudiano, G., Bravo, P., Quilico, A., Golding, B. T., and Rickards, R. W., *Tetrahedron Lett.,* p. 3567 (1966).
14. Gaudiano, G., Bravo, P., Quilico, A., Golding, B. T., and Rickards, R. W., *Gazz. Chim. Ital., 96,* 1470 (1966).
15. Struyk, A. P., Hoette, I., Drost, G., Waisvisz, J. M., van Eek, T., and Hoogerheide, J. C., in *Antibiotics Annual, 1957–1958,* p. 878. Medical Encyclopedia, Inc., New York (1958).
16. Newcomer, V. D., Sternberg, T. H., Wright, E. T., Reisner, R. M., McNall, E. G., and Sorensen, L. D., *Ann. N.Y. Acad. Sci., 89,* 240 (1960).

17. Welsh, A. L., *Ann. N.Y. Acad. Sci., 89,* 267 (1960).
18. Burns, J., and Holtman, D. F., *Antibiot. Chemother., 9,* 398 (1959).
19. Divekar, P. V., Bloomer, J. L., Eastham, J. F., Holtman, D. F., and Shirley, D. A., *Antibiot. Chemother., 11,* 377 (1961).
20. Golding, B. T., Rickards, R. W., Meyer, W. E., and Patrick, J. B., *Tetrahedron Lett.,* p. 3551 (1966).
21. Ceder, O., and Hansson, B., *Tetrahedron Lett.,* p. 3753 (1967).
22. Ceder, O., *Acta Chem. Scand., 18,* 126 (1964).
23. Davisson, J. W., Tanner, F. W., Jr., Finlay, A. C., and Solomons, I. A., *Antibiot. Chemother., 1,* 289 (1951).
24. Davisson, J. W., Tanner, F. W., Jr., Finlay, A. C., and Kane, J. H., *U.S. Patent 2,963,401* (1960).
25. *British Patent 718,021* (1954).
26. Cope, A. C., Axen, U., and Burrows, E. P., *J. Am. Chem. Soc., 88,* 4221 (1966).
27. Cope, A. C., Burrows, E. P., Derieg, M. E., Moon, S., and Wirth, W. D., *J. Am. Chem. Soc., 87,* 5452 (1965).
28. Gottlieb, D., and Pote, H. L., *Phytopathology, 50,* 817 (1960).
29. Rinehart, K. L., Jr., German, V. F., Tucker, W. T., and Gottlieb, D., *Liebigs Ann. Chem., 668,* 77 (1963).
30. Pandey, R. C., German, V. F., Nishikawa, Y., and Rinehart, K. L., Jr., *J. Am. Chem. Soc., 93,* 3738 (1971).
31. Rinehart, K. L., Jr., Tucker, W. P., and Pandey, R. C., *J. Am. Chem. Soc., 93,* 3747 (1971).
32. Nakazawa, K., *J. Agr. Chem. Soc. Japan, 29,* 650 (1955).
33. Horii, S., Shima, T., and Ouchida, A., *J. Antibiot. (Tokyo) Ser. A, 23,* 102 (1970).
34. Okami, Y., Utahara, R., Nakamura, S., and Umezawa, H., *J. Antibiot. (Tokyo) Ser. A, 7,* 98 (1954).
35. Brown, R., and Hazen, E. L., *Anbibiot. Chemother., 10,* 702 (1960).
36. Gottlieb, D., Ammann, A., and Carter, H. E., *Plant Dis. Rep., 39,* 219 (1955).
37. Ammann, A., Gottlieb, D., Brock, T. D., Carter, H. E., and Whitfield, G. B., *Phytopathology, 45,* 559 (1955).
38. Whitfield, G. B., Brock, T. D., Ammann, A., Gottlieb, D., and Carter, H. E., *J. Am. Chem. Soc., 77,* 4799 (1955).
39. Golding, B. T., Rickards, R. W., and Barber, M., *Tetrahedron Lett.,* p. 2615 (1964).
40. Ceder, O., and Ryhage, R., *Acta Chem. Scand., 18,* 558 (1964)..
41. Bergy, M. E., and Eble, T. E., *Biochemistry, 7,* 653 (1968).
42. Pandey, R. C., and Rinehart, K. L., Jr., *J. Antibiot. (Tokyo) Ser. A, 23,* 414 (1970).
43. Rickards, R. W., and Smith, R., *J. Antibiot. (Tokyo) Ser. A, 23,* 603 (1970).
44. Tytell, A. A., McCarthy, F. J., Fisher, W. P., Bolhofer, W. A., and Charney, J., in *Anbibiotics Annual, 1954–1955,* p. 716. Medical Encyclopedia, Inc., New York (1955).
45. Cope, A. C., Bly, R. K., Burrows, E. P., Ceder, O. J., Ciganek, E., Gillis, B. T., Porter, R. F., and Johnson, H. E., *J. Am. Chem. Soc., 84,* 2170 (1962).
46. Ball, S., Bessell, C. J., and Mortimer, A., *J. Gen. Microbiol., 17,* 96 (1957).
47. Umezawa, S., and Tanaka, Y., *J. Antibiot. (Tokyo) Ser. A, 11,* 26 (1958).
48. Ogawa, H., Ito, T., Inoue, S., and Nishio, M., *J. Antibiot. (Tokyo) Ser. A, 13,* 353 (1960).
49. Dhar, M. L., Thaller, V., and Whiting, M. C., *Proc. Chem. Soc. (London),* p. 310 (1960).
50. Burke, R. C., Swartz, J. H., Chapman, S. S., and Huang, W., *J. Invest. Dermatol., 23,* 163 (1954).
51. Uri, J., and Bekesi, I., *Nature, 181,* 908 (1958).
52. Schlegel, R., and Thrum, H., *Experientia, 24,* 11 (1968).
53. Wasserman, H. H., VanVerth, J. E., McCaustland, D. J., Borowitz, I. J., and Kamber, B., *J. Am. Chem. Soc., 89,* 1535 (1967).
54. Bognar, R., Brown, B. O., Lockley, W. J. S., Makleit, S., Toube, T. P., Weedon, B. C. L., and Zsupan, K., *Tetrahedron Lett.,* p. 471 (1970).
55. Gold, W., Stout, H. A., Pagano, J. S., and Donovick, R., in *Antibiotics Annual, 1955–1956,* p. 579. Medical Encyclopedia, Inc., New York (1956).
56. Vandeputte, J., Wachtel, J. L., and Stiller, E. T., in *Antibiotics Annual, 1955–1956,* p. 587. Medical Encyclopedia, Inc., New York (1956).
57. Artis, D., and Baum, G. L., *Antibiot. Chemother., 11,* 373 (1961).
58. Borowski, E., Schaffner, C. P., Lechevalier, H., and Schwartz, B. S., in *Antimicrobial Agents Annual, 1960,* p. 532. Plenum Press, New York (1961).
59. Hildick-Smith, G., Blank, H., and Sarkany, I., in *Fungus Diseases and Their Treatment,* p. 403. Little, Brown and Co., Boston, Massachusetts (1964).
60. Oroshnik, W., and Mebane, A. D., *Prog. Chem. Nat. Prod., 21,* 17 (1963).
61. Dutcher, J. D., Gold, W., Pagano, J. F., and Vandeputte, J., *U.S. Patent 2,908,611* (1959).
62. Mechlinski, W., Schaffner, C. P., Ganis, P., and Avitabile, G., *Tetrahedron Lett.,* p. 3873 (1970).
63. Borowski, E., Zielinski, J., Ziminski, T., Falkowski, L., Kolodziejczyk, P., Golik, J., Jereczek, E., and Adlercreutz, H., *Tetrahedron Lett.,* p. 3909 (1970).
64. Ganis, P., Avitabile, G., Mechlinski, W., and Schaffner, C. P., *J. Am. Chem. Soc., 93,* 4560 (1971).
65. Borowski, E., Falkowski, L., Golik, J., Zielinski, J., Ziminski, T., Mechlinski, W., Jereczek, E., Kolodziejczyk, P., Adlercreutz, H., Schaffner, C. P., and Neelakantan, S., *Tetrahedron Lett.,* p. 1987 (1971).
66. Taber, W. A., Vining, L. C., and Waksman, S. A., *Antibiot. Chemother., 4,* 455 (1954).

67. Vining, L. C., Taber, W. A., and Gregory, F. J., in *Antibiotics Annual, 1954–1955*, p. 980. Medical Encyclopedia, Inc., New York (1955).
68. Lechevalier, H., Borowski, E., Lampen, J. O., and Schaffner, C. P., *Antibiot. Chemother., 11*, 640 (1961).
69. Vining, L. C., and Taber, W. A., *Can. J. Chem., 34*, 1163 (1956).
70. Hosoya, S., Komatsu, N., Soeda, M., Yuwaguchi, T., and Sonoda, Y., *J. Antibiot. (Tokyo) Ser. B, 5*, 564 (1952).
71. Hosoya, S., Komatsu, N., Soeda, M., and Sonoda, Y., *Jap. J. Exp. Med., 22*, 505 (1952).
72. Nakano, H., *J. Antibiot. (Tokyo) Ser. A, 14*, 68 (1961).
73. Nakano, H., *J. Antibiot. (Tokyo) Ser. B, 15*, 39 (1962).
74. Hattori, K., Nakano, H., Seki, M., and Hirata, Y., *J. Antibiot. (Tokyo) Ser. A, 9*, 176 (1956).
75. Gopalkrishnan, K. S., Narasimhachari, N., Joshi, V. B., and Thirumalachar, M. J., *Nature (London), 218*, 597 (1968).
76. Thirumalachar, M. J., Rahalkar, P. W., Deshmukh, P. V., and Sukapure, R. S., *Hind. Antibiot. Bull., 8*, 6 (1965).
77. Pandey, R. C., Narasimhachari, N., Rinehart, K. L., Jr., and Millington, D. S., *J. Am. Chem. Soc., 94*, 4306 (1972).
78. Bohlmann, F., Dehmlow, E. V., Neuhahn, H. J., Brandt, R., and Reinicke, B., *Tetrahedron, 26*, 2191 (1970).
79. Bohlmann, F., Dehmlow, E. V., Neuhahn, H. J., Brandt, R., and Bethke, H., *Tetrahedron, 26*, 2199 (1970).
80. Dutcher, J. D., Boyack, G., and Fox, S., in *Antibiotics Annual, 1953–1954*, p. 191. Medical Encyclopedia, Inc., New York (1954).
81. Shenin, Yu. D., Kotienko, T. V., and Exempliarow, O. N., *Antibiotiki, 13*, 387 (1968).
82. Bognar, R., Farkas, I., Makleit, S., Rakosi, M., Soltesz, J., Somogyi, L., and Zsupan, K., *Antibiotiki, 10*, 1059 (1965).
83. Manwaring, D. G., Rickards, R. W., Gaudiano, G., and Nicolella, V., *J. Antibiot. (Tokyo) Ser. A, 22*, 545 (1969).
84. Oswald, E. J., Reedy, R. J., and Randall, W. A., in *Antibiotics Annual, 1955–1956*, p. 236. Medical Encyclopedia, Inc., New York (1956).
85. Perlman, D., *Prog. Ind. Microbiol., 6*, 1 (1967).
86. Mohan, R. R., Pianotti, R. S., Martin, G. F., Ringel, S. M., Schwartz, B. S., Bailey, E. G., McDaniel, L. E., and Schaffner, C. P., *Antimicrob. Agents Chemother.*, p. 462 (1963).
87. Lee, C. H., and Schaffner, C. P., *Tetrahedron Lett.*, p. 5837 (1966).
88. Lee, C. H., and Schaffner, C. P., *Tetrahedron, 25*, 2229 (1969).
89. Thirumalachar, M. J., Menon, S. K., and Bhatt, V. V., *Hind. Antibiot. Bull., 3*, 136 (1961).
90. Borowski, E., Malyshkina, M., Soloviev, S., and Ziminski, T., *Chemotherapia, 10*, 176 (1965–1966).
91. Thirumalachar, M. J., Rahalkar, P. W., Sukapure, R. S., and Gopalkrishnan, K. S., *Hind. Antibiot. Bull., 6*, 108 (1964).
92. Tzyganov, V. A., Goliakov, P. N., Besborodov, A. M., Namestnikova, V. P., Soloviev, S. N., Malyshkina, M. A., Khopko, G. V., and Bolshakova, L. O., *Antibiotiki, 4*, 21 (1959).
93. Craveri, R., Lugli, A. M., Sgarzi, B., and Giolitti, G., *Antibiot. Chemother., 10*, 306 (1960).

NON-POLYENE MACROLIDES

DR. NANCY N. GERBER

Name	Structure	Microorganism
Methymycin		*Streptomyces eurocidicus* sp.
Neomethymycin		*Streptomyces* sp.
Pikromycin (Picromycin, amaromycin)		*Streptomyces felleus venezuelae*
Antimycin A$_1$	$R = C_6H_{13}$	*Streptomyces kitasawaensis blastmyceticus* at least 5 other spp.
Antimycin A$_3$	$R = C_4H_9$	

Name	Structure	Microorganism

Narbomycin

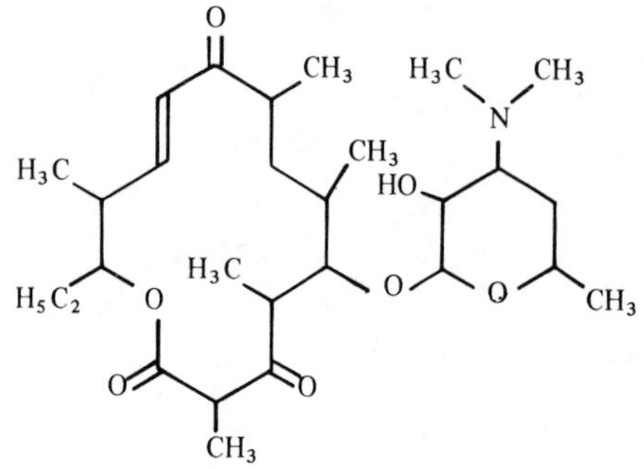

Streptomyces narbonensis

Chalcomycin

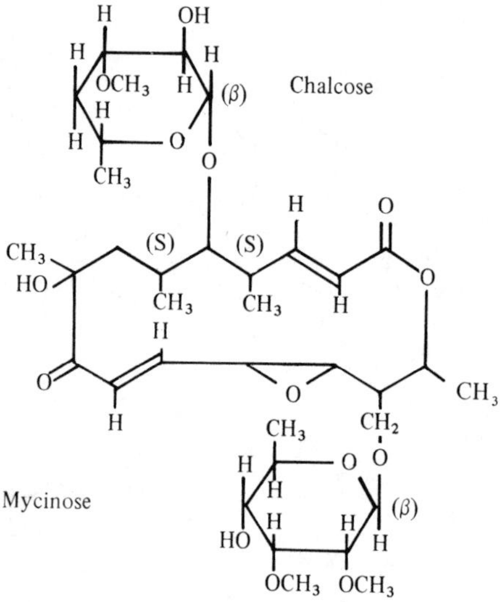

Chalcose

Mycinose

Streptomyces bikiniensis

Oleandomycin
(Amimycin, matromycin, romicil)

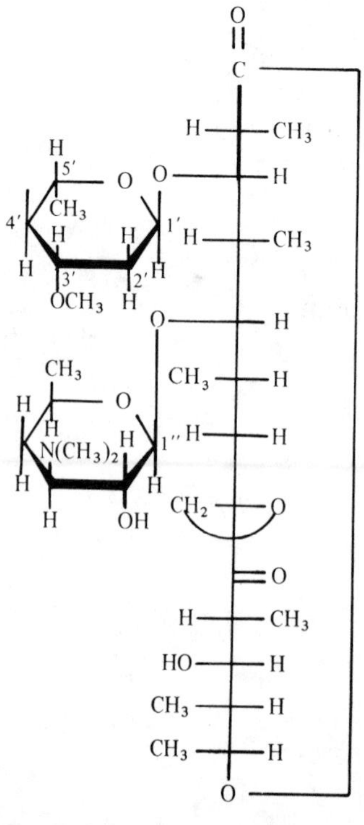

Streptomyces antibioticus olivochromogenes

Name	Structure	Microorganism
Erythromycin C		*Streptomyces erythreus*
Erythromycin B		*Streptomyces erythreus*
Erythromycin		*Streptomyces erythreus*

Name	**Structure**	**Microorganism**

Niddamycin
(3-Desacetylcarbomycin B)

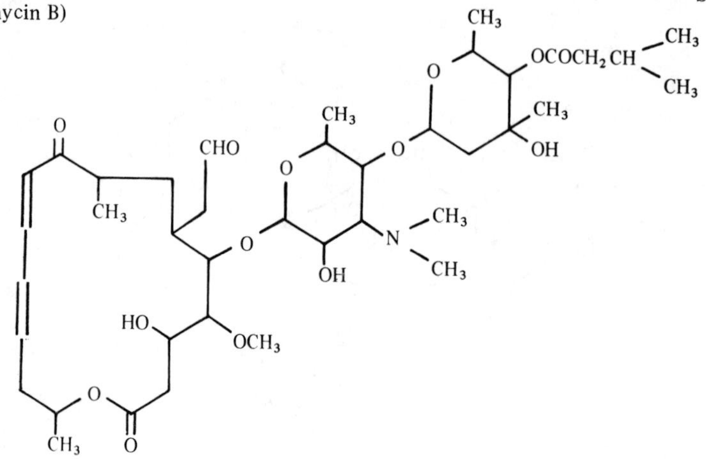

Streptomyces djakartensis

Lankamycin
(Lankavamycin)

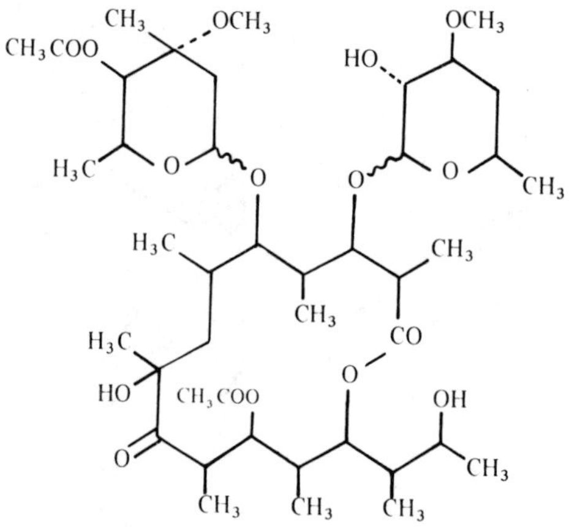

Streptomyces violaceoniger

Carbomycin
(Carbomycin A, magnamycin)

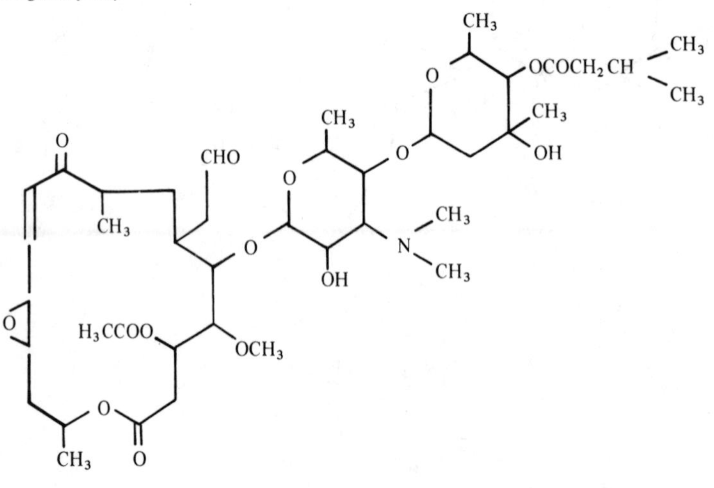

Streptomyces halstedii, hygroscopicus, albireticuli, macrosporus, tendae

Name	Structure	Microorganism

Carbomycin B
(Magnamycin B)

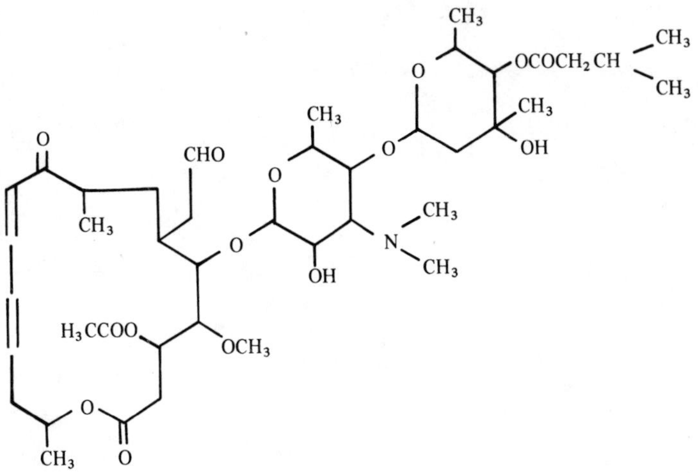

Streptomyces halstedii

Spiramycin I

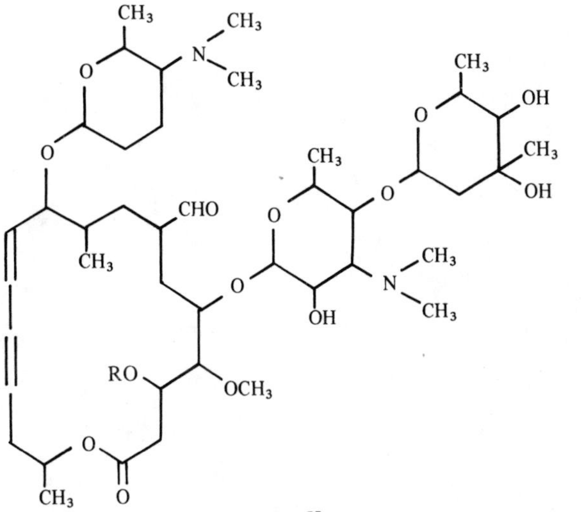

Streptomyces ambofaciens

R = H

Spiramycin II   R = OCH$_3$ *Streptomyces ambofaciens*

Spiramycin III   R = OC$_2$H$_5$ *Streptomyces ambofaciens*

REFERENCE

Umezawa, H. (Ed.), *Index of Antibiotics from Actinomycetes.* University of Tokyo Press, Tokyo, Japan (1967).

MICROBIAL TERPENES

DR. NANCY N. GERBER

More terpenes and terpenoids have been found in higher plants than in microorganisms. T. K. Devon and A. I. Scott, in *The Handbook of Naturally Occurring Compounds,* Vol. 2, *Terpenes* (Academic Press, 1972), list and show formulas of about 380 monoterpenes, 1,000 sesquiterpenes, 650 diterpenes, 13 sesterterpenes and 750 triterpenes from all sources. By comparison, 1 monoterpene, 61 sesquiterpenes, 23 diterpenes, 6 sesterterpenes and 40 triterpenes are found in Chapter 6, Terpenes and Steroids, of W. K. Turner's *Fungal Metabolites* (Academic Press, 1971). Microbial terpenes, however, often have novel and unexpected structures and are more conveniently utilized for biosynthetic studies. A few terpenoid substances have recently been shown to act as hormones in the development of fungi.

Both gibberellins and carotenoids are considered to be terpenes; they are discussed separately elsewhere in this volume, as are steroids and tetracyclic triterpenes. The pentacyclic triterpenes, with two exceptions, are listed among the lichen products.

Terpenes are biosynthesized from acetate via mevalonate. All structures that look as if they were formed from "isoprene units" (that is, mevalonate) are included, whether or not their biosynthesis has actually been investigated. Some compounds that are only partially terpenoid (for example, mycelianamide and bovinones) are also included; others are listed with the azaphilones, quinonemethides and γ-pyronemethides (atrovenetin, herqueinone, norherqueinone), piperazines (echinulin, neoechinulin), miscellaneous peptides (ilamycin), and ergot alkaloids.

TABLE 1
HEMITERPENES
(1 Isoprene Unit, 5 Carbon Atoms)

Compound	Properties	Source	Reference
Furan-3-carboxylic acid	Colorless crystals Melting point: 121°C	*Ceratocystis fimbriata* (on sweet potato)	4
2-Methyl-2-butene	Colorless liquid Boiling point: 38°C	*Puccinia graminis* (uredospores)	4

TABLE 2
MONOTERPENES
(2 Isoprene Units, Usually 10 Carbon Atoms)

Compound	Structure	Source	Ref.
Citronellol	$(CH_3)_2C=CHCH_2CH_2CH(CH_3)CH_2CH_2OR$ R = H	Fusel oil[a] *Ceratocystis variospora*	1, 3
Citronellyl acetate	Same as above, except: R = OAc	*Ceratocystis variospora*	3
a-Terpineol		Fusel oil[a]	1
Linalool	$(CH_3)_2C=CHCH_2CH_2C(CH_3)(OH)CH=CH_2$	Fusel oil[a] *Ceratocystis variospora*	1, 3

TABLE 2 (Continued)
MONOTERPENES

Compound	Structure	Source	Ref.
Geraniol	$(CH_3)_2C=CHCH_2CH_2C(CH_3)=CHCH_2OR$ R = H	Algae *Ceratocystis variospora*	2, 3
Geranyl acetate	Same as above, except: R = OAc	*Ceratocystis variospora*	3
d-Limonene		Algae	2
a-Pinene		Algae	2
Neral	$(CH_3)_2C=CHCH_2CH_2C(CH_3)=CHCHO$	*Ceratocystis variospora*	3
Geranial	$(CH_3)_2C=CHCH_2CH_2C(CH_3)=CHCHO$	*Ceratocystis variospora*	3
Methylisoborneol	OH	*Streptomyces* sp.	5
Ipomeanin	$COCH_2CH_2COCH_3$	*Ceratostomella fimbriata* (on sweet potato)	4
Batatic acid	$COCH_2CH_2CH(CH_3)COOH$	*Ceratostomella fimbriata* (on sweet potato)	4
Mycelianamide	OR	*Penicillium griseofulvum*	6

$$R = CH_2CH=C(CH_3)CH_2CH_2CH=C(CH_3)_2$$

a From black-rotted sweet potato.

TABLE 3
SESQUITERPENES
(3 Isoprene Units, Usually 15 Carbon Atoms)

Compound	Structure	Source	Ref.

Trichothecane Group

Trichodermin

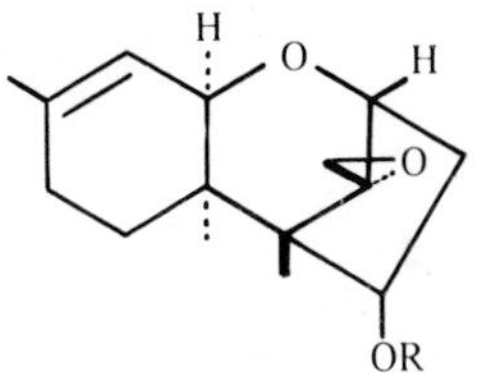

R = Ac

Trichoderma sp. — 6

Trichodermol (Roridin C) — Same as above, except: R = H — *Myrothecium roridum* — 6

Trichothecin

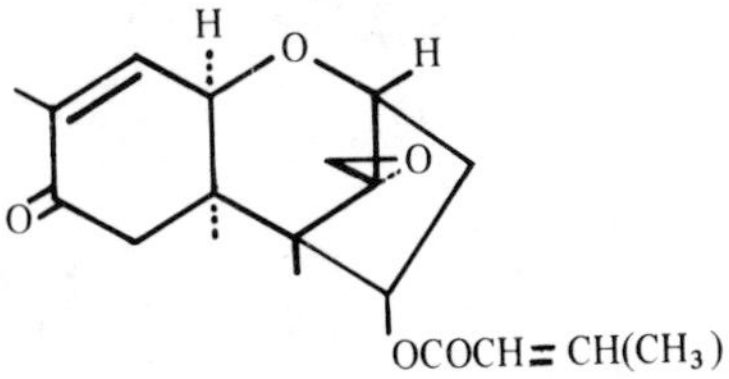

OCOCH=CH(CH$_3$)

Trichothecium roseum — 6

Crotocin (Antibiotic T)

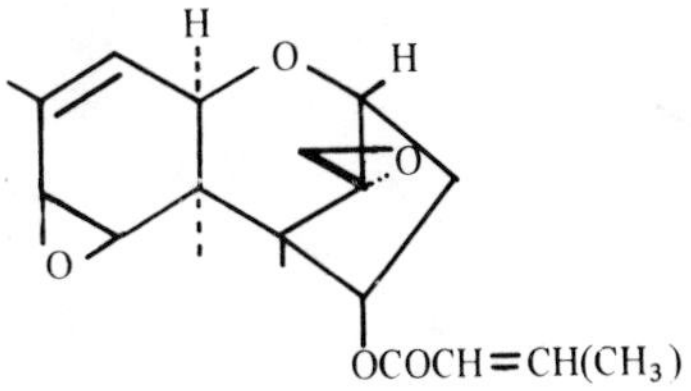

OCOCH=CH(CH$_3$)

Cephalosporium crotocinigenum
Trichothecium roseum — 6

Diacetoxyscirpenol

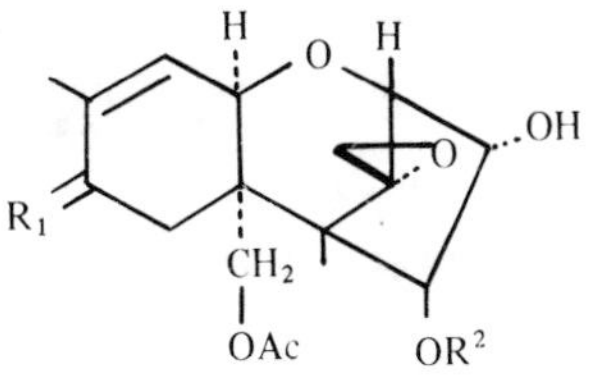

$R_1 = H_2$
$R_2 = Ac$

Fusarium scirpi (= F. equiseti) — 6

Toxin T-2 — Same as above, except: R_1 = H, OCOCH$_2$CH(CH$_3$)$_2$; R_2 = Ac — *Fusarium tricinctum* *Trichothecium lignorum* — 6

Toxin HT-2 — Same as above, except: R_1 = H, OCOCH$_2$CH(CH$_3$)$_2$; R_2 = H — *Fusarium tricinctum* — 6

Nivalenol

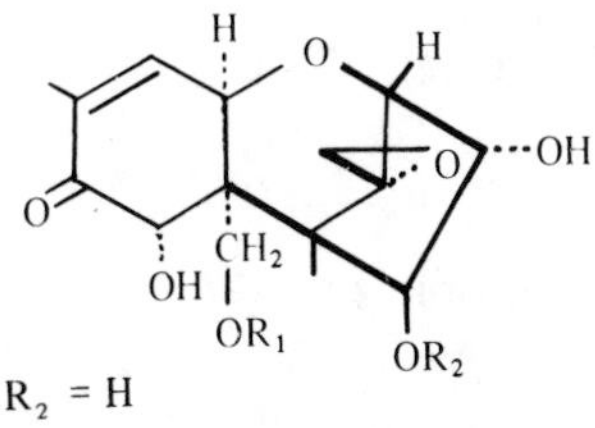

$R_1 = R_2 = H$

Fusarium nivale — 6

TABLE 3 (Continued)
SESQUITERPENES

Compound	Structure	Source	Ref.

Trichothecane Group (continued)

Compound	Structure	Source	Ref.
Nivalenol monoace- tate (Fusarenone)	Same as above, except: R_1 = H R_2 = Ac	*Fusarium* *nivale*	6
4β,15-Diacetoxy- 3α,7α-dihydroxy- 12,13-epoxytri- chothec-9-en-8- one	Same as above, except: $R_1 = R_2$ = Ac	*Fusarium* *scirpi*	6

| 4β,8α,15-Triace-
toxy-12,13-epoxy-
tricho-9-thene-
3α,7α-diol | | *Fusarium*
scirpi | 6 |

| Verrucarol | | *Myrothecium* sp. | 6 |

$R_1 = R_2$ = H

| Diacetoxyverru-
carol | Same as above, except:
$R_1 = R_2$ = Ac | *Myrothecium* sp. | 6 |
| Verrucarin A | Same as above, except: | *Myrothecium*
verrucaria | 6 |

$$R_1 R_2 = -COC-CCH_2 CH_2 OCOCH \overset{t}{=} CHCH \overset{c}{=} CHCO-$$

OH H (above), H Me (below)

| Verrucarin B | Same as above, except: | *Myrothecium*
verrucaria | 6 |

$$R_1 R_2 = -COCHCCH_2 CH_2 OCOCH \overset{t}{=} CHCH \overset{c}{=} CHCO-$$

O (above), Me (below)

| Verrucarin J | Same as above, except: | *Myrothecium*
verrucaria | 6 |

$$R_1 R_2 = -COCH = CCH_2 CH_2 OCOCH \overset{t}{=} CHCH \overset{c}{=} CHCO-$$

Me (below)

| 2'-Dehydroverru-
carin A | Same as above, except: | *Myrothecium*
verrucaria | 6 |

$$R_1 R_2 = -COCOCCH_2 CH_2 OCOCH \overset{t}{=} CHCH \overset{c}{=} CHCO-$$

H (above), Me (below)

TABLE 3 (Continued)
SESQUITERPENES

Compound	Structure	Source	Ref.
Trichothecane Group (continued)			

Roridin A — Same as above, except:

$$R_1 R_2 = \text{–COC–CCH}_2\text{CH}_2\text{OCHCH}\overset{t}{=}\text{CHCH}\overset{c}{=}\text{CHCO–}$$

with OH, H above; H, Me below; CH(OH)Me below

Myrothecium roridum — 6

Roridin D — Same as above, except:

$$R_1 R_2 = \text{–COCHCCH}_2\text{CH}_2\text{OCHCH}\overset{t}{=}\text{CHCH}\overset{c}{=}\text{CHCO–}$$

with O epoxide above; Me below; CH(OH)Me below

Myrothecium roridum — 6

Roridin E — Same as above, except:

$$R_1 R_2 = \text{–COCH=CCH}_2\text{CH}_2\text{OCHCH}\overset{t}{=}\text{CHCH}\overset{c}{=}\text{CHCO–}$$

with Me below; CH(OH)Me below

Myrothecium roridum — 6

Helminthosporal Group

Helminthosporal[a]

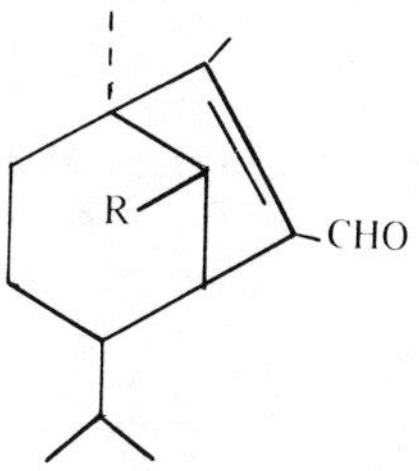

R = CHO

Helminthosporium sativum — 6

Helminthosporol[b] — Same as above, except:
R = CH$_2$OH

Helminthosporium sativum — 6

Prehelminthosporol

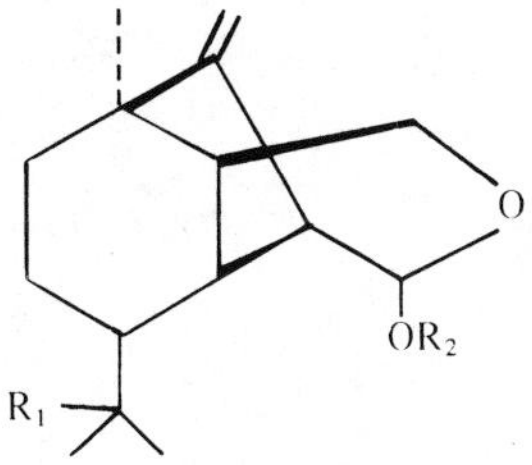

R$_1$ = R$_2$ = H

Helminthosporium sativum — 6

9-Hydroxyprehel- minthosporol — Same as above, except:
R$_1$ = OH
R$_2$ = H

Cochliobolus sativus — 6

Sativene

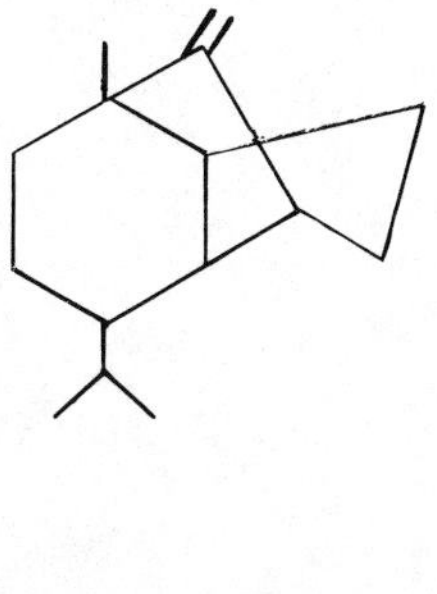

Helminthosporium sativum — 6

TABLE 3 (Continued)
SESQUITERPENES

Compound	Structure	Source	Ref.
Other Sesquiterpenes[c]			
Ipomeamaronol	CH$_2$COCH$_2$CH(CH$_3$)CH$_2$OH	Mold-damaged sweet potato	7
Deoxytrisporone	CH$_2$OH	*Choanephora trispora*	8
Fumagillin	OMe OCO(CH=CH)$_4$COOH	*Apergillus fumigatus*	6
Ovalicin (Graphinone)	OH OMe	*Pseudorotium ovalis* *Graphium* sp.	6
Cyclonerolidol	OH OH	*Trichothecium* sp.	6
Unnamed	OH OH	*Trichothecium* sp.	6
Geosmin	OH	Various Actinomycetes, mainly *Streptomyces*	14

TABLE 3 (Continued)
SESQUITERPENES

Compound	Structure	Source	Ref.

Other Sesquiterpenes (continued)

Cadin-4-ene-1-ol

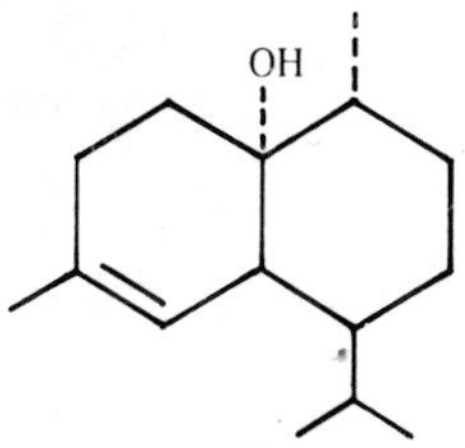

Streptomyces sp.   15

Selina-4-(14),7(11)-diene-9-ol

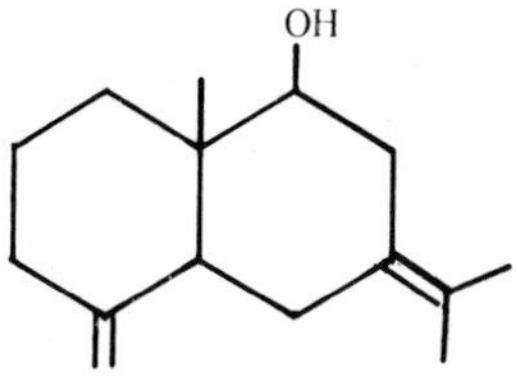

Streptomyces fradiae   16

Lactarorufin A

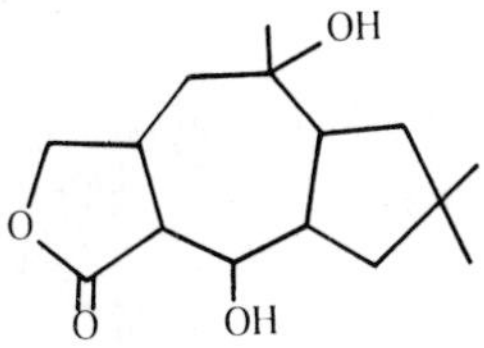

Not known   10

Unnamed

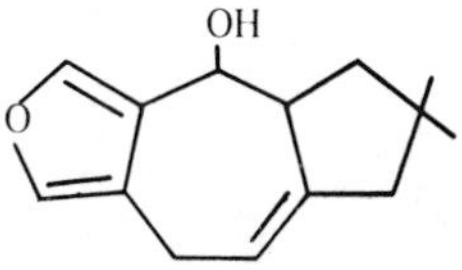

Fomitopsis insularis   9

Unnamed

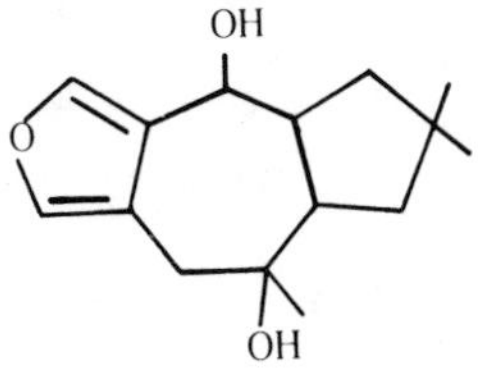

Fomitopsis insularis   9

Lactaroviolin[d]

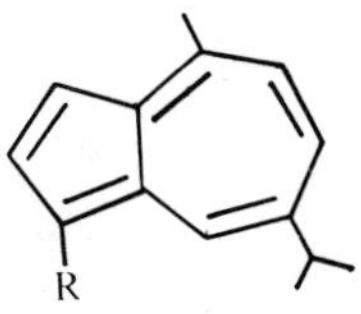

R = CHO

Lactarius deliciosus   6

Lactarazulene[d]   Same as above, except:
R = CH$_3$

Lactarius deliciosus   6

TABLE 3 (Continued)
SESTERTERPENES

Compound	Structure	Source	Ref.
Lactarofulvene[d]		*Lactarius deliciosus*	6
Unnamed	CH$_2$OR R = stearyl	*Lactarius deliciosus* (freshly collected)	12
Unnamed	Same as above, except: R = OH	*Lactarius deliciosus* (freshly collected)	12
Illudalic acid	OH, OHC, HO, O, O	*Clitocybe illudens*	6
Illudinine	OMe, N, COOH	*Clitocybe illudens*	6
Fomannosin	HOCH$_2$, O, O	*Fomes annosus*	6
Illudoic acid (tentative)	OH, O, HO, COOH	*Clitocybe illudens*	6

TABLE 3 (Continued)
SESQUITERPENES

Compound	Structure	Source	Ref.

Other Sesquiterpenes (continued)

Sirenin

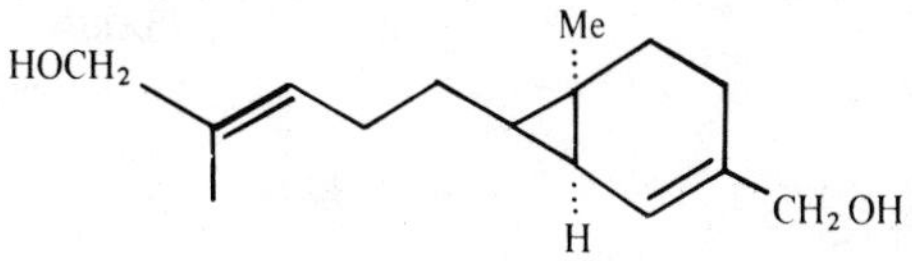

Allomyces sp.   6

Helicobasidin

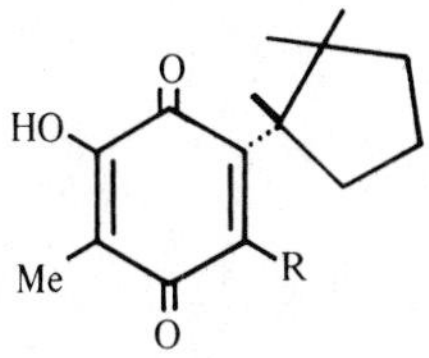

R = OH

Helicobasidium mompa   6

Deoxyhelicobasidin — Same as above, except: R = H — *Helicobasidium mompa* — 6

Unnamed

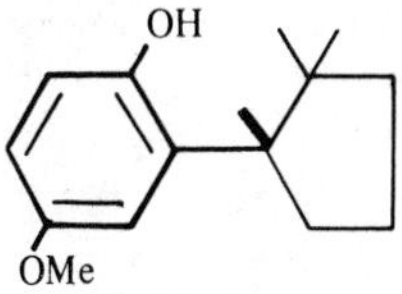

Helicobasidium mompa   6

Pebrolide

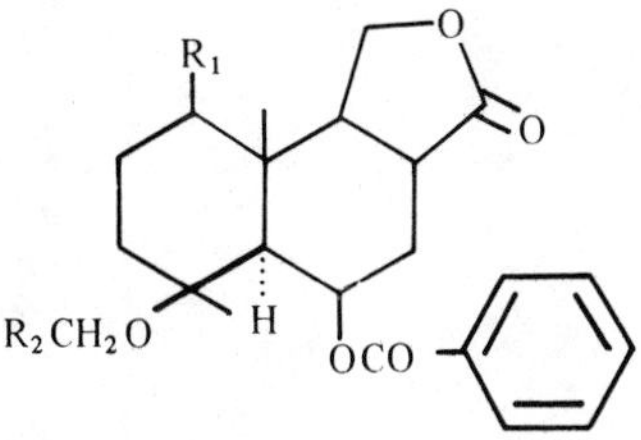

R_1 = OH
R_2 = Ac

Penicillium brevi-compactum   6

1-Deoxypebrolide — Same as above, except: R_1 = H, R_2 = Ac — *Penicillium brevi-compactum* — 6

Desacetylpebrolide — Same as above, except: R_1 = OH, R_2 = H — *Penicillium brevi-compactum* — 6

Coriolin

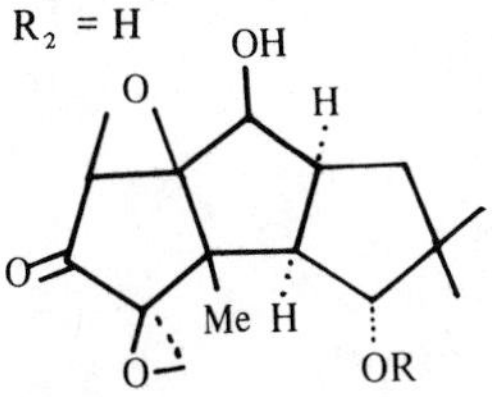

R = H

Coriolus consors   6, 13

Coriolin C — Same as above, except: R = $COCH(OH)C_6H_{13}$ — *Coriolus consors* — 6, 13

TABLE 3 (Continued)
SESQUITERPENES

Compound	Structure	Source	Ref.

Other Sesquiterpenes (continued)

Compound	Structure	Source	Ref.
Coriolin B	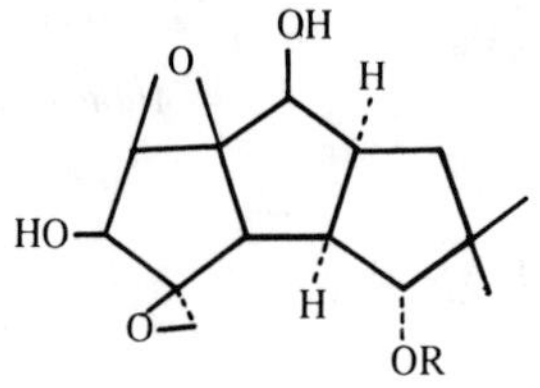	*Coriolus consors*	6, 13

$R = O_2 CC_7 H_{15}$

Compound	Structure	Source	Ref.
Illudol	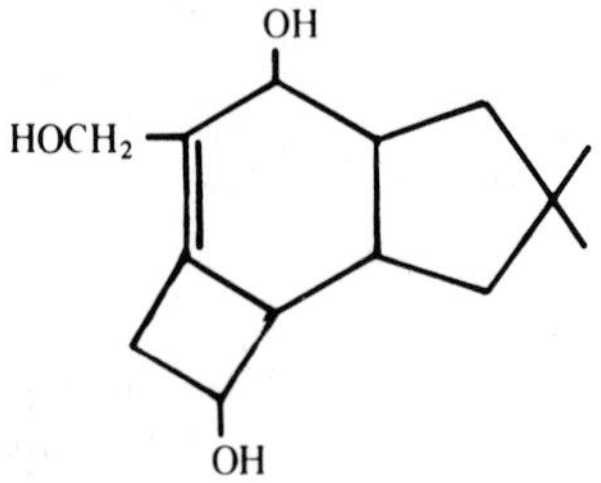	*Clitocybe illudens*	6
Illudin M	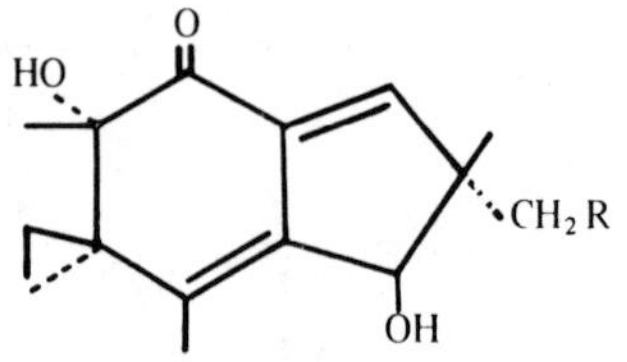	*Clitocybe illudens*	6

$R = H$

Illudin S (Lampterol)	Same as above, except: $R = OH$	*Clitocybe illudens* *Lampteromyces japonicus*	6
Dihydroilludin S	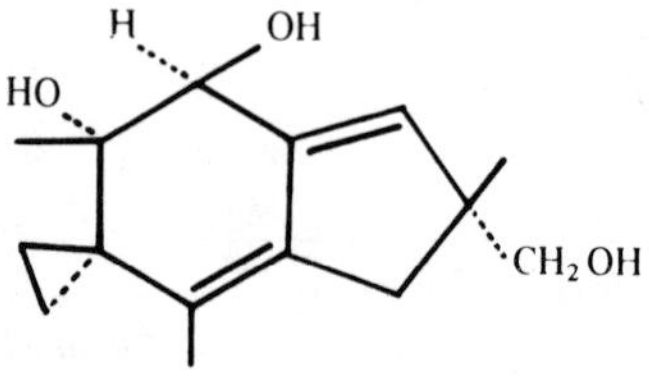	*Lampteromyces japonicus*	6
Marasmic acid	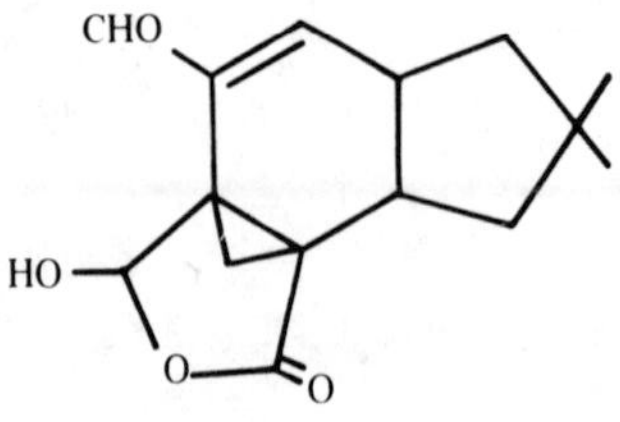	*Marasmius conigenus*	6

TABLE 3 (Continued)
SESQUITERPENES

Compound	Structure	Source	Ref.

Other Sesquiterpenes (continued)

Hirsutic acid C

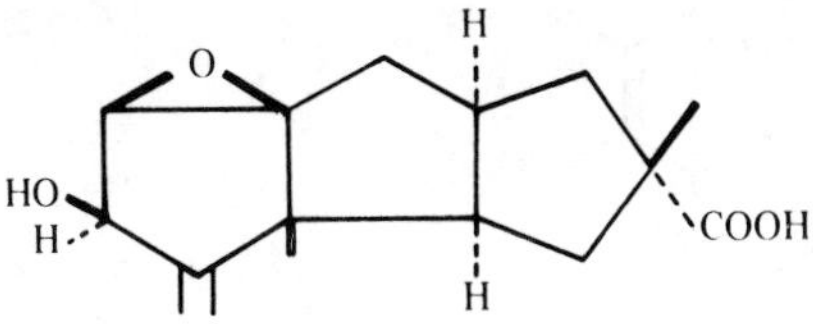

Stereum hirsutum — 6

Culmorin

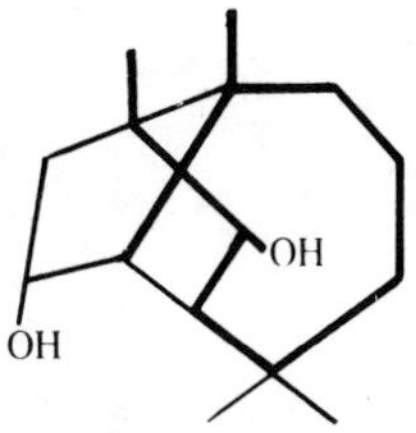

Fusarium culmorum — 6

[a] Known to be an artifact of the isolation procedure.
[b] May be an artifact.
[c] In order of increasing number of carbocyclic rings.
[d] May be an artifact due to enzyme action.

TABLE 4
COMPOUNDS WITH A SESQUITERPENE UNIT
ATTACHED TO AN AROMATIC NUCLEUS OF THE ORSELLINIC ACID TYPE

Compound	Structure	Source	Ref.

Grifolin

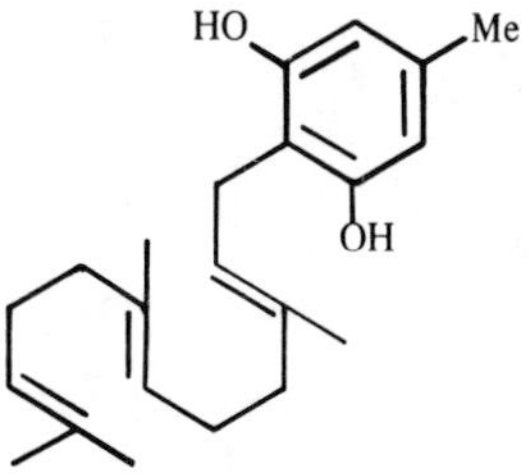

Grifola confluens — 6

Antibiotic LL-Z1272*a*

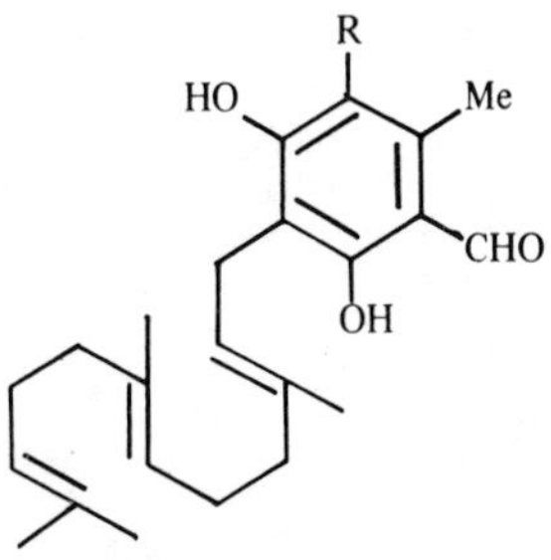

R = Cl

Fusarium sp. — 6

Antibiotic LL-Z1272*β*

Same as above, except:
R = H

Fusarium sp. — 6

TABLE 4 (Continued)
COMPOUNDS WITH A SESQUITERPENE UNIT
ATTACHED TO AN AROMATIC NUCLEUS OF THE ORSELLINIC ACID TYPE

Compound	Structure	Source	Ref.
Antibiotic LL-Z1272δ	R = Cl	*Fusarium* sp.	6
Antibiotic LL-Z1272ε	Same as above, except: R = H	*Fusarium* sp.	6
Antibiotic LL-Z1272γ (Ascochlorin)	R = H	*Fusarium* sp. *Ascochyta viciae*	6
Antibiotic LL-Z1272ζ	Same as above, except: R = OAc	*Fusarium* sp. *Ascochyta viciae*	6
Presiccanochromenic acid		*Helminthosporium siccans*	6
Siccanochromenic acid	R = CO_2H	*Helminthosporium siccans*	6

TABLE 4 (Continued)
COMPOUNDS WITH A SESQUITERPENE UNIT
ATTACHED TO AN AROMATIC NUCLEUS OF THE ORSELLINIC ACID TYPE

Compound	Structure	Source	Ref.
Siccanochromene A	Same as above, except: R = H	*Helminthosporium* *siccans*	6
Siccanochromene B		*Helminthosporium* *siccans*	6
Siccanochromene E		*Helminthosporium* *siccans*	6
Siccanin		*Helminthosporium* *siccans*	6
Tauranin		*Oospora* *aurantia*	6
Boviquinone-3		*Gomphidius* *rutilus*	17

TABLE 4 (Continued)
COMPOUNDS WITH A SESQUITERPENE UNIT
ATTACHED TO AN AROMATIC NUCLEUS OF THE ORSELLNIC ACID TYPE

Compound	Structure	Source	Ref.
Diboviquinone-3,4	$$X = -CH_2CH = C(CH_3) - CH_2 -$$	*Gomphidius rutilus*	17
Methylenediboviquinone-3,3	$$X = -CH_2CH = C(CH_3) - CH_2 -$$	*Gomphidius rutilus*	17

TABLE 5
DITERPENES (EXCLUDING GIBBERELLINS)
(4 Isoprene Units, Usually 20 Carbon Atoms)

Compound	Structure	Source	Ref.
13-Epi-(–)-manoyl oxide		*Gibberella fujikuroi*	6
Pleuromutilin		*Pleurotus mutilus passeckerianus* *Drosophila subatrata*	6

TABLE 5 (Continued)
DITERPENES (EXCLUDING GIBBERELLINS)

Compound	Structure	Source	Ref.
Antibiotic LL-Z1271α	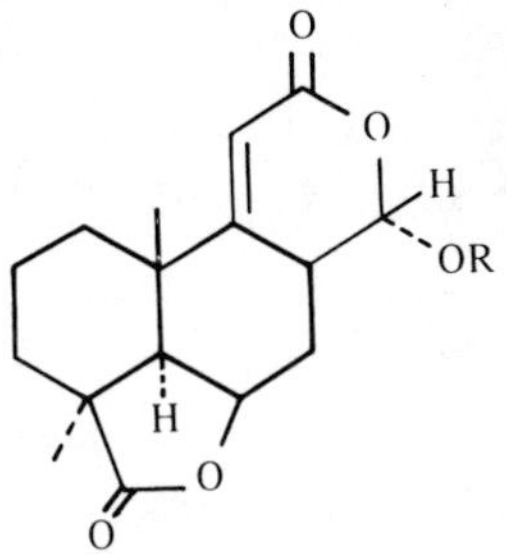	*Acrostalagmus* sp.	6
Antibiotic LL-Z1271γ	Same as above, except: R = H	*Acrostalagmus* sp.	6
Antibiotic LL-Z1271β	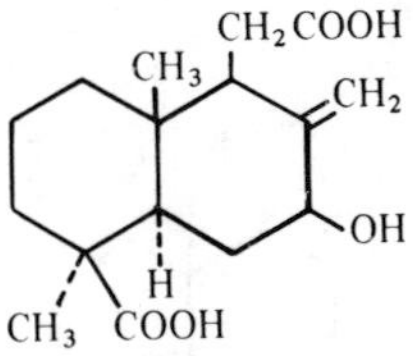	*Acrostalagmus* sp.	11
Virescenol A	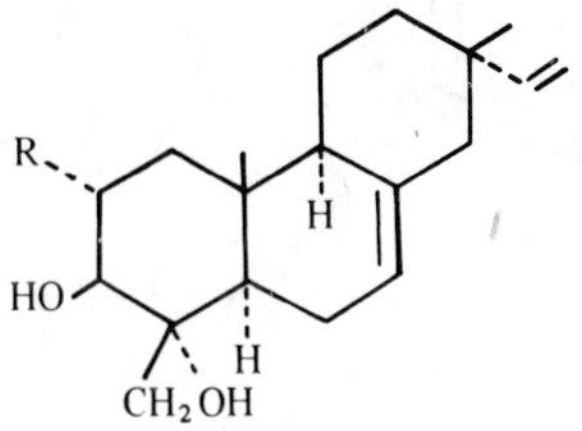	*Oospora virescens*	6
Virescenol B	Same as above, except: R = H	*Oospora virescens*	6
Virescenoside A	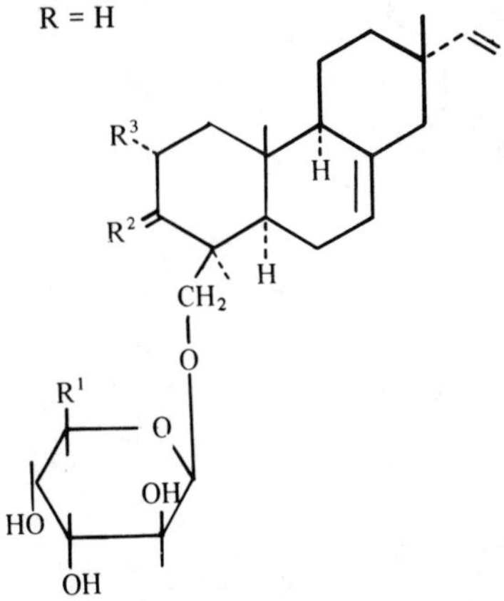	*Oospora virescens*	6

R = Me

R = OH

$R_1 = CH_2OH$
$R_2 = H, OH$
$R_3 = OH$

TABLE 5 (Continued)
DITERPENES (EXCLUDING GIBBERELLINS)

Compound	Structure	Source	Ref.
Virescenoside B	Same as above, except: R_1 = CH$_2$OH R_2 = H, OH R_3 = H	*Oospora virescens*	6
Virescenoside C	Same as above, except: R_1 = CH$_2$OH R_2 = O R_3 = H	*Oospora virescens*	6
Virescenoside F	Same as above, except: R_1 = COOH R_2 = H, OH R_3 = OH	*Oospora virescens*	6
Virescenoside G	Same as above, except: R_1 = COOH R_2 = H, OH R_3 = H	*Oospora virescens*	6
Virescenoside H		*Oospora virescens*	6
Rosenonolactone (Rosein I)	 R_1 = H R_2 = O	*Trichothecium roseum luteum cytosporium*	6
6β-Hydroxyrosenonolactone	Same as above, except: R_1 = OH R_2 = O	*Trichothecium roseum*	6
Rosenololactone	Same as above, except: R_1 = H R_2 = H, OH	*Trichothecium roseum cytosporium*	6
7-Deoxyrosenonolactone	Same as above, except: R_1 = H R_2 = H$_2$	*Trichothecium roseum*	6

TABLE 5 (Continued)
DITERPENES (EXCLUDING GIBBERELLINS)

Compound	Structure	Source	Ref.
Rosololactone (Rosein II, rosonolactone)	Same as above, except: R_1 = OH R_2 = H_2	*Trichothecium roseum*	6
Rosein III (11β-Hydroxyrosenonolactone)	HO	*Trichothecium roseum*	6
Isorosenolic acid	HO COOH	*Trichothecium roseum*	6

TABLE 6
COMPOUNDS WITH A DITERPENE UNIT ATTACHED TO AN AROMATIC NUCLEUS

Compound	Structure	Source	Ref.
Bovinone	HO ($CH_2CH=\overset{CH_3}{C}CH_2)_4$H OH	*Boletus bovinus*	6
Amitenone (Methylenediboviquinone-4,4)	HO OH HX$_4$ OH X$_4$H $X = -CH_2CH=\overset{CH_3}{C}-CH_2-$	*Suillus bovinus*	17, 18

TABLE 7
SESTERTERPENES
(5 Isoprene Units, 25 Carbon Atoms)

Compound	Structure	Source	Ref.
Ophiobolin A (Ophiobolin, ophiobalin, coch- liobolin, cochliobolin A)		*Cochliobolus (Ophiobolus) miyabeanus Helminthosporium turcicum zizaniae panici-miliacei leersii*	6
Ophiobolin B (Zizanin, zizanin B, ophiobolo- sin A, cochliobolin B)	 R = OH	*Helminthosporium turcicum zizaniae Cochliobolus miyabeanus heterostrophus*	6
Ophiobolin C (Zizanin A)	Same as above, except: R = H	*Helminthosporium turcicum zizaniae Cochliobolus miyabeanus heterostrophus*	6
Ophiobolin D (Cephalonic acid)		*Cephalosporium caerulens*	6
Ophiobolin F		*Cochliobolus heterostrophus*	6

TABLE 7 (Continued)
SESTERTERPENES

Compound	Structure	Source	Ref.
Geranylnerolidol		*Cochliobolus heterostrophus*	6
Fusicocum A		*Fusicocium amygdali*	6

TABLE 8
TRITERPENES[a]
(6 Isoprene Units, 30 Carbon Atoms)

Compound	Structure	Source	Ref.
Squalene		*Methylococcus capsulatus*	19
Hopene-b		A bacterium with optimal growth at 58–60°C and pH 2.6–2.8	20

TABLE 8 (Continued)
TRITERPENES[a]
(6 Isoprene Units, 30 Carbon Atoms)

Compound	Structure	Source	Ref.
Hopene-1		A bacterium with optimal growth at 58–60°C and pH 2.6–2.8	20

[a] Other triterpenes are listed with lichen products and with steroids and tetracyclic terpenes.

REFERENCES

1.   Taira, T., *Nippon Nogei Kagaku Kaishi, 37*, 49 (1963); *Chem. Abstr., 59*, 5729b.
2.   Katayama, T., *Kogoshima Daigaku Suisan Gakubu Kiyo, 13*, 58 (1964); *Chem. Abstr., 62*, 10822h.
3.   Collins, R. P., and Halim, A. F., *Lloydia, 33*, 481 (1970).
4.   Miller, M., *The Pfizer Handbook of Microbial Metabolites.* McGraw-Hill, New York (1961).
5.   Gerber, N. N., *J. Antibiotics (Tokyo), 22*, 508 (1969).
6.   Turner, W. B., *Fungal Metabolites,* Academic Press, New York (1971).
7.   Yang, D. T. C., Wilson, B. J., and Harris, T. M., *Phytochemistry, 10*, 1653 (1971).
8.   Grasselli, P., and Selva, A., *Chim. Ind. (Milan), 52*, 584 (1970); *Chem. Abstr., 73*, 131144e.
9.   Nozoe, S., Matsumoto, H., and Urano, S., *Tetrahedron Lett.*, p. 3125 (1971); *Chem. Abstr., 75*, 110447g.
10. Daniewski, W. M., and Kocor, M., *Bull. Acad. Pol. Sci. Ser. Sci. Chim., 19*, 553 (1971); *Chem. Abstr., 76*, 14738e.
11. Ellestad, G. A., Evans, R. H., and Kunstmann, M. P., *Tetrahedron Lett.*, p. 497 (1971); *Chem. Abstr., 74*, 142089n.
12. Vohac, K., Samek, Z., Heroul, V., and Sorm, F., *Collect. Czech. Chem. Comm., 35*, 1296 (1970).
13. Takahashi, S., Naganawa, H., Iinuma, H., Takita, T., Maeda, K., and Umezawa, H., *Tetrahedron Lett.*, p. 1955 (1971); *Chem. Abstr., 75*, 64012m.
14. Gerber, N. N., *Tetrahedron Lett.*, p. 2971 (1968).
15. Gerber, N. N., *Phytochemistry, 10*, 185 (1971).
16. Gerber, N. N., *Phytochemistry, 11*, 385 (1972).
17. Edwards, R. L., and Beaumont, P. C., *J. Chem. Soc. Sect. C*, p. 2582 (1971); *Chem. Abstr., 75*, 72462y.
18. Minami, K., Asawa, K., and Sawada, M., *Tetrahedron Lett.*, p. 5067 (1968).
19. Bird, C. W., Lynch, J. M., Pirt, F. J., Reid, W. W., Brooks, C. J. W., and Middleditch, B. S., *Nature, 230*, 473 (1971).
20. Minale, L., DeRosa, M., Gambacorta, A., and Bu'Lock, J. D., *J. Chem. Soc. Sect. D*, p. 619 (1971); *Chem. Abstr., 75*, 60086r.

GIBBERELLINS

DR. SONIA RUSSELL

Gibberellins are polar compounds with a gibbane skeleton that promote growth in higher plants. They have been isolated from *Gibberella fujikuroi* and higher plants by extraction with ethyl acetate.[1] Gibberellin esters[2,3] and glucosides[4] have been isolated by extraction with *n*-butanol. Pure gibberellins have been obtained from the extract by the use of countercurrent distribution,[5] adsorption chromatography,[6] partition chromatography,[7] and gas-liquid chromatography.[8] The purity of a gibberellin is established by thin-layer chromatography,[9] using sulfuric acid[10] as location reagent. Mass spectrometry[11] and nuclear magnetic resonance spectrometry[12] have been used in characterizing the compounds. Gibberellins have ill-defined melting points and characteristic infrared spectra. The gibberellane numbering system is shown below.

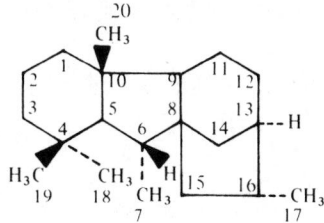

ABBREVIATIONS

br	=	broad
C	=	in chloroform
d	=	decomposing
Dc	=	deuterochloroform
di	=	dimorphous
Dx	=	in dioxane
E	=	epoxy
F	=	free acid
H	=	in deuteromethanol
K	=	in potassium bromide
L	=	lactone
Me	=	methyl ester

Mt	=	in methanol
n	=	needles
P	=	in pyridine
pr	=	prisms
Q	=	in deuteropyridine
R	=	in carbon tetrachloride
s	=	stable
S	=	stable form
sh	=	shoulder
T	=	in deuteroacetone
u	=	unsaturation
us	=	unstable

Compound Formula MW	Structure	Melting Point, °C	Specific Rotation[a]	Infrared Spectrum[b]	NMR Spectrum[c]	Reference
A_1 $C_{19}H_{24}O_6$ 348	7–COOH; 3,13–OH; 17=CH$_2$; 4–CH$_3$; $\gamma 4 \rightarrow 10L$	255–258, d	+36° (21°C, c = 1.74)	K: 1740, 1717	Me, Dc: 1.15, 2.67, 3.22, 3.85, 4.95, 5.04	6, 7, 13
A_2 $C_{19}H_{26}O_6$ 350	7–COOH; 3,17–OH; 4–CH$_3$; $\gamma 4 \rightarrow 10L$	255, d		3470 sh, 3380 br, 1752, 1708	Me, Dc: 1.03, 1.28 2.7, 3.22, 3.85	13, 14
A_3 $C_{19}H_{22}O_6$ 346	7–COOH; 3,13–OH; 17=CH$_2$; 4–CH$_3$ $\gamma 4 \rightarrow 10L$; $\Delta 1,2$ u	233–255, d	+86° (19°C, c = 2.12)	3390, 3305, 1746 Dx: 1784, 1736	Me, P: 1.54, 3.05, 3.69, 4.48, 5.02, 5.45, 6.11, 6.41	13, 15
A_4 $C_{19}H_{24}O_5$ 332	7–COOH; 3–OH; 17=CH$_2$; 4–CH$_3$; $\gamma 4 \rightarrow 10L$	di: 216; 255	–3° (20°C, c = 0.4)	3500, 2700, 2615, 1736, 1720, 1657	Me, Dc: 1.15, 2.71 3.22, 3.85, 4.85, 4.95	13.16
A_5 $C_{19}H_{22}O_5$ 330	7–COOH; 13–OH; 17=CH$_2$; 4–CH$_3$ $\gamma 4 \rightarrow 10L$; $\Delta 2,3$ u	260–261	–77°, Mt (27°C, c = 0.5)	3436, 2700 br, 1765, 1734, 1659, 1624, 893, 694		17
A_6 $C_{19}H_{22}O_6$ 346	7–COOH; 13–OH; 17=CH$_2$; 4–CH$_3$; 2,3 E; $\gamma 4 \rightarrow 10L$	di: 222–225; pr 206–209	–20°, Mt (26°C, c = 0.2)	pr: 3600, 3405, 3090, 1780, 1734, 1665, 910, 840		18
A_7 $C_{19}H_{22}O_5$ 330	7–COOH; 3–OH; 17=CH$_2$; 4–CH$_3$; $\gamma 4 \rightarrow 10L$; $\Delta 1,2$ u	di: n 169–172; pr 202, d	+20° (24°C, c = 0.5)	n: 3450, 1742, 1722, 1654	Me, Dc: 1.23, 2.71, 3.22, 4.13, 4.85, 4.95, 5.86, 6.30	13, 19
A_8 $C_{19}H_{24}O_7$ 364	7–COOH; 2,3,13–OH; 17=CH$_2$; 4–CH$_3$ $\gamma 4 \rightarrow 10L$	di: 210–215; 144.5–147.5	+13°, Mt (23°C, c = 2.05)	3500–3300, 2600, 1762, 1710, 1660, 880	F, P: 1.71, 3.17, 4.01, 4.22, 4.38, 5.03, 5.61	18, 20
A_9 $C_{19}H_{24}O_4$ 316	7–COOH; 17=CH$_2$; 4–CH$_3$; $\gamma 4 \rightarrow 10L$	208–211	–22° (17°C, c = 0.25)	3098, 1740, 1723, 1659, 893	Me, Dc: 1.08, 2.49, 2.72, 4.85, 4.95	13, 19
A_{10} $C_{19}H_{26}O_5$ 334	7–COOH; 17–OH; 4–CH$_3$; $\gamma 4 \rightarrow 10L$	245–246	+3°	3360, 2605, 1766, 1680	Me, Dc: 1.1, 1.38, 2.47, 2.77, 3.78	21

Compound Formula MW	Structure	Melting Point, °C	Specific Rotation[a]	Infrared Spectrum[b]	NMR Spectrum[c]	Reference
A_{11} $C_{19}H_{22}O_5$ 330	7–COOH; 17=CH_2; 4–CH_3; $\gamma 4 \rightarrow 2L$	242–244, d	+11° (c = 0.3)	3150, 3060, 1777, 1730, 1658, 896	Me, Dc: 1.17, 2.67, 2.88, 3.41, 5.1	22
A_{12} $C_{20}H_{28}O_4$ 332	4,7–COOH; 17=CH_2; 4,10–CH_3	245–248		3400, 3078, 2460, 1692 br, 1658, 882	F, Dx: 0.79, 1.13, 2.08, 4.06	23
A_{13} $C_{20}H_{26}O_7$ 378	4,7,10–COOH; 3–OH; 17=CH_2; 4–CH_3	194–196, d	–48° (17°C, c = 0.25)	3500 br, 1723–1700, 1660	Me, Dc: 1.10, 2.08 2.52, 3.61, 3.68 3.72, 3.85, 4.85, 5.00	24
A_{14} $C_{20}H_{28}O_5$ 348	4,7,10–COOH; 3–OH; 17=CH_2; 4–CH_3	232–233.5, d	–73° (17°C c = 0.4)	3540, 3350–2500, 1708, 1697, 1665, 879	F, P: 1.17, 1.93, 2.53, 3.05, 4.05, 4.62, 4.78 br	25
A_{15} $C_{20}H_{26}O_4$ 330	7–COOH; 17=CH_2; 4–CH_3; $\gamma 4 \rightarrow 10L$	274–276	+5°	3239 br, 1724, 1680, 1652, 889	Me, Dc: 1.15, 2.21 2.79, 3.67, 4.03, 4.42, 4.75	26
A_{16} $C_{19}H_{24}O_6$ 348	7–COOH; 1,3–OH; 17=CH_2; 4–CH_3; $\gamma 4 \rightarrow 10L$	157–165		Me: 3560, 3461, 1780, 1716, 1650, 901	F, P: 1.66, 3.59, 4.20, 4.63, 4.92, 7.2–8.2	27, 28
A_{17} $C_{20}H_{26}O_7$ 378	4,7,10–COOH; 13–OH; 17=CH_2; 4–CH_3	140–145		Me, C: 3500 br, 1723, 1663, 906	F, T: 1.16, 3.81, 4.76, 5.03	29
A_{18} $C_{20}H_{28}O_6$ 364	4,7–COOH; 3,13–OH; 17=CH_2; 4,10–CH_3	235–239		K: 3340, 3260, 2800–2400, 1695, 908	F, Q: 1.26, 2.00, 3.12, 4.26, 5.04, 5.53	30
A_{19} $C_{20}H_{26}O_6$ 362	4,7–COOH; 13–OH; 17=CH_2; 10–CHO; 4–CH_3	di: 124–126 us; 236–237 s		S: 3370, 1720, 1695, 887	Me, Dc: 0.27, 1.15, 2.43, 3.88, 4.98, 5.2	5, 26

Compound Formula MW	Structure	Melting Point, °C	Specific Rotation[a]	Infrared Spectrum[b]	NMR Spectrum[c]	Reference
A_{20} $C_{19}H_{24}O_5$ 332	7–COOH; 13–OH; 17=CH$_2$; 4–CH$_3$; γ4 → 10L	232–233		Me: 3540, 1785, 1756, 1740, 1726, 880	Me, Dc: 1.07, 2.5, 2.67, 3.66, 4.9, 5.2	31
A_{21} $C_{19}H_{22}O_7$ 362	4,7–COOH; 13–OH; 17=CH$_2$; γ4 → 10L	244–246		3280, 2520 br, 1779, 1730, 1682, 877	Me, Dc: 2.76, 3.11, 3.64, 3.66, 4.89, 5.2	32–34
A_{22} $C_{19}H_{22}O_6$ 346	7–COOH; 13–OH; 17=CH$_2$; Δ2,3 u; γ4 → 10L	213–214		3380, 3240, 1770, 1709, 1670, 1643, 892	Me, Dc: 2.84, 3.68, 3.76, 4.88, 5.22, 5.85	33, 34
A_{23} $C_{20}H_{26}O_7$ 378	4,7–COOH; 3,13–OH; 17=CH$_2$; 10–CHO; 4–CH$_3$	179–181		K: 3500–3350, 2800–2400, 1730–1690, 1660 sh, 900	F, Q: 1.95, 3.58, 3.98, 4.50, 5.00, 5.57	30
A_{24} $C_{20}H_{26}O_5$ 346	4,7–COOH; 17=CH$_2$; 10–CHO; 4–CH$_3$	198–203	–88° (23°C, c = 0.6)	3380 br, 3075, 2800 br, 1716, 1690, 875	Me, Dc: 1.11, 2.19, 3.81, 3.57, 3.66, 4.76, 4.84, 9.62	35
A_{25} $C_{20}H_{26}O_6$ 362	4,7,10–COOH; 17=CH$_2$; 4–CH$_3$	248–252	–69° (27°C, c = 0.7)	C: 2800 br, 1710 br, 880	Me, Dc: 1.10, 2.07, 3.51, 3.57, 3.63, 3.73, 4.73, 4.80	35
A_{26} $C_{19}H_{22}O_7$ 362	7–COOH; 2,3–OH; 12=C=O; 17=CH$_2$; 4–CH$_3$; γ4 → 10L	254–257		Me, C: 3500, 1780, 1733, 1716, 1655	Me, Dc: 1.21, 2.73, 3.37, 3.71, 3.75, 3.87, 5.05, 5.2	36
A_{27} $C_{20}H_{26}O_6$ 362	7–COOH; 2,3–OH; 17=CH$_2$; 4–CH$_3$; γ4 → 10L	163–165		Me: 3300, 1740, 1710, 1650	Me, Dc: 1.24, 3.69, 3.72, 3.86, 4.12, 4.46, 4.83, 4.95	36
A_{28} $C_{20}H_{26}O_8$ 394	4,7,10–COOH; 3,13–OH; 17=CH$_2$; 4–CH$_3$	224–225, d	–6.8° (12°C, c = 1.18)	K: 3430, 2800–2400, 1715 sh, 1705 sh, 1702	F, P: 2.02, 3.55, 4.73, 4.98, 5.06, 5.56	37
A_{29} $C_{19}H_{24}O_6$ 348	7–COOH; 2,13–OH; 17=CH$_2$; 4–CH$_3$; γ4 → 10L		Me: 3240, 1790, 1750, 1660	Me, T: 1.06, 2.56, 2.73, 3.72, 4.84, 5.18	38

Compound Formula MW	Structure	Melting Point, °C	Specific Rotation[a]	Infrared Spectrum[b]	NMR Spectrum[c]	Reference
A_{30} $C_{19}H_{22}O_6$ 346	7–COOH; 2,12–OH; 17=CH$_2$; 4–CH$_3$; Δ1,2 u; γ4 → 10L	188–191		3300, 1780, 1705, 1655	F, P: 1.6, 4.01, 4.48, 4.99, 5.02, 6.13, 6.43	39
A_{31} $C_{19}H_{22}O_5$ 330	7–COOH; 12–OH; 17=CH$_2$; 4–CH$_3$; Δ2,3 u; γ4 → 10L	227–231		3440, 1750, 1724, 1660	F, P: 1.44, 4.03 4.97, 5.07, 5.73	39
A_{32} $C_{19}H_{22}O_8$ 378	7–COOH; 3,12,13–OH; 17=CH$_2$; 4–CH$_3$; Δ1,2 u; γ4 → 10L				F, H: 2.44, 2.65, 3.16, 3.58, 3.97, 4.35, 6.36, 6.84	40
A_{33} $C_{19}H_{22}O_7$ 362	7–COOH; 2,12–OH; 3=C=O; 17=CH$_2$; 4–CH$_3$; γ4 → 10L	219–221		3470, 1794, 1733, 1696, 1658	Me, T: 1.11, 2.54, 2.94, 3.48, 3.74, 4.42, 4.48, 4.98, 5.08	41
A_{34} $C_{19}H_{24}O_6$ 348	7–COOH; 2,3–OH; 17=CH$_2$; 4–CH$_3$; γ4 → 10L	218–219		3340, 1655	Me, T: 1.15, 3.25, 3.62, 3.75, 4.84, 4.96	41
A_{35} $C_{19}H_{24}O_6$ 348	7–COOH; 3,11–OH; 17=CH$_2$; 4–CH$_3$; γ4 → 10L	248–252			Me, Dc: 1.15, 2.74, 3.26, 3.86, 4.22, 4.91, 5.04	42
A_{36} $C_{20}H_{26}O_6$ 362	4,7–COOH; 3–OH; 17=CH$_2$; 10–CHO; 4–CH$_3$	205–208		Me, R: 3635, 3540 br, 3070, 2760, 2720, 1735 br, 1658, 883	F, P: 1.86, 3.41, 3.51, 4.35, 4.8, 4.93	43
A_{37} $C_{20}H_{26}O_5$ 346	7–COOH; 3–OH; 17=CH$_2$; 4–CH$_3$; γ4 → 10L	228–232			F, P: 1.84, 3.28, 4.13,, 4.61, 4.78, 4.92	43
A_{38} $C_{20}H_{26}O_6$ 362	7–COOH; 3,13–OH; 17=CH$_2$; 4–CH$_3$; γ4 → 10L	237–239	+29.5°, Mt (20°C, c = 0.576)	K: 3390, 3270, 2800– 2500, 1728, 1696, 1667	Me, Dc: 1.19, 2.79, 3.7, 3.78, 4.09, 4.48, 4.91, 5.22	30

[a] Specific rotations were measured in ethanol unless otherwise stated; values are given in $[\alpha]_D$ at the temperature and concentration stated.

[b] Infrared spectra were determined on Nujol mulls unless otherwise stated; values are given in ν max/cm^{-1}

[c] Nuclear magnetic resonance spectra are chemical shifts (δ, ppm from tetramethylsiland as internal standard).

REFERENCES

1. Cross, B. E., Galt, R. H. B., Hanson, J. R., Curtis, P. J., Grove, J. F., and Morrison, A., *J. Chem. Soc.*, p. 2937 (1963).
2. Schreiber, K., Schneider, G., Sembdner, G., and Focke, I., *Phytochemistry, 5,* 1221 (1966).
3. Yokota, T., Takahashi, N., Murofushi, N., and Tamura, S., *Tetrahedron Lett.*, p. 2081 (1969).
4. Schrieber, K., Weiland, J., and Sembdner, G., *Phytochemistry, 9,* 189 (1970).
5. Murofushi, N., Iriuchijima, S., Takahashi, N., Tamura, S., Kato, J., Wada, Y., Watanabe, E., and Aoyama, T., *Agric. Biol. Chem., 30,* 917 (1966).
6. West, C. A., and Phinney, B. O., *J. Am. Chem. Soc., 81,* 2424 (1957).
7. Stodola, F. H., Nelson, G. E. N., and Spence, D. J., *Arch. Biochem. Biophys., 66,* 438 (1957).
8. Ikekawa, N., and Sumiki, Y., *Chem. Ind. (London),* p. 1728 (1963).
9. MacMillan, J., and Suter, P. J., *Nature, 197,* 790 (1963).
10. Jones, D. F., MacMillan, J., and Radley, M., *Phytochemistry, 2,* 307 (1963).
11. Wulfson, N. S., Zaretskii, V. I., Papernaja, I. B., Serebryakov, E. P., and Kucherov, V. F., *Tetrahedron Lett.*, p. 4209 (1965).
12. Sheppard, N., *J. Chem. Soc.*, p. 3040 (1960).
13. Hanson, J. R., *J. Chem. Soc.*, p. 5036 (1965).
14. Grove, J. F., *J. Chem. Soc.*, p. 3545 (1961).
15. Cross, B. E., *J. Chem. Soc.*, p. 4670 (1954).
16. Grove, J. F., MacMillan, J., Mulholland, T. P. C., and Turner, W. B., *J. Chem. Soc.*, p. 3049 (1960).
17. MacMillan, J., Seaton, J. C., Suter, P. J., *Tetrahedron, 11,* 60 (1960).
18. MacMillan, J., Seaton, J. C., and Suter, P. J., *Tetrahedron, 18,* 349 (1962).
19. Cross, B. E., Galt, R. H. B., and Hanson, J. R., *Tetrahedron, 18,* 451 (1962).
20. Yokota, T., Murofushi, N., Takahashi, N., and Tamura, S., *Agric. Biol. Chem., 35,* 573 (1971).
21. Hanson, J. R., *Tetrahedron, 22,* 701 (1966).
22. Brown, J. C., and Cross, B. E., *Tetrahedron, 23,* 4095 (1967).
23. Cross, B. E., and Norton, K., *J. Chem. Soc.*, p. 1570 (1965).
24. Galt, R. H. B., *J. Chem. Soc.*, p. 3143 (1965).
25. Cross, B. E., *J. Chem. Soc. Sect. C,* p. 501 (1966).
26. Hanson, J. R., *Tetrahedron, 23,* 733 (1967).
27. Vining, L. C., Personal communication (1973).
28. Galt, R. H. B., *Tetrahedron, 24,* 1337 (1968).
29. Pryce, R. J., and MacMillan, J., *Tetrahedron Lett.*, p. 4173 (1967).
30. Koshimizu, K., Ishii, H., Fukui, H., and Mitsui, T., *Phytochemistry, 11,* 2355 (1972).
31. Murofushi, N., Takahashi, N., Yokota, T., and Tamura, S., *Agric. Biol. Chem., 32,* 1239 (1968).
32. Takahashi, N., Murofushi, N., Yokota, T., Tamura, S., Kato, J., and Shiotani, Y., *Tetrahedron Lett.*, p. 4861 (1967).
33. Murofushi, N., Takahashi, N., Yokota, T., and Tamura, S., *Agric. Biol. Chem., 33,* 598 (1969).
34. Murofushi, N., Takahashi, N., Yokota, T., Kato, J., Shiotani, Y., and Tamura, S., *Agric. Biol. Chem., 33,* 592 (1969).
35. Harrison, D. M., and MacMillan, J., *J. Chem. Soc. Sect. C,* p. 631 (1971).
36. Takahashi, N., Yokota, T., Murofushi, N., and Tamura, S., *Tetrahedron Lett.*, p. 2077 (1969).
37. Fukui, H., Koshimizu, K., and Mitsui, T., *Phytochemistry, 10,* 671 (1971).
38. Yokota, T., Murofushi, N., and Takahshi, N., *Tetrahedron Lett.*, p. 1489 (1970).
39. Murofushi, N., Yokota, T., and Takahashi, N., *Agric. Biol. Chem., 34,* 1436 (1970).
40. Yamaguchi, I., Yokota, T., Murofushi, N., Ogawa, Y., and Takahashi, N., *Agric. Biol. Chem., 34,* 1439 (1970).
41. Murofushi, N., Yokota, T., and Takahashi, N., *Agric. Biol. Chem., 35,* 441 (1971).
42. Yamane, H., Yamaguchi, I., Murofushi, N., and Takahashi, N., *Agric. Biol. Chem., 35,* 1144 (1971).
43. Bearder, J. R., and MacMillan, J., *Agric. Biol. Chem., 36,* 342 (1972).

STEROIDS AND TETRACYCLIC TRITERPENES

DR. W. B. TURNER

FUNGAL STEROIDS AND TETRACYCLIC TRITERPENES

C_{27} COMPOUNDS

Structure	Compound	Organisms	References
	Zymosterol	Yeast	1, 2
	Cholesterol	*Penicillium funiculosum*	3
	7-Dehydrocholesterol	*Lactarius* sp.	4

C$_{28}$ STEROLS RELATED TO ERGOSTEROL

Structure	Compound	Position of Double Bonds	Organisms	References
	Campesterol	5	*Stachybotrys alternans*	5
	Fungisterol	7	Widespread	6
	22,23-Dihydroergosterol	5,7	*Leccinum aurantiacum*	7
			Suillus bovinus	7
			variegatus	7
			Yeasts	2, 8
	Brassicasterol	5,22	*Trichophyton rubrum*	9
	5,6-Dihydroergosterol	7,22	*Claviceps purpurea*	10, 11
			Fomes annosus	12
			applanatus	13
			fomentarius	12
			Leccinum aurantiacum	7
			Polyporus pinicola	6
			Suillus bovinus	7
			variegatus	7
			Yeasts	10, 11
	Episterol	7,24(28)	Yeasts	8
	Ascosterol	8,23	Yeasts	2, 8
	Fecosterol	8,24(28)	Yeasts	2, 8
	Ergosterol	5,7,22	Widespread	2, 8

C$_{28}$ STEROLS RELATED TO ERGOSTEROL (Continued)

Structure	Compound	Position of Double Bonds	Organisms	References
	Ergosta-5,7,24(28)-trien-3-ol	5,7,24(28)	*Phycomyces blakesleeanus*	14
	14-Dehydroergosterol	5,7,14,22	*Aspergillus niger*	15
	24(28)-Dehydroergosterol	5,7,22,24(28)	Yeasts	16,17
	Cerevisterol		*Aspergillus phalloides* Yeasts	18,19 18,19
	Ergosterol peroxide		*Aspergillus fumigatus* Others	20
	5α,8α-Epidioxyergosta-6,9(11),22-trien-3β-ol-12-one		*Fusarium moniliforme*	21

C$_{28}$ STEROLS RELATED TO ERGOSTEROL (Continued)

Structure	Compound	Organisms	References
	Ergosta-7,22-dien-3-one	*Fomes fomentarius*	22
	Ergosta-4,7,22-trien-3-one	*Fomes annosus*	12
	Ergosta-4,6,8(14),22-tetraen-3-one	*Fomes officinalis* *Penicillium rubrum*	23 24

THE C$_{30}$ TETRACYCLIC TRITERPENES

Structure	Compound	Substituents			Organisms	References

Mixture of lanosta-7,9-dien-3β-ol and lanosta-8-en-3β-ol — *Fomes pini* — 12

Obliquol — *Poria obliqua* — 25, 26

23-S-Hydroxylanosterol — *Scleroderma aurantium* — 27

Compound	R^1	R^2	R^3	Organisms	References
3α-Hydroxylanosta-8,24-dien-21-oic acid	H	OH	H	*Polyporus pinicola*	28
3β-Hydroxylanosta-8,24-dien-21-oic acid	OH	H	H	*Trametes odorata*	29
Pinicolic acid A	=O		H	*Polyporus pinicola*	30
15 α-Hydroxytrametenolic acid	OH	H	OH	*Lenzites trabea*	31

THE C_{30} TETRACYCLIC TRITERPENES (Continued)

Structure	Compound	Substituents		Organisms	References
		R^1	R^2		
	Lanosta-7,9(11),24-triene-3β,21-diol	H, β-OH	CH_2OH	*Polyporus pinicola*	6
	21-Hydroxylanosta-7,9(11),24-triene-3-one	O	CH_2OH	*Polyporus pinicola*	6
	3β-Lanosta-7,9(11),24-trien-21-oic acid	H, β-OH	CO_2H	*Poria cocos*	32
	Tyromycic acid			*Tyromyces albidus*	33
	Echinodol			*Echinodontium tinctorium*	34
	Senexdiolic acid			*Fomes senex*	35

THE C$_{30}$ TETRACYCLIC TRITERPENES (Continued)

Structure	Compound	Substituents	Organisms	References
	Senexonol	R = H, OH	*Fomes senex*	35
	Senexdione	R = O	*Fomes senex*	35
	Oxidosenexone		*Fomes senex*	35

THE C_{31} TETRACYCLIC TRITERPENES

Structure	Compound	Substituents			Organisms	References
		R^1	R^2	R^3		
	Eburicoic acid	H	H	H	*Fomes officinalis*	36, 37
					Lentinus dactyloides	36, 37
					Oospora astringens	38
					Polyporus anthracophilus	36, 37
					eucalyptorum	36, 37
					hispidus	39
					sulphureus	36, 37
					Poria cocos	39
	3-O-Acetyleburicoic acid	Ac	H	H	*Polyporus anthracophilus*	36
	Tumulosic acid	H	H	OH	*Polyporus australiensis*	39
					betulinus	39
					tumulosus	39
					Poria cocos	39
	Sulfurenic acid	H	OH	H	*Polyporus sulphureus*	40
	3-O-Acetyltumulosic acid	Ac	H	OH	*Trametes lilacino gilva*	41

THE C$_{31}$ TETRACYCLIC TRITERPENES (Continued)

Structure	Compound	Substituents				Organisms	References
		R^1	R^2	R^3	R^4		
	Polyporenic acid C*	=O		H	OH	*Polyporus benzoinus*	43, 44
						betulinus	43, 44
	6α-Hydroxypolyporenic acid C	=O		OH	OH	*Trametes feei*	45
	Dehydroeburicoic acid	OH	H	H	H	*Fomes officinalis*	39
						Lentinus dactyloides	39
	Dehydrotumulosic acid	OH	H	H	OH	*Polyporus australiensis*	39
						betulinus	39
						tumulosus	39
						Poria cocos	39
	Mixture of acetylglucosyltumulosic acid and acetylglucosyldehydrotumulosic acid					*Trametes dickinsii*	42

* A mixture of polyporenic acid C and dihydropolyporenic acid C has been isolated from *Trametes dickinsii*.

THE C$_{31}$ TETRACYCLIC TRITERPENES (Continued)

Structure	Compound	Substituents			Organisms	References
		R^1	R^2			
	Polyporenic acid A				*Polyporus betulinus*	46
	24-Methylenedihydrolanosterol	OH	H		*Phycomyces blakesleeanus*	47
	Eburicol	H	OH		*Fomes officinalis*	48
	Carboxyacetylquercinic acid			R = H	*Daedalea quercinus* *Trametes dickinsii*	49 50
	Carbomethoxyacetylquercinic acid			R = Me	*Trametes dickinsii*	50

MISCELLANEOUS TRITERPENES AND STEROIDS

Structure	Compound	Organisms	References
	24-Methylenelophenol	*Aspergillus fumigatus*	47
	4α-Methyl-24-methylene-24,25-dihydrozymosterol	*Saccharomyces cerevisiae*	51
	3α-Hydroxy-4,4,14α-trimethyl-Δ^8-5α-pregnen-20-one	*Fomes officinalis*	48
	Antheridiol	*Achlya bisexualis*	52, 53

THE FUSIDANES AND PROTOSTANES

Structure	Compound	Substituents	Organisms	References
	Fusidic acid (ramycin)		*Cephalosporium* sp.	54
			Fusidium coccineum	55
			Mucor ramannianus	56, 57
	Cephalosporin P₁	R = Ac	*Cephalosporium* sp.	58
	Desacetylcephalosporin P₁	R = H	*Cephalosporium acremonium*	59
	Helvolic acid[60]	R = OAc	*Acrocylindrium oryzae*	61
			Aspergillus fumigatus mut. *helvola*	62
			Cephalosporium caerulens	63
	Helvolinic acid	R = OH	*Cephalosporium caerulens*	64
	7-Desacetoxyhelvolic acid	R = H	*Cephalosporium caerulens*	64

THE FUSIDANES AND PROTOSTANES (Continued)

Structure	Compound	Substituents	Organisms	References
	3-Oxo-16β-acetoxyfusida-1,17(20)[16,21-*cis*],24-trien-21-oic acid		*Cephalosporium caerulens*	65
	3β-Hydroxy-4β-hydroxymethylfusida-17(20)[16,21-*cis*],24-diene	R = OH	*Cephalosporium caerulens*	66
	3β-Hydroxy-4β-methylfusida-17(20)[16,21-*cis*],24-diene	R = H	*Cephalosporium caerulens*	67
	3β-Hydroxyprotosta-13(17),24-diene (fusisterol)		*Cephalosporium caerulens* *Fusidium coccineum*	67 68

VIRIDIN AND RELATED COMPOUNDS

Structure	Compound	Substituents	Organisms	References
	Viridin	R = OMe	*Gliocladium virens*	69
	Desmethoxyviridin	R = H	*Apiospora camptospora*	70
			Unidentified sp.	70
	Viridiol	R = OMe	*Gliocladium virens*	71
	Desmethoxyviridiol	R = H	Unidentified sp.	70
	Wortmannin	R = OAc	*Penicillium wortmanni*	72
	Desacetoxywortmannin	R = H	*Aspergillus janus*	73
			Penicillium funiculosum	73

24-ETHYLSTEROLS

Structure	Compound	Position of Double Bonds	Organisms	References
	β-Sitosterol	5	*Stachybotrys alternans*	5
	Δ^{22}-Stigmasten-3β-ol	22	*Dictyostelium discoideum*	74
	Stigmasterol	5,22	*Debaromyces hansenii*	75
			Stachybotrys alternans	5
	Fucosterol	5,24(28)	Phycomycetes	76

ALGAL STEROLS

Compound	Position of Double Bonds	C-24 Substituents	Organisms	References
Cyanophyta (Blue-Green Algae)				
Cholesterol	5	—	*Anacystis nidulans*	77
			Fremyella diplosiphon	77
			Phormidium luridum var. *olivaceae*	78
β-Sitosterol (24α-ethylcholest-5-en-3β-ol)	5	α-Et	*Anacystis nidulans*	77
			Fremyella diplosiphon	77
24-Ethyl-Δ^7-cholesterol	7	Et	*Phormidium luridum* var. *olivaceae*	78
24-Ethyl-$\Delta^{7,22}$-cholestadienol	7,22	Et	*Phormidium luridum* var. *olivaceae*	78
24-Ethyl-$\Delta^{5,7,22}$-cholestatrienol	5,7,22	Et	*Phormidium luridum* var. *olivaceae*	78
24-Ethyl-$\Delta^{5,22}$-cholestadienol	5,22	Et	*Phormidium luridum* var. *olivaceae*	78
24-Ethyl-$\Delta^{5,7}$-cholestadienol	5,7	Et	*Phormidium luridum* var. *olivaceae*	78
Chlorophyta (Green Algae)				
Cholesterol	5	—	*Chaetomorpha crassa*	79
			Oocystis polymorpha	80
			Ulva pertusa	79
24-Methylenecholesterol	5	=CH$_2$	*Chaetomorpha crassa*	79

Compound	Position of Double Bonds	C-24 Substituents	Organisms	References
Chlorophyta (continued)				
22-Dihydrobrassicasterol	5	β-Me	*Chlorella*	
			ellipsoidea	81
			saccharophila	81
Campesterol (24α-methylcholest-5-en-3β-ol)	5	α-Me	*Chaetomorpha*	
			crassa	79
Haliclonasterol (24α-methyl-20-isocholest-5-en-3β-ol)	5	α-Me	*Monostroma*	
			nitidum	82
Δ⁷-Ergosterol	7	β-Me	*Chlorella* spp.	83
			vulgaris	84
			Oocystis	
			polymorpha	80
22-Dihydroergosterol	5,7	β-Me	*Chlorella*	
			ellipsoidea	81
			saccharophila	81
Brassicasterol	5,22	β-Me	*Chaetomorpha*	
			crassa	79
5-Dihydroergosterol	7,22	—	*Chlorella*	
			candida	83
			nocturna	83
			simplex	83
			sorokiniana	83
			vannielii	83
			Oocystis	
			polymorpha	80
Ergosterol	5,7,22	β-Me	*Chlorella*	
			candida	83
			nocturna	83
			pyranoidosa	85
			simplex	83
			sorokiniana	83
			vannielii	83
β-Sitosterol (24α-ethylcholest-5-3n-3β-ol)	5	α-Et	*Chaetomorpha*	
			crassa	79
Clionosterol	5	β-Et	*Chlorella*	
			ellipsoidea	81
			saccharophila	81
Fucosterol	5	=CHMe	*Chlorella*	
			ellipsoidea	81
			saccharophila	81
Δ⁷-Chondrillasterol	7	p-Et	*Chlorella*	
			vulgaris	84
			Oocystis	
			polymorpha	80

Compound	Position of Double Bonds	C-24 Substituents	Organisms	References
Chlorophyta (continued)				
Chondrillasterol	7,22	β-Et	*Chlorella vulgaris*	81, 84
			Oocystis polymorpha	80
			Scenedesmus obliquus	86
Poriferasterol	5,22	β-Et	*Chlorella ellipsoidea*	81
			saccharophila	81
Δ⁵-Avenasterol (stigmasta-5,11-dien-3β-ol)	5,11	a-Et	*Enteromorfa linza*	82
Phaeophyta (Brown Algae)				
Cholesterol	5	—	*Alaria crassifolia*	87
			Costaria costata	87
			Cystophyllum hakodatense	87
			Fucus evanescens	87
			Laminaria digitata	88
			faeroensis	88
			Pelvetia wrightii	87
			Sargassum confusum	87
			ringgoldianum	87
			thunbergii	87
24-Methylenecholesterol	5	=CH₂	As cholesterol	87, 88
			Dictyopteris divaricata	87
Chondrillasterol	7,22	β-Et	*Scenedesmus* sp.	89
Fucosterol	5	=CHMe	*Alaria crassifolia*	87, 90
			Ascophyllum nodosum	91, 92
			Chorda filum	91
			Cladostephus spongiosus	91
			Costaria costata	87, 93
			Cystophyllum hakodatense	87
			Dictyopteris divaricata	87
			Dictyota dichotoma	91

Compound	Position of Double Bonds	C-24 Substituents	Organisms	References
Phaeophyta (continued)				
Fucosterol (*cont.*)				
			Ectocarpus	
			tomentosus	91
			Eisenia	
			bicyclis	90
			Fucus	
			ceranoides	91
			evanescens	87
			serratus	92
			spiralis	92
			vesiculosus	92, 94
			Halidrys	
			siliquosa	91
			Heterochordaria	
			abietina	93
			Laminaria	
			angustata	93
			digitata	88, 91
			faeroensis	88
			japonica	93
			Myelophycus	
			caespitosus	90
			Padina	
			arborescens	93
			Pelvetia	
			canaliculata	92, 94
			wrightii	87
			Pilayella	
			littoralis	91
			Sargassum	
			confusum	87
			ringgoldanium	87
			thunbergii	87
			Sphacelaria	
			cirrhosa	91
			Stypocaulon	
			scoparium	91
			Undaria	
			pennatifida	93
Sargasterol (24-ethyliden-20-isocholesterol)	5	=CHMe	*Eisenia*	
			bicyclis	90
			Sargassum	
			ringgoldanium	95
Saringosterol	5	⟨OH / CH=CH₂⟩	As cholesterol	87, 88
			Dictyopteris	
			divaricata	87
Rhodophyta (Red Algae)				
Cholesterol	5	—	*Acanthopeltis*	
			japonica	96
			Ahnfeltia	
			phicata	97
			Coeloseira	
			pacifica	98

Compound	Position of Double Bonds	C-24 Substituents	Organisms	References
Rhodophyta (continued)				
Cholesterol (*cont.*)			*Chondrus*	
			crispus	97
			giganteus	98
			ocellatus	98
			Corallina	
			officinalis	97
			Cyrtymenia	
			sparsa	98
			Dilsea	
			carnosa	97
			Furcellaria	
			fastigiata	97
			Gelidium	
			amansii	96
			japonicum	96
			subcostatum	96
			Gigartina	
			stellata	97
			Gloidopeltis	
			furcata	98
			Grateloupia	
			elliptica	98
			Halosaccion	
			ramentaceum	99
			Iridophycus	
			cornucopiae	98
			Laurencia	
			pinnatifida	97
			Plocanium	
			vulgare	97
			Polyides	
			caprinus	97
			rotundus	99
			Polysiphonia	
			lanosa	97
			nigrescens	97
			Porphyra sp.	99
			purpurea	97
			Pterocladia	
			tenuis	100
			Rhodoglossum	
			pulcherum	96
			Rhodomela	
			confervoides	99
			larix	98
			Rhodymenia	
			palmata	97
			Tichocarpus	
			crinitus	98
22-Dehydrocholesterol	5,22	—	*Dilsea*	
			carnosa	97
			Halosaccion	
			ramentaceum	99
			Hypnea	
			japonica	101

Compound	Position of Double Bonds	C-24 Substituents	Organisms	References
Rhodophyta (continued)				
22-Dehydrocholesterol (*cont.*)			*Polyides*	
			caprinus	97
			rotundus	99
			Rhodomela	
			confervoides	99
Desmosterol	5,24	—	*Gigartina*	
			stellata	97
			Halosaccion	
			ramentaceum	99
			Laurencia	
			pinnatifida	97
			Polysiphonia	
			nigrescens	97
			Porphyra sp.	99
			purpurea	97
			Rhodomela	
			confervoides	99
			Rhodymenia	
			palmata	99
24-Methylenecholesterol	5	=CH$_2$	*Rytiphlea*	
			tinctoria	102
Campesterol (or its 24-isomer)	5	a-Me	*Rytiphlea*	
			tinctoria	102
β-Sitosterol	5	a-Et	*Halosaccion*	
			ramentaceum	99
			Polyides	
			rotundus	99
			Porphyra sp.	99
Stigmasterol	5,22	a-Et	*Halosaccion*	
			ramentaceum	99
Chrysophyta				
Cholesterol	5	—	*Synura*	
			petersenii	103
22-Dihydrobrassicasterol	5	β-Me	*Ochromonas*	
			danica	104
Brassicasterol	5,22	β-Me	*Ochromonas*	
			danica	104
Ergosterol	5,7,22	β-Me	*Ochromonas*	
			danica	104
β-Sitosterol	5	a-Et	*Synura*	
			petersenii	103
Clionasterol	5	β-Et	*Ochromonas*	
			danica	104

Compound	Position of Double Bonds	C-24 Substituents	Organisms	References
Chrysophyta (continued)				
Poriferasterol	5,22	β-Et	*Ochromonas danica*	104
			malhamensis	105
Stigmasterol	5,22	α-Et	*Ochromonas malhamensis*	106
			minuta	106
			sociabilis	106
7-Dehydroporiferasterol	5,7,22	β-Et	*Ochromonas danica*	104
Pyrrophyta				
Cholesterol	5	—	*Cryptomonas ovata*, var. *palustris*	107
Ergosterol	5,7,22	β-Me	*Chilomonas paramecium*	108, 109
Stigmasterol	5,22	α-Et	*Chilomonas paramecium*	108
			Cryptomonas ovata, var. *palustris*	107
Poriferasterol	5,22	β-Et	*Chilomonas paramecium*	109
Euglenophyta				
Cholesterol	5	—	*Euglena gracilis*	110
Δ^7-Cholestenol	7	—	*Euglena gracilis*	110
$\Delta^{5,7}$-Cholestadienol	5,7	—	*Euglena gracilis*	110
22-Dihydrobrassicasterol	5	β-Me	*Euglena gracilis*	110
Δ^7-Ergostenol	7	β-Me	*Euglena gracilis*	110
5-Dihydroergosterol	7,22	β-Me	*Euglena gracilis*	110
Ergosterol	5,7,22	β-Me	*Euglena gracilis*	111
Chalinasterol	5	$=CH_2$	*Euglena gracilis*	110

Compound	Position of Double Bonds	C-24 Substituents	Organisms	References
Euglenophyta (continued)				
Episterol	7	=CH$_2$	*Euglena gracilis*	110
Clionasterol	5	α-Et	*Euglena gracilis*	110
Δ^7-Chondrillastenol	7	α-Et	*Euglena gracilis*	110
$\Delta^{5,7}$-Chondrillastenol	5,7	α-Et	*Euglena gracilis*	110
Poriferasterol	5,22	α-Et	*Euglena gracilis*	110

PROTOZOAL STEROLS

Compound	Position of Double Bonds	C-24 Substituents	Organisms	References
Ergosterol	5,7,22	β-Me	*Acanthamoeba* (Neff)	112
			Acanthamoeba sp.	113
			Atasia ocellata	114
			Blastocrithidia culicis	115
			Crithidia fasciculata	115, 116
			oncopelti	114
			Haematococcus pluvialis	114
			Hartmanella rhysodes	113
			Leishmania tarentolae	115
			Mayorella palestinensis	113
			Prototheca zopfii	114
			Trypanosoma cruzi	116
			mega	114
			ranarum	115
			rhodesiense	114

Compound	Position of Double Bonds	C-24 Substituents	Organisms	References
22,23-Dihydroergosterol	5,7	β-Me	*Trypanosoma cruzi*	116
Stigmasterol	5,22	a-Et	*Acanthamoeba* sp.	113
			Hartmanella rhysodes	113
			Mayorella palestinensis	113
7-Dehydrostigmasterol	5,7,22	a-Et	*Acanthamoeba* (Neff)	112
7-Dehydroporiferasterol	5,7,22	β-Et	*Trypanosoma cruzi*	116
Spinasterol (or chondrillasterol)			*Atasia ocellata*	114
			Crithidia oncopelti	114
			Haematococcus pluvialis	114
7-Dehydroclionosterol	5,2	β-Et	*Trypanosoma cruzi*	116

BACTERIAL STEROLS

Compound	Position of Double Bonds	C-24 Substituents	Other Substituents	Organisms	References
Cholesterol	5	—	—	*Escherichia coli*	117
				Streptomyces olivaceus	118
4a-Methylcholest-8(9)-en-3β-ol	8(9)	—	4a-Me	*Methylococcus capsulatus*	119
4a-Methylcholest-8(9),24-dien-3β-ol	8(9),24	—	4a-Me	*Methylococcus capsulatus*	119
4,4-Dimethylcholest-8(9)-en-3β-ol (14-nordihydrolanosterol)	8(9)	—	4,4-di-Me	*Azotobacter chroococcum*	120
				Methylococcus capsulatus	119

Compound	Position of Double Bonds	C-24 Substituents	Other Substituents	Organisms	References
4,4-Dimethylcholest-8(9),24-dien-3β-ol(14-norlanosterol)	8(9),24	–	4,4-di-Me	*Azotobacter chroococcum* *Methylococcus capsulatus*	120 119
Lanosterol	8(9),24	–	4,4,14α-tri-Me	*Azotobacter chroococcum*	120
Campestrol	5	α-Me	–	*Azotobacter chroococcum*	120
Δ^7-Ergosten-3β-ol	7	β-Me	–	*Escherichia coli*	117
$\Delta^{7,22}$-Ergostadien-3β-ol	7,22	β-Me	–	*Azotobacter chroococcum*	120
Ergosterol	5,7,22	β-Me	–	*Azotobacter chroococcum*	120
β-Sitosterol	5	α-Et	–	*Escherichia coli*	117
Stigmasterol	5,22	α-Et	–	*Escherichia coli*	117

REFERENCES

1. Smedley-MacLean, I., *Biochem. J., 22,* 22 (1928).
2. Barton, D. H. R., and Cox, J. D., *J. Chem. Soc.,* p. 214 (1949).
3. Chen, Y. S., and Haskins, R. H., *Can. J. Chem., 41,* 1647 (1963).
4. Jayko, L. G., Baker, T. I., Stubblefield, R. D., and Anderson, R. F., *Can. J. Microbiol., 8,* 361 (1962).
5. Svishchuk, A. A., Seredyuk, L. S., Levchuk, Y. N., and Kolesnikova, S. G., *Khim. Prir. Soedin (Tashkent), 6,* 319 (1970); *Chem. Abstr., 73,* 106518 (1970).
6. Halsall, T. G., and Sayer, G. C., *J. Chem. Soc.,* p. 2031 (1959).
7. Shrivina, A. N., and Cherotchenko, Y. P., *Mikol. Fitopatol., 4,* 187 (1970).
8. Wieland, H., Rath, F., and Hesse, H., *Ann. Chem., 548,* 34 (1941).
9. Wirth, J. C., Beesley, T., and Miller, W., *J. Invest. Dermatol., 37,* 153 (1961).
10. Wieland, H., and Benend, W., *Ann. Chem., 554,* 1 (1943).
11. Barton, D. H. R., and Cox, J. D., *J. Chem. Soc.,* p. 1354 (1948).
12. Munro, H. D., and Musgrave, O. C., *J. Chem. Soc. Sect. C,* p. 685 (1971).
13. Pettit, G. R., and Knight, J. C., *J. Org. Chem., 27,* 2696 (1962).
14. Goulston, G., and Mercer, E. I., *Phytochemistry, 8,* 1945 (1969).
15. Barton, D. H. R., and Bruun, T., *J. Chem. Soc.,* p. 2728 (1951).
16. Breivak, O. N., Owades, J. L., and Light, R. F., *J. Org. Chem., 19,* 1734 (1954).
17. Petzoldt, K., Kühne, M., Blanke, E., Kieslich, K., and Kasper, E., *Ann. Chem., 709,* 203 (1967).
18. Wieland, H., and Coutelle, G., *Ann. Chem., 548,* 270 (1941).
19. Alt, G. H., and Barton, D. H. R., *J. Chem. Soc.,* p. 1356 (1954).
20. Wieland, P., and Prelog, V., *Helv. Chim. Acta, 30,* 1028 (1947).
21. Serebryakov, E. P., Simolin, A. V., Kucherov, V. F., and Rosynov, B. V., *Tetrahedron, 26,* 5215 (1970).
22. Pettit, G. R., and Knight, J. C., *J. Org. Chem., 27,* 2696 (1962).
23. Schulte, K. E., Rücker, G., and Fachmann, H., *Tetrahedron Lett.,* p. 4763 (1968).
24. White, J. D., and Taylor, S. I., *J. Amer. Chem. Soc., 92,* 5811 (1970).
25. Kier, L. B., *J. Pharm. Sci., 50,* 471 (1961).
26. Kier, L. B., and Brey, W. S., *J. Pharm. Sci., 52,* 465 (1963).
27. Entwistle, N., and Pratt, A. D., *Tetrahedron, 24,* 3949 (1968); *25,* 1449 (1969).
28. Beereboom, J. J., Fazakerley, H., and Halsall, T. G., *J. Chem. Soc.,* p. 3437 (1957).
29. Halsall, T. G., Hodges, R., and Sayer, G. C., *J. Chem. Soc.,* p. 2036 (1959).

30. Guider, J. M., Halsall, T. G., and Jones, E. R. H., *J. Chem. Soc.*, p. 4471 (1954).
31. Lawrie, W., McLean, J., and Watson, J., *J. Chem. Soc. Sect. C*, p. 1776 (1967).
32. Kanematsu, A., and Natori, S., *Chem. Pharm. Bull. (Tokyo), 18,* 779 (1970); *Yakugaku Zasshi, 90,* 475 (1970).
33. Gaudemer, A., Polonsky, J., Gmelin, R., Adam, H. K., and McCorkindale, N. J., *Bull. Soc. Chim. Fr.*, p. 1844 (1967).
34. Bond, F. T., Fullerton, D. S., Sciuchetti, L. A., and Catalfomo, P., *J. Amer. Chem. Soc., 88,* 3882 (1966).
35. Batta, A. K., and Rangaswami, S., *Curr. Sci., 39,* 416 (1970).
36. Gascoigne, R. M., Holker, J. S. E., Ralph, B. J., and Robertson, A., *J. Chem. Soc.*, p. 2346 (1951).
37. Lahey, F. N., and Strasser, P. H. A., *J. Chem. Soc.*, p. 873 (1951).
38. Yamamoto, I., *Agr. Biol. Chem., 25,* 400 (1961).
39. Cort, L. A., Gascoigne, R. M., Holker, J. S. E., Ralph, B. J., Robertson, A., and Simes, J. J. H., *J. Chem. Soc.*, p. 3713 (1954).
40. Fried, J., Grabowich, P., Sabo, E. F., and Cohen, A. I., *Tetrahedron, 20,* 2297 (1964).
41. Pinhey, J. T., Ralph, B. J., Simes, J. J. H., and Wootton, M., *Aust. J. Chem., 23,* 2141 (1970).
42. Inouye, H., Tokura, K., and Hayashi, T., *Tetrahedron Lett.*, p. 2811 (1970).
43. Bowers, A., Halsall, T. G., Jones, E. R. H., and Lemin, A. J., *J. Chem. Soc.*, p. 2548 (1953).
44. Bowers, A., Halsall, T. G., and Sayer, G. S., *J. Chem. Soc.*, p. 3070 (1954).
45. Simes, J. J. H., Wootton, M., Ralph, B. J., and Pinhey, J. T., *Chem. Commun.*, p. 1150 (1969); *Aust. J. Chem., 24,* 609 (1971).
46. Halsall, T. G., and Hodges, R., *J. Chem. Soc.*, p. 2385 (1954).
47. Goulston, G., Goad, L. J., and Goodwin, T. W., *Biochem. J., 102,* 15C (1967).
48. Epstein, W. W., and van Lear, G., *J. Org. Chem., 31,* 3434 (1966).
49. Adam, H. K., Bryce, T. A., Campbell, I. M., McCorkindale, N. J., Gaudemer, A., Gmelin, R., and Polonsky, J., *Tetrahedron Lett.*, p. 1461 (1967).
50. Inouye, H., and Tokura, K., *Z. Naturforsch. Teil B., 25,* 1194 (1970).
51. Barton, D. H. R., Harrison, D. M., and Widdowson, D. A., *Chem. Commun.*, p. 17 (1968).
52. McMorris, T. C., and Barksdale, A. W., *Nature, 215,* 320 (1967).
53. Arsenault, G. P., Biemann, K., Barksdale, A. W., and McMorris, T. C., *J. Amer. Chem. Soc., 90,* 5635 (1968).
54. Belgian Patent 619,287.
55. Godtfredsen, W. O., Jahnsen, S., Lorck, H., Roholt, K., and Tybring, L., *Nature, 193,* 987 (1962).
56. van Dijck, P. J., and de Somer, P., *J. Gen. Microbiol., 18,* 377 (1958).
57. Vanderhaeghe, H., van Dijck, P., and de Somer, P., *Nature, 205,* 710 (1965).
58. Burton, H. S., and Abraham, E. P., *Biochem. J., 50,* 168 (1951).
59. Chou, T. S., Eisenbraun, E. J., and Rapala, R. T., *Tetrahedron, 25,* 3341 (1969).
60. Iwasaki, S., Iqbal Sair, M., Igarashi, H., and Okuda, S., *Chem. Commun.*, p. 1119 (1970).
61. von Daehne, W., Lorch, H., and Godtfredsen, W. O., *Tetrahedron Lett.*, p. 4843 (1968).
62. Chain, E., Florey, H. W., Jennings, M. A., and Williams, T. I., *Brit. J. Exp. Pathol., 24,* 108 (1943).
63. Okuda, S., Iwasaki, S., Tsuda, K., Sano, Y., Hata, T., Udagawa, S., Nakayama, Y., and Yamaguchi, H., *Chem. Pharm. Bull. (Tokyo), 12,* 121 (1964).
64. Okuda, S., Nakayama, Y., and Tsuda, K., *Chem. Pharm. Bull. (Tokyo), 14,* 436 (1966).
65. Okuda, S., Sato, Y., Hattori, T., and Wakabayashi, M., *Tetrahedron Lett.*, p. 4847 (1968).
66. Okuda, S., Sato, Y., Hattori, T., and Igarashi, H., *Tetrahedron Lett.*, p. 4769 (1968).
67. Hattori, T., Igarashi, H., Iwasaki, S., and Okuda, S., *Tetrahedron Lett.*, p. 1023 (1969).
68. Arigoni, D., *Pure Appl. Chem., 17,* 331 (1968).
69. Brian, P. W., and McGowan, J. C., *Nature, 156,* 144 (1945).
70. Aldridge, D. C., Burrows, B. F., Galt, S., and Turner, W. B., Unpublished Data.
71. Moffatt, J. S., Bu'Lock, J. D., and Yuen, T. H., *Chem. Commun.*, p. 839 (1969).
72. MacMillan, J., Vantsone, A. E., and Yeboah, S. K., *Chem. Commun.*, p. 613 (1968).
73. Ger. Offen. 2,022,452.
74. Heftmann, E., Wright, B. E., and Liddel, C. U., *J. Amer. Chem. Soc., 81,* 6525 (1959); *Arch. Biochem. Biophys., 91,* 266 (1960).
75. Merdinger, E., and Devine, E. M., *J. Bacteriol., 89,* 1488 (1965).
76. McCorkindale, N. J., Hutchinson, S. A., Pursey, B. A., Scott, W. T., and Wheeler, R., *Phytochemistry, 8,* 861 (1969).
77. Reitz, R. C., and Hamilton, J. G., *Comp. Biochem. Physiol., 25,* 401 (1968).
78. de Souza, N. J., and Nes, W. R., *Science, 162,* 363 (1968).
79. Ikekawa, N., Morisaki, N., Tsuda, K., and Yoshida, T., *Steroids, 12,* 41 (1968).
80. Oreutt, D. M., and Richardson, B., *Steroids, 16,* 429 (1970).
81. Patterson, G. W., and Krauss, R. W., *Plant Cell Physiol., 6,* 211 (1965).
82. Tsuda, K., and Sakai, K., *Chem. Pharm. Bull. (Tokyo), 8,* 554 (1960).
83. Patterson, G. W., *Comp. Biochem. Physiol., 31,* 391 (1969).
84. Patterson, G. W., *Plant Physiol., 42,* 1457 (1967).

85.   Klosty, M., and Bergmann, W., *J. Amer. Chem. Soc., 74,* 1601 (1952).
86.   Bergmann, W., and Feeney, R. J., *J. Org. Chem., 15,* 812 (1950).
87.   Ikekawa, N., Morisaki, N., Tsuda, K., and Yoshida, T., *Steroids, 12,* 41 (1968).
88.   Patterson, G. W., *Comp. Biochem. Physiol., 24,* 501 (1968).
89.   Iwata, I., Nakata, H., Mizushima, M., and Sakurai, Y., *Agr. Biol. Chem., 25,* 319 (1961).
90.   Tsuda, K., Akagi, S., Kishida, Y., Hayatsu, R., and Sakai, K., *Chem. Pharm. Bull. (Tokyo), 6,* 724 (1958).
91.   Heilbron, I. M., *J. Chem. Soc.,* p. 79 (1942).
92.   Black, W. A. P., and Cornhill, W. J., *J. Sci. Food Agr., 2,* 387 (1951).
93.   Ito, S., Tamura, T., and Matsumoto, T., *Nippon Daigaku Kôgaku Kenkyûsho Ihô, 13,* 99 (1959); *Chem. Abstr., 53,* 13276d (1959).
94.   Heilbron, I., Phipers, R. F., and Wright, H. R., *J. Chem. Soc.,* p. 1572 (1934).
95.   Tsuda, K., Hayatsu, R., Kishida, Y., and Akagi, S., *J. Amer. Chem. Soc., 80,* 921 (1958).
96.   Tsuda, K., Akagi, S., and Kishida, Y., *Chem. Pharm. Bull. (Tokyo), 6,* 101 (1958).
97.   Gibbons, G. F., Goad, L. J., and Goodwin, T. W., *Phytochemistry, 6,* 677 (1967).
98.   Tsuda, K., Akagi, S., Kishida, Y., Hayatsu, R., and Sakai, K., *Chem. Pharm. Bull. (Tokyo), 6,* 724 (1958).
99.   Idler, D. R., Saito, A., and Wiseman, P., *Steroids, 11,* 465 (1968).
100.   Tsuda, K., Akagi, S., and Kishida, Y., *Science, 126,* 927 (1957).
101.   Tsuda, K., Sakai, K., Tanabe, K., and Kishida, Y., *J. Amer. Chem. Soc., 82,* 1442 (1960).
102.   Alcaide, A., Barbier, M., Potier, P., Magueur, A. M., and Teste, J., *Phytochemistry, 8,* 2301 (1969).
103.   Collins, R. P., and Kalnins, K., *Comp. Biochem. Physiol., 30,* 779 (1969).
104.   Gershengorn, M. C., Smith, A. R. H., Goulston, G., Goad, L. J., Goodwin, T. W., and Haines, T. H., *Biochemistry, 7,* 1698 (1968).
105.   Williams, B. L., Goodwin, T. W., and Ryley, J. F., *J. Protozool., 13,* 227 (1966).
106.   Avivi, L., Iaron, O., and Halevy, S., *Comp. Biochem. Physiol., 21,* 321 (1967).
107.   Collins, R. P., and Kalnins, K., *Can. J. Microbiol., 14,* 837 (1968).
108.   Avivi, L., Iaron, O., and Halevy, S., *Comp. Biochem. Physiol., 21,* 321 (1967).
109.   Williams, B. L., Goodwin, T. W., and Ryley, J. F., *J. Protozool., 13,* 227 (1966).
110.   Brandt, R. D., Pryce, R. J., Anding, C., and Ourisson, G., *Eur. J. Biochem., 17,* 344 (1970).
111.   Avivi, L., Iaron, O., and Halevy, S., *Comp. Biochem. Physiol., 21,* 321 (1967).
112.   Smith, F. R., and Korn, E. D., *J. Lipid Res., 8,* 405 (1968).
113.   Halevy, S., Avivi, L., and Katan, H., *J. Protozool., 13,* 480 (1966).
114.   Williams, B. L., Goodwin, T. W., and Ryley, J. F., *J. Protozool., 13,* 227 (1966).
115.   Halevy, S., and Avivi, L., *Ann. Trop. Med. Parasitol., 60,* 439 (1966).
116.   Korn, E. D., von Brand, T., and Tobie, E. J., *Comp. Biochem. Physiol., 30,* 601 (1969).
117.   Schubert, K., Rose, G., Tümmler, R., and Ikekawa, N., *Z. Physiol. Chem., 339,* 293 (1964).
118.   Schubert, K., Rose, G., and Hörhold, C., *Biochim. Biophys. Acta, 137,* 168 (1967).
119.   Bird, C. W., Lynch, J. M., Pirt, F. J., Reid, W. W., Brooks, C. J. W., and Middleditch, B. S., *Nature, 230,* 473 (1971).
120.   Schubert, K., Rose, G., Wachtel, H., Hörhold, C., and Ikekawa, N., *Eur. J. Biochem., 5,* 246 (1968).

AROMATIC COMPOUNDS

SIMPLE CARBOCYCLIC PHENOLS DERIVED BY THE ACETATE–POLYMALONATE PATHWAY

DR. JOHN FREDERICK GROVE

More complex phenols, e.g., depsides, are considered in other sections, as are phenolic compounds that also contain a heterocyclic ring or a quinone grouping. Compounds, e.g., gentisic acid, known to be derived by both shikimate and acetate-polymalonate pathways are included in this section, as are related compounds whose derivation is uncertain or unproven.

These simple carbocyclic phenols are further subdivided (Tables 1–6) into hydroxy derivatives of benzene, monocyclic aromatic carbonyl compounds, benzophenone, diphenylether, diphenyl, and naphthalene, respectively.

Many of the phenols, particularly those hydroxycarbonyl compounds capable of forming chelate complexes, show antimicrobial properties.

CODES FOR ORGANISMS IN TABLES 1 TO 6

Table 1

A = *Penicillium* sp.	H = *Aspergillus fumigatus*	P = *Fomes fastuosus*
B = *Penicillium patulum*	I = *Aspergillus terreus*	Q = *Fomes robiniae*
C = *Psathyrella conopilea*	J = *Gliocladium roseum*	R = *Lentinus degener*
D = *Phyllosticta* sp.	K = *Penicillium citrinum*	S = *Penicillium baarnense*
E = *Phoma* sp.	L = *Penicillium spinulosum*	T = *Helminthosporium siccans*
F = *Penicillium canadense*	M = *Stereum subpileatum*	U = *Alternaria zinniae*
G = *Penicillium martinsii*	N = *Aspergillus versicolor*	V = *Umbilicaria papulosa*
	O = *Penicillium divergens*	

Table 2

A = *Penicillium griseofulvum*	S = *Lentinus degener*	μ = *Epicoccum andropogonis*
B = *Pellicularia filamentosa*	T = *Phyllosticta* sp.	ν = *Epicoccum nigrum*
C = *Rhizoctonia solani*	U = *Aspergillus oniki*	ξ = *Paecilomyces victoriae*
D = *Penicillium patulum*	V = *Aspergillus fumigatus*	π = *Curvularia siddiqui*
E = *Streptomyces griseus*	W = *Aspergillus rugulosus*	ρ = *Aspergillus quadrilineatus*
F = *Claviceps paspali*	X = *Penicillium cyclopium*	σ = *Curvularia ellisii*
G = *Penicillium divergens*	Y = *Penicillium stipitatum*	τ = *Pseudomonas fluorescens*
H = *Aspergillus terreus*	Z = *Penicillium funiculosum*	υ = *Aspergillus clavatus*
I = *Polyporus tumulosus*	α = *Penicillium spinulosum*	φ = *Mortierella ramanniana*
J = *Penicillium citrinum*	β = *Penicillium baarnense*	χ = *Roccella fuciformis*
K = *Cyathus helanae*	γ = *Penicillium madriti*	ψ = *Sparassis ramosa*
L = *Boletus scaber*	δ = *Penicillium fennelliae*	ω = *Penicillium gladioli*
M = *Penicillium islandicum*	ε = *Aspergillus flavipes*	a = *Aspergillus oryzae*
N = *Fomes juniperinus*	ζ = *Chaetomium cochliodes*	b = *Aspergillus ustus*
O = *Penicillium brevi-compactum*	η = *Chaetomium globosum*	c = *Sclerotinia sclerotiorum*
P = *Daldinia concentrica*	θ = *Cladosporium fulvum*	d = *Sclerotinia libertiana*
Q = *Penicillium flexuosum*	ι = *Curvularia lunata*	e = *Penicillium notatum*
R = *Gliocladium roseum*	κ = *Gibberella fujikuroi*	f = *Pyricularia oryzae*
	λ = *Epicoccum purpurascens*	

Table 3

A = *Oospora sulphurea-ochracea*
B = *Penicillium patulum*
C = *Penicillium estinogenum*
D = *Aspergillus terreus*
E = *Penicillium frequentans*
F = *Aspergillus fumigatus*
G = *Aspergillus rugulosus*

Table 5

A = *Penicillium rubrum*
B = *Phlebia mellea*
C = *Phlebia albida*
D = *Alternaria tenuis*

Table 4

A = *Aspergillus rugulosus*
B = Sphaeropsidales (order)
C = *Oospora sulphurea-ochracea*
D = *Aspergillus terreus*
E = *Penicillium frequentans*

Table 6

A = *Daldinia concentrica*

TABLE 1
HYDROXYBENZENES

Formula	Systematic Name (Common Name)	Melting Point (°C)	Organism	Reference
$C_6H_4OCl_2$	2,4-Dichloro-1-hydroxybenzene	77–78*	A	1
$C_6H_6O_3$	1,2,3-Trihydroxybenzene (Pyrogallol)	134	B	2
$C_7H_4O_2Cl_4$	2,3,5,6-Tetrachloro-1-hydroxy-4-methoxybenzene (Drosophilin A)	118	C	3
$C_7H_7O_3Cl$	6-Chloro-1,4-dihydroxy-2-hydroxymethylbenzene	147–8	D,E,F	4,5,6
$C_7H_7O_3Cl$	5-Chloro-1,4-dihydroxy-2-hydroxymethylbenzene (Amudol)	146–7	G	7
C_7H_8O	1-Hydroxy-3-methylbenzene (m-Cresol)	11–12	H	8
$C_7H_8O_2$	1,3-Dihydroxy-5-methylbenzene (Orcinol)	109	H–L,V	9–13
$C_7H_8O_2$	1-Hydroxy-3-hydroxymethylbenzene (m-Hydroxybenzyl alcohol)	73	B	14
$C_7H_8O_2$	1,4-Dihydroxy-2-methylbenzene (Toluhydroquinone)	126–127	B,E	5,15
$C_7H_8O_2$	1-Hydroxy-4-methoxybenzene (Hydroquinone monomethylether)	53	M	16
$C_7H_8O_3$	1,2,5-Trihydroxy-3-methylbenzene (2,3,5-Trihydroxytoluene)	148	H,L	12
$C_7H_8O_3$	1,2,4-Trihydroxy-3-methylbenzene (Versicolin)	125–126	N	17,18

TABLE 1 (Continued)
HYDROXYBENZENES

Formula	Systematic Name (Common Name)	Melting Point (°C)	Organism	Reference
$C_7H_8O_3$	1,4-Dihydroxy-2-hydroxymethylbenzene (Gentisyl alcohol)	104–105	B,D,E,O	19–22
$C_7H_8O_4$	1,2,3,4-Tetrahydroxy-5-methylbenzene (2,3,4,5-Tetrahydroxytoluene)	170–171	H,L	12
$C_8H_6O_2Cl_4$	2,3,5,6-Tetrachloro-1,4-dimethoxybenzene (Drosophilin A methyl ether)	164–165	P,Q	23,24
$C_8H_6O_4Cl_3N$	3,5,6-Trichloro-1,4-dimethoxy-2-nitrobenzene	115–116	Q	24
$C_8H_{10}O_2$	1,5-Dihydroxy-2,3-dimethylbenzene	136	J	9
$C_8H_{10}O_3$	1,4-Dihydroxy-2-methoxy-5-methylbenzene (4-Methoxytoluhydroquinone)	125–126	R	25
$C_8H_{10}O_4$	1,3,4-Trihydroxy-2-methoxy-5-methylbenzene (Dihydrofumigatin)	108–109	H	26
$C_9H_{10}O_3$	Acetal of 1,4-dihydroxy-2-hydroxymethylbenzene (Gentisyl acetal)	103–105	E	5
$C_{10}H_{14}O_3$	5-Ethyl-1,2,3-trihydroxy-4,6-dimethylbenzene (Barnol)	145–146	S	27
$C_{10}H_{14}O_4$	1,4-Dihydroxy-2,3-dimethoxy-5,6-dimethylbenzene (Dihydroaurantiogliocladin)	84	J	28
$C_{11}H_{10}O_2$	1,3-Dihydroxy-4-(3'-methylbut-3'-en-1'-ynyl)benzene (Siccayne)	115–116	T	29
$C_{11}H_{16}O_3$	(-)1,5-Dihydroxy-2-methyl-3(2'-hydroxy-1'-methylpropyl)benzene	129	K	11
$C_{15}H_{22}O_4$	2,3-Dihydroxymethyl-1-methoxy-6-methyl-5-dimethylallyloxybenzene (Zinniol)	oil	U	30

*Literature mp = 45°C

TABLE 2

HYDROXYBENZENE ALDEHYDES, KETONES, CARBOXYLIC ACIDS, AND DERIVATIVES

Formula	Systematic Name (Common Name)	Melting Point (°C)	Organism	Reference
$C_7H_6O_3$	3-Hydroxybenzoic acid	200	A–C	1,2
$C_7H_6O_3$	2,5-Dihydroxybenzaldehyde (Gentisyl aldehyde)	99	D	3
$C_7H_6O_4$	2,3-Dihydroxybenzoic acid	204	E,F	4,5
$C_7H_6O_4$	2,4-Dihydroxybenzoic acid	218–219	A	1
$C_7H_6O_4$	2,5-Dihydroxybenzoic acid (Gentisic acid)	205	A,D,G–I	1,6–11
$C_7H_6O_4$	3,4-Dihydroxybenzoic acid (Protocatechuic acid)	199	A,J	1,12
$C_7H_6O_4$	3,5-Dihydroxybenzoic acid	232–233	A	1
$C_7H_6O_4$	2,4,5-Trihydroxybenzaldehyde (Chromocyathin)	223	K	13
$C_7H_6O_4$	3,4,5-Trihydroxybenzaldehyde	212(decomposes)	L	14
$C_8H_6O_4$	6-Formyl-2-hydroxybenzoic acid (6-Formylsalicylic acid)	137	D	15
$C_8H_6O_5$	3-Hydroxyphthalic acid	150(decomposes)	D,M	15,16
$C_8H_6O_5$	3,4,5-Trihydroxyphthalaldehyde (Fomecin B)	>230(decomposes)	N	17
$C_8H_6O_6$	3,5-Dihydroxyphthalic acid	188–190	H,O	7,18
$C_8H_8O_3$	2,6-Dihydroxyacetophenone	154–156	P	19
$C_8H_8O_3$	2-Hydroxy-6-methylbenzoic acid (6-Methylsalicylic acid)	173	A,H,Q–T	7,20–25
$C_8H_8O_3$	2,4-Dihydroxy-6-methylbenzaldehyde (Orsellinaldehyde)	181–182	H,W	26,27

TABLE 2 (Continued)

HYDROXYBENZENE ALDEHYDES, KETONES, CARBOXYLIC ACIDS, AND DERIVATIVES

Formula	Systematic Name (Common Name)	Melting point (°C)	Organism	Reference
$C_8H_8O_4$	2,4-Dihydroxy-6-hydroxymethylbenzaldehyde	148–150(decomposes)	W	27
$C_8H_8O_4$	2,4-Dihydroxy-6-methylbenzoic acid (Orsellinic acid)	176(decomposes)	A,H,J,V X–Z,α–κ	7,12, 28–36
$C_8H_8O_5$	2,3,4-Trihydroxy-6-hydroxymethylbenzaldehyde (Fomecin A)	>160(decomposes)	N	17
$C_9H_8O_5$	2,3,4-Trihydroxy-5-methylphthalaldehyde (Flavipin)	233–234(decomposes)	H,ϵ,λ–ν	37–39
$C_9H_8O_7$	4,5-Dihydroxy-3-methoxyphthalic acid	193–194(decomposes)	ζ	40
$C_9H_{10}O_4$	2,4-Dihydroxy-3,6-dimethylbenzoic acid	183–185	H	41
$C_9H_{10}O_4$	4,6-Dihydroxy-2,3-dimethylbenzoic acid	159–162(decomposes)	H,N,ϵ	7,29,41
$C_9H_{10}O_4$	3,5-Dimethoxybenzoic acid	186	π	42
$C_9H_{10}O_4$	6-Ethyl-2,4-dihydroxybenzoic acid	168–9	β	43
$C_{10}H_8O_6$	2,4-Dihydroxy-6-(1',2'-dioxopropyl) benzoic acid	125–135^1	O	44
$C_{10}H_{10}O_4$	5-Hydroxy-3-methoxy-4-methylphthalaldehyde (Quadrilineatin)	172(decomposes)	ρ	45
$C_{10}H_{10}O_5$	2-Acetyl-3,5-dihydroxyphenylacetic acid (Curvulinic acid)	238(decomposes)	π,σ	46,47
$C_{10}H_{10}O_5$	2,4-Dihydroxy-6-(2'-oxopropyl) benzoic acid	145–153(decomposes)	O	44
$C_{10}H_{10}O_5$	2,4-Diacetylphloroglucinol	168	τ	48
$C_{10}H_{10}O_6$	2,4-Dihydroxy-6-(1'-hydroxy-2'-oxopropyl) benzoic acid	202–204	O	44
$C_{10}H_{12}O_3$	2,4-Dihydroxy-3,5-dimethylacetophenone (Clavatol)	183	υ	49
$C_{10}H_{12}O_3$	2,6-Dihydroxybutyrophenone	117–118	P	19

TABLE 2 (Continued)

HYDROXYBENZENE ALDEHYDES, KETONES, CARBOXYLIC ACIDS, AND DERIVATIVES

Formula	Systematic Name (Common Name)	Melting Point (°C)	Organism	Reference
$C_{10}H_{12}O_4$	2,4-Dihydroxy-6-propylbenzoic acid	208–212(decomposes)[2]	O	50
$C_{10}H_{12}O_4$	2,4-Dihydroxy-3,5,6-trimethylbenzoic acid	193	φ	51
$C_{10}H_{12}O_4$	Ethyl-2,4-Dihydroxy-6-methylbenzoate (Ethyl orsellinate)	132	X	52
$C_{10}H_{12}O_4$	Methyl 2-Hydroxy-4-methoxy-6-methylbenzoate (Sparassol)	67–68	ψ	53
$C_{11}H_{10}O_5$	2,3-Diformyl-6-methoxy-5-methyl benzoic acid (Gladiolic acid)	160	ω	54
$C_{11}H_{10}O_6$	2,3-Diformyl—4-hydroxy-6-methoxy-5-methyl benzoic acid (Cyclopaldic acid)	224–225	X	55
$C_{11}H_{10}O_7$	4,6-Dihydroxy-3-methoxy-2(1',2'-dioxopropyl) benzoic acid (Dehydroustic acid)	121–122[1]	ξ	40
$C_{11}H_{12}O_4$	1,3,4-Trihydroxy-2-(1'-oxopent-3-enyl) benzene (Maltoryzine)	69	a	56
$C_{11}H_{12}O_5$	2-Formyl-3-hydroxymethyl-6-methoxy-5-methyl benzoic acid (Dihydrogladiolic acid)	135–136	ω	54
$C_{11}H_{12}O_5$	2-Acetyl-3-hydroxy-5-methoxyphenylacetic acid (O-Methylcurvulinic acid)	173	σ	47
$C_{11}H_{12}O_6$	2-Acetyl-3,5-dihydroxy-4-methoxylphenylacetic acid (Curvulic acid)	154	π	57
$C_{11}H_{12}O_6$	2-Formyl-4-hydroxy-3-hydroxymethyl-6-methoxy-5-methylbenzoic acid (Cyclopolic acid)	147–148	X	55

TABLE 2 (Continued)
HYDROXYBENZENE ALDEHYDES, KETONES, CARBOXYLIC ACIDS, AND DERIVATIVES

Formula	Systematic Name (Common Name)	Melting Point (°C)	Organism	Reference
$C_{11}H_{12}O_7$	4,6-Dihydroxy-3-methoxy-2(1'-hydroxy-2'-oxopropyl) benzoic acid (Ustic acid)	169—170(decomposes)	ξ, b	40,58
$C_{12}H_{14}O_5$	Ethyl 2-acetyl-3,5-dihydroxyphenylacetic acid (Curvulin)	145	π, σ	46,47
$C_{12}H_{14}O_5$	2-Acetyl-3,5-dimethoxyphenylacetic acid (Dimethylcurvulinic acid)	119	π	46
$C_{12}H_{14}O_5$	2-Acetonyl-4,6-dihydroxy-3,5-dimethyl benzoic acid (Sclerotinin B)	192—5(decomposes)	c	59
$C_{12}H_{16}O_4$	5-(2'-Formyloxy-1'-methylpropyl)-4-methylresorcinol	131—133	J	12
$C_{12}H_{16}O_5$	(-)2,6-Dihydroxy-4-(2'-hydroxy-1'-methylpropyl)-3-methylbenzoic acid	185(decomposes)	J	12
$C_{13}H_{14}O_4$	(+)3-Hydroxy-2-carboxy-α,4,5,6-tetramethylpropionic acid anhydride (Sclerin)	123	c,d	59,60
$C_{13}H_{16}O_5$	4,6-Dihydroxy-3,5-dimethyl-2-(2'-oxo-1'-methyl propyl) benzoic acid (Sclerotinin A)	205—208	c	59
$C_{13}H_{16}O_6$	Ethyl 2-acetyl-3,5-dihydroxy-4-methoxyphenyl acetate (Curvin)	123	π	57
$C_{14}H_{16}O_3$	1,3-Dihydroxy-6-(1'-oxohexa-2',4'-dienyl)-2,4-dimethyl benzene (Sorbicillin)	113—114	e	61
$C_{14}H_{16}O_4$	2-Hydroxy-6-(3',4'-dihydroxyhepta-1',5'-dienyl) benzaldehyde (*Pyricularia oryzae* toxin)	96—97	f	62

[1] Hydrate
[2] Literature m p = 179°C

177

TABLE 3
HYDROXYBENZOPHENONES

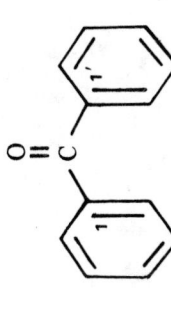

Formula	Systematic Name (Common Name)	Melting Point (°C)	Organism	Reference
$C_{16}H_{14}O_7$	6'-Carboxy-2,6,4'-trihydroxy-2'methoxy-4-methylbenzophenone (Desmethylsulochrin)	205–207	A	1
$C_{16}H_{15}O_6Cl$	3-Chloro-2,6,4'-trihydroxy-4,2'-dimethoxy-6'-methylbenzophenone (Griseophenone B)	205	B	2
$C_{16}H_{16}O_6$	2,6,4'-Trihydroxy-4,2'-dimethoxy-6'-methylbenzophenone (Griseophenone C)	183–184	B	2
$C_{17}H_{14}O_7Cl$	6'-Carbomethoxy-3,5-dichloro-2,6,4'-trihydroxy-2'-methoxy-4-methylbenzophenone (Dihydrogeodin)	229	C,D	3,4
$C_{17}H_{16}O_7$	6'-Carbomethoxy-2,6,4'-trihydroxy-2'-methoxy-4-methylbenzophenone (Sulochrin)	262	A,D,E	4,5,6
$C_{17}H_{17}O_5Cl$	3-Chloro-2,4'-dihydroxy-6,2'-dimethoxy-4,6'-dimethylbenzophenone	181–182	B	2,7
$C_{17}H_{17}O_6Cl$	3-Chloro-2,4'-dihydroxy-4,6,2'-trimethoxy-6'-methylbenzophenone (Griseophenone A)	212–214	B	2,7
$C_{18}H_{18}O_7$	6'-Carbomethoxy-2,4'-dihydroxy-6,2'-dimethoxy-4-methylbenzophenone (Methyl sulochrin)	198–199	F	8
$C_{25}H_{28}O_6$	6'-Formyl-2,6,2'-trihydroxy-3-dimethylallyl-5'-dimethylallyloxy-4'-methylbenzophenone (Arugosin A)	Oil	G	9
$C_{25}H_{28}O_6$	6'-Formyl-2,6,2'-trihydroxy-5-dimethylallyl-5'-dimethyl-allyloxy-4'-methylbenzophenone (Arugosin B)	Oil	G	9

TABLE 4
HYDROXYDIPHENYL ETHERS

Formula	Systematic Name (Common Name)	Melting Point (°C)	Organism	Reference
$C_4H_{14}O_3$	3,3'-Dihydroxy-5,5'-dimethyldiphenyl ether	Oil	A	1
$C_{15}H_{16}O_4$	2,3'-Dihydroxy-4-methoxy-6,5'-dimethyldiphenyl ether (Antibiotic LL-V125a)	122*	B	2,3
$C_{16}H_{14}O_8$	2,2'-Dicarboxy-4,3'-dihydroxy-6-methoxy-5'-methyldiphenyl ether (Methylosoic acid, dimethylasterric acid)	248(decomposes)	C,D	4
$C_{17}H_{14}O_8Cl_2$	2'-Carboxy-4',6'-dichloro-4-hydroxy-6-methoxy-2-methoxy-carbonyl-5'-methyldiphenyl ether (Geodin hydrate)	209	D	5
$C_{17}H_{16}O_8$	2'-Carboxy-4,3'-dihydroxy-6-methoxy-2-methoxycarbonyl-5'-methyldiphenyl ether (Dimethylosoic acid, asterric acid)	214	C,D,E	4,6
$C_{18}H_{18}O_8$	4,3'-Dihydroxy-6-methoxy-2,2'-dimethoxycarbonyl-5'-methyldiphenyl ether (Trimethylosoic acid, methyl asterrate)	190	C	4
$C_{19}H_{18}O_9$	3'-Acetoxy-2'-carboxy-4-hydroxy-6-methoxy-2-methoxy-carbonyl-5'-methyldiphenyl ether	200(decomposes)	C	4

*Hydrate

TABLE 5
HYDROXYDIPHENYLS

Formula	Systematic Name (Common Name)	Melting Point (°C)	Organism	Reference
$C_{14}H_{14}O_6$	2,3,6,2′,3′,6′-Hexahydroxy-4,4′-dimethyldiphenyl (Tetrahydrophoenicin)	247(decomposes)	A	1
$C_{14}H_{14}O_8$	2,3,5,6,2′,3′,5′,6′-Octahydroxy-4,4′-dimethyldiphenyl (Tetrahydroosporein)		B,C	2
$C_{15}H_{14}O_6$	2-Carboxy-3,4′,5′-trihydroxy-5-methoxy-2′-methyldiphenyl (Altenusin)	202—203	D	3

TABLE 6
HYDROXYNAPHTHALENES

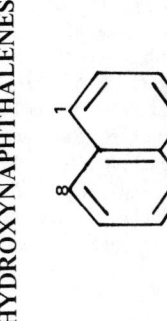

Formula	Systematic Name (Common Name)	Melting Point (°C)	Organism	Reference
$C_{11}H_{10}O_2$	1-Hydroxy-8-methoxynaphthalene	55—56	A	1
$C_{12}H_{12}O_2$	1,8-Dimethoxynaphthalene	158—161	A	1
$C_{20}H_{14}O_4$	4,5,4′,5′-Tetrahydroxydinaphthyl	225—230	A	1

REFERENCES

Table 1

1. Ando, K., Kato, A., and Susuki, S., *Biochem. Biophys. Res. Commun., 39,* 1104 (1970).
2. Bassett, E. W., and Tanenbaum, S. W., *Biochim. Biophys. Acta, 28,* 21 (1958).
3. Kavanagh, F., Hervey, A., and Robbins, W. J., *Proc. Natl. Acad. Sci. U.S.A., 38,* 555 (1952).
4. Sakamura, S., Ito, J., and Sakai, R., *Agric. Biol. Chem., 35,* 105 (1971).
5. Sequin-Frey, M., and Tamm, C., *Helv. Chim. Acta, 54,* 851 (1971).
6. McCorkindale, N. J., Roy, T. P., and Hutchinson, S. A., *Tetrahedron, 28,* 1107 (1972).
7. Kamal, A., Jarboe, C. H., Qureshy, I. H., Husain, S. A., Murtaza, N., and Noorani, R., *Pak. J. Sci. Ind. Res., 13,* 236 (1970).
8. Packter, N. M., *Biochem. J., 97,* 321 (1965).
9. Pettersson, G., *Acta Chem. Scand., 18,* 1202 (1964); *19,* 414 (1965).
10. Curtis, R. F., Harries, P. C., Hassall, C. H., and Levi, J. D., *Biochem. J., 90,* 43 (1964).
11. Curtis, R. F., Hassall, C. H., and Nazar, M., *J. Chem. Soc. C,* 85 (1968).
12. Simonart, P., and Verachtert, H., *Bull. Soc. Chim. Biol., 48,* 943 (1966).
13. Miller, E. V., Griffin, C. E., Schaefers, T., and Gordon, M., *Chem. Abstr., 63,* 8727 (1966).
14. Rebstock, M. C., *Arch. Biochem. Biophys., 104,* 156 (1964).
15. Scott, A. I., and Yalpani, M., *Chem. Commun.,* 945 (1967).
16. Bu'Lock, J. D., Hudson, A. T., and Kaye, B., *Chem. Commun.,* 817 (1967).
17. Dhar, A. K., and Bose, S. K., *J. Antibiot. Ser. A , 21,* 156 (1968).
18. Rickards, R. W., *J. Antibiot., 24,* 715 (1971).
19. Birkinshaw, J. H., Bracken, A., and Raistick, H., *Biochem. J., 37,* 726 (1943).
20. Barta, J., and Mecir, R., *Experientia, 4,* 227 (1948).
21. Closse, A., Mauli, R., and Sigg, H. P., *Helv. Chim. Acta, 49,* 204 (1966).
22. Sakamura, S., Chida, T., Ito, J., and Sakai, R., *Agric. Biol. Chem., 35,* 1810 (1971).
23. Singh, P., and Rangaswami, S., *Tetrahedron Lett.,* 1229 (1966).
24. Butruille, B., and Dominguez, X. A., *Tetrahedron Lett.,* 211 (1972).
25. Packter, N. M., *Biochem. J., 114,* 369 (1969).
26. Anslow, W. K., and Raistrick, H., *Biochem. J., 32,* 687 (1938).
27. Ljungerantz, I., and Mosbach, K., *Acta Chem. Scand., 18,* 638 (1964).
28. Steward, M. W., and Packter, N. M., *Biochem. J., 95,* 26C (1965).
29. Ishibashi, K., Nose, K., Shindo, T., Avai, M., and Mishima, H., *Chem. Abstr., 71,* 38498 (1969).
30. Starratt, A. N., *Can. J. Chem., 46,* 767 (1968).

Table 2

1. Simonart, P., and Wiaux, A., *Bull. Soc. Chim. Biol., 41,* 527 (1959).
2. Aoki, H., Sassa, T., and Tamura, T., *Nature, 200,* 575 (1963).
3. Tanenbaum, S. W., and Bassett, E. W., *Biochim. Biophys. Acta, 28,* 21 (1958).
4. Dyer, J. R., Heding, H., and Schaffner, C. P., *J. Org. Chem., 29,* 2802 (1964).
5. Arcamone, F., Chain, E. B., Ferretti, A., and Penella, P., *Nature, 192,* 552 (1961).
6. Crowden, R. K., and Ralph, B. J., *Aust. J. Chem., 14,* 475 (1961).
7. Curtis, R. F., Harries, P. C., Hassall, C. H., and Levi, J. D., *Biochem. J., 90,* 43 (1964).
8. Barta, J., and Mecir, R., *Experientia, 4,* 277 (1948).
9. Raistrick, H., and Simonart, P., *Biochem. J., 27,* 628 (1933).
10. Gatenbeck, S., and Lonnroth, I., *Acta Chem. Scand., 16,* 2298 (1962).
11. Crowden, R. K., *Can. J. Microbiol., 13,* 181 (1967).
12. Curtis, R. F., Hassall, C. H., and Nazar, M., *J. Chem. Soc. C,* 85 (1968).
13. Allbutt, A. D., Ayer, W. A., Brodie, H. J., Johri, B. N., and Taube, H., *Can. J. Microbiol, 17,* 1401 (1971).
14. Edwards, R. L., and Elsworthy, G. C., *J. Chem. Soc. C,* 410 (1967).
15. Bassett, E. W., and Tanenbaum, S. W., *Experientia, 14,* 38 (1958).
16. Gatenbeck, S., *Acta Chem. Scand., 11,* 555 (1957).
17. McMorris, T. C., and Anchel, M., *Can. J. Chem., 42,* 1595 (1964).
18. Oxford, A. E., and Raistrick, H., *Biochem. J., 26,* 1902 (1932).
19. Allport, D. C., and Bu'Lock, J. D., *J. Chem. Soc.,* 654 (1960).
20. Anslow, W. K., and Raistrick, H., *Biochem. J., 25,* 39 (1921).
21. Oxford, A. E., Raistrick, H., and Simonart, P., *Biochem. J., 29,* 1102 (1935).
22. Packter, N. M., *Biochem. J., 97,* 321 (1965).
23. Pettersson, G., *Acta Chem. Scand., 20,* 45 (1966).
24. Sasaki, M., Kaneko, Y., Oshita, K., and Takematsu, H., *Agric. Biol. Chem., 34,* 1296 (1970).

25. Sakamura, S., Chida, T., Ito, J., and Sakai, R., *Agric. Biol. Chem., 35,* 1810 (1971).
26. Curtis, R. F., Harries, P. C., Hassall, C. H., and Levi, J. D., *J. Chem. Soc. C,* 168 (1966).
27. Ballantine, J. A., Hassall, C. H., and Jones, B. D., *Phytochemistry, 7,* 1529 (1968).
28. Reio, L., *J. Chromatogr., 1,* 340 (1958).
29. Pettersson, G., *Acta Chem. Scand., 18,* 1202 (1964); *19,* 1724 (1965).
30. Simonart, P., and Verachtert, H., *Bull. Soc. Chim. Biol., 48,* 943 (1966).
31. Locci, R., Merlini, L., Hasini, G., and Locci, J. R., *G. Microbiol., 15,* 93 (1967).
32. Gaucher, G. M., and Shepard, M. G., *Biochem. Biophys. Res. Commun., 32,* 664 (1968).
33. Birkinshaw, J. H., and Gowlland, A., *Biochem. J., 84,* 342 (1962).
34. Mosbach, K., *Z. Naturforsch. B , 14,* 69 (1959).
35. Mosbach, K., *Acta Chem. Scand., 14,* 457 (1960).
36. Bentley, R., Ghaphery, J. A., and Keil, J. G., *Arch. Biochem. Biophys., 111,* 80 (1965).
37. Raistrick, H., and Rudman, P., *Biochem. J., 63,* 395 (1956).
38. Bamford, P. C., Norris, G. L. F., and Ward, G., *Trans. Br. Mycol. Soc., 44,* 354 (1961).
39. Eka, O. U., *Experientia, 25,* 924, 1278 (1970).
40. Vora, V. C., *J. Sci. Ind. Res. B. (India), 13,* 842 (1954).
41. Takenaka, S., Ojima, N., and Seto, S., *Chem. Commun.,* 391 (1972).
42. Qureshi, A. A., Rickards, R. W., and Kamal, A., *Tetrahedron, 23,* 3801 (1967).
43. Mosbach, K., *Acta. Chem. Scand., 18,* 1591 (1964).
44. Clutterbuck, P. W., Oxford, A. E.; Raistrick, H., and Smith, G., *Biochem. J., 26,* 1441 (1932).
45. Birkinshaw, J. H., Chaplen, P., and Lashoz-Oliver, R., *Biochem. J., 67,* 155 (1957).
46. Kamal, A., Ahmed, N., Khan, M. A., and Qureshi, I. H., *Tetrahedron 18,* 433 (1962).
47. Coombe, R. G., Jacobs, J. J., and Watson, T. R., *Aust. J. Chem., 21,* 783 (1968).
48. Reddi, T. K., and Borovkov, A. V., *Antibiotiki, 15,* 19 (1970).
49. Bergel, F., Morrison, A. L., Moss, A. R., and Rinderknecht, H., *J. Chem. Soc.,* 415 (1944).
50. Godin, P., *Antonie van Leeuwenhoek , 21,* 362 (1955).
51. Andres, W. W., Kunstmann, M. P., and Mitscher, L. A., *Experientia, 23,* 703 (1967).
52. Aberhart, D. J., and Overton, K. H., *J. Chem. Soc. C,* 704 (1969).
53. Falck, R., *Ber. Dtsch. Chem. Ges., 56,* 2555 (1923).
54. Grove, J. F., *Biochem. J., 50,* 648 (1952).
55. Birkinshaw, J. H., Raistrick, H., Ross, D. J., and Stickings, C. E., *Biochem. J., 50,* 610 (1952).
56. Iizuka, H., and Iida, M., *Nature, 196,* 681 (1962).
57. Kamal, A., Khan, H. A., and Qureshi, A. A., *Tetrahedron, 19,* 111 (1963).
58. Raistrick, H., and Stickings, C. E., *Biochem. J., 48,* 53 (1951).
59. Sassa, T., Aoki, H., Namiki, M., and Munakata, K., *Agric. Biol. Chem., 32,* 1432 (1968).
60. Tokaroyama, T., Kamikawa, T., and Kubota, T., *Tetrahedron, 24,* 2345 (1968).
61. Cram, D. J., and Tishler, M., *J. Am. Chem. Soc., 70,* 4238 (1948).
62. Iwasaki, S., Nozoe, S., Okuda, S., Sato, Z., and Kozaka, T., *Tetrahedron Lett.,* 3977 (1969).

Table 3

1. Natori, S., and Nishikawa, H., *Chem. Pharm. Bull. (Tokyo), 10,* 117 (1962).
2. Rhodes, A., Boothroyd, B., McGonagle, M. P., and Somerfield, S. A., *Biochem. J., 81,* 28 (1961).
3. Komatsu, E., *J. Agric. Chem. Soc. Jap., 31,* 905 (1957).
4. Curtis, R. F., Harris, P. C., Hassall, C. H., and Levi, J. D., *Biochem. J., 90,* 43 (1964).
5. Nishikawa, H., *Acta Phytochim. (Tokyo), 11,* 167 (1939).
6. Mahmoodian, A., and Stickings, C. E., *Biochem. J., 92,* 369 (1964).
7. McMaster, W. J., Scott, A. I., and Trippett, S., *J. Chem. Soc.,* 4628 (1960).
8. Turner, W. B., *J. Chem. Soc.,* 6658 (1965).
9. Ballantine, J. A., Francis, D. J., Hassall, C. H., and Wright, J. L. C., *J. Chem. Soc. C,* 1175 (1970).

Table 4

1. Ballantine, J. A., Hassall, C. H., and Jones, B. D., *Phytochemistry, 7,* 1529 (1968).
2. McGahren, W. J., Andres, W. W., and Kunstmann, *J. Org. Chem., 35,* 2433 (1970).
3. Cannon, J. R., Cresp, T. M., Metcalf, B. W., Sargent, M. V., and Elix, J. A., *Chem. Commun.,* 473 (1971).
4. Curtis, R. F., Hassall, C. H., Natori, S., and Nishikawa, H., *Chem. Ind. (Lond.),* 1360 (1961).
5. Curtis R. F., Harries, P. C., Hassall, C. H., and Levi, J. D., *Biochem. J., 90,* 43 (1964).
6. Mahmoodian, A., and Stickings, C. E., *Biochem. J., 92,* 369 (1964).

Table 5

1. Posternak, T., *Chem. Abstr., 34,* 4096 (1940).
2. Takeshita, H., and Anchel, M., *Science, 147,* 152 (1965).
3. Rogers, D., Williams, D. J., and Thomas, R., *Chem. Commun.,* 393 (1971).

Table 6

1. Allport, D. C., and Bu'Lock, J. D., *J. Chem. Soc.,* 654 (1960).

MICROBIAL QUINONES

DR. CHI-KIT WAT

A characteristic property of quinones is color, ranging from yellow to black. Quinones are reduced by mild reagents (e.g., sodium dithionite) to hydroquinones and revert upon aeration at alkaline pH. With few exceptions, quinones sublime when heated. They form colored oxonium salts in concentrated sulfuric acid; hydroxyquinones form intensely colored alkali salts; o-quinones form quinoxalines with o-phenylene-diamine. References to other chemical tests and spectroscopic properties are listed in Table 1.

TABLE 1

REFERENCES FOR CHEMICAL TESTS AND SPECTROSCOPY OF QUINONES

Types of Compounds	Chemical Tests	Spectroscopy

General	Redox reaction: 15
	Indole, pyrrole and ethylene–
	   diamine test: 16
	Ethyl–cyanoacetate (Craven's)
	   test: 17
		UV: 18
		IR: 19
		Mass: 20

Benzoquinines
  Monobenzoquinones
	Magnesium acetate test: 21	UV:
		   p − 22−25, 34
		   o − 26, 34
		IR: 25, 27−29
		Mass: 30, 31

Bibenzoquinones Terphenylquinones		UV: 32, 33 UV: 34, 35

Naphthoquinones	Novelli test: 36
	Boroacetate complex test: 37
	Magnesium acetate test: 21
	Nickel acetate test: 38
	Lead acetate test: 39
	Titanium chloride test: 40
		UV:
		   p − 38, 41−43
		   o − 41, 44
		IR: 45
		Mass:
		   p − 46−50
		   o − 51
		NMR: 52, 53
		ESR: 54

Perylenequinones	Boroacetate complex test: 37	UV: 55 IR: 56, 57 Mass: 58
Anthraquinones	Magnesium acetate test: 21 Zirconium nitrate test: 59 Boroacetate complex test: 37 Titanium chloride test: 40	UV: 41, 60−67 IR: 68, 69 Mass: 49, 50, 70
Anthracyclinones		UV: 60, 71 IR: 74 Mass: 72, 73 NMR: 74 Circular dichroism: 75

Quinones are commonly extracted from the culture medium with ethyl acetate or ether, and from mycelium and sporophore with acetone or alcohol. Their solubilities depend on the nature of substituent groups; e.g., monobenzoquinones and bibenzoquinones are only slightly soluble in water (exception: spinulosin hydrate) and petroleum ether, but soluble in other common organic solvents, whereas terphenylquinones are soluble only in pyridine or dioxane.

The distribution, biogenesis and identification of quinones are described by Thomson.[10] For a general discussion of antimicrobial and biological activity of quinones see References 1, 2, and 3.

The terpene quinones, i.e., the ubiquinones, the K vitamins, rhodoquinone and chlorobiumquinone, are purposely omitted. Their occurrence and chemical properties are given in References 4 to 10.

A detailed account of quinones isolated up to 1970 is given in Reference 10; other general information is given in References 2, 11, 12, and 13. Quinones from lichens are described in Reference 14.

STRUCTURES

Benzoquinone Derivatives

MONOBENZOQUINONES:

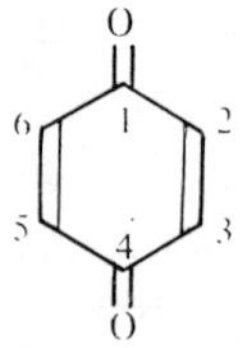

Benzoquinone (BQ) 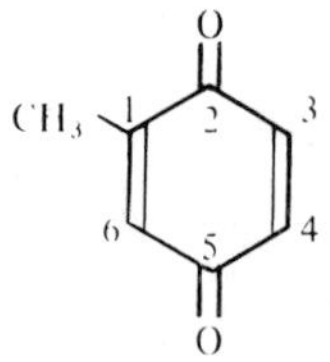   Toluquinone (TQ)

Naphthoquinone Derivatives

NAPHTHOQUINONE (NQ):

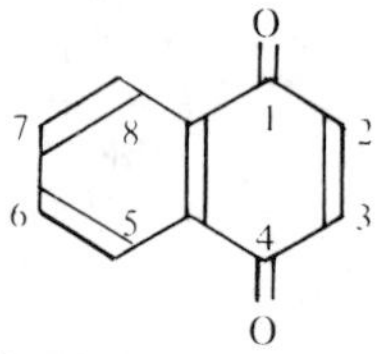

NAPHTHAZARIN (NZ):

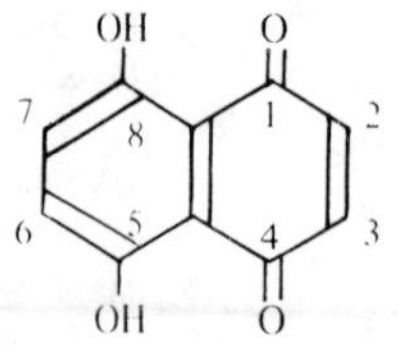

DIHYDROPYRANONAPHTHAZARIN:

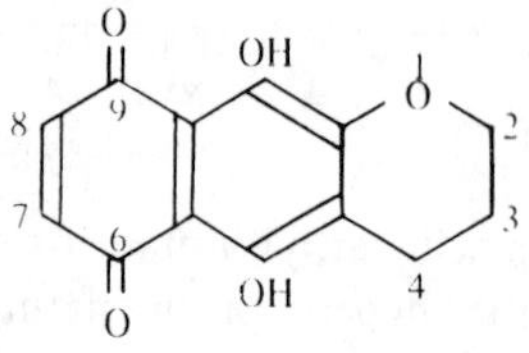

Naphthoquinone Derivatives

BIBENZOQUINONES (BBQ):

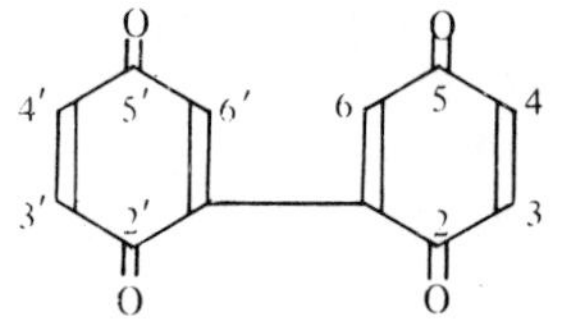

TERPHENYLQUINONES (TPQ):

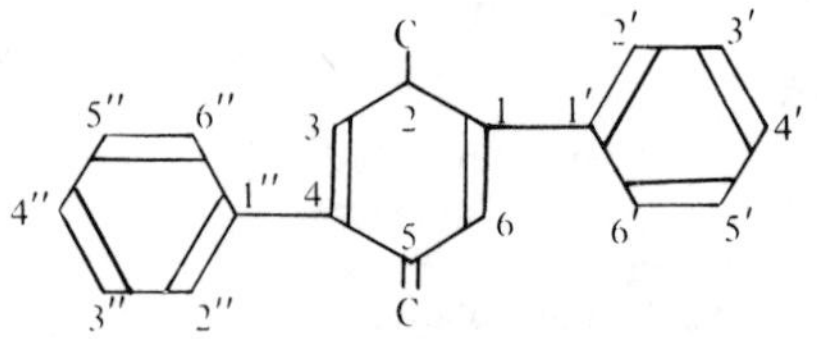

ISOPYRANONAPHTHAZARIN (IPNA):

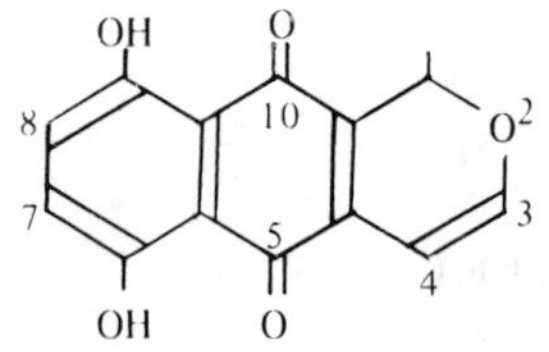

5,10-IPNZ

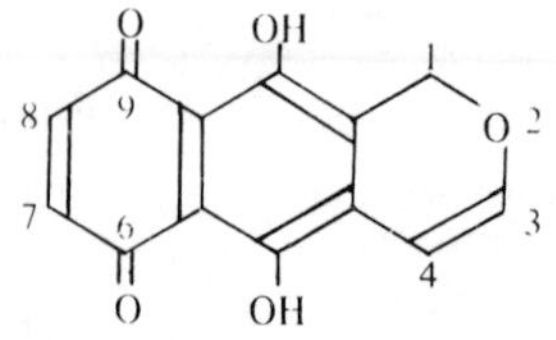

6,9-IPNZ

Anthraquinone Derivatives

ANTHRAQUINONES (AQ):

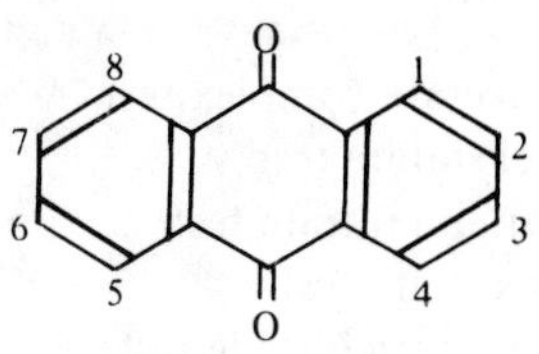

BENZ[a] ANTHRAQUINONES (BENZ-AQ):

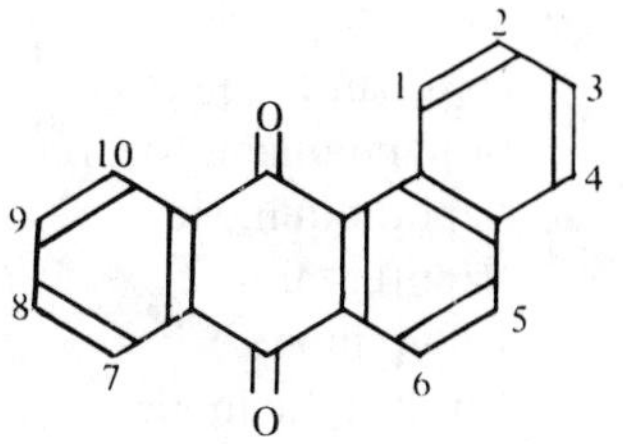

BIANTHRAQUINONES (BAQ):

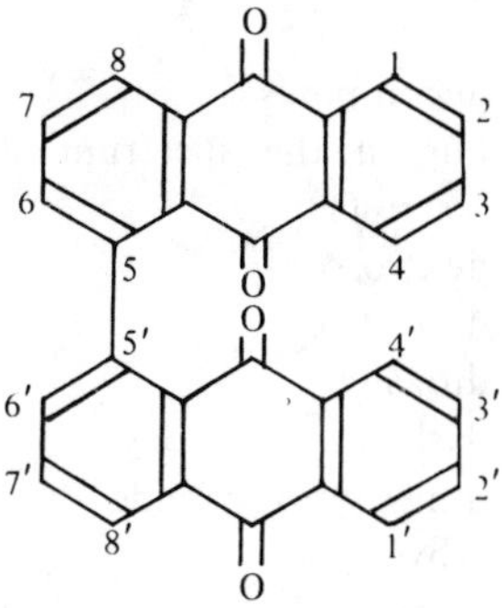

5,5'-Bianthraquinone

DIFUROANTHRAQUINONE (DFAQ):

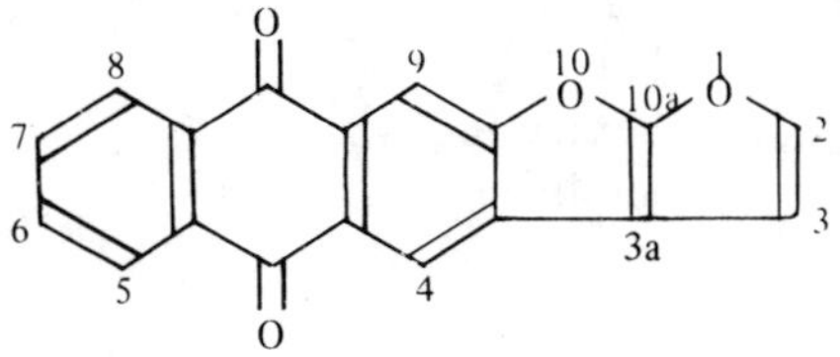

Anthracyclinones and Anthracyclines

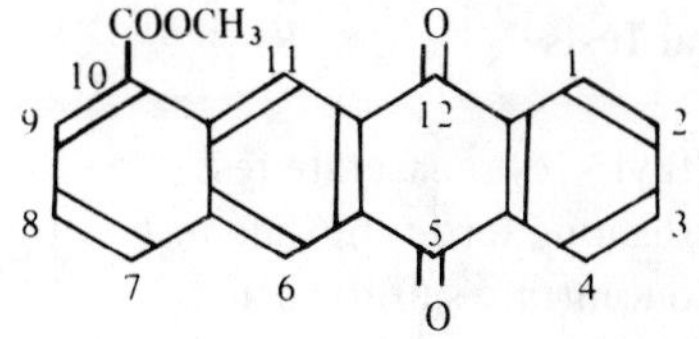

10-Carboxy-naphthacene-5,12-quinone
methyl ester (CNME)

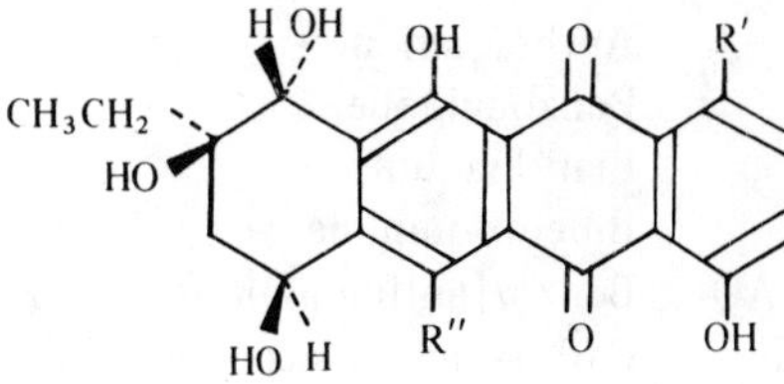

I

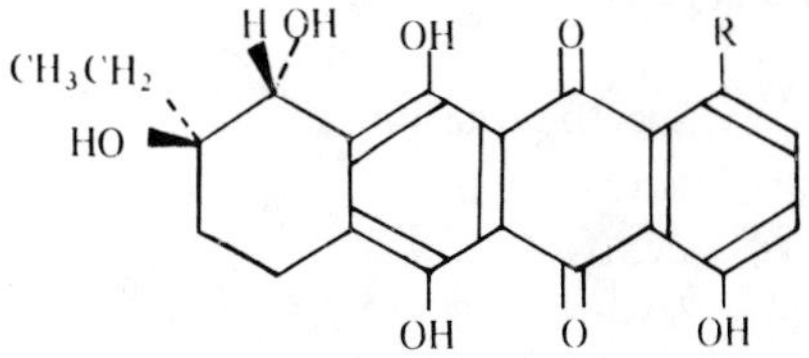

II

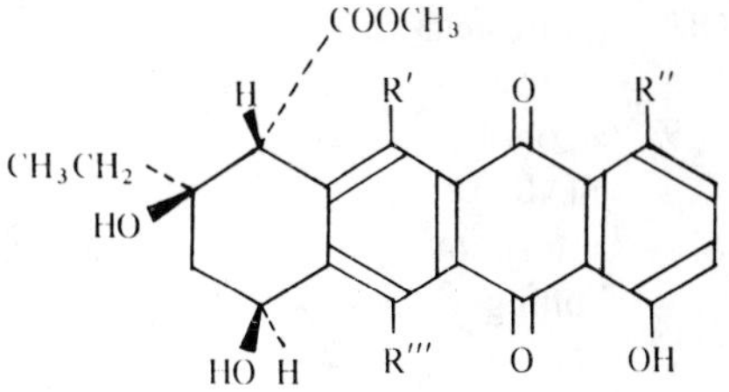

III

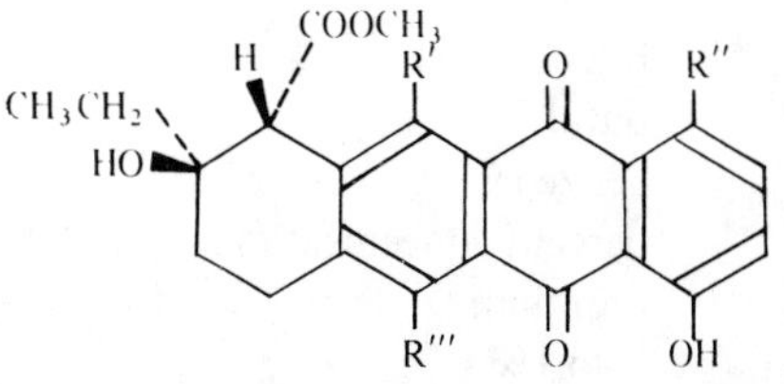

IV

SYMBOLS AND ABBREVIATIONS

Chemical Tests

A	Ethyl – cyanoacetate test	G	Boroacetate complex test
B	Ethanolic ferric chloride test	H	Nickel sulfate test
C	Concentrate sulfuric acid	I	Zirconium nitrate test
E	Magnesium acetate test	J	Lead acetate test
F	Thiosulfate test	K	Titanium trichloride test

Quinones and Their Derivatives[a]

AQ	Anthraquinone	DFAQ	Difuroanthraquinone
BQ	Benzoquinone	IPNZ	Isopyranonaphthazarin
BAQ	Bianthraquinone	NQ	Naphthoquinone
BBQ	Bibenzoquinone	NZ	Naphthazarin
Benz-AQ	Benz[a]anthraquinone	TQ	Toluquinone
CNME	Carboxynaphthacenequinone methyl ester	TPQ	Terphenylquinone

Miscellaneous

[a]	specific rotation	d	decomposes
>	above	D	line in the spectrum of sodium (subscript)
(+)	dextro-		
(−)	levo-	D	dextrose
λ_{max}	electronic absorption maximum: wavelength (nm) and log ϵ value (in parentheses); solvent at superscript	Der	derivative
		diox	dioxane
		dk	dark
		DMF	dimethylformamide
AcOH	acetic acid	Et	ethyl
ace	acetone	eth	diethyl ether
al	alcohol	EtOH	ethanol
alk	alkali	f	fairly
amor	amorphous	flr	fluorescence
b	boiling	gold	golden
bk	black	gr	green
bl	blue	gy	gray
BP	boiling point, °C	h	hot
br	brown	hx	hexane
bron	bronze	i	insoluble
bt	bright	i-	iso-
bu	butyl	lf	leaf
bz	benzene	lt	light
c	percent concentration	mag	magenta
Cd	cadmium	max	maximum
col	colorless	Me	methyl
con	concentrated	MeOH	methanol
cr	crystals	mod	moderately
cy	cyclohexane	Mol	molecular

[a] For numbering of each compound, see the section on structures above.

MP	melting point, °C	s	soluble
n	normal chain	sh	shoulder
nd	needles	sl	slightly
OAc	acetate	sol	solution
og	orange	solv	solvent
org	organic	st	stable
pa	pale	sub	sublimation
peth	petroleum ether	THF	tetrahydrofuran
pk	pink	tol	toluene
pl	plates	unst	unstable
ppt	precipitate	v	very
pr	prisms	visc	viscous
purp	purple	vt	violet
pw	powder	wr	warm
pyr	pyridine	Wt	weight
rh	rhomboid	ye	yellow
rn	reaction		

TABLE 2
BENZOQUINONE DERIVATIVES

No.	Compound (Common Name)	Formula (Mol. Wt.)	Organisms	Color and Crystalline Form (Melting Point, °C)	Significant Characteristics	Ref.
Monobenzoquinones						
1	2,5-Dihydroxy-3-geranyl-geranyl-BQ (Bovinone)	$C_{26}H_{36}O_4$ (412)	*Boletus (Suillus) bovinus* (Linn. ex Fr.) Kuntze	og nd (AcOH) (84—85)	λ_{max}^{EtOH}: 287; (4.31) Color rn: B, br ppt	95
2	2,5-Dimethoxy-BQ	$C_8H_8O_4$ (168)	*Polyporus fumosus* (Pers.) Fries	ye pr (EtOH, AcOH) (~250, d)	$\lambda_{max}^{CHCl_3}$: 278, 284, 370; (4.37, 4.38, 2.48) Red in con H_2SO_4 Possibly derived from a precursor in culture medium	93
3	2,3-Epoxy-6-(1-hydroxy-2,3-dimethylbut-2-enyl)-BQ (Panepoxydione)	$C_{11}H_{12}O_4$ (208)	*Panus conchatus* (Bull. *ex* Fr.) *rudis*	ye visc oil BP: 76°C/0.005 torr	$\lambda_{max}^{CH_2Cl_2}$: 261; (3.69) $[a]_D$: +173° (c = 0.1, MeOH); +223° (c = 0.1, $CHCl_3$)	94
4	5,6-Epoxy-2-hydroxymethyl-BQ (Phyllostine)	$C_8H_8O_4$ (168)	*Aspergillus fumigatus* Fres. *Penicillium spinulosum* Thom.	maroon nd (peth) (116)	λ_{max}^{EtOH}: 265, 450; (4.14, 2.96) Color rn: B, lt br; C, br—red	87
5	5-Hydroxy-3-hydroxymethyl-2,6-dimethyl-BQ (Shanorellin)	$C_9H_{10}O_4$ (182)	*Shanorella spirotricha* Benjamin	og-ye nd (bz—peth) (121)	$\lambda_{max}^{CHCl_3}$: 272, 406; (4.21, 2.79) Unst in strong alk sol; ruby red in con H_2SO_4	92
6	2-Hydroxymethyl-BQ (Gentisylquinone)	$C_{14}H_{14}O_6$ (278)	*Penicillium patulum* Bain.	ye nd (75—76)	λ_{max}^{EtOH}: 247, 432; (4.38, 1.32) Occurs as quinhydrone	86
7	15-(3-Methoxy-2,5-benzoquinonyl)-pentadec-2-enoic acid (Sarcodontic acid)	$C_{22}H_{32}O_5$ (376)	*Sarcodontia setosa* (Pers.) Donk	ye nd (ace) (122—124)	λ_{max}^{EtOH}: 264, 366; (4.82, 3.12)	251

TABLE 2 (Continued)
BENZOQUINONE DERIVATIVES

No.	Compound (Common Name)	Formula (Mol. Wt.)	Organisms	Color and Crystalline Form (Melting Point, °C)	Significant Characteristics	Ref.
Monobenzoquinones (continued)						
8	3,4-Epoxy-6-hydroxy-TQ (Terreic acid)	$C_7H_6O_4$ (154)	*Aspergillus terreus* Thom.	pa ye pl (bz), or nd (bz–peth) (127–127.5)	λ_{max}^{EtOH}: 214, 316; (4.03, 3.88) $[\alpha]_D^{22}$: -16.6° (c = 1, $CHCl_3$) Color rn: B, red	82
9	1,6-Epoxy-3-hydroxy-4-methoxy-TQ (Fumigatin oxide)	$C_8H_8O_5$ (184)	*Aspergillus fumigatus* Fres.	pa og lf (peth–bz) (74)	λ_{max}^{EtOH}: 216, 334; (4.05, 3.78) $[\alpha]_D^{14}$: +28.5 (c = 1.3, EtOH) s in $NaHCO_3$ (gr-br) and NaOH (ye) Color rn: C, ye→og→br; B, br with foaming; F, red	79, 84
10	3-Hydroxy-TQ	$C_7H_6O_3$ (138)	*Aspergillus fumigatus* Fres.		$\lambda_{max}^{CHCl_3}$: 270, 380 Color in aqu sol: pH<3, ye; pH 4–7, red; pH>8, irreversible fading	81
11	3-Hydroxy-4-methoxy-TQ (Fumigatin)	$C_8H_8O_4$ (168)	*Aspergillus fumigatus* Fres. *Penicillium spinulosum* Thom.	maroon nd (peth) (116)	λ_{max}^{EtOH}: 265, 450; (4.14, 2.96) Color rn: B, lt br; C, br→red	80
12	4-Hydroxy-3-methoxy-TQ	$C_8H_8O_4$ (168)	*Aspergillus fumigatus* Fres.		$\lambda_{max}^{CHCl_3}$: 283, 425 Color in aqu sol: pH<3, ye; pH 4–7, vt; pH>8, vt	
13	6-Hydroxy-4-methoxy-TQ	$C_8H_8O_4$ (168)	*Lentinus degener* Kalchbr.	br-og pr (sub) (202–203, d)	λ_{max}^{EtOH}: 290, 425; (4.20, 2.72) Der: mono-Me ether (ye nd, MP 125)	77
14	3-Hydroxy-4-methoxy-6-methyl-TQ	$C_9H_{10}O_4$ (182)	*Gliocladium roseum* Bain.	red-br pr (peth) (70)	λ_{max}^{EtOH}: 278, 442 (4.21, 2.65) Unst in strong alk sol; deep bl in con H_2SO_4	24

TABLE 2 (Continued)
BENZOQUINONE DERIVATIVES

No.	Compound (Common Name)	Formula (Mol. Wt.)	Organisms	Color and Crystalline Form (Melting Point, °C)	Significant Characteristics	Ref.
Monobenzoquinones (continued)						
15	3,4-Dihydroxy-TQ	$C_7H_6O_4$ (154)	*Aspergillus fumigatus* Fres.		$\lambda^{CHCl_3}_{max}$: 279, 385; Color in aqu sol: pH<3, ye; pH 4–7, vt; pH>8, irreversible fading	81
16	3,6-Dihydroxy-TQ	$C_7H_6O_4$ (154)	*Aspergillus fumigatus* Fres.		$\lambda^{CHCl_3}_{max}$: 282, 420; Color in aqu sol: pH<3, ye; pH 4–7, red-vt; pH>8, irreversible fading	81
17	3,6-Dihydroxy-4-methoxy-TQ (Spinulosin)	$C_8H_8O_5$ (184)	*Aspergillus fumigatus* Fres. *Penicillium cinerascens* Biourge *spinulosum* Thom.	purp-bk pl (203)	$\lambda^{CHCl_3}_{max}$: 297, 460; (4.35, 2.29) Color rm: B, br; C, bl	80, 83
18	3,4-Dihydroxy-6-methyl-TQ	$C_8H_8O_4$ (168)	*Gliocladium roseum* Bain.	og nd (peth) (182)	λ^{EtOH}_{max}: 277, 445; (4.07, 2.61) Unst in strong alk sol; cherry red in con H_2SO_4	24
19	1,3,6-Trihydroxy-4-methoxy-1,6-dihydro-TQ (Spinulosin hydrate)	$C_8H_{10}O_6$ (202)	*Aspergillus fumigatus* DH413 and Fres. 4399	pa ye nd (EtOH–bz) (181–182, d)	λ^{EtOH}_{max}: 225, 305; (3.78, 4.06) $[\alpha]^{14}_D$: −213° (c = 1, EtOH) d to spinulosin with ethanolic acid	84
20	4-Methoxy-TQ (Coprinin)	$C_8H_8O_3$ (154)	*Coprinus similis* Berk *et* Br. *Lentinus degener* Kalchbr.	ye pl (EtOH) (175)	$\lambda^{CHCl_3}_{max}$: 264, 360; (4.28, 2.86) Color rm: A, bl-vt→bl-gr→red-br; C, og-red	76, 77
21	6-(2-Carboxy-3-hydroxy-5-methoxyphenyl)-3-methoxy-TQ (Botrallin)	$C_{16}H_{14}O_7$ (318)	*Botrytis allii* Munn.	ye pr (165–185, d)	λ^{EtOH}_{max}: 238, 273, 370; (4.36, 4.20, 3.59) Color rn: B, vt-br; A, gr; NH_4OH, re→br→ye	253

TABLE 2 (Continued)
BENZOQUINONE DERIVATIVES

No.	Compound (Common Name)	Formula (Mol. Wt.)	Organisms	Color and Crystalline Form (Melting Point, °C)	Significant Characteristics	Ref.
Monobenzoquinones (continued)						
22	3,4-Dimethoxy-6-methyl-1,6-dihydro-TQ (Gliorosein)	$C_{10}H_{14}O_4$ (198)	*Gliocladium roseum* Bain.	col nd (48)	λ_{max}^{EtOH}: 289; (4.39) $[\alpha]_D^{22}$: +125° (c = 0.2, CCl$_4$) Rearranges to aurantiogliocladin in base	89, 91
23	3,4-Dimethoxy-6-methyl-TQ (Aurantiogliocladin)	$C_{10}H_{12}O_6$ (196)	*Gliocladium roseum* Bain.	og lf (peth) (63)	λ_{max}^{EtOH}: 275, 407; (4.58, 3.08) Color rn: C, vt	89, 90
24	Quinhydrone of aurantiogliocladin (Rubrogliocladin)	$C_{20}H_{26}O_8$ (394)	*Gliocladium roseum* Bain.	dk red nd (peth) (74)	λ_{max}^{EtOH}: 275, 407; (4.34, 2.81) Gives same hydroquinone (MP 84) as aurantiogliocladin	89, 90
25	(S)-6-Hydroxy-4-(1,2,2-trimethylcyclopentyl)-TQ (Deoxyhelicobasidin)	$C_{15}H_{20}O_3$ (248)	*Helicobasidium mompa* Tanaka	ye nd (hx) (194–195)	λ_{max}^{EtOH}: 274, 404; (4.11, 3.02) $[\alpha]_D^{24}$: -186.5° (c = 0.375, CHCl$_3$)	78
26	(S)-3,6-Dihydroxy-4-(1,2,2-trimethylcyclopentyl)-TQ (Helicobasidin)	$C_{15}H_{20}O_4$ (264)	*Helicobasidium mompa* Tanaka	og-red nd (MeOH, bz. –peth) (190–192)	λ_{max}^{EtOH}: 210, 297, 377, 430; (4.13, 4.15, 2.61, 2.47) $[\alpha]_D^{25}$: -123° (c = 1, CHCl$_3$) Color rn: E, purp; B, vt-br	85
27	(S)-4-(1,2,2,-Trimethylcyclopentan-4-one)-TQ (Lagopodin A)	$C_{15}H_{18}O_3$ (246)	*Coprinus lagopus* Fr.	ye nd (96–97)	$\lambda_{max}^{CHCl_3}$: 257, 310, 435; (4.25, 2.55, 1.55) $[\alpha]_D^{20}$: -10° (CHCl$_3$)	252
28	(7S-Lagopodin B)	$C_{15}H_{18}O_4$ (262)	*Coprinus lagopus* Fr.	ye cr (113–115)	λ_{max}^{EtOH}: 268, 412; (4.10, 2.98) $[\alpha]_D^{20}$: +17° (CHCl$_3$)	252

TABLE 2 (Continued)
BENZOQUINONE DERIVATIVES

Monobenzoquinones (continued)

No.	Compound (Common Name)	Formula (Mol. Wt.)	Organisms	Color and Crystalline Form (Melting Point, °C)	Significant Characteristics	Ref.
29	(Tauranin)	$C_{22}H_{30}O_4$ (358)	*Oospora aurantia* (Cooke) Sacc. *et Vogl.*	og-ye pr (150–160, d)	λ^{MeOH}_{max}: 266, 415; (4.07, 3.07) $[\alpha]^{21}_D$: –148° (c = 0.099, MeOH)	88

Bibenzoquinones

No.	Compound (Common Name)	Formula (Mol. Wt.)	Organisms	Color and Crystalline Form (Melting Point, °C)	Significant Characteristics	Ref.
1	6,6'-Dihydroxy-4,4'-dimethyl-BBQ (Phoenicin)	$C_{14}H_{10}O_6$ (274)	*Penicillium chermesinum* Biourge *phoeniceum* van Beyma *rubrum* O. Stoll	ye-br pl (EtOH) (231, d)	λ^{MeOH}_{max}: 266, 404; (4.43, 3.40) Ye-red sol at pH 1.6–3.5, red-vt at pH 4.9–5.0	96
2	3,3',6,6'-Tetrahydroxy-4,4'-dimethyl-BBQ (Oosporein)	$C_{14}H_{10}O_8$ (306)	*Acremonium* sp. *Beauveria bassiana* (Bals.) Vuill. *tenella* (Delacroix) Siem *Chaetomium* spp. *aurem* Chivers *trilaterale* Chivers *Oospora colorans* van Beyma	bron pl (MeOH–H_2O) (290–295, 260–275)	λ^{EtOH}_{max}: 216, 287, 372, 415; (3.51, 4.67, 2.94, 2.79) Color changes from ye to bl as pH is raised to 8	97, 254

TABLE 2 (Continued)
BENZOQUINONE DERIVATIVES

No.	Compound (Common Name)	Formula (Mol. Wt.)	Organisms	Color and Crystalline Form (Melting Point, °C)	Significant Characteristics	Ref.
Bibenzoquinones (continued)						
2	3,3',6,6'-Tetrahydroxy-4,4'-dimethyl-BBQ (cont.)		Phlebia albida Fries mellea Overholts Verticillium psalliotae Treschow			
3	3,4',6,6'-Tetrahydroxy-3',4'-dimethyl-BBQ (Iso-oosporein)	$C_{14}H_{10}O_8$ (306)	Unidentified soil fungus	purp (~250, d)	Der: tetra-Ac (MP 192); tetra-Me ether (og-red pr, MP 125)	98
Terphenylquinones						
1	3,6-Dihydroxy-TPQ (Polyporic acid)	$C_{18}H_{12}O_4$ (292)	Lopharia papyracea (Jungh.) Reid Peniophora filamentosa (B. et C.) Burt Polyporus nidulans Fr. rutilans (Pers.) Fr.	bron pl (tol) (305—307, d)	λ_{max}^{EtOH}: 205, 256, 262, 330, 465; (4.68, 4.63, 4.63, 4.06, 2.60) Der: di-OAc (MP 215); di-Me ether (MP 192)	100
2	5',5''-Dihydroxy-TPQ (Volucrisporin)	$C_{18}H_{12}O_4$ (292)	Volucrispora aurantiaca Haskins	red pl (pyr) (>300)	λ_{max}^{diox}: 281.5, 234, 329, 377; (4.00, 4.47, 3.90, 3.45) Der: di-OAc (ye nd, MP 223); di-Me ether (og cubes, MP 172 —174)	99
3	3,4',4'',6-Tetrahydroxy-TPQ (Atromentin)	$C_{18}H_{12}O_6$ (324)	Boletus (Suillus) bovinus (Linn ex Fr.) Kuntze Clitocybe subilludens Murr.	bron lf (AcOH) (>300)	λ_{max}^{diox}: 268, 385; (4.55, 3.66) Der: tetra-OAc (MP 242); leuco-hexa-OAc (MP 236)	101, 255

TABLE 2 (Continued)
BENZOQUINONE DERIVATIVES

No.	Compound (Common Name)	Formula (Mol. Wt.)	Organisms	Color and Crystalline Form (Melting Point, °C)	Significant Characteristics	Ref.
	Terphenylquinones (continued)					
3	3,4',4'',6-Tetrahydroxy-TPQ (*cont.*)		*Hydnellum diabolus* Banker *Paxillus atrotomentosus* (Batsch) Fr.			
4	3,4',4'',5'',6-Pentahydroxy-TPQ (Leucomelone)	$C_{18}H_{12}O_7$ (340)	*Polyporus leucomelas* Pers. *ex* Fr.	br lf (AcOH) (320, d)	Der: penta-OAc (ye nd, MP 226—227); leucohepta-OAc (lf, MP 204—205)	102
5	3,6-Dibenzoyloxy-4',4''-dihydroxy-TPQ (Aurantiacin)	$C_{32}H_{20}O_8$ (532)	*Hydnellum aurantiacum* (Batsch *ex* Fr.) Karst *caeruleum* (Pers.) Karst HA202 *pseudocaeruleum* ined. *scrobiculatum, var. zonatum* (Fr.) Harr.	dk red nd (diox) (285—295, 305—308)	λ_{max}^{diox}: 234, 402; (4.71, 3.70) Der: di-OAc (ye, MP 235—238); leucotetra-OAc (MP 246—253)	103
6	6-(4-Carboxy-buta-1,3-dienyl)-2',2''-dicarboxy-3-hydroxy-TPQ (Muscarufin)	$C_{25}H_{16}O_9$ (460)	*Amanita muscaria* (L. *ex* Fr.) S. F. Gray	og-red nd (275.5)	λ_{max}: 502 Color rm: alk, bl-red; C, purp-red Der: mono-OAc (og-ye, MP 197); leucotri-OAc (MP 184)	104

TABLE 2 (Continued)
BENZOQUINONE DERIVATIVES

No.	Compound (Common Name)	Formula (Mol. Wt.)	Organisms	Color and Crystalline Form (Melting Point, °C)	Significant Characteristics	Ref.
Terphenylquinones (continued)						
7	(Thelephoric acid)	$C_{18}H_8O_8$ (352)	(*Cantharellus multiple* Underw.) *Hydnellum aurantiacum* (Batsch ex Fr.) Karst *caeruleum* (Pers.) Karst *diabolus* Banker HA202 *scrobiculatum* (Secr.) Karst, var. *scrobiculatum* *scrobiculatum*, var. *zonatum* (Fr.) Harr. *suaveolens* (Scop. ex Fr.) Karst *Hydnum amarescens* Quél. *aspratum* Berk. *ferrugineum* Fr. *graveolens* Fr. *imbricatum* (L.) Fr. *nigrum* Fr. *scabrosum* Fr. *Phlebia strigoso-zonata* (Schw.) Lloyd *Polyozellus multiplex* (Underw.) Murr. *Polyporus versicolor* (L.) Fr.	dk vt pr (pyr)	λ_{max}^{EtOH}: 217, 264, 305, 390, 483; (4.33, 4.27, 4.30, 3.48, 4.83) Der: tetra-OAc (MP 330—335); tetra-Me ether (br-vt, MP 360, d)	103

TABLE 2 (Continued)
BENZOQUINONE DERIVATIVES

No.	Compound (Common Name)	Formula (Mol. Wt.)	Organisms	Color and Crystalline Form (Melting Point, °C)	Significant Characteristics	Ref.
Terphenylquinones (continued)						
7	(Telephoric acid) *(cont.)*		*Thelephora caryophyllea* Schaeff. *coralloides* Fr. *crustacea* Schum. *flabelliformis* Fr. *intybacea* Pers. *laciniata* Pers. *palmata* Scop. *terrestris* Ehrh.			
8	(Phlebiarubrone) 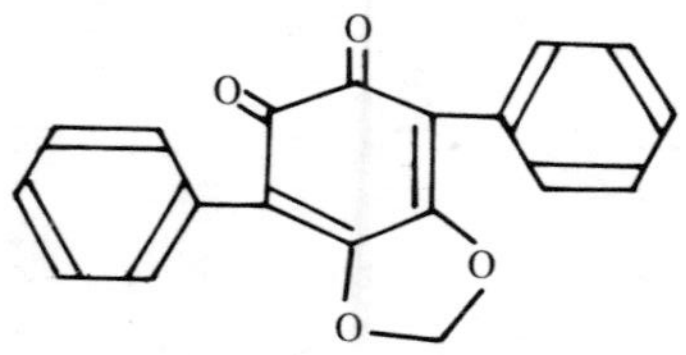	$C_{19}H_{12}O_4$ (304)	*Phlebia strigoso-zonata* (Schw.) Lloyd	red nd (AcOH) (248–250)	λ^{EtOH}_{max}: 268, 332, 465; (4.47, 3.64, 3.54) d by alk to polyporic acid and HCHO; Der: leuco-OAc (MP 227–229); *o*-phenylenediamine quinoxaline (dk red, MP ~305)	105
Methyl-*bis* benzoquinone						
1	(Amitenone) 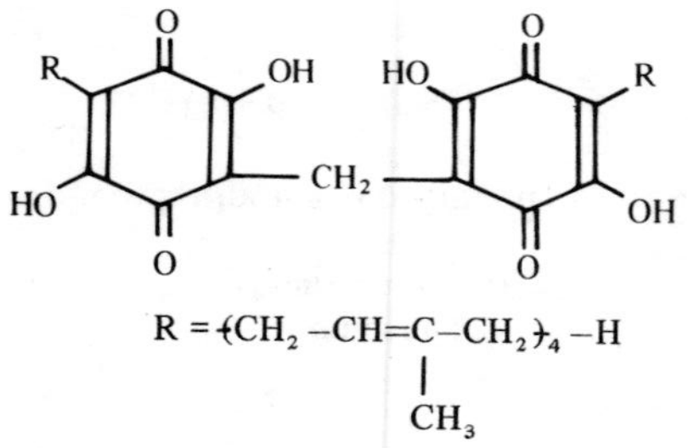	$C_{53}H_{72}O_8$ (836)	*Boletus (Suillus) bovinus* (Linn. *ex* Fr.) Kuntze	ye-og (187–188)	λ^{EtOH}_{max}: 288; (4.45) Der: tetra-Me ether (ye-be oil, BP 210, d)	106

$R = \{CH_2-CH=C-CH_2\}_4-H$
CH_3
(all *trans*)

TABLE 3
NAPHTHOQUINONE DERIVATIVES

No.	Compound (Common Name)	Formula (Mol. Wt.)	Organisms	Color and Crystalline Form (Melting Point, °C)	Significant Characteristics	Ref.
Naphthoquinones						
1	6-Methyl-NQ	$C_{11}H_8O_2$ (172)	*Marasmius graminum* (Libert) Berk.	gold-ye nd (AcOH–H_2O) (90–91)	λ_{max}^{EtOH}: 249, 255.5, 342.5; (4.34, 4.29, 3.49	109
2	2,5,7-Trihydroxy-NQ (Flaviolin)	$C_{10}H_6O_5$ (206)	*Aspergillus citricus* (Wehmer) Mosseray	dk red rh (contain diox and H_2O (~250, d)	λ_{max}^{EtOH}: 215, 262, 308, 403, 452; (4.43, 4.12, 3.93, 3.30, 3.35) sl s in bz, H_2O; s in AcOH, EtOH, eth, $CHCl_3$, diox Color rn: B, red-vt; C, red	107
3	5-Hydroxy-2,7-dimethoxy-NQ (Flaviolin-2,7-dimethyl ether)	$C_{12}H_{10}O_5$ (234)	*Streptomyces* sp.	og cr (diox–H_2O) (266–268)	$\lambda_{max}^{CHCl_3}$: 261, 302, 435; (4.18, 4.07, 3.63) Color rn: B, br; NaOH, red-purp	108
4	2,7-Dimethoxy-6-ethyl-5-hydroxy-NQ	$C_{14}H_{14}O_5$ (262)	*Hendersonula toruloidea* Nattrass	ye nd (187)	λ_{max}^{MeOH}: 221, 256(sh), 262, 305, 422, (4.48, 4.21, 4.22, 3.98)	256
5	8-Dichloroacetyl-5-hydroxy-2,7-dimethyl-NQ (Mollisin)	$C_{14}H_{10}O_4Cl_6$ (313)	*Mollisia caesia* Sacc. *fallens* Karst *sensu* Sydow	og-ye nd (peth) (202–203, d)	λ_{max}^{EtOH}: 259, 280(sh), 420; (4.26, 3.9, 3.52) Color rn: B, br-red; C, dk og; J, red sol; G, red; NaOH, vt-red	110
6	(Lambertellin)	$C_{14}H_8O_5$ (256)	*Lambertella* spp. *bicoriae* Whetzel *corni-maris* Hohn	og-red nd (ace), og pl (EtOH (253–254)	λ_{max}^{EtOH}: 284(sh), 290, 430; (4.08, 4.10, 3.68)	259

TABLE 3 (Continued)
NAPHTHOQUINONE DERIVATIVES

Naphthoquinones (continued)

No.	Compound (Common Name)	Formula (Mol. Wt.)	Organisms	Color and Crystalline Form (Melting Point, °C)	Significant Characteristics	Ref.
7	(Herbarin)	$C_{16}H_{16}O_6$ (304)	*Torula herbarum* (Pers.) Link *ex* Fr.	ye nd (MeOH) (192–193)	λ_{max}^{EtOH}: 216, 237(sh), 266, 285 (sh), 350(sh), 415; (4.83, 3.15, 4.18, 3.94, 3.15, 3.48) Color rn: C, red	283
8	7,9-Dimethoxy-3-methyl-isopyrano-5,10-NQ (Dehydroherbarin)	$C_{16}H_{14}O_5$ (286)	*Torula herbarum* (Pers.) Link *ex* Fr.	red nd (MeOH) (186–188)	λ_{max}^{EtOH}: 217, 250, 272, 335, 400, 485; (4.12, 3.85, 3.85, 3.28, 3.15, 3.15)	283
9	(Frenolicin)	$C_{18}H_{18}O_7$ (346)	*Streptomyces fradiae* (Waksman et Curtis) Waksman et Henrici	pa ye nd (161–162)	λ_{max}^{MeOH}: 234, 284(sh), 362; (4.26, 3.54, 3.72) $[\alpha]_d^{25}$: $-37.7°$ ($c = 1.5$, MeOH)	118
10	(Kalafungin)	$C_{16}H_{12}O_6$ (300)	*Streptomyces tanashiensis* Hata et *al.* (strain Kala)	og cr (EtOAc) (163–166)	λ_{max}^{MeOH}: 212, 256, 268(sh), 425; (4.62, 4.04, 4.02, 3.65) $[\alpha]_D^{25}$: $+159°$ ($c = 1.0$, CHCl$_3$)	10, 119, 224

TABLE 3 (Continued)
NAPHTHOQUINONE DERIVATIVES

No.	Compound (Common Name)	Formula (Mol. Wt.)	Organisms	Color and Crystalline Form (Melting Point, °C)	Significant Characteristics	Ref.
Naphthoquinones (continued)						
11	2,7-Dihydroxy-NZ (Mompain)	$C_{10}H_6O_6$ (222)	*Helicobasidium mompa* Tanaka	dk red lf (H_2O, diox) (>300, d)	EtOH: λ_{max} 228, 272, 318, 486, 517, 554; (4.43, 4.06, 3.93, 3.75, 3.80, 3.63) i in $CHCl_3$, bz; s in al Color rn: B, dk gr; E, vt-red; J, bl-vt ppt	78
12	2,7-Dimethoxy-NZ	$C_{12}H_{10}O_6$ (250)	*Streptomyces* sp. *Streptoverticillium* sp.	red-bk cr ($CHCl_3$–hx) (275–276)	$CHCl_3$: λ_{max} 285, 308, 480, 512, 550; (3.94, 3.97, 3.85, 3.93, 3.74) Color rn: B, bl; NaOH, purp	108, 243
13	2,7-Dimethyl-NZ	$C_{12}H_{10}O_4$ (194)	*Mollisia caesia* Sacc. *sensu* Sydow	red nd (125–126)	EtOH: λ_{max} 217, 280, 480, 510, 550; (4.55, 3.90, 3.77, 3.79, 3.58)	111
14	7-Acetonyl-2-methoxy-NZ (Norjavanicin)	$C_{14}H_{12}O_6$ (276)	*Fusarium martii* App. *et* Wr., var. *pisi* Jones	red nd (MeOH–$CHCl_3$) (200–204)		10, 257
15	7-Acetonyl-2-methoxy-6-methyl-NZ (Javanicin, solanione)	$C_{15}H_{14}O_6$ (290)	*Fusarium javanicum* Kds. *martii* App. *et* Wr., var. *pisi* Jones *solani* (Mart.) App. *et* Wr., var. *rosa* (strain D_2 purple)	red pl (EtQH) (208, d)	$CHCl_3$: λ_{max} 307, 510; (3.99, 3.86) Color rn: B, vt; J, vt sol	45, 112, 113
16	(+)-7-(2-Hydroxy-*n*-propyl-2-methoxy-6-methyl-NZ (Solaniol)	$C_{15}H_{16}O_6$ (292)	*Fusarium solani* (Mart.) App. *et* Wr. (strain D_2 purple)	dk red nd (190–194, d)	λ_{max}^{diox}: 227, 304, 472, 500, 536; (4.53, 3.97, 3.84, 3.91, 3.72) $[\alpha]_D^{25}$: +122° (c = 0.053, MeOH) Color rn: B, dk gr; J, vt	116

199

TABLE 3 (Continued)
NAPHTHOQUINONE DERIVATIVES

Naphthoquinones (continued)

No.	Compound (Common Name)	Formula (Mol. Wt.)	Organisms	Color and Crystalline Form (Melting Point, °C)	Significant Characteristics	Ref.
17	2-Acetonyl-8-methoxydihydropyranol [3,2-g]-NZ (Deoxyerythrostominone)	$C_{17}H_{16}O_7$ (332)	*Gnomonia erythrostoma* (Pers. ex Fr.) Auersw.	red rods (148–150)	λ^{EtOH}_{max}: 234, 275, 317, 477(sh), 505, 542; (4.49, 3.83, 3.88, 3.84, 3.91, 3.73) $[\alpha]^{25}_D$: +277° (ace)	260
18	2-Acetonyl4-hydroxy-8-methoxydihydropyrano-[3,2-g]-NZ (Erythrostominone)	$C_{17}H_{16}O_8$ (348)	*Gnomonia erythrostoma* (Pers. ex Fr.) Auersw.	red nd (184–186)	λ^{EtOH}_{max}: 231.5, 280, 315, 480(sh), 509, 546; (4.54, 3.89, 3.90, 3.86, 3.93, 3.72) $[\alpha]^{25}_D$: +231° (ace)	260
19	2-(2-Hydroxy-*n*-propyl)-8-methoxydihydropyrano-[3,2-g]-NZ (Deoxyerythrostominol)		*Gnomonia erythrostoma* (Pers. ex Fr.) Auersw.	red nd (139–141)	λ^{EtOH}_{max}: 234, 275, 317, 475(sh), 505, 541; (4.46, 3.85, 3.89, 3.84, 3.92, 3.73) $[\alpha]^{25}_D$: +271° (ace)	260
20	7-Hydroxy-3-methyl-6,9-IPNZ (O-Demethylanhydrofusarubin)	$C_{14}H_{10}O_6$ (274)	*Gibberella fujikuroi* (Saw.) Wr.	purp nd (ace) (202–204)	λ^{EtOH}_{max}: 237, 285, 353, 492(sh), 546; (4.40, 4.41, 3.81, 3.88, 3.99)	258
21	7-Methoxy-3-methyl-6,9-IPNZ (Anhydrofusarubin)	$C_{15}H_{12}O_6$ (288)	*Fusarium* spp.	vt-bk nb (bz, AcOH) (193–201, d)	λ^{EtOH}_{max}: 237.5, 291, 540; (4.32, 4.29, 4.00)	10, 114, 257
22	(Hydroxyjavanicin)	$C_{15}H_{14}O_7$ (306)	*Fusarium javanicum* Kds. *martii* App. et Wr., var. *pisi* Jones *solani* (Mart.) App. et Wr., var. *rosa* (strain D_2 purple)	red nd ($CHCl_3$–peth), pr (bz) (218, d)	$\lambda^{CHCl_3}_{max}$: 303, 505; (4.08, 3.96) Forms bl-vt hydroscopic K salt and red-br Cu salt. Fusarubin sulfuric acid ester (og rh or nd, oxidized form of fusarubinogen) isolated as ammonium salt	112, 114

TABLE 3 (Continued)
NAPHTHOQUINONE DERIVATIVES

Naphthoquinones (continued)

No.	Compound (Common Name)	Formula (Mol. Wt.)	Organisms	Color and Crystalline Form (Melting Point, °C)	Significant Characteristics	Ref.
23	(Marticin)	$C_{18}H_{16}O_9$ (376)	*Fusarium javanicum* Kds. *martii* App. *et* Wr., var. *pisi* Jones *solani* (Mart.) App. *et* Wr. *solani* (Mart.) App. *et* Wr., var. *minus* Wr.	dk red nd (200–201)	λ_{max}^{EtOH}: 227, 305, 497; (4.52, 3.96, 3.89) [α]$_D^{25}$: +132° (CHCl$_3$)	117
24	(Isomarticin)	$C_{18}H_{16}O_9$ (376)	*Fusarium javanicum* Kds. *martii* App. *et* Wr., var. *pisi* Jones *solani* (Mart.) App. *et* Wr. *solani* (Mart.) App. *et* Wr., var. *minus* Wr.	br pr (168–169)	λ_{max}^{EtOH}: 227, 306, 497; (4.49, 3.94, 3.88) [α]$_D^{25}$: +26° (CHCl$_3$) Partially converted to marticin in acid sol	117
25	(Granaticin)	$C_{22}H_{20}O_{10}$ (444)	*Streptomyces olivaceus* (Waksman) Waksman *et* Henrici *violaceoruber* (Waksman *et* Curtis) Waksman *sensu* Waksman *et* Kutzner	red cr (ace) (204–206, d)	λ_{max}^{EtOH}: 223, 286, 496(sh), 532, 576 Red in acid, bl in alk	245, 246

TABLE 3 (Continued)
NAPHTHOQUINONE DERIVATIVES

No.	Compound (Common Name)	Formula (Mol. Wt.)	Organisms	Color and Crystalline Form (Melting Point, °C)	Significant Characteristics	Ref.
Naphthoquinones (continued)						
26	(Granaticin B)	$C_{28}H_{30}O_{12}$ (558)	*Streptomyces violaceoruber* (Waksman *et* Curtis) Waksman *sensu* Waksman *et* Kutzner	red amor pw (EtOAc—peth)	λ^{MeOH}_{max}: 223, 285, 498(sh), 527, 566; (4.42, 3.68, 3.71, 3.76, 3.57) Acid hydrolysis→granaticin + L-rhodinose	246, 247
27			*Streptomyces* sp. No. B44-P1 *spectabilis* Dietz	red-og	A mixture of antibiotics (streptovaricin A, B, C, D, E, and G) present in the culture medium Labile to alk; ye in acid, red-amber at alk pH s in al, lower ketones, ace, $CHCl_3$; sl s in eth bz	121, 122
	(Streptovaricin, dalacin)					
	Streptovaricin A (B44P C) X = OH, Y - OAc	$C_{42}H_{53}NO_{16}$ (827)		og cr (eth) (194—196)	$[a]_D$: +610° ($CHCl_3$)	
	Streptovaricin B (B44P B)	$C_{42}H_{53}NO_{15}$ (811)		og cr (eth) (185—187)	$[a]_D$: +576° ($CHCl_3$)	
	Streptovaricin C (B44P A) X = H, Y = OH	$C_{40}H_{53}NO_{15}$ (771)		red-og amor (eth) (189—191)	λ^{MeOH}_{max} ($E^{1\%}_{1\,cm}$): 245, 430; (630, 190) $[a]_D$: +602° ($CHCl_3$)	

$O \longrightarrow C_{12}$ of granaticin

TABLE 3 (Continued)
NAPHTHOQUINONE DERIVATIVES

No.	Compound (Common Name)	Formula (Mol. Wt.)	Organisms	Color and Crystalline Form (Melting Point, °C)	Significant Characteristics	Ref.
Naphthoquinones (continued)						
27	(Streptovaricin, dalacin) (*cont.*)					
	Streptovaricin D	$C_{42}H_{55}NO_{15}$ (813)		amor (167–170)	$[a]_D$: +436° ($CHCl_3$)	
	Streptovaricin E	$C_{40}H_{55}NO_{13}$ (757)				
	Streptovaricin G	$C_{40}H_{51}NO_{15}$ (785)		amor (190–192)	$[a]_D$: +473° ($CHCl_3$)	
28	(Rubromycins)		*Actinomyces fluvoviolaceus violochromogenes* *Streptomyces collinus* (Lindenbein)		A mixture of antibiotics (a-, β-, γ-, and δ-rubromycin from mycelium) s in $CHCl_3$, ace; sl s in eth, al Color rn: K, bl-gr→red-vt	126, 262
	a-Rubromycin (Collinomycin)	$C_{27}H_{20}O_{12}$ (536)		ye-red nd (EtOH) (278–281, d)	$\lambda_{max}^{CHCl_3}$: 319, 352, 365, 415, 484; (4.32, 4.06, 4.05, 3.75, 3.90) Gives vt ppt of Na salt with Na_2CO_3	

TABLE 3 (Continued)
NAPHTHOQUINONE DERIVATIVES

No.	Compound (Common Name)	Formula (Mol. Wt.)	Organisms	Color and Crystalline Form (Melting Point, °C)	Significant Characteristics	Ref.
	Napthoquinones (continued)					
28	Rubromycins (*cont.*)					
	β-Rubromycin	$C_{27}H_{20}O_{12}$ (536)		red nd (AcOH, tol, bz) (225–227, d)	$\lambda_{max}^{CHCl_3}$: 316, 350, 364, 504; (4.36 4.10, 4.06, 3.82) Readily modified in acid to γ-rubromycin, and in pyr to α-rubromycin	
	γ-Rubromycin	$C_{26}H_{18}O_{12}$ (522)		red nd (AcOH, bz) (~235, d)	$\lambda_{max}^{CHCl_3}$: 317, 349, 364, 484, 513, 551; (4.35, 4.12, 4.06, 3.87, 3.91, 3.68)	

TABLE 3 (Continued)
NAPHTHOQUINONE DERIVATIVES

No.	Compound (Common Name)	Formula (Mol. Wt.)	Organisms	Color and Crystalline Form (Melting Point, °C)	Significant Characteristics	Ref.
Naphthoquinones (continued)						
29	(Griseorhodines)	$C_{25}H_{16}O_{12}$ (508)	*Streptomyces californicus* strains JA2640 and ATCC3312		A mixture of seven components (griseorhodines A_2, A, B, C_2, C, K, and L) from mycelium Griseorhodines A, B, and C possess the same chromophore, a trihydroxynaphthaquinone	123, 124

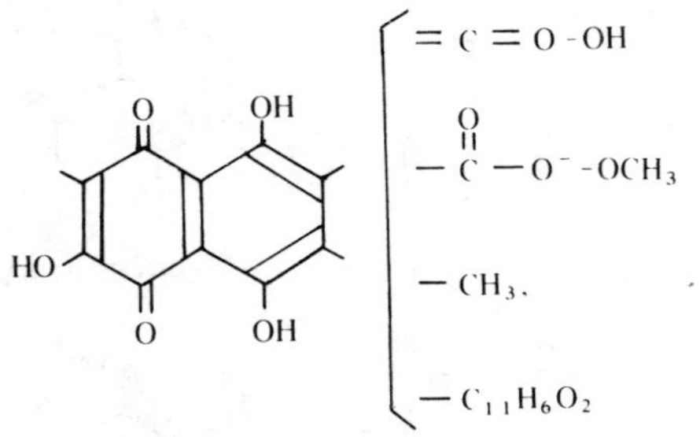

$\equiv C \equiv O - OH$

$- \overset{O}{\underset{\parallel}{C}} - O^- - OCH_3$

$- CH_3.$

$- C_{11}H_6O_2$

Griseorhodine A				dk red amor (diox) (280–282, d)	λ_{max}^{MeOH}: 231, 255(sh), 316, 360, 510 i in eth; sl s in al, ace, bz, $CHCl_3$; s in DMF, 2N NaOH (bl-vt), con H_2SO_4 (red)	
Griseorhodine B				dk red	λ_{max}^{MeOH}: 231, 255(sh), 314, 362, 511	
Griseorhodine C				dk red	λ_{max}^{MeOH}: 232, 254(sh), 316, 361, 511	

TABLE 3 (Continued)
NAPHTHOQUINONE DERIVATIVES

No.	Compound (Common Name)	Formula (Mol. Wt.)	Organisms	Color and Crystalline Form (Melting Point, °C)	Significant Characteristics	Ref.
Binaphthoquinones						
1	(Mycochrysone)	$C_{20}H_{12}O_7$ (364)	An inoperculate discomycete	red nd (ace—H_2O) (>195, d)	EtOH λ_{max}: 236, 303, 438: (4.49, 3.75, 3.80) [α]$_D^{22}$: -331° (ace) s in al, ace, NaHCO$_3$ (purp); f s in bz, CHCl$_3$, eth, H_2O Color rn: B, br; C, dk bl; G, purp; J, purp ppt; K, vt→vt-br	126
2	(Aureofusarin)		*Dactylium dendroides* (Bull.) Fr. *Fusarium culmorum* (W. G. Smith) Sacc. *decemcellulare* Brick. *graminearum* Schwabe *Hypomyces rosellus* (Alb. et Schw.) Tul.	ye pl (CHCl$_3$—MeOH) (>330)	CHCl$_3$ λ_{max}: 248, 269, 381, 422(sh); (4.69, 4.52, 3.99, 3.93) Color rn: E, red; alk, red→vt	120, 281

TABLE 3 (Continued)
NAPHTHOQUINONE DERIVATIVES

Binaphthoquinones (continued)

No.	Compound (Common Name)	Formula (Mol. Wt.)	Organisms	Color and Crystalline Form (Melting Point, °C)	Significant Characteristics	Ref.
3	[(+)-Actinorhodin]	$C_{32}H_{26}O_{14}$ (634)	*Streptomyces coelicolor* (Muller) Waksman *et* Henrici	red nd (diox, THF) (270, d)	An artifact; its precursor protoactinorhodin (pa red pr, MP 330, d) has been isolated from the mycelium and can be converted to actinorhodine by aeration in alk sol λ_{max} diox (precursor): 285, 523, 531 (sh); (4.25), 4.18, 4.05) s in pyr, THF, diox, alk (bl) Color rn: G, vt-bl; C, dk bl	127, 128, 129
4	(Xanthomegnin)	$C_{30}H_{22}O_{12}$ (574)	*Trichophyton megnini* Blanchard *rubrum* (Castellani) Sabouraud *violaceum* Sabouraud, apud Bodin	og pl (CHCl₃—bz); nd (AcOH—hx) (340, d)	λ_{max} diox: 228, 289(sh), 395; (4.73, 4.23, 4.00) $[\alpha]_D^{22}$: -155° (c = 1.0, CHCl₃) i in hydroxylic solv; sl s in ace, EtOAc, bz	130, 131
5	(Fuscofusarin)	$C_{30}H_{20}O_{11}$ (556)	*Fusarium culmorum* (W. G. Smith) Sacc.	br pw (MeOH) (>300)	λ_{max} EtOH: 225, 281, 346, 405; (4.50, 4.63, 3.89, 4.01) Color rn: E, br	261

TABLE 3 (Continued)
NAPHTHOQUINONE DERIVATIVES

Binaphthoquinones (continued)

No.	Compound (Common Name)	Formula (Mol. Wt.)	Organisms	Color and Crystalline Form (Melting Point, °C)	Significant Characteristics	Ref.
6	(Purpurogenone)	$C_{29}H_{20}O_{11}$ (544)	*Penicillium purpurogenum* Stoll	dk red pr (bz); red pr (ace) (310, d)	$\lambda_{max}^{CHCl_3}$: 252, 306, 387, 498, 528, 568; (4.18, 3.73, 3.80, 3.42, 3.47, 3.27) sl s in org solv except diox; forms a br, amor, sl s Na salt with $NaHCO_3$; NaOH, gr→bl→dk purp Color rn: B, gr	132
7	(Xylindein)	$C_{32}H_{24}O_{10}$ (568)	*Chlorociboria aeruginosa* (Oed.) Seaver *Lophiostoma viridarium* Cooke	lt ye to dk br rh lf (phenol) (>300)	$\lambda_{max}^{CHCl_3}$: 380, 405, 423, 603, 647; (4.19, 4.17, 4.06, 4.31, 4.46) Der: di-Me ether (bl-vt nd); di-OAc-di-Me ether (ye to ye-gr, MP 294−295)	134

TABLE 3 (Continued)
NAPHTHOQUINONE DERIVATIVES

Binaphthoquinones (continued)

No.	Compound (Common Name)	Formula (Mol. Wt.)	Organisms	Color and Crystalline Form (Melting Point, °C)	Significant Characteristics	Ref.
8	(Viopurpurin)	$C_{29}H_{20}O_{11}$ (544)	*Trichophyton violaceum* Sabouraud, apud Bodin	dk re (bz)	Der: tri-OAc (ye-og, MP 280–285); tri-Me ether (red, MP 173–174) λ_{max}^{EtOH} (tri-OAc): 217, 270, 277; (3.75, 3.97, 3.97)	133

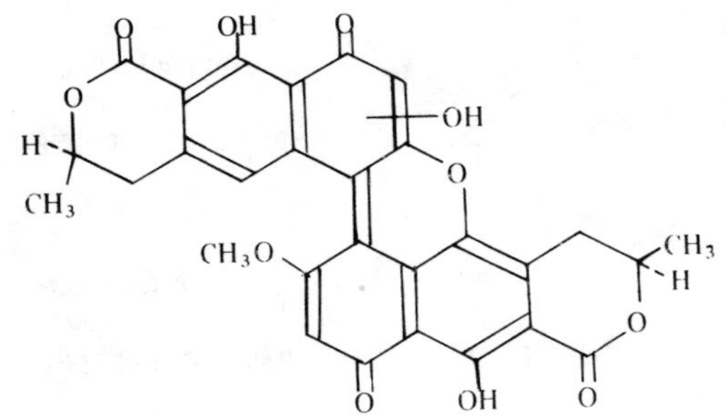

or

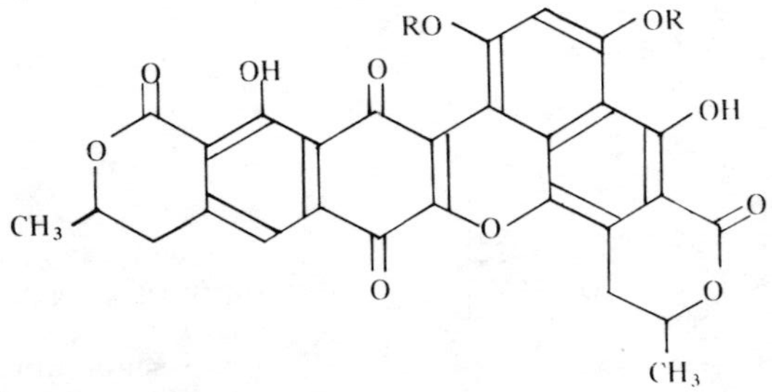

R=H or CH₃

TABLE 4
ANTHRAQUINONE DERIVATIVES

No.	Compound (Common Name)	Formula (Mol. Wt.)	Organisms	Color and Crystalline Form (Melting Point, °C)	Significant Characteristics	Ref.
Anthraquinones						
1	3-Methyl-1-hydroxy-AQ (Pachybasin)	$C_{15}H_{10}O_3$ (238)	*Aspergillus crystallinus* Kwon *et* Fennell *Pachybasium candidum* (Sacc.) Peyronel *Phoma foveata* Foister *Trichoderma viride* Pers. ex Fr. Two unidentified fungi	ye nd (MeOH or AcOH) (178)	λ_{max}^{EtOH}: 224, 252, 281, 403; (3.74, 4.01, 3.65, 3.02) Color rn: E, og-red	140, 264
2	3-Methyl-1,6-dihydroxy-AQ (Phomarin)	$C_{15}H_{10}O_4$ (254)	*Phoma fovesta* Foister	og nd (254—255)	λ_{max}^{EtOH}: 220, 245, 271.5, 280(sh), 292(sh), 337.5, 395(sh), 411, 425(sh); (4.45, 4.04, 4.46, 4.38, 4.18, 3.17, 3.71, 3.80, 3.71)	265
3	3-Methyl-1,8-dihydroxy-AQ (Chrysophanol)	$C_{15}H_{10}O_4$ (254)	*Aspergillus crystallinus* Kwon *et* Fennell *Chaetomium affine* Corda *Pachybasium candidum* (Sacc.) Peyronel *Penicillium islandicum* Sopp. *Phoma foveata* Foister *Sepedonium ampullosporum* Damon *Trichoderma viride* Pers. ex Fr.	og pl (bz) (196)	λ_{max}^{EtOH}: 225, 255, 277.5, 287.5, 430; (4.56, 4.34, 4.04, 4.07, 4.08) s in ace, AcOH, CHCl₃, h EtOH, h bz, NaOH (rose-pk); sl s in H₂O eth, peth Color rn: E, og; C, red	140,141, 142, 264, 265

TABLE 4 (Continued)
ANTHRAQUINONE DERIVATIVES

Anthraquinones (continued)

No.	Compound (Common Name)	Formula (Mol. Wt.)	Organisms	Color and Crystalline Form (Melting Point, °C)	Significant Characteristics	Ref.
4	2-Methyl-1,4,5-trihydroxy-AQ (Islandicin)	$C_{15}H_{10}O_5$ (270)	*Penicillium funiculosum* Thom. *islandicum* Sopp.	dk red pl (CHCl$_3$); lf (AcOH) (218)	λ_{max}^{EtOH}: 232, 252.5, 289, 390-92, 466—70, 492, 513, 527; (4.52, 4.28, 3.86, 3.32, 3.98, 4.09, 3.96, 3.92) Color rn: C, purp-red s in CHCl$_3$, ace; less s in AcOH (ye-og, gr flr), EtOH and eth; v sl s in peth; NaOH (vt →pa ye)	144
5	2-Methyl-1,7,8-trihydroxy-AQ (Cladofulvin)	$C_{15}H_{10}O_5$ (270)	*Cladosporium fulvum* Cooke	og nd (bz) (310, d)	λ_{max}^{EtOH}: 235, 270, 449; (4.43, 4.43, 4.05) Color rn: NaOH, purp-bl→vt→bl; C, purp-vt→vt; G, red	145
6	3-Methyl-1,5,8-trihydroxy-AQ (Helminthosporin)	$C_{15}H_{10}O_5$ (270)	*Deuterophoma tracheiphila* Petri *Helminthosporium catenarium* Drechsler *cynodontis* Marignoni *gramineum* Rabenhorst *tritici-vulgaris* Nishik *Phoma violacea* (Bertel) Eveleigh	dk red nd (EtOAc, CHCl$_3$) (226—227)	λ_{max}^{EtOH}: 231, 255, 289, 405—15, 480, 490, 510, 525; (4.53, 4.15, 3.79, 3.85, 3.99, 4.01, 3.89, 3.81) sl s in EtOH, eth, ace, AcOH (og-red, gr flr) Color rn: C, red; H, rose-pk	149
7	3-Methyl-1,6,8-trihydroxy-AQ (Emodin)	$C_{15}H_{10}O_5$ (270)	*Aspergillus fumigatus* Fr. *terreus* Thom. *Chaetomium affine* Corda *Cladosporium fulvum* Cooke	og pl (CHCl$_3$, EtOH); nd (AcOH) (257)	λ_{max}^{EtOH}: 222, 252, 265, 289, 437, 520—530; (4.55, 4.26, 4.27, 4.34, 4.10, 2.38) Occurs also as glycoside in *Dermocybe sanguinea* and *D. semisanguinea* Color rn: E, og; C, og	141, 145, 146, 147, 148, 265, 266, 268

TABLE 4 (Continued)
ANTHRAQUINONE DERIVATIVES

No.	Compound (Common Name)	Formula (Mol. Wt.)	Organisms	Color and Crystalline Form (Melting Point, °C)	Significant Characteristics	Ref.
Anthraquinones (continued)						
7	3-Methyl-1,6,8-trihydroxy-AQ(*cont.*)		*Cortinarius (Dermocybe) sanguinea* (Wulf.) Fr. *Dermocybe semisanguinea* (Fr.) *Penicilliopsis clavariaeformis* Solms-Laubach *Penicillium brunneum* Udagawa *frequentans* Westling *islandicum* Sopp. *Phoma foveata* Foister *Polystictus versicolor* (L.) Fr. *Talaromyces avellaneus* (Thom. et Turesson) C. R. Benjamin *Trichoderma viride* Pers. *ex* Fr. *Valsaria rubricosa* (Fr.) Sacc.			
8	2-Methyl-1,4,5,7-tetrahydroxy-AQ (Catenarin)	$C_{15}H_{10}O_6$ (286)	*Aspergillus amstelodami* (Mangin) Thom. *et* Church *Helminthosporium catenarium* Drechsler *gramineum* Rabenhorst *tritici-vulgaris* Nisikado *velutinum* Link *Penicillium islandicum* Sopp.	red pl (EtOH) (246)	λ_{max}^{EtOH}: 231, 255, 280, 298, 488.5, 508, 515—25; (4.51, 4.23, 4.24, 4.03, 4.16, 4.06, 3.99) Color rm: E, vt;C, red-vt	160, 161, 165

TABLE 4 (Continued)
ANTHRAQUINONE DERIVATIVES

Anthraquinones (continued)

No.	Compound (Common Name)	Formula (Mol. Wt.)	Organisms	Color and Crystalline Form (Melting Point, °C)	Significant Characteristics	Ref.
9	2-Methyl-1,4,5,8-tetrahydroxy-AQ (Cynodontin)	$C_{15}H_{10}O_6$ (286)	*Deuterophoma tracheiphila* Petri *Helminthosporium avenae* Eidam *cynodontis* Marignoni *euchlaenae* Zimm. *oryzae* Breda de Haan *victoriae* Meehan *et* Murphy *Phoma terrestris* Hansen *violacea* (Bertel) Eveleigh	bron pl (pyr) (260)	λ^{EtOH}_{max}: 241, 295, 471(sh), 483, 503(sh), 514, 539, 552; (4.56, 4.06, 4.06, 4.14, 4.31, 4.38, 4.37, 4.42) Color rn: AcOH, bl-red, crimson in thin layers; C, bl, red flr	162, 163
10	2-Methyl-1,4,7,8-tetrahydroxy-AQ	$C_{15}H_{10}O_6$ (286)	*Penicillium islandicum* Sopp.	red nd (tol) (255)	Color rn: E, bl; B, bl-vt; I, lt bl	164
11	2,6-Dimethyl-1,4,5,8-tetrahydroxy-AQ	$C_{16}H_{12}O_6$ (300)	*Cochliobus sativus* (Ito *et* Kurib.) Drechsler *et* Dastur. *spicifer* Nelson *Curvularia* spp. *lunata* (Wakker) Boedijin	red (bz) (263.5)	λ^{EtOH}_{max}: 241, 295, 454(sh), 470 (sh), 483, 504, 514, 540, 552; (4.58, 3.98, 3.56, 3.95, 4.09, 4.26, 4.34, 4.32, 4.37) Color rn: C, bl, red flr	168
12	3-Hydroxymethyl-1,6,8-trihydroxy-AQ (Citreorosein)	$C_{15}H_{10}O_6$ (286)	*Penicillium citreo-roseum* Dierckx *cyclopium* Westling *islandicum* Sopp. *Talaromyces avellaneus* (Thom. *et* Turesson) C. R. Benjamin	og nd (EtOH−H$_2$O) (288)	λ^{EtOH}_{max}: 222.5, 250.5, 266, 289.5, 436, 510−20; (4.46, 4.21, 4.21, 4.26, 4.04, 2.92) Color rn: C, red	153

TABLE 4 (Continued)
ANTHRAQUINONE DERIVATIVES

Anthraquinones (continued)

No.	Compound (Common Name)	Formula (Mol. Wt.)	Organisms	Color and Crystalline Form (Melting Point, °C)	Significant Characteristics	Ref.
13	2-Hydroxymethyl-1,4,5,7-tetrahydroxy-AQ (Tritisporin)	$C_{15}H_{10}O_7$ (302)	*Helminthosporium tritici-vulgaris* Nisikado	red pr (diox) (278—279)	λ_{max}^{EtOH}: 231, 255, 280, 300.5, 489.5, 510, 525; (4.50, 4.23, 4.26, 4.05, 4.18, 4.09, 4.01) sl s in common org solv Color rn: E, deep carmine, red flr; C, red-vt; B, dk vt; AcOH, og-ye, ye-gr flr	166
14	2-Hydroxymethyl-1,4,?,?-tetrahydroxy-AQ	$C_{14}H_{10}O_7$ (290)	*Gibberella fujikuroi* (Saw.) Wr.	red pl (CHCl₃ —peth) (325, d at 205)	s in CHCl₃, pyr, AcOH, con H₂SO₄, 10% NaOH (bl); v sl s in EtOH, bz	167
15	7-Hydroxymethyl-1,2,4,5,6-pentahydroxy-AQ (Asperthecin)	$C_{15}H_{10}O_8$ (318)	*Aspergillus nidulans* (Eidam) Wint. *nidulans* (Eidam) Wint. mut. *alba* Vuill. *quadrilineatus* Thom. et Raper *rugulosus* Thom. et Raper	br nd (MeOH) (>370, darkens)	λ_{max}^{EtOH}: 237.5, 262.5, 286.5, 318, 484, 510, 545—55; (4.41, 4.49, 4.27, 4.02, 4.24, 4.23, 3.66) v sl sol in CHCl₃, bz; sl s in al; mod s in ace, AcOH, diox; s in pyr (purp nd ppt), NaHCO₃ (purp-red) Color rn: B, C, E, purp	169
16	1-Hydroxy-3-methoxy-6-methyl-AQ	$C_{16}H_{12}O_4$ (268)	*Alternaria solani* (Ellis et Martin) Sorauer	og nd (184—185)	λ_{max}^{EtOH}: 226, 252, 261, 276, 409; (4.12, 4.25, 4.23, 4.23, 3.71)	143
17	1,6-Dihydroxy-8-methoxy-3-methyl-AQ (Questin)	$C_{16}H_{12}O_5$ (284)	*Aspergillus terreus* Thom. *Penicillium frequentans* Westling	ye-br or or-br nd (AcOH) (301—303)	λ_{max}^{EtOH}: 224, 248, 285, 425; (4.59, 4.12, 4.37, 3.97) Color rn: C, red; NaOH, dk rd	152, 267

TABLE 4 (Continued)
ANTHRAQUINONE DERIVATIVES

No.	Compound (Common Name)	Formula (Mol. Wt.)	Organisms	Color and Crystalline Form (Melting Point, °C)	Significant Characteristics	Ref.
Anthraquinones (continued)						
18	1,7-Dihydroxy-3-methoxy-6-methyl-AQ (Macrosporin)	$C_{16}H_{12}O_5$ (284)	*Alternaria cucumerina* (Ell. et Everh.) Elliott *solani* (Ellis et Martin) Sorauer *Macrosporium porri* Elliott	og-ye rh (AcOH) (300–302, d)	λ_{max}^{EtOH}: 225, 251(sh), 284, 305, 379; (4.23, 4.00, 4.45, 4.03, 3.74) sl s in eth, bz, $CHCl_3$, ; s in h ace, EtOH, AcOH Color rn: E, og-ye; C, og-red; B, dk br	143, 150
19	1,8-Dihydroxy-3-methoxy-6-methyl-AQ (Physcion, parietin)	$C_{16}H_{12}O_5$ (284)	*Aspergillus amstelodami* (Mangin) Thom. *et* Church *chevaleri* (Mangin) Thom. *et* Church *echinulatus* (Delarc.) Thom. *et* Church *niveo-glaucus* Thom. *et* Raper *ruber* (Bremer) Thom. *et* Raper *umbrosus* Bain. *et* Sart *Dermocybe sanguinea* (Wülf. *ex* Fr.) Wunsche *semisanguinea* (Fr.) *Penicillium herquei* Bain. *et* Sart. *Valsaria rubricosa* (Fr.) Sacc.	og-red nd (EtOAc) (204.5–205)	λ_{max}^{EtOH}: 257(sh), 266, 288, 431; (4.35, 4.36, 4.35, 4.20) Also occurs as glycoside in *Dermocybe sanguinea* and *D. semisanguinea* sl s in EtOH, ace; s in bz, NaOH (purp) Color rn: B, red-br; C, mag, on heating emodin	148, 151, 161, 266
20	1,2,8-Trihydroxy-3-methoxy-6-methyl-AQ (Dermoglaucin)	$C_{16}H_{12}O_6$ (300)	*Dermocybe sanguinea* (Wülf. *ex* Fr.) Wunsche *semisanguinea* (Fr.)	red cr (EtOAc) (236)	λ_{max}^{EtOH}: 248, 254, 282, 435, 580; (3.94, 3.97, 4.30, 3.86, 3.11) Also occurs as glycoside	148

TABLE 4 (Continued)
ANTHRAQUINONE DERIVATIVES

No.	Compound (Common Name)	Formula (Mol. Wt.)	Organisms	Color and Crystalline Form (Melting Point, °C)	Significant Characteristics	Ref.
Anthraquinones (continued)						
21	1,2,8-Trihydroxy-6-methoxy-3-methyl-AQ	$C_{16}H_{12}O_6$ (300)	*Alternaria solani* (Ellis et Martin) Sorauer	og cr (267–268)	λ^{EtOH}_{max}: 229, 280, 312, 425; (4.36, 4.43, 4.00, 4.03) Color rn: B, olive br	143
22	1,4,5-Trihydroxy-7-methoxy-2-methyl-AQ (Erythroglaucin)	$C_{16}H_{12}O_6$ (300)	Species in *Aspergillus glaucus* series	dk red pl or nd (AcOH) (205–206)	λ^{EtOH}_{max}: 231, 255, 275, 302.5, 460 (sh), 475(sh), 489, 511(sh), 523; (4.44, 4.17, 4.13, 3.92, 3.95, 4.01, 4.06, 3.95, 3.90) sl s in EtOH, eth; mod sol in bz, CHCl$_3$, EtOAc, ace, NaOH (purp→ppt) Color rn: AcOH, og-red, gr flr; C, vt-bl; B, br; G, purp	160, 165
23	1,2,4,8-Tetrahydroxy-3-methoxy-6-methyl-AQ (Dermocybin)	$C_{16}H_{12}O_7$ (316)	*Cortinarius sanguineus* (Wulf. ex Fr.) Wünsche *Dermocybe sanguinea* (Wulf. ex Fr.) Wünshe *semisanguinea* (Fr.)	red pr or nd (EtOH–H$_2$O) (228–229)	λ^{EtOH}_{max}: 219.5, 262.5, 279, 459, 486, 521; (4.43, 4.38, 4.33, 4.07, 4.14, 4.00) Also occurs as glycoside	146, 148
24	1,3-Dihydroxy-8-methoxy-6-hydroxymethyl-AQ (Carviolin, roseo-purpurin)	$C_{16}H_{12}O_6$ (300)	*Penicillium roseo-purpureum* Dierckx	ye nd (AcOH) (286)	Color: pH 5.6–7, ye→red-br; pH 10.2–12.5, red-br→red-purp	154
25	1,6-Dihydroxy-8-methoxy-3-hydroxymethyl-AQ (Questinol)	$C_{16}H_{12}O_6$ (300)	*Penicillium frequentans* Westling	og nd (MeOH) (280–282)	λ^{EtOH}_{max}: 224, 247, 286, 432; (4.57, 4.15, 4.34, 3.95)	267

TABLE 4 (Continued)
ANTHRAQUINONE DERIVATIVES

Anthraquinones (continued)

No.	Compound (Common Name)	Formula (Mol. Wt.)	Organisms	Color and Crystalline Form (Melting Point, °C)	Significant Characteristics	Ref.
26	1,8-Dihydroxy-6-methoxy-3-(2-hydroxy-n-propyl)-AQ (Nalgiovensin)	$C_{18}H_{16}O_6$ (328)	*Penicillium nalgiovensis* Laxa	og nd (CHCl₃); pl (AcOH) (199–200)	λ_{max}^{EtOH}: 225, 266, 287, 437; (4.53, 4.27, 4.22, 4.09) $[a]_{5461}^{20}$: +48.1° (c = 0.1786, CHCl₃) $[a]_{5790}^{20}$: +39.7° (c = 0.1786, CHCl₃) s in wr CHCl₃, AcOH, ace; less s in EtOH, NaOH (bl-red) Color rn: C, red, B, og-br	158, 159
27	6-Carboxy-1,3,8-trihydroxy-AQ (Emodic acid)	$C_{15}H_8O_7$ (300)	*Penicillium cyclopium* Westling; *Talaromyces avellanus* (Thom. et Turesson) C. R. Benjamin	og-red nd (AcOH or EtOH) (363–365)	λ_{max}^{EtOH}: 222, 248.5, 271.5, 293 (sh), 437; (4.40, 4.18, 4.18, 4.14, 4.02) sl s in org solv	153
28	2-Carboxy-3-methyl-1,6,8-trihydroxy-AQ (Endocrocin)	$C_{16}H_{10}O_7$ (314)	*Aspergillus amstelodami* (Mangin) Thom. et Church; *Claviceps purpurea* (Fr.) Tul.; *Dermocybe sanguinea* (Wulf. ex Fr.) Wünsche *semisanguinea* (Fr.); *Penicillium islandicum* Sopp.	ye-red lf (EtCOMe–CCl₄) (318, d)	λ_{max}^{MeOH}: 227, 274, 287, 311, 442; (4.43, 4.30, 4.21, 3.94, 3.99) v s in ace; mod s in EtOH, AcOH, EtOAc; sl s in eth, CHCl₃, bz; s in NaHCO₃ (br-red), Na₂CO₃, NaOH (vt-red) Color rn: E, og-red; B, red-br; C, red	148, 156, 157, 161
29	2-Carboxy-3-methyl-1,5,7,8-tetrahydroxy-AQ (Clavorubin)	$C_{16}H_{10}O_8$ (330)	*Claviceps purpurea* (Fr.) Tul.	dk red pw (MeCOEt–CCl₄) (232)	λ_{max}^{MeOH}: 265, 340, 496, 528, 564 s in MeOH, EtOAc; sl s in bz, cy, CCl₄; mod s in NaHCO₃ (red) Color rn: C, vt; B, red-br	157, 269

TABLE 4 (Continued)
ANTHRAQUINONE DERIVATIVES

Anthraquinones (continued)

No.	Compound (Common Name)	Formula (Mol. Wt.)	Organisms	Color and Crystalline Form (Melting Point, °C)	Significant Characteristics	Ref.
30	2-Carboxy-8-methoxy-3-methyl-1,6-dihydroxy-AQ (Dermolutein)	$C_{17}H_{12}O_7$ (328)	*Dermocybe sanguinea* (Wulf. ex Fr.) Wünsche *semisanguinea* (Fr.)	ye cr (~270, d)	λ^{EtOH}_{max}: 247, 286, 432; (4.32, 3.36, 3.95) Also occurs as glycoside Color rn: NH_3, og-red	148
31	2-Carboxy-8-methoxy-3-methyl-1,4,6-trihydroxy-AQ (Dermorubin)	$C_{17}H_{12}O_8$ (344)	*Dermocybe sanguinea* (Wulf. ex Fr.) Wünsche *semisanguinea* (Fr.)	dk red nd (EtOH–H_2O) (300, d)	λ^{EtOH}_{max}: 233, 279, 291(sh), 481; (4.54, 4.30, 4.14, 4.04) Also occurs as glycoside Color rn: NH_3, red-vt	148
32	2-Carboxy-5-chloro-8-methoxy-3-methyl-1,6-dihydroxy-AQ (5-Chloro-dermolutein)	$C_{17}H_{11}O_7Cl$ (362.5)	*Dermocybe sanguinea* (Wulf. ex Fr.) Wünsche *semisanguinea* (Fr.)	br nd (EtOH–H_2O) (~280, d)	λ^{EtOH}_{max}: 230, 253, 290, 436; 4.52, 4.23, 4.28, 4.00) Also occurs as glycoside Color rn: NH_3, red	148
33	2-Carboxy-5-chloro-8-methoxy-3-methyl-1,4,6-trihydroxy-AQ (5-Chloro-dermorubin)	$C_{17}H_{11}O_8Cl$ (378.5)	*Dermocybe sanguinea* (Wulf. ex Fr.) Wünsche *semisanguinea* (Fr.)	dk red nd (AcOH) (~300)	λ^{EtOH}_{max}: 255(sh), 283, 486; (4.14, 4.25, 4.09) Also occurs as glycoside Color rn: NH_3, bl-vt	148
34	2-Chloro-6-methyl-1,3,8-trihydroxy-AQ (2-Chloro-emodin)	$C_{15}H_9O_5Cl$ (304.5)	*Aspergillus fumigatus* Fr. (J-4) *Valsaria rubricosa* (Fr.) Sacc.	og-red nd (bz) (271–273)	λ^{EtOH}_{max}: 217.5, 258, 312, 324(sh), 437, 510; (4.49, 4.24, 4.22, 3.87, 3.72) s in ace, al; sl s in $CHCl_3$, bz Color rn: E, og-red	266, 268
35	6-Chloro-2-methyl-1,4,5,7-tetrahydroxy-AQ (Valsarin II)	$C_{15}H_9O_6Cl$ (320.5)	*Valsaria rubricosa* (Fr.) Sacc.		Color rn: DMF, vt→red-br; C, vt	266
36	2-Chloro-6-methyl-1,3,4,8-tetrahydroxy-AQ (Valsarin I)	$C_{15}H_9O_6Cl$ (320.5)	*Valsaria rubricosa* (Fr.) Sacc.	red nd (273–274)	λ^{cy}_{max}: 245, 257, 305, 485, 518 Color rn: DMF, vt→red-br; C, vt	266

TABLE 4 (Continued)
ANTHRAQUINONE DERIVATIVES

No.	Compound (Common Name)	Formula (Mol. Wt.)	Organisms	Color and Crystalline Form (Melting Point, °C)	Significant Characteristics	Ref.
Anthraquinones (continued)						
37	2-Chloro-6-hydroxymethyl-1,3,8-trihydroxy-AQ (2-Chloro-citreorosein)	$C_{15}H_9O_6Cl$ (320.5)	*Aspergillus fumigatus* Fr. (J-4)	og-red pw (EtOAc) (297)	λ^{EtOH}_{max}: 220, 249(sh), 306, 332(sh), 435, 453(sh), 520; (4.50, −, 4.16, −, 3.96, −, 3.62) s in ace, al, EtOAc; sl s in eth, CHCl$_3$; i in bz	268
38	2-Chloro-3-methoxy-6-(2-hydroxy-*n*-propyl)-1,8-dihydroxy-AQ (Nalgiolaxin)	$C_{18}H_{15}O_6Cl$ (362.5)	*Penicillium nalgiovensis* Laxa	ye pl or nd (EtOH) (248–248.5)	λ^{EtOH}_{max}: 226, 270, 310, 435, 452 (sh); (4.49, 4.49, 4.03, 4.08, 4.04) $[a]^{22}_{5790}$: +40.3° (c = 0.1072, CHCl$_3$) Color rn: B, og-br; C, red; NaOH, bl-red→purp→red ppt	158
39	2-Hexanoyl-1,3,6,8-tetra-hydroxy-AQ (Norsolorinic acid)	$C_{20}H_{18}O_7$ (370)	*Aspergillus versicolor* (Vuill.) Tiraboschi	red pr (256–257, d)	λ^{EtOH}_{max}: 235, 269, 284, 297, 314, 465; (4.39, 4.23, 4.27, 4.30, 4.36, 4.89) Color rn: E, red; B, red-br	270
40	2-(Hex-1-enyl)-1,3,6,8-tetrahydroxy-AQ (Averythin)	$C_{20}H_{16}O_6$ (344)	*Aspergillus versicolor* (Vuill.) Tiraboschi	red pr (MeOH) (229–231, d)	λ^{EtOH}_{max}: 223, 255(sh), 266, 294, 324, 453; (4.46, 4.12, 4.18, 4.45, 4.02, 3.95) s in most polar org sol, Na$_2$CO$_3$ (red-purp); sl s in NaHCO$_3$ (pk) Color rn: B, red-br; C, purp	242
41	2-(1-Hydroxy-*n*-hexyl)-1,3,6,8-tetrahydroxy-AQ (Averantin)	$C_{20}H_{20}O_7$ (372)	*Aspergillus versicolor* (Vuill.) Tiraboschi	og nd (233–234)	λ^{EtOH}_{max}: 223, 258(sh), 266, 287(sh), 294, 325, 454; (4.53, 4.18, 4.24, 4.47, 4.53, 4.01, 4.03) s in EtOH, h CHCl$_3$, h bz, NaOH (purp-red), Na$_2$CO$_3$ (pk-vt); sl s in NaHCO$_3$ (pk) Color rn: B, red; C, red	240, 241

TABLE 4 (Continued)
ANTHRAQUINONE DERIVATIVES

No.	Compound (Common Name)	Formula (Mol. Wt.)	Organisms	Color and Crystalline Form (Melting Point, °C)	Significant Characteristics	Ref.
Anthraquinones (continued)						
42	2-(1-Hydroxymethyl-3-hydroxy-*n*-propyl)-1,3,6,8-tetrahydroxy-AQ (Versiconol)	$C_{18}H_{16}O_8$ (360)	*Aspergillus versicolor* (Vuill.) Tiraboschi	og-red nd (265, d)	λ_{max}^{EtOH}: 224, 255, 265, 295, 322, 460; (4.65, 4.35, 4.35, 4.53, 4.14, 4.02) $[\alpha]_D^{25}$: −35.8° (c = 0.35, diox)	274
43	(Avermutin)	$C_{20}H_{18}O_7$ (370)	*Aspergillus flavus* Link *ex* Fries *versicolor* (Vuill.) Tiraboschi (a mutant)	og cr (CHCl₃) (271)	λ_{max}^{MeOH}: 223, 263, 282, 312, 452; (4.30, 4.16, 4.25, 4.00, 3.94) $[\alpha]_D^{22}$: +226° (diox)	272, 273
44	(Avermutin-8-methyl ether)	$C_{21}H_{20}O_7$ (384)	*Aspergillus versicolor* (Vuill.) Tiraboschi (a mutant)	og (252—254)		273
45	(Avermutin-6,8-dimethyl ether)	$C_{22}H_{22}O_7$ (398)	*Aspergillus versicolor* (Vuill.) Tiraboschi (a mutant)	og (219—220)		273
46	(Averufin)	$C_{20}H_{16}O_7$ (368)	*Aspergillus versicolor* (Vuill.) Tiraboschi	og-red laths (ace) (280—282, d)	λ_{max}^{EtOH}: 223, 256(sh), 266, 286(sh), 294, 324, 453; (4.54, 4.21, 4.26, 4.43, 4.52, 3.98, 4.03) $[\alpha]_D^{22}$: >1° (c = 0.3, EtOH) sl s in most org solv, Na_2CO_3 (purp) Color rn: C, red; B, og-br	170, 271

TABLE 4 (Continued)
ANTHRAQUINONE DERIVATIVES

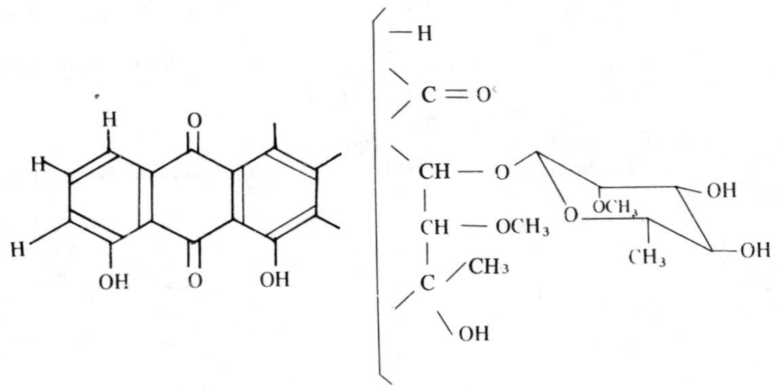

No.	Compound (Common Name)	Formula (Mol. Wt.)	Organisms	Color and Crystalline Form (Melting Point, °C)	Significant Characteristics	Ref.
Anthraquinones (continued)						
47	(Aranciamycin)	$C_{27}H_{28}O_{12}$ (544)	*Streptomyces echinatus* Corbaz *et al.*	ye-og pl (MeOH) (~240, d)	λ^{MeOH}_{max}: 241, 265, 440 $[\alpha]_D$: +149.5° (c = 0.5, MeOH)	10, 177
48	(Carviolacin)	$C_{20}H_{16}O_7$ (368)	*Penicillium carmino-violaceum* Biourge	lt br nd (EtOH–H_2O) (243, d)	Color rn similar to those of carviolin Der: tri-Me ether (ye nd, MP 214–215)	171
49	1,3,4-Trihydroxy-5-(or 8-) carboxy-AQ (Boletol)	$C_{15}H_8O_7$ (300)	*Boletus badius* Fr. *calopus* Fr. *erythropus* Fr. *luridus* Schaeff. ex Fries *satanus* Lenz.	red nd (eth-peth) (275–280, d)	Gives purpurin on heating of its sodium salt with soda lime s in EtOH, eth Its presence in the fungi is repudiated)	155, 284

Structure (compound 47):

— H
\C=O
CH—O
CH—OCH₃
C—CH₃ / OH

OCH₃
OH / CH₃
OH

(anthraquinone ring with OH, OH substituents)

TABLE 4 (Continued)
ANTHRAQUINONE DERIVATIVES

No.	Compound (Common Name)	Formula (Mol. Wt.)	Organisms	Color and Crystalline Form (Melting Point, °C)	Significant Characteristics	Ref.
9,10-Phenanthrenequinones						
1	(Piloquinone)	$C_{21}H_{20}O_5$ (352)	*Streptomyces pilosus* Ettlinger *et al.*	dk red nd (CHCl$_3$–eth) (176–178)	λ^{EtOH}_{max}: 233, 277, 286, 396(sh), 515; (4.50, 4.23, 4.19, 3.61, 3.74) Color rn: C, bl-gr→bl-vt	172
2	(4-Hydroxy-piloquinone)	$C_{21}H_{20}O_6$ (368)	*Streptomyces pilosus* Ettlinger *et al*	red-cr (MeOH–eth) (174–176)	λ^{EtOH}_{max}: 235, 250, 294, 500; (4.38, 4.22, 3.82, 3.89)	173
Benz [*a*] anthraquinones						
1	3-Methyl-1,7-dihydroxy-benz-AQ (Tetrangulol)	$C_{19}H_{12}O_4$ (304)	*Streptomyces rimosus* Sobin *et al.*	purp nd (EtOAc) (198–200)	λ^{MeOH}_{max}: 225, 250(sh), 315, 425, (4.70, 4.31, 4.39, 3.85)	174
2	3-Methyl-7-hydroxy-1-oxo-1,2,3,4-tetrahydro-benz-AQ (Ochromycinone)	$C_{15}H_{14}O_4$ (306)	*Streptomyces* spp.	ye cr (152–253)	λ^{EtOH}_{max}: 265, 405; (4.42, 3.55) $[\alpha]^{25}_D$: +204.5° (CHCl$_3$)	276
3	3-Methyl-3,7-dihydroxy-1-oxo-1,2,3,4-tetrahydro-benz-AQ (Tetrangomycin)	$C_{19}H_{14}O_5$ (322)	*Streptomyces rimosus* Sobin *et al.*	ye cr (EtOH) (182–184)	λ^{MeOH}_{max}: 206, 267, 330(sh), 400; (4.43, 4.50, 3.46, 3.72) $[\alpha]^{25}_D$: +41.8±3.50° (c = 0.861, CHCl$_3$) s to mod s in eth, EtOH, ace; s in CH$_2$Cl$_2$, CHCl$_3$, DMF Converted to tetrangulol in base	174

223

TABLE 4 (Continued)
ANTHRAQUINONE DERIVATIVES

No.	Compound (Common Name)	Formula (Mol. Wt.)	Organisms	Color and Crystalline Form (Melting Point, °C)	Significant Characteristics	Ref.
Benz [a] anthraquinones (continued)						
4	3-Methyl-3,6,7-trihydroxy-1-oxo-1,2,3,4-tetrahydro-benz-AQ (Rabelomycin)	$C_{19}H_{14}O_6$ (338)	*Streptomyces olivaceus* (Waksman) Waksman et Henrici	ye nd (bz—MeOH) (193, d)	λ^{MeOH}_{max}: 228, 267, 433; (4.25, 4.46, 3.90) $[a]_D$: $-102 \pm 10°$ ($c = 1$, $CHCl_3$) st in base; dehydrated in acid	175
5	(Aquayamycin)	$C_{25}H_{26}O_{16}$ (582)	*Streptomyces misawanensis* Hamada et Okami	og-ye cr (BuOAc) (189–190, contains BuOAc)	λ^{MeOH}_{max}, $E^{1\%}_{1cm}$: 220, 320, 430 (675, 140, 124) $[a]^{20}_D$: $+160°$ ($c = 1$, diox) s in EtOH, ace, diox, AcOH, NaOH (bl-vt); sl s in EtOAc, BuOAc, CHCl₃; i in eth, bz, hx Color rn: B, br; E, bl-vt; C, purp	176

Dihydroanthraquinones and Tetrahydroanthraquinones

No.	Compound (Common Name)	Formula (Mol. Wt.)	Organisms	Color and Crystalline Form (Melting Point, °C)	Significant Characteristics	Ref.
1	3-Methyl-1,6,8-trihydroxy-6,7-dihydro-AQ (Flavoskyrin)	$C_{15}H_{12}O_5$ (272)	*Penicillium islandicum* Sopp.	ye nd (diox—AcOH, ace—H₂O) (215, d)	$[a]$: $-295°$ (diox) s in diox; less s in ace, al, AcOH, CHCl₃; sl s in eth, bz, Na₂CO₃ (ye) Color rn: B, olive br; C, og→ purp-red; E, no rn, on standing →og (emodin)	178, 188
2	3-Methyl-1,4,6,8-tetrahydroxy-5,6-dihydro-AQ (5,6-Dihydrocatenarin)	$C_{15}H_{12}O_6$ (288)	*Penicillium islandicum* Sopp.	red cr (CHCl₃—MeOH) (~95–105)	λ^{EtOH}_{max}: 495, 530(sh), 570(sh) Color rn: E, turquoise	275

TABLE 4 (Continued)
ANTHRAQUINONE DERIVATIVES

No.	Compound (Common Name)	Formula (Mol. Wt.)	Organisms	Color and Crystalline Form (Melting Point, °C)	Significant Characteristics	Ref.
Dihydroanthraquinones and Tetrahydroanthraquinones (continued)						
3	3-Methyl-1,4,6,8-tetrahydroxy-5,6,7,8-tetrahydro-AQ (5,6,7,8-Tetrahydrocatenarin)	$C_{15}H_{14}O_6$ (290)	*Penicillium islandicum* Sopp.	red cr (CHCl$_3$ – MeOH) (~130, d)	λ_{max}^{EtOH}: 485, 514, 554; (3.78, 3.82, 3.61) Color rn: C, purp; E, purp	275
4	3-Methoxy-6-methyl-1,6,7-trihydroxy-5,6,7,8-tetrahydroxy-AQ (Altersolanol B)	$C_{16}H_{16}O_6$ (304)	*Alternaria solani* (Ellis et Martin) Sorauer	red-br pl (EtOH) (228–230)	λ_{max}^{EtOH}: 217, 265, 285, 421; (4.52, 4.14, 3.97, 3.58)	143
5	3-Methoxy-6-methyl-1,4,6,7,8-pentahydroxy-5,6,7,8-tetrahydro-AQ (Bostrycin)	$C_{16}H_{16}O_8$ (336)	*Bostryoconema alpestre cetati*	red cr (pyr–H$_2$O) (222–224, d)	λ_{max}^{EtOH}: 228, 303, 472, 505, 542; (4.42, 3.48, 3.71, 3.78, 3.59) sl s in org solv, except pyr	180
6	3-Methoxy-6-methyl-1,5,6,7,8-pentahydroxy-5,6,7,8-tetrahydro-AQ (Altersolanol A)	$C_{16}H_{16}O_8$ (336)	*Alternaria solani* (Ellis et Martin) Sorauer	(218–218.5, d)	λ_{max}^{EtOH}: 219, 240, 268, 285(sh), 422; (4.57, 3.96, 4.15, 3.84, 3.65) $[a]_D^{23}$: -292° (c = 0.25, pyr)	143, 179
Difuroanthraquinones						
1	4,5,7-Trihydroxy-3a, 10a-dihydro-DFAQ (Versicolorin A)	$C_{18}H_{10}O_7$ (338)	*Aspergillus versicolor* (Vuill.) Tiraboschi	og-ye nd (ace) (289, d)	λ_{max}^{EtOH}: 222, 255, 267, 290, 326, 450; (4.45, 4.13, 4.26, 4.40, 3.83, 3.85) $[a]_D^{18}$: -354° (c = 0.75, diox) s in ace, al, diox, EtOAc, eth; sl s in bz, CHCl$_3$ Color rn: E, red; C, purp; B, red-br	181

TABLE 4 (Continued)
ANTHRAQUINONE DERIVATIVES

Difuroanthraquinones (continued)

No.	Compound (Common Name)	Formula (Mol. Wt.)	Organisms	Color and Crystalline Form (Melting Point, °C)	Significant Characteristics	Ref.
2	4,5,7-Trihydroxy-2,3,3a,10a-tetrahydro-DFAQ (Versicolorin B)	$C_{18}H_{12}O_7$ (340)	*Aspergillus versicolor* (Vuill.) Tiraboschi	og-ye nd (ace) (298, d)	λ_{max}^{EtOH}: 223, 255, 266, 291, 324, 450; (4.38, 4.13, 4.29, 4.38, 4.11, 3.94) $[\alpha]_D^{25}$: $-223°$ ($c = 0.42$, diox) s in ace, al, diox, EtOAc, eth; sl s in bz, $CHCl_3$ Color rn: E, red; C, purp; B, red-br	181
3	(Versicolorin C)	$C_{18}H_{12}O_7$ (340)	*Aspergillus flavus* Link *ex* Fries *versicolor* (Vuill.) Tiraboschi	og-red nd (ace) (>310)	λ_{max}^{EtOH}: 223, 225, 267, 292, 326, 450; (4.46, 4.20, 4.31, 4.46, 4.00, 4.03) $[\alpha]_D^{25}$: $0°$ ($c = 0.44$, diox) s in ace, al, diox, EtOAc, eth; sl s in bz, $CHCl_3$ Color rn: E, red; C, purp; B, red-br Probably a racemate of versicolorin B	181
4	3,3a,4,5,8-Pentahydroxy-2,3,3a,10a-tetrahydro-DFAQ (Dothistromin)	$C_{18}H_{12}O_9$ (372)	*Dothistroma pini* Hulbary	red	λ_{max}^{EtOAc}: 478, 490, 509, 523 Color rn: E, dk purp-red; alk, purp	183
5	5,7-Dimethoxy-4-hydroxy-2,3,3a,10a-tetrahydro-DFAQ (Aversin)	$C_{20}H_{16}O_7$ (368)	*Aspergillus versicolor* (Vuill.) Tiraboschi	gold nd (ace) (217)	λ_{max}^{EtOH}: 224, 251, 285, 313, 363, 440; (4.56, 4.13, 4.53, 3.95, 3.70, 3.89) $[\alpha]_D^{20}$: $-222°$ ($c = 0.248$, $CHCl_3$) Color rn: C, purp; B, red-br	10, 182

TABLE 4 (Continued)
ANTHRAQUINONE DERIVATIVES

No.	Compound (Common Name)	Formula (Mol. Wt.)	Organisms	Color and Crystalline Form (Melting Point, °C)	Significant Characteristics	Ref.
Aza-anthraquinones						
1	(Phomazarin)	$C_{19}H_{17}O_8N$ (387)	*Phoma terrestris* Hansen	og nd (196, d)	λ_{max}^{EtOH}: 235, 274, 325(sh), 390(sh), 439; (4.30, 4.48, 3.83, 3.78, 3.88)	184
2	(Bostrycoidin)	$C_{15}H_{11}O_5N$ (342)	*Fusarium bostrycoides* Wr. et Rkg., *solani* (Mart.) App. et Wr.	br (ace) or red pl (sub) (243—244)	λ_{max}^{EtOH}: 251, 320, 475, 497, 525	184
Bianthraquinones						
1	(+)-3,3'-Dimethyl-1,1',8, 8'-tetrahydroxy-BAQ (Dianhydrorugulosin)	$C_{30}H_{18}O_8$ (506)	*Penicillium islandicum* Sopp.	og-red pr (321)	λ_{max}^{diox}: 282(sh), 439; (4.28, 4.33) Reductive cleavage→chrysophan-ol	186
2	(+)-3,3'-Dimethyl-1,1',4, 8,8'-pentahydroxy-BAQ (Roseoskyrin)	$C_{30}H_{18}O_9$ (522)	*Penicillium islandicum* Sopp.	(>300)	Reductive cleavage→chrysophan-ol + islandicin	186
3	(+)-3,3'-Dimethyl-1,1',6, 8,8'-pentahydroxy-BAQ (Auroskyrin)	$C_{30}H_{18}O_9$ (522)	*Penicillium islandicum* Sopp.	(>300)	Reductive cleavage→chrysophan-ol + emodin	186

TABLE 4 (Continued)
ANTHRAQUINONE DERIVATIVES

No.	Compound (Common Name)	Formula (Mol. Wt.)	Organisms	Color and Crystalline Form (Melting Point, °C)	Significant Characteristics	Ref.
Bianthraquinones (continued)						
4	(+)-3,3'-Dimethyl-1,1',4, 4',8,8'-hexahydroxy-BAQ (Iridoskyrin)	$C_{30}H_{18}O_{10}$ (538)	*Penicillium islandicum* Sopp.	red rods or pl (bz) (358–360, d)	λ_{max}^{diox}: 286, 502; (4.26, 4.39) s in $CHCl_3$, ace, wr bz, N–NaOH (purp-bl); sl s in eth; v s in EtOH Color rn: E, vt-red; C, bl, B, vt	186, 187, 191
5	(+)-2,3'-Dimethyl-1,1',4', 6',8,8'-hexahydroxy-BAQ (Rhodoislandin A)	$C_{30}H_{18}O_{10}$ (538)	*Penicillium islandicum* Sopp.	(>300)	Reductive cleavage→chrysophan-ol + catenarin	186
6	(+)-3,3'-Dimethyl-1,1',4, 6',8,8'-hexahydroxy-BAQ (Rhodoislandin B)	$C_{30}H_{18}O_{10}$ (538)	*Penicillium islandicum* Sopp.	(>300)	Reductive cleavage→emodin + islandicin	186
7	(+)-3,3'-Dimethyl-1,1',6, 6',8,8'-hexahydroxy-BAQ (Skyrin, endothianin)	$C_{30}H_{18}O_{10}$ (538)	*Endothia fluens* (Sow.) Shear *et* Stevens *gyrosa* (Scw.) Fr. *longirostris* Earle *parasitica* (Murr.) P. J. *et* H. W. And. *tropicalis* Shear *et* Stevens *Penicilliopsis clavariaeformis* Solms-Laubach *Penicillium brunneum* Udagawa *islandicum* Sopp. *piceum* Raper *et* Fen-nel *rugulosum* Thom. *tardum* Thom. *variabile* Sopp. *wortmanni* Klocker	red-og rods (eth) (>380)	λ_{max}^{EtOH}: 257, 300, 462; (4.69, 4.50, 4.37) sl s in diox, ace; v sl s in eth, al, $CHCl_3$, AcOH; s in h pyr (red-br sol→red rh cr ppt Color rn: E, og; C, red→red-gr; B, br-red	147, 186, 188, 189, 194, 198

TABLE 4 (Continued)
ANTHRAQUINONE DERIVATIVES

No.	Compound (Common Name)	Formula (Mol. Wt.)	Organisms	Color and Crystalline Form (Melting Point, °C)	Significant Characteristics	Ref.
	Bianthraquinones (continued)					
7	(+)-3,3'-Dimethyl-1,1',6,6',8,8'-hexahydroxy-BAQ (cont.)		*Preussia multispora* (Saito et Minoura) Cain *Sepedonium ampullosporum* Damon			
8	(+)-3,3'-Dimethyl-1,1',4,4',6,8,8'-heptahydroxy-BAQ (Punicoskyrin)	$C_{30}H_{18}O_{11}$ (554)	*Penicillium islandicum* Sopp.	(>300)	Reductive cleavage→islandicin + catenarin	186
9	(+)-3,3'-Dimethyl-1,1',4,6,6',8,8'-heptahydroxy-BAQ (Aurantioskyrin)	$C_{30}H_{18}O_{11}$ (554)	*Penicillium islandicum* Sopp.	(>300)	Reductive cleavage →emodin + catenarin	186
10	(+)-3,3'-Dimethyl-1,1',4,4',6,6',8,8'-octahydroxy-BAQ (Dicatenarin)	$C_{30}H_{18}O_{12}$ (570)	*Penicillium islandicum* Sopp.	(>300)	Der: octa-OAc (pa ye nd, MP 295 −297)	186,190
11	6,6'-Dimethoxy-3,3'-dimethyl-1,1',4,4',8,8'-hexahydroxy-BAQ (Fusaroskyrin)	$C_{32}H_{22}O_{12}$ (598)	*Fusarium* sp.	dk red nd (>300)	$\lambda_{max}^{CHCl_3}$: 256, 276, 305-310, 505 s in ace Color rn: E, bl	190
12	(+)-3,3'-Dihydroxymethyl-1,1',6,6',8,8'-hexahydroxy-BAQ (Skyrinol)	$C_{30}H_{18}O_{12}$ (570)	*Penicillium islandicum* Sopp.	red cr (>360)	λ_{max}^{diox}: 258, 290, 448 s in 5% NaHCO$_3$ Color rn: C, purp→gr; E, og-red	191

TABLE 4 (Continued)
ANTHRAQUINONE DERIVATIVES

No.	Compound (Common Name)	Formula (Mol. Wt.)	Organisms	Color and Crystalline Form (Melting Point, °C)	Significant Characteristics	Ref.
Bianthraquinones (continued)						
13	(+)-3-Hydroxymethyl-3'-methyl-1,1',6,6',8,8'-hexahydroxy-BAQ (Oxyskyrin)	$C_{30}H_{18}O_{11}$ (554)	*Endothia parasitica* Anderson *et* Anderson *Penicillium islandicum* Sopp.	og-red nd (ace−bz) (>360, darkens >270)	λ_{max}^{EtOH}: 257, 300, 462: (4.69, 4.50, 4.37) sl s in ace; partially s in NaHCO₃; completely s in Na₂CO₃ (purp) Color rn: E, og-red; C, purp→red-gr	191, 192
14	[(−)-Rubroskyrin]	$C_{30}H_{22}O_{12}$ (574)	*Penicillium islandicum* Sopp.	dk red pl (EtOH) (281, d)	$\lambda_{max}^{CHCl_3}$: 275, 415, 435, 530, 540; (4.31, 4.10, 4.10, 3.97, 3.97) $[\alpha]_D$: levorotatory (color of sol prevents exact measurement) v sl s in bz, eth; sl s in CHCl₃; mod s in EtOAc, AcOH, EtOH, diox; f s in ace, carbonate, NaOH (gr→br-ye) Color rn: E, gr; C red→bl; B, purp-br	187, 193, 195, 196

TABLE 4 (Continued)
ANTHRAQUINONE DERIVATIVES

No.	Compound (Common Name)	Formula (Mol. Wt.)	Organisms	Color and Crystalline Form (Melting Point, °C)	Significant Characteristics	Ref.
Bianthraquinones (continued)						
15	[(+)-Rugulosin, radicalisin]	$C_{30}H_{22}O_{10}$ (542)	*Endothia fluens* (Sow.) Shear *et* Stevens *gyrosa parasitica* And. *et* And. *Penicillium brunneum* Udagawa *rugulosum* Thom. *tardum* Thom. *variabile* Sopp. *wortmanni* Klöcker *Sepedonium ampullosporum* Damon	ye pr or rods (eth, ace, MeOH) (290, d)	$[\alpha]_D^{19}$: +492° (c = 0.5, diox); +432° (c = 0.5, ace) s in ace, al, EtOAc, AcOH, pyr, NaHCO₃ (ye); v sl s in h bz, CHCl₃ Color rn: C, ye; B, gr-br	193, 194, 197, 198

TABLE 4 (Continued)
ANTHRAQUINONE DERIVATIVES

No.	Compound (Common Name)	Formula (Mol. Wt.)	Organisms	Color and Crystalline Form (Melting Point, °C)	Significant Characteristics	Ref.
Bianthraquinones (continued)						
16	[(−)-Luteoskyrin]	$C_{30}H_{22}O_{12}$ (574)	Mycelia Sterilia *Penicillium islandicum* Sopp.	ye pr (281, d)	$[\alpha]_D^{23.4}$: $-857 \pm 19°$ ($c = 0.107$, ace) s in $NaHCO_3$, Na_2CO_3, ace; less s in $CHCl_3$, bz, NaOH (og-ye) Color rn: E, no rn; C, og-red→ red-purp	193

TABLE 4 (Continued)
ANTHRAQUINONE DERIVATIVES

No.	Compound (Common Name)	Formula (Mol. Wt.)	Organisms	Color and Crystalline Form (Melting Point, °C)	Significant Characteristics	Ref.
	Bianthraquinones (continued)					
17	Julimycins		*Streptomyces shiodaensis* nov. sp.		A series of twenty antibiotics extracted from mycelium with ace and from medium with EtOAc; each compound consists of two of the nine component units (Q_1 to Q_9), coupled to each other at β,β'-position of C_7 and$_{7'}$ (e.g., Julimycin B-II = Q_1 –Q_1)	199

TABLE 4 (Continued)
ANTHRAQUINONE DERIVATIVES

No.	Compound (Common Name)	Formula (Mol. Wt.)	Organisms	Color and Crystalline Form (Melting Point, °C)	Significant Characteristics	Ref.
	Bianthraquinones (continued)					
17	Julimycins (cont.)					

$Q_5 =$

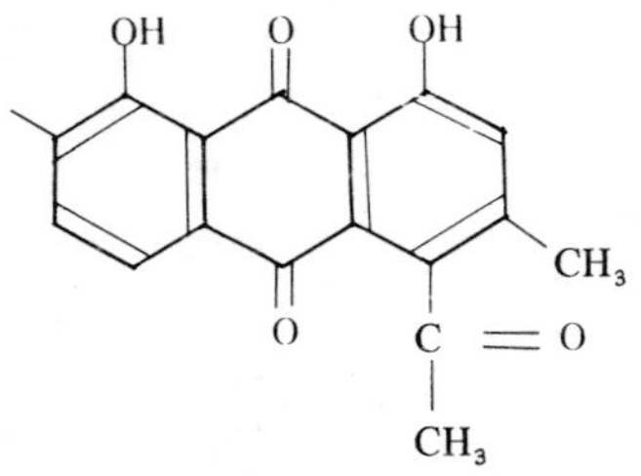

$Q_6 =$

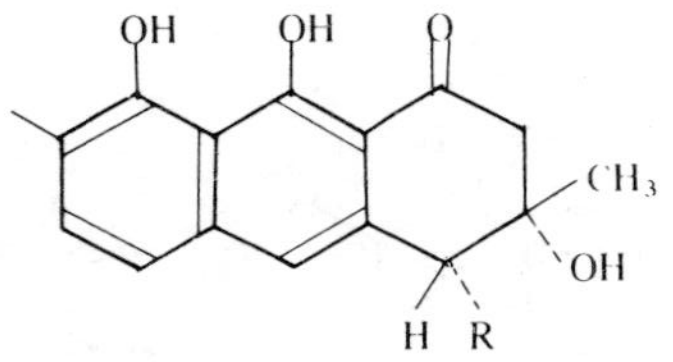

$Q_7 =$

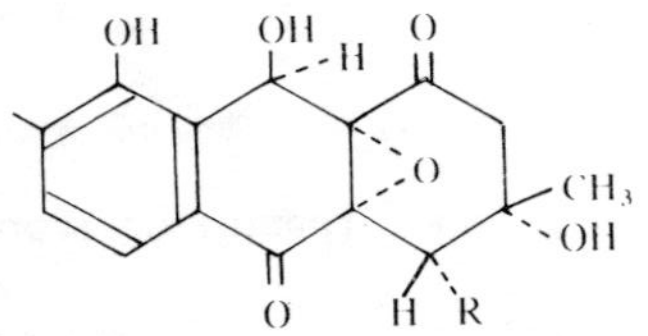

$Q_8 =$

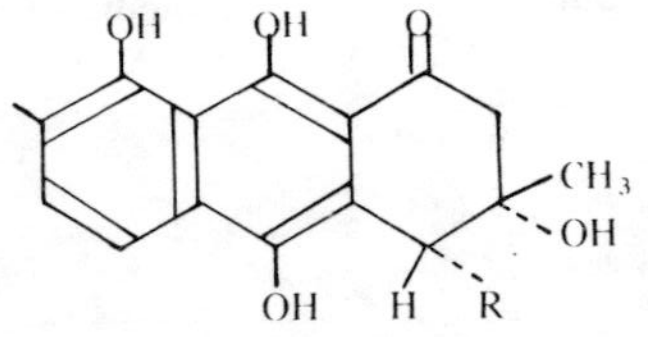

TABLE 4 (Continued)
ANTHRAQUINONE DERIVATIVES

No.	Compound (Common Name)	Formula (Mol. Wt.)	Organisms	Color and Crystalline Form (Melting Point, °C)	Significant Characteristics	Ref.
Bianthraquinones (continued)						
17	Julimycins (*cont.*)					
	$Q_9 =$					
	$R = $					
	$Q_1 - Q_1$ (Julimycin B-II)	$C_{38}H_{34}O_{14}$ (714)		dk red pr (MeOH) (215—220, d, darkens at 170)	λ_{max}^{MeOH}: 234, ~275(sh), 456; (4.70, —, 4.07) $[\alpha]_D^{24.5}$: $-82.9 \pm 20°$ ($c = 0.06$, MeOH) st in weak acid and neutral state, v unst in alk s in diox, DMF, THF, pyr; sl s in al, ace, EtOAc, CHCl$_3$; i in eth, bz, hx, H$_2$O Color rn: B, purp-gy; E, gy-bl; C, red-purp	200, 201, 202
	$Q_1 - Q_2$ (Julichrome $Q_{1.2}$)	$C_{38}H_{32}O_{13}$ (696)		red pw (166—170)	$\lambda_{max}^{CHCl_3}$: 266, 460; (4.60, 4.27) Color rn: Mg(OAc)$_2$, purp	119, 238

TABLE 4 (Continued)
ANTHRAQUINONE DERIVATIVES

No.	Compound (Common Name)	Formula (Mol. Wt.)	Organisms	Color and Crystalline Form (Melting Point, °C)	Significant Characteristics	Ref.
Bianthraquinones (continued)						
17	Julimycins (*cont.*)					
	$Q_1 - Q_3$ (Julichrome $Q_{1.3}$)	$C_{38}H_{36}O_{15}$ (732)		red pr (MeOH) (190–210)	λ_{max}^{MeOH}: 221, 260(sh), 273(sh), 450; (4.57, 4.33, 4.32, 3.81) v hygroscopic Color rn: $Mg(OAc)_2$, vt	203, 204
	$Q_1 - Q_4$ (Julichrome $Q_{1.4}$)	$C_{38}H_{34}O_{15}$ (730)		red pr (MeOH) (200–220, d)	λ_{max}^{MeOH}: 230, 247(sh), 390, 433 (sh); (4.53, 4.51, 3.96, 3.89) Color rn: $Mg(OAc)_2$, br	203
	$Q_1 - Q_5$ (Julichrome $Q_{1.5}$, julimycin B-I)	$C_{36}H_{28}O_{11}$ (636)		red pr (ace) (>300)	$\lambda_{max}^{CHCl_3}$: 265, 460; (4.59, 4.30) Color rn: $Mg(OAc)_2$, purp	199, 238
	$Q_1 - Q_6$ (Julichrome $Q_{1.6}$)	$C_{38}H_{36}O_{13}$ (700)		dk red lf (EtOAc) (191–197)	Color rn: $Mg(OAc)_2$, gr	119
	$Q_1 - Q_7$ (Julichrome $Q_{1.7}$)	$C_{38}H_{36}O_{15}$ (732)		red pr (bz) (>290)	λ_{max}^{MeOH}: 220, 247(sh), 275(sh), 450; (4.51, 4.32, 4.22, 3.76)	205
	$Q_1 - Q_9$ (Julichrome $Q_{1.9}$)	$C_{38}H_{34}O_{16}$ (746)		red pw ($CHCl_3$–peth) (172–177, d)	λ_{max}^{MeOH}: 226, 284, 446; (4.50, 4.28, 374) Color rn: $Mg(OAc)_2$, vt	205
	$Q_2 - Q_2$ (Julichrome $Q_{2.2}$)	$C_{38}H_{30}O_{12}$ (678)		ye-og pr (ace) (250–251, d)	Color rn: $Mg(OAc)_2$, red	199, 238
	$Q_2 - Q_3$ (Julichrome $Q_{2.3}$)	$C_{38}H_{34}O_{14}$ (714)		ye pw (bz) (165–190, d)	λ_{max}^{diox}: 230, 268, 442; (4.54, 4.54, 4.15) Color rn: $Mg(OAc)_2$, red	203

TABLE 4 (Continued)
ANTHRAQUINONE DERIVATIVES

No.	Compound (Common Name)	Formula (Mol. Wt.)	Organisms	Color and Crystalline Form (Melting Point, °C)	Significant Characteristics	Ref.
Bianthraquinones (continued)						
17	Julimycins (cont.)					
	Q_2-Q_5 (Julichrome $Q_{2.5}$)	$C_{36}H_{26}O_{10}$ (618)		red pr (CHCl$_3$–MeOH) (>300)	$\lambda_{max}^{CHCl_3}$: 268, 292(sh), 462; (4.70, 4.34, 4.49) Color rn: Mg(OAc)$_2$, red	199, 238
	Q_3-Q_3 (Julichrome $Q_{3.3}$)	$C_{38}H_{38}O_{16}$ (750)		col pr (ace) (>300)	λ_{max}^{MeOH}: 295, 333, 400; (3.85, 3.75, 3.37) Color rn: Mg(OAc)$_2$, ye	205
	Q_3-Q_4 (Julichrome $Q_{3.4}$)	$C_{38}H_{36}O_{16}$ (748)		pa ye amor pw (>280)	λ_{max}^{MeOH}: 208.5, 267, 370; (4.48, 4.25, 3.94) Color rn: Mg(OAc)$_2$, negative (color somewhat deepened)	203
	Q_3-Q_5 (Julichrome $Q_{3.5}$)	$C_{36}H_{30}O_{12}$ (654)		og nd (MeOH) (228, d)	λ_{max}^{MeOH}: 229, 265, 444; (4.54, 4.48, 4.07) Color rn: Mg(OAc)$_2$, red	199, 238
	Q_3-Q_8 (Julichrome $Q_{3.8}$, julimycin B-III)	$C_{38}H_{38}O_{15}$ (734)		ye pr (CHCl$_3$) (195–200)	$[a]_D^{24}$: +47.2 ± 4.5° ($c = 0.178$, MeOH) Color rn: Mg(OAc)$_2$, negative	205
	Q_4-Q_5 (Julichrome $Q_{4.5}$)	$C_{36}H_{28}O_{12}$ (652)		og pr (MeOH) (241–246)	λ_{max}^{MeOH}: 229, 262, 439; (4.60, 4.63, 4.25) Color rn: Mg(OAc)$_2$, dk red	199, 238
	Q_5-Q_5 (Julichrome $Q_{5.5}$)	$C_{34}H_{22}O_{8}$ (558)		ye pr (CHCl$_3$) (>300)	$\lambda_{max}^{CHCl_3}$: 266.5, 292(sh), 464; (4.72, 4.36, 4.46) Color rn: Mg(OAc)$_2$, ye→red→ye	199, 238
	Q_5-Q_6 (Julichrome $Q_{5.6}$)	$C_{36}H_{30}O_{10}$ (622)		og amor pw	Color rn: Mg(OAc)$_2$, mag	199
	Q_6-Q_6 (Julichrome $Q_{6.6}$)	$C_{38}H_{38}O_{12}$ (686)		ye pr (bz) (240–243)	Color rn: Mg(OAc)$_2$, negative	199

TABLE 4 (Continued)
ANTHRAQUINONE DERIVATIVES

No.	Compound (Common Name)	Formula (Mol. Wt.)	Organisms	Color and Crystalline Form (Melting Point, °C)	Significant Characteristics	Ref.
	Bianthraquinones (continued)					
17	Julimycins (*cont.*)					
	Q_8–Q_8 (Julichrome $Q_{8.8}$)	$C_{38}H_{38}O_{14}$ (718)		ye pr (CHCl$_3$) (178–180)	Color rn: Mg(OAc)$_2$, negative	205

TABLE 5
ANTHRACYCLINONES AND ANTHRACYCLINES

No.	Compound (Common Name)	Formula (Mol. Wt.)	Organisms	Color and Crystalline Form (Melting Point, °C)	Significant Characteristics	Ref.
	Aklavinones					
1	(7S,9R,10R)-III; R' = R'' = H, R''' = OH (Aklavinone)	$C_{22}H_{20}O_8$ (412)	*Streptomyces galilaeus* Ettlinger *et al.* unidentified sp.	og nd (EtOH) (171–172)	λ_{max}^{MeOH}: 229, 258, 278(sh), 288, 430; (4.64, 4.43, 4.08, 4.06, 4.12) $[\alpha]_D^{19}$: +142° (CHCl$_3$) $[\alpha]_D^{28}$: +213° (diox)	206, 207, 208, 210
2	(9R, 10R)-IV; R' = R'' = H, R''' = OH (7-Deoxy-aklavinone, galirubinone D)	$C_{22}H_{20}O_7$ (396)	*Streptomyces galilaeus* Ettlinger *et al.*	ye-red cr (EtOAc) (224–225)	λ_{max}^{MeOH}: 229, 259, 290, 431; (4.56, 4.47, 4.01, 4.13)	206, 207, 210
3	9-Ethyl-4,6-dihydroxy-CNME (Bisanhydroaklavinone, galirubinone B$_1$)	$C_{22}H_{16}O_6$ (376)	*Streptomyces galilaeus* Ettlinger *et al.* unidentified sp.	gold-ye nd (EtOAc) (234–236)	λ_{max}^{hx}: 242, 262, 279, 290, 440, 462, 474; (4.65, 4.68, 4.28, 4.29, 4.28, 4.16, 4.19) Red in con H$_2$SO$_4$	206

TABLE 5 (Continued)
ANTHRACYCLINONES AND ANTHRACYCLINES

No.	Compound (Common Name)	Formula (Mol. Wt.)	Organisms	Color and Crystalline Form (Melting Point, °C)	Significant Characteristics	Ref.
Aklavinone Glycosides						
1	Aglycone: aklavinone Sugar: $C_8H_{17}O_4N$ (Aklavin)	$C_{30}H_{37}O_{11}N$ (487)	*Streptomyces* unidentified sp.	og	Acid hydrolysis→aklavinone and basic sugar linked to the 2° OH group of aglycone λ_{max} MeOH: 228, 258, 288, 427 HCl salt: v s in H_2O, EtOH, pyr, diox, $CHCl_3$; s in ace; mod s in EtOAc. Color rn: E, purp-bl; C, red-purp	209
2	Aglycone: aklavinone (Galirubin B)		*Streptomyces galilaeus* Ettlinger et al.	og or red amor	Acid hydrolysis→aklavinone and two sugars λ_{max} MeOH: 230, 259, 291(sh), 431	206
Citromycinones						
1	(7S,9R,10R)-I; R' = R'' = H (α-Citromycinone)	$C_{20}H_{18}O_7$ (370)	*Streptomyces purpuracens* Lindenbein	ye nd ($CHCl_3$—peth) (135—137)	λ_{max}^{cy}: 418, 436; (4.04, 4.04) Obtained from acid hydrolysate of anthracyclines in culture	210, 211
2	I; R' = R'' = H; H instead of OH at C-7 or C-10 (λ-Citromycinone)	$C_{20}H_{18}O_6$ (354)	*Streptomyces purpuracens* Lindenbein	ye nd ($CHCl_3$—peth) (207, d)	λ_{max}^{cy}: 421, 438; (4.08, 4.09) Obtained from acid hydrolysate of anthracyclines in culture	211
Pyrromycinones						
1	(7S,9R,10R)-III; R' = H, R'' = R''' = OH (ε-Pyrromycinone, rutilantinone)	$C_{22}H_{20}O_9$ (428)	*Streptomyces* unidentified sp.	red nd (bz) (213—214)	λ_{max}^{cy}: 498, 518, 533 $[\alpha]_{Cd}^{20}$: +143 ± 7° (c = 1.0, $CHCl_3$)	206, 210, 212, 216

TABLE 5 (Continued)
ANTHRACYCLINONES AND ANTHRACYCLINES

No.	Compound (Common Name)	Formula (Mol. Wt.)	Organisms	Color and Crystalline Form (Melting Point, °C)	Significant Characteristics	Ref.
Pyrromycinones (continued)						
2	(9R,10R)-IV: R' = H, R'' = R''' = OH (ζ-Pyrromycinone, galirubinone C)	$C_{22}H_{20}O_8$ (412)	*Streptomyces galilaeus* Ettlinger *et al.* unidentified sp.	red nd (bz) (216)	λ_{max}^{cy}: 483, 494, 516, 528 $[\alpha]_{Cd}^{20}$: +74 ± 6° (c = 1.0, $CHCl_3$) Brick red, gr flr, in org solv; vt, red flr, in con H_2SO_4, piperidine, and alk	206, 210, 212, 213
3	9-Ethyl-1,4,6-trihydroxy-CNME (η-Pyrromycinone, galirubinone B_2, dianhydrorutilantinone, ciclacidine)	$C_{22}H_{16}O_7$ (392)	*Streptomyces copoamus* sp.nov. *galilaeus* Ettlinger *et al.* unidentified sp.	dk red nd (bz) (236—237)	λ_{max}^{cy}: 484, 494, 506, 518, 529 sl s in bz, diox, $CHCl_3$; f s in pyr bl, red flr, in con H_2SO_4	206, 212, 214, 215
4	1,4,6-Trihydroxy-9-methyl-CNME (η_1-Pyrromycinone)	$C_{21}H_{14}O_7$ (378)	*Streptomyces* unidentified sp.	red cr (bz—hx) (290—292)	λ_{max}^{cy}: 252, 275, 482, 492, 504, 515, 526 v sl s in org solv	277
ε-Pyrromycinone Glycosides						
1	(Pyrromycin)	$C_{30}H_{35}O_{11}N$ ·HCl (621.5)	*Streptomyces* unidentified sp.	HCl salt: red cr (ace) (162—165, d)	$\lambda_{max}^{CHCl_3}$ (HCl salt): 498, 518, 533 $[\alpha]_{Cd}^{20}$ (HCl salt): +132 ± 27° (c = 0.4, MeOH) s in H_2O, al, pyr; f s in $CHCl_3$; sl s in bz (br-red sol with intense gr flr) vt→bl in con H_2SO_4 Acid hydrolysis→ε-pyrromycinone (main product) + η-pyrromycinone + rhodosamine	216

TABLE 5 (Continued)
ANTHRACYCLINONES AND ANTHRACYCLINES

No.	Compound (Common Name)	Formula (Mol. Wt.)	Organisms	Color and Crystalline Form (Melting Point, °C)	Significant Characteristics	Ref.
ε-Pyrromycinone Glycosides (continued)						
2	Aglycone: ε-pyrromycinone Sugars: rhodosamine + 2-deoxy-L-fucose + unidentified sugar (Cinerubin A, ryemycin B$_2$)		*Streptomyces antibioticus* (Waksman et Woodruff) Waksman et Henrici *galilaeus* Ettlinger et al. *niveoruber* Ettlinger et al.	dk red pw (bz—MeOH) (155—158, resolidifies at 160—180, remelts >249)	λ_{max}^{EtOH}: 235, 259, 294, 473, 487, 497, 517, 533; (4.75, 4.42, 4.00, 4.13, 4.19, 4.22, 4.10, 4.04)	13, 217
3	Aglycone: ε-pyrromycinone Sugars: rhodosamine + 2-deoxy-L-fucose + unidentified sugar (Cinerubin B, ryemycin B$_1$)		*Streptomyces antibioticus* (Waksman et Woodruff) Waksman et Henrici *galilaeus* Ettlinger et al. *niveoruber* Ettlinger et al.	og-red rodlets (168—178, resolidifies to long needles, remelts at 240—243, melts at ~180—181)	λ_{max}^{EtOH}: 235, 258, 294, 473, 488, 497, 519, 432; (4.73, 4.40, 3.98, 4.12, 4.18, 4.21, 4.11, 4.05)	13, 217
4	Aglycone: ε-pyrromycinone (Galirubin A)		*Streptomyces galilaeus* Ettlinger et al.	og or red amor	λ_{max}^{MeOH}: 233, 258, 290, 495	206
5	Aglycone: ε-pyrromycinone (Rutilantins A, B, and C)		Unidentified Actinomycetes	dk red		278, 279

TABLE 5 (Continued)
ANTHRACYCLINONES AND ANTHRACYCLINES

Rhodomycinones

No.	Compound (Common Name)	Formula (Mol. Wt.)	Organisms	Color and Crystalline Form (Melting Point, °C)	Significant Characteristics	Ref.
1	(7S,9R,10R)-I; R′ = H, R, R″ = OH, reverse configuration at C-7 (α-Rhodomycinone)	$C_{20}H_{18}O_8$ (386)	*Actinomyces* unidentified spp. *Streptomyces purpuracens* Lindenbein (non-isorhodomycin-producing strain)	red pr (CHCl$_3$) (217–220, d)	λ_{max}: similar to β-rhodomycinone; s in pyr (ye-red); f s in CHCl$_3$, bz, ace; sl s in cy. Obtained by acid hydrolysis of rhodomycin mixture	210, 218, 223, 262
2	(7S,9R,10R)-I; R′ = OH, R″ = H (α$_2$-Rhodomycinone)	$C_{20}H_{18}O_8$ (386)	*Streptomyces purpuracens* Lindenbein	red lf (AcOH) (207–209)	λ_{max}^{cy}: 482, 494, 515, 529. Obtained by acid hydrolysis of anthracycline mixture	210, 211
3	(7S,9R,10R)-I; R′ = H, R″ = OH (β-Rhodomycinone)	$C_{20}H_{18}O_8$ (386)	*Actinomyces* unidentified spp. *Streptomyces purpuracens* Lindenbein (non-isorhodomycin-producing strain)	red nd (AcOH) (224–225, d)	λ_{max}^{cy}: 481, 492, 514, 527; s in ace, al, AcOH, bz, pyr; sl s in peth. Color rn: piperidine, bl; C, bl-vt, red flr; G, vt, red flr. Obtained by acid hydrolysis of rhodomycin mixture	210, 219, 220, 223, 262, 280
4	(9R,10R)-II; R = H, CH$_3$ instead of CH$_3$CH$_2$ at C-9 (β$_1$-Rhodomycinone)	$C_{19}H_{16}O_7$ (356)	*Streptomyces purpuracens* Lindenbein (non-isorhodomycin-producing strain)	red pr (CHCl$_3$) (~260, d)	λ_{max}^{cy}: 495, 517, 531; sl s in org solv. Obtained by acid hydrolysis of rhodomycin mixture	210, 218

TABLE 5 (Continued)
ANTHRACYCLINONES AND ANTHRACYCLINES

No.	Compound (Common Name)	Formula (Mol. Wt.)	Organisms	Color and Crystalline Form (Melting Point, °C)	Significant Characteristics	Ref.
5	(9R,10R)-II; R = H (γ-Rhodomycinone)	$C_{20}H_{18}O_7$ (370)	*Actinomyces* unidentified spp. *Streptomyces purpuracens* Lindenbein (non-isorhodomycin-producing strain)	red nd (bz), pl (AcOH–H_2O) (230–240, d)	λ_{max}^{cy}: 483, 494, 517, 529 s in pyr; f s in ace, al, bz, AcOH; sl s in cy Obtained by acid hydrolysis of rhodomycin mixture	210, 219, 220, 224, 262
6	(9R)-II; R = H, H instead of OH at C-10 (10-Deoxy-γ-rhodomycinone)	$C_{20}H_{18}O_6$ (354)	*Streptomyces purpuracens* Lindenbein	red nd (CHCl$_3$–peth) (232)	λ_{max}^{cy}: 467, 485, 496, 519, 532 sl s in cy, pyr; f s in CHCl$_3$, bz, ace	210, 218
7	(7S,9R,10R)-III; R' = R'' = OH, R''' = H (δ-Rhodomycinone)	$C_{22}H_{20}O_9$ (428)	*Streptomyces purpuracens* Lindenbein	red pr (AcOH) (195–197)	λ_{max}^{cy}: 483, 395, 517, 530	210, 221
8	(7S,9R,10R)-III; R' = R''' = OH, R'' = H (ε-Rhodomycinone)	$C_{22}H_{20}O_9$ (428)	*Streptomyces purpuracens* Lindenbein unidentified sp.	red nd (cy) (210)	λ_{max}^{cy}: 483, 493, 515, 529 f s in al ace, CHCl$_3$; sl s in peth	210, 220, 222, 280
9	(9R,10R)-IV; R' = R''' = OH, R'' = H (ζ-Rhodomycinone)	$C_{22}H_{20}O_8$ (412)	*Streptomyces purpuracens* Lindenbein unidentified sp.	red nd (275)	Der: tri-OAc (pa ye nd, MP 222, d)	74, 210, 220, 280
10	9-Ethyl-4,6,7,9,11-pentahydroxy-7,8,9,10-tetrahydro-CNME (θ-Rhodomycinone)	$C_{22}H_{20}O_9$ (428)	*Streptomyces* unidentified sp.	red-br nd (THF) (220, d)	λ_{max}^{EtOH}: 234, 258, 296, 480, 493, 513, 527; (4.60, 4.35, 3.82, 4.10, 4.13, 4.01, 3.95) $[a]_D^{25}$: +191.5° (c = 0.223, THF)	280

TABLE 5 (Continued)
ANTHRACYCLINONES AND ANTHRACYCLINES

No.	Compound (Common Name)	Formula (Mol. Wt.)	Organisms	Color and Crystalline Form (Melting Point, °C)	Significant Characteristics	Ref.
Rhodomycinone Glycosides: β-Rhodomycins						
1	(β-Rhodomycin I, rhodomycin B, S-583-A-II)	$C_{28}H_{33}O_{10}N$ (543)	*Streptomyces purpuracens* Lindenbein	HCl salt: red amor pw (185–190, d)	Acid hydrolysis→β-rhodomycinone + rhodosamine λ_{max}^{MeOH} (HCl salt): 235, 254, 294, 495, 528(sh), 562(sh) $[\alpha]_D^{23}$: +273 ± 62° (c = 0.011, MeOH) s in H_2O Amphoteric, red in acid, bl in alk	225, 226, 227

TABLE 5 (Continued)
ANTHRACYCLINONES AND ANTHRACYCLINES

Rhodomycinone Glycosides: β-Rhodomycins (continued)

No.	Compound (Common Name)	Formula (Mol. Wt.)	Organisms	Color and Crystalline Form (Melting Point, °C)	Significant Characteristics	Ref.
2	(β-Rhodomycin II, rhodomycin A, S-583-A-III)	$C_{36}H_{48}O_{12}N_2$ (700)	*Streptomyces purpuracens* Lindenbein	HCl salt: red amor pw (205)	Acid hydrolysis→β-rhodomycinone + 2 rhodosamine λ^{MeOH}_{max} (di-HCl salt): 236, 255, 295, 497, 530, 564 $[\alpha]^{23}_D$: +418 ± 74° (c = 0.011, MeOH) s in H_2O Amphoteric, red in acid, bl in alk	225, 226, 227, 228
3	Aglycone: β-rhodomycinone Sugars: (Rhodosamine) + other sugars (S-583-B)		*Streptomyces purpuracens* Lindenbein	HCl salt: red amor pw (CHCl₃ –eth) (184–190, d)	$[\alpha]^{24}_D$: +122 ± 50° (c = 0.0106, $CHCl_3$) HCl salt: s in H_2O, al, DMF, $CHCl_3$; sl s in ace; i in bz, eth, hx, EtOAc Basic, red in acid, bl-purp in alk	225

TABLE 5 (Continued)
ANTHRACYCLINONES AND ANTHRACYCLINES

No.	Compound (Common Name)	Formula (Mol. Wt.)	Organisms	Color and Crystalline Form (Melting Point, °C)	Significant Characteristics	Ref.
Rhodomycinone Glycosides: γ-Rhodomycins						
1	Aglycone: γ-rhodomycin-one Sugar: rhodosamine (γ-Rhodomycin I)		*Streptomyces purpuracens* Linden-bein	red	Basic, H_2O-s antibiotic; gives on hydrolysis with $0.1N$ HCl for 2 hours at $70-75°C$ one mole of γ-rhodomycinone and one or more sugars	229, 230
2	Aglycone: γ-rhodomycin-one Sugars: 2-rhodosamine (γ-Rhodomycin II)		*Streptomyces purpuracens* Linden-bein	red	Same as above	229, 230
3	Aglycone: γ-rhodomycin one Sugars: 2-rhodosamine + 2-deoxy-L-fucose (γ-Rhodomycin III)		*Streptomyces purpuracens* Linden-bein	red	Same as above	229, 230
4	Aglycone: γ-rhodomycinone Sugars: 2-rhodosamine + 2-deoxy-L-fucose + rhodinose (γ-Rhodomycin IV)		*Streptomyces purpuracens* Linden-bein	red	Same as above	229, 230
Isorhodomycinones						
1	(7S, 9R, 10R)-I; R' = R'' = OH (β-Isorhodomycinone)	$C_{20}H_{18}O_9$ (402)	*Streptomyces purpuracens* Linden-bein	dk red nd	$\lambda_{max}^{CHCl_3}$: 491, 514, 524, 552, 564 s in pyr (bl-red), piperidine (bl); mod s in $CHCl_3$, AcOH, ace, bz; i in cy Obtained by acid hydrolysis of anthracycline mixture	210, 223, 231

TABLE 5 (Continued)
ANTHRACYCLINONES AND ANTHRACYCLINES

No.	Compound (Common Name)	Formula (Mol. Wt.)	Organisms	Color and Crystalline Form (Melting Point, °C)	Significant Characteristics	Ref.
Isorhodomycinones (continued)						
2	(9R,10R)-II; R = OH (γ-Isorhodomycinone)	$C_{20}H_{18}O_8$ (386)	*Streptomyces purpuracens* Lindenbein	red amor	$\lambda_{max}^{CHCl_3}$: 491, 514, 525, 552, 564	210, 231
3	(7S,9R,10R)-III; R' = R'' = R''' = OH (ε-Isorhodomycinone)	$C_{22}H_{20}O_{10}$ (444)	*Streptomyces purpuracens* Lindenbein unidentifed sp.	dk red lf (MeOH−bz) (227−229, d)	λ_{max}^{diox}: 242, 298, 523, 535, 550, 560; (4.69, 3.91, 4.30, 4.19, 4.28, 3.28) s in ace, al, AcOH, bz, CHCl₃; i in peth	210, 222, 232, 233, 280
4	(9R,10R)-IV; R' = R'' = R''' = OH (ζ-Isorhodomycinone)	$C_{22}H_{20}O_9$ (428)	*Streptomyces purpuracens* Lindenbein unidentifed sp.	dk red nd (258−260)	λ_{max}^{diox}: 244, 298−300, 521, 548, 560	233, 280
Isorhodomycinone Glycoside						
1	(β-Isorhodomycin II, isorhodomycin A)	$C_{36}H_{48}O_{13}N_2$ (716)	*Streptomyces purpuracens* Lindenbein	HCl salt: dk red (EtOH−HCl) (220)	λ_{max}: similar to β-isorhodomycinone; $[\alpha]^{18}_{606-760}$ (HCl salt): +268 ± 30° (c = 0.1, MeOH)	226, 228

TABLE 5 (Continued)
ANTHRACYCLINONES AND ANTHRACYCLINES

Miscellaneous Anthracyclinones and Anthracyclines

No.	Compound (Common Name)	Formula (Mol. Wt.)	Organisms	Color and Crystalline Form (Melting Point, °C)	Significant Characteristics	Ref.
1	(7S,9S,1'R,3'S,4'S,5'S-Daunomycin, rubidomycin, rubomycin C)	$C_{27}H_{29}O_{10}N$ (527)	*Streptomyces coeruleorubidus* Gause et al. *peuceticus* Grein et al.	HCl salt: red nd (188–190, d)	λ_{max}^{MeOH}, $E_{1\ cm}^{1\%}$ (HCl salt): 234, 252, 290, 480, 495, 532; (665, 462, 153, 214, 218, 112) HCl salt: s in H_2O, al; i in eth, bz, $CHCl_3$ Color rn: acid, pk; alk, bl Acid hydrolysis→daunomycinone + daunosamine; daunomycinone: $C_{21}H_{18}O_8$, Mol. Wt. 398, MP 213–214, $[a]_D = +193°$ (c = 0.1, diox); daunosamine; $C_6H_{13}O_3N$, MP (HCl) 168, $[a] = -54.5°$ (H_2O)	234, 235, 239
2	(7S,9S,1'R-Adriamycin)	$C_{27}H_{29}O_{11}N$ (435)	*Streptomyces peuceticus* Grein et al. (a mutant)	HCl salt: og-red nd (204–205)	HCl salt: s in H_2O, MeOH, aq al; i in ace, eth, bz, $CHCl_3$, peth Acid hydrolysis→adriamycinone + daunosamine; adriamycinone: $C_{21}H_{18}O_9$, Mol. Wt. 414, MP 223–224, $[a]_D^{23} = +188°$ (c = 0.1, diox); daunosamine: see above	236

TABLE 5 (Continued)
ANTHRACYCLINONES AND ANTHRACYCLINES

Miscellaneous Anthracyclinones and Anthracyclines (continued)

No.	Compound (Common Name)	Formula (Mol. Wt.)	Organisms	Color and Crystalline Form (Melting Point, °C)	Significant Characteristics	Ref.
3	(Ruticulomycin A and B)		*Streptomyces rubrireticuli* (Waksman) Waksman *et* Henrici		Ruticulomycin A differs from B in the sugar portion	237
	(Ruticulomycin A)			dk red pl (ace—hx) (183—184, d)	λ_{max}^{MeOH}, $E_{1\ cm}^{1\%}$: 235, 258, 295, 475; (650, 300, 120, 170)	
	(Ruticulomycin B)			red rosettes (ace—eth) (179—180, d)	λ_{max}^{MeOH}, $E_{1\ cm}^{1\%}$: 235, 258, 295, 475; (690, 300, 120, 190)	
4	(Ryemycin A_1 and A_2)		*Streptomyces ryensis*			13
	(Ryemycin A_1)	$C_{22}H_{20}O_8$ (412)		red (224—226, d)	λ_{max}^{MeOH}, $E_{1\ cm}^{1\%}$: 233, 259, 490, 512, 532; (736.6, 532.0, 235.3, 197.8, 163.7 $[\alpha]_D^{24}$: +121.5 ± 10° Weakly acidic; og in acid, vt in alk	

TABLE 5 (Continued)
ANTHRACYCLINONES AND ANTHRACYCLINES

Miscellaneous Anthracyclinones and Anthracyclines (continued)

No.	Compound (Common Name)	Formula (Mol. Wt.)	Organisms	Color and Crystalline Form (Melting Point, °C)	Significant Characteristics	Ref.
4	(Ryemycin A$_1$ and A$_2$) (cont.)					
	(Ryemycin A$_2$)	C$_{22}$H$_{20}$O$_7$ (396)		ye (234–235, d)	λ_{max}^{MeOH}, E$_{1\ cm}^{1\%}$: 228, 259, 289, 432; (887.9, 698.7, 247.5, 29.9) $[\alpha]_D^{24}$: +65.9 ± 10° Weakly acidic	
5	(Isoquinocycline A, PA-371 γ)	C$_{33}$H$_{33}$O$_{10}$N$_2$ ·HX	*Streptomyces aureofaciens* Duggar			

TABLE 5 (Continued)
ANTHRACYCLINONES AND ANTHRACYCLINES

No.	Compound (Common Name)	Formula (Mol. Wt.)	Organisms	Color and Crystalline Form (Melting Point, °C)	Significant Characteristics	Ref.
Miscellaneous Anthracyclinones and Anthracyclines (continued)						
5	(Isoquinocycline A, PA-371 γ) (cont.)					
	(Isoquinocycline A, PA-371γ)			ye	λ_{max} 0.1N HCl (HCl salt): 231, 260, 292, 425–35; (4.74, 4.28, 4.03, 4.11, 4.12) $[a]_D^{25}$ (HCl salt): +27.2° (c = 1.0, AcOH) Acid hydrolysis→isoquinocycline + anhydro-sugar; isoquinocyclinone HCl: $C_{25}H_{22}O_6N_2 \cdot$HCl, λ_{max} 0.1N HCl = 230, 259, 291, 425, 435 (4.71, 4.53, 4.99, 4.10, 4.11), $[a]_D^{25}$ = +27.2° (c = 1.0, AcOH); anhydro-sugar: $C_8H_{14}O_4$, Mol. Wt. 174, MP 153–154, $[a]_D^{25}$ = –144° (c = 2.0, H_2O)	249, 250
	(Isoquinocycline B, PA-371γ)			ye	$[a]_{Hg}^{25}$: +24° (MeOH) Basic	13, 248

TABLE 5 (Continued)
ANTHRACYCLINONES AND ANTHRACYCLINES

No.	Compound (Common Name)	Formula (Mol. Wt.)	Organisms	Color and Crystalline Form (Melting Point, °C)	Significant Characteristics	Ref.
Miscellaneous Anthracyclinones and Anthracyclines (continued)						
6	Aglycone: nogalarol; Sugar: nogalose (Nogalamycin)	$C_{39}H_{49}O_{17}N$ (803)	*Streptomyces nogalater*, var. *nogalater* Bhuyan *et* Dietz	og-red (195—196, d)	λ_{max}^{EtOH}: 236, 258, 292, 390, 480; (4.72, 4.39, 3.99, 3.61, 4.19) $[\alpha]_D^{25}$: +479° (CHCl$_3$) Acid hydrolysis→nogalarol + nogalose + nogalarene; nogalarol: $C_{29}H_{31}O_{13}N$, Mol. Wt. 501, MP >200, d, λ_{max}^{MeOH} = 234, 258, 288, 475 (4.73, 4.39, 3.99, 4.19); nogalose: $C_{10}H_{20}O_5$, Mol. Wt. 220, MP 115—121, $[\alpha]_D^{25}$ = +15.5° (H$_2$O), $[\alpha]_D^{25}$ = −10.6° (MeOH)	282
7	Tetramid quinones	$C_{19}H_{13}O_8N$ (377)	*Streptomyces aureofaciens* Duggar	og	From mutants of tetracline-producing strains	10

(Protetrone)

TABLE 5 (Continued)
ANTHRACYCLINONES AND ANTHRACYCLINES

No.	Compound (Common Name)	Formula (Mol. Wt.)	Organisms	Color and Crystalline Form (Melting Point, °C)	Significant Characteristics	Ref.
Miscellaneous Anthracyclinones and Anthracyclines (continued)						
7	Tetramid quinones (*cont.*)					
	("Tetramid-green")	$C_{20}H_{13}O_7N$ (379)		red		
	("Tetramid-blue")	$C_{21}H_{16}O_7N_2$ (408)			bl in alk	

TABLE 6
PERYLENEQUINONES AND PENTAPHENQUINONE

No.	Compound (Common Name)	Formula (Mol. Wt.)	Organisms	Color and Crystalline Form (Melting Point, °C)	Significant Characteristics	Ref.
1	(3,10-Dihydroxyperylene-4,9-quinone)	$C_{20}H_{10}O_4$ (314)	*Daldinia concentrica* (Bolt) Ces. *et* de Not.	dk red nd (sub) (>350, d)	$\lambda_{max}^{C_2H_2Cl_4}$: 265, 340, 419, 444, 493, 526, 467; (4.33, 3.73, 4.32, 4.50, 3.77 4.00, 4.17) sl s in all solv	135
2	(Elsinochromes)		*Elsinoe annonae* Bitanc. *et* Jenkins *australis* Bitanc. *et* Jenkins *fawcettii* Bitanc. *et* Jenkins *jasmini* Bitanc. *et* Jenkins *mattiroliana* Arn. *et* Bitanc. *phaseoli* Jenkins *Sphaceloma* sp. *lagoa-santense* Bitanc. *et* Jenkins *perseae* Jenkins *populi* (Sacc.) Jenkins *randii* Jenkins *et* Bitanc.		s in common org solv	136, 137, 138

TABLE 6 (Continued)
PERYLENEQUINONES AND PENTAPHENQUINONE

No.	Compound (Common Name)	Formula (Mol. Wt.)	Organisms	Color and Crystalline Form (Melting Point, °C)	Significant Characteristics	Ref.
2	(Elsinochromes) *(cont.)*					
	$R^1 = R^2 = Ac$ (Elsinochrome A, phycaron A)	$C_{30}H_{24}O_{11}$ (560)		dk red nd (CHCl$_3$ – EtOH) (248)	EtOH λ_{max}: 219, 262, 280, 330, 430, 455, 525, 563; (4.76, 4.59, 4.55, 3.74, 4.40, 4.44, 4.11, 4.21) Color rn: G, dk purp; C, purp-red	
	$R^1 = Ac, R^2 = CH(OH)CH_3$ (Elsinochrome B, phycaron B)	$C_{30}H_{26}O_{10}$ (546)		red-br cr (bz–hx) (208)	EtOH λ_{max}: 262, 339, 455, 525, 563; (4.54, 3.70, 4.33, 4.12, 4.17)	
	$R^1 = R^2 = CH(OH)CH_3$ (Elsinochrome C, phycaron C)	$C_{30}H_{28}O_{10}$ (548)		red-br cr pw (bz) (293)	EtOH λ_{max}: 272, 339, 355, 455, 523, 563; (4.49, 3.67, 3.62, 4.26, 4.05, 4.13)	
	(Elsinochrome D)	$C_{30}H_{24}O_{10}$ (544)	*Elsinoe annonae* Bitanc. *et* Jenkins	og (CHCl$_3$–hx) (159–161)	EtOH λ_{max}: 224, 253, 267, 348, 463, 526, 560; (4.70, 4.50, 4.46, 3.66, 4.28, 4.05, 3.15)	138

TABLE 6 (Continued)
PERYLENEQUINONES AND PENTAPHENQUINONE

No.	Compound (Common Name)	Formula (Mol. Wt.)	Organisms	Color and Crystalline Form (Melting Point, °C)	Significant Characteristics	Ref.
3	(Cladochrome A)	$C_{38}H_{42}O_{14}$ (722)	*Cladosporium cucumerinum* Ell. *et* Arth.	red cr (eth–peth) (197–199)	EtOH: λ_{max} 270, 335–340, 478, 540 (sh), 585 dk gr in alk	263
4	(Cercosporin)	$C_{29}H_{26}O_{10}$ (434)	*Cercosphora kikuchii* (Matsumoto *et* Tomoyasu) Gardner	red pr (CHCl$_3$–bz) (241)	λ_{max}^{MeOH}: 223, 260, 271, 275, 325, 472, 564 $[\alpha]_{700}^{20}$: +470° (c = 0.5, CHCl$_3$) v s in pyr, diox, CHCl$_3$, al, ace, NaOH (gr); s in eth, bz Color rn: B, red; E, gr; C, purp-bl	10, 139
5	(Phenocyclinone)	$C_{35}H_{24}O_{14}$ (668)	*Streptomyces coelicolor* (Muller) Waksman, Kutzner *et* Waksman (strain PR26)	red amor	$\lambda_{max}^{CHCl_3}$: 292, 500; (4.58, 4.28) s in ace, CHCl$_3$, alk (bl) Color rn: C, vt; G, vt	244

REFERENCES

1. Sexton, W. A., in *Chemical Constitution and Biological Activity*, p. 238. E. & F. N. Spon Ltd., London, England (1963).
2. Waksman, S. A., and Lechevalier, H. A., *The Actinomycetes*, Vol. 3, Antibiotics of Actinomycetes. The Williams and Wilkins Co., Baltimore (1962).
3. Newton, B. A., *Advan. Pharmacol. Chemother., 8*, 149 (1970).
4. Morton, R. A., Ed., *Biochemistry of Quinones*. Academic Press, New York (1965).
5. International Symposium on Recent Advances in Research on Vitamins K and Related Quinones (Vitamins K, Ubiquinones or Coenzymes Q, Plastoquinones) in Honor of Professor Henrik Dam, *Vitamins Hormones, 24*, 293 (1966).
6. Ramasarma, T., Lipid Quinones. *Advan. Lipid Res., 6*, 107 (1968).
7. Morton, R. A., *Biol. Rev., 46*, 47 (1971).
8. Morimoto, H., Imada, I., Watanabe, M., and Nakao, Y., *Liebigs Ann. Chem., 729*, 158 (1969).
9. Imamoto, S., and Senoh, S., *Tetrahedron Lett.,* No. 13, 1237 (1967).
10. Thomson, R. H., *Naturally Occurring Quinones*. Academic Press, New York (1971).
11. Shibata, S., Natori, S., and Udagawa, S., *List of Fungal Products*. C. C Thomas, Springfield, Illinois (1964).
12. Miller, M. W., *Pfizer's Handbook of Microbial Metabolites*. McGraw-Hill Book Co., Inc., New York, Toronto, and London (1961).
13. Umezawa, H., *Index of Antibiotics from Actinomycetes*. University of Tokyo Press, Tokyo, and University Park Press, State College, Pennsylvania (1967).
14. Culberson, C. F., *Chemical and Botanical Guide to Lichen Products*. The University of North Carolina Press, Chapel Hill, North Carolina (1969).
15. Feigl, F., *Spot Tests in Organic Analysis*, 6th ed. p. 123, translated by R. E. Oesper. Elsevier Publishing Co., Amsterdam, London, New York, and Princeton (1960).
16. Karius, H., and Mapstone, G. E., *Chem. Ind. (London),* p. 266 (1956).
17. Craven, R., *J. Chem. Soc.,* p. 1605 (1931).
18. Morton, R. A. (Ed.), in *Biochemistry of Quinones*, p. 23. Academic Press, New York (1965).
19. Josien, M. L., Fuson, N., Lebas, J. M., and Gregory, T. M., *J. Chem. Phys., 21*, 331 (1953).
20. Budzikiewicz, H., Djerassi, C., and Williams, D. H., *Mass Spectrometry of Organic Compounds*, p. 527. Holden-Day, Inc., San Francisco, Cambridge, London, and Amsterdam (1967).
21. Shibata, S., Takito, M., and Tanaka, O., *J. Amer. Chem. Soc., 72*, 2789 (1950).
22. Braude, E. A., *J. Chem. Soc.,* p. 409 (1945).
23. Flaig, W., Salfeld, J. C., and Baume, E., *Liebigs Ann. Chem., 618*, 117 (1958).
24. Pettersson, G., *Acta Chem. Scand., 18*, 2303 (1964).
25. Natori, S., Nishikawa, H., and Ogawa, H., *Chem. Pharm. Bull. (Tokyo), 12*, 236 (1964).
26. Teuber, H. J., and Staiger, G., *Chem. Ber., 88*, 802 (1955).
27. Yates, P., Ardao, M. I., and Fieser, L. F., *J. Amer. Chem. Soc., 78*, 650 (1956).
28. Flaig, W., and Salfeld, J. C., *Liebigs Ann. Chem., 626*, 215 (1959).
29. Bycroft, B. W., and Roberts, J. C., *J. Org. Chem., 28*, 1429 (1963).
30. Bowie, J. H., Cameron, D. W., Giles, R. G. F., and Williams, D. H., *J. Chem. Soc. Sect. B*, p. 335 (1966).
31. Aplin, R. T., and Pike, W. T., *Chem. Ind. (London),* p. 2009 (1966).
32. Endtman, H., Granath, M., and Schultz, G., *Acta Chem. Scand., 8*, 1442 (1954).
33. Musso, H., and Bormann, D., *Chem. Ber., 98*, 2774 (1965).
34. Flaig, W., Ploetz, T., and Kullmer, A., *Z. Naturforsch., 10B*, 668 (1955).
35. Gripenberg, J., *Acta Chem. Scand., 12*, 1762 (1958).
36. Novelli, A., *Science, 93*, 358 (1941).
37. Dimroth, O., and Faust, T., *Chem. Ber., 54B*, 3020 (1921).
38. Macbeth, A. K., Price, J. R., and Winzor, F. L., *J. Chem. Soc.,* p. 325 (1935).
39. Arnstein, H. R. V., and Cook, A. H., *J. Chem. Soc.,* p. 1021 (1947).
40. Weygand, F., and Csendes, E., *Chem. Ber., 85*, 45 (1952).
41. Spruit, C. J. P., *Rec. Trav. Chim. Pays-Bas, 68*, 309 (1949).
42. Singh, I., Ogata, R. T., Moore, R. E., Chang, C. W. J., and Scheuer, P. J., *Tetrahedron, 24*, 6053 (1968).
43. Hill, R. R., and Mitchell, G. H., *J. Chem. Soc. Sect. B*, p. 61 (1969).
44. Birch, A. J., and Donovan, *Aust. J. Chem., 6*, 373 (1953).
45. Weiss, S., and Nord, F. F., *Arch. Biochem., 22*, 288 (1949).
46. Bowie, J. H., Cameron, D. W., and Williams, D. H., *J. Amer. Chem. Soc., 87*, 5094 (1965).
47. Becher, D., Djerassi, C., Moore, R. E., Singh, H., and Scheuer, P. J., *J. Org. Chem., 31*, 3650 (1966).
48. Elwood, T. A., Dudley, K. H., Tesarek, J. M., Rogerson, P. F., and Bursey, M. M., *Org. Mass Spectra, 3*, 841 (1970).
49. Bowie, J. H., White, P. Y., and Hoffmann, P. J., *Tetrahedron, 25*, 1629 (1969).
50. Bowie, J. H., Hoffmann, P. J., and White, P. Y., *Tetrahedron, 26*, 1163 (1970).
51. Oliver, R. W. A., and Rashman, R. M., *J. Chem. Soc. Sect. B*, p. 1141 (1968).

52. Moore, R. E., and Scheuer, P. J., *J. Org. Chem., 31,* 3272 (1966).
53. Brockmann, H., and Zeeck, A., *Chem. Ber., 101,* 4211 (1968).
54. Piette, L. H., Okamura, M., Rabold, G. P., Ogata, R. T., Moore, R. E., and Scheuer, P. J., *J. Phys. Chem., 71,* 29 (1967).
55. Calderbank, A., Johnson, A. W., and Todd, A. R., *J. Chem. Soc.,* p. 1285 (1954).
56. Johnson, A. W., Quale, J. R., Robinson, T. S., Sheppard, N., and Todd, A. R., *J. Chem. Soc.,* p. 2633 (1951).
57. Brown, B. R., and Todd, A. R., *J. Chem. Soc.,* p. 1280 (1954).
58. Bowie, J. H., and Cameron, D. W., *J. Chem. Soc. Sect. B,* p. 684 (1966).
59. Feigl, F., *Spot Tests in Organic Analysis,* 6th ed, p. 210, translated by R. E. Oesper. Elsevier Publishing Co., Amsterdam, London, New York, and Princeton (1966).
60. Hartmann, H., and Lorenz, E., *Z. Naturforsch., 7A,* 360 (1952).
61. Spruit, C. J. P., *Rec. Trav. Chim. Pays-Bas, 68,* 325 (1949).
62. Briggs, L. H., Nicholls, G. A., and Paterson, R. M. L., *J. Chem. Soc.,* p. 1718 (1952).
63. Peters, R. H., and Summer, H. H., *J. Chem. Soc.,* p. 2101 (1953).
64. Birkinshaw, *Biochem. J., 59,* 485 (1955).
65. Yoshimoto, T., and Minami, K., *Nippon Mokuzai Gakkaishi, 9,* 175 (1963).
66. Yoshimoto, T., *Nippon Kagaku Zasshi, 84,* 733 (1963).
67. Brockmann, H., and Budde, G., *Chem. Ber., 86,* 432 (1953).
68. Flett, M. St. C., *J. Chem. Soc.,* p. 1441 (1948).
69. Bloom, H., Briggs, L. H., and Cleverley, B., *J. Chem. Soc.,* p. 178 (1959).
70. Beynon, J. H., and Williams, A. E., *Appl. Spectrosc., 14,* 156 (1960).
71. Brockmann, H., and Müller, W., *Chem. Ber., 92,* 1164 (1959).
72. Brockmann, H. Jr., Budzikiewicz, H., Djerassi, C., Brockmann, H., and Niemeyer, J., *Chem. Ber., 98,* 1260 (1965).
73. Reed, R. I., and Reid, W. K., *Tetrahedron, 19,* 1817 (1963).
74. Brockmann, H., *Fortschr. Chem. Org. Naturst., 21,* 121 (1963).
75. Brockmann, H. Jr., and Legrand, M., *Tetrahedron, 19,* 395 (1963).
76. Anchel, M., Hervey, A., Kavanagh, F., Polatnick, J., and Robbins, W. J., *Proc. Nat. Acad. Sci. U.S.A., 34,* 498 (1948).
77. Pettersson, G., *Acta Chem. Scand., 20,* 45 (1966).
78. Natori, S., Inouye, Y., and Nishikawa, H., *Chem. Pharm. Bull. (Tokyo), 15,* 380 (1967).
79. Yamamoto, Y., Nitta, K., Tango, K., Saito, T., and Tsuchimuro, M., *Chem. Pharm. Bull. (Tokyo), 13,* 935 (1965).
80. Anslow, W. K., and Raistrick, H., *Biochem. J., 32,* 687 (1938).
81. Pettersson, G., *Acta Chem. Scand., 17,* 1771 (1963).
82. Sheehan, J. C., Lawson, W. B., and Gaul, R. J., *J. Amer. Chem. Soc., 80,* 5536 (1958).
83. Birkinshaw, J. H., and Raistrick, H., *Phil. Trans. Roy. Soc. London Ser. B, 220,* 254 (1931).
84. Yamamoto, Y., Shinya, M., and Oohata, Y., *Chem. Pharm. Bull. (Tokyo), 18,* 561 (1970).
85. Natori, S., Nishikawa, H., and Ogawa, H., *Chem. Pharm. Bull. (Tokyo), 12,* 236 (1964).
86. Engel, B. G., and Brzeski, W., *Helv. Chim. Acta, 30,* 1472 (1947).
87. Sakamura, S., Ito, J., and Sakai, R., *Agr. Biol. Chem., 34,* 153 (1970).
88. Kawashima, K., Nakanishi, K., Tada, M., and Nishikawa, H., *Tetrahedron Lett.,* No. 20, 1227 (1964).
89. Brian, P. W., Curtis, P. J., Howland, S. R., Jefferys, E. G., and Raudnitz, H., *Experientia, 7,* 266 (1951).
90. Vischer, E. B., *J. Chem. Soc.,* p. 815 (1953).
91. Grove, J. F., *J. Chem. Soc.,* p. 985 (1966).
92. Wat, C. K., Tse, A., Bandoni, R. J., and Towers, G. H. N., *Phytochemistry, 7,* 2177 (1968).
93. Bu'Lock, J. D., *J. Chem. Soc.,* p. 575 (1955).
94. Kis, Z., Closse, A., Sigg, H. P., Hruban, L., and Snatzke, G., *Helv. Chim. Acta, 53,* 1577 (1970).
95. Beaumont, P. C., and Edwards, R. L., *J. Chem. Soc. Sect. C,* p. 2398 (1969).
96. Posternak, T., *Helv. Chim. Acta, 21,* 1326 (1938).
97. Cooke, J. C., and Collins, R. P., *Lloydia, 33,* 269 (1970).
98. Shigematsu, N., *J. Inst. Polytech. Osaka City Univ. Ser. C 5,* 100 (1956).
99. Divekar, P. V., Read, G., Vining, L. C., and Haskins, R. H., *Can. J. Chem., 37,* 1970 (1959).
100. Jirawongse, F., Ramstad, E., Wolinsky, J., *J. Pharm. Sci., 51,* 1108 (1962).
101. Kogl, F., and Becker, H., *Ann. Chem., 465,* 211, 243 (1928).
102. Akagi, M., *Pharm. Soc. Jap. J., 62,* 129 (1942).
103. Sullivan, G., Brady, L. R., and Tyler, V. E. Jr., *Lloydia, 30,* 84 (1967).
104. Kogl, F., and Erxleben, H., *Ann. Chem., 479,* 11 (1930).
105. McMorris, T. C., and Anchel, M., *Tetrahedron Lett.,* No. 5, 335 (1963).
106. Minami, K., Asawa, K., and Sawada, M., *Tetrahedron Lett.,* No. 49, 5067 (1968).
107. Astill, B. D., and Roberts, J. C., *J. Chem. Soc.,* p. 2063 (1962).
108. Gerber, N. N., and Wieclawek, B., *J. Org. Chem., 31,* 1496 (1966).
109. Bendz, G., *Acta Chem. Scand., 5,* 489 (1951).
110. Overeem, J. C., and van der Kerk, J. M., *Rec. Trav. Chim. Pays-Bas, 83,* 995 (1964).

111.   Bentley, R., and Gatenbeck, S., *Biochemistry, 4,* 1150 (1965).
112.   Arnstein, H. R. V., and Cook, A. H., *J. Chem. Soc.,* p. 1021 (1947).
113.   Hardegger, E., Steiner, K., Widmer, E., and Pfiffner, A., *Helv. Chim. Acta, 47,* 2027 (1964).
114.   Ruelius, H. W., and Gauhe, A., *Ann. Chem., 569,* 38 (1950).
115.   Ruelius, H. W., and Gauhe, A., *Ann. Chem., 570,* 121 (1951).
116.   Arsenault, G. P., *Tetrahedron, 24,* 4745 (1968).
117.   Kern, H., and Naef-Roth, S., *Phytopathol. Z., 53,* 45 (1965).
118.   Ellestad, G. A., Whaley, H. A., and Patterson, E. L., *J. Amer. Chem. Soc., 88,* 4109 (1966).
119.   Bergy, M. E., *J. Antibiot. (Tokyo) Ser. A, 21,* 454 (1968).
120.   Morishita, E., Takeda, T., and Shibata, S., *Chem. Pharm. Bull. (Tokyo), 16,* 411 (1968).
121.   Yamazaki, H., *J. Antibiot. (Tokyo) Ser. A, 21,* 204 (1968).
122.   Rinehart, K. L. Jr., Mathur, H. H., Sasaki, K., Martin, P. K., and Coverdale, C. E., *J. Amer. Chem. Soc., 90,* 6241 (1968).
123.   Eckardt, K., in *International Congress on Antibiotics, Prague 1964, Antibiotic Advances in Research, Production and Clinical Use,* p. 414, M. Herold, Ed. Butterworth, London, England (1966).
124.   Eckardt, K., *Chem. Ber., 98,* 24 (1965).
125.   Brockmann, H., and Zeeck, A., *Chem. Ber., 103,* 1709 (1970).
126.   Read, G., Rashid, A., and Vining, L. C., *J. Chem. Soc. Sect. C,* p. 2059 (1969).
127.   Brockmann, H., and Loeschcke, V., *Chem. Ber., 88,* 778 (1955).
128.   Brockmann, H., Pini, H., and von Plotho, O., *Chem. Ber., 83,* 161 (1950).
129.   Zeeck, A., and Christiansen, P., *Liebigs Ann. Chem., 724,* 172 (1969).
130.   Wirth, J. C., Beesley, T. E., Anand, S. R., *Phytochemistry, 4,* 505 (1965).
131.   Just, G., and Day, W. C., *Can. J. Chem., 41,* 74 (1963).
132.   King, T. J., Roberts, J. C., and Thompson, D. J., *Chem. Commun.,* p. 1499 (1970).
133.   Ng, A. S., Just, G., and Blank, F., *Can. J. Chem., 47,* 1223 (1969).
134.   Blackburn, G. M., Neilson, A. H., and Todd, L., *Proc. Chem. Soc.,* p. 327 (1962).
135.   Anderson, J. M., and Murray, J., *Chem. Ind. (London),* p. 376 (1956).
136.   Weiss, U., Ziffer, H., Batterham, T. J., Blumer, M., Hackeng, W. H. L., Copier, H., and Salemink, C. A., *Can. J. Microbiol., 11,* 57 (1965).
137.   Lousberg, R. J. J. Ch., Salemink, C. A., Weiss, U., and Batterham, T. J., *J. Chem. Soc. Sect. C,* p. 1219 (1969).
138.   Lousberg, R. J. J. Ch., Salemink, C. A., and Weiss, U., *J. Chem. Soc. Sect. C,.* p. 2159 (1970).
139.   Kuyama, S., *J. Org. Chem., 27,* 939 (1962).
140.   Shibata, S., and Takido, M., *Chem. Pharm. Bull. (Tokyo), 3,* 156 (1955).
141.   Arkley, V., Dean, F. M., Jones, P., Robertson, A., and Tetaz, J., *Croatica Chem. Acta, 29,* 141 (1957).
142.   Howard, B. H., and Raistrick, H., *Biochem. J., 46,* 49 (1950).
143.   Stoessl, A., *Can. J. Chem., 47,* 767 (1969).
144.   Howard, B. H., and Raistrick, H., *Biochem. J., 44,* 227 (1949).
145.   Agosti, G., Brikinshaw, J. H., and Chaplen, P., *Biochem. J., 85,* 528 (1962).
146.   Kögl, K., and Postowsky, J. J., *Ann. Chem., 444,* 1 (1925).
147.   Shibata, S., and Udagawa, S. I., *Chem. Pharm. Bull. (Tokyo), 11,* 402 (1963).
148.   Steglich, W., Lösel, W., and Austel, V., *Chem. Ber., 102,* 4104 (1969).
149.   Charles, J. H. V., Raistrick, H., Robinson, R., and Todd, A. R., *Biochem. J., 27,* 499 (1933).
150.   Suemitsu, R., Nakajima, M., and Hiura, M., *Agr. Biol. Chem., 25,* 100 (1961).
151.   Ashley, J. N., Raistrick, H., and Richards, T., *Biochem. J., 33,* 1291 (1939).
152.   Platel, A., Ueno, Y., and Fromageot, P., *Bull. Soc. Chim. France, 50,* 678 (1968).
153.   Anslow, W. K., Breen, J., and Raistrick, H., *Biochem. J., 34,* 159 (1940).
154.   Posternak, T., *Helv. Chim. Acta, 23,* 1046 (1940).
155.   Kögl, F., and Oeijs, W. B., *Ann. Chem., 515,* 23 (1935).
156.   Gatenbeck, S., *Acta Chem. Scand., 13,* 386 (1959).
157.   Franck, B., and Reschke, T., *Chem. Ber., 93,* 347 (1960).
158.   Raistrick, H., and Ziffer, J., *Biochem. J., 49,* 563 (1951).
159.   Birch, A. J., and Massy-Westropp, R. A., *J. Chem. Soc.,* p. 2215 (1957).
160.   Anslow, W. K., and Raistrick, H., *Biochem. J., 34,* 1124 (1940).
161.   Shibata, S., and Natori, S., *Chem. Pharm. Bull. (Tokyo), 1,* 160 (1953).
162.   Raistrick, H., Robinson, R., and Todd, A. R., *Biochem. J., 27,* 1170 (1933).
163.   Wright, D. E., and Schofield, K., *Nature (London), 188,* 233 (1960).
164.   Gatenbeck, S., *Acta Chem. Scand., 13,* 705 (1959).
165.   Chandrasenan, K., Neelakantan, S., and Seshadri, T. R., *Proc. Indian Acad. Sci., 51A,* 296 (1960).
166.   Neelakantan, S., Pocker, A., and Raistrick, H., *Biochem. J., 64,* 464 (1956).
167.   Nakamura, Y., Shimomura, T., and Ono, J., *Nippon Nogei Kagaku Kaishi, 31,* 669 (1957).
168.   Bohlmann, F., Leuders, W., and Plettner, W., *Arch. Pharm., 294,* 521 (1961).
169.   Birkinshaw, J. H., and Gourlay, R., *Biochem. J., 81,* 618 (1961).

170. Pusey, D. F. G., and Roberts, J. C., *J. Chem. Soc.*, p. 3542 (1963).
171. Hind, H. G., *Biochem. J., 34,* 37, 577 (1940).
172. Polonsky, J., Johnson, C. B., Cohen, P., and Lederer, E., *Bull. Soc. Chim. France,* p. 1909 (1963).
173. Lounasmaa, M., and Zylber, J., *Bull. Soc. Chim. France,* p. 3100 (1969).
174. Kuntsmann, M. P., and Mitscher, L. A., *J. Org. Chem., 31,* 2920 (1966).
175. Liu, W. C., Parker, W. L., Slusarchyk, D. S., Greenwood, G. L., Graham, S. F., and Meyers, E., *J. Antibiot. (Tokyo) Ser. A, 23,* 437 (1970).
176. Sezaki, M., Kondo, S., Maeda, K., Umezawa, H., and Ohno, M., *Tetrahedron, 26,* 5171 (1970).
177. Keller-Schierlein, W., Sauerbier, J., Vogler, U., and Zähner, H., *Helv. Chim. Acta, 53,* 779 (1970).
178. Shibata, S., Ikekawa, T., and Kishi, T., *Chem. Pharm. Bull. (Tokyo), 8,* 889 (1960).
179. Stoessl, A., *Can. J. Chem., 47,* 777 (1969).
180. Noda, T., Take, T., Watanabe, T., and Abe, J., *Tetrahedron, 26,* 1339 (1970).
181. Hamasaki, T., Hatsuda, Y., Terashima, N., and Renbutsu, M., *Agr. Biol. Chem. (Tokyo), 31,* 11 (1967).
182. Bullock, E., Kirkaldy, D., Roberts, J. C., and Underwood, J. G., *J. Chem. Soc.,* p. 829 (1963).
183. Bear, C. A., Waters, J. M., and Waters, T. N., *Chem. Commun.,* p. 1705 (1970).
184. Birch, A. J., Butler, D. N., and Richards, R. W., *Tetrahedron Lett.,* No. 28, 1853 (1964).
185. Arsenault, G. P., *Tetrahedron Lett.,* No. 45, 4033 (1965).
186. Ogihara, Y., Kobayashi, N., and Shibata, S., *Tetrahedron Lett.,* No. 15, 1881 (1968).
187. Howard, B. H., and Raistrick, H., *Biochem. J., 57,* 212 (1954).
188. Howard, B. H., and Raistrick, H., *Biochem. J., 56,* 56 (1954).
189. Tanaka, O., and Kaneko, C., *Chem. Pharm. Bull. (Tokyo), 3,* 284 (1955).
190. Fujise, S., Hishida, S., Shibata, M., and Matsueda, S., *Chem. Ind. (London),* p. 1754 (1961).
191. Shibata, S., Takido, M., and Nakajima, T., *Chem. Pharm. Bull. (Tokyo), 3,* 286 (1955).
192. Shibata, S., Takido, M., Ohta, A., and Kurosu, T., *Chem. Pharm. Bull. (Tokyo), 5,* 573 (1957).
193. Sankawa, U., Seo, S., Kobayashi, N., Ogihara, Y., and Shibata, S., *Tetrahedron Lett.,* No. 53, 5557 (1968).
194. Briggs, L. H., and LeQuesne, P. W., *J. Chem. Soc.,* p. 2290 (1965).
195. Shibata, S., and Kitagawa, I., *Chem. Pharm. Bull. (Tokyo), 4,* 309 (1956).
196. Shibata, S., and Kitagawa, I., *Chem. Pharm. Bull. (Tokyo), 8,* 884 (1960).
197. Shibata, S., Murakami, T., Tanaka, O., Chihara, G., and Sumimoto, M., *Chem. Pharm. Bull. (Tokyo), 3,* 274 (1955).
198. Breen, J., Dacre, J. C., Raistrick, H., and Smith, G., *Biochem. J., 60,* 618 (1955).
199. Tsuji, N., Nagashima, K., Kimura, T., and Kyotani, H., *Tetrahedron, 25,* 2999 (1969).
200. Shoji, J., Kimura, Y., and Katagiri, K., *J. Antibiot. (Tokyo) Ser. A, 17,* 156 (1964).
201. Tsuji, N., *Tetrahedron, 24,* 1765 (1968).
202. Tsuji, N., and Nagashima, K., *Tetrahedron, 24,* 4233 (1968).
203. Tsuji, N., and Nagashima, K., *Tetrahedron, 25,* 3007 (1969).
204. Tsuji, N., and Nagashima, K., *Tetrahedron, 25,* 3017 (1969).
205. Tsuji, N., and Nagashima, K., *Tetrahedron, 26,* 5201 (1970).
206. Eckardt, K., *Chem. Ber., 100,* 2561 (1967).
207. Brockmann, H., and Niemeyer, J., *Chem. Ber., 101,* 2409 (1968).
208. Gordon, J. J., Jackman, L. M., Ollis, W. D., and Sutherland, I. O., *Tetrahedron Lett.,* No. 8, 28 (1960).
209. Strelitz, F., Flon, H., Weiss, U., and Asheshov, I. N., *J. Bacteriol., 72,* 90 (1956).
210. Brockmann, H., Brockmann, H., Jr., and Niemeyer, J., *Tetrahedron Lett.,* No. 45, 4719 (1968).
211. Brockmann, H., and Niemeyer, J., *Chem. Ber., 101,* 1341 (1968).
212. Brockmann, H., and Lenk, W., *Chem. Ber., 92,* 1880 (1959).
213. Brockmann, H., and Lenk, W., *Naturwissenschaften, 47,* 135 (1960).
214. Brockmann, H., and Brockmann, H., Jr., *Naturwissenschaften, 47,* 135 (1960).
215. Brockmann, H., Brockmann, H., Jr., Gordon, J. J., Keller-Schlierlein, W., Lenk, W., Ollis, W. D., Prelog, V., and Sutherland, I. O., *Tetrahedron Lett.,* No. 8, 25 (1960).
216. Brockmann, H., and Lenk, W., *Chem. Ber., 92,* 1904 (1959).
217. Ettlinger, L., Gäumann, E., Hütter, R., Keller-Schierlein, W., Kradolfer, F., Neipp, L., Prelog, V., Reusser, P., and Zähner, H., *Chem. Ber., 92,* 1867 (1959).
218. Brockmann, H., Niemeyer, J., Brockmann, H., Jr., and Budzikiewicz, H., *Chem. Ber., 98,* 3785 (1965).
219. Brockmann, H., Boldt, P., and Niemeyer, J., *Chem. Ber., 96,* 1356 (1963).
220. Brockmann, H., and Wimmer, E., *Chem. Ber., 98,* 2797 (1965).
221. Brockmann, H., and Brockmann, H., Jr., *Chem. Ber., 96,* 1771 (1963).
222. Brockmann, H., and Brockmann, H., Jr., *Chem. Ber., 94,* 2681 (1961).
223. Brockmann, H., and Niemeyer, J., *Chem. Ber., 100,* 3578 (1967).
224. Röhrl, M., and Hoppe, W., *Chem. Ber., 103,* 3502 (1970).
225. Shoji, J., Kozuki, S., Nishimura, H., Mayama, M., Motokawa, K., Tanaka, Y., and Otsuka, H., *J. Antibiot. (Tokyo) Ser. A, 21,* 643 (1968).
226. Brockmann, H., and Patt, P., *Chem. Ber., 88,* 1455 (1955).
227. Brockmann, H., and Spohler, E., *Naturwissenschaften, 48,* 716 (1961).
228. Brockmann, H., Waehneldt, T., and Niemeyer, J., *Tetrahedron Lett.,* No. 6, 415 (1969).

229. Brockmann, H., and Waehneldt, T., *Naturwissenschaften, 48,* 717 (1961).
230. Brockmann, H., and Waehneldt, T., *Naturwissenschaften, 50,* 43 (1963).
231. Brockmann, H., Niemeyer, J., and Rode, W., *Chem. Ber., 98,* 3145 (1965).
232. Brockmann, H., and Franck, B., *Chem. Ber., 88,* 1792 (1955).
233. Brockmann, H., and Boldt, P., *Chem. Ber., 94,* 2174 (1961).
234. DiMarco, A., Silverstrini, R., Gaetani, M., Soldati, M., Orezzi, P., Dasdia, T., Scarpinato, B. M., and Valentini, L., *Nature (London), 201,* 706 (1964).
235. Arcamone, F., Cassinelli, G., Franceschi, G., and Orezzi, P., *Tetrahedron Lett.,* No. 30, 3353 (1968).
236. Arcamone, F., Franceschi, G., Penco, S., and Selva, A., *Tetrahedron Lett.,* No. 13, 1007 (1969).
237. Mitscher, L. A., McCrae, W., Andres, W. W., Lowery, J. A., and Bohonos, N., *J. Pharm. Sci., 53,* 1139 (1964).
238. Tsuji, N., and Nagashima, K., *Tetrahedron, 26,* 5719 (1970).
239. Iwamoto, R. H., Lim, P., and Bhacca, N. S., *Tetrahedron Lett.,* No. 36, 3891 (1968).
240. Birkinshaw, J. H., and Hammady, I. M. M., *Biochem. J., 65,* 162 (1957).
241. Birkinshaw, J. H., Roberts, J. C., and Roffey, P., *J. Chem. Soc. Sect. C,* p. 855 (1966).
242. Roberts, J. C., and Roffey, P., *J. Chem. Soc.,* p. 3666 (1965).
243. Tresner, H. D., Hayes, J. A., and Borders, D. B., *Appl. Microbiol., 21,* 562 (1971).
244. Brockmann, H., and Christiansen, P., *Chem. Ber., 103,* 708 (1970).
245. Cobaz, R., Ettlinger, L., Gäumann, E., Kalvoda, J., Keller-Schierlein, W., Kradolfer, F., Manukian, B. K., Neipp, L., Prelog, V., Reusser, P., and Zähner, H., *Helv. Chim. Acta, 40,* 1262 (1957).
246. Keller-Schierlein, W., Brufani, M., and Barcza, S., *Helv. Chim. Acta, 51,* 1257 (1968).
247. Barcza, S., Brufani, M., Keller-Schierlein, W., and Zähner, H., *Helv. Chim. Acta, 49,* 1736 (1966).
248. Cosulich, D. B., Mowat, J. H., Broschard, R. W., Patrick, J. B., and Meyer, W. E., *Tetrahedron Lett.,* No. 7, 453 (1963).
249. Tulinsky, A., *J. Amer. Chem. Soc., 86,* 5368 (1964).
250. Webb, J. S., Broschard, R. W., Cosulich, D. B., Mowat, J. H., and Lancaster, J. E., *J. Amer. Chem. Soc., 84,* 3183 (1962).
251. Krepinsky, J., Herout, V., Sorm, F., Vystrcil, A., Prokes, R., and Jommi, G., *Coll. Czech. Chem. Comm., 30,* 2626 (1965).
252. Bollinger, P., *Thesis,* Eidgenössische Technische Hochschule, Zürich,(1965).
253. Overeem, J. C., and van Dijkman, A., *Rec. Trav. Chim., 87,* 940 (1968).
254. Takeshita, H., and Anchel, M., *Science, 147,* 152 (1965).
255. Sullivan, G., and Guess, W. L., *Lloydia, 32,* 72 (1969).
256. Howe, R., and Moore, R. H., *Experientia, 25,* 474 (1969).
257. Chilton, W. S., *J. Org. Chem., 33,* 4299 (1968).
258. Cross, B. E., Myers, P. L., and Webster, G. R. B., *J. Chem. Soc. Sect. C,* p. 930 (1970).
259. Armstrong, J. J., and Turner, W. B., *J. Chem. Soc.,* p. 5927 (1965).
260. Cross, B. E., and Edinberry, M. N., *Chem. Commun.,* p. 209 (1970).
261. Takeda, T., Morishita, E., and Shibata, S., *Chem. Pharm. Bull. (Tokyo), 16,* 2213 (1968).
262. Yakubov, G. Z., Khokhlova, M. Yu., Artamonova, O. I., Sergeeva, L. N., and Kalmykova, G. Ya., *Microbiology, 35,* 875 (1966).
263. Overeem, J. C., Sijpesteijn, A. K., and Fuchs, A., *Phytochemistry, 6,* 99 (1967).
264. Cserjesi, A. J., and Smith, R. S., *Mycopathol. Mycol. Appl., 35,* 91 (1968).
265. Bick, I. R. C., and Rhee, C., *Biochem. J., 98,* 112 (1966).
266. Bohman, G., *Acta Chem. Scand., 23,* 2241 (1969).
267. Mahmoodian, A., and Stickings, C. E., *Biochem. J., 92,* 369 (1964).
268. Yamamoto, Y., Kiriyama, N., and Arahata, S., *Chem. Pharm. Bull. (Tokyo), 16,* 304 (1968).
269. Franck, B., and Zimmer, I., *Chem. Ber., 98,* 1514 (1965).
270. Hamasaki, T., Renbutsu, M., and Hatsuda, Y., *Agr. Biol. Chem., 31,* 1513 (1967).
271. Roffey, P., and Sargent, M. V., *Chem. Commun., 24,* 913 (1966).
272. Holker, J. S. E., Kagal, S. A., Mulhevin, L. J., and White, P. M., *Chem. Commun., 24,* 911 (1966).
273. Heathcote, J. G., and Dutton, M. F., *Tetrahedron, 25,* 1497 (1969).
274. Hatsuda, Y., Hamasaki, T., Ishida, M., and Yoshikawa, S., *Agr. Biol. Chem., 33,* 131 (1969).
275. Bu'Lock, J. D., and Smith, J. R., *J. Chem. Soc. Sect. C,* p. 1941 (1968).
276. Bowie, J. H., and Johnson, A. W., *Tetrahedron Lett.,* No. 16, 1449 (1967).
277. Hegyi, J. R., and Gerber, N. N., *Tetrahedron Lett.,* No. 13, 1587 (1968).
278. Ollis, W. D., and Sutherland, I. O., in *Chemistry of Natural Phenolic Compounds,* p. 212. W. D. Ollis, Ed. Pergamon Press, London, England (1961).
279. Ollis, W. D., Sutherland, I. O., and Gordon, J. J., *Tetrahedron Lett.,* No. 16, 17 (1959).
280. Bowie, J. H., and Johnson, A. W., *J. Chem. Soc.,* p. 3927 (1964).
281. Gray, J. S., Martin, G. C. J., and Rigby, W., *J. Chem. Soc. Sect. C,* p. 2580 (1967).
282. Wiley, P. F., MacKellar, F. A., Caron, E. L., and Kelly, R. B., *Tetrahedron Lett.,* p. 663 (1968).
283. Nagarajan, R., Narasimhachari, N., Kadkol, M. V., and Gopalkrishnan, K. S., *J. Antibiot. (Tokyo) Ser. A, 24,* 249 (1971).
284. Bräm, A., and Eugster, C. H., *Helv. Chim. Acta, 52,* 165 (1969).

TROPOLONES

DR. L. C. VINING

Microbial tropolones are restricted to fungi and can be isolated from culture filtrates by precipitation with acid or by solvent extraction. They crystallize from polar solvents, vary from yellow to almost colorless, and have relatively high melting points. All are insoluble in petroleum ether, slightly soluble in ether or ethyl acetate, and soluble in ethanol, dioxane or pyridine. Most are insoluble in water and give green to brown colors with ferric chloride.[1]

Ultraviolet absorption spectra are usually distinctive and show a very intense band in the 200 to 300 nm region, with weaker absorption maxima at 300 to 400 nm.[2] To avoid variations due to mixtures of ionic species, pH should be controlled. Infrared absorption maxima in the 1550 to 1650 cm^{-1} region are associated with the tropolone ring.

Numbering systems are shown below.

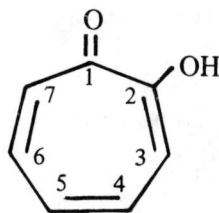

Tropolone

6-Hydroxy-1,3,4,7 tetrahydro-cyclohepta [c] pyran-7-one

Common Name (Systematic Name) Formula (Molecular Weight)	Organisms	Significant Characteristics	Reference
Stipitatic acid (6-Hydroxytropolone-4-carboxylic acid $C_8H_6O_5$ (182)	Penicillium stipitatum	Melting point: 302–304°C (decomposes) λ_{max}^{water}: 262, 332, 360 nm (log ε 4.49, 3.71, 3.61) Red with FeCl$_3$ Sublimes (high vacuum) at 190–200°C	3, 4
Stipitatonic acid (6-Hydroxytropolone-4,5-dicarboxylic acid anhydride) $C_9H_4O_6$ (208)	Penicillium stipitatum	Melting point: 237–237.5°C (decomposes) $\lambda_{max}^{dioxane}$: 258, 355, 382 nm (log ε 4.45, 3.98, 3.97) ν_{max}^{KBr}: 1745, 1825 cm^{-1} Brown with FeCl$_3$ Gives stipitatic acid on reflux in water	5
Ethyl stipitatate (6-Hydroxytropolone-4-carboxylic acid ethyl ester) $C_{10}H_{10}O_5$ (210)	Penicillium stipitatum	Melting point: 250–251°C $\lambda_{max}^{95\% ethanol}$: 262, 372 nm ν_{max}: 1770 cm^{-1} Red to green with FeCl$_3$ Sublimes (high vacuum) at 190°C	6, 7

Common Name (Systematic Name) Formula (Molecular Weight)	Organisms	Significant Characteristics	Reference
Puberulic acid (3,4-Dihydroxytropolone-5-carboxylic acid) $C_8H_6O_6$ (198)	*Penicillium* *puberulum* *johannioli* *aurantio-virens* *cyclopium-viridicatum*	Melting point: 318–320°C (decomposes) λ_{max}^{water}: 270, 350 nm (log ϵ 4.45, 3.86) Red-brown with $FeCl_3$	8, 9
Puberulonic acid (3,4-Dihydroxytropolone-5,6-dicarboxylic acid anhydride) $C_9H_4O_7$ (224)	*Penicillium* *puberulum* *johannioli* *aurantio-virens* *cyclopium-viridicatum*	Melting point: 298°C (decomposes) λ_{max}: 274.5, 317 nm (log ϵ 4.52, 3.98) ν_{max}^{KBr}: 1770, 1830 cm^{-1} Red-brown with $FeCl_3$ Gives puberulic acid on reflux with water	9
Sepedonin (3,6,9-Trihydroxy-3-methyl-1,3,4,7-tetrahydro-cyclohepta [*c*] pyran-7-one $C_{11}H_{12}O_5$ (224)	*Sepedonium* *chrysospermum*	Melting point: 190–205°C (decomposes) $\lambda_{max}^{ethanol}$: 253, 327 nm (log ϵ 4.57, 3.81) Deep green with $FeCl_3$ Gives anhydride (melting point 205°C) on reflux in acid	10

REFERENCES

1. Nozoe, T., Tropolones and Tropoids, in *Fortschritte der Chemie Organischer Naturstoffe*, Vol. 13, p. 232, L. Zechmeister, Ed. Springer Verlag, Vienna, Austria (1956).
2. Aulin-Erdtman, G., *Acta Chem. Scand., 5,* 301 (1951).
3. Birkinshaw, J. H., Chambers, A. R., and Raistrick, H., *Biochem. J., 36,* 242 (1942).
4. Corbett, R. E., Johnson, A. W., and Todd, A. R., *J. Chem. Soc.,* p. 147 (1950).
5. Segal, W., *J. Chem. Soc.,* p. 2847 (1959).
6. Tannenbaum, S. W., Basset, E. W., and Kaplan, M., *Arch. Biochem. Biophys., 81,* 169 (1959).
7. Bentley, R., and Keil, J. G., *J. Biol. Chem., 238,* 3806 (1963).
8. Birkinshaw, J. H., and Raistrick, H., *Biochem. J., 26,* 441 (1932).
9. Corbett, R. E., Hassall, C. H., Johnson, A. W., and Todd, A. R., *J. Chem. Soc.,* p. 1 (1950).
10. Divekar, P. V., Raistrick, H., Dobson, T. A., and Vining, L. C., *Can. J. Chem., 43,* 1835 (1965).

TETRACYCLINES, INCLUDING BIOSYNTHETIC PRECURSORS

DR. ROBERT THOMAS AND DR. RODNEY A. BASSETT

The tetracycline antibiotics are a clinically important group of *Streptomyces* metabolites. Studies of their chemistry,[1] mode of action[2] and biosynthesis[3-6] have received a great deal of attention; the relevant references are cited in numerous review articles and previous lists of the tetracyclines and their chemical derivatives.[1-10]

A compilation of the naturally occurring tetracyclines, including established precursors and shunt metabolites, is presented with selected characteristic properties (where available) and at least one key reference for each compound. Except for tetracycline precursors that have not as yet been detected as microbial metabolites, the parent organism is also listed.

STRUCTURES

XI

Me///, OH

O

OH

CONH$_2$

OH   O   OH   O

XII

O   NMe$_2$

OH

CONH$_2$

OH   O   OH   OH

ORGANISMS

A = *Streptomyces aureofaciens*   B = *Streptomyces rimosus*   C = *Nocardia sulphurea*

CHARACTERISTICS OF TETRACYCLINES AND RELATED PRODUCTS

Name	Organism	Structure	Characteristics	References
Tetracycline	A, B	I R_1 = H R_2 = Me R_3 = H	Melting point (trihydrate): 170–175°C; decomposes UV maxima (0.1N HCl): 220, 268, 355 $[a]_D^{25}$: –239° (c = 1.0 in MeOH)	10
5-Hydroxytetracycline (oxytetracycline, terramycin)	B	I R_1 = OH R_2 = Me R_3 = H	Melting point (dihydrate): 181–182°C; decomposes UV maxima (pH 4): 247, 275, 353 $[a]_D^{25}$: –197° (c = 1.0 in MeOH)	10
7-Chlorotetracycline (chlortetracycline, aureomycin)	A	I R_1 = H R_2 = Me R_3 = Cl	Melting point: 168–169°C; decomposes UV maxima (0.1N HCl): 230, 262.5, 367.5 UV maxima (0.1N NaOH): 255, 258, 345 $[a]_D^{25}$: –275° (MeOH)	10
7-Bromotetracycline	A	I R_1 = H R_2 = Me R_3 = Br	Melting point: 170–172°C; decomposes UV maxima (0.1N HCl): 268, 345 $[a]_D^{20}$: –196° (0.1N HCl)	10
6-Demethyltetracycline	A	I R_1 = H R_2 = H R_3 = H	Melting point: 203–209°C; decomposes UV maxima: cf. tetracycline $[a]_D^{25}$: –259° (0.1N H$_2$SO$_4$)	10, 18
7-Chloro-6-demethyl-tetracycline	A	I R_1 = H R_2 = H R_3 = Cl	Melting point: 174–178°C; decomposes UV maxima: cf. chlorotetracycline $[a]_D^{25}$: –258° (0.1N H$_2$SO$_4$)	10, 18
5-Hydroxy-7-chloro-tetracycline	B	I R_1 = OH R_2 = Me R_3 = Cl	UV maxima (MeOH/0.01N HCl): 229, 266, 372 $[a]_D^{25}$: –237° (MeOH/HCl)	11
2-Acetyl-2-decarboxa-midooxytetracycline	B	II R_1 = OH R_2 = H	Melting point: 200–203°; decomposes UV maxima (MeOH/0.01N HCl): 240 (sh), 277, 316 (sh), 357 $[a]_D^{25}$: –46.6° (0.1N HCl)	12, 14

Name	Organism	Structure	Characteristics	References
2-Acetyl-2-decarboxa-midotetracycline	A	II R_1 = H R_2 = H	Melting point: 179–186°C; decomposes UV maxima (MeOH/0.01N HCl): 220, 240 (sh), 277, 316 (sh), 332 (sh), 360 $[a]_D^{25}$: –125° (MeOH/0.5N HCl)	13, 14
2-Acetyl-2-decarboxa-midochlorotetracy-cline	A	II R_1 = H R_2 = Cl	UV maxima (MeOH/0.01N HCl): 232, 279, 310 (sh), 337 (sh), 373	13, 14
Chelocardin (2-acetyl-2-decarboxamido-4-dedimethylamino-4-epiamino-9-methyl-5a,6-anhydrotetra-cycline	C	III	UV maxima (MeOH): 226, 276, 437	15
4-Dedimethylamino-4-ketoanhydrotetracy-cline		IV		4, 20
4-Dedimethylamino-4-aminoanhydrotetra-cycline	B	V R_1 = H R_2 = H R_3 = Me R_4 = H		16
4-Dedimethylamino-4-amino-7-chloro-6-demethylanhydro-tetracycline	A	V R_1 = H R_2 = H R_3 = H R_4 = Cl	UV maxima: 223, 269, 302, 314, 329, 424	16, 17
6-Demethylanhydrotet-racycline		V R_1 = Me R_2 = Me R_3 = H R_4 = H		29
Anhydrotetracycline		V R_1 = Me R_2 = Me R_3 = Me R_4 = H	Melting point: 225–226°C $[a]_D^{25}$: +25° (Cellosolve)	21, 29
7-Chloroanhydrotetra-cycline		V R_1 = Me R_2 = Me R_3 = Me R_4 = Cl	Melting point: 210°C; decomposes at 220–235°C UV maxima (0.1N HCl): 227, 277, 445 $[a]_D^{25}$: +16° (ethoxyethanol)	21, 29, 30
7-Chloro-6-demethyl-anhydrotetracycline	A	V R_1 = Me R_2 = Me R_3 = H R_4 = Cl	Melting point: 205–210°C; decomposes UV maxima (0.1N HCl): 223, 272, 430 $[a]_D^{25}$: +105° (methyl Cellosolve)	19, 20
4-Dedimethylamino-4-methylaminoanhydro-tetracycline	B	V R_1 = H R_2 = Me R_3 = Me R_4 = H		16

Name	Organism	Structure	Characteristics	References
4-Dedimethylamino-4-methylamino-7-chloroanhydrotetracycline	A	V R_1 = H R_2 = Me R_3 = Me R_4 = Cl		16
4-Dedimethylamino-4-4-methylamino-7-chloro-6-demethylanhydrotetracycline	A	V R_1 = H R_2 = Me R_3 = H R_4 = Cl		16
4-Dedimethylamino-4-methylamino-6-demethylanhydrotetracycline	B	V R_1 = H R_2 = Me R_3 = H R_4 = H		16
4-Dedimethylamino-4-amino-6-demethyl-anhydrotetracycline	A, B	V R_1 = H R_2 = H R_3 = H R_4 = H		16
4-Dedimethylamino-4-amino-7-chloroanhydrotetracycline	A	V R_1 = H R_2 = H R_3 = Me R_4 = Cl		16, 17
5-Hydroxyanhydrotetracycline		VI		29
4-Dedimethylamino-4-ethylmethylaminotetracycline	B	VII		22
Pretetramid		VIII R_1 = H R_2 = H	UV maxima (sulfuric acid/boric anhydride): 232, 264, 281, 333, 394, 490	23
6-Methylpretetramid		VIII R_1 = H R_2 = Me	UV maxima (sulfuric acid/boric acid): 263, 278, 341, 400, 512	24
4-Hydroxy-6-methylpretetramid	A	VIII R_1 = OH R_2 = Me	Melting point: 260–310°C; decomposes UV maxima (sulfuric acid/boric acid): 280, 316, 467, 520	24
5a,11a-Dehydrotetracycline	A	IX R = H		25
7-Chloro-5a,11a-dehydrotetracycline	A	IX R = Cl	UV maxima: 221, 251, 375 $[a]_D^{25}$: +15.5° (0.03 N HCl)	26
Protetrone	A	X	Melting point: 186–190°C; decomposes UV maxima (0.1N HCl/MeOH): 255, 276 (sh), 286 (sh), 432	27
Tetramid green	A	XI		3
Tetramid blue	A	XII		28

REFERENCES

1. Clive, D. J., *Q. Rev. Chem. Soc. Lond., 22,* 435 (1968).
2. Franklin, T. J., in *Biochemical Studies of Antimicrobial Drugs,* p. 192, B. A. Newton and P. E. Reynolds, Eds. Cambridge University Press, New York (1966).
3. McCormick, J. R. D., in *Biogenesis of Antibiotic Substances,* p. 73, Z. Vanek and Z. Hostalek, Eds. Academic Press, New York (1965).
4. McCormick, J. R. D., in *Antibiotics,* Vol. 2, p. 113, D. Gottlieb and P. D. Shaw, Eds. Springer-Verlag, New York (1967).
5. Mitscher, L. A., *J. Pharm. Sci, 57,* 1633 (1968).
6. Turley, R. H., and Snell, J. F., in *Biosynthesis of Antibiotics,* Vol. 1, p. 35, J. F. Snell, Ed. Academic Press, New York (1966).
7. Barratt, J. M., *J. Pharm. Sci., 52,* 309 (1963).
8. Blackwood, R. K., and English, A. R., *Adv. Appl. Microbiol., 13,* 237 (1970).
9. Evans, R. M., in *The Chemistry of Antibiotics Used in Medicine,* p. 103. Pergamon Press, London, England (1965).
10. Miller, M. W., *The Pfizer Handbook of Microbial Metabolites.* McGraw-Hill, New York (1961).
11. Mitscher, L. A., Martin, J. H., Miller, P. A., Shu, P., and Bohonos, N., *J. Am. Chem. Soc., 88,* 3647 (1966).
12. Hochstein, F. A., Schach von Wittenau, M., Tanner, F. W., Jr., and Murai, K., *J. Am. Chem. Soc., 82,* 5934 (1960).
13. Miller, M. W., and Hóchstein, F. A., *J. Org. Chem., 27,* 2525 (1962).
14. Keiner, J., Huttenrauch, R., and Poethke, W., *Arch. Pharm. (Weinheim), 300,* 840 (1967).
15. Mitscher, L. A., Juvarkar, J. V., Rosenbrook, Wm., Jr., Andres, W. W., Schenck, J., and Egan, R. S., *J. Am. Chem. Soc., 92,* 6070 (1970).
16. Miller, P. A., Saturnelli, A., Martin, J. H., Mitscher, L. A., and Bohonos, N., *Biochem. Biophys. Res. Commun., 16,* 285 (1964).
17. McCormick, J. R. D., Jensen, E. R., Johnson, S., and Sjolander, N. O., *J. Am. Chem. Soc., 90,* 2201 (1968).
18. McCormick, J. R. D., Sjolander, N. O., Hirsch, U., Jensen, E. R., and Doerschuk, A. P., *J. Am. Chem. Soc., 79,* 4561 (1957).
19. Webb, J. S., Broschard, R. W., Cosulich, D. B., Stein, W. J., and Wood, C. F., *J. Am. Chem. Soc., 79,* 4563 (1957).
20. McCormick, J. R. D., and Jensen, E. R., *J. Am. Chem. Soc., 91,* 206 (1969).
21. Waller, C. W., Hutchings, B. L., Broschard, R. W., Goldman, A. A., Stein, W. J., Wolf, C. F., and Williams, J. H., *J. Am. Chem. Soc., 74,* 4981 (1952).
22. Neidleman, S. L., Weisenborn, F. L., and Bouchard, J., *Biotechnol. Bioeng., 5,* 83 (1963).
23. McCormick, J. R. D., Rosenthal, J., Johnson, S., and Sjolander, N. O., *J. Am. Chem. Soc., 85,* 1694 (1963).
24. McCormick, J. R. D., and Jensen, E. R., *J. Am. Chem. Soc., 87,* 1794 (1965).
25. Miller, P. A., Hash, J. H., Lincks, M., and Bohonos, N., *Biochem. Biophys. Res. Commun., 18,* 325 (1965).
26. McCormick, J. R. D., Miller, P. A., Growich, J. A., Sjolander, N. O., and Doerschuk, A. P., *J. Am. Chem. Soc., 80,* 5572 (1958).
27. McCormick, J. R. D., and Jensen, E. R., *J. Am. Chem. Soc., 90,* 7127 (1968).
28. McCormick, J. R. D., and Gardner, W. E., *U.S. Patent 3,074,975* (1963).
29. McCormick, J. R. D., Miller, P. A., Johnson, S., Arnold, N., and Sjolander, N. O., *J. Am. Chem. Soc., 84,* 3023 (1962).

AROMATIC MICROBIAL METABOLITES
OTHER THAN RECOGNIZED POLYKETIDES

DR. ROBERT THOMAS AND DR. RODNEY A. BASSETT

The aromatic compounds listed here are of established or probable shikimate origin, although a few are biosynthetically derived from other sources, such as the polyisoprenoid viridin (Structure LX). Each product is listed with the parent organism, selected physical properties, and at least one key reference.

A significant characteristic of the majority of shikimate-derived metabolites is the substitution of the aromatic nucleus by a single carbon side chain, as in methyl cinnamate (Structure XXI), although multiple carbon substituents are also known (for example, novobiocin, Structure LII). It is also noteworthy that, in contrast to the polyketides, shikimate-derived metabolites are rarely polycarbocyclic, although such structures are known (for example, the naphthoquinones of the vitamin K group). An example of a naphthalene derivative of probable polyketide origin is 1,8-dimethoxynaphthalene (Structure LVIII).

STRUCTURES

I.   Pyrogallol

II.   Drosophilin A
          (R = H)

IIa.   Drosophilin A methyl ether
          (R = Me)

III.   *o*-Aminophenol

IV.   Benzoic acid
          ($R_1 = R_2 = R_3 = R_4 = H$)

V.   *p*-Hydroxybenzoic acid
          ($R_1 = R_2 = R_4 = H, R_3 = OH$)

VI.   Protocatechuic acid
          ($R_1 = R_4 = H, R_2 = R_3 = OH$)

VII.   Gentisic acid
          ($R_2 = R_3 = H, R_1 = R_4 = OH$)

VIII.   2,3-Dihydroxybenzoic acid
          ($R_3 = R_4 = H, R_1 = R_2 = OH$)

IX.   Gallic acid
          ($R_1 = H, R_2 = R_3 = R_4 = OH$)

X.   *p*-Methylanisate

XI.   *p*-Hydroxybenzaldehyde
($R_1 = R_2 = R_3 = H$)

CHO

R_3 R_1

OR_2

XIa.   3,4,5-Trihydroxybenzaldehyde
($R_2 = H$, $R_1 = R_3 = OH$)

XII.   *p*-Anisaldehyde
($R_1 = R_3 = H$, $R_2 = Me$)

XIII.   Gentisyl alcohol

CH_2OH

OH

HO

XIV.   3-Chloro-2,5-dihydroxybenzyl alcohol

CH_2OH

OH

HO   Cl

XV.   Phenylacetic acid

$CH_2 \cdot COOH$

XVI.   *p*-Hydroxyphenylacetic acid

$CH_2 \cdot COOH$

OH

XVII.   Homoprotocatechuic acid

$CH_2 \cdot COOH$

OH

OH

XVIII.   2,5-Dihydroxyphenylglyoxylic acid
($R_2 = R_3 = H$, $R_1 = R_4 = OH$)

$CO \cdot COOH$

R_1

R_4 R_2

R_3

XVIIIa.   3,4-Dihydroxyphenylglyoxylic acid
($R_1 = R_4 = H$, $R_2 = R_3 = OH$)

XIX.   2,4,5-Trihydroxyphenylglyoxylic acid
($R_2 = H$, $R_1 = R_3 = R_4 = OH$)

XX.   *trans*-Cinnamic acid

$CH = CH \cdot COOH$

XXI.   Methyl *trans*-cinnamate

$CH = CH \cdot COOMe$

XXII.   Methyl-*p*-coumarate
($R_1 = Me$, $R_2 = R_3 = H$)

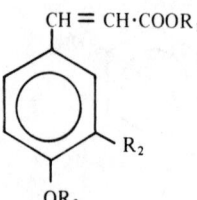

$CH = CH \cdot COOR_1$

R_2

OR_3

XXIII.   Methyl-*p*-methoxycinnamate
           (R_2 = H, R_1 = R_3 = Me)

XXIIIa.   Caffeic acid
           (R_1 = R_3 = H, R_2 = OH)

XXIIIb.   Methyl isoferulate
           (R_1 = R_3 = Me, R_2 = OH)

XXIV.   *trans*-Cinnamic acid amide

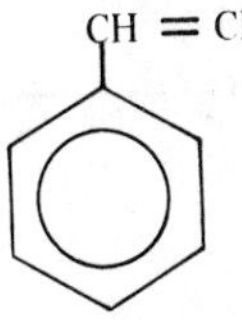

XXV.   Chloramphenicol

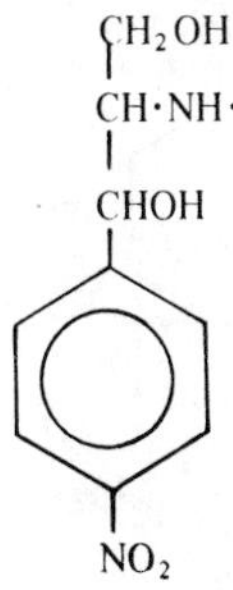

XXVI.   2,3-Dihydroxybenzoylglycine

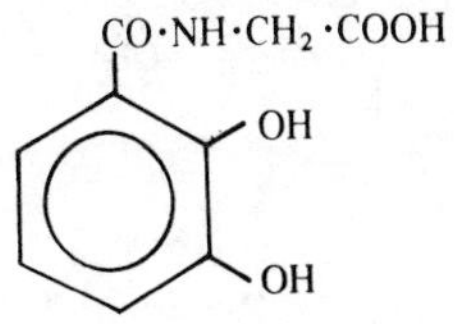

XXVII.   4-Carboxy-2-oxo-3-phenylhept-3-
           enedioic acid

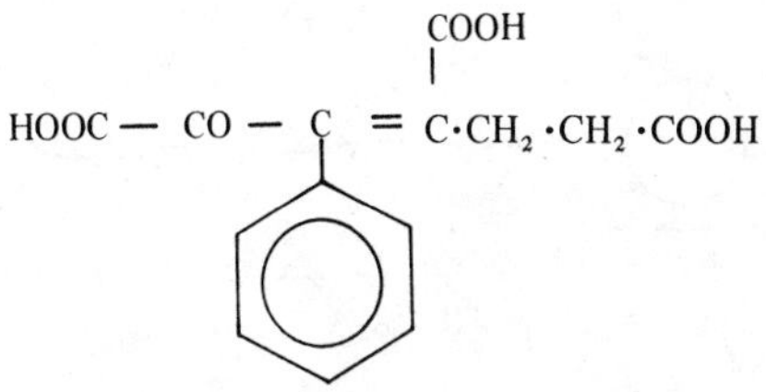

XXVIII.   Pulvic anhydrĭde
           (R_1 = R_2 = R_3 = R_4 = H)

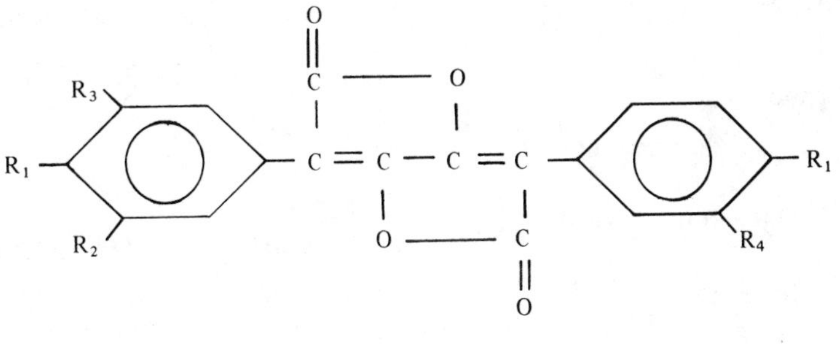

XXIX.   Variegatic acid lactone
           (R_3 = H, R_1 = R_2 = R_4 = OH)

XXX.   Xerocomic acid lactone
           (R_3 = R_4 = H, R_1 = R_2 = OH)

XXXI.   Gomphidic acid lactone
           (R_4 = H, R_1 = R_2 = R_3 = OH)

XXXII.   Calycin

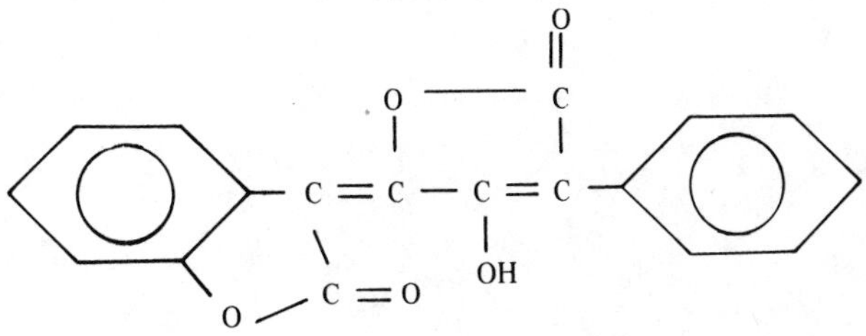

XXXIII.   Vulpinic acid

XXXIV.   Pinastric acid

XXXV.   Leprapic acid
(R = H)

XXXVI.   Leprapic acid methyl ether
(R = Me)

XXXVII.   Epanorin

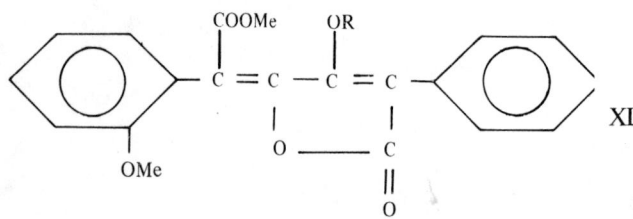

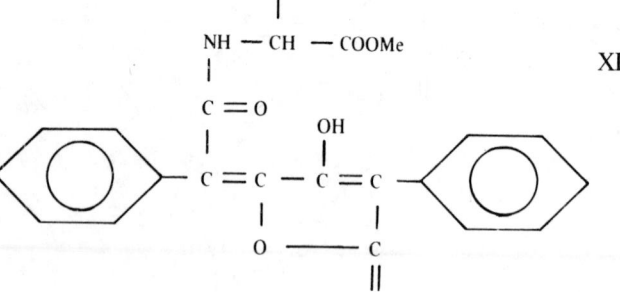

XXXVIII.   Rhizocarpic acid

XXXIX.   *p*-Aminobenzoic acid

XL.   Anthranilic acid

XLI.   Phenylacetamide

XLII.   β-Phenylethylamine

XLIII.   Phenylethyl alcohol

XLIV.   Phthallic acid
         (R = H)

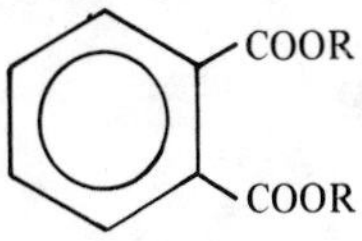

$COOR$
$COOR$

XLV.   Dimethylphthalate
         (R = Me)

XLVI.   2-(2-Hydroxy-5-methoxyphenyl)-
         ethanol

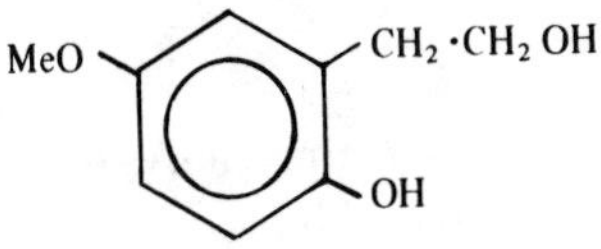

MeO $CH_2 \cdot CH_2 OH$
       OH

XLVII.   5-Methoxycoumarone

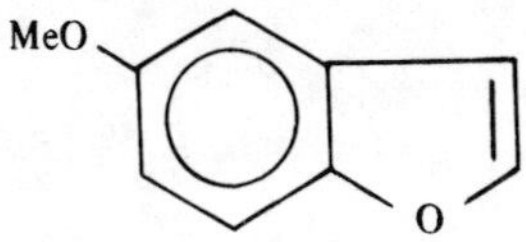

MeO   O

XLVIII. Tuberin

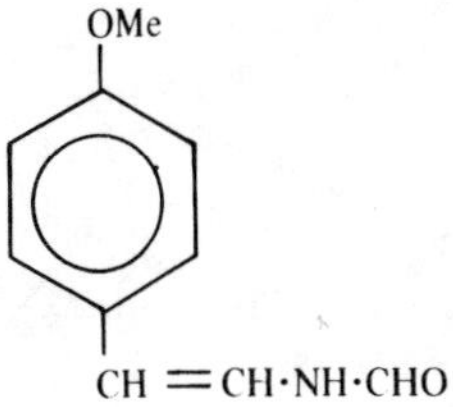

OMe
$CH = CH \cdot NH \cdot CHO$

XLIX.   Dihydrochalcone

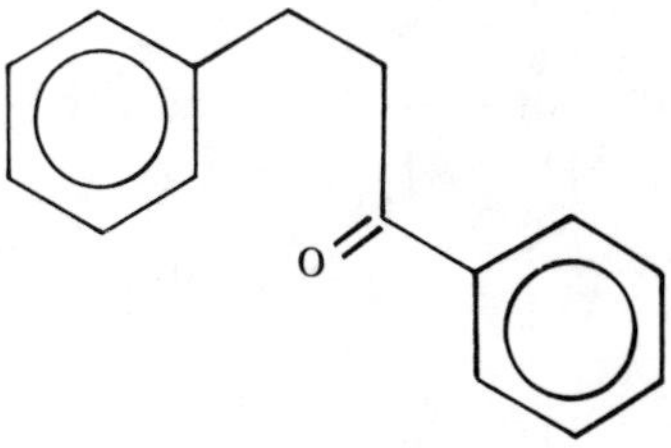

O

L.   Aureothin

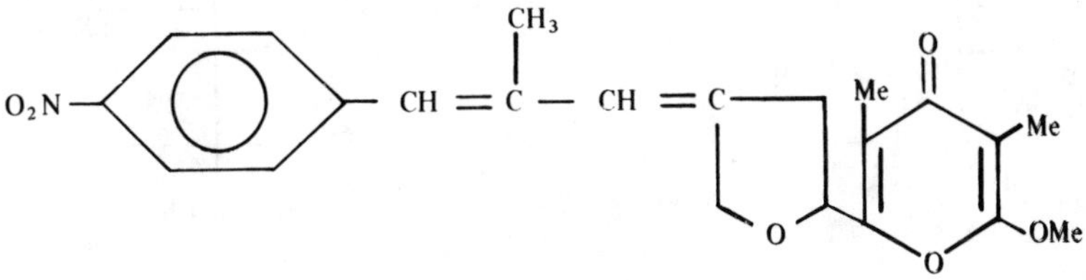

CH_3
O_2N — $CH = C$ — $CH = C$ — Me   O   Me
                                  O   OMe
                              O

LI.   Protoleucomelone (probable structure)

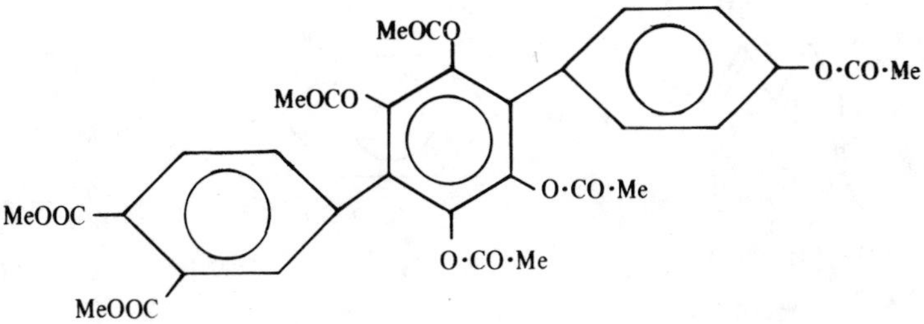

MeOCO
MeOCO $O \cdot CO \cdot Me$
MeOOC $O \cdot CO \cdot Me$
MeOOC $O \cdot CO \cdot Me$

LII.   Novobiocin

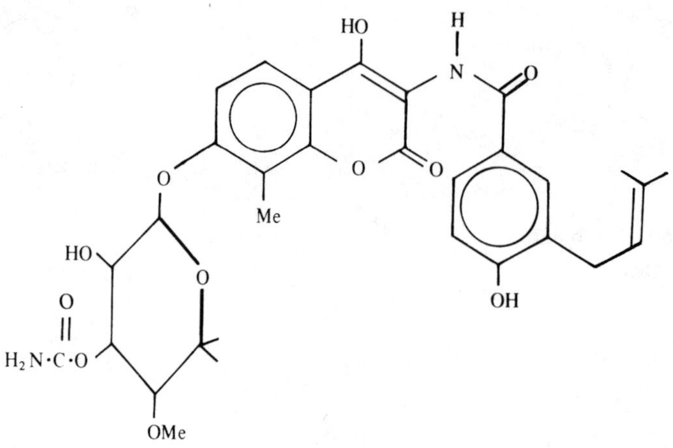

LIII.   Coumermycin-A_1
            (R_1 = H)

LIV.   Coumermycin-A_2 (R
            (R_1 = Me)

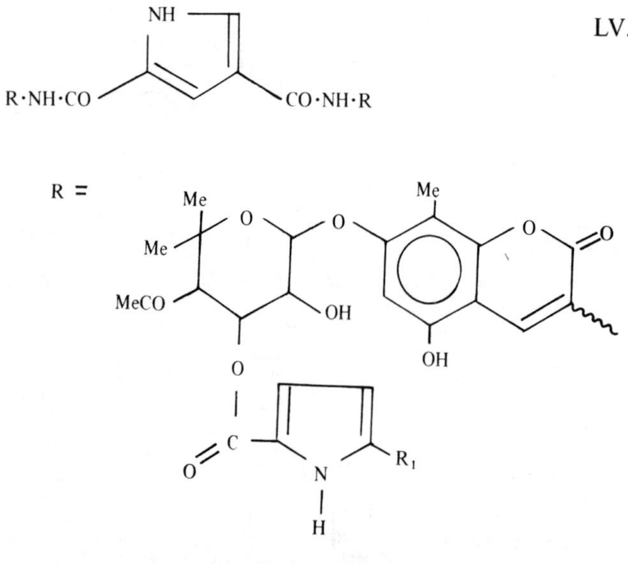

LV.   Homomycin

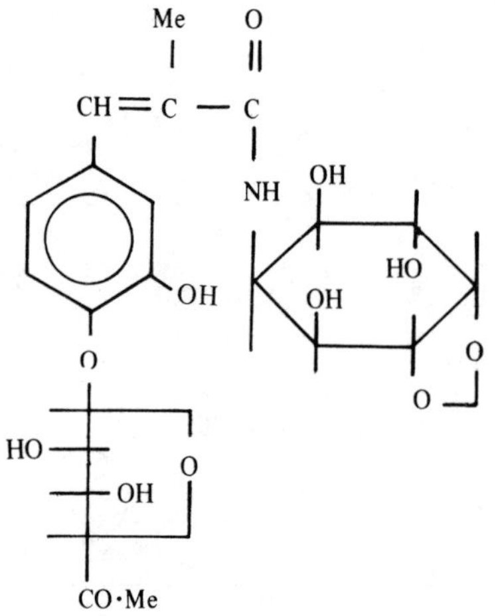

R =

LVI.   Mycelianamide

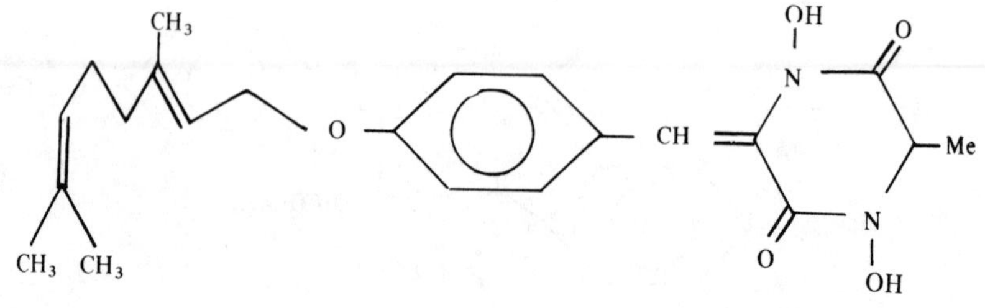

LVII.   8-Methoxy-1-naphthol
$(R_1 = H, R_2 = Me)$

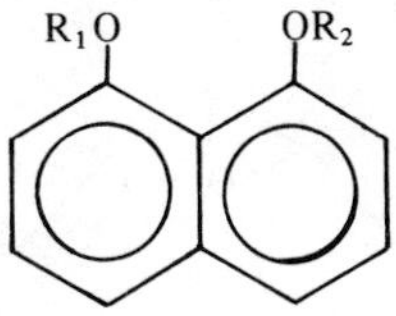

LVIII.   Dimethoxynaphthalene
$(R_1 = R_2 = Me)$

LIX.   Illudinine

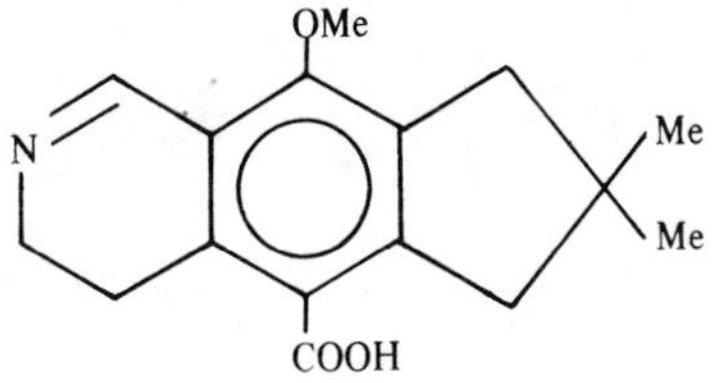

LX.   Viridin
$(R = MeO)$

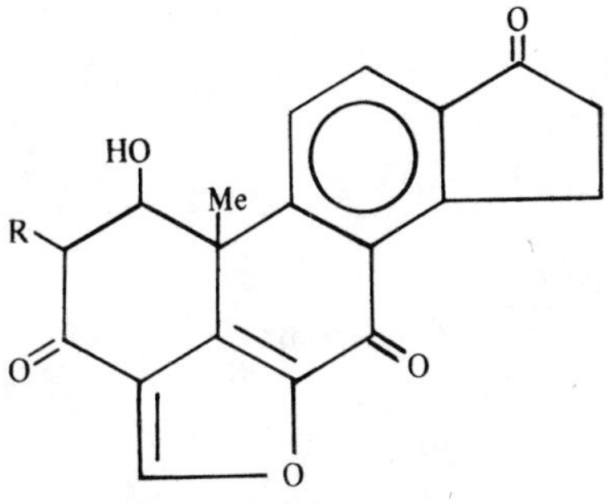

LXI.   Desmethoxyviridin
$(R = H)$

LXII.   Viridiol

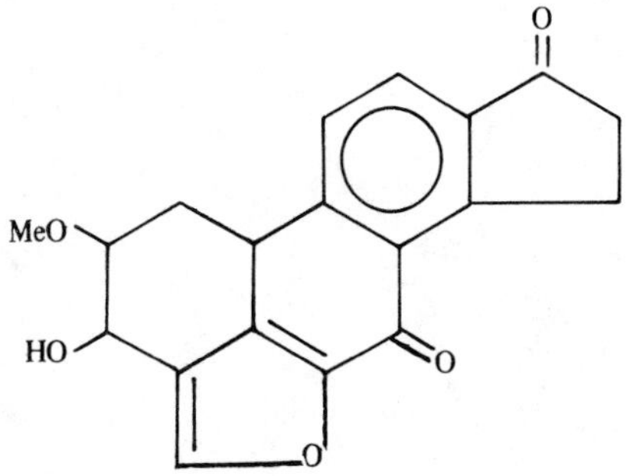

LXIII.   *p*-Hydroxymethylbenzene-
diazonium ion

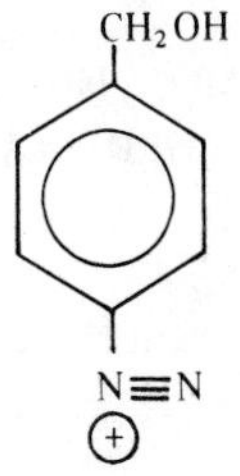

LXIV.   N-Methylnitrosoaminobenzaldehyde

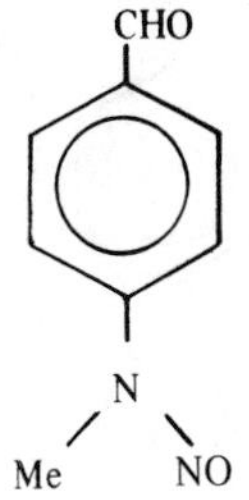

LXV.   2-Pyruvoylaminobenzamide

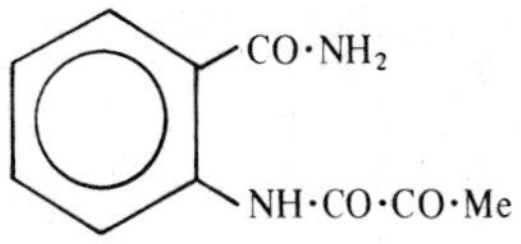

LXVI.   Agaritine

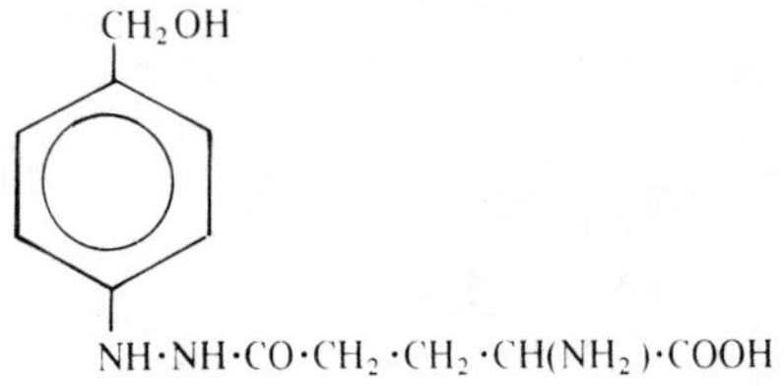

LXVII.   Viridicatin
$(R_1 = R_2 = H)$

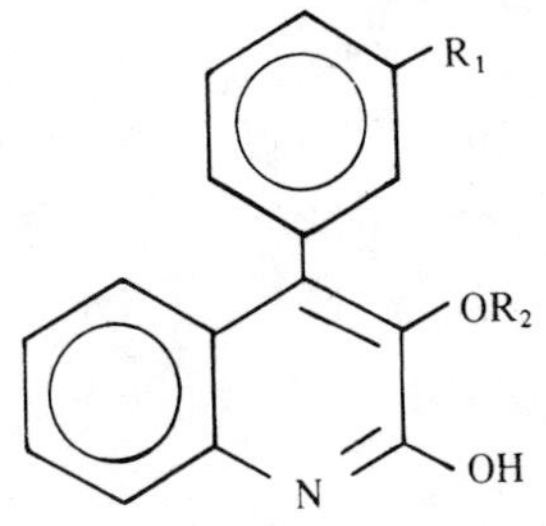

LXVIII.   Viridicatol
(R₁ = OH, R₂ = H)

LXIX.   3-O-Methylviridicatin
(R₁ = H, R₂ = Me)

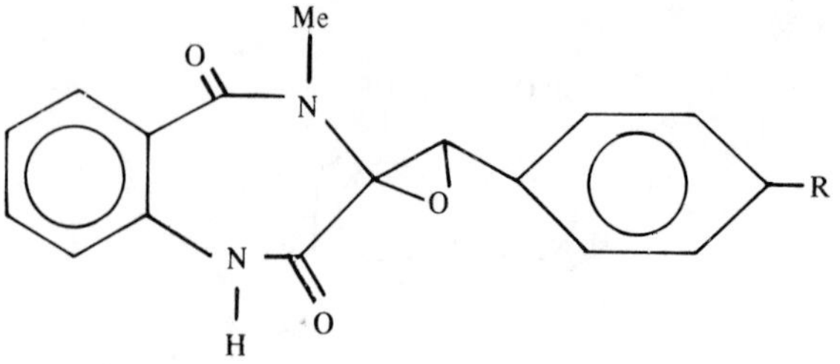

LXXI.   Cyclopenol
(R = OH)

LXXII.   Aspergellone

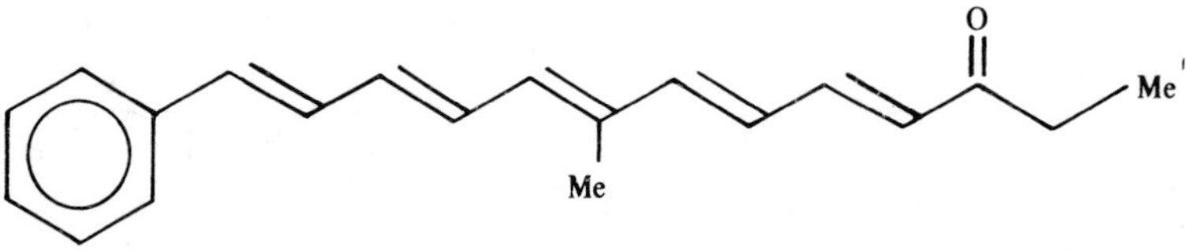

LXX.   Cyclopenin
(R = H)

LXXIII.   Funicone

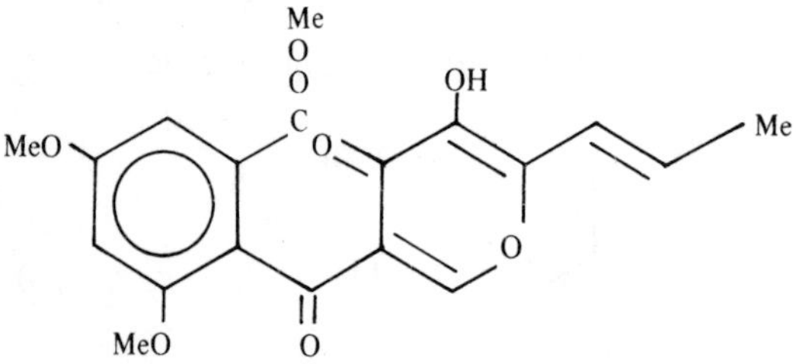

LXXIV.   Anisomycin

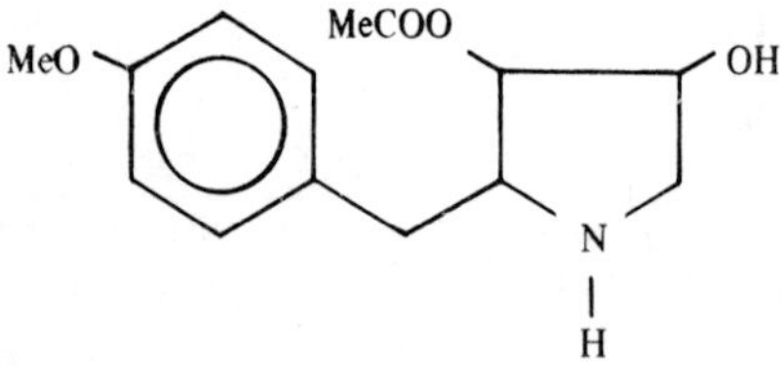

LXXV.   Tryptanthrin

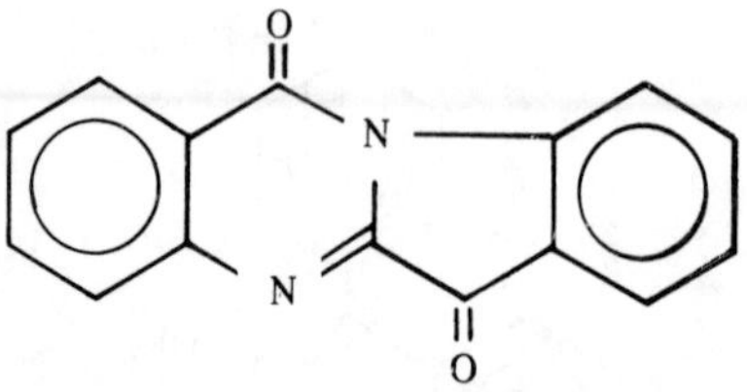

LXXVI.   Hispidin

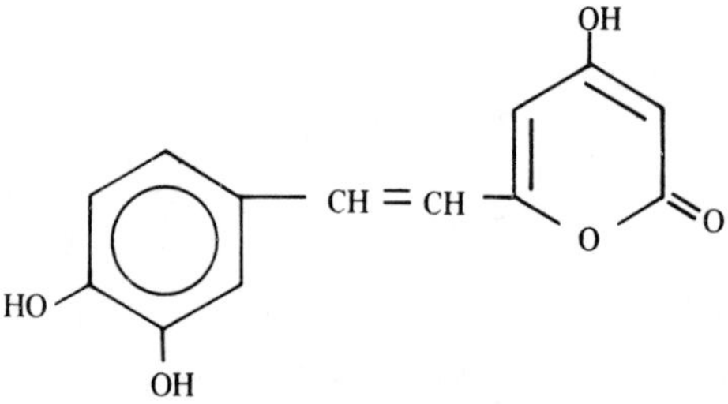

LXXVII.   Chlorflavonin

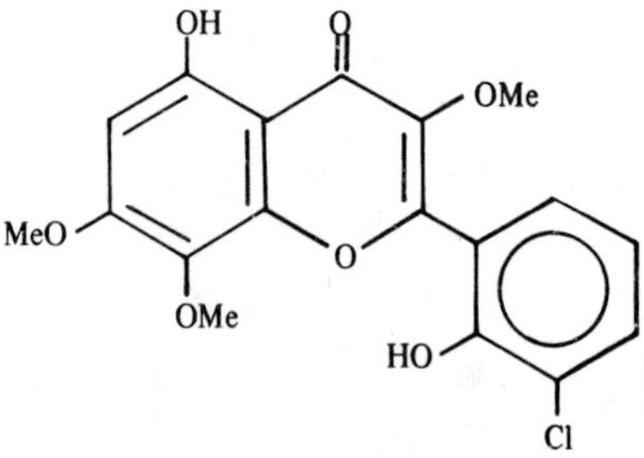

CHARACTERISTICS

Name Structure Formula	Organism	Characteristics	References
Pyrogallol I $C_6H_6O_3$	*Penicillium* *patulum*	Colorless crystals; turn brown in air Melting point: 133°C UV maxima (methanol), nm: 244, 267, 275	1
Drosophilin A (*p*-methoxy-tetrachlorophenol) II $C_7H_4O_2Cl_4$	*Drosophila* *substrata* *Psathyrella* *conopilea*	Yellow crystals Melting point: 118°C UV maximum (ethanol), nm: 310	2
Drosophilin A methyl ether IIa $C_8H_6O_2Cl_4$	*Fomes* *fastuosus*	Colorless needles Melting point: 164−165°C	3
o-Aminophenol III C_6H_7ON	*Streptomyces* spp.	Colorless crystals, amphoteric Melting point: 170−175°C UV maximum (0.1N HCl), nm: 270	4
Benzoic acid IV $C_7H_6O_2$	Yeast	Colorless plates Melting point: 122.5°C UV maxima (water), nm: 226, 269	5
p-Hydroxybenzoic acid V $C_7H_6O_3$	*Penicillium* *patulum* *Polyporus* *tumulosus*	Colorless crystals Melting point: 213°C UV maxima (0.1N HCl), nm: 207, 255 UV maximum (0.05N NaOH), nm: 278	1, 10
Protocatechuic acid VI $C_7H_6O_4$	*Phycomyces* *blakesleeanus*	White or tan crystalline powder; darkens in air Melting point: 200°C; decomposes UV maxima (0.1N HCl), nm: 217, 259, 293	6, 7
Gentisic acid VII $C_7H_6O_4$	*Penicillium* *griseofulvum* *jenseni* *divergens* *Polyporus* *tumulosus*	Colorless crystals Melting point: 199°C UV maximum (water), nm: 320	8−11
2,3-Dihydroxybenzoic acid VIII $C_7H_6O_4$	*Streptomyces* *rimosus* *Claviceps* *paspali*	Prisms Melting point: 206°C UV maxima (ethanol), nm: 248, 317	12, 13
Gallic acid IX $C_7H_6O_5$	*Phycomyces* *blakesleeanus*	Colorless or pale-tan crystals (mono-hydrate from water) Melting point: 225−250°C; decomposes UV maximum (methanol), nm: 272	6, 7
Methyl anisate X $C_9H_{10}O_3$	*Lentinus* *lepideus* *Trametes* *suaveolens*	Colorless crystals Melting point: 48−49°C Boiling point: 256°C UV maximum (ether), nm: 252	14, 15

Name Structure Formula	Organism	Characteristics	References
p-Hydroxybenzaldehyde XI $C_7H_6O_2$	*Streptomyces rimosus*	Colorless prisms Melting point: 114°C UV maxima (methanol), nm: 221, 285, 292 (inflection)	12
3,4,5-Trihydroxybenzalde-hyde XIa $C_7H_6O_4$	*Boletus scaber (Leccinum scaber)*	Tan needles Melting point: 212°C; decomposes PMR, τ: 2.94 (2p), 0.24 (1 p)	88
Anisaldehyde XII $C_8H_8O_2$	*Trametes suavolens Lentinus lepideus Daedalea juniperina*	Oily liquid, n_D^{13} = 1.5764 Boiling point: 248°C Melting point (2,4-dinitrophenylhydrazone): 251°C UV maximum (ethanol), nm: 277	14–16
Gentisyl alcohol XIII $C_7H_8O_3$	*Penicillium patulum divergens*	Colorless crystals Melting point: 100–101°C (triacetyl-gentisyl alcohol, 120–125°C)	9, 11, 17
3-Chloro-2,5-dihydroxy-benzylalcohol XIV $C_7H_7O_3Cl$	*Phyllosticta* spp.		18
Phenylacetic acid XV $C_8H_8O_2$	*Streptomyces rimosus*	Plates Melting point: 77°C UV maxima (ethanol), nm: 229, 248, 253, 256, 259, 268	12
p-Hydroxyphenylacetic acid XVI $C_8H_8O_3$	*Hypochnus sasakii (Corticium sasakii) Polyporus tumulosus*	Colorless crystals Melting point: 148°C; sublimes UV maximum (water), nm: 275	10, 19
Homoprotocatechuic acid XVII $C_8H_8O_4$	*Polyporus tumulosus*	Colorless plates (3,4-dimethoxyphenyl-acetic acid, lustrous plates) Melting point: 128.5°C (3,4-dimethoxy-phenylacetic acid, 98°C) UV maximum (water), nm: 280	10, 20
2,5-Dihydroxyphenylgly-oxylic acid XVIII $C_8H_6O_5$	*Polyporus tumulosus*	Yellow needles Melting point: 141°C UV maxima (water), nm: 230, 263, 360	21
3,4-Dihydroxyphenylgly-oxylic acid XVIIIa $C_8H_6O_5$	*Polyporus tumulosus*	Dull-yellow needles Melting point: 159–160°C UV maxima (water), nm: 234, 285, 310	10
2,4,5-Trihydroxyphenylgly-oxylic acid XIX $C_8H_6O_6$ (pyridine salt, $C_8H_6O_6 \cdot C_6H_5N \cdot H_2O$)	*Polyporus tumulosus*	Bright-red prisms (pyridine salt, greenish-yellow prisms) Melting point: 193°C (pyridine salt, 188°C); decomposes	20

Name Structure Formula	Organism	Characteristics	References
trans-Cinnamic acid XX $C_9H_8O_2$	*Ceratostomella fimbriata* *Stereum subpileatum*	Colorless crystals Melting point: 133°C UV maxima (ethanol), nm: 216, 222, 273	22, 23
Methyl *trans*-cinnamate XXI $C_{10}H_{10}O_2$	*Lentinus lepideus*	Clear, pale-yellow oil, $n_D^{13} = 1.5766$, or white crystals Boiling point (oil): 94–110°C (2–3 mm) Melting point (crystals): 35–37°C UV maximum (3% MeCN), nm: 279	14
Methyl *p*-coumarate XXII $C_{10}H_{10}O_3$	*Lentinus lepideus*	Colorless crystals Melting point: 137–139°C UV maximum (water), nm: 310	24, 25
Methyl *p*-methoxycinnamate XXIII $C_{11}H_{12}O_3$	*Lentinus lepideus*	Colorless crystals Melting point: 88°C	14
Caffeic acid (3,4-dihydroxy- cinnamic acid) XXIIIa $C_9H_8O_4$	*Boletus scaber*	Derivative, 3,4-diacetoxycinnamic acid: Colorless needles Melting point: 190°C	88
Methyl isoferulate XXIIIb $C_{11}H_4O_4$	*Lentinus lepideus*	Melting point: 227–229°C UV maxima (water), nm: 290, 320	89
trans-Cinnamic acid amide XXIV C_9H_9ON	*Streptomyces* spp.	Colorless crystals Melting point: 147–149°C	26
Chloramphenicol (chloromy- cetin) XXV $C_{11}H_{12}O_5N_2Cl_2$	*Streptomyces venezuelae phaeochromogenes*	Colorless crystals Melting point: 149.7°C UV maxima (water), nm: 278, 298 $[\alpha]_D^{25}$ (ethylacetate): −25.5°	27–29
2,3-Dihydroxybenzoylgly- cine XXVI $C_9H_9O_5N$	*Bacillus subtilis*	Colorless needles Melting point: 210–211°C UV maxima (methanol), nm: 250, 314	30
4-Carboxy-2-oxo-3-phenyl- hept-3-enedioic acid XXVII $C_{14}H_{12}O_7$	*Chaetomium indicum*	Colorless prisms Melting point: 170°C; decomposes UV maximum (ethanol), nm: 288 UV minimum (ethanol), nm: 237	31
Pulvic anhydride XXVIII $C_{18}H_{10}O_4$	*Stricta aurata* *Candellariella vitellina*	Yellow needles Melting point: 222–224°C	32, 33
Variegatic acid lactone XXIX $C_{18}H_{10}O_8$	*Suillus (Boletus) variegatus* *Boletus satamus luridus erythropus*	Red needles Melting point: 235°C; softens UV maxima (10% ethanol), nm: 272, 386	90, 91

Name Structure Formula	Organism	Characteristics	References
Xerocomic acid lactone XXX $C_{18}H_{10}O_7$	*Xerocomus chrystenteron* *Gomphidus glutinosus*	Orange crystals Melting point: 302°C UV maxima (ethanol), nm: 261, 411 UV maxima (water), nm: 256, 380	92, 93
Gomphidic acid lactone XXXI $C_{18}H_{10}O_8$	*Gomphidus glutinosus*	IR (KBr) of tetraacetyl gomphidic acid lactone: see reference	93
Calycin XXXII $C_{18}H_{10}O_5$	*Lepraria candelaris* *Stricta aurata crocata*	Orange-red crystals Melting point: 248.5°C UV maxima (cyclohexane), nm: 208, 237, 253, 430	33—35
Vulpinic acid XXXIII $C_{19}H_{14}O_5$	*Evernia vulpinia* *Cyphelium chrysocephalum* *Cetraria juniperina* *Candellariella vitellina*	Yellow crystals Melting point: 148—149°C UV maximum, nm: 371	33, 36
Pinastric acid (chrysocetraric acid) XXXIV $C_{20}H_{16}O_6$	*Lepraria flava* *Cetraria pinastri tubulosa juniperina*	Orange needles Melting point: 200—203°C	34, 37
Leprapic acid (leprapinic acid) XXXV $C_{20}H_{16}O_6$	*Lepraria citrina chlorina* *Biatora lucida*	Golden plates, deep-yellow color with H_2SO_4 Melting point: 159—160°C UV maxima (methanol), nm: 270, 316 (inflection)	38—40
Leprapinic acid methyl ether XXXVI $C_{21}H_{18}O_6$	*Lepraria chlorina* *Biatora lucida*	Colorless needles Melting point: 150—152°C UV maxima (methanol), nm: 229 (inflection), 261, 336	40, 41
Epanorin XXXVII $C_{25}H_{25}O_6N$	*Lecanora epanora*	Yellow needles Melting point: 135—136°C $[\alpha]_D^{26}$ (chloroform, c = 6.48): −1.86° ± 0.2°	42
Rhizocarpic acid XXXVIII $C_{28}H_{23}O_6N$	*Rhizocarpon geographicum viridiatrum* *Calicium hyperellum*	Yellow needles Melting point: 177—178°C $[\alpha]_D^{20}$ (chloroform, c = 1.22): +110.4° ± 2.1°	42
p-Aminobenzoic acid XXXIX $C_7H_7O_2N$	*Hansenula anomala* *Mycotorula lipolytica*	Yellow needles Melting point: 186°C UV maxima (ethanol), nm: 220, 288	43

Name Structure Formula	Organism	Characteristics	References
Anthranilic acid XL $C_7H_7O_2N$	*Corynebacterium diphtheriae* *Pseudomonas* spp.	Leaflets Melting point: 144°C UV maxima (ethanol), nm: 249, 326	44
Phenylacetamide XLI C_8H_9ON	*Streptomyces rimosus*	Crystals Melting point: 154–156°C UV maxima (ethanol), nm: 222, 247, 252, 258, 264	12
β-Phenylethylamine XLII $C_8H_{11}N$	*Boletus edulis* *Claviceps purpurea* *Polyporus sulphureus* *Marasmins personatus* *Phlegmacium* spp.	Liquid Boiling point: 196–198°C	45, 46
Phenylethyl alcohol XLIII $C_8H_{10}O$	*Gibberella fujikuroi*	Oil; distills at 35°C (0.1 mm) Melting point (dinitrobenzoate): 107– 108°C	47
Phthallic acid XLIV $C_8H_6O_4$	*Gibberella fujikuroi*	Prisms Melting point: 212–213°C	47
Dimethylphthalate XLV $C_{10}H_{10}O_4$	*Giberella fujikuroi*	UV maximum, nm: 277 $E_{1\ cm}^{1\%}$: 57.7	47
2-(2-Hydroxy-5-methoxy-phenyl)ethanol XLVI $C_9H_{12}O_3$	*Stereum subpileatum*		48
5-Methoxycoumarone XLVII $C_9H_8O_2$	*Stereum subpileatum*	Colorless leaflets Melting point: 34°C UV maxima (ethanol), nm: 221, 247, 293, 300.5	48, 49
Tuberin XLVIII $C_{10}H_{11}O_2N$	*Streptomyces amakusaensis*	Colorless plates Melting point: 132–133°C UV maxima, nm: 219, 285 $[\alpha]_D$: 0°	50, 51
Dihydrochalcone XLIX $C_{15}H_{14}O$	*Phallus impudicus*		52
Aureothin L $C_{22}H_{23}O_6N$	*Streptomyces thiolenteus*	Yellow crystals Melting point: 158°C UV maxima (ethanol), nm: 257, 346	53–55
Protoleucomelone LI $C_{32}H_{28}O_{14}$	*Polyporus leucomelas*	Colorless crystals Melting point: 203–205°C	56

Name Structure Formula	Organism	Characteristics	References
Novobiocin LII $C_{31}H_{36}O_{11}N_2$	*Streptomyces spheroides niveus griseus*	Pale-yellow crystals Melting point: 152—156°C; decomposes (polymorphic, 174—178°C) UV maxima (0.1N HCl), nm: 324, 390 $[\alpha]_D^{26}$ (ethanol, c = 1): −63°	57
Coumermycin-A$_1$ LIII $C_{55}H_{59}O_{20}N_5$	*Streptomyces rishiriensis spinicoumarensis*	Melting point: 258—260°C; decomposes UV maxima (ethanol), nm: 280, 336 $[\alpha]_D^{25}$, sodium salt (75% acetone, c = 1) −141°	58
Coumermycin-A$_2$ LIV $C_{53}H_{55}O_{20}N_5$	*Streptomyces rishiriensis spinicoumarensis*	UV maxima (ethanol), nm: 267, 340	58, 59
Homomycin LV $C_{23}H_{29}O_{12}N$	*Streptomyces noboritoensis*	Melting point: >160°C; decomposes UV maxima (water), nm: 214, 270, 291 $[\alpha]_D^{18}$ (water, c = 1): −146°	60
Mycelianamide LVI $C_{22}H_{28}O_5N_2$	*Penicillium griseofulvum*	Colorless leaflets Melting point: 170—172°C; decomposes $[\alpha]_D^{19}$ (chloroform, c = 0.869): −217°	61, 62
8-Methoxy-1-naphthol LVII $C_{11}H_{10}O_2$	*Daldinia concentrica*	Oil UV maxima (ethanol), nm: 298, 315, 330	63
1,8-Dimethoxynaphthalene LVIII $C_{12}H_{12}O_2$	*Daldinia concentrica*	Colorless crystals Melting point: 158—161°C UV maxima (ethanol), nm: 227, 298, 310, 316, 325, 330	63
Illudinine LIX $C_{16}H_{17}O_3N$	*Clitocybe illudens*	Melting point: 228—229°C; decomposes UV maxima (methanol), nm: 232, 300, 332	64
Viridin LX $C_{20}H_{16}O_6$	*Apiospora camptospora Gliocladium virens Trichoderma viride*	Prisms Melting point: 245°C; decomposes UV maxima, nm: 242, 300 $[\alpha]_D^{19}$: −224°	65—69
Desmethoxyviridin LXI $C_{19}H_{14}O_5$	*Apiospora camptospora Gliocladium virens*		67—69 68, 69
Viridiol LXII $C_{20}H_{18}O_6$	*Gliocladium virens Trichoderma viride*	Melting point: 198—201°C; decomposes UV maxima (ethanol), nm: 250, 317	70
p-Hydroxymethylbenzene-diazonium ion LXIII $C_7H_7ON_2$	*Agaricus bisporus*	Color reaction and chromatography	71

Name Structure Formula	Organism	Characteristics	References
N-Methylnitrosoaminobenz-aldehyde LXIV $C_8H_8N_2O_2$	*Clitocybe suaveoleus*	Melting point: 81–82°C UV maximum, nm: 304 UV minimum, nm: 254	72
2-Pyruvoylaminobenzamide LXV $C_{10}H_{10}O_3N_2$	*Penicillium chrysogenum notatum*	Needles Melting point: 181–184°C UV maxima, nm: 211, 247, 302	73
Agaritine LXVI $C_{12}H_{17}O_4N_3$	*Agaricus bisporus*	Melting point: 203–208°C; decomposes UV maximum, nm: 237.5 $[\alpha]_D^{23}$ (water, c = 1.10): +6°	74, 75
Viridicatin LXVII $C_{15}H_{11}O_2N$	*Penicillium viridicatum cyclopium puberulum*	Colorless needles Melting point: 268°C UV spectrum: see Reference 76	76, 77
Viridicatol LXVIII $C_{15}H_{11}O_3N$	*Penicillium viridicatum cyclopium*	Melting point: 280°C UV maxima (methanol), nm: 226, 284, 304 (sh), 316, 329.5	79, 80
3-O-Methylviridicatin LXIX $C_{16}H_{13}O_2N$	*Penicillium viridicatum cyclopium puberulum*	Crystals Melting point: 248–249°C UV maxima (ethanol), nm: 223, 281, 313, 324, 337	76, 78
Cyclopenin LXX $C_{17}H_{14}O_3N_2$	*Penicillium cyclopium*	Colorless needles Melting point: 183–184°C $[\alpha]_D^{20}$ (methanol, c = 1.20): –290°	77, 79, 80
Cyclopenol LXXI $C_{17}H_{14}O_4N_2$	*Penicillium cyclopium viridicatum*	Prisms Melting point: 215°C; decomposes $[\alpha]_D^{20}$ (methanol, c = 1.3): –309°	79, 80
Aspergellone (asperenone) LXXII $C_{20}H_{22}O$	*Aspergillus awamori niger*	Crystals Melting point: 128–130°C UV maxima, nm: 371.5, 392, 415	81–83
Funicone LXXIII $C_{19}H_{18}O_8$	*Penicillium funiculosum*	White crystals Melting point: 176–178°C, UV maxima (95% ethanol), nm: 245, 310	84
Anisomycin (flagecidin) LXXIV $C_{14}H_{19}O_4N$	*Streptomyces griseolus roseochromo-genes*	White needles Melting point: 140–141°C UV maxima (ethanol), nm: 224, 277, 283 $[\alpha]_D^{25}$ (chloroform, c = 1): –45° ± 3°	85, 86
Tryptanthrin LXXV $C_{15}H_8O_2N_2$	*Candida lipolytica*	Yellow crystals Melting point: 266–267°C	87
Hispidin LXXVI $C_{13}H_{10}O_5$	*Polyporus hispidus schweinitzi*	Yellow needles (hispidin hydrate) Melting point: 256–258°C UV maxima (ethanol), nm: 223, 248, 367	94, 95

Name Structure Formula	Organism	Characteristics	References
Chlorflavonin LXXVII $C_{18}H_{15}O_7Cl$	*Aspergillus candidus*	Crystals Melting point: 212°C	96, 97

REFERENCES

1. Bassett, E. W., and Tanenbaum, S. W., *Biochim. Biophys. Acta, 28,* 21, 247 (1958).
2. Kavanagh, F., Hervey, A., and Robbins, W. J., *Proc. Natl. Acad. Sci., U.S.A., 38,* 555 (1952).
3. Singh, P., and Rangaswami, S., *Tetrahedron Lett.,* p. 1229 (1966).
4. Anzai, K., Isono, K., Okuna, K., and Suzuki, S., *J. Antibiot. (Tokyo) Ser. A, 13,* 125 (1960).
5. Kuhn, R., and Schwarz, K., *Chem. Ber., 74,* 1617 (1941).
6. Schröter, H. B., *Angew. Chem., 68,* 158 (1956).
7. Halam, E., Haworth, R. D., and Knowles, P. E., *J. Chem. Soc.,* p. 1854 (1961).
8. Raistrick, H., and Simonart, P., *Biochem. J., 27,* 628 (1933).
9. Barta, J., Mecir, R., *Experientia, 4,* 277 (1948).
10. Crowden, R. K., and Ralph, B. J., *Aust. J. Chem., 14,* 475 (1961).
11. Brack, A., *Helv. Chim. Acta, 30,* 1 (1947).
12. Catlin, E. R., Hassall, C. H., and Pratt, P. C., *Biochem. Biophys. Acta, 156,* 109 (1968).
13. Arcamone, F., Chain, E. B., Ferretti, A., and Penella, P., *Nature, 192,* 552 (1961).
14. Birkinshaw, J. H., and Findlay, W. P. K., *Biochem. J., 34,* 82 (1940).
15. Birkinshaw, J. H., Bracken, A., and Findlay, W. P. K., *Biochem. J., 38,* 131 (1944).
16. Birkinshaw, J. H., and Chaplen, P., *Biochem. J., 60,* 255 (1955).
17. Engel, B. G., and Brzeski, W., *Experimentia, 30,* 1472 (1947).
18. Sakamura, S., Ito, J., and Sakai, R., *Agric. Biol. Chem., 35,* 105 (1971).
19. Chen, Y. S., *Bull. Agric. Chem. Soc. Jap., 22,* 136 (1958).
20. Ralph, B. J., and Robertson, A., *J. Chem. Soc.,* p. 3380 (1950).
21. Moir, G. F. J., and Ralph, B. J., *Chem. Ind. (London),* p. 1143 (1954).
22. Kubota, T., and Naya, K., *Chem. Ind. (London),* p. 1427 (1954).
23. Birkinshaw, J. H., Chaplen, P., and Findlay, W. P. K., *Biochem. J., 66,* 188 (1957).
24. Eberhardt, G., *J. Am. Chem. Soc., 78,* 2832 (1956).
25. Shimazono, H., Nord, F. F., *Arch. Biochem. Biophys., 78,* 263 (1958).
26. Sekizawa, Y., *Biochem. J. (Jap.), 45,* 9, (1958).
27. Ehrlich, J., Bartz, R., Smith, R. M., Joslyn, D. A., and Burkholder, P. R., *Science, 106,* 417 (1947).
28. Controulis, J., Rebstock, M. C., and Crooks, H. M., *J. Am. Chem. Soc., 71,* 2463 (1949).
29. McGarth, R., Vining, L. C., Sala, F., and Westlake, D. W. S., *Can. J. Biochem., 46,* 587 (1968).
30. Ito, T., and Neilands, J. B., *J. Am. Chem. Soc., 80,* 4645 (1958).
31. Johnson, D. H., Robertson, A., and Walley, W. B., *J. Chem. Soc.,* p. 2429 (1953).
32. Hesse, O., *J. Prakt. Chem., 170,* 334 (1900).
33. Mosbach, K., *Acta Chem. Scand., 21,* 2331 (1967).
34. Asano, M., and Kameda, Y., *Chem. Ber., 68,* 1568 (1935).
35. Akermark, B., *Acta Chem. Scand., 15,* 1695 (1961).
36. Karrer, P., Gehrckens, A., and Heuss, W., *Helv. Chim. Acta, 9,* 446 (1926).
37. Asahino, Y., and Shibata, S., in *Chemistry of Lichen Substances,* p. 43. Japan Society for the Advancement of Science, Tokyo, Japan (1954).
38. Mittel, O. P., and Seshadri, T. R., *J. Chem. Soc.,* p. 3053 (1955).
39. Mittel, O. P., and Seshadri, T. R., *J. Chem. Soc.,* p. 1734 (1956).
40. Agarwal, S. C., and Seshadri, T. R., *Tetrahedron, 21,* 3205 (1965).
41. Grover, P. K., and Seshadri, T. R., *J. Sci. Ind. Res. (India) Sect. B, 18,* 238 (1959).
42. Frank, R. L., Cohen, S. M., and Coker, J. N., *J. Am. Chem. Soc., 72,* 4454 (1950).
43. Preston, W. H., Yeasts in Feeding, *Symposium, Milwaukee* (1948).
44. Takeda, R., and Nakanishi, I., *J. Ferment. Technol.,* p. 37 (1959).
45. Reuter, C., *Z. Physiol. Chem., 78,* 167 (1912).
46. List, P. H., *Planta Med., 6,* 424 (1958).
47. Cross, B. E., Galt, R. H. B., Hanson, J. R., Curtis, P. J., Grove, J. F., and Morrison, A., *J. Chem. Soc.,* p. 2937 (1963).

48. Bu'Lock, J. D., Hudson, A. T., and Kaye, B., *Chem. Commun.*, p. 814 (1967).
49. Birkinshaw, J. H., Chaplen, P., Findlay, W. P. K., *Biochem. J.*, *66*, 188 (1957).
50. Ohkuma, K., Anzai, K., and Suzuki, S., *J. Antibiot. (Tokyo) Ser. A*, *15*, 115 (1962).
51. Anzai, K., *J. Antibiot. (Tokyo) Ser. A*, *15*, 117, 123 (1962).
52. List, P. H., and Freund, B., *Planta Med. Suppl.*. p. 123 (1968).
53. Maeda, K., *J. Antibiot. (Tokyo) Ser. A.*, *6*, 137 (1953).
54. Hirata, Y., Nakata, H., and Yamada, Y., *J. Chem. Soc. (Jap.)*, *79*, 1390 (1958).
55. Hirata, Y., Nakata, H., and Yamada, Y., *J. Chem. Soc., (Jap.)*, *81*, 340 (1960).
56. Akagi, M., *J. Pharm. Sci. (Jap.)*, *62*, 129 (1942).
57. Hoeksema, H., Caron, E. L., and Hinman, J. W., *J. Am. Chem. Soc.*, *78*, 2019 (1956).
58. Kawaguchi, H., Tsukiura, H., Okanishi, M., Ohmori, T., Fujisawa, K., and Koshiyama, H., *J. Antibiot. (Tokyo) Ser. A*, *18*, 1 (1965).
59. Kawaguchi, H., Miyaka, T., and Tsukiura, H., *J. Antibiot. (Tokyo) Ser. A*, *18*, 220 (1965).
60. Namiki, M., Isno, K., and Suzuki, S., *J. Antibiot. (Tokyo) Ser. A*, *10*, 160 (1957).
61. Oxford, A. E., and Raistrick, H., *Biochem. J.*, *42*, 323 (1948).
62. Birch, A. J., *Proc. Chem. Soc.*, p. 233 (1957).
63. Allport, D. C., and Bu'Lock, J. D., *J. Chem. Soc.*, p. 654 (1960).
64. Nair, M. S. R., Takeshita, H., McMorris, T. C., and Anchel, M., *J. Org. Chem.*, *34*, 240 (1969).
65. Brain, P. W., and McGowan, J. C., *Nature*, *156*, 144 (1945).
66. Grove, J. F., *J. Chem. Soc. Sect. C*, p. 549 (1969).
67. Turner, T. W., in *Fungal Metabolites*, p. 267. Academic Press, London, England, and New York (1971).
68. Grove, J. F., Moffatt, J. S., and Vischer, E. B., *J. Chem. Soc.*, p. 3803 (1965).
69. McCloskey, P., *J. Chem. Soc.*, p. 3811 (1965).
70. Moffatt, J. S. Bu'Lock, J. D., and Yuen, T. H., *Chem. Commun.*, p. 839 (1969).
71. Levenberg, B., *Biochem. Biophys. Acta*, *63*, 212 (1962).
72. Herrmann, H., *Z. Physiol. Chem.*, *326*, 16 (1961).
73. Sutter, P. J., and Turner, W. B., *J. Chem. Soc.*, p. 2240 (1967).
74. Daniels, E. G., Kelly, R. B., and Hinman, J. W., *J. Org. Chem.*, *27*, 3229 (1962).
75. Levenberg, B., *J. Biol. Chem.*, *239*, 2256 (1964).
76. Cunningham, K. G., and Freeman, G. G., *Biochem. J.*, *53*, 328 (1953).
77. Bracken, A., Pocker, A., and Raistrick, H., *Biochem. J.*, *57*, 587 (1954).
78. Austin, D. J., and Meyers, M. B., *J. Chem. Soc.*, p. 1197 (1964).
79. Birkinshaw, J. H., Luckner, M., Mohammed, Y. S., Mothes, K., and Stickings, C. E., *Biochem. J.*, *89*, 196 (1963).
80. Mohammed, Y. S., and Luckner, M., *Tetrahedron Lett.*, p. 1953 (1963).
81. Yu, J., Tamura, G., Takahasi, N., and Arima, K., *Agric. Biol. Chem.*, *31*, 831 (1967).
82. Jefferson, W. E., *Biochemistry*, *6*, 3479 (1967).
83. Patterden, G., *Tetrahedron Lett.*, p. 4049 (1969).
84. Merlini, L., Nasini, G., and Selva, A., *Tetrahedron*, *26*, 2739 (1970).
85. Sobin, B. A., and Tanner, F., *J. Am. Chem. Soc.*, *76*, 4053 (1954).
86. Beereboom, J. J., Butler, K., Pennington, F. C., and Solomon, I. A., *J. Org. Chem.*, *30*, 2334 (1965).
87. Brufani, M., Fedeli, W., Mazza, F., Gerhard, A., and Keller-Schierlein, W., *Experientia*, *27*, 1249 (1971).
88. Edwards, R. L., and Elsworthy, G. C., *J. Chem. Soc. Sect. C.* p. 410 (1967).
89. Shimazono, H., *Arch. Biochem. Biophys.*, *83*, 206 (1959).
90. Edwards, R. L., and Elsworthy, G. C., *Chem. Commun.*, p. 373 (1967).
91. Beaumont, P. C., Edwards, R. L., and Elsworthy, G. C., *J. Chem. Soc. Sect. C*, p. 2968 (1968).
92. Steglich, W., Furtner, W., and Prox, A., *Z. Naturforsch. Teil B*, *23*, 1044 (1968).
93. Steglich, W., Furtner, W., and Prox, A., *Z. Naturforsch. Teil B*, *24*, 941 (1969).
94. Edwards, R. L., Lewis, D. G., and Wilson, D. V., *J. Chem. Soc.*, p. 4995 (1961).
95. Bu'Lock, J. D., Leeming, P. R., and Smith, H. G., *J. Chem. Soc.*, p. 2085 (1962).
96. Bu'Lock, J. D., in *Essays in Biosynthesis and Microbial Development*, p. 2. John Wiley and Sons, London, England (1967).
97. Richards, M., Bird, A. E., and Munden, J. E., *J. Antibiot. (Tokyo) Ser. A*, *22*, 388 (1969).

NITROGEN-CONTAINING COMPOUNDS

SIMPLE AMINES AND AMIDES

DR. NANCY N. GERBER

Many of the simple amines result from the decarboxylation of common amino acids.

Name	Structure	Microorganism	Reference
Ammonia	NH_3	Widely distributed	1
Methylamine	CH_3NH_2	Widely distributed	1, 2
Ethylamine	$CH_3CH_2NH_2$	*Claviceps purpurea, Polyporus sulfureus*	1, 2
Dimethylamine	CH_3NHCH_3	*Phallus impudicus, Clathrus ruber, Russula aurata*	1, 2
Ethanolamine	$HOCH_2CH_2NH_2$	Widely distributed	1, 2
Aminoacetone	$CH_3COCH_2NH_2$	*Staphylococcus aureus*	1
Trimethylamine	$(CH_3)_3N$	Many fungi	1, 2
Cycloserine	$CH_2ONHCOCHNH_2$	*Streptomyces* sp.	3
n-Propylamine	$CH_3CH_2CH_2NH_2$	*Claviceps purpurea, Polyporus sulfureus*	1, 2
Isopropylamine	$(CH_3)_2CHNH_2$	*Claviceps purpurea*	1, 2
Methylaminoethanol	$HOCH_2CH_2NHCH_3$	*Neurospora crassa* mutant	1, 2
n-Butylamine	$CH_3CH_2CH_2CH_2NH_2$	Bacteria	4
Isobutylamine	$(CH_3)_2CHCH_2NH_2$	*Claviceps purpurea*	1, 2
1-Amino-2-methyl-2-propanol	$(CH_3)_2C(OH)CH_2NH_2$	*Neurospora crassa*	1, 2
Putrescine	$H_2N(CH_2)_4NH_2$	*Boletus edulis, B. luteus, B. elegans, Amanita muscaria*	1, 2
Pantherine (Agarin)	$ON=C(OH)CH=CCH_2NH_2$	*Amanita pantherina, A. muscaria*	3
Histamine	$HNCH=NCH=CCH_2CH_2NH_2$	*Claviceps purpurea, Coprinus comatus*	1, 2
Amylamine	$CH_3(CH_2)_4NH_2$	Bacteria	4
Isoamylamine	$(CH_3)_2CHCH_2CH_2NH_2$	Many fungi	1, 2
Cadaverine	$H_2N(CH_2)_5NH_2$	*Brevibacterium linens*	5
Triethylamine	$(CH_3CH_2)_3N$	*Brevibacterium linens*	5
n-Hexylamine	$CH_3(CH_2)_5NH_2$	*Claviceps purpurea*	1, 2

Name	Structure	Microorganism	Reference
Dimethylhistamine	$HNCH{=}NCH{=}CCH_2CH_2N(CH_3)_2$	*Coprinus comatus*	1, 2
Acetylcholine	$(CH_3)_3N^+CH_2CH_2OCOCH_3$ OH^-	*Streptobacterium plantarum*	1, 2
3-Butenyltrimethyl-ammonium chloride	$CH_2{=}CHCH_2CH_2N^+(CH_3)_3Cl^-$	*Amanita muscaria*	3
Spermidine	$H_2N(CH_2)_3NH(CH_2)_4NH_2$	Yeast, *Neurospora crassa*	1, 2
β-Phenylethylamine	$PhCH_2CH_2NH_2$	Fungi	1, 2, 4
Tyramine	$4\text{-}HOPhCH_2CH_2NH_2$	*Coprinus comatus, Claviceps purpurea*	1, 2
2-Aminoacetophenone	$2\text{-}H_2NPhCOCH_3$	*Pseudomonas aeruginosa*	5
Muscarine	$OCH(CH_3)CH(OH)CH_2CHCH_2N^+(CH_3)_3$	*Amanita muscaria*	1, 2, 3
Muscaridine	$CH_3CH(OH)CH(OH)(CH_2)_3N^+(CH_3)_3$	*Amanita muscaria*	1, 2, 3
Spermine	$H_2N(CH_2)_3NH(CH_2)_4NH(CH_2)_3NH_2$	Yeast, *Neurospora crassa*	1, 2
Necrosamine	$CH_3(CH_2)_{14}CH(NH_2)CH(NH_2)CH_2CH_2CH_3$	*Escherichia coli*	1
Slaframine		*Rhizoctonia legumini-cola*	3
Thiourea	$(H_2N)_2C{=}S$	*Verticillium albo-atrum, Botrytis cinerea*	1, 2
Guanidine	$(H_2N)_2C{=}NH$	*Boletus edulis, Hydnum aspratum*	1, 2
Cellocidin (Aquamycin)	$H_2NCOC{\equiv}CCONH_2$	*Streptomyces chibaensis, S. reticuli*	1
1-Ethoxy-1,2-ethylene-dicarboxamide	$H_2COC(OCH_2CH_3){=}CHCONH_2$	*Streptomyces* sp.	1

REFERENCES

1. Miller, M. W., *The Pfizer Handbook of Microbial Metabolites.* McGraw-Hill, New York (1961).
2. Shibata, S., Natori, S., and Udagawa, S., *List of Fungal Products.* Charles C Thomas, Springfield, Illinois (1964).
3. Turner, W. B., *Fungal Metabolites.* Academic Press, New York (1971).
4. Bast, E., *Arch. Mikrobiol.*, *79,* 7 (1971).
5. Tokita, F., and Hosono, A., *Milchwissenschaft, 23,* 690 (1968).
6. Mann, S., *Arch. Mikrobiol., 54,* 184 (1966).

AMINO ACIDS

DR. NANCY N. GERBER

The common amino acids from proteins are discussed in Volume II of the Handbook of Microbiology. The following list of microbial amino acids, undoubtedly incomplete, was compiled primarily from References 1, 2 and 3. Amino acids with a heterocyclic ring system (e.g., pyridine and indole) are listed elsewhere in this volume.

Name	Formula	Microorganism	Reference
Glycine	H_2NCH_2COOH	Widely distributed	1
Hadacidin	$(OHC)(OH)NCH_2COOH$	*Penicillium frequentans*	2, 3
Sarcosine	CH_3NCH_2COOH	*Cladonia sylvatica*	1
L-Alanine	$CH_3CH(NH_2)COOH$	Widely distributed	1
β-Alanine	$H_2NCH_2CH_2COOH$	Widely distributed	1
2,3-Diaminopropionic acid	$H_2NCH_2CH(NH_2)COOH$	*Chainia olivacea*	4
L-Serine	$HOCH_2CH(NH_2)COOH$	Widely distributed	1
L-Aspartic acid	$HOOCCH_2CH(NH_2)COOH$	Widely distributed	1
L-Asparagine	$H_2NCOCH_2CH(NH_2)COOH$	Widely distributed	1
d-Diaminosuccinic acid	$HOOCCH(NH_2)CH(NH_2)COOH$	*Streptomyces rimosus*	1
O-Carbamyl-D-serine	$H_2NCOOCH_2CH(NH_2)COOH$	*Streptomyces polychromogenes*	1
Allantoic acid (Allantonic acid)	$(H_2NCONH)_2CHCOOH$	*Coprinus miraceus* *Collybia dryophila*	1, 2
L-(+)-α-Aminobutyric acid	$CH_3CH_2CH(NH_2)COOH$	*Escherichia coli,* *Corynebacterium diphtheriae*	1
γ-Aminobutyric acid	$H_2NCH_2CH_2CH_2COOH$	Widely distributed	1
L-Threonine	$CH_3CHOHCH(NH_2)COOH$	Widely distributed	1
S-Methyl-L-cysteine	$CH_3SCH_2CH(NH_2)COOH$	*Neurospora crassa*	1, 2
N-Methyl-O-methyl-L-serine	$CH_3OCH_2CH(NCH_3)COOH$	*Mycobacterium butyricum,* *M. avium*	5
Azaserine	$N_2CHCOOCH_2CH(NH_2)COOH$	Streptomycete sp.	1
L-Proline	$HNCH_2CH_2CH_2CHCOOH$	Widely distributed	1
L-Glutamic acid	$HOOCCH_2CH_2CH(NH_2)COOH$	Widely distributed	1
L-Glutamine	$H_2NCOCH_2CH_2CH(NH_2)COOH$	Widely distributed	1

Name	Formula	Microorganism	Reference
L-Valine	$(CH_3)_2 CHCH(NH_2)COOH$	Widely distributed	1
Betaine	$(CH_3)_3 N^+CH_2 COO^-$	*Aspergillus oryzae, Patella vulgata, Claviceps purpurea,* other fungi	1, 2
L-Methionine	$CH_3 SCH_2 CH_2 CH(NH_2)COOH$	Widely distributed	1
L-Ornithine	$H_2 NCH_2 CH_2 CH_2 CH(NH_2)COOH$	Widely distributed	1
Choline sulfate	$(CH_3)_3 N^+CH_2 CH_2 OSO_3^-$	*Aspergillus sydowi, Penicillium chrysogenum,* lichens, yeasts	1, 2
4-Imidazoylacetic acid	$N=CHNHCH=CCH_2 COOH$	*Polyporus sulfureus*	1, 2
Muscazone	$OCONHCH_2 CHCH(NH_2)COOH$	*Amanita muscaria*	3
Ibotenic acid	$ON=COHCH=CCH(NH_2)COOH$	*Amanita muscaria, A. pantherina, A. strobiliformis*	3
Tricholomic acid	$ONHCOCH_2 CHCH(NH_2)COOH$	*Tricholoma muscarium*	3
3-Amino-L-proline	$HNCH_2 CH_2 C(NH_2)CHCOOH$	*Morchella esculenta*	3
Imidazoleacetol	$HNCH_2=N^+HCH=CCH_2 COCH_2 OH$ Cl^-	*Neurospora crassa* and *Escherichia coli* mutants	1, 2
L-Histidine	$HNCH=NCH=CCH_2 CH(NH_2)COOH$	*Claviceps purpurea*	1
6-Diazo-5-oxo-L-norleucine	$N_2 CHCOCH_2 CH_2 CH(NH_2)COOH$	*Streptomyces* sp.	1
Imidazoleglycerol	$HNCH=N^+HCH=CCHOHCHOHCH_2 OH$ Cl	*Neurospora crassa* mutant	1, 2
L-Histidinol	$HNCH=N^+HCH=CCH_2 CH(NH^+_3)CH_2 OH$ $2Cl^-$	*Escherichia coli* mutant	1
L-Leucine	$(CH_3)_2 CHCH_2 CH(NH_2)COOH$	Widely distributed	1
L-Isoleucine	$CH_3 CH_2 CH(CH_3)CH(NH_2)COOH$	Widely distributed	1
L-a-Aminoadipic acid	$HOOC(CH_2)_3 CH(NH_2)COOH$	*Aspergillus oryzae*	1, 2
L-Lysine	$H_2 N(CH_2)_4 CH(NH_2)COOH$	*Ustilago maydis*	1
L-Arginine	$H_2 NC(=NH)NHCH_2 CH_2 CH_2 CH$	Widely distributed	1
β-Methylene-L-(+)-norvaline	$CH_3 CH_2 C(=CH_2)CH(NH_2)COOH$	*Lactarius helvus*	3
δ-Oxy-L-lysine	$H_2 NCH_2 CHOHCH_2 CH_2 CH(NH_2)COOH$	*Mycobacterium phlei*	1
1-Amino-2-nitro-cyclopentane carboxylic acid	$O_2 NCHCH_2 CH_2 CH_2 C(NH_2)COOH$	*Aspergillus wentii*	3

Name	Formula	Microorganism	Reference
a-Aminoheptanoic acid	$CH_3(CH_2)_4 CH(NH_2)COOH$	*Claviceps purpurea*	6
L-Carnitine	$(CH_3)_3 N^+ CH_2 CHOHCH_2 COO^-$	*Neurospora crassa*	2
Stachydrine	$(CH_3)_2 N^+ CH_2 CH_2 CH_2 \overline{CHCOO^-}$	*Aspergillus oryzae*, other fungi	1, 2
3-Butenyltrimethyl-ammonium chloride	$CH_2=CHCH_2 CH_2 N^+(CH_3)_3 Cl^-$	*Amanita muscaria*	3
2,6-Diaminopimelic acid	$HOOCCH(NH_2)(CH_2)_3 CH(NH_2)COOH$	*Corynebacterium diphtheriae*, *Mycobacterium tuberculosis*, *Bacillus anthracis*, *Escherichia coli* mutants	1
β-Methyllanthionine	$HOOCCH(NH_2)CH_2 SCH(CH_3)CH(NH_2)COOH$	Yeast, *Phallus impudicus*	1, 2, 7
Furanomycin (Threomycin)	$\overline{OCH(CH_3)}CH=CHCCH(NH_2)COOH$	*Streptomyces threomyceticus*	8
Anthranilic acid	$2\text{-}NH_2 PhCOOH$	*Corynebacterium diphtheriae*	1
p-Aminobenzoic acid	$4\text{-}NH_2 PhCOOH$	*Hansenula anomala*, *Mycotorula lipolytica*	1
3-Hydroxyanthranilic acid	$3\text{-}OH\text{-}2\text{-}NH_2 PhCOOH$	Degradation of tryptophan	3
2-Amino-3-hydroxy-2,3-dihydrobenzoic acid	$H_2 NCHCHOHCH=CHCH=CCOOH$	*Streptomyces aureofaciens*	3
N-Formyl anthranilic acid	$2\text{-}OHCNHPhCOOH$	*Streptomyces* sp., *Neurospora crassa*	9
a-Amino octanoic acid	$CH_3(CH_2)_5 CH(NH_3)COOH$	*Aspergillus* atypical	14
Pencolide	$COC(CH_3)=CHCONC(COOH)=CHCH_3$	*Penicillium multicolor*	2
L-Phenylalanine	$PhCH_2 CH(NH_2)COOH$	Widely distributed	1
L-Tyrosine	$4\text{-}OH\text{-}PhCH_2 CH(NH_2)COOH$	Widely distributed	1
3-(3,4-Dihydroxy-phenyl)-L-alanine	$3\text{-}OH\text{-}4\text{-}OHPhCH_2 CH(NH_2)COOH$	Actinomycetes	13
Hercynine	$HNCH=NCH=CCH_2 CH(COO^-)N(CH_3)_3$	*Amanita muscaria*, *Agaricus campestris*, *Boletus edulis*, *Polyporus sulfureus*	1, 2
Laminine	$(CH_3)_3 N^+(CH_2)_4 CH(NH_2)COOH$	*Lamanaria augustata*	15
Ergothioneine	$HNC(SH)=NCH=CCH_2 CH(COO^-)N(CH_3)_3$	*Claviceps purpurea*, *Coprinus comatus*, *Mycobacterium tuberculosis*	1, 2
2-Pyruvoylamino-benzamide	$2\text{-}CH_3 COCONHPhCONH_2$	*Penicillium chrysogenum* *P. notatum*	3

Name	Formula	Microorganism	Reference
Fusarinine	$HOCH_2CH_2CH(CH_3)=CHCONOH(CH_2)_3-$ $CH(NH_2)COOH$	*Fusarium* sp.	11
2-Methylenecyclo- heptene-1,3-diglycine		*Lactarius helvus*	3
Agaritine	$4-HOCH_2\,PhNHNHCOCH_2\,CH_2\,C(NH_2\,)COOH$	*Agaricus bisporus*	2, 3
N-Jasmonoylisoleucine and N-dihydrojasmonoyl- isoleucine		*Gibberella fujikuroi*	12

REFERENCES

1. Miller, M. W., *The Pfizer Handbook of Microbial Metabolites.* McGraw-Hill, New York (1961).
2. Shibata, S., Natori, S., and Udagawa, S., *List of Fungal Products.* Charles C Thomas, Springfield, Illinois, (1964).
3. Turner, W. B., *Fungal Metabolites.* Academic Press, New York (1971).
4. Lechevalier, M. P., Lechevalier, H. A., and Heintz, C. E., *Int. J. Syst. Bacteriol., 23,* 157 (1973).
5. Viekas, E., Rojas, A., and Lederer, E., *Compt. Rend., 261,* 4258 (1965).
6. Steiner, M., and Hartmann, T., *Biochem. Z., 340,* 436 (1964).
7. List, P. H., and Reinhard, C., *Arch. Pharm., 295,* 564 (1962).
8. Katagiri, K., Tori, K., Kimura, Y., Yoshida, T., Nagasaki, T., and Minato, H., *J. Med. Chem., 10,* 1149 (1967).
9. Teuscher, E., *Pharmazie, 21,* 320 (1966).
10. Shiman, R., and Neilands, J. B., *Biochemistry, 4,* 2233 (1965).
11. Emery, T., *Biochemistry, 4,* 1410 (1965); *7,* 184 (1968).
12. Cross, B., and Webster, G. R. B., *J. Chem. Soc. Sect. C.,* p. 1839 (1970).
13. Amao, S., Nii, M., Kobayashi, T., Hoshi, K., and Ishibashi, K., *Annu. Rep. Sanhyo Res. Lab. Jap., 23,* 245 (1971).
14. Thadee, S., Allard, C., and Xuong, N. D., *Compt. Rend., 260,* 3502 (1965).
15. Takemoto, T., Daigo, K., and Takagi, N., *Yakugaku Zasshi, 84,* 1176 (1964).

CYCLODEPSIPEPTIDE ANTIBIOTICS

D. J. HOOK

All the cyclodepsipeptides show infrared absorption at 1735 to 1750 cm^{-1} (lactone) and at *ca.* 1650 and 1680 cm^{-1} (amide). They are insoluble in water or saturated hydrocarbons but soluble in most organic solvents. (There are two exceptions to this general rule: telomycin is soluble in water but not in most organic solvents, and destruxin A is soluble in water). They can be extracted by organic solvents from culture filtrates adjusted to the appropriate pH as well as from the mycelium. Most are amorphous powders, although some crystallize as needles. They are all colorless substances, active against Gram-positive but not against Gram-negative bacteria.

Cyclodepsipeptides are often produced by the culture as families of chemically related compounds. Separation of the congeners may be difficult or impossible, and even when it is possible, some are only obtained in small amounts, sufficient for identification of their structure but not for determination of their physical properties. The triostin, quinoxaline and sporidesmolide groups are examples of this situation. When the components cannot be separated, their structure can sometimes be inferred, but the physical properties reported are those for the mixture. The stendomycin complex[1] provides an example of this situation.

For convenience the compounds have been divided into three groups: peptide lactones, true cyclodepsipeptides, and miscellaneous compounds. In peptide lactones, the lactone linkage is through a hydroxy amino acid. They are either totally composed of amino acids or have a hetero moiety attached outside the lactone ring. Two classes in this group, the triostins and the quinomycins, have been modified by cross-bridging. The best methods for identifying these compounds are mass spectroscopy and amino acid analysis; melting points are not useful because they are modified according to the solvent of crystallization and because of the tendency of these compounds to decompose. UV absorption may indicate the chromophore present, but it is not useful for identifying individual compounds, especially in the same family.

The cyclodepsipeptides contain both amino and hydroxy acids. Two classes can be distinguished: those containing alternating hydroxy and amino acid residues, and those consisting largely of amino acids, with the lactone linkage through hydroxy or hydroxy amino acids. Again melting points are not good criteria for identification; mass spectroscopy or hydroxy and amino acid composition are preferred.

The third group consists of cyclodepsipeptides that contain amino and hydroxy acids, occasionally with other moieties, and have undergone some modification in structure. Again mass spectroscopy is the preferred method of identification, although melting points may be useful here.

PEPTIDE LACTONES

Systematic Names and Structures

Cycloheptamycin   N-(N-formyl-L-valyl)-threonyl-O-methyl-D-tyrosyl-D-alanyl-N-methylalloisoleucyl-β-hydroxy-L-norvalyl-5-methoxytryptophanyl lactone

Doricin   N-(3-hydroxypicolinyl)-L-threonyl-D-α-aminobutyryl-L-prolyl-β-dimethylamino-N-methyl-L-phenylalanyl-L-aspartyl-L-phenylglycyl lactone

Etamycin   N-(3-hydroxypicolinyl) L-threonyl-D-leucyl-4-allohydroxy-D-prolylsarcosyl-N-3,4-trimethyl-L-norvalyl-L-alanyl-L-2-phenylsarcosyl lactone

Griselimycin   N-(N-acetyl-N-methyl-L-valyl-4-methyl-L-prolyl)-N-methyl-L-threonyl-L-leucyl-4-methyl-L-prolyl-L-leucyl-N-methyl-L-valyl-L-prolyl-N-methyl-D-leucyl lactone

Ostreogrycin B3   N-(3-hydroxypicolinyl)-L-threonyl-D-*a*-aminobutyryl-L-prolyl-*p*-dimethylamino-N-methyl-L-phenylalanyl-5-hydroxy-4-oxopipecolyl-L-phenylglycyl lactone

Staphylomycin S   N-(3-hydroxypicolinyl)-L-threonyl-D-*a*-aminobutyryl-L-prolyl-N-methyl-L-phenylalanyl-4-oxopipecolyl-L-phenylglycyl lactone

Telomycin   N-(N-β-L-aspartyl-L-seryl)-L-threonyl-allo-L-threonyl-L-alanylglycyl-*trans*-3-hydroxy-L-prolylerythro-3-hydroxyl-L-leucyl-β-methyl-L-tryptophanyl-Δ^2-dehydrotryptophanyl-*cis*-3-hydroxyl-L-prolyl lactone

Vernamycin B$_a$ N-(3-hydroxypicolinyl)-L-threonyl-D-*a*-aminobutyryl-L-prolyl-*p*-dimethylamino-N-methyl-L-phenylalanyl-4-oxopipecolyl-L-phenylglycyl lactone

Vernamycin B$_\beta$ N-(3-hydroxypicolinyl)-L-threonyl-D-*a*-aminobutyryl-L-prolyl-*p*-methylamino-N-methyl-L-phenylalanyl-4-oxopipecolyl-L-phenylglycyl lactone

Vernamycin B$_\gamma$ N-(3-hydroxypicolinyl)-L-threonyl-D-alanyl-L-prolyl-*p*-dimethylamino-N-methyl-L-phenylalanyl-4-oxopipecolyl-L-phenylglycyl lactone

Quinomycins

	R_1	R_2
Quinomycin A:	H	CH_3
Quinomycin B:	CH_3	CH_3
Quinomycin C:	CH_3	H

Triostins

	R_1	R_2	R_3	R_4
Triostin A:	H	H	H	H
Triostin C:	CH_3	CH_3	CH_3	CH_3

Identifying Properties

Common Name (Alternate Names) Molecular Formula	Significant Characteristics	Organism(s)	References
Cycloheptamycin $C_{48}H_{68}O_{12}N_8$	Melting point: 256–258°C UV maximum (EtOH): 276.5 (ϵ = 7550) $[a]_D^{20}$ (CHCl$_3$, c = 1): +37°	*Streptomyces* sp.	28
Doricin $C_{43}H_{52}O_{11}N_8$	Melting point: 170–190°C UV maximum: 305 $[a]_D^{20}$ (MeOH, c = 1): –92°	*Streptomyces loidensis*	2
Etamycin (Viridogresin, 6613, F1370A, K179) $C_{44}H_{62}O_{11}N_8$	Melting point: 163–170°C, decomposes UV maximum (EtOH): 305 (log ϵ = 3.91)[3] UV maximum (MeOH): 304 ($E_{1\ cm}^{1\%}$ = 86)[4] UV maximum (CHCl$_3$, c = 1): 305 ($E_{1\ cm}^{1\%}$ = 95)[25] $[a]_D^{25}$ (CHCl$_3$, c = 5): +62°[3] $[a]_D^{25}$ (MeOH, c = 2): +7.7°[4]	*Streptomyces griseus griseoroseus*	3,4,21,24,25
Griselimycin (RP 11072) $C_{57}H_{96}O_{12}N_{10}$	Melting point: 226–228°C $[a]_D^{22}$ (MeOH, c = 1): –108°	*Streptomyces coelicus griseus*	5,6
Ostreogrycin B3 $C_{44}H_{52}O_{10}N_8$	Melting point: 215°C UV maximum: 305 $[a]_D^{25}$ (MeOH, c = 1): –57°	*Streptomyces ostreogriseus*	7

Common Name (Alternate Names) Molecular Formula	Significant Characteristics	Organism(s)	References
Quinomycin A (Echinomycin) $C_{50}H_{60}O_{12}N_{12}S_2$	Melting point: 218–221°C, decomposes UV maxima (MeOH): 243 (log ϵ = 4.81), 320 (log ϵ = 4.07)[10] UV maxima (MeOH): 243 ($E_1^{1\%}{}_{cm}$ = 694), 320 ($E_1^{1\%}{}_{cm}$ = 130)[11] $[a]_D^{25}$ (CHCl$_3$, c = 0.86): –310°[8] $[a]_D^{25}$ (CHCl$_3$, c = 1): –315°[11]	*Streptomyces* sp. 1752	8,9,10,11
Quinomycin B $C_{52}H_{64}O_{12}N_{12}S_2$	Melting point: 215–219°C, decomposes;[10] 221°C, decomposes[12] UV maxima (MeOH): 243 (log ϵ = 4.85), 320 (log ϵ = 4.12)[10] UV maxima (MeOH): 243 ($E_1^{1\%}{}_{cm}$ = 678), 320 ($E_1^{1\%}{}_{cm}$ = 118)[12] $[a]_D^{26}$ (CHCl$_3$, c = 1.308): –250°[8] $[a]_D^{26}$ (CHCl$_3$, c = 1.308): –300°[12]	*Streptomyces* sp. 732	8,10,12
Quinomycin C $C_{54}H_{68}O_{12}N_{12}S_2$	Melting point: 215–218°C, decomposes UV maxima (MeOH): 243 (log ϵ = 4.82), 320 (log ϵ = 4.08) $[a]_D^{25}$ (CHCl$_3$, c = 0.97): –250°	*Streptomyces* sp. 732	8,10
Staphylomycin S (PA 114 B-2) $C_{43}H_{62}O_{10}N_7$	Melting point: 240–242°C UV maximum (EtOH): 305 (log ϵ = 3.85)[14] UV maximum (MeOH): 304 ($E_1^{1\%}{}_{cm}$ = 86)[15] $[a]_D^{20}$ (EtOH, c = 1): –28°	*Streptomyces* sp. related to *S. virginiae*	14,15
Telomycin $C_{59}H_{79}O_9N_{13}$	Melting point: 230–240°C, decomposes UV maxima (EtOH:H$_2$O = 2:1): 290 (ϵ = 11,890), 339 (ϵ = 22,000)[13] UV maximum (MeOH): 340 ($E_1^{1\%}{}_{cm}$ = 127)[17] $[a]_D^{28}$ (MeOH/H$_2$O 2:1, c = 1): –133°	*Streptomyces* sp.	13,16,17
Triostin A $C_{50}H_{62}O_{12}N_{12}S_2$	Melting point: 245–248°C UV maxima (MeOH): 243 (log ϵ = 4.85), 320 (log ϵ = 4.11) $[a]^{23.5}$ (CHCl$_3$, c = 0.97): –157°	*Streptomyces* sp. S-2-210 *Streptomyces* sp. related to *S. aureus*	10,20
Triostin C $C_{54}H_{70}O_{12}N_{12}S_2$	Melting point: 260°C, decomposes; 210–214°C[20] UV maxima (MeOH): 243 (log ϵ = 4.85), 320 (log ϵ = 4.11)[10] UV maxima (MeOH): 242.5 ($E_1^{1\%}{}_{cm}$ = 622), 322.5 ($E_1^{1\%}{}_{cm}$ = 115)[20] $[a]_D^{24}$ (CHCl$_3$, c = 1.121): –143.9°[18] $[a]_D^{22}$ CHCl$_3$, c = 1): –133.4°[22]	*Streptomyces* sp. S-2-210 *Streptomyces* sp. related to *S. aureus*	10,18,20,22

Common Name (Alternate Names) Molecular Formula	Significant Characteristics	Organism(s)	References
Vernamycin B_α (Ostreogrycin B, mikamycin B, synergistin B-1, streptogramin B1) $C_{45}H_{54}O_{10}N_8$	Melting point: 266–268°C; 263°C[27] UV maximum (EtOH): 305 (ϵ = 8580)[22] UV maximum (MeOH): 305 ($E_1^{1\%}$ $_{cm}$ = 101)[23] UV maximum (MeOH): 305 (ϵ = 9560)[27] $[\alpha]_D^{20}$ (MeOH, c = 1): –66.8°[22] $[\alpha]_D^{23}$ (MeOH, c = 1): –63°[23] $[\alpha]_D^{20}$ (MeOH, c = 1): –60.3°[27]	*Streptomyces loidensis mitakaensis ostreogriseus*	19,22,23,24,27
Vernamycin B_β (Ostreogrycin B2, pristinamycin IB) $C_{44}H_{52}O_{10}N_8$	Melting point: 192–193°C UV maximum (MeOH): 305 ($E_1^{1\%}$ $_{cm}$ = 84)[23] $[\alpha]_D^{22}$ (EtOH, c = 1): –44.5°[23] $[\alpha]_D^{23}$ (MeOH, c = 1): –53°[23]	*Streptomyces loidensis*	19,23,27
Vernamycin B_γ (Ostreogrycin B1, pristinamycin IC)	Melting point: 198°C UV maximum (EtOH): 302 ($E_1^{1\%}$ $_{cm}$ = 98) $[\alpha]_D^{22}$ (EtOH, c = 0.5): –62.5°	*Streptomyces pristinaespiralis*	29

TRUE CYCLODEPSIPEPTIDES

Systematic Names

Angolide	cyclo-(D-alloisoleucyl-L-α-hydroxyisovaleryl-L-isoleucyl-L-α-hydroxyisovaleryl)
Beauvericin	cyclo-(N-methyl-L-phenylalanyl-D-α-hydroxyisovaleryl)$_3$
Destruxin A	D-2-hydroxy-4-pentenoyl-L-prolyl-L-isoleucyl-N-methyl-L-valyl-N-methyl-L-alanyl-β-alanyl lactone
Destruxin B	D-α-hydroxy-γ-methylvaleryl-L-propyl-L-isoleucyl-N-methyl-L-valyl-N-methyl-L-alanyl-β-alanyl lactone
Enniatin A	cyclo-(N-methyl-L-isoleucyl-D-α-hydroxyisovaleryl)$_3$
Enniatin B	cyclo-(N-methyl-L-valyl-D-α-hydroxyisovaleryl)$_3$
Isariin	D-β-hydroxydodecanoyl-L-valyl-L-alanyl-D-leucyl-L-valylglycyl lactone
Peptidolipin NA	D-β-hydroxyeicosanoyl-L-threonyl-L-valyl-D-alanyl-L-prolyl-D-alloisoleucyl-L-alanyl lactone
Pithomycolide	cyclo-(D-β-hydroxy-β-phenylpropionyl-D-β-hydroxy-β-phenylpropionyl-L-α-hydroxyisovaleryl-N-methyl-L-alanyl-L-alanyl)
Serratamolide	cyclo-(D-β-hydroxydecanoyl-L-seryl)$_2$

Sporidesmolide I   cyclo-(L-*a*-hydroxyisovaleryl-D-valyl-D-leucyl-L-*a*-hydroxyisovaleryl-
                      L-valyl-N-methyl-L-leucyl)

Sporidesmolide II   cyclo-(L-*a*-hydroxyisovaleryl-D-alloisoleucyl-D-leucyl-L-
                      a-hydroxyisovaleryl-L-valyl-N-methyl-L-leucyl)

Sporidesmolide III   cyclo-(L-*a*-hydroxyisovaleryl-D-valyl-D-leucyl-L-*a*-
                      hydroxyisovaleryl-L-valyl-L-leucyl)

Surfactin   3-hydroxy-13-methyltetradecanoyl-L-glutamyl-L-leucyl-D-
                      leucyl-L-valyl-L-aspartyl-D-leucyl-L-leucyl lactone

Valinomycin   cyclo-(L-lactyl-L-valyl-D-*a*-hydroxyisovaleryl-D-valyl)$_3$

Viscosin   N-(D-β-hydroxydecanoyl-L-leucyl-D-glutamyl)-D-allothreonyl-D-
                      valyl-L-leucyl-D-seryl-L-leucyl-D-seryl-L-isoleucyl lactone

Identifying Properties

Common Name Molecular Formula	Significant Characteristics	Organism(s)	References

Angolide
$C_{22}H_{38}O_6N_2$
	Melting point: 261–262°C, decomposes
	$[a]_D^{22}$ (CHCl$_3$, c = 1): –83°
		Pithomyces
		  cynodantis
		  sacchari
			30,67

Beauvericin $C_{45}H_{57}O_9N_3$	Melting point: 93–94°C $[a]_D^{25}$ (MeOH, c = 1): +65.8°	*Beauveria* *bassiana*	45,57,73
Destruxin A $C_{29}H_{47}O_7N_5$	Melting point: 188°C $[a]_D^{15}$ (MeOH, c = 2.25): –224.8°	*Metarrhizium* *anisopliae*[a]	31,72,78
Destruxin B $C_{30}H_{51}O_7N_5$	Melting point: 234°C, decomposes $[a]_D^{23}$ (MeOH, c = 0.5): –228°	*Metarrhizium* *anisopliae*[a]	31,43,44

Enniatin A
$C_{36}H_{63}O_9N_3$
	Melting point: 121–122°C
	$[a]_D^{18}$ (CHCl$_3$, c = 1): –91°
		Fusarium
		  orthoceras var.
		    enniatinum
		  scirpi
			32,35
			
			33,36,58

Enniatin B $C_{33}H_{57}O_9N_3$	Melting point: 173–175°C $[a]_D^{19}$ (CHCl$_3$, c = 1): –107°	*Fusarium* sp.	32,34,35,37,59,60
Isariin $C_{33}H_{59}O_7N_5$	Melting point: 250°C	*Isaria* *cretacea*	38,39,68
Peptidolipin NA $C_{50}H_{89}O_{11}N_7$	Melting point: 232–233°C $[a]_D$ (CHCl$_3$, c = 3.39): +42.7°	*Nocardia* *asteroides*	75,76,77
Pithomycolide $C_{30}H_{36}O_2N_8$	Melting point: 242–244°C $[a]_D$ (CHCl$_3$, c = 0.095): –60°	*Pithomyces* *chatarum*	40

Common Name (Alternate Names) Molecular Formula	Significant Characteristics	Organism(s)	References
Serratamolide $C_{26}H_{46}O_8N_2$	Melting point: 159–160°C $[a]_D^{25}$ (EtOH, c = 2.5): = +4.8° $[a]_D^{25}$ (CHCl$_3$, c = 2.3): +7.9°	*Serratia marcescens*	41,71
Sporidesmolide I $C_{33}H_{58}O_8N_4$	Melting point: 261–263°C $[a]_D$ (AcOH, c = 1.5): –98° $[a]_D$ (CHCl$_3$, c = 1.5): –217°	*Pithomyces chatarum*	42
Sporidesmolide II[b] $C_{24}H_{60}O_8N_4$	Melting point: 226–228°C $[a]_D^{20}$ (CHCl$_3$, c = 0.6): –195°	*Pithomyces chatarum*	62,63,74
Sporidesmolide III $C_{32}H_{56}O_8N_4$	Melting point: 277–278°C; 294–295°C $[a]_D^{18}$ (AcOH, c = 1.6): –80°	*Pithomyces chatarum*	46,81
Sporidesmolide IV $C_{34}H_{60}O_8N_4$	Melting point: 232–233°C $[a]_D$ (CHCl$_3$, c = 2): –212°	*Pithomyces maydicus*	47,64,65
Surfactin $C_{53}H_{93}O_{13}N_7$	Melting point: 140°C $[a]_D^{27}$ (CHCl$_3$, c = 1): +40° $[a]_D^{27}$ (MeOH, c = 1): –39°	*Bacillus subtilis*	49,50,52,53
Valinomycin $C_{54}H_{90}O_{18}N_6$	Melting point: 190°C $[a]_D^{20}$ (C$_6$H$_6$, c = 1.6): +31° $[a]_D^{22}$ (C$_6$H$_6$, c = 1): –31.3°	*Streptomyces* sp. *fulvissimus*	54,55,56,61,79,80
Viscosin $C_{54}H_{97}O_{17}N_9$	Melting point: 270–273°C $[a]_D^{29}$ (EtOH, c = 1): –168.3°	*Pseudomonas viscosa*	48,69,70

[a] Originally *Oospora destructor*.
[b] Synthetic.

MISCELLANEOUS COMPOUNDS

Structures

Griseoviridin

Ostreogrycin A

Ostreogrycin G

Pyridomycin

Identifying Properties

Common Name Molecular Formula	Significant Characteristics	Organism(s)	References
Griseoviridin $C_{22}H_{27}O_7N_3S$	Melting point: 228–230°C UV maximum(EtOH): 220.5 (ϵ = 44,000) $[a]_D^{27}$ (MeOH, c = 0.5): –237°	*Streptomyces* sp. *griseus*	82,83,84,85,86
Ostreogrycin A $C_{28}H_{35}O_7N_3$	Melting point: 203–205°C UV maximum (EtOH, 95%): 228 (log ϵ = 4.15) $[a]_D^{20}$ (EtOH, c = 0.34): –218°	*Streptomyces* *ostreogriseus*	87,89
Ostreogrycin G $C_{28}H_{37}O_7N_3$	Melting point: 122–127°C, decomposes UV maximum (EtOH): 215 (log ϵ = 4.53) $[a]_D^{21}$ (EtOH, c = 1.36): +78°	*Streptomyces* *ostreogriseus*	88
Pyridomycin $C_{27}H_{32}O_8N_4$	Melting point: 214–216°C UV maximum (EtOH): 303 ($E_{1\,cm}^{1\%}$ = 209) $[a]_D^{16}$ (CHCl$_3$, c = 1): –90.3°	*Streptomyces* *pyndomyceticus*	90,91,92

REFERENCES

1. Bodanszky, M., Izdebski, J., and Muramatsu, I., *J. Amer. Chem. Soc., 91,* 2351 (1969).
2. Bodanszky, M., Sheehan, J. T., *Antimicrob. Agents Chemother.* p. 38 (1963).
3. Sheehan, J. C., Zachau, H. G., and Lawson, W. B., *J. Amer. Chem. Soc., 80,* 3349 (1958).
4. Heinemann, B., Gourevitch, A., Leir, J., Johnson, D. L., Kaplan, M. A., Varas, D., and Hooper, I. R., *Antibiotics Annual 1954–1955,* p. 728 (1955).
5. Terlain, B., and Thomas, J.-P., *C. R. Acad. Sci. (Paris), 269C,* 1546 (1969).
6. Noufflard-Guy-Loe, H., and Bertreaux, S., *Rev. Tuberc. Pneumol., 29,* 301 (1965).
7. Cox, B. R., Eastwood, F. W., Snell, B. K., and Lord Todd, *J. Chem. Soc. Sect. D,* p. 1623 (1970).
8. Yoshida, T., Katagiri, K., and Yokozawa, S., *J. Antibiot. (Tokyo) Ser. A 14,* 330 (1961).
9. Keller-Schierlein, W., Mihailovic, N. Lj., and Prelog, V., *Helv. Chim. Acta, 42,* 305 (1969).
10. Otsuka, H., and Shoji, J., *Tetrahedron, 23,* 1535 (1967).
11. Katagiri, K., Shoji, J., and Yoshida, T., *J. Antibiot. (Tokyo) Ser. A 15,* 273 (1962).

12. Yoshida, T., and Katagiri, K., *J. Antibiot. (Tokyo) Ser. A 15,* 272 (1962).
13. Sheehan, J. C., Mania, D., Nakamura, S., Stock, J. A., and Maeda, K., *J. Amer. Chem. Soc., 90,* 462 (1969).
14. Vanderheaghe, H., and Parmentier, G., *J. Amer. Chem. Soc., 82,* 4414 (1960).
15. Vanderheaghe, H., Van Dijck, P., Parmentier, G., and de Somer, P., *Antibiot. Chemother., 7,* 606 (1957).
16. Sheehan, J. C., Drummond, P. E., Gardner, J. N., Maeda, K., Mania, D., Nakamura, S., Sen, A. K., and Stock, J. A., *J. Amer. Chem. Soc., 85,* 2867 (1963).
17. Misiek, M., Fardig, O. B., Gourevitch, A., Johnson, D. L., Hooper, I. R., and Lein, J., *Antibiotics Annual, 1957–1958,* p. 852 (1958).
18. Shoji, J., and Katagiri, K., *Tetrahedron, 29,* 2931 (1963).
19. Bodanszky, M., and Odetti, M. A., *Antimicrob. Agents Chemother.,* p. 360 (1963).
20. Otsuka, H., and Shoji, J., *J. Antibiot. (Tokyo) Ser. A 14,* 335 (1961).
21. Arnold, R. B., Johnson, A. W., and Manger, A. B., *J. Chem. Soc.,* p. 4466 (1958).
22. Eastwood, F. W., Snell, B. K., and Todd, S. A., *J. Chem. Soc.,* p. 2286 (1960).
23. Prend'homme, J., Belloc, A., Charpentier, Y., and Tarridec, P., *C. R. Acad. Sci. (Paris), 260C,* 1309 (1965).
24. Bartz, W. R., Standiford, J., Mold, J. D., Johannessen, D. W., Ryder, A., Maretzki, A., and Haskell, T. H., *Antibiotics Annual, 1954–1955,* p. 777 (1955).
25. Haskell, T. H., Maretzki, A., and Bartz, Q. R., *Antibiotics Annual 1954–1955,* p. 784 (1955).
26. Jolles, G., Terlain, B., and Thomas, J. P., *Nature, 207,* 199 (1965).
27. Watarabe, K., *J. Antibiot. (Tokyo) Ser. A 14,* 1 (1961).
28. Godtfredson, W. O., and Vangedar, S., *Tetrahedron, 26,* 4931 (1970).
29. Prend'homme, J., Tarridec, P., and Belloc, A., *Bull. Soc. Chim. Fr.,* p. 585 (1968).
30. Russell, D. W., *J. Chem. Soc. Sect. C,* p. 4664 (1965).
31. Suzuki, A., Takahashi, N., and Tamura, S., *Org. Mass Spectra, 4,* 175 (1970).
32. Plattner, P. A., Nager, U., and Boller, A., *Helv. Chim. Acta, 31,* 594 (1948).
33. Plattner, P. A., and Nager, U., *Helv. Chim. Acta, 31,* 2192 (1948).
34. Plattner, P. A., and Nager, U., *Helv. Chim. Acta, 31,* 665 (1948).
35. Shemyakin, M. M., Ovchinnikov, Yu. A., Ivarov, V. T., and Kiryushkin, A. A., *Tetrahedron, 19,* 581 (1963).
36. Quitt, P., Studer, R. O., and Vogler, K., *Helv. Chim. Acta, 46,* 1715 (1963).
37. Plattner, P. A., Vogler, K., Studer, R. O., Quitt, P., and Keller-Schierlein, W., *Helv. Chim. Acta, 46,* 927 (1963).
38. Wolstenhome, W. A., and Vining, L. C., *Tetrahedron Lett.,* p. 7285 (1966).
39. Vining, L. C., and Taber, W. A., *Can. J. Chem., 40,* 1579 (1962).
40. Briggs, L. H., Colebrook, L. D., Davis, B. R., and LeQuense, P. W., *J. Chem. Soc.,* p. 5626 (1964).
41. Wasserman, H. H., Keggi, J. J., and McKeon, J. E., *J. Amer. Chem. Soc., 84,* 2978 (1962).
42. Russell, D. W., *J. Chem. Soc.,* p. 753 (1962).
43. Kuyama, S., and Tamura, S., *Agr. Biol. Chem., 29,* 168 (1965).
44. Tamura, S., Kuyama, S., Kodaira, Y., and Higashikawa, S., *Agr. Biol. Chem., 28,* 137 (1964).
45. Kiryushkin, A. A., Ovchinnikov, Yu. A., and Shemyakin, M. M., *Khim. Prir. Soedin. (Tashk.), 1,* 58 (1965).
46. Russell, D. W., MacDonald, C. G., and Shannon, J. S., *Tetrahedron Lett.,* p. 2759 (1964).
47. Bishop, E., and Russell, D. W., *J. Chem. Soc. Sect. C,* p. 634 (1967).
48. Hiramoto, M., Okada, K., and Nagai, S., *Tetrahedron Lett.,* p. 1087 (1970).
49. Kakinuma, A., Ouchida, A., Shina, T., Sugiro, H., Isono, M., Tamura, G., and Arima, K., *Agr. Biol. Chem., 33,* 1669 (1969).
50. Kakinuma, A., Hori, M., Sugiro, H., Yoshida, I., Isono, M., Tamura, G., and Arima, K., *Agr. Biol. Chem., 33,* 1523 (1969).
51. Kakinuma, A., Sugiro, H., Isono, M., Tamura, G., and Arima, K., *Agr. Biol. Chem., 33,* 973 (1969).
52. Kakinuma, A., Hori, M., Isono, M., Tamura, G., and Arima, K., *Agr. Biol. Chem., 33,* 971 (1969).
53. Arima, K., Kakinuma, A., and Tamura, G., *Biochem. Biophys. Res. Commun., 31,* 488 (1968).
54. Brown, R., Brennan, J., and Kelley, C., *Antibiot. Chemother., 12,* 482 (1962).
55. Brockmann, H., and Schmidt-Kastner, G., *Chem. Ber., 88,* 57 (1955).
56. Brockmann, H., and Geeren, H., *Justus Liebigs Ann. Chem., 603,* 216 (1957).
57. Hamill, R. L., Higgers, C. E., Booz, N. E., and Gorman, M., *Tetrahedron Lett.,* p. 4255 (1969).
58. Shemyakin, M. M., Ovchinnikov, Yu. A., Kiryushkin, A. A., and Ivanov, V. T., *Izv. Akad. Nauk SSSR Ser. Khim.,* p. 1148 (1963).
59. Shemyakin, M. M., Ovchinnikov, Yu. A., Kiryushkin, A. A., and Ivanov, V. T., *Izv. Akad. Nauk SSSR Ser. Khim.,* p. 579 (1963).
60. Shemyakin, M. M., Ovchinnikov, Yu. A., Kiryushkin, A. A., and Ivanov, V. T., *Tetrahedron Lett.,* p. 885 (1963).
61. Shemyakin, M. M., Aldarova, N. A., Vinogradoya, E. I., and Feigina, M. Yu., *Tetrahedron Lett.,* p. 1921 (1963).
62. Bertaud, W. S., Probine, M. C., Shannon, J. S., and Taylor, A., *Tetrahedron, 21,* 677 (1965).
63. Bertaud, W. S., Morice, I. M., Russell, D. W., and Taylor A., *J. Gen. Microbiol., 32,* 385 (1963).
64. Bishop, E., and Russell, D. W., *Biochem. J., 92,* 19p (1964).
65. Bishop, E., Griffiths, H., Russell, D. W., Ward, V., and Gartside, R. N., *J. Gen. Microbiol., 38,* 289 (1965).
66. Fones, W. S., *J. Amer. Chem. Soc., 76,* 1377 (1954).

67.   MacDonald, C. G., and Shannon, J. S., *Tetrahedron Lett.,* 3113 (1964).
68.   Kiryushkin, A. A., Ovchinnikov, Yu. A., Rozynov, B. V., and Vul'fson, N. S., *Khim. Prir. Soedin. (Tashk.), 2,* 203 (1966).
69.   Hiramoto, M., Okada, K., Nagai, S., and Kawamoto, H., *Biochem. Biophys. Res. Commun. 35,* 702 (1969).
70.   Ohno, T., Tajima, S., and Toki, K., *Bull. Agr. Chem. Soc. Jap., 27,* 665 (1953).
71.   Wasserman, H. H., Keggi, J. J., and McKeon, J. E., *J. Amer. Chem. Soc., 83,* 4107 (1961).
72.   Suzuki, A., Kuyama, S., Kodaira, Y., and Tamura, S., *Agr. Biol. Chem., 30,* 517 (1966).
73.   Ovchinnikov, Yu. A., Ivanov, V. T., and Mikhaleva, I. I., *Tetrahedron Lett.,* p. 159 (1971).
74.   Shemyakin, M. M., Ovchinnikov, Yu. A., Ivanov, V. T., Kiryushkin, A. A., and Khalilulira, K. Kh., *Zh. Obshch. Khim., 35,* 1399 (1965).
75.   Guirard, M., Michel, G., and Lederer, E., *C. R. Acad. Sci. (Paris), 259,* 1267 (1964).
76.   Guirard, M., and Michel, G., *Biochim. Biophys. Acta, 125,* 75 (1966).
77.   Barber, M., Wolstenholme, W. A., Guirard, M., Michel, G., Das, B. C., and Lederer, E., *Tetrahedron Lett.,* p. 1331 (1963).
78.   Kodaira, Y., *Agr. Biol. Chem., 26,* 137 (1962).
79.   Ohnishi, M., and Urry, D. W., *Biochem. Biophys. Res. Commun., 36,* 194 (1969).
80.   Shemyakin, M. M., Aldarova, N. A., Virogradova, E. I., and Feigina, M. Yu., *Tetrahedron Lett.,* p. 1921 (1963).
81.   Ovchinnikov, Yu. A., Kiryushkin, A. A., and Shemyakin, M. M., *Zh. Obshch. Khim., 36,* 620 (1966).
82.   Ames. D. E., and Bowman, R. E., *J. Chem. Soc.,* p. 4265 (1955).
83.   Ames, D. E., and Bowman, R. E., *J. Chem. Soc.,* p. 2925 (1955).
84.   Ames, D. E., Bowman, R. E., Cavalla, J. F., and Evans, D. D., *J. Chem. Soc.,* p. 4260 (1955).
85.   Fallona, M. C., McMorris, T. C., De Mayo, P., Money, T., and Stoessl, A., *J. Amer. Chem. Soc., 84,* 4162 (1962).
86.   Fallona, M. C., De Mayo, P., and Stoessl, A., *Can. J. Chem., 42,* 394 (1964).
87.   Delpierre, G. R., Eastwood, F. W., Gream, G. E., Kingston, D. G. I., Lord Todd, and Williams, D. H., *J. Chem. Soc. Sect C,* p. 1653 (1966).
88.   Kingston, D. G. I., Sarin, P. S., Lord Todd, and Williams, D. H., *J. Chem. Soc. Sect C,* p. 1856 (1966).
89.   Kingston, D. G. I., Lord Todd, and Williams, D. H., *J. Chem. Soc, Sect. C,* p. 1669 (1966).
90.   Ogawara, H., Maeda, K., Koyama, G., Naganawa, H., and Unezawa, H., *Chem. Pharm. Bull. (Tokyo) 16,* 679 (1968).
91.   Koyama, G., Iitaka, Y., Maeda, K., and Unezawa, H., *Tetrahedron Lett.,* p. 3587 (1967).
92.   Maeda, K., Kosaka, H., Okani, Y., and Unezawa, H., *J. Antibiot. Ser. A 6,* 140 (1953).

OTHER PEPTIDES

DR. NANCY N. GERBER

Name	Structure	Microorganism	Ref.
DL-Fumarylyl alanine	$HOOCCH=CHCONHCH(CH_3)COOH$	*Penicillium resticulosum*	1
N-Succinyl-L-glutamic acid	$HOOCCH_2CH_2CH(COOH)NHCO-CH_2CH_2COOH$	*Bacillus megatherium*	1
Lycomarasmine	$H_2NCOCH_2CH(COOH)NHCOCH_2-NHC(CH_3)(OH)COOH$	*Fusarium lycopersici*	1, 2
Aspergillomarasmine B (Lycomarasmic acid)	$HOOCCH_2CH(COOH)NHCH_2-CH(COOH)NHCH_2COOH$	*Aspergillus flavus oryzae*	2
Anhydroaspergillomarasmine B		*Aspergillus flavus oryzae*	2
d-Pantothenic acid	$HOCH_2C(CH_3)_2CH(OH)CONHCH_2-CH_2COOH$	Penicillin liquors Yeasts	*1*
Aspergillomarasmin A	$HOOCCH_2CH(COOH)NHCH_2CH-(COOH)NHCH_2CH(NH_2)COOH$	*Aspergillus flavus oryzae*	2
Toxin of tobacco wild-fire disease		*Pseudomonas tabaci*	1
Glutathione	$HOOCCH(NH_2)CH_2CH_2CONH-CH(CH_2SH)CONHCH_2COOH$	Yeasts	1
N-Succinyl-L-diaminopimelic acid	$HOOCCH(NH_2)(CH_2)_3CH(COOH)-NHCOCH_2CH_2COOH$	*Escherichia coli* mutant	1
Pantetheine	$HOCH_2C(CH_3)_2CH(OH)CONH-CH_2CH_2CONHCH_2CH_2SH$	Yeasts Many microorganisms	1
Glutathione-cysteine disulfide	$H_2NCH(COOH)CH_2CH_2CONHCH-(CONHCH_2COOH)CH_2SSCH_2-CH(NH_2)COOH$	*Saccharomyces cerevisiae*	1
Serratamic acid	$CH_3(CH_2)_6CH(OH)CH_2CONH-CH(CH_2OH)COOH$	*Serratia* sp.	1
δ-(*a*-Aminoadipyl)-cysteinylvaline	$HOOCCH(NH_2)(CH_2)_3CONHCH-(CH_2SH)CONHCH(isopr)COOH$	*Penicillium chrysogenum*	1

Name	Structure	Microorganism	Ref.
Duazomycin B	$N_2CHCOCH_2CH_2CH(COOH)NHCOCH-$ $(N_2CHCOCH_2CH_2)NHCOCH_2CH_2CH-$ $(NH_2)COOH$	*Streptomyces ambofaciens*	3
Netropsin	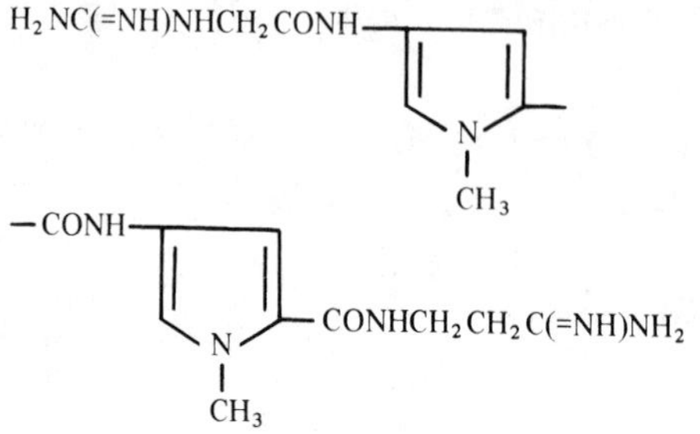 $H_2NC(=NH)NHCH_2CONH-$ (see structure)	*Streptomyces* spp.	3
Pantethine	$[HOCH_2C(CH_3)_2CH(OH)CONHCH_2-$ $CH_2CONHCH_2CH_2S]_2$	Yeasts Many microorganisms	1
Eulicin	$H_2NC(=NH)NH(CH_2)_8CH(OH)CH$ $(CH_2CH_2CH_2NH_2)NHCO(CH_2)_8$ $NHC(=NH)NH_2$	*Streptomyces* sp.	3
Nocardamin	$NHCH_2CH_2CON(OH)(CH_2)_4CH_2$	*Nocardia* sp.	1, 3
Aspochracein	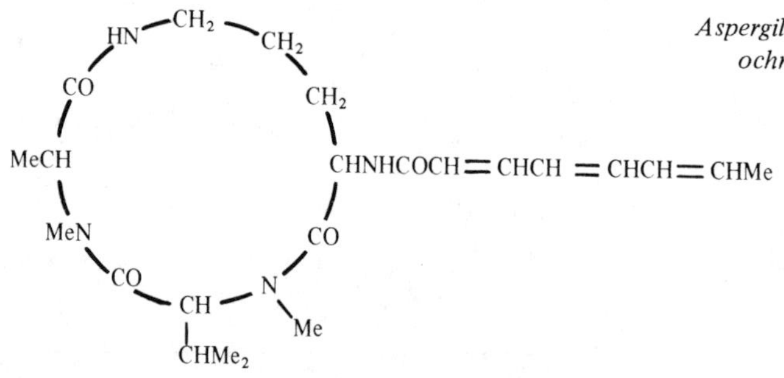 (see structure)	*Aspergillus ochraceus*	2
Islanditoxin	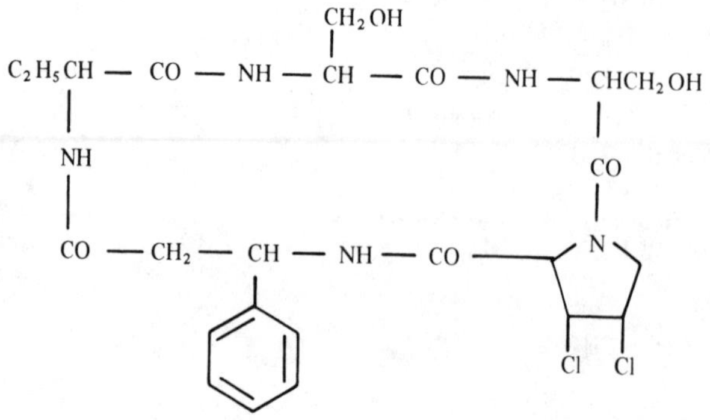 (see structure)	*Penicillium islandicum*	2

Netropsin structure:

$H_2NC(=NH)NHCH_2CONH-$ [pyrrole ring with N—CH_3 and CH_3]

$-CONH-$ [pyrrole ring with N—CH_3] $-CONHCH_2CH_2C(=NH)NH_2$

Aspochracein structure (ring):

HN—CH_2—CH_2
CO—CH_2
MeCH
MeN—CO
CO—CH—N—Me
$\overset{|}{CHMe_2}$
CHNHCOCH=CHCH=CHCH=CHMe

Islanditoxin structure:

C_2H_5CH — CO — NH — $\overset{CH_2OH}{CH}$ — CO — NH — $CHCH_2OH$
$\overset{|}{NH}$
CO — CH_2 — CH — NH — CO — [pyrrolidine ring with CO, N, Cl, Cl]
[benzene ring]

Name	Structure	Microorganism	Ref.

Bottromycin A₁ — *Streptomyces* sp. — 3

$$CH_3$$
$$H_3C-C-CH_3$$

$$R_1-NH-CH-CO-NH-CH-CO-[pyrrolidine\ ring]-C-NH-CH_2-\cdot$$

with H_3C CH_3 / CH group and R_2 substituent, $\parallel$ NH

$$\cdot-CO-NH-CH-CO-NH-CH-CH_2-COOCH_3$$

with phenyl–CH–CH₃ group and isothiazole (N=C–S) ring

Bottromycin A₂ — *Streptomyces* sp. — 3

$$A_1 : R_1 = CH_3-\underset{\underset{CH_3}{|}}{\overset{\overset{CH_3}{|}}{C}}-CO-,\ R_2=CH_3$$

$$A_2 : R_1 = \underset{CH_3}{\overset{CH_3}{>}}CH-CH=CH-CO-,\ R_2=CH_3$$

$$B : R_1 = CH_3-\underset{\underset{CH_3}{|}}{\overset{\overset{CH_3}{|}}{C}}-CO-,\ R_2=H$$

Bottromycin B — *Streptomyces* sp. — 3

Malformin A — *Aspergillus niger* other *A.* spp. — 2

Me Et groups on CH, cyclic peptide with S–S–CH₂ bridge, Me₂CHCH₂–CH D, DCH, CH₂, CHMe₂, CH D, L CH, etc.

Viomycin — *Streptomyces* sp. — 3

$$CH_2-O-\overset{\overset{O}{\parallel}}{C}-CH_2-\underset{}{\overset{\overset{NH_2}{|}}{CH}}-CH_2-CH_2-CH_2-NH_2$$

$$[ring]-C-NH-CH-\overset{O}{\underset{\parallel}{C}}-NH-CH-\overset{O}{\underset{\parallel}{C}}-NH-C-NH_2$$

with $HOCH_2$ and H_2N, ring N=C, C=O, CH, CH₂, NH groups

Name	Structure	Microorganism	Ref.
Ilamycin (Rufomycin A)	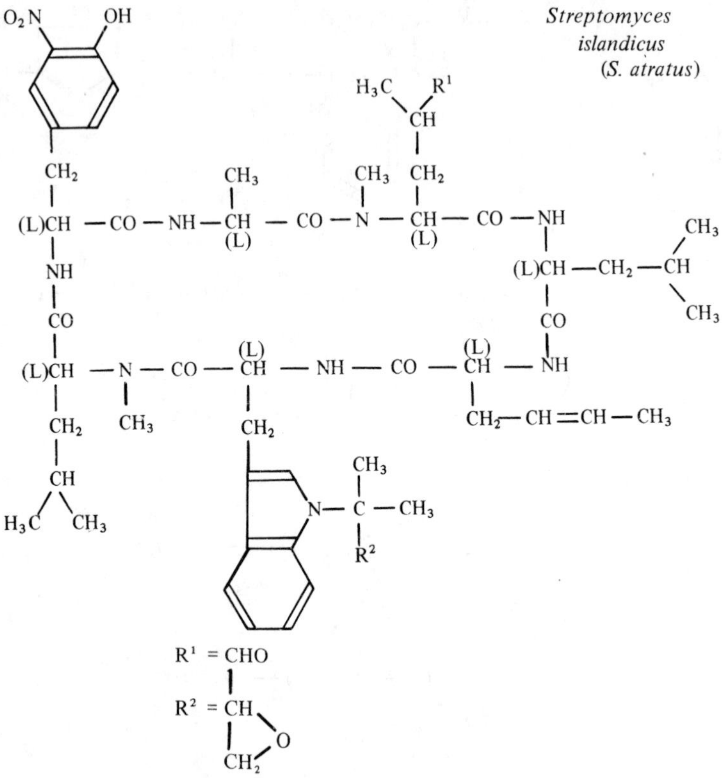	*Streptomyces islandicus* (*S. atratus*)	3

O_2N OH

H_3C R^1
         CH

CH_2 CH_3 CH_3 CH_2
(L)CH — CO — NH — CH — CO — N — CH — CO — NH
             (L)   (L)
NH   (L)CH — CH$_2$ — CH   CH$_3$
CO   CO   CH$_3$
(L)CH — N — CO — CH — NH — CO — CH — NH
        CH$_3$ (L)   (L)
CH$_2$ CH$_2$ CH$_2$ — CH = CH — CH$_3$
CH
H$_3$C   CH$_3$

N — C — CH$_3$
    CH$_3$
    R^2

R^1 = CHO

R^2 = CH
        CH$_2$ — O

| Ilamycin B$_1$ (Rufomycin B$_1$) | 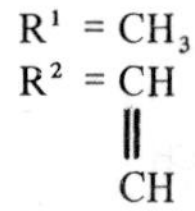 $R^1 = CH_3$ $R^2 = CH$ $\parallel$ CH | *Streptomyces islandicus* (*S. atratus*) | |

| Ilamycin B$_2$ (Rufomycin B$_2$) | $R^1 = CH_3$ $R^2 = CH$—O CH$_2$ | *Streptomyces islandicus* (*S. atratus*) | |

| Fungisporine | 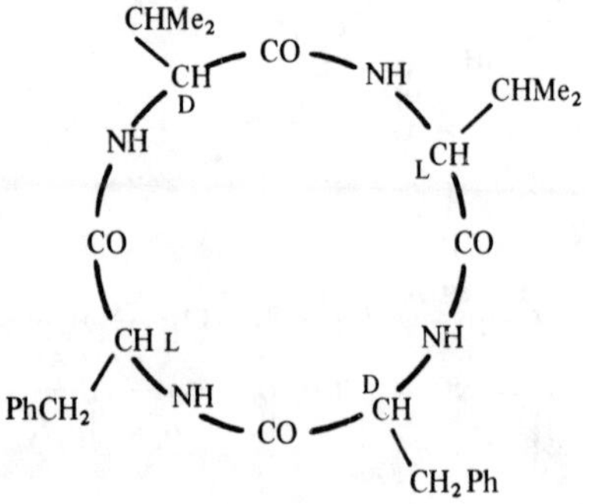 | *Penicillium* sp. *Aspergillus* sp. | 2 |

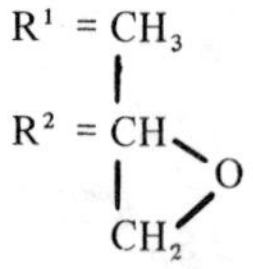

CHMe$_2$
CH — CO — NH
D   CHMe$_2$
NH   CH L
CO   CO
CH L   NH
PhCH$_2$ NH — CO — CH D
                    CH$_2$Ph

Name	Structure	Microorganism	Ref.
Amanita toxins (about 10 are known)	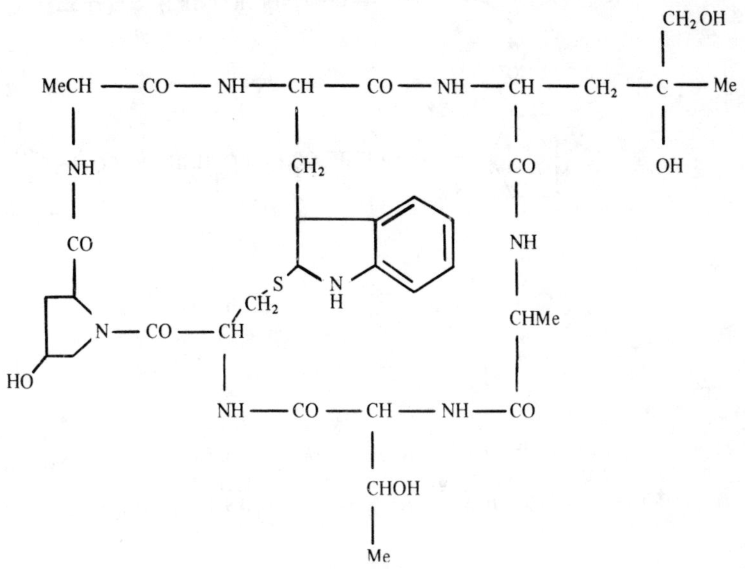	*Amanita* sp.	2

Phalloidin group

CH$_2$OH

MeCH — CO — NH — CH — CO — NH — CH — CH$_2$ — C — Me

NH   CH$_2$ CO   OH

CO   NH

CH$_2$ S   N-H

HO   N — CO — CH   CHMe

NH — CO — CH — NH — CO

CHOH

Me

Phalloidin

Amanitin group   CH$_2$OH   *Amanita* sp.   2

CHOH

CHMe

NH — CH — CO — NH — CH — CO — NH — CH$_2$ — CO

CO   CH$_2$ NH

    Me

HC — CH

O — S   N-H   OH   CH$_2$Me

HO   N   CH$_2$ CO

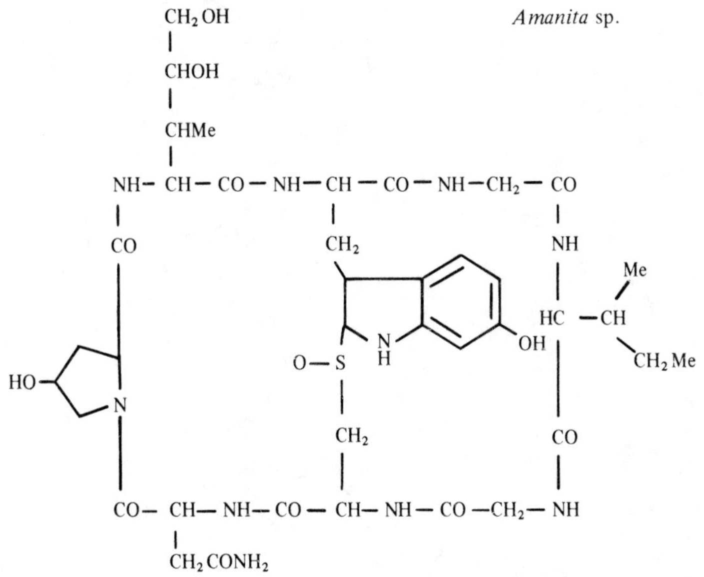

CO — CH — NH — CO — CH — NH — CO — CH$_2$ — NH

CH$_2$CONH$_2$

α-Amanitin

Name	Structure	Microorganism	Ref.

Antamanid (an antitoxin)

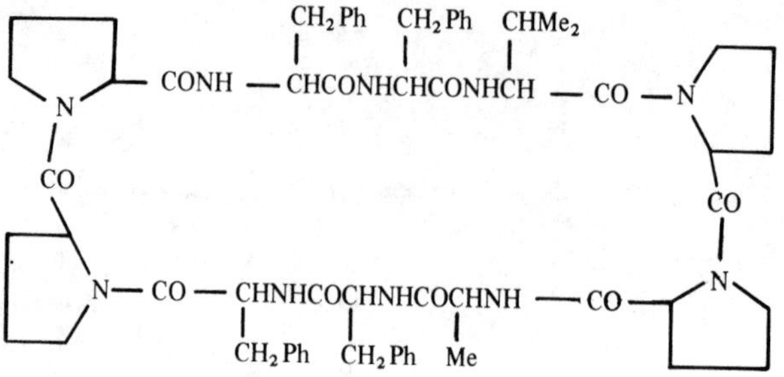

2

REFERENCES

1. Miller, M. W., *The Pfizer Handbook of Microbial Metabolites.* McGraw-Hill, New York (1961).
2. Turner, W. B., *Fungal Metabolites.* Academic Press, New York (1971).
3. Umezawa, H. (Ed.), *Index of Antibiotics from Actinomycetes.* University of Tokyo Press, Tokyo, Japan (1967).

HETEROCYCLIC COMPOUNDS

MICROBIAL PTERIDINES[a]

DR. C. VAN BAALEN

The pteridine ring system can be related to the monocyclic nitrogen heterocycles, pyrimidine and pyrazine (it is fused pyrimido-4,5-b-pyrazine), or to the purine ring system from which it can be formally derived by the addition of one carbon to the five-membered ring of purine. The ring is generally numbered as shown below.

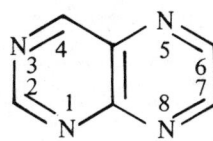

Nearly all naturally occurring pteridines are 2-amino-4-hydroxy pteridines, and they differ only in the substituents and state of oxidation of the pyrazine ring.

Pteridines can be extracted from whole or broken cells with water or under alkaline (NH$_4$OH) conditions. *In vivo* pteridines are usually in the reduced form, and upon extraction the reduced forms oxidize and decompose to give a multiplicity of products. Extraction with 5% acetic acid together with a small amount of solid manganese dioxide mildly oxidizes the reduced forms and usually gives a single aromatic product.

Purification and identification can generally be done by paper chromatography. The aromatic pteridines can be readily detected by their strong fluorescence (emission maximum about 410–500 nm when irradiated with wavelengths of 365 nm or lower).

A generally useful criterion for aromatic pteridines is their typical UV-absorption spectrum, which in a number of cases has two absorption maxima in alkaline solution which generally shift to lower wavelengths in acid solution.

CODE FOR ORGANISMS

A = *Azotomonas insolita*
B = *Anacystis nidulans*
C = *Rhodospirillium rubrum*
D = *Rhodomicrobium vannielii*
E = *Rhodopseudomonas palustris*
F = *Corynebacterium* sp.
G = *Pseudomonas roseus fluorescens*
H = *Synechocystis* sp.
I = *Pseudomonas* sp.
J = *Azotobacter agilis*
K = *Bacillus subtilis*
L = *Ochromonas danica*
M = *Tetrahymena pyriformis*

N = *Anabaena variabilis*
O = *Nostoc muscorum* G
P = *Anabaena cylindrica*
Q = *Synechococcus* sp.
R = *Nostoc muscorum* A
S = *Escherichia coli*
T = *Mycobacterium avium*
U = *Mycobacterium smegmatis*
V = *Aspergillus niger*
W = *Mycobacterium lacticola*
X = *Mycobacterium tuberculosis*
Y = *Eremothecium ashbyii*

[a] Simple pteridines, i.e., those not containing the *p*-aminobenzoyl residue.

2-Amino-4-hydroxy 6-Substituted Pteridines

Trivial Name	Substituents at 6	Substituents at Other Positions	Organisms	UV Maxima, nm 0.1N HCl	UV Maxima, nm 0.1N NaOH	Reference
Pterin	H		A, B, C, D, E	331	253, 358	1, 2, 3
Monapterin	CH$_2$OH		F, G	248, 320	254, 365	3, 4, 5
	CH$_2$O·glucosyl		H	248, 320	254, 365	2
Neopterin	CHOH·CHOH·CH$_2$OH (D-*erythro*)		A, I	248, 320	254, 365	1, 6
	CHOH·CHOH·CH$_2$O·glucuronyl		J, K	248, 320	254, 365	7, 8
Biopterin	CHOH·CHOH·CH$_3$ (L-*erythro*)		A, L	248, 320	254, 365	1, 9
Ciliapterin	CHOH·CHOH·CH$_3$ (L-*threo*)		M	248, 320	254, 365	10
	CHOH·CHOH·CH$_3$	Monosaccharide or uronic acid at 1' or 2'	B, D, E, N, O, P, Q, R	248, 320	254, 365	2
	CHOH·CHOH·CH$_2$O·P:O(OH)$_2$		S	248, 320	254, 365	11
	(CHOH)$_5$·CH$_2$OH	Glucose at 2'	T, U	248, 320	254, 365	12, 13
	CH$_2$·COOH	OH at 7	V	—	—	14
Erythopterin	OH	CH$_2$·CO·COOH at 7	W	238, 306, 319, 340^s, 450	238, 274^s, 317, 334^s, 470	15
Xanthopterin	OH		C, X	235, 263, 358	256, 392	3, 6, 16
	OSO$_3$H		A, D, S	235, 290, 324	262, 370	1, 3
Isoxanthopterin	H	OH at 7	A	288, 342	254, 340	1
	NH$_2$		B, O	272, 375	262, 393	17

2,4-Dihydroxypteridines (Lumazines)

Trivial Name	Substituents at 6	Substituents at Other Positions	Organisms	UV Maxima, nm		Reference
				0.1N HCl	0.1N NaOH	
Compound V		8-ribityl-7-hydroxy-6-methyl	Y	281, 327	289, 345	18
Compound G		8-ribityl-6,7-dimethyl	Y	258, 275, 407	230, 280, 313	19

Note:
s = shoulder.

REFERENCES

1. Goto, M., Forrest, H. S., Dickerman, L. H., and Urushibara, T., *Arch. Biochem. Biophys., 111,* 8 (1965).
2. Hatfield, D. L., Van Baalen, C., and Forrest, H. S., *Plant Physiol., 36,* 240 (1961).
3. Kobayashi, K., and Forrest, H. S., *Biochim. Biophys. Acta, 141,* 642 (1967).
4. Krumdieck, C. L., Baugh, C. M., and Shaw, E. N., *Biochim. Biophys. Acta, 90,* 573 (1964).
5. Viscontini, M., Pouteau-Thouvenot, M., Buchler-Moor, R., and Schroder, M., *Helv. Chim. Acta, 47,* 1948 (1964).
6. Guroff, G., and Rhoads, C. A., *J. Biol. Chem., 244,* 142 (1969).
7. Kobayashi, K., and Forrest, H. S., *Comp. Biochem. Physiol., 33,* 201 (1970).
8. Suzuki, A., and Goto, M., *J. Biochem. (Tokyo), 63,* 798 (1968).
9. Nathan, H. A., and Funk, H. B., *Amer. J. Clin. Nutr., 7,* 375 (1959).
10. Kidder, G., and Dewey, V. C., *J. Biol. Chem., 243,* 826 (1968).
11. Goto, M., and Forrest, H. S., *Biochem. Biophys. Res. Commun., 6,* 180 (1961).
12. Goto, M., Kobayashi, K., Sato, H., and Korte, F., *Justus Liebigs, Ann. Chem., 689,* 221 (1965).
13. Korte, F., and Goto, M., *Tetrahedron Lett.,* p. 55 (1961).
14. Kaneko, Y., *Nippon Nogei Kagaku Kaishi, 31,* 122 (1957).
15. Tschescue, R., and Vester, F., *Chem. Ber., 86,* 454 (1953).
16. Crowe, M. O'L., and Walker, A., *Brit. J. Exp. Pathol., 35,* 18 (1954).
17. Van Baalen, C., and Forrest, H. S., *J. Amer. Chem. Soc., 81,* 1770 (1959).
18. McNutt, W. S., *J. Amer. Chem. Soc., 82,* 217 (1960).
19. Masuda, T., Kishi, T., Asai, M., and Kuwada, S., *Chem. Pharm. Bull., 7,* 361 (1959).

MICROBIAL PYRROLES

DR. EDWARD E. GARCIA

Pyrroles may be extracted from fermentation broths with ethyl acetate, chloroform, or acetone. Most pyrroles are crystalline compounds and have moderately low melting points. Those pyrroles containing a nitro group are generally yellow in color.

The presence of the pyrrole ring can be demonstrated by the violet color produced by the Ehrlich reaction. The infrared spectrum is characterized by a sharp band at approximately 3300 cm^{-1} (N–H), while the aromatic protons of the pyrrole ring appear in the NMR spectrum at approximately δ 6.2 to 7 ppm. Pyrroles are generally unstable in acid. Some pyrroles have antifungal activity.

CODE FOR ORGANISMS

A = *Aerobacter aerogenes*
B = *Paracolobactrum aerogenoides*
C = *Spirillum serpens*
D = *Chromobacterium violaceum*
E = *Nocardia corallina*
F = *Pseudomonas pyrrocinia*
G = *Pseudomonas pyrrolnitrica*
H = *Pseudomonas aeruginosa*
I = *Pseudomonas mephitica*
J = *Pseudomonas ovalis*
K = *Pseudomonas shuylkilliensis*
L = *Pseudomonas aureofaciens*

M = *Serratia marcescens*
N = *Pseudomonas bromoutilis*
O = *Myrothecium verrucaria*
P = *Streptomyces rishiriensis*
Q = *Streptomyces spinicoumarensis*
R = *Streptomyces spinichromogenes*
S = *Streptomyces hazeliensis*
T = *Streptomyces distallicus*
U = *Streptomyces hygroscopicus*
V = *Streptomyces albocinerescens*
W = *Streptomyces roseochromogenes*

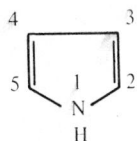

(Common Name)	Producing Organisms	Significant Characteristics	References
2-Carboxylic acid	A, B, C, D, E	Melting point: 202–203°C (decomposes) UV maximum: 260 nm	1, 2
2-Methyl-3-amyl	M	Not isolated (measured as prodigiosin formed)	3
3-Chloro-4-(2-nitro-3-chlorophenyl) (Pyrrolnitrin)	F, G, H, I, J, K, L	Melting point: 125°C UV maximum: 252 nm Exposure to sunlight causes color to change from yellow to red	4, 5, 6, 7
2,3-Dichloro-4-(2-nitrophenyl) (Isopyrrolnitrin)	G	Melting point: 105–108°C	8
3-Chloro-4-(2-nitro-3-chloro-6-hydroxy-phenyl) (Oxypyrrolnitrin)	G	Melting point: 215–216°C UV maximum: 290 nm	9

(Common Name)	Producing Organisms	Significant Characteristics	References
3-Chloro-4-(2-nitrophenyl)	G	Melting point: 105–108°C	10
2,3-Dichloro-4-(2-nitro-3-chlorophenyl)	L	Melting point: 143–144°C	11
3-Chloro-4-(2-amino-3-chlorophenyl)	L	Melting point: 88–90°C	11
3-Bromo-4-(2-nitro-3-bromophenyl)	G	Melting point: 72–74°C (decomposes)	12
2,3-Dibromo-4-(2-nitro-3-bromophenyl)	G	Melting point: 86–87°C (decomposes)	12
2,3-Dibromo-4-(2-nitro-5(or 4)-bromophenyl)	G	Melting point: 84–85°C (decomposes)	12
3-Chloro-4-(2-nitro-3-chloro-4-fluorophenyl)	L	Melting point: 155°C UV maximum: 248 nm	13
3-Chloro-4-(2-nitro-3-methylphenyl)	L	Melting point: 102–104°C UV maximum: 245 nm	13
2-(3,5-Dibromo-2-hydroxyphenyl)-3,4,5-tribromo	N	Melting point: 130–170° (decomposes)	14
2-(2,6-Dihydroxybenzoyl)-4,5-dichloro (Pyoluteorin)	H	Melting point: 177–186°C	15, 16
3-Acetyl-4-hydroxymethyl (Verrucarin E)	O	Melting point: 89–92°C UV maxima: 198, 249, 270 (inflection) nm	17

(Coumermycin A$_1$)	P, Q, R, S	Melting point: 258–260°C UV maxima: 280, 336 nm	18–21
(Coumermycin A$_2$)	P, S	Melting point: 258–260°C UV maxima: 267, 340 nm	18–21

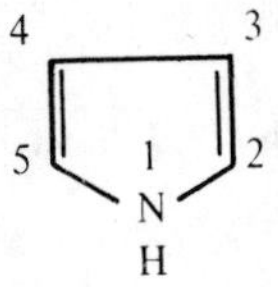

4 3
5 2
1
N
H

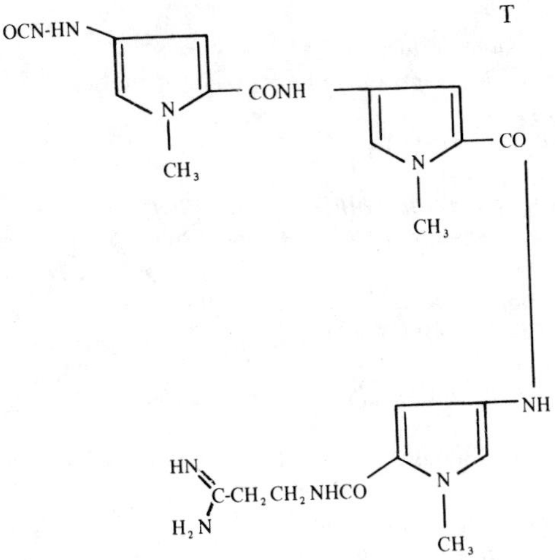

T

Melting point (HCl): 184–187°C   22, 23

(Distamycin A)

(Distamycin B)

(Distamycin C)

UV maxima: 237, 240, 303, 304 nm

UV maxima: 230, 318, 333, 350, 380, 405 nm

UV maxima: 236, 300–350 nm

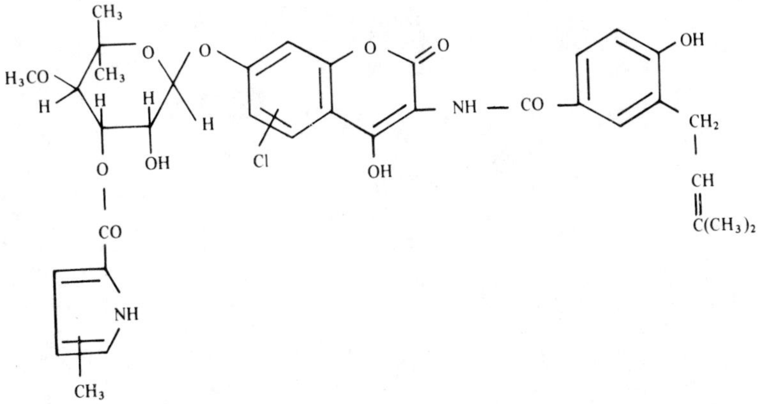

(Antibiotic 18631 RP)   U, V, W

Melting point: 206°C
UV maxima (CHCl$_3$): 275, 307, 337 nm

24

REFERENCES

1. Coupe, W. A., *Appl. Microbiol., 11,* 145 (1963).
2. Sax, K. J., Holmlund, C. E., Feldman, L. I., Evans, R. H., Jr., Blank, R. H., Shay, A. J., Schultz, J. S., and Dann, M., *Steroids, 5,* 345 (1965).
3. Goldschmidt, M. C., and Williams, R. P., *J. Bacteriol., 96,* 609 (1968).
4. Arima, K., Imanaka, H., Kousaka, M., Fukuda, A., and Tamura, G., *J. Antibiot. (Tokyo) Ser. A, 18,* 201 (1965); *Chem. Abstr., 64,* 4864 (1966).
5. Arima, K., Tamura, G., Imanaka, H., Kousaka, M., and Fukuda, A., *Japanese Patent 21,750* (1965); *Chem. Abstr., 68,* P38175c (1968).
6. *Netherlands Application 6,503,853* (1965); *Chem. Abstr., 65,* PC18434 (1966).
7. Lively, D. H., Gorman, M., Haney, M. E., and Mabe, J. A., *Antimicrob. Agents Chemother.,* p. 462 (1966).
8. Hashimoto, M., and Hattori, K., *Bull. Chem. Soc. Jap., 39,* 410 (1966).
9. Hashimoto, M., and Hattori, K., *Chem. Pharm. Bull. (Tokyo), 14,* 1314 (1966).
10. Hashimoto, M., and Hattori, K., *Chem. Pharm. Bull. (Tokyo), 16,* 1144 (1968).
11. Hamill, R. L., Elander, R., Mabe, J., and Gorman, M., *Antimicrob. Agents Chemother.,* p. 388 (1967).
12. Ajiska, M., Kariyone, K., Jomon, K., Yazawa, H., and Arima, K., *Agr. Biol. Chem., 33,* 294 (1969); *Chem. Abstr., 70,* 103817 (1969).
13. Gorman, M., Hamill, R. L., Elander, R., and Mabe, J., *Biochem. Biophys. Res. Commun., 31,* 294 (1968).
14. Burkholder, P. R., Pfister, R. M., and Leitz, F. H., *Appl. Microbiol., 14,* 649 (1966).
15. Takeda, R., *J. Ferment. Technol., 36,* 281 (1958).
16. Birch, A. J., Hodge, P., Rickards, R. W., Takeda, R., and Watson, T. R., *J. Chem. Soc.,* p. 2641 (1964).
17. Pfaffli, P., and Tamm, C., *Helv. Chim. Acta, 52,* 1911 (1969).
18. Kawaguchi, H., Tsukiura, H., Okanishi, M., Miyaki, T., Ohmori, T., Fujisawa, K., and Koshiyama, H., *J. Antibiot. (Tokyo) Ser. A, 18,* 1 (1965).
19. Kawaguchi, H., Naito, T., and Tsukiura, H., *J. Antibiot. (Tokyo) Ser. A, 18,* 11 (1965).
20. Kawaguchi, H., Miyaki, T., and Tsukiura, H., *J. Antibiot. (Tokyo) Ser. A, 18,* 220 (1965).
21. Berger, J., Schocher, A. J., Batcho, A. D., Pecherer, B., Keller, O., Maricq, J., Karr, A. H., Vaterlaus, B. P., Furlenmeier, A., and Spiegelberg, H., *Amtimicrob. Agents Chemother.,* p. 778 (1965).
22. DiMarco, A., Gaetani, M., Orezzi, P., Scotti, T., and Arcamone, F., *Cancer Chemother. Rep., 18,* 15 (1962).
23. Arcamone, F., Penco, S., Orezzi, P., Nicolella, V:, and Pirelli, A., *Nature, 203,* 1064 (1964).
24. *Netherlands Patent 69,02381* (1969); *B.A.S.I.C., 39,* 229 (1969).

PURINES, PYRIMIDINES, AND PYRIMIDOTRIAZINES

DR. NANCY N. GERBER

The nucleoside antibiotics are reviewed in Volume II of the *Handbook of Microbiology* (nucleosides are purines or pyrimidines linked to sugars). The purines and pyrimidines known to be components of nucleic acids are also listed and described in Volume II.

Tautomerism is an important property of hydroxypurines and pyrimidines. For convenience we have listed only the hydroxy form, although the keto form may predominate.

Purine

Pyrimidine

Name	Structure	Source	Ref.
Adenine	6-Aminopurine	*Coprinus comatus* *Boletus edulis* *Polyporus sulfureus*	1
Guanine	2-Amino-6-hydroxypurine	*Coprinus comatus* *Boletus edulis*	1
Hypoxanthine	6-Hydroxypurine	*Amanita muscaria* *Boletus edulis* *Agaricus nebularis* *Polyporus sulfureus*	1
Xanthine	2,6-Dihydroxypurine	*Amanita muscaria*	1
Uric Acid	2,6,8-Trihydroxypurine	*Aspergillus oryzae*	1
Heteroxanthine	2,6-Dihydroxy-7-methylpurine	Yeast	1
Kinetin	6-Fufurylaminopurine	Yeast extracts	1
8-Azaguanine	2-Amino-6-hydroxy N in place of C at position 8	*Streptomyces albus* *Streptomyces morookaensis*	2
Planomycin (Fervenulin)		*Streptomyces rubrireticuli* *Streptomyces fervens*	2
Xanthothricin (Toxoflavin)		*Streptomyces* sp. *Pseudomonas cocovenenans*	1, 2

Name	Structure	Source	Ref.
Cytosine	6-Amino-2-hydroxypyrimidine	*Agaricus nebularis*	1
Uracil	2,6-Dihydroxypyrimidine	*Agaricus nebularis*	1
4,5-Diaminouracil	4,5-Diamino-2,6-dihydroxy-pyrimidine	*Eremothecium ashbyii*	1
Polyoxin C	2-OH, 3-$C_6H_{10}NO_5$, 5-CH_2OH, 6-OH-pyrimidinine	*Streptomyces cacaoi*	2
Amicetin		*Streptomyces vinaceus-drappus* *Streptomyces fasiculatus* *Streptomyces sacromyceticus*	2

Bamicetin		*Streptomyces plicatus*	2

Blasticidin S		*Streptomyces griseochromogenes* *Streptomyces globifer* *Streptomyces morookaensis*	2

Name	Structure	Source	Ref.
Cytomycin		*Streptomyces griseochromogenes*	2
Gougerotin		*Streptomyces gougerotii*	2
Plicacetin		*Streptomyces plicatus*	2

REFERENCES

1. Miller, M. W., *The Pfizer Handbook of Microbial Metabolites.* McGraw-Hill, New York (1961).
2. Umezawa, H. (Ed.), *Index of Antibiotics from Actinomycetes.* University of Tokyo Press, Tokyo, Japan (1967).

PRODIGIOSIN-LIKE PIGMENTS

DR. NANCY N. GERBER

Although prodigiosin-like pigments are soluble in non-polar solvents, polar solvents such as acetone and ethanol are often required to remove them from cells. Once dissolved, the pigments are readily recognized by their red color in acid solutions, which changes to yellow in neutral or basic solution. They are unstable to acid or light, especially in solution; typically, at room temperature all of the original pigment in a dilute solution disappears in a few days or weeks.

The visible-absorption spectra are quite characteristic, with a strong maximum between 525 and 555 nm, depending on the structure and the solvent (which must be slightly acid). A shoulder always appears on this main band at about 35 nm lower wave length.

Extracts containing prodigiosin-like pigments may be conveniently examined and compared by 10- to 15-minute thin-layer chromatography on small pieces of Eastman chromagram in chloroform-ethyl acetate mixtures.

STRUCTURES

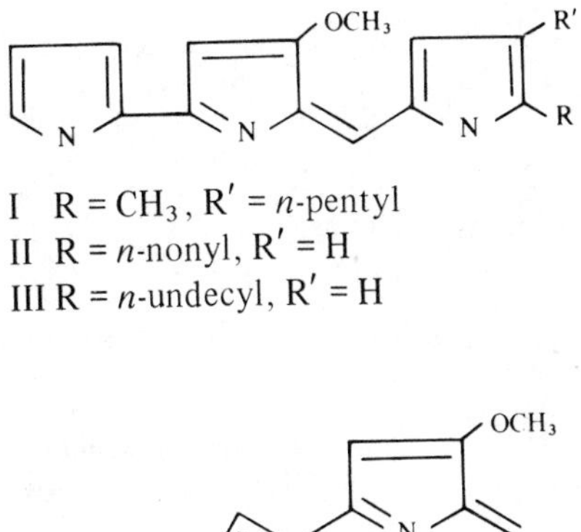

I R = CH$_3$, R$'$ = n-pentyl
II R = n-nonyl, R$'$ = H
III R = n-undecyl, R$'$ = H

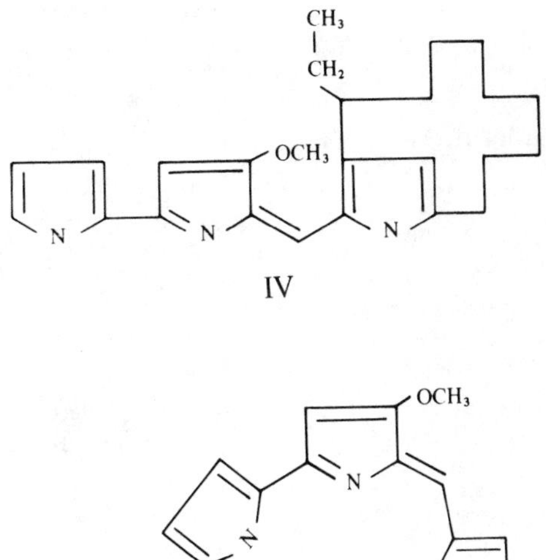

IV

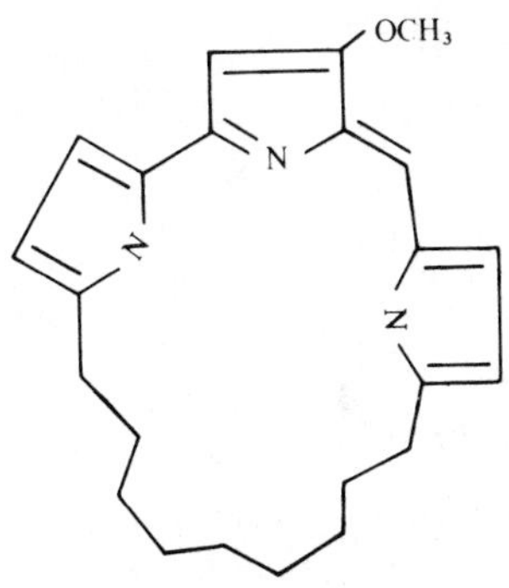

V

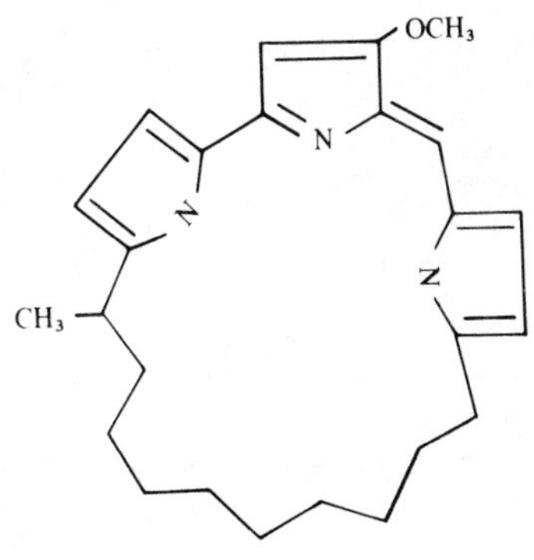

VI

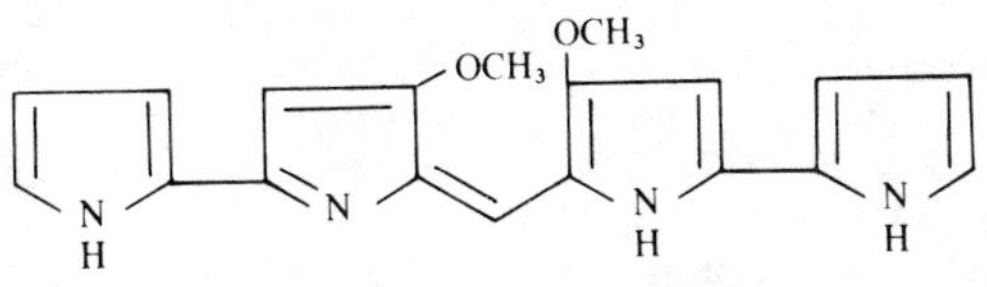

VII

IDENTIFYING CHARACTERISTICS

Name	Structure	Microorganism	Absorption Maximum and Solvent	Ref.
Prodigiosin 2-Methyl-3-*n*-pentylprodiginine 2-Methyl-3-amyl-6-methoxy-prodigiosene[a]	I	*Serratia marcescens*	537, acidified ethanol	3, 4
Norprodigiosin	I, with —OH instead of —OCH$_3$	*Serratia marcescens* mutant	No data available	8
Nonylprodiginine Nonylprodigiosin	II	*Actinomadura (Nocardia) madurae*	525, acidified ethanol	2
Undecylprodiginine Undecylprodigiosin	III	*Streptomyces longisporusruber Actinomadura (Nocardia) pelletieri*	525, acidified ethanol	5, 6
Metacycloprodigiosin	IV	*Streptomyces longisporusruber*	530, hydrochloride salt in methanol	7
Cyclononylprodiginine	V	*Actinomadura (Nocardia) madurae*	542, acidified ethanol 551, chloroform	9
Cyclomethyldecyl prodiginine	VI	*Actinomadura (Nocardia) pelletieri*	539, acidified ethanol 549, hexane	10
Unnamed blue pigment	VII	*Serratia marcescens* mutant	588, hydrochloride salt	11

[a] For an explanation of nomenclature, see References 1 and 2. *Chemical Abstracts* indexes prodigiosin as follows: 2,2′-Bipyrrole, 4-methoxy-5-[(5-methyl-4-pentyl-2H-pyrrol-2-ylidine)methyl].

REFERENCES

1. Hearn, W. R., Elson, M. K., Williams, R. H., and Medina-Castro, J., *J. Org. Chem., 35,* 142 (1970).
2. Gerber, N. N., *Appl. Microbiol., 18,* 1 (1969).
3. Castro, A. J., Corwin, A. H., Waxham, F. J., and Bulby, A. L., *J. Org. Chem., 24,* 455 (1959).
4. Williams, R. P., and Hearn, W. R., in *Antibiotics,* pp. 410–432, D. Gottlieb and P. D. Shaw, Eds. Springer-Verlag, Berlin, Germany, and New York (1967).
5. Wasserman, H. H., Rodgers, G. C., and Keith, D. D., *Chem. Commun.,* p. 825 (1966).
6. Harashima, K., Tsuchida, N., Tanaka, T., and Nagatsu, J., *Agric. Biol. Chem., 31,* 481 (1967).
7. Wasserman, H. H., Rodgers, G. C., and Keith, D. D., *J. Am. Chem. Soc., 91,* 1263 (1969).
8. Hearn, W. R., Worthington, R. E., Burgus, R. C., and Williams, R. P., *Biochem. Biophys. Res. Commun., 17,* 517 (1964).
9. Gerber, N. N., *Tetrahedron Lett.,* p. 809 (1970).
10. Gerber, N. N., *J. Antibiot. (Tokyo), 24,* 636 (1971).
11. Wasserman, H. H., Friedland, D. J., and Morrison, D. A., *Tetrahedron Lett.,* p. 641 (1968).

MICROBIAL PHENAZINES

DR. NANCY N. GERBER

Phenazines, with the exception of internal salts, can generally be extracted from whole broth by chloroform, are easily crystallized, and have relatively high melting points. Many sublime. Colors range from pale yellow to blue and purple.

When treated with sodium hydrosulfite, phenazines lacking a 1-COOH group fade to colorless dihydro derivatives. Phenazines with a 1-COOH group become more intensely colored. N-oxide groups are irreversibly removed by this reagent. Many phenazines show a deepening of color when treated with concentrated hydrochloric acid.

Electronic absorption spectra are often quite characteristic, with an intense peak in the range from 250 to 290 nm and a weaker peak between 350 and 400 nm. If a phenolic or amino group is present, an additional band above 400 nm appears.

Many phenazines have antibiotic activity.

The properties given in the table below are, in the opinion of the compiler, the most significant for easy recognition. In the table, the symbol "PZ" is substituted for "phenazine". The Chemical Abstracts numbering system is shown below; occasionally, in older literature, one encounters an alternate system.

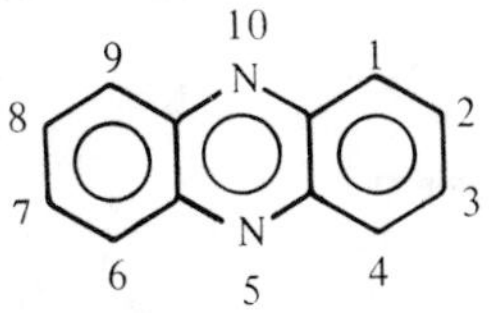

Chemical Abstracts System

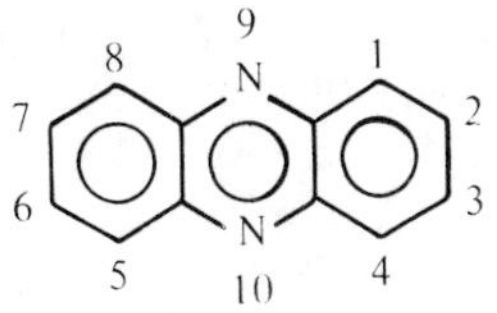

Alternate System

CODE

Organisms

A = *Pseudomonas aeruginosa*

B = *Streptomyces thioluteus*

C = a member of a group of novel *Nocardiaceae*, now classified as *Actinomadura dassonvillei*

D = *Pseudomonas aureofaciens*

E = *Streptomyces misakiensis*

F = an unidentified bacterium, a non-motile Gram-negative rod

G = *Pseudomonas chloraphis*

H = *Microbispora (Waksmania) aerata*

I = *Brevibacterium iodinum*, erroneously called *Pseudomonas* in Reference 11

J = *Streptosporangium amethystogenes*

K = *Sorangium* sp.

L = *Streptomyces griseoluteus*

M = *Streptomyces lomondensis*

N = *Streptomyces luteoreticuli*

Solubility

i = insoluble; r = readily; s = soluble; sp = sparingly

IDENTIFYING PROPERTIES

Systematic Name (Common Name)	Organism	Significant Characteristics	Reference
1-PZol (Hemipyocyanine)	A, B	Melting point: 155–157°C UV maxima: 264, 352, 360, 384, 425	1, 2
1-Methoxy PZ	N		24
1-PZol-10-oxide	C	Melting point: 165–167°C UV maxima: 279, 326, 334, 368, 380, 387, 468	3
5-Methyl-1-hydroxy-phenazinium betaine; 5-Methyl-1(5H) phenazinone (Pyocyanine)	A	Internal salt; blue in water or dilute base and extracted by chloroform; red in dilute acid and not extracted by chloroform	2, 4
2-PZol	D	UV maxima in dilute hydrochloric acid: 217, 263, 388 UV maxima in dilute sodium hydroxide: 275, 367	5
PZ-1-carboxylic acid (Tubermycin B)	A, C, D, E, F, G	Melting point: 243°C UV maxima: 250, 364 Solubility: sp s in dilute bicarbonate solution	6–10
PZ-1-carboxylic acid, methyl ester	N		24
$C_{17}H_{16}N_2O_2$ (Tubermycin A)	E	Melting point: 174°C UV maxima: 254, 370	8
PZ-1-carboxamide (Oxychlororaphine)	A	Melting point: 237°C	2, 9
1,6-PZdiol	C, F, H, I; see also Ref. 22	Melting point: 265°C (sealed tube) UV maxima: 273, 374, 445 Blue in ferric chloride test	3, 10, 11
6-Methoxy-1-PZol	B	Melting point: 192°C UV maxima: 271, 371, 438	1
1,6-Dimethoxy-PZ	B, N	Melting point: 252°C UV maxima: 270, 370, 432 Intense yellow-green fluorescence	1
1,6-PZdiol-5-oxide	B, C, F, H, I	Melting point: 245°C UV maxima: 283, 487	3, 10, 13
1,6-PZdiol-5,10-dioxide (Iodinin)	B, C, F, H, I, J, K; see also Ref. 22	Shiny purple crystals UV maxima: 290, 350, 530 Solubility: i in ethanol, sp s in most solvents	2, 3, 10, 11, 14–17
6-Methoxy-1-PZol-5,10-dioxide (Myxin)	K	UV maxima in dilute hydrochloric acid: 283, 340, 505 Solubility: s in most solvents Easily loses one oxide function	15–17
1,8-PZdiol	F	Melting point: 230°C UV maxima: 268, 383, 432 Solubility: s in bicarbonate solution C = O in IR	10

Systematic Name (Common Name)	Organism	Significant Characteristics	Reference
1,8-PZdiol-10-oxide	F	Melting point: 235–240°C; decomposes UV maxima: 285, 405, 540 UV maxima, acidic: 283, 403, 470 UV maxima, basic: 295, 440, 550	10
8-Amino-1-PZol	F	Melting point, methyl ether: 180°C, then 243°C UV maxima: 275, 376, 480 UV maxima, basic: 277, 372, 495	10
PZ-1,6-dicarboxylic acid	F	Melting point: > 300°C UV maxima: 250, 370 Solubility: i in ethanol and chloroform	10
6-Hydroxymethyl-PZ-1-carboxylic acid	L	Melting point, methyl ester: 197–201°C	18
2-Hydroxy-PZ-1-carboxylic acid	D, F	UV maxima: 252, 363 UV maxima, basic: 287, 363, 490	6, 10, 19
9-Hydroxy-PZ-1-carboxylic acid	F	Melting point: > 270°C UV maxima: 268, 370 Solubility: i in chloroform	10
6-Methyl-9-methoxy-PZ-1-carboxylic acid	L	Melting point, methyl ester: 124–126°C UV maxima: 267, 346, 363	18
2,9-Dihydroxy-PZ-1-carboxylic acid	F	Melting point: > 270°C Solubility: r s in bicarbonate solution, i in chloroform	10
6-Hydroxymethyl-PZ-9-methoxy-PZ-1-carboxylic acid (Griseoleutic acid)	L	Melting point, methyl ester: 189°C UV maxima: 269, 369	18

(Griseolutein A)

	L	Melting point: 194–197°C Melting point, methyl ester: 149°C UV maxima: 267, 368	18

(Griseolutein B)

	L	Darkens at 150°C, decomposes at approximately 220°C UV maxima: relatively weak benzenoid: 281, 342	20

Systematic Name (Common Name)	Organism	Significant Characteristics	Reference
5-Methyl-7-amino-PZ-1-carboxylic acid (Aeruginosin A)	A (red strains)	Internal salt; not extracted by solvents; unstable in base UV maxima: 280, 395, 515	21
5-Methyl-7-amino-1-carboxy-PZ-3-sulfonic acid (Aeruginosin B)	A (red strains)	Internal salt; not extracted by solvents; unstable in base UV maxima: similar to aeruginosin A	12, 21
6-Formyl-4,7,9-trihydroxy-PZ-1-carboxylic acid, methyl ester (Lomofungin; Lomondomycin)	M	Melting point: > 320°C Melting point, triacetate: 166–168°C UV maxima: 257, 364	23

REFERENCES

1. Gerber, N. N., *J. Org. Chem., 32,* 4055 (1967).
2. Swan, G. A., and Felton, D. G. I., *Phenazines.* Interscience Publications, John Wiley and Sons, New York (1957).
3. Gerber, N. N., *Biochemistry, 5,* 3824 (1966).
4. MacDonald, J. C., in *Antibiotics,* Vol. 2, p. 52, D. Gottlieb and P. D. Shaw, Eds. Springer-Verlag, Berlin, Germany, and New York (1967).
5. Levitch, M. E., and Reitz, P., *Biochemistry, 5,* 689 (1966).
6. Toohey, J. I., Nelson, C. D., and Krothov, G., *Can. J. Bot., 43,* 1055 (1965).
7. Haynes, W. C., *et al., J. Bacteriol., 72,* 412 (1956).
8. Isono, K., Anzai, K., and Suzuki, S., *J. Antibiot. (Tokyo) Ser. A, 11,* 264 (1958).
9. Takeda, R., and Nakashini, I., *Hakko Kogaku Zasshi, 38,* 9 (1959).
10. Gerber, N. N., *J. Heterocycl. Chem., 6,* 297 (1969).
11. Gerber, N. N., and Lechevalier, M. P., *Biochemistry, 3,* 598 (1964).
12. Herbert, R. B., and Holliman, F. G., *Proc. Chem. Soc.,* p. 19 (1964).
13. Gerber, N. N., and Lechevalier, M. P., *Biochemistry, 4,* 176 (1965).
14. Prauser, H., and Eckardt, K., *Z. Allg. Microbiol., 7,* 409 (1967).
15. Edwards, O. E., and Gillespie, D. C., *Tetrahedron Lett.,* p. 4867 (1966). (Incorrect structure)
16. Weigele, M., and Leimgruber, W., *Tetrahedron Lett.,* p. 715 (1967).
17. Sigg, H. P., and Toth, A., *Helv. Chim. Acta, 50,* 716 (1967).
18. Yagishita, K., *J. Antibiot. (Tokyo) Ser. A, 13,* 83 (1960).
19. Olsen, E. S., and Richards, J. H., *J. Org. Chem., 32,* 2887 (1967).
20. Nakamura, S., Maeda, K., and Umezawa, H., *J. Antibiot. (Tokyo) Ser. A, 17,* 33 (1964).
21. Holliman, F. G., *S. Afr. Ind. Chem., 15,* 233 (1961).
22. Suzuki, T., Uno, K., and Deguchi, T., *Agr. Biol. Chem., 35,* 92 (1971); *Chem. Abstr., 76,* 139047h; *Japanese Patent 7,204,998.*
23. Tipton, C. D., and Rinehart, K. L., Jr., *J. Am. Chem. Soc., 92,* 1425 (1970).
24. Yamagishi, S., Koyama, Y., Fukakusa, Y., Kyomura, N., Ohishi, J., Hamamichi, N., and Arai, T., *Yakugaku Zasshi, 91,* 351 (1971); *Chem. Abstr., 75,* 1565y.

PHENOXAZINONES AND ACTINOMYCINS

Phenoxazinones

DR. NANCY N. GERBER

Most microbial phenoxazinones, including actinomycins, are orange-red. Their spots on paper or thin-layer chromatograms change to deep red in the presence of HCl fumes. The ultraviolet and visible absorption spectra commonly show 2 peaks, one at 240 nm and the other in the 430–460 nm range.

Most of these compounds crystallize readily when pure. Their solubility in organic solvents ranges from good (2-amino-3H-phenoxazin-3-one) through moderate (actinomycins) to poor (cinnabarin); all are poorly soluble in water. Their physical properties are determined to a considerable extent by internal hydrogen bonding.

Name	Structure	Microorganism	Ref.
2-Amino-3H-phenoxazin-3-one (Questiomycin B)	$R_1 = R_2 = R_3 = H$	*Streptomyces* sp. *Streptomyces thioluteus* *Microbispora aerata* *Brevibacterium iodinum* *Actinomadura dassonvillei*	1, 2, 3
2-Acetamido-3H-phenoxazin-3-one	$R_1 = R_2 = H$, $R_3 = -COCH_3$	*Streptomyces thioluteus* *Microbispora aerata* *Brevibacterium iodinum* *Actinomadura dassonvillei*	3
2-Ethanolamine-3H-phenoxazin-3-one	$R_1 = R_2 = H$, $R_3 = -CH_2CH_2OH$	*Streptomyces thioluteus*	2
2-Amino-1-carboxy-3H-phenoxazin-3-one	$R_1 = R_3$ H, $R_2 = COOH$	*Nocardia* sp. later classified as *Actinomadura dassonvillei*	3
Cinnabarin (Polystictin)	$R_1 = CH_2OH$, $R_2 = COOH$, $R_3 = H$	*Polyporus cinnabarinus*	1
Cinnabarinic acid	$R_1 = R_2 = COOH$, $R_3 = H$	*Polyporus cinnabarinus*	1
Tramesanguin	$R_1 = COOH$, $R_2 = CHO$, $R_3 = H$	*Polyporus cinnabarinus*	1

REFERENCES

1. Turner, W. B., *Fungal Metabolites*. Academic Press, New York (1971).
2. Gerber, N. N., *J. Org. Chem., 32,* 4055 (1967).
3. Gerber, N. N., *Biochemistry, 5,* 3824 (1966).

Actinomycins

DR. EDWARD KATZ

The actinomycins constitute a family of chromopeptide antibiotics that differ solely in the peptide portion of the molecule. The chromophore, actinocin, is 2-amino-4,6-dimethylphenoxazinone (3)-1,9-dicarboxylic acid; attached to the chromophore are two pentapeptide lactone rings, which may be identical or differ in amino acid composition. An example of an actinomycin structure is shown in Figure 1.

The biological activity of the actinomycins depends on their inhibition of DNA-dependent RNA synthesis, an effect exerted through their ability to complex with double-helical DNA. These antibiotics have been extremely useful biochemical tools for investigations of cellular metabolic processes, particularly macromolecular biosyntheses and viral replication.

Actinomycin is of considerable clinical importance in the treatment of Wilm's tumor, gestational choriocarcinoma and mixed metastatic carcinoma of the testes. Wilm's tumor therapy with actinomycin in combination with radiotherapy and surgery has resulted in long-term remissions and cures in 60 to 90% of patients.

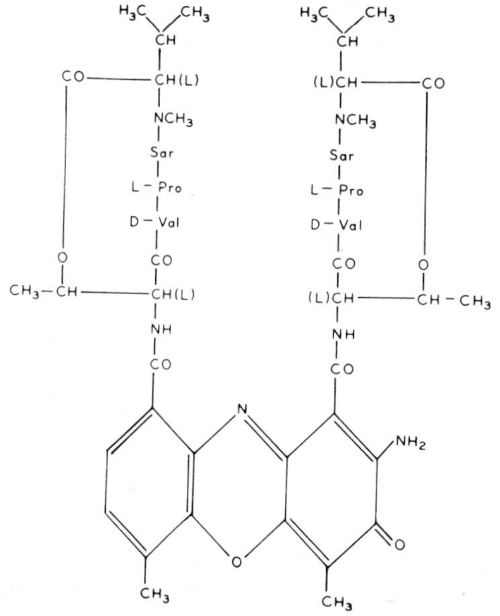

FIGURE 1.   Structure of Actinomycin IV (D, C_1).

ABBREVIATIONS

ahypro = allo-(*cis*)-4-hydroxy-L-proline; hypro = *trans*-4-hydroxy-L-proline; oxopro = 4-oxo-L-proline; pro = L-proline; 5-mepro = 5-methylproline; 4-oxo-5-mepro = 4-oxo-5-methylproline; 3-hy-4-oxo-5-mepro = 3-hydroxy-4-oxo-5-methylproline; pip = L-pipecolic acid; oxopip = 4-oxo-L-pipecolic acid; hypip = *trans*-4-hydroxy-L-pipecolic acid; thr = L-threonine; methr = N-methylthreonine; val = D-valine; alleu = D-alloisoleucine; sar = sarcosine; meval = N-methyl-L-valine; meala = N-methylalanine; mealleu = N-methyl-L-alloisoleucine.

SYMBOLS

symbolizes the chromophore; the horizontal lines represent the connections between the peptide lactones and the α and β rings respectively; the fork at the bottom represents the quinoid-ring functions.

denotes uncertainty regarding the assignment of an amino acid to the peptide lactone attached to the benzene or quinoid ring of the actinomycin chromophore.

TABLE 1
NATURALLY OCCURRING ACTINOMYCINS

Synonyms	Formula
I, A_1, B_1, $X_{O\beta}$	thr — val — pro — sar — meval — O thr — val — hypro — sar — meval — O
$X_{O\delta}$	thr — val — pro — sar — meval — O thr — val — ahypro — sar — meval — O
II, A_{II}, B_{II}, F_8	thr — val — sar — sar — meval — O thr — val — sar — sar — meval — O
III, A_{III}, B_{III}, F_9, $X_{O\gamma}$	thr — val — pro — sar — meval — O thr — val — sar — sar — meval — O
IV, A_{IV}, B_{IV}, D, C_1, I_1, X_1	thr — val — pro — sar — meval — O thr — val — pro — sar — meval — O
X_{1a}	thr — val — oxopro — sar — meval — O thr — val — sar — sar — meval — O
V, A_V, B_V, X_2	thr — val — pro — sar — meval — O thr — val — oxopro — sar — meval — O
VI, C_2, I_2	thr — val — pro — sar — meval — O thr — aIleu — pro — sar — meval — O
C_{2a}	thr — aIleu — pro — sar — meval — O thr — val — pro — sar — meval — O
VII, C_3	thr — aIleu — pro — sar — meval — O thr — aIleu — pro — sar — meval — O
Z_1	thr — val — 4-oxo-5-mepro — sar — meval — O methr — val — 3-hy-4-oxo-5-mepro — sar — meala — O
Z_5	thr — val — 4-oxo-5-mepro — sar — meval — O thr — val — 5-mepro — sar — meala — O
Z_2, Z_3, Z_4	thr — val — 4-oxo-5-mepro — sar — meval — O methr — val — 5-mepro — sar — meala — O
Z_{03}	thr — val — 4-hy-5-mepro — sar — meval — O methr — val — 5-mepro — sar — meala — O
Z_{01}	thr — val — 3-hy-4-oxo-5-mepro — sar — meval — O methr — val — 4-hy-5-mepro — sar — meala — O

TABLE 2
ACTINOMYCINS PRODUCED BY CONTROLLED BIOSYNTHESIS

Synonyms   Formula

Addition of Pipecolic Acid to *Streptomyces antibioticus*

Pip 2

thr — val — pip — sar — meval — O
thr — val — pip — sar — meval — O

Pip 1β

thr — val — pip — sar — meval — O
thr — val — pro — sar — meval — O

Pip 1α

thr — val — oxopip — sar — meval — O
thr — val — pip — sar — meval — O

Pip 1δ

thr — val — oxopip — sar — meval — O
thr — val — pro — sar — meval — O

Pip 1γ

thr — val — hypip — sar — meval — O
thr — val — pip — sar — meval — O

Pip 1ε

thr — val — hypip — sar — meval — O
thr — val — pro — sar — meval — O

Addition of Sarcosine to *Streptomyces chrysomallus*

F_1

thr — alleu — sar — sar — meval — O
thr — val — sar — sar — meval — O

F_2

thr — alleu — sar — sar — meval — O
thr — val — pro — sar — meval — O

F_3

thr — alleu — sar — sar — meval — O
thr — alleu — sar — sar — meval — O

F_4

thr — alleu — sar — sar — meval — O
thr — alleu — pro — sar — meval — O

Addition of Isoleucines (L-Alloisoleucine, D-Isoleucine) to *Streptomyces chrysomallus*

E_1

thr — alleu — pro — sar — mealleu — O
thr — alleu — pro — sar — meval — O

E_2

thr — alleu — pro — sar — mealleu — O
thr — alleu — pro — sar — mealleu — O

REFERENCES

1. Brockmann, H., Die Actinomycine, *Fortschr. Chem. Org. Naturst., 18,* 1 (1960).
2. Reich, E., and Goldberg, I. H., Actinomycin and Nucleic Acid Function, *Prog. Nucleic Acid Res., 3,* 183 (1964).
3. Katz, E., Actinomycin, in Antibiotics, Vol. 2, Biosynthesis, pp. 276—341, D. Gottlieb and P. D. Shaw, Eds. Springer-Verlag, New York (1967).
4. Waksman, S. A., (Ed.), *Actinomycin Nature, Formation and Activities.* Interscience Publications, John Wiley and Sons, New York (1968).
5. Friedman, P. A., and Cerami, A., Actinomycin, in *Cancer Medicine,* J. F. Holland and E. Frei, III, Eds. Lea and Febiger, Philadelphia, Pennsylvania (1973).
6. Stock, J. A., Actinomycins, in *Experimental Chemotherapy,* Vol. 4, pp. 243—267, R. J. Schnitzer and F. Hawking, Eds., Academic Press, New York (1966).
7. Farber, S., and Mitus, A., Chemotherapy of Wilm's Tumor, in *Chemotherapy of Cancer,* pp. 277—285, W. H. Cole, Ed. Lea and Febiger, Philadelphia, Pennsylvania (1970).
8. Meienhofer, J., and Atherton, E., Structure-Activity Relationships in the Actinomycins, in *Advances in Applied Microbiology,* Vol. 16, pp. 203—300, D. Perlman, Ed. Academic Press, New York (1973).

PYRIDINE, QUINOLINE, AND INDOLE COMPOUNDS

DR. NANCY N. GERBER

The basic ring systems and numbering are shown below. Natural products with a pyridine structure and their biosyntheses were recently reviewed.[4]

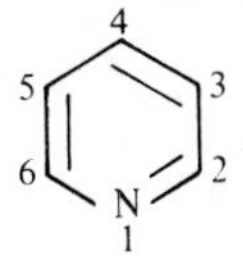

Pyridine

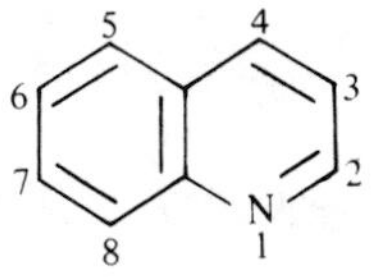

Quinoline

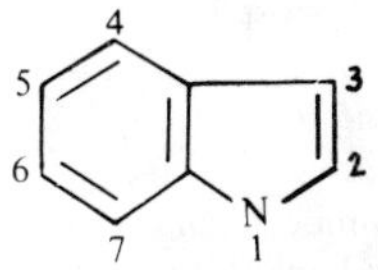

Indole

Name	Substituent and Location	Microorganism	Reference
Nicotinic acid	3–COOH	Widespread	3
a-Pycolinic acid	2–COOH	*Piricularia oryzae*	2
3-Hydroxymethyl pyridine	3–CH$_2$OH	*Mycobacterium platypoecilus, M. tuberculosis, M. bovis, M. phlei, M. smegmatis, M. fortuitum, M. avium*	4
Trigonelline	1–CH$_3$, 3–COOH internal salt (betaine)	*Polyporus sulfureus*	1, 3
Homarine	1–CH$_3$, 2–COOH internal salt	*Polyporus sulfureus*	1, 2, 3
Pyridine-3,4-dicarboxylic acid (Cinchomeronsäure)	3–COOH, 4–COOH	*Escherichia coli* mutant	4
2,6-Dipicolinic acid	2–COOH, 6–COOH	Bacterial spores *Penicillium citreoviride*	1, 2, 3
Ethylhydrogen-2,6-dipicolinate	2–COOEt, 6–COOH	*Bacillus cereus* spores	1
Fusaric acid	2–COOH, 5–(CH$_2$)$_3$CH$_3$	*Fusarium* sp.	1, 2, 3
Dehydrofusaric acid	2–COOH, 5–(CH$_2$)$_2$CH=CH$_2$	*Fusarium* sp.	1, 2, 3
Hydroxyfusaric acid	2–COOH, 5–(CH$_2$)$_2$CHOHCH$_3$	*Fusarium* sp.	3
Pyrimine	2–$\overset{\displaystyle O}{\overset{\|}{C}}CH_2CH_2$C(NH$_2$)COOH	*Pseudomonas* GH	4
Nigrifactin	2–CH=CHCH=CHCH=CHCH$_3$ 3,4,5,6 tetrahydro	*Streptomyces* sp.	4

Name	Substituent and Location	Microorganism	Reference
Piericidin A and B A: R = H B: R = CH$_3$	2–CH$_2$CH=C(CH$_3$)CH=CHCH$_2$C= (CH)$_3$CHCH(CH$_3$)CH(OR)CH(CH$_3$)=CHCH$_3$ 3–CH$_3$, 4–OH, 5–OMe, 6–OMe	*Streptomyces mobaraensis*	4
Pyridoxine	3–CH$_2$OH, 4–CH$_2$OH, 5–OH, 6–CH$_3$	Widespread	1
Pyridoxamine	3–CH$_2$OH, 4–CH$_2$NH$_2$, 5–OH, 6–CH$_3$	Widespread	1
Pyridoxal phosphate	3–CH$_2$–O Ⓟ 4–CHO, 5–OH, 6–CH$_3$	Widespread	1
Coenzyme III	1-Ribose– Ⓟ⌇Ⓟ 3–CONH$_2$	Yeast	1
Nicotinamide adenine dinucleotide (NAD)	1–Ribose– Ⓟ⌇Ⓟ –adenosine, 3–CONH$_2$	Widespread	4
Nicotinamide adenine dinucleotide phosphate (NADP)	1–Ribose– Ⓟ⌇Ⓟ⌇Ⓟ –adenosine, 3–CONH$_2$	Widespread	1
Caerulomycin		*Streptomyces caerulens*	4
Indigoidine		*Corynebacterium* *insidiosum, Arthro-* *bacter atrocyaneus,* *Pseudomonas indigo-* *ferus, Arthrobacter* *polychromogenes*	5
2-*n*-Heptyl-4-oxy- quinoline	2–(CH$_2$)$_6$CH$_3$, 4–OH	*Pseudomonas aeruginosa*	1
2-*n*-Heptyl-3-oxy- 4-quinoline	2–(CH$_2$)$_6$CH$_3$, 3–OH, 4=O	*Pseudomonas aeruginosa*	1
2-*n*-Heptyl-4-oxy- quinoline N-oxide	1 → 0, 2–(CH$_2$)$_6$CH$_3$, 4–OH	*Pseudomonas aeruginosa*	1
2-(2-Heptenyl)-3- methyl-4-quinolinol	2–CH$_2$CH=CH(CH$_2$)$_3$CH$_3$, 3–CH$_3$, 4–OH	*Pseudomonas* sp.	6
2-(*n*-Δ^1-Nonenyl)-4- oxyquinoline	2–CH=CH(CH$_2$)$_6$CH$_3$, 4–OH	*Pseudomonas aeruginosa*	1
2-*n*-Nonyl-4-oxy- quinoline	2–(CH$_2$)$_8$CH$_3$, 4–OH	*Pseudomonas aeruginosa*	1
2-*n*-Nonyl-4-oxy- quinoline N-oxide	1 → 0, 2–(CH$_2$)$_8$CH$_3$, 4–OH	*Pseudomonas aeruginosa*	1
2-*n*-Undecyl-4-oxy- quinoline N-oxide	1 → 0, 2–(CH$_2$)$_{10}$CH$_3$, 4–OH	*Pseudomonas aeruginosa*	1
Viridicatin	2–OH, 3–OH, 4–Ph	*Penicillium viridicatum*	1, 2, 3
Viridicatol	2–OH, 3–OH, 4–(3–OHPh)	*Penicillium viridicatum*	2, 3
3-O-Methyl- viridicatin	2–OH, 3–OCH$_3$, 4–Ph	*Penicillium viridicatum*	3

Name	Substituent and Location	Microorganism	Reference
Indole	None	*Escherichia coli* mutants, yeasts, *Treponema* sp.	1, 9
Indole-3-aldehyde	3–CHO	*Lasiodiplodia theobromae*	3
Indole-3-carboxylic acid	3–COOH	*Lasiodiplodia theobromae*	3
3-Methylindole (Skatole)	3–CH$_3$	*Tricholoma* sp., *Leprota buchnallii*	9
Indole-3-acetic acid	3–CH$_2$COOH	*Rhizopus suinus*, *R. nigricans*, *Aspergillus niger*, *Penicillium notatum*, *Absidia ramosa*, *Boletus edulis*, yeasts	1
L-Tryptophan	3–CH$_2$CH(NH$_2$)COOH	Widely distributed	1
Indoleisopropionic acid and its ester	3–CH(CH$_3$)COOR R = H R = D-mannitol	*Claviceps* sp.	2
Serotonin	3–CH$_2$CH$_2$NH$_2$, 5–OH	*Panaeolus campanulatus*	1
Bufotenin	3–CH$_2$CH$_2$N(CH$_3$)$_2$, 5–OH	*Amanita mappa*	3
Psilocin	3–CH$_2$CH$_2$N(CH$_3$)$_2$	*Psilocybe* sp.	3
Psilocybin	3–CH$_2$CH$_2$N(CH$_3$)$_2$, 4–OH$_2$PO$_3$	*Psilocybe* sp.	3
Baeocystin	3–CH$_2$CH$_2$NHCH$_3$, 4–OH$_2$PO$_3$	*Psilocybe baeocystis*	3
Norbaeocystin	3–CH$_2$CH$_2$NH$_2$, 4–OH$_2$PO$_3$	*Psilocybe baeocystis*	3
5-Chloro-6-methoxy-1-methyl isatin	1–CH$_3$, 2=O, 3=O, 5–Cl, 6–OMe	*Micromonospora carbonacea*	7
Indigo (Indigotin)		*Schizophyllum commune*, *Agaricus campestris*	1, 2, 3
Violacein R = OH		*Chromobacterium violaceum*	1, 8
Deoxyviolacein R = H		*Chromobacterium violaceum*	1, 8
Tetrahydro-1-methyl-β-carboline carboxylic acid		*Amanita muscaria*	3

REFERENCES

1. Miller, M. W., *The Pfizer Handbook of Microbial Metabolites.* McGraw-Hill, New York (1961).
2. Shibata, S., Natori, S., and Udagawa, S., *List of Fungal Products.* Charles C Thomas, Springfield, Illinois (1964).
3. Turner, W. B., *Fungal Metabolites.* Academic Press, New York (1971).
4. Gross, D., *Prog. Chem. Org. Nat. Prod., 28,* 109 (1970).
5. Kuhn, R., Starr, M. P., Kuhn, D. A., Bauer, H., and Knackmuss, H. J., *Arch. Mikrobiol., 51,* 71 (1965).
6. Hashimoto, M., and Hattori, K., *Chem. Pharm. Bull. (Tokyo), 15,* 718 (1967).
7. Reimann, H., and Jaret, R., *Chem. Ind. (London),* p. 2173 (1967).
8. DeMoss, R. D., in *Antibiotics,* Vol. 2, p. 77, D. Gottlieb and P. D. Shaw, Eds., Springer-Verlag, New York (1967).
9. Hilber, O., *Z. Pilzkd., 34,* 153 (1968).

MICROBIAL PYRAZINES

DR. J. C. MACDONALD

Microbial pyrazines are listed in Table 1, together with producing organisms and properties, which include melting points and absorption maxima in appropriate solvents. Soluble pyrazines can often be detected from the ultraviolet spectrum of the culture filtrate.

The largest number of microbial pyrazines is substituted 2(1H)pyrazinones. These, depending on their structures, can generally be classified into three groups, as analogues of either aspergillic acid, hydroxyaspergillic acid, or flavacol. Chemical properties and ultraviolet spectra for compounds of a group are similar, and often infrared spectra from 1600 to 4000 cm^{-1} are also similar. Separation and identification of members of a group generally require chromatography and proton magnetic-resonance spectra.

STRUCTURES

Structure 1: emimycin

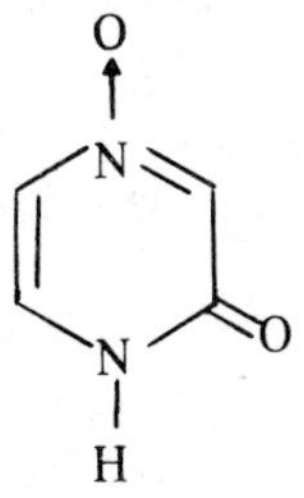

Structure 2: pulcherriminic acid

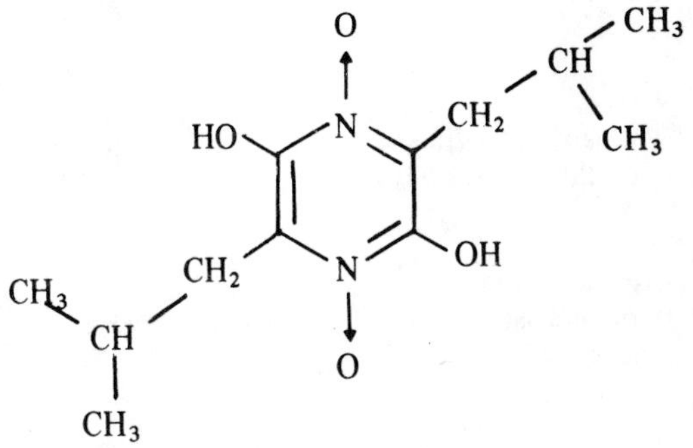

Structure 3: 3,6-substituted-1-hydroxy-2(1H) pyrazinones

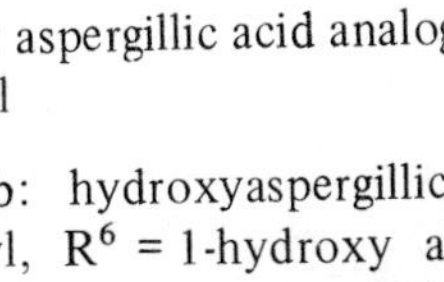

Structure 4: 3,6-substituted-2(1H)pyrazinones, R^3 = alkyl, R^6 = alkyl or 1-hydroxy-alkyl

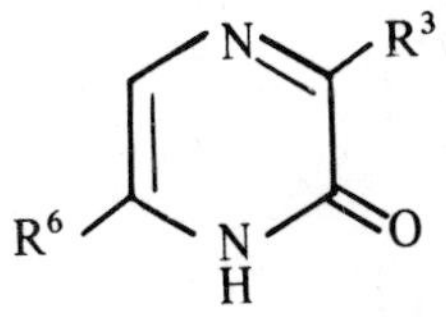

Structure 3a: aspergillic acid analogues, R^3 = alkyl, R^6 = alkyl

Structure 3b: hydroxyaspergillic acid analogues, R^3 = alkyl, R^6 = 1-hydroxy alkyl,

i.e., $C-$, where R = alkyl, R' = H or CH$_3$

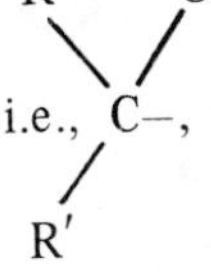

TABLE 1
PYRAZINE DERIVATIVES

Code

A1	*Aspergillus flavus*	B1	*Bacillus subtilis*	C1	*Corynebacterium glutamicum* (mutant)
A2	*Aspergillus sojae*	B2	*Bacillus natto*	C2	*Candida pulcherrima*
A3	*Aspergillus sclerotium*	B3	*Bacillus cereus*	M1	*Micrococcus violagabriellae*
A4	*Aspergillus oryzae*			S1	*Streptomyces* species

Compound and Structure	Organisms and Location of Compound	Properties of Compound	References
2,3,5,6-Tetramethyl pyrazine	B1, B2, C1 Filtrate	Melting point: 87°C UV maximum: 300 nm in 0.1M perchloric acid Forms a picrate	1, 2
2(1H)pyrazinone-4-oxide(emimycin) Structure 1	S1 Filtrate	Decomposes above 200°C UV maxima: 223, 276, 331 nm in methanol Forms silver salt	3
Pulcherriminic acid Structure 2	B3 Filtrate	Decomposes at 162–164°C UV maxima, sodium salt: 243, 282, 410 nm in 2M sodium hydroxide Forms insoluble ferric salt	4
Pulcherrimin (ferric salt of pulcherriminic acid)	B1, B3, C2, M1 Cells	Dark-red pigment, insoluble in water and organic solvents but easily converted to sodium salt (see above)	4, 5, 6, 7, 8
Aspergillic acid Structure 3a; 3-isobutyl-6-*sec*-butyl	A1, A2 Filtrate	Melting point: 99°C UV maxima: 235, 328 nm in ethanol; 232, 336 nm in pH 7–7.3 buffer Forms green cupric salt, gives red color with ferric chloride; has antibiotic activity	9, 10, 11
Aspergillic acid analogues	Filtrate	Properties similar to those of aspergillic acid	
Neoaspergillic acid Structure 3a; 3,6-diisobutyl	A1, A3	Melting point: 129°C	12, 13
Structure 3a; 3,6-di-*sec*-butyl	A1, A2	Oil	11, 12
Structure 3a; 3-isobutyl-6-isopropyl	A1, A2	Melting point: 99°C	12, 14
Structure 3a; 3,6-diisopropyl	A1	Melting point: 74–75°C	12
Structure 3a; 3-propyl-6-*sec*-butyl	A1	Melting point: 80–81°C	12
Structure 3a; 3,6-dipropyl	A1	Melting point: 106–107°C	12
Structure 3a; 3-isobutyl-6-propyl	A1	Melting point: 130–131°C	12

Compound and Structure	Organisms and Location of Compound	Properties of Compound	References
Hydroxyaspergillic acid Structure 3a; 3-isobutyl-6-(1-hydroxy-1-methyl propyl	A1, A2, A4 Filtrate	Melting point: 148−150°C UV maxima: 235, 328 nm in ethanol; 230, 336 nm in pH 7.3 buffer Forms green cupric salt, gives red color with ferric chloride; has weak antibiotic activity	10, 14, 15
Hydroxyaspergillic acid analogues	Filtrate	Properties similar to those of hydroxy-aspergillic acid	
Neohydroxyaspergillic acid Structure 3b; 3-isobutyl-6-(1-hydroxy-2-methyl propyl)	A1, A3	Melting point: 170−171°C	12, 13
Muta-aspergillic acid Structure 3b; 3-isobutyl-6-(1-hydroxy-1-methyl ethyl)	A4	Melting point: 173−174°C	15
Structure 3b; 3-sec-butyl-6-(1-hydroxy-1-methyl propyl)	A1, A2	Melting point: 120−121°C	12, 14
Flavacol Structure 4; 3,6-diisobutyl	A1, A2 Filtrate	Melting point: 147−149°C UV maxima: 208, 230, 326 nm in ethanol No copper salts, no color with ferric chloride; fluoresces under ultraviolet light	16, 17
Flavacol analogues	Filtrate	Properties similar to those of flavacol UV maxima: 208−210, 227−230, 312−326 nm in ethanol	17
Deoxyaspergillic acid Structure 4; 3-isobutyl-6-sec-butyl	A2	Melting point: 102°C	9, 18
Structure 4; 3-isobutyl-6-(1-hydroxy-1-methyl propyl)	A2	Melting point: 107°C	17
Structure 4; 3,6-di-sec-butyl	A2	Melting point: 128°C	17
Structure 4; 3-sec-butyl-6-(1-hydroxy-1-methyl propyl)	A2	Melting point: 120−121°C	18
Structure 4; 3-isobutyl-6-isopropyl	A2	Melting point: 109−111°C	18
Structure 4; 3-isobutyl-6-(1-hydroxy-1-methyl ethyl)	A2	Melting point: 135°C	17

REFERENCES

1. Kosuge, T., and Kamiya, H., *Nature, 193,* 776 (1962).
2. Demain, A. L., Jackson, M., and Trenner, N. R., *J. Bacteriol., 94,* 323 (1967).
3. Terao, M., *J. Antibiot. (Tokyo) Ser. A, 16,* 182 (1963).
4. Kupfer, D. G., Uffen, R. L., and Canale-Parola, E., *Arch. Mikrobiol., 56,* 9 (1967).
5. Uffen, R. L., and Canale-Parola, E., *Z. Allg. Mikrobiol., 9,* 231 (1969).
6. MacDonald, J. C., *Can. J. Microbiol., 13,* 17 (1967).
7. MacDonald, J. C., *Can. J. Microbiol., 12,* 65 (1966).
8. MacDonald, J. C., *Can. J. Chem., 41,* 165 (1963).
9. Dutcher, J. D., *J. Biol. Chem., 171,* 321 (1947).
10. Dutcher, J. D., *J. Biol. Chem., 232,* 785 (1958).
11. Yokotsuka, T., Asao, Y., and Sasaki, M., *J. Agr. Chem. Soc. Jap., 42,* 346 (1968).
12. MacDonald, J. C., *Can. J. Biochem., 48,* 1165 (1970).
13. Micetich, R. G., and MacDonald, J. C., *J. Chem. Soc.,* p. 1507 (1964).
14. Sasaki, M., Asao, Y., and Yokotsuka, T., *J. Agr. Chem. Soc. Jap., 42,* 351 (1968).
15. Nakamura, S., *Agr. Biol. Chem., 25,* 665 (1961).
16. Dunn, G., Newbold, G. T., and Spring, F. S., *J. Chem. Soc.,* p. 2586 (1949).
17. Sasaki, M., Kikuchi, T., Asao, Y., and Yokotsuka, T., *J. Agr. Chem. Soc. Jap., 41,* 154 (1967).
18. Sasaki, M., Asao, Y., and Yokotsuka, T., *J. Agr. Chem. Soc. Jap., 42,* 288 (1968).

MICROBIAL PIPERAZINES

DR. J. C. MACDONALD

Except for anhydroaspergillomarasmine B, all the microbial piperazines can be considered to be derivatives of 2,5-diketopiperazine. Most of the compounds have to be isolated before they can be identified. Simple diketopiperazines are named in Table 1 as cyclic dipeptides, but cyclic dipeptides containing proline may be present in peptone and may not be fermentation products.[1]

PROPERTIES OF PIPERAZINE DERIVATIVES

Melting Points

In general, these are not good criteria of purity, because the compounds often sublime or decompose near their melting points.

Infrared Absorption

Selected peaks of the infrared spectra are shown with the preparation method used. All the compounds have absorption in the carbonyl region, and therefore the absorption frequencies in this region are given. In addition, the highest absorption frequency in the spectrum (usually due to N–H or O–H vibrations) is given if available.

Ultraviolet Absorption

Maxima for visible or ultraviolet spectra are shown with the solvent used. The diketopiperazine ring itself does not absorb light strongly, and maxima are due to substituents that absorb.

Optical Rotation

Optical rotation is shown together with solvent and concentration. Most of the 2,5-diketopiperazines have asymmetric carbon atoms at the 3- and/or 6- positions of the ring and have optical activity. Asymmetric carbon atoms are also present in some substituents.

Other Characteristics

Derivatives of lysergic or isolysergic acid, such as those of Structure 17 (see below), give a blue color with p-aminobenzaldehyde in sulfuric acid solution.[20]

STRUCTURES

Structures of more complex compounds are shown below, with the omission of ring hydrogen atoms.

Structure 1:

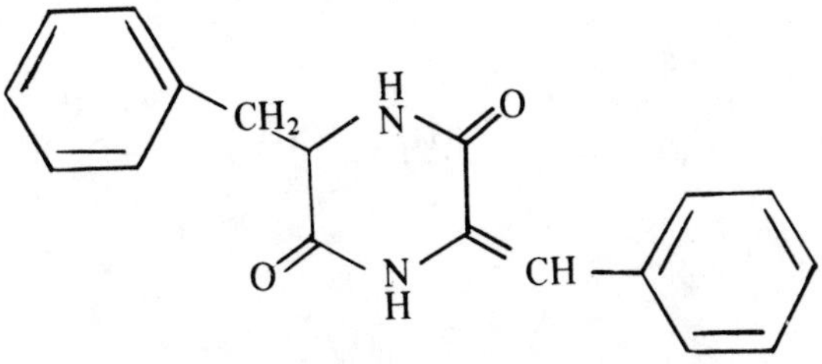

Structure 2:

Structure 2a: R = R′ = phenyl

Structure 2b: R = isopropyl, R′ = phenyl

Structure 2c: R = p-methoxyphenyl, R′ = phenyl

Structure 3:

Structure 3a: rhodotorulic acid, R = R′ = CH_3- O -OH$(CH_2)_3$—

Structure 3b: dimerum acid, R = R′ = $HOCH_2$—$\overset{\overset{\displaystyle CH_3}{|}}{C}$=CH—$\overset{\overset{\displaystyle O}{||}}{C}$—$\overset{\overset{\displaystyle OH}{|}}{N}$—$(CH_2)_3$—

Structure 3c: echinulin, R = CH_3, R′ =

Structure 4: gliotoxin

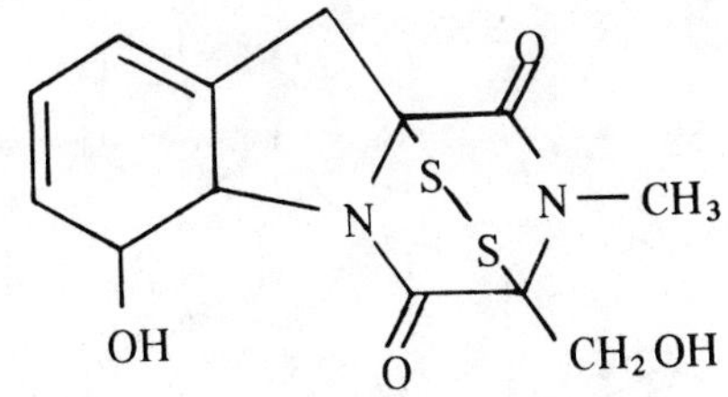

Structure 5: dehydrogliotoxin

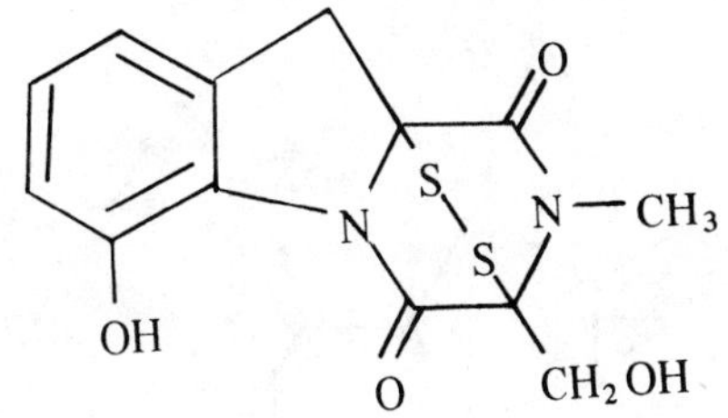

Structure 6:

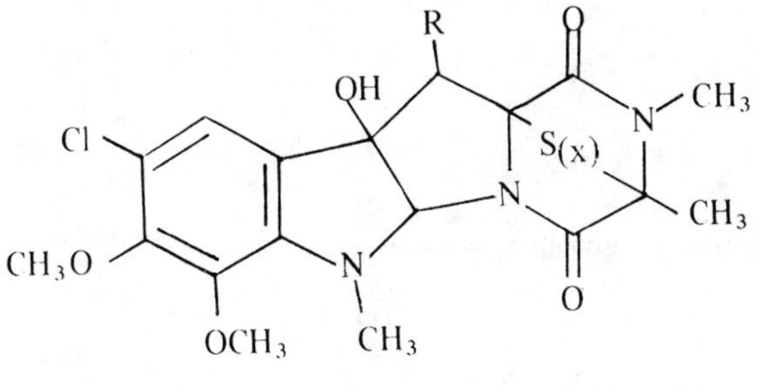

Structure 6a: sporidesmin, R = OH, x = 2

Structure 6b: sporidesmin B, R = H, x = 2

Structure 6c: sporidesmin E, R = OH, x = 3

Structure 7: sporidesmin D

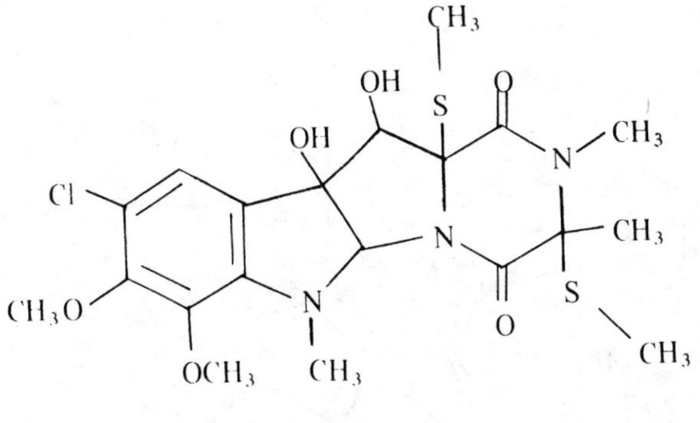

Structure 8: sporidesmin F

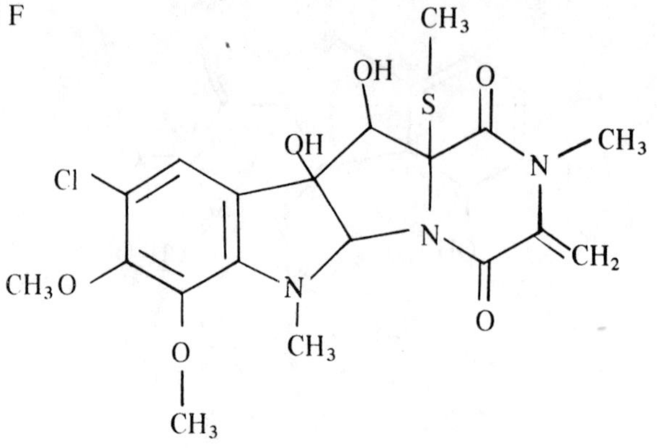

Structure 9:

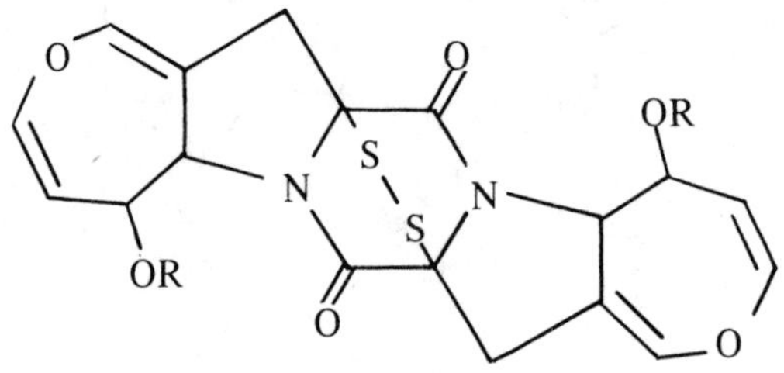

Structure 9a: aranotin, R = H

Structure 9b: acetylaranotin, R = acetyl

Structure 10: bisdethio-di(methylthio)-acetylaranotin

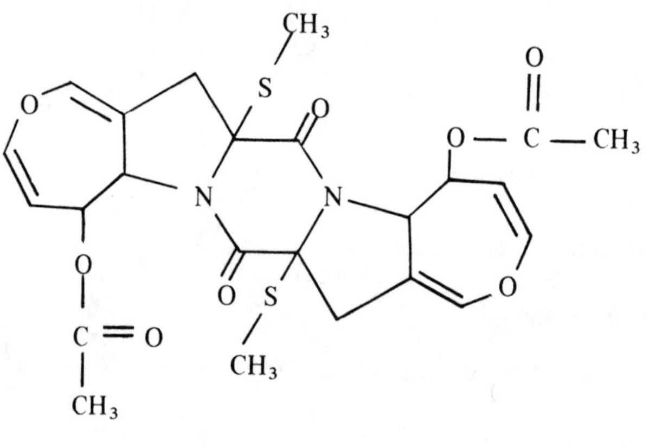

Structure 11: apoaranotin

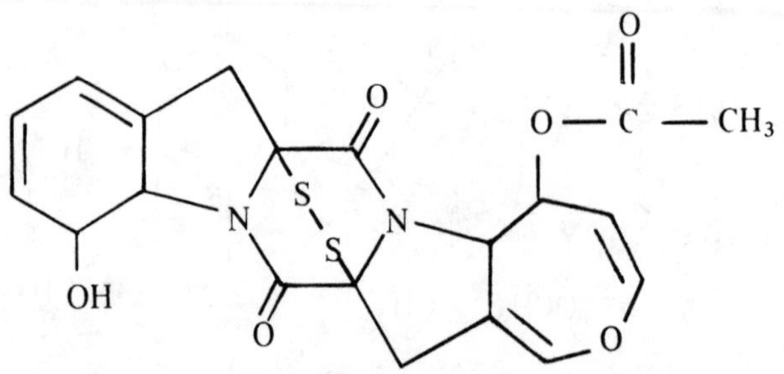

Structure 12: bisdethio-di(methylthio)-acetylapoaranotin

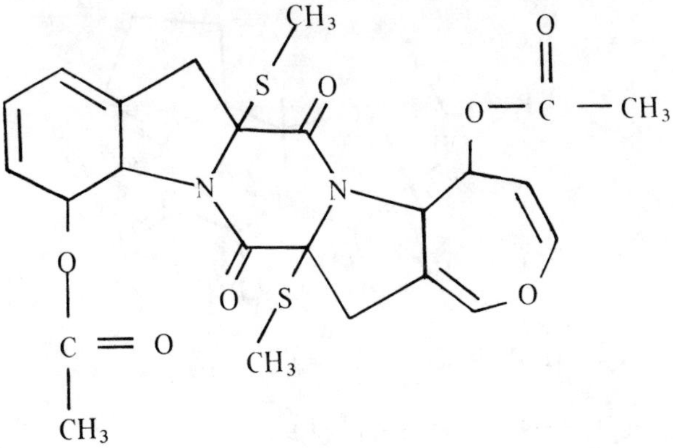

Structure 13: mycelianamide

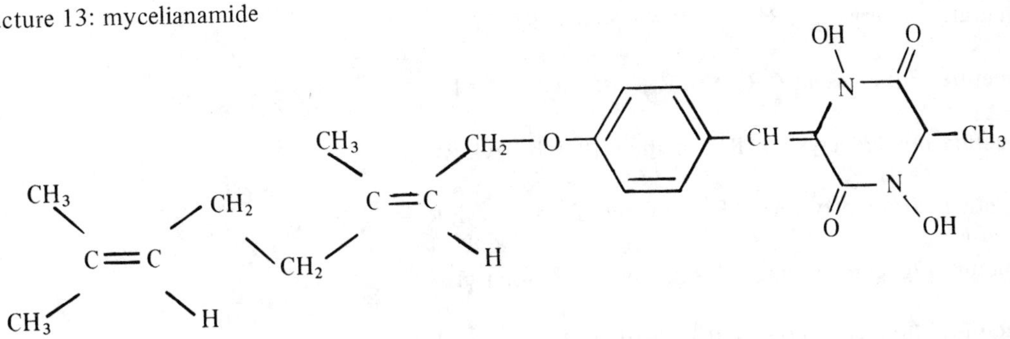

Structure 14: brevianamide A

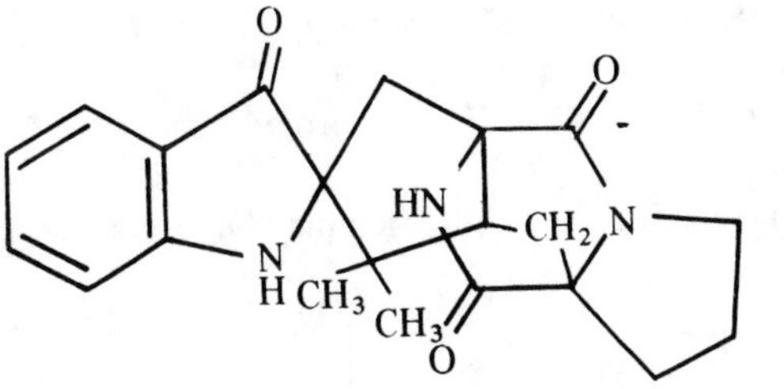

Structure 15: brevianamide E

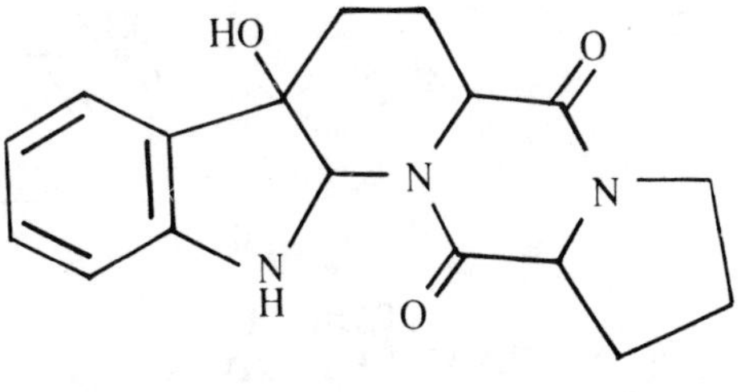

Structure 16: anhydroaspergillomarasmine B

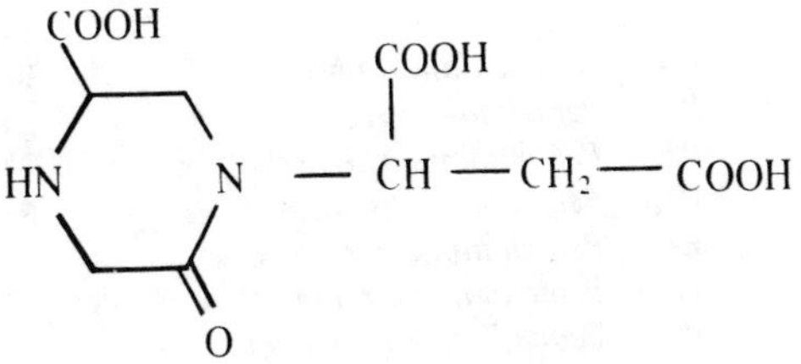

Structure 17: amides of lysergic acid (L) or of isolysergic acid (I).

Structure 17a: ergotamine, R^1 = H, R^2 = benzyl, R^3 = L

Structure 17b: ergotaminine, R^1 = H, R^2 = benzyl, R^3 = I

Structure 17c: ergosine, R^1 = H, R^2 = isobutyl, R^3 = L

Structure 17d: ergosinine, R^1 = H, R^2 = isobutyl, R^3 = I

Structure 17e: ergocrystine, R^1 = methyl, R^2 = benzyl, R^3 = L

Structure 17f: ergocrystinine, R^1 = methyl, R^2 = benzyl, R^3 = I

Structure 17g: ergocryptine, R^1 = methyl, R^2 = isobutyl, R^3 = L

Structure 17h: ergocryptinine, R^1 = methyl, R^2 = isobutyl, R^3 = I

Structure 17i: ergocornine, R^1 = methyl, R^2 = isopropyl, R^3 = L

Structure 17j: ergocorninine, R^1 = methyl, R^2 = isopropyl, R^3 = I

Structure 18: picrorocellin (substituents on nitrogen may be reversed)

TABLE 1
PIPERAZINE DERIVATIVES

Code

A1	*Aspergillus fumigatus*	G1	*Gliocladium fimbriatum*
A2	*Aspergillus glaucus* group	P1	*Penicillium nigrans*
A3	*Aspergillus chevalieri*	P2	*Penicillium terlikowskii*
A4	*Aspergillus terreus*	P3	*Penicillium obscurum*
A5	*Aspergillus flavus*	P4	*Penicillium cinerascens*
A6	*Aspergillus oryzae*	P5	*Penicillium patulum*
Ar	*Arachniotus aureus*	P6	*Penicillium griseofulvin*
C	*Claviceps* species	P7	*Penicillium brevi-compactum*
F1	*Fusarium* species	Pi	*Pithomyces chartarum*
		R1	*Rosellinia necatrix*

Rh	Species of *Rhodotorula*, *Rhodosporidium, Leucosporidium, Sporidiobolus, Sporobolomyces*
Ro	*Roccella fuciformis*
S1	*Streptomyces griseus*
S2	*Streptomyces noursei*
S3	*Streptomyces albus*, var. *fungatus*
S4	*Streptomyces thioluteus*
T1	*Trichoderma viride*

TABLE 1 (Continued)
PIPERAZINE DERIVATIVES

Compound and Structure	Organisms and Location of Compound	Properties of Compound	References
Cyclo-L-leucyl-L-prolyl	A1, R1, S1 Filtrate	Melting point: 158–161°C IR peaks: 3260, 1670, 1635 cm^{-1} in Nujol® $[a]_D^{20}$: –141° in ethanol (c = 3.33)	2
Cyclo-L-prolyl-L-valyl	R1 Filtrate	Melting point: 191–193°C IR peaks: 3260, 1660 cm^{-1} in Nujol® $[a]_D^{20}$: –161° in ethanol (c = 1)	2
Cyclo-L-prolyl-L-phenylalanyl	R1	Melting point: 127–128°C IR peaks: 3300, 1705, 1690, 1660, 1640, 1610 cm^{-1} in Nujol® UV maximum: 258 nm in methanol $[a]_D^{20}$: –99.8° in ethanol (c = 1)	2
Cyclo-di-(L-phenylalanyl)	P1, S2 Cells	Melting point: 311–312°C IR peaks: 3189, 1669, 1665 cm^{-1} in potassium bromide $[a]_D^{25}$: –242° in pyridine (c = 0.059)	3
3-Benzyl-6-benzylidine-2,5-di-oxopiperazine Structure 1	S2	Melting point: 288–290°C IR peaks: 3189, 1661, 1626 cm^{-1} in potassium bromide UV maximum: 296 nm in ethanol $[a]_D^{24}$: –490° in pyridine (c = 0.08)	3
3,6-Dibenzylidene-2,5-dioxo-piperazine Structure 2a	S2	Melting point: 298–300°C IR peaks: 3169, 1674, 1620 cm^{-1} in potassium bromide UV maxima: 234, 338 nm in ethanol $[a]_D$: 0 in pyridine	3
3-Benzylidene-6-isobutylidine-2,5-dioxopiperazine (albo-noursin) Structure 2b	S2, S3	Melting point: 272°C IR peaks: 3184, 1680, 1638 cm^{-1} in potassium bromide UV maxima: 234, 318 nm in ethanol $[a]_D$: 0 in pyridine	3
3-Anisylidene-6-benzylidene-2,5-dioxopiperazine Structure 2c	S4	Melting point: 275–277°C IR peaks: 3100, 1660, 1590 cm^{-1} in Nujol® UV maxima: 360 nm in chloroform; 393 nm in dilute sodium hydroxide	4
Echinulin Structure 3c	A2 Cells	Melting point: 240–246°C IR peaks: 3460, 1670 cm^{-1} in potassium bromide UV maxima: 228, 278 nm in ethanol $[a]_D^{20}$: –24° in chloroform	5
Rhodotorulic acid Structure 3a	Rh Filtrate	Decomposes at 217–218°C IR peaks: 3190, 1682, 1594 cm^{-1} in potassium bromide UV maximum, ferric complex: 480 nm in aqueous acid	6
Dimerum acid Structure 3b	F1 Filtrate	IR peaks, ferric complex: 3300, 1680, 1600 cm^{-1} in potassium bromide UV maxima, ferric complex: 210, 430 nm in pH 7.3 buffer	7

TABLE 1 (Continued)
PIPERAZINE DERIVATIVES

Compound and Structure	Organisms and Location of Compound	Properties of Compound	References
Gliotoxin Structure 4	T1, G1, P2, P3, P4, A1, A3 Filtrate	Decomposes at 196–221°C IR peaks: 3573, 3454 cm^{-1} in carbon tetrachloride; 1660 cm^{-1} in potassium bromide UV maximum: 267 nm in ethanol $[a]_D^{26}$: –255° in chloroform (c = 0.1) Has antibiotic activity	8, 9, 10, 11
Dehydrogliotoxin Structure 5	P2 Filtrate	Melting point: 185–189°C IR peaks: 3580 cm^{-1} in carbon tetrachloride; 1660, 1600 cm^{-1} in potassium bromide UV maxima: 215, 272, 300 nm in ethanol Has antibiotic activity	8
Sporidesmin Structure 6a	Pi	Benzene solvate decomposes at 110–120°C IR peaks, benzene solvate: 3558, 1715, 1664 cm^{-1} in Nujol® UV maxima, benzene solvate: 219, 254, 302 nm in ethyl ether $[a]_D^{23}$, benzene solvate: –33.5° in methanol (c = 1.1) Very toxic	12
Sporidesmin B Structure 6b	Pi	Melting point: 183°C IR peaks: *ca.* 3500 cm^{-1} in potassium bromide; 1697, 1666 cm^{-1} in Nujol® UV maxima: 218, 256, 307 nm in ethyl ether $[a]_D^{21}$: –27° in methanol (c = 1) Very toxic	12
Sporidesmin D Structure 7	Pi	Melting point, ethanol solvate: 105–107°C IR peaks, ether solvate: 3450, 1680, 1665, 1605 cm^{-1} in potassium bromide UV maxima, ethanol solvate: 216, 252, 300 nm in methanol $[a]_D^{23}$, ethanol solvate: +58° in chloroform (c = 0.11)	13
Sporidesmin E Structure 6c	Pi	Melting point, ether solvate: 180–185°C IR peaks, ether solvate: 3325, 1690, 1655 cm^{-1} in potassium bromide UV maxima, ether solvate: 217, 252, 295 nm in methanol $[a]_D^{20}$, ether solvate: –132° in chloroform (c = 0.06)	14
Sporidesmin F Structure 8	Pi	Melting point: 65–75°C IR peaks: 3430, 1690, 1615, 1605 cm^{-1} in potassium bromide UV maxima: 216, 250, 298 nm in methanol	13
Aranotin Structure 9a	Ar Filtrate	Decomposes at 198–200°C Has antiviral activity	11, 15, 16

TABLE 1 (Continued)
PIPERAZINE DERIVATIVES

Compound and Structure	Organisms and Location of Compound	Properties of Compound	References
Acetylaranotin Structure 9b	Ar, A4 Filtrate	Decomposes at 201–230°C IR peaks: 1740, 1665 cm^{-1} in Nujol® UV maxima: end absorption, shoulders at 220, 270 nm in methanol $[a]_D^{26}$: –550° in chloroform Has antiviral activity	11, 15, 16
Bisdethio-di(methylthio)-ace-tylaranotin Structure 10	Ar, A4 Filtrate	Decomposes at 214–236°C IR peaks: 1744, 1678 cm^{-1} in potassium bromide UV maxima: end absorption, shoulders at 220, 255 nm in methanol $[a]_D^{26}$: –282° in chloroform	11, 15, 16
Apoaranotin Structure 11	Ar Filtrate	Decomposes at 200–205°C IR peaks: 3350, 1725, 1660 cm^{-1} in Nujol® UV maximum: 265 nm in ethanol $[a]_D^{26}$: –492° in chloroform Has antiviral activity	11
Bisdethio-di(methylthio)-ace-tylapoaranotin Structure 12	Ar	Decomposes at 105–107°C $[a]_D^{26}$: –175° in chloroform	11
Mycelianamide Structure 13	P1, P5, P6 Cells	Decomposes at 158–159°C IR peaks: 3230, 1675, 1645, 1625, 1610 cm^{-1} in potassium bromide* UV maxima: 234, 324 nm in ethanol $[a]_D^{20}$: –160° in chloroform ($c = 0.9$) Forms ferric complex	17
Brevianamide A Structure 14	P7 Filtrate	Sublimes at 220–250°C IR peaks: 3420, 1715-1680, 1625 cm^{-1} in chloroform UV maxima: 235, 256, 404 nm in ethanol $[a]_D^{25}$: +413° in ethanol	18
Brevianamide E Structure 15	P7 Filtrate	Ir peaks: 3600, 1690, 1680 cm^{-1} in chloroform UV maxima: 239, 296 nm in ethanol $[a]_D^{25}$: –30° in ethanol	18
Anhydroaspergillomarasmine B Structure 16	A5, A6 Filtrate	Decomposes at 270–292°C $[a]_D^{20}$: –103° in pH 7 phosphate buffer ($c = 1$)	19
Ergotamine Structure 17a	C	Decomposes at 212–214°C $[a]_D^{20}$: –160° in chloroform ($c = 1.0$)	20
Ergotaminine Structure 17b	C	Decomposes at 241–243°C $[a]_D^{20}$: +369° in chloroform ($c = 0.5$)	20
Ergosine Structure 17c	C	Decomposes at 220–230°C $[a]_D^{20}$: –183° in chloroform ($c = 1$)	20

TABLE 1 (Continued)
PIPERAZINE DERIVATIVES

Compound and Structure	Organisms and Location of Compound	Properties of Compound	References
Ergosinine Structure 17d	C	Decomposes at 228°C $[a]_D^{20}$: +420° in chloroform ($c = 1$)	20
Ergocrystine Structure 17e	C	Acetone solvate decomposes at 160–175°C $[a]_D^{20}$: –183° in chloroform ($c = 1$)	20
Ergocrystinine Structure 17f	C	Decomposes at 226°C $[a]_D^{20}$: +366° in chloroform ($c = 1$)	20
Ergocryptine Structure 17g	C	Decomposes at 212–214°C $[a]_D^{20}$: –190° in chloroform ($c = 1$)	20
Ergocryptinine Structure 17h	C	Decomposes at 240–242°C $[a]_D^{20}$: +408° in chloroform ($c = 1$)	20
Ergocornine Structure 17i	C	Decomposes at 182–184°C $[a]_D^{20}$: –188° in chloroform ($c = 1$)	20
Ergocorninine Structure 17j	C	Decomposes at 228°C $[a]_D^{20}$: +409° in chloroform ($c = 1$)	20
Picrorocellin Structure 18	Ro	Melting point: 190–220°C, heating-rate dependent $[a]_D$: +12.5° in chloroform	21

REFERENCES

1. Tamura, S., Suzuki, A., Aoki, Y., and Otake, N., *Agr. Biol. Chem., 28,* 650 (1964).
2. Chen, Y., *J. Agr. Chem. Soc. Jap., 24,* 372 (1960).
3. Brown, R., Kelley, C., and Wiberley, S. E., *J. Org. Chem., 30,* 277 (1965).
4. Gerber, N. N., *J. Org. Chem., 32,* 4055 (1967).
5. Quilico, A., *Research Progress in Organic Biological and Medicinal Chemistry,* Vol. 1, p. 225, V. Gallo and S. Santamaria, Eds. North-Holland Publishing Co., Amsterdam (1964).
6. Atkin, C. L., Nielands, J. B., and Phaff, H. J., *J. Bacteriol., 103,* 722 (1970).
7. Diekmann, H., *Arch. Mikrobiol., 73,* 65,(1970).
8. Lowe, G., Taylor, A., and Vining, L. C., *J. Chem. Soc. Sect. C,* p. 1799 (1966).
9. Johnson, J. R., Bruce, W. F., and Dutcher, J. D., *J. Amer. Chem. Soc., 65,* 2005 (1943).
10. Wilkinson, S., and Spilsburg, J. F., *Nature, 206,* 619 (1965).
11. Neuss, N., Nagarajan, R., Molloy, B. B., and Huckstep, L. L., *Tetrahedron Lett.,* No. 72, 4467 (1968).
12. Ronaldson, J. W., Taylor, A., White, E. P., and Abraham, R. J., *J. Chem. Soc.,* p. 3172 (1963).
13. Jamieson, W. D., Rahman, R., and Taylor, A., *J. Chem. Soc. Sect. C,* p. 1564 (1969).
14. Rahman, R., Safe, S., and Taylor, A., *J. Chem. Soc. Sect. C,* p. 1665 (1969).
15. Miller, P. A., Trown, P. W., Fulmor, W., Morton, G. O., and Karliner, J., *Biochem. Biophys. Res. Commun., 33,* 219 (1968).
16. Nagarajan, R., Huckstep, L. L., Lively, D. H., Delong, D. C., March, M. M., and Neuss, N., *J. Amer. Chem. Soc., 90,* 2980 (1968).
17. Gallina, C., Romeo, H., Tarzina, G., and Torotorella, V., *Gazz. Chim. Ital., 94,* 1301 (1964).
18. Birch, A. J., and Wright, J. J., *Tetrahedron, 26,* 2329 (1970).
19. Haenni, A. L., Robert, M., Vetter, W., Roux, L., Barbier, M., and Lederer, E., *Helv. Chim. Acta, 48,* 729 (1965).
20. Stoll, A., and Hofmann, A., The Ergot Alkaloids, in *The Alkaloids,* Vol. 8, Ch. 21, R. H. F. Manske, Ed. Academic Press, New York (1965).
21. Forster, M. O., and Saville, W. B., *J. Chem. Soc., 121,* 816 (1922).

PORPHYRINS, CHLOROPHYLLS, AND VITAMIN B$_{12}$

DR. NANCY N. GERBER

PORPHYRINS

Compound	Structure	Microorganisms	Ref.
Protoporphyrin	$R_1 = R_3 = R_5 = R_8 = CH_3$ $R_2 = R_4 = -CH=CH_2$ $R_6 = R_7 = CH_2CH_2COOH$	Yeasts *Rhodopseudomonas spheroides* Other photosynthetic bacteria	1
Hematin	Same as protoporphyrin, except with Fe^+ in the center bound to all four pyrrole nitrogens (no hydrogens on the pyrrole nitrogens)	*Saccharomyces anamensis*	1
Coproporphyrin I	$R_1 = R_3 = R_5 = R_7 = CH_3$ $R_2 = R_4 = R_6 = R_8 = CH_2CH_2COOH$	*Saccharomyces cerevisiae* *Saccharomyces anamensis* Other yeasts *Aspergillus oryzae* Photosynthetic bacteria	1
Coproporphyrin III	$R_1 = R_3 = R_5 = R_8 = CH_3$ $R_2 = R_4 = R_6 = R_7 = CH_2CH_2COOH$	*Mycobacterium tuberculosis* *Rhodopseudomonas spheroides* *Corynebacterium diphtheriae*	1
Uroporphyrin III	$R_1 = R_3 = R_5 = R_8 = CH_2COOH$ $R_2 = R_4 = R_6 = R_7 = CH_2CH_2COOH$	*Rhodopseudomonas spheroides*	1

CHLOROPHYLLS

Compound	Structure	Microorganisms	Ref.
Chlorophyll a	$R_1 = CH=CH_2$ $R_2 = COOMe$ $R_3 = C_{20}H_{39}$ (phytyl) $R_4 = CH_3$ $R_5 = H$	Algae	2
Chlorophyll b	$R_1 = CH=CH_2$ $R_2 = COOMe$ $R_3 = C_{20}H_{39}$ (phytyl) $R_4 = CHO$ $R_5 = H$	Algae	2
Chlorophyll d	$R_1 = CHO$ $R_2 = COOMe$ $R_3 = C_{20}H_{39}$ (phytyl) $R_4 = CH_3$ $R_5 = H$	Rhodophyta (red algae)	2
Chlorophyll c	Structure not known	Phaeophyta (brown algae)	2
Protochlorophyll a	Same as chlorophyll a, except with C=C at carbons 7 and 8	Some algae grown in the dark	2
Bacteriochlorophyll a (formerly bacterio-chlorophyll)	3,4-Dihydro $R_1 = COCH_3$ $R_2 = COOMe$ $R_3 = C_{20}H_{39}$ (phytyl) $R_4 = CH_3$ $R_5 = H$	Most known purple bacteria *Thiorhodaceae athiorhodaceae*	3
Bacteriochlorophyll b	Not known	*Rhodopseudomonas* sp.	3
Bacteriochlorophyll c (Chlorobium chloro-phyll 660)	$R_1 = CH(OH)CH_3$ $R_2 = H$ $R_3 = C_{15}H_{25}$ (farnesyl) $R_4 = R_5 = CH_3$	*Chlorobium lumicola* Other *Chlorobium* spp. *Pelodictyon clathratiforme* *Pelodictyon abgregamonas ethylicum*	3
Bacteriochlorophyll d (Chlorobium chloro-phyll 650)	$R_1 = CH(OH)CH_3$ $R_2 = H$ $R_3 = C_{15}H_{25}$ (farnesyl) $R_4 = CH_3$ $R_5 = H$	Various *Chlorobium* spp. *Pelodictyon aggregatum* *Pelodictyon clathratiforme*	3

VITAMIN B$_{12}$

Compound	Microorganisms	Ref.
Vitamin B$_{12}$	*Streptomyces griseus*	1
	Streptomyces antibioticus	
	Streptomyces roseochromogenes	
	Streptomyces olivaceus	
	Mycobacterium smegmatis	
	Lactobacillus arabinosus	
	Propionibacteria	

REFERENCES

1. Miller, M. W., *The Pfizer Handbook of Microbial Metabolites.* McGraw-Hill, New York (1961).
2. Lechevalier, H. A., and Pramer, D., *The Microbes.* J. B. Lippincott, Philadelphia, Pennsylvania (1972).
3. Pfennig, N., *Annu. Rev. Microbiol., 21,* 305 (1967).

COMPOUNDS WITH SULFUR IN THE RING

DR. NANCY N. GERBER

The most important compounds ever isolated from fungi are the penicillins.[1] The penicillins and the closely related cephalosporins were recently comprehensively reviewed.[2]

Penicillins are isolated from a number of Penicillia and Aspergilli and from a Cephalosporium. Penicillin N has also been obtained from a *Streptomyces* species.

Substances with sulfur atoms bridged across dioxopiperazine rings, such as gliotoxin and sporidesmin, are listed with the piperazines (see pp. 345–354).

TABLE 1

Compound	R	References
6-Aminopenicillanic acid	H–	1
Penicillin F	$CH_3CH_2CH=CHCH_2CO–$	1
Dihydropenicillin F	$CH_3(CH_2)_4CO–$	1
Penicillin G	$PhCH_2CO–$	1
Penicillin K	$CH_3(CH_2)_6CO–$	1
Penicillin N (Cephalosporin N, synnematin B)	D-α-Aminoadipoyl	1
Isopenicillin N	L-α-Aminoadipoyl	1
Penicillin X	p-OHPhCH$_2$CO–	1

TABLE 2

Compound	R	Reference
7-Aminocephalosporanic acid	H–	1
Cephalosporin C	D-α-Aminoadipoyl	1

TABLE 3

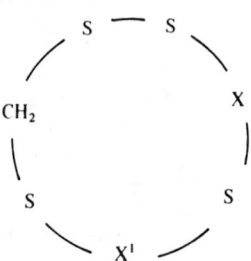

Compound	R, R¹	Microorganisms	Reference
Holomycin	CH_3, H	*Streptomyces griseus*	1
Thiolutin	CH_3, CH_3	*Streptomyces albus thioluteus*	1, 4
Aureothricin	CH_3CH_2, CH_3	*Streptomyces celluloflavus thioluteus*	1, 4
Isobutyropyrrothine	$(CH_3)_2CH$, CH_3	*Streptomyces* sp.	1
Holothin	R¹ = H; side chain NH_2 only	*Streptomyces griseus*	5

TABLE 4

Compound	X, X¹	Microorganisms	Reference
Lenthionine (1,2,3,5,6-Pentathiapane)	CH_2, S	*Lentinus edodes*	1
1,2,4,6-Tetrathiapane	CH_2, CH_2	*Lentinus edodes*	1
1,2,3,4,5,6-Hexathiapane	S, S	*Lentinus edodes*	1

TABLE 5

Compound	Structure	Microorganisms	Reference
Lipoic acid	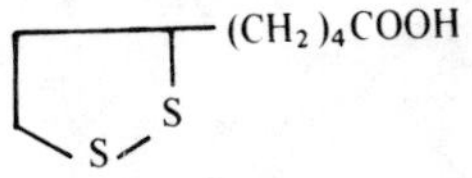	Yeast	3
Junipal	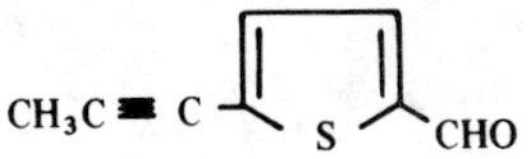	*Daedalea juniperina*	3
Actithiazic acid (Acidomycin, mycobacidin)	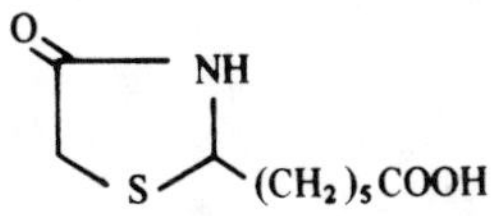	*Streptomyces virginiae cinnamonensis lavendulae*	3
Biotin	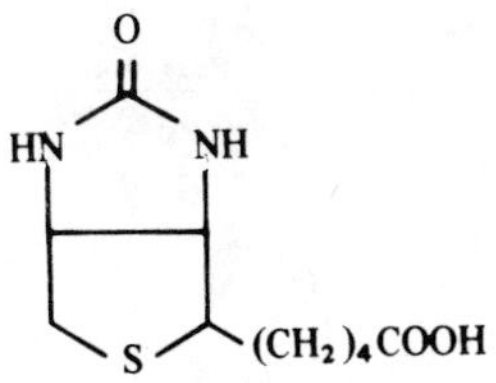	*Torula* and other yeasts, molds, and bacteria	3
Biotin-1-sulfoxide	(→O on sulfur)	*Aspergillus niger*	3
Biocytin	Side chain: $-(CH_2)_4CONH(CH_2)_4CH(NH_2)COOH$	Yeast	3
Thiamin	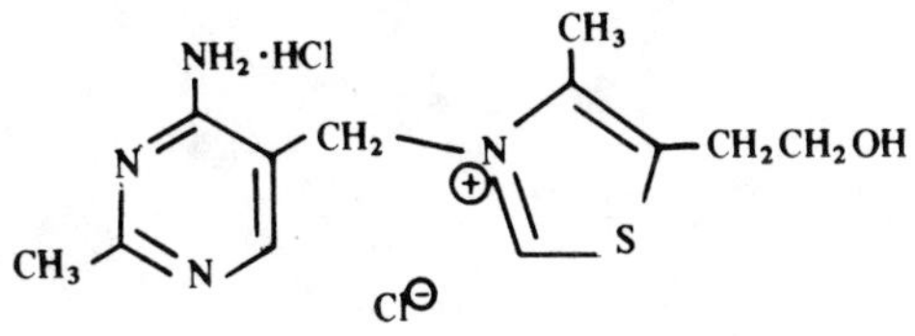	Most yeasts, molds, and bacteria	3

REFERENCES

1. Turner, W. B., *Fungal Metabolites.* Academic Press, New York (1971).
2. Flynn, E. H. (Ed.), *Cephalosporins and Penicillins, Chemistry and Biology.* Academic Press, New York (1972).
3. Miller, M. W., *The Pfizer Handbook of Microbial Metabolites.* McGraw Hill, New York (1961).
4. Waksman, S. A., and Lechevalier, H. A., *The Actinomycetes,* Vol. 3. Williams and Wilkins, Baltimore, Maryland (1962).
5. Umezawa, H. (Ed.), *Index of Antibiotics from Actinomycetes.* University of Tokyo Press, Tokyo, Japan (1967).

GRISANS

DR. JOHN FREDERICK GROVE

The naturally occurring grisans are all metabolic products of fungi and can be extracted from the culture filtrates by chloroform. They are high-melting, sparingly soluble, highly crystalline, colorless or yellow solids and are conveniently considered as two groups — those related to griseofulvin (Structure II; R = Cl), shown in Table 1, and those related to geodin (Structure VI; R = Me, R' = H, R'' = Cl), shown in Table 2. Geodoxin (Structure VII), although not a grisan, is included in Table 2 because of its relationship to geodin.

Grisans containing a ring C enone or dienone chromophore adjacent to the asymmetric spiran center show large values for the optical rotation at the sodium D line. The ultraviolet absorption spectra are characteristic of the substituted coumaranone chromophore λ_{max} 285-291 nm (log $\epsilon \sim 4.3$), and the contribution of the ring C enone or dienone is generally masked. Fluorescence spectra have been useful in identification and estimation.

Griseofulvin is in veterinary and clinical use for the systemic treatment of mycoses, particularly those involving dermatophytes. The characteristic distortion and helical curling of fungal hyphae, first observed with griseofulvin, is found only in analogues with the (d,d) configuration (see below) at the asymmetric centers. Some analogues of geodin show weak fungistatic and bacteriostatic activity.

The chemistry and physical properties of griseofulvin and its analogues have been reviewed,[1] as have studies of the mode of action and biosynthesis.[2]

NOMENCLATURE

The systematic nomenclature is based on the trivial name grisan for the spiro coumarancyclohexane tricyclic system (Structure I), numbered as shown. The absolute configuration of griseofulvin has been determined as (2S, 6'R), but since correlation of configuration in a group of closely related compounds is not always readily evident on the chirality system of Cahn, Ingold and Prelog, the configuration at the two asymmetric centers 2 and 6' has been arbitrarily assigned d or l. The spiran center is placed first and (2S, 6'R)-griseofulvin follows the prefix (d,d). All naturally occurring analogues of griseofulvin have the (d,d) configuration or, when only the spiran center is present, (d). The stereochemical relationship of the geodin group of grisans to (d)[(-)]-dehydrogriseofulvin has not been established. Bisdechlorogeodin occurs in both enantiomeric forms. Naturally occurring erdin was racemic.

STRUCTURES

ORGANISMS

a *Penicillium griseofulvum* Dierckx
b *Penicillium patulum* Bain
c *Penicillium raciborskii* Zal.
d *Penicillium albidum* Sopp.
e *Penicillium melinii* Thom
f *Penicillium janczewskii* Zal.
    [= *P. nigricans* (Bain.) Thom]
g *Penicillium raistrickii* Smith
h *Penicillium brefeldianum* Dodge
i *Penicillium janthinellum* Biourge

j *Penicillium martensii* Biourge
k *Penicillium sclerotigenum* Yamamoto
l *Nigrospora oryzae* (Berk. and Br.) Petch
m *Nigrospora saccharii*
n *Aspergillus terreus* Thom
o *Penicillium estinogenum*
p *Oospora sulphurea-ochracea* v. Beyma
q *Penicillium frequentans* Westling
r *Aspergillus fumigatus* Fresenius

TABLE 1
GRISANS RELATED TO GRISEOFULVIN

Formula (Structure)	Systematic Name (Common Name)	Melting Point, °C	$[a]_D$	Organism(s)	Reference
$C_{16}H_{16}O_6$ [a] (V)	4,6-Dimethoxy-2'-methylgrisan-3,4',6'-trione (Dechlorogriseofulvic acid)	245–248 (decomposes)	+420°	b	3
$C_{17}H_{15}O_6Cl$ (III)	7-Chloro-4,6,2'-trimethoxy-6'-methylgrisa-2',6'-dien-3,4'-dione [(−)-Dehydrogriseofulvin]	276	−26°	b, j	3
$C_{17}H_{17}O_6Br$ [b] (II; R = Br)	7-Bromo-4,6,2'-trimethoxy-6'-methylgris-2'-ene-3,4'-dione	204–205	+300°	a, f	4
$C_{17}H_{17}O_6Cl$ (II; R = Cl)	7-Chloro-4,6,2'-trimethoxy-6'-methylgris-2'-ene-3,4'-dione (Griseofulvin)	222	+340°	a–m	5, 6, 7
$C_{17}H_{18}O_6$ (II; R = H)	4,6,2'-Trimethoxy-6'-methylgris-2'-ene-3,4'-dione (Dechlorogriseofulvin)	179–181	+390°	a, f	8
$C_{17}H_{19}O_6Cl$ (IV)	7-Chloro-4,6,2'-trimethoxy-6'-methylgrisan-3,4'-dione (Dihydrogriseofulvin)	206–208	−20°	j	9

[a] Possible artifact.

[b] Semisynthetic, by addition of bromide to a chloride-deficient medium.

TABLE 2
GRISANS RELATED TO GEODIN

Formula (Structure)	Systematic Name (Common Name)	Melting Point, °C	$[\alpha]_D$	Organism(s)	Reference
$C_{16}H_{10}O_7Cl_2$ (VI; R = R' = H, R'' = Cl)	5,7-Dichloro-4-hydroxy-2'-methoxy-6-methyl-3,4'-dioxogrisa-2',6'-dien-6'-carboxylic acid (Erdin)	211–212	0°	n	10, 11
$C_{17}H_{12}O_7Cl_2$ (VI; R = Me, R' = H, R'' = Cl)	Methyl 5,7-dichloro-4-hydroxy-2'-methoxy-6-methyl-3,4'-dioxogrisa-2',6'-dien-6'-carboxylate (Geodin)	235 (decomposes)	+140°	n, o	10, 11
$C_{17}H_{14}O_7$ (VI; R = Me, R' = R'' = H)	Methyl 4-hydroxy-2'-methoxy-6-methyl-3,4'-dioxogrisa-2',6'-dien-6'-carboxylate[a] [(−)-Bisdechlorogeodin]			n, p	12
$C_{17}H_{14}O_7$ (VI; R = Me, R' = R'' = H)	Methyl 4-hydroxy-2'-methoxy-6-methyl-3,4'-dioxogrisa-2',6'-dien-6'-carboxylate[a] [(+)-Bisdechlorogeodin]	170–173	+188°	q	13
$C_{18}H_{16}O_7$ (VI; R = R' = Me, R'' = H)	Methyl 4,2'-dimethoxy-6-methyl-3,4'-dioxogrisa-2',6'-dien-6'-carboxylate (Trypacidin)	239–240	−160°	r	14
$C_{17}H_{12}O_8Cl_2$ (VII)	(Geodoxin)	216–217 (decomposes)	0°	n	15

[a] Not optically pure.[13]

REFERENCES

1. Grove, J. F., *Fortschr. Chem. Org. Naturst., 22,* 203 (1964).
2. Huber, F. M., and Grove, J. F., in *Antibiotics,* Vol. 1, p. 181, Vol. 2, pp. 123, 440, D. Gottlieb and P. D. Shaw, Eds. Springer Verlag, Berlin, Germany (1967).
3. McMaster, W. J., Scott, A. I., and Trippett, S., *J. Chem. Soc.,* p. 4628 (1960).
4. MacMillan, J., *J. Chem. Soc.,* p. 2585 (1954).
5. Oxford, A. E., Raistrick, H., and Simonart, P., *Biochem. J., 33,* 240 (1939).
6. Grove, J. F., and McGowan, J. C., *Nature, 160,* 574 (1947).
7. Grove, J. F., MacMillan, J., Mulholland, T. P. C., and Rogers, M. A. T., *J. Chem. Soc.,* p. 3977 (1952).
8. MacMillan, J., *J. Chem. Soc.,* p. 1697 (1953).
9. Holbrook, A., Bailey, F., and Bailey, G. M., *J. Pharm. Pharmacol., 15,* 274T (1963).
10. Raistrick, H., and Smith, G., *Biochem. J., 30,* 1315 (1936).
11. Barton, D. H. R., and Scott, A. I., *J. Chem. Soc.,* p. 1767 (1958).
12. Natori, S., and Nishikawa, H., *Chem. Pharm. Bull., 10,* 117 (1962).
13. Stickings, C. E., and Mahmoodian, A., *Biochem. J., 92,* 359 (1964).
14. Balan, J., Kjaer, A., Kovac, S., and Shapiro, R. H., *Acta Chem. Scand., 19,* 528 (1965).
15. Hassall, C. H., and McMorris, T. C., *J. Chem. Soc.,* p. 2831 (1959).

FURANS, BENZOFURANS, AND DIBENZOFURANS

DR. NANCY N. GERBER

Some complex molecules containing a furan or dihydrofuran ring are listed elsewhere in this book, for example, aflatoxins, monocerins, trichothecins, and some quinones.

Furan

Benzofuran

Compound	Structure	Microorganism	Reference
Furan-3-carboxylic acid	3-COOH	*Ceratostomella fimbriata* (on sweet potato)	1, 2
5-Hydroxymethyl furan-2-carboxylic acid	$2-COOH, 5-CH_2OH$	Various aspergilli	1, 2, 3
Ipomeanine	$3-COCH_2CH_2COCH_3$	*Ceratostomella fimbriata* (on sweet potato)	1, 2
Batatic acid	$3-COCH_2CH_2CH(CH_3)COOH$	*Ceratostomella fimbriata* (on sweet potato)	1, 2
Ipomeamarone	$3-CHCH_2CH_2C(CH_3)COCH_2CH_2CH(CH_3)_2$ ⌊————O	*Ceratostomella fimbriata* (on sweet potato)	1, 2
2(Buta-1,3-dienyl)-3-hydroxy-4-(penta-1,3-dienyl) tetrahydrofuran	No double bonds in ring $2-CH=CHCH=CH_2$, $3-OH$, $4-CH=CHCH=CH-CH_3$	*Chaetomium coarctatum*	3
Muscarine	No double bonds in ring $2-CH_2N^{\oplus}(CH_3)_3$, $4-OH, 5-CH_3$	*Amanita muscaria* *Inocybe* spp. *Clitocybe* spp.	3
5-Methoxybenzofuran	$5-OCH_3$	*Stereum subpileatum*	3
5-Formylbenzofuran	5-CHO	*Stereum subpileatum*	3
5-Hydroxymethylbenzofuran	$5-CH_2OH$	*Stereum subpileatum*	3
Unnamed	$5-CH(OH)CH(OH)CH_3$	*Stereum subpileatum*	3
Unnamed	$5-COCH(OH)CH_3$	*Stereum subpileatum*	3
Unnamed	$5-CHCHCH_3$ \\O⁄	*Stereum subpileatum*	3
2-(6-Hydroxy-2-methoxy 3,4-methylene-dioxyphenyl)-benzofuran		Yeast	1

Compound	Structure	Microorganism	Reference
Curvulol	R₁ = CH₃, R₂ = OCH₃ R₃ = H	*Curvularia siddiqui*	3

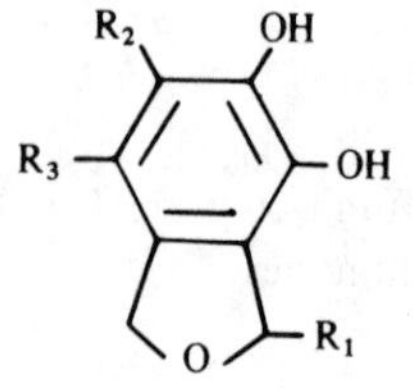

$R_1 = CH_3$,
$R_2 = OCH_3$
$R_3 = H$

Compound	Structure	Microorganism	Reference
Unnamed	$R_1 = H, R_2 = OH, R_3 = CH_3$	*Aspergillus terreus*	3
d- and *l*-Usnic acid	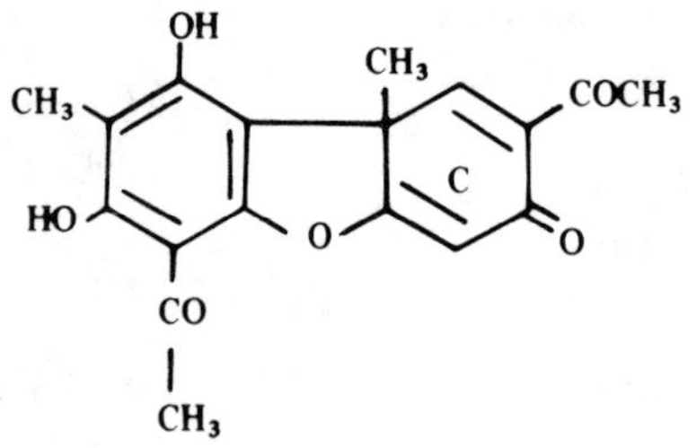	Most yellow lichens	1
Iso-usnic acid	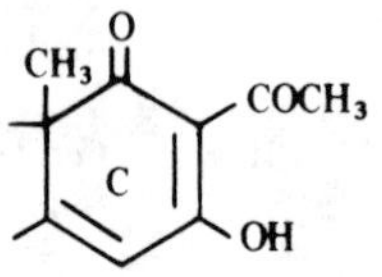	*Cladonia* sp.	3
Pannaric acid	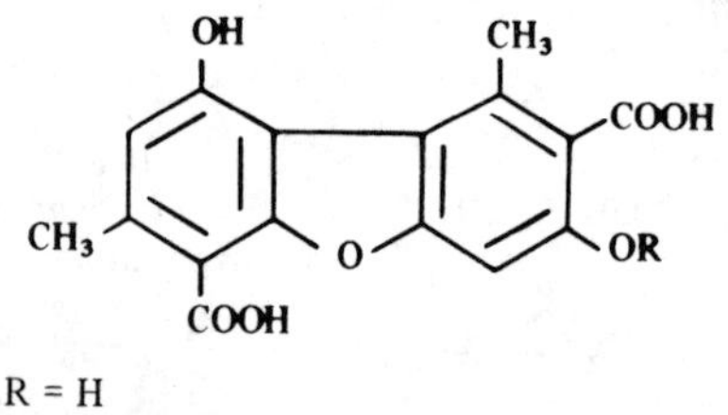 R = H	*Crocynea membranacea* = *Pannaria lanuginosa*	1
Schizopeltic acid	R = CH₃	*Schizopelta californica*	3
Strepsilin	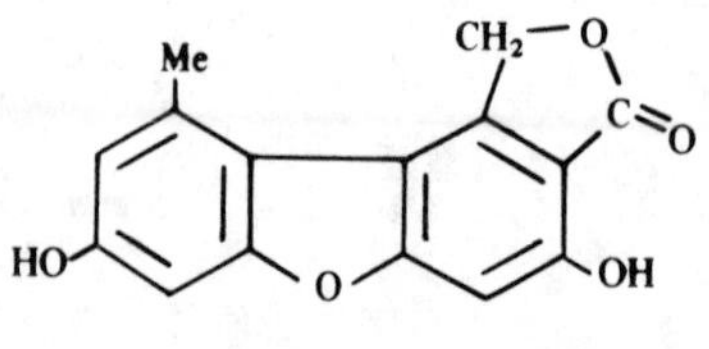	*Cladonia strepsilis*	1, 3
Porphyrilic acid	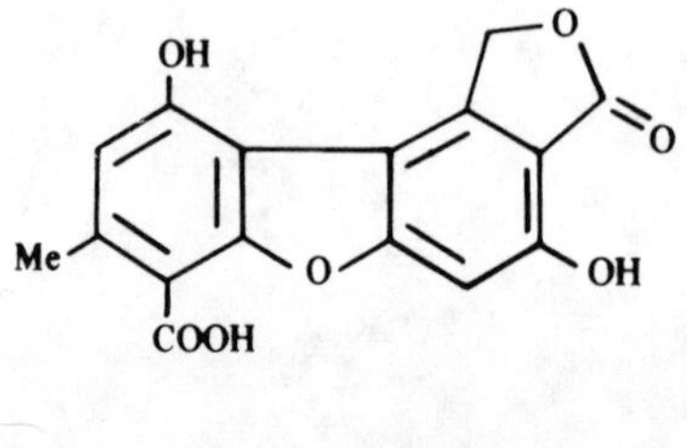	*Haematomma coccineum* *Haematomma porphyrium*	1, 3

Compound	Structure	Microorganism	Reference
Didymic acid		*Cladonia* sp.	1

REFERENCES

1. Miller, M. W., *The Pfizer Handbook of Microbial Metabolites.* McGraw-Hill, New York (1961).
2. Shibata, S., Natori, S., and Udagawa, S., *List of Fungal Products.* Charles C Thomas, Springfield, Illinois, (1964).
3. Turner, W. B., *Fungal Metabolites.* Academic Press, New York (1971).

FIVE-MEMBERED-RING (γ) LACTONES, INCLUDING TETRONIC ACID DERIVATIVES

DR. NANCY N. GERBER

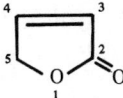

Certain phenyl-substituted γ-lactones and dilactones of the vulpinic acid type are not included here, but some will be found in the chapter on lichen products. An extensive listing of other lactones begins on page 51.

Compound	Structure	Microorganism	Reference
Ascorbic acid (Vitamin C)	3–OH 4–OH 5–CHOHCH$_2$OH	*Serratia marcescens* *Aspergillus niger*	1, 2
Penicillic acid	4–OCH$_3$ 5–OH, 5–C(CH$_3$)=CH$_2$	*Penicillium* spp. *Aspergillus* spp.	1, 2
4,5-Dihydroxyoct-6-enoic acid 1,4-lactone	No C=C in ring 5–CH(OH)CH=CHCH$_3$	*Nigrospora* sp.	3
Unnamed	3–(CH$_2$)$_3$CH$_3$ 5–CH$_3$	*Streptomyces griseus odorifer*	4
Unnamed	3–CH$_2$CH$_2$CH(CH$_3$)$_2$ 5–CH$_3$	*Streptomyces griseus odorifer*	4
Canadensic acid	3–C(=CH$_2$)COOH 5–(CH$_2$)$_3$CH$_3$	*Penicillium canadense*	3
Isocanadensic acid	3–CH$_3$ 4–COOH 5–(CH$_2$)$_4$CH$_3$	*Penicillium canadense*	3
Hydroxyisocanadensic acid	3–CH$_3$ 4–COOH 5–CH(OH)(CH$_2$)$_3$CH$_3$	*Penicillium canadense*	3
Dihydroisocanadensic acid	No C=C in ring 3–CH$_3$ 4–COOH 5–(CH$_2$)$_4$CH$_3$	*Penicillium canadense*	3
Nemotin	No C=C in ring 5–CH=C=CH–C≡C–C≡CH	*Poria corticola tenuis*	1, 2
Unnamed	3–(CH$_2$)$_3$CH(CH$_3$)$_2$ 5–CH$_3$	*Streptomyces griseus odorifer*	4

Compound	Structure	Microorganism	Reference
Unnamed	$3-(CH_2)_5CH_3$ $5-CH_3$	*Streptomyces griseus odorifer*	4
Unnamed	$3-(CH_2)_4CH(CH_3)_2$ $5-CH_3$	*Streptomyces griseus odorifer*	4
Minioluteic acid	No C=C in ring $3-(CH_2)_9CH_3$ $4-COOH, 4-OH$ $5-COOH$	*Penicillium minioluteum*	3
Nephrosterinic acid	No C=C in ring $3=CH_2$ $4-COOH$ $5-(CH_2)_{10}CH_3$	Lichens *Nephromopsis endocrocea (= Cetraria endocrocea)*	1, 3
Nephrosteranic acid	No C=C in ring $3-CH_3$ $4-COOH$ $5-(CH_2)_{10}CH_3$	Lichens *Nephromopsis endocrocea*	1, 3
Spiculisporic acid	No C=C in ring $5-COOH, 5-CH(COOH)CH_2(CH_2)_8CH_3$	*Penicillium* sp.	3
Acaranoic acid	No C=C in ring $3-CH_2COOH$ $5-(CH_2)_{10}CH_3$	*Acarospora chlorophana*	3
Acarenoic acid	$3-CH_2COOH$ $5-(CH_2)_{10}CH_3$	*Acarospora chlorophana*	3
Lichesterinic acid	$3-CH_3$ $4-COOH$ $5-(CH_2)_{12}CH_3$	*Cetraria islandica* *Nephromopsis stracheyi* Lichens	1, 3
Protolichesterinic acid	No C=C in ring $3=CH_2$ $4-COOH$ $5-(CH_2)_{12}CH_3$	Lichens *Cetraria islandica* *Parmelia sinodensis* *Cladonia papillaria*	1, 3
Nephromopsic acid	No C=C in ring $3-CH_3$ $4-COOH$ $5-(CH_2)_{12}CH_3$	*Nephromopsis stracheyi* Lichens	1, 3
Isochracein		*Hypoxylon coccineum*	3

Compound	Structure	Microorganism	Reference
3,5-Dimethyl-6-hydroxy-phthalide		*Penicillium gladioli*	1, 2, 3
Gladiolic acid	 R = CHO	*Penicillium gladioli*	1, 2
Dihydrogladiolic acid	R = CH_2OH	*Penicillium gladioli*	1, 2
Mycophenolic acid	 R = $CH_2CH=CH(CH_3)CH_2CH_2COOH$	*Penicillium brevi-compactum*	3
Mycophenolic acid diol lactone	 R = $CH_2CH(OH)$-	*Penicillium brevi-compactum*	3
Ethyl mycophenolate	R = $CH_2CH=CH(CH_3)CH_2CH_2COOEt$	*Penicillium brevi-compactum*	3
Mychromenic acid		*Penicillium brevi-compactum*	3
Variegatic acid		*Suillus variegatus*	5

REFERENCES

1. Miller, M. W., *The Pfizer Handbook of Microbial Metabolites.* McGraw-Hill, New York (1961).
2. Shibata, S., Natori, S., and Udagawa, S., *List of Fungal Products.* Charles C Thomas, Springfield, Illinois (1964).
3. Turner, W. B., *Fungal Metabolites.* Academic Press, New York (1971).
4. Gerber, N. N., *Tetrahedron Lett.,* p. 771 (1973).
5. Edwards, R. L., and Elsworthy, G. C., *Chem. Commun.,* p. 373 (1967).

SIX-MEMBERED-RING LACTONES (δ-LACTONES):
a-PYRONES AND ISOCOUMARINS

DR. NANCY N. GERBER

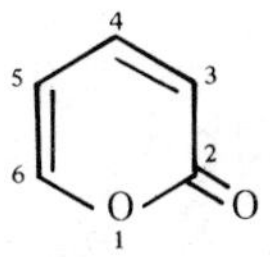

a-Pyrone 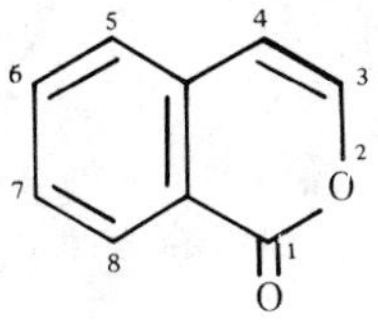   Isocoumarin

Aflatoxins are not included in this group. An extensive listing of other lactones begins on page .

Compound	Structure	Microorganism	Reference
a-Pyrones			
Triacetic acid lactone	4$-$OH 6$-$CH$_3$	*Penicillium* *stipitatum*	1
Methyltriacetic acid lactone	3$-$CH$_3$ 4$-$OH 6$-$CH$_3$	*Penicillium* *stipitatum*	1, 2
6-Allyl-5,6-dihydro-5-hydroxypyran-2-one	5,6$-$Dihydro 5$-$OH 6$-$CH=CHCH$_3$	*Nigrospora* sp.	1
Tetraacetic acid lactone	4$-$OH 6$-$CH$_2$COCH$_3$	*Penicillium* *stipitatum*	1
Hydroxyramulosin	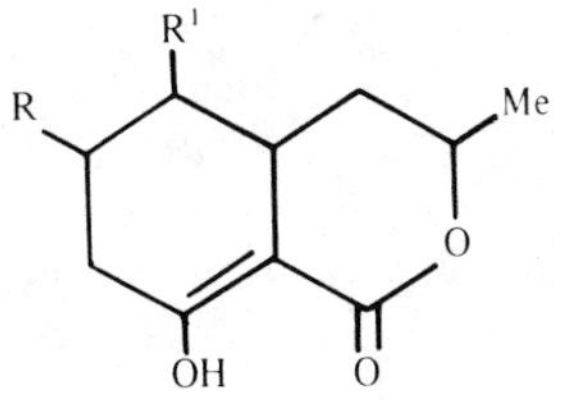 R = OH R^1 = H	*Pestalotia* *ramulosa*	1

Alternaric acid	5,6$-$Dihydro
	4$-$OH
	5$-$COCH$_2$C(=CH$_2$)CH$_2$CH=CH$-$
	  C(OH)(COOH)CH(OH)CH(CH$_3$)$-$
	  CH$_2$Me
	6$-$Me
		Alternaria
		solani
			1

Portentol	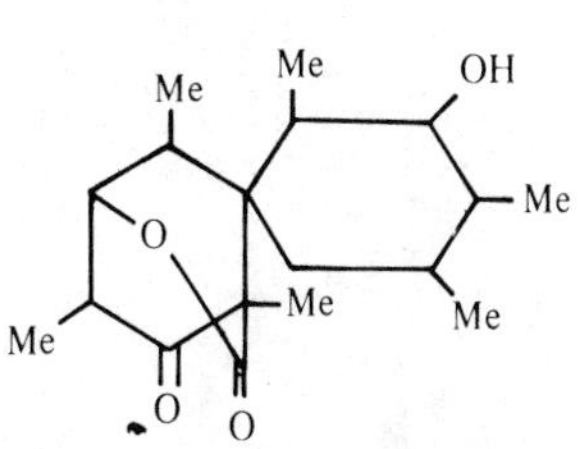	*Roccella* *fuciformis*	1

Compound	Structure	Microorganism	Reference

a-Pyrones (continued)

Hispidin	4–OH
	6—CH=CH— (phenyl) OH, OH
		Polyporus
		  hispidus
		  schweinitzii
			1, 2

Luteoreticulin	3–CH$_3$ 4–OCH$_3$ 6–C(CH$_3$)=CHC(CH$_3$)=CH—(phenyl)—NO$_2$	*Streptomyces* *luteoreticula*	5
Citreoviridin	4–OCH$_3$ 5–CH$_3$ 6–CH=CHCH=CHCH=CHC(CH$_3$)=CH— (ring: OH, CH$_3$, OH, CH$_3$, O, CH$_3$)	*Penicillium* *citreo-viride*	1, 2

Isocoumarins

Mellein	3,4–Dihydro 3–Me 8–OH	*Aspergillus* *melleus*	1
4-Hydroxymellein	3,4–Dihydro 3–CH$_3$ 4–OH 8–OH		
cis-isomer		*Lasiodiplodia* *theobromae*	1
trans-isomer		*Aprospora* *camptospora*	1

6-Methoxymellein	3,4–Dihydro
	3–CH$_3$
	6–OCH$_3$
	8–OH
		Sporormia
		  bipartis
		  affinis
			1

5-Methylmellein	3,4–Dihydro 3–CH$_3$ 5–CH$_3$ 8–OH	*Fusicoccum* *amygdali*	1
Isochracein	3,4–Dihydro 3–CH$_2$Me 8–OH	*Hypoxylon* *coccineum*	1
8-Hydroxy-3-methyliso- coumarin	3–CH$_3$ 8–OH	*Marasmius* *ramealis*	1, 2, 3
8-Hydroxy-6-methoxy-3- methylisocoumarin	3–CH$_3$ 6–OMe 8–OH	*Ceratocystis* *fimbriata*	1

Compound	Structure	Microorganism	Reference
Isocoumarins (continued)			
3,4-Dihydro-6,8-dihydroxy-3-methylisocoumarin	3,4–Dihydro 3–CH$_3$ 6–OH 8–OH	*Aspergillus terreus*	1
Reticulol	3,4–Dihydro 3–CH$_3$ 6–OH 7–OCH$_3$ 8–OH	*Streptomyces rubrireticuli*	1
5-Chloro-8-hydroxy-6-methoxy-3-methyl-3,4-dihydroisocoumarin	3,4–Dihydro 3–CH$_3$ 5–Cl 6–OMe 8–OH	*Periconia macrospinosa*	1
7-Chloro-8-hydroxy-6-methoxy-3-methyl-3,4-dihydroisocoumarin	3,4–Dihydro 3–CH$_3$ 6–OMe 7–Cl 8–OH	*Sporormia affinis*	1
5,7-Dichloro-8-hydroxy-6-methoxy-3,4-dihydroisocoumarin	3,4–Dihydro 3–CH$_3$ 5–Cl 6–OCH$_3$ 7–Cl 8–OH	*Sporormia affinis*	1
Oospolactone	3–CH$_3$ 4–CH$_3$ 8–OH	*Oospora astringens*	1
Oospoglycol	4–CHOHCH$_2$OH 8–OH	*Oospora astringens*	1, 2
Oosponol	4–COCH$_2$OH 8–OH	*Oospora astringens*	1
4-Acetyl-6,8-dihydroxy-5-methylisocoumarin	4–COCH$_3$ 5–CH$_3$ 6–OH 8–OH	*Oospora astringens*	1
Diaporthin	3–CH$_2$CHOHCH$_3$ 6–OCH$_3$ 8–OH	*Endothia parasitica*	4
Ochratoxin A	3,4–Dihydro 3–CH$_3$ 5–Cl 7–CONHCH(COOH)CH$_2$Ph 8–OH	*Aspergillus ochraceus* *Penicillium viridicatum*	1
Ochratoxin B	3,4–Dihydro 3–CH$_3$ 7–CONHCH(COOH)CH$_2$Ph 8–OH	*Aspergillus ochraceus*	1

Compound	Structure	Microorganism	Reference

Isocoumarins (continued)

Compound	Structure	Microorganism	Reference
Ochratoxin C	3,4–Dihydro 3–CH$_3$ 5–Cl 7–CONHCH(COOEt)CH$_2$Ph 8–OH	*Aspergillus ochraceus*	1
Sclerotinin B	3,4–Dihydro 3–CH$_3$, 3–OH 5–CH$_3$ 6–OH 7–CH$_3$ 8–OH	*Sclerotinia sclerotiorum*	1
Sclerotinin A	3,4–Dihydro 3–CH$_3$, 3–OH 4–CH$_3$ 5–CH$_3$ 6–OH 7–CH$_3$ 8–OH	*Sclerotinia sclerotiorum* *Penicillium citrinum*	1
Dihydrocitrinone	3,4–Dihydro 3–CH$_3$ 4–CH$_3$ 5–CH$_3$ 6–OH 7–COOH 8–OH	*Penicillium citrinum* *Aspergillus terreus*	
Decarboxydihydrocitrinone	3,4–Dihydro 3–CH$_3$ 4–CH$_3$ 5–CH$_3$ 6–OH 8–OH	*Penicillium citrinum*	1
Fumarin	3–CH$_2$CH$_2$CH=CHCH$_3$ 5–*n*-butyl 6–OH 8–OH	*Fusarium* sp.	1
Canescin	3–CH$_3$ 6–OH 8–OH	*Penicillium canesiens* *Aspergillus malynus*	1
Monocerin	 R^1 = H R^2 = H$_2$	*Helminthosporium monoceras*	1

Compound	Structure	Microorganism	Reference
Isocoumarins (continued)			
Hydroxymonocerin	R^1 = OH R^2 = H_2	*Helminthosporium monoceras*	1
Monocerone	R^1 = H R^2 = O	*Helminthosporium monoceras*	1
Monocerolide	Same as above, except with =O instead of side chain	*Helminthosporium monoceras*	1
Duclauxin		*Penicillium duclauxi*	1
Xenoclauxin		*Penicillium duclauxi*	1
Cryptoclauxin		*Penicillium duclauxi*	1

Compound	Structure	Microorganism	Reference

Isocoumarins (continued)

Xylindein	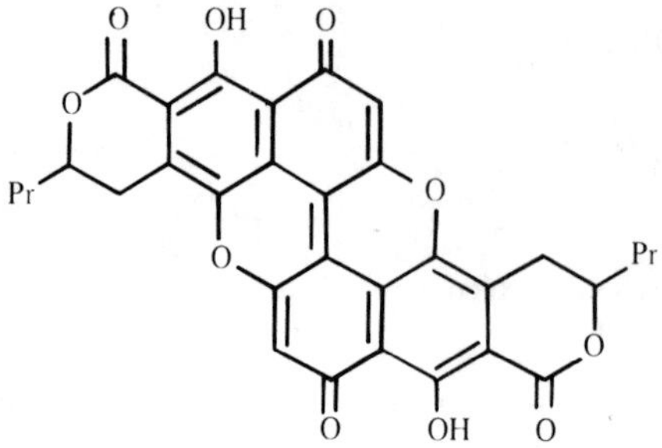	*Chlorociboria aeruginosa*	1
Novobiocin	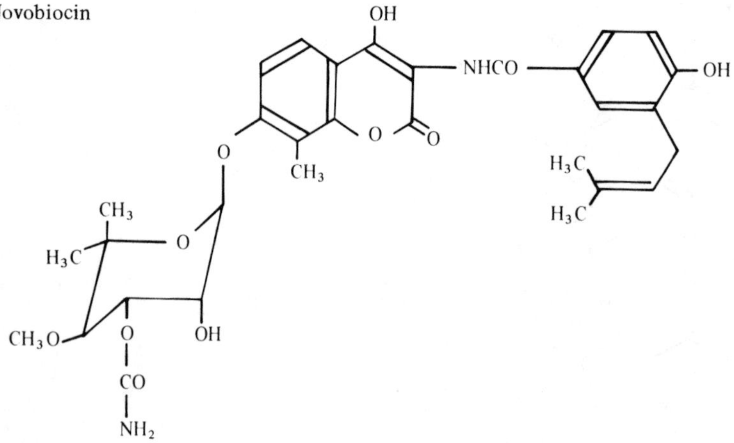	*Streptomyces spheroides niveus griseus*	3
Fuscin	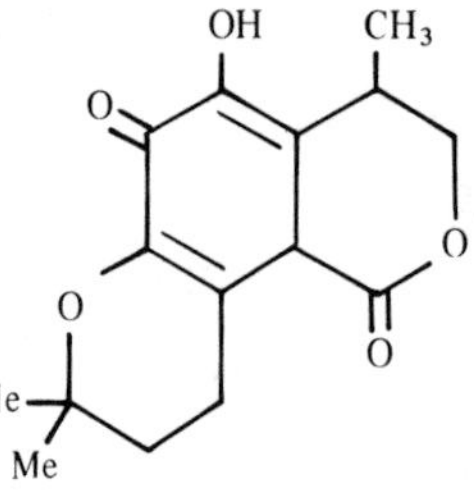	*Oidiodendron fuscum*	1, 2, 3
Alternariol	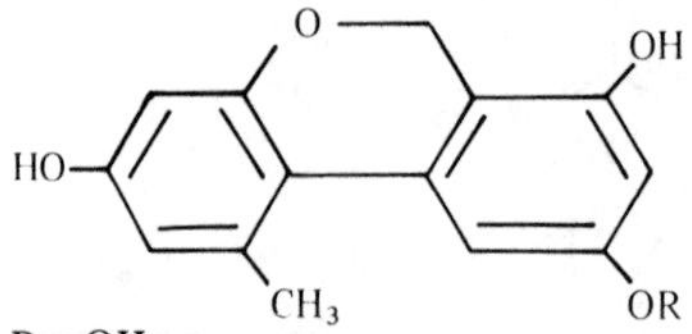	*Alternaria tenuis*	1, 2, 3

R = OH

Alternariol methyl ether   R = CH₃

Compound	Structure	Microorganism	Reference

Isocoumarins (continued)

Chartreusin

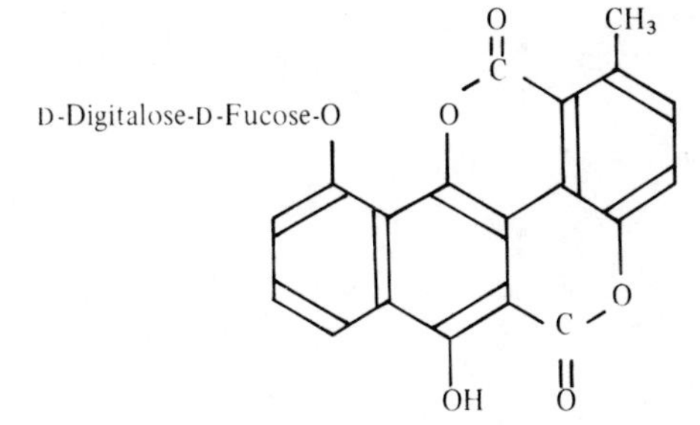

D-Digitalose-D-Fucose-O

Streptomyces chartreusis other spp.

6

Coumermycin A₁

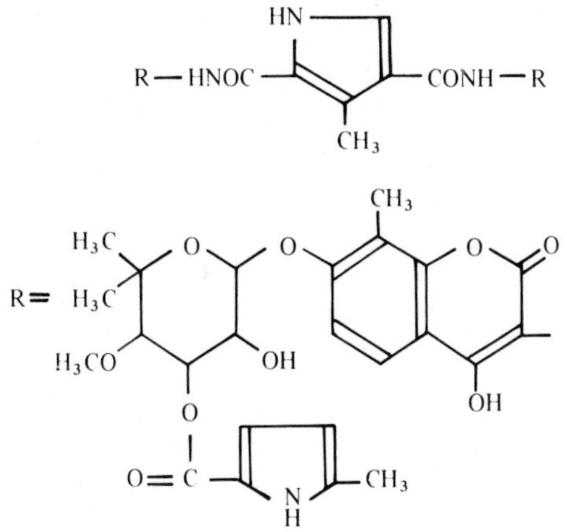

Streptomyces rishiriensis spinicoumarensis spinichromogenes

6

Coumermycin A₂

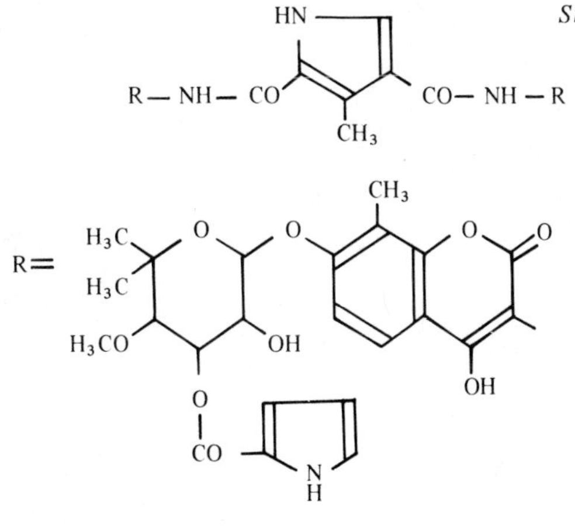

Streptomyces rishiriensis spinicoumarensis spinichromogenes

6

REFERENCES

1. Turner, W. B., *Fungal Metabolites.* Academic Press, New York (1971).
2. Shibata, S., Natori, S., and Udagawa, S., *List of Fungal Products.* Charles C Thomas, Springfield, Illinois (1964).
3. Miller, M. W., *The Pfizer Handbook of Microbial Metabolites.* McGraw-Hill, New York (1961).
4. Hardegger, E., Rieder, W., Walser, A., and Kugler, F., *Helv. Chim. Acta, 49,* 1283 (1966).
5. Koyama, Y., Fukakusa, Y., Kyomura, N., and Yamagishi, S., *Tetrahedron Lett.,* p. 355 (1969).
6. Umezawa, H. (Ed.), *Index of Antibiotics from Actinomycetes.* University of Tokyo Press, Tokyo, Japan (1967).

GAMMA-PYRONES, BENZPYRONES, XANTHONES (DIBENZPYRONES), NAPHTHAPYRONES AND DIMERS

DR. NANCY N. GERBER

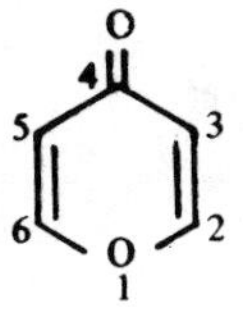

γ-Pyrone

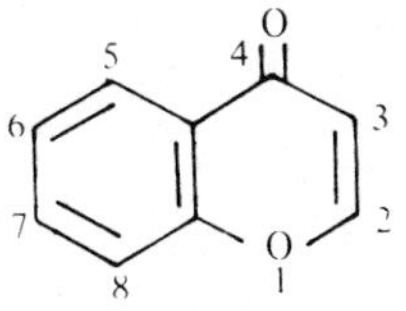

Benzpyrone

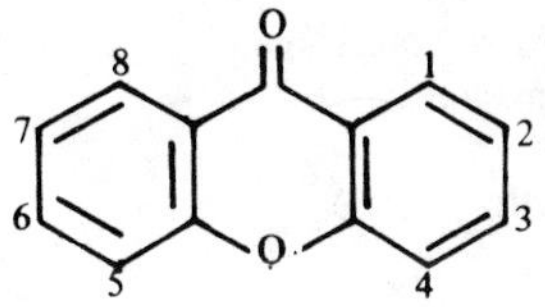

Xanthone

Compound	Structure	Microorganisms	References
Pyrones			
Comenic acid	2−COOH, 5−OH	*Gluconoacetobacter liquefaciens*	1
Rubiginic acid	2−COOH, 3−OH, 5−OH	*Gluconoacetobacter liquefaciens*	1
Kojic acid	2−CH$_2$OH, 5−OH	Various aspergilli	1, 2, 3
Isokojic acid	2−CH$_2$OH, 6−OH	*Gluconoacetobacter roseum* (on fructose)	1
Rubiginol	3−OH, 5−OH	*Gluconoacetobacter liquefaciens*	1
Versicolin	2,3-Dihydro, 2−CHO, 3−CH$_3$	*Aspergillus versicolor*	3
Aureothin	2−OCH$_3$, 3−CH$_3$, 5−CH$_3$, 6−CHCH$_2$ C=CHC(CH$_3$)=CH−pNO$_2$ Ph | | O — CH$_2$	*Streptomyces thioluteus*	1, 3
Radicinin	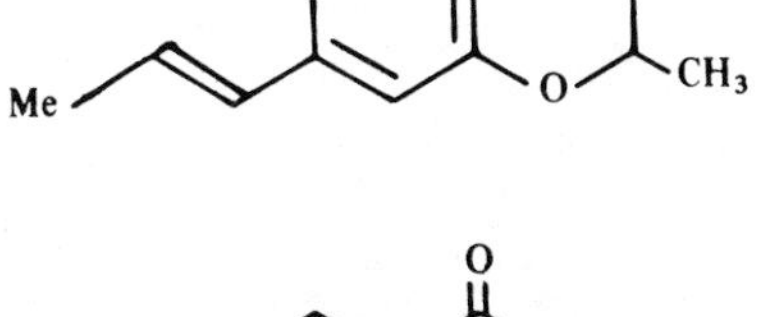	*Stemphylium radicinum*	1, 2, 3
Citromycetin	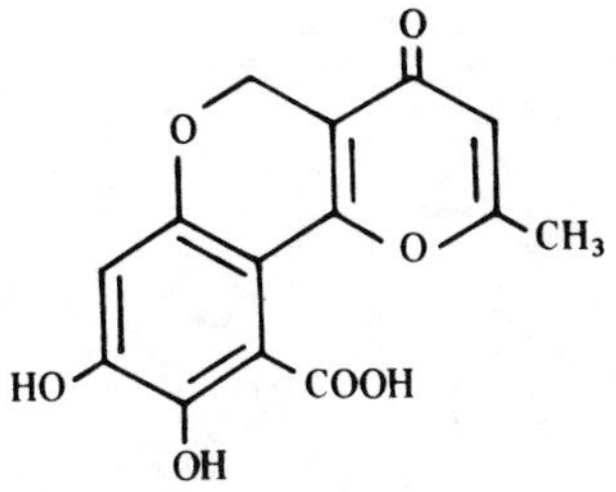	Various penicillia	2, 3

Compound	Structure	Microorganisms	References
Benzpyrones			
5-Hydroxy-2-methyl-chromone	$2-CH_3$, $5-OH$	*Daldinia concentrica*	1, 2, 3
5-Hydroxy-2-methyl-chromanone	2,3-Dihydro, $2-CH_3$, $5-OH$	*Daldinia concentrica*	1, 2, 3
Eugenetin	 $R^1 = CH_3$, $R^2 = H$, $R^3 = CH_3$	*Lecanora rupicola*	3
Sordidone (Rupicolon)	$R^1 = H$, $R^2 = Cl$, $R^3 = CH_3$	*Lecanora sordida rupicola*	3
Eugenitol	$R^1 = R^2 = H$, $R^3 = CH_3$	*Lecanora rupicola*	3
Lepraric acid (Fuciformic acid)	$R^1 = CH_3$, $R^2 = H$, $R^3 = -CH_2 OOCCH=C(CH_3)CH_2 COOH$	*Lepraria latebrarum* *Rocella fuciformis*	3
Chlorflavonin	· $3-OCH_3$, $5-OH$, $7-OCH_3$, $8-OCH_3$	*Aspergillus candidus*	3
Fulvic acid		*Penicillium flexuosum brefeldianum griseofulvum*	1, 2
Ergochromes (2,2′ or 4,4′ dimers of the units shown			

Compound	Structure	Microorganisms	References
Benzpyrones (continued)			
Ergochrome AA (4,4′) (Secalonic acid)		*Claviceps purpurea* *Parmelia entocherochrosa*	3
Ergochrome BB (4,4′)		*Claviceps purpurea*	3
Ergochrome CC (2,2′) (Ergoflavin)		*Claviceps purpurea*	3
Ergochrome AB (4,4′)		*Claviceps purpurea*	3
Ergochrome AC (2,2′)		*Claviceps purpurea*	3
Ergochrome BC (2,2′)		*Claviceps purpurea*	3
Ergochrome AD (2,2′)		*Claviceps purpurea*	3
Ergochrome BD (2,2′)		*Claviceps purpurea*	3
Ergochrome CD (2,2′)		*Claviceps purpurea*	3
Ergochrome DD (2,2′)		*Claviceps purpurea*	3
Xanthones			
Pinselin	$1-OH$, $3-CH_3$, $7-OH$, $8-COOCH_3$	*Penicillium amarum*	2, 3
Pinselic acid	$1-OH$, $3-CH_3$, $7-OH$, $8-COOH$	*Penicillium amarum*	2, 3
Ravenelin	$1-OH$, $3-CH_3$, $4-OH$, $8-OH$	*Helminthosporium ravenelii*	1, 2, 3
Norlichexanthone	$1-OH$, $3-OH$, $6-OH$, $8-CH_3$	*Lecanora reuteri*	3
Griseoxanthone C	$1-OH$, $3-OH$, $6-OCH_3$, $8-CH_3$	*Penicillium patulum*	3
Lichexanthone	$1-OH$, $3-OCH_3$, $6-OCH_3$, $8-CH_3$	Several lichens	3
2-Chloronorliche-xanthone	$1-OH$, $2-Cl$, $3-OH$, $6-OH$, $8-CH_3$	*Lecanora straminea*	3
2,4-Dichloronor-lichexanthone	$1-OH$, $2-Cl$, $3-OH$, $4-Cl$, $6-OH$, $8-CH_3$	*Lecanora straminea*	3

Compound	Structure	Microorganisms	References

Xanthones (continued)

Compound	Structure	Microorganisms	References
2,7-Dichloronor-lichexanthone	1−OH, 2−Cl, 3−OH, 6−OH, 7−Cl, 8−CH$_3$	*Lecanora straminea*	3
Arthetholin	1−OH, 2−Cl, 3−OH, 4−Cl, 6−OH, 7-Cl, 8−CH$_3$	*Lecanora straminea*	3
2,5,7-Trichloronor-lichexanthone	1−OH, 2−Cl, 3−OH, 5-Cl, 6−OH, 7−Cl, 8−CH$_3$	Several lichens	3
Thiophanic acid	1−OH, 2−Cl, 3−OH, 4−Cl, 5−Cl, 6−OH, 7−Cl, 8−CH$_3$	Several lichens	3
Thiophaninic acid	1−OH, 2−Cl, 3−OH, 4−Cl, 6−OCH$_3$, 8−CH$_3$	*Pertusarta* sp.	3
3-*o*-Methyl-2,5-di-chloronorlichexan-thone	1−OH, 2−Cl, 3−OCH$_3$, 5−Cl, 6−OH, 8−CH$_3$	*Lecanora contractula*	3
3-*o*-Methyl-2,5,7-trichloronorliche-xanthone	1−OH, 2−Cl, 3−OCH$_3$, 5−Cl, 6−OH, 7−Cl, 8-CH$_3$	Serveral lichens	3
Thuringione	1-OH, 2−Cl, 3−OCH$_3$, 4−Cl, 6−OH, 7−Cl, 8−CH$_3$	*Lecidea carpathica*	3
2,7-Dichloroliche-xanthone	1−OH, 2−Cl, 3−OCH$_3$, 6−OCH$_3$, 7−Cl, 8−CH$_3$	*Buellia glaziovana*	3
2,5-Dichloroliche-xanthone	1−OH, 2−Cl, 3−OCH$_3$, 5−Cl, 6-OCH$_3$, 8−CH$_3$	*Lecidea populicola*	3
Sterigmatocystin	R^1 = R^2 = R^3 = H	*Aspergillus versicolor*	3
5-Methoxysterigma-tocystin	R^1 = OCH$_3$, R^2 = R^3 = H	*Aspergillus versicolor*	3
O-Methylsterigma-tocystin	R^1 = R^3 = H, R^2 = Me	*Aspergillus flavus*	3
Aspertoxin (3-Hydroxy-6,7-di-methoxydifuroxan-thone)	R^1 = H, R^2 = CH$_3$, R^3 = OH	*Aspergillus flavus*	3
Fonsecin		*Aspergillus fonsecaeus*	2, 3

Compound	Structure	Microorganisms	References

Xanthones (continued)

Rubrofusarin

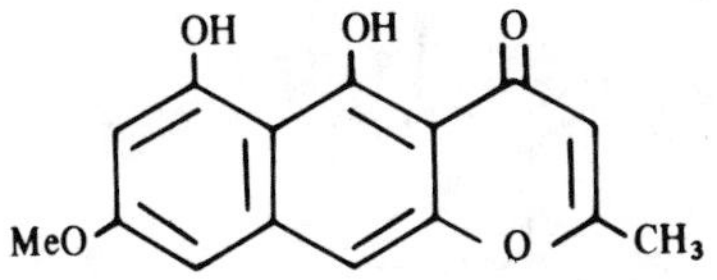

Fusarium culmorum graminearum

1, 2, 3

Flavasperone
(Asperxanthone)

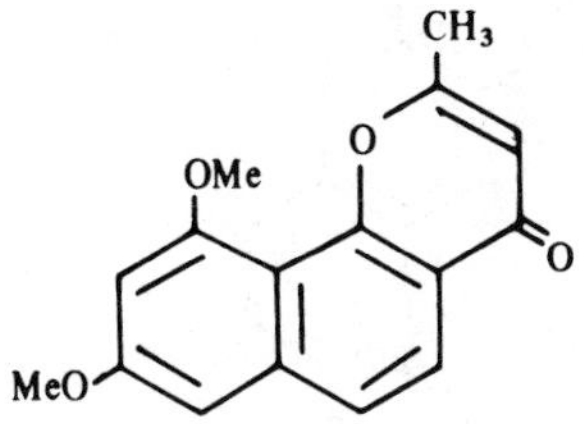

Aspergillus niger

1, 2, 3

Ustilaginoidin A

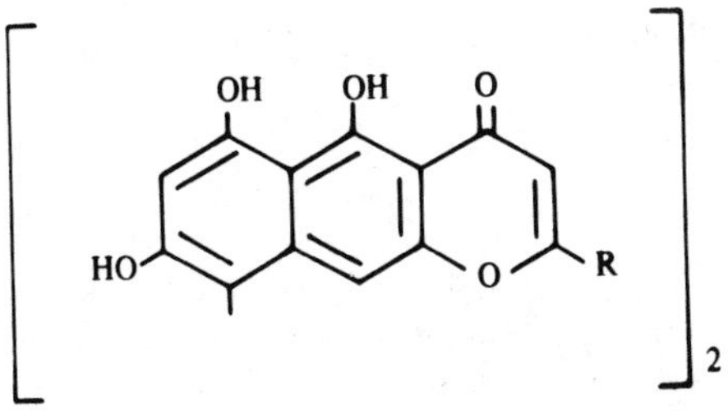

R = CH₃

Ustilaginoidea virens

3

Ustilaginoidin C

R = CH₂OH

Ustilaginoidea virens

3

Ustilaginoidin B

R = CH₃, R = CH₂OH

Ustilaginoidea virens

3

Cephalochromin

Ustilaginoidin A without C=C in the pyrone ring

Cephalosporium sp.

3

Aurofusarin

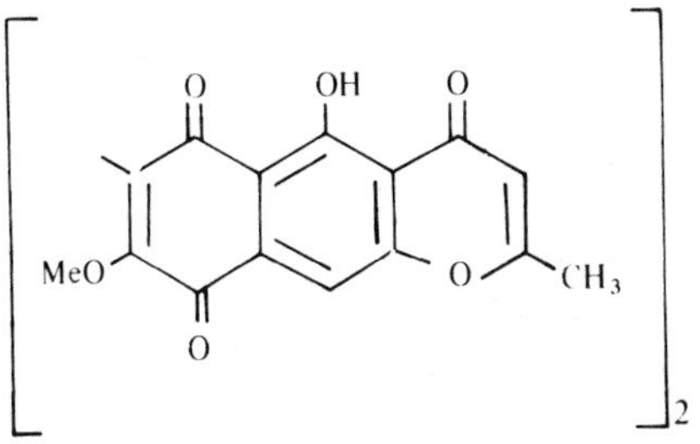

Fusarium sp.
Hypomyces rosellus

3

Fuscofusarin

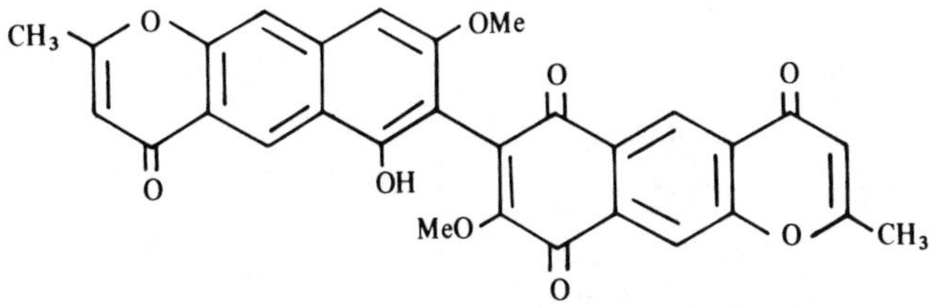

Fusarium culmorum

3

Compound	Structure	Microorganisms	References

Xanthones (continued)

Aurasperone A

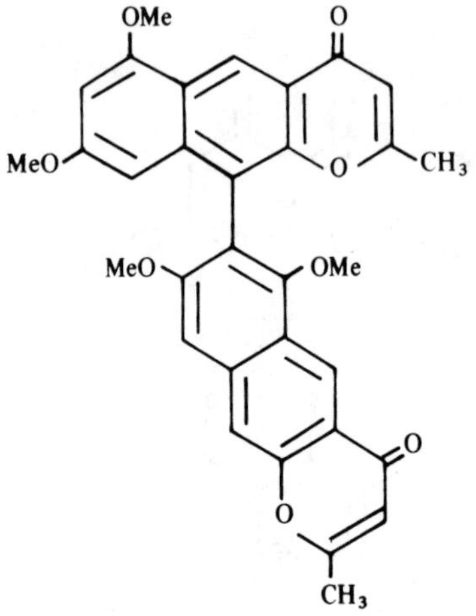

Aspergillus
niger
awamori

3

Aurasperone B   Pyrone double bonds hydrated as in fonsecin   3

REFERENCES

1. Miller, M. W., *The Pfizer Handbook of Microbial Metabolites.* McGraw-Hill, New York (1961).
2, Shibata, S., Natori, S., and Udagawa, S., *List of Fungal Products.* Charles C Thomas, Springfield, Illinois (1964).
3. Turner, W. B., *Fungal Metabolites.* Academic Press, New York (1971).

CYCLIC ANHYDRIDES AND CYCLOHEXIMIDES (GLUTARIMIDES)

DR. NANCY N. GERBER

Name	Structure	Microorganism	Reference
Itaconitin	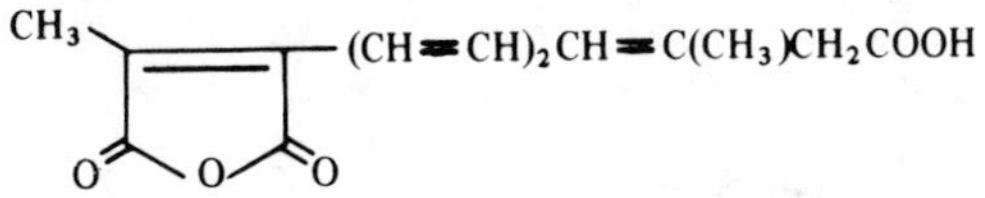	*Aspergillus itaconicus*	1
Teleocidin B	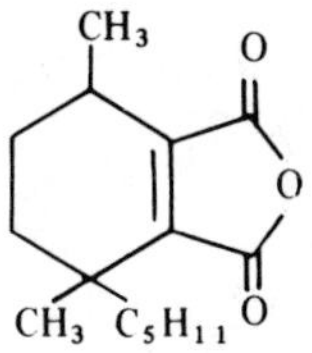	*Streptomyces kitasotoensis*	2
Stipitatonic acid	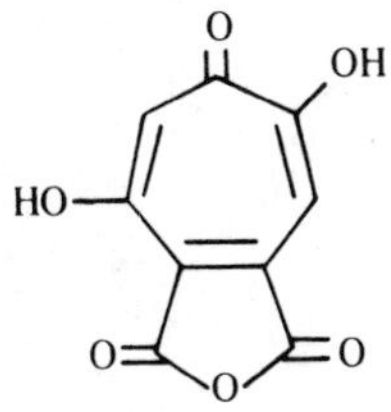	*Penicillium stipitatum*	1
Puberulonic acid	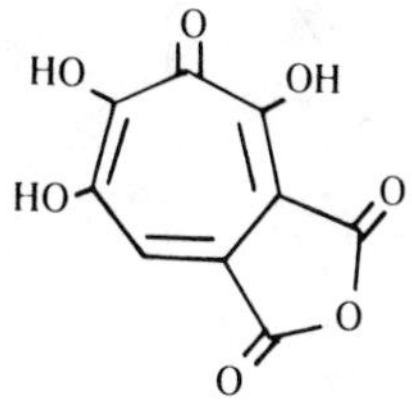	*Penicillium* spp.	
Sclerin	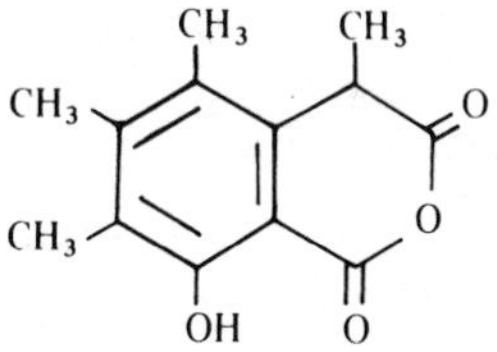	*Sclerotinia libertiana sclerotiorum*	1
Glauconic acid	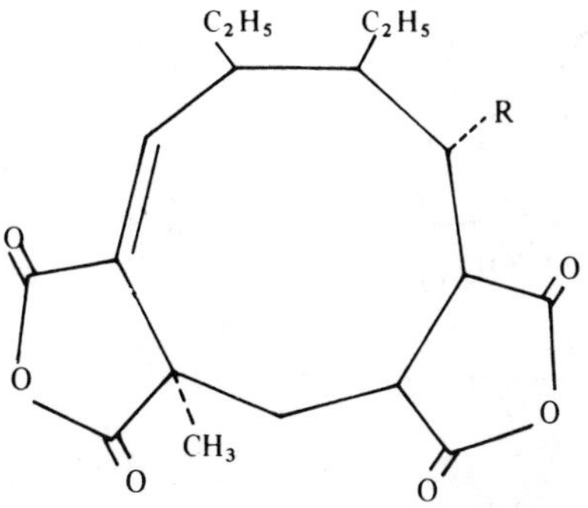	*Penicillium glaucum*	1

CH_3 $(CH = CH)_2 CH = C(CH_3)CH_2COOH$

R = OH

Name	Structure	Microorganism	Reference
Glaucanic acid	R = H	*Penicillium glaucum*	1

Byssochlamic acid

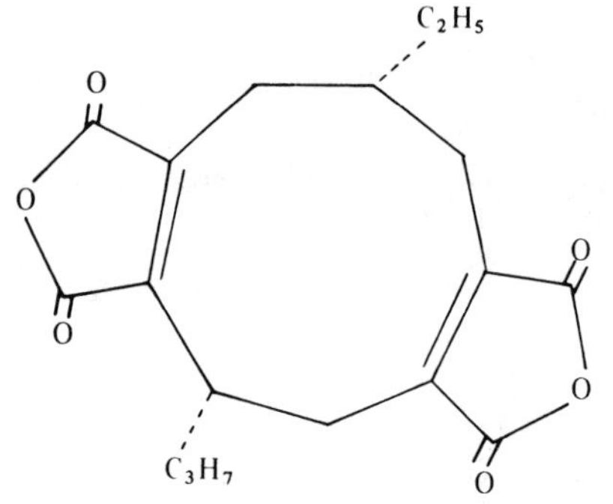

Byssochlamys fulva

Rubratoxin A

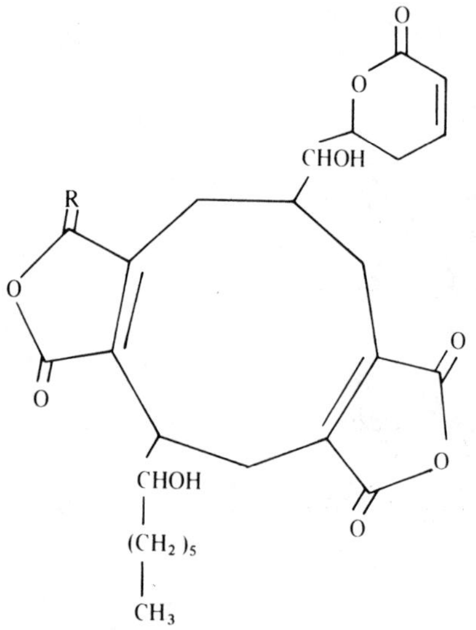

R = H, OH

Penicillium rubrum

Rubratoxin B   R = O

"Naphthalic anhydride"

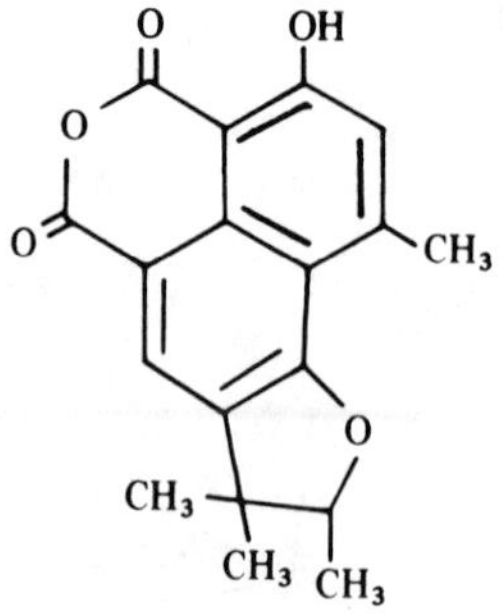

Penicillium herquei   1

Xenoclauxin   See *delta*-lactones, p.   *Penicillium duclauxi*   1

Name	Structure	Microorganism	Reference
Actiphenol		*Streptomyces* sp.	2, 3
Cycloheximide		*Streptomyces griseus noursei*	2, 3
Isocycloheximide	Side chain is opposite configuration	*Streptomyces griseus noursei*	
Inactone	Double bond between carbons 2 and 3		2, 3
Naramycin B	Same as cycloheximide, but 6-methyl is opposite configuration	*Streptomyces* sp.	2, 3
Streptovitacin A	4-OH cycloheximide	*Streptomyces griseus*	2, 3
Streptovitacin B	5-OH cycloheximide	*Streptomyces griseus*	2, 3
Streptovitacin C_2	6-OH cycloheximide	*Streptomyces griseus*	2, 3
4-Acetoxycyclo-heximide E-73	4-OOCCH$_3$ cycloheximide	*Streptomyces albulus*	2, 3
Streptimidone	CH$_2$=CHC(CH$_3$)=CHCR(CH$_3$)- R = H	*Streptomyces* sp.	2, 3
Protomycin	R = −C(CH$_3$)$_2$OH	*Streptomyces reticuli*	2

REFERENCES

1. Turner, W. B., *Fungal Metabolites.* Academic Press, New York (1971).
2. Umezawa, H. (Ed.), *Index of Antibiotics from Actinomycetes.* University of Tokyo Press, Tokyo (1967).
3. Miller, M. W., *The Pfizer Handbook of Microbial Metabolites.* McGraw-Hill, New York (1961).

AZAPHILONES, QUINONEMETHIDES, AND γ-PYRONEMETHIDES

DR. NANCY N. GERBER

The generic name "azaphilones" is now thought to be too restrictive.[1]

Compound	Structure	Microorganism	Reference
Pulvilloric acid		*Penicillium pulvillorum*	1
Ascochitine		*Ascochyta fabae*	1
Mitorubrin	R = Me	*Penicillium rubrum*	1
Mitorubrinol	R = CH$_2$OH	*Penicillium rubrum*	1
Mitorubrinic acid	R = COOH	*Penicillium funiculosum*	1
Citrinin	R = COOH	*Penicillium* various species *Crotalaria crispata*	1, 2, 3
Decarboxycitrinin	R = H	*Penicillium citrinum*	1
Sclerotiorin	R = CH=CHC(CH$_3$)=CHCH(CH$_3$)CH$_2$CH$_3$	*Penicillium sclerotiorum*	1, 2
7-*epi*-Sclerotin	Same as above, except with opposite arrangement at carbon 7	*Penicillium hirayamae*	1, 2

Compound	Structure	Microorganism	Reference
Rotiorin	 R^1 = COMe R^2 = CH=CHC(CH$_3$)=CH(CH$_3$)CH$_2$Me	*Penicillium* *sclerotiorum*	1, 2
Rubropunctatin	R^1 = CO(CH$_2$)$_4$Me R^2 = CH=CHCH$_3$	*Monascus* *rubropunctatus*	
Monascorubrin	 R^1 = CH(CH$_2$)$_6$Me R^2 = CO(CH$_2$)$_6$Me	*Monascus* *purpureus*	
Monascin (Monascoflavin)	R^1 = CO(CH$_2$)$_4$Me R^2 = CH=CHCH$_3$	*Monascus* sp.	
Atrovenetin		*Penicillium* *atrovenetum* *herquei*	1
Herqueinone	 R^1 or R^2 = Me	*Penicillium* *herquei*	1
Norherqueinone	R^1 = R^2 = H	*Penicillium* *herquei*	1

REFERENCES

1. Turner, W. B., *Fungal Metabolites.* Academic Press, New York (1971).
2. Miller, M. W., *The Pfizer Handbook of Microbial Metabolites.* McGraw-Hill, New York (1961).

MICROBIAL 1,4-BENZODIAZEPINES

DR. W. LEIMGRUBER

The antibiotic anthramycin (I) and the structurally related "yellow pigment" (II) are extracted from the fermentation broth with l-butanol. Cyclopenin (III) and cyclopenol (IV) are absorbed from fermentation broths at pH 2 with charcoal, from which they can be eluted with methanol. Except for the "yellow pigment," these compounds are very labile substances; they are stable only under essentially neutral conditions.

The ultraviolet spectra of cyclopenin and cyclopenol show only an anthranilamide chromophore, whereas those of anthramycin and the "yellow pigment" exhibit absorption bands that are attributable to the chromophore of the conjugated side chain.

I: $R_1 = CH_3$, $R_2 = R_3 = OH$

II: $R_1 = R_2 = R_3 = H$

III: $R = C_6H_5$

IV: $R = m\text{-}HOC_6H_4$

Compound (Formula)	Organism	Significant Characteristics	References
Anthramycin ($C_{16}H_{17}N_3O_4$)	*Streptomyces refuineus* var. *thermotolerans*, NRRL 3143	Yellow prisms from actone–water Melting point: 188–194°C $\lambda_{max}^{CH_3CN}$: 235 ($\epsilon = 18{,}200$), 333 ($\epsilon = 31{,}800$) $[a]_D^{25}$ (DMF, $c = 1$): +930°	1, 2
"Yellow pigment" ($C_{15}H_{15}N_3O_2$)	*Streptomyces refuineus* var. *thermotolerans*, NRRL 3143	Yellow needles from methanol–benzene Melting point: 280–282°C $\lambda_{max}^{2\text{-}PrOH}$: 238 ($\epsilon = 18{,}100$), 324 ($\epsilon = 33{,}100$), 335 ($\epsilon = 32{,}950$) $[a]_D^{25}$ (DMSO, $c = 1$): +883°	1, 2
Cyclopenin ($C_{17}H_{14}N_2O_3$)	*Penicillium cyclopium* Westling *viridicatum* Westling	Melting point: 183–184°C λ_{max}: 211 ($\epsilon = 37{,}200$), 290 ($\epsilon = 2060$) $[a]_{546}^{20}$ (MeOH, $c = 1$): −291°	3, 4, 5, 6
Cyclopenol ($C_{17}H_{14}N_2O_4$)	*Penicillium cyclopium* Westling *viridicatum* Westling	Melting point: 215°C λ_{max}: 285 ($\epsilon = 3740$) $[a]_{546}^{20}$ (MeOH, $c = 1$): −309°	3, 4, 5, 6

REFERENCES

1. Leimgruber, W., Stefanovic, V., Schenker, F., Karr, A., and Berger, J., *J. Amer. Chem. Soc.*, 87, 5791 (1965).
2. Leimgruber, W., Batcho, A. D., and Schenker, F., *J. Amer. Chem. Soc.*, 87, 5793 (1965).
3. Bracken, A., Pocker, A., and Raistrick, H., *Biochem. J., 57*, 587 (1954).
4. Luckner, M., and Mothes, K., *Tetrahedron Lett.*, p. 1035 (1962).
5. Mohammed, Y. S., and Luckner, M., *Tetrahedron Lett.*, p. 1953 (1963).
6. Martin, P. K., Rapoport, H., Smith, H. W., and Wong, J. L., *J. Org. Chem., 34*, 1359 (1969).

MITOMYCINS

G. S. BEZANSON

With the exception of mitiromycin, which inhibits only Gram-positive bacteria, mitomycins are broad-spectrum antibiotics. Produced by actinomycetes, they are extracted from broth by chloroform and are obtained as red to purple crystals from acetone. Generally they are soluble in chloroform, acetone, ethyl acetate, and ethanol; they are slightly soluble in water, and insoluble in petroleum ether and carbon tetrachloride. Aqueous solutions have indicator properties (yellow → red → purple as the pH increases). Mitomycins are labile in acid or base and light-sensitive.

STRUCTURES[1,2]

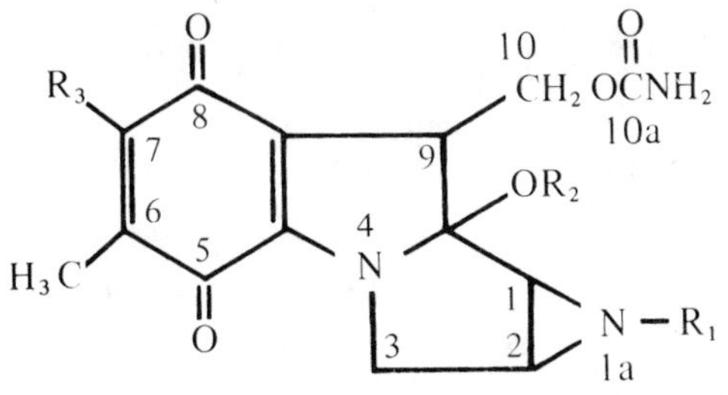

	R_1	R_2	R_3
Mitomycin A	H	CH_3	OCH_3
Mitomycin B	CH_3	H	OCH_3
Mitomycin C	H	CH_3	NH_2
Porfiromycin	CH_3	CH_3	NH_2

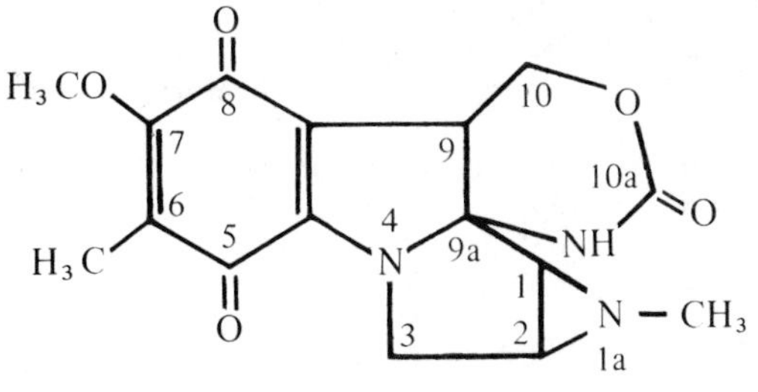

Mitiromycin

CODE FOR ORGANISMS

A = *Streptomyces caespitosus*
B = *Streptomyces verticillatus*

C = *Streptomyces reticuli* var. *shimofusaenis*
D = *Streptoverticillium ardus*

SIGNIFICANT CHARACTERISTICS

Mitomycins have characteristic absorption spectra, as shown in the table below. Individual members may be separated by paper or thin-layer chromatography.[3,4] Mass spectra provide a valuable means of identification and give useful information on structures.[5]

Compound (Mol. Wt.)	Organism	Melting Point, °C	Absorption Maxima, nm ($E_{1\ cm}^{1\%}$)	References
Mitomycin A (349)	A, B	159–161 (decomposes)	215 (234), 316–318 (122), 530 (18.8), in water	6, 7
Mitomycin B (349)	A, B	182–184 (decomposes)	220 (118), 320 (55), 550 (9.9), in water	6, 7
Mitomycin C (334)	A, B	> 360 (decomposes)	216 (742), 360 (742), 560 (0.06), in methanol	7, 8
Porfiromycin (348)	B, D	202–204 (decomposes)	216 (665), 240 (sh), 358 (638), 550 (6.5), in methanol	7, 9, 10
Mitiromycin (331)	A	124–126	218 (448), 323 (278), 530 (34), in methanol	7
G-253 B$_1$ (~ 330)	C	161–162 (decomposes)	218 (410), 250 (sh), 366 (355), 370 (4), in water	4
G-253 B$_2$ (~ 330)	C	164–165 (decomposes)	218 (820), 250 (sh), 366 (410), 572 (5.5), in water	4
G-253 C$_1$ (~ 320)	C	> 300	218 (730), 250 (sh), 366 (690), 574 (7), in water	4

REFERENCES

1. Webb, J. S., Cosulich, D., Mowat, J., Patrick, J., Broshard, R., Meyer, W., Williams, R., Wolf, C., Fulmor, W., Pidacks, C., and Lancaster, J., *J. Amer. Chem. Soc., 84,* 3185 (1962).
2. Morton, G. O., Van Lear, G., and Fulmor, W., *J. Amer. Chem. Soc., 92,* 2588 (1970).
3. Kirsch, E. J., in *Antibiotics,* Vol. 2, Biosynthesis, pp. 66–76, D. Gottlieb and P. D. Shaw, Eds. Springer-Verlag, New York (1967).
4. Nomura, S., Yamamoto, H., Umesawa, I., Matsumae, A., and Hata, T., *J. Antibiot. (Tokyo) Ser. A , 20,* 55 (1967).
5. Van Lear, G. E., *Tetrahedron, 26,* 2587 (1970).
6. Hata, T., Sano, Y., Sugawara, R., Matsumae, A., Kanamori, K., Shima, T., and Hoshi, T., *J. Antibiot. (Tokyo) Ser. A, 9,* 141 (1956).
7. Lefemine, D. V., Dann, M., Barbatachi, F., Hausmann, W., Zbinovsky, V., Monnikendam, P., Adam, J., and Bohonos, N. S., *J. Amer. Chem. Soc., 84,* 3184 (1962).
8. Wakaki, S., Marumo, H., Tomioka, K., Shimizu, G., Kato, K., Kamada, H., Kudo, S., and Fujimoto, Y., *Antibiot. Chemother., 8,* 228 (1958).
9. Bohonos, N., Dann, M., Hausmann, W., Zbinovsky, V., and Brackus, E., *United States Patent 3,219,530* (1961).
10. DeBoer, C., Dietz, A., Lummis, N., and Savage, G., in *Antimicrobial Agents Annual,* p. 17. Plenum Press, New York (1960).

ERGOT ALKALOIDS

DR. L. C. VINING

Ergot alkaloids were originally recognized as a series of physiologically active bases isolated from ergot (*Claviceps purpurea*). However, the group is expanded here to include all microbial metabolites biogenetically derived from 4-dimethylallyltryptophan.[1] Their known distribution is restricted to fungi and higher plants.[2]

   Except for a few internal salts, ergot alkaloids can be extracted from alkaline broth with chloroform. Cultures usually produce mixtures that can be separated by chromatography on alumina.[3] Thin-layer chromatography on silica gel is useful for checking identity and purity of samples.[4] As indoles, the ergot alkaloids all react with Ehrlich's reagent (dimethylaminobenzaldehyde in strong acid) to give green, blue, or purple colors.[5] Those with a simple indole chromophore have the characteristic absorption spectrum and a maximum near 280 nm; those with a double bond conjugated to position 4 of the indole ring have absorption maxima near 310 nm and a strong blue fluorescence under ultraviolet light. Mass spectrometry is useful for identification and structure elucidation.[6] Alkaloids of well-defined structure are listed in the tables. In natural ergoline derivatives the C-5 hydrogen is β (as in Structure 1) and the stereochemistry of most alkaloids is known.[7,8,9]

STRUCTURAL FORMULAS

Structure 1: ergoline

Structure 2: chanoclavines

Structure 3: rugulovasines

Structure 4: cyclopiazonic acids

Structure 5: bissecodehydrocyclopiazonic acid

Structure 6: clavicipitic acid

Structure 7: depsipeptide moiety

SYMBOL CODE

Melting point given in °C; (d) = melting with decomposition. Optical rotation: measured in pyridine solution unless another solvent is specified; c = g/100 ml.

Organisms:

A = *Claviceps purpurea*
B = *Claviceps paspali*
C = *Claviceps gigantea*
D = *Claviceps* sp. from *Elymus mollis*
E = *Claviceps* sp. from *Agropyrum semicostatum*
F = *Claviceps* sp. from *Pennisetum typhoideum*
G = *Sphaecelia sorghi*
H = *Aspergillus flavus*
I = *Aspergillus fumigatus*

J = *Rhizopus nigricans*
K = *Rhizopus arrhizus*
L = *Penicillium concavo-rugulosum*
M = *Penicillium rugulosum*
N = *Penicillium chermesinum*
O = *Penicillium cyclopium*
P = *Corticium caeruleum*
Q = *Pellicularia filamentosa*
R = *Lenzites trabea*

TABLE 1
CLAVINE GROUP I
Derivatives of 6-Methylergolene

Common Name Formula (M. W.)	Ring D Substitution			Melting Point	$[a]_D$ $(c, °C)$	Organisms	References
	C-8	C-9	C-10(H)				
Festuclavine $C_{16}H_{20}N_2$ (252)	β-CH_3		a	239–240 (d)	–170° (0.13, 14)	C, I	10, 11, 12
Pyroclavine $C_{16}H_{20}N_2$ (252)	a-CH_3		a	204	–90° (0.2, 20)	C, E	11, 13
Costaclavine $C_{16}H_{20}N_2$ (252)	β-CH_3		β	182	+44° (0.2, 20)	E, N	13, 14
Dihydrolysergol I $C_{16}H_{20}N_2O$ (268)	β-CH_2OH		a	281–283	–93° (0.4, 20)	C	11
Fumigaclavine A $C_{18}H_{22}N_2O_2$ (298)	β-CH_3	$OCOCH_3$		84–85	–57° at 546 nm	I	12, 15
Fumigaclavine B $C_{16}H_{20}N_2O$ (268)	β-CH_3	OH		244–245	–113° at 546 nm (0.6, 22)	I	12, 15
Fumigaclavine C $C_{22}H_{30}N_2O_2$ (354)				191–193	–132° (1.0, 22)	I	15

TABLE 2
CLAVINE GROUP II
Derivatives of $\Delta^{8,9}$-6-Methylergolene

Common Name Formula (M. W.)	Ring D Substitution			Melting Point	$[a]_D$ $(c, °C)$	Organisms	References
	C-8	C-9	C-10 (H)				
Agroclavine $C_{16}H_{18}N_2$ (238)	CH_3		a	205 (d)	−155° in chloroform (0.9, 20)	A, C, E, F, H, J, K	11, 15, 16, 17, 18
Elymoclavine $C_{16}H_{18}N_2O$ (254)	CH_2OH		a	245−247 (d)	−152° (0.9, 20)	D, F, H, I	15, 16, 17
Elymoclavine-O-β-D-fructoside $C_{22}H_{28}N_2O_6$ (416)	CH_2O Fru		a			F	19
Molliclavine $C_{16}H_{18}N_2O_2$ (270)	CH_2OH	OH		253 (d)	+30° (0.2, 17)	D	20
$\Delta^{8,9}$-Lysergic acid $C_{16}H_{16}N_2O_2$ (268)	COOH		a	245−247 (d)	−208° in 0.1N NaOH (0.4)	A, B	21, 22

TABLE 3
CLAVINE GROUP III
Derivatives of $\Delta^{9,10}$-6-Methylergolene

Common Name Formula (M. W.)	Ring D Substitution		Melting Point	$[a]_D$ $(c, °C)$	Organisms	References
	C-8a	C-8β				
Lysergine $C_{16}H_{18}N_2$ (238)	H	CH_3	275 (d)	+70° (0.2, 18)	E	23
Lysergene $C_{16}H_{16}N_2$ (236)	= CH_2		244 (d)	+461° (0.2, 18)	D	23
Lysergol $C_{16}H_{18}N_2O$ (254)	H	CH_2OH	245 (d)	+49° (0.2, 18)	D	23
Isolysergol $C_{16}H_{18}N_2O$ (254)	CH_2OH	H	139.5−140	+224° (0.2, 20)	F	26
Lysergic acid $C_{16}H_{16}N_2O_2$ (268)	H	COOH	240 (d)	+40° (0.5, 20)	A, B	21, 22
Isolysergic acid $C_{16}H_{16}N_2O_3$ (268)	COOH	H	218 (d)	+281° (1, 20)		3
Setoclavine $C_{16}H_{18}N_2O$ (254)	OH	CH_3	229−234	+174° (1.1, 20)	D, E, F	24, 25
Isosetoclavine $C_{16}H_{18}N_2O$ (254)	CH_3	OH	234−237 (d)	+107° (0.5, 20)	E	13, 25
Penniclavine $C_{16}H_{18}N_2O_2$ (270)	OH	CH_2OH	222 (d)	+151° (0.5, 20)	D	17, 24
Isopenniclavine $C_{16}H_{18}N_2O_2$ (270)	CH_2OH	OH	163−165	+146° (0.5, 20)	F	25

TABLE 4
CLAVINE GROUP IV
Modified in Rings C or D of 6-Methylergolene

Common Name Formula (M. W.)	Structure	Melting point	$[a]_D$ $(c, °C)$	Organisms	References
Norsetoclavine $C_{15}H_{16}N_2O$ (240)	8-(a-Hydroxy-β-methyl-$\Delta^{9,10}$-ergolene)	163–165		F	27
Chanoclavine I $C_{16}H_{20}N_2O$ (256)	Structure 2; R′ = H, R″ = OH, β-H at C-5, a-H at C-10	222	−240° (0.5, 20)	B, C, F, L, I	11, 15, 18 25, 27, 28 29, 30
Chanoclavine II $C_{16}H_{20}N_2O$ (256)	Structure 2; R′ = H, R″ = OH, H's at C-5/C-10 *cis*	174; racemic form 179	−332° (0.5, 20)	F	28
Isochanoclavine I $C_{16}H_{20}N_2O$ (256)	Structure 2; R′ = OH, R″ = H, β-H at C-5, a-H at C-10	181	−216° (0.5, 20)	F	28
Rugulovasine A $C_{16}H_{16}N_2O_2$ (268)	Structure 3; optical isomer	138 (d)	−3.0° at 436 nm (1.0, 22)	L, M, P, Q, R	29, 31
Rugulovasine B $C_{16}H_{16}N_2O_2$ (268)	Structure 3; optical isomer	187 (d)	+1.4° at 436 nm (1.0, 22)	L, M, P, Q, R	29, 31
Cyclopiazonic acid $C_{20}H_{20}N_2O_3$ (336)	Structure 4; R = O	245–246	$\Delta\epsilon_{228}$ nm: −40.3 in methanol (0.0015, 20)	O	32
Cyclopiazonic acid imine $C_{20}H_{21}N_3O_2$ (335)	Structure 4; R = NH	277–278		O	33
Bissecodehydrocyclopiazonic acid $C_{20}H_{22}N_2O_3$ (338)	Structure 5	168–169	$\Delta\epsilon_{280}^{20°}$ nm: −5.0	O	33
Clavicipitic acid $C_{16}H_{18}N_2O_2$ (270)	Structure 6	262 (d)		F	34
4-Dimethylallyltryptophan $C_{16}H_{20}N_2O_2$ (272)		Racemic form 210 (d)		F	35

TABLE 5
LYSERGAMIDE GROUP
Amide Derivatives of D-Lysergic, D-Isolysergic, or D-Dihydrolysergic Acid

Common Name Formula (M. W.)	Amide Substituent		Melting Point	$[a]_D$ (c, °C)	Organisms	References
	Config- uration	Structure				
Isoergine $C_{16}H_{17}N_3O_2$ (267)	β	$-NH_2$	135 (d)	+11° (1, 20)	B	22, 36
Ergine $C_{16}H_{17}N_3O_2$ (267)	α		242 (d)	+448° in chloroform (0.9, 20)	B	22, 36
Lysergic acid methyl-carbinolamide $C_{18}H_{21}N_3O_2$ (311)	β	CH₃ | −NH−C−H | OH	135	+29° in dimethylformamide (1, 20)	B	22, 36
Ergometrine $C_{19}H_{23}N_3O_2$ (325)	β	CH₂OH | −NH−C−H | CH₃	162 (d)	−44° in chloroform (20)	A, B	3, 22
Ergometrinine $C_{19}H_{23}N_3O_2$ (325)	α		196 (d)	+414° in chloroform (20)	A, B	3, 22
Lysergyl-L-valyl methyl ester $C_{22}H_{27}N_3O_3$ (381)	β	CH(CH₃)₂ | −NH−C−H | COOCH₃	Amorphous base 80–85; acid maleate 185 (d)	−65° (1.4, 20)	A	37
Ergosecaline $C_{24}H_{28}N_4O_4$ (436)	β	CH₃ | −NH−C−O−CO | CO−NH−CH | CH(CH₃)₂			A	18
Ergosecalinine $C_{24}H_{28}N_4O_4$ (436)	α		217 (d)	+417° (0.2, 18)	A	18
Ergosine $C_{30}H_{37}N_5O_5$ (547)	β	Structure 7; R' = Me, R'' = i-Bu	228 (d)	−179° in chloroform (20)	A, J	3, 38
Ergosinine $C_{30}H_{37}N_5O_5$ (547)	α		228 (d)	+420° in chloroform (20)	A, J	3, 38
Ergotamine $C_{33}H_{35}N_5O_5$ (581)	β	Structure 7; R' = Me, R'' = Bz	212–214 (d)	−160° in chloroform (20)	A	3
Ergotaminine $C_{33}H_{35}N_5O_5$ (581)	α		241–243 (d)	+385° in chloroform (20)	A	3
Ergocornine $C_{31}H_{39}N_5O_5$ (559)	β	Structure 7; R' = i-Pr, R'' = i-Pr	182–184 (d)	−188° in chloroform (20)	A	3
Ergocorninine $C_{31}H_{39}N_5O_5$ (559)	α		228 (d)	+409° in chloroform (20)	A	3

TABLE 5 (Continued)
LYSERGAMIDE GROUP
Amide Derivatives of D-Lysergic, D-Isolysergic, or D-Dihydrolysergic Acid

Common Name Formula (M. W.)	Amide Substituent		Melting Point	$[a]_D$ $(c, °C)$	Organisms	References
	Config- uration	Structure				
Ergocristine $C_{35}H_{39}N_5O_5$ (610)	β	Structure 7; R' = i-Pr, R'' = Bz	165–170 (d)	–183° in chloro- form (20)	A	3
Ergocristinine $C_{35}H_{39}N_5O_5$ (610)	a		226 (d)	+366° in chloro- form (20)	A	3
Ergostine $C_{34}H_{37}N_5O_5$ (596)	β	Structure 7; R' = Et, R'' = Bz	204–208 (d)	–38° (1, 20)	A	39
Ergostinine $C_{34}H_{37}N_5O_5$ (596)	a		215–216 (d)	+429° (1, 20)	A	39
α-Ergokryptine $C_{32}H_{41}N_5O_5$ (575)	β	Structure 7; R' = i-Pr R'' = i-Bu	212–214 (d)	–187° in chloro- form (20)	A	3, 38, 40
α-Ergokryptinine $C_{32}H_{41}N_5O_5$ (575)	a		240–242 (d)	+408° in chloro- form (20)	A	3, 38, 40
β-Ergokryptine $C_{32}H_{41}N_5O_5$ (575)	β	Structure 7; R' = i-Pr, R'' = sec-Bu	174–177 (d)	–91° (2, 20)	A	40
β-Ergokryptinine $C_{32}H_{41}N_5O_5$ (575)	a		220 (d)	+492° (1, 20)	A	40
Dihydroergosine $C_{30}H_{39}N_5O_5$ (549)	β	(Structure 7; R' = Me, R'' = i-Bu)-amide of 9, 10-dihydro- lysergic acid	212 (d)	–52° (1, 20)	G	41

REFERENCES

1. Voigt, R., *Pharmazie, 23,* 285, 353, 419 (1968).
2. Kelleher, W. J., *Advan. Appl. Microbiol., 11,* 211 (1970).
3. Stoll, A., in *Fortschr. Chem. Organ. Naturst., 9,* 114 (1952).
4. Agurell, S., *Acta Pharm. Suec., 2,* 357 (1965).
5. Michelon, L. E., and Kelleher, W. J., *Lloydia, 26,* 192 (1963).
6. Barber, M., Weisbach, J. A., Douglas, B., and Dudek, G. O., *Chem. Ind.,* p. 1072 (1965).
7. Schreier, E., *Helv. Chim. Acta, 41,* 1984 (1958).
8. Stadler, P. A., and Hofmann, A., *Helv. Chim. Acta, 45,* 2005 (1962).
9. Acklin, W., Fehr, T., and Arigoni, D., *Chem. Commun.,* p. 799 (1966).
10. Abe, M., and Yamatodani, S., *J. Agr. Chem. Soc. Jap., 33,* 1031 (1959).
11. Agurell, S. L., and Ramstad, E., *Acta Pharm. Suec., 2,* 231 (1965).
12. Spilsbury, J. F., and Wilkinson, S., *J. Chem. Soc.,* p. 2085 (1961).
13. Abe, M., Yamatodani, S., Yamano, T., and Kusumoto, M., *Bull. Agr. Chem. Soc. Jap., 20,* 59 (1956).
14. Agurell, S. L., *Experientia, 20,* 25 (1964).
15. Yamano, T., Kishino, K., Yamatodani, S., and Abe, M., *Annu. Rep. Takeda Res. Labs., 21,* 95 (1962).
16. Abe, M., Yamano, T., Kozu, Y., and Kusumoto, M., *J. Agr. Chem. Soc. Jap., 25,* 458 (1952).
17. Stoll, A., Brack, A., Kobel, H., Hofmann, A., and Brunner, R., *Helv. Chim. Acta, 37,* 1815 (1954).

18. Abe, M., Yamano, T., Yamatodani, S., Kozu, Y., Kusumoto, M., Komatsu, H., and Yamada, S., *Bull. Agr. Chem. Soc. Jap., 23,* 246 (1959).

19. Floss, H. G., Gunther, H., Mothes, U., and Becker, I., *Z. Naturforsch., 22B,* 399 (1967).

20. Abe, M., and Yamatodani, S., *Bull. Agr. Chem. Soc. Jap., 19,* 161 (1955).

21. Castagnoli, N., and Mantle, P. G., *Nature, 211,* 859 (1966).

22. Kobel, H., Schreier, E., and Rutschmann, J., *Helv. Chim. Acta, 47,* 1052 (1964).

23. Abe, M., Yamatodani, S., Yamano, T., and Kusumoto, M., *Agr. Biol. Chem., 25,* 594 (1961).

24. Abe, M., Yamatodani, S., Yamano, T., and Kusumoto, M., *Bull. Agr. Chem. Soc. Jap., 19,* 92 (1955).

25. Hofmann, A., Brunner, R., Kobel, H., and Brack, A., *Helv. Chim. Acta, 40,* 1358 (1957).

26. Agurell, S., *Acta Pharm. Suec., 3,* 7 (1966).

27. Ramstad, E., Chan Lin, W-N., Shough, H. R., Goldner, K. J., Parikh, R. P., and Taylor, E. H., *Lloydia, 30,* 441 (1967).

28. Stauffacher, D., and Tscherter, H., *Helv. Chim. Acta, 47,* 2186 (1964).

29. Abe, M., Ohmomo, S., Ohashi, T., and Tabuchi, T., *Agr. Biol. Chem., 33,* 469 (1969).

30. Gröger, D., *Pharmazie, 20,* 523 (1965).

31. Yamatodani, S., Asahi, Y., Matsukura, A., Ohmomo, S., and Abe, M., *Agr. Biol. Chem., 34,* 485 (1970).

32. Holzapfel, C. W., *Tetrahedron, 24,* 2101 (1968).

33. Holzapfel, C. W., Hutchison, R. D., and Wilkins, D. C., *Tetrahedron, 26,* 5239 (1970).

34. Robbers, J. E., and Floss, H. G., *Tetrahedron Lett.,* p. 1857 (1969).

35. Agurell, S., and Lindgren, J.-E., *Tetrahedron Lett.,* p. 5127 (1968).

36. Arcamone, F., Chain, E. B., Ferretti, A., Minghetti, A., Pennella, P., Tonolo, A., and Vero, L., *Proc. Royal Soc. Ser. B, 155,* 26 (1961).

37. Schlientz, W., Brunner, R., and Hofmann, A., *Experientia, 19,* 397 (1963).

38. Sallam, L., El-Refai, A-M., and Naim, N., *Jap. J. Microbiol., 13,* 218 (1969).

39. Schlientz, W., Brunner, R., Stadler, P. A., Frey, A. J., Ott, H., and Hofmann, A., *Helv. Chim. Acta, 47,* 1921 (1964).

40. Schlientz, W., Brunner, R., Rüegger, A., Berde, B., Stürmer, E., and Hofmann, A., *Pharm. Acta Helv., 43,* 497 (1968).

41. Mantle, P. G., and Waight, E. S., *Nature, 218,* 581 (1968).

THE POLYETHER ANTIBIOTICS

DR. J. W. WESTLEY AND DR. J. BERGER

The polyether antibiotics are produced by various streptomycetes. Although compounds of this type were first isolated[1-3] twenty years ago, it was not until 1967 that the first structure, that of monensin,[4] was solved. A later report,[5-7] stating that monensin and three other polyether antibiotics – X-206,[1] nigericin,[2,3] and dianemycin[8] – were orally effective in poultry coccidiosis, created considerable interest in this class of antibiotics. By the end of 1970, twelve distinct compounds of the polyether type had been reported (Tables 1 and 2).

The antibiotics are characterized by good *in vitro* activity against many Gram-positive bacteria and mycobacteria, but are in general inactive against Gram-negative bacteria. They may also have certain specific antifungal or insecticidal activity, but parenteral toxicity has prevented their use in humans.

The structures of these compounds (Table 3) have been elucidated almost entirely by X-ray analysis of their heavy atom salts. Microanalysis reveals a high oxygen content, but the compounds contain few hydroxyl groups. The oxygen atoms are accounted for by a number of cyclic ether functions. In addition, the compounds are all monocarboxylic acids ranging in molecular weight from 590 to 1000. They are characterized by a large number of C-alkyl groups.

TABLE 1
ORGANISMS PRODUCING POLYETHER ANTIBIOTICS

Antibiotic	Producing Organism	References
X-206	*Streptomyces* sp. X-206	1, 22
X-537A	*Streptomyces* sp. X-537A	1, 17–19
Nigericin*	*Streptomyces* sp. X-464	1, 13
[X-464,	Unidentified *Streptomyces*	2, 3, 11
Polyetherin A (E-749C)	*Streptomyces hygroscopicus* ATCC 21368	12, 28
Helixin C	Actinomycete isolate A 158	5–7, 14
K-178	*Streptomyces albus*	5–7, 15
Azalomycin M]	*Streptomyces hygroscopicus* ATCC 13810	5–7, 16
Grisorixin	*Streptomyces griseus*	21
Dianemycin	*Streptomyces hygroscopicus* NRRL 3444	5–8, 22
Monensin	*Streptomyces cinnamonensis* ATCC 15413 or NRRL B1588	4–7
Factor B	*Streptomyces cinnamonensis*	5–7
Factor C	*Streptomyces cinnamonensis*	5–7
Factor D	*Streptomyces cinnamonensis*	5–7
Antibiotic A-204A	*Streptomyces albus* NRRL 3384	25–27
Antibiotic A-204B	*Streptomyces albus*	25–27
Antibiotic K-358	*Streptomyces albus* K-358	24

* X-464, nigericin and polyetherin A have been conclusively demonstrated to be identical; the other three compounds listed are paper chromatographically[5] not separable.

TABLE 2
PHYSICAL CONSTANTS OF THE POLYETHER ANTIBIOTICS

Antibiotic	Molecular Formula	Molecular Weight	Melting Point, °C	$[a]_{Me}^{D}$*	$[a]_{Chl}^{D}$*	pKa (66% DMF)
X-206						
Free acid	$C_{45}H_{78}O_{13}$	828	133–145	+17.73°	−1.87°	7.6
Sodium salt	$C_{45}H_{77}O_{13}Na$	850	189–190	+14.96°	+21.0°	
X-537A						
Free acid	$C_{34}H_{54}O_{8}$	590	100–109	−7.55°	−39.82°	5.13
Sodium salt	$C_{34}H_{53}O_{8}Na$	612	168–171	−30°	−84.64°	
Nigericin						
Free acid	$C_{40}H_{68}O_{11}$	724	183–185	+9.21°	+35.2°	7.6
Sodium salt	$C_{40}H_{67}O_{11}Na$	746	245–255	+7.8°		
Grisorixin						
Free acid	$C_{40}H_{68}O_{10}$	708	75–80	+16° (acetone)		7.05
Sodium salt	$C_{40}H_{67}O_{10}Na$	730	242–246			(ethanol–H_2O, 1:1)
Dianemycin						
Free acid	$C_{47}H_{78}O_{14} \cdot H_2O$	956	156–157	+39.9°		6.6
Sodium salt	$C_{47}H_{77}O_{14}Na$	978	212	+37.1°		
Monensin						
Free acid	$C_{36}H_{62}O_{11}$	670	103–105	+47.7°		6.65
Sodium salt	$C_{36}H_{61}O_{11}Na$	692	267–269	+57.3°		
Factor B	$C_{35}H_{60}O_{11}$	656	227–228			
Factor C	$C_{37}H_{64}O_{11}$	684	212–214			
Factor D	$C_{37}H_{64}O_{11}$	684	251–252			
A-204A						
Sodium salt	$C_{48}H_{82}O_{16}Na$	960 (X-ray)	178–179	+54.95°		6.1
A-204B						
Sodium salt	$C_{50}H_{86}O_{17}Na$		177–179	+42.3°		6.3
K-358						
Sodium salt	$C_{34}H_{56}O_{9}Na$	600	246 (decomposes)			

* Me = methanol; Chl = chloroform.

In the crystalline state, the molecules exist in a cyclic conformation, with the two ends held together by a hydrogen bond between the carboxyl group and a tertiary hydroxyl on the terminal tetrahydropyran ring. The oxygen functions are concentrated in the center of the molecule, and the hydrophobic alkyl groups are all on the surface. This accounts for the unusual solubility properties of the antibiotic salts. They are virtually insoluble in water, but soluble in solvents such as benzene, ether, and chloroform.

The characteristic conformation of the antibiotic salts probably accounts for their biological activity. They possess the ability to transport cations across membranes. They are, however, distinct from the other classes of ionophorous antibiotics (the valinomycin-type depsipeptides, the macrotetralide nonactins, and the peptide gramicidins), which are all neutral cyclic compounds. Their mode of action[9,10] is presumed to be either that of simple carriers that confer lipid solubility on the ions transported, or a channel mechanism whereby stacks of molecules stretch across the membrane to form a hydrophilic tunnel.

About one year after the structure of monensin was revealed, simultaneous papers on the X-ray analysis of nigericin[11] and polyetherin A[12] appeared. Soon afterwards, both nigericin and polyetherin A, together with X-464, were conclusively shown[13] to be identical. Three other compounds — helixin C,[14] K-178,[15] and azalomycin M[16] — are not separable from nigericin by paper chromatography.[5-7] Nigericin[2,3] and antibiotic X-464,[1] now shown to be identical, were amongst the original compounds of this class isolated in 1950.

The structure of antibiotic X-537A, which was also isolated at that time, was solved[17,18] in 1970 and is

TABLE 3
STRUCTURES OF POLYETHER ANTIBIOTICS

Antibiotic | Structure

X-206

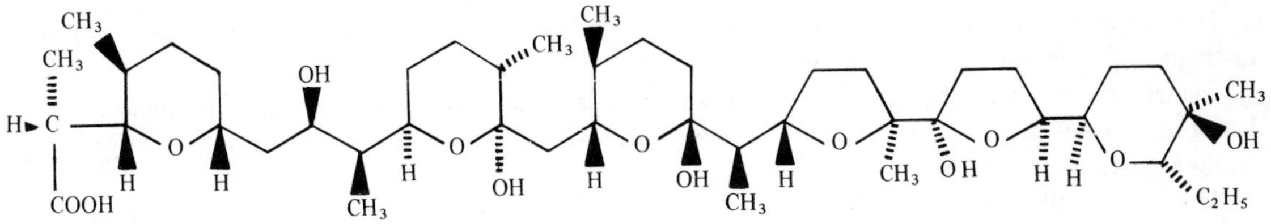

X-537A

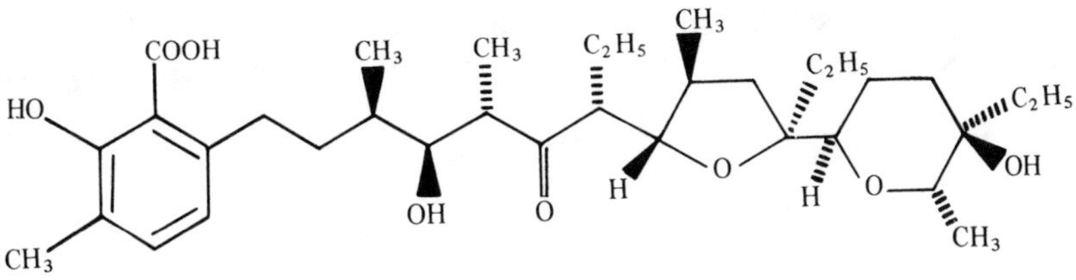

X-464
  Nigericin (R = OH)
  Grisorixin (R = H)

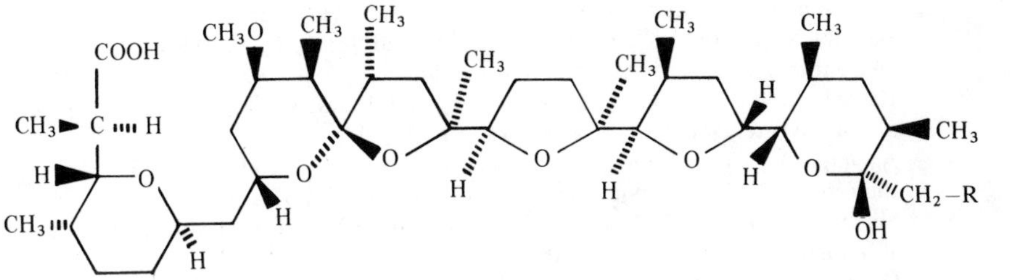

Monensin group

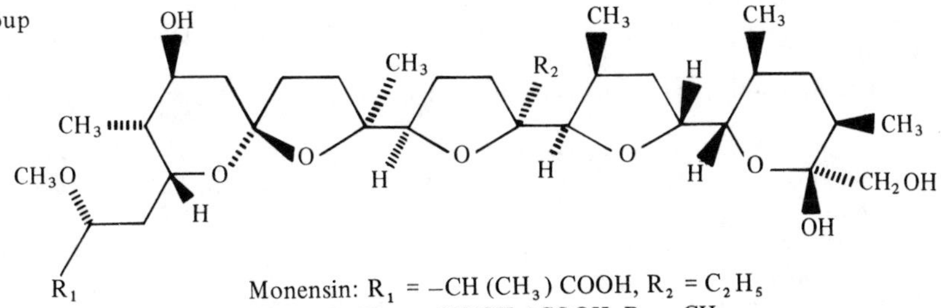

Monensin: $R_1 = -CH(CH_3)COOH$, $R_2 = C_2H_5$
Factor B: $R_1 = -CH(CH_3)COOH$, $R_2 = CH_3$
Factor C: $R_1 = (CH_2)_3COOH$, $R_2 = CH_3$

Dianemycin

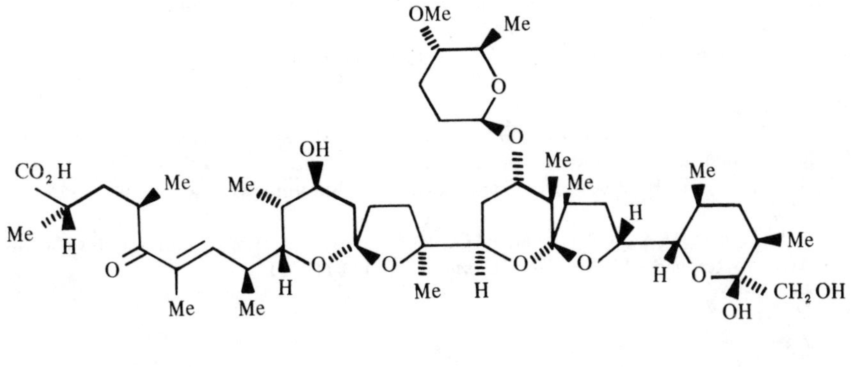

so far unique in the polyether class in possessing an aromatic chromophore. The other unique structural feature of this antibiotic was the presence of three C-ethyl groups, which prompted an investigation of its biosynthesis. This in turn revealed[19,20] the first illustration of the incorporation of a complete butyric acid unit to form a C-ethyl group.

The next structure to appear in the literature was grisorixin,[21] which was found to differ by only a single oxygen atom from nigericin.

The structure of antibiotic X-206[1] was recently solved[22] by X-ray analysis of its silver salt and is distinguished from the other polyether antibiotics by its three lactol rings.

Dianemycin has been shown[23] to contain an α-β-unsaturated carbonyl group and a terminal 1,2-glycol. Unsaturation has also been indicated[24] in antibiotic K-358 by a strong UV absorption at 215 nm and a weaker maximum at 240 nm.

The latest antibiotics in this class to be reported[25-27] were A-204A and A-204B, which were isolated as the complex A-204. They behaved identically to X-206[1,22] on TLC, but NMR revealed that A-204A has four or five methoxyls and A-204B has three, whereas X-206 contains no methoxyls. Although extremely toxic[25-27] (oral LD_{50} = 8 mg/kg in both rats and mice), A-204 was reported to be effective in the treatment of coccidial infections in poultry, as shown earlier[5-7] for monensin, nigericin, X-206, and dianemycin.[29]

REFERENCES

1. Berger, J., Rachlin, A. I., Scott, W. E., Sternbach, L. H., and Goldberg, M. W., The Isolation of Three New Crystalline Antibiotics from *Streptomyces, J. Amer. Chem. Soc., 73,* 5295 (1951).
2. Harned, R. L., Hidy, P. H., Corum, C. J., and Jones, K. L., Nigericin, A New Crystalline Antibiotic from an Unidentified *Streptomyces, Antibiot. Chemother., 1,* 594 (1951).
3. Harned, R. L., Hidy, P. H., Corum, C. J., and Jones, K. L., A New Crystalline Antibiotic from an Unidentified Streptomycete, *Proc. Indiana Acad. Sci., 59,* 38 (1950).
4. Agtarap, A., Chamberlin, J. W., Pinkerton, M., and Steinrauf, L., The Structure of Monensic Acid, A New Biologically Active Compound, *J. Amer. Chem. Soc., 89,* 5737 (1967).
5. Shumard, R. F., and Callender, M. E., Monensin, A New Biologically Active Compound, VI, Anticoccidial Activity, *Antimicrob. Agents Chemother.,* p. 369 (1968).
6. Gorman, M., Chamberlin, J. W., and Hamill, R. L., Monensin, A New Biologically Active Compound, V, Compounds Related to Monensin, *Antimicrob. Agents Chemother.,* p. 363 (1968).
7. Gorman, M., and Hamill, R. L., Nigericin for Treating Coccidiosis, *U.S. Patent 3,555,150* (1971).
8. Lardy, H. A., Johnson, D., and McMurray, W. C., Antibiotics Are Tools for Metabolic Studies, *Arch. Biochem. Biophys., 78,* 587 (1958).
9. Lardy, H., Influence of Antibiotics and Cyclic Polyethers on Ion Transport in Mitochondria, *Fed. Proc., 27,* 1278 (1968).
10. Pressman, B. C., Ionophorous Antibiotics as Models for Biological Transport, *Fed. Proc., 27,* 1283 (1968).
11. Steinrauf, L. K., Pinkerton, M., and Chamberlin, J. W., The Structure of Nigericin, *Biochem. Biophys. Res. Commun., 33,* 29 (1968).
12. Kubota, T., Matsutani, S., Shiro, M., and Koyama, H., The Structure of Polyetherin A , *Chem. Commun.,* p. 1541 (1968).
13. Stempel, A., Westley, J. W., and Benz, W., The Identity of Nigericin, Polyetherin A, and X-464, *J. Antibiot. (Tokyo), 22,* 384 (1969).
14. Smeby, R. R., Leben, C., Keitt, G. W., and Strong, F. M., Production and Purification of the Antibiotic Helixin, *Phytopathology, 42,* 506 (1952).
15. Horvath, I., Lovrekovich, I., and Varga, J. M., Antibiotics Produced by *Streptomyces,* III, A New Antibiotic, K-178 – Biological Studies, *Z. Allg. Mikrobiol., 4,* 236 (1964).
16. Okazaki, H., and Arai, M., *Japanese Patent 13,791.* Sankyo Co. Ltd. (1966).
17. Westley, J. W., Evans, R. H., Williams, T., and Stempel, A., Structure of Antibiotic X-537A, *Chem. Commun.,* p. 71 (1970).
18. Johnson, S. M., Herrin, J., Liu, S. J., and Paul, I. C., The Crystal and Molecular Structure of the Barium Salt of an Antibiotic Containing a High Proportion of Oxygen, *J. Amer. Chem. Soc., 92,* 4428 (1970).
19. Westley, J. W., Evans, R. H., Pruess, D. L., and Stempel, A., Biosynthesis of Antibiotic X-537A, *Chem. Commun.,* p. 1467 (1970).
20. Westley, J. W., Pruess, D. L., and Pitcher, R. G., Incorporation of [1-13C] Butyrate into Antibiotic X-537A: 13C Nuclear Magnetic Resonance Study, *Chem. Commun.,* p. 161 (1972).

21. Gachon, P., Kergomard, A., Veschambre, H., Esteve, C., and Staron, T., Grisorixin, a New Antibiotic Related to Nigericin, *Chem. Commun.,* p. 1421 (1970).

22. Blount, J. F., and Westley, J. W., X-Ray Crystal and Molecular Structure of Antibiotic X-206, *Chem. Commun.,* p. 927 (1971).

23. Hamill, R. L., Hoelm, M. M., Pittenger, G. E., Chamberlin, J., and Gorman, M., Dianemycin, an Antibiotic of the Group Affecting Ion Transport, *J. Antibiot. (Tokyo), 22,* 161 (1969).

24. Gimesi, J., Horvath, I., and Szentirmai, A., Antibiotics Produced by *Streptomyces,* V, A New Antibiotic, K-358, *Z. Allg. Mikrobiol., 4,* 269 (1964).

25. Hamill, R. L., Hoehn, M. M., and Gorman, M., A-204, a New Biologically Active Compound – Discovery and Isolation, *Tenth Interscience Conference on Antimicrobial Agents and Chemotherapy, October 18–21, 1970, Chicago, Illinois,* Paper 15. American Society of Microbiology, Bethesda, Maryland (1971).

26. Whitney, J. G., Wicker, K. J., and Boeck, L. D., A-204, a New Biologically Active Compound – Fermentation Studies, *Tenth Interscience Conference on Antimicrobial Agents and Chemotherapy, October 18–21, 1970, Chicago, Illinois,* Paper 16. American Society of Microbiology, Bethesda, Maryland (1971).

27. Eli Lilly and Co., *Belgian Patent 728,382* (1969).

28. Shoji, J., Kozuki, S., Matsutani, S., Kubota, T., Nishimura, H., Mayama, M., Matokawa, K., Tanaka, Y., Shimaoka, N., and Otsuka, H., Studies on Polyetherin A, I, Isolation and Characterization, *J. Antibiot. (Tokyo), 21,* 402 (1968).

29. Czerwinski, E. W., and Steinrauf, L. K., Structure of the Antibiotic Dianemycin, *Biochem. Biophys. Res. Commun., 45,* 1284 (1971).

MISCELLANEOUS
COMPOUNDS

MICROBIAL CHLORINE-CONTAINING METABOLITES

DR. G. M. STRUNZ

The production of halogen-containing metabolites by microorganisms involves, in general, the oxidative metabolism of ionic halide, derived from the nutrient source. Of the microbial halometabolites that have been reported, those containing chlorine form the largest group.

Chlorometabolite-producing microorganisms, grown *in vitro,* vary in their responses to changes in the amount of chloride available.[1] In a number of cases, enrichment of the culture medium with chloride ion has been reported to enhance the yield of chlorometabolite produced.[4,12] Whereas available chloride is, by definition, required for the biosynthesis of a chlorometabolite, the corresponding deschloro-analogue is not necessarily produced when a chlorometabolite-synthesizing organism is deprived of chloride.[1] Factors that promote or inhibit the production of halometabolites have been reported for a number of microbial systems.[1,180]

Physical and chemical characteristics of known chlorometabolites have been compiled in the accompanying table from a survey of the literature available up to the early part of 1970.[1,2]* As in other recent compilations of data on microbial metabolites,[1,2] the table includes information on metabolites of lichens. An effort has been made to indicate those literature references concerned principally with synthesis or biosynthetic studies, using the designation "syn" or "biosyn" respectively. Citations not so designated refer to reports of the isolation of metabolites and elucidation of their structures. Taxonomic data have, in general, been quoted from the original literature, and have not been checked for authenticity.

Among the listed metabolites, the abundance of structures apparently derived, at least in part, from polyketide precursors may be noted.

Many of the compounds tabulated possess antibiotic properties.

* The formidable literature search associated with this compilation could not have been undertaken without the kind assistance of the library staffs of Ayerst Research Laboratories, Montreal, P.Q., and Hoffmann-La Roche Inc., Nutley, N.J. The author is deeply indebted to Dr. M. Götz, Ayerst Research Laboratories, and Dr. A. Brossi and Dr. P. F. Sorter, Hoffmann-La Roche, for making possible this generous cooperation.

Compound (Alternate Name) Formula	Structure	Organisms	Properties	References
Anthraquinone, 2-chloro-1,3,8-trihydroxy-6-methyl (AO-1) $C_{15}H_9ClO_5$		*Nephroma laevigatum* Ach. non auct. nonn. *Anaptychia obscurata* (Nyl.) Vain (syn. *A. heterochroa*) *Aspergillus fumigatus* *Lasallia papulosa* (Ach.) Llano	**Yamamoto et al.[4]** Orange-red needles from benzene Melting point: 271–273°C UV maxima (ethanol), nm (log ε): 217.5 (4.49), 258 (4.24), 312 (4.22), 324 (inflection), 437 (3.87), 510 (3.72) Gives fragilin with diazomethane **Yosioka et al.[5]** Orange needles from methanol Melting point: 286–287°C UV maxima (ethanol), nm (log ε): 273 (4.35), 282 (4.35), 307 (4.18), 431 (4.05), 460 (3.96), 521 (3.42) IR maxima (potassium bromide), cm^{-1}: 3333, 1663, 1611 Color reactions: red with concentrated sulfuric acid, pink-red with magnesium acetate Reduction (Zn/HOAc) gives emodin	3–6, 7 (syn)
Anthraquinone, 2-chloro-1,3,8-trihydroxy-6-hydroxymethyl $C_{15}H_9ClO_6$		*Aspergillus fumigatus*	Orange-red powder from ethyl acetate Melting point: 297°C UV maxima (ethanol), nm (log ε): 220 (4.50), 249 (inflection), 306 (4.16), 332 (inflection), 435 (3.96), 453 (inflection), 520 (3.62)	4
Anthraquinone, 2-chloro-1,3,4,8-tetrahydroxy-6-methyl (Papulosin) $C_{15}H_9ClO_6$		*Lasallia papulosa* (Ach.) Llano, var. *rubiginosa* (Pers.) *Valsaria rubricosa* (Fr.) Sacc.	Red-brown plates from methyl ethyl ketone Melting point: 268–269°C (sublimes) UV maxima (methanol), nm: 239, 264, 307, 351, 465, 491, 520, 572 Possibly identical with one of the valsarins[9]	6, 8, 9

Compound (Alternate Name) Formula	Structure	Organisms	Properties	References
Anthraquinone, 2-chloro-1,3,5,8-tetrahydroxy-6-methyl $C_{15}H_9ClO_6$		*Lasallia papulosa* (Ach.) Llano, var. *rubiginosa* (Pers.) *Valsaria rubricosa* (Fr.) Sacc.	Possibly identical with one of the valsarins[9]	8
Anthraquinone, 2,4,-dichloro-1,3,8-trihydroxy-6-methyl (AO-2) $C_{15}H_8Cl_2O_5$		*Anaptychia obscurata* (Nyl.) Vain. (syn. *A. heterochroa*)	Orange needles from benzene Melting point: 267–269°C IR maxima (potassium bromide), cm^{-1}: 3317, 1666, 1623 Color reactions: red with concentrated sulfuric acid, pink with magnesium acetate	5
Anthraquinone, 2-chloro-1,3,-dihydroxy-8-methoxy-6-methyl $C_{16}H_{11}ClO_5$		*Nephroma laevigatum* Ach. non auct. nonn. (syn. *N. lusitanicum* Schaer.)		3
Anthraquinone, 2-chloro-1,8-dihydroxy-3-methoxy-6-methyl (Fragilin) $C_{16}H_{11}ClO_5$		*Sphaerophorus fragilis* (L.) Ach. non *globosus* Vain.	Crystals from chloroform Melting point: 267–268°C UV maxima (chloroform), nm (ε): 271.5 (36,500), 312.5 (14,000), 434.5 (15,000) IR maxima (potassium bromide), cm^{-1}: 1680, 1630	3, 4 (syn), 7 (syn), 10
Anthraquinone, 2-chloro-1-hydroxy-3,8-dimethoxy-6-methyl $C_{17}H_{13}ClO_5$		*Nephroma laevigatum* Ach. non auct. nonn.		3

Compound (Alternate Name) Formula	Structure	Organisms	Properties	References
Anthraquinone, 2-chloro-1,8-dihydroxy-6(2-hydroxypropyl)-3-methoxy (Nalgiolaxin) $C_{18}H_{15}ClO_6$		*Penicillium nalgiovensis* Laxa.	Yellow plates or needles Melting point: 248°C $[a]_{5790}^{22}$: +40.3° (c = 0.1072 in chloroform)	11, 12
9-Anthrone, 2-chloro-1,3,8-trihydroxy-6-methyl $C_{15}H_{11}ClO_4$		*Aspergillus fumigatus*	Pale-yellow plates Melting point: 227–229°C (decomposes) UV maxima (ethanol), nm (log ϵ): 226 (4.44), 273 (4.01), 360 (4.16) *p*-Nitrosodimethylaniline test: deep green Color reactions: violet with ammonium hydroxide or sodium carbonate Oxidation gives corresponding anthraquinone	4
Ascochlorin (LL-Z1272 γ, c.f. *Fusarium* sp. metabolites) $C_{23}H_{29}ClO_4$		*Ascochyta viciae Fusarium* sp.	Melting point: 153–154°C (decomposes),[13] 172–173°C[14] $[a]_D^{25}$: –31° (c = 0.99 in methanol) UV maxima (methanol), nm (ϵ): 230 (35,700), 293 (11,370), 347 (10,150)	13–15
Butanoic acid, 2-amino-4,4-dichloro- (2-Amino-4,4-dichlorobutyric acid) $C_4H_7Cl_2NO_2$	(L configuration)	*Streptomyces armentosus*, var. *armentosus*	**After Recrystallization in Aqueous Methanol** $[a]_D^{25}$: +6.7° (c = 0.74 in water); +26.2° (c = 0.74 in aqueous hydrochloric acid, pH 1.0) IR maxima (*inter alia*), cm^{-1}: 2950–2585, 1612, 1588, 785 Reduction with sodium borohydride in dilute alkali gave homoserine	16

Compound (Alternate Name) Formula	Structure	Organisms	Properties	References
Caldariomycin $C_5H_8Cl_2O_2$		*Caldariomyces fumago*	Colorless needles Melting point: 121°C $[\alpha]_{5461}^{20}: +59.2°$ ($c = 0.338$ in water) Acetone-dried mycelial powders of *C. fumago* catalyze the conversion of β-ketoadipic acid to δ-chlorolevulinic acid, an intermediate in the biosynthesis of caldariomycin[20],[22] The haloperoxidase of *C. fumago* has been used to chlorinate 16-ketoprogesterone and 16-keto-A-norprogesterone[27]	17, 18–19 (syn), 20–26 (biosyn), 27
Chloramphenicol (Chloromycetin; levomycetin; D-threo-N-(1,1'-dihydroxy-1-p-nitrophenylisopropyl)-dichloroacetamide) $C_{11}H_{12}Cl_2N_2O_5$		*Streptomyces omiyaensis venezuelae*	Colorless crystals Melting point: 150°C $[\alpha]_D^{25}: -25.5°$ (ethyl acetate); $+19°$ (ethanol) UV maximum (water), nm: 278	28, 29, 30–31 (syn), 32 (biosyn), 33
Chloramphenicol, monobromo analogue (N-Monobromomonochloroacetyl-p-nitrophenyl serinol) $C_{11}H_{12}BrClN_2O_5$		*Streptomyces sp.*		34
Chlorflavonin (3'-Chloro-2',5-dihydroxy-3,7,8-trimethoxy flavone) $C_{18}H_{15}ClO_7$		*Aspergillus candidus*	Yellow powder from sublimation Melting point: 212°C UV maxima (ethanol), nm (log ϵ): 266 (4.44), 305 (inflection, 3.85), 350 (3.85) Ethanolic ferric chloride test: deep green Color reactions: deep yellow with aqueous sodium hydroxide, yellow-orange with concentrated sulfuric acid	35–37

Compound (Alternate Name) Formula	Structure	Organisms	Properties	References
Chloroatranorin $C_{19}H_{17}ClO_8$		*Parmelia furfuracea* (L.) Ach. *physodes* Ach. *chryptochlorophaea perlata* (Huds.) Ach. *pseudoreticulata* Tav. *Evernia prunastri* (L.) Ach. *Siphula polyschides* Krempelh. *dactyliza* Nyl. *Lecanora rupicola* (L.) Zahlbr. *Lecidea carpathica* (Koerb.) Szat. Wide occurrence	Colorless prisms Melting point: 208°C	38–43; 44 (biosyn)
Chloroisorotiorin, (–)-7-epi-5- $C_{23}H_{23}ClO_5$		*Penicillium hirayamae* Udagawa	Red pigment	45, 46 (syn)
Cryptosporiopsin [(1S,5S)-2-Cyclopentene-1-carboxylic acid; 2-*trans*-allyl-3,5-dichloro-1-hydroxy-4-oxo methyl ester; c.f. *Sporormia affinis* metabolites] $C_{10}H_{10}Cl_2O_4$		*Sporormia affinis* Sacc., Bomm et Rouss. *Cryptosporiopsis* sp. *Periconia macrospinosa* (metabolite detected chromatographically)	Colorless needles from cyclohexane Melting point: 138–139°C $[\alpha]_D^{25}$: +61.5 ± 2.9° (c = 1.023 in methanol); +129° (c = 1.35 in chloroform) UV maximum (methanol), nm (ϵ): 289 (22,525)	47–51, 52 (syn)
Cylindrochlorin $C_{23}H_{27}ClO_4$		*Cylindrocladium* sp.	Molecular weight: 402 UV maxima, nm (ϵ): 234 (49,400), 291 (13,800), 347 (8,500) IR maxima, cm^{-1}: 1710, 1680, 1638	53

Compound (Alternate Name) Formula	Structure	Organisms	Properties	References
Diploicin $C_{16}H_{10}Cl_4O_5$		*Buellia canescens* (Dicks.) De Not. [*Diplocia canescens* (Dicks.) Massal.] *Lecidea carpathica* (Koerb.) Szat.	Colorless crystals Melting point: 232°C	33, 54–57, 58 (syn)
Drosophilin A (*p*-Methoxytetrachlorophenol) $C_7H_4Cl_4O_2$		*Psathyrella subatrata* (Batsch. *ex* Fr.) Quel. [*Drosophila subatrata* (Batsch. *ex* Fr.) Quel.] *Fomes fastuosus*	Crystals from hexane or dilute alcohol Melting point: 114°C UV maximum, nm: 301	33, 59, 60 (syn), 61, 62
(±)-Erdin $C_{16}H_{10}Cl_2O_7$		*Aspergillus terreus* Thom.	Needles from chloroform–methanol Melting point: 210–212°C UV maximum (ethanol), nm (ϵ): 284 (21,000) Erdin occurs naturally as racemate; methyl ester, geodin, present in the same culture is (+) enantiomer	33, 63, 64
Exfoliatin $C_{27}H_{40}ClO_{19}$		*Streptomyces exfoliatus* Umezawa	Isolated as monohydrate Colorless needles from alcohol Melting point: 172°C Positive ferric chloride and Molisch tests; negative Fehling test	65
Ferramido chloromycin $C_{127}H_{201}Cl_2N_{24}O_{70}SFe_{10}$ $C = 45.6\%$, $H = 6.0\%$, $Cl = 2.02\%$, $N = 9.9\%$, $S = 0.99\%$, $Fe = 11.3\%$	Polypeptide; acid hydrolysis gives cysteine, lysine, glycine, alanine, valine and leucine plus an unidentified chlorine-containing component	*Streptomyces* AS13	Minute brownish-gray plates; brown at 57°C, deep brown at 82°C, charred at 99°C	66

Compound (Alternate Name) Formula	Structure	Organisms	Properties	References
Flavoobscurin A [1,3,8,1′,3′,8′-hexahydroxy-2,2′,4′-trichloro-6,6′-dimethylbisanthronyl-(10,10′)] $C_{30}H_{19}Cl_3O_8$		*Anaptychia obscurata* Vain.	Greenish-yellow needles from glacial acetic acid Melting point: > 360°C (darkens above 195°C) $[a]_D$: +30° (c = 0.1 in acetone) UV maxima (ethanol), nm (log ε): 233 (inflection, 4.59), 281 (4.27), 370 (4.33) IR maxima (potassium bromide), cm^{-1}: 3470, 3400, 1626 Negative color test with magnesium acetate	67
Flavoobscurin B₁ and B₂ [1,3,8,1′,3′,8′-hexahydroxy-2,4,2′,4′-tetrachloro-6,6′-dimethylbisanthronyl-(10,10′)] $C_{30}H_{18}Cl_4O_8$		*Anaptychia obscurata* Vain.	Flavoobscurins B₁ and B₂ are considered to be rotational or stereoisomers about the C_{10}—C_{10} bond; pure B₁ or B₂, on refluxing in acetone, gives rise to an equilibrium mixture of B₁ and B₂ **B₁** Melting point: > 360°C (darkens above 210°C) $[a]_D$: +35° (c = 0.1 in acetone) UV maxima (ethanol), nm (log ε): 233 (inflection, 4.60), 286 (4.26), 377 (4.35) IR maxima (potassium bromide), cm^{-1}: 3425, 1628 **B₂** Melting point: > 360°C (darkens above 215°C) $[a]_D$: +7° (c = 0.1 in acetone) UV maxima (ethanol), nm (log ε): 232 (inflection, 4.61), 276 (4.27), 383 (4.32) IR maxima (potassium bromide), cm^{-1}: 3465, 3385, 1631	67

Compound (Alternate Name) Formula	Structure	Organisms	Properties	References
Flavoobscurin B_1 and B_2 (cont.)			**$B_1 + B_2$** Yellow needles from acetone—chloroform Melting point: $> 360°C$ (darkens above $245°$ C) $[a]_D$: $+19°$ ($c = 0.1$ in acetone) UV maxima (ethanol), nm (log ϵ): 232 (inflection, 4.57), 278 (4.23), 380 (4.31) IR maxima (potassium bromide), cm^{-1} : 3435, 3400, 1632	
Fumigachlorin $C_{16}H_{25}Cl_2NO_4$		*Sartorya fumigata*, var. *spinosa*	Stable at acidic and neutral pH; slightly unstable at alkaline pH	68
Fusarium sp. metabolite LL-Z1272 a $C_{23}H_{31}ClO_3$		*Fusarium* sp.	Melting point: 72.5–73°C UV maxima (methanol), nm (ϵ): 228 (11,150), 293 (10,400), 345 (7,800)	14
Fusarium sp. metabolite LL-Z1272 δ, c.f. Asco-chlorin $C_{23}H_{31}ClO_4$		*Fusarium* sp.	Melting point: 129.5–130.5°C $[a]_D^{25}$: $+6°$ ($c = 1.0$ in methanol) UV maxima (methanol), nm (ϵ): 231 (23,000), 293 (12,000), 346 (9,150)	14
Fusarium sp. metabolite LL-Z1272 ζ $C_{25}H_{31}ClO_6$		*Fusarium* sp.	Melting point: 156.5–157°C $[a]_D^{25}$: $-15°$ ($c = 1.0$ in methanol) UV maxima (methanol), nm (ϵ): 239 (39,800), 293 (11,700), 347 (9,600)	14
Gangaleodin $C_{18}H_{14}Cl_2O_7$		*Lecanora gangaleoides* Nyl.	Colorless needles Melting point: 214–215°C	69–71

Compound (Alternate Name) Formula	Structure	Organisms	Properties	References
(+)-Geodin $C_{17}H_{12}Cl_2O_7$		*Aspergillus terreus* Thom., Strain No. 45 *Penicillium paxilli*, var. *echinulatum*	Prisms from chloroform—ether Melting point: 228—231°C $[\alpha]_D$: +140° (c = 0.80 in chloroform) UV maximum (ethanol), nm (ϵ): 284 (19,000)	33, 63, 64, 72, 73 (biosyn), 92,
Geodoxin $C_{17}H_{12}Cl_2O_8$		*Aspergillus terreus* Thom., Strain Ac 100	Yellow needles from chloroform—ether Melting point: 216—217°C (decomposes) $[\alpha]_D$: 0° (c = 1.0 in chloroform) UV maxima, nm (log ϵ): 270 (3.95), 345 (3.73)	74, 75—76 (bio-syn)
Griseofulvin (Fulvicin, Grisovin) $C_{17}H_{17}ClO_6$		*Penicillium griseofulvum* Dierckx. *janczewskii* Zal. [= *P. nigricans* (Bain.) Thom.] *patulum albidum* Sopp. *raciborskii* Zal. *melinii* Thom. *urticae* Bain. *raistrickii* *Carpenteles brefeldianum* Dodge (Shear)	Colorless needles or prisms from ethanol or benzene Melting point: 220—221°C $[\alpha]_D^{21}$: +337° (c = 1.0 in acetone) UV maxima, nm: 286, 325	33, 77—89, 90—91 (biosyn), 92, 93—97 (syn), 98 (review)
Indole, 3-chloro- C_8H_6ClN		*Pseudomonas pyrrocinia*	Synthetic Melting point: 96°C (decomposes)	99, 100

Compound (Alternate Name) Formula	Structure	Organisms	Properties	References
Islanditoxin $C_{24}H_{31}Cl_2N_5O_7$		*Penicillium islandicum* Sopp.	Colorless amorphous solid Melting point: 250–251°C (decomposes) $[a]_D^{21}$: $-47.7°$ ($c = 1.99$ in glacial acetic acid) UV maximum, nm (ϵ): 257 (292)	101–104
Isocoumarin, 5-chloro-3,4-dihydro-8-hydroxy-6-methoxy-3-methyl- $C_{11}H_{11}ClO_4$		*Periconia macrospinosa*	Prisms from acetone–light petroleum Melting point: 123–124°C $[a]_D^{25}$: $-68°$ ($c = 0.51\%$ in methanol) UV maxima, nm (ϵ): 219 (23,000), 265 (11,700), 312 (5,780)	51
Isocoumarin, 7-chloro-3,4-dihydro-8-hydroxy-6-methoxy-3-methyl- $C_{11}H_{11}ClO_4$		*Sporormia affinis* Sacc., Bomm *et* Rouss.	Crystals from ethyl acetate–hexane Melting point: 170–171°C $[a]_D^{25}$: $-71.3 \pm 5.9°$ ($C = 0.505$ in methanol) UV maxima (methanol), nm (ϵ): 305 (680), 272 (12,600), 224 (25,400) UV maxima (methanolic 0.1N sodium hydroxide), nm (ϵ): 340 (6,420), 272 (25,000), 230 or 224 (29,700)	105
Isocoumarin, 5,7-dichloro-3,4-dihydro-8-hydroxy-6-methoxy-3-methyl- $C_{11}H_{10}Cl_2O_4$		*Sporormia affinis* Sacc., Bomm *et* Rouss.	Crystals from ethyl acetate–hexane Melting point: 225–226°C $[a]_D^{25}$: $-142.0 \pm 2.8°$ ($c = 1.067$ in methanol) UV maxima (methanol), nm (ϵ): 310 (5,540), 260 (7,890), 224 (24,830) UV maxima (methanolic 0.1N sodium hydroxide), nm (ϵ): 310 (28,060), 244 (17,300)	105

Compound (Alternate Name) Formula	Structure	Organisms	Properties	References
Mollisin (8-Dichloroacetyl-5-hydroxy-2,7-dimethyl-1,4-naphthaquinone) $C_{14}H_{10}Cl_2O_4$		*Mollisia caesia* Sacc. sensu Sydow *fallens* Karst.	Orange-yellow needles Melting point: 202–203°C (decomposes) UV maxima (ethanol), nm (log ϵ): 259 (4.26), 280 (3.9), 420 (3.52)	106–108, 109 (biosyn)
Monamycins		*Streptomyces jamaicensis*		110

	R_1	R_2	R_3
Monamycin G_1 $C_{33}H_{54}ClN_7O_8$	H	H	CH_3
Monamycin G_2 $C_{33}H_{54}ClN_7O_8$	H	CH_3	H
Monamycin G_3 $C_{33}H_{54}ClN_7O_8$	CH_3	H	H
Monamycin H_1 $C_{34}H_{56}ClN_7O_8$	CH_3	H	CH_3
Monamycin H_2 $C_{34}H_{56}ClN_7O_8$	H	CH_3	CH_3
Monamycin I $C_{35}H_{58}ClN_7O_8$	CH_3	CH_3	CH_3

Compound (Alternate Name) Formula	Structure	Organisms	Properties	References
Monorden (Radicicol) $C_{18}H_{17}ClO_6$		*Monosporium bonorden Nectria radicicola* Gerlach and Nilsson	Crystals from chloroform, ethanol, or benzene Melting point: 193.5°C (195°C) $[\alpha]_D^{20}$: +203° (in chloroform) UV maxima (ethanol), nm (ϵ): 264 (13,200), 272 (13,100), 315 (inflection, 2,800)	33, 111–114
Nidulin $C_{20}H_{17}Cl_3O_5$		*Aspergillus nidulans* (Eidam) Wint.	Colorless rods from light petroleum Melting point: 180°C UV maxima, nm (ϵ): 267 (9,010), 323 (inflection, 1,200)	33, 115–117, 118 (biosyn)
Nidulin, nor– (Ustin) $C_{19}H_{15}Cl_3O_5$		*Aspergillus nidulans* NRRL No. 2006	Hexagonal plates or prisms from benzene–light petroleum Melting point: 185–186°C UV maxima, nm (ϵ): 266 (8,120), 323 (inflection, 746)	116, 117, 119, 120
Nidulin, dechloronor– (Ustin II) $C_{19}H_{16}Cl_2O_5$		*Aspergillus nidulans* NRRL No. 2006	Needles from methanol Melting point: 212–214°C UV maximum (ethanol), nm (ϵ): 264 (9,810)	117, 121
Nordin $C_{18}H_{16}Cl_2O_8$		*Penicillium paxilli*, var. *echinulatum*	Needles Melting point: 134–136°C Contains three methoxy groups Occurs with geodin (estin)	122

Compound (Alternate Name) Formula	Structure	Organisms	Properties	References
Ochratoxin A $C_{20}H_{18}ClNO_6$		*Aspergillus ochraceus* Wilh., Strain K-804	Crystals from xylene Melting point: 169°C $[\alpha]_D$: −118° (c = 1.1 in chloroform) UV maxima (ethanol), nm (ϵ): 215 (34,000), 333 (2,400)	33, 123—125
Ochratoxin C $C_{22}H_{22}ClNO_6$	Ethyl ester of Ochratoxin A	*Aspergillus ochraceus* Wilh., Strain K-804	Isolated amorphous $[\alpha]_D$: −100° (c = 1.2 in ethanol) UV maxima (ethanol), nm (ϵ): 214 (30,000), 333 (7,000) IR maxima (chloroform), cm^{-1}: 1730, 1680	124
Pannarin $C_{18}H_{15}ClO_6$		*Pannaria languinosa* Korb. *fulvescens* Nyl. *lurida* Nyl. *Lecanora hercynica* Hoelt and Ullrich	Colorless prisms Melting point: 216°C	126—128
Penicillium islandicum Sopp. toxic chlorine-containing peptide	Constituent amino acids: L-serine, L-α-amino-*n*-butyric acid, L-β-amino, β-phenylpropionic acid, and a chlorine-containing component (2 Cl)	*Penicillium islandicum* Sopp.	Melting point: 251°C	129
Dehydrochlorinated-peptide amide obtained by treatment of the above with concentrated ammonium hydroxide $C_{24}H_{32}N_6O_7$			Melting point: 270—273°C (decomposes) UV maximum (ethanol), nm (ϵ): 268 (14,200)	129

Compound (Alternate Name) Formula	Structure	Organisms	Properties	References
Periconia macrospinosa metabolite (Methyl-2-allyl-3,5-dichloro-1,4-dihydroxy-cyclopent-2-enoate) $C_{10}H_{12}Cl_2O_4$		*Periconia macrospinosa*	Prisms from ether–light petroleum Melting point: 121–122°C $[\alpha]_D^{25}$: –90° (c = 0.56% in methanol) UV maxima, nm (ϵ): 247 (22,500), 256 (inflection, 16,000)	51
Pyoluteorin $C_{11}H_7Cl_2NO_3$		*Pseudomonas aeruginosa*	Melting point: 174–175°C (decomposes) UV maxima (ethanol), nm (ϵ): 255 (4,200), 310 (13,000)	130, 131, 132–133 (syn)
Pyrrole, 3-chloro-4-(*o*-nitrophenyl)- $C_{10}H_7ClN_2O_2$		*Pseudomonas pyrrolnitrica*	Pale-yellow crystals from benzene–*n*-hexane Melting point: 105–108°C	134
Pyrrole, 3-chloro-4-(2-amino-3-chlorophenyl)- $C_{10}H_8Cl_2N_2$		*Pseudomonas aureofaciens*		135
Pyrrole, 3-chloro-4-(3-chloro-2-nitrophenyl)- (Pyrrolnitrin) $C_{10}H_6Cl_2N_2O_2$		*Pseudomonas pyrrocinia pyrrolnitrica aureofaciens*	Crystals from benzene Melting point: 125°C	135 (biosyn), 136, 137, 138 (syn), 139, 140
Pyrrole, 2,3-dichloro-4-(*o*-nitrophenyl)- (Isopyrrolnitrin) $C_{10}H_6Cl_2N_2O_2$		*Pseudomonas* sp.	Yellow crystals Melting point: 105–108°C	141

Compound (Alternate Name) Formula	Structure	Organisms	Properties	References
Pyrrole, 3-chloro-4-(3-chloro-6-hydroxy-2-nitrophenyl)- (Oxypyrrolnitrin) $C_{10}H_6Cl_2N_2O_3$		*Pseudomonas pyrrocinia*	Pale-yellow crystals from ether–hexane Melting point: 215–216°C (decomposes) UV maximum (ethanol), nm (log ϵ): 290 (3.54)	142
Pyrrole, 2,3-dichloro-4-(3-chloro-2-nitrophenyl)- $C_{10}H_5Cl_3N_2O_2$		*Pseudomonas aureofaciens*	Synthetic; crystals from benzene–hexane Melting point: 140–141°C	135, 143 (syn)
Sclerotiorin $C_{21}H_{23}ClO_5$		*Penicillium sclerotiorum van Beyma multicolor* G.M.P. *implicatum* Biourge	Fine yellow needles from alcohol Melting point: 206–207°C $[\alpha]_D^{21}$: +500° (c = 1 in chloroform) UV maxima (95% ethanol), nm (log ϵ): 224 (4.06), 287 (4.11), 365 (4.50), 449 (4.11)	144–146, 147 (bio-syn), 148, 149, 150 (syn)
(–)-Sclerotiorin [7-epi-(+)-Sclerotiorin] $C_{21}H_{23}ClO_5$		*Penicillium hirayamae* Udagawa	Melting point: 204–205°C $[\alpha]_D^{26}$: –480° (c = 0.01 in ethanol) UV maxima (90% ethanol), nm (log ϵ): 215 (4.11), 286 (4.23), 368 (4.39)	150 (syn), 151–153
Sordidone (8-Chloro-5,7-dihydroxy-2,6-dimethylchromone) $C_{11}H_9ClO_4$		*Lecanora sordida* (Pers.) Th. Fr. *rupicola* (L.) Zahlbr. *carpinea* (L.) Ach. em. Vain.	Melting point: 260–262°C UV maxima (ethanol), nm (ϵ): 263 (16,000), 296 (5,750), 332 (3,650) IR maxima (Nujol), cm^{-1}: 1660, 1624, 1586	154–155 (syn)
Sporidesmin $C_{18}H_{20}ClN_3O_6S_2$		*Pithomyces chartarum* (Berk. *et* Curt.) Ellis	**Benzene Solvate** ($C_{18}H_{20}ClN_3O_6S_2 \cdot C_6H_6$) Melting point: 110–126°C (decomposes) $[\alpha]_D^{23}$: –33.5° (c = 1.1 in methanol); +6.9° (c = 1.4 in chloroform)	156–160

Compound (Alternate Name) Formula	Structure	Organisms	Properties	References
Sporidesmin (cont.)			UV maxima, nm (log ϵ): 218.5 (4.60), 254 (4.12), 302 (3.45) **After Recrystallization in Aqueous Methanol** Colorless needles, solvent-free (faint green sheen) Melting point: 179°C $[\alpha]_D^{20}: -45°$ ($c = 0.98$ in methanol)	
Sporidesmin B $C_{18}H_{20}ClN_3O_5S_2$		*Pithomyces chartarum* (Berk. *et* Curt.) Ellis	**After Recrystallization in Aqueous Acetone** Colorless needles Melting point: 183°C $[\alpha]_D^{21}: -27°$ ($c = 1.0$ in methanol); $+12°$ ($c = 0.75$ in chloroform) UV maxima, nm (log ϵ): 218 (4.50), 256 (4.08), 307 (3.41)	156, 157, 159
Sporidesmin C (isolated as diacetate) $C_{22}H_{24}ClN_3O_8S_3$		*Pithomyces chartarum* (Berk. *et* Curt.) Ellis	**Diacetate** Needles from isopropanol Melting point: 230−240°C (decomposes) $[\alpha]_D: -215°$ ($c = 0.46$ in chloroform) UV maxima (methanol), nm (ϵ): 222 (22,300), 255 (inflection, 9,900), 309 (2,900)	161
Sporidesmin D $C_{20}H_{26}ClN_3O_6S_2$		*Pithomyces chartarum* (Berk. *et* Curt.) Ellis	**Ether Solvate** ($C_{20}H_{26}ClN_3O_6S_2 \cdot C_4H_{10}O$) Melting point: 110−120°C **Ethanol Solvate** ($C_{20}H_{26}ClN_3O_6S_2 \cdot C_2H_5OH$) Melting point: 105−107°C $[\alpha]_D^{23}: +58°$ ($c = 0.11$ in chloroform) UV maxima (methanol), nm (log ϵ): 216 (4.45), 252 (4.00), 300 (3.28) Biologically inactive	159, 162

Compound (Alternate Name) Formula	Structure	Organisms	Properties	References
Sporidesmin E $C_{18}H_{20}ClN_3O_6S_3$		*Pithomyces chartarum* (Berk. *et* Curt.) Ellis	**Ether Solvate** ($C_{18}H_{20}ClN_3O_6S_3 \cdot C_4H_{10}O$) Melting point: 180–185°C $[\alpha]_D^{20}$: −132° ($c = 0.064$ in chloroform) UV maxima (methanol), nm (log ϵ): 217 (4.52), 252 (4.22), 295 (3.50) Treatment with triphenyl phosphine gives sporidesmin; treatment with sodium borohydride and methyl iodide gives sporidesmin D Shows cytotoxic activity against HeLa cells at 0.1 ng per ml ("the most cytotoxic mold metabolite described so far")	160, 163
Sporidesmin F $C_{19}H_{22}ClN_3O_6S$		*Pithomyces chartarum* (Berk. *et* Curt.) Ellis	Isolated as amorphous solid Melting point: 65–75°C UV maxima (methanol), nm (log ϵ): 216 (4.46), 250 (4.14), 298 (3.30) Little or no biological activity	162
Sporormia affinis metabolite [(1S,5S)-2-Cyclopentene-1-carboxylic acid; *2-trans*-allyl-5-chloro-1-hydroxy-4-oxo-methyl ester; c.f. Cryptosporiopsin] $C_{10}H_{11}ClO_4$		*Sporormia affinis* Sacc. Bomm *et* Rouss.	Colorless crystals from ether–hexane Melting point: 91.5–92.5°C $[\alpha]_D^{25}$: +105 ± 3.0° ($c = 1.00$ in ethyl acetate) UV maximum (methanol), nm (ϵ): 277 (23,040) Material turns yellow on storing	47
Sporormia affinis metabolite [(1S,5R)-2-Cyclopentene-1-carboxylic acid; *2-trans*-allyl-5-chloro-1-hydroxy-4-oxo methyl ester] $C_{10}H_{11}ClO_4$		*Sporormia affinis* Sacc., Bomm *et* Rouss.	Colorless crystals from ether–ethyl acetate–hexane Melting point: 83.5–84.5°C $[\alpha]_D^{25}$: +322 ± 2.9° ($c = 1.00$ in ethyl acetate) UV maximum (methanol), nm (ϵ): 277 (22,010)	47

Compound (Alternate Name) Formula	Structure	Organisms	Properties	References
Streptomyces ambofaciens metabolite (Antibiotic MSD-819; 6-chloro-2-quinoxaline-carboxylic acid, 1,4-dioxide) $C_9H_5ClN_2O_4$		*Streptomyces ambofaciens*		164
Terrecin Formula not known $C = 51.89\%$, $H = 3.51\%$, $Cl = 19.1\%$, $N = 3.8\%$		*Aspergillus terreus*	Pale-yellow prisms from acetone and benzene Melting point: 219–220°C Soluble in alkali; precipitated from solution by addition of acid Ferric chloride test: deep violet	165
Tetracycline, 4-aminodedimethylaminoanhydrodemethylchlor- $C_{19}H_{15}ClN_2O_7$		*Streptomyces aureofaciens*, mutant 1E1407	Isolated as hydrate hydrochloride UV maxima, nm (ε): 424 (8,600), 329 (3,520), 314 (inflection, 3,720), 302 (inflection, 5,470), 269 (53,400), 223 (35,000)	166, 167
Tetracycline, anhydrodemethylchlor- $C_{21}H_{19}ClN_2O_7$		*Streptomyces aureofaciens*, mutant 1E6113	Paper chromatography: orange fluorescing spot at R_f 0.48 in pH 8.3 EDTA–butanol system Identical with the compound derived from dehydration of 7-chloro-6-demethyl tetracycline Melting point: 205–210°C (decomposes) $[\alpha]_D^{25}$: +105° ($c = 0.467\%$ in methyl Cellosolve) UV maxima (0.1N hydrochloric acid), nm (ε): 223 (30,200), 272 (52,500), 430 (8,050)	168, 169

Compound (Alternate Name) Formula	Structure	Organisms	Properties	References
Tetracycline, 7-chloro-6-demethyl- (Demethylchlortetracycline) $C_{21}H_{21}ClN_2O_8$		*Streptomyces aureofaciens*, Duggar mutant	Isolated as crystalline sesquihydrate Melting point: 174–178°C (decomposes) $[\alpha]_D^{25}: -258°$ ($c = 0.5\%$ in $0.1N$ sulfuric acid)	33, 170
Tetracycline, 7-chloro-5a(11a)-dehydro- $C_{22}H_{21}ClN_2O_8$		*Streptomyces aureofaciens*, Duggar mutant S-1308	**Hydrochloride** $[\alpha]_D^{25}: +15.5°$ ($c = 0.65\%$ in $0.03N$ hydrochloric acid) UV maxima, nm (e): 221 (27,500), 251 (23,000), 375* (4,300)	171
Tetracycline, 7-chloro- (Aureomycin; biomycin) $C_{22}H_{23}ClN_2O_8$		*Streptomyces aureofaciens*	Fine yellow crystals Melting point: 168–169°C $[\alpha]_D^{23}: -275.0°$ (in methanol) UV maxima ($0.1N$ hydrochloric acid), nm: 230, 262.5, 367.5 UV maxima ($0.1N$ sodium hydroxide), nm: 255, 285, 345	33, 172–180
Tumidulin (Methyl-3,5-dichlorolecanorate) $C_{17}H_{14}Cl_2O_7$		*Ramalina ceruchis* (Ach.) De Not. *tumidula* (Tayl.) Hun. *et* Follm. *flaccescens* Nyl. *peruviana* Ach. *chilensis* Bert. *cacacearum* Follm.	Crystals from dry benzene (in darkness) Melting point: 177–177.5°C (decomposes) UV maxima (95% ethanol), nm (log e): 259 (4.2), 318 (4.1) IR maxima (potassium bromide), cm⁻¹: 3400, 1715, 1685, 1620, 1600, 1575 Insoluble in sodium bicarbonate Positive ferric chloride test	181 (syn), 182, 183
Vicanicin $C_{17}H_{14}Cl_2O_5$		*Teloschistes flavicans*	Colorless needles Melting point: 248–250°C UV maxima, nm: 270, 324 (inflection)	184

*Very broad

435

Compound (Alternate Name) Formula	Structure	Organisms	Properties	References
Xanthen-9-one, 2-chloro-1,3,6-trihydroxy-8-methyl- (2-Chloronorlichexanthone) $C_{14}H_9ClO_5$		*Lecanora straminea*	Yellow crystals Melting point: 240–244°C	185
Xanthen-9-one, 2,4-dichloro-1,3,6-trihydroxy-8-methyl- (2,4-Dichloronorlichexanthone) $C_{14}H_8Cl_2O_5$		*Lecanora straminea*	Yellow crystals Melting point: 286–287°C	185, 186 (syn)
Xanthen-9-one, 2,7-dichloro-1,3,6-trihydroxy-8-methyl- (2,7-Dichloronorlichexanthone) $C_{14}H_8Cl_2O_5$		*Lecanora straminea*	Yellow crystals Melting point: 273–274°C (decomposes)	187
Xanthen-9-one, 2,4,7-trichloro-1,3,6-trihydroxy-8-methyl- (Arthothelin) $C_{14}H_7Cl_3O_5$		*Arthothelium pacificum* Follm. *Lecidea quernea* (Dicks.) Arn. *Lecanora reuteri* Shaer *straminea*	Melting point: 275–276°C Soluble in sodium bicarbonate Treatment with one equivalent diazomethane gives thuringion	186 (syn), 188–190
Xanthen-9-one, 2,4,5,7-tetrachloro-1,3,6-trihydroxy-8-methyl- (Thiophanic acid) $C_{14}H_6Cl_4O_5$		*Lecanora rupicola* (L.) Zahlbr. (= *L. sordida* Th. Fr.) *straminea*	Yellow prisms from benzene Melting point: 243° or 249.5°C UV maxima, nm (ε): 248 (8,200), 272 (5,300), 364 (6,200) IR maxima (potassium bromide), cm⁻¹: 3585, 3440, 1635, 1568	186 (syn), 190, 191, 192 (syn)

Compound (Alternate Name) Formula	Structure	Organisms	Properties	References
Xanthen-9-one, 2,4,-dichloro-1,3-dihydroxy-6-methoxy-8-methyl- (Thiophaninic acid) $C_{15}H_{10}Cl_2O_5$		*Pertusaria flavicans*	Yellow prisms on sublimation Melting point: 278–279°C	193–195
Xanthen-9-one, 2,4,7-trichloro-1,6-dihydroxy-3-methoxy-8-methyl- (Thuringion) $C_{15}H_9Cl_3O_5$		*Lecidea carpathica* (Koerb.) Szat.	Yellow needles from ethyl acetate–ethanol Melting point: 278–279°C UV maxima (methanol), nm (log ϵ): 246 (4.53), 314 (4.18), 356 (4.01)	190 (syn), 196

REFERENCES

1. Petty, M. A., An Introduction to the Origin and Biochemistry of Microbial Halometabolites, *Bacteriol. Rev., 25,* 111 (1961).
2. Miller, M. W., *The Pfizer Handbook of Microbial Metabolites.* McGraw-Hill Book Co., Inc., Blakiston Division, New York (1961).
3. Bendz, G., Bohman, G., and Santesson, J., *Acta Chem. Scand., 21,* 2889 (1967).
4. Yamamoto, Y., Kiriyama, N., and Arahata, S., *Chem. Pharm. Bull. (Tokyo), 16,* 304 (1968).
5. Yamauchi, H., Morimoto, K., and Kitagawa, I., *Tetrahedron Lett.,* p. 1149 (1968).
6. Fox, C. H., Maass, W. S. G., and Forrest, T. P., *Tetrahedron Lett.,* p. 919 (1969).
7. Sargent, M. V., Smith, D. O'N., and Elix, J. A., *J. Chem. Soc. Sect. C,* p. 307 (1970).
8. Bohman, G., *Acta Chem. Scand., 23,* 2241 (1969).
9. Briggs, L. H., and Castaing, D. R., *Bull. Nat. Inst. Sci. India, 28,* 71 (1965).
10. Bruun, T., Hollis, D. P., and Ryhage, R., *Acta Chem. Scand., 19,* 839 (1965).
11. Raistrick, H., and Ziffer, J., *Biochem. J., 49,* 563 (1951).
12. Birch, A. J., and Stapleford, K. S. J., *J. Chem. Soc. Sect. C,* p. 2570 (1967).
13. Tamura, G., Suzuki, S., Takatsuki, A., Ando, K., and Arima, K., *J. Antibiot. (Tokyo), 21,* 539 (1968).
14. Ellestad, G. A., Evans, R. H., Jr., and Kunstmann, M. P., *Tetrahedron, 25,* 1323 (1969).
15. Nawata, Y., Ando, K., Tamura, G., Arima, K., and Iitaka, Y., *J. Antibiot. (Tokyo), 22,* 511 (1969).
16. Argoudelis, A. D., Herr, R. R., Mason, D. J., Pyke, T. R., and Zeiserl, J. F., *Biochemistry, 6,* 165 (1967).
17. Clutterbuck, P. W., Mukhopadhyay, S. L., Oxford, A. E., and Raistrick, H., *Biochem. J., 34,* 664 (1940).
18. Beckwith, J. R., and Hager, L. P., *J. Org. Chem., 26,* 5206 (1961).
19. Burgsthaler, A. W., Lewis, T. B., and Abdel-Rahman, M. O., *J. Org. Chem., 31,* 3516 (1966).
20. Shaw, P. D., and Hager, L. P., *J. Amer. Chem. Soc., 81,* 1011 (1959).
21. Shaw, P. D., and Hager, L. P., *J. Amer. Chem. Soc., 81,* 6527 (1959).
22. Shaw, P. D., Beckwith, J. R., and Hager, L. P., *J. Biol. Chem., 234,* 2560 (1959).
23. Shaw, P. D., and Hager, L. P., *J. Biol. Chem., 234,* 2563 (1959).
24. Shaw, P. D., and Hager, L. P., *J. Biol. Chem., 236,* 1626 (1961).
25. Beckwith, J. R., Clark, R., and Hager, L. P., *J. Biol. Chem., 238,* 3086 (1963).
26. Beckwith, J. R., and Hager, L. P., *J. Biol. Chem., 238,* 3091 (1963).
27. Neidleman, S. L., Diassi, P. A., Junta, B., Palmere, R. M., and Pan, S. C., *Tetrahedron Lett.,* p. 5337 (1966).
28. Ehrlich, J., Bartz, Q. R., Smith, R. M., Joslyn, D. A., and Burkholder, P. R., *Science, 106,* 417 (1947).
29. Rebstock, M. C., Crooks, H. M., Jr., Controulis, J., and Bartz, Q. R., *J. Amer. Chem. Soc., 71,* 2458 (1949).
30. Long, L. M., and Troutman, H. D., *J. Amer. Chem. Soc., 71,* 2469 (1949).
31. Controulis, J., Rebstock, M. C., and Crooks, H. M., Jr., *J. Amer. Chem. Soc., 71,* 2463 (1949).
32. McGrath, R., Vining, L. C., Sala, F., and Westlake, D. W. S., *Can. J. Biochem., 46,* 587 (1968).
33. Stecher, P. G. (Ed.), *The Merck Index,* 8th ed. Merck and Co., Inc., Rahway, New Jersey (1968).
34. Smith, C. G., *J. Bacteriol., 75,* 577 (1958).
35. Richards, M., Bird, A. E., and Munden, J. E., *J. Antibiot. (Tokyo), 22,* 388 (1969).
36. Bird, A. E., and Marshall, A. C., *J. Chem. Soc. Sect. C,* p. 2418 (1969).
37. Munden, J. E., Butterworth, D., Hanscomb, G., and Verrall, M. S., *Appl. Microbiol., 19,* 718 (1970).
38. Curd, F. H., Robertson, A., and Stephenson, R. J., *J. Chem. Soc.,* p. 130 (1933).
39. Koller, G., and Pöpl, K., *Monatsh. Chem., 64,* 106 (1934).
40. Koller, G., and Pöpl, K., *Monatsh. Chem., 64,* 126 (1934).
41. St. Pfau, A., *Helv. Chim. Acta, 17,* 1319 (1934).
42. Huneck, S., and Follmann, G., *Z. Naturforsch. Teil B, 20,* 1138 (1965).
43. Culberson, C. F., *J. Pharm. Sci., 54,* 1815 (1965).
44. Yamazaki, M., Matsuo, M., and Shibata, S., *Chem. Pharm. Bull. (Tokyo), 13,* 1015 (1965).
45. Udagawa, S., *Chem. Pharm. Bull. (Tokyo), 11,* 366 (1963).
46. Gray, R. W., and Whalley, W. B., *Chem. Commun.,* p. 762 (1970).
47. McGahren, W. J., van den Hende, J. H., and Mitscher, L. A., *J. Amer. Chem. Soc., 91,* 157 (1969).
48. Stillwell, M. A., Wood, F. A., and Strunz, G. M., *Can. J. Microbiol., 15,* 501 (1969).
49. Strunz, G. M., Court, A. S., Komlossy, J., and Stillwell, M. A., *Can. J. Chem., 47,* 2087 (1969).
50. Strunz, G. M., Court, A. S., Komlossy, J., and Stillwell, M. A., *Can. J. Chem., 47,* 3700 (1969).
51. Giles, D., and Turner, W. B., *J. Chem. Soc. Sect. C,* p. 2187 (1969).
52. Strunz, G. M., and Court, A. S., *Experientia, 26,* 1054 (1970).
53. Kato, A., Ando, K., Tamura, G., and Arima, K., *J. Antibiot. (Tokyo), 23,* 168 (1970).
54. Zopf, W., *Justus Liebigs Ann. Chem., 336,* 46 (1904).
55. Nolan, T. J., *Sci. Proc. Roy. Dublin Soc., 21,* 67 (1934).
56. Spillane, P. A., Keane, J., and Nolan, T., *Sci. Proc. Roy. Dublin Soc., 21,* 333 (1936).
57. Nolan, T. J., Algar, J., McCann, E. P., Manahan, W. A., and Nolan, N., *Sci. Proc. Roy. Dublin Soc., 24,* 319 (1948).
58. Brown, C. J., Clark, D. E., Ollis, W. D., and Veal, P. L., *Proc. Chem. Soc.,* p. 393 (1960).

59. Kavanagh, F., Hervey, A., and Robbins, W. J., *Proc. Nat. Acad. Sci. U.S.A., 38,* 555 (1952).
60. Anchel, M., *J. Amer. Chem. Soc., 74,* 2943 (1952).
61. Singh, P., and Rangaswami, S., *Tetrahedron Lett.,* p. 1229 (1966).
62. Bureš, E., and Hutter, J., *Čas. Česk. Lékárnictva, 11,* 29, 57 (1931); *Chem. Abstr., 25,* 5153 (1931).
63. Raistrick, H., and Smith, G., *Biochem. J., 30,* 1315 (1936).
64. Barton, D. H. R., and Scott, A. I., *J. Chem. Soc.,* p. 1767 (1958).
65. Umezawa, H., Takahashi, S., Takeuchi, T., Maeda, K., and Okami, Y., *Jap. J. Med. Sci. Biol., 5,* 311 (1952); *Chem. Abstr., 47,* 4949a (1953).
66. Shimi, I. R., and Shoukry, S., *J. Antibiot. (Tokyo) Ser. A, 19,* 110 (1966).
67. Yosioka, I., Yamauchi, H., Morimoto, K., and Kitagawa, I., *Tetrahedron Lett.,* p. 3749 (1968).
68. Atsumi, K., Takada, M., Mizuno, K., and Ando, T., *J. Antibiot. (Tokyo), 23,* 223 (1970).
69. Hardiman, J., Keane, J., and Nolan, T. J., *Sci. Proc. Roy. Dublin Soc., 21,* 141 (1935).
70. Nolan, T. J., and Keane, J., *Sci. Proc. Roy. Dublin Soc., 22,* 199 (1940).
71. Davidson, V. E., Keane, J., and Nolan, T. J., *Sci. Proc. Roy. Dublin Soc., 23,* 143 (1943).
72. Delmotte, P., Delmotte-Plaquée, J., and Bastin, R., *J. Pharm. Belg., 11,* 200 (1956).
73. Rhodes, A., McGonagle, M. P., and Somerfield, G. A., *Chem. Ind. (London),* p. 611 (1962).
74. Hassall, C. H., and McMorris, T. C., *J. Chem. Soc.,* p. 2831 (1959).
75. Hassall, C. H., and Lewis, J. R., *J. Chem. Soc.,* p. 2312 (1961).
76. Hassall, C. H., The Biosynthesis of Geodoxin and Related Compounds, in *Biogenesis of Antibiotic Substances,* p. 51, Z. Vanek and Z. Hostalek, Eds. Academic Press, New York, London (1965).
77. Oxford, A. E., Raistrick, H., and Simonart, P., *Biochem. J., 33,* 240 (1939).
78. Brian, P. W., Curtis, P. J., and Hemming, H. G., *Trans. Brit. Mycol. Soc., 29,* 173 (1946).
79. McGowan, J. C., *Trans. Brit. Mycol. Soc., 29,* 188 (1946).
80. Grove, J. F., and McGowan, J. C., *Nature, 160,* 514 (1947).
81. Brian, P. W., Curtis, P. J., and Hemming, H. G., *Trans. Brit. Mycol. Soc., 32,* 30 (1949).
82. Grove, J. F., Ismay, D., Macmillan, J., Mulholland, T. P. C., and Thorold-Rogers, M. A., *Chem. Ind. (London),* p. 219 (1951).
83. Grove, J. F., Macmillan, J., Mulholland, T. P. C., and Thorold-Rogers, M. A., *J. Chem. Soc.,* p. 3949 (1952).
84. Grove, J. F., Ismay, D., Macmillan, J., Mulholland, T. P. C., and Thorold-Rogers, M. A., *J. Chem. Soc.,* p. 3958 (1952).
85. Grove, J. F., Macmillan, J., Mulholland, T. P. C., and Zealley, J., *J. Chem. Soc.,* p. 3967 (1952).
86. Grove, J. F., Macmillan, J., Mulholland, T. P. C., and Thorold-Rogers, M. A., *J. Chem. Soc.,* p. 3977 (1952).
87. Mulholland, T. P. C., *J. Chem. Soc.,* p. 3987 (1952).
88. Mulholland, T. P. C., *J. Chem. Soc.,* p. 3994 (1952).
89. Brian, P. W., Curtis, P. J., and Hemming, H. G., *Trans. Brit. Mycol. Soc., 38,* 305 (1955).
90. Birch, A. J., Massy-Westropp, R. A., Rickards, R. W., and Smith, H., *Proc. Chem. Soc.,* p. 98 (1957).
91. Birch, A. J., Massy-Westropp, R. A., Rickards, R. W., and Smith, H., *J. Chem. Soc.,* p. 360 (1958).
92. Macmillan, J., *J. Chem. Soc.,* p. 1823 (1959).
93. Day, A. C., Nabney, J., and Scott, A. I., *Proc. Chem. Soc.,* p. 284 (1960); *J. Chem. Soc.,* p. 4067 (1961).
94. Brossi, A., Baumann, M., Gerecke, M., and Kyburz, E., *Helv. Chim. Acta, 43,* 1444, 2071 (1960).
95. Kuo, C. H., Hoffsommer, R. D., Slates, H. L., Taub, D., and Wendler, N. L., *Chem. Ind. (London),* p. 1627 (1960).
96. Stork, G., and Tomasz, M., *J. Amer. Chem. Soc., 84,* 310 (1962); *86,* 471 (1964).
97. Taub, D., Kuo, C. H., Slates, H. L., and Wendler, N. L., *Tetrahedron, 19,* 1 (1963).
98. Grove, J. F., *Quart. Rev. Chem. Soc. London, 17,* 1 (1963).
99. Lively, D. H., Gorman, M., Haney, M. E., and Mabe, J. A., *Antimicrob. Agents Chemother.,* p. 462 (1966); *Chem. Abstr., 67,* 79836s (1967).
100. Pappalardo, J., and Vitali, T., *Gazz. Chim. Ital., 88,* 1147 (1958).
101. Marumo, S., and Sumiki, Y., *J. Agr. Chem. Soc. Jap., 29,* 305 (1955).
102. Marumo, S., Miyao, K., Matsuyama, A., and Sumiki, Y., *J. Agr. Chem. Soc. Jap., 29,* 913 (1955).
103. Marumo, S., *Bull. Agr. Chem. Soc. Jap., 19,* 258 (1955).
104. Marumo, S., *Bull. Agr. Chem. Soc. Jap., 23,* 428 (1959).
105. McGahren, W. J., and Mitscher, L. A., *J. Org. Chem., 33,* 1577 (1968).
106. Gremmen, J., *Antonie van Leeuwenhoek, J. Microbiol. Serol., 22,* 58 (1956).
107. van der Kerk, G. J. M., and Overeem, J. C., *Rec. Trav. Chim. Pays-Bas, 76,* 425 (1957).
108. Overeem, J. C., and van der Kerk, G. J. M., *Rec. Trac. Chim. Pays-Bas, 83,* 955 (1964).
109. Bentley, R., and Gatenbeck, S., *Biochemistry, 4,* 1150 (1965).
110. Bevan, K., Davies, J. S., Hall, M. J., Hassall, C. H., Morton, R. B., Phillips, D. A. S., Ogihara, Y., and Thomas, W. A., *Experientia, 26,* 122 (1970).
111. Delmotte, P., and Delmotte-Plaquée, J., *Nature, 171,* 344 (1953).
112. McCapra, F., Scott, A. I., Delmotte, P., Delmotte-Plaquée, J., and Bhacca, N. S., *Tetrahedron Lett.,* p. 869 (1964).
113. Mirrington, R. N., Ritchie, E., Shoppee, C. W., Taylor, W. C., and Sternhell, S., *Tetrahedron Lett.,* p. 365 (1964).
114. Mirrington, R. N., Ritchie, E., Shoppee, C. W., Sternhell, S., and Taylor, W. C., *Aust. J. Chem., 19,* 1265 (1966).
115. Dean, F. M., Robertson, A., Roberts, J. C., and Raper, K. B., *Nature, 172,* 344 (1953).

116. Dean, F. M., Roberts, J. C., and Robertson, A., *J. Chem. Soc.,* p. 1432 (1954).

117. Dean, F. M., Deorha, D. S., Erni, A. D. T., Hughes, D. W., and Roberts, J. C., *J. Chem. Soc.,* p. 4829 (1960).

118. Beach, W. F., and Richards, J. H., *J. Org. Chem., 28,* 2746 (1963).

119. Kurung, J. M., *Science, 102,* 11 (1945).

120. Doering, W. D., Dubos, R. J., Noyce, D. S., and Dreyfus, R., *J. Amer. Chem. Soc., 68,* 725 (1946).

121. Dean, F. M., Erni, A. D. T., and Robertson, A., *J. Chem. Soc.,* p. 3545 (1956).

122. Komatsu, E., *Japanese Patent 4799* (1953); *Chem. Abstr., 48,* 11010 (1954).

123. Scott, de B., *Mycopathol. Mycol. Appl., 25,* 213 (1965).

124. van der Merwe, K. J., Steyn, P. S., and Fourie, L., *J. Chem. Soc.,* p. 7083 (1965).

125. Natori, S., Sakaki, S., Kurata, H., Udagawa, S., Ichinoe, M., Saito, M., and Umeda, M., *Chem. Pharm. Bull. (Tokyo), 18,* 2259 (1970).

126. Yosioka, I., *J. Pharm. Soc. Jap. (Yakugaku Zasshi), 61,* 332 (1941).

127. Asahina, Y., and Shibata, S., *The Chemistry of Lichen Substances.* Japanese Society for the Promotion of Science, Tokyo (1954).

128. Huneck, S., *Z. Naturforsch. Teil B, 21,* 80 (1966).

129. Sato, M., and Tatsuno, T., *Chem. Pharm. Bull. (Tokyo), 16,* 2182 (1968).

130. Takeda, R., *J. Amer. Chem. Soc., 80,* 4749 (1958).

131. Birch, A. J., Hodge, P., Rickards, R. W., Takeda, R., and Watson, T. R., *J. Chem. Soc.,* p. 2641 (1964).

132. Bailey, K., and Rees, A. H., *Chem. Commun.,* p. 1284 (1969).

133. Bailey, D. M., and Johnson, R. E., *Tetrahedron Lett.,* p. 3555 (1970).

134. Hashimoto, M., and Hattori, K., *Chem. Pharm. Bull. (Tokyo), 16,* 1144 (1968).

135. Hamill, R. L., Elander, R., Mabe, J., and Gorman, M., *Antimicrob. Agents Chemother.,* p. 388 (1967); *Chem. Abstr., 70,* 2415h (1969).

136. Arima, K., Imanaka, H., Kousaka, M., Fukuta, A., and Tamura, G., *Agr. Biol. Chem., 28,* 575 (1964).

137. Imanaka, H., Kousaka, M., Tamura, G., and Arima, K., *J. Antibiot. (Tokyo) Ser. A, 18,* 207 (1965).

138. Nakano, H., Umio, S., Kariyone, K., Tanaka, K., Kishimoto, T., Noguchi, H., Ueda, I., Nakamura, H., and Morimoto, Y., *Tetrahedron Lett.,* p. 737 (1966).

139. Gorman, M., and Lively, D. H., *Antibiotics, 2,* 433 (1967); *Chem. Abstr., 68,* 48256u (1968).

140. Morimoto, Y., Hashimoto, M., and Hattori, K., *Tetrahedron Lett.,* p. 209 (1968).

141. Hashimoto, M., and Hattori, K., *Bull. Chem. Soc. Jap., 39,* 410 (1966).

142. Hashimoto, M., and Hattori, K., *Chem. Pharm. Bull. (Tokyo), 14,* 1314 (1966).

143. Hattori, K., and Hashimoto, M., *Japanese Patent 16,135* (1968); *Chem. Abstr., 70,* 57621c (1969).

144. Curtin, T. P., and Reilly, J., *Biochem. J., 34,* 1419 (1940).

145. Birkinshaw, J. H., *Biochem. J., 52,* 283 (1952).

146. Fielding, H. C., Robertson, A., Traners, R. B., and Whalley, W. B., *J. Chem. Soc.,* p. 1814 (1958).

147. Birch, A. J., Fitton, P., Pride, E., Ryan, A. J., Smith, H., and Whalley, W. B., *J. Chem. Soc.,* p. 4576 (1958).

148. Dean, F. M., Staunton, J., and Whalley, W. B., *J. Chem. Soc.,* p. 3004 (1959).

149. Yamamoto, Y., and Nishikawa, N., *J. Pharm. Soc. Jap. (Yakugaku Zasshi), 79,* 297 (1959).

150. Chong, R., King, R. R., and Whalley, W. B., *Chem. Commun.,* p. 1512 (1969).

151. Udagawa, S., *Chem. Pharm. Bull. (Tokyo), 11,* 366 (1963).

152. Gregory, E. M., and Turner, W. B., *Chem. Ind. (London),* p. 1625 (1963).

153. Ellestad, G. A., and Whalley, W. B., *J. Chem. Soc.,* p. 7260 (1965).

154. Arshad, M., Devlin, J. P., Ollis, W. D., and Wheeler, R. E., *Chem. Commun.,* p. 154 (1968).

155. Huneck, S., and Santesson, J., *Z. Naturforsch. Teil B, 24,* 750 (1969).

156. Ronaldson, J. W., Taylor, A., White, E. P., and Abraham, R. J., *J. Chem. Soc.,* p. 3172 (1963).

157. Hodges, R., Ronaldson, J. W., Taylor, A., and White, E. P., *Chem. Ind. (London),* p. 42 (1963).

158. Fridrichsons, J., and Mathieson, A. McL., *Tetrahedron Lett.,* p. 1265 (1962).

159. Rahman, R., and Taylor, A., *Chem. Commun.,* p. 1032 (1967).

160. Brewer, D., Rahman, R., Safe, S., and Taylor, A., *Chem. Commun.,* p. 1571 (1968).

161. Hodges, R., and Shannon, J. S., *Aust. J. Chem., 19,* 1059 (1966).

162. Jamieson, W. D., Rahman, R., and Taylor, A., *J. Chem. Soc. Sect. C,* p. 1564 (1969).

163. Rahman, R., Safe, S., and Taylor, A., *J. Chem. Soc. Sect. C,* p. 1665 (1969).

164. Stapley, E. O., Hendlin, D., Matta, J. M., Hernandez, S., Miller, A. K., Jackson, M., and Wallick, H., *Antimicrob. Agents Chemother.,* p. 249 (1968–1969).

165. Iwata, K., and Yosioka, I., *J. Antibiot. (Tokyo) Ser. B, 3,* 193 (1950).

166. McCormick, J. R. D., Jensen, E. R., Johnson, S., and Sjölander, N. O., *J. Amer. Chem. Soc., 90,* 2201 (1968).

167. Miller, P. A., Saturnelli, A., Martin, J. H., Mitscher, L. A., and Bohonos, N., *Biochem. Biophys. Res. Commun., 16,* 285 (1964).

168. McCormick, J. R. D., and Jensen, E. R., *J. Amer. Chem. Soc., 91,* 206 (1969).

169. Webb, J. S., Broschard, R. W., Cosulich, D. B., Stein, W. J., and Wolf, C. F., *J. Amer. Chem. Soc., 79,* 4563 (1957).

170. McCormick, J. R. D., Sjölander, N. O., Hirsch, U., Jensen, E. R., and Doerschuk, A. P., *J. Amer. Chem. Soc., 79,* 4561 (1957).

171. McCormick, J. R. D., Miller, P. A., Growich, J. A., Sjölander, N. O., and Doerschuk, A. P., *J. Amer. Chem. Soc., 80,* 5572 (1958).
172. Broschard, R. W., Dornbush, A. C., Gordon, S., Hutchings, B. L., Kohler, A. R., Krupka, G., Kushner, A., Lefemine, D. V., and Pidacks, C., *Science, 109,* 199 (1949).
173. Duggar, B. M., *U.S. Patent 2,482,055* (1949).
174. Stephens, C. R., Conover, L. H., Hochstein, F. A., Regna, P. P., Pilgrim, F. J., Brunings, K. J., and Woodward, R. B., *J. Amer. Chem. Soc., 74,* 4976 (1952).
175. Waller, C. W., Hutchings, B. L., Broschard, R. W., Goldman, A. A., Stein, W. J., Wolf, C. F., and Williams, J. H., *J. Amer. Chem. Soc., 74,* 4981 (1952).
176. Stephens, C. R., Conover, L. H., Pasternack, R., Hochstein, F. A., Moreland, W. T., Regna, P. P., Pilgrim, F. J., Brunings, K. J., and Woodward, R. B., *J. Amer. Chem. Soc., 76,* 3568 (1954).
177. Doerschuk, A. P., McCormick, J. R. D., Goodman, J. J., Szumski, S. A., Growich, J. A., Miller, P. A., Bitler, B. A., Jensen, E. R., Petty, M. A., and Phelps, A. S., *J. Amer. Chem. Soc., 78,* 1508 (1956).
178. Dobryinin, V. N., Gurevich, A. I., Karapetyan, M. G., Kolosov, M. N., and Shemyakin, M. M., *Tetrahedron Lett.,* p. 901 (1962).
179. Donohue, J., Dunitz, J. D., Trueblood, K. N., and Webster, M. S., *J. Amer. Chem. Soc., 85,* 851 (1963).
180. Goodman, J. J., and Matrishin, M., *Nature, 219,* 291 (1968).
181. Bendz, G., Santesson, J., and Wachtmeister, C. A., *Acta Chem. Scand., 19,* 1188 (1965).
182. Huneck, S., and Follmann, G., *Z. Naturforsch. Teil B, 20,* 611 (1965).
183. Huneck, S., *Chem. Ber., 99,* 1106 (1966).
184. Neelakantan, S., Seshadri, T. R., and Subramanian, S. S., *Tetrahedron Lett. No. 9,* p. 1 (1959).
185. Santesson, J., *Ark. Kemi, 30,* 461 (1969).
186. Santesson, J., and Sundholm, G., *Ark. Kemi, 30,* 427 (1969).
187. Santesson, J., *Ark. Kemi, 30,* 455 (1969).
188. Huneck, S., and Follmann, G., *Z. Naturforsch. Teil B, 22,* 461 (1967).
189. Santesson, J., *Acta Chem. Scand., 22,* 1698 (1968).
190. Santesson, J., *Ark. Kemi, 30,* 449 (1969).
191. Huneck, S., *Tetrahedron Lett.,* p. 3547 (1966).
192. Jayalakshmi, V., Neelakantan, S., and Seshadri, T. R., *Curr. Sci., 37,* 196 (1968).
193. Hesse, O., *J. Prakt. Chem., 58,* 465 (1898).
194. Zopf, W., *Justus Liebigs Ann. Chem., 317,* 110 (1901).
195. Santesson, J., and Wachtmeister, C. A., *Ark. Kemi, 30,* 445 (1969).
196. Huneck, S., and Santesson, J., *Z. Naturforsch. Teil B, 24,* 756 (1969).

BROMINE-CONTAINING COMPOUNDS

DR. NANCY N. GERBER

Microbial metabolites containing bromine are relatively rare. They are isolated either from marine organisms or from microorganisms that normally furnish chlorine-containing compounds. In the latter cases, the media have been prepared with bromide instead of chloride salts.

Name	Structure	Organism	Reference
-Dibromo-4-hydroxy-enzyl alcohol	3,5-diBr-4-OH-PhCH$_2$OH	Odonthalia dentata, Rhodomela confervoides (both red alga)	1
-Dibromo-4,5-dihydroxy-enzyl alcohol	3,5-diBr-4,5-diOH-PhCH$_2$OH		
Bromo-3,4-dihydroxy-benzaldehyde	5-Br-3,4-diOHPhCHO	Polysiphonia morrowii	1
-Dibromobenzyl alcohol-4,5-disulfate, dipotassium salt	2,3-diBr-4,5-diKOSO$_3$PhCH$_2$OH	Polysiphonia lanosa	1
-Dibromo-4,5-dihydroxy-benzaldehyde	2,3-diBr-4,5-diOHPhCHO	Rhodomela larix, Odonthalia corymbifera	1
-Dibromo-4,5-dihydroxy-benzylmethyl ether	2,3-diBr-4,5-diOHPhCH$_2$OCH$_3$		1
5-Dibromo-1-hydroxy-4-keto-2,5-cyclohexadiene-1-acetamide		Verongia fistularis, V. cauliformis (sponge)	2
eroplysinin		Ianthella ardis (Caribbean sponge)	3
cifenol		Laurencia pacifica (red alga)	4
pirolaurenone		Laurencia glandulifera	5
aurinterol		Laurencia intermidia	6

Name	Structure	Organism	Referen
Isolaurinterol		*Laurencia intermidia*	7
Laurencin		*Laurencia glandulifera*	8
Bromo analogs of chloramphenicol	$4-NO_2PhCHOHCH(NCOCHHal_2)COOH$ $Hal_2 = Br_2$ or $BrCl$	*Streptomyces* sp.	9
Bromo analog of griseofulvin		*Penicillium* sp.	10
Bromotetracycline		*Streptomyces aureofaciens*	11
Bromo analog of caldariomycin		*Caldariomyces fumago*	12
	$R_1 = R_2 = R_3 = R_5 = R_6 = Br; R_4 = OH$	*Pseudomonas bromoutilis*	13
Bromo analogs of pyrrolnitrin	(1) $R_2 = R_5 = Br; R_4 = NO_2; R_1 = R_3 = R_6 = H$ (2) $R_1 = R_2 = R_5 = Br; R_4 = NO_2; R_3 = R_6 = H$ (3) $R_1 = R_2 = R_6 = Br; R_4 = NO_2; R_3 = R_5 = H$	*Pseudomonas pyrrolnitrica* and several other pyrrolnitrin-producing strains	14

REFERENCES

1. Craigie, J. S., and Gruenig, D. E., *Science, 157,* 1058 (1967).
2. Sharma, G. P., and Burkholder, P. R., *J. Antibiot. (Tokyo), Ser. A, 20,* 200 (1967); *Tetrahedron Lett.,* p. 4147 (1967).
3. Fulmor, W., Van Lear, G. E., Morton, G. O., and Mills, R. D., *Tetrahedron Lett.,* p. 4551 (1970).
4. Sims, J. J., Fenical, W., Wing, R. M., and Radlick, P., *J. Am. Chem. Soc., 93,* 3774 (1971).
5. Suzuki, M., Kurosawa, E., and Irie, T., *Tetrahedron Lett.,* p. 4995 (1970).
6. Irie, T., Suzuki, M., and Kurosawa, E., *Tetrahedron Lett.,* p. 1837 (1966).
7. Irie, T., Suzuki, M., Kurosawa, E., and Masamune, T., *Tetrahedron, 26,* 3271 (1970).
8. Irie, T., Suzuki, M., and Masamune, T., *Tetrahedron Lett.,* p. 1091 (1965).
9. Smith, C. G., *J. Bacteriol., 75,* 577 (1958).
10. MacMillan, J., *J. Chem. Soc.,* p. 2585 (1954).
11. Doerschuck, A. P., et al., *J. Am. Chem. Soc., 81,* 3069 (1959).
12. Patterson, E. L., Andres, W. W., and Mitscher, L. A., *Appl. Microbiol., 15,* 528 (1967).
13. Burkholder, P. R., Pfister, R. M., and Leitz, F. H., *Appl. Microbiol., 14,* 649 (1966).
14. Ajisaka, M., Kariyone, K., Jamm, K., Yazawa, H., and Arima, K., Progress in Antimicrobial and Anticancer Chemotherapy, in *Proceedings of the 6th International Congress of Chemotherapy, Tokyo, 1969,* Vol. 1, p. 77.

NITRO COMPOUNDS

DR. W. B. TURNER

FUNGAL COMPOUNDS

Structure	Compound	Organisms	Reference
$O_2N \cdot CH_2 \cdot CH_2 \cdot CO_2H$	β-Nitropropionic acid	*Aspergillus*	
		avenaceus	1
		flavus	2, 3
		oryzae	4
		wentii	5
		Penicillium	
		atrovenetum	6
	1-Amino-2-nitrocyclopentane-carboxylic acid	*Aspergillus wentii*	5
	3-Nitro-4-hydroxyphenylacetic acid	*Rhizoctonia solani*	7
	O-Glucosyl-2-nitrophenol	*Rhizoctonia solani*	8

BACTERIAL COMPOUNDS

Structure	Compound	Organisms	Reference
	4-Hydroxy-3-nitrophenylalanine (a component of the peptide antibiotics ilamycin and rufomycin)		
	Chloramphenicol (D(−)-*threo*-2-dichloroacetamide-1-*p*-nitrophenyl-1,3-propanediol, Chloromycetin, Levomycetin)	*Streptomyces venezuelae*	9, 10
	Luteoreticulin	*Streptomyces luteoreticuli*	11

Structure	Compound	Organisms	Reference

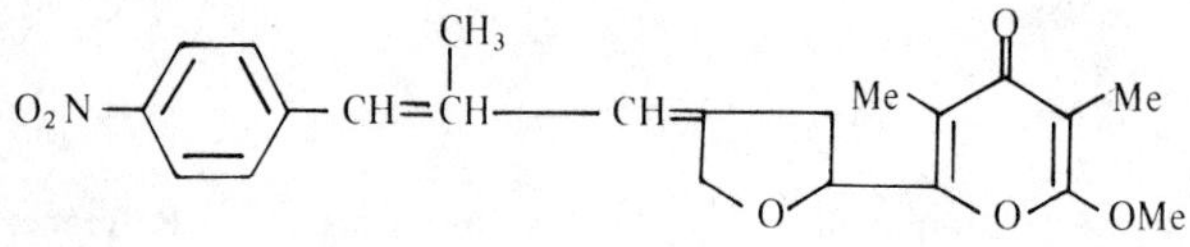

Aureothin

Aureothin — *Streptomyces thioluteus* — 12, 13

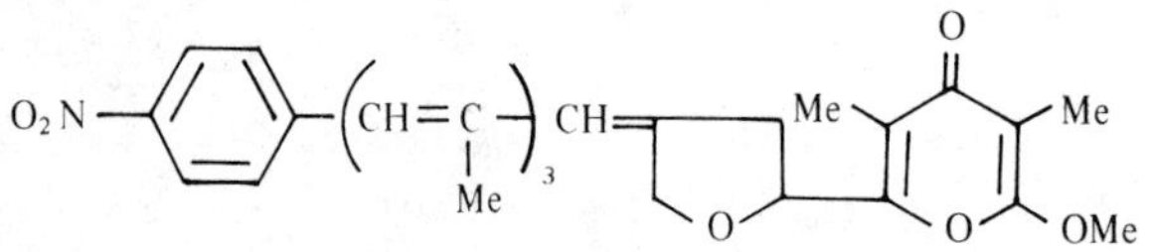

Neoaureothin

Neoaureothin — *Streptoverticillium orinoci* — 14

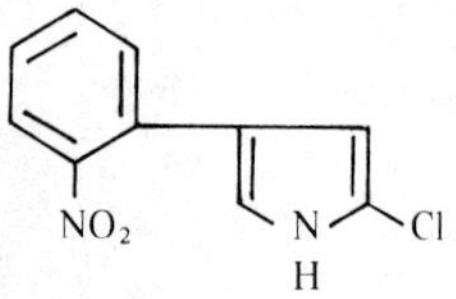

3-Chloro-4-(*o*-nitrophenyl)-pyrrole — *Pseudomonas pyrrolnitrica* — 15

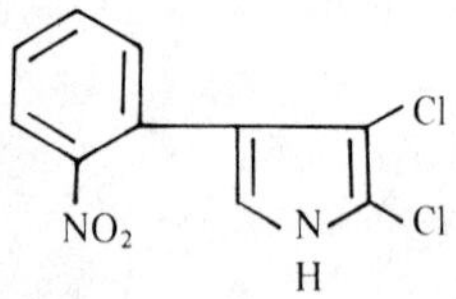

Isopyrrolnitrin [2,3-dichloro-4-(*o*-nitrophenyl)pyrrole] — *Pseudomonas pyrrolnitrica* — 16

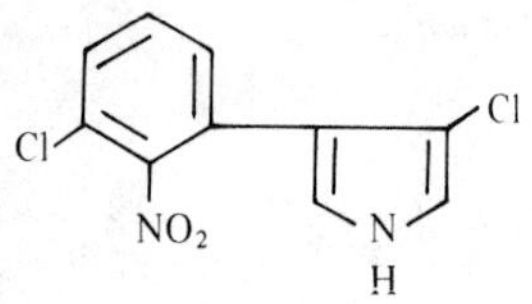

Pyrrolnitrin — *Pseudomonas aureofaciens* 17 *pyrrocinia* 18, 19

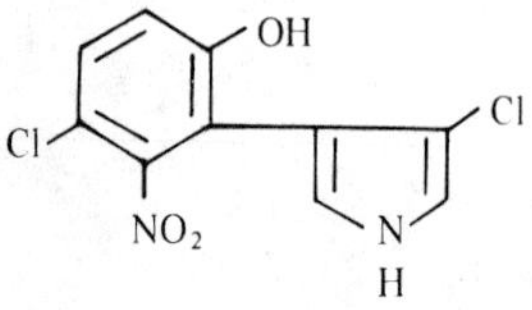

Oxypyrrolnitrin — *Pseudomonas pyrrolnitrica* — 20

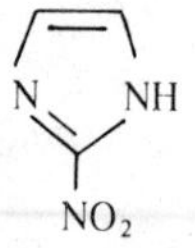

Azomycin (2-nitroimidazole) — *Nocardia mesenterica* 21, 22 *Streptomyces eurocidicus* 21, 22

$O_2N \cdot NH \cdot CH_2CO_2H$

Nitraminoacetic acid — *Streptomyces noursei* 805-MC$_3$ — 23

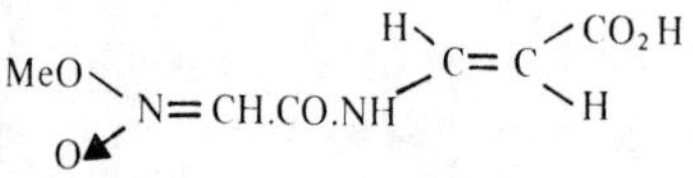

Enteromycin (seligocidin) — *Streptomyces achromogenes* 24 *albireticuli* 25, 26

Structure	Compound	Organisms	Reference
MeO–O N=CH.CO.NH, H\C=C/CONH$_2$, /C=C\ H	Enteromycin carboxamide	*Streptomyces* sp.	27, 28
MeO–O N=CH, H\C=C/CO$_2$H, /C=C\ H	Antibiotic U-22956	*Streptomyces fervens*, var. *melrosporus*	24

REFERENCES

1. Brookes, D., Tidd, B. K., and Turner, W. B.; *J. Chem. Soc.,* p. 5385 (1963).
2. Bush, M. T., Goth, A., and Dickison, H. L., *J. Pharmacol. Exp. Ther., 84,* 262 (1945).
3. Bush, M. T., Touster, O., and Brockman, J. E., *J. Biol. Chem., 188,* 685 (1951).
4. Nakamura, S., and Shimoda, C., *J. Agr. Chem. Soc. Jap., 28,* 909 (1954).
5. Burrows, B. F., and Turner, W. B., *J. Chem. Soc. Sect. C,* p. 255 (1966).
6. Raistrick, H., and Stössl, A., *Biochem. J., 68,* 647 (1958).
7. Aoki, H., Sassa, T., and Tamura, T., *Nature, 200,* 575 (1963).
8. Sherwood, R. T., U.S. Patent 3,179,653. *Chem. Abstr., 63,* 1170 (1965).
9. Ehrlich, J., Bartz, Q. R., Smith, R. M., Joslyn, D. A., and Burkholder, P. R., *Science, 106,* 416 (1947).
10. Rebstock, M. C., Crookes, H. M., Jr., Controulis, J., and Bartz, Q. R., *J. Amer. Chem. Soc., 71,* 2458 (1949).
11. Koyama, Y., Fukakusa, Y., Kyomura, N., Yamagishi, S., and Arai, T., *Tetrahedron Lett.,* p. 355 (1969).
12. Maeda, K., *J. Antibiot. (Tokyo), 6,* 137 (1953).
13. Hirata, Y., Nakata, H., Yamada, K., Okuhara, K., and Naito, T., *Tetrahedron, 14,* 252 (1961).
14. Cardani, C., Ghiringhelli, D., Selva, A., Arcamone, F., Camerino, B., and Cassinelli, G., *Chim. Ind. Milan, 52,* 793 (1970).
15. Hashimoto, M., and Hattori, K., *Chem. Pharm. Bull. (Tokyo), 16,* 1144 (1968).
16. Hashimoto, M., and Hattori, K., *Bull. Chem. Soc. Jap., 39,* 310 (1966).
17. Lively, D. H., Gorman, M., Haney, M. E., and Mabe, J. A., *Antimicrob. Agents Chemother.,* p. 462 (1966).
18. Arima, K., Imanaka, H., Kousaka, M., Fukuta, A., and Tamura, G., *Agr. Biol. Chem., 28,* 575 (1964); *J. Antibiot. (Tokyo), 18,* 201 (1965).
19. Imanaka, H., Kousaka, M., Tamura, G., and Arima, K., *J. Antibiot. (Tokyo), 18,* 207 (1965).
20. Hashimoto, M., and Hattori, K., *Chem. Pharm. Bull. (Tokyo), 14,* 1314 (1966).
21. Maeda, K., Osato, T., and Umezawa, H., *J. Antibiot. (Tokyo), 6,* 182 (1953).
22. Nakamura, S., *Pharm. Bull. (Tokyo), 3,* 379 (1955).
23. Miyazaki, Y., Kono, Y., Shimazu, A., Takeuchi, S., and Yonehara, H., *J. Antibiot. (Tokyo), 21,* 279 (1968).
24. Wiley, P. F., Herr, R. H., MacKellar, F. A., and Argedoulis, A. D., *J. Org. Chem., 30,* 2330 (1965).
25. Nakamura, S., Maeda, Y., and Umezawa, H., *J. Antibiot. (Tokyo), 7,* 57 (1954).
26. Mizunu, K., *Bull. Chem. Soc. Jap., 34,* 1419, 1425, 1631, 1633 (1961).
27. DeVoe, S. E., McCrae, W., and Mitscher, L. A., *Antimicrob. Agents Chemother.,* p. 105 (1964).
28. Mitscher, L. A., McCrae, W., and DeVoe, S. E., *Tetrahedron, 21,* 267 (1965).

SIDEROCHROMES [IRON(III)-TRIHYDROXAMATES]

DR. H. DIEKMANN

Desferri-siderochromes are found in bacteria, streptomycetes and fungi in higher concentrations only when the microoorganisms are grown on iron-deficient media. They are hydrophilic and can be extracted from the filtrate by benzyl alcohol or a mixture of chloroform and phenol (1:1, vol:w). Some have been crystallized. The iron–hydroxamate complex is characterized by a red-brown or yellow-brown color; the desferri compounds are white after recrystallization.

Trivalent iron, which is rather selectively bound, may be removed from the complex by treatment with 8-hydroxyquinoline or by extraction with ether after addition of $6N$ hydrochloric acid. Peptide bonds are split by heating to $100°C$ for six hours with hydriodic acid in a sealed tube. Under these conditions the –NHOH is reduced to a NH_2 group. The hydroxamate group of the desferri compounds may be split by periodic acid, in some cases by $1N$ hydrochloric acid and heating to $80°C$. The hydroxamate bonding can be reduced to a peptide bond by zinc and glacial acetic acid or by hydrogen in the presence of raney nickel.

All siderochromes have a broad peak in the range from 420–450 nm and additional characteristic bands according to special substituents. The IR spectra are not very characteristic. The NMR spectra of the desferri compounds have helped a great deal in the elucidation of the structures.

The siderochromes may be classified according to their biological activities. Sideramines are growth factors and sideromycins are antibiotics, but in a few instances a sideramine may be inhibitory to a micro organism and a sideromycin may accelerate the growth of a microorganism.

Much valuable information on siderochromes and related compounds not included in this chapter (e.g., rhodotorulic acid, dimerum acid, the fusarinines, the mycobactins, terregens-factor, aerobactin, aspergillic acid, etc.) is given in the articles by Keller-Schierlein, Prelog and Zähner[1] and by Neilands,[2] and on the sideromycins, especially, in some chapters of the books on antibiotics by Gottlieb and Shaw[3] and by Korzybski et al.[4]

STRUCTURES

Structure 1: ferrioxamines A_1 ,B,D_1 ,G; ferrimycin A_1

Structure 2: ferrioxamines D_2 , E

Structure 3: coprogen; coprogen B

Structure 4: fusigen

Structure 5: ferrichrome; ferrichrysin; ferricrocin; ferrirubin; ferrirhodin; ferrichrome A; albomycin-grisein group

SYMBOLS AND ABBREVIATIONS

Solvents

(w) = water
(e) = ethanol

Dissociation Constants and Partition Coefficients

K   partition coefficient in the system butanol:benzyl alcohol:0.001N hydrochloric acid:saturated sodium chloride in water = 10:5:15:3
K′   partition coefficient in the system 18:18:30:10
pK dissociation constant in water
pK* apparent dissociation constant in 80% methyl Cellosolve®

Culture Collection

ATCC   American Type Culture Collection, Rockville, Maryland, U.S.A.
CBS   Centraalbureau for Schimmelcultures, Baarn, The Netherlands
ETH   Eidgenössische Technische Hochschule, Zürich, Switzerland
NRRL   Northern Utilization Research and Development Division, U.S. Department of Agriculture, Peoria, Illinois, U.S.A.
Tü   Institut für Biologie, Lehrbereich Mikrobiologie, Universität Tübingen, Germany

Microorganisms

a   *Streptomyces pilosus* Ettlinger *et al.*, ETH 21748

b   *Micromonospora fusca* Jensen, ETH 27556

c   *Nocardia asteroides* (Eppinger) Blanchard, ETH 27472

d   *Streptomyces aureofaciens* Duggar, ETH 22765

e   *Streptomyces galilaeus* Ettlinger *et al.*, ETH 18822

f   *Streptomyces glaucescens* (Preobrazhenskaya) Pridham *et al.*, ETH 22794

g   *Streptomyces griseus* (Krainsky) Waksman *et* Henrici, ETH 7419

h   *Streptomyces griseus* (Krainsky) Waksman *et* Henrici, ETH 10112

i   *Streptomyces griseoflavus* (Krainsky) Waksman *et* Henrici, ETH 9578

j   *Streptomyces lavendulae* (Waksman *et* Curtis) Waksman *et* Henrici, ETH 21510

k   *Streptomyces olivaceus* (Waksman) Waksman *et* Henrici, ETH 6445

l   *Streptomyces olivaceus* (Waksman) Waksman *et* Henrici, ETH 7346

m   *Streptomyces olivaceus* (Waksman) Waksman *et* Henrici, ETH 7437

n   *Streptomyces polychromogenus* Hagemann *et al.*, ETH 21837

o   *Arthrobacter simplex* (Jensen) Lochhead, Weihenstephan 6946

p   *Nocardia* sp.

q   *Nocardia brasiliensis* (Lindenberg) Pinoy, ETH 27413

r   *Streptomyces albus* (Rossi-Doria) Waksman *et* Henrici, ETH 24457

s   *Streptomyces antibioticus* (Waksman *et* Woodruff) Waksman *et* Henrici, ETH 27083

t   *Streptomyces galileus* Ettlinger *et al.*, ETH 27028

u   *Streptomyces fulvissimus* (Jensen) Waksman *et* Henrici, ETH 27362

v   *Streptomyces griseoflavus* (Krainsky) Waksman *et* Henrici, ETH 27081

w   *Streptomyces viridochromogenes* (Krainsky) Waksman *et* Henrici, ETH 27454

x   *Streptosporangium roseum* Couch, ETH 27525

y   *Chainia* sp., ETH 28904

z   *Chromobacterium violaceum* (Schroeter) Bergonzini

aa	*Streptomyces griseoflavus* (Krainsky) Waksman *et* Henrici, ETH 15311
ab	*Streptomyces lavendulae* (Waksman *et* Curtis) Waksman *et* Henrici, ETH 14677
ac	*Streptomyces aureofaciens* Duggar, ETH 22083
ad	*Streptomyces aureofaciens* Duggar, ETH 22931
ae	*Streptomyces olivochromogenes* (Waksman) Waksman *et* Henrici
af	*Streptomyces albaduncus* Tsukiura *et al.*, strain 13246
ag	*Penicillium* sp.
ah	*Penicillium urticae* Bainier, ETH 4277
ai	*Penicillium notatum* Westling, ETH 1613
aj	*Penicillium camemberti* Thom, ETH M98
ak	*Penicillium chrysogenum* Thom, ETH 4360
al	*Penicillium citrinum* Thom, ETH M3614
am	*Neurospora crassa* Shear *et* Dodge, ETH 3627
an	*Neurospora crassa* Shear *et* Dodge, strain Em 5297a
ao	*Ascodesmis sphaerospora* Obrist, CBS 125.61
ap	*Leptosphaerulina australis* McAlpine, ETH 4619
aq	*Myrothecium roridum* Tode *ex* Fries, CBS 372.50
ar	*Myrothecium striatisporum* Preston, CBS 277.48
as	*Myrothecium verrucaria* Ditmar *ex* Fries, CBS 328.52
at	*Fusarium* sp., Tü 511
au	*Nectria cinnabarina* (Tode) Fries, Tü 568
av	*Fusarium dimerum* Penzig, var. *pusillum* Wr., CBS 254.50
aw	*Gibberella saubinetii* (Mont.) Sacc., ETH M427
ax	*Fusarium* sp., Tü 554
ay	*Fusarium cubense* Smith, ETH M862

az	*Cylindrocarpon radicicola* Wollenweber, ETH M2617
ba	*Aspergillus fumigatus* Fresenius, Tü 149
bb	*Gibberella fujikuroi* (Saw.) Wollenweber, ETH M82
bc	*Trichothecium roseum* Link, CBS 142.65
bd	*Fusarium roseum* Link, ATCC 12822
be	*Ustilago sphaerogena* Burrill, ATCC 12421
bf	*Aspergillus niger* v. Tieghem, ETH M3573
bg	*Penicillium resticulosum* Birkinshaw, ETH M3583
bh	*Ustilago maydis* (DeCandolle) Corda
bi	*Aspergillus melleus* Yukawa, ETH M2853
bj	*Aspergillus terreus* Thom, ETH 4785
bk	*Aspergillus oryzae* (Ahlberg) Cohn
bl	*Aspergillus versicolor* (Vuill.) Tiraboshi, ETH M3636
bm	*Aspergillus fumigatus* Fresenius, ETH M57
bn	*Aspergillus nidulans* (Eidam.) Wint., ETH M2734
bo	*Aspergillus humicola* Chaudhuri *et* Sachav, ETH 3635
bp	*Microsporum gypseum* (Bodin) Guiart *et* Grigorakis
bq	*Penicillium variabile* Sopp, ETH M2733
br	*Spicaria* sp., ETH M4622
bs	*Paecilomyces varioti* Bainier, ETH M4646
bt	*Onygena piligena* Fr., CBS 298.49
bu	*Actinomyces subtropicus* Kudrina *et* Kochetkova
bv	*Streptomyces griseus* (Krainsky) Waksman *et* Henrici, ATCC 3478
bw	*Streptomyces* sp., strain LA 1787
bx	*Streptomyces* sp., strain LA 5352
by	*Streptomyces griseus* (Krainsky) Waksman *et* Henrici, ETH 10073
bz	*Streptomyces griseus,* var. X-2455, NRRL 3456

LIST OF SIDEROCHROMES

The siderochromes in the following table are arranged according to their chemical structures. In the case of the trihydroxamates, the name of the compound denotes the iron (III) complex; the name of the iron-free compound is formed by the prefix desferri.

Compound Formula (M.W.)	Structure	Organisms	Significant Characteristics	References
Ferrioxamine A$_1$ $C_{24}H_{43}O_8N_6Fe$ (599.5)	Structure 1; a = 4, R' = CH$_3$, R'' = H	a	Red-brown, amorphous K = 0.11; pK* = 9.89 $\lambda_{max}^{(w)}$: 430–440 nm; log ϵ: 3.35	5, 6
Ferrioxamine B $C_{25}H_{45}O_8N_6Fe$ (613.5)	Structure 1; a = 5, R' = CH$_3$, R'' = H	a–o	As hydrochloride: red-brown, amorphous K = 0.23; pK* = 9.74 $\lambda_{max}^{(w)}$: 430–440 nm; log ϵ: 3.40	7–11
Ferrioxamine D$_1$ $C_{27}H_{47}O_9N_6Fe$ (655.6)	Structure 1; a = 5, R' = CH$_3$, R'' = COCH$_3$	a	Red-brown needles from methanol–ether Melting point: 195–200°C K = 1.80 $\lambda_{max}^{(w)}$: 430–440 nm; log ϵ: 3.46	10, 12
Ferrioxamine D$_2$ $C_{26}H_{45}O_9N_6Fe$ (641.6)	Structure 2; b = 4	a	Crystals from methanol–ether Melting point: 220–223°C K = 0.86	5, 6
Ferrioxamine E (nocardamine–Fe^{III} complex) $C_{27}H_{45}O_9N_6Fe$ (653.6)	Structure 2; b = 5	a, p–z	Needles from methanol K = 1.59 $\lambda_{max}^{(w)}$: 430–440 nm; log ϵ: 3.44	7, 8, 13–15
Ferrioxamine G $C_{27}H_{47}O_{10}N_6Fe$ (671.6)	Structure 1; a = 5, R' = CH$_2$CH$_2$COOH, R'' = H	a	Red-brown, amorphous K = 0.40; pK* = 5.79, 10.53 $\lambda_{max}^{(w)}$: 440 nm; log ϵ: 3.32	15, 16
Ferrimycin A$_1$ $C_{41}H_{65}O_{14}N_{10}Fe$ (977.9)	Structure 1; a = 5, R' = CH$_3$, R'' =	e, i, j, aa, ab	As dihydrochloride: red-brown, amorphous, unstable above pH 5 K = 0.372; pK* = 4.11, 7.92, 11.4 $\lambda_{max}^{(w)}$: 229, 319, 430 nm; log ϵ_{430}: 3.41	17–19
Ferrimycin A$_2$		i	K = 0.175; pK* = 4.04, 7.91, 11.5 λ_{max}: 231, 319, 435 nm	17, 18

Compound Formula (M.W.)	Structure	Organisms	Significant Characteristics	References
Antibiotics of "group 2"[a]				
Antibioticum A 22765		d, ac, ad		17, 21
Succinimycin	Found: C = 45.03–46.22% H = 6.63–7.0% N = 8.11–8.90% Fe = 4.3–4.8%	ae	As acetate: reddish-orange, amorphous, stable at pH 6–7 λ_{max}: 430 nm	22
Danomycin	Found: C = 48.82% H = 7.05% N = 7.81% Fe = 3.1%	af	Reddish-orange Melting point: 135–138°C λ_{max}: 270, 325, 430 nm	23
Coprogen $C_{35}H_{53}O_{13}N_6Fe$ (821.7)	Structure 3; R = $COCH_3$	ag–ap	Red-brown, amorphous K' = 2.0 $\lambda_{max}^{(e)}$: 217, 250, 440 nm; log ϵ_{440}: 3.47	24, 25
Coprogen B $C_{33}H_{51}O_{12}N_6Fe$ (779.7)	Structure 3; R = H	am, aq–av	Red-brown, amorphous K' = 0.36; pK = 7.2 $\lambda_{max}^{(w)}$: 220, 250, 440 nm; log ϵ_{440}: 3.44	26
Fusigen (fusarinine C) $C_{33}H_{51}O_{12}N_6Fe$ (779.7)	Structure 4	ai, ak, au, aw–bd	Red-brown, amorphous pK = 7.1 $\lambda_{max}^{(w)}$: 250, 439 nm; log ϵ_{430}: 3.50	27, 28
Fusigen B (fusarinine B) $C_{33}H_{53}O_{13}N_6Fe$ (797.7)	One ester group of fusigen is hydrolyzed	ai, ak, au, aw–bd	Red-brown, amorphous r_e = 0.68 relative to fusigen in electrophoresis at pH 5.0	27, 28
Ferrichrome $C_{27}H_{42}O_{12}N_9Fe$ (740.6)	Structure 5; R = CH_3, R' = R'' = R''' = H	an, be–bh	Red-brown needles from methanol Decomposes at 240–242°C K' = 1.19 $\lambda_{max}^{(e)}$: 214, 425 nm; log ϵ_{425}: 3.51	29–33
Ferrichrysin $C_{29}H_{46}O_{14}N_9Fe$ (800.6)	Structure 5; R = CH_3, R' = R'' = CH_2OH, R''' = H	bi–bk	Orange-brown needles from methanol Decomposes at 260°C K' = 0.63 $\lambda_{max}^{(e)}$: 430 nm; log ϵ: 3.48	32, 34
Ferricrocin $C_{28}H_{44}O_{13}N_9Fe$ (770.6)	Structure 5; R = CH_3, R', R'', R''' = 2 H, 1 CH_2OH	bl–bp	Orange-brown crystals from ethanol Decomposes at 250°C K' = 0.55 $\lambda_{max}^{(e)}$: 430 nm; log ϵ: 3.42	32, 34

Compound Formula (M.W.)	Structure	Organisms	Significant Characteristics	References
Ferrirubin $C_{41}H_{64}O_{17}N_9Fe$ (1010.9)	Structure 5; $R' = R'' =$ CH_2OH, $R''' = H$, $R =$ 	bi, bq–bs	Dark-red prisms from metha- nol $K' = 5.1$ $\lambda_{max}^{(e)}$: 216, 254, 450 nm; log ϵ_{450}: 3.53	32, 35
Ferrirhodin $C_{41}H_{64}O_{17}N_9Fe$ (1010.9)	Structure 5; $R' = R'' =$ CH_2OH, $R''' = H$, $R =$ 	ax, bl, bn bt	Crystals from methanol– ether Decomposes at 270°C $K' = 7.3$ $\lambda_{max}^{(e)}$: 217, 262, 445 nm; log ϵ_{445}: 3.58	32, 35
Ferrichrome A $C_{41}H_{58}O_{20}N_9Fe$ (1052.8)	Structure 5; $R' = R'' =$ CH_2OH, $R''' = H$, $R =$ 	be, bh	Crystals from methanol– water $K' = 6.4$ $\lambda_{max}^{(e)}$: 217, 254, 445 nm; log ϵ_{445}: 3.62	35–37
δ_2-Albomycin $C_{39}H_{62}O_{20}N_{12}SFe$ (?)	Structure 5: $R = CH_3$, $R'' =$ $R''' = CH_2OH$, $R' =$ 	bu	Main component of albomy- cin mixture, red, amor- phous λ_{max}: 425 nm	38, 39
δ_1-Albomycin	Structure 5; $R = CH_3$, $R'' =$ $R''' = CH_2OH$, $R' =$ 	bu	Decomposition product of δ_2-albomycin	38, 39

Compound Formula (M.W.)	Structure	Organisms	Significant Characteristics	References
ϵ-Albomycin	Structure 5; R = CH$_3$, R'' = R''' = CH$_2$OH, R' = NH	bu	Decomposition product of δ_2-albomycin	38, 39

Other antibiotics of the albomycin-grisein group[b]

Grisein		bv	May be a mixture of com-ponents, as albomycin	40–42
Antibioticum A 1787		bw		43
Antibiotic LA-5352		bx		44
Antibioticum 10073		by		17
Antibiotic Ro 5-2667	Identical with δ_2-albomycin	bz		45, 46
Antibiotic Ro 7-7730	Identical with δ_1-albomycin	bz		45, 46
Antibiotic Ro 7-7731	Identical with ϵ-albomycin	bz		45, 46

[a] For chromatographic separation from the ferrimycins and the albomycin-grisein group see References 17 and 20.
[b] For chromatographic separation from other sideromycins see Reference 17.

REFERENCES

1. Keller-Schierlein, W., Prelog, V., and Zähner, H., *Fortschr. Chem. Org. Naturst., 22,* 279 (1964).
2. Neilands, J. B., *Struct. Bonding, 1,* 59 (1966).
3. Bhuyan, B. K., in *Antibiotics,* Vol. 1, p. 153, D. Gottlieb and P. D. Shaw, Eds; Nüesch, J., and Knüsel, F., *ibid.,* p. 499. Springer Verlag, Berlin, Heidelberg, New York (1967).
4. Korzybski, T., Kowszyk-Gindifer, Z., and Kurylowicz, W., in *Antibiotics,* p. 361. Pergamon Press, Oxford, England (1967).
5. Bickel, H., Bosshardt, R., Gäumann, E., Reusser, P., Vischer, E., Voser, W., Wettstein, A., and Zähner, H., *Helv. Chim. Acta, 43,* 2118 (1960).
6. Keller-Schierlein, W., Mertens, P., Prelog, V., and Walser, A., *Helv. Chim. Acta, 48,* 710 (1965).
7. Zähner, H., Bachmann, E., Hütter, R., and Nüesch, J., *Pathol. Microbiol., 25,* 708 (1962).
8. Müller, A., and Zähner, H., *Arch. Mikrobiol., 62,* 257 (1968).
9. Bickel, H., Hall, G. E., Keller-Schierlein, W., Prelog, V., Vischer, E., and Wettstein, A., *Helv. Chim. Acta, 43,* 2129 (1960).
10. Prelog, V., and Walser, A., *Helv. Chim. Acta, 45,* 631 (1962).
11. Bickel, H., Keberle, H., and Vischer, E., *Helv. Chim. Acta, 46,* 1385 (1963).

12. Keller-Schierlein, W., and Prelog, V., *Helv. Chim. Acta, 44,* 709 (1961).

13. Stoll, A., Renz, J., and Brack, A., *Schweiz. Z. Pathol. Bakteriol., 14,* 225 (1951).

14. Keller-Schierlein, W., and Prelog, V., *Helv. Chim. Acta, 44,* 1981 (1961).

15. Keller-Schierlein, W., and Prelog, V., *Helv. Chim. Acta, 45,* 590 (1962).

16. Prelog, V., and Walser, A., *Helv. Chim. Acta, 45,* 1732 (1962).

17. Bickel, H., Gäumann, E., Keller-Schierlein, W., Prelog, V., Vischer, E., Wettstein, A., and Zähner, H., *Experientia, 16,* 129 (1960).

18. Bickel, H., Gäumann, E., Nussberger, G., Reusser, P., Vischer, E., Voser, W., Wettstein, A., and Zähner, H., *Helv. Chim. Acta, 43,* 2105 (1960).

19. Bickel, H., Mertens, P., Prelog, V., Seibl, J., and Walser, A., *Tetrahedron, Suppl. 8,* 171 (1966).

20. Benz, F., and Bickel, H., cited in *Antibiotics,* Vol. 1, p. 504, D. Gottlieb and P. D. Shaw, Eds. Springer Verlag, Berlin, Heidelberg, New York (1967).

21. Zähner, H., Hütter, R., and Bachmann, E., *Arch. Mikrobiol., 36,* 325 (1960).

22. Haskell, T. H., Bunge, R. H., French, J. C., and Bartz, O. R., *J. Antibiot. (Tokyo), 16,* 67 (1963).

23. Tsukiura, H., Okanishi, M., Ohmori, T., Koshiyama, H., Miyaki, T., Kitazima, H., and Kawaguchi, H., *J. Antibiot. (Tokyo), 17,* 39 (1964).

24. Pidacks, C., Whitehill, A. R., Pruess, L. M., Hesseltine, C. W., Hutchings, B. L., Bohonos, N., and Williams, J. H., *J. Amer. Chem. Soc., 75,* 6064 (1953).

25. Keller-Schierlein, W., and Diekmann, H., *Helv. Chim. Acta, 53,* 2035 (1970).

26. Diekmann, H., *Arch. Mikrobiol., 73,* 65 (1970).

27. Diekmann, H., and Zähner, H., *Eur. J. Biochem., 3,* 213 (1967).

28. Sayer, J. M., and Emery, T. F., *Biochemistry, 7,* 184 (1968).

29. Neilands, J. B., *J. Amer. Chem. Soc., 74,* 4846 (1952).

30. Emery, T., and Neilands, J. B., *J. Amer. Chem. Soc., 83,* 1626 (1961).

31. Rogers, S., Warren, R. A. J., and Neilands, J. B., *Nature, 200,* 167 (1963).

32. Zähner, H., Keller-Schierlein, W., Hütter, R., Hess-Leisinger, K., and Deér, A., *Arch. Mikrobiol., 45,* 119 (1963).

33. Keller-Schierlein, W., and Maurer, B., *Helv. Chim. Acta, 52,* 603 (1969).

34. Keller-Schierlein, W., and Deér, A., *Helv. Chim. Acta, 46,* 1907 (1963).

35. Keller-Schierlein, W., *Helv. Chim. Acta, 46,* 1920 (1963).

36. Garibaldi, J. A., and Neilands, J. B., *J. Amer. Chem. Soc., 77,* 2429 (1955).

37. Zalkin, A., Forrester, J. D., and Templeton, D. H., *J. Amer. Chem. Soc., 88,* 1810 (1966).

38. Turková, J., Mikeš, O., and Šorm, F., *Experientia, 19,* 633 (1963).

39. Turková, J., Mikeš, O., and Šorm, F., in *Antibiotics: Advances in Research, Production and Clinical Use,* p. 424, M. Herold, and Z. Gabriel, Eds. Butterworth, London, England (1966).

40. Reynolds, D. M., Schatz, A., and Waksman, S. A., *Proc. Soc. Exp. Biol. Med., 64,* 50 (1947).

41. Kuehl, F. A., Bishop, M. N., Chaiet, L., and Folkers, K., *J. Amer. Chem. Soc., 73,* 1770 (1951).

42. Stapley, E. O., and Ormond, E. E., *Science, 125,* 587 (1957).

43. Thrum, H., *Naturwissenschaften, 44,* 561 (1957).

44. Sensi, P., and Timbal, M. T., *Antibiot. Chemother., 9,* 160 (1959).

45. Maehr, H., and Berger, J., *Biotechnol. Bioeng., 11,* 1111 (1969).

46. Maehr, H., and Pitcher, R. G., *J. Antibiot. (Tokyo), 24,* 830 (1971).

COMPOUNDS PRODUCED EXCLUSIVELY BY LICHENS

DR. H. A. LECHEVALIER

Lichens form a large number of compounds; many of these are not known to be produced by other organisms. Most lichen products are probably synthesized by the fungus partner from carbohydrates supplied by the alga.

The following table lists compounds believed to be produced exclusively by lichens. The information was taken from the compilation of Chicita F. Culberson, *Chemical and Botanical Guide to Lichen Products* (1969) and is presented here by permission of the University of North Carolina Press, Chapel Hill, NC. Readers should consult this text for further information.

Compound Formula	Structure and Chemical Properties	Producing Organisms
Siphulitol (1-Deoxy-D-*glycero*-D-*talo*-hepitol) $C_7H_{16}O_6$	 Melting point: 122—123°C (methanol) $[a]_D^{20}$: −8° (water, c = 1.5)	*Siphula ceratites*
3-O-β-D-Glucopyranosyl-D-mannitol $C_{12}H_{24}O_{11}$	 Melting point (dihydrate): 97—100°C (ethanol—water) $[a]_D^{20}$: −6° (water, c = 2.0)	*Peltigera aphthosa horizontalis*
Peltigeroside (3-O-β-D-Galactofuranosyl-D-mannitol) $C_{12}H_{24}O_{11}$	 Melting point (monohydrate): 161—163°C (96% ethanol) $[a]_D^{20}$: (monohydrate) −61° (water, c = 2.0)	*Peltigera horizontalis*

Compound Formula	Structure and Chemical Properties	Producing Organisms
Umbilicin (2-O-β-D-Galactofuranosyl-D-arab-itol) $C_{11}H_{22}O_{10}$	Melting point: $138-139°C$ (ethanol) $[a]_D^{20}: -81°$ (water, c = 2.0)	Agyrophora rigida Cetraria islandica Haematomma ventosum Lasallia pustulata
Isolichenin $(C_6H_{12}O_6)_n$	Soluble in cold water; I_2 KI + blue, but less intense than starch	Cetraria islandica Peltigera canina Roccella montagnei Xanthoria parietina
Lichenin $(C_6H_{12}O_6)_n$	Molecular weight $\sim 20,000-40,000$; insoluble in cold water, soluble in hot water $[a]_D^{20}: -8.3°$ (2N sodium hydroxide)	Alectoria jubata Cetraria islandica Hubbsia lumbricoides Lecanora muralis Ramalina calicaris sinensis Usnea longissima Xanthoria parietina

Compound Formula	Structure and Chemical Properties	Producing Organisms
Pustulan $(C_6H_{12}O_6)_n$	$[a]_D$: $-46°$ (water, c = 2.0)	*Lasallia pustulata* *Umbilicaria hirsuta*
Picroroccellin $C_{20}H_{22}N_2O_4$	Substitution on nitrogens may be reversed Melting point: $190-220°C$, depending on heating rate (ethanol) $[a]_D$: $+12.5°$ (chloroform)	*Roccella fuciformis*
Acaranoic acid $C_{17}H_{30}O_4$	Melting point: $154-155°C$ (ether) $[a]_D^{25}$: $-30°$ (chloroform, c = 0.29)	*Acarospora chlorophana oxytona*
Acarenoic acid $C_{17}H_{28}O_4$	Melting point: $144-146.5°C$ (benzene) $[a]_D^{25}$: $-39°$ (chloroform, c = 0.25)	*Acarospora chlorophana oxytona*

Compound Formula	Structure and Chemical Properties	Producing Organisms
(−)-Caperatic acid $C_{21}H_{38}O_7$	CH₃OOC, CH₂COOH, OH, COOH, H, $CH_3(CH_2)_{12}$ Melting point: 132–133.5°C (methanol–water) $[\alpha]_D^{10}: -3.85°$ (chloroform)	*Cetraria stracheyi* *Parmelia caperata cryptochlorophaea* *Usnea orientalis*
(−)-Lichesterinic acid $C_{19}H_{32}O_4$	HOOC, CH₃, O, $CH_3(CH_2)_{12}$ Melting point (plates): 123–124°C (acetic acid or ethanol) $[\alpha]_D^{15}: -30°$ (chloroform)	*Cetraria ericetorum* *Nephromopsis stracheyi*
(−)-Nephromopsinic acid $C_{19}H_{34}O_4$	HOOC, CH₃, O, $CH_3(CH_2)_{12}$ Melting point (leaflets): 137°C (ethanol) $[\alpha]_D^{12}: -85.1°$ (chloroform)	*Cetraria stracheyi*
(+)-Nephrosteranic acid $C_{17}H_{30}O_4$	HOOC, CH₃, O, $CH_3(CH_2)_{10}$ Melting point (plates): 95°C (petroleum ether) $[\alpha]_D^{21}: +38.4°$ (chloroform)	*Cetraria endocrocea*
(+)-Nephrosterinic acid $C_{17}H_{28}O_4$	HOOC, CH₂, O, $CH_3(CH_2)_{10}$ Melting point (leaflets): 96°C (dilute acetic acid) $[\alpha]_D^{10}: +10.8°$ (chloroform)	*Cetraria endocrocea*

Compound Formula	Structure and Chemical Properties	Producing Organisms
(+)-Norrangiformic acid $C_{20}H_{36}O_6$	 Melting point (from hydrolysis of rangiformic acid): 114–122°C $[\alpha]_D^{18}: +12.9°$	*Cladonia* *mitis*
(−)-Protolichesterinic acid $C_{19}H_{32}O_4$	 Melting point: 106°C (acetic acid) $[\alpha]_D^{27}: -12.7°$ (chloroform) $[\alpha]_D^{22}: -15°$ (chloroform)	*Cetraria* *ericetorum* *stracheyi*
(+)-Protolichesterinic acid $C_{19}H_{32}O_4$	 Melting point: 107.5°C (benzene or acetic acid) $[\alpha]_D^{19.5}: +12.1°$ (chloroform)	*Cetraria* *cucullata* *ericetorum* *islandica* *Parmelia* *cirrhata* *fraudans* *Pycnothelia* *papillaria*
(−)-*allo*-Protolichesterinic acid $C_{19}H_{32}O_4$	 Melting point: 107°C (acetic acid) $[\alpha]_D^{18}: -102°$ (chloroform)	*Cetraria* *ericetorum* *islandica*

Compound Formula	Structure and Chemical Properties	Producing Organisms
(+)-Rangiformic acid $C_{21}H_{38}O_6$	Melting point (colorless needles): 104–105°C (acetic acid) $[a]_D^{20}$: +18° (ethanol, c = 1.92)	*Cladonia mitis rangiformis retipora* *Lecanora polytropa*
(+)-Roccellaric acid $C_{19}H_{34}O_4$	Melting point (colorless prisms): 110–111°C (methanol) $[a]_D^{20}$: +35° (chloroform, c = 1.73)	*Roccellaria mollis*
(+)-Roccellic acid $C_{17}H_{32}O_4$	Melting point: 129–130°C (ethyl acetate) $[a]_D^{26}$: +16.8°	*Dirina lutosa* *Lecanora cenisea rupicola* *Lepraria membranacea* *Roccella boergesenii fuciformis fucoides gayana hypomecha montagnei portentosa* *Roccellina condensata*

Compound Formula	Structure and Chemical Properties	Producing Organisms
9,10,12,13-Tetrahydroxyheneicosa-noic acid $C_{21}H_{42}O_6$	$$\overset{HO}{}\overset{OH}{}\overset{HO}{}\overset{OH}{}$$ $CH_3(CH_2)_7CHCHCH_2CHCH(CH_2)_7COOH$	*Haematomma ventosum*
Ventosic acid (9,10,12,13-Tetrahydroxydocosanoic acid) $C_{22}H_{44}O_6$	$$\overset{HO}{}\overset{OH}{}\overset{HO}{}\overset{OH}{}$$ $CH_3(CH_2)_8CHCHCH_2CHCH(CH_2)_7COOH$ White amorphous powder; nearly insoluble in sodium hydroxide solution Melting point: 183—185°C (dioxane, ethanol, glacial acetic acid)	*Cetraria nivalis* *Cladonia alpestris* *Haematomma ventosum* *Stereocaulon tomentosum* *Usnea pectinata*
Acetylportentol $C_{19}H_{28}O_6$	 Melting point: 215—216°C $[\alpha]_D$: −35°	*Roccella fuciformis*
Portentol $C_{17}H_{26}O_5$	 Melting point: 240—241°C, d (ethanol); 244—245°C (chloroform—ethanol) $[\alpha]_D$: +21°	*Dirina repanda* *Roccella fuciformis portentosa* *Roccellina condensata*

Compound Formula	Structure and Chemical Properties	Producing Organisms
Methyl β-orcinolcarboxylate $C_{10}H_{12}O_4$ $C_{12}H_{16}O_7$	Melting point (prismatic rods): 143–144°C (benzene)	*Parmelia tinctorum*
– (+)-Montagnetol $C_{10}H_{12}O_4$	Melting point: 135–136°C $[\alpha]_D$: +16.0° (water); +12.6° (acetone); racemized by boiling in water for 3 hours	*Roccella montagnei*
– Anziaic acid $C_{24}H_{30}O_7$	Melting point (needles): 124°C, d (ethanol–water)	*Anzia japonica*
– Confluentic acid $C_{28}H_{36}O_8$	Melting point: 157°C (methanol)	*Enterographa crassa* *Herpothallon sanguineum* *Lecidea confluens macrocarpa tumida*
– Diploschistesic acid $C_{16}H_{14}O_8$	Sodium hypochlorite + blue Melting point: 174°C, d (acetone–water–acetic acid)	*Diploschistes scruposus*

Compound Formula	Structure and Chemical Properties	Producing Organisms
Divaricatic acid $C_{21}H_{24}O_7$	Melting point (needles): 137–138°C (benzene)	*Anzia* *japonica* *Evernia* *divaricata* *esorediosa* *Haematomma* *ventosum* *Lecidea* *kochiana*
Erythrin $C_{20}H_{22}O_{11}$	Melting point: 156–157°C $[\alpha]_D: +8.0°$	*Chiodecton* *cretaceum* *Combea* *mollusca* *Dirina* *approximata* *catalinariae* *ceratoniae* *chilena* *limitata* *repanda* *Dirinastrum* *chilenum* *Ingaderia* *pulcherrima* *Roccella* *africana* *babingtonii* *boergesenii* *fimbriata* *flaccida* *fuciformis* *fucoides* *linearis* *montagnei* *phycopsis* *podocarpa* *portentosa* *teneriffensis* *tinctoria*

Compound Formula	Structure and Chemical Properties	Producing Organisms
Evernic acid $C_{17}H_{16}O_7$	Melting point (needles): 172–174° C, d(acetone)	Evernia prunastri Ramalina pollinaria Usnea misaminensis
Glomelliferic acid $C_{25}H_{30}O_8$	Melting point: 143–144° C (benzene)	Lecidea leucophaea Parmelia glomellifera isidiotyla pulla
Gyrophoric acid $C_{24}H_{20}O_{10}$	Melting point: 220–225° C, d (acetone); 212–214° C, d (acetone–methanol)	Acarospora fuscata montana Dactylina arctica ramulosa Dolichocarpus chilensis Lasallia papulosa pustulata Lecanora fuscoatra Lecidea granulosa griseoatra Ochrolechia tartarea Parmelia sancti-angelii Rinodina oreina Umbilicaria deusta esculenta mammulata

Compound Formula	Structure and Chemical Properties	Producing Organisms
Hiascic acid $C_{24}H_{20}O_{11}$	Melting point: 190.5°C (ethanol)	*Cetraria delisei*
Imbricaric acid $C_{23}H_{28}O_7$	Melting point: 125.5—126.0°C (benzene—petroleum ether)	*Cetrelia alaskana cetrarioides* *Haematomma polycarpum puniceum* *Parmelia locarnensis*
Lecanoric acid $C_{16}H_{14}O_7$	Melting point: 177—187.5°C, d, shrinking at 183°C, bubbling at 184.5°C (ethanol—water)	*Byssocaulon niveum* *Chiodecton sphaerale* *Dermatiscum thunbergii* *Dirina capensis veruculosa* *Hubbsia lumbricoides* *Parmelia dilatata glabra gossweileri meizospora subargentifera subrudecta tinctorum* *Pseudevernia intensa*

Compound Formula	Structure and Chemical Properties	Producing Organisms
Lecanoric acid (cont.)		*Roccella* *africana* *arboricola* *canariensis* *gayana* *immutata* *linearis* *montagnei* *podocarpa* *portentosa* *Roccellina* *condensata* *vicentina* *Stereocaulon* *corticulatum* *Stricta* *fuliginosa*
Methyl 3,5-dichlorolecanorate (Tumidulin) $C_{17}H_{14}Cl_2O_7$	 Melting point: 177–177.5°C (benzene); 174–175°C (methanol–water)	*Ramalina* *cactacearum* *ceruchis* *chilensis* *flaccescens* *inanis* *peruviana* *tumidula*
Microphyllinic acid $C_{29}H_{36}O_9$	 Melting point: 116°C (benzene–petroleum ether)	*Cetrelia* *japonica*

Compound Formula	Structure and Chemical Properties	Producing Organisms

Obtusatic acid $C_{18}H_{18}O_7$	Melting point: 208–209°C (acetone)	*Ramalina*
		  commixta
		  ligulata
		  obtusata
		  pollinaria

Olivetoric acid $C_{26}H_{32}O_8$	Melting point (needles): 151°C (benzene)	*Cetraria*
		  ciliaris
		Cetrelia
		  olivetorum
		Cornicularia
		  pseudosatoana
		Pseudevernia
		  furfuracea

Perlatolic acid $C_{25}H_{32}O_7$	Melting point: 107–108°C (benzene–petroleum ether)	*Cetrelia*
		  cetrarioides
		Cladonia
		  impexa

Planaic acid $C_{27}H_{36}O_7$	Melting point: 110–111°C (methanol)	*Lecidea*
		  lithophila
		  plana

Sphaerophorin $C_{23}H_{28}O_7$	Melting point: 137°C (benzene)	*Sphaerophorus*
		  fragilis
		  globosus
		  melophorus
		  melanocarpus

Compound Formula	Structure and Chemical Properties	Producing Organisms
Tenuiorin $C_{26}H_{24}O_{10}$	Melting point (plates): 178–180°C, solidifies, then d at 238°C (benzene)	*Lobaria linita Peltigera canina polydactyla*
Umbilicaric acid $C_{25}H_{22}O_{10}$	Melting point (plates): 185–189°C (ethanol); 203° (ethanol–water)	*Actinogyra polyrrhiza Umbilicaria deusta hyperborea polyphylla*
Boninic acid $C_{25}H_{32}O_8$	Melting point (plates): 134.5°C (benzene–petroleum ether)	*Ramalina boninensis*
Cryptochlorophaeic acid $C_{25}H_{32}O_8$	Melting point: 182–184°C (benzene)	*Cladonia cryptochlorophaea Parmelia cryptochlorophaea Ramalina paludosa*
Homosekikaic acid $C_{24}H_{30}O_8$	Melting point: 133–137°C (benzene–petroleum ether)	*Cladonia dissimilis nemoxyna pityrea submultiformis subpityrea*

Compound Formula	Structure and Chemical Properties	Producing Organisms
Merochlorophaeic acid $C_{24}H_{30}O_8$	Melting point: 164–166°C (benzene–hexane)	*Cladonia merochlorophaea pseudorangiformis*
Paludosic acid $C_{23}H_{28}O_8$	Melting point (fine colorless needles): 158.5–159.5°C (benzene)	*Ramalina paludosa*
Ramalinolic acid $C_{23}H_{28}O_8$	Melting point: 163–164°C (benzene)	*Physcia aegialita Ramalina calicaris geniculata nervulosa usnea*
Scrobiculin $C_{22}H_{26}O_8$	Melting point: 135.5–136°C (benzene–petroleum ether)	*Lobaria amplissima scrobiculata*
Sekikaic acid $C_{22}H_{26}O_6$	Melting point (colorless prisms): 150–151°C (benzene)	*Physcia aegialita Ramalina boulhautiana calicaris chilensis*

Compound Formula	Structure and Chemical Properties	Producing Organisms
Sekikaic acid (*cont.*)		*Ramalina (cont.)* *farinacea* *geniculata* *nervulosa*
Alectoronic acid $C_{28}H_{32}O_9$	 Melting point: 193°C (benzene); hydrate 120–121°C (ethanol–water), resolidifies at 140°C, melts at 193°C	*Alectoria* *japonica* *sarmentosa* *Cetraria* *halei* *pseudocomplicata* *Cetrelia* *nuda* *Lecanora* *atra*
a-Collatolic acid $C_{29}H_{34}O_9$	 Melting point: 124–125°C (benzene–petroleum ether); hydrate 90–95°C (ethanol–water)	
Diploicin $C_{16}H_{10}Cl_4O_5$	 Melting point: 232°C (ethanol or benzene)	*Buellia* *canescens*
Grayanic acid $C_{23}H_{26}O_7$	 Melting point: 186–189°C, d (50% ethanol)	*Cladonia* *cylindrica* *grayi*

Compound Formula	Structure and Chemical Properties	Producing Organisms
Iobaric acid $C_{25}H_{28}O_8$	 Melting point: 196—197°C (acetone)	*Anzia hypoleucoides* *Lecanora badia* *Stereocaulon antarcticum exutum paschale sorediiferum*
4-O-Methylphysodic acid $C_{27}H_{32}O_8$	 Melting point: 151—152°C	*Parmelia livida*
Norlobaridone $C_{23}H_{26}O_6$	 Melting point: 188°C (ethanol–water or benzene); 188—190°C (ether–petroleum ether)	*Parmelia ecaperata nairobiensis scabrosa*
Physodic acid $C_{26}H_{30}O_8$	 Melting point: 205°C, d (acetone–carbon disulfide or methanol–water)	*Cetraria ciliaris* *Hypogymnia physodes* *Parmelia livida* *Pseudevernia furfuracea*

477

Compound Formula	Structure and Chemical Properties	Producing Organisms
Variolaric acid $C_{16}H_{10}O_7$	 Melting point: 296°C (80% aqueous acetone)	*Ochrolechia* *parella* *Pertusaria* *lactea*
Picrolichenic acid $C_{25}H_{30}O_7$	 Melting point (prisms): 190°C, d (acetic acid–water); 184–187°C (benzene) $[\alpha]_D^{20}$: 0° (chloroform, C = 5)	*Pertusaria* *amara*
Atranorin $C_{19}H_{18}O_8$	 Melting point: 196°C (acetone)	*Anaptychia* *ciliaris* *neoleucomelaena* *palmulata* *speciosa* *Anzia* *hypoleucoides* *japonica* *opuntiella* *Buellia* *canescens* *Cetrelia* *nuda* *Crocynia* *neglecta* *Evernia* *prunastri* *Haematomma* *coccineum*

Compound Formula	Structure and Chemical Properties	Producing Organisms
Atranorin *(cont.)*		*Heterodermia*
		hypoleuca
		Himantormia
		lugubris
		Hypogymnia
		physodes
		Lasallia
		pustulata
		Lecanora
		gangaleoides
		melanaspis
		rupicola
		Lecidea
		granulosa
		insularis
		pertingens
		stigmatea
		Lepraria
		neglecta
		Menegazzia
		terebrata
		Ochrolechia
		tartarea
		Parmelia
		acetabulum
		camtschadalis
		cirrhata
		dilatata
		entotheiochroa
		galbina
		gossweileri
		leucotyliza
		livida
		pseudoreticulata
		sancti-angelii
		saxatilis
		subrudecta
		tinctorum
		zollingeri
		Pertusaria
		dealbata
		Physcia
		setosa

Compound Formula	Structure and Chemical Properties	Producing Organisms

Atranorin *(cont.)*

Pseudevernia
  intensa
Pseudocyphellaria
  nitida
Ramalina
  druidarum
Stereocaulon
  antarcticum
  corticulatum
  exutum
  nanodes
  paschale
  ramulosum
  sorediiferum
  tomentosum
Thamnolecania
  gerlachei
Usnea
  canariensis
Xylographa
  vitiligo

Baeomycesic acid
$C_{19}H_{18}O_8$

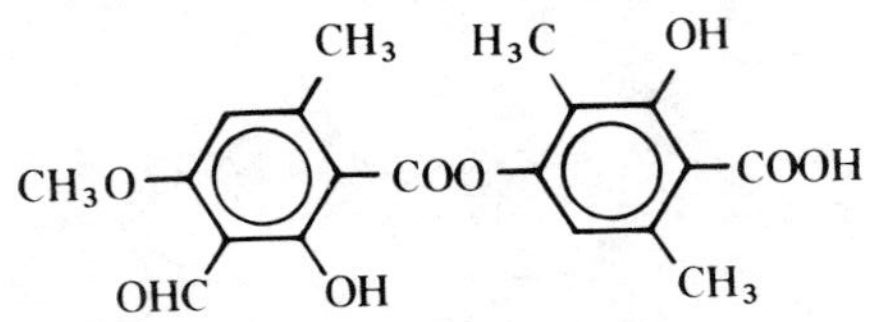

Melting point: 222–223°C, d (acetone–water)

Baeomyces
  roseus
Cladonia
  strepsilis
Thamnolia
  subuliformis

Barbatic acid
$C_{19}H_{20}O_7$

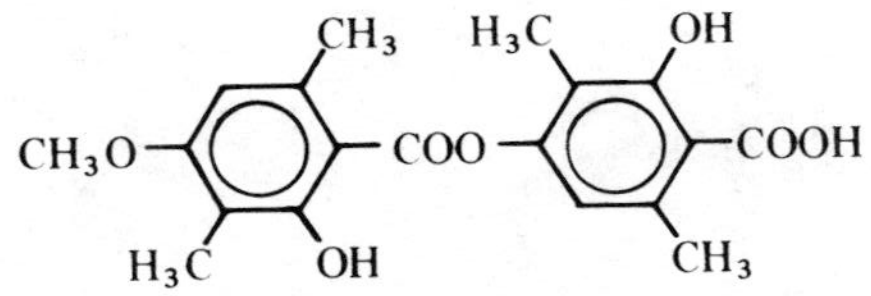

Melting point: 187°C, d (benzene)

Cladonia
  amaurocraea
  floerkeana
Usnea
  barbata
  ceratina
  compacta
  longissima
  ludicra

Compound Formula	Structure and Chemical Properties	Producing Organisms

Barbatic acid (*cont.*)

Usnea (cont.)
  misaminensis
  orientalis
  perplcctans
  venosa

Chloroatranorin
$C_{19}H_{17}ClO_8$

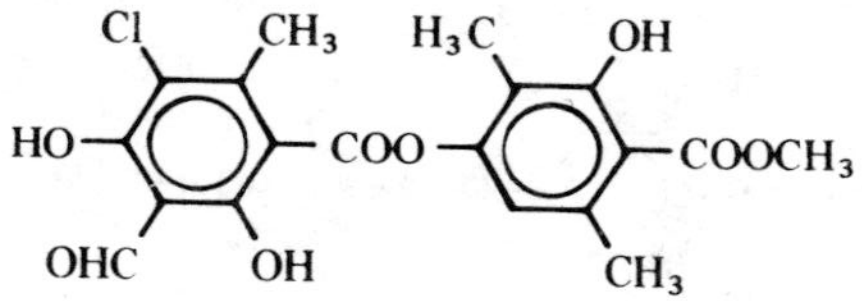

Melting point: 208—208.5°C (acetone)

Anaptychia
  neoleucomelaena
Buellia
  canescens
Cetrelia
  cetrarioides
  japonica
Evernia
  prunastri
Hypogymnia
  physodes
Lecanora
  gangaleoides
  rupicola
Parmelia
  cryptochlorophaea
  pseudoreticulata
  tinctorum
Pseudevernia
  furfuracea
  intensa
Ramalina
  druidarum
Usnea
  canariensis

4-O-Demethylbarbatic acid
$C_{18}H_{18}O_7$

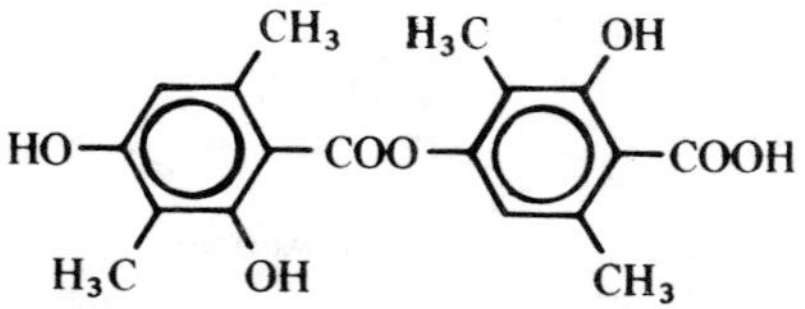

Melting point (needles): 176—177°C, d (methanol—water)

Ramalina
  subdecipiens

Compound Formula	Structure and Chemical Properties	Producing Organisms
Diffractaic acid $C_{20}H_{22}O_7$	Melting point: 189–190°C (benzene)	*Alectoria ochroleuca* *Parmelia insueta mesogenes* *Usnea deminuta diffracta*
Squamatic acid $C_{19}H_{18}O_9$	Melting point: 219°C, d (acetic acid); 228°C, d (acetone)	*Cladonia bellidiflora crispata pseudodidyma squamosa uncialis* *Dermatiscum thunbergii* *Sphaerophorus meiophorus* *Thamnolia subuliformis*
Decarboxythamnolic acid $C_{18}H_{16}O_9$	May form from thamnolic acid during extraction or chromatography Melting point: 215°C (acetone)	*Cladonia polydactyla* *Haematomma ventosum* *Siphula decumbens* *Thamnolia vermicularis*
Haemathamnolic acid $C_{19}H_{16}O_{10}$	Melting point (pale-yellow laths): 202–204°C, d (acetone)	*Pertusaria rhodesiaca*

Compound Formula	Structure and Chemical Properties	Producing Organisms
Hypothamnolic acid $C_{19}H_{18}O_{10}$	Melting point: 225—227°C, d (80% acetone—water)	*Cladonia pseudostellata*
Thamnolic acid $C_{19}H_{16}O_{11}$	Melting point: 223°C, d (acetone)	*Cladonia gorgonina polydactyla Haematomma ventosum Parmeliopsis placorodia Pertusaria corallina dealbata Siphula decumbens Thamnolia subuliformis vermicularis Usnea eulychniae*
Fumarprotocetraric acid $C_{22}H_{16}O_{12}$	Melting point: 250—260°C, d, discolors above 230°C (acetone)	*Cetraria islandica Cladonia cryptochlorophaea endiviaefolia pityrea subtenuis Dendrographa leucophaea minor*

Compound Formula	Structure and Chemical Properties	Producing Organisms
Gangaleodin $C_{18}H_{14}Cl_2O_7$	Melting point: 214–215°C (ethanol–acetone)	*Lecanora gangaleoides*
Hypoprotocetraric acid $C_{18}H_{16}O_7$	Melting point: 242–243°C, d, turning pink near 230°C (acetone–petroleum ether); 250–251°C, d (methanol–water)	*Ramalina druidarum farinacea hypoprotocetrarica siliquosa tumidula*
Norstictic acid $C_{18}H_{12}O_9$	Melting point: 286–287°C, d (acetone–water)	*Buellia sororioides Cladonia subcariosa Diploschistes ocellatus Lecanora radiosa Lecidea pantherina Lobaria isidiosa oregana pulmonaria scrobiculata Medusulina chilena Parmelia acetabulum Pertusaria pseudocorallina*

Compound Formula	Structure and Chemical Properties	Producing Organisms
Norstictic acid *(cont.)*		*Ramalina*
		chilensis
		Roccella
		canariensis
		mossamedana
		Stereocaulon
		dactylophyllum
		Umbilicaria
		torrefacta
		Usnea
		aspera
		ludiera
		Xylographa
		hians
		vitiligo
Pannarin C$_{18}$H$_{15}$ClO$_6$	Melting point: 216—217°C (acetone)	*Bombyliospora*
		japonica
		Lecanora
		hercynica
		Pannaria
		fulvescens
		lurida
		pityrea
		rubiginosa
Physodalic acid C$_{20}$H$_{16}$O$_{10}$	Melting point: 230—260°C, d (acetic acid or ethyl ether)	*Dactylina*
		chinensis
		ramulosa
		Hypogymnia
		physodes
		Parmelia
		ferax
		gerlachei

Compound Formula	Structure and Chemical Properties	Producing Organisms
Protocetraric acid $C_{18}H_{14}O_9$	Melting point: 245–250° C, d, carbonizes (acetone)	*Parmelia* *camtschadalis* *caperata* *dilatata* *zollingeri* *Ramalina* *farinacea* *Roccella* *portentosa* *Roccellinastrum* *spongioideum* *Usnea* *elongata* *lacerata* *perplectans*
Psoromic acid $C_{18}H_{14}O_8$	Melting point (needles): 265° C (ethanol)	*Alectoria* *sulcata* *Chiodecton* *stalactinum* *Cladonia* *aberrans* *Darbishirella* *gracillima* *Everniopsis* *trulla* *Ingaderia* *pulcherrima* *Pentagenella* *fragillima* *Ramalina* *tigrina* *Roccellodea* *nigerrima* *Roccellographa* *cretacea* *Squamarina* *crassa* *Usnea* *aspera*

Compound Formula	Structure and Chemical Properties	Producing Organisms

Salazinic acid $C_{18}H_{12}O_{10}$	The "salazinic acid" of Zopf (*Ann. Chem., 295,* 222–300, 1897) is now known as norstitic acid
	Melting point (needles): 260–280° C, d, carbonizes (80% acetone—water)
		Lobaria
		  pulmonaria
		Parmelia
		  camtschadalis
		  cetrata
		  cirrhata
		  meizospora
		  pseudoreticulata
		  saxatilis
		  scabrosa
		  taractica
		Ramalina
		  crassa
		Usnea
		  aureola
		  compacta
		  florida
		  japonica
		  ludicra
		  orientalis
		  perplectans
		  rubicunda
		  venosa

Stictic acid $C_{19}H_{14}O_{9}$	Melting point: 270–272° C (acetone)	*Baeomyces*
		  placophyllus
		Lecanora
		  jussuffii
		Lobaria
		  isidiosa
		  oregana
		  pulmonaria
		  scrobiculata
		  subretigera
		Menegazzia
		  terebrata

Compound Formula	Structure and Chemical Properties	Producing Organisms
Stictic acid (*cont.*)		*Parmelia* *conspersa* *Ramalina* *curnowii* *Stereocaulon* *foliosum* *nabewariense* *tomentosum* *Umbilicaria* *torrefacta* *Usnea* *orientalis* *rubicunda*
Vicanicin $C_{18}H_{16}Cl_2O_5$	 Melting point: 248–250° C	*Teloschistes* *flavicans*
Virensic acid $C_{18}H_{14}O_8$	 Melting point: 245–247° C	*Alectoria* *tortuosa*
Barbatolic acid $C_{18}H_{14}O_{10}$	 Melting point (needles): 206–207° C, d, softens at 190° C (acetic acid or dioxane)	*Alectoria* *implexa* *Himantormia* *lugubris* *Usnea* *barbata*

Compound Formula	Structure and Chemical Properties	Producing Organisms
Pyxiferin $C_{13}H_8O_8$	Melting point (red crystals): 300°C (chloroform)	*Pyxine coccifera*
Didymic acid $C_{22}H_{26}O_5$	Melting point: 172—173°C (ligroin)	*Cladonia corallifera floerkeana*
Pannaric acid $C_{16}H_{12}O_7$	Melting point (anhydrous crystals): 243—245°C (acetone—water); forms a dihydrate	*Lepraria membranacea*
Porphyrilic acid $C_{16}H_{10}O_7$	Melting point: 280—283°C, d, darkens above 270°C	*Haematomma coccineum porphyrium Lecidea silacea*
Schizopeltic acid $C_{19}H_{18}O_7$	$R = CH_3$, $R' = H$ or $R = H$, $R' = CH_3$. Melting point (colorless needles): 228—230° (acetone—water); 233—235°C (methanol)	*Reinkella parishii Roccellina luteola Schizopelte californica*

Compound Formula	Structure and Chemical Properties	Producing Organisms
Strepsilin $C_{15}H_{10}O_5$	 Melting point: 324°C (acetic acid)	*Cladonia strepsilis*
(+)-Isousnic acid $C_{18}H_{16}O_7$	 Melting point (yellow prisms): 150−152°C (benzene−methanol) $[\alpha]_D^{21}$: +500° (dioxane)	*Cladonia arbuscula mitis submitis*
(−)-Isousnic acid $C_{18}H_{16}O_7$	Melting point: 150−152°C $[\alpha]_D$: −490° (dioxane)	*Cladonia pleurota*
(+)-Usnic acid $C_{18}H_{16}O_7$	 Melting point (yellow prisms or needles): 203−204°C (benzene or chloroform−ethanol) $[\alpha]_D^{20}$: +495°C (chloroform)	*Alectoria ochroleuca Cladonia arbuscula mitis Evernia divaricata prunastri*

Compound Formula	Structure and Chemical Properties	Producing Organisms
(+)-Usnic acid *(cont.)*		*Lecanora*
		badia
		handelii
		polytropa
		sulphurea
		Lobaria
		scrobiculata
		Nephroma
		gyelnikii
		Parmelia
		dilatata
		scabrosa
		taractica
		Ramalina
		boulhautiana
		capitata
		chilensis
		fraxinea
		inanis
		linearis
		subamplicata
		terebrata
		tigrina
		tumidula
		Usnea
		aspera
		aureola
		compacta
		dasypoga
		eulychniae
		florida
		hirta
		implicita
		lacerata
		longissima
		misaminensis
		orientalis
		pusilla
		rubicunda

Compound Formula	Structure and Chemical Properties	Producing Organisms
(−)-Usnic acid $C_{18}H_{16}O_7$	Melting point (yellow prisms or needles): 203°C (benzene or chloroform−ethanol) $[\alpha]_D^{20}$: −495° (chloroform)	*Alectoria japonica ochroleuca sarmentosa Cetraria nivalis stracheyi Cladonia aberrans alpestris deformis endiviaefolia impexa reticulata Dactylina arctica chinensis endochrysea madreporiformis ramulosa Everniopsis trulla Haematomma coccineum Lecanora melanophthalma Ramalina calicaris reticulata Usnea barbata*
Lepraric acid $C_{18}H_{18}O_8$	Melting point (colorless crystals): 155−156.5°C, d (chloroform−benzene) Melting point (plates): 161−162°C (methanol)	*Lepraria latebrarum Roccella fuciformis teneriffensis Sagenidium molle*

Compound Formula	Structure and Chemical Properties	Producing Organisms
Lepraric acid (*cont.*)		
Siphulin C₂₄H₂₆O₇		*Siphula ceratites*
Sordidone C₁₁H₉ClO₄	Melting point: 180°C, d	*Lecanora rupicola sordida*
Arthothelin C₁₄H₇Cl₃O₅	Melting point (cream-colored crystals): 260–262°C	*Arthothelium pacificum Lecanora pinguis straminea Lecidea guernea*
2,4-Dichloronorlichexanthone C₁₄H₈Cl₂O₅	Melting point (yellow prisms): 275–276°C (ethyl acetate)	*Lecanora straminea*

Compound Formula	Structure and Chemical Properties	Producing Organisms
2,7-Dichloronorlichexanthone $C_{14}H_8Cl_2O_5$		*Lecanora straminea*
Lichexanthone $C_{16}H_{14}O_5$	 Melting point (yellow needles): 187–190°C (acetone)	*Graphina confluens* *Lecidea stigmatea* *Parmelia formosana* *Pyxine caesiopruinosa*
Norlichexanthone $C_{14}H_{10}O_5$		*Lecanora reuteri straminea*
Thiophanic acid $C_{14}H_6Cl_4O_5$	 Melting point (yellow prisms): 243°C (benzene)	*Lecanora rupicola*
Thiophaninic acid $C_{15}H_{10}Cl_2O_5$	 Melting point (lemon-yellow prisms): 269–271°C (ethyl acetate)	*Pertusaria flavicans flavicunda*

Compound Formula	Structure and Chemical Properties	Producing Organisms
Thuringione $C_{15}H_9Cl_3O_5$	Melting point: 278–279°C (ethyl acetate)	*Lecanora pinguis* *Lecidea carpathica*
Vinetorin $C_{15}H_{11}ClO_5$	Melting point: 243–245°C	*Lecanora* sp.
1,3-Dihydroxy-8-methoxy-2-chloro-6-methylanthraquinone $C_{16}H_{11}ClO_5$		*Nephroma laevigatum*
Fallacinal $C_{16}H_{10}O_6$	Melting point (orange-red needles): 250–252°C (chloroform-ethanol)	*Teloschistes flavicans* *Xanthoria fallax*
Fragilin $C_{16}H_{11}ClO_5$	Melting point (yellow): 267–268°C (chloroform; vacuum sublimation)	*Acroscyphus sphaerophoroides* *Nephroma laevigatum* *Sphaerophorus fragilis* *globosus* *melanocarpus*

Compound Formula	Structure and Chemical Properties	Producing Organisms
1-Hydroxy-3,8-dimethoxy-2-chloro-6-methylanthraquinone $C_{17}H_{13}ClO_5$		*Nephroma laevigatum*
Norsolorinic acid $C_{20}H_{18}O_7$		*Solorina crocea*
Parietinic acid $C_{16}H_{10}O_7$	Melting point (red plates): 269–270°C (ethanol)	*Xanthoria aureola contortuplicata parietina*
Solorinic acid $C_{21}H_{20}O_7$	Melting point: ∿300°C with sublimation	*Solorina crocea*
Teloschistin $C_{16}H_{12}O_6$	Melting point (orange-red crystals): 201°C (acetic acid) Melting point (orange needles): 244–246°C (benzene; purified through the acetate)	*Teloschistes flavicans Xanthoria fallax*

Compound Formula	Structure and Chemical Properties	Producing Organisms
1,3,8-Trihydroxy-2,4-dichloro-6-methylanthraquinone $C_{15}H_8Cl_2O_5$	 Melting point (orange needles): 267—269°C (benzene)	*Heterodermia obscurata*
Xanthorin $C_{16}H_{12}O_6$	 Melting point (red): 253°C (toluene)	*Laurera purpurina* *Xanthoria elegans*
7β-Acetoxy-22-hydroxyhopane $C_{23}H_{54}O_3$	 Melting point: 247—248°C (benzene) $[\alpha]_D^{20}$: +26° (chloroform, c = 0.71)	*Pseudocyphellaria nitida* *Stricta billardierii*
16β-O-Acetylleucotylic acid $C_{32}H_{52}O_5$	 Isolated as the methyl ester (melting point 176°C) $[\alpha]_D$: +95° (chloroform)	*Parmelia entotheiochroa*

Compound Formula	Structure and Chemical Properties	Producing Organisms
6α-O-Acetylleucotylin $C_{32}H_{54}O_4$	Melting point: 225°C $[\alpha]_D$: +36° (chloroform)	*Parmelia entotheiochroa*
6-Deoxy-16β-O-Acetylleucotylin $C_{32}H_{54}O_3$	Melting point: 228°C $[\alpha]_D$: +52° (chloroform)	*Parmelia entotheiochroa*
6-Deoxyleucotylin $C_{30}H_{52}O_2$	Melting point: 268°C $[\alpha]_D$: +68° (chloroform)	*Parmelia entotheiochroa*
6α,16β-Di-O-acetylleucotylin $C_{34}H_{56}O_5$	Melting point: 232°C $[\alpha]_D$: +109°	*Parmelia entotheiochroa*

Compound Formula	Structure and Chemical Properties	Producing Organisms
15a,22-Dihydroxyhopane $C_{30}H_{52}O_2$	 Melting point: 249°C (benzene) $[a]_D^{20}$: +34° (chloroform, c = 1.3)	*Pseudocyphellaria intricata* *Stricta billardierii*
Leucotylic acid $C_{30}H_{50}O_4$	 Melting point: 260°C $[a]_D^{20}$: +330° (chloroform, c = 0.15)	*Parmelia entotheiochroa leucotyliza*
Leucotylin $C_{30}H_{52}O_3$	 Melting point: 335–336°C (methanol) $[a]_D^{20}$: +56.5° (chloroform, c = 0.566)	*Lecanora muralis* *Parmelia entotheiochroa homogenes leucotyliza*

499

Producing Organisms

*Pyxine
endocrysina*

*Cladonia
deformis*

*Anapthychia
  neoleucomelaena
  speciosa
Cladonia
  deformis
Heterodermia
  hypoleuca
  leucomela
  obscurata
Lecanora
  handelii
  muralis
  polytropa
  sulphurea*

Structure and Chemical Properties

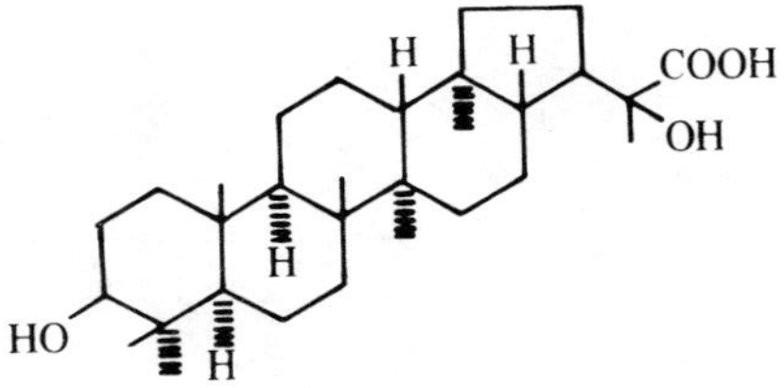

Melting point: 254–255°C
$[\alpha]_D$: +62° (ethanol, c = 0.3)

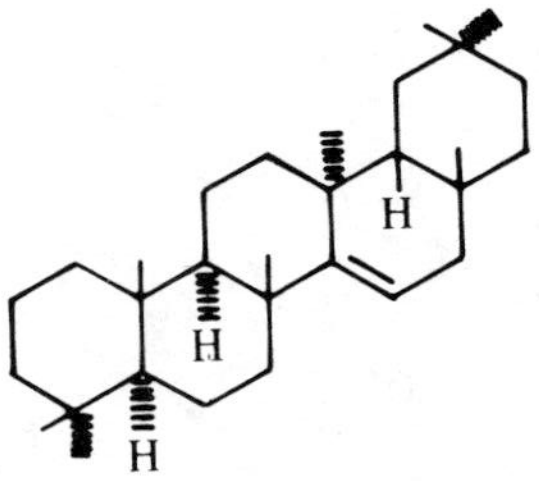

Melting point: 238–239°C
$[\alpha]_D$: +3° (chloroform)

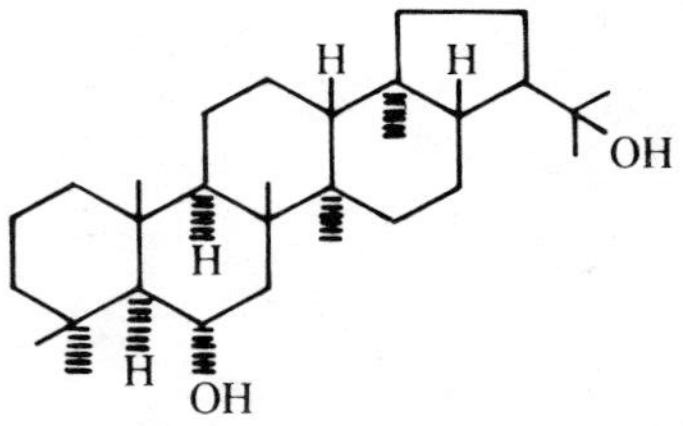

Melting point: variable; ∼245–253°C, ∼223–227°C
$[\alpha]$: +54° (chloroform, c = 0.50)

**Compound
Formula**

Pyxinic acid
$C_{30}H_{50}O_4$

Taraxerene
$C_{30}H_{50}$

Zeorin
$C_{30}H_{52}O_2$

Compound Formula

Zeorin *(cont.)*

Calycin
$C_{18}H_{10}O_5$

Structure and Chemical Properties

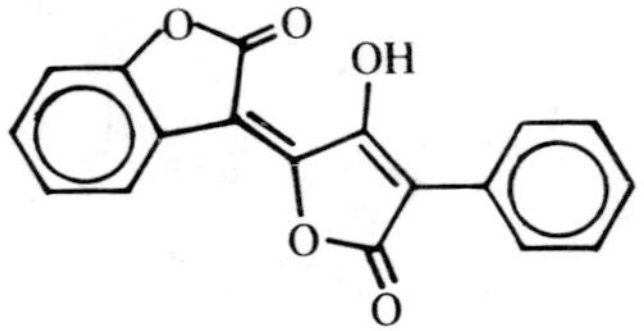

Melting point (red crystals): 249–249.5°C (acetic acid)

Producing Organisms

Nephroma
  antarcticum
  arcticum
  gyelnikii
  laevigatum
  parile
Parmelia
  entotheiochroa
  galbina
  leucotyliza
Peltigera
  dolichorrhiza

Buellia
  rhodesiaca
Candelaria
  concolor
Candelariella
  aurella
  medians
  vitellina
Chrysothrix
  nolitangere
Lecidea
  lucida
Lepraria
  candelaris
  chlorina
Pseudocyphellaria
  aurata
  crocata
  durvillei
  hirsuta
  nitida
Stricta
  aurata
  colensoi
  coronata

Compound Formula	Structure and Chemical Properties	Producing Organisms
Epanorin $C_{25}H_{25}NO_6$		*Lecanora epanora*
Leprapinic acid $C_{20}H_{16}O_6$	 Melting point (yellow needles): $135-136°C$ (methanol) $[\alpha]_D^{26}: -1.86° \pm 0.2°$ (chloroform, $c = 6.48$)	*Lecidea lucida* *Lepraria citrina*
Leprapinic acid methyl ether $C_{21}H_{18}O_6$	 Melting point (golden-yellow plates): $164-165°C$ (methanol)	*Lepraria chlorina*
Pinastric acid $C_{20}H_{16}O_6$	 Melting point (colorless needles): $150-152°C$ (methanol)	*Cetraria juniperina pinastri* *Lepraria candelaris* *Temnospora fulgens*
	 Melting point (orange rectangular plates): $202-204°C$ (benzene)	

**Compound
Formula**

Structure and Chemical Properties

Producing Organisms

Pulvinic dilactone
$C_{18}H_{10}O_4$

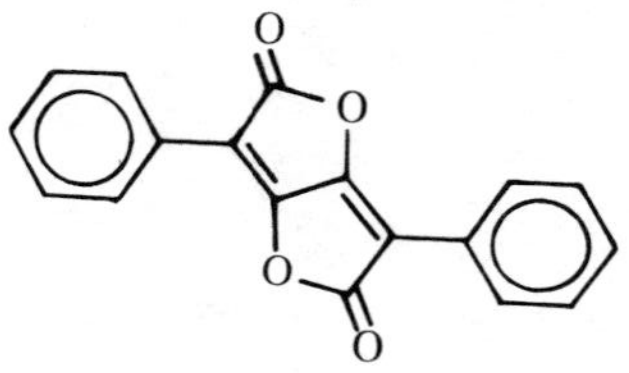

Melting point (yellow needles): 222—224°C (glacial acetic acid); 227°C (benzene)

*Candelaria
    concolor
Candelariella
    aurella
    vitellina
Pseudocyphellaria
    aurata
    crocata
    durvillei
    hirsuta
    nitida
Sticta
    colensoi
    coronata
Thelocarpon
    epibolum
    laureri*

Rhizocarpic acid
$C_{28}H_{23}NO_6$

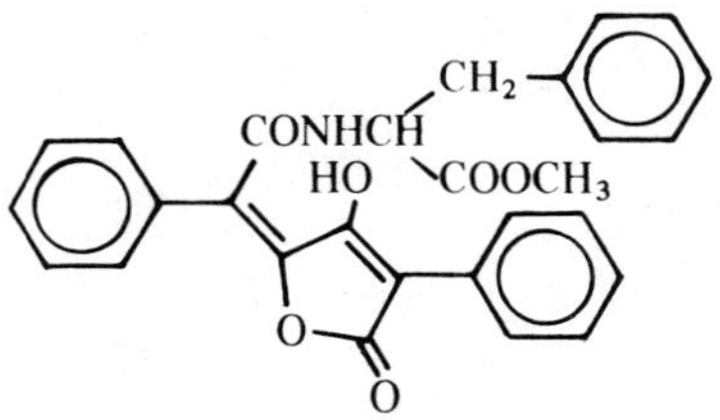

Melting point (yellow needles): 177—178°C (ethanol)
$[\alpha]_D^{20}$: +110.4° ± 2.1° (chloroform, c = 1.22)

*Acarospora
    chlorophana
    gobiensis
    oxytona
    schleicheri
Coniocybe
    furfuracea
Dermatiscum
    thunbergii
Lecanora
    hercynica
Rhizocarpan
    geographicum*

MISCELLANEOUS LARGE-RING COMPOUNDS

DR. NANCY N. GERBER

Compound	Structure	Microorganism	Reference
Brefeldin A (Cyanein, decumbin)		*Penicillium brefeldianum* other species *Nectria radicicola* *Curvularia lunati subulata*	1
Curvularin		*Curvularia lunata*	1
Dehydrocurvularin	Double bond in conjugation with ketone carbonyl	*Curvularia* sp.	1
Pyrenophorin	R = O	*Pyrenophora avenae* *Stemphylium radicinum*	1
Pyrenophorol	R = H, OH	*Byssochlamys nivea*	1
Colletodiol		*Colletotrichum capsici*	1
Lasiodiplodin	R = Me	*Lasiodiplodia theobromae*	1

Compound	Structure	Microorganism	Reference
Desmethyllasiodiplodin	R = H	*Lasiodiplodia theobromae*	1
Monorden (Radicicol)		*Monosporium bonorden* *Nectria radicicola*	1
Zearalenone		*Gibberella zea* *Fusarium moniliforme*	1
Cytochalasin A	R = O	*Helminthosporium dematioideum*	1
Cytochalasin B (Phomin)	R = H, OH	*Helminthosporium dematioideum* *Phoma* sp.	1

Compound	Structure	Microorganism	Reference
Cytochalasin C	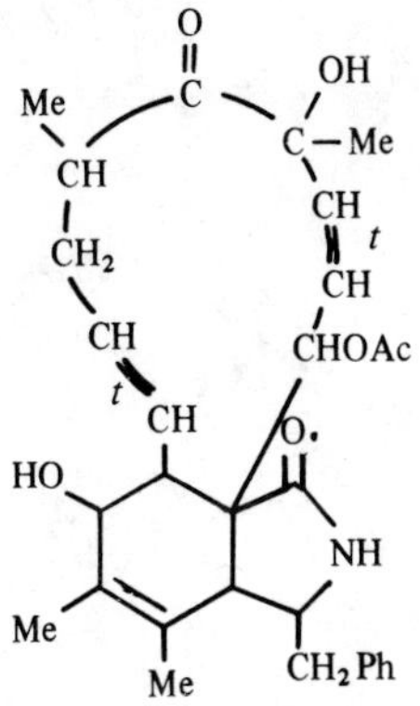	*Metarrhizium anisopliae*	1

O
Me—CH—C—C—Me (OH)
CH₂
CH — CH ‖ CH (t)
CH (t) — CHOAc
CH — O
HO — NH
Me — CH₂Ph
Me

| Cytochalasin D (Zygosporin A) | 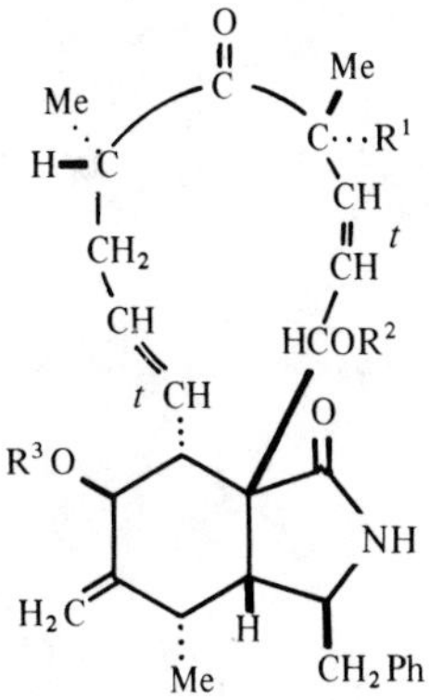 | *Metarrhizium anisopliae* *Zygosporium masonii* | 1 |

O
Me—H·C—C—C···R¹ (Me)
CH₂
CH — CH ‖ CH (t)
t CH — HCOR²
R³O — O
H₂C — NH
Me — H — CH₂Ph

R¹ = OH
R² = Ac
R³ = H

| Acetylcytochalasin D (Zygosporin F) | R¹ = OH
R² = Ac
R³ = Ac | *Zygosporium masonii* | 1 |

Compound	Structure	Microorganism	Reference
Desacetylcytochalasin D (Zygosporin D)	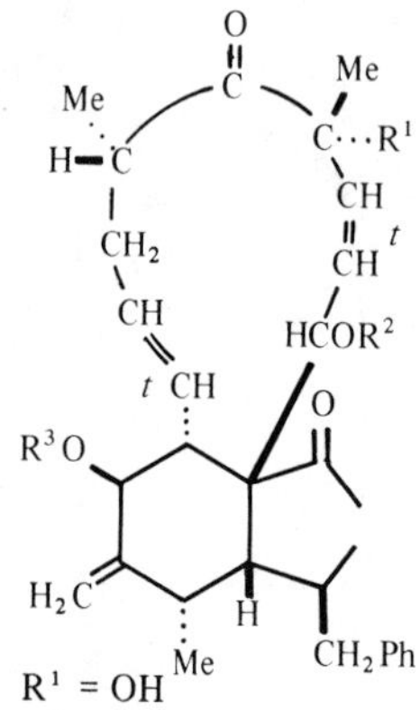	*Zygosporium masonii*	1
Zygosporin E		*Zygosporium masonii*	1
Zygosporin G	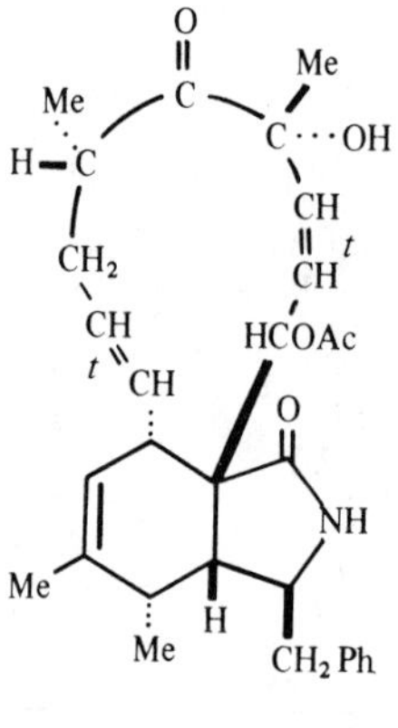	*Zygosporium masonii*	1
Rifamycin B	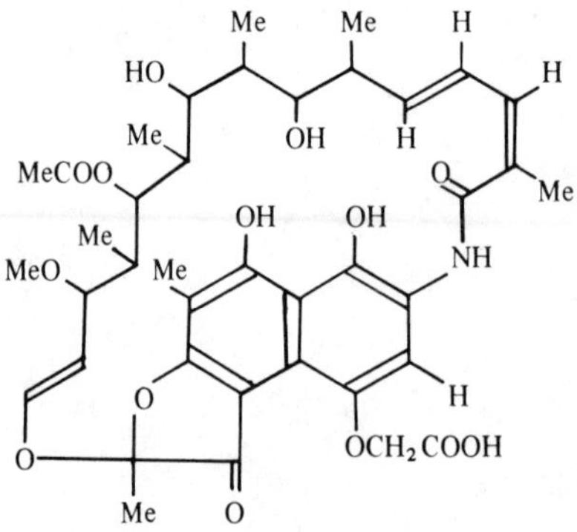	*Streptomyces mediterranei*	2

$R^1 = OH$
$R^2 = H$
$R^3 = H$

$R^1 = H$
$R^2 = H$
$R^3 = Ac$

Compound	Structure	Microorganism	Reference

Rifamycin O

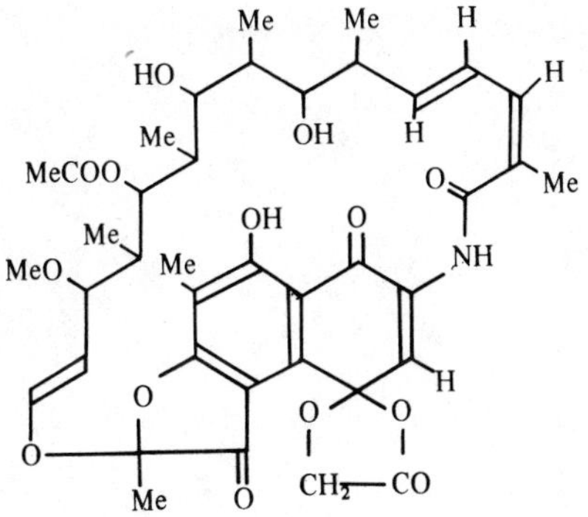

Streptomyces sp. 2

REFERENCES

1. Turner, W. B., *Fungal Metabolites,* Academic Press, New York (1971).
2. Umezawa, H. (Ed.), *Index of Antibiotics from Actinomycetes.* University of Tokyo Press, Tokyo, Japan (1967).

MISCELLANEOUS PRODUCTS

DR. NANCY N. GERBER

Compound	Structure	Microorganism	Ref.
Elaiomycin	$\overset{\displaystyle O}{\underset{\displaystyle \uparrow}{}}$ $CH_3(CH_2)_5CH=CHN=NCH(CH_2OCH_3)CH(OH)CH_3$	*Streptomyces gelaticus* sp.	1
Gyromitrin	$CH_3CH=NN(CH_3)CHO$	*Gyromitra esculenta*	2
Enteromycin group			
U-15774	$HON=CHCONHCH=CHCONH_2$	*Streptomyces achromogenes*	1
U-22956	$\overset{\displaystyle O}{\underset{\displaystyle \uparrow}{}}$ $CH_3ON=CHCH=CHCOOH$	*Streptoverticillium fervens*	1
Enteromycin	$\overset{\displaystyle O}{\underset{\displaystyle \uparrow}{}}$ $CH_3ON=CHCONHCH=CHCOOH$	*Streptomyces achromogenes albireticuli*	1
Enteromycin carboxamide	$\overset{\displaystyle O}{\underset{\displaystyle |}{}}$ $CH_3ON=CHCONHCH=CHCOONH_2$	*Streptomyces* sp.	1
Noformicin	$HN=$⟨ring⟩$N-CONHCH_2CH_2C(=NH_2)NH_2$	*Nocardia formica*	1
Actinonin	$HO-$⟨ring⟩$N-COCH(CHMe_2)NHCOCH(C_5H_{11})-$ $-CH_2CONH(OH)$	*Streptomyces* sp.	1
Azomycin	⟨imidazole ring⟩ $N-NO_2$	*Nocardia mesenterica* *Streptomyces eurocidicus*	1
Dethiobotin	$HN-NH$ ⟨ring⟩ $CH_3 \quad (CH_2)_3COOH$	*Penicillium chrysogenum*	3
p-Hydroxymethylbenzene-diazonium	$N\equiv\overset{\oplus}{N}-$⟨benzene⟩$-CH_2OH$	*Agaricus bisporus*	2
N-Methylnitrosoamino-benzaldehyde	$CH_3N(NO)-$⟨benzene⟩$-CHO$	*Clitocybe suaveolens*	2

Compound	Structure	Microorganism	Ref.
6-Methoxybenzoxazoli-done		*Ustilago maydis* (spores)	3
Erythroskyrine		*Penicillium islandicum*	2
Phomazarine		*Phoma terrestris*	2
Nybomycin		*Streptomyces* sp.	1
Celesticetin		*Streptomyces caelestis*	1
Lincomycin		*Streptomyces lincolnensis*	1

Compound	Structure	Microorganism	Ref.
Lincomycin group			
U-11921	Same as lincomycin, but with $-C_2H_5$ on S	*Streptomyces lincolnensis*	1
U-11973	N-Demethyl lincomycin	*Streptomyces lincolnensis*	1
U-21699	Same as lincomycin, but with C_2H_5 instead of C_3H_7	*Streptomyces lincolnensis*	1

REFERENCES

1. Umezawa, H. (Ed.), *Index of Antibiotics from Actinomycetes.* University of Tokyo Press, Tokyo, Japan (1967).
2. Turner, W. B., *Fungal Metabolites.* Academic Press, New York (1971).
3. Miller, M. W., *The Pfizer Handbook of Microbial Metabolites.* McGraw-Hill, New York (1961).

TOXINS

BACTERIAL PROTEIN TOXINS

DR. JOHN P. ARBUTHNOTT

INTRODUCTION

The concept of exotoxins as harmful diffusible products of pathogenic bacteria was introduced early in the history of microbiology. By the mid-1890's the existence of three potent exotoxins, namely those of diphtheria, tetanus and botulism, had been demonstrated experimentally. In each case immunity to the toxin conferred immunity against the disease. These pioneering studies firmly established the importance of extracellular toxins in microbial pathogenicity. In the ensuing eighty years, microbiologists assembled a formidable list of agents that can be classed as exotoxins, although the roles of individual toxins in pathogenicity are fully understood in only a few cases. Medical considerations apart, there is an increasing awareness of the importance of bacterial toxins as tools for biochemical and pharmacological research.

Although most bacterial toxins are proteins or contain significant amounts of polypeptide, it must be pointed out that the chemical characterization of many is incomplete. Toxins vary widely in biological potency and in the nature of the lesion or harmful effect produced. At the present time there is no consensus regarding the degree of toxicity that any factor must possess to merit inclusion as a bacterial toxin. Moreover, since the most sensitive measurement of activity usually involves biological assay, the problem of biological variation arises, and discrepancies between the findings of different research groups are not uncommon. Not all toxins have been purified to a high degree, and differences in purity and/or physical state may well account for apparently conflicting reports in the literature. In view of these limitations, material included in any summary of bacterial toxins must be selected at the author's discretion.

Clearly a precise and all-embracing definition of the term "exotoxin" is not easy to formulate. Differentiation between exotoxin and endotoxin on the basis of site of location of the toxin, so much a part of classical medical bacteriology, no longer holds. Raynaud and Alouf[1] have described the main classes of bacterial toxins. A modification of this system has been adopted for the presentation of summaries, although at present such a scheme must be tentative.

Class A. Cell-Associated Toxins. This group includes protein toxins that appear to be intracytoplasmic, since they are released only on autolysis or extraction of cells. In addition to the protein toxins of *Shigella dysenteriae, Pasteurella pestis* and *Bordetella pertussis,* the insect toxin of *Bacillus thuringiensis* and the recently discovered enterotoxin of *Clostridium perfringens* type A are included here. The lipopolysaccharide endotoxin of Gram-negative bacterial cell envelopes is not within the scope of this chapter.

Class B. Extracellular Toxins. In this, the main class, are included toxins that to date are considered extracellular. However, our knowledge of the mechanism of secretion for these toxins is at best fragmentary and future work may necessitate reclassification.

Class C. Toxins Located Both Intracellularly and Extracellularly. The two most potent toxic agents known, namely the neurotoxins of *Clostridium tetani* and *Clostridium botulinum,* fall within this group.

Space does not permit a comprehensive list of references; those listed are intended to provide the reader with key source material. Details of purification procedures have not been included; an excellent review of the purification of bacterial toxins has been compiled by Alouf and Raynaud.[2] All biological potencies are given as values per milligram of protein.

ABBREVIATIONS

LD_{50}	Dose causing death in 50% of tested animals	MW	Molecular weight
MLD	Minimal lethal dose	pI	Isoelectric point
HU	Haemolytic unit	NADH	Reduced nicotinamide adenine dinucleotide
RBC	Red blood cells	NAD	Nicotinamide adenine dinucleotide
MRD	Skin-minimal-reactive dose (diphtheria toxin)	CoQ	Coenzyme Q
		SLO	Streptolysin O
$S_{20,w}$	Sedimentation constant, given in Svedberg units reduced to water	SLS	Streptolysin S

PROPERTIES OF BACTERIAL TOXINS

Class A. Cell-Associated Toxins

Toxin	Significant Features	References
Shigella dysenteriae		
"Neurotoxin"	Neurotoxin is distinct from endotoxin: 4×10^5 LD_{50}/mg (rabbit), 4×10^3 LD_{50}/mg (mouse). Iron concentration is critical for production. Extraction requires heating the cells to $56-59°C$, then extracting with $1N$ KOH. The toxin is a protein. $S_{20,w} = 4.8$ S, MW = 82,000. It causes neurological symptoms, and histology suggests action on blood vessels. Neurotoxin is rapidly fixed *in vivo*.	3–7
Bordetella pertussis		
Heat-labile toxin	This toxin is produced by autolysis or extraction of cells. It is lethal and dermonecrotic; 840 LD_{50}/mg (mouse), 8,300 necrotizing doses/mg (rabbit). The toxin is extracted by $CaCl_2$, by freezing and thawing, or by disintegration of cells. It is presumed to be a protein. Pneumotropic, it causes inflammation and necrosis in the respiratory tract. It is antigenic; the antibody does not protect against whooping cough. The organism produces numerous other biologically active antigens.	8–11
Pasteurella pestis		
Murine toxins	Dried cells are lysed with acetone; the toxins are extracted with 2.5% NaCl for 24 hours, then purified by paper-curtain and glass-bead electrophoresis. 2,000 LD_{50}/mg. The toxins are proteins. They can be separated by acrylamide electrophoresis into components A and B, which are immunologically non-identical. MW of A = 240,000, $S_{20,w} = 10.8$ S; MW of B = 120,000, $S_{20,w} = 7.8$ S. Murine toxins are denatured by urea and inhibited by sulfhydryl reagents. They are poorly antigenic; the antibody is not directly correlated with protection. Rabbit, dog, and monkey are resistant to murine toxins. The toxins inhibit electron transport of mitochondria *in vitro* at the level of NADH-CoQ reductase. They cause swelling and pre-	12–15

Toxin	Significant Features	References
Pasteurella pestis Murine toxins (*cont.*)	vent ion accumulation in mitochondria. Their effect on blood pressure is similar to that of endotoxin. Their role in pathogenesis is not yet clear.	
Clostridium perfringens Enterotoxin	This toxin is isolated from sporulating cultures of type A food-poisoning strains. It causes diarrhea in man and experimental animals, fluid accumulation in ligated intestinal loops of rabbits and lambs, and erythema in guinea pig and rabbit skin. Extracted from sporulating cells by sonication, the toxin appears in filtrates after 24 to 48 hours and has been purified by gel filtration and ion exchange chromatography. It is a protein. MW = 36,000, pI = 4.3. 3.5×10^2 MLD/mg (mouse), 4×10^2 erythema units/mg (guinea pig). Enterotoxin is antigenic.	16, 17
Bacillus thuringiensis Crystalline insect toxin	A protein. This toxin is produced as a bipyramidal crystalline inclusion within the bacterial cell during sporulation and is released on lysis of the cell in the terminal stages of sporulation. The crystals are insoluble in water, but soluble in alkaline solution, pH > 11.0. The toxin is activated by proteolytic enzymes of insect larval gut. MW of the protoxin = $> 200,000$; MW of the active toxin = $5,000-55,000$. 2×10^4 LD_{50}/mg/g of lepidopterous larvae.	18, 19

Class B. Extracellular Toxins

Toxin	Significant Features	References
Corynebacterium diphtheriae Diphtheria toxin	Concentration of iron, within narrow limits, controls toxin production. Only strains lysogenic for phage β are toxigenic. The roles of iron and phage are not fully understood. The toxin has been crystallized in three laboratories. It is a protein. $S_{20,w}$ = 4.2 S, MW = $64,500-70,000$. The amino acid composition is known. The toxin is lethal and erythrogenic; 1.5×10^4 MLD/mg (guinea pig), 1×10^8 MRD/mg. The toxin does not affect electron transport, as earlier suggested. Active toxin consists of two polypeptides linked by a disulfide bond. One part of the molecule is concerned with penetrating susceptible cells; the other blocks incorporation of amino acids into cellular protein and inhibits incorporation of ^{14}C-amino acid into polypeptides in a cell-free system. NAD is essential for inhibition. The toxin specifically inactivates aminoacyl transferase II by transferring the ADP-ribose portion of NAD. It appears to act primarily on cardiac tissue of susceptible animals. It is antigenic; the antibody protects against the disease.	20–27

Toxin	Significant Features	References

Staphylococcus aureus

a-Toxin — Production is enhanced by 20–30% CO_2 or by incorporation of a diffusate of yeast extract. The toxin has been purified to a high degree in a number of laboratories. Electrofocusing gives pure product in high yield. a-Toxin is a protein. $S_{20,w}$ = 3.0 S, MW = 33,000–36,000. The toxin forms a biologically inactive aggregate, $S_{20,w}$ = 12.0 S, on contact with cell membranes and membrane lipids. Four biologically active forms can be isolated by electrofocusing; the pI of the main component is 8.55. a-Toxin is hemolytic, lethal and dermonecrotic; 1 x 10^3 LD_{50}/mg (mouse), 2 x 10^4 HU/mg (rabbit). It acts on cell membranes and possesses surface-active properties. It is antigenic; the antibody does not protect against infection. — References 28–32

β-Toxin — Production is stimulated by CO_2. The purified toxin is unstable. β-Toxin is a protein. Reported values of MW range from 18,000 to 59,000; pI = 9.4. It is a "hot-cold" hemolysin acting mainly on sheep RBC; production requires Mg^{2+}. The lethal effect of the toxin is still in question. Enzymic in nature, β-toxin hydrolyses sphingomyelin, releasing N-acyl sphingosine. It is antigenic. Its role in pathogenicity has not been established. — References 33–36

δ-Toxin — This hemolytic toxin is distinct from a- and β-toxins. It is active against the RBC of many species. Hydroxyapatite is particularly useful in its purification. δ-Toxin is a protein. pI = 9.8–10.0. A wide range of MW has been reported, probably because it forms aggregates. Its antigenicity and enzymic nature are disputed. Its biological role is unknown. — References 37, 38

Enterotoxins — Six serologically distinct enterotoxins have been identified: A, B, C_1, C_2, D, and E. The toxin is responsible for symptoms of staphylococcal food poisoning. It causes emesis in young rhesus monkeys. An *in vitro* precipitin test is used for assay. Protein hydrolysate media + niacin favor production; 200–500 μg/ml of toxin B are found in culture filtrates. Enterotoxins are produced by more than 50% of strains. They are synthesized on the outer surface of the cell. Cation exchange chromatography is the main method of purification. Toxins A, B, C_1, and C_2 are well characterized. They are proteins. $S_{20,w}$ = 2.8–3.0 S, MW = 30,000–34,700, pIs = 6.8, 8.6, 8.6, and 7.0 respectively. The amino acid sequence of toxin B has been determined; it has a single polypeptide chain and one disulfide bridge. Toxin B contains 200 emetic doses/mg (monkey); the presumed minimal dose for man is 1 μg. The site of emetic action is in the abdominal viscera; the sensory stimulus reaches the vomiting center via the vagus and sympathetic nerves. Symptoms of food poisoning may arise from indirect action on the nervous system. Some properties of enterotoxin are similar to those of the lipopolysaccharide endotoxin of Gram-negative bacteria. Specific antisera are produced in rabbits by using purified toxins; some cross reaction occurs with heterologous antisera. — References 39–42

Toxin	Significant Features	References
Leukocidin	Leukocidin is a complex of two proteins, F and S, which together are toxic to polymorphonuclear leukocytes and macrophages of man and rabbits; no other cell type is affected. It is distinct from other cytolytic toxins of *Staphylococcus aureus*. F and S components are separated by ion exchange chromatography; both components have been crystallized. Both F and S are proteins; $S_{20,w}$ of F = 3.0 S, MW = 32,000; $S_{20,w}$ of S = 3.3 S, MW = 38,000; pI = 9.0. Primarily, the toxin causes increased permeability of leukocytes to cations. The F component interacts with the hydrophobic region of phospholipids; the S component does not. Both F and S act on triphosphoinositide of the leukocyte cell membrane. Evidence suggests complex interaction between F and S components and the leukocyte membrane phospholipid. Their roles in pathogenicity have not been established. Both F and S are antigenic.	43–46
Epidermolytic toxin (exfoliatin)	Exfoliatin is produced by phage group II staphylococci, which are isolated from impetigo and scalded skin syndrome. It is a protein. MW = 24,000–33,000. Epidermolytic toxin is distinct from a- and δ-toxins. It causes extensive splitting of the epidermis in neonatal mice.	47–49

Streptococcus pyogenes

Streptolysin O (SLO)	This oxygen-labile hemolysin is produced by strains of groups A, C and G. It has properties in common with other oxygen-labile hemolysins, namely pneumolysin, tetanolysin, *Clostridium perfringens* θ-toxin, and the cereolysin of *Bacillus cereus*. All are neutralized by hyperimmune horse anti-SLO, activated by reducing agents, and inhibited by cholesterol. MW of SLO = 80,000, $2-5 \times 10^5$ HU/mg protein, 1×10^3 LD_{50}/mg/20 g animal tissue (mouse, rabbit, guinea pig). SLO is rapidly lethal. It is a potent cardiotoxic agent, exerting a two-phase action on a perfused heart; it is also dermonecrotic, and toxic for tissue cells *in vitro*. SLO binds to cholesterol in cell membranes. It is inhibited non-specifically by certain serum proteins. The level of specific anti-SLO antibody is useful in the diagnosis of streptococcal disease.	50–54
Streptolysin S (SLS)	This oxygen-stable hemolysin is produced by strains of groups A, C and G. Formation of SLS is induced by serum, yeast RNA, serum albumin, and detergents. The toxin is produced in washed metabolizing cell suspensions and in growing cultures. It exists as a complex with non-specific carriers. RNA-SLS is a carrier-hemolysin complex of polypeptide and oligonucleotide (molar ratio 0.3:1). MW = 12,000–20,000, $S_{20,w}$ = 2.4 S, 2×10^6 HU/mg; the estimated MW of the polypeptide portion is 2,800. Of the twelve amino acids present, none are aromatic. The mechanism of release of SLS is extremely complex. The toxin can exist in cell-bound form and may occupy an intermediate position between Class A and Class B toxins. It is inhibited by phospholipid, aniline dyes, and certain	55–59

Toxin	Significant Features	References
Streptococcus pyogenes Streptolysin S (SLS) (*cont.*)	proteolytic enzymes. Cytotoxic for many cell types, SLS inhibits mitochondrial respiration and disrupts lysosomes; membrane phospholipids probably are the target site. The toxin is non-antigenic. Its role in pathogenicity is not known.	
Erythrogenic toxins	At least three serologically distinct toxins (A, B, and C) are known. They are responsible for the erythematous rash of scarlet fever. A and C toxins are highly purified. Toxin A is 80% protein and 20% hyaluronic acid. $S_{20,w}$ = 1.8–2.7 S, MW = 27,000, 1 x 10^9 skin test doses/mg (man). Evidence suggests that erythrogenic, pyrogenic, and endotoxin-stimulating activities are due to the same entity. The toxins have been isolated from extracts of streptococcal lesions. The mechanism of erythrogenic reaction is complex; it probably involves elements of direct toxic action and delayed-type hypersensitivity.	60–62
Clostridium perfringens		
a-Toxin	This is the main lethal toxin of *Clostridium perfringens* type A strains that cause gas gangrene and food poisoning in man. Its cultural requirements are complex; the peptide components of peptone stimulate production. Many purification procedures have been published; the toxin is difficult to separate from other extracellular products; electrofocusing produces purified toxin in high yield. The toxin exists in two forms: a_A, the main form, pI = 5.5, and a_B, pI = 5.25. A variety of MWs have been reported: 31,000, 53,000, and 106,000. The toxin is lethal. It is a "hot-cold" hemolysin, to which sheep RBC is the most sensitive. The toxin causes turbidity of egg-yolk emulsions and opacification of serum. It is an enzyme (phospholipase C), and requires divalent cations. It hydrolyzes phosphatidyl choline and sphingomyelin, but its action on other phospholipids remains in question. a-Toxin is cytotoxic; it impairs all respiration. Its role in gas gangrene is controversial. The toxin is antigenic.	63–66
β-Toxin	Produced by type B and C strains, which cause enterotoxemia in animals, β-toxin is lethal. It has not been characterized. It is assayed by the lethal effect in mice in the presence of appropriate antiserum.	67
ε-Toxin	ε-Toxin is produced by type C and D strains. It is excreted as a low-activity "protoxin", which is activated by proteolytic enzymes. Toxicity is expressed as MLD after activation. No marked alteration of structure occurs on activation. $S_{20,w}$ of ε-protoxin = 2.48–2.85 S, MW = 25,000–40,500; $S_{20,w}$ of ε activated toxin = 2.8 S, MW = 38,000; 4.5 x 10^5 MLD/mg protein. The toxin affects the permeability of the small intestine and accumulates in brain tissue; it causes edema and necrosis of nervous tissue. Active immunization protects against disease.	67–70

Toxin	Significant Features	References
ι-Toxin	This toxin is produced by type E strains and is activated by proteolytic enzymes. Necrotic and lethal, it causes increased permeability of capillary blood vessels. The toxin is assayed by the lethal effect in mice in the presence of type A antiserum to neutralize α-toxin. It has not been characterized.	67, 71
θ-Toxin	An oxygen-labile hemolysin, θ-toxin is activated by reducing agents. It is inhibited by cholesterol and anti-streptolysin O. It can be purified by electrofocusing (pI = 6.5). MW = 61,000. θ-Toxin is cardiotoxic. Its role in pathogenicity is unknown.	64, 67, 72
Vibrio cholerae		
Cholera enterotoxin	Enterotoxin is distinct from endotoxin. It acts on gut epithelium, causing excessive loss of electrolyte and water into the gut lumen. It is released into culture filtrates during logarithmic growth. MW of the purified active toxin = 90,000; it is homogeneous by several criteria. A naturally occurring toxoid is also found in the filtrates: MW of the toxoid = 60,000. Several animal models are used in the assay of the toxin. These are ligated segments of small intestine of rabbits, oral administration to suckling rabbits, isolated segments of small bowel in dogs, and isolated ileal mucosa of dogs. The toxin may activate adenylcyclase of the intestinal epithelial cell membrane. It is inactivated by the ganglioside of gut epithelium. The permeability factor (PF) causes increased permeability of the small blood vessels of skin. PF is probably identical to the enterotoxin. PF activity provides convenient alternative assay.	73–76
Bacillus anthracis		
Anthrax toxin	A complex toxin consisting of three antigenic components: edema factor (EF, I), protective antigen (PA, II), and lethal factor (LF, III). Full toxicity requires all three factors. The toxic complex is lethal and dermonecrotic. It is assayed by the lethal action in rats (Fischer 344 albino strain) or by edema in guinea pigs and rabbits. The toxin was originally isolated from the plasma of infected guinea pigs, but is now produced *in vitro;* HCO_3^- and a sugar source are important in production. The components are proteins or lipoproteins; they probably form aggregates. The primary effect of the toxin complex may be on the respiratory center of the central nervous system. Mice, guinea pigs, rabbits, and monkeys are relatively resistant to the lethal effect. Each of the toxin components, alone or in combination, induces varying degrees of immunity to the toxin complex and to anthrax spores in experimental animals.	77–80
Bacillus cereus		
Lethal toxin, phospholipase C, hemolysin	The lethal toxin, phospholipase C, and the hemolysin have been separated by gel filtration. Phospholipase C	81–83

Toxin	Significant Features	References
Bacillus cereus Lethal toxin, phospholipase C, hemolysin (*cont.*)	acts on phosphatidyl ethanolamine and phosphatidyl choline; it has no requirement for divalent cations. The three factors differ in thermal inactivation. The lethal toxin acts rapidly and differs from anthrax toxin in toxicity for mice. These toxins are immunologically distinct. $S_{20,w}$ of the hemolysin (cereolysin) = 3.7 S, MW = 50,000, 2×10^6 HU/mg; it resembles other oxygen-labile hemolysins.	

Class C. Toxins Located Intracellulary and Extracellulary

Toxin	Significant Features	References
Clostridium tetani Tetanus toxin (tetanospasmin)	This toxin is responsible for symptoms of tetanus. The guinea pig is the most susceptible species. The toxin is assayed by its lethal effect in mice. It has been purified in several laboratories and was crystallized by Pillemer and coworkers. It probably exists as monomer and dimer. Both forms are proteins. Respective $S_{20,w}$ for monomer and dimer = 3.9 S and 7.8 S, respective MW = 67,000 and 148,000, 3.0×10^7 MLD/mg (mouse). Maximal release of the toxin occurs after autolysis. It has been estimated that 5 to 10% of bacterial dry weight is toxin. General tetanus and local tetanus have been reproduced in experimental animals. The toxin is fixed specifically by ceramidyloligosaccharides (gangliosides); optimal fixation is achieved by mixtures of ganglioside and cerebroside (1:3). Tetanospasmin suppresses synaptic inhibition in the spinal cord. It causes increased activity of α-motoneurones but has no effect on γ-motoneurones. Peripheral action may involve blocking of neuromuscular transmission. Tetanus toxin is antigenic; active immunization prevents tetanus.	84–89
Clostridium botulinum Botulinum toxins	Six serologically distinct neurotoxins (A, B, C, D, E, and F) are known. Type A, a crystalline toxin, is a protein. $S_{20,w}$ = 17.3 S, MW = 900,000. It is not homogeneous and can be resolved into α and β components. The α component is lethal. $S_{20,w}$ = 7.0 S, MW = 150,000. The β component is a hemagglutinin. $S_{20,w}$ = 13 S, MW = 500,000. Antibody to the α component neutralizes toxicity. The β component can be further resolved to β_1, β_2, and β_3. Crystalline type A toxin possesses 4×10^7 MLD/mg (mouse). Type B toxin MW = 10,000–500,000; it contains 4×10^7 MLD/mg (mouse). Type C toxin contains 5×10^6 MLD/mg (mouse). Type E toxin is potentiated by proteolytic enzymes; it contains 1.3×10^7 MLD/mg (mouse) after activation. MW of Type E = 200,000; a small-molecular-weight peptide (MW = 18,600) has been isolated from some Type E toxin preparations. Botulinum toxins cause presynaptic inhibition of transmission at neuromuscular junction, paralyzing cholinergic fibers at the point of acetylcholine release. Antagonism of Ca^{2+} transport has been suggested to explain the mechanism of action. Death results from suffocation following paralysis of the diaphragm. The toxins are antigenic; active immunization protects.	90–95

REFERENCES

1. Raynaud and Alouf, in *Microbial Toxins,* Vol. 1, p. 67, Ajl, Kadis and Montie, Eds. Academic Press, New York (1970).
2. Alouf and Raynaud, in *Microbial Toxins,* Vol. 1, p. 119, Ajl, Kadis and Montie, Eds. Academic Press, New York (1970).
3. Boivin and Mesrobeanu, *C.R. Acad. Sci., 204,* 302 (1937).
4. van Heyningen and Gladstone, *Br. J. Exp. Pathol., 34,* 202 (1953).
5. Bridgewater, Morgan, Rowson and Wright, *Br. J. Exp. Pathol., 36,* 447 (1955).
6. Cavanagh, Howard and Whitby, *Br. J. Exp. Pathol., 37,* 272 (1956).
7. van Heyningen, in *Microbial Toxins,* Vol. 2A, p. 255, Kadis, Montie and Ajl, Eds. Academic Press, New York (1971).
8. Robbins and Pillemer, *Proc. Soc. Exp. Biol. Med., 74,* 75 (1950).
9. Munoz, Ribi and Larson, *J. Immunol., 83,* 496 (1959).
10. Banerjea and Munoz, *J. Bacteriol., 84,* 269 (1962).
11. Munoz, in *Microbial Toxins,* Vol. 2A, p. 271, Kadis, Montie and Ajl, Eds. Academic Press, New York (1971).
12. Ajl, Rust, Hunter, Woebke and Bent, *J. Immunol., 80,* 435 (1958).
13. Montie, Montie and Ajl, *Biochim. Biophys. Acta, 130,* 406 (1966).
14. Montie and Ajl, in *Microbial Toxins,* Vol. 3, p. 1, Montie, Kadis and Ajl, Eds. Academic Press, New York (1970).
15. Kadis and Ajl, in *Microbial Toxins,* Vol. 3, p. 39, Montie, Kadis and Ajl, Eds. Academic Press, New York (1970).
16. Hauschild and Hilsheimer, *Can. J. Microbiol., 17,* 1425 (1971).
17. Stark and Duncan, *Infect. Immun., 4,* 89 (1971).
18. Cooksey, in *Microbial Control of Insects and Mites,* p. 247, Burges and Hussey, Eds. Academic Press, New York (1971).
19. Lecadet, in *Microbial Toxins,* Vol. 3, p. 437, Montie, Kadis and Ajl, Eds. Academic Press, New York (1970).
20. Pappenheimer and Johnson, *Br. J. Exp. Pathol., 17,* 335 (1936).
21. Barksdale, *C.R. Acad. Sci., 240,* 1831 (1955).
22. Pope and Stevens, *Br. J. Exp. Pathol., 39,* 139 (1958).
23. Kato, Nakamura, Uchida, Koyama and Katsura, *Jap. J. Exp. Med., 30,* 129 (1960).
24. Relyveld and Raynaud, *Ann. Inst. Pasteur, 107,* 618 (1964).
25. Strauss and Hendee, *J. Exp. Med., 109,* 145 (1959).
26. Collier and Pappenheimer, *J. Exp. Med., 120,* 1007, 1019 (1964).
27. Honjo, Nishizuka, Hayaishi and Kato, *J. Biol. Chem., 243,* 3553 (1968).
28. Bernheimer and Schwartz, *J. Gen. Microbiol., 30,* 455 (1963).
29. Arbuthnott, Freer and Bernheimer, *J. Bacteriol., 94,* 1170 (1967).
30. Freer, Arbuthnott and Bernheimer, *J. Bacteriol., 95,* 1153 (1968).
31. Wadström, *Biochim. Biophys. Acta, 168,* 228 (1968).
32. Arbuthnott, in *Microbial Toxins,* Vol. 3, p. 189, Montie, Kadis and Ajl, Eds. Academic Press, New York (1970).
33. Doery, Magnusson, Gulasekharam and Pearson, *J. Gen. Microbiol., 40,* 283 (1965).
34. Wiseman and Caird, *Can. J. Microbiol., 13,* 369 (1967).
35. Wiseman, in *Microbial Toxins,* Vol. 3, p. 237, Montie, Kadis and Ajl, Eds. Academic Press, New York (1970).
36. Wadström and Möllby, *Biochim. Biophys. Acta, 242,* 288, 308 (1971).
37. Wiseman and Caird, *Proc. Soc. Exp. Biol. Med., 128,* 428 (1968).
38. Kreger, Kim, Zaboretsky and Bernheimer, *Infect. Immun., 3,* 449 (1971).
39. Bergdoll, Surgalla and Dack, *J. Immunol., 83,* 334 (1959).
40. Shantz, Roessler, Wagman, Spero, Dunnery and Bergdoll, *Biochemistry, 4,* 1011 (1965).
41. Sugiyama and Hayama, *J. Infect. Dis., 115,* 330 (1965).
42. Bergdoll, in *Microbial Toxins,* Vol. 3, p. 265, Montie, Kadis and Ajl, Eds. Academic Press, New York (1970).
43. Woodin, *Biochem. J., 75,* 158 (1960).
44. Woodin and Wieneke, *Biochem. J., 99,* 479 (1966).
45. Woodin and Wieneke, *Biochem. J., 105,* 1029 (1967).
46. Woodin, in *Microbial Toxins,* Vol. 3, p. 327, Montie, Kadis and Ajl, Eds. Academic Press, New York (1970).
47. Arbuthnott, Kent, Lyell and Gemmell, *Br. J. Dermatol., 85,* 145 (1971).
48. Kapral and Miller, *Infect. Immun., 4,* 541 (1971).
49. Arbuthnott, Kent, Lyell and Gemmell, *Br. J. Dermatol., 86, Suppl. 8,* 35 (1972).
50. Halbert and Auerbach, *J. Exp. Med., 113,* 131 (1961).
51. Alouf and Raynaud, *C.R. Acad. Sci. Sect. D, 264,* 2524 (1967).
52. Reitz, Prager and Feigen, *J. Exp. Med., 128,* 1401 (1968).
53. Alouf and Raynaud, *Ann. Inst. Pasteur, 115,* 97 (1968).
54. Halbert, in *Microbial Toxins,* Vol. 3, p. 69, Montie, Kadis and Ajl, Eds. Academic Press, New York (1970).
55. Bernheimer, in *Streptococcal Infections,* p. 19, McCarty, Ed. Columbia University Press, New York (1954).

56.   Koyama and Egami, *J. Biochem. (Tokyo)*, *55*, 629 (1964).
57.   Elias, Heller and Ginsburg, *Isr. J. Med. Sci.*, *2*, 302 (1966).
58.   Bernheimer, *J. Bacteriol.*, *93*, 2024 (1967).
59.   Ginsburg, in *Microbial Toxins*, Vol. 3, p. 99, Montie, Kadis and Ajl, Eds. Academic Press, New York (1970).
60.   Stock, *J. Biol. Chem.*, *142*, 777 (1942).
61.   Watson, *J. Exp. Med.*, *111*, 255 (1960).
62.   Watson and Kim, in *Microbial Toxins*, Vol. 3, p. 173, Montie, Kadis and Ajl, Eds. Academic Press, New York (1970).
63.   Macfarlane and Knight, *Biochem. J.*, *35*, 884 (1941).
64.   Haberman, *Arch. Exp. Pathol. Pharmakol.*, *235*, 513 (1959).
65.   Shemanova, Vlasova, Tsvetkov, Logunov and Levin, *Biokhimija*, *33*, 130 (1968).
66.   Bernheimer, Grushoff and Avigad, *J. Bacteriol.*, *95*, 2439 (1968).
67.   Hauschild, in *Microbial Toxins*, Vol. 2A, p. 159, Kadis, Montie and Ajl, Eds. Academic Press, New York (1971).
68.   Griner, *Am. J. Vet. Res.*, *22*, 429 (1961).
69.   Thompson, *J. Gen. Microbiol.*, *31*, 79 (1963).
70.   Habeeb, *Can. J. Biochem.*, *42*, 545 (1964).
71.   Craig and Miles, *J. Pathol. Bacteriol.*, *81*, 481 (1961).
72.   Smyth, Ph.D. Thesis, University of Glasgow, Bearsden, Glasgow, Scotland (1972).
73.   De and Chatterje, *J. Pathol. Bacteriol.*, *66*, 559 (1953).
74.   Finkelstein and LoSpalluto, *J. Infect. Dis.*, *121*, *Suppl.*, S63 (1970).
75.   Carpenter, *Am J. Med.*, *50*, 1 (1971).
76.   Craig, in *22nd Symposium of the Society for General Microbiology*, p. 129, Smith and Pearce, Eds. Cambridge University Press, New York (1972).
77.   Smith and Keppie, in *5th Symposium of the Society for General Microbiology*, p. 126, Howie and O'Hea, Eds. Cambridge University Press, New York (1955).
78.   Stanley and Smith, *J. Gen. Microbiol.*, *26*, 49 (1961).
79.   Fish, Mahlandt, Dobbs and Lincoln, *J. Bacteriol.*, *95*, 907 (1968).
80.   Lincoln and Fish, in *Microbial Toxins*, Vol. 3, p. 361, Montie, Kadis and Ajl, Eds. Academic Press, New York (1970).
81.   Johnson and Bonventre, *J. Bacteriol.*, *94*, 306 (1967).
82.   Bernheimer and Grushoff, *J. Gen. Microbiol.*, *46*, 143 (1967).
83.   Bonventre and Johnson, in *Microbial Toxins*, Vol. 3, p. 415, Montie, Kadis and Ajl, Eds. Academic Press, New York (1970).
84.   Pillemer, Wittler, Burrel and Grossberg, *J. Exp. Med.*, *88*, 205 (1948).
85.   Mangalo, Bizzini, Turpin and Raynaud, *Biochim. Biophys. Acta*, *168*, 583 (1969). ·
86.   van Heyningen and Miller, *J. Gen. Microbiol.*, *24*, 107 (1961).
87.   Curtis, *J. Physiol.*, *145*, 175 (1959).
88.   van Heyningen and Mellanby, in *Microbial Toxins*, Vol. 2A, p. 69, Kadis, Montie and Ajl, Eds. Academic Press, New York (1971).
89.   Diamond and Mellanby, *J. Physiol.*, *210*, 186P (1970).
90.   Putnam, Lamanna and Sharp, *J. Biol. Chem.*, *165*, 735 (1946).
91.   Wagman, *Arch. Biochem. Biophys.*, *100*, 414 (1963).
92.   DasGupta, Boroff and Rothstein, *Biochem. Biophys. Res. Commun.*, *22*, 750 (1966).
93.   Sakaguchi and Sakaguchi, *J. Bacteriol.*, *78*, 1 (1959).
94.   Ambache, *J. Physiol.*, *108*, 127 (1949).
95.   Boroff and DasGupta, in *Microbial Toxins*, Vol. 2A, p. 1, Kadis, Montie and Ajl, Eds. Academic Press, New York (1971).

MYCOTOXINS

DR. ALEX CIEGLER

The mycotoxins are a large group of unrelated toxic secondary metabolites produced by fungi. The disease initiated by a mycotoxin is a mycotoxicosis, as distinguished from a mycosis, and does not involve an invasion of the host by the fungus. The earliest recognized form of a mycotoxicosis, ergotism, has been known to man for much of his recorded history. Much of the early work on mycotoxins was carried out in Japan and Russia, but mycotoxicoses acquired world-wide interest with the discovery of the aflatoxins in 1960 in England.

Certain useful diagnostic features characterize outbreaks of mycotoxicoses: (a) the diseases are not transmissible; (b) drug and antibiotic treatment have little or no effect on the disease; (c) in field outbreaks, the trouble is often seasonal; (d) the outbreak is usually associated with a specific food or feedstuff; and (e) examination of the suspected food or feed reveals signs of fungal activity.

A large number of reviews have appeared on the subject. The most recent of those are listed below.

1. Goldblatt, L. A., *Aflatoxin*. Academic Press, New York, 1969.

2. Lillehoj, E. B., Ciegler, A., and Detroy, R. W., Fungal Toxins, in *Essays in Toxicology*, pp. 1–154, F. Blood, Ed. Academic Press, New York, 1970.

3. Ciegler, A., Kadis, S., and Ajl, S., *Microbial Toxins*, Vols. 6, 7, and 8. Academic Press, New York, 1971.

TABLE 1
CHEMICAL STRUCTURES OF MYCOTOXINS

Name	Structure
Aflatoxin B$_1$	
Aflatoxin B$_2$	
1-acetoxy-Aflatoxin B$_2$	

TABLE 1 (Continued)
CHEMICAL STRUCTURES OF MYCOTOXINS

Name	Structure
1-ethoxy-Aflatoxin B$_2$	
1-methoxy-Aflatoxin B$_2$	
2-methoxy-Aflatoxin B$_2$	
Aflatoxin B$_{2a}$	
Aflatoxin B$_3$	
Aflatoxin G$_1$	
Aflatoxin G$_2$	

TABLE 1 (Continued)
CHEMICAL STRUCTURES OF MYCOTOXINS

Name	Structure
2-ethoxy-Aflatoxin G_2	
Aflatoxin G_{2a}	
Aflatoxin $G-M_1$	
Aflatoxin M_1	
Aflatoxin M_2	
Aflatoxin P_1	

TABLE 1 (Continued)
CHEMICAL STRUCTURES OF MYCOTOXINS

Name   Structure

Aflatoxin R_0

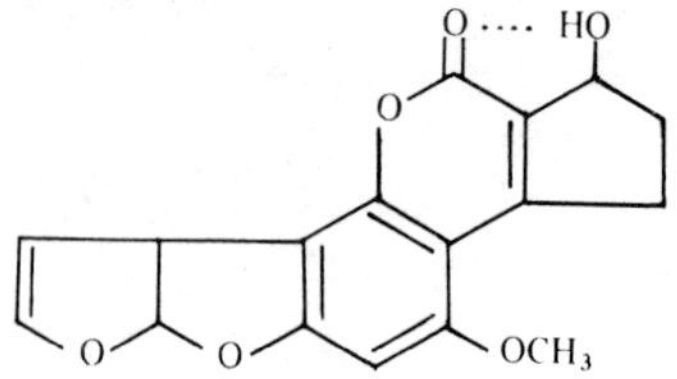

Agarin   *see* Muscimol

Agroclavine

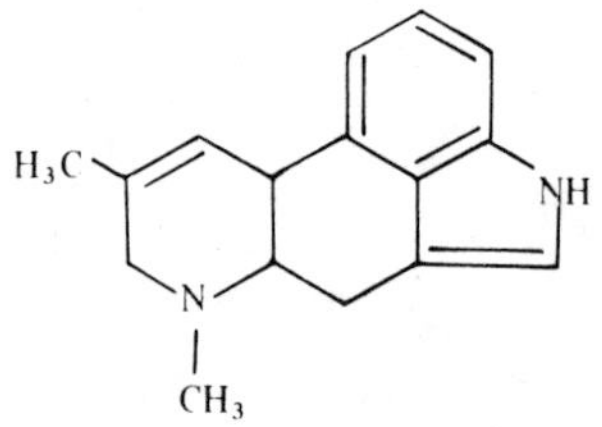

Agrocybin

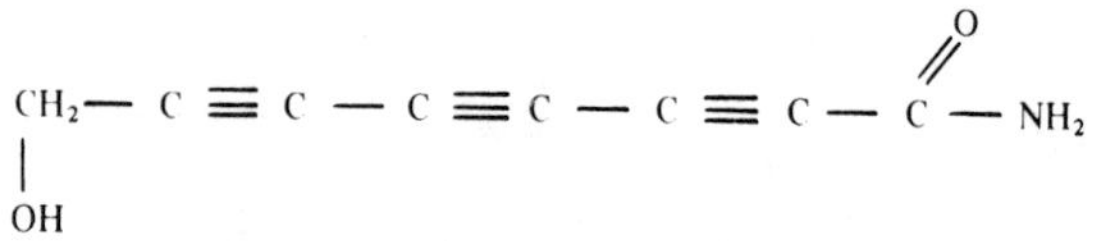

$$CH_2 - C \equiv C - C \equiv C - C \equiv C - C - NH_2$$
$$\overset{|}{OH}$$

Alpha-toxin   *see* Ibotenic acid

Alternaria longipes toxin   not known

Alternaria mali toxin   not known

Amanin   *see* β-Amanitine

Amanita toxins   *see* α-, β-, γ-, and ε-Amanitine

α-Amanitine

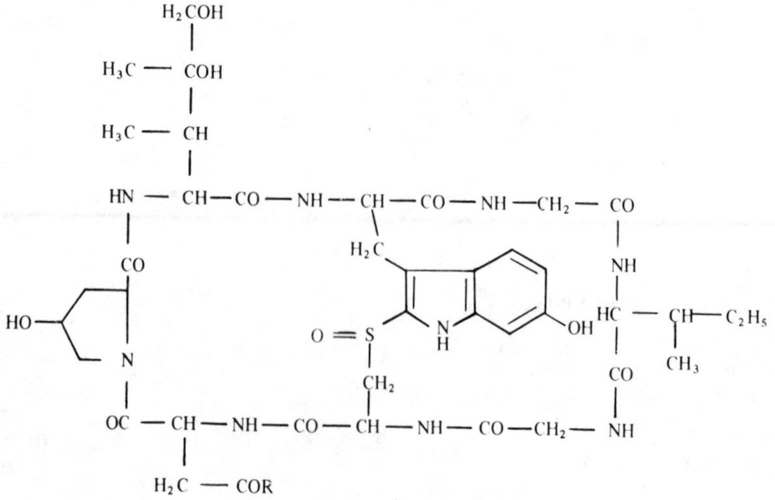

R = NH₂

TABLE 1 (Continued)
CHEMICAL STRUCTURES OF MYCOTOXINS

Name	Structure

β-Amanitine

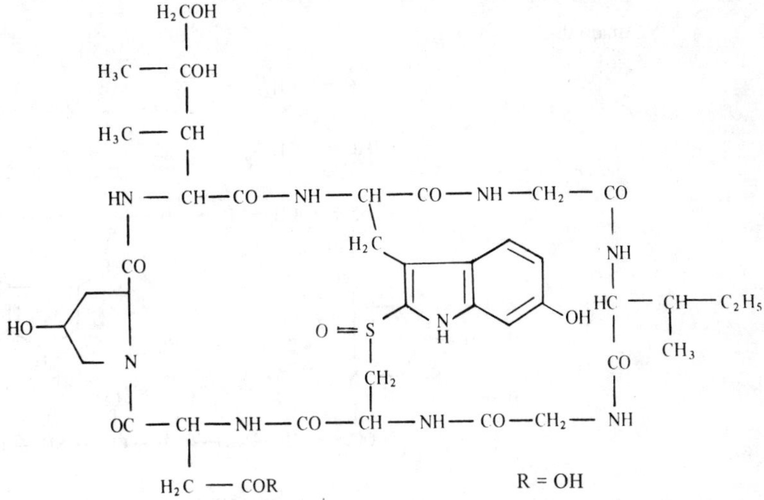

$$H_2COH$$
$$|$$
$$H_3C — COH$$
$$|$$
$$H_3C — CH$$
$$|$$
$$HN — CH—CO—NH—CH—CO—NH—CH_2 — CO$$

$$H_2C$$ NH

$$CO$$

$$HC —— CH—C_2H_5$$

$$O = S \quad N \quad OH \quad CH_3$$
$$H$$

$$HO$$ $$CO$$

$$N$$

$$CH_2$$

$$OC — CH—NH—CO—CH—NH—CO—CH_2 — NH$$

$$H_2C — COR$$ R = OH

γ-Amanitine

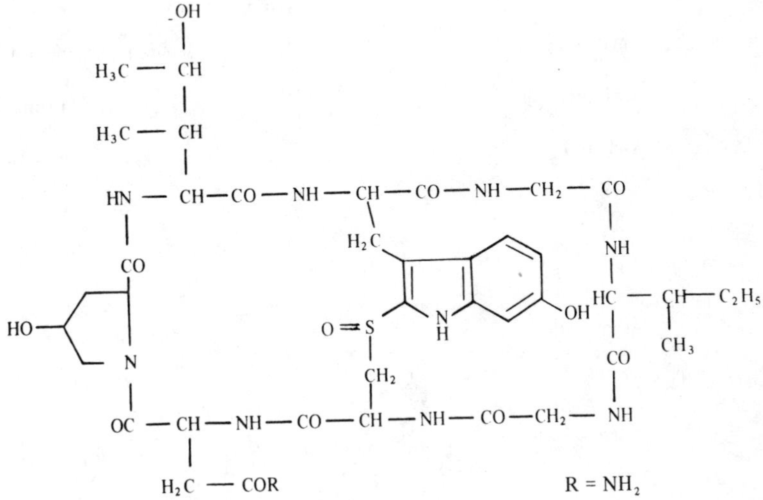

$$OH$$
$$|$$
$$H_3C — CH$$
$$|$$
$$H_3C — CH$$
$$|$$
$$HN — CH—CO—NH—CH—CO—NH—CH_2 — CO$$

$$H_2C$$ NH

$$CO$$

$$HC —— CH—C_2H_5$$

$$O = S \quad N \quad OH \quad CH_3$$
$$H$$

$$HO$$ $$CO$$

$$N$$

$$CH_2$$

$$OC — CH—NH—CO—CH—NH—CO—CH_2 — NH$$

$$H_2C — COR$$ R = NH_2

ε-Amanitine

$$OH$$
$$|$$
$$H_3C — CH$$
$$|$$
$$H_3C — CH$$
$$|$$
$$HN — CH—CO—NH—CH—CO—NH—CH_2 — CO$$

$$H_2C$$ NH

$$CO$$

$$HC —— CH—C_2H_5$$

$$O = S \quad N \quad OH \quad CH_3$$
$$H$$

$$HO$$ $$CO$$

$$N$$

$$CH_2$$

$$OC — CH—NH—CO—CH—NH—CO—CH_2 — NH$$

$$H_2C — COR$$ R = OH

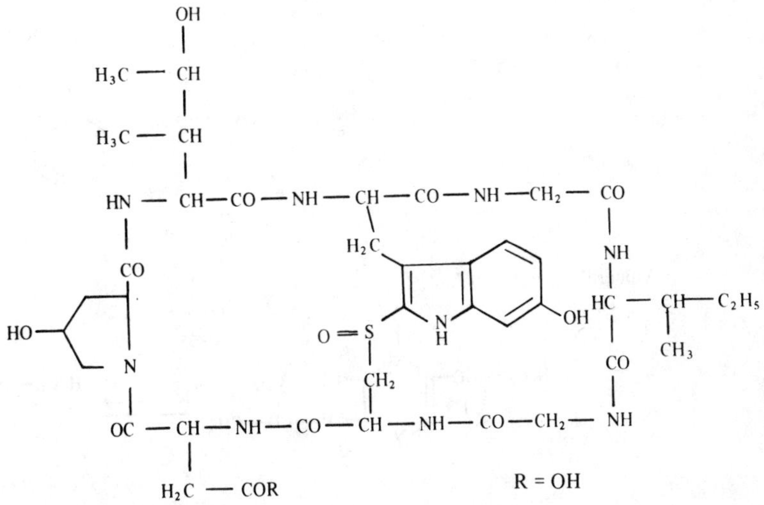

TABLE 1 (Continued)
CHEMICAL STRUCTURES OF MYCOTOXINS

Name	Structure

Amanullin

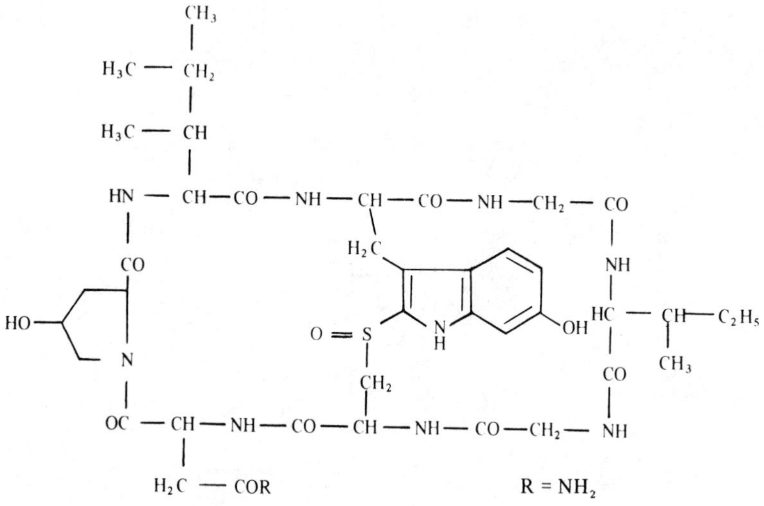

R = NH$_2$

Amatoxins *see* α-, β-, γ-, and ε-Amanitine

Antimycin *see* Citrinin

Ascladiol

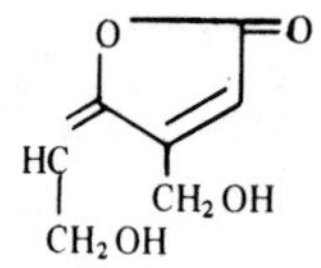

Ascotoxin *see* Decumbin

Aspercolorin

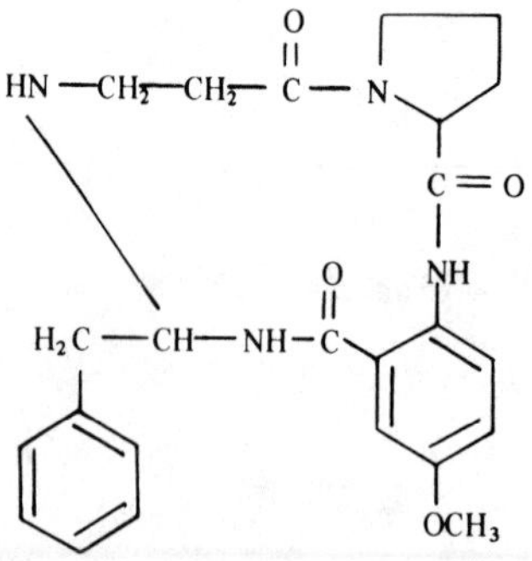

Aspergillic acid

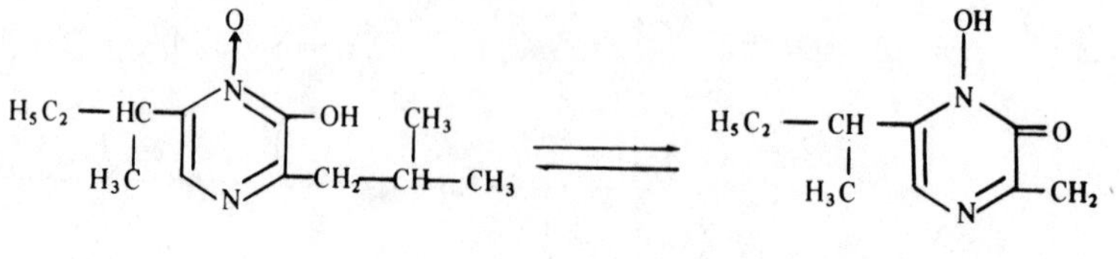

531

TABLE 1 (Continued)
CHEMICAL STRUCTURES OF MYCOTOXINS

Name	Structure

Aspertoxin

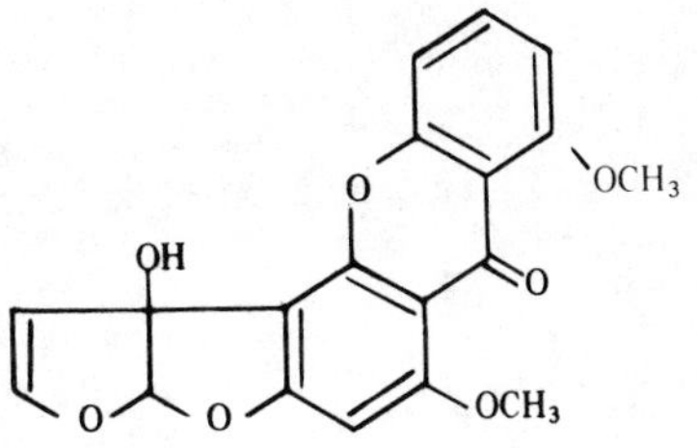

B-24 toxin   *see* Diacetoxyscirpenol

Beta-toxin   *see* Muscimol

Bufotenine

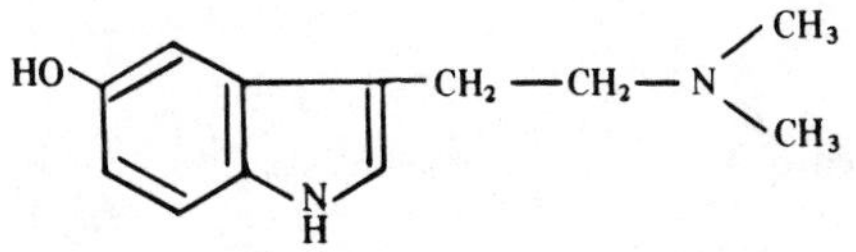

Butenolide

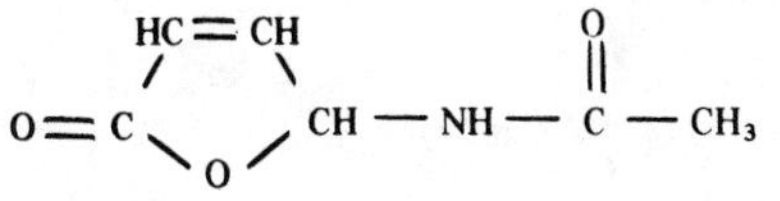

Byssochlamic acid

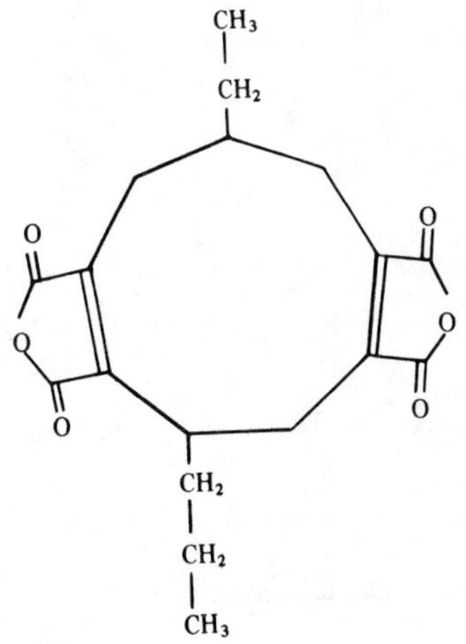

Chaetomium globosum toxin   not known

Citreoviridin

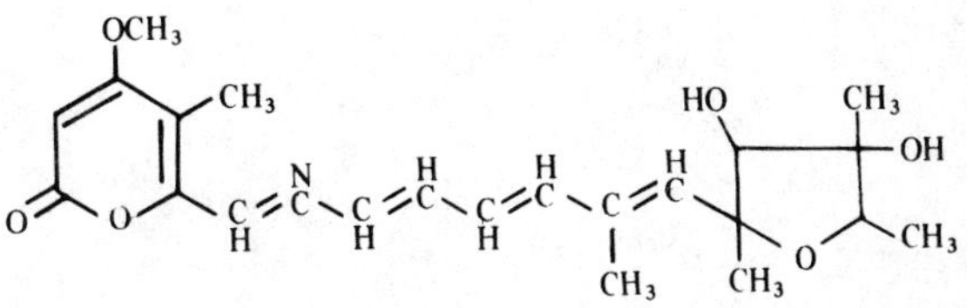

TABLE 1 (Continued)
CHEMICAL STRUCTURES OF MYCOTOXINS

Name	Structure

Citrinin

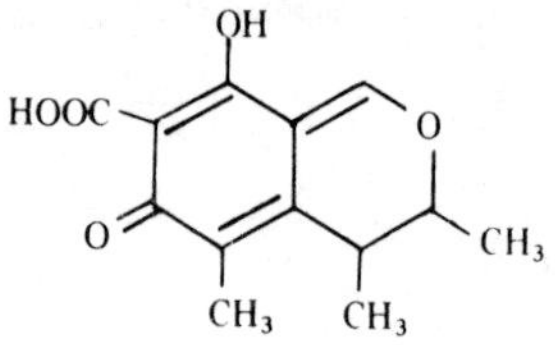

Citromycetin

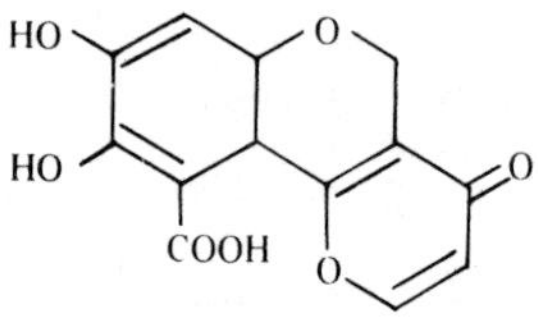

Clavacin *see* Patulin

Clavatin *see* Patulin

Claviformin *see* Patulin

Cochliobolus toxin not known

Cyclopiazonic acid

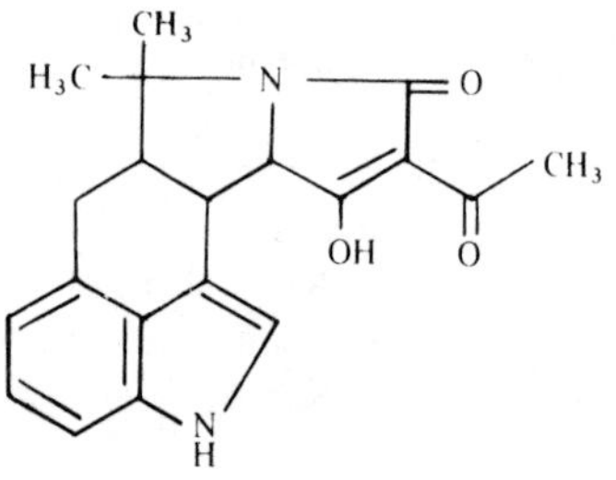

Decumbin

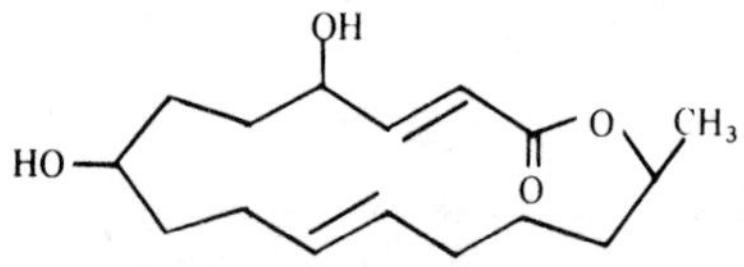

Dendrodochine not known

Diacetoxyscirpenol

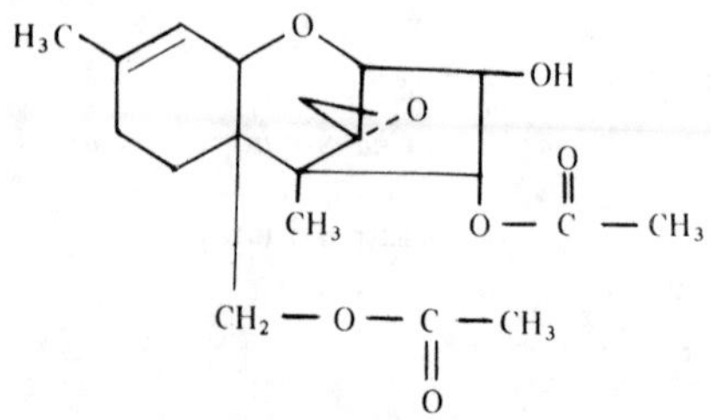

Emetic toxin not known

TABLE 1 (Continued)
CHEMICAL STRUCTURES OF MYCOTOXINS

Name	Structure

Emodin

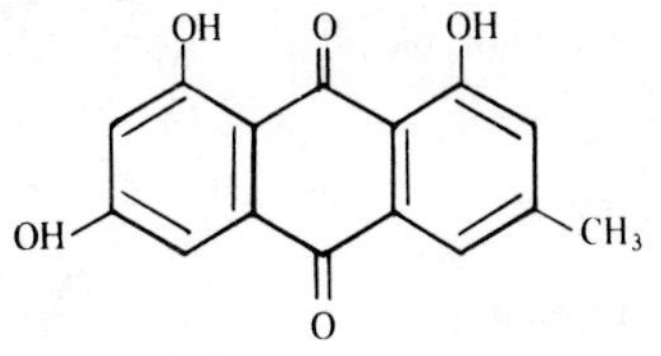

Ergobasine   *see* Ergometrine

Ergoclinine   *see* Ergometrine

Ergocornine

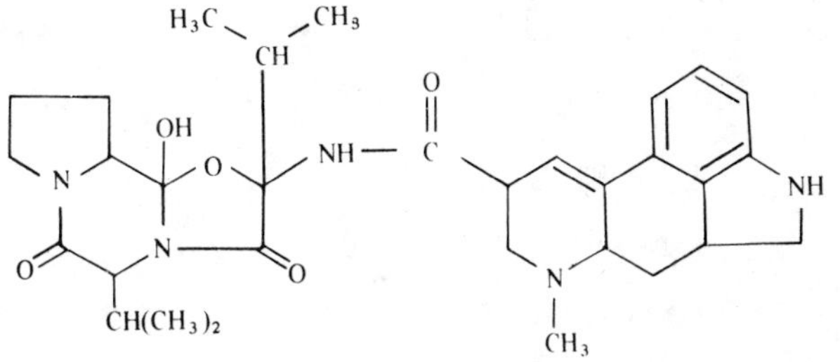

Ergocristine

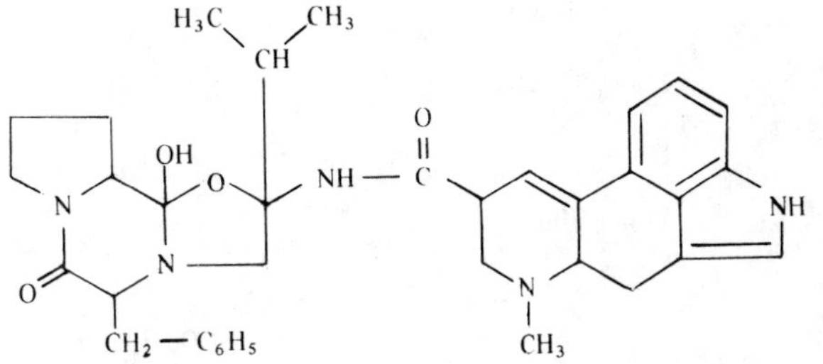

Ergocryptine

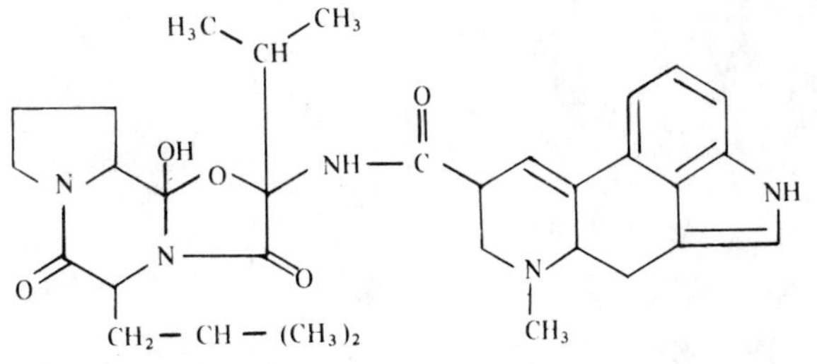

Ergometrine

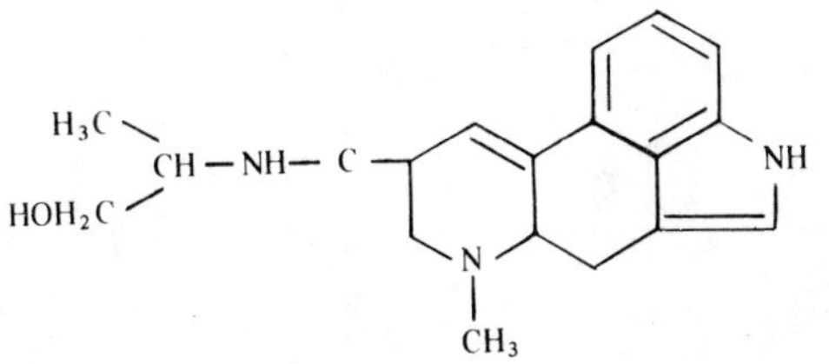

Ergonovine   *see* Ergometrine

Ergostetrine   *see* Ergometrine

TABLE 1 (Continued)
CHEMICAL STRUCTURES OF MYCOTOXINS

Name	Structure
Ergot alkaloids	*see* Agroclavine, Ergocornine, Ergocristine, Ergocryptine, Ergometrine, and Ergotamine

Ergotamine

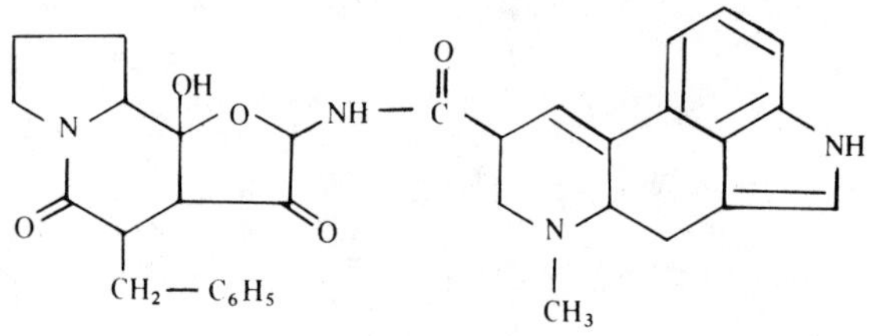

$$CH_2 - C_6H_5$$

Ergotocine	*see* Ergometrine
Ergotrate	*see* Ergometrine
Expansin	*see* Patulin
F-2 toxin	*see* Zearalenone
Fescue toxin	*see* Butenolide
Frangulic acid	*see* Emodin
Frequentic acid	*see* Citromycetin

Fumagillin

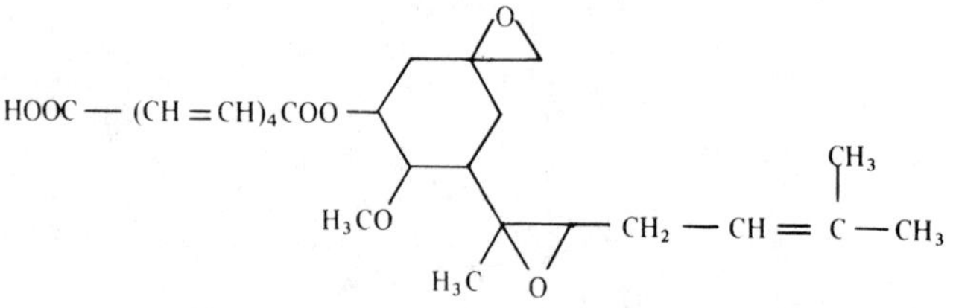

$$HOOC - (CH = CH)_4 COO -$$

Fumigacin	*see* Helvolic acid
Fumigatoxin	not known
Fusarenon	not known
Fusarium culmorum toxin	not known

Gliotoxin

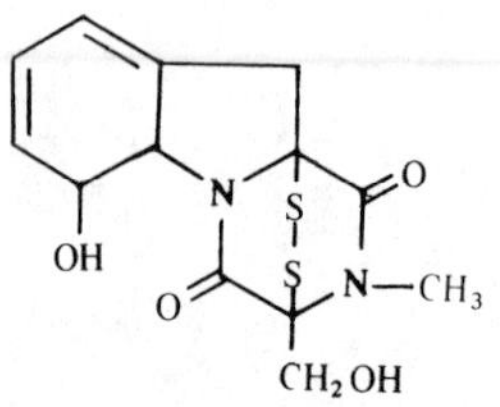

Gyromitrin

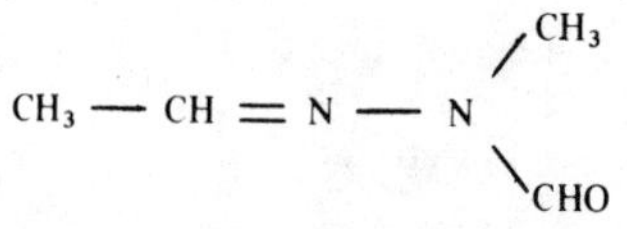

$$CH_3 - CH = N - N \begin{array}{c} CH_3 \\ CHO \end{array}$$

TABLE 1 (Continued)
CHEMICAL STRUCTURES OF MYCOTOXINS

Name	Structure

Helvolic acid

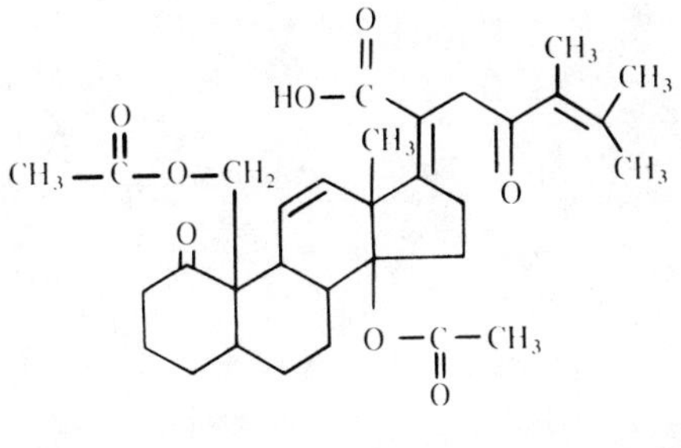

Hiptagenic acid

see β-Nitropropionic acid

HT-2 toxin

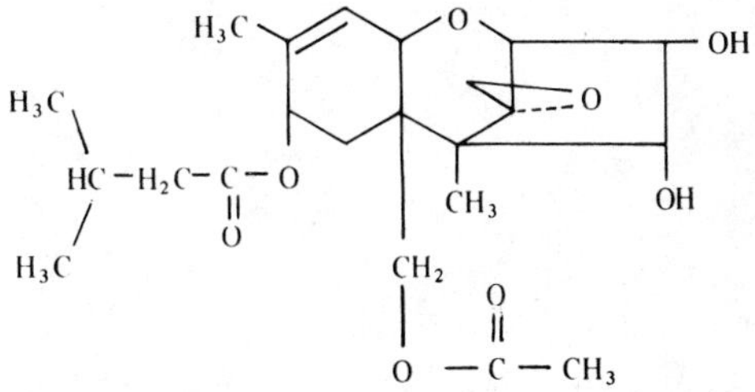

Ibotenic acid

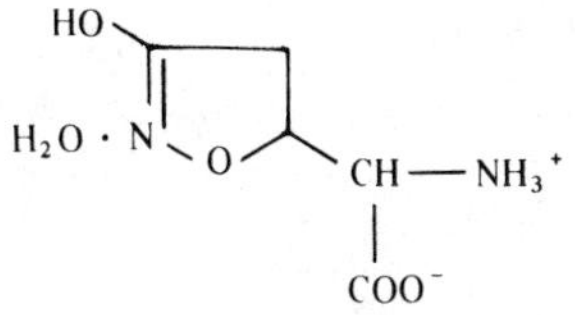

Islanditoxin

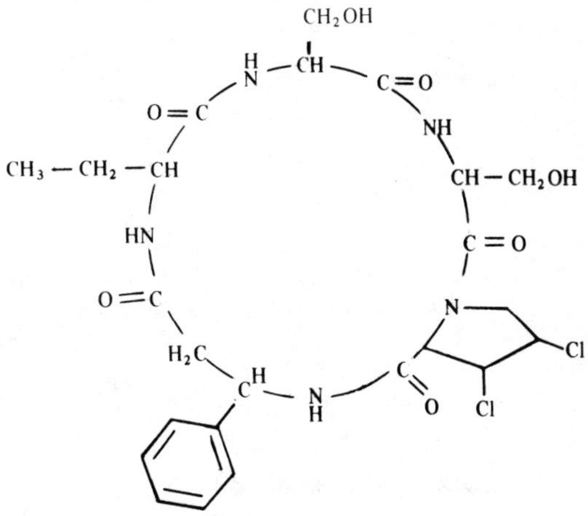

Kojic acid

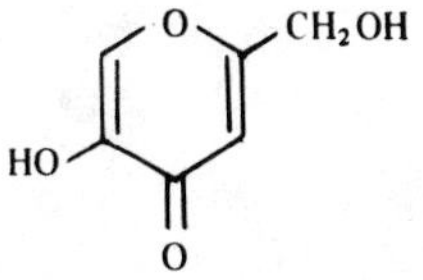

TABLE 1 (Continued)
CHEMICAL STRUCTURES OF MYCOTOXINS

Name	Structure
Luteoskyrin	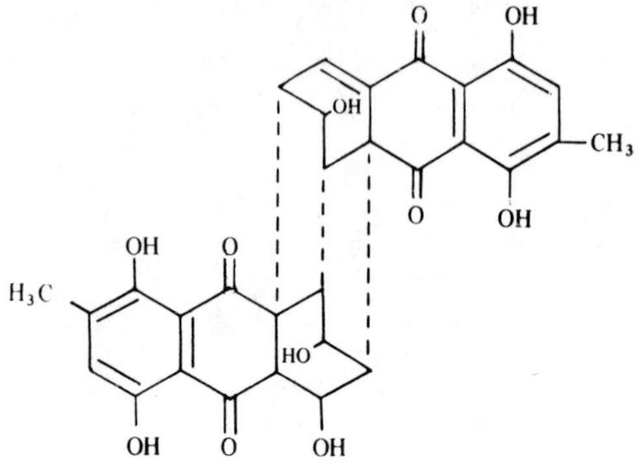
Lysergic acid	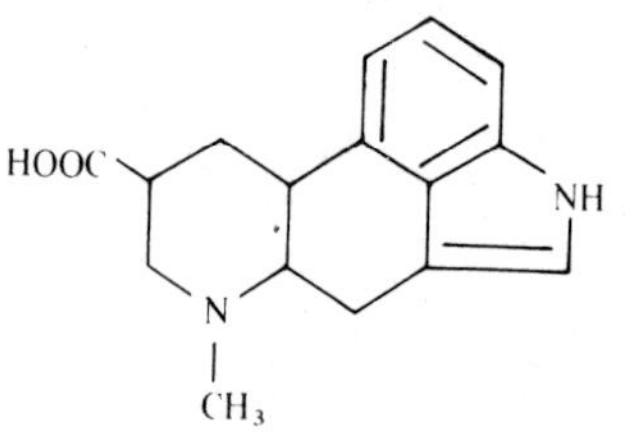
Maltoryzine	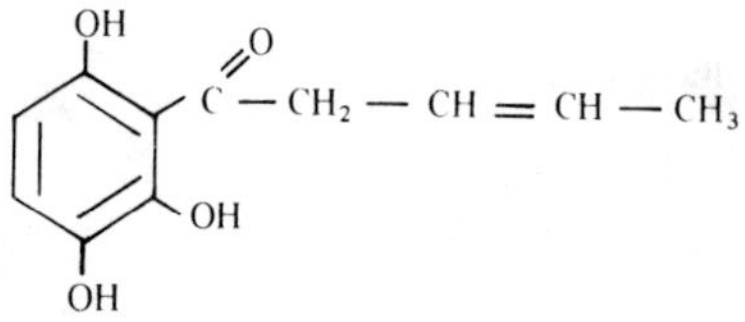
Mappine	*see* Bufotenine
Mescaline	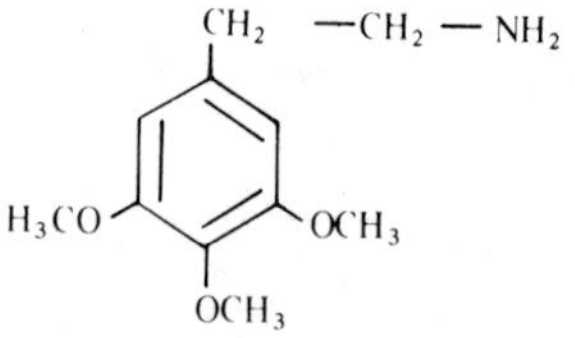
8-Methoxypsoralen	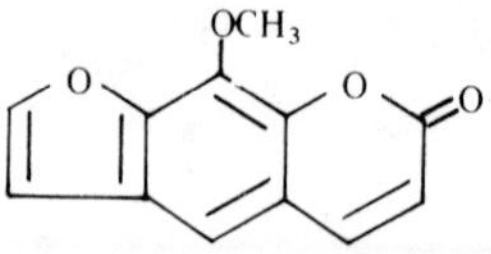
O-Methylsterigmatocystin	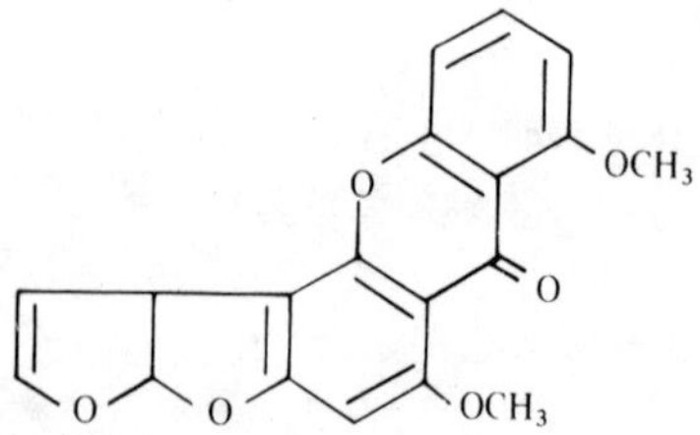

TABLE 1 (Continued)
CHEMICAL STRUCTURES OF MYCOTOXINS

Name	Structure
Muscarine	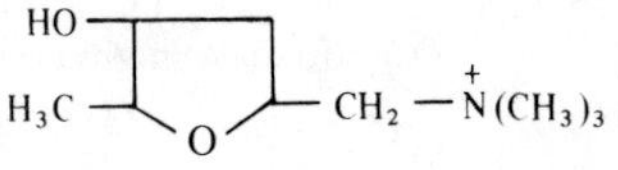
Muscazone	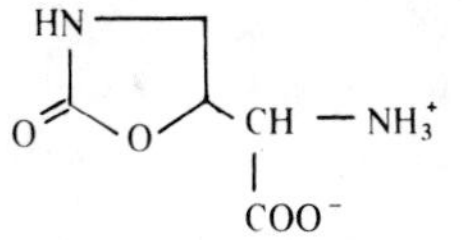
Muscimol	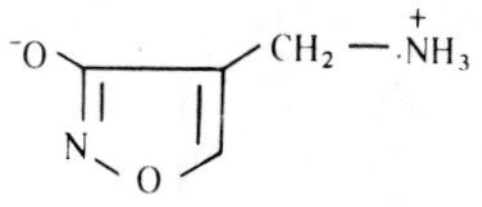
Mycoin	*see* Patulin
Mycophenolic acid	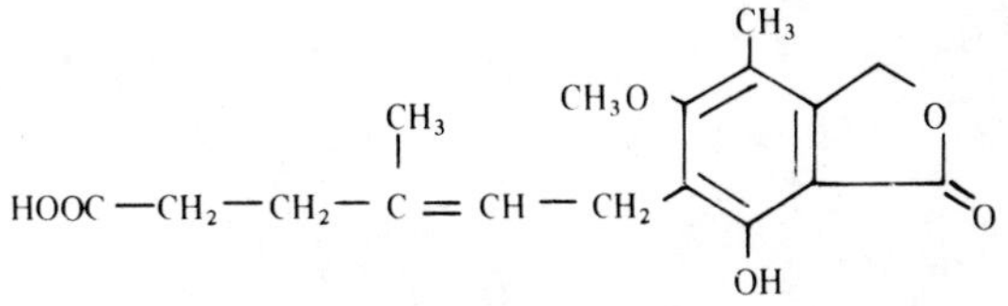
Nidulotoxin	not known
β-Nitropropionic acid	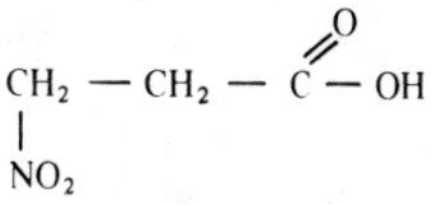
Nivalenol	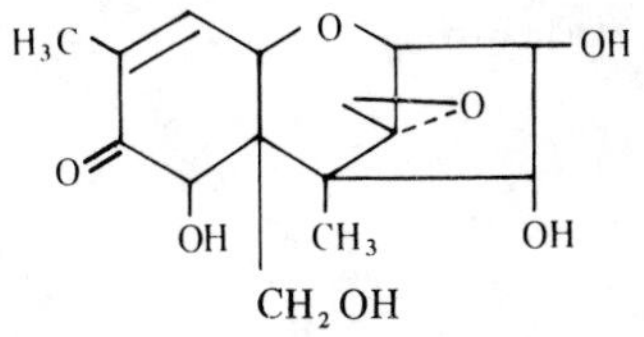
Ochratoxin A	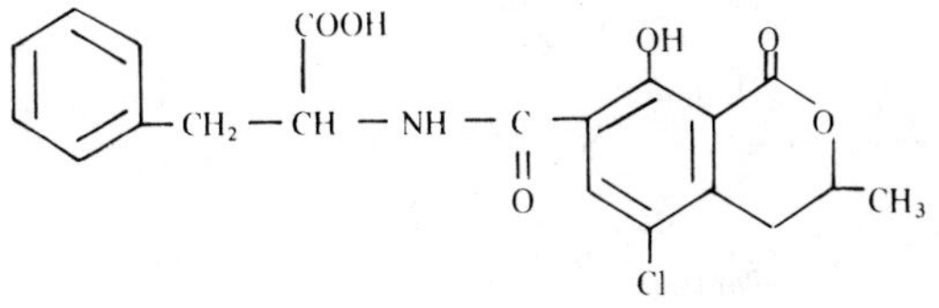
Ochratoxin A ethyl ester	*see* Ochratoxin C

Muscarine:

HO— , H$_3$C— , O , CH$_2$—$\overset{+}{N}$(CH$_3$)$_3$

Muscazone:

HN , O=, O , CH—NH$_3^+$, COO$^-$

Muscimol:

$^-$O— , N , O , CH$_2$—$\overset{+}{N}$H$_3$

Mycophenolic acid:

HOOC—CH$_2$—CH$_2$—$\overset{\overset{\displaystyle CH_3}{|}}{C}$=CH—CH$_2$, CH$_3O , CH_3$, O , O , OH

β-Nitropropionic acid:

CH$_2$—CH$_2$—$\overset{\overset{\displaystyle O}{\|}}{C}$—OH , NO$_2$

Nivalenol:

H$_3$C , O , OH , O , OH , CH$_3$, OH , CH$_2$OH

Ochratoxin A:

CH$_2$—CH—NH—$\overset{\overset{\displaystyle COOH}{|}}{\underset{\underset{\displaystyle O}{\|}}{C}}$, OH , O , O , CH$_3$, Cl

TABLE 1 (Continued)
CHEMICAL STRUCTURES OF MYCOTOXINS

Name	Structure
Ochratoxin A methyl ester	
Ochratoxin B	
Ochratoxin B ethyl ester	
Ochratoxin B methyl ester	
Ochratoxin C	
Orellanine	not known
Oxalic acid	$(COOH)_2 \cdot 2H_2O$
Pantherine	*see* Muscimol
Parasiticol	*see* Aflatoxin B_3
Patulin	

TABLE 1 (Continued)
CHEMICAL STRUCTURES OF MYCOTOXINS

Name	Structure
Penicidin	*see* Patulin

Penicillic acid

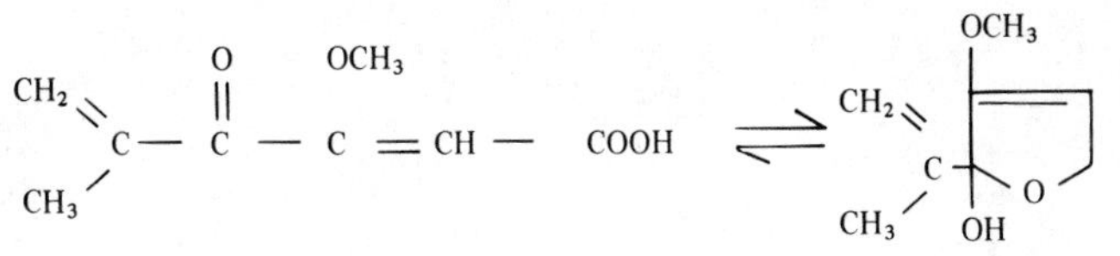

$$CH_2 = \underset{CH_3}{\overset{}{C}} - \overset{O}{\overset{\|}{C}} - \overset{OCH_3}{\overset{}{C}} = CH - COOH \rightleftharpoons$$

Penicillium puberulum toxin   not known

Phallicidine

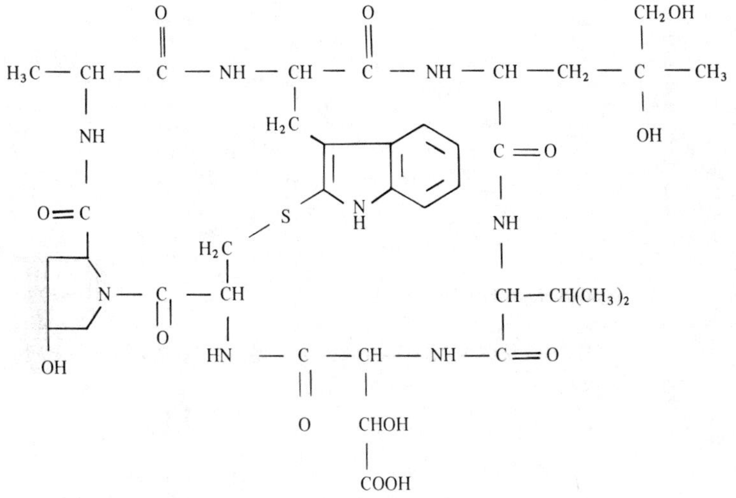

Phallin B*

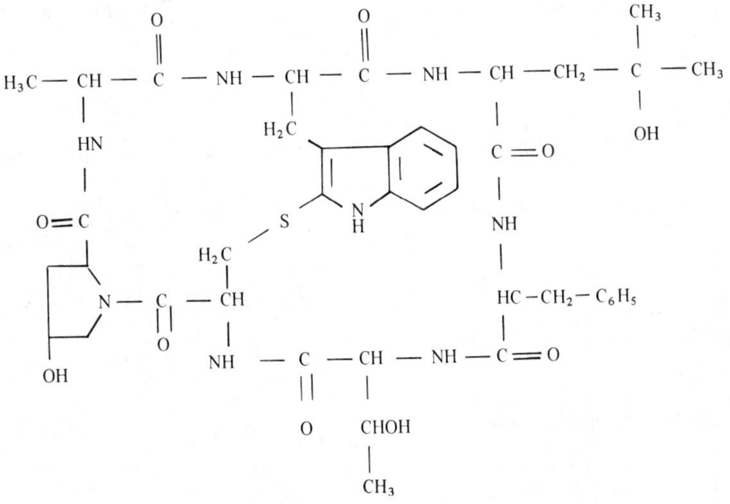

---

*   Structure tentative.

TABLE 1 (Continued)
CHEMICAL STRUCTURES OF MYCOTOXINS

Name	Structure

Phalloidin

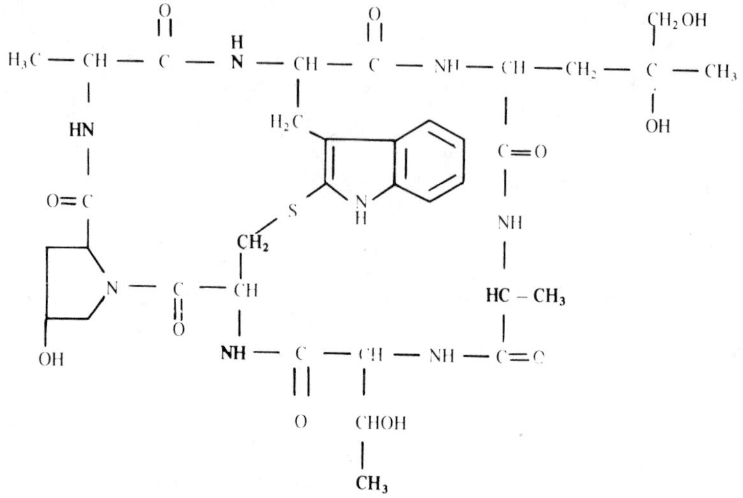

Phalloin

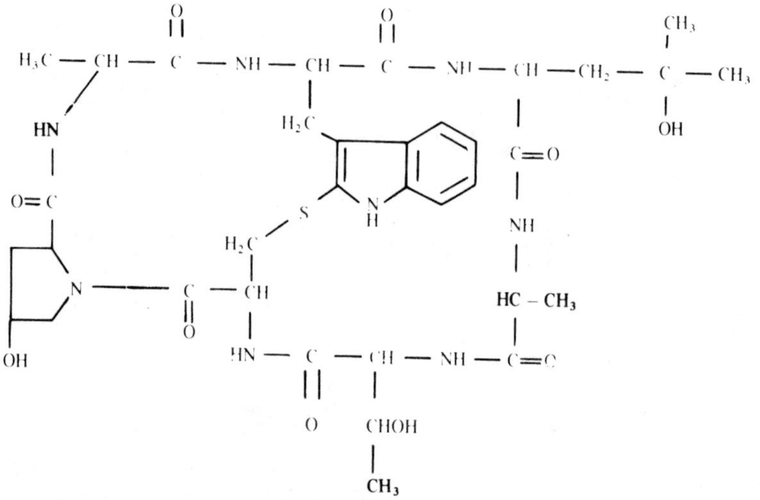

Phallotoxins   *see* Phallacidine, Phallin B, Phalloidin, and Phalloin

Pigment A

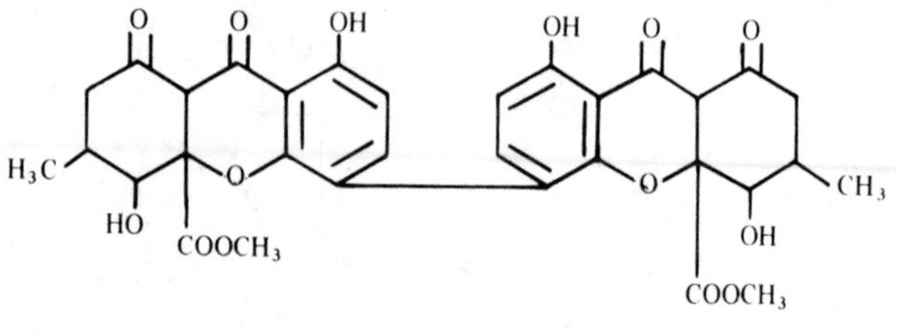

Prämuscimol   *see* Ibotenic acid

TABLE 1 (Continued)
CHEMICAL STRUCTURES OF MYCOTOXINS

Name	Structure

Psilocybin

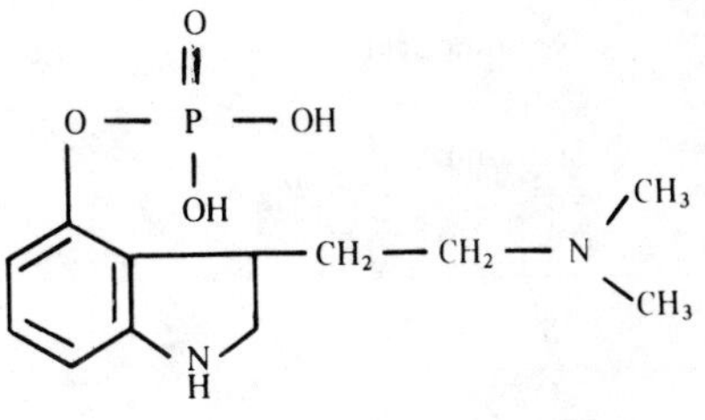

Psoralens

see 8-Methoxypsoralen and 4,5′,8-Trimethylpsoralen

Rubratoxin A

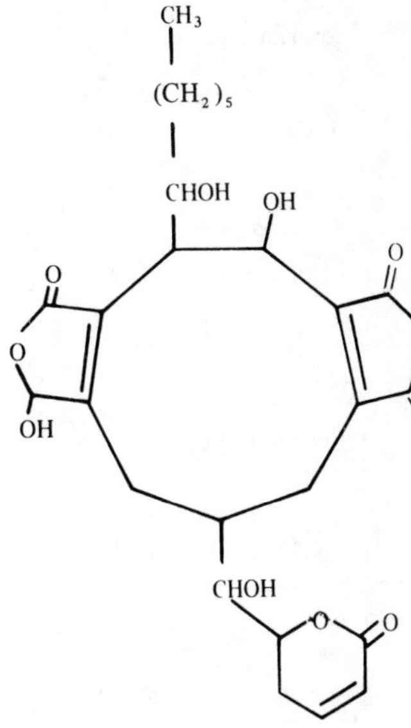

Rubratoxin B

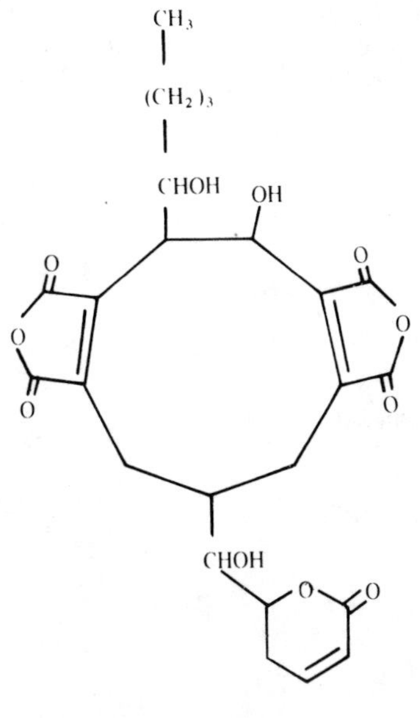

TABLE 1 (Continued)
CHEMICAL STRUCTURES OF MYCOTOXINS

Name	Structure

Name

Structure

Secalonic acid

see Pigment A

Slaframine

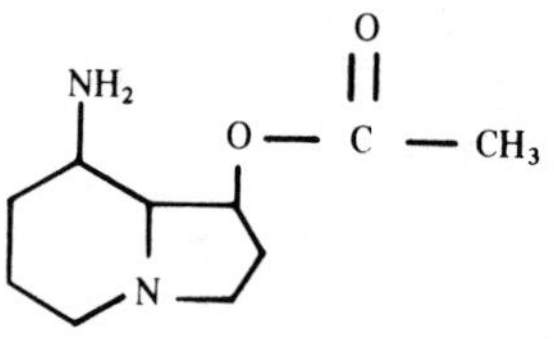

Slobber factor

see Slaframine

Sporidesmin A

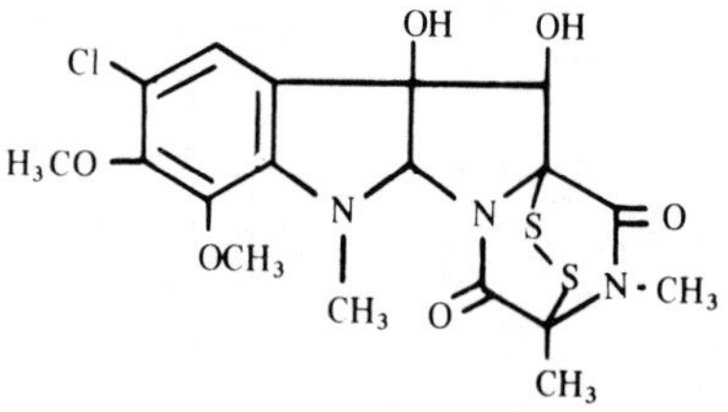

Sporidesmin B

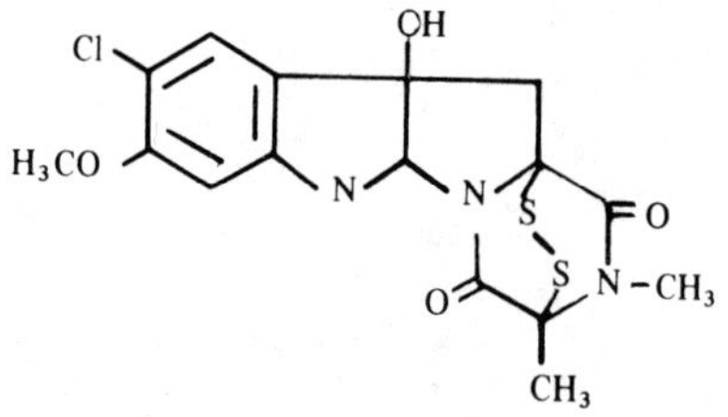

Sporidesmin C

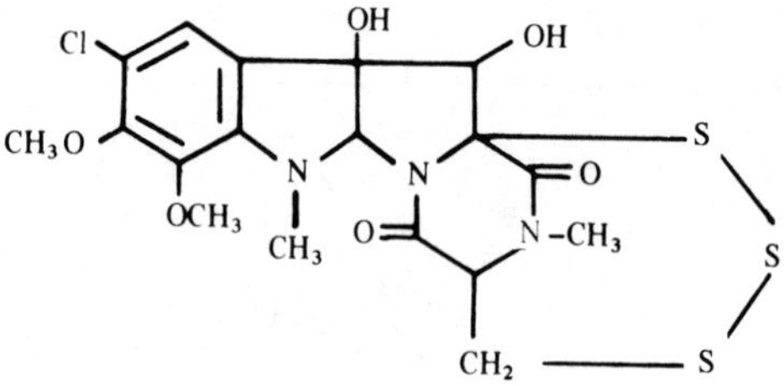

Sporidesmin D

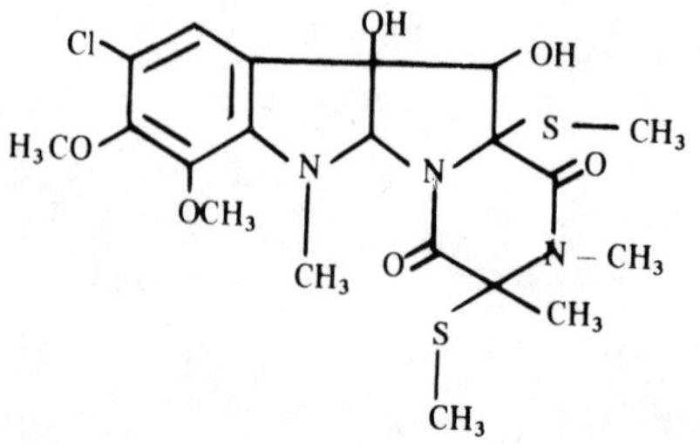

TABLE 1 (Continued)
CHEMICAL STRUCTURES OF MYCOTOXINS

Name	Structure

Sporidesmin E

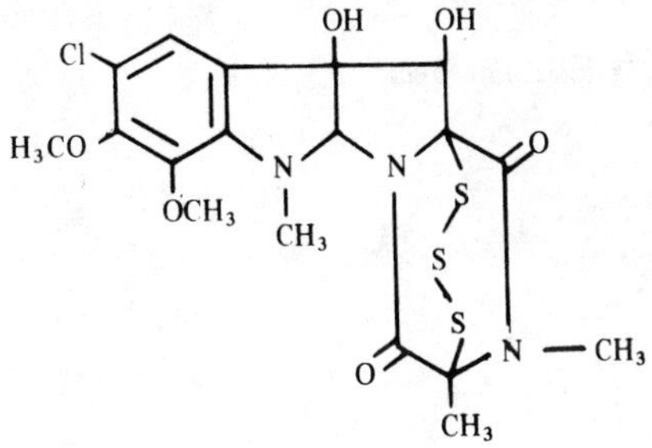

Stachybotrys toxin — not known

Stemphone — not known

Sterigmatocystin

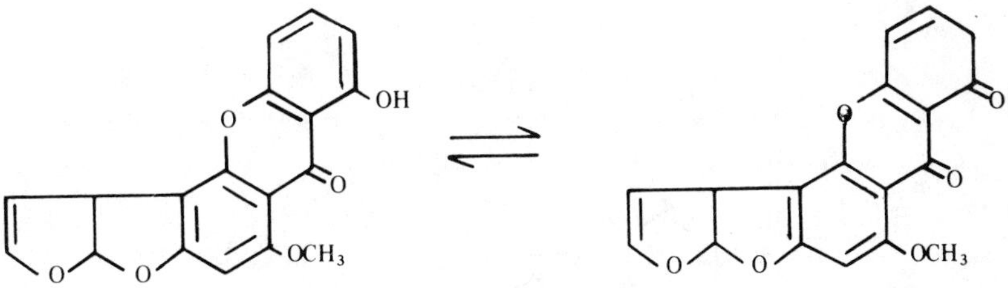

T-2 toxin

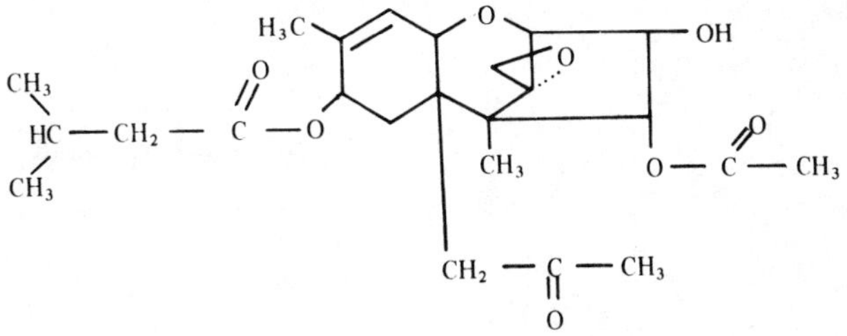

Terreic acid

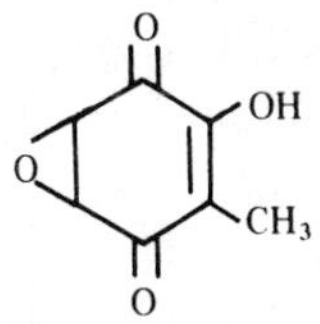

Tremorgen — not known

Tremortin A (Penitrem A) — not known

Tremortin B (Penitrem B) — not known

Trichodermin

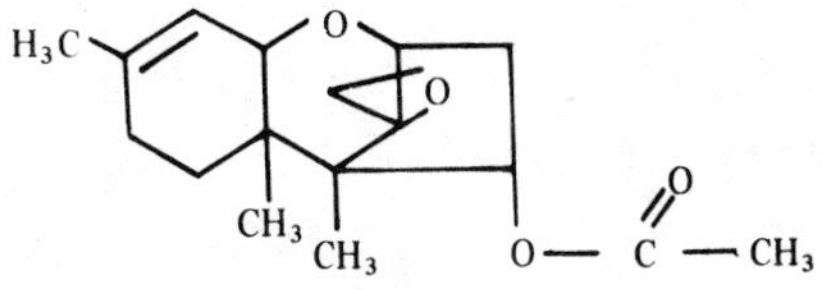

TABLE 1 (Continued)
CHEMICAL STRUCTURES OF MYCOTOXINS

Name	Structure

Tricholomic acid

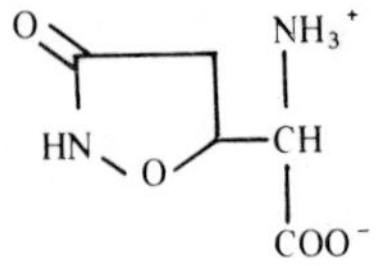

Trichothecin

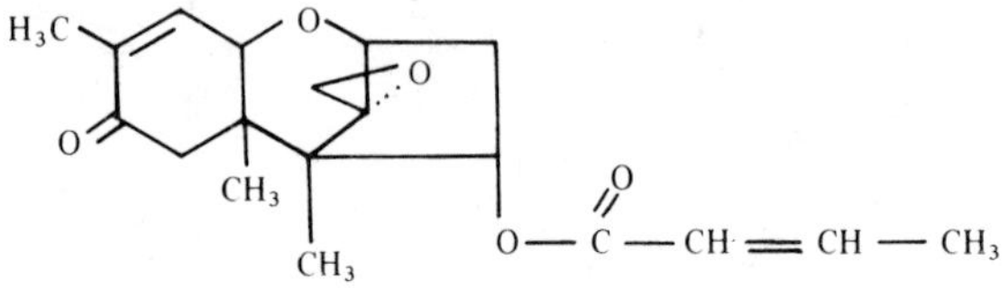

4,5′,8-Trimethylpsoralin

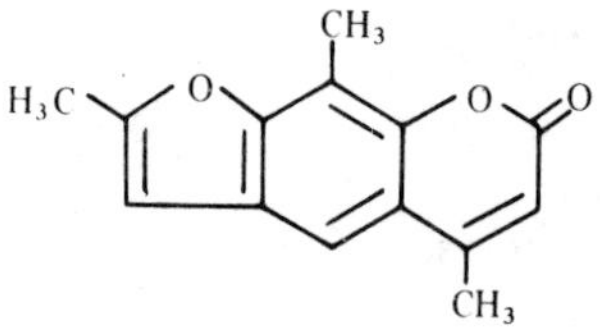

Verrucarin A

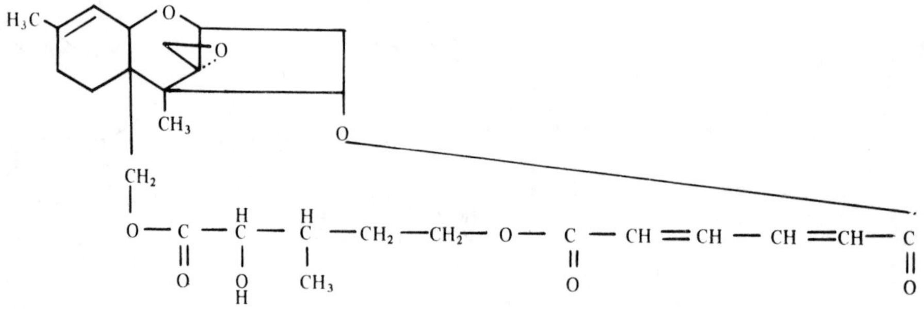

Verrucarin B

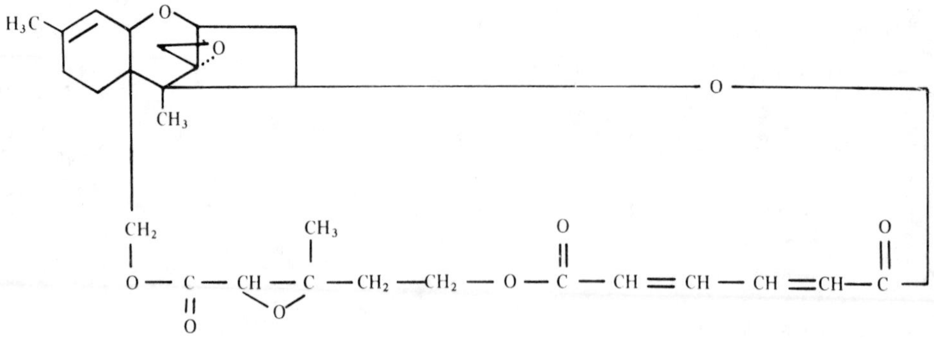

Viriditoxin

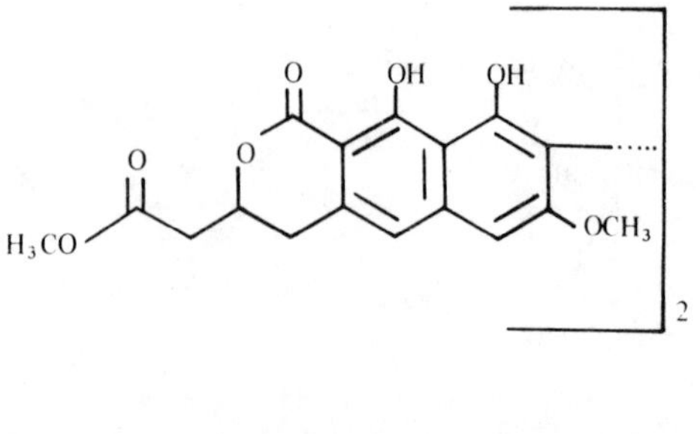

TABLE 1 (Continued)
CHEMICAL STRUCTURES OF MYCOTOXINS

Name	Structure

Xanthocillin

Zearalenone

TABLE 2
CHEMICAL AND PHYSICAL CHARACTERISTICS OF MYCOTOXINS

Name	Formula	Molecular Weight	Melting Point, °C
Aflatoxin B_1	$C_{17}H_{12}O_6$	312	268–269
Aflatoxin B_2	$C_{17}H_{14}O_6$	314	286–289
1-acetoxy-Aflatoxin B_2	$C_{19}H_{16}O_8$	372	225
1-ethoxy-Aflatoxin B_2	$C_{19}H_{18}O_7$	358	247
1-methoxy-Aflatoxin B_2	$C_{18}H_{16}O_7$	360	220–223
2-methoxy-Aflatoxin B_2	$C_{18}H_{16}O_7$	360	245–247
Aflatoxin B_{2a}	$C_{17}H_{14}O_7$	330	240
Aflatoxin B_3	$C_{16}H_{14}O_6$	302	217 (233, 234)
Aflatoxin G_1	$C_{17}H_{12}O_7$	328	244–246
Aflatoxin G_2	$C_{17}H_{14}O_7$	330	237–240
2-ethoxy-Aflatoxin G_2	$C_{19}H_{18}O_8$	374	203
Aflatoxin G_{2a}	$C_{17}H_{14}O_8$	346	190
Aflatoxin $G-M_1$	$C_{17}H_{12}O_8$	344	276
Aflatoxin M_1	$C_{17}H_{12}O_7$	328	299
Aflatoxin M_2	$C_{17}H_{14}O_7$	330	293
Aflatoxin P_1	$C_{16}H_{10}O_6$	298	No data
Aflatoxin R_0	$C_{17}H_{16}O_6$	314	230–234
Agarin (*see* Muscimol)			
Agroclavine	$C_{16}H_{18}N_2$	238.32	198–203 (decomposes)
Agrocybin	$C_8H_5O_2N$	147.13	140 (conflagrates)
Alpha-toxin (*see* Ibotenic acid)			
Alternaria longipes toxin	Not known	No data	No data
Alternaria mali toxin	Not known	No data	No data
Amanin (*see* β-Amanitine)			
Amanita toxins (*see* α-, β-, γ-, and ε-Amanitine)			
α-Amanitine	$C_{40}H_{56}O_{13}N_{10}S$	917	254–255
β-Amanitine	$C_{40}H_{55}O_{14}N_9S$	918	300
γ-Amanitine	$C_{39}H_{54}O_{12}N_{10}S$	890	No data
ε-Amanitine	$C_{39}H_{53}O_{13}N_9S$	888	No data
Amanullin	$C_{39}H_{54}O_{11}N_{10}S$	874	No data

TABLE 2 (Continued)
CHEMICAL AND PHYSICAL CHARACTERISTICS OF MYCOTOXINS

Name	Formula	Molecular Weight	Melting Point, °C
Amatoxins (*see* α-, β-, γ-, and ϵ-Amanitine)			
Antimycin (*see* Citrinin)			
Ascladiol	$C_7H_8O_4$	156	65–66
Ascotoxin (*see* Decumbin)			
Aspercolorin	$C_{23}H_{27}O_4N_4$	423.48	No data
Aspergillic acid	$C_{12}H_{20}O_2N_2$	224.3	97–99
Aspertoxin	$C_{19}H_{14}O_7$	354.07	325–327
B-24 toxin (*see* Diacetoxy-scirpenol)			
Beta-toxin (*see* Muscimol)			
Bufotenine	$C_{12}H_{16}ON_2$	204.26	146–147
Butenolide	$C_6H_7O_3N$	141	116.5–118.5
Byssochlamic acid	$C_{18}H_{20}O_6$	332.34	163.5
Chaetomium globosum toxin	Not known	No data	No data
Citreoviridin	$C_{23}H_{30}O_6$	402	107–111
Citrinin	$C_{13}H_{14}O_5$	250.24	175
Citromycetin	$C_{14}H_{10}O_7$	290.22	290–300 (decomposes)
Clavacin (*see* Patulin)			
Clavatin (*see* Patulin)			
Claviformin (*see* Patulin)			
Cochliobolus toxin	Not known	No data	No data
Cyclopiazonic acid	$C_{20}H_{20}O_3N_2$	336.15	245–246
Decumbin	$C_{16}H_{24}O_4$	280.35	203
Dendrodochine	Not known	No data	No data
Diacetoxyscirpenol	$C_{19}H_{26}O_7$	366.40	161–162
Emetic toxin	Not known	No data	No data
Emodin	$C_{15}H_{10}O_5$	270.23	256–257
Ergobasine (*see* Ergometrine)			
Ergoclinine (*see* Ergometrine)			
Ergocornine	$C_{31}H_{39}O_5N_5$	561.66	182–184 (decomposes)
Ergocristine	$C_{35}H_{39}O_5N_5$	609.74	155–157 (decomposes)
Ergocryptine	$C_{32}H_{41}O_5N_5$	575.69	212 (decomposes)
Ergometrine	$C_{19}H_{23}O_2N_3$	325	162–163 (decomposes)
Ergonovine (*see* Ergometrine)			
Ergostetrine (*see* Ergometrine)			
Ergot alkaloids (*see* Agrocla-vine, Ergocornine, Ergo-cristine, Ergocryptine, Er-gometrine, and Ergotamine)			
Ergotamine	$C_{33}H_{35}O_5N_5$	581.65	212–214 (decomposes)
Ergotocine (*see* Ergometrine)			
Ergotoxine (a 1:1:1 mixture of ergocornine, ergocristine, and ergocryptine)			
Ergotrate (*see* Ergometrine)			
Expansine (*see* Patulin)			
F-2 toxin (*see* Zearalenone)			
Fescue toxin (*see* Butenolide)			
Frangulic acid (*see* Emodin)			
Frequentic acid (*see* Citromy-cetin)			
Fumagillin	$C_{26}H_{34}O_7$	458.53	194–195
Fumigacin (*see* Helvolic acid)			
Fumigatoxin	Not known	No data	No data

TABLE 2 (Continued)
CHEMICAL AND PHYSICAL CHARACTERISTICS OF MYCOTOXINS

Name	Formula	Molecular Weight	Melting Point, °C
Fusarenon	$C_{15}H_{20}O_7$	312.31	No data
Fusarium culmorum toxin	Not known	No data	No data
Gliotoxin	$C_{13}H_{14}O_4N_2S_2$	326.39	221 (decomposes)
Gyromitrin	$C_4H_8ON_2$	100.12	No data
Helvolic acid	$C_{33}H_{44}O_8$	568.68	215
Hiptagenic acid (*see* β-Nitropropionic acid)			
HT-2 toxin	$C_{21}H_{32}O_8$	412.47	No data
Ibotenic acid	$C_5H_8O_5N_2$	176.13	145 (decomposes)
Islanditoxin	$C_{25}H_{36}O_8N_5Cl_2$	605.49	No data
Kojic acid	$C_6H_4O_4$	142.11	153—154
Luteoskyrin	$C_{30}H_{22}O_{12}$	574.48	278
Lysergic acid	$C_{16}H_{16}O_2N_2$	268.32	240 (decomposes)
Maltoryzine	$C_{11}H_{14}O_4$	210.22	69 (decomposes)
Mappine (*see* Bufotenine)			
Mescaline	$C_{11}H_{17}O_3N$	211.25	35—36
8-Methoxypsoralen	$C_{12}H_8O_4$	206.18	145—148
O-Methylsterigmatocystin	$C_{19}H_{14}O_6$	328.30	265 (decomposes)
Muconomycin A (*see* Verrucarin A)			
Muscarine	$C_9H_{20}O_2N^+$	174.26 (ion)	No data
Muscazone	$C_5H_6O_4N_2$	158.11	190 (decomposes)
Muscimol	$C_4H_6O_2N_2$	114.10	172—174 (decomposes)
Mycoin (*see* Patulin)			
Mycophenolic acid	$C_{17}H_{20}O_6$	320.35	141
Nidulotoxin	Not known	No data	No data
β-Nitropropionic acid	$C_3H_5O_4N$	124	68
Nivalenol	$C_{15}H_{20}O_7$	312	222—223
Ochratoxin A	$C_{20}H_{18}O_6NCl$	403.83	169
Ochratoxin A ethyl ester (*see* Ochratoxin C)			
Ochratoxin A methyl ester	$C_{21}H_{20}O_6NCl$	417	No data
Ochratoxin B	$C_{20}H_{19}O_6N$	369.45	221
Ochratoxin B ethyl ester	$C_{22}H_{23}O_6N$	397	102—103 (from ether)
Ochratoxin B methyl ester	$C_{21}H_{21}O_6N$	383.48	134—135 (from benzene)
Ochratoxin C	$C_{22}H_{22}O_6NCl$	431	No data
Orellanine	Not known	No data	No data
Oxalic acid	$C_2H_2O_4 \cdot 2H_2O$	126.07	101—102
Pantherine (*see* Muscimol)			
Parasiticol (*see* Aflatoxin B_3)			
Patulin	$C_7H_6O_4$	154.12	111.0
Penicidin (*see* Patulin)			
Penicillic acid	$C_8H_{10}O_4$	170.16	83—84
Penicillium puberulum toxin	Not known	No data	No data
Phallicidine	$C_{36}H_{50}O_{12}N_8S$	786	No data
Phallin B	$C_{41}H_{52}O_9N_8S$	832.95	No data
Phalloidin	$C_{35}H_{48}O_{11}N_8S$	788.89	280—282
Phalloin	$C_{35}H_{48}O_{10}N_8S$	772	250—280 (decomposes)
Phallotoxins (*see* Phallicidin, Phallin B, Phalloidin, and Phalloin)			
Pigment A	$C_{32}H_{30}O_{14}$	638.59	244—250
Prämuscimol (*see* Ibotenic acid)			
Psilocybin	$C_{12}H_{17}O_4N_2P$	284.27	220—228 (crystals from water)

TABLE 2 (Continued)
CHEMICAL AND PHYSICAL CHARACTERISTICS OF MYCOTOXINS

Name	Formula	Molecular Weight	Melting Point, °C
Psoralens (*see* 8-Methoxy-psoralen and 4,5′,8-Tri-methylpsoralen)			
Rubratoxin A	$C_{26}H_{32}O_{11}$	520	No data
Rubratoxin B	$C_{26}H_{30}O_{11}$	518	185–186 (decomposes)
Secalonic acid (*see* Pigment A)			
Slaframine	$C_{10}H_{18}O_2N_2$	198.26	183–184 (slaframine dipicrate)
Slobber factor (*see* Slafra-mine)			
Sporidesmin A	$C_{18}H_{20}O_6N_3ClS_2$	473.95	179; benzene solvate, 110–120 (decomposes)
Sporidesmin B	$C_{18}H_{20}O_5N_3ClS_2$	457.95	183
Sporidesmin C	$C_{18}H_{20}O_6N_3ClS_3$	506.05	No data
Sporidesmin D	$C_{19}H_{23}O_6N_3ClS_2$	478.99	No data
Sporidesmin E	$C_{18}H_{20}O_6N_3ClS_3$	506.05	180–185
Stachybotrys toxin	Not known	No data	No data
Stemphone	$C_{30}H_{42}O_8$	530	160.5–161.5
Sterigmatocystin	$C_{19}H_{14}O_6$	338	265 (decomposes)
T-2 toxin	$C_{24}H_{34}O_9$	466	151–152
Terreic acid	$C_7H_6O_4$	154	127–127.5
Tremorgen	Not known	No data	No data
Tremortin A	$C_{37}H_{44}O_6NCl$	633	210–230 (decomposes)
Tremortin B	$C_{37}H_{45}O_5N$	583	185–195 (decomposes)
Trichodermin	$C_{17}H_{24}O_4$	292.36	46
Tricholomic acid	$C_5H_8O_4$	132.11	207
Trichothecin	$C_{19}H_{24}O_5$	332.38	118
4,5′,8-Trimethylpsoralen	$C_{14}H_{12}O_3$	228.24	226–228
Verrucarin A (Penitrem A)	$C_{27}H_{34}O_9$	501.53	212–215
Verrucarin B (Penitrem B)	$C_{27}H_{32}O_9$	500	176–179 (hexahydro derivative)
Viriditoxin	$C_{34}H_{30}O_{14}$	662	245
Xanthocillin	$C_{18}H_{12}O_2N_2$	288	200 (decomposes)
Zearalenone	$C_{18}H_{22}O_5$	318	164–165

TABLE 3
SOURCES AND EFFECTS OF MYCOTOXINS

Producing Fungi or Other Sources	Substrates (Opt. Temp., °C)	Animals Affected	Toxin Effects	LD_{50}	References
Aflatoxin B$_1$					
Aspergillus flavus parasiticus	Peanuts, corn, wheat, rice, nuts, coconut, cottonseed meal, various foods, milk (25–30)	Most mammals, man, birds, fish	Hepatotoxic, carcinogenic	Dog: 1.0 mg/kg; Duckling: 0.3–0.4 mg/kg; Guinea pig: 1.4 mg/kg; Monkey (i.p.): 2.5–5.0 mg/kg; Rabbit, weanling (i.p.): 0.5 mg/kg; Rat, adult female: 180 mg/kg; Rat, adult male: 72 mg/kg	1, 2, 3
Aflatoxin B$_2$					
Aspergillus flavus parasiticus	Same as aflatoxin B$_1$ (25–30)	Same as aflatoxin B$_1$	Same as aflatoxin B$_1$	Duckling: 1.7 mg/kg	1, 2, 3, 4
1-acetoxy-Aflatoxin B$_2$					
Synthesis		No data	No data	No data	5
1-ethoxy-Aflatoxin B$_2$					
Synthesis					6
1-methoxy-Aflatoxin B$_2$					
Synthesis		No data	No data	No data	7
2-methoxy-Aflatoxin B$_2$					
Synthesis		No data	No data	No data	7

TABLE 3 (Continued)
SOURCES AND EFFECTS OF MYCOTOXINS

Producing Fungi or Other Sources	Substrates (Opt. Temp., °C)	Animals Affected	Toxin Effects	LD_{50}	References
Aflatoxin B$_{2a}$					
Acid catalysis of aflatoxin B$_1$		No data	Almost nontoxic	No data	5, 8
Aflatoxin B$_3$					
Aspergillus flavus parasiticus	Rice, wheat (No data)	No data	No data	Chick embryo: 5–10 µg/egg	9, 10
Aflatoxin G$_1$					
Aspergillus flavus parasiticus	Same as aflatoxin B$_1$ (15–20)	Same as aflatoxin B$_1$	Same as aflatoxin B$_1$	Duckling: 0.78 mg/kg	1, 2, 3
Aflatoxin G$_2$					
Aspergillus flavus parasiticus	Same as aflatoxin B$_1$ (No data; probably 15–20)	Duckling; no other data available	Probably the same as aflatoxin G$_1$	Duckling: 3.45 mg/kg	1, 2, 3
1-ethoxy-Aflatoxin G$_2$					
Synthesis		No data	No data	No data	6
Aflatoxin G$_{2a}$					
Acid catalysis of aflatoxin G$_1$		No data	No data	No data	5, 8
Aflatoxin G-M$_1$					
Aspergillus flavus	Aflatoxin G$_1$ is converted in the animal body to aflatoxin G-M$_1$	No data	No data	No data	11

TABLE 3 (Continued)
SOURCES AND EFFECTS OF MYCOTOXINS

Producing Fungi or Other Sources	Substrates (Opt. Temp., °C)	Animals Affected	Toxin Effects	LD_{50}	References
Aflatoxin M_1					
Aspergillus flavus parasiticus	Same as aflatoxin B_1; found in milk and urine of animals fed aflatoxin B_1 (Probably the same as aflatoxin B_1)	Duckling; no other data available	Carcinogenic; probably the same as aflatoxin B_1	Duckling: 0.32 mg/kg	12, 13
Aflatoxin M_2					
Aspergillus flavus parasiticus	No data (No data)	Duckling; no other data available	No data	Duckling: 1.23 mg/kg	14
Aflatoxin P_1					
Isolated from the urine of rhesus monkeys		No data	No data	No data	15
Aflatoxin R_0					
Transformation of aflatoxin B_1 by *Dactylium dendroides*		Duckling; no other data available	Bile duct hyperplasia	Duckling: 0.48—0.64 mg/kg	16
Agarin (*see* Muscimol)					
Agroclavine					
Claviceps purpurea	Grasses, grains	Pig, mouse	Agalactia	No data	17

TABLE 3 (Continued)
SOURCES AND EFFECTS OF MYCOTOXINS

Producing Fungi or Other Sources	Substrates (Opt. Temp., °C)	Animals Affected	Toxin Effects	LD_{50}	References
Agrocybin					
Agrocybe dura		Man	Extremely toxic		18
Alpha-toxin (*see* Ibotenic acid)					
Alternaria longipes toxin					
Alternaria longipes	Corn (25—28)	Chick	Anorexia, diarrhea, loss of muscular control, gizzard erosion, proventriculous hemorrhage, coma, death	No data	19
Alternaria mali toxin					
Alternaria mali	Mycological media (26)	Mouse	Hyperkeratosis of the forestomach, pulmonary hemorrhage, vacuolated liver cord cells	No data	20
Amanin (*see* β-Amanitine)					
Amanita toxins (*see* α-, β-, γ-, and ε-Amanitine)					

TABLE 3 (Continued)
SOURCES AND EFFECTS OF MYCOTOXINS

Producing Fungi or Other Sources	Substrates (Opt. Temp., °C)	Animals Affected	Toxin Effects	LD_{50}	References
α-Amanitine					
Amanita bisporigera phalloides (Fr.) Secr. *tenuifolia verna (virosa)*		Man, monkey, guinea pig, rabbit, rat, mouse, horse, goat, sheep, pigeon	Man: salivation, vomiting, bloody stools, cyanosis, muscular twitching, convulsions; can be fatal	Mouse: 0.3 mg/kg	21, 22, 23, 24
Galerina spp.			Also: fatty degeneration of the liver, liver necrosis, kidney necrosis		
β-Amanitine					
Same as α-amanitine		Same as α-amanitine	Same as α-amanitine	Mouse: 0.4 mg/kg	21, 22, 23, 24
γ-Amanitine					
Same as α-amanitine		Same as α-amanitine	Same as α-amanitine	Mouse: 0.15 mg/kg	21, 22, 23, 24
ε-Amanitine					
Same as α-amanitine		Same as α-amanitine	Same as α-amanitine	Mouse: 0.5 mg/kg	21, 22, 23, 24
Amanullin					
Same as α-amanitine			Nontoxic		21, 22, 23, 24

TABLE 3 (Continued)
SOURCES AND EFFECTS OF MYCOTOXINS

Producing Fungi or Other Sources	Substrates (Opt. Temp., °C)	Animals Affected	Toxin Effects	LD_{50}	References
Amatoxins (*see* α-, β-, γ-, and ϵ-Amanitine)					
Antimycin (*see* Citrinin)					
Ascladiol					
Aspergillus clavatus	Wheat flour (No data)	No data	No data	No data	25
Ascotoxin (*see* Decumbin)					
Aspercolorin					
Aspergillus versicolor	Corn (No data)	Duckling	No data	No data	26
Aspergillic acid					
Aspergillus flavus	No data (No data)	Mouse	Lethal	Mouse (LD_{100}): 150 mg/kg, i.p. 250 mg/kg, oral	27
Aspertoxin					
Aspergillus flavus	Shredded wheat (No data)	Chick embryo	Beak malformation, general edema, loss of muscle tone, hemorrhage from umbilical vessels	Chick embryo: 0.7 µg/egg	28, 29
B-24 toxin (*see* Diacetoxyscirpenol)					
Beta-**toxin** (*see* Muscimol)					

TABLE 3 (Continued)
SOURCES AND EFFECTS OF MYCOTOXINS

Producing Fungi or Other Sources	Substrates (Opt. Temp., °C)	Animals Affected	Toxin Effects	LD_{50}	References
Bufotenine					
Amanita citrina mappa muscaria pantherina		Man	Hallucinogenic, respiratory arrest, ataxia, cardiovascular effects		30
Butenolide					
Fusarium nivale	Tall fescue (*Festuca arundinacea*) (15)	Cattle (?), mouse	Gangrene of the extremities	Mouse: 43.6 mg/kg	31
Byssochlamic acid					
Byssochlamys fulva	Fruit, fruit juices (30)	Mouse	Hemorrhage (?)	No data	32
***Chaetomium globosum* toxin**					
Chaetomium globosum	Corn, commercial feed (20–25)	Rat	CNS damage, hemoglobinuria, hemorrhagic enteritis, subdural hemorrhaging	No data	33
Citreoviridin					
Penicillium citreo-viride ochrosalmoneum toxicarium	Rice (No data)	Rat	Neurotoxic, damage to CNS, adrenal cortex and liver, kidney paralysis, respiratory failure	No data Rat (LD_{100}): 8–30 mg/kg	34, 35

TABLE 3 (Continued)
SOURCES AND EFFECTS OF MYCOTOXINS

Producing Fungi or Other Sources	Substrates (Opt. Temp., °C)	Animals Affected	Toxin Effects	LD_{50}	References
Citrinin					
Aspergillus candidus niveus terreus Penicillium citreo-viride citrinum fellutanum implicatum jenseni lividum viridicatum	Rice, corn, wheat, barley (30)	Cattle (?), swine, laboratory animals	Nephrotoxic	Mouse: 35 mg/kg	36, 37
Citromycetin					
Citromyces spp. *Penicillium frequentans roseo-purpureum spinulosum*	Rice (No data)	Man	Nephrotoxic	No data	35
Clavacin (*see* Patulin)					
Clavatin (*see* Patulin)					
Claviformin (*see* Patulin)					
Cochliobolus toxin					
Cochliobolus carbonum	Cereals (25)	Mouse	Respiratory difficulty, ataxia, pilo erection, death	No data	38

TABLE 3 (Continued)
SOURCES AND EFFECTS OF MYCOTOXINS

Producing Fungi or Other Sources	Substrates (Opt. Temp., °C)	Animals Affected	Toxin Effects	LD_{50}	References
Cyclopiazonic acid					
Penicillium cyclopium	Corn (26)	Duckling, rat	Lethal	No data	39
Decumbin					
Ascochyta imperfecta *Penicillium decumbens*	Corn, potato medium (23–26)	Rat, goldfish	Anorexia, diarrhea, cyanosis, respiratory difficulty, nasal bleeding	Rat (oral): 275 mg/kg	40, 41
Dendrochine					
Dendrodochium toxicum	Wheat, straw, oats, Sudan grass (25)	Man, domestic animals	Skin necrosis, damage to the hemopoietic system and alimentary tract	No data	42
Diacetoxyscirpenol					
Fusarium diversisporum equiseti sambucinum scirpi tricinctum *Gibberella intricans*	Corn (8)	Cattle, laboratory animals	Internal hemorrhage, skin necrosis	Rat: 0.75 mg/kg (i.p.) 7.3 mg/kg (oral)	43

TABLE 3 (Continued)
SOURCES AND EFFECTS OF MYCOTOXINS

Producing Fungi or Other Sources	Substrates (Opt. Temp., °C)	Animals Affected	Toxin Effects	LD$_{50}$	References
Emetic toxin					
Fusarium avenaceum culmorum equiseti graminearum moniliforme nivale poae roseum scirpi	Corn, barley, wheat (No data)	Man, dog, pigeon, farm animals	Nausea, emesis	No data	44
Emodin					
Chaetomium affine Cortinarius sanguineus Penicillium brunneum	Rice (No data)	No data	No data	No data	35

Ergobasine (*see* Ergot alkaloids)

Ergoclinine (*see* Ergot alkaloids)

Ergocornine (*see* Ergot alkaloids)

Ergocristine (*see* Ergot alkaloids)

Ergocryptine (*see* Ergot alkaloids)

Ergometrine (*see* Ergot alkaloids)

Ergonovine (*see* Ergot alkaloids)

Ergostetrine (*see* Ergot alkaloids)

TABLE 3 (Continued)
SOURCES AND EFFECTS OF MYCOTOXINS

Producing Fungi or Other Sources	Substrates (Opt. Temp., °C)	Animals Affected	Toxin Effects	LD_{50}	References
Ergot alkaloids					
Aspergillus fumigatus *Claviceps purpurea* *Rhizopus arrhizus*	Rye, wild grasses (25)	Man, sheep cattle	Vasoconstriction, uterus contraction, adrenalin and serotonin antagonism, damage to CNS (reduced activity of vasomotor center)		17, 45
Ergotamine (*see* Ergot alkaloids)					
Ergotocine (*see* Ergot alkaloids)					
Ergotrate (*see* Ergot alkaloids)					
Expansine (*see* Patulin)					
F-2 toxin (*see* Zearalenone)					
Fescue toxin (*see* Butenolide)					
Frangulic acid (*see* Emodin)					
Frequentic acid (*see* Citromycetin)					
Fumagillin					
Aspergillus fumigatus	Cereal grains (No data)	Man, mouse	Abdominal pain, skin rash	Mouse (s.c.): 800 mg/kg	46
Fumigacin (*see* Helvolic acid)					

TABLE 3 (Continued)
SOURCES AND EFFECTS OF MYCOTOXINS

Producing Fungi or Other Sources	Substrates (Opt. Temp., °C)	Animals Affected	Toxin Effects	LD_{50}	References
Fumigatoxin					
Aspergillus fumigatus	Laboratory media (27)	Mouse	Paralysis of the extremities, cyanosis	Mouse (LD_{100}): 150 µg/kg	47
Fusarenon					
Fusarium nivale	Wheat, rice (27)	Mouse	Damage to hematopoietic tissue, proliferating cells of the intestinal epithelium, testes	Mouse: 4.2 mg/kg	48
***Fusarium culmorum* toxin**					
Fusarium culmorum	Barley, corn (No data)	Pig, dairy cattle	Vomiting, anorexia, loss of milk production	No data	49
Gliotoxin					
Aspergillus chevaleri fumigatus Gliocladium fimbriatum Penicillium cinerascens jenseni obscurum restrictum terlikowski Trichoderma viride	Grain (25)	No data on occurrence as a mycotoxin, but closely related to toxic sporidesmins		Mouse (s.c.): 45–65 mg/kg	50

TABLE 3 (Continued)
SOURCES AND EFFECTS OF MYCOTOXINS

Producing Fungi or Other Sources	Substrates (Opt. Temp., °C)	Animals Affected	Toxin Effects	LD_{50}	References
Gyromitrin					
Helvella esculenta gigas infula underwoodii		Man, guinea pig, rabbit	Vomiting, diarrhea, jaundice, liver damage, convulsions, delirium, heart failure	No data	51
Helvolic acid					
Aspergillus fumigatus	Cereal grains (No data)	Mouse	Fatty livers	Mouse (i.p.): 400 mg/kg	52
Hiptagenic acid (*see β-Nitropropionic acid*)					
HT-2 toxin					
Fusarium tricinctum	Corn (24)	Rat	Dermatitis	No data	53
Ibotenic acid					
Amanita muscaria pantherina pantherina-gemmata strobiliformis		Man, fly	Man: psychotomimetic Fly: lethal		54
Islanditoxin					
Penicillium islandicum	Rice (Room temperature)	Laboratory animals, man (?)	Hepatotoxic, hemorrhage	Mouse: 0.475 mg/kg (s.c.) 6.55 mg/kg (oral) 0.3 mg/kg (i.v.)	55

TABLE 3 (Continued)
SOURCES AND EFFECTS OF MYCOTOXINS

Producing Fungi or Other Sources	Substrates (Opt. Temp., °C)	Animals Affected	Toxin Effects	LD_{50}	References
Kojic acid					
Aspergillus flavus glaucus oryzae tamarii	Corn (35)	Mouse, dog	Convulsant	Mouse (i.p.): 250 mg/kg	27
Luteoskyrin					
Penicillium islandicum	Rice (No data)	Mouse, rat	Hepatotoxic	Mouse: 6.65 mg/kg (i.v.) 147.0 mg/kg (s.c.) 221.0 mg/kg (oral)	56, 57
Lysergic acid					
Claviceps paspali	Mycological media (25)	Man	Psychotomimetic (principal moiety of D-lysergic acid diethylamide)		58
Maltoryzine					
Aspergillus oryzae var. *microspora*	Malt sprouts (30)	Mouse, dairy cattle	Mouse: muscular narcotism	Mouse (i.p.): 3 mg/kg	59
Mappine (*see* Bufotenine)					
Mescaline					
Lophophora williamsii		Man, guinea pig, rat, mouse, frog	Psychotomimetic for man, convulsions, respiratory arrest, cardiac arrest	Rat (i.p.): 370 mg/kg	60

TABLE 3 (Continued)
SOURCES AND EFFECTS OF MYCOTOXINS

Producing Fungi or Other Sources	Substrates (Opt. Temp., °C)	Animals Affected	Toxin Effects	LD_{50}	References
8-Methoxypsoralen					
Sclerotinia sclerotiorum	Celery (No data)	Man	Dermatitis	No data	61
O-Methylsterigmatocystin					
Aspergillus flavus	Laboratory media (Room temperature)	May be nontoxic			62
Muconomycin A (*see* Verrucarin A)					
Muscarine					
Amanita muscaria *Clitocybe cerussata dealbata rivulosa truncicola* *Inocybe lacera napipes patouillardii picrosma* *Oinphalotus olearius*		Man, cat, frog	Sialorrhea, lacrimation, diaphoresis, nausea, vomiting, diarrhea, bradycardia, circulatory collapse		63, 64

TABLE 3 (Continued)
SOURCES AND EFFECTS OF MYCOTOXINS

Producing Fungi or Other Sources	Substrates (Opt. Temp., °C)	Animals Affected	Toxin Effects	LD_{50}	References
Muscazone					
Amanita muscaria		Man	Sialorrhea, lacrimation, diaphoresis, nausea, vomiting, diarrhea, bradycardia, circulatory collapse		54
Muscimol					
Amanita muscaria pantherina pantherina-gemmata		Man	Psychotomimetic		54
Mycoin (*see* Patulin)					
Mycophenolic acid					
Penicillium brevi-compactum stoloniferum urticae	Corn (No data)	Mouse	Affects leukocytes	Mouse (LD_{100}): 500 mg/kg (i.v.) 2,000 mg/kg (oral)	65, 66
Nidulotoxin					
Aspergillus nidulans	Laboratory media (30)	Chick, duck embryo	Growth retardation	Chick embryo: 1 μg/embryo	67

TABLE 3 (Continued)
SOURCES AND EFFECTS OF MYCOTOXINS

Producing Fungi or Other Sources	Substrates (Opt. Temp., °C)	Animals Affected	Toxin Effects	LD_{50}	References
β-Nitropropionic acid					
Aspergillus flavus oryzae *Penicillium atrovenetum*	Laboratory media (24)	Dairy cattle	Vasodilation	No data	27, 68
Nivalenol					
Fusarium nivale	Wheat, rice (27)	Mouse	Cell degeneration of bone marrow, lymph nodes, intestines, testes, thymus	Mouse (i.p.): 4 mg/kg	69
Ochratoxin A					
Aspergillus alliaceus melleus ochraceus ostianus sclerotiorum sulfureus *Penicillium viridicatum*	Cereals (25)	Rat, duckling, chick	Tubular necrosis of the kidney, mild liver degeneration, enteritis	Duckling (oral): 3.0 mg/kg	70
Ochratoxin A ethyl ester (*see* Ochratoxin C)					
Ochratoxin A methyl ester					
Aspergillus ochraceus	Corn, wheat (No data)	Same as ochratoxin A	Same as ochratoxin A	Duckling: 135–170 µg/bird	71

TABLE 3 (Continued)
SOURCES AND EFFECTS OF MYCOTOXINS

Producing Fungi or Other Sources	Substrates (Opt. Temp., °C)	Animals Affected	Toxin Effects	LD_{50}	References
Ochratoxin B					
Same as ochratoxin A	Corn, other cereals (No data)		Comparatively nontoxic		72
Ochratoxin B ethyl ester					
Aspergillus ochraceus	Corn (No data)		Nontoxic		71
Ochratoxin B methyl ester					
Aspergillus ochraceus	Corn (No data)		Nontoxic		71
Ochratoxin C					
Aspergillus ochraceus	Corn (No data)	Same as ochratoxin A	Same as ochratoxin A	Duckling: 135—170 µg/bird	71
Orellanine					
Cortinarius orellanus		Man, laboratory animals	Gastric disturbances, oliguria, anuria, convulsions, renal lesions	Cat: 4.9 mg/kg; Guinea pig: 8.0 mg/kg; Mouse: 8.3 mg/kg	73
Oxalic acid					
Aspergillus niger *Penicillium oxalicum* Others	Mycological and natural products (No data)	No data as mycotoxin	Calcium depletion in rats, neurological coma, gastric irritation	Mouse (i.p.): 150 mg/kg	27

TABLE 3 (Continued)
SOURCES AND EFFECTS OF MYCOTOXINS

Producing Fungi or Other Sources	Substrates (Opt. Temp., °C)	Animals Affected	Toxin Effects	LD_{50}	References
Pantherine (*see* **Muscimol**)					
Parasiticol (*see* **Aflatoxin B₃**)					
Patulin					
Aspergillus clavatus giganteus terreus *Byssochlamys nivea* *Penicillium claviforme expansum urticae*	Moldy feed (25)	Cattle, laboratory animals	Edema, carcinogenic	Mouse, Rat (i.v.): 15–35 mg/kg	74
Penicidine (*see* **Patulin**)					
Penicillic acid					
Aspergillus melleus ochraceus quercinus sulfureus *Penicillium baarnense cyclopium madriti martensii palitans puberulam roqueforti thomii*	Corn, cereals (5–10)	Laboratory animals	Digitalis-like action on heart, dilator action on systemic blood vessels, antidiuretic action	Mouse: 110 mg/kg (s.c.) 250 mg/kg (i.v.)	75, 76

TABLE 3 (Continued)
SOURCES AND EFFECTS OF MYCOTOXINS

Producing Fungi or Other Sources	Substrates (Opt. Temp., °C)	Animals Affected	Toxin Effects	LD_{50}	References
Penicillium puberulum toxin					
Penicillium puberulum	Corn, wheat, millet, oats, peanuts (28)	Mouse, duckling	Incoordination, ataxia, convulsions, apnea	No data	77
Phallicidine					
Amanita bisporigera phalloides (Fr.) Secr. *tenuifolia verna (virosa) Galerina* spp.		Man, monkey, guinea pig, rabbit, rat, mouse, horse, goat, sheep, pigeon	Man: salivation, vomiting, bloody stools, cyanosis, muscular twitching, convulsions; can be fatal / Also: fatty degeneration of the liver, liver necrosis, kidney necrosis	Mouse: 2.5 mg/kg	21, 22, 23, 24
Phallin B					
Same as phallicidine		Same as phallicidine	Same as phallicidine	Mouse: 15 mg/kg	21, 22, 23, 24
Phalloidin					
Same as phallicidine		Same as phallicidine	Same as phallicidine	Guinea pig: 3.5 mg/kg / Mouse: 2.0 mg/kg	21, 22, 23, 24
Phalloin					
Same as phallicidine		Same as phallicidine	Same as phallicidine	Mouse: 1.8 mg/kg	21, 22, 23, 24

TABLE 3 (Continued)
SOURCES AND EFFECTS OF MYCOTOXINS

Producing Fungi or Other Sources	Substrates (Opt. Temp., °C)	Animals Affected	Toxin Effects	LD$_{50}$	References
Phallotoxins (*see* Phallicidine, Phallin B, Phalloidin, and Phalloin)					
Pigment A					
Claviceps purpurea *Penicillium oxalicum*	Cereals (No data)	Rat, mouse, duckling	Lethal	No data	78, 79
Prämuscimol (*see* Ibotenic acid)					
Psilocybin					
Conocybe siligenoides *Panaeolus companulatus* *Psilocybe aztecorum caerulescens cordispora cubensis hoogshageni mexicana mixaeensis semperviva wassonii yungensis zapotecorum* *Strophana cubensis*		Man	Hallucinogenic	Mouse: 280 mg/kg	60
Psoralens (*see* 8-Methoxypsoralen and 4,5′,8-Trimethylpsoralen)					

TABLE 3 (Continued)
SOURCES AND EFFECTS OF MYCOTOXINS

Producing Fungi or Other Sources	Substrates (Opt. Temp., °C)	Animals Affected	Toxin Effects	LD_{50}	References
Rubratoxin A					
Penicillium purpurogenum rubrum	Grains, laboratory media (25)	Swine, cattle, laboratory animals	Hepatotoxic, hemorrhage	Mouse (i.p.): 6.3 mg/kg	80
Rubratoxin B					
Same as rubratoxin in A	Same as rubratoxin A (25)	Same as rubratoxin A	Same as rubratoxin A, HeLa cells	Mouse (i.p.): 3.0 mg/kg	80, 81
Secalonic acid (*see* Pigment A)					
Slaframine					
Rhizoctonia leguminicola	Red clover, hay (25)	Ruminants, laboratory animals	Excessive salivation, bloat, anorexia, lacrimation, diarrhea	No data	82
Slobber factor (*see* Slaframine)					
Sporidesmin A					
Pithomyces chartarum	Pasture grasses, cereals (24)	Sheep, cattle	Hepatotoxic, facial eczema, edema	Sheep (oral): 0.5–1.0 mg/kg	83
Sporidesmin B					
Same as sporidesmin	Same as sporidesmin (No data)	Same as sporidesmin	Same as sporidesmin	No data	83
Sporidesmin C					
Same as sporidesmin	Same as sporidesmin (No data)	No data	No data	No data	84

TABLE 3 (Continued)
SOURCES AND EFFECTS OF MYCOTOXINS

Producing Fungi or Other Sources	Substrates (Opt. Temp., °C)	Animals Affected	Toxin Effects	LD_{50}	References
Sporidesmin D					
Same as sporidesmin	Rye (No data)	No data	No data	No data	85
Sporidesmin E					
Same as sporidesmin	Pasture grasses (No data)	Same as sporidesmin	Same as sporidesmin; also toxic to HeLa cells	No data	86, 87
Stachybotrys toxin					
Stachybotrys atra	Straw, feed (25)	Man, horse	Hemorrhage, necrosis of mucous membranes	No data	88
Stemphone					
Stemphylium sarcinaeforme	Red clover (25)	Chick embryo, fish larvae	Lethal	Chick embryo (LD_{33}): 100 µg/embryo	89
Sterigmatocystin					
Aspergillus nidulans versicolor *Bipolaris* spp.	Corn (No data)	Rat, mouse	Carcinogenic, hepatotoxic, nephrotoxic, diarrhea	Mouse (oral): 800 mg/kg Rat: 65 mg/kg (i.p.) 120 mg/kg (oral)	90, 91
T-2 toxin					
Fusarium tricinctum, strain T-2	Corn (8)	Cattle, trout	Hemorrhage, skin necrosis, shedding of intestinal mucosa in trout	Mouse (LD_{100}): 100–150 mg/kg	92

TABLE 3 (Continued)
SOURCES AND EFFECTS OF MYCOTOXINS

Producing Fungi or Other Sources	Substrates (Opt. Temp., °C)	Animals Affected	Toxin Effects	LD_{50}	References
Terreic acid					
Aspergillus terreus	Cereal grains (No data)	Man, mouse	Irritation of human mucous membranes	Mouse (i.v.): 71–119 mg/kg	93
Tremorgen					
Aspergillus flavus	Cereals (No data)	Mouse	Tremors (sustained)	No data	94
Tremortin A (Penitrem A)					
Penicillium cyclopium crustosum gramulatum martensii palitans puberulum	Feedstuffs (25)	Cattle, sheep, horse, mouse, rat	Tremors, convulsions, ataxia, anorexia, pilo erection	Mouse (i.p.): 1.2 mg/kg	95, 96, 97
Tremortin B (Penitrem B)					
Same as tremortin A	Same as tremortin A (25)	Same as tremortin A	Same as tremortin A	Mouse (i.p.): 5 mg/kg	96
Trichodermin					
Fusarium spp. *Trichoderma viride*	Corn (Low temperature ?)	Mouse, cattle (?)	Hemorrhage (?)	Mouse: 500–1,000 mg/kg (s.c.) > 1,000 mg/kg (oral)	98

TABLE 3 (Continued)
SOURCES AND EFFECTS OF MYCOTOXINS

Producing Fungi or Other Sources	Substrates (Opt. Temp., °C)	Animals Affected	Toxin Effects	LD_{50}	References
Tricholomic acid					
Tricholoma muscarium		Fly	Leg paralysis, intense vacuole formation in the intestinal epithelium	No data	99
Trichothecin					
Fusarium spp. *Trichothecium roseum*	Corn (8)	Cattle (?)	Hemorrhage (?)	No data	98
4,5',8-Trimethylpsoralen					
Sclerotinia sclerotiorum	Celery (No data)	Man	Dermatitis	No data	101
Verrucarin A					
Myrothecium roridum verrucaria	Laboratory media (25 or lower)	Man	Dermatitis	Mouse (i.v.): 1.5 mg/kg	100, 101
Verrucarin B					
Same as verrucarin A	Same as verrucarin A; corn (?) (Same as verrucarin A)	Same as verrucarin A	Same as verrucarin A	Mouse (i.v.): 7.0 mg/kg	100, 101
Viriditoxin					
Aspergillus viridi-nutens	Laboratory media, corn (28)	Mouse	Lethal	Mouse (i.p.): 2.8 mg/kg	102

TABLE 3 (Continued)
SOURCES AND EFFECTS OF MYCOTOXINS

Producing Fungi or Other Sources	Substrates (Opt. Temp., °C)	Animals Affected	Toxin Effects	LD_{50}	References
Xanthocillin					
Aspergillus chevalieri *Penicillium notatum*	Cereal grains (25)	Laboratory animals	Hepatotoxic	No data	103
Zearalenone					
Fusarium graminearum	Corn, commercial feed (12—27)	Swine, rat, sheep	Enlarged vulvae and mammary glands, vaginal prolapse, abortion	No data	104

REFERENCES

1. Detroy, R. W., Lillehoj, E. B., and Ciegler, A., Aflatoxin and Related Compounds, in *Microbial Toxins,* Vol. 6 (Fungal Toxins). Academic Press, New York (in press).
2. Goldblatt, L. A., *Aflatoxin.* Academic Press, New York (1969).
3. Lillehoj, E. B., Ciegler, A., and Detroy, R. W., Fungal Toxins, in *Essays in Toxicology,* p. 1, F. Blood, Ed. Academic Press, New York (1970).
4. Hartley, R. D., Nesbitt, B. F., and Kelly, J. O., *Nature, 198,* 1056 (1963).
5. Dutton, M. F., and Heathcote, J. G., *Chem. Ind.,* p. 418 (1968).
6. Dutton, M. F., and Heathcote, J. G., *Chem. Ind.,* p. 983 (1969).
7. Waiss, A. C., and Wiley, M., *Chem. Commun.,* p. 512 (1969).
8. Lillehoj, E. B., and Ciegler, A., *Appl. Microbiol., 17,* 516 (1969).
9. Heathcote, J. G., and Dutton, M. F., *Tetrahedron, 25,* 1497 (1969).
10. Stubblefield, R. D., Shotwell, O. L., Shannon, G. M., Weisleder, D., and Rohwedder, W. K., *J. Agr. Food Chem., 18,* 391 (1970).
11. Dutton, M. F., and Heathcote, J. G., *J. S. Afr. Chem. Inst., 12,* S107 (1969).
12. Buchi, G., and Weinreb, S. M., *J. Amer. Chem. Soc., 91,* 5408 (1969).
13. Sinnhuber, R. O., Lee, D. J., Wales, J. H., Landers, M. K., and Kegl, A. C., *Fed. Proc., 29,* 568 Ab (1970).
14. Purchase, I. F. H., *Food Cosmet. Toxicol., 5,* 339 (1967).
15. Dalezios, J., and Wogan, G. N., *Science,* 584 (1971).
16. Detroy, R. W., and Hesseltine, C. W., *Nature 219,* 967 (1968).
17. Mantle, P. G., *J. Stored Prod. Res., 5,* 237 (1969).
18. Bu'Lock, *J. Chem. Ind.,* p. 990 (1954).
19. Doupnik, B., and Sobers, E. K., *Appl. Microbiol., 16,* 1596 (1968).
20. Slifkin, M. K., and Spalding, J., *Toxicol. Appl. Pharmacol., 17,* 375 (1970).
21. Dessey, G., and Francioli, M., *Boll. Inst. Zier Milan, 17,* 779 (1938).
22. Tyler, V. E., Brady, L. R., Benedict, R. G., Khanna, J. M., and Malone, M. H., *Lloydia, 26,* 154 (1963).
23. Wieland, O., *Clin. Chem., 11,* 323 (1965).
24. Wieland, T., *Fortschr. Chem. Org. Naturst., 25,* 214 (1967).
25. Tanabe, H., and Suzuki, T., in *U.J.N.R. Conference on Toxic Microorganisms, Honolulu, Hawaii, Oct. 7-10, 1968* (unnumbered publication), p. 127. U.S. Department of the Interior, Washington, D.C. (1970).
26. Aucamp, P. J., and Holzapfel, C. W., *J. S. Afr. Chem. Inst., 22,* S35 (1969).
27. Wilson, B. J., *Bacteriol. Rev., 30,* 478 (1966).
28. Rodricks, J. V., Lustig, E., Campbell, A. D., and Stoloff, L., *Tetrahedron Lett.,* p. 2875 (1968).
29. Rodricks, J. V., Henery-Logan, K. R., Campbell, A. D., Stoloff, L., and Verrett, M. J., *Nature, 217,* 668 (1968).
30. Fischer, R., *Nature, 220,* 411 (1968).
31. Yates, S. G., Tookey, H. L., Ellis, J. J., and Burkhardt, H. J., *Phytochemistry, 7,* 139 (1968).
32. de Scott, B., *Mycopathol. Mycol. Appl., 25,* 213 (1965).
33. Christensen, C. M., Nelson, G. H., Mirocha, C. J., Bates, F., and Dorworth, C. E., *Appl. Microbiol., 14,* 774 (1966).
34. Sakabe, N., Goto, T., and Hirata, Y., *Tetrahedron Lett.,* p. 1825 (1964).
35. Uraguchi, K., *J. Stored Prod. Res., 5,* 227 (1969).
36. Friis, P., Hasselager, E., and Krogh, P., *Acta Pathol. Microbiol. Scand., 77,* 559 (1969).
37. Kinosita, R., and Shikata, T., in *Mycotoxins in Foodstuffs,* p. 111, G. N. Wogan, Ed. M.I.T. Press, Cambridge, Massachusetts (1965).
38. Hamilton, P. B., Nelson, R. R., and Harris, B. S. H., *Appl. Microbiol., 16,* 1719 (1968).
39. Holzapfel, C. W., *Tetrahedron, 24,* 2101 (1968).
40. Singleton, V. L., Bohonos, N., and Ullstrup, A. J., *Nature, 181,* 1072 (1958).
41. Suzuki, Y., Tanaka, H., Aoki, H., and Tamura, T., *Agr. Biol. Chem. (Japan), 34,* 395 (1970).
42. Bilai, V. I., *Mikrobiologiya (Engl. Transl.), 30,* 834 (1962).
43. Bamburg, J. R., Marasas, W. F., Riggs, N. V., Smalley, E. B., and Strong, F. M., *Biotechnol. Bioeng., 10,* 445 (1968).
44. Prentice, N., and Dickson, A. D., *Biotechnol. Bioeng., 10,* 413 (1968).
45. Stoll, A., and Hofmann, A., The Ergot Alkaloids, in *The Alkaloids,* Vol. 8, p. 725, R. H. F. Manske, Ed. Academic Press, New York (1965).
46. Tarbell, D. S., Carman, R. M., Chapman, D. D., Huffman, K. B., and McCorkindale, N. J., *J. Amer. Chem. Soc., 82,* 1005 (1962).
47. Iwata, K., Nagai, T., and Okudaira, M., *J. S. Afr. Chem. Inst., 22,* S131 (1969).
48. Saito, M., and Okuba, K., in *U.J.N.R. Conference on Toxic Microorganisms, Honolulu, Hawaii, Oct. 7–10, 1968* (unnumbered publication), p. 82. U.S. Department of the Interior, Washington, D.C. (1970).
49. Fisher, E. E., Kellock, A. W., and Wellington, N. A. M., *Nature, 215,* 322 (1967).
50. Bell, M. R., Johnson, J. R., Wilde, B. S., and Woodward, R. B., *J. Amer. Chem. Soc., 80,* 1001 (1958).
51. List, P. H., and Luft, P., *Tetrahedron Lett.,* p. 1893 (1967).
52. Okuda, S., Iwasaki, S., Sair, M. I., Machida, Y., Inoue, A., and Tsuda, K., *Tetrahedron Lett.,* p. 2295 (1967).

53. Bamburg, J. R., and Strong, F. M., *Phytochemistry, 8,* 2405 (1969).
54. Benedict, R. G., Tyler, V. E., and Brady, L. R., *Lloydia, 29,* 333 (1966).
55. Miyake, M., and Saito, M., in *Mycotoxins in Foodstuffs,* p. 133, G. N. Wogan, Ed. M.I.T. Press, Cambridge, Massachusetts (1965).
56. Saito, M., Enomoto, M., Ishiko, T., Shikata, T., and Miyake, M., *Jap. J. Exp. Med., 31,* 435 (1961).
57. Shibata, S., Ogihara, Y., Koboyashi, N., Seo, S., and Kitagawa, J., *Tetrahedron Lett.,* p. 3179 (1968).
58. Arcamone, F., Bonnino, C., Chain, E. B., Ferretti, A., Minghetti, A., Pennella, P., Tonola, A., and Vero, L., *Proc. Royal Soc. (London) Ser. B, 155,* 26 (1951).
59. Izuka, H., and Mitsugi, I., *Nature, 190,* 681 (1962).
60. Hoffer, A., and Osmond, H., *The Hallucinogens.* Academic Press, New York (1967).
61. Scheel, L. D., Perone, V. B., Larkin, R. L., and Kupel, R. E., *Biochemistry, 2,* 1127 (1963).
62. Burkhardt, H. J., and Forjacs, J., *Tetrahedron, 24,* 717 (1968).
63. Wieland, T., *Science, 159,* 946 (1968).
64. Wilkinson, S., *Quart. Rev. Chem. Soc. London, 15,* 153 (1961).
65. Burton, H. S., *Brit. J. Exp. Pathol., 30,* 151 (1949).
66. Logan, W. R., and Newbold, G. T., *J. Chem. Soc.,* p. 1946 (1957).
67. Lafont, P., Lafont, J., and Frayssinet, C., *Experientia, 26,* 61 (1970).
68. Raistrick, H., and Stössl, A., *Biochem. J., 68,* 647 (1958).
69. Tatsuno, T., Fujimoto, Y., and Morita, Y., *Tetrahedron Lett.,* p. 2823 (1969).
70. Theron, J. J., van der Merwe, K. J., Liebenberg, N., Joubert, H. J. B., and Nel, W., *J. Pathol. Bacteriol., 91,* 521 (1966).
71. Steyn, P. S., and Holzapfel, C. W., *J. S. Afr. Chem. Inst., 20,* 186 (1967).
72. Purchase, I. F. H., and Nel, W., in *Biochemistry of Some Foodborne Microbial Toxins,* p. 153, R. I. Mateles and G. N. Wogan, Eds. M.I.T. Press, Cambridge, Massachusetts (1967).
73. Desvignes, A., *Aliment. Vie, 53,* 155 (1965).
74. Ciegler, A., Detroy, R. W., and Lillehoj, E. B., in *Microbial Toxins,* Vol. 6 (Fungal Toxins). Academic Press, New York (in press).
75. Kurtzman, C. P., and Ciegler, A., *Appl. Microbiol., 20,* 204 (1970).
76. Ciegler, A., and Kurtzman, C. P., *Appl. Microbiol., 20,* 761 (1970).
77. Wilson, B. J., Harris, T. M., and Hayes, A. W., *J. Bacteriol., 93,* 1737 (1967).
78. Steyn, P. S., *J. S. Afr. Chem. Inst., 22,* S20 (1969).
79. Stoll, A., Renz, J., and Brack, A., *Helv. Chim. Acta, 35,* 2022 (1952).
80. Moss, M. O., Wood, A. B., and Robinson, F. V., *Tetrahedron Lett.,* p. 367 (1969).
81. Hayes, A. W., and Wilson, B. J., *Appl. Microbiol., 16,* 1163 (1968).
82. Aust, S. D., Broquist, H. P., and Rinehart, K. L., Jr., *Biotechnol. Bioeng., 10,* 403 (1968).
83. Taylor, A., in *Biochemistry of Some Foodborne Microbial Toxins,* p. 69, R. I. Mateles and G. N. Wogan, Eds. M.I.T. Press, Cambridge, Massachusetts (1967).
84. Hodges, R., and Shannon, J. S., *Aust. J. Chem., 19,* 1059 (1966).
85. Rahman, R., and Taylor, A., *Chem. Commun.,* p. 1032 (1967).
86. Brewer, D., Rahman, R., Safe, S., and Taylor, A., *Chem. Commun.,* p. 1571 (1968).
87. Raham, R., Safe, S., and Taylor, A., *J. Chem. Soc. Sect. C,* p. 1665 (1969).
88. Forjacs, J., in *Mycotoxins in Foodstuffs,* p. 87, G. N. Wogan, Ed. M.I.T. Press, Cambridge, Massachusetts (1965).
89. Scott, P. M., and Lawrence, J. W., *Can. J. Microbiol., 14,* 1015 (1968).
90. Holzapfel, C. W., Purchase, I. F. H., Steyn, P. S., and Gouws, L., *S. Afr. Med. J., 40,* 1100 (1966).
91. Lillehoj, E. B., and Ciegler, A., *Mycopathol. Mycol. Appl., 35,* 373 (1968).
92. Bamburg, J. R., Riggs, N. V., and Strong, F. M., *Tetrahedron Lett., 24,* 3329 (1968).
93. Sheenan, J., Lawson, W., and Gaul, R., *J. Amer. Chem. Soc., 80,* 5536 (1958).
94. Wilson, B. J., and Wilson, C. H., *Science, 144,* 177 (1964).
95. Ciegler, A., *Appl. Microbiol., 18,* 128 (1969).
96. Hou, C. T., Ciegler, A., and Hesseltine, C. W., *Can. J. Microbiol., 17,* 599 (1971).
97. Wilson, B. J., Wilson, C. H., and Hayes, A. W., *Nature, 220,* 77 (1968).
98. Bamburg, J. R., Marasas, W. F., Riggs, N. V., Smalley, E. B., and Strong, F. M., *Biotechnol. Bioeng., 10,* 445 (1968).
99. Takemoto, T., *Jap. J. Pharm. Chem., 33,* 252 (1961).
100. von Härri, E., Loeffler, W., Sigg, H. P., Stahelin, H., Stoll, C., Tamm, C., and Wiesinger, D., *Helv. Chim. Acta, 45,* 839 (1962).
101. Bohner, B., Fetz, E., von Härri, E., Sigg, H. P., Stoll, C., and Tamm, C., *Helv. Chim. Acta, 48,* 1079 (1965).
102. Lillehoj, E. B., and Ciegler, A. (in preparation).
103. Hagedorn, I., and Tönjes, H., *Pharmazie, 11,* 409 (1956).
104. Mirocha, C. J., Christensen, C. M., and Nelson, G. H., *Appl. Microbiol., 15,* 497 (1967).

MICROBIAL INSECTICIDES

DR. MARTIN H. ROGOFF

MICROBIAL PREPARATIONS AVAILABLE COMMERCIALLY OR IN TEST QUANTITIES

Insecticidal Agent	Target Insect	Product Name	Supplier
Bacteria			
Bacillus			
lentimorbus	Japanese beetle	Japidemic	Ditman Corp., USA
popilliae	Japanese beetle	Doom	Fairfax Biological Laboratories, USA
sphaericus	Non-aedine mosquitoes		International Minerals and Chemical Corp., USA
thuringiensis	Various lepidopteran larvae	Agritol	Merck and Co., USA
		Bakthane L-69	Rohm and Haas Co., USA
		Baktospeine	Pechiney Progil Laboratory, Roger Bellon, France
		Bakthurin	Chemapol, Biokrma, Czechoslovakia
		Biospor 2802	Farbwerke Hoechst, Germany
		Dendrobacillin	Moskovs. zavod bakt., USSR
		Dipel	Abbott Laboratories, USA
		Entobakterin 3	All-Union Institute for Plant Protection, USSR
		Parasporin	Grain Processing Corp., USA
		Sporeine	Laboratoire L.I.B.E.C., France
		Thuricide	International Minerals and Chemical Corp., USA
Fungi			
Beauveria			
bassiana	Lepidopteran larvae	—	International Minerals and Chemical Corp., USA
		Biotrol FBB	Nutrilite Products, Inc., USA
Metarrhizium			
anisopliae	Broad spectrum	—	International Minerals and Chemical Corp., USA
Viruses			
Polyhedroses	Cotton bollworm	Biotrol VHZ	Nutrilite Products, Inc., USA
		Virex	Hays-Sammons, USA
		Viron H	International Minerals and Chemical Corp., USA
	Sawfly	Polyvirocide	Indiana Farm Bureau Co-op Association, USA
	Beet armyworm	*Spodoptera* virus	International Minerals and Chemical Corp., USA
		Biotrol VSE	Nutrilite Products, Inc., USA
	Cabbage looper	*Trichoplusia* virus	Biological Control Supplies, USA
		Trichoplusia virus	International Minerals and Chemical Corp., USA
		Biotrol VTN	Nutrilite Products, Inc., USA
	Yellow-striped armyworm	Biotrol VPO	Nutrilite Products, Inc., USA
Nematodes			
Nematode DD-136	Codling moth	Biotrol NCS	Nutrilite Products, Inc., USA

MAJOR INSECT-PATHOGENIC MICROORGANISMS
POTENTIALLY USEFUL IN INSECT CONTROL

The organisms listed in the following table have been propagated and tested on at least laboratory scale.

Insect Pathogen, Species or Type	Host (Common Name)	References
Bacteria		
Azotobacter spp.	(Cabbage worm)	9
Bacillus		
cazauban	(Wax moth)	71
cereus	(Larch sawfly)	37
lentimorbus	(Japanese beetle)	21
popilliae	(Japanese beetle)	36
sphaericus	(Non-aedine mosquitoes)	41
thuringiensis	Broad spectrum of lepidopteran larvae	55, 60, 71
	(Mosquitoes)	58
Cloaca		
cloacae acridiorum	(Grasshoppers)	14
Clostridium		
brevifaciens	(Tent caterpillar)	15
malacasomae	(Tent caterpillar)	15
Pseudomonas		
aeruginosa	(Wax moth)	66
	(Grasshoppers)	16
Serratia		
marcescens	Broad spectrum	16
Streptococcus spp.	(Wax moth)	17
Fungi		
Aschersonia spp.	(Whiteflies, scale insects)	1, 2, 3, 10, 71
Aspergillus		
flavus	(Cecropia moth)	72
fumigatus	(Desert locust)	54
Beauveria spp.	(Whitegrubs, potato beetle, corn borer, chinch bugs)	23, 63
Cephalosporium		
licanii	(Wood lice)	29
Coelomomyces spp.	(Mosquitoes)	71
Cordyceps spp.	Broad spectrum	48
Entomophthora spp.	(Browntail moth, cicada, mealy bugs, flies)	71, 79
	(Blackgrass bug)	34
	(Pea aphid, green peach aphid)	26
	(Mosquitoes)	44
	(Pine aphid)	51
	(Alfalfa aphid)	35
	(Imported applesucker)	25
	(Wireworms)	24
Fusarium spp.	(Desert locust)	7
	(White flies)	71
Hirsutella		
thompsonii	(Citrus rust mite)	47
Metarrhizium		
anisopliae	Broad spectrum	1, 2, 3, 52
	(Pasture insects)	5
Oospora		
destructor	(Silkworms)	73
Spicaria spp.	Broad spectrum	8
	(European corn borer)	77
	(Cabbage looper)	30
Verticillium spp.	(White flies)	71

Insect Pathogen, Species or Type	Host (Common Name)	References
Protozoa		
Crithidia		
fasciculata	(Mosquitoes, adult)	1, 2, 3, 45
Farinocystis		
tribolii	(Flour beetles)	1, 2, 3, 81
Herpetomonas		
muscarum	(Flies)	45
Malamoeba		
locusti	(Grasshoppers)	76
Mattesia		
dispora	(Flour moth, wax moth)	81
grandis	(Boll weevil)	50
Nosema		
bombycis	Broad lepidopteran spectrum	1, 2, 3, 74
infesta	(Lawn moth)	33
lymantreae	(Fall webworm)	80
mesnili	(Cabbageworm)	81
muscularis	(Gypsy moth)	81
Plistophora spp.	Broad lepidopteran spectrum	81
Tetrahymena		
pyriformis	Broad spectrum	45
Thelohania spp.	(Mosquitoes)	40
hyphantriae	(Tent caterpillar)	81
Viruses		
Polyhedroses	*Antheraea*	
	eucalypti (Emperor gum moth)	1, 2, 3, 39
	pernyi (Chinese oak silkworm)	39
	Barathra	
	brassicae (Cabbage armyworm)	57
	Bombyx	
	mori (Silkworm)	39
	Choristoneura	
	fumiferana (Spruce budworm)	69
	Colias	
	eurytheme (Alfalfa caterpillar)	53
	Dendrolimus	
	spectabilis (Pine caterpillar)	39
	Diprion	
	hercyniae (European spruce sawfly)	12
	Galleria	
	melonella (Greater wax moth)	6
	Heliothis	
	phloxiphaga	39
	virescens (Tobacco budworm)	18
	zea (Cotton bollworm)	39
	Hemerocampa	
	leucostigma (White-masked tussock moth)	39
	Kotochalia	
	junodi (Wattle bagworm)	56
	Laphygma	
	frugiperda (Fall webworm)	28
	Malacasoma	
	fragile (Great basin tent caterpillar)	19
	Neodiprion	
	sertifer (European pine sawfly)	39
	swainei (Swaine jack pine sawfly)	67

Insect Pathogen, Species or Type	Host (Common Name)	References

Viruses (continued)

Polyhedroses (*cont.*)	*Pectinophora*	
	gossypiella (Pink bollworm)	39
	Peridroma	
	margaritosa (Variegated cutworm)	70
	Phryganida	
	californica (Oakworm)	64
	Plusia	
	gamma (Gamma noctuid)	78
	Porthetria	
	dispar (Gypsy moth)	62
	Prodenia	
	eridania (Southern armyworm)	39
	litura (Cotton leafworm)	4
	ornithogalli (Yellow-striped armyworm)	39
	praefica (Western yellow-striped armyworm)	71
	Samia	
	walkeri	39
	Spodoptera	
	exigua (Beet armyworm)	39
	Thaumetopoea	
	pityocampa (Pine processionary caterpillar)	32
	Trichoplusia	
	ni (Cabbage looper)	65
Granuloses	*Argyrotaenia*	
	velutinana (Red-banded leafroller)	31
	Carpocapsa	
	pomonella (Codling moth)	75
	Eucosma	
	griseana (Larch bud moth)	39
	Hyphantria	
	cunea (Fall webworm)	28
	Pieris	
	brassicae (Cabbageworm)	22
	rapae (Imported cabbageworm)	49
	Spodoptera spp. (Armyworms)	39
Non-inclusion	*Panonychus*	
	citri (Citrus red mite)	68
	Sericesthis iridescent virus (Grain insects)	39
	Tipula iridescent virus (Range crane fly)	68

TOXIC MATERIALS PRODUCED BY INSECT PATHOGENS
REPORTED AS RELATED TO OBSERVED PATHOLOGY IN THE HOST

Species	Toxic Material	References
Bacteria		
Bacillus		
cereus	a-Exotoxin (Phospholipase C)	42
thuringiensis	a-Exotoxin (Phospholipase C)	38
	β-Exotoxin; a substitution compound of adenine, containing adenine, phosphate, ribose, glucose and allomucic acid; a probable inhibitor of RNA polymerization	13
	δ-Endotoxin; the proteinaceous parasporal body or crystal; a protoxin yielding toxic peptides (mol wt < 5000) on enzymatic activation	20
	Water-soluble toxin; an unidentified compound (mol wt > 30,000) that causes pathology similar to that of δ-endotoxin	27
Pseudomonas		
aeruginosa	Proteolytic enzyme	43
	Cell-wall component	82
Serratia		
marcescens	Possibly a proteolytic enzyme	46
Fungi		
Aspergillus spp.	Aflatoxins	46
	Kojic acid	59
Beauveria		
bassiana	Beauvericin; a depsipeptide	46
	Proteinase	46
Cordyceps spp.	Possibly cordycepin (3-deoxyadenosine)	59
Entomophthora spp.	Proteinaceous toxins, possibly with proteolytic activity	46, 59
Metarrhizium		
anisopliae	Destruxins A and B; cyclic depsipeptides	46

REFERENCES

General References

1. Steinhaus, E. A. (Ed.), *Insect Pathology, An Advanced Treatise*, Vols. 1–2. Academic Press, New York (1963).
2. DeBach, P. (Ed.), *Biological Control of Insect Pests and Weeds*. Reinhold Publishing Co., New York (1964).
3. Burges, H. D., and Hussey, N. W. (Eds.), *Microbial Control of Insects and Mites*. Academic Press, New York (1971).

Specific References

4. Abul Nasr, S., *J. Insect Pathol., 1,* 112 (1959).
5. Adamik, L., *Folia Microbiol., 10,* 255 (1965).
6. Aizawa, K., *J. Insect Pathol., 4,* 122 (1962).
7. Akbar, K., Hague, H., and Abhas, H. M., *FAO (Food Agr. Organ. U. N.) Plant Prot. Bull., 6,* 59 (1958).
8. Aoki, K., Samamoto, K., and Nakarata, Y., *J. Sericult. Sci. Jap., 24,* 231 (1955).
9. Atger, P., Dusanssoy, G., and Bourguignon, S., *Rev. Pathol. Veg. Entomol. Agr. Fr., 43,* 191 (1964).
10. Berger, E. W., *Quart. Bull. State Board Fla., 43,* 191 (1921).

11.   Biliotti, E., Grison, F., Maury, R., and Vago, C., *C. R. Séances Acad. Agr. Fr., 45,* 407 (1959).

12.   Bird, F. T., and Burk, J. M., *Can. Entomol., 93,* 228 (1961).

13.   Bond, R., Boyce, C., Rogoff, M., and Shieh, T., in *Microbial Control of Insects and Mites,* p. 275, H. D. Burges and W. N. Hussey, Eds. Academic Press, New York (1971).

14.   Bucher, G. E., *J. Insect Pathol., 2,* 172 (1960).

15.   Bucher, G. E., *Can. J. Microbiol., 7,* 641 (1961).

16.   Bucher, G. E., in *Insect Pathology, An Advanced Treatise,* Vol. 2, p. 117, E. A. Steinhaus, Ed. Academic Press, New York (1963).

17.   Cameron, G. R., *J. Pathol. Bacteriol., 38,* 441 (1934).

18.   Chamberlin, F., and Dutky, S. R., *J. Econ. Entomol., 51,* 560 (1958).

19.   Clark, E. C., and Thompson, C. G., *J. Econ. Entomol., 47,* 268 (1954).

20.   Cooksey, K. W., in *Microbial Control of Insects and Mites,* p. 247, H. D. Burges and W. N. Hussey, Eds. Academic Press, New York (1971).

21.   Costilow, R. N., Sylvester, J. C., and Pepper, R. E., *Appl. Microbiol., 14,* 161 (1966).

22.   David, W., and Gardiner, B., *J. Invertebr. Pathol., 7,* 285 (1965).

23.   Dunn, P. H., and Mechalas, B. J., *J. Insect Pathol., 5,* 451 (1963).

24.   Durnovo, Z. P. (Abstr.), *Rev. Appl. Entomol., 23,* 573 (1935).

25.   Durtan, A. G., *Agr. Gaz. Can., 10,* 16 (1923).

26.   Evlakhiva, A., and Woronina, E., in *Proceedings of the 12th International Congress of Entomology, London, 1965,* Vol. 2, p. 867.

27.   Fast, P. G., *J. Invertebr. Pathol., 17,* 301 (1971).

28.   Franz, J. M., *Annu. Rev. Entomol., 6,* 183 (1961).

29.   Ganhao, J., *Broteria, 25,* 71 (1956).

30.   Getzin, L. W., *J. Insect Pathol., 3,* 2 (1961).

31.   Glass, E. H., *J. Econ. Entomol., 51,* 454 (1958).

32.   Grison, P., Vago, C., and Maury, R., *Rev. Forest. Fr., 5,* 353 (1959).

33.   Hall, I. M., *Hilgardia, 22,* 535 (1954).

34.   Hall, I. M., *J. Insect Pathol., 1,* 48 (1959).

35.   Hall, I. M., and Halfhill, J. C., *J. Econ. Entomol., 52,* 30 (1959).

36.   Haynes, W. C., and Rhodes, L. J., *J. Invertebr. Pathol., 13,* 161 (1969).

37.   Heimpel, A. M., *Can. J. Zool., 33,* 311 (1955).

38.   Heimpel, A. M., *Annu. Rev. Entomol., 12,* 287 (1967).

39.   Ignoffo, C. M., *Curr. Top. Microbiol. Immunol., 42,* 129 (1968).

40.   Kellen, W. R., in *Proceedings of the 12th International Congress of Entomology, London, 1965,* Vol. 3, p. 728.

41.   Kellen, W. R., Clark, T. B., Hindegren, J. E., Ho, B. C., Rogoff, M. H., and Singer, S., *J. Invertebr. Pathol., 7,* 442 (1965).

42.   Krieg, A., *J. Invertebr. Pathol., 17,* 297 (1971).

43.   Kucera, M., and Lysenko, O., *J. Invertebr. Pathol., 17,* 203 (1971).

44.   Laird, M., and Colless, R., in *Proceedings of the 11th International Congress of Entomology, Vienna, 1962,* Vol. 2, p. 867.

45.   Lipa, J., in *Insect Pathology, An Advanced Treatise,* Vol. 2, p. 335, E. A. Steinhaus, Ed. Academic Press, New York (1963).

46.   Lysenko, O., and Kucera, M., in *Microbial Control of Insects and Mites,* p. 205, H. D. Burges and W. N. Hussey, Eds. Academic Press, New York (1971).

47.   McCoy, C. W., and Kanavel, R. F., *J. Invertebr. Pathol., 14,* 386 (1969).

48.   McEwen, F. L., in *Insect Pathology, An Advanced Treatise,* Vol. 2, p. 273, E. A. Steinhaus, Ed. Academic Press, New York (1963).

49.   McEwen, F. L., and Hervey, G., *J. Insect Pathol., 1,* 86 (1959).

50.   McLaughlin, R. E., *J. Invertebr. Pathol., 7,* 464 (1965).

51.   MacLeod, D. M., and Soper, R. S., in *Proceedings of the 12th International Congress of Entomology, London, 1965,* Vol. 3, p. 724.

52.   Madelin, M. F., in *Insect Pathology, An Advanced Treatise,* Vol. 2, p. 233, E. A. Steinhaus, Ed. Academic Press, New York (1963).

53.   Martignoni, M. E., and Milstead, J. E., *J. Insect Pathol., 4,* 113 (1962).

54.   Mirra, A. P., *Curr. Sci., 2,* 225 (1952).

55.   Norris, J. R., *J. Appl. Bacteriol., 33,* 192 (1970).

56.   Ossowski, L., *J. Insect Pathol., 2,* 350 (1960).

57.   Ponsen, M. B., *Meded. Rijksfac. Landbouwwetensch. Gent, 31* (3), 553 (1966).

58.   Reeves, E. L., and Garcia, C., Jr., *Abstracts of the 4th International Colloquium on Insect Pathology, College Park, Maryland, 1970.*

59.   Roberts, D. W., Paper presented at Symposium on Microbial Insecticides. American Chemical Society, New York (1969).

60. Rogoff, M. H., *Adv. Appl. Microbiol.*, *8*, 299 (1966).
61. Rogoff, M. H. and Yousten, A., *Ann. Rev. Microbiol.*, *23*, 357 (1969).
62. Rollinson, L., Lewis, F., and Waters, M. A., *J. Invertebr. Pathol.*, *7*, 515 (1965).
63. Scharffenberg, B., *Anz. Schaedlingskd.*, *30*, 390 (1957).
64. Schmidt, L., and Phillips, G., *Poljoprivredno Sumorski Fakultet Zavod za Entomologija Zagreb*, *1*, 1 (1958).
65. Semel, M., *J. Econ. Entomol.*, *54*, 698 (1961).
66. Sen Gupta, K., *Curr. Sci. (India)*, *33*, 21 (1964).
67. Smirnoff, V., *J. Insect. Pathol.*, *3*, 29 (1961).
68. Smith, K. M., in *Insect Pathology: An Advanced Treatise*, Vol. 1, p. 457, E. A. Steinhaus, Ed. Academic Press, New York (1963).
69. Stairs, G., and Bord, F. T., *Can. Entomol.*, *94*, 966 (1962).
70. Steinhaus, E. A., *J. Insect Pathol.*, *2*, 327 (1960).
71. Steinhaus, E. A., in *Biological Control of Insect Pests and Weeds*, p. 515, P. DeBach, Ed. Reinhold Publishing Co., New York (1964).
72. Sussman, A. S., *Mycologia*, *44*, 493 (1952).
73. Tamura, S., Kuyama, S., Kodama, Y., and Hishagawa, S., in *Proceeding of the 12th International Congress on Entomology*, Vol. 3, p. 749 (1965).
74. Tanada, Y., in *Insect Pathology: An Advanced Treatise*, Vol. 2, p. 423, E. A. Steinhaus, Ed. Academic Press, New York (1963).
75. Tanada, Y., *J. Insect Pathol.*, *6*, 378 (1963).
76. Taylor, A. B., and Ring, R. L., *Trans. Am. Microsc. Soc.*, *56*, 172 (1937).
77. Toumanoff, C., *Sci. Rep. Int. Corn Borer Invest.*, *1*, 74 (1928).
78. Vago, C., and Cayrol, R., *Ann. Inst. Natl. Rech. Agron. Ser. C Ann. Epiphyt.*, *6*, 421 (1955).
79. Von Celi, B., Hertefuss, R., and Fuchs, W., *Z. Pflanzenkr. Pflanzenpathol Pflanzenschutz*, *72*, 201 (1965).
80. Weiser, J., *Cesk. Parasitol.*, *4*, 359 (1957).
81. Weiser, J., in *Insect Pathology: An Advanced Treatise*, Vol. 2, p. 291, E. A. Steinhaus, Ed. Academic Press, New York (1963).
82. Ziprin, R., and Hartman, P. A., *J. Invertebr. Pathol.*, *17*, 265 (1971).

ENZYME INHIBITORS OF MICROBIAL ORIGIN

DR. H. UMEZAWA, DR. T. TAKEUCHI AND DR. T. AOYAGI

STRUCTURES

A–1: Antipain

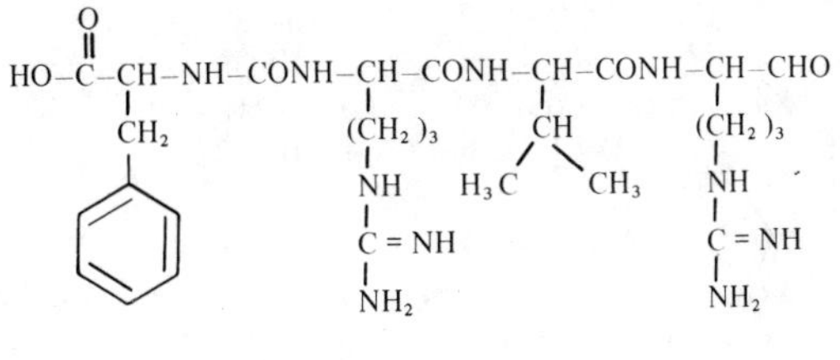

$$HO-\underset{\underset{O}{\parallel}}{C}-\underset{\underset{CH_2}{|}}{CH}-NH-CONH-\underset{\underset{(CH_2)_3}{|}}{CH}-CONH-\underset{\underset{CH}{|}}{CH}-CONH-\underset{\underset{(CH_2)_3}{|}}{CH}-CHO$$

A–2: Aquayamycin

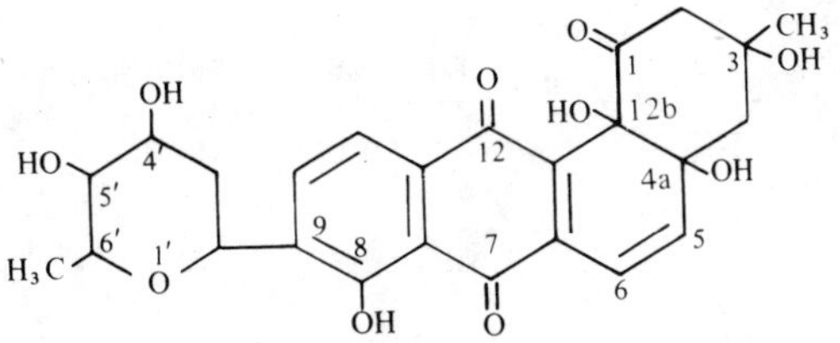

D–1: Dopastin

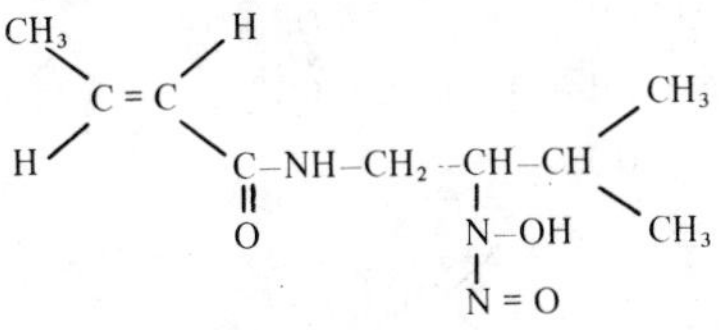

F–1: Fusaric acid

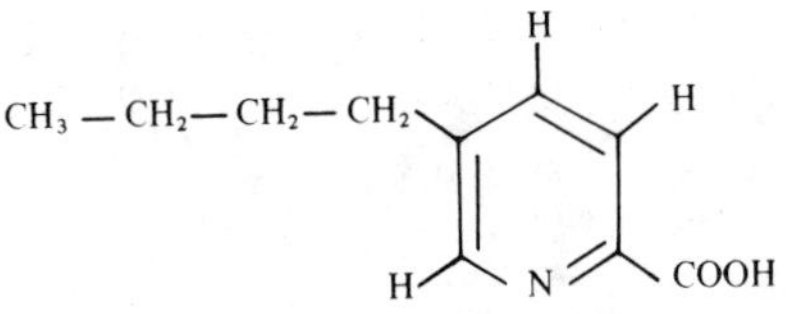

$$CH_3-CH_2-CH_2-CH_2-$$

L–1: Leupeptin

$$R-CO-NH-CH-CO-NH-CH-CO-NH-CH-CHO$$
(L)   (L)   (DL)

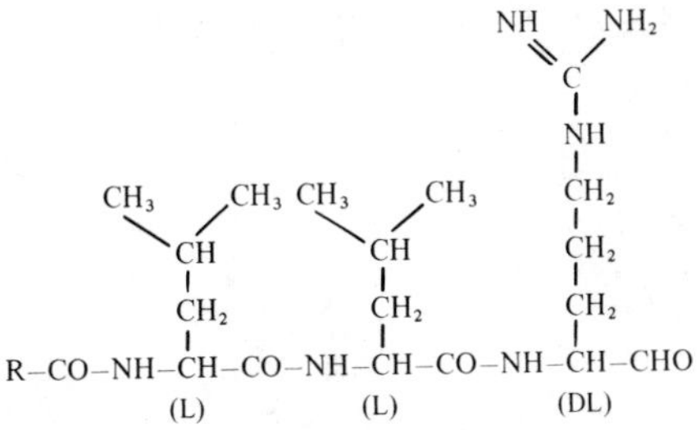

Ac–L $R = CH_3CO-$

Pr–L $R = CH_3CH_2CO-$

O–1: Oosponol

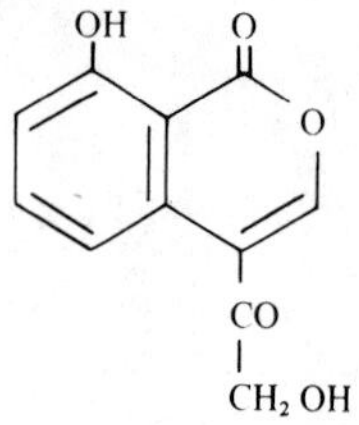

O–2: Oudenone

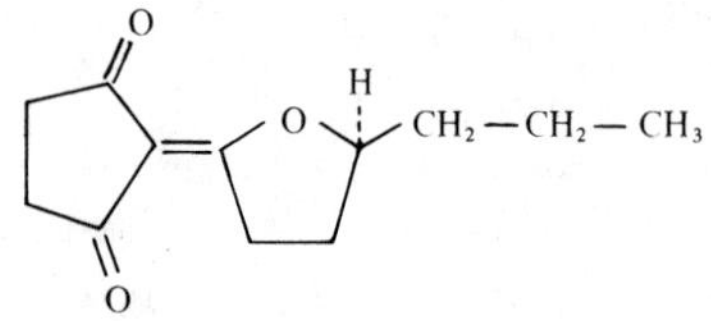

$CH_2-CH_2-CH_3$

P–1: Panosialin

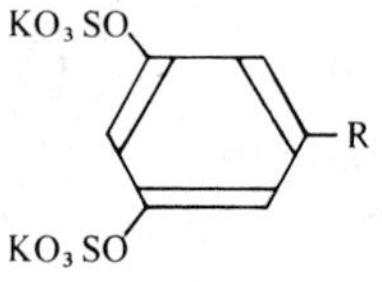

$R = -(CH_2)_{12}-CH(CH_3)_2$
$-(CH_2)_{14}-CH_3$
$-(CH_2)_{13}-CH(CH_3)_2$

P–2: Pepstatin

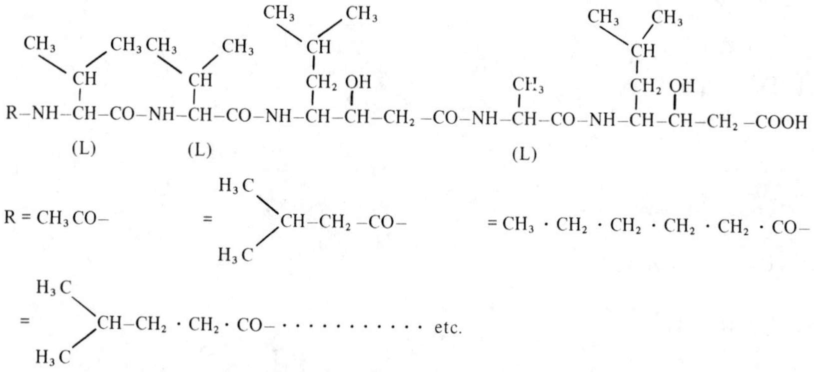

$R = CH_3CO-$

$= \;\; (CH_3)_2CH-CH_2-CO-$

$= CH_3 \cdot CH_2 \cdot CH_2 \cdot CH_2 \cdot CH_2 \cdot CO-$

$= \;\; (CH_3)_2CH-CH_2 \cdot CH_2 \cdot CO- \cdots\cdots\cdots$ etc.

S–1: S–PI

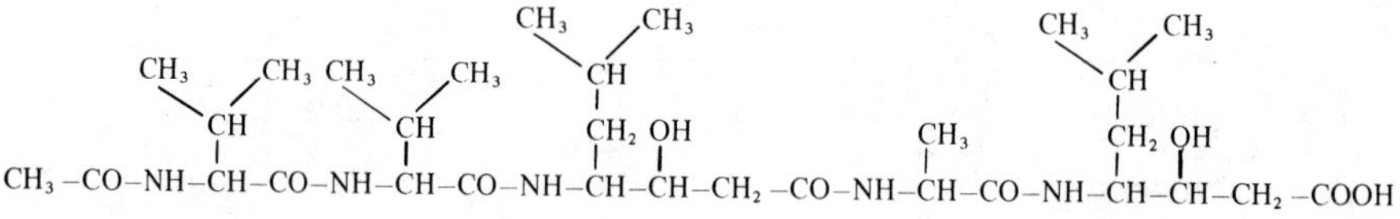

BIOLOGICAL PROPERTIES

Name Structure Formula	Producer	Inhibitory Activity	Toxicity[a]	Remarks	References
Antipain A-1 $C_{27}H_{44}N_{10}O_6$	*Streptomyces michiganensis yokusakaensis* var. *antipainicus*	Papain, trypsin, thrombokinase	M	Inhibition of carrageenine edema and blood coagulation	1, 2
Aquayamycin A-2 $C_{25}H_{26}O_{10} \cdot C_6H_{12}O_2$	*Streptomyces misawanensis*	Dopamine β-hydroxylase, tyrosine hydroxylase; antibacterial	T		3–8
Chrothiomycin $C_{27}H_{31-33}O_{13}NS$	*Streptomyces pluricolorescens*	Dopamine β-hydroxylase, tyrosine hydroxylase; antibacterial (weak)	T		9
Chymostatin	*Streptomyces hygroscopicus lavendulae*	Chymotrypsin, papain	L (ip, or)	Inhibition of carrageenin edema	10
Dopastin D-1 $C_9H_{17}N_3O_3$	*Pseudomonas* sp.	Dopamine β-hydroxylase	L	Anti-hypertension	11
Fusaric acid F-1 $C_{10}H_{13}NO_2$	*Fusarium oxysporum*	Dopamine β-hydroxylase, polyphenol oxylase, tyrosine hydroxylase; antibacterial	M	Anti-hypertension	12–15
Leupeptin L-1 Ac-L·HCl $C_{20}H_{38}O_4N_6 \cdot HCl \cdot H_2O$ Pr-L·HCl $C_{21}H_{40}O_4N_6 \cdot HCl \cdot H_2O$	*Streptomyces albireticuli chartreusis lavendulae nobortoensis roseochromogenes roseus thioluteus*	Kallikrein, plasmin, papain, thrombokinase, trypsin	M (rat)	Inhibition of carrageenin edema, blood coagulation and burn	16–21

Name Structure Formula	Producer	Inhibitory Activity	Toxicity[a]	Remarks	References
Oosponol O-1 $C_{11}H_8O_5$	*Gloeophyllum striatum Oospora astringens*	Dopamine β-hydroxylase	M (ip)	Anti-hypertension, increase of capillary permeability	22–24
Oudenone O-2 $C_{12}H_{16}O_3$	*Oudemansiella radicata*	Tyrosine hydroxylase	L	Anti-hypertension	25, 26
Panosialin P-1 $C_{21-22}H_{34-36}O_8S_2K_2$	*Streptomyces pseudoverticillus rimosus forma panosialinus*	Various enzymes	M (ip)	Detergent	27, 28
Pepstatin P-2 $C_{34}H_{63}N_5O_9$	*Streptomyces longisporoflavus argenteolus var. toyonakensis testaceus*	Pepsin, gastricsin, cathepsin-D, and other acid proteases	L	Anti-peptic ulcer, inhibition of carrageenin edema	29–32
S-PI S-1 $C_{31}H_{57}N_5O_9$	*Streptomyces naniwaensis*	Pepsin and other acid proteases			33–36
Surfactin $C_{53}H_{93}N_7O_{13}$	*Bacillus subtilis*	Blood coagulation		Surface-active	37, 38

[a] L = low toxicity ($LD_{50} > 200$ mg/kg, iv in mice); M = medium toxicity (LD_{50} 200 to 50 mg/kg, iv in mice); T = toxic ($LD_{50} < 50$ mg/kg, iv in mice); iv = intravenous; ip = intraperitoneal; or = oral.

REFERENCES

1. Suda, H., Aoyagi, T., Hamada, M., Takeuchi, T., and Umezawa, H., *J. Antibiot. (Tokyo), 25,* 263 (1972).
2. Tatsuta, K., Tsuchiya, T., and Umezawa, S., *J. Antibiot. (Tokyo), 25,* 267 (1972).
3. Sezaki, M., Hara, T., Ayukawa, S., Takeuchi, T., Okami, Y., Hamada, M., Nagatsu, T., and Umezawa, H., *J. Antibiot. (Tokyo), 21,* 91 (1968).
4. Ayukawa, S., Takeuchi, T., Sezaki, M., Hara, T., and Umezawa, H., *J. Antibiot. (Tokyo), 21,* 350 (1968).
5. Nagatsu, T., Ayukawa, S., Takeuchi, T., and Umezawa, H., *J. Antibiot. (Tokyo), 21,* 354 (1968).
6. Sezaki, M., Kondo, S., Maeda, K., Umezawa, H., and Ohno, M., *Tetrahedron, 26,* 5171 (1970).
7. Hayaishi, O., Okuno, S., Fujisawa, H., and Umezawa, H., *Biochem. Biophys. Res. Commun., 39,* 643 (1970).
8. Nozaki, M., Okuno, S., and Fujisawa, H., *Biochem. Biophys. Res. Commun., 44,* 1109 (1971).
9. Ayukawa, S., Hamada, M., Kojiri, K., Takeuchi, T., Hara, T., Nagatsu, T., and Umezawa, H., *J. Antibiot. (Tokyo), 22,* 303 (1969).
10. Umezawa, H., Aoyagi, T., Morishima, H., Kunimoto, S., Matsuzaki, M., Hamada, M., and Takeuchi, T., *J. Antibiot. (Tokyo), 23,* 425 (1970).
11. Iinuma, H., Takeuchi, T., Kondo, S., Matsuzaki, M., Ohno, M., and Umezawa, H., *J. Antibiot. (Tokyo), 25,* 497 (1972).
12. Suda, H., Takeuchi, T., Nagatsu, T., Matsuzaki, M., Matsumoto, I., and Umezawa, H., *Chem. Pharm. Bull. (Tokyo), 17,* 2377 (1969).
13. Nagatsu, T., Hidaka, H., Kuzuya, H., Takeya, K., Umezawa, H., Takeuchi, T., and Suda, H., *Biochem. Pharmacol., 19,* 35 (1970).
14. Hidaka, H., Nagatsu, T., Takeya, K., Takeuchi, T., Suda, H., Kojiri, K., Matsuzaki, M., and Umezawa, H., *J. Antibiot. (Tokyo), 22,* 228 (1969).
15. Terasawa, F., and Kameyama, M., *Jap. Circ. J., 35,* 339 (1971).
16. Aoyagi, T., Takeuchi, T., Matsuzaki, M., Kawamura, K., Kondo, S., Hamada, M., Maeda, K., and Umezawa, H., *J. Antibiot. (Tokyo), 22,* 283 (1969).
17. Kondo, S., Kawamura, K., Iwanaga, J., Hamada, M., Aoyagi, T., Maeda, K., Takeuchi, T., and Umezawa, H., *Chem. Pharm. Bull. (Tokyo), 17,* 1896 (1969).
18. Kawamura, K., Kondo, S., Maeda, K., and Umezawa, H., *Chem. Pharm. Bull. (Tokyo), 17,* 1902 (1969).
19. Aoyagi, T., Miyata, S., Nanbo, M., Kojima, F., Matsuzaki, M., Ishizuka, M., Takeuchi, T., and Umezawa, H., *J. Antibiot. (Tokyo), 22,* 558 (1969).
20. Maeda, K., Kawamura, K., Kondo, S., Aoyagi, T., Takeuchi, T., and Umezawa, H., *J. Antibiot. (Tokyo), 24,* 402 (1971).
21. Shimizu, B., Saito, A., Ito, A., Tokawa, K., Maeda, K., and Umezawa, H., *J. Antibiot. (Tokyo),* in press.
22. Umezawa, H., Iinuma, H., Ito, M., Matsuzaki, M., Takeuchi, T., and Tanabe, O., *J. Antibiot. (Tokyo), 25,* 239 (1972).
23. Yamamoto, I., Nitta, K., and Yamamoto, Y., *Agric. Biol. Chem., 26,* 486 (1962).
24. Nitta, K., Imai, J., Yamamoto, I., and Yamamoto, Y., *Agric. Biol. Chem., 27,* 817 (1963).
25. Umezawa, H., Takeuchi, T., Iinuma, H., Suzuki, K., Ito, M., and Matsuzaki, M., *J. Antibiot. (Tokyo), 23,* 514 (1970).
26. Ohno, M., Okamoto, M., Kawabe, N., Umezawa, H., Takeuchi, T., Iinuma, H., and Takahashi, S., *J. Am. Chem. Soc., 93,* 1285 (1971).
27. Aoyagi, T., Yagisawa, M., Kumagai, M., Hamada, M., Okami, Y., Takeuchi, T., and Umezawa, H., *J. Antibiot. (Tokyo), 24,* 860 (1971).
28. Kumagai, M., Suhara, Y., Aoyagi, T., and Umezawa, H., *J. Antibiot. (Tokyo), 24,* 870 (1971).
29. Umezawa, H., Aoyagi, T., Morishima, H., Matsuzaki, M., Hamada, M., and Takeuchi, T., *J. Antibiot. (Tokyo), 23,* 259 (1970).
30. Morishima, H., Takita, T., Aoyagi, T., Takeuchi, T., and Umezawa, H., *J. Antibiot. (Tokyo), 23,* 263 (1970).
31. Aoyagi, T., Kunimoto, S., Morishima, H., Takeuchi, T., and Umezawa, H., *J. Antibiot. (Tokyo), 24,* 687 (1971).
32. Ikezawa, H., Aoyagi, T., Takeuchi, T., and Umezawa, H., *J. Antibiot. (Tokyo), 24,* 488 (1971).
33. Murao, S., and Satoi, S., *Agric. Biol. Chem., 34,* 1265 (1970).
34. Fukumura, M., Satoi, S., Kuwana, N., and Murao, S., *Agric. Biol. Chem., 35,* 1310 (1971).
35. Murao, S., and Satoi, S., *Agric. Biol. Chem., 35,* 1477 (1971).
36. Satoi, S., and Murao, S., *Agric. Biol. Chem., 35,* 1482 (1971).
37. Arima, K., Kakinuma, A., and Tamura, G., *Biochem. Biophys. Res. Commun., 31,* 488 (1968).
38. Kakinuma, A., and Arima, K., *Annu. Rep. Takaeda Res. Lab., 28,* 140 (1969).

ENZYMES

MICROBIAL PROTEASES

DR. DAISUKE TSURU

Numerous bacteria and fungi are known to produce various kinds of proteases intracellularly and/or extracellularly. Some protease-forming bacteria secrete two types of endopeptidases (proteinases); one type is most active at neutral pH and the other in the alkaline region. Some bacteria also have the ability to form intracellular and extracellular exopeptidases. Some fungi have been found to produce three types of endopeptidases (acid, neutral, and alkaline proteases) together with exopeptidases. Streptomycetes and yeasts have also been shown to be good producers of exopeptidases as well as endopeptidases. Yeast proteases usually accumulate in the cells and are liberated by autodigestion or physical and chemical treatments.

In Tables 1 to 4, some enzymatic and physicochemical properties of microbial proteases are summarized; Tables 5, 6 and 7 list the amino acid compositions of several highly purified proteases.

The following organisms, not listed in the tables, have also been known to produce endopeptidases, but the details of their enzymatic properties remain obscure:[32,114,164-173]

Achromobacter sp. (collagenase);[241]

Ajellomyces dermatitidis (elastase);[220]

Altescheria hoydii (elastase);[220]

Arthroderma benhamii (elastase);[220]

Ascochyta visa (milk-clotting enzyme);[238]

Aspergillus usami, A. carbonarius, A. japonicus, A. awamori, A. nakazawaii, A. fumigatus (optimal pH 6.7 and 10.0);[165]

Chaetomium globosum (milk-clotting enzyme);[242]

Coriolus consors (milk-clotting enzyme);[238]

Fomitopsis pinicola (milk-clotting enzyme);[238]

Fusarium moniliforme (milk-clotting enzyme);[242]

Byssochlamyces fulva;[167]

Penicillium brevi-compactum, P. citrinum (milk-clotting protease),[211] *P. expansum, P. italicum, P. spinulosum, P. stipitatum, P. wortmanii, P. vermiculatum;*

Rhizopus formosensis, R. tonkinensis, R. javanicus, R. delemar, R. candidus, R. nodosus, R. batatae, R. oligosporus,[173] *R. peka;*

Malbranchae pulchella;[219]
several species of *Mucor: M. hiemalis,*[185] *M. gypseum,*[114] *M. audouini,*[114] *M. rubrum,*[114] *M. canis;*[114]

Monascus anka;

Bacillus larvae,[168,169] *B. brevis, B. mycoides, B. anthracis, B. anthracoides, B. thuringiensis;*

Bacteroides melaninogenicus (collagenase),[239] *B. amylophilus;*[186]

Chondrococcus columnaris;

Clostridium acetobutyricum, C. butyricum, C. putrificum, C. chauvoei, C. lentoputrescens, C. capitovale, C. bifermentans (collagenase), *C. septicum, C. oedematiens, C. welchii* (collagenase), *C. sporogenes* (optimal pH 6—6.2), *C. perfringens* (collagenase), *C. tetani* (collagenase);

Flavobacterium elastolyticum (elastase);

Halobacterium salinarium (collagenase);[187]

Micrococcus freudenreichii, M. caseinolyticus, M. lysodeikticus; Mycobacterium tuberculosis (collagenase);[229]

Lactobacillus casei;

Narrizzia fulva (elastase);[220]

Pseudomonas cholerae, P. pseudomallei, P. pyocyanea;

Sarcina flava;

Streptomyces rimosus (keratinase), *S. liquefaciens, S. scabies;*

Vibrio cholerae.

Various species of yeast and yeast-like fungi have shown caseinolytic activity:[188,210]

Aureobasidium pullulans;

Bullera alba;

Candida lipolytica,[210] *C. punicea, C. pseudotropicalis, C. curiosa, C. aquatica;*

certain species of *Cephalosporium;*

Debaromyces hansenii;

Endomycopsis capsularis, E. fibuliger, E. selenospora;

Geotrichum candidum;

Kluyveromyces aestuarii, K. africans, K. drosophilarum, K. fragilis, K. polysporus;

Protomyces enouyei, P. inundatus, P. pachydermus;

Scytalidium lignicolum (M-133);[209]

Sporidiobolus johnsonii;

Sporobolomyces pararoseus, S. proseus, S. salmonicolor;

Sporotrichum schenkii;

Taphrina sp.;

Trichosporon cutaneum.

Exopeptidases have also been reported to be formed by the following organisms:[67,91,172]

Aspergillus niger, A. saitoi, A. oryzae, A. aureus, A. awamori, A. sojae,[198-200] *A. parasiticus;*[216]

Mucor miehei, M. pusillus;

Penicillium janthinellum, P. dalae, P. funiculosum;[201-203]

Talaromyces duponti (MW = 400,000), *T. emersonii, T. auranticus;*[172]

Halobacterium salinarium (dipeptidase);

Humicola lanigunosa;

Escherichia coli;

Leuconostoc mesenteroides;

Mycobacterium phlei (dipeptidase);[213]

Pseudomonas sp.,[204,205] *P. stutzer* (Glu-specific and Asp-specific Cpase);

Streptomyces rectus, S. peptidofaciens;[208]

thermophilic Actinomycetes.

A characteristic peptidase, pyrrolidonecarboxylyl peptidase, was found in two bacterial species:[174]

Bacillus subtilis;

Pseudomonas fluorescens.

A number of cell-wall-lytic peptidases have recently been found in the following bacteria:

Achromobacter lyticus;[234]

Aeromonas;[180]

Bacillus thuringiensis;[189]

Escherichia coli;[190]

Flavobacterium;[191,192]

Micrococcus;[133,134]

Myxobacter;[121,122]

Pseudomonas aeruginosa;[231,232]

Staphylococcus epidermidis ALE;[233]

several species of *Streptomyces;*[193-196,235,236]

thermophilic Actinomycetes: *Micropolyspora* sp. No. 434,[223] *Thermoactinomyces vulgaris,*[237] *Thermomonospora fusca.*[237]

TABLE 1
FUNGAL PROTEASES

Organism	Type of Protease	Optimal pH (Stable pH)	Inhibitor[a]	$E_{1\ cm}^{1\%}$ [b]	$S_{20,w}$	pI	MW x 10^{-3}	Remarks	Reference
Acremonium									
kiliense	Alkaline	10.5 (5–10)					28		212
Acrocylindrium	Acid	2–3 (1.5–5.0)			3.2				25
Alternaria									
tenuissima	Acid	3–5			2.35	10.5	23–25		166, 181
Aspergillus									
candidus IAM 2015	Alkaline	10.0–11.0 (5–11)	DFP, PPI	7.1	3.0	4.8	23		14
flavus	Alkaline[c]	8.5–10 (5–10)	DFP, PPI		3.0		23		13
melleus	Semi-alkaline	8.0 (6–9)	DFP		2.8		33		20
niger macrosporus	Acid A	2.0 (2–5)							23
	Acid B	2.8 (2–5)							23
ochraceus	Neutral	7.5 (5–9)	EDTA, HM						15, 16
oryzae	Acid I	3.0 (2.5–6)	SLS	11.5	3.16		34.6	Showed trypsinogen activation activity	1–5
	Acid II	3.0 (2.6–6.0)	SLS	5.9	3.93		66.8	Showed trypsinogen activation activity; sugar content, 60%	1–5

TABLE 1 (Continued)
FUNGAL PROTEASES

Organism	Type of Protease	Optimal pH (Stable pH)	Inhibitor[a]	$E_1^{1\%}$ₘ[b]	$S_{20,w}$	pI	MW x 10⁻³	Remarks	Reference
Aspergillus oryzae (cont.)									
	Alkaline[d]	8.5–10.0 (5–8)	DFP, PPI	9.0	3.05		18–20		5–10
	Cpase I[e]	4.0 (3–8)	SH reagent				120		24, 200
	Cpase II	3.0 (5–6)	Hg²⁺				105		24, 200
	Cpase III	3.0 (3–6)	Cu²⁺				61		24, 200
	Cpase IV	4.0 (5–7)	Pb²⁺				45		24, 200
	Apase I[f]	8.5 (7–9)	MC				26.5		24, 200
	Apase II	8.0 (6–9)	Cysteine				61.0		24, 200
	Apase III	8.0 (5–7)	HM						200
oryzae microsporus	Alkaline	8.5–9.5 (5–8)	DFP, PPI	8.3	2.80	6.0	22		11, 12
	Semi-alkaline	8.2–8.4 (5–8)	DFP, PPI	10.2	2.90	5.4	24		11, 12
	Neutral	6–7 (4–10)	EDTA, HM	13.2	3.8	4.2	38	Zn-protease	11, 12
saitoi	Acid[g]	2.5–3.0 (2.5–6.0)	SLS	13.15	3.33	3.4	34.9	Showed trypsinogen activation activity	21, 22
sojae	Alkaline	8.5–10 (5–9)	DFP, PPI	9.0	2.82	5.1	25.5		17
	Neutral I	6.5–7.5 (5–9)	EDTA			4.7	42	Zn-protease	18
	Neutral II	6.5–7.5 (5–9)	EDTA			4.2	20	Zn-protease	18
sulphureus	Alkaline	7.5–9.5 (6–11)	DFP, NBS		2.78		23		215

TABLE 1 (Continued)
FUNGAL PROTEASES

Organism	Type of Protease	Optimal pH (Stable pH)	Inhibitor[a]	$E_{1\,cm}^{1\%}b$	$S_{20,w}$	pI	MW x 10^{-3}	Remarks	Reference
Aspergillus sydowi	Semi-alkaline	8.0 (6–9)	DFP						19
Cephalosporium	Alkaline	11.0 (5–11)	DFP, PI	11.1	2.9	10.5	22.5		217, 218
Cladosporium	Acid	2.5–2.9 (2.5–7.0)	SLS	10.7		4.6	38	Pepsin-like	26
Endothia parasitica	Acid	<5 (3.8–4.5)				4.6	34–39	High milk-clotting activity	27
Fusarium solani M77	Alkaline	8.5–9.5 (6–10)						Collagenase-like	29
sp.	Alkaline	10–11 (5–10)			3.1	10.5	32		28
Gliocladium roseum	Alkaline I	11 (6–8)	HM						30
	Alkaline II	10 (6–10)	HM						30
Irpex lactis	Acid	2.5 (3–5)				5.2	34	High milk-clotting activity	31

TABLE 1 (Continued)
FUNGAL PROTEASES

Organism	Type of Protease	Optimal pH (Stable pH)	Inhibitor[a]	$E_{1cm}^{1\%}b$	$S_{20,w}$	pI	MW x 10^{-3}	Remarks	Reference
Mucor									
miehei	Acid[h]	4.0 (3–8)		11.8	3.35	4.2	38.2	High milk-clotting activity	34
pusillus Lint	Acid[i]	4.0 (4–6)	HM	10.0	2.39	3.5	30	High milk-clotting activity	32, 33
Paecilomyces									
varioti	Acid	3.0 (3–6.5)	PCMB, Hg²⁺		2.75	3.8	27.3		35
Penicillium									
caseicolum	Neutral	6.5 (4–8)	EDTA, HM						43
chrysogenum	Neutral	7.8 (4–7)	HM						42
cyaneo-fulvum	Alkaline	10 (5–10)							44
cyclopium	Acid	3.5 (3–7)	SLS		3.09	5.0	30.0		40, 41
janthinellum	Acid[j]	3–4 (2.2–6.6)	SLS	13.5	3.1	3.8	32.0	Showed trypsinogen activation activity	36, 37
	Cpase[k]	4.7 (4–6)	Fe³⁺, Cd²⁺						37
Penicillium									
notatum	Acid	3.8–4.2 (3–7)						Active toward glycoprotein	38
	Alkaline	7.5–9.5 (4–10)			1.6	4.92	20.0		175

TABLE 1 (Continued)
FUNGAL PROTEASES

Organism	Type of Protease	Optimal pH (Stable pH)	Inhibitor[a]	$E_{1cm}^{1\%b}$	$S_{20,w}$	pI	MW x 10^{-3}	Remarks	Reference
Penicillium									
purpurogenum	Acid	3.0–3.5 (3–6)	SLS						39
rubrum	Acid	3.0–3.5 (3–6)	SLS						39
Phymatotrichum									
omnivorum	Semi-acid Neutral	5–5.5 7–8[l]	EDTA, DFP DFP, MC				33.0 33.4	Aromatic amino acid-specific esterase	45
	Cpase	7.5 (7–10.5)	MC				31.4	Aromatic C-terminal-specific Cpase	46
Rhizopus									
chinensis	Acid[m]	2.9–3.3 (2.8–6.5)	SLS, HM	12.9	2.83	5.2	35.0	Showed trypsinogen activation activity; high milk-clotting activity	47, 48
niveus	Acid	2.9–3.3 (2.8–6.5)	SLS, HM	12.1	2.7	5.2	39.0		49
Scopulariopsis									
	Neutral	7.5 (4–8)	EDTA, HM						50
	Alkaline	11 (4–8)	HM						51
Trametes									
sanguinea	Acid[n]	2.3–2.5 (2–6)	HM	11.4	3.15	3.5	34		52

TABLE 1 (Continued)
FUNGAL PROTEASES

Organism	Type of Protease	Optimal pH (Stable pH)	Inhibitor[a]	$E_1^{1\%}$ cm[b]	$S_{20,w}$	pI	MW x 10^{-3}	Remarks	Reference
Trichophyton									
granulosum	Alkaline	9.5–9.8 (4.7–8.0)	EDTA, HM	22.15			34.0	Keratinase	53
schoenleinii	Collagenase	6.5	EDTA, Cys				20.0		177

[a] Cys = cysteine; DFP = diisopropyl fluorophosphate; EDTA = ethylenediaminetetraacetic acid; HM = heavy metals; MC = metal chelator; NBS = N-bromosuccinimide; PCMB = p-chloromercuribenzoate; PI = pepsin inhibitor; PPI = protease inhibitor obtained from potato; SLS = sodium lauryl sulfate.
[b] Optical density at 280 nm of 1% enzyme solution at 1 cm optical path.
[c] Aspergillopeptidase C.
[d] Aspergillopeptidase B.
[e] Cpase = carboxypeptidase.
[f] Apase = aminopeptidase.
[g] Aspergillopeptidase A.
[h] Rennilase.
[i] Mucorrennet.
[j] Penicillopepsin.
[k] Penicillocarboxypeptidase.
[l] Optimal pH of estrase activity.
[m] Rhizopepsin.
[n] Rapidase.

TABLE 2
ACTINOMYCETIC PROTEASES

Organism	Type of Protease	Optimal pH (Stable pH)	Inhibitor[a]	$E_1^{1\%}$ cm[b]	$S_{20,w}$	pI	MW $\times 10^{-3}$	Remarks	Reference
Streptomyces									
caespitosus	Neutral	7.8 (5–9)	EDTA	15.5	1.95		15.0		54
erythreus	Alkaline	9 (5–10)	DFP, TLCK					Trypsin-like	55
fradiae	Ia	9.5 (5–9)	DFP, PPI					Thermostable	56, 57
	Ib	11 (6–12)	DFP, PPI					Thermostable	56, 57
	II	9–10 (5–9)	DFP, PPI		2.4	9	27.0	Keratinase	56–59
	III	8 (6–9)	DFP, PPI					Elastase	56, 57
	IV	8–9 (5–9)	DFP, PPI					Trypsin-like	56, 57
	Apase[c]	10 (7–11)	DFP, PPI						60
	Cpase[d]	7–8 (4–8)							60
griseus K-1	Alkaline A	8–10 (5–10)	DFP	9			16.0		61–64
	Alkaline B	8.5–9.5 (5–10)	DFP, TLCK	14.95			18.6	Trypsin-like	61–64
	Alkaline C	9–10 (3–10)	DFP, PPI	13.6			13.5		61–64
	Neutral[e]	7.5–8 (6–9)	EDTA, HM	11			20.0		65.
griseolus	Neutral	7.5–8.5 (3–10)	HM, MC, I_2						221

TABLE 2 (Continued)
ACTINOMYCETIC PROTEASES

Organism	Type of Protease	Optimal pH (Stable pH)	Inhibitor[a]	$E_{1cm}^{1\%}b$	$S_{20,w}$	pI	MW x 10^{-3}	Remarks	Reference
Streptomyces									
griseus ATCC 3463 (*Actinomadura*)	Apase	8 (6–9)	EDTA, HM						66
kinoluteus	Neutral	6.5 (5–9)	HM, MC						221
maderatus sp.n.	Alkaline	8 (5–10)	DFP, HM		2.57	8.6		Intracellular	222
madurae	Neutral	7.5 (4.5–8.5)	Cys, EDTA				35.0	Collagenase-like	67, 240
	Semi-alkaline	8 (5–10)	DFP, HM		2.57	8.6			182
naraensis	Neutral	7.5 (6–9)	EDTA, HM		3.3	4.2	37.0	Zn-protease	68
proteolyticus	Neutral	6.8–7.0 (5–8)	EDTA, HM						69
rectus	Alkaline A	10.6–10.8 (5–10)	DFP, PCMB	18.2	3.0	9.5	21.5	Thermostable	70, 71
	Alkaline B	10.6–10.8 (5–10)	DFP		3.0	9.5	21.5	Thermostable	70, 71
sioyaensis	Apase	8–9 (4.5–9)	EDTA, HM					Co^{2+} activation	67
	Cpase	8 (4.5–9)	HM					Ca^{2+} activation	67
verticillatus	Neutral I	6.5–8.5 (6–9)	HM, MC						221
	Neutral II	7–8 (6–8)	HM, MC						221

TABLE 2 (Continued)
ACTINOMYCETIC PROTEASES

Organism	Type of Protease	Optimal pH (Stable pH)	Inhibitor[a]	$E_{1\ cm}^{1\ \%}$ [b]	$S_{20,w}$	pI	MW $\times 10^{-3}$	Remarks	Reference
Streptomyces									
S-1033	Neutral	7–8 (5–8)	EDTA, HM		1.95		15.0		67
sp-41	Alkaline	10 (5–9)	DFP, PPI						72
Thermonospora									
fusca	Alkaline	8–9	I$_2$					Thermostable	73
Thermoactinomyces									
vulgaris	Alkaline	8–9	I$_2$					Thermostable; bacteriolytic	73

a Cys = cysteine; DFP = diisopropyl fluorophosphate; EDTA = ethylnediaminetetraacetic acid; HM = heavy metals; MC = metal chelator; PCMB = p-chloromercuribenzoate; PPI = protease inhibitor obtained from potato; TLCK = tosyl-L-lysylchloromethylketone.

b Optical density at 280 nm of 1% enzyme solution at 1 cm optical path.

c Apase = aminopeptidase.

d Cpase = carboxypeptidase.

e Pronase; the pronase preparations commercially available now are mixtures of alkaline and neutral proteases and several kinds of exopeptidases.

TABLE 3
BACTERIAL PROTEASES

Organism	Type of Protease	Optimal pH (Stable pH)	Inhibitor[a]	$E_1^{1\%}_{cm} b$	$S_{20,w}$	pI	MW x 10^{-3}	Remarks	Reference
Aeromonas									
hydrophyla	Staphylolytic	9.0			1.1	9.5		Bacteriolytic	180
proteolytica	Semi-alkaline	8.0						Zn-protease	74, 224
	Apase[c]	8.5—9.0 (8)	MC		3.39		29.5	Zn-protease	75
Arthrobacter	Semi-alkaline	7—9 (2—11)	DFP	44.4	2.7		22.0		76—79
Bacillus									
cereus	Neutral	7 (5—9)	MC				6.30	Zn-protease	225
	Alkaline	10.5—11.0 (6—11)	DFP						80
licheniformis	Alkaline[d]	10.3—10.8 (5—11)	DFP, PPI				27.8		81, 82
	Apase	8.5—9.5							81
	Neutral	6.5—7.5 (5—10)	DFP						81
megaterium	Neutral[e]	7 (6—9)	EDTA, HM					Zn-protease	83, 84
mesentericus	Semi-alkaline	8.5 (6—9)	EDTA, HM				32.0		85, 86
pasteurianus	Alkaline	11—12 (10—12)						Intracellular	87
polymyxa	Semi-alkaline	8.0—8.5 (7—9)	EDTA, MIA				35.9	Zn-protease; high milk-clotting activity	88

TABLE 3 (Continued)
BACTERIAL PROTEASES

Organism	Type of Protease	Optimal pH (Stable pH)	Inhibitor[a]	$E_{1cm}^{1\%}$,b	$S_{20,w}$	pI	MW $\times 10^{-3}$	Remarks	Reference
Bacillus									
pumilis	Alkaline[d]	10.3–10.8 (5–11)	DFP						82
stearothermophilus	Neutral	6.9–7.2 (6–9.5)	EDTA, HM					Thermostable	89
	Apase I	7.5–8.0[f] (7–8)		10.2			400.0	Co²⁺ activation	90, 91
	Apase II						70–80	Co²⁺ activation	90, 91
	Apase III						70–80	No metal need	90, 91
subtilis	Alkaline[g]	10.3–10.8 (5–11)	DFP, PPI	8.7	2.85	9.8	27.6		104, 105
amyloliquefaciens	Apase I	8.2 (5–9)	HM			4.2	257.0	Intracellular; Co²⁺ or Mn²⁺ activation	92, 93
	Apase II	7.5 (5–9)	HM				115.0	Intracellular; Co²⁺ or Mn²⁺ activation	92, 93
	Neutral[h]	6.5–7.5 (6–10)	EDTA, HM	13.6	3.24	8.95	40.0	Zn-protease	94, 95
	Alkaline[i]	10.2–10.7 (5–11)	DFP, PPI	11.7	2.77	7.8	27.7		96–99
alkalophilic	Alkaline	10–12 (4–12)	DFP		3.5	9.4	30.0		226
subtilis var. *amylosacchariticus*	Apase I	8.2 (5–9)	HM			3.7	90.0	Intracellular; Co²⁺ or Mn²⁺ activation	92, 93
	Apase II	7.5 (5–9)	HM			4.2	105.0	Intracellular; Co²⁺ or Mn²⁺ activation	92, 93
	Neutral[j]	7.0 (6–10)	EDTA, HM	13.8	3.02	8.5	34.0	Zn-protease	100–102
	Alkaline[k]	10.5 (5–11)	DFP, PPI	11.9	2.89	8.5	25–28		99, 103

TABLE 3 (Continued)
BACTERIAL PROTEASES

Organism	Type of Protease	Optimal pH (Stable pH)	Inhibitor[a]	$E_{1cm}^{1\%}$ b	$S_{20,w}$	pI	MW x 10^{-3}	Remarks	Reference
Bacillus									
subtilis var. *natto*	Alkaline	10.3–10.8 (5–11)	DFP	8.8	2.8	9.75	27.0		109
subtilis WI	Alkaline[l]	8	DFP, PPI		1.8		17–30		106
subtilis, transformable	Acidic[m]	8.0 (5.5–10.5)	DFP				30.7		107, 108
	Basic[n]	7.5 (5.5–10.5)	DFP				36.0		107, 108
thermoproteolyticus	Neutral[o]	7.5–8.0 (6–9.5)	EDTA, HM	17.7	3.6		37–39	Zn-protease	110–112
Clostridium									
botulinum Type B	Apase	7–8 (5–8)						Fe^{3+} activation	113
	Neutral	6.2–7				4.62	34.0	Trypsin-like	227
histolyticum	Collagenase[p]	7.5–8.0 (6–9)	EDTA, HM		5.2		95.0	Zn-protease	114–116
	Collagenase	6.5–7.0 (5–9)	EDTA, HM		5.1		79.0		114–116
	Clostripain[q]	7–8 (6–)	DFP, PCMB		4.43	4.8	50.0	Trypsin-like; plasminogen activity	117, 118
	Apase	7 (6–8)	EDTA, HM						119
parabutylinum	Apase	7–8 (5–8)							113
Escherichia									
coli	Apase	8–10.5 (6.5–10.5)	Zn, MC		12.0			Heat-stable	228

TABLE 3 (Continued)
BACTERIAL PROTEASES

Organism	Type of Protease	Optimal pH (Stable pH)	Inhibitor[a]	$E_{1cm}^{1\%}$	$S_{20,w}$	pI	MW $\times 10^{-3}$	Remarks	Reference
Escherichia									
freundii	Alkaline	10 (4.5–9.5)	EDTA			5.0	42.0	Thermolabile	120
Micrococcus sp.	Lysostaphin	7.5				> 9.0	30.0	Bacteriolytic	183, 184
Myxobacter AL-I	Semi-alkaline	9 (7–10)	EDTA, HM	15.8	2.4		14.3	Bacteriolytic	121, 122
Proteus									
mirabilis[r]	Alkaline	9–10 (6.5–11)	PCMB, HgCl$_2$						125, 126
vulgaris[r]	Alkaline	9–10 (6.5–11)	PCMB, HgCl$_2$						123–125
Pseudomonas									
aeruginosa	Alkaline	8.5 (5–10)	EDTA, HM	14.71	3.99	4.1	48.4	Collagenase-like	127–130
	Elastase	8 (5–10)	EDTA, HM	14.52	3.38	5.9	39.5		131
fluorescens	Semi-alkaline	7–8.5 (5–10)	EDTA, HM						132
marinoglutinosa	Collagenase	7.5 (6–9)	EDTA, HM				74		230
myxogenes	Semi-alkaline	7–8.5 (5–10)	EDTA, HM			5.5	77.0		133
sp. No. 548	Alkaline	10.5 (7–9)	EDTA					Maximum production of enzyme at 5°C	134

TABLE 3 (Continued)
BACTERIAL PROTEASES

Organism	Type of Protease	Optimal pH (Stable pH)	Inhibitor[a]	$E_{1cm}^{1\%}$ b	$S_{20,w}$	pI	MW x 10^{-3}	Remarks	Reference
Serratia									
marcescens	Alkaline	9 (6—8)	EDTA						136
sp.	Semi-alkaline	8 (6—8)	EDTA				30.0		135
	Semi-alkaline	7—8 (6—8)	EDTA	13.0	3.8	5—5.5	60.0	Zn-protease	137
	Apase I	8	EDTA				> 100.0		135
	Apase II	8	EDTA				> 100.0		135
Sorangium									
	a-Lytic	8.5	DFP		2.5	> 9	19.0	Bacteriolytic	138—140
	β-Lytic	8.5		20.5	2.2		20.0	Bacteriolytic	138—140
Staphylococcus									
aureus	Neutral[s]	7.5 (6—8)	ε-ACA			6.7	22.5	Plasminogen activation	179, 214
	Collagenase	7.5 (Labile)				5.3	44.0		141, 142
Streptococcus									
faecalis	Neutral	7.6 (5—9)	EDTA, SHR						143, 144
lactis	Semi-alkaline	8.5 (7)	PCMB					Thermostable	145
group A	Neutral	7.6 (4.5—8.5)	PCMB			8.2	32.0	SH-protease; accumulated as proenzyme	146—148
group C	Neutral[t]	7.3—7.6 (5—9)		9.45		4.7	47—48	Plasminogen activation	149, 150, 178

TABLE 3 (Continued)
BACTERIAL PROTEASES

a ε-ACA = ε-aminocaproic acid; DFP = diisopropyl fluorophosphate; EDTA = ethylenediaminetetraacetic acid; HM = heavy metals; MC = metal chelator; MIA = monoiodoacetate; PCMB = p-chloromercuribenzoate; PPI = protease inhibitor obtained from potato; SHR = SH-reagent.
b Optical density at 280 nm of 1% enzyme solution at 1 cm optical path.
c Apase = aminopeptidase.
d Subtilisin, Carlsberg type.
e Megateriopeptidase.
f Substrate: LNA (L-leucine-p-nitroanilide).
g Subtilopeptidase A = subtilisin Carlsberg = bacillopeptidase A.
h Neutral subtilopeptidase amyloliquefaciens.
i Subtilopeptidase B = subtilisin BPN′ = subtilisin, type novo = bacillopeptidase B.
j Neutral subtilopeptidase amylosacchariticus.
k Subtilisin amylosacchariticus = bacillopeptidase D.
l Bacillopeptidase C.
m Bacillopeptidase E.
n Bacillopeptidase F.
o Thermolysin.
p Clostripeptidase A.
q Clostripeptidase B.
r $P.$ $mirabilis$ and $P.$ $vulgaris$ seemed to be identical strains.
s Staphylokinase.
t Streptokinase.

TABLE 4
YEAST PROTEASES

Organism	Type of Protease	Optimal pH (Stable pH)	Inhibitor$a$	$E^{1\%b}_{1\,cm}$	$S_{20,w}$	pI	MW x 10^{-3}	Remarks	Reference
Candida									
albicans	Acid	3.2 (3–5)			3.2		42.0		151
Endomycopsis									
fibliget	Acid	2.3–2.6 (4.5–7.0)	SLS, PI						152
Rhodotorula									
glutinis	Acid	2–3 (2.5–6.5)	PI	14.0		4.5	30.0	Thermostable	152

TABLE 4 (Continued)
YEAST PROTEASES

Organism	Type of Protease	Optimal pH (Stable pH)	Inhibitor[a]	$E_{1\,cm}^{1\%}$[b]	$S_{20,w}$	pI	MW x 10^{-3}	Remarks	Reference
Saccharomyces									
carlsbergensis	Acid	3.0 (6.0–6.5)						Glycoprotein	160
	Neutral	7.0 (6.0–6.5)						Glycoprotein	160
cerevisiae	A	2–3 (5)				3.5	50.0	Glycoprotein	153–156
	B	9 (Labile)	PCMB, DFP						153–156
	C[c]	5–6 (7)	PCMB, DFP	14.8	4.23	3.6	61.0	Exists as proenzyme (pro-C)	153–156
	Pro-C								
	Cpase[d]	6 (7)	EDTA, HM		5.27		79.0		153–156 157–158
brewer's yeast	Cpase, a	6.0–6.2 (5–7)	MC, PCMB						159
	Neutral, β	6.0 (5–7)	DFP, PCMB						159
Torulopsis									
ingeniosa	Acid	2.5–3.0 (3.5–7.0)	PI						152

[a] DFP = diisopropyl fluorophosphate; EDTA = ethylenediaminetetraacetic acid; HM = heavy metals; MC = metal chelator; PCMB = p-chloromercuribenzoate; PI = pepsin inhibitor; SLS = sodium lauryl sulfate.

[b] Optical density at 280 nm of 1% enzyme solution at 1 cm optical path.

[c]

[d] Cpase = carboxypeptidase.

TABLE 5
AMINO ACID COMPOSITION OF MOLD PROTEASES

Organism	Alternaria tenuissima	Aspergillus flavus	Aspergillus melleus	Aspergillus oryzae	Aspergillus oryzae microsporus	Aspergillus oryzae microsporus
Type of Protease	Acid	Alkaline	Semi-alkaline	Alkaline	Alkaline	Alkaline
Amino Acids						
Lysine	1	11	5	11–12	12	15
Histidine	1	3–4	2	4	4	3
Arginine	10	2	2	2	3	3
Aspartic Acid	18	21	36	21	21–22	28
Threonine	33	11	25	11	13	14
Serine	35	20	37	19	23	22
Glutamic Acid	12	12–13	18	12	13	14
Proline	4	4–5	9	4	5–6	5
Glycine	35	20	45	19	21	24
Alanine	22	23	46	23	23	28
½-Cystine	6	0	0	0	0	0
Valine	19	15	33	15	16	17
Methionine	0	1	0	0	1	2
Isoleucine	7	9–10	15	9–10	10–11	11
Leucine	7	9	21	9	10	11
Tyrosine	13	5	11	5	4–5	7
Phenylalanine	6	5	7	5	6	7
Tryptophan	4	2	2	2	2	4
Amide-NH₂	23	17	11	15		11
Total Amino Acids	253	173–177	314	171–173	187–191	215
Molecular Weight	25,000	18,000	33,000	17,800	19,700	22,000
N-Terminal		Glycine		Glycine	Glycine	
C-Terminal		Alanine		Alanine	Alanine	
Reference	181	13	20	8	161	11

TABLE 5 (Continued)
AMINO ACID COMPOSITION OF MOLD PROTEASES

Organism	Aspergillus oryzae microsporus	Aspergillus oryzae microsporus	Aspergillus saitoi	Aspergillus sojae	Cladosporium	Endothia parasitica
Type of Protease	Semi-alkaline	Neutral	Acid	Alkaline	Acid	Acid
Amino Acids						
Lysine	16	15	11	14	6	12
Histidine	6	7	3	5	0	3
Arginine	5	13	1	3	3	2
Aspartic Acid	28	48	34—35	31	34	26
Threonine	14	24	25	18	46	50
Serine	25	25	42—43	28	50	44
Glutamic Acid	15	25	22	19	19	15
Proline	8	18	10	6	11	13
Glycine	25	38	31—32	27	36	38
Alanine	29	31	20	32	29	29
½-Cystine	1	1—2	1	2	1	2
Valine	18	13	22	18	18	22
Methionine	2	5	0	2	2	1
Isoleucine	12	15	11—12	14	14	17
Leucine	12	25	19—20	14	11	19
Tyrosine	8	20	17—18	8	14	19
Phenylalanine	7	11	13	7	20	13
Tryptophan	4	6	1	2	2	3
Amide-NH$_2$		20	31—32	20		31
Total Amino Acids	235	340—341	283—289	250	316	328
Molecular Weight	24,000	38,000	35,000	26,000	38,000	34,000
N-Terminal			Serine	Glycine		
C-Terminal			Alanine	Alanine		
Reference	11	11	21	17	26	172

TABLE 5 (Continued)
AMINO ACID COMPOSITION OF MOLD PROTEASES

Organism	Mucor miehei	Mucor pusillus	Paecilomyces varioti	Penicillium janthinellum	Rhizopus chinensis	Trichophyton granulosum
Type of Protease	Acid	Acid	Acid	Acid	Acid	Alkaline
Amino Acids						
Lysine	9	11–12	28	5	12 (13)	9
Histidine	2	1–2	7	3	0 (0)	6
Arginine	6	4	3	0	9 (9)	14
Aspartic Acid	42	44	45	36	43 (43)	46
Threonine	28	21	27	28	32 (30)	22
Serine	34	22	39	42	26 (27)	19
Glutamic Acid	24	20	23	28	21 (21)	16
Proline	17	14	20	12	14 (13)	17
Glycine	32	34	33	39	39 (46)	34
Alanine	26	16–17	26	23	23 (22)	20
½-Cystine	4	2	1	0	4 (4)	0
Valine	24	24	18	22	19 (20)	9
Methionine	6	3	1	0	3 (2)	Trace
Isoleucine	18	12	15	12–13	21 (21)	8
Leucine	19	15	21	20	23 (21)	15
Tyrosine	19	13	16	14	14 (15)	14
Phenylalanine	20	19	14	19	16 (17)	13
Tryptophan	3	2–3	3	4–5	5	25
Amide NH$_2$	42			43	32	
Total Amino Acids	333	277–281	343	307–309	324	294
Molecular Weight	39,000	30,000	37,000	32,000	35,000 (35,000)	34,000
N-Terminal			Leucine	Alanine	Alanine (Alanine)	
C-Terminal			Alanine	Alanine		
Reference	34	33	35	36, 37	48 (37)	53

TABLE 6
AMINO ACID COMPOSITION OF BACTERIAL PROTEASES

Organism	Bacillus subtilis	Bacillus subtilis var. amyloliquefaciens	Bacillus subtilis var. amyloliquefaciens	Bacillus subtilis var. amylosacchariticus	Bacillus subtilis var. amylosacchariticus	Bacillus subtilis var. amylosacchariticus
Type of Protease	Alkaline[a]	Alkaline[b]	Neutral[c]	Alkaline[d]	Alkaline[d]	Neutral[c]
Amino Acids						
Lysine	9	11	20	7	8	15
Histidine	5	6	7	5	6	5
Arginine	4	2	10	4	4	9
Aspartic Acid	28	28	61	24	27	43
Threonine	19	13	37	15	17	31
Serine	32	37	42	38	44	30
Glutamic Acid	12	15	33	14	16	23
Proline	9	14	14	12	13	11
Glycine	35	33	37	30	35	27
Alanine	41	37	35	31	32	27
½-Cystine	0	0	0	0	0	0
Valine	31	30	24	23	27	16
Methionine	5	5	5	4	4	2
Isoleucine	10	13	18	14	16	12
Leucine	16	15	26	14	16	22
Tyrosine	13	10	29	11	12	23
Phenylalanine	4	3	14	2	3	9
Tryptophan	1	3	4	3	3	3
Amide-NH₂	28	28	35	17		32
Total Amino Acids	274	275	416	251	283	308
Molecular Weight	27,600	27,700	44,000	25,000	28,500	34,000
N-Terminal	Alanine	Alanine	Alanine	Alanine	Alanine	Alanine
C-Terminal	Glutamine	Glutamine	Leucine			
Reference	97	97	95	109	97	101

TABLE 6 (Continued)
AMINO ACID COMPOSITION OF BACTERIAL PROTEASES

Organism	Bacillus subtilis var. natto	Bacillus subtilis, transformable	Bacillus subtilis, transformable	Bacillus thermoproteolyticus	Bacillus thermoproteolyticus	Clostridium histolyticum
Type of Protease	Alkaline	Acidic[e]	Basic[f]	Neutral[g]	Neutral[g]	Collagenase A[h]
Amino Acids						
Lysine	9	11	9	12	14	65 ± 2
Histidine	5	5	5	9	10	11 ± 1
Arginine	4	4	5	10	12	19
Aspartic Acid	30	44	28	43	56	105
Threonine	20	16	17	23	32	48
Serine	32	23	33	23	34	48
Glutamic Acid	12	30	13	20	27	60
Proline	8	23	11	8	10	17
Glycine	35	39	32	36	45	57
Alanine	41	35	30	28	36	41
½-Cystine	0	0	0	0	0	0
Valine	30	20	22	24	27	39
Methionine	3	4	4	2	3	8
Isoleucine	9	13	14	18	22	37
Leucine	16	15	16	17	19	46
Tyrosine	12	8	12	29	35	46
Phenylalanine	4	9	5	10	13	29
Tryptophan	1			5	4	
Amide NH₂	24		38	41		
Total Amino Acids	270		315	399		
Molecular Weight	27,000	36,000	30,700	37,000	42,700	95,000
N-Terminal	Alanine					
C-Terminal						
Reference	109	108	108	111	112	116

TABLE 6 (Continued)
AMINO ACID COMPOSITION OF BACTERIAL PROTEASES

Organism	Clostridium histolyticum	Pseudomonas aeruginosa	Pseudomonas aeruginosa	Sorangium	Sorangium	Streptococci Group A
Type of Protease	Collagenase B[i]	Alkaline	Elastase	α-Lytic	β-Lytic	Neutral
Amino Acids						
Lysine	50	16	12	2	3	17
Histidine	9	6	7	1	8	8
Arginine	15	7	16	12	5	9
Aspartic Acid	86	66	49	15	21	39
Threonine	33	24	20	18	13	11
Serine	35	42	26	20	22	26
Glutamic Acid	51	35	20	13	10	28
Proline	18	11	12	4	8	14
Glycine	44	65	39	32	24	37
Alanine	32	58	32	24	13	22
½-Cystine	0	0	5	6	4	1
Valine	29	25	20	19	5	20
Methionine	10	0	9	2	4	5
Isoleucine	28	17	10	8	4	12
Leucine	40	37	15	10	9	17
Tyrosine	35	21	24	4	13	19
Phenylalanine	22	20	19	6	6	12
Tryptophan		6	5	2	5	
Amide-NH$_2$				22		
Total Amino Acids		454	340	198	177	
Molecular Weight	79,000	48,000	39,500	20,000	20,000	32,000
N-Terminal						
C-Terminal						
Reference	116	128	131	176	139, 176	147, 148

a Bacillopeptidase A (subtilisin Carlsberg).
b Bacillopeptidase B (subtilisin BPN').
c Neutral subtilopeptidase.
d Bacillopeptidase D (subtilisin amylosacchariticus).
e Bacillopeptidase F.
f Bacillopeptidase E.
g Thermolysin.
h Collagenase I.
i Collagenase II.

TABLE 7
AMINO ACID COMPOSITION OF PROTEASES FROM BACTERIA, STREPTOMYCES AND YEASTS

Organism	Arthrobacter	Bacillus stearothermophilus	Myxobacter AL-1	Serratia sp.	Streptomyces griseus	Rhodotorula glutinis	Saccharomyces cerevisiae
Type of Protease	Semi-alkaline	Apase I	Semi-alkaline	Semi-alkaline	Trypsin-like	Acid	C
Amino Acids							
Lysine	9	205	2	20	7	7	17
Histidine	2	84	6	14	1	1	7
Arginine	1	179	2	10	8	4	8
Aspartic Acid	24	347	16	90	18	25	54
Threonine	19	234	11	35	17	25	17
Serine	27	118	17	39	15	38	27
Glutamic Acid	12	410	7	48	17	16	36
Proline	12	172	5	11	8	12	23
Glycine	29	380	27	66	28	34	32
Alanine	23	371	10	52	26	28	20
½-Cystine	4	0	2	0	6	0	7
Valine	12	332	3	27	18	18	24
Methionine	4	52	2	0	3	1	4
Isoleucine	5	246	3	25	8	7	19
Leucine	8	292	7	30	11	18	30
Tyrosine	15	73	8	20	8	11	24
Phenylalanine	10	120	4	32	6	13	21
Tryptophan	5	60	3	10		5	11
Glucosamine	0	0	0	0	1	0	7
Amide-NH$_2$	29	275					55
Total Amino Acids	221	3,675	136	592		263	388
Molecular Weight	22,000	400,000	14,300	60,000	20,000	30,000	61,000
Reference	77, 78	90, 172	121, 122	137	162	152	163

REFERENCES

1. Kurono, K., *Hakko Kyokai Zasshi, 19,* 2 (1934).
2. Nunokawa, Y., Namba, Y., and Koromoyama, Y., *J. Agric. Chem. Soc. Jap., 36,* 879 (1962).
3. Matsushima, K., and Shimada, K., *J. Agric. Chem. Soc. Jap., 36,* 193 (1962).
4. Nakanishi, K., *J. Biochem., 46,* 1263 (1964).
5. Tsujita, Y., *Abstr. Annu. Meet. Agric. Chem. Soc. Jap., 43,* 205 (1968).
6. Crewther, W. G., and Lennox, F. G., *Aust. J. Biol. Sci., 6,* 428 (1953).
7. Nunokawa, Y., *Hakko Kyokai Zasshi, 21,* 379 (1963).
8. Subramanian, A. R., and Kalnitsky, G., *Biochemistry, 3,* 1861, 1868 (1964).
9. Bergkvist, R., *Acta Chem. Scand., 17,* 1521, 1541 (1963).
10. Matsushima, K., *J. Agric. Chem. Soc. Jap., 32,* 215 (1958).
11. Misaki, T., Ph.D. Thesis. Osaka City University, Osaka, Japan (1969).
12. Misaki, T., Yamada, M., Okazaki, T., and Sawada, J., *Agric. Biol. Chem., 34,* 1383 (1970).
13. Turkova, J., Mikes, O., Gancerin, K., and Boublick, M., *Biochim. Biophys. Acta, 178,* 100 (1969); *257,* 257 (1972).
14. Nasuno, S., and Obara, T., *Agric. Biol. Chem., 36,* 1791, 1797 (1972).
15. Kishida, T., and Yoshimura, Y., *J. Biochem., 55,* 95 (1964).
16. Miake, S., Yoshimura, Y., Hashimoto, Y., and Ueda, H., *Sci. Rep. Hyogo Univ. Ser. Agric., 3,* 147 (1958).
17. Hayashi, K., Fukushima, D., and Mogi, K., *Agric. Biol. Chem., 31,* 642, 1171, 1237 (1967); *32,* 988 (1968).
18. Sekine, H., *Agric. Biol. Chem., 36,* 2143 (1972).
19. Danno, G., and Yoshimura, S., *Agric. Biol. Chem., 31,* 1151, 1159 (1967).
20. Ito, M., and Sugiura, E., *Yakugaku Zasshi, 88,* 1576, 1583, 1591 (1968).
21. Ichishima, E., and Yoshida, F., *Biochim. Biophys. Acta, 99,* 360 (1968); *47,* 257 (1972).
22. Ichishima, E., *Hakko Kyokai Zasshi, 22,* 393 (1964).
23. Koaze, Y., Goi, H., Ezawa, K., and Hara, T., *Agric. Biol. Chem., 28,* 216 (1964).
24. Nakadai, T., Nasuno, S., and Iguchi, N., *Abstr. Annu. Meet. Agric. Chem. Soc. Jap., 46,* 204 (1971).
25. Uchino, F., Kurono, G., and Doi, S., *Agric. Biol. Chem., 31,* 428 (1967).
26. Funagoshi, S., Oda, K., and Murao, S., *Abstr. Annu. Meet. Agric. Chem. Soc. Jap., 46,* 60 (1971).
27. Sardinas, J. L., *Appl. Microbiol., 16,* 248 (1968).
28. Isono, M., Tomoda, K., and Miyata, K., *Abstr. Annu. Meet. Agric. Soc. Jap., 45,* 132 (1970).
29. Meguro, M., Yamaguchi, T., and Watanabe, T., *Abstr. Annu. Meet. Agric. Chem. Soc. Jap., 44,* 57 (1969).
30. Kishida, T., and Yoshimura, S., *Agric. Biol. Chem., 30,* 1183 (1966).
31. Kawai, M., *Agric. Biol. Chem., 35,* 1517 (1971).
32. Arima, K., Iwasaki, S., and Tamura, G., *Agric. Biol. Chem., 31,* 540, 546, 1421, 1427 (1967).
33. Yu, J., Tamura, G., and Arima, K., *Biochim. Biophys. Acta, 171,* 138 (1969).
34. Ottesen, M., and Rickert, W., *C.R. Trav. Lab. Carlsberg, 37,* 310 (1970); *38,* 1 (1970).
35. Sawada, J., *Agric. Biol. Chem., 27,* 677 (1963); *28,* 348 (1964); *30,* 393 (1966).
36. Hofmann, T., and Show, R., *Biochim. Biophys. Acta, 92,* 543 (1964).
37. Sódek, J., and Hofmann, T., in *Methods in Enzymology,* Vol. 19, p. 376, G. E. Perlmann and L. Lorand, Eds. Academic Press, New York (1970).
38. Marshall, W. E., Martin, R., and Porath, J., *Biochim. Biophys. Acta, 151,* 414 (1968).
39. Yoshimura, S., Kishida, T., and Danno, G., *J. Agric. Chem. Soc. Jap., 38,* 128 (1964).
40. Shimada, K., and Matsushima, K., *Abstr. Annu. Meet. Agric. Chem. Soc. Jap., 43,* 205 (1968).
41. Shimada, K., Ph.D. Thesis, Osaka City University, Osaka, Japan (1968).
42. Yoshimura, S., *Sci. Rep. Hyogo Univ. Ser. Agric., 13,* 1 (1962).
43. Matsuoka, H., and Tsugo, T., *J. Agric. Chem. Soc. Jap., 37,* 444 (1963).
44. Ankel, H., and Martin, M., *Biochem. J., 91,* 431 (1964).
45. Bergum, A. A., and Prescott, J. M., *Arch. Biochem. Biophys., 111,* 391 (1965).
46. Prescott, J. M., and Boston, J. D., *Arch. Biochem. Biophys., 121,* 555 (1967).
47. Fukumoto, J., Tsuru, D., and Yamamoto, T., *Agric. Biol. Chem., 31,* 710 (1967); *33,* 1419 (1969).
48. Tsuru, D., Hattori, A., Tsuji, H., and Fukumoto, J., *J. Biochem., 67,* 415 (1970).
49. Tsuru, D., Hattori, A., Yamamoto, T., and Fukumoto, J., Unpublished Data.
50. Miake, S., Yoshimura, S., and Iwaoka, K., *Sci. Rep. Hyogo Univ. Agric., 3,* 129, 135 (1958).
51. Yoshimura, S., and Danno, G., *J. Agric. Chem. Soc. Jap., 38,* 178 (1964).
52. Tomoda, K., and Shimazono, H., *Agric. Biol. Chem., 28,* 770, 774 (1964).
53. Day, W.-C., Toncic, D., Stratman, S. L., Leeman, U., and Harmon, S. R., *Biochim. Biophys. Acta, 167,* 597 (1969).
54. Yokote, Y., Kawasaki, K., Nakashima, J., and Noguchi, Y., *J. Agric. Chem. Soc. Jap., 43,* 125 (1970).
55. Inoue, H., Yoshida, N., and Sasaki, A., *Biochim. Biophys. Acta, 284,* 451 (1972).
56. Morihara, K., Oka, T., and Tsujuki, H., *Biochim. Biophys. Acta, 139,* 382 (1967).
57. Morihara, K., and Tsuzuki, H., *Arch. Biochem. Biophys., 128,* 971 (1968).
58. Nickerson, W. J., *Biochim. Biophys. Acta, 77,* 73, 87 (1963).
59. Novel, J. J., and Nickerson, W. J., *J. Bacteriol., 77,* 251 (1959).

60. Morihara, K., Oka, T., and Tsuzuki, H., *Proc. Symp. Enz. Chem. Jap., 18,* 238 (1966).
61. Narahashi, Y., and Fukuyama, J., *J. Biochem., 66,* 743 (1969).
62. Hiramatsu, A., and Ouchi, T., *J. Biochem., 54,* 462 (1963).
63. Wahlby, S., and Engström, L., *Biochim. Biophys. Acta, 151,* 394, 402 (1968).
64. Simons, E. R., and Blout, E. R., *Biochim. Biophys. Acta, 92,* 197 (1964).
65. Nomoto, M., and Narahashi, Y., *J. Biochem., 46,* 653, 839, 1481 (1959).
66. Ouchi, T., and Hiramatsu, A., *Proc. Symp. Enz. Chem. Jap., 13,* 152 (1961).
67. Morihara, K., and Tsuzuki, H., *Annu. Rep. Shionogi Res. Lab., 13,* 9 (1963).
68. Hiramatsu, A., *J. Biochem., 62,* 353, 364 (1967).
69. Tytell, A. A., Charney, J., Bolhofer, W. A., and Curran, C., *Fed. Proc., 13,* 312 (1954).
70. Mizusawa, K., Ichishima, I., and Yoshida, F., *Agric. Biol. Chem., 28,* 884 (1964); *30,* 35 (1966).
71. Mizusawa, K., and Yoshida, F., *Abstr. Annu. Meet. Agric. Chem. Soc. Jap., 46,* 203 (1971).
72. Iwasa, T., Sugimoto, H., Motai, H., and Yokotsuka, T., *J. Agric. Chem. Soc. Jap., 38,* 84, 90 (1964).
73. Desai, A. J., and Dhala, S. A., *J. Bacteriol., 100,* 149 (1969).
74. Prescott, J. M., and Williams, C. R., *Fed. Proc., 19,* 337 (1960).
75. Prescott, J. M., and Wilkes, S. H., *Arch. Biochem. Biophys., 117,* 328 (1966); *J. Biol. Chem., 246,* 1756 (1971).
76. Hofsten, B. V., and Tjeder, C., *Biochim. Biophys. Acta, 110,* 576 (1965).
77. Hofsten, B. V., Kley, H. V., and Eaker, D., *Biochim. Biophys. Acta, 110,* 585 (1965).
78. Hofsten, B. V., and Reinhammer, B., *Biochim. Biophys. Acta, 110,* 599 (1965).
79. Wahlby, S., Engström, L., Bjave, U., and Hofsten, B. V., *Acta Chem. Scand., 20,* 1993 (1966).
80. Furukawa, Y., Fujii, Y., and Takahashi, H., *Agric. Biol. Chem., 32,* 822, 907 (1968).
81. Hall, F. F., Kunkel, H. O., and Prescott, J. M., *Arch. Biochem. Biophys., 114,* 145 (1966).
82. Keay, L., and Moser, P. W., *Biochem. Biophys. Res. Commun., 34,* 600 (1969).
83. Millet, J., and Archer, R., *Eur. J. Biochem., 9,* 456 (1969).
84. Millet, J., and Archer, R., *Biochim. Biophys. Acta, 151,* 302 (1958).
85. Miake, S., Hamaguchi, Y., and Takahashi, K., *Seikagaku, 26,* 105 (1954).
86. Hamaguchi, Y., and Takahashi, K., *Seikagaku, 29,* 777, 780, 908 (1957).
87. Horikoshi, H., *Abstr. Annu. Meet. Agric. Chem. Soc. Jap., 46,* 201 (1971).
88. Irie, T., Kanazawa, Y., Yoshikawa, M., and Imai, T., *Abstr. Annu. Meet. Agric. Soc. Jap., 44,* 59 (1969).
89. O'Brien, R. T., and Campbell, I. L., Jr., *Arch. Biochem. Biophys., 70,* 432 (1957).
90. Roncori, G., and Zuber, H., *Int. J. Protein Res., 1,* 45 (1969).
91. Zuber, H., in *Proceedings of the International Symposium on the Structure–Function Relationship of Proteolytic Enzymes,* p. 188. Royal Society of London, England (1970).
92. Minamiura, N., Yamamoto, T., and Fukumoto, J., *Agric. Biol. Chem., 30,* 186 (1966).
93. Minamiura, N., Matsumura, Y., Yamamoto, T., and Fukumoto, J., *Agric. Biol. Chem., 33,* 653 (1969); *35,* 975 (1971).
94. Fukumoto, J., Yamamoto, T., and Ichikawa, K., *J. Agric. Chem. Soc. Jap., 32,* 233 (1958).
95. Tsuru, D., McConn, J. D., and Yasunobu, K. T., *Biochem. Biophys. Res. Commun., 15,* 369 (1964); *J. Biol. Chem., 239,* 3706 (1964); *240,* 2415 (1965).
96. Matsubara, H., Hagihara, B., Nakai, M., Komaki, T., Yonetani, T., and Okunuki, K., *J. Biochem., 45,* 251 (1958).
97. Smith, E. L., Markland, F. S., and Glazer, A. N., in *Proceedings of the International Symposium on the Structure–Function Relationship of Proteolytic Enzymes,* p. 160. Royal Society of London, England (1970).
98. Ottesen, M., and Spector, A., *C.R. Trav. Lab. Carlsberg, 32,* 63 (1960).
99. Matsubara, H., Kasper, C. B., Brown, D. M., and Smith, E. L., *J. Biol. Chem., 240,* 1125 (1965).
100. Tsuru, D., Kira, H., Yamamoto, T., and Fukumoto, J., *Agric. Biol. Chem., 30,* 651, 856, 1164 (1966).
101. Tsuru, D., Yoshimoto, T., Yoshida, T., Kira, H., and Fukumoto, J., *Int. J. Protein Res., 2,* 75, 257 (1970).
102. Tsuru, D., Yoshida, T., and Fukumoto, J., *J. Biochem., 67,* 867 (1970).
103. Tsuru, D., Kira, H., Yamamoto, T., and Fukumoto, J., *Agric. Biol. Chem., 30,* 1261 (1966); *31,* 330 (1967).
104. Guntelberg, A. B., and Ottesen, M., *C.R. Trav. Lab. Carlsberg, 29,* 36 (1954).
105. Johansen, G., and Ottesen, M., *C.R. Trav. Lab. Carlsberg, 34,* 199 (1964).
106. Rappaport, H. P., Riggsby, W. S., and Holden, D. A., *J. Biol. Chem., 240,* 78, 87 (1965).
107. Boyer, H. W., and Carlton, B. C., *Arch. Biochem. Biophys., 128,* 442 (1968).
108. Hageman, J. H., and Carlton, B. C., *Arch. Biochem. Biophys., 139,* 67 (1970).
109. Yoshimoto, T., Fukumoto, J., and Tsuru, D., *Int. J. Protein Res., 3,* 285 (1971).
110. Endo, T., *J. Ferment. Technol., 40,* 346 (1962).
111. Ota, Y., Ogura, Y., and Wada, A., *J. Biol. Chem., 241,* 5919 (1966); *242,* 509 (1967).
112. Matsubara, H., Molecular Mechanism of Temperature Adaptation, *Proc. Meet. Am. Assoc. Adv. Sci.,* p. 283 (1969).
113. Millonig, R. C., *J. Bacteriol., 72,* 301 (1956).
114. Mandle, I., *Adv. Enzymol., 23,* 163 (1961).
115. Mandle, I., *Biochemistry, 3,* 1737 (1964).
116. Yoshida, E., and Noda, H., *Biochim. Biophys. Acta, 105,* 562 (1965).
117. Mitchell, W. M., and Harrington, W. F., *J. Biol. Chem., 243,* 4683 (1968).
118. Mitchell, W. M., and Schmidt, J. J., *Biochim. Biophys. Acta, 175,* 207 (1969); *178,* 194 (1969).

119. Mandle, I., Ferguson, L. T., and Zaffuto, S. F., *Arch. Biochem. Biophys., 69,* 565 (1957).
120. Nakashima, M., and Yoshida, F., *Abstr. Annu. Meet. Agric. Soc. Jap., 46,* 59 (1971); *47,* 252 (1972).
121. Jackson, R. L., and Wolfe, R. S., *J. Biol. Chem., 243,* 879 (1967).
122. Ensign, J. C., and Wolfe, R. S., *J. Bacteriol., 90,* 395 (1965); *91,* 524 (1966).
123. Burgum, A. A., and Prescott, J. M., *Arch. Biochem. Biophys., 111,* 391 (1965).
124. Bensusan, H. B., Derow, M. A., and Walker, B. S., *Arch. Biochem. Biophys., 49,* 293 (1954).
125. Mills, G. L., and Wilkins, J. M., *Biochim. Biophys. Acta, 30,* 63 (1958).
126. Hampson, S. E., Mills, G. L., and Spencer, T., *Biochim. Biophys. Acta, 73,* 476 (1963).
127. Morihara, K., *J. Bacteriol., 88,* 745 (1964).
128. Morihara, K., and Tsuzuki, H., *Biochim. Biophys. Acta, 73,* 113, 125 (1963); *92,* 351, 361 (1964).
129. Schaellmann, G., and Fisher, E., Jr., *Biochim. Biophys. Acta, 122,* 557 (1966).
130. Ogino, S., Wada, H., and Matsui, N., *Agric. Biol. Chem., 34,* 1126 (1970).
131. Morihara, K., Tsuzuki, H., Oka, T., Inoue, H., and Ebata, M., *J. Biol. Chem., 240,* 3295 (1965).
132. Mashmann, E., *Ergeb. Enz. Forsch., 9,* 155 (1943).
133. Morihara, K., *Bull. Agric. Chem. Soc. Jap., 20,* 243 (1956); *21,* 11 (1957); *23,* 49, 60 (1959).
134. Kato, N., Nagasawa, T., Adachi, S., Tani, Y., and Ogata, K., *Agric. Biol. Chem., 36,* 1185 (1972).
135. Rydén, A.-C., and Hofsten, B. V., *Acta Chem. Scand., 22,* 2803 (1968).
136. Quade, A. B., and Wrether, W. G., *Biochim. Biophys. Acta, 191,* 762 (1969).
137. Miyata, K., Maejima, K., Tomoda, K., and Isono, M., *Agric. Biol. Chem., 34,* 310, 1457 (1970); *35,* 460 (1971).
138. Whitaker, D. R., *J. Ferment. Technol., 44,* 875 (1966).
139. Whitaker, D. R. *et al., Can. J. Biochem., 43,* 1935, 1955, 1961 (1965); *45,* 917 (1967).
140. Smillie, L. B., and Whitaker, D. R., *J. Am. Chem. Soc., 89,* 3350, 3352 (1967).
141. Duthie, E. S., *J. Gen. Microbiol., 10,* 427 (1954).
142. Duthie, E. S., and Haughton, G., *Biochem. J., 70,* 125 (1958).
143. Shugart, L. R., and Beck, R. W., *J. Bacteriol., 88,* 586 (1964).
144. Bleiweis, A. S., and Zimmerman, L. N., *J. Bacteriol., 88,* 653 (1964).
145. Williamson, W. T., Tore, S. B., and Speck, M. L., *J. Bacteriol., 87,* 49 (1964).
146. Elliott, S. D., *J. Exp. Med., 92,* 201 (1950).
147. Liu, T.-Y., Neumann, N. P., and Elliott, S. D., *J. Biol. Chem., 238,* 251 (1963).
148. Liu, T.-Y., Stein, W. H., Moore, S., and Elliott, S. D., *J. Biol. Chem., 240,* 1138, 1143 (1965); *242,* 4029 (1967).
149. Tillett, W. S., and Gardner, R. L., *J. Exp. Med., 58,* 485 (1932).
150. Brown, K. O., Jacobs, G., and Laskowski, M., *J. Biol. Chem., 194,* 445 (1952).
151. Remond, H., Fasold, H., and Stalb, F., *Biochim. Biophys. Acta, 167,* 399 (1968).
152. Kamada, S., Koyama, T., Maeyama, K., Arai, M., Oda, K., and Murao, S., *Agric. Biol. Chem., 36,* 1095, 1103 (1972); *J. Agric. Chem. Soc. Jap., 46,* 167, 171 (1972).
153. Hata, T., Hayashi, R., and Doi, E., *Agric. Biol. Chem., 31,* 150, 160, 357, 1103 (1967).
154. Lenny, J. F., *J. Biol. Chem., 221,* 919 (1956).
155. Hayashi, R., Oka, Y., Doi, E., and Hata, T., *Agric. Biol. Chem., 31,* 1103 (1967); *32,* 359, 367 (1968); *33,* 196 (1969).
156. Hayashi, R., Aibara, S., and Hata, T., *Biochim. Biophys. Acta, 212,* 359 (1970).
157. Grassmann, W., and Dyckerhoff, H., *Z. Phys. Chem., 179,* 41 (1928).
158. Johnson, M. J., *J. Biol. Chem., 137,* 575 (1940).
159. Felix, F., and Brouillet, N., *Biochim. Biophys. Acta, 122,* 127 (1966).
160. Maddex, I. S., and Hough, I. S., *Biochem. J., 117,* 843 (1970).
161. Nordwig, A., and John, W. F., *Eur. J. Biochem., 3,* 519 (1968).
162. Turasek, L., Fackre, D., and Smillie, L. B., *Biochem. Biophys. Res. Commun., 37,* 99 (1969).
163. Hayashi, R., Hibara, S., and Hata, T., *Agric. Biol. Chem., 35,* 658 (1971).
164. Pollock, M. R., in *The Bacteria,* Vol. 4, p. 121, I. C. Gunsalus and R. Y. Stanier, Eds. Academic Press, New York (1962).
165. Jönsson, G. A., and Martin, S. M., *Agric. Biol. Chem., 28,* 734 (1964).
166. Jönsson, G. A., and Martin, S. M., *Agric. Biol. Chem., 29,* 787 (1965).
167. Knight, S. G., *Can. J. Microbiol., 12,* 420 (1966).
168. Patel, N. G., and Cutknop, L. K., *J. Econ. Entomol., 54,* 773 (1961).
169. Underkofler, L. A., and Charles, R. L., *Dev. Ind. Microbiol., 1,* 125 (1960).
170. Waldvogel, F. A., and Swartz, M. N., *J. Bacteriol., 98,* 662 (1969).
171. Berger, J., Johnson, M. J., and Peterson, W. H., *J. Bacteriol., 36,* 521 (1938).
172. Perlmann, G. E., and Lorand, L. (Eds.), *Methods in Enzymology,* Vol. 19. Academic Press, New York (1970).
173. Wang, H. L., Ruttle, D. I., and Hesseltine, C. W., *Can. J. Microbiol., 15,* 99 (1969).
174. Doolittle, R. F., in *Methods in Enzymology,* Vol. 19, p. 555, G. E. Perlmann and L. Lorand, Eds. Academic Press, New York (1970).
175. Below, M., and Porath, J., in *Methods in Enzymology,* Vol. 19, p. 576, G. E. Perlmann and L. Lorand, Eds. Academic Press, New York (1970).

176. Whitaker, D. R., in *Methods in Enzymology,* Vol. 19, p. 599, G. E. Perlmann and L. Lorand, Eds. Academic Press, New York (1970).

177. Rippon, J. W., *J. Bacteriol., 95,* 43 (1968).

178. Taylor, F. B., Jr., and Tomar, R. H., in *Methods in Enzymology,* Vol. 19, p. 807, G. E. Perlmann and L. Lorand, Eds. Academic Press, New York (1970).

179. Lack, C. H., and Glanville, K. L. A., in *Methods in Enzymology,* Vol. 19, p. 706, G. E. Perlmann and L. Lorand, Eds. Academic Press, New York (1970).

180. Coles, N. W., Gilbo, C. M., and Broad, A. J., *Biochem. J., 111,* 7 (1969).

181. Jönsson, A. G., *Arch. Biochem. Biophys., 129,* 62 (1969).

182. Reusser, F., *Arch. Biochem. Biophys., 112,* 156 (1965).

183. Schindler, C. A., and Schuhardt, V. T., *Proc. Natl. Acad. Sci. U.S.A., 51,* 414 (1964).

184. Browder, H. P., Zygmunt, W. A., Yong, J. R., and Tavormina, P. A., *Biochem. Biophys. Res. Commun., 19,* 383 (1965).

185. Wang, H. L., *J. Bacteriol., 93,* 794 (1967).

186. Blackburn, T. H., *J. Gen. Microbiol., 53,* 27, 37 (1968).

187. Norberg, P., and Hofsten, B. V., *J. Gen. Microbiol., 55,* 251 (1969).

188. Ahearn, D. G., Meyers, S. P., and Nichols, R. A., *Appl. Microbiol., 16,* 1370 (1968).

189. Kingan, S. L., and Ensign, J. C., *J. Bacteriol., 96,* 629 (1968).

190. Weidel, W., and Pelzer, H., *Adv. Enzymol., 26,* 193 (1964).

191. Kato, K., Hirata, T., Murayama, Y., Suginaka, H., and Kotani, S., *Biken J., 11,* 1 (1968).

192. Kato, K., and Strominger, J. L., *Biochemistry, 7,* 2754 (1968).

193. Gillespie, D. C., and Cook, F. D., *Can. J. Microbiol., 11,* 169 (1965).

194. Ghuysen, J. M., et al., *Bacteriol. Rev., 32,* 425 (1968).

195. Petit, J. F., Muñoz, E., and Ghuysen, J. M., *Biochemistry, 5,* 2764 (1966).

196. Yoshimoto, T., Nakanishi, K., Fukumoto, J., and Tsuru, D., *Agric. Biol. Chem., 35,* 1775 (1971).

197. Matsubara, H., and Feder, J., in *The Enzymes,* Vol. 3, p. 721, P. D. Boyer, Ed. Academic Press, New York (1971).

198. Ichishima, E., *Biochim. Biophys. Acta, 258,* 274 (1972).

199. Ichishima, E., Sonoki, S., Hirai, K., Torii, Y., and Yokoyama, S., *J. Biochem., 72,* 1045 (1972).

200. Nakadai, T., Nasuno, S., and Iguchi, N., *Agric. Biol. Chem., 36,* 1461, 1473, 1481 (1972); *37,* 757, 767, 775 (1973).

201. Show, R., *Biochim. Biophys. Acta, 92,* 558 (1964).

202. Jones, S. R., and Hofmann, T., *Can. J. Biochem., 36,* 1297 (1972).

203. Yokoyama, S., and Ichishima, E., *Agric. Biol. Chem., 36,* 1259 (1972).

204. Levy, C. C., and Goldman, P., *J. Biol. Chem., 242,* 2933 (1967).

205. Goldman, P., and Levy, C. C., *Proc. Natl. Acad. Sci. U.S.A., 58,* 1299 (1967).

206. McCullough, J. L., Chabner, B. A., and Bertino, J. R., *J. Biol. Chem., 246,* 7207 (1971).

207. Chabner, B. A., and Bertino, J. R., *Biochim. Biophys. Acta, 276,* 234 (1972).

208. Uwajima, T., Yoshikawa, N., and Terada, O., *Agric. Biol. Chem.,* (1972); *37,* 1517 (1973).

209. Murao, S., Oda, K., and Matsushita, Y., *Agric. Biol. Chem., 36,* 1647 (1972).

210. Mitsugi, K., Takami, T., Tobe, S., Kimura, M., Nakase, T., and Komagata, K., *Agric. Biol. Chem., 35,* 1633 (1971).

211. Abdel-Fattoh, A. F., Mabrouk, S. S., and El-Hawwong, N. M., *J. Gen. Microbiol., 70,* 151 (1972).

212. Heyningen, S. V., *Biochem. J., 125,* 1159 (1971); *Eur. J. Biochem., 27,* 436 (1972); *28,* 432 (1972).

213. Plancat, M. T., and Han, K. K., *Eur. J. Biochem., 28,* 327 (1972).

214. Drapeau, G. R., Boily, Y., and Houmard, J., *J. Biol. Chem., 247,* 8720 (1972).

215. Danno, G., *Agric. Biol. Chem., 34,* 264 (1970).

216. Johnson, M. J., and Peterson, W. H., *J. Biol. Chem., 112,* 25 (1936).

217. Yagi, J., Yano, T., Jomon, K., Sakai, H., and Ajisaka, M., *J. Ferment. Technol., 50,* 592, 810, 816, 823, 829 (1972).

218. Pisano, M. A., Oleniacz, W. S., Mason, R. T., and Fleischman, A. I., *Appl. Microbiol., 11,* 111 (1963).

219. Ong, P. S., and Gaucher, G. M., in *Abstracts of the 4th International Fermentation Symposium (Kyoto),* p. 47 (1972).

220. Rippon, J. W., *Science, 157,* 947 (1967); *Appl. Microbiol., 22,* 471 (1971).

221. Nakamura, S., Hamada, M., and Umezawa, H., *Chem. Pharm. Bull. (Tokyo), 17,* 2044 (1969); *17,* 714 (1969); *18,* 2112, 2577 (1970).

222. Reusser, F., *Arch. Biochem. Biophys., 112,* 156 (1965).

223. Okazaki, H., *J. Ferment. Technol., 50,* 580 (1972).

224. Griffin, T. B., and Prescott, J. M., *J. Biol. Chem., 245,* 1348 (1970).

225. Keay, L., in *Abstracts of the 4th International Fermentation Symposium (Kyoto),* p. 51 (1972).

226. Horikoshi, K., *Agric. Biol. Chem., 35,* 1407 (1971).

227. Dasgupta, B. R., and Sugiyama, H., *Biochim. Biophys. Acta, 268,* 719 (1972).

228. Vogt, V. M., *J. Biol. Chem., 245,* 4760 (1970).

229. Takahashi, S., *J. Bacteriol., 61,* 258 (1967).

230. Mizutani, T., Hamada, K., Yamaguchi, M., Tsuji, H., Misaki, T., and Sawada, J., *Agric. Biol. Chem., 35,* 1651 (1971).

231. Tipper, D. J., *Biochemistry, 8,* 2192 (1969).
232. Zyskind, J. W., *Science, 147,* 1458 (1965).
233. Suginaka, H., *Biken J., 11,* 13 (1968).
234. Takahashi, T., Yamazaki, Y., and Isono, M., in *Abstracts of the 4th International Fermentation Symposium (Kyoto),* p. 252 (1972).
235. Mori, Y., Kato, K., Matsubara, T., and Kotani, S., *Biken J., 3,* 139 (1960).
236. Yoshimoto, T., and Tsuru, D., *J. Biochem., 72,* 379 (1972).
237. Dessai, A. J., and Phala, S. A., *J. Bacteriol., 100,* 149 (1969).
238. Kawai, M., and Mukai, N., *Agric. Biol. Chem., 34,* 159, 164 (1970).
239. Waldvogel, F. A., *J. Bacteriol., 90,* 662 (1969).
240. Rippon, J. W., *Biochim. Biophys. Acta, 159,* 147 (1968).
241. Thomson, J. A., Woods, D. R., and Welton, R. L., *J. Gen. Microbiol., 70,* 315 (1972).
242. Knight, S. G., *Can. J. Microbiol., 12,* 420 (1966).

MICROBIAL LIPASES

DR. YASUHIDE OTA

Microbial lipases (EC 3.1.1.3) can hydrolyze triglycerides and methyl alcohol esters of long-chain fatty acids. Some can also split the ester bond of methyl butyrate, usual substrate for carboxylesterase (EC 3.1.1.1), or the ester bond of phospholipids.

Microbial lipases can be purified from the extract of cells or from broth free of cells by precipitation with organic solvents (e.g., acetone), precipitation with ammonium sulfate, gel filtration, column chromatography with ion exchangers, and so on. All lipases listed in the table were highly purified by the above methods; some were obtained in crystalline form.

Many lipases contain carbohydrate, such as mannose, in the enzyme molecule. Lipase from *Mucor javanicus* contains phospholipid, which stabilizes the enzyme and changes the properties of the enzyme.

Two lipases require activators for the reaction; lipase from *Candida paralipolytica* is activated by bile salts or long-chain fatty acids (in combination with Ca^{2+}), and lipase from *Mucor javanicus* is activated by human serum or bovine serum albumin.

Organism	Production Site	Optimal pH and Temperature	Stable pH Range and Heat Loss	Hydrolysis of Monoacid Triglyceride[a]	Positional Specificity	Molecular Weight	Reference
Pseudomonas							
fragi	Extracellular	8.0; 54°C	— 27.6% (63°C, 30 min)	C_{12}	a, a'	—	1, 2
species							
I	Extracellular	7.0; 40°C	4.0–9.0	—	—	—	3
II	Extracellular	7.0; 40°C	— 4.0–9.0 —	—	—	—	3
Propionibacterium							
shermanii	Intracellular	7.2; 47°C	5.5–8.0 (35°C, 5 min) 75% (45°C, 10 min)	C_3	—	—	4
Staphylococcus							
aureus	Extracellular	8.3; 45°C	— 80% (65°C, 30 min)	—	a, β, a'	—	5
Candida							
cylindracea							
I	Extracellular	5.2; 47.5°C	1.7–8.5 (5°C, 12 hr) 82% (65°C, 20 min)	C_{12}	a, β, a'	52,600	6, 7
II	Extracellular	5.2; 57.5°C	1.1–9.0 (5°C, 12 hr) 64% (65°C, 20 min)	C_{12}	a, β, a'	55,900	6, 7

Organism	Production Site	Optimal pH and Temperature	Stable pH Range and Heat Loss	Hydrolysis of Monoacid Triglyceride[a]	Positional Specificity	Molecular Weight	Reference
Candida (cont.)							
paralipolytica							
Purified	Extracellular	8.0; —	3.5–9.0 (5°C, 22 hr) 45% (45°C, 20 min)	C$_8$	—	—	8, 9
Modified	—	7.0; —	3.5–9.0 (5°C, 22 hr) —	—	—	—	8, 9
Torulopsis							
ernobii	Extracellular	6.5; 45°C	3.0–9.0 (37°C, 1 hr) Stable (65°C, 10 min)	C$_4$	—	42,500	10, 11
Aspergillus							
niger	Extracellular	5.6; 25°C	2.2–6.8 (30°C, 24 hr) 50% (60°C, 15 min)	—	α, β, α'	—	12, 13
Penicillium							
crustosum							
I	Extracellular	9.0; —	6.0–9.0 (30°C, 24 hr) 80% (60°C, 15 min)	—	—	29,300	14
II	Extracellular	9.0; —	6.0–9.0 (30°C, 24 hr) 80% (60°C, 15 min)	—	—	32,200	14
III	Extracellular	9.0; —	— —	—	—	11,000	14

Organism	Production Site	Optimal pH and Temperature	Stable pH Range and Heat Loss	Hydrolysis of Monoacid Triglyceride[a]	Positional Specificity	Molecular Weight	Reference
Mucor							
javanicus	Extracellular	7.0; 40°C	5.0–6.5 (40°C, 3 hr) 60% (60°C, 10 min)	—	a, a'	—	15
lipolyticus							
F-1	Extracellular	–; –	—	—	—	>200,000	16, 17
F-2	Extracellular	–; –	—	—	—	59,000	16, 17
F-3A	Extracellular	9.0; –	—	—	—	25,400	16, 17
F-3B	Extracellular	8.0; –	—	—	—	29,000	16, 17
pusillus	Extracellular	5.0–5.5; 50°C	– –	—	—	—	18
Rhizopus							
arrhizus	Extracellular	7.0; 37°C	5.0–7.0 (30°C, 2 hr)	—	a, a'	43,000	19–21
delemar							
A	Extracellular	5.6; 35°C	3.8–8.0 (30°C, 24 hr) Stable (65°C, 15 min)	—	—	—	22, 23
C	Extracellular	6.0; 35°C	4.0–7.0 (30°C, 24 hr) 60% (55°C, 15 min)	—	—	—	22, 23

Organism	Production Site	Optimal pH and Temperature	Stable pH Range and Heat Loss	Hydrolysis of Monoacid Triglyceride[a]	Positional Specificity	Molecular Weight
Rhizopus (cont.) species						
I	Intracellular	6.0–6.5; —	4.0–7.0 (30°C, 20 hr) 50% (60°C, 15 min)	C_4	—	24
II	Intracellular	6.0–6.5; —	4.0–7.0 (30°C, 20 hr) 50% (60°C, 15 min)	C_8	—	24

REFERENCES

1. Lu, J. Y., and Liska, B. J., *Appl. Microbiol., 18* 108 (1969).
2. Mencher, J. R., and Alford, J. A., *J. Gen. Microbiol., 48,* 317 (1967).
3. Narasaki, T., Saiki, T., Tamura, G., and Arima, K., *Agric. Biol. Chem., 31,* 993 (1967).
4. Oterholm, A., Ordal, Z. J., and Witter, L. D., *Appl. Microbiol., 20,* 16 (1970).
5. Vadehra, D. V., and Harmon, L. G., *Appl. Microbiol., 15,* 480 (1967).
6. Tomizuka, N., Ota, Y., and Yamada, K., *Agric. Biol. Chem., 30,* 576 (1966).
7. Tomizuka, N., Thesis. The University of Tokyo, Tokyo, Japan (1969).
8. Ota, Y., Nakamiya, T., and Yamada, K., *Agric. Biol. Chem., 34,* 1368 (1970).
9. Ota, Y., Nakamiya, T., and Yamada, K., *Agric. Biol. Chem., 36,* 1895 (1972).
10. Motai, H., Ichishima, E., and Yoshida, F., *Nature, 210,* 308 (1966).
11. Yoshida, F., Motai, H., and Ichishima, E., *Biochim. Biophys. Acta, 154,* 586 (1968).
12. Fukumoto, J., Iwai, M., and Tsujisaka, Y., *J. Gen. Appl. Microbiol., 9,* 353 (1963).
13. Iwai, M., Tsujisaka, Y., and Fukumoto, J., *J. Ferment. Assoc. Jap., 22,* 419 (1964).
14. Oi, S., Sawada, A., and Satomura, Y., *Agric. Biol. Chem., 31,* 1357 (1967).
15. Saiki, T., Takagi, Y., Suzuki, T., Narasaki, T., Tamura, G., and Arima, K., *Agric. Biol. Chem., 33,* 414 (1969).
16. Nagaoka, K., and Yamada, Y., *Agric. Biol. Chem., 33,* 986 (1969).
17. Nagaoka, K., and Yamada, Y., *Abstr. Annu. Meet. Agric. Chem. Soc. Jap.,* p. 242 (1971).
18. Somkuti, G. A., Babel, F. J., and Somkuti, A. C., *Appl. Microbiol., 17,* 606 (1969).
19. Laboureur, P., and Labrousse, M., *Bull. Soc. Chim. Biol., 48,* 747 (1966).
20. Semeriva, M., Benzonana, G., and Desnuelle, P., *Bull. Soc. Chim. Biol., 49,* 71 (1967).
21. Semeriva, M., Benzonana, G., and Desnuelle, P., *Biochim. Biophys. Acta, 191,* 598 (1969).
22. Fukumoto, J., Iwai, M., and Tsujisaka, Y., *J. Gen. Appl. Microbiol., 10,* 257 (1964).
23. Tsujisaka, Y., and Iwai, M., *Abstr. Annu. Meet. Agric. Chem. Soc. Jap.,* p. 279 (1967).
24. Oi, S., Yamazaki, O., Sawada, A., and Satomura, Y., *Agric. Biol. Chem., 33,* 729 (1969).

DEXTRANASES

DR. ROBERT G. BROWN

Dextranases (a-1,6-glucan-6-glucanohydrolase EC 3.2.1.11) are enzymes capable of hydrolyzing the a-1,6-glucosidic linkages of the bacterial polysaccharide dextran. The term dextran is applied to a group of glucans produced by cultivating selected bacteria in media containing sucrose. Dextrans from 96 strains of bacteria, representing five different genera, have been characterized and classified.[10] Bacterial dextrans contain 1,6-, 1,4- and 1,3-glucosidic linkages in different proportions, and this forms the basis for the above classification scheme. The a-1,6-glucosidic linkage accounts for at least 50% of the linkages and may be present to an extent of 95% or more. The other known occurrence of this linkage is the 5 to 10% found in starch, glycogen and similar polysaccharides. The major interest in dextranase centered on the possibility of using the enzyme to degrade dextran to a desired molecular size, suitable for use as synthetic blood volume expanders.[3] Recent interest in dextranase involves its potential use in preventing formation of microbial dental plaques.[5]

Both endo- and exo- types of dextranase may be obtained from a wide variety of sources. Endodextranases have been found in fungi,[3,8,11,17] whereas both endo- and exo- dextranases are produced by bacteria;[4,7,9,16] indeed, dextranase activity has also been detected in mammalian tissues.[14,15] It is not known if this dextran-splitting activity is due to a specific dextranase or the "debranching" enzymes involved in the cleavage of the a-1,6-linkages of glycogen. Most known mammalian dextranases are exodextranases.

Microbial dextranases are inducible enzymes; therefore, media containing dextran are generally used for their production. However, excellent yields of dextranase may be obtained by using other inducers, for example, isomaltose dipalmitate[13] or ketodextran.[2]

TABLE 1
FUNGAL DEXTRANASES

Organism	Optimal pH	Temperature Stability	Mode of Action	Reference
Aspergillus		—	endo	17, 18
Penicillium				
funiculosum	5.5–6.5	Inactivated (70°C, 10 min)	endo	2, 17
lilacinum	7.0	—	—	2, 17
luteum	5	Inactivated (60°C, 15 min)	endo	6
verruculosum	5.0–6.0	—	—	17
Spicaria				
violacea		—	—	2
Verticillium		—	—	12

TABLE 2
BACTERIAL DEXTRANASES

Organism	Optimal pH	Temperature Stability	Mode of Action	Reference
Bacteroides	5.0–5.5	–	endo	16
	7.0–7.5	–	exo	16
Lactobacillus				
bifidus	5.5–6.5	–	endo	4
Cellvibrio				
fulvus	5.3	–	endo	9

REFERENCES

1. Bailey, R. W., and Clarke, R. T. J., *Biochem. J., 72,* 49 (1959).
2. Brown, R. G., *Can. J. Microbiol., 16,* 841 (1970).
3. Charles, A. F., and Farrell, L. N., *Can. J. Microbiol., 3,* 239 (1957).
4. Clarke, R. T. J., *J. Gen. Microbiol., 20,* 549 (1959).
5. Fitzgerald, R. J., and Jordan, H. V., Polysaccharide-Producing Bacteria and Caries, in *Art and Science of Dental Caries Research,* pp. 79–89, R. S. Harris, Ed. Academic Press, New York (1968).
6. Fukumoto, J., Isuji, H., and Tsuru, D., *J. Biochem., 69,* 1113 (1971).
7. Hehre, E. J., and Sery, T. W., *J. Bacteriol., 63,* 424 (1952).
8. Hultin, E., and Nordstrom, L., *Acta Chem. Scand., 3,* 1405 (1949).
9. Ingelman, B., *Acta Chem. Scand., 2,* 803 (1948).
10. Jeanes, A., Haynes, W. C., Wilham, C. A., Rankin, J. C., Meloin, E. H., Austin, M. J., Cluskey, J. E., Fisher, B. E., Tsuchiya, H. M., and Rest, C. E., *J. Am. Chem. Soc., 76,* 5041 (1954).
11. Kobayashi, T., *J. Agric. Chem. Soc. Jap., 28,* 352 (1954).
12. Nordstrom, L., and Hultin, E., *Sven. Kem. Tidskr., 60,* 283 (1948).
13. Reese, E. T., Lola, J. E., and Parrish, F. W., *J. Bacteriol., 100,* 1151 (1969).
14. Rosenfeld, E. L., and Lukomskaya, I. S., *Clin. Chim. Acta, 2,* 105 (1957).
15. Rosenfeld, E. L., *Biokhimiya* (Translation), *21,* 77 (1956).
16. Sery, T. W., and Hehre, E. J., *J. Bacteriol., 71,* 373 (1956).
17. Tsuchiya, H. M., Jeanes, A., Bricker, H. M., and Wilham, C. A., *J. Bacteriol., 64,* 513 (1952).
18. Whiteside-Carlson, V., and Carlson, W. W., *Science, 115,* 43 (1952).

β-GLUCANASES

DR. ROBERT BROWN

β-Glucan is a general term for polysaccharides containing β-D-glucopyranosidic residues; however, the term is also used more specifically in reference to polymers containing predominantly β-(1 → 3)-linked glucopyranosidic units. Other terms, such as pachyman,[43] callose,[4,24] laminarin,[12] and paramylon,[42] are employed to denote a variety of related polysaccharides; however, in all cases the polymers contain a preponderance of β-(1 → 3)-glucosidic linkages. Many glucans contain β-(1 → 6)- in addition to β-(1 → 3)-linked glucose residues. For example, laminarin and glucan from the cell walls of a variety of fungi contain both linkages.[8,10] Because of this, β-(1 → 3)-glucanases (β-1,3-glucan 3-glucanhydralase, E.C. 3.2.1.6) and β-(1 → 6)-glucanases are considered together under the title β-glucanases. Indeed, there is a close association between the two enzymic activities. For example, an exo-β-glucanase from a number of yeast species hydrolyzed both β-(1 → 3) and β-(1 → 6)-glucosidic linkages;[1] an exo-β-glucanase from a basidiomycete, on the other hand, cleaved only β-(1 → 3)- linkages, but could bypass β-(1 → 6)- linkages, yielding gentiobiose quantitatively from glucans containing both β-(1 → 3)- and β-(1 → 6)-glucosidic bonds.[31] In addition, induction of β-(1 → 6)-glucanase production in *Penicillium lilacinum* with pustulan, a glucan containing only β-(1 → 6) linkages, resulted in simultaneous induction of β-(1 → 3)-glucanase activity, which was attributable to three different enzymes.

β-(1 → 3)-Glucanases are widely distributed in nature and have been reported from bacteria,[5,17,26] fungi,[1,2,7,13,14] plants,[3,11,20,25,27] marine and litter invertebrates,[32,38] euglenoids,[6] sea urchins,[18,28] molluscs,[23,37] fish[33] and the sand dollar.[41] β-(1 → 6)-Glucanases have a more restricted distribution; they have been found only in bacteria,[5,29,30,36,39] fungi[1,7,9,40] and plants.[20]

TABLE 1
FUNGAL β-(1→3)-GLUCANASES

Organism	Optimal pH	Temperature Stability	Mode Of Action	Linkage Hydrolyzed	Reference
Aspergillus *niger*	5.2–5.6		Exo	1→3 1→3	13 14
oryzae	5.4		Endo	1→3	13
Basidiomycete	5.0	Stable at 55°C for 16 minutes	Exo	1→3	13, 22
Cercaspora *salina*	4.0, 5.0		Endo	1→3	13
Claviceps *fusiformis*			Exo	1→3	16
Diplodina sp.	5.5–5.7		Exo	1→3	13
Hanseniaspora *uvarum* *valbyensis*	3.5–4.5 3.5–4.5		Endo Endo	1→3 1→3	2 2
Helicoma sp.	5.0		Exo	1→3	13

TABLE 1 (Continued)
FUNGAL β-(1→3)-GLUCANASES

Organism	Optimal pH	Temperature Stability	Mode of Action	Linkage Hydrolyzed	Reference
Irpex					
lacteus	4.0, 5.5		Intermediate	1→3	13
Mortierella					
alpina	4.9, 6.2		Exo	1→3	13
Myrothecium					
verricaria	4.9, 6.2		Intermediate	1→3	13
Penicillium					
claviforme	5.4		Intermediate	1→3	13
lilacinum	6.0			1→3	9
stipitatum	4.9		Intermediate	1→3	13
violaceum	5.5		Endo	1→3	13
Phytophthora					
palmivora	5–6		Exo	1→3	21
Rhizopus					
arrhizus	4.5–5.0		Exo and endo	1→3	19
nodosus	4.7		Endo	1→3	13
Saccharomyces					
cerevisiae	5.5		Exo	1→3 and 1→6	7
	4.7, 5.5, 6.5		Exo and endo	1→3	15
ellipsoides	5.5	Inactivated at 70°C in 15 minutes		1→3	35
lactis	5.6		Exo	1→3	40
Schizophyllum					
commune	3.8, 4.8	Two components: (a) 50% inactivation at 50°C in 3.5 minutes (b) 50% inactivation at 50°C in 8.8 hours		1→3	34
Sporotrichum					
pruinosum	4.5		Exo	1→3	13
Trichoderma					
album	5.4, 6.2		Exo	1→3	13
Yeasts			Exo	1→3 and 1→4	1

TABLE 2
BACTERIAL β-(1→3)-GLUCANASES

Organism	Optimal pH	Mode of Action	Linkage Hydrolyzed	Reference
Arthrobacter sp.		Endo	1→3	17
Bacillus circulans			1→3	39
Cytophaga johnsonii		Endo	1→3	5
Streptomycete	6.0		1→3	36

TABLE 3
β-(1→6)-GLUCANASES

Organism	Optimal pH	Mode of Action	Linkage Hydrolyzed	Reference
Bacillus circulans			1→6	39
Cytophaga johnsonii		Endo	1→6	5
Penicillium lilacinum	6.0	Endo	1→6	9
Saccharomyces cerevisiae	5.5	Exo	1→6 and 1→3	7
lactis	5.6	Exo	1→6 and 1→3	40
Streptomyces griseus	6.7		1→6	29, 30
niveoruber	6.7		1→6	29, 30
parvus	6.7		1→6	29, 30
Streptomycete	6.0		1→6	36
Yeasts		Exo	1→6 and 1→3	1

REFERENCES

1. Abd-el-al, A. T. H., and Phaff, H. J., *Biochem. J., 109,* 347 (1968).
2. Abd-el-al, A. T. H., and Phaff, H. J., *Can. J. Microbiol., 15,* 697 (1969).
3. Abeles, F. B., and Forrence, L. E., *Plant Physiol., 45,* 395 (1970).
4. Aspinall, G. O., and Kessler, G., *Chem. Ind. (London),* p. 1573 (1957).
5. Bacon, J. S. D., Gordon, A. H., Jones, D., Taylor, I., and Webley, D., *Biochem. J., 120,* 67 (1970).
6. Barras, D. R., and Stone, B. A., *Biochim. Biophys. Acta, 191,* 329 (1969).
7. Brock, T. D., *Biochim. Biophys. Res. Commun., 19,* 623 (1965).
8. Brown, R. G., and Lindberg, B., *Acta Chem. Scand., 21,* 2379 (1967).
9. Brown, R. G., *Can. J. Microbiol., 18,* 1543 (1972).

10. Bull, A. T., *Adv. Enzymol., 28,* 325 (1966).
11. Chen, S., and Luchsingen, W. W., *Arch. Biochem. Biophys., 106,* 71 (1964).
12. Chesters, C. G. C., and Bull, A. T., *Biochem. J., 86,* 28 (1963).
13. Chesters, C. G. C., and Bull, A. T., *Biochem. J., 86,* 38 (1963).
14. Clarke, A. E., and Stone, B. A., *Biochem. J., 96,* 793 (1965).
15. Cortat, M., Matile, P., and Wiemken, A., *Arch. Mikrobiol., 82,* 189 (1972).
16. Dickerson, A. G., Mandle, P. G., and Szczyrbak, C. A., *J. Gen. Biol., 60,* 941 (1970).
17. Doi, K., Doi, A., and Fukui, A., *J. Biochem., 70,* 711 (1971).
18. Epel, D., Weaver, A. M., Muchmore, A. V., and Schimke, R. T., *Science, 163,* 294 (1969).
19. Garcia-Ballestra, J. P., *Microbiol. Esp., 24,* 257 (1971).
20. Heyn, A. N., *Arch. Biochem. Biophys., 132,* 442 (1969).
21. Holten, V. Z., and Bartnicki-Garcia, S., *Biochim. Biophys. Acta, 276,* 221 (1972).
22. Huotari, F. I., Nelson, T. E., Smith, F., and Kirkwood, S., *J. Biol. Chem., 243,* 952 (1968).
23. Isakov, V. V., Sova, V. V., Denisenko, V. G., Sakharovsky, V. G., and Elyakova, L. A., *Biochim. Biophys. Acta, 268,* 184 (1972).
24. Kessler, G., Feingold, D. S., and Hassid, W. Z., *Plant Physiol., 35,* 505 (1960).
25. Manners, D. J., and Marshall, J. J., *Phytochemistry, 12,* 547 (1973).
26. Marshall, J. J., *Carbohydr. Res., 26,* 274 (1973).
27. Moore, A. E., and Stone, B. A., *Biochim. Biophys. Acta, 258,* 238 (1972).
28. Muchmore, A. V., Epel, D., Weaver, A. M., and Schimke, R. T., *Biochim. Biophys. Acta, 178,* 551 (1969).
29. Nakamura, N., and Tanabe, O., *Nature, 196,* 774 (1962).
30. Nakamura, N., and Tanabe, O., *Nature, 200,* 1337 (1963).
31. Nelson, T. E., Johnson, J., Jantzen, E., and Kirkwood, S., *J. Biol. Chem., 244,* 5972 (1969).
32. Nielson, C. O., *Nature, 199,* 1001 (1963).
33. Piavaux, A., and Dandrifosse, G., *Arch. Int. Physiol. Biochim., 80,* 51 (1972).
34. Schneberger, G. L., and Luchsinger, W. W., *Can. J. Microbiol., 13,* 969 (1967).
35. Shimoda, C., and Yanagishima, N., *Physiol. Plant., 24,* 46 (1971).
36. Skujins, J. J., Potgieter, H. J., and Alexander, M., *Arch. Biochem. Biophys., 111,* 358 (1965).
37. Sova, V. V., and Elyakova, L. A., *Biochim. Biophys. Acta, 258,* 219 (1972).
38. Sova, V. V., Elyakova, L. A., and Vaskovsky, V. E., *Comp. Biochem. Physiol., 32,* 459 (1970).
39. Tanaka, H., Ogasawara, N., Nakajima, T., and Tamari, K., *J. Gen. Appl. Microbiol., 16,* 39 (1970).
40. Tingle, M. A., and Halvorson, H. O., *Biochim. Biophys. Acta, 250,* 165 (1971).
41. Vacquier, B. C., *Dev. Biol., 26,* 1 (1971).
42. Vogel, K., and Barber, A. A., *J. Protozool., 15,* 657 (1968).
43. Warsi, S. A., and Whelan, W. J., *Chem. Ind. (London),* p. 1573 (1957).

PENTOSANASES

DR. F. J. SIMPSON*

Pentosanase is a general name applied to enzymes that hydrolyze glycosidic bonds of polysaccharides composed of pentose sugars. These include the homopentosans, such as the arabans and xylans, the heteropentosans, such as araboxylans, and the hemicelluloses that contain short chains of pentoses attached to chains of other aldoses and alduronic acids. The arabans of terrestrial plants are usually low-molecular-weight polymers containing both $a(1\rightarrow5)$- and $a(1\rightarrow3)$-linked arabofuranose units, whereas the xylans contain mainly the $\beta(1\rightarrow4)$-xylopyranoside linkage and range from 80 to 200 units in size. Xylans possessing $\beta(1\rightarrow3)$-xylopyranoside bonds are found in marine algae. The structure of a $\beta(1\rightarrow4)$-xylan is analogous to that of cellulose; the configurations of the hydroxyl groups are the same, although in the xylan the hydroxymethyl group at C_5 is replaced with H. Thus it is not surprising to find that some celluloses are capable of hydrolyzing xylans.[62,96] A similar relationship exists between $\beta(1\rightarrow3)$-xylan and laminaran.

The enzymes hydrolyzing the glycosidic links between xylose units in a polysaccharide are commonly referred to as xylanases, and those hydrolyzing the glycosidic links between arabinose units as arabanases. Both endo- (liquefying) and exo- (saccharifying) types of pentosanases exist.[79] Calcium and magnesium ions have an activating effect on some pentosanases.[4,59] The pentosanases of fungi and some bacteria are usually induced enzymes and are not produced in the absence of a suitable substrate. At present the detection and assay of pentosanases depend on the availability of naturally occurring pentosans and hemicelluloses.[79-82] Access to more substrates of known structure and to pure enzymes will increase our ability to accurately determine specificity. The terminology of these enzymes then likely will undergo changes.

Pentosanases have been found in plants, particularly malts,[90,94] in insects and wood-boring larvae,[1,91] in snails,[77,92] in protozoa,[17,24] and in soil,[66,95] and are produced by both aerobic and anaerobic bacteria. The latter are largely responsible for degradation of pentosans in the rumen and intestines.[1,5,9,15,17,24,62,65,66] Microorganisms reported to produce pentosanases are listed in Tables 1 and 2.

The terms "pentosidase", "arabinosidase", and "xylosidase" are used for glycosidases that hydrolyze bonds between a pentose and an aglycone or another sugar. The relationship of these enzymes to pentosanases is not clear, for definitive studies on specificity are lacking and the terms "arabinosidase" and "xylosidase" are not always used with discrimination. For this reason, some sources of pentosidases are listed in Tables 3 and 4. Plants,[85,87,90] animals,[45,88-90] and gastropods[86] also possess pentosidases.

Some properties of purified pentosanases are listed in Table 5.

*Tables were compiled with the assistance of Mr. G. Murphy.

TABLE 1
SOURCES OF PENTOSANASES
Arabanases

Source	Substrate	pH	Reference
Bacteria			
Clostridium			
felsineum	Beet araban	5.6	97
Fungi			
Alternaria			
solani	Araban		38
Aspergillus			
awamori	Beet araban	3.6	40
niger	Araban, beet araban	3.0–6.0	21, 22, 39, 40, 54, 74, 95
oryzae	Beet araban	3.6	40
Botrytis			
allii	Araban		38
cinerea	Beet araban, arabanose	3.0	18
fabae	Araban	3.8–4.8	38
tulipae	Araban		38
Cladosporium			
cucumerinum	Araban		38
Colletotrichum			
lindemuthianum	Araban		38
Coniothyrium			
diplodiella	Beet araban	2.0–8.0	18, 41
Fusarium			
oxysporum f. *lupini*	Araban		38
oxysporum f. *lycopersici*	Araban		38
oxysporum f. *pisi*	Araban		38
Gloeosporium			
kaki	Beet araban, arabanose	4.0–6.0	18

TABLE 1 (Continued)
SOURCES OF PENTOSANASES
Arabanases

Source	Substrate	pH	Reference
Fungi (continued)			
Glomerella			
cingulata	Araban	4.8	38
Monilia			
fructigena	Araban		38
Mycosphaerella			
pinodes	Araban		38
Sclerotinia			
libertiana	Beet araban, arabanose	3.0	18
sclerotiorum	Araban	3.6−5.8	38
Trichoderma			
viride	Araboxylan		74
Protozoa			
Epidinium			
ecaudatum	Araboxylan, xylodextrans		17

TABLE 2
SOURCES OF PENTOSANASES
Xylanases

Source	Substrate	pH	Reference
Bacteria			
Bacillus sp.	Xylan		4, 42
Bacillus			
firmus	Xylan	7.0−7.2	57
licheniformis	Wheat flour pentosan		78
megaterium	Wheat flour pentosan		78
polymyxa	Wheat flour pentosan	6.5	64, 78
pumilus	Wheat flour pentosan		64, 78
subtilis	Xylan	6.0	42, 57, 64 78

TABLE 2 (Continued)
SOURCES OF PENTOSANASES
Xylanases

Source	Substrate	pH	Reference
Bacteria (continued)			
Bacillus (cont.)			
xylophagus			62
Bacteroides			
amylagenes			65
Butyrivibrio			
fibrisolvens	Hemicellulose fractions		44
Cellvibrio			
fulvus	Wheat straw xylan	7.0–7.5	62
Clostridium sp.	Wheat straw xylan	6.0–6.5	62
Micromonospora			
chalcea	Wheat straw xylan	6.0–6.5	62
Streptomyces sp.	Xylan, wheat araboxylan	4.5–7.3	28, 48, 49, 56
Streptomyces			
albogriseolus	Xylan		32, 62
albus	Wheat straw xylan	6.0–6.5	62
olivaceus	Xylan		32
xylophagus	Xylan		31, 32, 43
Fungi			
Alternaria spp.	Wheat flour pentosan		78
Alternaria			
kikuchiana			2
Aspergillus			
amstelodami	Xylan		10
batatae	Xylan	4.5–5.0	6
foetidus	Corn cob xylan	3.4–4.5	93
fumigatus	Wheat flour pentosan		78

TABLE 2 (Continued)
SOURCES OF PENTOSANASES
Xylanases

Source	Substrate	pH	Reference
Fungi (continued)			
Aspergillus (cont.)			
niger	Xylan, phenyl-β-D-xylose	4.2–5.0	10, 19, 25, 50, 51, 78
oryzae	Xylan		8, 37, 43, 78
Chaetomium			
globosum	Wheat xylan, $\beta(1\rightarrow3)$-xylan	3.5–6.5	3, 12, 14, 61, 62, 67, 78
trilateralo	Rice straw xylan, corn cob xylan, wheat straw xylan		32, 33, 53
Chrysosporium			
lignorum	Xylan		26, 55
Cochliobolus			
miyabeanus	Rice straw, xylan	6–7	7
Collybia			
yelutipes			14, 60
Coniophora			
cerebella	Xylan, 4-O-methylglucoronoxylan	5	14, 35, 60, 67
puteana	Holocellulose		73
Diplodia			
viticola	Grape xylan	3.8	47
Fomes			
annosus			55
igniarus			14
marginatus			14, 60, 67
Fusarium spp.	Wheat flour pentosan		78
Fusarium			
moniliforme	Xylan		10

TABLE 2 (Continued)
SOURCES OF PENTOSANASES
Xylanases

Source	Substrate	pH	Reference
Fungi (continued)			
Fusarium (cont.)			
roseum	Xylan	6.3	10
Gibberella			
saubinetti			2
Gloeophyllum			
saepiarium	Xylan		67
Gloeosporium			
kawakamii			2
oryzae			2
Irpex			
lacteus	Xylan		10, 13, 58
Lenzites			
saepiaria			14
Merulius			
lacrymans	Xylan		67
silvester			14
Myrothecium			
verrucaria	Xylan		10, 96
Penicillium spp.	Xylan, corn cob xylan		29, 30
Penicillium			
cyclopium	Xylan		76
digitatum	Potato cell wall preparation		70
expansum			2
funiculosum	Xylan		10, 63, 76
janthinellum			62, 63
lilacinum	Xylan		10
luteum	Xylan		10

TABLE 2 (Continued)
SOURCES OF PENTOSANASES
Xylanases

Source	Substrate	pH	Reference
Fungi (continued)			
Penicillium (cont.)			
pinophilum	Xylan		10
rugulosum	Xylan		76
verruculosum	Xylan		76
viridicatum	Xylan		10
Phellinus			
igniarius			60, 67
Piricularia			
oryzae	Rice straw xylan	6.4–8.1	2, 59
Polyporus			
betulinus			14
Rhizopus			
tritici			2
Schizophyllum			
commune			69, 71
Sclerotinia			
cinerea			2
rolfsii	Xylan	4.0–5.0	34
Stereum			
sanguinolentum			55
Trametes			
versicolor		5.0	14, 60, 67
Trichoderma			
koningi	Xylan		11
lignorum	Corn cob xylan		29
viride	Xylan	4.2–5.6	10, 19, 23, 25, 27, 63

TABLE 2 (Continued)
SOURCES OF PENTOSANASES
Xylanases

Source	Substrate	pH	Reference
Fungi (continued)			
Trichothecium			
roseum	Hemicellulose		75
Protozoa			
Entodinium sp.	Hemicellulose fraction		17
Eremoplastron			
bovis	Hemicellulose fraction		17
Eudiplodinium			
medium	Xylan	7.5	24
Eupidinium			
ecaudatum	Hemicellulose fraction		17
Polyplastron			
multivesiculatum	Wheat flour pentosan	6.3	15

TABLE 3
SOURCES OF PENTOSIDASES
Arabinosidases

Source	Substrate	pH	Reference
Fungi			
Phytophthora			
palmivora	*p*-Nitrophenyl-α-L-arabinofurano-side	4.0	83, 84
Sclerotinia			
fructigena	Pectin		72
Protozoa			
Epidinium			
ecaudatum	Arabinoxylan		17

TABLE 4
SOURCES OF PENTOSIDASES
Xylosidases

Source	Substrate	pH	Reference
Algae			
Cladophora			
rupestris		5.9	52
Rhodymenia			
palmata		5.9	52
Bacteria			
Bacteroides			
amylogenes	Xylan, xylobiose	6.8	65
Butyrivibrio sp.	Phenyl β-xyloside		9
Fungi			
Aspergillus sp.			
Aspergillus			
niger	Indolyl β-D-xylopyranoside	4.4–5.4	36
oryzae	β-Xylosides	5–6	20
parasiticus	Indolyl β-D-xylopyranoside	4.4–5.4	36
Chrysosporum			
lignorum			55
Collybia			
velutipes			16
Coniophora			
cerebella			35
Fomes			
annosus			55
marginatus			16
Stereum			
sanguinolentum			55
Trametes			
versicolor			16
Trichoderma			
viride	Indolyl β-D-xylopyranoside	4.4–5.4	36

TABLE 5
SOME PROPERTIES OF PURIFIED PENTOSANASES

Source	Degree of Purity	Optimal pH	Temperature Stability	Mode of Action	Bond Broken	Products	Reference
Arabanases							
Aspergillus niger	Homogenous protein	4.0	Stable at 60°C for 10 minutes	exo?		L-Arabinose	21, 39
Coniothyrium diplodiella	Highly purified	3.6–3.8		exo?		L-Arabinose, D-galactose	18
Xylanases							
Aspergillus batatae							
Xylanase A	Crystalline			endo		Oligosaccharides, xylotriose, xylobiose, xylo-arabinose	6
Xylanase B	Crystalline			exo	β-1,4'-	D-Xylose	6
Bacillus sp.	Crystalline			endo	β-1,4'-	Oligosaccharides, xylotriose, xylobiose	4
Chaetomium globosum	Highly purified	3.5		exo	β-1,3'-	Xylobiose, xylose	12, 68
Piricularia oryzae							
Xylanase A	Highly purified	6.4	Stable at 40°C for 2 hours; inactivated at 57°C in 2 hours			Xylobiose, xylose	59

TABLE 5 (Continued)
SOME PROPERTIES OF PURIFIED PENTOSANASES

Source	Degree of Purity	Optimal pH	Temperature Stability	Mode of Action	Bond Broken	Products	Reference
Xylanases (continued)							
Piricularia oryzae (cont.)							
Xylanase B	Highly purified	8.1	Stable at 50°C for 2 hours; inactivated at 70°C in 2 hours			Xylotriose, xylobiose	59
Streptomyces xylophagus	Homogenous protein preparation	6.2	Stable at 40°C for 10 minutes; inactivated at 70°C in 10 minutes	endo	β-1,4'-	Oligosaccharides, xylobiose, xylose	43
Cellulose[a]	Highly purified	4.5	Inactivated at 80°C in 1 hour	endo	β-1,4'-	Oligosaccharides, xylose	46

[a] Commercial preparation, Miles Laboratories.

REFERENCES

1. Yamafuji and Inaoka, *J. Agric. Chem. Soc. Jap., 23,* 502 (1950).
2. Ozawa, *Rep. Ohara Inst. Agric. Res., 40,* 110 (1952).
3. Sørensen, *Physiol. Plant., 5,* 183 (1952).
4. Inaoka and Soda, *Nature, 178,* 202 (1956).
5. Inaoka and Soda, *Mem. Ehime Univ. Sect. 6 (Agric.), 1,* 1 (1955).
6. Fukui and Sato, *Bull. Agric. Chem. Soc. Jap., 21,* 392 (1957).
7. Asada, *Nihon Shokubutsu Byori Gakkai-ho, 21,* 191 (1956).
8. Fukui, *J. Gen. Appl. Microbiol., 4,* 39 (1958).
9. Howard, Jones and Purdom, *Biochem. J., 74,* 173 (1960).
10. Gascoigne and Gascoigne, *J. Gen. Microbiol., 22,* 242 (1960).
11. Toyama, *Hakko Kogaku Zasshi, 38,* 81 (1960).
12. Fukui, Suzuki, Kitahara and Miwa, *J. Gen. Appl. Microbiol., 6,* 270 (1960).
13. Nishizawa, Morimoto, Handa, Shibata and Ikawa, *Koso Kagaku Shinpojiumu, 15,* 26 (1961).
14. Lyr, *Z. Allg. Mikrobiol., 3,* 25 (1963).
15. Akkada, Eadie and Howard, *Biochem. J., 89,* 268 (1963).
16. Lyr, *Z. Allg. Mikrobiol., 4,* 249 (1964).
17. Bailey and Gaillard, *Biochem. J., 95,* 758 (1965).
18. Kaji, Tagawa and Motoyama, *Nippon Nogei Kagaku Kaishi, 39,* 352 (1965).
19. Fujii and Toyama, *Hakko Kogaku Zasshi, 42,* 105 (1964).
20. Legler, *Z. Physiol. Chem., 345,* 197 (1966).
21. Kaji, Tagawa and Matsubara, *Agric. Biol. Chem., 31,* 1023 (1967).
22. Tagawa and Terui, *Hakko Kogaku Zasshi, 46,* 693 (1968).
23. Nomura, Yasui, Kiyooka and Kobayashi, *Hakko Kogaku Zasshi, 46,* 634 (1968).
24. Naga and El-Shazly, *J. Gen. Microbiol., 53,* 305 (1968).
25. Sugiura, Ogiso, Hatano and Tanaka, *Gifu Yakka Daigaku Kiyo, 17,* 117 (1967).
26. Eriksson and Rzedowski, *Arch. Biochem. Biophys., 129,* 683 (1969).
27. Nomura, Yasui, Kiyooka and Kobayashi, *Hakko Kogaku Zasshi, 47,* 313 (1969).
28. Kusakabe, Yasui and Kobayashi, *Nippon Nogei Kagaku Kaishi, 43,* 145 (1969).
29. Bilai, Pidoplichko, Strizhevskaya and V'yun, *Eksp. Mikol.,* p. 8 (1968).
30. Strizhevskaya, *Fermenty Med. Pishch. Prom-st. Sel. Khoz., 244* (1968).
31. Kawaminami and Iizuka, *Agric. Biol. Chem., 33,* 1787 (1969).
32. Iizuka and Kawaminami, *Agric. Biol. Chem., 33,* 1257 (1969).
33. Kawaminami and Iizuka, *Hakko Kogaku Zasshi, 48,* 161 (1970).
34. Van Etten and Bateman, *Phytopathology, 59,* 968 (1969).
35. King and Fuller, *Biochem. J., 108,* 571 (1968).
36. Esterly, Standen and Pearson, *J. Histochem. Cytochem., 16,* 489 (1968).
37. Kundu, Mukherjee, Das and Das Gupta, *Indian J. Exp. Biol., 6,* 129 (1968).
38. Fuchs, Jobsen and Wouts, *Nature, 206,* 714 (1965).
39. Kaji and Tagawa, *Nippon Nogei Kagaku Kaishi, 38,* 580 (1964).
40. Kaji, Taki, Shimazaki and Shinkai, *Kagawa Daigaku Nogakubu Gakuzyutu Hokoku, 15,* 34 (1963).
41. Kaji and Tagawa, *Kagawa Daigaku Nogakubu Gakuzyutu Hokoku, 16,* 143 (1965).
42. Takahashi and Hashimoto, *Hakko Kogaku Zasshi, 41,* 116 (1963).
43. Iizuka and Kawaminami, *Agric. Biol. Chem., 29,* 520 (1965).
44. Gaillard, Bailey and Clarke, *J. Agric. Sci., 64,* 449 (1965).
45. Robinson and Abrahams, *Biochim. Biophys. Acta, 132,* 212 (1967).
46. Hrazdina and Neukom, *Biochim. Biophys. Acta, 128,* 402 (1966).
47. Strobel, *Phytopathology, 53,* 592 (1963).
48. Goldschmid and Perlin, *Can. J. Chem., 41,* 2272 (1963).
49. Perlin and Reese, *Can. J. Biochem. Physiol., 41,* 1842 (1963).
50. Sasaki and Inaoka, *Mem. Ehime Univ. Sect. 6 (Agric.), 12,* 149 (1967).
51. Sasaki and Inaoka, *Mem. Ehime Univ. Sect. 6 (Agric.), 12,* 157 (1967).
52. Manners and Mitchell, *Biochem. J., 103,* 43P (1967).
53. Iizuka and Kawaminami, *Jap. Patent 11,* 999 (1967).
54. Ahlgren, Eriksson and Vesterberg, *Acta Chem. Scand., 21,* 937 (1967).
55. Ahlgren and Eriksson, *Acta Chem. Scand., 21,* 1193 (1967).
56. Inaoka, *Mem. Ehime Univ. Sect. 6 (Agric.), 6,* 105 (1961).
57. Inaoka, *Mem. Ehime Univ. Sect. 6 (Agric.), 6,* 91 (1961).
58. Nishizawa, Morimoto, Handa and Hashimoto, *Arch. Biochem. Biophys., 96,* 152 (1962).
59. Sumizu, Yoshikawa and Tanaka, *J. Biochem., 50,* 538 (1961).
60. Lyr, *Enzymologia, 23,* 231 (1961).

61. Sørensen, *Physiol. Plant., 5,* 183 (1952).
62. Sørensen, *Acta Agric. Scand., Suppl. 1,* 1 (1957).
63. Reese and Mandels, *Appl. Microbiol., 7* (1959).
64. Simpson, *Can. J. Microbiol., 2,* 28 (1956).
65. Walker, *Aust. J. Biol. Sci., 20,* 799 (1967).
66. Sørensen, *Nature, 176,* 74 (1955).
67. Lyr, *Arch. Mikrobiol., 33,* 266 (1959).
68. Fukui, *Tampakushitsu Kakusan Koso, 6,* 90 (1961).
69. Varadi and Jurasek, *Drev. Vysk., 1,* 7 (1970).
70. Cole and Wood, *Phytochemistry, 9,* 695 (1970).
71. Varadi and Jurasek, *Drev. Vysk., 3,* 89 (1969).
72. Fielding and Byrde, *J. Gen. Microbiol., 58,* 73 (1969).
73. Jurasek and Sopko, *Drev. Vysk., 2,* 71 (1962).
74. Kulp, *Cereal Chem., 45,* 339 (1968).
75. Zhdanova and Salmanova, *Tr. Vses. Nauchno-Issled. Inst. Pivo-Bezalkogol'n Vinodel'chesk Prom., 12,* 23 (1965).
76. Strizhevskaya, *Eksp. Mikol.,* p. 24 (1968).
77. Myers and Northbote, *Biochem. J., 69,* 54P (1958).
78. Simpson, *Can. J. Microbiol., 1,* 131 (1954).
79. Sugiura, Ito and Aoyama, *Yakugaku Zasshi, 90,* 1258 (1970).
80. Dingle, Reid and Solomons, *J. Sci. Food Agric., 4,* 149 (1953).
81. Husemann, Loës and Lösterle, *Makromol. Chem., 6,* 163 (1951).
82. Inaoka, *Mem. Ehime Univ. Sect. 6 (Agric.), 6,* 111 (1961).
83. Akinrefon, *New Phytol., 67,* 543 (1968).
84. Akinrefon, *J. Gen. Microbiol., 51,* 67 (1968).
85. Manners and Taylor, *Carbohyd. Res., 7,* 497 (1968).
86. Fukuda, Muramatsu, Egami, Takahashi and Yasuda, *Biochim. Biophys. Acta, 159,* 215 (1968).
87. Shibata and Nishizawa, *Arch. Biochem. Biophys., 109,* 516 (1965).
88. Fisher, Higham, Kent and Pritchard, *Biochem. J., 98,* 46 (1966).
89. Patel and Tappel, *Biochim. Biophys. Acta, 191,* 86 (1969).
90. Fisher, Whitehouse and Kent, *Nature, 213,* 204 (1967).
91. Allman and Duspiva, *Experientia (Basel), 22,* 231 (1966).
92. Ehrenstein, *Helv. Chim. Acta, 9,* 332 (1926).
93. Whistler and Masak, *J. Am. Chem. Soc., 77,* 1241 (1955).
94. Bass, Meredith and Anderson, *Cereal Chem., 30,* 313 (1953).
95. Tagawa and Kaji, *Kagawa Daigaku Nogakubu Gakuzyutu Hokuku, 15,* 45 (1963).
96. Bishop and Whitaker, *Chem. Ind., 119* (1955).
97. Kaji, Anabuhu, Taki and Oyama, *Kagawa Daigaku Nogakubu Gakuzyutu Hokuku, 15,* 40 (1963).

AMYLASES OF MICROBIAL ORIGIN

DR. JAMES H. LITTLEJOHN

The following table is a list of some of the amylases of microbial origin. It includes those enzymes capable of degrading starch or other polysaccharides whose glucose monomers are basically bound by an a-1,4 linkage with a-1,6 branching.. This table does not list every microorganism capable of producing amylases, due to the vast number cited and vagueness of the data reported.

Organism	Enzyme	pH	Temperature	Activators	Inhibitors	Comments	References
Aerobacter							
aerogenes	Pullulanase	5.0	≤47.5°C			Hydrolyzes α-1,6 linkage only if α-1,4 linkages are present; MW = 145,000; hydrolyzes amylopectin and glycogen; cell bound and soluble isoenzymes	1
	Isoamylase	6.1	40°C				2
Aspergillus							
awamorii	α-Amylase	4.5—6.2	40°C			Dextrogenic amylase at acid pH; no saccharogenic amylase; *A. awamorii*, *A. usamii*, and *A. niger* have amylase different from that of *A. oryzae*; acid-stable and acid-labile α-amylases are present	3—5
	β-Amylase	3.5—7.0	50°C				6—8
	Glucoamylase	4.5—4.7	55—75°C				9
candidus	α-Amylase	4.4—5.3		Ca++			10
niger	α-Amylase	4.7—6.0	65°C			Produces glucose > maltose (high temperature) > at pH panose + maltotriose; isoelectric point (IEP) at pH 3.7; acid-stable α-amylase IEP at pH 3.4	11, 12
	Glucoamylase	3.8	50°C			Stable at pH 1.8—8.2, inactivated at 55°C, pH 4.5, for 10 minutes; IEP at pH 5.5	13
shirousamii	α-Amylase			Na+, K+, NH4+, Ca++, Mg++			14
usamii	α-Amylase Debranching						14, 15
							15
	Glucoamylase	5.0	55°C				15
oryzae	α-Amylase	5.5—5.9	50—57°C			IEP at pH 4.2	11, 16—18
	β-Amylase	4.8	30°C			Stable at pH 5.4—8.6, 15°C; produces maltotriose > maltose > glucose	11, 16—18
	Glucoamylase	4.8	50°C			60% hydrolysis of starch; MW = 70,000; total reducing sugar produced was 34.5%, and glucose 21.6%	19

Organism	Enzyme	pH	Temperature	Activators	Inhibitors	Comments	References
Azotobacter							
chroococcum	Amylase					Prevalent in older cultures	20
Bacillus							
amyloliquefaciens	α-Amylase	5.7–6.0	55–60°C	CaCl$_2$, NaCl, Na$_2$SO$_4$	Glycine	Maximum enzyme production in the stationary phase	21–23
coagulans	α-Amylase	6.5–8.0	60–70°C	Cl$^-$		Thermophilic; stable at pH 5.0–9.5	24
	α-Amylase	6.5–8.0	45–55°C	Cl$^-$		Mesophilic; stable at pH 5.0–9.5	24
diastaticus	α-Amylase	5.8	70°C			Thermophilic enzyme, resistant to low pH; 25% activity remained at pH 3.0	25
macerans	α-Amylase	5.5–5.6	55°C			IEP at pH 4.5	26, 27
mesentericus	α-Amylase	6.0–7.5	70°C	Cl$^-$, Br$^-$, NO$_3^-$		pH range 4.0–10.0	21, 28
polymyxa	α-Amylase	6.2–7.5	45°C			Able to degrade Schardinger dextrins; able to by-pass α-1,6 linkages of amylopectin; the main product is β-maltose	29–31
stearothermophilus	α-Amylase	4.6–5.1	70°C	Cl$^-$, Na$^+$, Ca^{++}, Mg^{++}, Sr^{++}	EDTA	Thermophilic; IEP at pH 4.8; MW = 15,600; N-ethylmalemide and p-hydroxy-mercuribenzoate does not inhibit the enzyme	24, 27, 32, 33
	α-Amylase	4.6–5.1	37°C	Ca^{++}, Cl$^-$		Mesophilic; 98% activity loss at 55°C for 30 minutes	34
subtilis	α-Amylase	6.0	50°C	Ca^{++}, Zn^{++}	6M urea, pH 8, 50°C	pH range 6–11; MW = 49,000; sulfhydryl groups are necessary for activity; maximum enzyme production in the exponential phase	27, 35–39
Candida							
sitophila	α-Amylase						40
	β-Amylase						40

Organism	Enzyme	pH	Temperature	Activators	Inhibitors	Comments	References
Candida (cont.)							
tropicalis	α-Amylase	4.0	65°C			Single protein	41
japonica	α-Amylase	5.0–6.0	55°C			Percent hydrolysis increases with chain length	42
Clostridium							
acetobutylicum	α-Amylase	4.6	40°C		10–16% by $10^{-4}\,M$ and $10^{-3}\,M$ EDTA respectively $10^{-3}\,M$ CuSO$_4$ gave 95% inhibition		43
	β-Amylase	4.8	37°C			Enzyme not activated by Cl$^-$, PO$_4^{\equiv}$, KCN, MgCl$_2$, NaF or AsO$_4$	44
butyricum	α-Amylase	4.6–6.1	48°C			Main products are maltose, maltotriose, and glucose; not activated by Ca^{++} or Cl$^-$	45
novyi	α-Amylase	6.9–7.0		Resorcinol NaCl	Hydroquinone		46
pasteurianum	α-Amylase	5.5–7.7	70°C		NaNO$_3$ at pH 5.0	Saccharification at 60°C, pH 6	47
perfringens	α-Amylase	6.0–7.0					48
Endomyces sp.	Glucoamylase	4.8	55°C			MW = 55,000; hydrolyzes 80% of the glycolytic linkages in starch	49
Endomycopsis							
capsularis	α-Amylase Glucoamylase	4.5 4.5	40–50°C 40–50°C			pH range 4.2–8.0 at 20°C	50, 51 50, 51
fibuliger	α-Amylase Glucoamylase	4.8–5.0 4.8–5.0				The saccharogenic enzyme is separated from the dextrinogenic enzyme by overnight incubation at 30°C in 0.5M citrate buffer	52 52
Escherichia							
intermedia	Isoamylase	5.5–6.0	47°C		Hg^{++}, Cu^{++}, Fe^{+++}	Hydrolyzes α-1,6 linkages in starch and glycogen; hydrolyzes pullulan	53

Organism	Enzyme	pH	Temperature	Activators	Inhibitors	Comments	References
Physarum							
polycephalum	α-Amylase	6.9	20°C	NaCl		50% activity loss at 65°C for 5 minutes	54
Pseudomonas							
sp.	Isoamylase	3.0—4.0	40°C			pH range 3–6; IEP at pH 4.4; MW = 95,000; hydrolyzes amylopectin, but not pullulan	55
saccharophila	α-Amylase	5.3—5.8	40°C			pH range 4.5—8.0	56, 57
	β-Amylase	5.2					58
Pullularia							
pullulans	α-Amylase	4.2—4.8	45°C		$10^{-1} M$ EDTA	pH range 3.5—5.0	59
	Glucoamylase	4.6—4.8	54°C			pH range 3.5—5.0; IEP at pH 5.8	59
Rhizopus							
delemar	Glucoamylase	5.5	45°C		pH 4.5 at 45°C for 10 minutes inactivates	pH range 3.0—7.8; IEP at pH 7.0; both α-1,4 and α-1,6 bonds are hydrolyzed; trace of α-amylases	13, 60, 61
javanicus	α-Amylase	6.9	20°C	$Na_2S_2O_3$	$ZnSO_4$, $Ca(OAc)_2$	Total reducing sugar 70.5%, glucose 58.3%	60, 62, 63
	β-Amylase	4.8	50°C			The major amylase produced	60, 62
	Glucoamylase						60, 62, 63
niveus	Glucoamylase	5.0	55°C			Products: glucose > isomaltose > maltose	64, 65
peka	α-Amylase	6.9	20°C			Trace amount produced	60
	Glucoamylase					The major amylase produced	60
tonkinensis	α-Amylase	6.9	20°C			Trace amount produced	60
	Glucoamylase					The major amylase produced	60
Saccharomyces							
diastaticus	β-Amylase	4.6	25°C			Produces maltose and residual dextrin; negligible action on β-limit dextrin	66
	Debranching	6.4	32°C			Heat-inactivated at 55°C	66
	Glucoamylase	4.6	32°C			Extracellular enzyme; inactivated at pH 6.4 and 55°C	67

Organism	Enzyme	pH	Temperature	Activators	Inhibitors	Comments	References
Saccharomyces (cont.)							
cerevisiae	β-Amylase						66
	Isoamylase	6.0	25°C		$BO_3^=$, MgCl	Hydrolyzes interchain linkages in amylopectin and glycogen; pullulan is not hydrolyzed	67–69
	Glucoamylase					Intracellular enzyme	67
Streptococcus							
bovis	α-Amylase	6.0	39°C			Products: maltotriose > maltose = 1.0:0.63	70
mitis	Pullulanase	5.4	30°C			Inactivated above 40°C; best action on α-1,6 linkages in α-maltosyldextrinyl-1,6-maltodextrin	71
Streptomyces							
aureofaciens	α-Amylase	5.0–6.2	30–40°C			Contains high level of acidic amino acids	72
diastochromogens and other species	α-Amylase	5.0	70°C			Activity decreases above 70°C; characteristics are more like those of plant enzymes than those of bacterial and animal enzymes	73
Thermoactinomyces							
vulgaris	α-Amylase	5.9–7.0	60°C			More active at pH 5.9 in the absence of substrate; rapidly inactivated at pH 7.0	74

REFERENCES

1. Bender, H., and Wallenfels, K., *Methods Enzymol., 8,* 555 (1966).
2. Fujio, Y., Shiosaka, M., and Ueda, S., *J. Ferment. Technol., 48,* 8 (1970).
3. Minoda, Y., Mogi, K., and Nakamura, T., *Nippon Nogei Kagaku Kaishi, 29,* 115 (1955).
4. Hayashida, S., *Bull. Agric. Chem. Soc. Jap., 21,* 386 (1957).
5. Watanabe, K., and Fukimbara, T., *J. Ferment. Technol., 45,* 226 (1967).
6. Tokuoka, Y., *J. Ferment. Technol., 26,* 198 (1947).
7. Dobrolinskaya, G. M., and Rodzevich, V. I., *Ferment. Spirt. Prom-st., 32,* 9 (1966).
8. Grigorov, V. S., *Prikl. Biokhim. Mikrobiol., 4,* 384 (1968).
9. Lulla, B. S., and Johar, D. S., *J. Sci. Ind. Res., 15C,* 233 (1956).
10. Takaoka, K., *Symp. on Enzyme Chem. Jap., 8,* 48 (1953); *J. Agric. Chem. Soc. Jap., 27,* 111 (1953).
11. Burger, M., and Beran, K., *Chem. Listy, 50,* 133 (1956).
12. Stinson, E. E., *Iowa State J. Sci., 29,* 509 (1955).
13. Tsujisaka, Y., Fukumoto, J., Yamamoto, T., *Nature, 181,* 770 (1958).
14. Okazaki, H., *Proc. Int. Sym. Enzyme Chem. Tokyo and Kyoto, 2,* 494 (1957).
15. Komaki, T., Matsuba, Y., Okamoto, N., and Sato, T., *Dempun Kogyo Gakkaishi, 4,* 98 (1956–57).
16. Matsuyama, M., *J. Ferment Technol., 28,* 461 (1950).
17. Roy, D. K., *Ann. Biochem. Exp. Med. (Calcutta), 16,* 111 (1956).
18. Oshima, K., *J. Coll. Agric. Hokkaido Imp. Univ., 19,* 135 (1928).
19. Morita, Y., Shimizu, K., Oga, M., and Korenaga, T., *Agric. Biol. Chem., 30,* 114 (1966).
20. Zinoveva, K. G., *Mikrobiol. Zh. Akad. Nauk Ukr. R.S.R., 19,* 22 (1957).
21. Hagihara, B., *Symp. Enzyme Chem. Jap., 7,* 105 (1952).
22. Fukumoto, J., *J. Agric. Chem. Soc. Jap., 19,* 487 (1943).
23. Tsura, D., and Fukumoto, J., *Koso Kagaku Shimpojiumu, 18,* 30 (1962).
24. Cambell, L. L., Jr., *Arch. Biochem. Biophys., 54,* 154 (1955).
25. Kislukhina, O. V., *Prikl. Biokhim. Mikrobiol., 2,* 544 (1966).
26. Samec, M., *Akad. Znanosti Umetnosti Ljubljani Kem. Lab. Kem. Studije,* p. 3 (1947).
27. Pazur, H. H., in *Starch Chemistry and Technology,* p. 133. Academic Press, New York (1965).
28. Velcheva, P., *Izv. Mikrobiol. Inst. Bulgar. Akad. Nauk, 12,* 83 (1960).
29. Robyt, J. F., and French, D., *Arch. Biochem. Biophys., 104,* 338 (1964).
30. Rose, D., *Arch. Biochem., 16,* 349 (1948).
31. Dunn, C. G., Fuld, G. L., Yamada, K., Mas Urioste, J., and Casey, P. R., *Appl. Microbiol., 7,* 212 (1959).
32. Endo, S., *Hakko Kogaku Zasshi, 37,* 353 (1959).
33. Manning, G. B., and Cambell, L. L., Jr., *J. Biol. Chem., 236,* 2953 (1961).
34. Isono, K., *Biochem. Biophys. Res. Comm., 41,* 852 (1970).
35. Stein, A., and Fischer, E. H., *Biochim. Biophys. Acta, 39,* 287 (1960).
36. Di Carlo, F. J., and Redfern, S., *Arch. Biochem., 15,* 333 (1947).
37. Yamanaka, T., Higashi, T., Horio, T., and Okunuki, K., *J. Biochem. (Tokyo), 44,* 637 (1957).
38. Welker, N. E., and Cambell, L. L., *J. Bacteriol., 94,* 1131 (1967).
39. Toda, H., and Narita, K., *J. Biochem. (Tokyo), 63,* 302 (1968).
40. Ishii, R., and Akagi, S., *Hakko Kogaku Zasshi, 26,* 276 (1948).
41. Sawai, T., *J. Biochem. (Tokyo), 45,* 49 (1958).
42. Ebertova, H., *Folia Microbiol., 8,* 333 (1963).
43. Scott, D., and Hendrick, L. R., *J. Bacteriol., 63,* 795 (1952).
44. Tomoeda, M., Horitsu, H., and Kameyama, M., *Gifu Daigaku Nogakubu Kenkyu Hokoku, 16,* 151 (1962).
45. Hobson, P. N., and Macphearson, M. J., *Biochem. J., 52,* 671 (1952).
46. Shemanova, G. F., and Blagoveshchenskii, V. F., *Biokhimiya, 22,* 799 (1957).
47. Proskuryakov, N. I., and Dmitrisvskaya, N. V., *Dokl. Akad. Nauk SSSR, 67,* 699 (1949).
48. Volchok, A. K., Ivanov, V. I., and Lobanova, A. V., *Biokhimiya, 20,* 522 (1955).
49. Toshio, F., and Ziro, N., *Agric. Biol. Chem., 33,* 884 (1969).
50. Ebertova, H., *Folia Microbiol., 11,* 14 (1966).
51. Ebertova, H., *Folia Microbiol., 11,* 422 (1966).
52. Hattori, Y., and Takeuchi, I., *Rika Gaku Kenkyusho Hokoku, 37,* 37 (1961).
53. Ueda, S., and Nanri, N., *Appl. Microbiol., 15,* 492 (1967).
54. Zelden, M. H., and Ward, J. M., *Nature, 198,* 389 (1963).
55. Harada, T., Yokobayashi, K., and Misaki, A., *Appl. Microbiol., 16,* 1439 (1968).
56. Markovitz, A., *Univ. Wash. Seattle Diss. Abstr., 15,* 941 (1955).
57. Markovitz, A., Klein, H. P., and Fischer, E. H., *Biochim. Biophys. Acta, 19,* 267 (1956).
58. Thayer, P. S., *J. Bacteriol., 66,* 656 (1953).
59. Yamada, N., *Nippon Nogei Kagaku Kaishi, 37,* 712 (1963).
60. Sakamoto, M., and Shuzui, K., *Hakko Kogaku Zasshi, 35,* 238 (1957).

61. Hiromi, K., Kawai, M., and Ono, S., *J. Biochem., 59,* 476 (1966).
62. Otani, Y., Takahashi, S., and Yamamoto, E., *Hakko Kogaku Zasshi, 36,* 241 (1958).
63. Liu, P., and Chen, S., *Chung Kuo Nung Yeh Hua Hsueh Hui Chih, 1,* 24 (1963).
64. Kawamura, S:, Watanabe, T., and Matsuda, K., *Tohoku J. Agric. Res., 20,* 137 (1969).
65. Komaki, T., Matsuba, Y., Okamoto, N., and Sato, T., *Dempun Kogyo Gakkaishi, 4,* 89 (1956–57).
66. Hopkins, R. H., in *European Brew. Conv. Proc. 5th Cong. Baden-Baden,* p. 52 (1955).
67. Hopkins, R. H., and Kulka, D., *Arch. Biochem. Biophys., 69,* 45 (1957).
68. Liu, P., and Chen, S., *Chung Kuo Nung Yeh Hua Hsueh Hui Chih, 5,* 87 (1967).
69. Bathgate, G. N., and Manners, D. J., *Biochem. J., 107,* 443 (1968).
70. Walker, G. J., *Biochem. J., 94,* 289 (1965).
71. Walker, G. J., *Biochem. J., 108,* 33 (1968).
72. Vecher, A. S., Babitskaya, V. G., and Ryabushko, T. A., *Mikrobiologiya, 38,* 999 (1969).
73. Ohtsuki, T., and Kawazu, M., *Nihon Kin Gakki Kaiho, 3,* 1 (1962).
74. Kuo, M. J., and Hartman, P. A., *Can. J. Microbiol., 13,* 1157 (1967).

MICROBIAL CELLULASES

DR. V. R. SRINIVASAN

Cellulose is abundantly distributed in nature, but relatively few groups of microorganisms are known to produce extracellular enzymes capable of directly degrading native crystalline cellulose. Many more organisms are known to produce enzymes active against partially substituted soluble cellulose or against regenerated or acid-swollen cellulose, whose crystalline structure has been modified. These latter organisms lack the specific factor (C_1 or swelling factor) necessary for degradation of native cellulose, Certain organisms, especially fungi, release extracellular enzymes to the surrounding medium. Other microorganisms, especially bacteria, seem capable of degrading cellulose only if their cells are intimately in contact with the cellulose fibers. Very little is known about the mode of action of the C_1 enzymes, which may also be called the "swelling factor".

A — *Trichoderma viride*

B — *Cellvibrio fulvus*

C — Many organisms

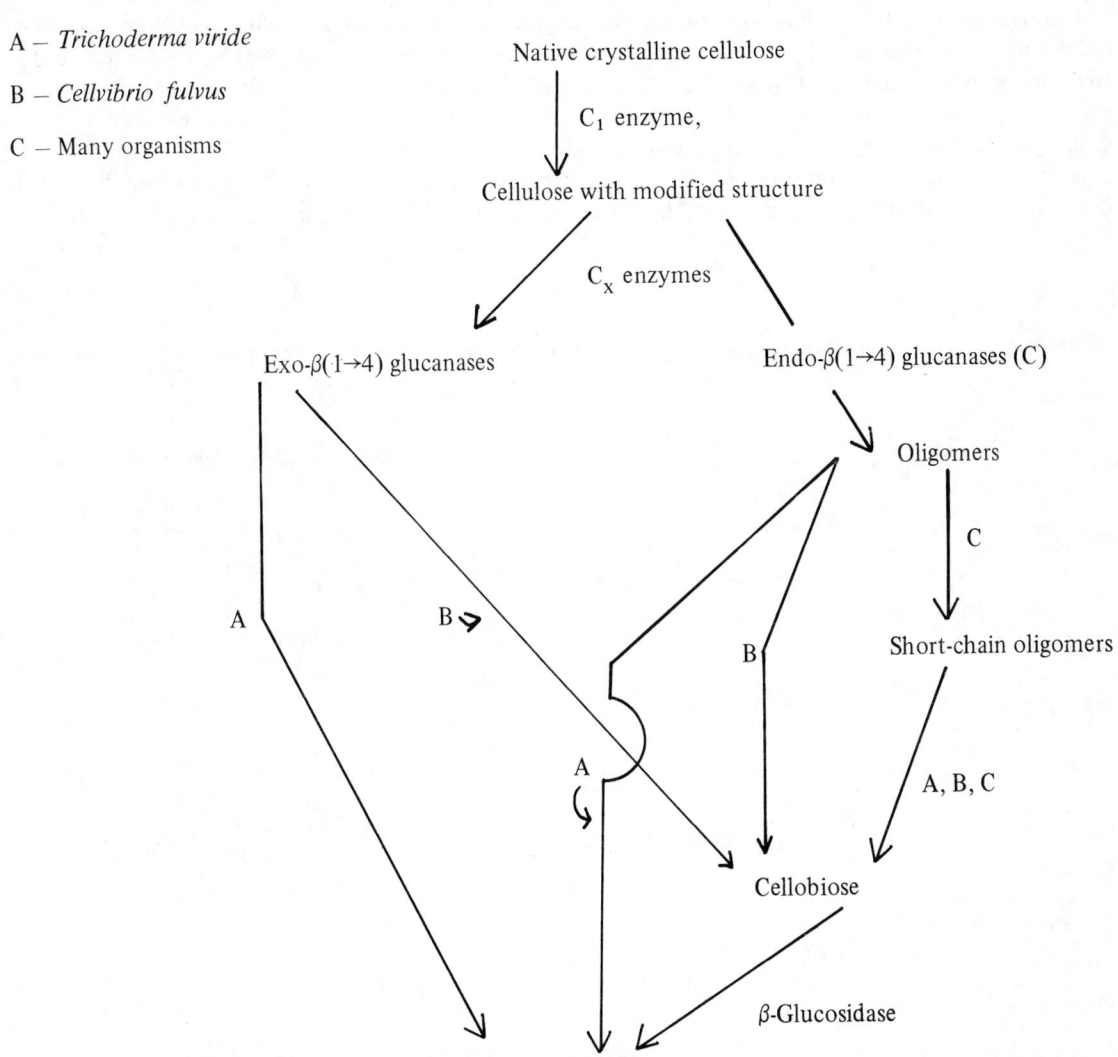

The C_x enzymes are also active on regenerated or partially substituted cellulose, such as carboxymethylcellulose (CMC).

Cellulase activity has generally been measured by two different methods.

1. Increase of reducing sugar by enzymatic activity on soluble or insoluble substrate. It should be emphasized that in tests for "true" cellulase activity a crystalline cellulose should be used as a substrate, e.g., filter paper, Avicel or adsorbent cotton (see Reference 10, p. 393).

2. Decrease in viscosity of 0.5 or 1.0% (w/v) solution of carboxymethylcellulose or hydroxyethyl cellulose. The degree of substitution of such celluloses markedly affects the rate of degradation. Since no standard substrate has been adopted, it is hard to make meaningful comparisons of activities of cellulases from different sources. A method has been presented that allows the use of a "conversion factor" for comparing the cellulolytic activities on carboxymethylcelluloses (CMC) of different degrees of polymerization (DP).[10]

Contemporary research has focused on the degradation of native crystalline cellulose, including lignocellulose. The enzymic hydrolysis of cellulose and the application of cellulases have been reviewed in two monographs edited by Reese and his collaborators.[10,25] A comprehensive table on cellulases and their properties, studied up to 1960, is presented in *The Biological Degradation of Cellulose* by Gascoigne and Gascoigne, Butterworth Scientific Publications, London, England.

The following table summarizes the cellulases from fungal and bacterial sources and lists their characteristics. The information presented here was obtained mainly from studies conducted after 1960.

Cellulase Source	Substrates	Fractionation Methods	Properties	Reference
Aspergillus				
fumigatus	CMC		Optimal temperature for activity 55°C at pH 4.5	18
niger	Filter paper; CMC		Optimal activity at pH 4.5, 40°C; stable between 37 and 50°C	34
niger 439	CMC		Optimal activity between pH 3.0 and 3.4, 40–60°C	14
Chaetomium				
globosum	CMC		Optimal activity at pH 4.0, 40°C; MW = 15,800, by ultracentrifugation	37
Coniophora				
cerebella	CMC		Optimal activity between pH 3.5 and 4.5, 40–50°C	13
Chrysosporium				
lignorum	CMC	Gel filtration on Sephadex G-100; DEAE-Sephadex A-50	Optimal activity at pH 5.0, 40°C	8

Cellulase Source	Substrates	Fractionation Methods	Properties	Reference
Fusarium				
moniliforme C_1 enzyme	Filter paper		Optimal activity at pH 5.0, 40°C	19
oxysporum C_1 enzyme	Filter paper		Optimal activity at pH 5.0, 40°C	19
solani C_1 enzyme	Filter paper		Optimal activity at pH 5.0, 40°C	19
solani C_1 and C_x enzymes	Dewaxed cotton; native cotton; CMC		Activity at pH 4.8, 40–50°C	40
splendens C_1 enzyme	Filter paper		Optimal activity at pH 5.0, 40°C	19
Humicola				
insolens	CMC			24
Mucor				
pusillus NRRL 2543	Acid-swollen cellulose; CMC	Gel filtration on Sephadex G-75	Optimal activity at pH 5.0, 37°C	32
Myrothecium				
verrucaria	Unmercerized cotton; mercerized cotton	Ethanol fractionation	Optimal activity at pH 5.6, 30°C; a labile enzyme, A, required for extensive degradation of fibrous material, and a stable enzyme, B; MWs of the three major cellulolytic components: 55,000, 30,000, and 5,300, by gel filtration	26–28
		Ammonium sulfate precipitation; gel filtration on Sephadex G-25 and G-75 Ion-exchange chromatography with Amberlite CG-50	MW = 49,000, by Archibald's method	
Neurospora spp.	Filter paper; CMC		Optimal activity between pH 4.5 and 6.0, 40°C; optimal temperature for activity 50°C, K_m for CMC = 1.07 g/liter, pH 4.5	15
Penicillium				
notatum	Hydroxyethyl cellulose	Gel filtration on Sephadex G-75; ion-exchange chromatography on DEAE-Sephadex A-25; zone electrophoresis on Sephadex G-25	Stable at pH 4–8.5 at room temperature; optimal activity at pH 4.5–7.0, 25°C; MW = 34,500	20, 23

Cellulase Source	Substrates	Fractionation Methods	Properties	Reference
Piricularia				
oryzae			Optimal temperature for activity 35–40°C; two distinct pH optima: 5.4 and 6.4	12
Polyporus				
schweinitzii	Regenerated cellulose; CMC	Ammonium sulfate fractionations; gel filtration on Sephadex G-100; ion-exchange gel chromatography on DEAE-Sephadex A-25; 1550-fold purification	Stable at pH 2.6–7.5; optimal activity at pH 4.0; activity decreased sharply above pH 6.0; optimal temperature for activity 60°C; MW = 45,000 ± 4,500, on Sephadex G-100	1
versicolor	CMC	Gel filtration on Sephadex G-100; zone electrophoresis	Four components were active on CMC at pH 5.0; two of these had MWs 11,400 and 51,000	21, 22
Sclerotium				
rolfsii	CMC	Resolved into three components by Sephadex G-75; Biogel P-100 and starch gel electrophoresis at pH 4.5	Optimal activity at pH 4.0; MW of the three components: 11,000–13,000, 22,000–26,000, and 34,000–40,000	2
Stachybotrys				
atra	CMC; phosphoric acid-swollen cellulose; cotton linters	Partial separation by chromatography		9
Stereum				
sanguinolentum	CMC	P-150 polyacrylamide columns	$K_m = 3 \times 10^{-3}$ g/liter; MW = 21,200, by ultracentrifugation	3, 5
Streptomyces				
antibioticus	CMC	Starch block electrophoresis	Three components	7
sp. 0143	CMC	P-100 polyacrylamide gel columns	Optimal activity at pH 5.9, 37°C; stable at pH 5.0 at room temperature for several days	31
Trichoderma				
koningi	Cotton yarn; CMC	Ammonium sulfate precipitation; gel filtration on Sephadex G-75; ion-exchange gel chromatography on DEAE-Sephadex and SE-Sephadex	C_1 or S-factor shows two peaks of activity, in the pH ranges 3.2–3.4 and 4.8–5.8; C_x is active at pH 3.1–3.5 and pH 5.0; MW = 50,000, by ultracentrifugation	11, 39

Cellulase Source	Substrates	Fractionation Methods	Properties	Reference
Trichoderma (cont.)				
viride	Cotton yarn; gauze; Avicel; filter paper; CMC	Gel filtration on Sephadex G-75; ion-exchange gel chromatography on DEAE-Sephadex	Exo- and endo-β 1-4 glucanases; activity at pH 4.5, 40°C; three components, MWs = 12,000, 48,000–62,000, and 61,000	17, 36
Acetobacter				
xylinum	CMC			38
Cellvibrio strains	CMC		Optimal activity at pH 7.0; no activity below pH 6.0	4
Cellvibrio				
gilvus	CMC	Starch gel electrophoresis; zone electrophoresis	Optimal activity at pH 5.9–6.0	33
Pseudomonas				
fluorescens var. *cellulosa*	Filter paper; cellodextrins, CMC	Zone electrophoresis at pH 8.6 on cellulose acetate films; gel filtration on Sephadex G-100; ion-exchange gel chromatography on DEAE-Sephadex A-50	Two components, A and B, had optimal activity at pH 8.0; a third component, C, was active at pH 7.0; all three components were stable between pH 7.0 and 8.0; inactivated at 60°C for 10 minutes; MW of components A and C approximately 50,000	35
Ruminococcus				
albus	Avicel; acid-swollen cellulose; CMC		Both C_1 and C_x activities are present; activity at pH 5.8, 47°C	16
Sprocytophaga				
myxococcoides	Filter paper; CMC		Optimal activity at neutral or alkaline pH	6

REFERENCES

1. Bailey, P. J., Liese, W., Roesch, R., Gunda, K., and Afting, E. G., *Biochim. Biophys. Acta, 185,* 392 (1969).
2. Bateman, D. F., and Rogowiez, J., *Phytopathology, 59,* 37 (1969).
3. Bucht, B., and Eriksson, K.-E., *Arch. Biochem. Biophys., 124,* 149 (1968).
4. Berg, B., Hofsten, B. V., and Goransson, B., *Appl. Microbiol., 16,* 1424 (1968).
5. Bjorndal, H., and Eriksson, K.-E., *Arch. Biochem. Biophys., 124,* 149 (1968).
6. Charpentier, M., *Ann. Inst. Pasteur, 109,* 171 (1965).
7. Enger, M. D., and Sleeper, B. P., *J. Bacteriol., 89,* 23 (1965).
8. Eriksson, K.-E., and Rzedowski, W., *Arch. Biochem. Biophys., 129,* 689 (1969).

9. Gilligan, W., and Reese, E. T., *Can. J. Microbiol., l,* 90 (1954).
10. Hajny, G. J., and Reese, E. T., *Cellulases and Their Applications.* American Chemical Society, Washington, D. C. (1969).
11. Iwasaki, T., Ikeda, R., Hayashi, K., and Funatsu, M., *J. Biochem.(Tokyo), 57,* 478 (1965).
12. Jothianandan, D., and Shanmugasundaram, E. R. B., *Enzymologia, 35,* 11 (1968).
13. King, N. J., *Biochem. J., 100,* 784 (1966).
14. Korkulanin, A., Hrasak, D., and Johanides, V., *Mikrobiologiya, 5,* 163 (1968).
15. Kuroda, H., and Mochizaki, T., *J. Ferment. Technol., 45,* 341 (1969).
16. Leatherwood, J. M., *Appl. Microbiol., 13,* 771 (1965).
17. Li, L. H., Flora, R. M., and King, K. W., *Arch. Biochem. Biophys., 111,* 439 (1965).
18. Loginova, L. G., and Tashpulatov, Zh., *Mikrobiologiya, 34,* 258 (1965).
19. Miyauchi, K., Iida, H., and Kurata, H., *Jap. Patent 69, 08, 068 Fpp.* (1969).
20. Pettersen, G., *Arch. Biochem. Biophys., 126,* 776 (1968).
21. Pettersen, G., Cowling, E. B., and Porath, J., *Biochim. Biophys. Acta, 67,* 1 (1963).
22. Pettersen, G., and Porath, J., *Biochim. Biophys. Acta, 67,* 9 (1963).
23. Pettersen, G., and Eacker, D. L., *Arch. Biochem. Biophys., 124,* 154 (1968).
24. Ramabhadran, R., *Indian J. Exp. Biol., 7,* 186 (1969).
25. Reese, E. T., *Advances in Enzymic Hydrolysis of Cellulose and Related Materials.* Pergamon Press, New York (1963).
26. Reese, E. T., Smakula, E., and Perlin, A. S., *Arch. Biochem. Biophys., 85,* 171 (1959).
27. Selby, K., *Biochem. J., 79,* 562 (1961).
28. Selby, K., Maitland, C. C., and Thompson, K. V. A. *Biochem. J., 88,* 288 (1963).
29. Selby, K., and Maitland, C. C., *Biochem. J., 94,* 578 (1965).
30. Selby, K., and Maitland, C. C., *Biochem. J., 104,* 716 (1967).
31. Sietsma, J. H., Eveleigh, D. E., and Haskins, R. H., *Antonie Van Leeuwenhoek J. Microbiol. Serol., 34,* 331 (1968).
32. Somkuti, G. A., Babel, F. J., and Somkuti, A. C., *Appl. Microbiol., 17,* 888 (1969).
33. Storwick, W. O., and King, K. W., *J. Biol. Chem., 235,* 303 (1960).
34. Suguira, M., Tanaka, H., Kato, S., Nagase, K., and Doke, M., *Yakuzaigaku, 26,* 69 (1966); *Chem. Abstr., 69,* 8356lg (1966).
35. Suzuki, H., Yamane, K., and Nisizawa, K., *Adv. Chem. Ser., 95,* 60 (1969).
36. Toyama, N., and Ogawa, M. K., *Hakko Kogaku Zasshi, 44,* 741 (1966); *Chem. Abstr., 69,* 49902j (1966).
37. Watanabe, T.-H, *Hakko Kogaku Zasshi, 46,* 303 (1968); *Chem. Abstr., 69,* 49364k (1968).
38. Werner, R., *J. Polymer Sci. Part C,* p. 4429 (1965).
39. Wood, T. M., *Biochem. J., 109,* 217 (1968).
40. Wood, T. M., and Phillips, D. R., *Nature, 222,* 986 (1969).

CHARACTERIZATION OF NUCLEOPHOSPHODIESTERASES

DR. JOSEPH L. POTTER

The following table summarizes some of the properties of the better-defined nucleophosphodiesterases. In many cases the described properties may be dependent upon rather restricted experimental conditions. It is now apparent that pH, nature of the metal activator, and configuration of the substrate may influence the specificity of the enzyme. Specificity may also change during different phases of the reaction. The survey is not exhaustive in scope; several excellent reviews are listed below, which should be consulted.

Laskowski, M., Sr., in *Handbook of Biochemistry,* 2nd ed., p. H-118, H. A. Sober, Ed. The Chemical Rubber Co., Cleveland, Ohio (1970).

Barnard, E. A., *Annu. Rev. Biochem., 36,* 677 (1969).

Lehman, I. R., *Annu. Rev. Biochem., 36,* 645 (1967).

Laskowski, M., Sr., *Advan. Enzymol., 28,* 165 (1967).

Deoxyribonucleases[a]

Enzyme	Products	Preferential Attack	Secondary Attack and/or Resistance	Comments	Ref.
1. Streptococcal nuclease A (streptodornase)	5'-Mononucleotides and 5'-terminated polynucleotides	Endonucleolytic; faster on native DNA; preference for pX-pG; slight preference for poly dA	Dinucleotides are resistant; tri-d poly A is resistant; Pyp-Pu poorly attacked	Streptodornase refers to one of at least four DNases produced by S. pyogenes and extensively purified by Lederle; requires M²⁺ and is not inhibited by RNA	1–6
2. Streptococcal nucleases B, C, and D	5'-Terminated products	Endonucleolytic; attack native DNA faster B: pX-pG; poly dC C: no distinct preferences D: pX-pA; poly dA	Will attack d(pA)₄, d(pA)₃, and d(pA)₂ poly G is not attacked by B, C, or D	All require M²⁺; B and D possess RNase activity and are inhibited by RNA	1–6
3. Escherichia coli endonuclease I	5'-Terminated products	Attacks duplex DNA preferentially; no distinct preference for bases	Oligonucleotides of chain length 7 or less are resistant; pA-pX is relatively resistant; inhibited by RNA	Predominantly double-strand breaks	7–9
4. Escherichia coli endonuclease II	Single-strand breaks with 5'-phosphate termini	Most active on double-stranded, alkylated DNA; single-strand cleavages in non-alkylated DNA and single- and double-strand cleavages in alkylated DNA	T₄ phage DNA is resistant; not inhibited by RNA	Does not require Mg²⁺	10, 11
5. Escherichia coli endonuclease III (restriction enzyme)	Duplexes containing little or no single-strand molecules and without single-chain breaks	DNA from cells lacking the modification allele	Duplexes with only one modified strand are not attacked	ATP, Mg²⁺, and S-adenosyl methionine are required	12–14

Enzyme	Products	Preferential Attack	Secondary Attack and/or resistance	Comments	Ref.
6. *Haemophilus influenzae* endonuclease (restriction enzyme)	5'-Terminated products	Specific for pPupApC at the 5'-break; 3' end of the break is complementary		Requires Mg^{2+}, but no ATP or S-adenoxyl methionine; double-strand breaks	15, 16
7. *Escherichia coli* endonuclease I—tRNA complex	Singly nicked circular CoIE₁ or φX174-RFI DNA	Covalently closed circular DNA	No strand specificity; inhibited by EDTA	Active in 0.5M NaCl and Mg^{2+}	17
8. *Escherichia coli* exonuclease I	5'-Mononucleotides	Attack starts at 3'-OH terminus; denatured DNA is attacked faster	Terminal di-nucleotide is resistant	Active on phage DNA containing hydroxymethyl cytosine; A-T polymer is attacked	18, 19
9. *Escherichia coli* exonuclease II	5'-Mononucleotides	Attack starts at 3'-OH terminus; duplex DNA is attacked more rapidly	5'-OH or 5'-phosphate terminus is also attacked, but more slowly	Associated with DNA polymerase	20, 21
10. *Escherichia coli* exonuclease III (phosphatase exonuclease)	5'-Mononucleotides	Attack starts at 3'-OH; if terminated in 3'-phosphate, Pi is cleaved first; preference for duplex DNA	Small oligonucleotides and single-stranded DNA are resistant		22, 23
11. *Escherichia coli* exonuclease IV (oligonucleotide diesterase)	5'-Mononucleotides	Oligonucleotides; pancreatic DNase digest is attacked 20 times faster than denatured DNA		Exonucleases IV A and IV B are physically separable, but have similar substrate specificity	24
12. Yeast endonuclease A	5'-Terminated oligonucleotides	Single-stranded DNA	Attacks duplex DNA at 1/750 the rate	Requires Mg^{2+} or Mn^{2+}; not activated by Ca^{2+}	25
13. *Proteus mirabilis* endonuclease I	Acid-soluble products	Single- or double-stranded DNA	In the presence of tRNA, degrades factor E, φX174, and SV40 DNA to 8s fragments; the latter are resistant to further degradation		26

Enzyme	Products	Preferential Attack	Secondary Attack and/or Resistance	Comments	Ref.
14. *Micrococcus lysodeickticus* endonuclease	5'-Terminated di- to pentanucleotides and some longer fragments	Attacks duplex DNA at 40 times the rate for single-strand; attacks φX174 DNA		Requires ATP or dATP for activity; dCTP is inactive	27
15. Pneumococcal endonuclease	5'-Terminated products	Acts on single- or double-stranded DNA			28
16. Pneumococcal exonuclease	5'-Mononucleotides and single-stranded DNA	Duplex DNA is the preferred substrate	Single-stranded DNA is very resistant	Has phosphatase activity against 3'-phosphates	28
17. *Haemophilus influenzae* exonuclease	5'-Mononucleotides and single-stranded DNA	Most active on sonically treated DNA or limited DNase I digests	Single-stranded DNA is very resistant	Has phosphatase activity against 3'-phosphate; requires M^{2+}	29
18. *Serratia marcescens* endonuclease	5'-Terminated products, 2% mononucleotides and di- to tetranucleotides	Attacks single and double strands and DNA and RNA at the same time	No apparent base preference	Requires M^{2+}; a single protein has both DNase and RNase activities	30
19. *Bacillus subtilis* exonuclease	3'-Terminated mononucleotides	Native and denatured DNA and RNA; attacks 3'-oligoribonucleotides from the 3'-terminus	Slower attack on non-terminated oligoribonucleotides and enhancement of attack from the 5'-end; glucosylated T_4 DNA is only partially degraded	Some endonucleolytic activity; probably a mixture of enzymes	31, 32
20. *Neurospora crassa* endonuclease	5'-Terminated products	Highly specific for denatured DNA; also attacks RNA; prefers G or dG residues	Inhibited by EDTA; removal of 5'-phosphate inhibits; 5'-terminated penta- and smaller oligonucleotides are resistant	Activated synergistically by $Mg^{2+} + Co^{2+}$; RNase and DNase are probably the function of a single protein	33

Enzyme	Products	Preferential Attack	Secondary Attack and/or Resistance	Comments	Ref.
21. T₂ phage-induced exonuclease A (oligonucleotide diesterase)	5'-Mononucleotides	Attacks DNase I digests 100 times more rapidly than denatured DNA		A similar enzyme is T₄ phage-induced	34
22. T₂ and T₄ phage-induced exonuclease	5'-Mononucleotides		Inhibited or masked in the presence of nucleoside triphosphates	Physically associated with phage-induced polymerase	35, 36
23. T₂ phage-induced endonuclease	5'-Terminated oligonucleotides	Preference for native DNA	Inhibited by tRNA		37
24. T₄ phage-induced endonuclease A	5'-Terminated products	Prefers duplex DNA; dG at 3'-terminus and dG and dA at 5'-terminus		About 1% of diester bonds are broken	38
25. T₄ phage-induced endonuclease II	5'-Terminated products; average chain length is 1000 units	dC and dG at 3'-terminus, and all 4 bases at 5'-terminus; prefers native DNA	Not inhibited by tRNA, no attack on T₄ DNA or glucosylated DNA	Single-strand breaks in native DNA are produced	39
26. T₄ phage-induced endonuclease IV	5'-Terminated oligonucleotides; chain length is 150 units	Specific for single-stranded DNA; attacks circular single-stranded phage fd DNA	No attack on single-stranded DNA containing HMC; not inhibited by tRNA		40, 41
27. T₅ phage-induced DNase	5'-Terminated oligonucleotides, average chain length 4 to 5, and 5'-mononucleotides	Attacks single- and double-stranded DNA; more mononucleotides are produced on native DNA		Both endo- and exonucleolytic	42–44

Enzyme	Products	Preferential Attack	Secondary Attack and/or Resistance	Comments	Ref.
28. T$_7$ phage-induced endonuclease I	5'-Terminated products	Prefers single-stranded DNA that is 50% degraded to acid soluble fragments of chain length 10; preference for pyrimidine residues	Produces single- and double-strand breaks in duplex DNA; limit product is duplex DNA, 125 units in length; high concentration of RNA inhibits	Acts on *Escherichia coli*, T$_4$, T$_7$, and λ phage DNA	45, 46
29. Phage SP3-induced exonuclease	90% dinucleotides, 10% trinucleotides	Prefers denatured DNA; initiates hydrolysis at 3'-OH terminus		No polymerase activity; requires Mg^{2+}	47
30. Phage λ-induced exonuclease	5'-Mononucleotides	Prefers native DNA 350X to denatured; initiates attack at 5'-terminus, with preference for a 5'-phosphate vs a 5'-OH	Attacks single-stranded DNA poorly	The enzyme attaches best to the terminus of a molecule, and less well to nicks; it remains firmly bound until the susceptible portion of the molecule is degraded; isolated in crystalline form	48–50
31. *Micrococcus lysodeickticus* endonuclease for UV-irradiated DNA	DNA containing about 10 single-strand breaks per molecule	Attacks near pyrimidine dimers in UV-irradiated single- or double-stranded DNA		Does not require Mg^{2+}	51, 52
32. *Micrococcus lysodeickticus* exonuclease for UV-irradiated DNA	Nucleotide fragments containing thymine dimers	Attacks a free end generated by the endonuclease	Does not attack duplex DNA		51, 52
33. *Lactobacillus acidophilus* exonuclease	3'-Mononucleotides	Initiates attack from the 5'-terminus; attacks both ribo- and deoxyribonucleic acid polymers; absolute requirement for a free 5'-OH terminus	Isolated with endogenous inhibitor	Has two pH optima; at pH 4 and 5, the rate is independent of chain length; at pH 7 and 8, the rate decreases with increasing chain length	53

Enzyme	Products	Preferential Attack	Secondary Attack and/or Resistance	Comments	Ref.
34. *Herpes simplex*-induced exonuclease	5'-Mononucleotides	Attack begins at 3'-terminus; attacks pancreatic DNase I digests, or denatured DNA more rapidly	Sensitive to heating at 45°C	Requires M²⁺; has no RNase activity	54, 55
35. Poxvirus-induced DNases		Alkaline exonuclease prefers duplex DNA; acid exonuclease prefers single-stranded DNA; neutral DNase prefers single-stranded DNA			56–60
36. Polyhedral virus-induced DNase	5'-Mononucleotides		Only 2% of activity at pH 7.0	Alkaline pH is optimal	61–63
37. Micrococcal nuclease	3'-Mono- and poly-nucleotides	Prefers denatured DNA to duplex DNA, prefers Np-Ap+Np-Tp in native DNA and random attack in denatured DNA	First endonucleolytic, then exonucleolytic; three distinct phases of digestion, with different requirements for Ca²⁺	Requires Ca²⁺; endo- and exonucleolytic autoacceleration effects	64–69
Ribonucleases					
38. *Bacillus pumillis, Mucor genevensis, Aspergillus* sp. ribonucleases	3'-Terminated fragments	Specific for G residues			70
39. *Bacillus cereus, Lenzites tenius, Monascus pilosus, Aspergillus* sp. ribonucleases	3'-Terminated mononucleotides	No distinct preference for bases	Can attack trinucleotides from either end		70

Enzyme	Products	Preferential Attack	Secondary Attack and/or Resistance	Comments	Ref.
40. *Escherichia coli* RNase I (endonuclease)	3'-Terminated mononucleotides	Cyclic A and cyclic C are attacked faster; all bonds are attacked	Inhibited by Mg^{2+}	Associated with ribonucleoprotein	71–73
41. *Escherichia coli* RNase II (exonuclease)	5'-Terminated mononucleotides	Specific for single-stranded RNA; no base preference	Short-chain oligonucleotides are resistant; inhibited by M urea	Requires K^+ and M^{2+}	74–76
42. *Escherichia coli* RNase III (endonuclease)	3'-Terminated mononucleotides and polynucleotides	Specific for duplex RNA; digests the RNA strand of DNA-RNA hybrids	No attack on single-strand RNA, or on double-stranded DNA	Requires M^+ and M^{2+} loosely bound to ribosomes	77
43. *Escherichia coli* RNase IV (endonuclease)	3'-Terminated or cyclic phosphate-terminated fragment	Specific for site about 1/3 the distance in from the 5'-terminus	Attacks phage R17 and phage M52 RNAs		78–80
44. *Escherichia coli* RNase V (exonuclease)		Heated rRNA, heated or fragmented phage R17 RNA	Native rRNA and intact phage R17 RNA	Probably initiates attack from the 5'-terminus	81, 82
45. *Ustilago sphaerogena* RNase	3'-Terminated mono-, di-, and trinucleotides	Pu-Pu and Pu-Py bonds		Intermediate cyclic phosphate is formed	83, 84
46. Ribonuclease U_4 (RNA-induced) exonuclease	3'-Mononucleotides	Initiates attack from the 5'-terminus		Cyclic phosphates are formed as intermediates; the level and type of RNases produced is dependent on the nature of the inducer	85
47. Ribonuclease T_1, *Aspergillus oryzae*	3'-Mono, di, tri-, and tetra-nucleotides	Gp-Xp is preferred		2',3'-Cyclic phosphate is formed as an intermediate	86–88
48. Ribonuclease T_2, *Aspergillus oryzae*	3'-Mononucleotides	Preference for A residues		2',3'-Cyclic phosphate is formed as an intermediate	89, 90

Enzyme	Products	Preferential Attack	Secondary Attack and/or Resistance	Comments	Ref.
49. Ribonuclease H	Single strands	Specific for DNA-RNA hybrids	Poly rU; poly rA is completely resistant		91
50. *Lactobacillus plantarum* exonuclease II	5'-Mononucleotides	Specific for single-stranded RNA; degrades the single chain to completion prior to attack on another chain	Oligonucleotides are resistant	Requires M^{2+} and M^+	92
51 Streptococcal endonuclease B	5'-Mononucleotides	Poly A and poly C; requires 6-NH_2 group	Does not attack poly I or poly U; degrades deaminated RNA very slowly	Requires M^{2+}; DNase and RNase reside in a single protein; shows auto retardation	1–6
52. Streptococcal endonuclease D	5'-Mononucleotides	Poly C; slower attack than on poly U or poly A	Does not attack poly I	Requires M^{2+}; DNase and RNase reside in a single protein; shows autoretardation	1–6
53. *Bacillus subtilis* RNase	3'-Terminated mono- and dinucleotides	Prefers Gp-Gp and Gp-Ap linkages	Inhibited by $0.1M$ sodium phosphate at pH 8.5; other base sequences are attacked much more slowly; terminal sequence Xp-Gp is resistant	A cyclic phosphate intermediate is formed	93
54. *Lactobacillus casei* phosphodiesterase	5'-Terminated mononucleotides	18 to 24 residue polynucleotides accumulate early in the digestion	*Escherichia coli* and *L. casei* ribosomal and soluble RNAs are attacked	Requires K^+; exo- and endo-nucleolytic	94
55. *Azotobacter agilis* endonuclease	5'-Terminated di-, tri-, and tetra-nucleotides	Attacks longer chains more rapidly; M^{2+} inhibits and urea stimulates activity against yeast RNA	Poly U and poly C are attacked slowly, poly G is resistant; terminal phosphodiester bonds are resistant	The enzyme attacks DNA to produce 5'-terminated products	95

Enzyme	Products	Preferential Attack	Secondary Attack and/or Resistance	Comments	Ref.
56. *Salmonella typhimurium* endonuclease	2',3'-Cyclic nucleotides	Prefers single-stranded RNA; poly A, poly C, and poly U are attacked faster than transfer RNA	Poly I is not attacked; cyclic nucleotides are slowly converted to 3'-mononucleotides		96
57. *Streptomyces aureofaciens* endonuclease	3'-Terminated products	Specific for Gp-Xp; can degrade poly G or poly I	No attack on poly A, poly C, or poly U	A cyclic phosphate intermediate is formed	97
58. *Streptomyces erythreus* ribonuclease	3'-Terminated products	Prefers Gp-Xp	A, C, and U cyclic phosphates are resistant		98

[a] The abbreviations follow the recommendations of the IUPAC-IUB Combined Commission on Biochemical Nomenclature as summarized in *Biochemistry*, 5, 1445 (1966). M^+ = metal$^+$ and M^{2+} = metal^{++}, without specific designation of the metal.

REFERENCES

1. Potter and Laskowski, Sr., *J. Biol. Chem., 234,* 1263 (1959).
2. Georgatsos, Unterholzner and Laskowski, Sr., *J. Biol. Chem., 237,* 2626 (1962).
3. Winter and Bernheimer, *J. Biol. Chem., 239,* 215 (1964).
4. Yasmineh, Gray and Wannamaker, *Biochemistry, 7,* 91 (1968).
5. Yasmineh and Gray, *Biochemistry, 7,* 105 (1968).
6. Stone and Burton, *Biochem. J., 83,* 492 (1962).
7. Lehman, Roussos and Pratt, *J. Biol. Chem., 237,* 819 (1962).
8. Lehman, Roussos and Pratt, *J. Biol. Chem., 237,* 829 (1962).
9. Studier, *J. Mol. Biol., 11,* 373 (1965).
10. Friedberg and Goldthwait, *Proc. Natl. Acad. Sci. U.S.A., 62,* 934 (1969).
11. Friedberg and Goldthwait, *Cold Spring Harbor Symp. Quant. Biol., 33,* 271 (1968).
12. Meselson and Yuan, *Nature, 217,* 1110 (1968).
13. Roulland-Doussoix and Boyer, *Biochim. Biophys. Acta, 195,* 219 (1969).
14. Arker and Linn, *Annu. Rev. Biochem., 38,* 467 (1969).
15. Smith and Wilcox, *J. Mol. Biol., 51,* 379 (1970).
16. Kelly and Smith, *J. Mol. Biol., 51,* 393 (1970).
17. Goebel and Helinski, *Biochemistry, 9,* 4793 (1970).
18. Lehman, *J. Biol. Chem., 239,* 1479 (1960).
20. Richardson, Schildkraut, Aposhian and Kornberg, *J. Biol. Chem., 239,* 222 (1964).
21. Lehman and Richardson, *J. Biol. Chem., 239,* 233 (1964).
22. Richardson and Kornberg, *J. Biol. Chem., 239,* 242 (1964).
23. Richardson, Lehman and Kornberg, *J. Biol. Chem., 239,* 251 (1964).
24. Jorgensen and Koerner, *J. Biol. Chem., 241,* 3090 (1966).
25. Pinon, *Biochemistry, 9,* 2839 (1970).
26. Goebel and Helinski, *J. Biol. Chem., 246,* 3851, 3857 (1971).
27. Anai, Hirahashi, Yamanaka and Takogi, *J. Biol. Chem., 245,* 775, 781 (1970).
28. Lacks and Greenberg, *J. Biol. Chem., 242,* 3108 (1967).
29. Gunther and Goodgal, *J. Biol. Chem., 245,* 5341 (1970).
30. Nestle and Roberts, *J. Biol. Chem., 244,* 5213, 5219 (1969).
31. Kerr, Chien and Lehman, *J. Biol. Chem., 242,* 2700 (1967).
32. Okazaki, Okazaki and Sakable, *Biochem. Biophys. Res. Commun., 22,* 611 (1966).
33. Linn and Lehman, *J. Biol. Chem., 240,* 1287 (1965); *241,* 2694 (1966).
34. Gleson and Koerner, *J. Biol. Chem., 239,* 2935 (1964).
35. Goulian, Lucas and Kornberg, *J. Biol. Chem., 243,* 627 (1968).
36. Short and Koerner, *Proc. Natl. Acad. Sci. U.S.A., 54,* 595 (1965).
37. Base and Nossal, *Fed. Proc., 23,* 272 (1964).
38. Ando, Takogi, Kosawa and Ikeda, *J. Biochem. (Tokyo), 66,* 1 (1969).
39. Hurwitz, Belker, Gefter and Gold, *J. Cell. Physiol. Suppl., 70,* 181 (1967).
40. Sadowski, Ginsberg, Yudelevich, Ferner and Hurwitz, *Cold Spring Harbor Symp. Quant. Biol., 33,* 165 (1968).
41. Sadowski and Hurwitz, *J. Biol. Chem., 244,* 6182, 6192 (1969).
42. Pfefferkorn and Amos, *Virology, 6,* 299 (1958).
43. Paul and Lehman, *J. Biol. Chem., 241,* 3441 (1966).
44. Crawford, *Virology, 7,* 359 (1959).
45. Sadowski, *J. Biol. Chem., 246,* 209 (1971).
46. Center and Richardson, *J. Biol. Chem., 245,* 6285, 6292 (1970).
47. Trilling and Aposhian, *Proc. Natl. Acad. Sci. U.S.A., 54,* 622 (1965); *60,* 214 (1968).
48. Carter and Radding, *J. Biol. Chem., 246,* 2502 (1971); *238,* 3390 (1963).
49. Little, Lehman and Karsen, *J. Biol. Chem., 242,* 672, 679 (1967).
50. Weissbach and Korn, *J. Biol. Chem., 237,* 3312 (1962); *238,* 3390 (1963).
51. Takagi, Sekiguchi, Okubo, Nakayama, Shimada, Yasuda, Nishimoto and Yashihara, *Cold Spring Harbor Symp. Quant. Biol., 33,* 219 (1968).
52. Grossman, Kaplan, Kushner and Mabler, *Cold Spring Harbor Symp. Quant. Biol., 33,* 229 (1968).
53. Fiers and Khorana, *J. Biol. Chem., 238,* 2780, 2789 (1963).
54. Keir and Gold, *Biochem. J., 87,* 25 (1963).
55. Russell, Gold, Keir, Omura, Watson and Wildy, *Virology, 22,* 103 (1964).
56. Jungwuth and Joklik, *Virology, 27,* 80 (1965).
57. McAuslan, *Biochem. Biophys. Res. Commun., 19,* 15 (1965).
58. Eron and McAuslan, *Biochem. Biophys. Res. Commun., 22,* 518 (1966).
59. McAuslan and Kates, *Proc. Natl. Acad. Sci. U.S.A., 55,* 1581 (1966).
60. McAuslan, Herde, Pett and Ross, *Biochem. Biophys. Res. Commun., 20,* 586 (1965).
61. Yamofuji and Yoshihara, *Nature, 196,* 1340 (1962).

62. Yamofuji, Yoshihara and Hirayami, *Enzymologia, 19,* 53 (1958).
63. Mukai and Yamafuji, *Enzymologia, 23,* 214 (1961).
64. Sulkowski and Laskowski, Sr., *J. Biol. Chem., 237,* 2620 (1962).
65. Rushizky, Knight, Roberts and Dekker, *Biochem. Biophys. Res. Commun., 2,* 153 (1960).
66. Reddi, *Biochim. Biophys. Acta, 47,* 47 (1961).
67. Cunningham, Catlin and Privat de Ganlhe, *J. Am. Chem. Soc., 78,* 4642 (1956).
68. Sulkowski and Laskowski, Sr., *J. Biol. Chem., 241,* 4386 (1966).
69. Von Hippel and Felsenfeld, *Biochemistry, 3,* 27 (1964).
70. Rushizky, Greco, Hartley and Sober, *J. Biol. Chem., 239,* 2165 (1964).
71. Spahr and Hollingworth, *J. Biol. Chem., 236,* 823 (1961).
72. Spahr, in *Procedures in Nucleic Acid Research,* p. 64, Cantoni and Davies, Eds., Harper and Row, New York (1966).
73. Heppel, *Science, 156,* 1451 (1967).
74. Spahr, *J. Biol. Chem., 239,* 3716 (1964).
75. Singer and Tolbert, *Biochemistry, 4,* 139 (1965).
76. Spahr and Schlessinger, *J. Biol. Chem., 238,* 225 (1963).
77. Robertson, Webster and Zinder, *J. Biol. Chem., 243,* 82 (1968).
78. Spahr and Gesteland, *Proc. Natl. Acad. Sci. U.S.A., 59,* 876 (1968).
79. Min Jou, Fiers, Goodman and Spahr, *J. Mol. Biol., 42,* 143 (1969).
80. Min Jou and Fiers, *J. Mol. Biol., 40,* 187 (1969).
81. Kuwano, Apirion and Schlessinger, *Science, 168,* 1225 (1970).
82. Kuwano, Kwan, Apirion, and Schlessinger, *Proc. Natl. Acad. Sci. U.S.A., 64,* 693 (1969).
83. Arima, Uchida and Egami, *Biochem. J., 106,* 601, 609 (1968).
84. Uchida, Arima and Egami, *J. Biochem. (Tokyo), 67,* 91 (1970).
85. Blank, Holloman and Dekker, *Fed. Proc. (Abstr.), 30,* 1093 (1971).
86. Sato and Egami, *J. Biochem. (Tokyo), 44,* 753 (1957).
87. Takahashi, *J. Biochem. (Tokyo), 49,* 1 (1961).
88. Rushizky and Sober, *J. Biol. Chem., 237,* 2883 (1962).
89. Naoi-Toda, Sato, Asano and Egami, *J. Biochem. (Tokyo), 46,* 757 (1959).
90. Rushizky and Sober, *J. Biol. Chem., 238,* 371 (1963).
91. Hausen and Stein, *Eur. J. Biochem., 14,* 278 (1970).
92. Logan and Singer, *J. Biol. Chem., 243,* 6161 (1968).
93. Rushizky, Greco, Hartley and Sober, *Biochem. J., 2,* 787, 794 (1963).
94. Keir, Mathog and Carter, *Biochemistry, 3,* 1188 (1964).
95. Stevens and Hilmoc, *J. Biol. Chem., 235,* 3016, 3023 (1968).
96. Chakabartly and Burmer, *J. Biol. Chem., 243,* 1133 (1968).
97. Zelinkova, Baiova and Zelinka, *Biochim. Biophys. Acta, 235,* 335, 343 (1971).
98. Tanaka, in *Procedures in Nucleic Acid Research,* Vol. 2, p. 14, Cantoni and Davies, Eds., Harper and Row, New York (1971).

MICROBIAL RNASES

DR. TSUNEKO UCHIDA and DR. FUJIO EGAMI*

Source and Name of Enzyme	pH Optima	Heat Stability	Molecular Weight	Specificity†	Other Information	Reference
Bacteria						
Azotobacter						
agilis	7.5	Yes, in acid		tra, endo, →N>p, non-s	Intracellular RNase; solubilization and purification; effectors; activation by Ca²⁺	1
Bacillus						
cereus	7.9		30,000–40,000	→Np, non-s	Purification; inactivation by EDTA	2
pumilus	7.9		10,000–15,000	→Np, g-s	Purification; no inactivation by EDTA	2
subtilis H (amyloliquefaciens)	7.5–8.5	Yes	10,700	tra, endo, →N>p →Np, non-s	Crystallization; relative specificity; amino acid composition^a	3–5
subtilis Marburg	5.0	No		tra, endo, →N>p →Np, non-s, hydrolysis very slow	Partial purification	6
subtilis (intracellular)	5.8			tra, endo, →N>p, non-s	Inhibition by EDTA; effects of nucleotides^a	7–10

* With the collaboration of Dr. E. Ohmura.

† Abbreviations: tra = transphosphorylation; hyd = direct hydrolysis; > p = 2′,3′-cyclic phosphate; endo = endonucleolytic; exo = exonucleolytic; non-s = non-specific; g-s = guanine-specific; pur-s = purine-specific; pyr-s = pyrimidine-specific.

Source and Name of Enzyme	pH Optima	Heat Stability	Molecular Weight	Specificity	Other Information	Reference
Bacteria (continued)						
Clostridium						
acetobutylicum						
RNase II	4.5			endo, →Np	Enzyme formation; purification; effectors	11
Escherichia						
coli						
RNase I	8.1	Yes, in acid		tra, endo, →N>p →Np, non-s	Purification; basic protein; relative specificity; intracellular localization; activation by Mg^{2+}; compare with related enzymes	12–14
RNase II	7–8	No	65,000	hyd, exo, →pN, non-s	[a]	15–17
RNase III	7.6–9.7			endo, double-stranded specific →Np and pN	Requirement for Mg^{2+} and K$^+$ or NH$_4^{+}$[a]	18–20
RNase IV					[a]	21–23
RNase V					[a]	21–23
Lactobacillus						
casei						
RNase II	8.1–8.2 (7.4)			endo, exo, →pN	Purification; activation by K$^+$	24
plantarum						
RNase II	8.6			hyd, exo, →pN	Purification; monovalent and divalent cations required; single-stranded preferred	25

Source and Name of Enzyme	pH Optima	Heat Stability	Molecular Weight	Specificity	Other Information	Reference
Bacteria (continued)						
Myxobacterium						
avium	7.5	Yes		pur-s or pur-preferential	Purification	26
Proteus						
mirabilis						
RNase II	7.0			tra, →N>p, non-s; CpN bonds are fairly resistant	Purification	27
Salmonella						
typhimurium	7.0	Not very stable		tra, endo, →N>p →Np, non-s; hydrolysis very slow; poly I is resistant	Intracellular enzyme; purification; Na$^+$, K$^+$, and Mg^{2+} have no effect; tRNA is fairly resistant	28
Thiobacillus						
thioparus						
IIA	7.0	Yes		endo, →Np, pyr-s	Purification	29, 30
IIB	9.5	Yes		endo, →Np, pyr-s		
IA-2	5.5	Yes				
Actinomycetes						
Actinomyces						
aureoventicillatus	7.6—7.8		13,000	tra, endo, →N>p →Np, g-s	Purification; specificity for methylated guanylates; splitting rate of GpNp: GpUp > GpCp	31, 32

Source and Name of Enzyme	pH Optima	Heat Stability	Molecular Weight	Specificity	Other Information	Reference
Actinomycetes (continued)						
Streptomyces						
albogriseolus	7.0—8.5	Yes, in acid		tra, endo, →N>p →Np, g-s	Purification; effectors	33
erythreus	7.3—7.4	Yes		tra, endo, →N>p →Np, g-s	Purification; relative specificity for methylated guanylates	34
Yeasts						
Endomyces	4.5			→Np		35
Rhodotorula						
glutinis						
I	7			→pN, non-s	Relation between enzyme formation and culture conditions	36—38
II	4			→Np, non-s		
III	7			→Np, non-s		
Saccharomyces						
cerevisiae (baker's yeast)	7.5	No		hyd, exo, →Np, non-s	Intracellular enzyme; purification; effectors; activated by phosphate and inhibited by Zn^{2+}	39
Fungi[b]						
Acrocylindrium sp.	8.0	Yes		tra, endo, →N>p, g-s		40

Source and Name of Enzyme	pH Optima	Heat Stability	Molecular Weight	Specificity	Other Information	Reference
Fungi (continued)						
Aspergillus						
niger	3.0–3.5		*c*	→Np, non-s	Purification, inhibitors	41–43
oryzae						
RNase T$_1$	7.4–7.5	Yes	11,000	tra, endo, →N>p →Np, g-s	*a*	72
RNase T$_2$	4.5	Yes	36,000	tra, endo, →N>p →Np, non-s	*a*	72
saitoi						
RNase M	4.0	Yes	30,000	tra, endo, →N>p →Np, non-s	Purification; inhibition by Zn^{2+} and Cu^{2+}; inhibition by nucleotides; effectors; kinetic data; photooxidation	44–48
Chalaropsis						
species						
RNase Ch				tra, endo, →N>p →Np, g-s		49
Lenzites						
tenuis	7.9		30,000–40,000	→Np, non-s	Inhibited by EDTA	2
Monascus						
pilotus	4.5		30,000–40,000	→Np, non-s	Not inhibited by EDTA	2
Mucor						
genevensis	7.9		12,000	→Np, g-s	Scarcely inhibited by EDTA	2

Source and Name of Enzyme	pH Optima	Heat Stability	Molecular Weight	Specificity	Other Information	Reference
Fungi[b] **(continued)**						
Neurospora						
crassa	7.5	Yes			Genetics; inhibited by EDTA	54, 55
RNase N_1	7.0	Yes	11,000	tra, endo, →N>p →Np, g-s	Formation[a]	50—53
RNase N_2	8.0		36,000	tra, endo, →N>p →Np, non-s	Formation	51, 52
RNase N_3	6—7		A little larger than N_1	tra, endo, →N>p →Np, g-s	Intracellar enzyme; formation	51,52
Physarum						
polycephalum						
RNase PP_1	6.7	Yes	40,000	tra, endo, →N>p →Np, g-s	Inhibitors; purification	56
RNase PP_2	4.5	Yes	40,000	tra, endo, →N>p →Np, non-s	Inhibitors; purification	56
RNase PP_3	5.5	Yes	10,000	tra, endo, →N>p →Np, non-s	Inhibitors; purification	56
RNase PP_4	4.0	No		hyd, endo, →pN	Inhibitors; purification	57
Pleospora	7.5			tra, endo, →N>p →Np, pur-s	Induced formation; inhibition by phosphate; Michaelis constant	58-60
Rhizopus RNase	5.0	Fairly stable	*d*	endo, →Np, non-s, pur-preferential	Crystallization; inhibition by Zn^{2+} and Cu^{2+}; no inhibition by EDTA	61

Source and Name of Enzyme	pH Optima	Heat Stability	Molecular Weight	Specificity	Other Information	Reference
Fungi[b] **(continued)**						
Trichoderma						
koningi						
I and II	4.5		25,000	tra, endo, →N>p →Np, non-s	Similar to RNase T$_2$	62
III	4.5		10,000	tra, endo, →N>p →Np, pur-preferential	Activated by Mg^{2+}	62
Ustilago						
sphaerogena						
RNase U$_1$	8.0–8.5	Yes	11,000	tra, endo, →N>p →Np, g-s	Induced formation[a]	63–66
RNase U$_2$	4.5	Yes	10,000	tra, endo, →N>p →Np, pur-s	[a]	63, 64, 67
RNase U$_3$	4.5	Yes	10,000	tra, endo, →N>p →Np, pur-s		63, 64
RNase U$_4$	8.0–8.5	No	Much larger than U$_1$ and U$_2$	hyd, exo, →Xp, non-s	Induced formation	63, 64
zea	8.9	No		tra, endo, →N>p →Np, g-s	Induced formation	68
Protozoa						
Euglena						
gracilis	4.5	No		non-s		69

Source and Name of Enzyme	pH Optima	Heat Stability	Molecular Weight	Specificity	Other Information	Reference
Protozoa (continued)						
Paramecium						
aurelia						
I	6.5	Yes (mixture of I and II)		non-s	Inhibitors; inhibited by Hg^{2+}; resistant to other metal ions	70
II	5.5					
Tetrahymena						
pyriformis						
I, II, and III	5.0			tra, endo, $\rightarrow N > p$ $\rightarrow Np$; relative specificity for I, II and III differs	$4M$ urea increases the activity	71

[a] Further information regarding this enzyme can be found in Reference 72.
[b] Including a slime mold.
[c] $S_{20,w} = 2.59$ S.
[d] $S_{20,w} = 2.42$ S.

Data taken from: Uchida, T., and Egami, F., in *The Enzymes*, 3rd ed., Vol. 4, pp. 244–248, P. D. Boyer, Ed. (1971). Reproduced by permission of Academic Press, New York.

REFERENCES

1. Shiio, I., Ishii, K., and Shimizu, S., *J. Biochem. (Tokyo), 59,* 363 (1966).
2. Rushizky, G. W., Greco, A. E., Hartley, R. W., Jr., and Sober, H. A., *J. Biol. Chem., 239,* 2165 (1964).
3. Nishimura, S., in *Procedures in Nucleic Acid Research,* Vol. 1, p. 56, G. L. Cantoni and D. R. Davies, Eds. Harper and Row, New York (1966).
4. Whitfeld, P. R., and Witzel, H., *Biochem. Biophys. Acta, 72,* 362 (1963).
5. Rushizky, G. W., Greco, A. E., Hartley, R. W., Jr., and Sober, H. A., *Biochemistry, 2,* 787 (1963).
6. Nikai, M., Minami, I., Yamasaki, T., and Tsugita, A., *J. Biol. Chem., 57,* 96 (1965).
7. Nishimura, S., and Maruo, B., *Biochem. Biophys. Acta, 40,* 355 (1960).
8. Yamasaki, M., and Arima, K., *Biochem. Biophys. Acta, 139,* 202 (1967).
9. Yamasaki, M., and Arima, K., *Biochem. Biophys. Res. Commun., 37,* 430 (1969).
10. Saito, M., Furuichi, Y., Takeishi, K., Yoshida, M., Yamasaki, M., Arima, K., Hayatsu, H., and Ukita, T., *Biochem. Biophys. Acta, 195,* 299 (1969).
11. Tomoyeda, M., Horitsu, H., and Kumagai, K., *Res. Bull. Fac. Agric., Gifu Univ., 28,* 153 (1969).
12. Spahr, P. F., and Hollingworth, B. R., *J. Biol. Chem., 236,* 823 (1961).
13. Anraku, Y., and Mizuno, D., *Biochem. Biophys. Res. Commun., 18,* 462 (1965).
14. Anraku, Y., and Mizuno, D., *J. Biochem. (Tokyo), 61,* 70 (1967).
15. Spahr, P. F., *J. Biol. Chem., 239,* 3716 (1964).
16. Singer, M. F., and Tolbert, G., *Biochemistry, 4,* 1319 (1965).
17. Nossal, N. G., and Singer, M. F., *J. Biol. Chem., 243,* 915 (1968).
18. Robertson, H. D., Webster, R. E., and Zinder, N. D., *Virology, 32,* 718 (1967).
19. Robertson, H. D., Webster, R. E., and Zinder, N. D., *J. Biol. Chem., 243,* 82 (1968).
20. Libonati, M., *Boll. Soc. Ital. Biol. Sper., 44,* 786, 789 (1968).
21. Spahr, P. F., and Gesteland, R. F., *Proc. Natl. Acad. Sci. U.S.A., 59,* 876 (1968).
22. Kuwano, M., Ning Kwan, C., Apirion, D., and Schlessinger, D., *Proc. Natl. Acad. Sci. U.S.A., 64,* 693 (1968).
23. Hamida, F. B. and William, F. R., *Bull. Soc. Chim. Biol., 51,* 1545 (1969).
24. Keir, H. M., Mathog, R. H., and Carter, C. E., *Biochemistry, 3,* 1188 (1964).
25. Logan, D. M., and Singer, M. F., *J. Biol. Chem., 243,* 6161 (1968).
26. Tsugita, A., and Matsui, K., *Seikagaku, 41,* 588 (1969).
27. Center, M. S., and Behal, F. J., *Biochem. Biophys. Acta, 151,* 698 (1968).
28. Chakraburtty, K., and Burma, D. P., *J. Biol. Chem., 243,* 1133 (1968).
29. Ostrowski, W., and Walczak, Z., *Acta Biochim. Pol., 8,* 345 (1961).
30. Walczak, Z., and Ostrowski, W., *Acta Biochim. Pol., 11,* 241 (1964).
31. Abrosimova-Amelyanchik, N. H., Tatarskaya, R. I., Venkstern, T. V., Aksel'rod, V. D., and Bayev, A. A., *Biokhimiya, 30,* 1269 (1965); *Biochemistry (English Transl.), 30,* 1086 (1965).
32. Tatarskaya, R. I., Abrosimova-Amelyanchik, N. H., and Aksel'rod, V. D., *Biokhimiya, 31,* 1017 (1966); *Biochemistry (English Transl.), 31,* 882 (1966).
33. Yoneda, M., *J. Biochem. (Tokyo), 55,* 469 (1964).
34. Tanaka, K., in *Procedures in Nucleic Acid Research,* Vol. 1, p. 14, G. L. Cantoni and D. R. Davies, Eds. Harper and Row, New York (1966).
35. Hattori, T., and Nakamura, S., *Seikagaku, 38,* 563 (1966).
36. Nakao, Y., and Ogata, K., *Agric. Biol. Chem., 27,* 116 (1963).
37. Nakao, Y., and Ogata, K., *Agric. Biol. Chem., 27,* 499 (1963).
38. Nakao, Y., and Ogata, K., *Agric. Biol. Chem., 27,* 507 (1963).
39. Ohtaka, Y., Uchida, K., and Sakai, T., *J. Biochem. (Tokyo), 54,* 322 (1963).
40. Suhara, I., Kusaka, F., and Ohmura, E., *Koso Kagaku Shimpoziumu, 16,* 115 (1964).
41. Eto, Y., Goto, Y., and Tomoyeda, M., *Nippon Nogei Kagaku Taikai Abstr.,* p. 30 (1969).
42. Azuma, Y., Horitsu, H., and Tomoyeda, M., *Nippon Nogei Kagaku Taikai Abstr.,* p. 30 (1969).
43. Horitsu, H., Okamoto, K., Azuma, Y., and Tomoyeda, M., *Nippon Nogei Kagaku Taikai Abstr.,* p. 30 (1969).
44. Imazawa, M., Irie, M., and Ukita, T., *J. Biochem. (Tokyo), 64,* 595 (1968).
45. Irie, M., *J. Biochem. (Tokyo), 62,* 509 (1967).
46. Irie, M., *J. Biochem. (Tokyo), 65,* 133 (1969).
47. Irie, M., *J. Biochem. (Tokyo), 66,* 569 (1969).
48. Irie, M., *J. Biochem. (Tokyo), 66,* 907 (1969).
49. Hash, J. H., and Elsevier, E., *Science, 162,* 681 (1968).
50. Takai, N., Uchida, T., and Egami, F., *Biochem. Biophys. Acta, 128,* 218 (1966).
51. Takai, N., Uchida, T., and Egami, F., *Seikagaku, 39,* 473 (1967).
52. Takai, N., Uchida, T., and Egami, F., *Seikagaku, 39,* 285 (1967).
53. Kasai, K., Uchida, T., Egami, F., Yoshida, K., and Nomoto, M., *J. Biochem. (Tokyo), 66,* 389 (1969).
54. Suskind, S. R., and Bonner, D. M., *Biochem. Biophys. Acta, 43,* 173 (1960).
55. Ishikawa, T., Toh-e, A., Uno, I., and Hasunuma, K., *Genetics, 63,* 75 (1969).

686 *Handbook of Microbiology*

56.   Hiramaru, M., Uchida, T., and Egami, F., *J. Biochem. (Tokyo), 65,* 697 (1969).
57.   Hiramaru, M., Uchida, T., and Egami, F., *J. Biochem. (Tokyo), 65,* 701 (1969).
58.   Cuchillo, C. M., Ventura, J. M., Concustell, E., and Villar-Palasi, V., *Rev. Esp. Fisiol., 23,* 81 (1967).
59.   Cuchillo, C. M., Ventura, J. M., Concustell, E., and Villar-Palasi, V., *Rev. Esp. Fisiol., 23,* 87 (1967).
60.   Cuchillo, C. M., Ventura, J. M., Concustell, E., and Villar-Palasi, V., *Rev. Esp. Fisiol., 23,* 93 (1967).
61.   Tomoyeda, M., Eto, Y., and Yoshino, T., *Arch. Biochem. Biophys., 131,* 191 (1969).
62.   Hamada, M., and Irie, M., *Seikagaku, 41,* 587 (1969).
63.   Arima, T., Uchida, T., and Egami, F., *Biochem. J., 106,* 601 (1968).
64.   Arima, T., Uchida, T., and Egami, F., *Biochem. J., 106,* 609 (1968).
65.   Glitz, D. G., and Dekker, C. A., *Biochemistry, 3,* 1391 (1964).
66.   Glitz, D. G., and Dekker, C. A., *Biochemistry, 3,* 1399 (1964).
67.   Uchida, T., Arima, T., and Egami, F., *J. Biochem. (Tokyo), 67,* 91 (1970).
68.   Yanagida, M., Uchida, T., and Egami, F., *Nippon Nogei Kagaku Kaishi, 38,* 531 (1964).
69.   Fellig, J., and Wiley, C. E., *Science, 132,* 1835 (1960).
70.   Gross, G., Skoczylas, B., and Tunski, W., *Acta Protozool., 4,* 59 (1966).
71.   Lazarus, L. H., and Scherbaum, O. H., *Biochem. Biophys. Acta, 142,* 368 (1967).
72.   Uchida, T., and Egami, F., in *The Enzymes,* 3rd ed., Vol. 4, p. 205, P. D. Boyer, Ed. Academic Press, New York (1971).

NUCLEIC ACID-METABOLIZING ENZYMES ASSOCIATED WITH VIRIONS OF ANIMAL VIRUSES

DR. SATOSHI MIZUTANI

All enzyme activities related to nucleic acid metabolism found so far in the virions of animal viruses are included in the following table. Other enzyme activities, such as protein kinase, protease, etc., are not included. The enzyme activities listed may not be necessary for virus replication.

Enzyme	Virus	Inhibitor[a]	Reference

DNA polymerase	Leukovirus group[b]
	  Avian
	  Murine
	  Feline
	  Viper
	  Murine mammary tumor
	  Primate
		Actinomycin D, rifampicin derivatives, ethidium bromide, poly-U, etc.	1

	Visna		1
	Syncytium-forming		1
	Picodnavirus group Kilham rat	p-Chloromercuribenzoate	2

	Poxvirus group
	  Vaccinia virus
	  Frog virus 3
			6
			6

RNA polymerase			

DNA-dependent	Poxvirus group
	  Vaccinia virus
	  Yaba tumor
			3, 4
			5

RNA-dependent	Diplornavirus group
	  Reovirus
	  Cytoplasmic polyhydrosis
		Mercaptoethanol	7, 8
			9

	Orthomyxovirus group
	  Influenza
		Polyanions
		  Polyvinylsulfate
		  Polydextransulfate
			10, 11

	Paramyxovirus group
	  Newcastle disease
	  Sendai virus
			12
			13

	Rhabdovirus group Vesicular stomatitis		14
Nuclease			
Deoxyribonuclease			
Exonuclease	Leukovirus group Rous sarcoma		1 15
	Poxvirus group Vaccinia virus	Inorganic phosphates, tRNA	19

Enzyme	Virus	Inhibitor[a]	Reference
Nuclease (*cont.*)			
Deoxyribonuclease (*cont.*)			
Endonuclease	Leukovirus group		1
	Rous sarcoma		16
	Avian myeloblastosis		17, 18
	Rauscher leukemia		17
	Poxvirus group		
	Vaccinia virus	Inorganic phosphates, tRNA	19
	Yaba tumor		5
	Adenovirus group		
	Adenovirus type 2	Actinomycin D, tRNA, dCMP, etc.	20
	Papovavirus group		
	Polyoma virus		21
	Simian virus 40		37
Ribonuclease	Leukovirus group		1
	Rous sarcoma		22
	Avian myeloblastosis		23
	Orthomyxovirus group		38
	Influenza		
	Fowl plague		
	Paramyxovirus group		38
	Newcastle disease		
	Sendai virus		
Ribonuclease H	Leukovirus group		
	Avian myeloblastosis	2,5-Dimethyl-4-N-benzyldemethyl	24, 39
	Rous sarcoma	rifampicin	40
	Murine leukemia, etc.		
Polynucleotide ligase	Leukovirus group		
	Rous sarcoma		15
	Avian myeloblastosis		17
	Rauscher leukemia		17
Nucleotide phosphohy- drase	Leukovirus group		
	Rous sarcoma		25
	Avian myeloblastosis		25–27
	Rauscher leukemia		27
	Diplornavirus group		
	Reovirus	Inorganic pyrophosphates, nucleo- side diphosphates	28, 29
	Orthomyxovirus group		
	Influenza		27
	Rhabdovirus group		
	Vesicular stomatitis		27
	Poxvirus group		
	Vaccinia virus	*p*-Hydroxymercuribenzoate	30, 31
	Yaba tumor		5
	Frog virus 3	Inorganic phosphates	32

Enzyme	Virus	Inhibitor[a]	Reference
Nucleotide phospho-transferase			
Nucleoside triphosphate phosphotransferase	Leukovirus group		27
	Avian myeloblastosis		
	Rauscher leukemia		
	Orthomyxovirus group		27
	Influenza		
	Rhabdovirus group		27
	Vesicular stomatitis		
Nucleoside diphosphate phosphotransferase	Leukovirus group		
	Rous sarcoma		25
	Avian myeloblastosis		33
Aminoacyl tRNA synthetase	Leukovirus group		
	Avian myeloblastosis		34
RNA-methylase	Leukovirus group		
	Avian myeloblastosis		35
Ribonucleic acid terminal transferase	Leukovirus group		
	Rous sarcoma	Spermine	36
	Rous-associated virus 1	Spermidine	
	Murine sarcoma	Sulfhydryl reagents	
	Murine leukemia	EDTA	

[a] Only inhibitors actually tested in each system are listed.
[b] Enzyme activities found in the virions of the leukovirus group are reviewed in detail in Reference 1.

REFERENCES

1. Temin, H. M., and Baltimore, D., DNA Synthesis and RNA Tumor Viruses, *Adv. Virus Res., 17,* 129 (1972).
2. Salzman, L. A., *Nature New Biol., 231,* 174 (1971).
3. Kates, J. R., and McAuslan, B. R., *Proc. Natl. Acad. Sci. U.S.A., 57,* 314 (1967).
4. Munyon, W., Paoletti, E., and Grace, J. T., Jr., *Proc. Natl. Acad. Sci. U.S.A., 58,* 2280 (1967).
5. Schwartz, J., and Dales, S., *Virology, 45,* 797 (1971).
6. Tan, K. B., and McAuslan, B. R., *J. Virol., 9,* 70 (1972).
7. Borsa, J., and Graham, A. F., *Biochem. Biophys. Res. Commun., 33,* 896 (1968).
8. Shatkin, A. J., and Sipe, J. D., *Proc. Natl. Acad. Sci. U.S.A., 61,* 1462 (1968).
9. Lewandowski, L. J., Kalmakoff, J., and Tanada, Y., *J. Virol., 4,* 857 (1969).
10. Chow, N., and Simpson, R. W., *Proc. Natl. Acad. Sci. U.S.A., 68,* 752 (1971).
11. Penholt, E., Miller, H., Doyle, M., and Blatti, S., *Proc. Natl. Acad. Sci. U.S.A., 68,* 1369 (1971).
12. Huang, A. S., Baltimore, D., and Bratt, M. A., *J. Virol., 7,* 389 (1971).
13. Robinson, W. S., *J. Virol., 8,* 81 (1971).
14. Baltimore, D., Huang, A. S., and Stampfer, M., *Proc. Natl. Acad. Sci. U.S.A., 66,* 572 (1970).
15. Mizutani, S., Temin, H. M., Kodama, M., and Wells, R. D., *Nature New Biol., 230,* 232 (1971).
16. Mizutani, S., Boettiger, D., and Temin, H. M., *Nature, 228,* 424 (1970).
17. Hurwitz, J., and Leis, J. P., *J. Virol., 9,* 116 (1972).
18. Riman, A. M. J., *Neoplasma, 18,* 575 (1971).
19. Pogo, B. G. T., and Dales, S., *Proc. Natl. Acad. Sci. U.S.A., 63,* 820 (1969).
20. Burlingham, B. T., and Doerfler, W., *J. Mol. Biol., 60,* 45 (1971).
21. Cuzin, M. M. F., Blangy, D., and Rouget, P., *C. R. Acad. Sci. (Paris), 273,* 2650 (1971).
22. Quintrell, N., Fanshier, L., Evans, B., Levinson, W., and Bishop, J. M., *J. Virol., 8,* 17 (1971).

23. Rosenbergova, M., Lacour, F., and Huppert, J., *C. R. Acad. Sci. (Paris), 260,* 5145 (1965).
24. Möiling, K., Bolognesi, D. P., Bauer, H., Büsen, W., Plassman, H. W., and Hausen, P., *Nature New Biol., 234,* 240 (1971).
25. Mizutani, S., and Temin, H. M., *J. Virol., 8,* 409 (1971).
26. Mommaerts, A. B., Eckert, E. A., Beard, D., Sharp, D. G., and Beard, J. W., *Proc. Soc. Exp. Biol. Med., 72,* 450 (1952).
27. Roy, P., and Bishop, D. H. L., *Biochim. Biophys. Acta, 235,* 191 (1971).
28. Borsa, J., Grover, J., and Chapman, J. D., *J. Virol., 6,* 295 (1970).
29. Kapular, A. M., Mendelsohn, N., Klett, H., and Acs, G., *Nature, 225,* 1209 (1970).
30. Gold, P., and Dales, S., *Proc. Natl. Acad. Sci. U.S.A., 60,* 845 (1968).
31. Pogo, B. G. T., and Dales, S., *Proc. Natl. Acad. Sci. U.S.A., 63,* 1297 (1969).
32. Vilagines, R., and McAuslan, B. R., *J. Virol., 7,* 619 (1971).
33. Miller, L. K. and Wells, R. D., *Proc. Natl. Acad. Sci. U.S.A., 68,* 2298 (1971).
34. Erikson, E., and Erikson, R. L., *J. Virol., 9,* 231, (1972).
35. Gantt, R. R., Stromberg, K. J., and DeOca, F. M., *Nature, 234,* 35 (1971).
36. Nakata, Y., and Sakamoto, Y., in *Abstracts of the Japanese Cancer Congress XXX, Tokyo,* p. 76. The Japanese Cancer Association, Tokyo, Japan (1971).
37. Kidwell, W. R., Saral, R., Martin, R. G., and Ozer, H. L., *J. Virol., 10,* 410 (1972).
38. Rosenbergová, M., Matisova, E., and Pristasova, S., *Acta Virol., 15,* 515 (1971).
39. Baltimore, D., and Smoler, D. F., *J. Biol. Chem., 247,* 7282 (1972).
40. Grandgenett, D. P., Gerard, D. F., and Green, M., *J. Virol., 10,* 1136 (1972).

ANTIBIOTICS, MICROBIAL INHIBITORS, AND FERMENTATION PRODUCTS

ANTIFUNGAL ANTIBIOTICS AND SOME SELECTED ANTIFUNGAL SUBSTANCES USED IN CHEMOTHERAPY

DR. E. DROUHET

INTRODUCTION

Among the thousand antifungal agents effective *in vitro,* few have any *in vivo* activity. *In vivo* antifungal activity is related to the permeability of parasite and host cells to the antifungal agents. The classical concept of drug—parasite—host interaction is of particular importance in the control of human, animal or plant fungal infections.

Before the era of antibiotics, the majority of antifungal agents were chemical substances, strong fungicides[1] and surface antiseptics, inactivating vital enzymes by heavy metal ions, with high toxic effects on bacteria, protozoa, fungi, animals and plants. Some organic substances, such as sulfur compounds, fatty acids, benzoic and phenol derivatives, quaternary ammonium compounds, etc., have shown more specific antifungal activities, but are limited to topical use.

Chemotherapy of mycoses dates from 1903, when the beneficial effect of the first oral chemothera-peutic agent, potassium iodide, was reported by de Beurmann and Raymond in treating sporotrichosis, the only mycosis susceptible to this drug. The mechanism of action is still obscure, since potassium iodide lacks antifungal activity *in vitro* or in animal experiments.

During the next 50 years, the only antifungal chemotherapeutic agents found were sulfonamides, with some controversial value in histoplasmosis and suppressive activity in paracoccidioidomycosis, and aromatic diamidines, with activity in blastomycosis but producing noteworthy neurotoxic side-effects.

Discovery of antibacterial antibiotics and the associated increased prevalence of superinfections with such fungi as *Candida* stimulated a vigorous search for antifungal antibiotics. Waksman *et al.*[2] recognized two groups of antifungal antibiotics since 1952, one active against fungi and bacteria but too toxic for therapy, and the other active against fungi only. Discovery by Hazen and Brown[3] of nystatin, active in *Candida* infections, opened up a new era in the therapy of mycoses. After candicidin[4] and trichomycin,[5] more than sixty antibiotics produced by Actinomycetes and grouped under the name of polyenes[6] have been isolated and described.[7,8] All are active *in vitro* against saprophytic or pathogenic fungi, yeasts, or filamentous fungi responsible for superficial or deep-seated mycoses. Despite similarity in chemical structure, therapeutic effectiveness in experimental and clinical infections varies considerably, due to differences in such factors as solubility, diffusibility, toxicity, inactivation by serum components, etc.

Among these polyene antibiotics, only amphotericin B[9,10] is absorbed, at least partially and without toxic effect, when administered orally. Moreover, it has a remarkable therapeutic effect on some of the deep human mycoses when given parenterally. Recently hamycin, administered orally, showed antifungal activity in patients with blastomycosis.[11] When studied by various authors, the mode of action of polyenes was found to be an alteration of cell permeability, leading to leakage of potassium and metabolites essential for life[12,13] or to respiratory troubles blocking ATP regeneration.[14] Among other antibiotics produced by Actinomycetes (cycloheximide, eulicin), only saramycetin, an oligopeptide also known as X-5079 C and as RO2-7758,[15] has been active in some deep-seated mycoses (histoplasmosis, blastomycosis).

As shown by Gentles[16] in 1958, among the antibiotics produced by fungi, griseofulvin, discovered in 1939 but at first applied only in plant fungous infections,[17] is unique in being absorbed when given by mouth; it is active against both superficial and deep infections due to dermatophytes. The only antifungal antibiotic of bacterial origin, pyrrolnitrin, was recently described,[18] but its activity on pathogenic fungi is less than that of amphotericin B, hamycin or saramycetin.

Among new chemical agents, 5-fluorocytosine[19] has a marked effect on *Candida albicans, Cryptococcus neoformans* and other fungi when given by the oral route. Clotrimazole[20] has a broad antifungal spectrum *in vitro* and was reported effective in some experimental and clinical mycoses.

The anatomy and physiology of the fungi are so different from those of the bacteria and actinomycetes that virtually all antibacterial and antiactinomycotic agents are not antifungal. The composition of the fungal cell wall, the material surrounding it, and the cytoplasmic membranes within it all play important roles in the permeability of fungi to antifungal agents. One cell wall component, chitin, is absent in bacteria and actinomycetes, but present in considerable amounts in saprophytic or pathogenic fungi. According to Brian,[21] griseofulvin, a "cell-wall antibiotic" particularly active on dermatophytes, interferes with biosynthesis of the chitinous elements of the hyphal wall. The material surrounding the fungal cell wall is either a polysaccharide, as in *Cryptococcus neoformans,* or of lipoid nature (sterols and phospholipids), as in *Candida albicans, Blastomyces dermatitidis* and other fungi. Fungi, protozoa and higher algae that are sensitive to antifungal polyenes[12] contain sterols, whereas bacteria and blue-green algae do not. This emphasizes the importance not only of the water-solubility but also of the lipo-solubility of antifungal substances. It is noteworthy that the principal antibiotics of therapeutic value, such as nystatin, amphotericin B and griseofulvin, and some recent chemical antifungal drugs, such as 5-fluorocytosine and clotrimazole [1-(-chlortrityl)-imidazole], are almost insoluble in water.

The first step in antifungal action is uptake of the drug by the fungal cells. Microorganisms highly susceptible to griseofulvin, such as dermatophytes, assimilate this antibiotic more than do those only moderately susceptible, such as *Aspergillus.* Resistant organisms, such as yeast and bacteria, do not absorb it at all. The yeasts take up polyenes in great quantity and are sensitive to such drugs as nystatin and amphotericin B; bacteria do not assimilate these antibiotics and are resistant.

It is not essential to chemotherapeutic activity that an antifungal agent be fungicidal. Methods such as the phenol coefficient, the antifungal references of the past, which measure the *in vitro* fungicidal action of antifungal agents, are not relevant. Griseofulvin and saramycetin, both excellent antifungal agents, have little or no fungicidal activity. Amphotericin B is fungicidal *in vitro*, but only fungistatic at therapeutically obtainable levels. Tolnaftate, active on dermatophytes *in vitro* and *in vivo*, is not fungicidal at all; 5-fluorocytosine and clotrimazole are fungistatic at therapeutically obtainable levels and fungicidal only at concentrations about 100 times greater. Prolonged fungistatic effect at a level non-toxic for host cells assists in progressive reduction in the number of parasites and finally permits cure of the mycosis by various immunologic and non-specific resistance factors of the host.

In vivo antifungal activity is related to the selective difference in permeability of parasite and host cells to the antifungal agent: an effective drug is more toxic to the pathogen than to the host. If a chemotherapeutic agent acts by inhibiting an enzymatic system of fundamental importance to the pathogen, that system must be absent in the host or, if present, less susceptible to the drug than that of the pathogen.

General data on antifungal antibiotics and other substances and on their mode of action are summarized in various reports (see References 12 to 14 and 23 to 32).

PRINCIPAL SYSTEMIC ANTIFUNGAL AGENTS USED IN MODERN THERAPY

The principal antifungal agents can be divided into two groups on the basis of their origin: (1) antibiotics produced by various microorganisms, and (2) chemical agents obtained by synthesis (Table 1). In the first group, many hundreds of antibiotics have been isolated during the last thirty years, but few have had the type and degree of selective toxicity that would permit their use as chemotherapeutic agents for human and animal mycoses. In the second or chemical group, only "old drugs", such as potassium iodide, sulfonamides and aromatic diamidines, and two more recently discovered agents, 5-fluorocytosine and clotrimazole, can be used by oral or parenteral administration.

The various antibiotics so far isolated from cultures of different microorganisms may be divided after Waksman *et al.*[31,32] into three broad groups on the basis of their respective antimicrobial spectrum.

1. Antibiotics with antibacterial but not antifungal activity: penicillin, streptomycin, tetracyclines, chloramphenicol, macrolides, etc. These antibiotics enhance the development *in vivo* of endogenous fungi, such as *Candida* and *Geotrichum*, but are not active *in vitro*, despite some conflicting results. Recent reports[33] present evidence that these antibiotics enhance the invasiveness of *Candida albicans* not only by a direct effect on the intestinal flora and on the fungus itself, but also by depressing the host defense mechanisms.

TABLE 1
PRINCIPAL SYSTEMIC ANTIFUNGAL AGENTS USED IN
THE THERAPY OF DEEP MYCOSES

Agent	Route of Administration	Principal Mycoses Affected *in vivo*
Antibiotics		
Polyenes		
Nystatin	Oral, local	Candidosis
Amphotericin B	Oral, intravenous	Deep mycoses (candidosis, cryptococcosis, histo-plasmosis, blastomycosis, etc.)
Hamycin	Oral	North American blastomycosis
Griseofulvin	Oral	Dermatophytosis
Saramycetin (X-5079 C)	Subcutaneous	North American blastomycosis, histoplasmosis
Chemical Agents		
Old drugs		
Potassium iodide	Oral	Sporotrichosis
Sulfonamides	Oral	South American blastomycosis, histoplasmosis
Aromatic diamidines	Oral, intravenous	North American blastomycosis
New drugs		
5-Fluorocytosin	Oral	Candidosis, cryptococcosis, chromomycosis, etc.
Clotrimazole	Oral	Candidosis, cryptococcosis, aspergillosis, etc.

2. Antibiotics with both antibacterial and antifungal activity. These may be produced by bacteria (tyrothricin, pyocianin, eumycin, toximycin, bacillomycin, and others), by fungi (clavacin, gliotoxin, patulin, trichothecin, and others), and by Actinomycetes (streptothricin, mycothricin, pleocidin). These antibiotics are powerful inhibitors *in vitro*, but they have no therapeutic application because of their high toxicity *in vivo*. In this category, but less toxic, is eulicin,[34] an oligopeptide antibiotic that contains two guanido groups and a peptide linkage. Eulicin inhibits some pathogenic fungi of deep mycoses in very small amounts *in vitro* (<0.1 μg/ml), and *Nocardia asteroides* at 2.3 μg/ml. Eulicin is active in mice against experimental infections caused by *Blastomyces dermatitidis* and *Cryptococcus neoformans*, but ineffective in experimental candidosis and histoplasmosis.

3. Antibiotics with selective antifungal activity but not active on bacteria. These antibiotics are produced by Actinomycetes (cycloheximides, polyenes, saramycetin) or by fungi (griseofulvin, cyanein, variotin) and have important therapeutic activities (Table 2). Some secondary properties are shown in Table 3. Chemical and biological properties of the principal antifungal antibiotics are summarized in Table 4.

TABLE 2
PRINCIPAL ANTIBIOTICS ACTIVE AGAINST
PATHOGENIC FUNGI

Antibiotic	Producing Organism	Reference
Actinomycetes-Produced		
Cycloheximide	*Streptomyces griseus*	35
Polyenes		
Nystatin	*Streptomyces noursei*	3
Ascosin	*Streptomyces canescus*	42
Trichomycin	*Streptomyces hachijaenis*	5
Candicidin	*Streptomyces griseus*	4
Filipin	*Streptomyces filipinensis*	43
Amphotericin B	*Streptomyces nodosus*	9
Pimaricin	*Streptomyces natalensis*	44
Levorin	*Actinomyces levoris*	45
Hamycin	*Streptomyces pimprina*	46
Durhamycin	*Streptomyces durhamensis*	47
Polypeptides		
Eulicin	*Streptomyces parvus*	48
Saramycetin	*Streptomyces* nov. sp.	49
Scopafungin	*Streptomyces hygroscopicus*	50
Fungal Sources		
Griseofulvin	*Penicillium griseofulvum*	17
Variotin	*Paecilomyces varioti*	51
Cyanein	*Penicillium cyaneum*	52, 53
Bacterial Source		
Pyrrolnitrin	*Pseudomonas pyrrocinia*	18

TABLE 3
SECONDARY PROPERTIES OF SOME ANTIFUNGAL ANTIBIOTICS

Antifungal Antibiotic	Secondary Activities
Nystatin	Antileishmanial (*in vitro*)
Amphotericin B	Antileishmanial (*in vitro* and *in vivo*) and antiamebal
Trichomycin	Antitrichomonal
Streptovitacins A and B	Carcinolytic
Fumagillin	Carcinolytic, amoebicidal, and antitrichomonal
Thiolutin	Antiprotozoal and antibacterial
Oligomycin	Antinematodal
Flavensomycin	Insecticidal

TABLE 4

CHEMICAL AND BIOLOGICAL PROPERTIES OF PRINCIPAL ANTIFUNGAL BIOTICS AGAINST PATHOGENIC FUNGI

Antibiotic	Synonym	Chemical Data	Biological Data and Practical Application	References
Cycloheximide	Acti-dione Naramycin A	$C_{15}H_{23}NO_4$, MW 281.34; white powder, soluble in water (2%), chloroform, ether, acetone, ethanol, methanol; stable at pH 3–5; UV absorption maximum at 287 nm	Selective against some yeasts (especially *Cryptococcus neoformans*), molds, phytopathogenic fungi; not active against human-pathogenic fungi; toxic, causes skin irritation. Laboratory use: selective isolation of pathogenic fungi. Agricultural use: fungicide	35–40
Nystatin	Fungicidin Mycostatin Moronal Nystan	$C_{46-47}H_{73-75}NO_{18}$ (tetraene); yellow powder, amphoteric, insoluble in water, soluble in dimethylformamide (DMF), dimethylsulfoxide (DMSO), ethanol, methanol, propylene glycol (>10 mg per ml); inactivated by acid, alkali, heat, light; UV absorption maxima at 235, 291, 304, 319 nm in ethanol	Large spectrum of yeasts and filamentous fungi, saprophytic or pathogenic, especially *Candida*; non-toxic by oral route, no (or poor) absorption. Medical use: treatment of candidosis, oral or topical. Veterinary use: poultry, swine; growth promotant, treatment of mycotic infection (mycotic cattle mastitis). Laboratory use: tissue culture	3, 14, 25–27, 29, 56, 63
Pimaricin	Tennecetin Notamycin Pimafucin Myprozin	$C_{33}H_{47}NO_{13}$ (tetraene), MW 665.75; soluble in water (0.005–0.01%), methanol (0.2%) propylene glycol (2%); thermostable; UV absorption maxima at 279, 290, 303, 318 nm	Spectrum similar to that of nystatin. Uses: similar to those of nystatin	
Filimarisin	Filipin 14-Deoxylagosine	$C_{35}H_{58}O_{11}$ (pentaene), MW 654.85; yellow, soluble in DMF, 95% ethanol, methanol, insoluble in water; UV absorption maxima at 355, 338, 322 nm in methanol	Large antifungal spectrum; toxic, LD_{50} in mice, i.p., 17 mg/kg. Medical use: topical. Experimental pathology: rapidly produces arthritis in rabbits	43, 49

TABLE 4 (Continued)
CHEMICAL AND BIOLOGICAL PROPERTIES OF PRINCIPAL ANTIFUNGAL BIOTICS AGAINST PATHOGENIC FUNGI

Antibiotic	Synonym	Chemical Data	Biological Data and Practical Application	References
Candicidin	Candeptin Vanobid	Heptaene, antibiotic complex; fraction A: reddish brown, soluble in water, ethanol, insoluble in acetone, UV absorption maxima at 340, 360, 380, 403 nm in ethanol; fraction B: greenish, soluble in butanol, insoluble in water, ethanol, acetone, UV absorption maxima at 340, 362, 381, 404 nm in ethanol	Active *in vitro* on yeasts, certain filamentous fungi, including human and plant pathogens; active in experimental candidosis, blastomycosis, sporotrichosis; LD_{50} of crude candicin in mice, i.p., 79 mg/kg, s.c., 663 mg/kg Medical use: mucocutaneous candidosis Agricultural use: protection against phytopathogenic fungi	4, 7
Candidin		$C_{46}H_{73}NO_{16}$ (heptaene), MW 869.10; golden-yellow needles, insoluble water, soluble in DMF, pyridine, somewhat soluble in 60% ethanol, acetone, dioxane	Active *in vitro* on *Candida albicans, Sporotrichum schenkii, Blastomyces dermatitidis* and various dermatophytes; LD_{50} in mice i.p., 7—36 mg/kg	7
Amphotericin B	Fungizone	$C_{46}H_{73}NO_{20}$ (heptaene), MW 960.10; deep yellow, soluble in water at pH 2 or 11 (0.1 mg/ml), DMF (2—4 mg/ml), DMSO (30—40 mg/ml), insoluble in water at pH 6—7; dispersible colloidal complex with sodium desoxycholate in aqueous media (i.v. preparation for medical use)	Large spectrum of yeasts and filamentous fungi, especially agents of deep mycoses in humans and animals; highly active in experimental mycoses, oral diffusion in mice, poor diffusion in man; LD_{50} in mice, i.p., 280 mg/kg Medical use: powerful agent in deep mycoses (histoplasmosis, blastomycosis, coccidioidomycosis, etc.) by i.v. route (1 mg/kg every 2 days; oral and topical application in candidosis)	9, 10, 14, 27, 29, 58, 59, 69, 75
Hamycin	Primemycin	Heptaene; yellow powder, amphoteric, insoluble in water, benzene, chloroform, soluble in basic solvents, aqueous lower alcohols	Large spectrum of yeasts and filamentous fungi; active in experimental mycoses (debatable), some oral absorption; LD_{50} in mice, i.v., 6.16 mg/kg Medical use: oral treatment in North American blastomycosis	11, 46, 73

TABLE 4 (Continued)

CHEMICAL AND BIOLOGICAL PROPERTIES OF PRINCIPAL ANTIFUNGAL BIOTICS AGAINST PATHOGENIC FUNGI

Antibiotic	Synonym	Chemical Data	Biological Data and Practical Application	References
Levorin		Heptaene, antibiotic complex, fraction A_2 quite identical to candicidin; yellow powder, insoluble in water, poorly soluble in 95% alcohol, soluble in ether, chloroform, acetone, benzene; UV absorption maxima at 400, 378, 358, 340 nm in methanol	Large spectrum of yeasts and filamentous fungi; LD_{50} in mice, i.p., 9 mg/kg, i.v., 43 mg/kg, LD_{50} in rabbits, i.v., 8 mg/kg Medical use: oral, vaginal, topical treatment of candidosis	45, 58
Ascosin		Heptaene complex; soluble in ethanol, methanol, dioxane; inactivated by strong acids and bases; UV absorption maxima of fraction A at 340, 358, 377, 399 nm, fraction B at 340, 358, 376, 398 nm in ethanol	Active *in vitro* on yeasts and certain filamentous fungi; active in certain experimental mycoses (histoplamosis, cryptococcosis); highly toxic, LD_{50} in mice, 8.6 mg/kg, i.v., 12.5 mg/kg Medical use: topical treatment of tinea capitis	42
Trichomycin	Cabimicina	Heptaene, 3 biological components (A, B, C); yellow, water-soluble sodium salt; UV absorption maxima at 286, 346, 364, 384, 405 nm in ethanol	Active *in vitro* on yeasts, filamentous fungi, protozoa; active in some experimental mycoses; oral tolerance >300 mg/kg, LD_{50} in mice, i.p., i.v., 2.2 mg/kg Medical use: superficial candidosis	5, 65, 66
Eulicin		Polypeptide; hygroscopic white powder; trihydrochloride $C_{24}H_{52}N_8O_2 \cdot 3HCl$, soluble in water; helianthate $C_{24}H_{52}N_8O_2 \cdot 3C_{14}H_{15}N_3O_3S$, crystals from methanol	Active *in vitro* on yeasts, filamentous fungi, *Nocardia asteroides*; active in experimental blastomycosis and cryptococcosis; LD_{50} in mice, i.p., 17 mg/kg, i.v., 3 mg/kg, s.c., 46 mg/kg	48

TABLE 4 (Continued)

CHEMICAL AND BIOLOGICAL PROPERTIES OF PRINCIPAL ANTIFUNGAL BIOTICS AGAINST PATHOGENIC FUNGI

Antibiotic	Synonym	Chemical Data	Biological Data and Practical Application	References
Saramycetin	X-5079 C Ro 2-7758	Polypeptide, 12–14% S, containing 5 amino acids (cysteic acid, aspartic acid, glycine, threonine, proline); white powder, sodium salt soluble in water	Selective activity *in vitro* on *Histoplasma capsulatum, Blastomyces dermatitidis*, some molds; active in experimental histoplasmosis, blastomycosis, coccidioidomycosis, sporotrichosis; LD_{50} in mice, i.p., i.v., s.c., 1000 mg/kg Medical use: subcutaneous histoplasmosis, blastomycosis, sporotrichosis, other mycoses	15, 78
Griseofulvin	Curling factor Fulcin Neo-Fulcin Fulvincin Grifulvin Grisovin Lamoryl Grisactin Sporostatin Spirofulvin Likuden Poncyl	$C_{17}H_{17}ClO_6$, MW 352.5; almost insoluble in water (10 μg/ml), petroleum ether, slightly soluble in ethanol, methanol, acetone, chloroform, soluble in DMF; thermostable; UV absorption maxima at 324, 291, 252, 236 nm	Active *in vitro* on all dermatophytes and on numerous plant-pathogenic fungi (*Botrytis allii*); active in experimental dermatophytosis; absorbed by oral route, serum levels, penetration in dermis, horny layers and incorporation into keratin Medical use: all ringworm infections, antiinflammatory effects in rheumatic arthritis Veterinary use: same as medical use Agricultural use: plant protection	16, 17, 24, 29, 41, 76
Variotin		$C_{15}H_{27}NO_4$; crystallized as needles, MP 41.5–42.5° C; UV absorption maximum at 320 nm, closely related to that of ω-substituted triene carboxylic acids	← Active *in vitro* on dermatophytes (*Microsporum, Trichophyton, Epidermophyton*), *Blastomyces dermatitidis*; no absorption by oral route	51
Pyrrolnitrin		$C_{10}H_6Cl_2N_2O_2$, MW 257.09; pale-yellow crystals from hot cyclohexane; slightly soluble in water, petroleum, soluble in ethanol, methanol; inactivated by serum	Active *in vitro* on *Trichophyton, Microsporum, Epidermophyton, Histoplasma capsulatum, Blastomyces dermatitidis, Cryptococcus neoformans, Candida albicans, Penicillium*; inactive by oral route in experimental mycoses Medical use: topical treatment of dermatophytosis	18

TABLE 4 (Continued)

CHEMICAL AND BIOLOGICAL PROPERTIES OF PRINCIPAL ANTIFUNGAL BIOTICS AGAINST PATHOGENIC FUNGI

Antibiotic	Synonym	Chemical Data	Biological Data and Practical Application	References
Cyanein	Blefeldin A	Slightly soluble in water, insoluble in ethanol	Active *in vitro* on yeasts (*Candida albicans, C. tropicalis, C. parakrusei, Trichosporum capitatum, Asperigillus fumigatus, Coccidioides immitis*), MIC 12.5—25 μg/ml; other fungi ± sensitive; morphological modifications in *T. cutaneum, A. fumigatus, Paecilomyces viridis*; antinematode, cytostatic, antimitotic activities, cancerostatic for mouse tumor; LD_{50} in mice, i.m., 40 mg/kg	52, 53

CYCLOHEXIMIDE

Cycloheximide (Acti-dione) is a water-soluble and thermostable diketone containing one hydroxyl and one ketone group. Isolated from different species of *Streptomyces*,[35] it was one of the first antifungal agents of medical interest. The high inhibitory action on some yeasts and filamentous saprophytic fungi (*Aspergillus, Mucor, Penicillium*) led to the use of cycloheximide as a selective antibiotic for the isolation of dermatophytes and of pathogenic dimorphic fungi, such as *Histoplasma capsulatum, Blastomyces dermatitidis* and *Coccidioides immitis,* at 25°C or room temperature.[36] The yeast phases of the dimorphic fungi are inhibited by cycloheximide at 37°C.[37] The selective inhibition of several *Candida* species, particularly *C. tropicalis* and *C. parakrusei,* is helpful for the taxonomy of *Candida.*[38] The high sensitivity of *Cryptococcus neoformans* to cycloheximide (minimum inhibitory concentration <0.2 µg/ml) led to utilization of this antibiotic, by intramuscular and intrathecal routes, in the treatment of cryptococcosis in man before the discovery of amphotericin B.

A small change in the structure of cycloheximide, that is, the introduction of a single hydroxyl group, produces the streptovitacins. These compounds are cytotoxic as well as antifungal and antitrichomonal. Another cycloheximide streptimidone is active against both protozoa and fungi (Figure 1).

FIGURE 1.   Biological activities of the cycloheximides.

Cycloheximide inhibits protein and nucleic acid synthesis in yeasts[39] by an immediate effect on the synthesis of DNA and protein at the stage of polypeptide formation from aminoacyl-soluble substances.[40] Cycloheximide thus causes the accumulation of rapidly labeled RNA. Inhibition can be eliminated by washing the organism free from the antibiotic. Under certain conditions, cycloheximide will prevent glycolysis in *Saccharomyces* and *Candida* strains. Cycloheximide is useful in clarifying the mechanisms of protein and nucleic acid formation in fungi.[41]

POLYENE ANTIFUNGAL ANTIBIOTICS

Chemical Structure and Properties

More than sixty antifungal antibiotics elaborated by the Actinomycetes have been described, characterized by their multipeaked ultraviolet-absorption spectra and by chromophore groups formed by four, five, six, or seven conjugated double bonds.[32] These compounds have been grouped by Oroshnik *et al.*[6,8] as tetraenes (nystatin, pimaricin, rimocidin), pentaenes (filipin, eurocidin, lagosin, durhamycin), hexaenes (flavacid, endomycin), and heptaenes (ascosin, candicidin, candidin, trichomycin, amphotericin B, hamycin), as shown on pages 94 to 96. Unquestionably, a number of these polyenes will eventually be found to be identical. The tetraene tennecetin, for example, proved to be a rediscovery of pimaricin;[54] and hamycin is similar, if not identical, to trichomycin on spectral and chromatographic analysis[55] and levorin A is similar to candicidin. Data on the chemistry of polyene antibiotics are reviewed by Oroshnik and Mebane[8] (see also pages 97 to 103).

The polyenes have a large macrocyclic lactone ring and, for that reason, were named polyenic macrolides by analogy with the antibacterial, non-polyenic macrolides, such as erythromycin. The active structure of these compounds is the macrolide ring with its rigid lipophilic and flexible hydrophilic parts. A major advance in elucidating the structure of the polyene antibiotics was made by Dutcher et al.[56] when they demonstrated the presence of a carbohydrate moiety — mycosamine — first in nystatin and amphotericin B, and later in other tetraenes and heptaenes, including pimaricin, candidin, and trichomycin.

Pimaricin is the first polyene antibiotic for which a complete structure has been suggested.[57] The structures of lagosin, filipin and amphotericin B[58] have also been elucidated, but those of other polyenes, such as nystatin and trichomycin, are still only partly known.

Chemically, the polyenes are very poorly soluble in the common organic solvents and in water, but reasonably soluble in highly polar solvents, for example, pyridine, dimethylformamide, and dimethyl-sulfoxide. The solubilities of amphotericin B and nystatin in dimethylformamide are 4 mg/ml and 110,000–150,000 units/ml respectively. In solution, loss of potency of both these antibiotics is evident within one hour. The solubility of amphotericin B in dimethylsulfoxide is 30–40 mg/ml. In combination with biliary salts such as sodium deoxycholate, amphotericin B is readily soluble in 5% glucose; this colloidal preparation is used for perfusion in therapy in man.[59] The polyenes are quite unstable in aqueous, acid or alkaline media, but in the dry state and in the absence of heat and light they remain stable for indefinite periods.[8] Polyenes are retained by Seitz filters. Preparations for laboratory use should be fresh, or handled so that their biological activity is known, if reproducible results are to be obtained.

Biological Properties

All polyene antibiotics are produced by soil Actinomycetes of the immense genus *Streptomyces*.

The biological properties of polyenes can be summarized as follows: (a) they show broad antifungal activity against organisms ranging from yeasts to filamentous fungi and from saprophytic to pathogenic fungi, but there are great differences between the sensitivities of different species of fungi to the different polyenes; their activity is fungistatic and fungicidal, even to resting cells; (b) many algae and some protozoa (for example, *Leishmania brasiliensis, L. donovani, L. tropicalis, Trypanosoma cruzei, Trichomonas vaginalis, Entamoeba histolytica*,[60] and *Naegleria*[61]) are sensitive to the polyenes, but no significant activity is shown against bacteria, actinomycetes, viruses or animal cells; (c) resistance to these antibiotics is rare, both clinically and in the laboratory; it develops slowly and never reaches high levels; (d) polyenes can be tolerated orally in relatively large doses; when given parenterally, they show significant toxicity, but they can be used therapeutically when administered in limited amounts.

Fungistatic Activity

Numerous studies of *in vitro* activity of polyene antibiotics have been conducted by a number of investigators, chiefly for nystatin,[3,27,62,63] amphotericin B,[9,27,64] candicidin,[4] trichomycin,[55,66] pimaricin, and hamycin.[11,46] To compare these antibiotics, it is necessary to study them under the same experimental conditions. These conditions are summarized in Table 5. For *Candida* spp., *Cryptococcus neoformans,* and for the dimorphic fungi causing systemic mycoses, the minimal inhibitory concentration of tetraene antibiotics, such as nystatin and pimaricin, is of the order of several micrograms (1.56–12.5 µg/ml), whereas that of the heptaenes is one tenth those amounts, and usually less than 1 µg/ml.

The activity of amphotericin B on the causal agents mycetoma, chromomycoses, phycomycoses and adiaspiromycoses is shown in Table 6. Of the causal organisms of mycetoma, *Madurella mycetomi* is most sensitive, with a minimal inhibitory concentration of 0.78–1.56 µg/ml for the majority of strains. The other causal agents are also sensitive (but less so than *M. mycetomi*), as are some of the fungi that cause chromomycosis. Since a concentration of 1.5–2 µg/ml can be achieved after intravenous administration, one might expect therapeutic activity in human disease.

TABLE 5
ANTIFUNGAL SPECTRUM OF PRINCIPAL ANTIFUNGAL ANTIBIOTICS: MINIMAL INHIBITORY CONCENTRATIONS[a]

Organisms	Antifungal Antibiotics				Chemical Agents	
	Nystatin	Amphotericin B	Saramycetin	Griseofulvin	5-Fluorocytosine	Clotrimazole
Pathogenic yeasts						
Candida						
albicans	3.12	0.04–1.56	>1000	>100	0.09–3.12	0.09–1.56
tropicalis	6.25	0.10–1.56	>1000	>100	3.12–6.25	0.09–1.56
paracrusei	6.25	0.04–1.56	>1000	>100	0.09–3.12	0.04–0.39
Cryptococcus						
neoformans	1.56	0.01–1.56	>1000	>100	0.09	1.56
Dimorphic fungi causing systemic mycoses						
Histoplasma						
capsulatum	1.56	0.04–1.56	1.3–2.1	>100	>100	3.12
Blastomyces						
dermatitidis	1.56	0.04–0.78		>100	25	0.18
Coccidioides						
immitis	1.56	0.78	>1000	>100	>100	0.39
Sporotrichum						
schenckii	12.5	0.14		>100	>100	
Opportunistic fungi (molds)						
Aspergillus						
fumigatus	3.12	3.12	0.78–10	100	6.25 (0.39–1.56)	0.78–1.56
Cephalosporium sp.	6.25	3.12	1.56	100	12.5–100	3.12–6.25
Geotrichum						
candidum	6.25	6.25		100	6.25	12.5
Dermatophytes						
Microsporum						
audouinii	6.25	12.5	>1000	0.62	>100	0.78
Trichophyton						
mentagraphytes	12.5	12.5	>1000	0.62	>100	1.56
Epidermophyton						
floccosum	3.12	3.12	>1000	2.5	>100	1.56
Actinomycetes	>100	>100	>1000	>100	>100	>100
Bacteria	>100	>100	>1000	>100	>100	>100

[a] Minimal Inhibitory Concentrations (MIC), given in μg/ml, represent 100% inhibition after culture in Sabouraud liquid medium (24 hours for yeasts and 5 days for other fungi); values in parentheses represent 50% inhibition.

TABLE 6
IN VITRO ACTIVITY OF FOUR ANTIFUNGAL AGENTS OF THERAPEUTIC INTEREST ON SOME FUNGI OF DEEP MYCOSES[a]

Mycoses	Pathogenic Fungi	Amphotericin B	Griseofulvin	5-Fluorocytosine	Clotrimazole
Mycetoma	*Madurella*				
	mycetomi	0.78–6.25	2.5–100	100 (0.04–3.12)	0.39–0.78
	Leptosphaeria				
	senegalensis	3.12–6.25	100 (12.5)	6.25–100 (0.09–50)	1.56–3.12
	Pyrenochaeta				
	romeroi	12.5	>100 (50)	12.5–100 (1.56–25)	0.39–12.5
	Neotestudina				
	rosatii	50	>100 (3.12)	>100 (0.78–100)	3.12
	Monosporium				
	apiospermum				6.25
Chromomycoses	*Phialophora*				
	pedrosoi	0.10	>100	3.12–25 (0.78–3.12)	1.56–6.25
	jeanselmii	50	>100	1.25–100 (0.78–3.12)	3.12–12.5
	verrucosa	6.25	>100	0.78–12.5 (0.18–0.78)	3.12–6.25
	compacta	100 (1.56)	>100	0.78–12.5 (0.18–0.78)	3.12
	gougerotii	50	>100	(0.78–6.25)	1.56–3.12
	Cladosporium				
	carrionnii	100 (25)	>100	6.25–25 (0.18–0.78)	0.18–0.36
	trichoides	1.56–12.5	>100 (1.56)	3.12 (0.09–0.78)	1.56–6.25
	werneckii	>100	>100	>100 (100)	0.18
Phycomycoses	*Mucor*				
	corymbifera	0.18	>100	>100	0.78
	Absidia				
	italiana	>100	>100	>100	0.18
	Basidiobolus				
	ranarum	12.5	50 (3.12–12.5)	>100	6.25
	meristoporus	12.5	50 (3.12–12.5)	>100	1.56–3.12
Adiaspiromycoses	*Emmonsia*				
	crescens	0.10	>100	>100	0.39–0.78
	parva	6.25	>100	12.5	0.78

[a] Minimal Inhibitory Concentrations (MIC), given in μg/ml, represent 100% inhibition after 5 days culture in Sabouraud liquid medium; values in parentheses represent 50% inhibition.

Values for the minimum inhibitory concentration *in vitro* for trichomycin or hamycin may be less than the values for amphotericin B, but the toxicity of these polyenes *in vivo* is also greater than that of amphotericin B.

Fungicidal Activity

The polyene antibiotics have not only fungistatic but also fungicidal activity.[3],[62] Nystatin is absorbed by yeasts more rapidly and in larger amounts at an acid pH than at a pH of 6 or above.[13] A concentration of four to ten times the minimum inhibitory concentration killed 99.9% of organisms in a few hours; the surviving cells were found to be as sensitive to nystatin and other polyenes as the parent culture. The pH of the medium influences the fungicidal effect: at pH 4.1 only 1.5 μg/ml of amphotericin B are needed to kill 90% of 4×10^7 *Candida albicans* yeasts, whereas at pH 7.1 a concentration of 50 μg/ml is needed to kill 75% of the cells.[14]

Tissue Culture

Polyenes can be tolerated in tissue cultures in certain amounts. Nystatin (10–20 μg/ml or 50 units) and amphotericin B (2.5 μg/ml) have been shown to be significantly useful for the control of fungal contaminants in tissue-culture studies.[67],[68] Pimaricin, thermostable, can be used at 10–20 μg/ml.

Erythrocytes

Rat and human erythrocytes suspended in isotonic saline are rapidly lysed by low concentrations of amphotericin B, ascosin, filipin, etruscomycin, candidin, trichomycin and hamycin.[30] The extent of lysis is dependent on the length of incubation time and on the ratio of the concentrations of antibiotic and erythrocytes. Lysis was also inhibited by sucrose. These results are analogous to those obtained by Kinsky[30] with *Neurospora crassa* protoplasts and with very low concentrations of serum. Polyene-binding by a component in the serum may explain why amphotericin B, for example, is effective in the treatment of systemic fungal infections and does not usually give rise to serious side effects, although hemolytic anemia is a common side effect with prolonged administration of amphotericin B. The "sterol hypothesis" emphasized by Kinsky[30] provides a basis for these phenomena. Polyene-binding to sterols may produce alterations in membrane lipid structure, which could result in conformational charges in the carrier involved in cation transport.

Resistance

It is of critical importance in the treatment of the systemic mycoses — as these are chronic conditions, in which therapy is administered over a long period of time — to determine the degree to which resistant strains develop. But, in contrast to the relatively high level of iatrogenically induced resistance of bacteria to antimicrobial antibiotics, it is rather difficult to induce resistance to antifungal antibiotics, and particularly to the polyene antibiotics.

The careful screening of *Candida* spp. isolated from the first clinical cases of disseminated and localized *Candida* infections, treated with nystatin[26],[63] and, more recently, others showed the absence of resistant strains from clinical material.[27-29],[67] Repeated efforts to develop resistant strains of *Candida in vitro* show that it is possible to obtain "less sensitive" polyene strains of "low-level" resistance,[68] generally after 30 to 58 transfers.[69] Some strains resistant to nystatin, but not to pimaricin,[70] were induced. Strains whose resistance to candidin was increased 150-fold and to amphotericin B 4-, 16-, 45- and 60-fold were developed by subculturing the organism in gradually increasing concentrations of antibiotic in broth or by repeated transfers on gradiant plates. Cross-resistance was observed for candicidin and amphotericin, but not for fungimycin and nystatin.[71]

Mode of Action

Various reviews report different aspects of the mechanism of action.[12-14,30] Lampen and his group,[12,13] working on *Saccharomyces cerevisiae,* and Kinsky,[30] working on *Neurospora crassa,* emphasized alteration of cell permeability and leakage of potassium and other essential metabolites as a result of polyene-binding to cell membrane sterols. Susceptible fungi bound substantial quantities of polyenes, whereas unsusceptible ones did not. Bacterial protoplasts and cells do not bind polyenes, and various workers have stated that the chemical basis for specificity is the presence of sterols in the cell membrane of susceptible organisms. Added sterols protect fungi against polyenes, and formation of polyene—sterol complexes can be demonstrated *in vitro.* Ghosh and Chatterjee,[72] working with *Leishmania donovani,* added support to the sterol-binding hypothesis. Binding of the polyene to the membrane sterol alters membrane structure sufficiently to make the organism unable to concentrate metabolites essential for its maintenance and growth from the medium. There is also a leakage of small ions. Since binding of polyenes is irreversible, the cells die.

Drouhet *et al.,*[14] working on *Candida albicans,* found that one of the primary effects of polyenes is stimulation of oxygen uptake, whereas inhibition of respiration is due to the subsequent death of the cells; the effect is greater on endogenous respiration. Stimulation of respiration was observed with the polyenes nystatin and ascosin. Stimulation of oxygen consumption, rapid oxidation of carbohydrate reserves, and leakage of essential metabolites are successive effects. It has been shown that amphotericin B penetrates cells so rapidly that *C. albicans,* kept in contact with this antibiotic for only 15 minutes, retains antibiotic even when washed six successive times; in these antibiotic-containing cells rapid inhibition, rather than stimulation of oxygen uptake, is observed.

Stimulation of oxygen uptake can be compared to the action of 2:4 dinitrophenol (DNP), which uncouples oxidative phosphorylation during cell respiration. Dinitrophenol induces degradation of ATP to ADP with a brief increase in oxygen uptake, but with cessation of vital synthesis when ATP regeneration is blocked. At fungistatic concentrations of DNP, stimulation of the respiration of *Candida albicans* was similar to that described for yeasts, molds, bacteria and mitochondrial preparations of animal cells and was compared to the modifications obtained with polyenes. DNP stimulated cellular oxidation in various warm-blooded and cold-blooded animals, producing hyperthermia as polyene antibiotics do; intravenous infusion of therapeutic doses of nystatin (2,000,000 units/day) or amphotericin B (1 mg/kg/day) produces a transitory rise of body temperature from $37°-40°C$.

Amphotericin B produces inhibition of the synthesis of proteins, ribonucleic acid, carbohydrate and polyphosphate reserves.[14] The disturbance of phosphorus metabolism is related to the stimulation of endogenous or exogenous oxidations. The action of polyenes on respiration is distinct from that of other antifungal antibiotics, such as cycloheximide and griseofulvin.

Activity *in vivo*

Biological data on tolerance, absorption and distribution in tissue and body fluids of polyenes used in clinical trials by oral or parenteral administration are reported in Table 7. Details on the biology and pharmacology of antifungal antibiotics can be found in general reports.[7,25-29,31,32]

Oral Administration

Polyene antibiotics usually can be tolerated orally in relatively large doses.

Nystatin. In doses of 4 to 8g ($8-16 \times 10^6$ units), nystatin is remarkably non-toxic, due to poor absorption from the gastrointestinal tract. In 24 hours, 32% of orally administered nystatin is recovered from the feces, and less than 1% is recovered from the urine. These data explain the effectiveness of nystatin in *Candida* infections of the alimentary tract and its ineffectiveness in deep infections. Even though nystatin is not absorbed into the blood stream, it may sterilize the intestinal tract and decrease the likelihood of dissemination of *Candida.* In experimental deep mycoses, the parenteral administration of

TABLE 7
ANTIFUNGAL ACTIVITY *IN VITRO*, TOXICITY IN ANIMALS, AND BLOOD LEVELS AT THERAPEUTIC DOSES OF SOME PRINCIPAL ANTIFUNGAL AGENTS USED IN THERAPY

Antifungal Agents	MIC[a]	LD$_{50}$ for Laboratory Animals, mg/kg			Plasma Levels, µg/ml, after Therapeutic Doses	
		Oral	Intra-peritoneal	Intra-venous	Oral	Intravenous
Polyene Antibiotics						
Tetraenes						
Nystatin	1.56—6.25	>4000—8000	45	3	0 or trace	10—50 after 200,000 units in man
Pimaricin	3.25—12.5	1500	250	5—10	0	
Heptaenes						
Amphotericin B	0.04—1	>8000	280	4—66	0.1—1 after 100 mg/kg	0.2—3 after 1 mg/kg in man
Candicidin	0.05—0.2	98—400	2.1—7			
Trichomycin	0.04—1	>100	2.2	2.2	0	2—10 in animals
Hamycin	0.04—1	100—300	17.8[b]	9[b]	0.05 after 50 mg/kg	
Chemical Agents						
5-Fluorocytosine	0.09—3.12	>2000	>1000	>500	10—40 after 50 mg/kg	
Clotrimazole	0.04—1.56	500—2000			2.4 after 50 mg/kg	

[a] Minimal Inhibitory Concentrations, given in µg/ml, for *Candida albicans* and other *Candida* spp. in Sabouraud agar.
[b] Colloidal.

nystatin is of value. In natural disease in man, and despite blood levels obtained, the associated toxicity precludes use of the parenteral route.

Pimaricin. When given orally, pimaricin, like nystatin, is not absorbed by the oral route and is, therefore, ineffective in the deep mycoses.

Candicidin, Ascosin, and Trichomycin. These heptaenes, closely related in their ultraviolet-light absorption spectra, are more toxic by the oral route than the polyenes described in Table 7. They are not absorbed from the intestinal tract and are not active by mouth against the systemic mycoses.

Hamycin. Hamycin is alleged to be absorbed from the gastrointestinal tract and to be active therapeutically in experimental cryptococcosis, histoplasmosis and blastomycosis. The toxicity and biologic activity of different batches of hamycin vary considerably. According to Drouhet,[73] no therapeutic effect was observed with hamycin given orally in daily doses of 25—50 mg/kg in experimental systemic candidosis, cryptococcosis, aspergillosis, histoplasmosis and blastomycosis. Shadomy *et al.*,[74] using a sensitive method, obtained serum levels ranging from 0.01 to 3.5 µg per ml in patients initially treated with 20 mg/kg and then with 40 mg/kg. However, Utz *et al.* cured only two out of seven patients with blastomycosis treated for several weeks.[11]

Amphotericin B. This was the first polyene to be effective when orally administered in experimental and human deep mycoses. It had no toxic effects on laboratory animals even in doses of 8 g/kg. In spite of the limited amounts that can be detected in plasma (0.03—0.08 µg/ml) and in urine (0.1% of the daily dose), this antibiotic prevents death in animals experimentally infected with *Candida albicans, Cryptococcus* *neoformans, Aspergillus fumigatus, Histoplasma capsulatum, H. duboisii, Blastomyces dermatitidis, Coccidioides immitis, Rhizopus oryzae* and *Sporotrichum schenckii*. The protective and curative doses are reported by different authors to be about 100 mg/kg/day. In man, the daily oral administration of 10 and 16 g induced plasma levels of 0.7—3.2 and 3—5 μg/ml respectively.[75] Some authors failed to detect serum levels after oral doses of 3—7 g. After daily doses of 4.5 or 8 g, serum levels of 0.11, 0.14 and 0.20 μg/ml were obtained in adults. With a daily total oral dose of 5 g, the 24-hour urinary excretion of antibiotic ranged from 160 to 300 μg. Therapeutic effects have been observed in some patients with systemic mycoses (cryptococcosis, histoplasmosis, coccidioidomycosis and blastomycosis), but relapses occurred frequently. Less favorable results in man may reflect doses lower than those used in animals.

Parenteral Administration

Administered parenterally, the polyene antibiotics are rather toxic, although the therapeutic index is high. Amphotericin B is the best tolerated antibiotic in this group.

A colloidal preparation with deoxycholate produced serum levels in excess of the minimal inhibitory concentrations for most pathogenic fungi that cause systemic mycotic infections, and these levels persisted. The intravenous administration of amphotericin B in doses of 1 mg/kg produced serum concentrations up to 3.0 μg/ml, measured 40 hours after infusion. After a dose of 1 mg/kg intravenously, Louria detected cerebrospinal-fluid levels ranging from 0.015 to 0.075 μg/ml.[75] Toxic side-effects (hyperthermia, anemia, alterations in the serum electrolytes, particularly a fall in potassium, and impairment of kidney function) are observed.

GRISEOFULVIN

Biochemical and Biological Properties

Griseofulvin was discovered by Oxford, Raistrick and Simonart[17] in 1939 as a metabolic product of *Penicillium griseofulvum*. The molecular structure — 7-chloro-2',4,6-trimethoxy-6'-methylspiro-[benzofuran-2(3H),1'-[2]cyclohexene]-3,4'-dione — is illustrated in Figure 2. Griseofulvin was identified in 1947 as the "curling factor", isolated by Brian,[21] and responsible for a profound disturbance at terminal growing hyphae characterized by branching of the filaments. Although it was shown that the *in vitro* growth of fungi pathogenic for man was impaired by griseofulvin, no *in vivo* studies were made until Gentles[16] in 1958 reported its successful oral use in experimental infections with *Microsporum canis* and *Trichophyton mentagrophytes* in guinea pigs.

FIGURE 2.   Griseofulvin.

Antifungal Activity

Roth, Sallman and Blank,[76] while studying the sensitivity of dermatophytes to griseofulvin, found the minimal inhibitory concentration to be as low as 0.14-0.44 μg/ml at 3 days incubation, and 1.4-2.4 μg/ml at 7 days incubation. These concentrations also produced morphologic alterations in the fungi. At a concentration of 30 μg/ml, the drug had no inhibitory activity against various bacteria, yeasts, and such fungi as *Candida albicans, Cryptococcus neoformans, Phialophora compacta, P. verrucosa, Aspergillus fumigatus, Blastomyces dermatitidis, Histoplasma capsulatum, Sporotrichum schenckii* and *Geotrichum candidum*. Some of the causal agents for mycetoma are sensitive to griseofulvin.

Mode of Action

This is not yet known, although a great deal of work has been done by Lampen *et al.*[12] A recent review[24] discusses the various proposed hypotheses. The mode of action is closely connected with the filamentous type of fungal growth, since yeast forms are not affected. Griseofulvin is a "cell wall antibiotic", producing distortion, irregular swelling and spiral curling, which reflects an alteration, in orientation of newly deposited structural elements of the hyphal wall. Young germinative tubes appear to be more sensitive than mature forms.

Electron-microscopic studies of cells in the presence of small amounts of griseofulvin show both unusually large and irregularly shaped nuclei and a large number of unusually small nuclei. Similar nuclear abnormalities near the hyphal tips of treated dermatophytic fungi have been observed under the light microscope. Thus, griseofulvin appears to affect mitosis in fungi in a way similar to that in mammalian and plant tissue. These effects on mitosis resemble those of colchicine, also responsible for alterations in fibrillar structure and chemical composition of cell walls. It is not yet clear if colchicine, unrelated structurally to griseofulvin, acts by a similar mechanism.

The structural similarities between griseofulvin and the purine ribosides suggested to McNall[77] that the former may act as a competitive analogue at one or more steps in the biosynthesis of nucleic acids. Independent studies by Lampen *et al.*[12] have shown that griseofulvin does, in fact, inhibit nucleic acid synthesis. They proved that organisms sensitive to the antibiotic are capable of taking it up in large amounts, whereas resistant organisms are not. Unlike polyenes, uptake of griseofulvin needs active cells. Uptake is essential for antibiotic action, and the degree of susceptibility of a fungus is correlated with the extent to which a complex is formed with the organism, particularly with the nucleic acid fraction. Griseofulvin, labeled with tritium in the 4-methoxy position, is taken up into the ribonucleic acid fraction and bound exclusively to adenylate residues.[12] Uptake depends on metabolic energy and seems to need the synthesis of a new transport system.

The action of griseofulvin is essentially fungistatic rather than fungicidal. A fungicidal action of the spores and mycelium of dermatophytes has been reported by some authors, but these results were not confirmed by other investigators.

Resistance

It is difficult to obtain wild strains resistant to griseofulvin. In the laboratory, such strains revert rapidly to the original sensitivity.

In Vivo Data

An oral dose of 100 mg/kg gives serum levels of 1–2 μg/ml. Griseofulvin is concentrated selectively in the lungs; after administration of a dose of 100 mg/kg, the lung concentrations, after 1 and 2 hours, are 22.5 and 30 μg/g. Griseofulvin is also found in liver, kidney, bone and brain in amounts of 1 μg/ml. The beneficial results observed in sporotrichosis and mycetoma due to *Nocardia brasiliensis* can be explained by the anti-inflammatory activity of griseofulvin, shown to be comparable to one third to one tenth of that produced by cortisone acetate.

SARAMYCETIN (X-5079 C, RO2-7758)

In vitro Data

Saramycetin[15] is the first non-polyenic agent to show activity against certain mycotic infections. The microbial spectrum includes *Histoplasma capsulatum*, *Blastomyces dermatitidis* and *Coccidioides immitis*, but not dermatophytes, non-pathogenic fungi, yeasts, bacteria, protozoa or viruses. It is fungistatic rather than fungicidal and decays rapidly (25% of the activity is lost in 2 days, 82% is lost in 8 days).

In vivo Data

Saramycetin administered subcutaneously to patients was detected in serum at concentrations between 2.0 and 13.5 μg/ml. Activity at doses in the range from 35 mg/kg was demonstrated in experimental histoplasmosis, blastomycosis and coccidioidomycosis[15] and in human histoplasmosis and blastomycosis.[11]

Mode of Action

This has not yet been studied, but the discrepancy between the low activity *in vitro* and the high activity *in vivo* suggests that an active antifungal component is liberated *in vivo*.

VARIOTIN, PYRROLNITRIN, AND CYANEIN

The properties of variotin (Figure 3), pyrrolnitrin (Figure 4), and cyanein (Figure 5) are given in Table 4.

FIGURE 3.   Variotin.

FIGURE 4.   Chemical structure of pyrrolnitrin.

FIGURE 5.   Cyanein.

NEW ANTIFUNGAL AGENTS FOR ORAL ADMINISTRATION

5-FLUOROCYTOSINE (5-FC)

5-Fluorocytosine is a fluorpyrimidine (Figure 6). Unlike other compounds of this series, including 5-fluorouracil, it was found to lack significant cytostatic activity and toxicity in mammals, but showed marked and selective antifungal activity against *Candida albicans* and *Cryptococcus neoformans*, as reported by Grunberg *et al.*[15] in regard to animals and by Utz *et al.*[79,80] in regard to man.

FIGURE 6.   Structure of 5-fluorocytosine (5-FC) and 5-fluorouracil (5-FU).

In vitro **Data**

The first study of Grunberg *et al.*[15] in peptone complex medium showed distinct activity against *Cryptococcus neoformans, Candida albicans,* and *Aspergillus fumigatus,* but not against *Histoplasma capsulatum, Blastomyces dermatitidis,* and *Coccidioides immitis.* Studies of Drouhet in a Sabouraud liquid medium[28] and Scholer's experiments in a semisynthetic medium[81] showed sensitivity of other pathogenic fungi to 5-FC, including. agents of subcutaneous mycoses (mycetoma, chromoblastomycosis). Shadomy[82,83] emphasized that 5-FC is relatively inactive when tested in broth containing beef and yeast extracts and peptones; this inactivity was attributed to competitive inhibition by cytosine as well as to the inactivation of the drug by peptones. In a synthetic medium, minimal fungistatic and fungicidal concentrations for *Cryptococcus neofromans* were in the range of 0.46–3.9 μg/ml and 3.9–15.6 μg/ml respectively. Corresponding values for *Candida albicans* were 0.46–3.9 μg/ml and 15.6 μg/ml or more respectively.

Synergic action of amphotericin B and 5-FC was reported.[89]

Mode of Action

As expected on the basis of its structure, and demonstrated by the cytosine reversal of the inhibitory effects, 5-FC was considered an antagonist of cytosine, with which it competes metabolically. Such a direct function of 5-FC as an antimetabolite may be distinguished from a more indirect effect due to release of 5-FU from systems possessing the necessary deaminase. The action of 5-FC is abolished competitively by cytosine and not competitively by uridine.[81] In preliminary experiments, in which ^{14}C-labeled 5-FC at fungistatic concentrations was incubated *in vitro* with *Candida albicans,* Koechlin *et al.*[84] demonstrated incorporation into RNA as fluorouracil riboside. The mechanism by which this is effected remains to be investigated.

Resistance

Naturally occurring and induced strains resistant to 1,000–2,000 μg/ml were observed for *Cryptococcus neoformans* and *Candida albicans.* Several phenotypes of spontaneous resistant mutants of *C. albicans* were observed,[85,86] similar to those obtained by induced resistance in *Saccharomyces cerevisiae.*[86]

In vivo **Data**

The low toxicity of 5-FC (LD_{50} in mice: > 2000 mg/kg orally or subcutaneously) and absorption by the oral route permitted therapeutic evaluation of this drug in systemic experimental mycoses. 5-Fluorocytosine exerted a marked effect in mouse infections with *Candida albicans* and *Cryptoccoccus neoformans,* a moderate effect with *Aspergillus fumigatus,* but no effect with *Histoplasma capsulatum* and *Blastomyces dermatitidis.* Doses between 100 and 500 mg/kg/day resulted in prolonged survival for lethal experimental infections, but not in complete fungal sterilization of organs.

In rats there was partial deamination after oral administration, presumably by action of intestinal flora, resulting in excretion of the known metabolites of 5-FU.

In man 5-FC diffuses rapidly into the body fluids (blood, spinal fluid and peritoneal dialysis fluids).[88] Apparent fungistatic concentrations of 10–40 μg/ml persisted in the blood for 6 to 10 hours after a single dose of 2 g; the half-life ranged from 4 to 8 hours. Since the half-life of 5-FC in mice had been estimated by microbiologic assay to be only about 1 hour, the assumption that equivalent doses will be more effective in man than in mice can be justified.

The therapeutic trial in man studied by Utz *et al.*[79,80] showed remarkable activity of 5-FC in all forms of cryptococcosis, even in meningitis, but doses larger than 100 mg/kg are necessary, due to the frequent emergence of resistant strains. Remarkable results were obtained in *Candida* septicemia with renal insufficiency.[87]

CLOTRIMAZOLE [(1-*o*-CHLORTRITYL)-IMIDAZOLE, BAY b 5097]

In vitro Data

This new compound of the tritylimidazol group (Figure 7), at concentrations varying between 1 and 10 µg/ml, had a large antifungal spectrum, including dermatophytes, pathogenic yeasts (*Candida albicans, Cryptococcus neoformans*), pathogenic opportunistic filamentous fungi (*Aspergillus, Mucor*), deep mycotic agents (*Histoplasma capsulatum, Penicillium brasiliensis, Coccidioides immitis*), and subcutaneous mycoses agents (*Phialophora pedrosoi, Cladosporium carrionii*). The antifungal activity is fungistatic rather than fungicidal, according to studies by Plempel *et al.*[20] No resistant strains were observed after 20 to 40 subcultures in the presence of the drug. No bacterial activity was observed.

FIGURE 7. Clotrimazole [1-(*o*-chlortrityl)-imidazole)].

Mode of Action

The mode of action of this promising agent is not yet known. An active substance is observed in human serum after oral administration, but other metabolites are found in animal sera. In the stomach, this product is partially hydrolyzed, giving imidol and 2-chlortritylcarbinal; no trace of biologic activity was seen in urine.

In vivo Data

The oral toxicity in animals is low: LD_{50} was 500–2000 mg/kg, and the chronic toxicity studies produced only liver alterations. By the oral route, serum levels between 10 and 20 µg/ml were obtained after daily doses of 30–228 mg/kg/day; peak levels were attained 5 hours after ingestion of the drug.

The therapeutic activity observed in laboratory animals on such experimental mycoses as candidosis, cryptococcosis, aspergillosis, histoplasmosis and trichophytosis with oral doses of about 100 mg/kg was partially confirmed in such human mycoses as systemic candidosis and aspergillosis. This drug has no effect by the parenteral route.

REFERENCES

1. Horsfall, J. G., *Principles of Fungicidal Action.* Chronica Botanica Co., Waltham, Massachusetts (1956).
2. Waksman, S. A., Romano, A. H., Lechevalier, H. A., and Raubitschek, F., *Bull. W. H. O., 6,* 163 (1952).
3. Hazen, E. L., and Brown, R., *Proc. Soc. Exp. Biol. Med., 76,* 93 (1951).
4. Lechevalier, H., Acker, R. F., Corke, C. T., Haenseler, C. M., and Waksman, S. A., *Mycologia, 45,* 155 (1953).
5. Hosoya, S., Komatsu, N., Soeda, M., and Yamaguchi, T., *J. Antibiot. (Tokyo) Ser. A, 5,* (1952).
6. Oroshnik, W., Vining, L. C., Mebane, A. D., and Taber, W. A., *Science, 121,* 147 (1955).
7. Waksman, S. A., and Lechevalier, H. A., *Antibiotics of Actinomycetes.* Williams and Wilkins, Baltimore, Maryland (1962).
8. Oroshnik, W., and Mebane, A. D., *Fortschr. Chem. Org. Naturst., 21,* 18 (1963).

9.   Gold, W., Stout, H. A., Paganol, J. F., and Donovick, R., in *Antibiotics Annual 1955–1956,* p. 579, H. Welch and F. Marti-Ibañez, Eds. Medical Encyclopedia, Inc., New York (1956).

10. Vandeputte, J., Nachtel, J. L., and Stiller, E. T., in *Antibiotics Annual 1955–1956,* p. 587, H. Welch and F. Marti-Ibañez, Eds. Medical Encyclopedia, Inc., New York (1956).

11. Utz, J. P., Witorsch, P., Williams, T. W., Jr., Emmons, C. W., Shadomy, H. J., and Piggott, W., *Am. Rev. Respir. Dis., 95,* 506 (1967).

12. Lampen, J. O., McLellan, W. L., Jr., and El-Nakeeb, M. A., in *Antimicrobial Agents and Chemotherapy,* p. 1006, J. C. Sylvester, Ed. American Society for Microbiology, Ann Arbor, Michigan (1965).

13. Lampen, J. O., in *Biochemical Studies of Antimicrobial Drugs,* p. 110, N. E. Reynolds and B. A. Newton, Eds. Cambridge University Press, London, England (1966).

14. Drouhet, E., Hirth, L., and Lebeurier, G., Some Aspects of the Mode of Action of Polyene Antifungal Antibiotics, *Ann. N.Y. Acad. Sci., 89,* 134 (1960).

15. Grunberg, E., Berger, J., and Titsworth, E., *Am. Rev. Respir. Dis., 64,* 504 (1961).

16. Gentles, J. C., *Nature (London), 182,* 476 (1958).

17. Oxford, A. E., Raistrick, H., and Simonard, P., *Biochem. J., 33,* 240 (1939).

18. Arima, K., Imanaka, H., Kousaka, M., Fukuda, A., and Tamura, G., *J. Antibiot. (Tokyo) Ser. A, 18,* 201 (1965).

19. Grunberg, E., Titsworth, E., and Bennett, M., in *Antimicrobial Agents and Chemotherapy,* p. 566, J. C. Sylvester, Ed. American Society for Microbiology, Ann Arbor, Michigan (1963).

20. Plempel, M., Bartman, K., Buchel, L. K., *et al., Dtsch. Med. Wochenschr., 26,* 1356 (1969).

21. Brian, P. W., *Ann. Bot. (London), 13,* 59 (1949).

22. Albert, A. (Ed.), *Selective Toxicity,* 2nd ed. Methuen, London, England (1960).

23. Woods, D. D., and Tuckner, R. G., in *The Strategy of Chemotherapy, (Eighth Symposium of the Society of General Microbiology),* p. 1, S. T. Cowan and E. Rowatt, Eds. Cambridge University Press, London, England (1958).

24. Bent, K. J., and Moore, R. H., in *Biochemical Studies of Antimicrobial Drugs,* p. 82, N. E. Reynolds and B. A. Newton, Eds. Cambridge University Press, London, England (1966).

25. Brown, R., in *Experimental Chemotherapy,* p. 418, R. J. Schnitzer and F. Hawkins, Eds. Academic Press, New York (1964).

26. Drouhet, E., in *Fungous Diseases and Their Treatment,* p. 193, R. W. Riddel and G. T. Stewart, Eds. Butterworth, London, England (1958).

27. Drouhet, E., in *Ciba Foundation Symposium on Systemic Mycoses,* p. 206, G. E. W. Wolstenholme and R. Porter, Eds. Churchill, London, England (1968).

28. Drouhet, E., in *Modern Treatment,* Vol. 7, p. 539, J. P. Utz, Ed. Hoeber Medical Division, Harper and Row, New York (1970).

29. Hildick-Smith, G., Blank, H., and Sarkany, I., *Fungus Diseases and Their Treatment.* Little, Brown and Co., Boston, Massachusetts (1964).

30. Kinsky, S. C., in *Antibiotics,* Vol. 1, Mechanism of Action, p. 122, D. Gottlieb and P. D. Shaw, Eds. Springer Verlag, Berlin, Germany (1967).

31. Waksman, S. A., Lechevalier, H. A., and Schaffner, C. P., *Bull. W. H. O., 33,* 219 (1965).

32. Waksman, S. A., and Lechevalier, H. A., *Antibiotics of Actinomycetes.* Williams and Wilkins, Baltimore, Maryland (1962).

33. Seelig, S. M., *Bacteriol Rev., 30,* 442 (1966).

34. West, M. K., Verwey, W. F., and Miller, A. K., in *Antibiotics Annual 1955–1956,* p. 231, H. Welch and F. Marti-Ibañez, Eds. Medical Encyclopedia, Inc., New York (1956).

35. Whiffen, A. J., *J. Bacteriol., 56,* 283 (1948).

36. Georg, L. K., Ajello, L., and Papageorge, C., *J. Lab. Clin. Med., 44,* 422 (1954).

37. McDonough, E. S., Ajello, L., Georg, L. K., and Brinkmans, S., *J. Lab. Clin. Med., 58,* 116 (1960).

38. Drouhet, E., Segretain, G., and Mariat, F., in *Techniques en Parasitologie et en Mycologie,* p. 353, Y. L. Golvan and E. Drouhet, Eds. Flammarion-Médecine-Sciences, Paris, France (1972).

39. Kerridge, D. J., *J. Gen. Microbiol., 19,* 497 (1958).

40. Siegel, M. R., and Sisler, H. D., *Biochim. Biophys. Acta, 87,* 70 (1964).

41. Lampen, J. O., McLellan, W. L., Jr., and El-Nakeeb, M. A., in *Antimicrobial Agents and Chemotherapy,* p. 1009, J. C. Sylvester, Ed. American Society for Microbiology, Ann Arbor, Michigan (1965).

42. Hickey, R. J., Corum, C. J., Hidy, P. H., Lohen, I. R., Nager, U. F. B., and Kropp, E., *Antibiot. Chemother., 2,* 472 (1952).

43. Gottlieb, D., Carter, H. E., Sloneker, J. H., and Amman, A., *Science, 128,* 361 (1958).

44. Struyk, A. P., Hoette, I., Drost, G., Waisvisz, J. M., van Eek, T., and Hoogerheide, J. C., in *Antibiotics Annual 1957–1958,* p. 880, H. Welch and F. Marti-Ibañez, Eds. Medical Encyclopedia Inc., New York (1958).

45. Tsyganov, V. A., Golyakov, P. N., Bezborodov, A. M., Namnestikova, V. P., Khopko, T. V., Soloviev, S. N., Malyshkina, M. A., and Bolshakova, L. O., *Antibiotiki, 4,* 21 (1959).

46. Thirumalachar, M. J., Menon, S. K., and Bhatt, V. V., *Hind. Antibiot. Bull., 3,* 136 (1961).

47. Gordon, M. A., and Lapa, E. W., *Appl. Microbiol., 14,* 754 (1966).

48. Charney, J., Machlowitz, R. A., McCarthy, F. J., Ratkowski, G. A., Tytell, A. A., and Fisher, W. P., in *Antibiotics Annual 1955–1956,* p. 228, H. Welch and F. Marti-Ibañez, Eds. Medical Encyclopedia Inc., New York (1956).

49.   Pras, M., and Weissman, G., *Drug Trade News*, p. 40 (1966).

50.   Johnson, L. E., and Dietz, A., *Appl. Microbiol.*, *22*, 303 (1971).

51.   Takeuchi, S., Yonehara, H., and Umezawa, H., *J. Antibiot. (Tokyo) Ser. A, 12*, 195 (1959).

52.   Nemec, P., and Betina, V., *Biologia (Bratisl.), 14*, 601 (1959).

53.   Betina, V., Drouhet, E., and Segretain, G., *Ann. Inst. Pasteur (Paris), 109*, 933 (1965).

54.   Divekar, P. V., Bloomer, J. L., Eastham, J. F., Holtman, D. F., and Shirley, D. A., *Antibiot. Chemother., 11*, 377 (1961).

55.   Divekar, P. V., Vora, V. C., and Khan, A. W., *J. Antibiot. (Tokyo) Ser. A, 19*, 63 (1966).

56.   Dutcher, J. D., Youwa, M. D., Sherman, J. M., Hibbits, W., and Walters, D. J., in *Antibiotics Annual 1956–1957*, p. 866, H. Welch and F. Marti-Ibañez, Eds. Medical Encyclopedia Inc., New York (1957).

57.   Patrick, J. B., and Williams, R. P., *J. Amer. Chem. Soc., 80*, 6689 (1958).

58.   Borowski, E., Zielniski, J., Ziminski, T., Falkowski, L., Koldziejczyk, P., Golik, J., and Jereczek, E., *Tetrahedron Lett.*, p. 3909 (1970).

59.   Bartner, E., Zinnes, H., Moe, R. A., and Kulesza, J. S., in *Antibiotics Annual 1957–1958*, p. 53, H. Welch and F. Marti-Ibañez, Eds. Medical Encyclopedia Inc., New York (1958).

60.   Actor, P., Wind, S., and Pagano, S., *Proc. Soc. Exp. Biol. Med., 110*, 409 (1962).

61.   Carter, R. F., *J. Clin. Pathol., 22*, 470 (1969).

62.   Donovick, R., Pansy, R. E., Stout, H. A., Stander, H., Weinstein, M. J., and Gold, W., in *Therapy of Fungous Diseases*, p. 176, T. H. Sternberg and V. D. Newcomer, Eds. Little, Brown and Co., Boston, Massachusetts (1955).

63.   Drouhet, E., *Ann. Inst. Pasteur (Paris), 88*, 298 (1955).

64.   Drouhet, E., *Sem. Hop. Paris, 37*, 101 (1961).

65.   Magara, M., Yokouti, E., Senda, T., and Amino, E., *Antibiot. Chemother., 6*, 433 (1954).

66.   Nekano, H., *J. Antibiot. (Tokyo) Ser. A, 14*, 68 (1961).

67.   Drouhet, E., *Br. Med. J., 1*, 699 (1964).

68.   Stout, H. A., and Pagano, J. F., in *Antibiotics Annual 1955–1966*, p. 704, H. Welch and F. Marti-Ibañez, Eds. Medical Encyclopedia Inc., New York (1955).

69.   Littman, M. L., Horowitz, P. L., and Swadley, J. G., *Am. J. Med., 24*, 568 (1958).

70.   McNall, E. G., Halde, P., Newcomer, V. D., and Sternberg, T. H., *Antibiotics Annual*, p. 131, H. Welch and F. Marti-Ibañez, Eds. Medical Encyclopedia Inc., New York (1957-58).

71.   Hebeca, E. K., and Solotrovosky, M., *J. Bacteriol., 89*, 1153 (1965).

72.   Ghosh, B. K., and Chatterjee, A. N., *Ann. Biochem. Exp. Med. (Calcutta), 23*, 309 (1963).

73.   Drouhet, E., in *5th International Congress of Chemotherapy, Vol. A 111*, p. 87. Verlag der Wiener Medizinischen Akademie, Vienna, Austria (1967).

74.   Shadomy, S., *et al., J. Bacteriol., 97*, 481 (1969).

75.   Louria, D. B., *Antibiot. Med. Clin. Ther., 5*, 295 (1958).

76.   Roth, F. J., Jr., Sallman, B., and Blank, H., *J. Invest. Dermatol., 33*, 403 (1959).

77.   McNall, E. G., *Arch. Dermatol., 81*, 657 (1960).

78.   Utz, J. P., Andriole, V. T., and Emmons, C. W., *Am. Rev. Respir. Dis., 84*, 514 (1961).

79.   Utz, J. P., Tynes, B. J., Shadomy, J., Duma, R. J., Kannan, M. M., and Mason, K. N., *Antimicrobial Agents and Chemotherapy*, p. 344, J. C. Sylvester, Ed. American Society for Microbiology, Ann Arbor, Michigan (1968).

80.   Utz, J. P., *N. Engl. J. Med., 286*, 776 (1972).

81.   Scholer, H. J., *Mykosen, 13*, 179 (1970).

82.   Shadomy, S., *Infect. Immun., 2*, 484 (1970).

83.   Shadomy, S., Kirchoff, C. B., and Ingroff, A. E., *Antimicrob. Agents Chemother., 3*, 9 (1973).

84.   Koechlin, B. A., Rubio, F., and Palmer, R. F., *Biochem. Pharmacol., 15*, 435 (1972).

85.   Normark, S., and Schönebeck, J., *Antimicrob. Agents Chemother., 2*, 114 (1972).

86.   Drouhet, E., Mercier-Soucy, L., and Montplaisir, S., *Bull. Soc. Fr. Mycol. Med., 2*, 135 (1973).

87.   Jund, R., and Lacroute, F., *J. Bacteriol., 607*, 102 (1970).

88.   Drouhet, E., Babinet, P., Chapusot, J. P., and Kleinknecht, D., *Biomed. Expr. Eur. J. Clin. Biol. Res., 19*, 408 (1973).

89.   Medoff, G., Kobayashi, G. S., Kwan, C. N., Schlessinger, D., and Venkov, P., *Proc. Nat. Acad. Sci. U.S.A., 69*, 196 (1972).

ANTIBIOTICS PRODUCED BY ACTINOMYCETES

DR. YOSHIRO OKAMI

The following table is a compilation of the properties of the antibiotics known to be produced by Actinomycetales, with the exception of members of the family Mycobacteriaceae. Search of the literature was completed at the end of 1970, but a few antibiotics described in 1971 are included. Trademarks are omitted. In cases where an antibiotic is also produced by organisms other than actinomycetes, the alternate producers are listed as well.

The generic names of the producing actinomycetes are abbreviated as follows:

Acm. = *Actinomyces*
Acp. = *Actinoplanes*
Cha. = *Chainia*
Mib. = *Microbispora*
Mim. = *Micromonospora*
Mip. = *Micropolyspora*
Noc. = *Nocardia*
Stm. = *Streptomyces*
Sts. = *Streptosporangium*
Stv. = *Streptoverticillium*
Tha. = *Thermoactinomyces*
Thm. = *Thermomonospora*
Thp. = *Thermoactinopolyspora*

Chemical structures, if known or partially known, are found at the end of this table. Antibiotic activity against various groups of microorganisms is indicated as follows:

S = strong activity
W = weak activity

Unless otherwise indicated, toxicity data are based on the intravenous route in mice.

L   = low toxicity (LD_{50} : >200 mg/kg)
.M.   = medium toxicity (LD_{50} : 200–50 mg/kg)
T   = toxic (LD_{50} : <50 mg/kg)
LM   = intermediate toxicity between low and medium
MT   = intermediate toxicity between medium and toxic

sc   = subcutaneous
ip   = intraperitoneal
or   = oral

Under remarks, some indication of the locus of the mode of action is given when the information is available.

C   = interference with cell wall synthesis
M   = interference at the membrane level
N   = inhibition of nucleic acid (DNA, RNA) synthesis
Oxid. Phsph. = inhibition of oxido-phosphorylation
P   = inhibition of protein synthesis
R   = inhibition of respiration

Other symbols and abbreviations used in the table are the following:

*	=	see remarks.
G+	=	Gram-positive bacteria
G–	=	Gram-negative bacteria
Myco	=	mycobacteria
Protoz	=	protozoa
Tox	=	toxicity

Information concerning antibiotics and some reference samples of antibiotics are available from the International Center of Information on Antibiotics, 32 Bd de la Constitution, B4000, Liège, Belgium.

Name	Synonym	Producer	Formula	Struct. Ref. No.	G +	G -	Myco	Fungi	Virus	Protoz	Tumor	Others	Tox.	Remarks	Ref.
Aabomycin A		Stm. hygroscopicus subsp. aabomyceticus	$C_{39-40}H_{65-67}O_{11}N$					S	*S				L (ip)	*TMV & NDV	A-1, A-2, A-3
Abikoviromycin	Latumcidin, virocidin	Stm. abikoensis Stm. rubescens	$C_{10}H_{11}ON—H_2SO_4$	A-1	W				S	S			T		A-4, A-5
Ablastmycin		Stm. aburaviensis var. ablastmyceticus	$C_{18}H_{31}N_5O_{10}$					S							A-6
Aburamycin		Stm. aburaviensis	C: 55.57, H: 7.54		S		W				S		T	Aureolic acid group	1, 2
Aburamycin isomer	M5-18903				S		S							Aureolic acid group	1, 2
Acetomycin		Stm. ramulosus	$C_{10}H_{14}O_5$	A-2	W	W	S			S			M (sc)		1, 2
Acetopyrrothin	Thiolutin													See Thiolutin	1, 2
2-Acetyl-2-decarboxamidooxytetracycline	Terramycin X	Stm. rimosus (strain 7478)	$C_{23}H_{25}O_9N \cdot HCl$	A-3	S	S								Tetracycline group	1, 2, A-7, A-8, A-9
4-Acetoxy-cycloheximide	E-73	Stm. sp.	$C_{17}H_{25}O_6N$	A-4				S			S		T (ip)	Glutarimide group Mode act.: P	1, A-10, A-11, A-12, A-13
2-Acetyl-2-decarboxamido-tetracycline		Stm. psammoticus	$C_{23}H_{45}NO_8 \cdot H_2O$											See Tetracycline	2, A-14
Acetylenedicarboxamide	Cellocidin Aquamycin													See Aquamycin	A-15
Achromoviromycin		Stm. achromogenes			W				S				L		1, 2
Achromycin	Puromycin													See Puromycin	1, 2
Acidomycin	Actithiazic acid													See Actithiazic acid	1, 2

Name	Synonym	Producer	Formula	Struct. Ref. No.	G +	G −	Myco	Fungi	Virus	Protoz	Tumor	Others	Tox.	Remarks	Ref.
Acrylamidine	D-274-2	Stm. eurythermus D-274-2	$C_3H_7N_2Cl$	A-5				*S					T	*Candida	A-16
Actidione	Cycloheximide													See Cycloheximide	1, 2, 3
Actiduins		Stm. sp.	contg. N & S		S										1, 2
Actinin	Mycetin	Stm. felis			S	S					S			Peptide group	1, 2
Actinobolin	Actinovolin	Stm. griseoviridis var. atrofaciens	$C_{13}H_{20}O_6N_2$	A-6	W	W					S		L	Mode act.: N	1, A-17, A-18, A-19
Actinochrysin (C₁, C₂, C₃)					S								T	See Actinomycin C	1, 2
Actinoflavin	Actinomycin J				S								T	See Actinomycin J	1, 2
Actinoflocin	Kikuchi's 3rd substance	Stm. sp. E 212 res Stm. albus			*S						W		T (ip)	*Streptococcus hemolyticus Actinomycin group	1, 2
Actinogan		Stm. sp.	C: 44.61, H: 5.85, N: 2.00								S			Peptide group (glycoprotein)	1, 2
Actinoidin		Proactinomyces (Noc.) actinoides 9765			*S								T	Peptide group *Aerobacter aerogenes	1, 2
Actinoleukin	MA-537A₁	Stm. abikoensis, Stm. aureus, Stm. sp.	$C_{29-30}H_{40-42}O_{7-8}N_6S$		S						S		T (ip)	Quinoxaline group Mode act.: N	1, 2, A-20
Actinolysin		Stm. albicans									*S			*Actinomycetes	1, 2
Actinomycelin		Stm. sp. Stm. antibioticus			S										1, 2

Name	Synonym	Producer	Formula	Struct. Ref. No.	G +	G −	Myco	Fung	Virus	Protoz	Tumor	Others	Tox.	Remarks	Ref.
Actinomycetin		*Stm. albus* *Stm. sp.*			S		S							Probably peptide group (endopeptidase)	1, 2, A-21
Actinomycin A ($A_I – A_V$)	Actinomycin	*Stm. antibioticus*		A-7	S						S		T	Mode act.: N	1, 2, 3, A-22, A-23, A-24, A-25
Actinomycin B	Actinomycin	*Stm. antibioticus*		A-7	S								T		1, 2
Actinomycin C	Actinochry-sin, HBF-386, Actinomycin S-67	*Stm. chrysomallus*		A-7	S						S		T	Mode act.: N	1, 2
Actinomycin C_1		*Stm. chrysomallus* *Stm. sp.* BOP 476		A-7	S						S		T		1, A-26
Actinomycin C_2		*Stm. chrysomallus* *Stm. sp.* BOP 476		A-7	S						S		T		1, 2
Actinomycin C_3		*Stm. chrysomallus* *Stm. sp.* BOP 476		A-7	S						S		T		1, 2
Actinomycin D		*Stm. chrysomallus* *Stm. antibioticus*		A-7	S						S		T	Mode act.: N	1, 2
Actinomycin E_1		*Stm. sp.*	$C_{64}H_{96}O_{16}N_{12}$	A-7	S						S		T		1, 2
Actinomycin E_2			$C_{65}H_{98}O_{16}N_{12}$	A-7	S						S		T		1, 2
Actinomycin F				A-7	S						S		T		1, 2
Actinomycin F_0				A-7	S						S		T		1, 2
Actinomycin F_1		*Stm. chrysomallus* *Stm. sp.* BOP 476	$C_{58}H_{88}O_{16}N_{12}$	A-7	S						S		T		1, 2

Name	Synonym	Producer	Formula	Struct. Ref. No.	G +	G −	Myco	Fungi	Virus	Protoz	Tumor	Others	Tox.	Remarks	Ref.
Actinomycin F₂			$C_{60}H_{90}O_{16}N_{12}$	A-7	S						S		T		1, 2
Actinomycin F₃			$C_{59}H_{88}O_{16}N_{12}$	A-7	S						S		T		1, 2
Actinomycin F₄			$C_{61}H_{92}O_{16}N_{12}$	A-7	S						S		T		1, 2
Actinomycin F₅₋₉				A-7	S						S		T		1, 2
Actinomycin FS				A-7	S						S		T		1, 2
Actinomycin H		*Stm. sp.* 1784	C: 56.81, H: 6.60, N: 12.36	A-7	S						S		T		1, 2
Actinomycin I	Actinomycin A₁, B₁ Xab Actinomycin J (in german)	*Stm. parvullus* *Stm. antibioticus* *Stm. sp.*		A-7	S						S		T	Mode act.: N	1, 2
Actinomycin J	Actinoflavin	*Stm. flaveolus* *Stm. sp.* *Stm. flavus*		A-7	S						S		T	Mode act.: N	1, 2
Actinomycin K		*Stm. sp.*		A-7	S						S		T		1, 2
Actinomycin L		*Stm. sp.* No. 2104L		A-7	S						S		T	Mode act.: N	1, 2
Actinomycin M		*Stm. antibioticus*		A-7	S						S		T		1, 2
Actinomycin P	PA-126P₂	*Stm. aureofaciens* ATCC 13336		A-7	S						S		T		1
Actinomycin S (S₀ – S₃)		*Stm. sp.* *Stm. flaveolus*		A-7	S						S		T		1,2
Actinomycin S₂		*Stm. flaveolus*	C: 59.20, H: 6.90, N: 13.47	A-7	S						S		T		2, A-27

Name	Synonym	Producer	Formula	Struct. Ref. No.	G+	G-	Myco	Fungi	Virus	Protoz	Tumor	Others	Tox.	Remarks	Ref.
Actinomycin S$_3$		Stm. flaveolus	C: 57.20, H: 6.65, N: 13.21	A-7	S						S		T		2, A-26
Actinomycin U (U$_1$ – U$_4$)		Stm. chrysomallus		A-7	S						S		T		1
Actinomycin X		Stm. fradiae		A-7	S						S		T		1, 2
Actinomycin Z				A-7	S						S		T		1, 2
Actinomycin I	Actinomycin A$_1$, B$_1$, X$_{0\beta}$			A-7	S						S		T		1, 2
Actinomycin II	Actinomycin A$_{II}$, B$_{II}$	Stm. antibioticus		A-7	S						S		T		1, 2
Actinomycin III	Actinomycin A$_{III}$, B$_{III}$, X$_{0\gamma}$			A-7	S						S		T		1, 2
Actinomycin IV	Actinomycin A$_{IV}$, B$_{IV}$, C$_1$, I$_1$, X$_I$			A-7	S						S		T		1, 2
Actinomycin V	Actinomycin A$_V$, B$_V$, X$_2$			A-7	S						S		T		1, 2
Actinomycin VI	C$_2$			A-7	S						S		T		1, 2
Actinomycin VII	C$_3$			A-7	S						S		T		1, 2
Actinon(e)		Stm. sp. res. Stm. antibioticus						*S					L	*Saccharomyces & Trichophyton	1, 2
Actinone B		Stm. sp. res. Stm. antibioticus						S							1
Actinonin		Stm. felis	C$_{19}$H$_{35}$O$_5$N$_3$	A-8	W	W							ML		1, 2

Name	Synonym	Producer	Formula	Struct. Ref. No.	G+	G-	Myco	Fungi	Virus	Protoz	Tumor	Others	Tox.	Remarks	Ref.
Actinorhodin	Coelicomycin	*Stm. coelicolor*	$C_{32}H_{26}O_{14}$	A-9	W									pH Indicator	1, 2
Actinorubin		*Stm. sp.* res. *Stm. erythreus*	$C_6H_{14}O_2N_3$ or $C_9H_{22}O_4N_5$		S	S	S						M1 (ip)	Streptothricin group	1, 2
Actinospectacin	Spectinomycin M-141 Trobicin	*Stm. spectabilis* NRRL 2494 *Stm. flavopersicus*	$C_{14}H_{24}O_7N_2$	A-10									T	Mode act.: P	1, 2, 3
Actinovorin	Actinobolin													See Actinobolin	1
Actinoxanthin(e)		*Acm. globisporus* 1131			S						S		T (ip)	Peptide group	1, 2, A-28
Actiphenol	C-73	*Stm. sp.* ETH 7792, *Stm. albus*, *Stm. noursei*, *Stm. pulveraceus*	$C_{15}H_{17}O_4N$	A-11				S						Glutarimide group Mode act.: P	1, 2, A-29, A-30, A-31
Actithiazic acid	Acidomycin, Cinnamomycin, Mycobacidin, PA-95 Thiazolidone antibiotic	*Stm. virginiae*, *Stm. acidomyceticus*, *Stm. roseochromogenes*	$C_9H_{15}O_3NS$	A-12	S		S						L	Inhibit of synth. Biotin	1, 2, A-32, A-33, A-34, A-35
Acumycin		*Stm. griseoflavus*	$C_{38}H_{61}O_{12}N$		S	S	S							Macrolide group	1, 2
Adriamycin		*Stm. peucetius* var. *caesius* (IMI 131.502 IMRU 3.920)	$C_{27}H_{29}NO_{11}$	A-13	W	W	S				S				A-36
Ahygroscopin		*Acm. ahygroscopicus* Yen 508	$C_{12}H_{14}N_5O_4$					S						Co-product: Anisomycin, similar to Toyocamycin	A-37

Name	Synonym	Producer	Formula	Struct. Ref. No.	G+	G-	Myco	Fungi	Virus	Protoz	Tumor	Others	Tox.	Remarks	Ref.
Akimycin		Stm. lavendurae E20-27	C: 22.39%, H: 4.06% N: 22.15%, S: 21.80%		S	S	S						T	Streptothricin group	1
Akitamycin	Toyamycin	Stm. akitaensis	C: 57.26 – 57.11 H: 7.68 – 7.65, N: 1.64 – 1.84					S		S			T	Polyene (tetraene) group	1
Aklavin		Stm. sp.	$C_{30}H_{37}O_{11}N$	A-14	S			S	S				M	Anthracycline group	1, 2
Alanosine		Stm. alanosinicus	$C_3H_7N_3O_4$	A-15				*S	S	S	S		L	*Yeasts	A-38
Alazopeptine		Stm. griseoplanus	$C_{15}H_{21}O_6N_7 \cdot H_2O$								S		M (rat)	(Azaamino acid group) Mode act.: N	1, 2, A-39
Albimycin	Hondamycin	Stm. sp. res. Stm. griseochromogenes & Stm. albochromogenes Stm. griseochromogenes var. albicans	$C_{18}H_{30}O_5$					*S					T (ip)	*Inactive against yeasts	1, A-40
Albocycline	TA2407	Stm. brunneogriseus Stm. roseocitreus Stm. roseochromogenes var. albocyclini	$C_{18}H_{28}O_4$		S								L (ip)	Similar to Cineromycin B	A-41

Name	Synonym	Producer	Formula	Struct. Ref. No.	G +	G −	Myco	Fungi	Virus	Protoz	Tumor	Others	Tox.	Remarks	Ref.
Albofungin		Stm. albus Stm. albus var. fungatus			S			S					T		1, 2
Albomycetin		Stm. albus Stm. sp. res. Stm. albus	$C_{32}H_{54}O_9N$		S							S	ML	Probably macrolide group	1, 2
Albomycin complex		Stm. subtropicus	$C_{40}H_{61}O_{21}N_{10}SFe$	A-16	S	S								Sideromycin group Mode act.: N, M	1, 3, A-42, A-43
Albonurusin	B-73	Acm. albus var. fungatus, Stm. noursei	$C_{15}H_{16}O_2N_2$	A-17							S				1, 2
Alboverticillin		Stm. alboverticillatus	$C_{13}H_{29}O_{5-6}N_5$		S		S						M	Peptide group	1, 2
Aldgamycin E	LL-AL471-E	Var. of Stm. lavendulae	O–Me: 4.45, C–Me: 8.65 C: 58.75, H: 7.92		S		W							Macrolide group	1, A-44
Aldgamycin E-like		Stm. sp. No. B-10,000	C: 58.80, H: 7.99		S								M	Macrolide group	A-45
Aliomycin		Stm. acidomyceticus						S		S	S		MT (ip)	Polyene (pentaene) group	1, 2
Allomycin	Amicetin	Stm. sindenensis	$C_{29}H_{44}O_9N_6$		S		S						T	Pyrimidine group	1, 2
Almarcetin		Stm. albus NRRL B-3141			W	W	W							Peptide group	1
Alomycin		Stm. sp. 181						*S						*Candida	1, 2
Althiomycin	116a Matamycin Prob.	Stm. althioticus	$C_{27}H_{28}O_{10}N_8S_3$		S	S	S	S					L (ip)	Mode act.: P	1, 2, A-46
Alveomycin		Stm. sp. Sd 094	Fe: 1.5		S	S	M							Sideromycin group	1

727

Name	Synonym	Producer	Formula	Struct. Ref. No.	G+	G-	Myco	Fungi	Virus	Protoz	Tumor	Others	Tox.	Remarks	Ref.
Alubimycin	Albimycin													See Albimycin	A-40
Amaromycin	Amaromycin I Picromycin	*Stm. flaveochromogenes*	$C_{25}H_{39}O_7N$	A-18	S								ML	Macrolide group	1, 2, A-47
Amethobottromycin	(Bottromycin B)													See Bottromycin B	A-48
Amicetin	Allomycin, Sacromycin, D-13	*Stm. vinaceus-drappus, Stm. fasciculatus, Stm. sacromyceticus*	$C_{29}H_{42}O_9N_6$	A-19	S		S						M	Pyrimidine group Mode act.: P	1, 2, 3, A-49, A-50
Amicetin B	Plicacetin, Antibiotic C, R-285	*Stm. plicatus Stm. sp.* (R-285) ATCC 13064T	$C_{25}H_{35}O_7N_5$	A-19	S		S						ML	Pyrimidine group	1, 2
Amicetin C		*Stm. vinaceus-drappus*	$C_{56}H_{94}O_{19}N_{10}S$		S								L		1
Amidinomycin	Myxoviromycin	*Stm. sp.* 2171-I₃ related to *Stm. flavochromogenes Stm. kasugaensis*	$C_9H_{18}ON_4$	A-20	*S				S				T	Depsipeptide group *B. subtilis*	1, 2, A-51
Amidomycin		*Stm. sp.*	$C_{40}H_{68}O_{12}N_4$ or $C_{20}H_{34}O_6N_2$					S						Probably peptide	1, 2
2-Amino-3-dimethylaminopropionic acid	L-4-Azaleucin	*Stm. neocaliberis var. neocaliberis*	$C_5H_{12}N_2O_2$	A-21	S	S								Active only in a synthetic medium	A-52
D-4-Amino-3-isoxazolidone	Cycloserine Oxamycin	*Stm. lavendulae Stm. garyphalus Stm. orchidaceus Pseudomonas fluorescens*		C-32										See Cycloserine	1, 2

Name	Synonym	Producer	Formula	Struct. Ref. No.	G+	G-	Myco	Fungi	Virus	Protoz	Tumor	Others	Tox.	Remarks	Ref.
2-Amino-4,4-dichlorobutyric acid	Armentomycin	*Stm. armentosus* var. *armentosus*	$C_4H_7Cl_2NO_2$	A-22		S							MT		A-52
2-Amino-4-methyl-5-hexenoic acid		*Stm. sp.* UC5159	$C_7H_{13}NO_2$	A-23	S										A-53
Aminomycin (Brockmann)	Valinomycin	*Stm. sp.* IHR 582	$C_{19}H_{32}O_6N_2$		S		S	*S						Depsipeptide group *Piricularia oryzae	1, 2
Aminomycin	Fungimycin Perimycin NC-1968	*Stm. sp.*												See Perimycin	1
Aminosidin	Catenulin, Crestomycin Farmiglucin Hydroxymycin, Paromomycin, Zygomycin A, FI 1600, FI 5853	*Stm. chrestomyceticus* *Stm. fradiae* var. *italicus*	$C_{23}H_{45}O_{14}N_5 \cdot \frac{5}{2}H_2SO_4 \cdot 2H_2O$		S	S	S						ML	Aminocyclitol (Neomycin) group	1, 2, A-54, A-55
															A-56
Amphomycin		*Stm. canus* *Stm. violaceus*, *Stm. sp. res. Stm. laverdulae*	C: 54.4, H: 7.19 N: 14.2		S								M	Peptide group	1, 2
Amphotericin A		*Stm. nodosus* *Stm. sp.*	$C_{46}H_{73}O_{19}N$					S					L (ip)	Polyene (tetraene) group	1, 2
Amphotericin B		*Stm. nodosus* *Stm. sp.*	$C_{47}H_{73}O_{18}N$	A-24				S					M (ip)	Polyene (heptaene) group Mode act.: C, M	1, 2, 3, A-56

Name	Synonym	Producer	Formula	Struct. Ref. No.	G +	G -	Myco	Fungi	Virus	Protoz	Tumor	Others	Tox.	Remarks	Ref.
Amycin(e)		Stm. lavendulae	−HCl: C: 41.18, H: 7.58, N: 17.59, Cl: 8.42		S	S	S	W					M		1, 2
Amylocyarin		Stm. coelicolor						S						Pigment group	1, 2
Angolamycin		Stm. erythermus	$C_{49-51}H_{87-91}O_{18}N$		S								M	Macrolide group Mode act.: P	1, 2, 3, A-57, A-58, A-59
Angustmycin A	Decoynin	Stm. hygroscopicus var. angustmyceticus	$C_{11}H_{13}O_4N_5 \cdot H_2O$	A-25			S							Purine group Mode act.: N	1, 2. A-60, A-61, A-62, A-63, A-64
Angustmycin C	Psicofuranine	Stm. hygroscopicus var. angustmyceticus	$C_{11}H_{15}O_5N_5$	A-26										Purine group Mode act.: N	1, 2
Anisomycin	PA106, Flagecidin	Stm. griseolus Stm. roseochromogenes	$C_{14}H_{19}O_4N$	A-27	W			*S		S			M	*Candida albicans Mode act.: P	1, 2, A-65, A-66
Anthelmycin		Stm. longissimus ATCC 14562	$C_{25}H_{44}O_{16}N_5$		W	W							T	Anti-Ascaris	1, 2
Anthelvencin A, B		Stm. venezuelae	$C_{19}H_{27}O_3N_9$		S	S	S	S					MT		1
Anthracidin A		Stm. sp. 190 res. Stm. hygroscopicus	$C_{20}H_{29}O_5N$		*S	S					S			*B. anthracis	1
Anthracidin B		Stm. sp. 190 res. Stm. hygroscopicus	$C_{20}H_{27}O_6N$		*S	S								*B. anthracis	1
Anthracyclines				A28-A33	S A28-A33									Generic name for glycoside of Anthracyclinone(s) (Rhodomycins, Pyrromycins, Rutilantin, Aklavin, cinerubins)	2, A-67
Anthracyclinon(e)s				A-28, A-33	S A-33										2

Name	Synonym	Producer	Formula	Struct. Ref. No.	G+	G-	Myco	Fungi	Virus	Protoz	Tumor	Others	Tox.	Remarks	Ref.
Anthracydine A		*Stm. sp.* 190	$C_{20}H_{29}O_5N$		W	W		W							A-68
Anthracydine B		*Stm. sp.* 190	$C_{20}H_{27}O_6N$		W	W		W							A-68
Anthramycin	Refuin	*Stm. refuineus var. thermotolerans*	$C_{16}H_{17}O_4N_3$	A-34	S	W					S				1
Antifongin 4915		*Stm. paucisporogenes*						S						Polyene (heptaene) group	1
Antimycin A complex		*Stm. kitasawaensis* *Stm. griseus* *Stm. sp.*						S						Mode act.: R	1, 2, 3, A-69, A-70, A-71, A-72
Antimycin A$_1$		*Stm. kitasawaensis* *Stm. griseus*	$C_{28}H_{40}O_9N_2$	A-35				S					T	Above	Above
Antimycin A$_2$a		*Stm. kitasawaensis* *Stm. griseus*	$C_{25}H_{34}O_9N_2$	A-35										Above	Above
Antimycin A$_2$b			$C_{25}H_{30}O_9N_2$	A-35										Above	
Antimycin A$_3$	Blastmycin	*Stm. kitasawaensis* *Stm. sp.*	$C_{26}H_{36}O_9N_2$	A-35				S					T	Above	Above
Antimycin A$_4$		*Stm. kitasawaensis* *Stm. sp.*	$C_{26}H_{39}O_9N_2$	A-35				S					T	Above	Above
Antimycoins	C-381 Fungicidin	*Stm. aureus*						S					M (ip)	Polyene (tetraene) group Mode act.: M	1, 2, 3

Name	Synonym	Producer	Formula	Struct. Ref. No.	G +	G −	Myco	Fungi	Virus	Protoz	Tumor	Others	Tox.	Remarks	Ref.
Antiphlei Factor		*Stm. aureus*					S						L		2, A-73. A-74
Antipiriculin	Antimycins	*Stm. kitasawaensis*	$C_{28}H_{40}O_9N_2$				S	S					T		1, 2
Antiprotozoin		*Stm. hygroscopicus var. indica*	C: 55.3, H: 8.5 N: 4.4, P: 0.48, Ash: 3.2				S	S		S					1
Antisideromycin (A₁, A₂, A₃, & B)		*Stm. griseoflavus* *Stm. lavendulae* *Stm. galilaeus* *Stm. pilosus, Stm. polychromogenes* *Stm. viridochromogenes* *Stm. aureofaciens* *Stm. olivaceus* *Stm. griseus* *Stm. glaucescens*	A₁: Fe 4.5-5.5		S			S							A-75
Antivirubin		*Acm. longispororuber*			W				S	S			T	pH Indicator	1, 2
Aquamycin	Acetylene dicarboxyamide Cellocidin Lenamycin	*Stm. reticuli*	$C_4H_4O_2N_2$	A-36	S	S	S				S		T		A-15
Aquayamycin		*Stm. misawanensis*	$C_{30-31}H_{34-40}O_{12}$		S	W	W	W			S		T	Hydroxyquinone group	A-76
9-β-D-Arabinofuranosyl adenine		*Stm. antibioticus*		A-37					S						A-77
Argomycin	Pikromycin													See Pikromycin	1, 2

Name	Synonym	Producer	Formula	Struct. Ref. No.	G +	G -	Myco	Fungi	Virus	Protoz	Tumor	Others	Tox.	Remarks	Ref.
Aristeromycin	Aristelomycin	*Stm. citricolor*	$C_{11}H_{15}O_3N_5$	A-38		*S[1]		*S[2]					L	*1 *Xanthomonas oryzae* *2 *Pricularia oryzae Alternaria Kikuchiana*	A-78, A-79
Arizinomycin	3-methyl-2-carbethoxy-2(2H)-arizirine-carboxylic acid	*Stm. aureus*	$C_4H_5NO_2$	A38/9	S	S							T (ip)		A-79/80
Armentomycin	2-Amino-4-4-dichlorobutyric acid			A-22										See, 2-Amino-4,4-dichlorobutyric acid	A-52
Arsimycin		*Stm. arsitiensis* *Stm. roseus* NCIB8996	$C_{27}H_{47}O_{11}N_7S$		*S									Peptide group *B. subtilis	1
Arvomycin		*Stm. sp.* J-4 *Stm. sp.* res. *Stm. fungicidicus*	$C_{24}H_{26}O_5N_3$		S		S								1
Ascaricidin		*Stm. sp.*										*S		*Anti-Ascaris	1, 2
Ascarinase		*Stm. sp.* U-13 belonging to *Stm. hachijoensis*	C: 41.64, H: 7.62, N: 11.32									*S	T	*Anti-Ascaris	1
Ascomycin		*Stm. sp.* KKS317, res., *Stm. hygroscopicus*					W	S					M (ip)		1, 2
Ascosin A		*Stm. canescus*						*S					T	Polyene (heptaene) group *Yeasts	1, 2
Ascosin B		*Stm. canescus*						*S						Polyene (heptaene) *Yeasts	1, 2

Name	Synonym	Producer	Formula	Struct. Ref. No.	G+	G-	Myco	Fungi	Virus	Protoz	Tumor	Others	Tox.	Remarks	Ref.
Aspartocin		*Stm. griseus* var. *spiralis* *Stm. violaceus* *Stm. violaceus* var. *aspartocicus*	C: 53.58–53.18, H: 7.39–7.58, N: 13.14–13.58, S: 0.36–0.49, Cl: 0.00–0.14		S		W						T	Peptide group	1, 2
Audricurin	Curamycin													See Curamycin	1
Aurant(h)ins		*Stm. sp.* res. *Stm. aurantiacus*		A-39	S						S		T	Actinomycin group Mode act.: N	1, 2, 3
Aurenin															A-80
Aureofacin		*Stm. aureofaciens*						S					T (ip)	Polyene(heptaene) group	1, 2
Aureofungin		*Stm. cinnamomeus* var. *terricola*	C: 60.4, H: 7.9, N: 2.2					S					T	Polyene(heptaene) group	1
Aureolic acid		*Stm. sp.*	$(C_{56-60}H_{96-104}O_{29-31})_2 Mg$		S								T		1, 2
Aureothricin	Farcinicin	*Stm. thioluteus* *Stm. farcinicus* *Stm. celluloflavus* *Stm. cyanoflavus* *Stm. sp.* 2336	$C_9H_{10}O_2N_2S_2$	A-40	S	S		S					M (sc)	Pyrrothine group	1, 2
Aurimycin		*Stm. lavendulae*			S	S								Aminocyclitol (Neomycin) group	1
Aurovertin			$C_{25}H_{32}O_9$											Mode act.: R, O.	A-81
Avilamycin	(Exfoliatin)	*Stm. viridochromogenes* NRRL 2860	$C_{31}H_{47}O_{17}Cl$		S								L (sc)		1

Name	Synonym	Producer	Formula	Struct. Ref. No.	G +	G –	Myco	Fungi	Virus	Protoz	Tumor	Others	Tox.	Remarks	Ref.
Axenomycin A, B		*Stm. lisandri*	A: C 61.74, H 8.52, O 28.87 B: C 62.8, H 8.58, O 27.8							S					A-82
Ayamycin A		*Stm. sp.* 0-80 res. *Stm. flaveolus*	$C_{28}H_{38}O_{10}$		S						S		T	pH Indicator	1, 2
Ayamycin A_1		*Stm. sp.* 0-80 res. *Stm. flaveolus*			S						S			pH Indicator	1, 2
Ayamycin A_2		*Stm. sp.* 0-80 res. *Stm. flaveolus*	$C_{28}H_{38}O_{10}$		S						S			pH Indicator	1, 2
Ayamycin A_3		*Stm. sp.* 0-80 res. *Stm. flaveolus*			S						S			pH Indicator	1, 2
Ayamycin B	(Luteomycin)	*Stm. sp.* 0-80 res. *Stm. flaveolus*			S						S			pH Indicator	1, 2
Ayfactin A	AYF, (Aureofacin)	*Stm. aureofaciens* *Stm. viridifaciens*	$C_{25}H_{35-6}O_7N$					S					T (ip)	Polyene(heptaene) group	1, 2
Ayfactin B		*Stm. aureofaciens* *Stm. viridifaciens*	$C_{25}H_{35-6}O_7N$					S						Polyene(heptaene) group	1, 2
Azacolutin	F17-C	*Stm. cinnamomeus* forma *azacoluta*						S						Polyene(heptaene) group	1
5-Azacytidine	U-18496	*Streptoverticillium* *S. ladakanus* var. *ladakanus* UC 2654	$C_8H_{12}N_4O_5$	A-41	S	S					S				A-83, A-84

Name	Synonym	Producer	Formula	Struct. Ref. No.	G+	G-	Myco	Fungi	Virus	Protoz	Tumor	Others	Tox.	Remarks	Ref.
8-Azaguanine	Pathocidin													See Pathocidin	1
L-4-Azaleucin	2-Amino-3-dimethyl-aminopropionic acid	*Stm. neocaliberis* var. *neocaliberis*	$C_5H_{12}N_2O_2$		S	S									A-52
Azalomycin B		*Stm. hygroscopicus* var. *azalomyceticus*	$C_{14}H_{24}O_5$		S		W						L (ip)		1, 2
Azalomycin M		*Stm. hygroscopicus* var. *azalomyceticus*	$C_{41}H_{70}O_{11}$		S		S	S		S			T (ip)		2, A-85
Azalomycin F		*Stm. hygroscopicus* var. *azalomyceticus*	$C_{30}H_{50}O_{10}N_2$		S		S	S		S			T (ip)	Mode act.: Damage on cell surface of *C. albicans*	1, 2, A-86, A-87
Azalomycin F₃		*Stm. hygroscopicus* var. *azalomyceticus*	C: 60.41, H: 8.80, N: 3.54		S		S	S		S					1, 2, A-86
Azalomycin F₄		*Stm. hygroscopicus* var. *azalomyceticus*	C: 61.51, H: 8.90, N: 3.74		S		S	S		S					1, 2, A-86
Azalomycin F₅		*Stm. hygroscopicus* var. *azalomyceticus*	C: 62.27, H: 8.94, N: 3.88		S		S	S		S					1, 2, A-86
Azaserine		*Stm. fragilis*	$C_5H_7O_4N_3$	A-42	S	S	W	W		S	S		M (ip)	Azaamino acid group Mode act.: N	1, 2, 3, A-88, A-89, A-90, A-91, A-92, A-93, A-94, A-95, A-96

Name	Synonym	Producer	Formula	Struct. Ref. No.	G+	G−	Myco	Fungi	Virus	Protoz	Tumor	Others	Tox.	Remarks	Ref.
Azomultin		*Stm. noboritoensis var. azomultinus* 63210B	$C_{13}H_{22}N_6O_4 \cdot H_2SO_4$		S	S		S					M		A-97
Azomycin	2-Nitroimidazol, 11A	*Nocardia mesenterica* *Stm. eurocidicus*	$C_3H_3O_2N_3$	A-43	S	S	S	W		S			MT		1, 2
Azotomycin	Duazomycin B													See Duazomycin B	1
B-Mycin	Bottromycin A													See Bottromycin A	1
Bagacidin		*Stm. fradiae* NRRL 2598	–HCl: C: 39.5, H: 7.2, N: 15.5, Cl: 17.2		S	S		*S						Pyrimidine group *Candida*	1
Bamicetin	Antibiotic D N-noramicetin	*Stm. plicatus*	$C_{28}H_{40}O_9N_6$	B-1	S		S							Pyrimidine group	1, 2
Bandamycin A		*Stm. goshikiensis*	C: 58.83, H: 8.25		S	S	W						L	Macrolide group	1, 2
Bandamycin B		*Stm. goshikiensis*	C: 59.96, H: 8.11		S	S	W						L	Macrolide group	1, 2
Bihoromycin		*Stm. filipinensis var. bihoroensis*	$C_{41}H_{76}O_{13}$		S			*[1]S	*[2]S				*[3]T (ip)	*[1]*Candida albicans Piricularia oryzae* *[2]TMV virus *[3]Also Phyto-toxic	B-1
Biomycin (USSR)	Chlortetracycline													See Tetracycline	1
Biovetin															B-2
Blamycin		*Stm. sp.* 658	$C_{32}H_{55}NO_{11}$		*[1]S		S	*[2]S					T (or)	*[1]*Sarcina lutea* *[2]*Piricularia oryzae*	B-3, B-4
Blasticidin A (B, C)		*Stm. griseochromogenes* 2A-327	$C_{51-52}H_{99-101}O_{23}N$ ½ Ca		S	*[1]S		*[2]S					T (ip)	*[1]*Pseudomonas solanacearum, Proteus vulgaris, Shigella sonnei* *[2]Plant pathogens	1, 2, B-5

Name	Synonym	Producer	Formula	Struct. Ref. No.	G+	G-	Myco	Fungi	Virus	Protoz	Tumor	Others	Tox.	Remarks	Ref.
Blasticidin S		Stm. griseochromogenes, Stm. globifer, Stm. morookaensis, Stm. sp.	$C_{17}H_{26}O_5N_8$	B-2	W	W	W	*S					T	*Piricularia oryzae, Mode act.: P	1, 2, 3, B-6, B-7
Blastmycin	Antimycin A₃	Stm. blastmyceticus		A-35										See Antimycin A₃, Mode act.: R.	1, 2, B-8, B-9
Bleomycins		Stm. verticillus NIHJ 424		B-3	S	S	S	W			S		MT (ip)	Peptide group, Mode act.: N	1, 3, B-10, B-11, B-12, B-13, B-14, B-15, B-16, B-17, B-18
Bluensin	Bluensomycin Glebomycin U-12898													See Bluensomycin	1
Bluensomycin	Glebomycin U-12898 Bluensin	Stm. bluensis var. bluensis	$C_{21}H_{39}O_{14}N_5 \cdot 2HCl$	B-4	S	S							L (ip)	Aminocyclitol group, Mode act.: P	1, 2, 3, B-19, B-20
Boromycin		Stm. antibioticus	$C_{45}H_{74}BNO_{15}$	B-4/5	S		W	*S					M (or)	*Cand. alb., Rhodotorula rubra, Paecilom. varioti	B-20/21
Borrelidin		Stm. rochei, Stm. sp. C2898 related to Stm. griseus	$C_{28}H_{43}O_6N$	B-5	S				S	S	S		MT		1, 2, B-21, B-22
Boseinycin	Ac₆ 569	Stm. sp. Ac₆ 569	$C_{24}H_{46}N_9O_8 \cdot 4HCl$		S	S		S						Streptothricin-like	B-23, B-24

Name	Synonym	Producer	Formula	Struct. Ref. No	G+	G−	Myco	Fungi	Virus	Protoz	Tumor	Others	Tox.	Remarks	Ref.
Botrycidin		*Stm. aureofaciens* Tü 342	C: 66.35, H: 9.18, O: 24.90, MW: 701					S*						*Botrytis cinerea* Co-product: venturicidins	B-25
Bottromycin complex		*Stm. canadensis* MA-959 (ATCC 17776)										*S		Contain metho-bottromycin *Mycoplasma Co-product: Netropsin	B-26, B-27
Bottromycin	Bottromycin A (or NA$_1$), B-Mycin	*Stm. bottropensis*	$C_{38}H_{57-61}O_{7-8}N_7S$		S	S	S						M	Peptide group	1, 2
Bottromycin A$_1$	A$_1$ (formerly A): Bottro-mycin	*Stm. sp.* 3668-L2	$C_{41}H_{62}O_7N_8S$	B-6	S	S	S						M		1
Bottromycin A$_2$	MA-251C$_1$	*Stm. sp.* 3668-L2	$C_{42}H_{62}O_7N_8S$	B-7	S	S	S								1
Bottromycin B		*Stm. sp.* 3668-L2	$C_{40}H_{60}O_7N_8S$	B-8	S	S									1
Bottromycin B$_2$		*Stm. sp.* 3668-L2	$C_{41}H_{60}O_7N_8S$		S	W	S								B-28, B-29
Bottromycin C$_2$		*Stm. sp.* 3668-L2	$C_{43}H_{64}N_8O_7S$		S	W	S								B-28, B-29
Bovinocidin	β-nitropropionic acid	*Stm. sp.*	$C_3H_5O_4N$	B-9			W						MT		1, 2
Bramycin		*Stm. sp.* 658 related to *Stm. diastatochromogenes*	$C_{32}H_{55}O_{11}N$					S					T		1, B-30, B-31, B-32
Bromotetracycline				T-4										See Tetracycline	

Name	Synonym	Producer	Formula	Struct. Ref. No.	G +	G -	Myco	Fungi	Virus	Protoz	Tumor	Others	Tox.	Remarks	Ref.
Bruneomycin	Streptonigrin	*Acm. albus.* var. *bruneomycini*, *Acm. sp.*			S	S					S			Mode act.: N	1, 3, B-33
Bryamycin	Thiostrepton	*Stm. hawaiiensis*	C: 51.7–52.2, H: 5.56–5.66 N: 15.3–18.1, S: 9.2–10.1		S								L (ip)		1, 2
Bulgerin		*Stm. aburaviensis* var. *tuffformis*	$C_{17}H_{24}O_{10}N_4$					S							B-34
Bulging factor		*Stm. sp.*						S							1, 2
Bundlin A		*Stm. griseofuscus*	C: 64.35, H: 7.00, N: 2.99		S			W					L (or)		1, 2 1, 2
Bundlin B	T2636A	*Stm. griseofuscus*	C: 63.24, H: 7.20, N: 2.30		S								L (or)		1, 2, NA-99
Cacaomycetin		*Stm. sp.* related to *Stm. cacaoi*			W	W	W	W					L		1, 2
Caerulomycin		*Stm. caeruleus* *Stm. tenebrarius*	$C_{12}H_{11}O_2N_3$	C-1	S	W	W	S						Co-product: tenebrimycin	1, C-1, C-2
Camphomycin		*Stm. rutgersensis* var. *castelarense*			S	W	S	W					L		1, 2
Canarius	U-13714	*Stm. canarius* var. *canarius*	$C_{13}H_{21}O_8N_7$					S	S				T (ip)		1
Cancidin A	(Gancidin A)										S		MT		1
Cancidin W	Gancidin W										W				1
Candicidin A, B, C		*Stm. griseus* *Stm sp.*	A: C: 62.9, H: 9.6, N: 4.7 B: C: 57.8, H: 9.9, N: 7.3					S					ML (sc)	Polyene (heptaene) group Mode act.: M membrane	1, 2, 3

Name	Synonym	Producer	Formula	Struct. Ref. No.	G+	G-	Myco	Fungi	Virus	Protoz	Tumor	Others	Tox.	Remarks	Ref.
Candidin		*Stm. viridoflavus*	$C_{46}H_{73}O_{16}N$					S					TM (ip)	Polyene (heptaene) group Mode act.: M	1, 2, 3
Candimycin		*Stm. ehimensis*	C: 57.17, H: 8.81, N: 1.70				S						T (sc)	Polyene (heptaene) group	1, 2
Capacidin		*Stm. sp.*	$C_{54}H_{35}O_{18}N_2$				S	S					T (ip)	Polyene (pentaene) group	1, 2
Capreomycin I	Capromycin	*Stm. capreolus*	$C_{25-28}H_{50-53}O_{10}N_{14}$		W	W	S	W					L	Peptide group Mode act.: P	1, 2, C-3
Capreomycin II		*Stm. capreolus*	$C_{24-25}H_{51-52}O_{9-10}N_{12-14}Cl_4$		W	W	S	W					L		1, 2, C-3
O-Carbamyl-D-Serine		*Stm. sp.* *Stm. narbonensis* *Stm. fradiae, Stm. polychromogenes*	$C_4H_8O_4N_2$	C-2	S		S						L	Amino acid group Mode act.: C	1, 2, 3, C-4, C-5
Carbomycin	Carbomycin A, M-4209 Magnamycin	*Stm. halstedii, hygroscopicus, tendae, albireticuli macrosporeus*	$C_{42}H_{67}O_{16}N$	C-3	S	W	S						L	Macrolide group Mode act.: P	1, 2, 3
Carbomycin B	Magnamycin B	*Stm. halstedii*	$C_{42}H_{67}O_{15}N$	C-4	S	W	S						L		1, 2
Cis-β-Carboxy-acrylamidine	U-20,904	*Acm. sp.*	$C_4H_6N_2O_2$	C-5							S		T (ip)		1, C-6
3-Carboxy-2,4-pentadienal lactol	A 415-Z3, PA-147	*S. sp. A415-Z3*	$C_6H_5O_3Ca\frac{1}{2} \cdot H_2O$	C-6	S						W		ML		1
Carcinomycin	Ganmmycin	*Stm. carcinomyceticus* *Stm. ganmmycicus*	C: 40.35, H: 6.23, N: 10.93								S		L	Macromolecular peptide group	1
Cardicin		*Noc. sp.*			S		S	S					T		1, 2

Name	Synonym	Producer	Formula	Struct. Ref. No.	G+	G-	Myco	Fungi	Virus	Protoz	Tumor	Others	Tox.	Remarks	Ref.
Caryomycin		*Stm. filamentosus*			S	W	S								1, 2
Carzinocidin		*Stm. kitasawaensis*	C: 37.2, H: 6.1 N: 12.2, S: 3.5								S		T	Macromolecular peptide group	1, 2
Carzinomycin	Carcinomycin													See Carcinomycin	1
Carzinophilin A		*Stm. sahachiroi*	$C_{50}H_{58}O_{18}N_5$	C-7	S	W	S				S		T	Mode act.: N	1, 2, C-7
Carzinostatin A, B		*Stm. sp.* closely related to *Stm. albus*									S			Mode act.: N	1, 2
Catenulin	Paromomycin	*Stm. catenulae* *Stm. sp.*	$-H_2SO_4$ C: 31.45, H: 6.15 N: 7.92, SO_4: 28.12		S	S	S						ML	Aminocyclitol group Mode act.: P	1, 2
Celesticetin	D-52	*Stm. caelestis*	$C_{24}H_{36}O_9N_2S$	C-8	S								ML (ip)	Lincomycin group	1, 2, 3
Cellocidin	Acetylene dicarboxamide; aquamycin, lenamycin	*Stm. chibaensis*	$C_4H_4O_2N_2$	C-9	S	S					S		T		1, 2
Cellostatin		*Stm. cellostaticus*	C: 52.1, H: 4.52, N: 13.15		W	W		W		W	S		T		1, 2
Cephalomycin		*Stm. tanashiensis* var. *cephalomyceticus*	C: 55.39, H: 6.66, N: 9.93				W		S				T	Peptide group	1, 2
Cerevioccidin		*Stm. sp.*	$C_{22}H_{39}O_4N_5$					S					M		1, 2
Cerulomycin		*Acm. coerulescens*	No N & S						S				M		2
Cervicarcin		*Stm. ogaensis*	$C_{19}H_{20}O_9$ (½H_2O)	C-10							S		MT		1, 2
Chainin		*Chainia minutisclerotica*	$C_{33}H_{54}O_{11}$					S				S		Polyene (pentaene) group	C-8

Name	Synonym	Producer	Formula	Struct. Ref. No.	G +	G −	Myco	Fungi	Virus	Protoz	Tumor	Others	Tox.	Remarks	Ref.
Chalcomycin		*Stm. bikiniensis*	$C_{35}H_{56}O_{14}$	C-11	S	W	S							Neutral macrolide Mode act.: P	1, 2, 3
Chantalmycin		*Stm. sp.*			S		S	S						Polyene group	1
Chartreusin	X465A, 747, 1293	*Stm. chartreusis* *Stm. sp.*	$C_{32}H_{32}O_{14}$	C-12	S		S						L (ip)		1, 2
Chelocardin		*Noc. sulphusea*	$C_{22}H_{21}NO_7$		S	S							M	Tetracycline group	1
Chloramphenicol	Levomycin (USSR)	*Stm. venezuelae* *Stm. phaeochromogenes var. chloromyceticus* *Stm. omiyaensis*	$C_{11}H_{12}O_5N_2Cl_2$	C-13	S	S	S		S				ML	Mode act.: P	1, 2, 3, C-9, C-10
7-Chloro-5a(11a)-dehydrotetracycline				T-5										See Tetracycline	
7-Chloro-6-Demethyl-Tetracycline														See Tetracycline	
Chlortetracycline				C-14											
Chlorothricin		*Stm. antibioticus* 99	$C_{50}H_{63}O_{16}Cl$	C-15	S									See Tetracycline	C-11
Chrestomycin	Aminosidin													See Aminosidin	1
Chromin		*Stm. sp.*	C: 58.19, H: 7.81, N: 2.29					S					T (ip)	Polyene (tetraene) group	1, 2
Chromocyclomycin		*Stm. sp.* La-7017	$C_{48}H_{64}O_{21}$	C-16											C-12.
Chromomycin complex		*Stm. griseus* *Stm. olivochromogenes*	A_4: $C_{48}H_{68}O_{22}$		S						S		T (ip)	Aurelic acid group	1

Name	Synonym	Producer	Formula	Struct. Ref. No.	G+	G-	Myco	Fungi	Virus	Protoz	Tumor	Others	Tox.	Remarks	Ref.
Chromomycin A_2		Stm. griseus / Stm. olivochromogenes	$C_{59}H_{86}O_{26}$	C-17	S						S		T (ip)	Aureolic acid group	1, 2
Chromomycin A_3	Toyomycin	Stm. griseus / Stm. olivochromogenes	$C_{57}H_{82}O_{26}$	C-18	S						S		T (ip)	Aureolic acid group; Mode act.: N	1, 2, C-14, C-15, C-16, C-17, C-18, C-19
Chromomycin A_4		Stm. grisins / Stm. olivochromogenes	$C_{48}H_{68}O_{22}$	C-18											
Chrothiomycin		Stm. pluricolorescens	$C_{27}H_{31-33}O_{13}NS$		W								TM	Catecholaminoxy-genase inhibitor	C-20
Chrysomallin		Acm. chrysomallus 2703									S	S	T	Actinomycin group	1
Chrysomycin		Stm. sp.	$C_{22}H_{20}O_7$						S				MT	Aureolic acid group	1, 2
Ciclacidin(a)		Stm. capoamus		C-22							S	S	MT	pH Indicator	1, C-21
Cicla-micin(a)		Stm. c?poamus 4670-IA-37				S	S				S	S	MT	Co-product: Ciclacidin(a)	1
Cineromycin A		Stm. cinero-chromogenes	$C_{37}H_{42}O_9\,NCl$ or $C_{37}H_{42}O_9\,NClCH_3OH$			S	S	S	S	W	S	S	T		1
Cineromycin B		Stm. cinero-chromogenes	$C_{17}H_{26}O_4$			S	S	S	S	W	S	S	ML (ip)		1
Cinerubin A	Ryemycine B_2	Stm. antibioticus / Stm. galilaeus / Stm. niveoruber	$C_{44}H_{59}O_{18}\,N \pm CH_2$			S	S	S	S	W	S	S	T	Anthracycline group; Mode act.: N	1, 2, 3, C- 22, C-23
Cinerubin B	Ryemycine B_1	Stm. antibioticus / Stm. galilaeus / Stm. niveoruber	$C_{44}H_{59}O_{18}\,N \pm CH_2$			S	S	S	S	W	S	S	T	Anthracycline group; Mode act.: N	1
Cinnamonin	Actithiazic acid													See Actithiazic acid	1

Name	Synonym	Producer	Formula	Struct. Ref. No.	G +	G -	Myco	Fungi	Virus	Protoz	Tumor	Others	Tox.	Remarks	Ref.
Cinnamycin		Stm. cinnamomeus			W		S						T (ip)	Peptide group	1, 2
Cirolerosus	U-12241	Stm. bellus var. cirolerosus NRRL 3107			S	S					S				1
Cirramycin A		Stm. cirratus	$C_{27}H_{45}O_8N$		S	W							M	Macrolide group	1, 2, C-24
Cirramycin A₁		Stm. cirratus 12090	$C_{13}H_{51}O_{10}N$	C-23	S	S		S				*S	L	Macrolide group *Mycoplasma	C- 25, C-26
Cirramycin B		Stm. cirratus	$C_{36}H_{59}O_{12}N$		S	W							M	Macrolide group	1, 2
Citromycin		Stm. sp. IN-1483 Stm. sp. IN-2035	–HCl: C: 35.58 H: 6.36, N: 17.42 O: 26.96, Cl: 11.13		S	S	S						M	Streptothricin-like	C-27
a-Citromycinon		Stm. purpurascens	$C_{20}H_{18}O_7$	C-24	W										C-28
γ-Citromycinon		Stm. purpurascens	$C_{20}H_{18}O_6$	C-24	W										C-28
Cladomycin		Stm. lilacinus			S	S	W						L		1, 2
Coccomycin		Stm. sp.			S						S				1
Coelicolorin		Stm. coelicolor			W								L (ip)	pH Indicator	1, 2
Coelicomycin	Actinorhodin													See Actinorhodin	1
Coerulomycin		Acm. coerulescens			S				W					Different from caerulomycin	1
Coformycin		Noc. interforma Stm. lavendulae Stm. kaniharaensis	$C_{11}H_{16}O_5N_4$			S							T (Delayed)	Only active when combined with Formycin A	C-29
Colimycin (USSR)	Neomycin													See Neomycin	1, 2
Collinomycin		Stm. collinus	C: 59.41, H: 4.01, OMe: 16.2		S									pH Indicator	1, 2

Name	Synonym	Producer	Formula	Struct. Ref. No.	G+	G-	Myco	Fungi	Virus	Protoz	Tumor	Others	Tox.	Remarks	Ref.
Congocidin(e)	Netropsin	*Stm. ambofaciens*	$C_{18}H_{26}O_3N_{10}$		S	S							ML (sc)		1, 2
Copiamycin		*Stm. hygroscopicus var. chrystallogenes*	C: 58.30, H: 8.90, N: 3.94					S					T (ip)		1
Costreptomycin		*Stm. griseus* LS-1												Aminocyclitol group; Hydrolysis gives streptomycin	1
Coumermycin A$_1$		*Stm. rishiriensis* / *Stm. spinicoumarensis* / *Stm. spinichromogenes*	$C_{55}H_{59}O_{20}N_5$	C-26	S	S	S						ML (sc)		1
Coumermycin A$_2$		*Stm. rishiriensis* / *Stm. spinicoumarensis* / *Stm. spinichromogenes*	Na: $C_{53}H_{54}O_{20}N_5$	C-27	S	W	S								1
Cranomycin		*Stm. viridogriseus*	$C_{22}H_{29-31}O_6N_3$		S	W	S	S					T (ip)		1, 2
Cremeomycin	(U-23643)	*Acm. cremeus*	$C_8H_6N_2O_4$		S	S		S							C-30
Cremomycin		*Stm. sp.*			S	S							M		1
Croceomycin	Resistomycin	*Stm. arabicus* 6762	$C_{22}H_{18}O_6$		S		S						T (ip)		1

Name	Synonym	Producer	Formula	Struct. Ref. No.	G +	G −	Myco	Fungi	Virus	Protoz	Tumor	Others	Tox.	Remarks	Ref.
Cryptocidin(e)		*Stm. sp.* 963			S	W		S					T	Polyene(hexaene) group	1, 2
Crystallinic acid	Novobiocin	*Stm. griseus*, *Stm. spheroides*, *Stm. griseoflavus*, *Stm. niveus*	$C_{31}H_{36}O_{11}N_2$		S								ML (ip)		1
Crystallomycin		*Stm. violaceoniger var. crystallomycini*	N: 12.24		S								M (ip)		1, 2
Curamycin		*Stm. curacoi*	$C_{53-55}H_{82-86}O_{32-33}Cl_2$	C-28	S				S						1
Cyanomycin	4738-B, Pyocyanine	*Stm. cyanoflavus*	$C_{15}H_{12}O_2N_2$		S	S	W						T	pH Indicator	1, 2
1-substituted-3-cyclohexene		*Stm. antibioticus*		C-29	S	S	S								C-31
Cycloheximide	Naramycin A, Actidione	*Stm. griseus*, *Stm. albulus*, *Stm. naraensis*, *Stm. pulveraceus*, *Stm. noursei*, *Stm. chrysomallus*	$C_{15}H_{23}O_4N$	C-30				S	S	S			M (ip)	Glutarimide group Mode act.: P, (C), N	1, 2, 3, C-32, C-33
Cycloheptamycin		*Stm. sp.* (MZ11158)	$C_{48}H_{68}N_8O_{12}$	C-31	S		S								C-34
Cycloserine	Orientomycin	*Stm. garyphalus*, *Stm. lavendulae*, *Stm. nagasakiensis*, *Stm. orchidaceus*, *Stm. roscochromogenes*, *Stm. sp.* E-733	C: 35.4, H: 5.98, N: 26.9	C-32	S	W	S						L	Mode act.: C	1, 2, 3, C-35, C-36
Cytomycin		*Stm. griseochromogenes*	$C_{17}H_{23}O_5N_7$	C-33						S			L (ip)	Pyrimidine group	1, 2

Name	Synonym	Producer	Formula	Struct. Ref. No.	G+	G−	Myco	Fungi	Virus	Protoz	Tumor	Others	Tox.	Remarks	Ref.
Cytovirin		Stm. olivochromogenes $(C_4H_5ON_3)_n$ var. cytovirinus	$C_{17}H_{24-28}O_5N_8$	C-34					*S					*Plant virus	1
Dactinomycin	Actinomycin D													See Actinomycin	1
Danomycin		Stm. albaduncus	$C_{73}H_{125}O_{37}N_{10}Fe$		S		W	W					L	Sideromycin group	1, 2
Danubomycin		Stm. griseus	$C_{22}H_{24-28}O_{4-6}N$		S	S	W	W						pH Indicator	1
Daunomycin	F. I 1762 Daunorubicin Rubidomycin	Stm. peucetius IMI 101335 CIB 9475 IM 3868	$C_{27}H_{29}O_{10}N \cdot HCl$	D-1	W					S	S		T	Anthracycline group Mode act.: N	1, 2, 3, D-1, D-2
Daunorubicin	Daunomycin Rubidomycin													See Daunomycin	
Decoyinine	Angustomycin A	Stm. hygroscopicus var. decoyicus Stm. endus. or. Stm. hygroscopicus CBS	$C_{11}H_{13}O_4N_5$	D-2	S	S	S	S						Purine group	1, 2
a-dehydrobiotin		Stm. lydicus	$C_{10}H_{14}N_2DS$	D-3	S	S	S	S							D-3
Dehydrochlortetra-cycline														See Tetracycline	1
Demethylchlortetra-cycline														See Tetracycline	1, 2
L'-Demethyl clindamycin		Addition of clindamycin to fermentations of Stm. punipalus	$C_{17}H_{31}N_2O_5SCl$	D-4											D-4
4-N-demethyl-4-N-ethyloxytetracy-cline														See Tetracycline	

Name	Synonym	Producer	Formula	Struct. Ref. No.	G +	G −	Myco	Fungi	Virus	Protoz	Tumor	Others	Tox.	Remarks	Ref.
4-N-demethyl-4-N-ethyltetracycline														See Tetracycline	
4-N-demethyl-4-N-ethyl-6-demethyl-tetracycline														See Tetracycline	
4-N-demethyl-4-N-ethylhalotetracycline														See Tetracycline	
Demethyltetracycline				T-6										See Tetracycline	
N-demethyl-lincomycin														See lincomycin	
N-demethyl-streptomycin														See Streptomycin	
1-demethyl-thio-1-hydroxylincomycin														See lincomycin	
Demetric acid	NSC B-152222	*Stm. umbrosus* var. *suragaoensis* ATCC 19104	$C_{18}H_{18}O_2$	D-5	S	S	S	S		S	S		M (ip)		1, D-5
Denamycin		*Stm. sp.* 8756-CC$_2$	C: 68.26, H: 8.71, O: 22.70		S						S		L (ip)	Mode act.: N	D-6
Denofungin		*Stm. hygroscopicus*						S							D-7
Deoxynybomycin		*Stm. hyalinum*	$C_{16}H_{14}N_2O_3$	D-6	S	S	S								D-8
Dermostatin		*Stm. viridogriseus*	C: 64.6, H: 8.6, O: 27.0					S					MT (ip)		1, 2
Desdamethine	U-11,994	*Stm. caelestis*	$C_8H_{14}N_2OS$		S	S								Related to desdanine	D-9
Desdanine		*Stm. caelestis*	$C_7H_{10}N_2O$			S									D-9

Name	Synonym	Producer	Formula	Struct. Ref. No.	G +	G -	Myco	Fungi	Virus	Protoz	Tumor	Others	Tox.	Remarks	Ref.
Desertomycin		*Stm. flavofungini*	$C_{33}H_{60-62}O_{14}N$		S	S		S	S		S		T		1
Desideus		*Stm. griseus var. desideus*	$C_{40}H_{64}O_{12}$		S	S									1
Destomycin A		*Stm. rimofaciens*	$C_{20}H_{37}O_{13}N_3$	D-7	S	S	S	W					MT	Aminocyclitol group (related to hygromycin B)	1
Destomycin B		*Stm. rimofaciens*			S	S	S	W					MT	Aminocyclitol group	1
Detoxin C_1, D_1 (A,B,C,D,E,F,G,H)		*Stm. caespitosus var. detoxicus* 7072 GC_1, *Stm. mobaraensis*	C_1: $C_{29-30}H_{44-46}O_9N_4$ D_1: C: 59.09, H: 7.46, O: 23.49, N: 8.98		S									Synergist of Blasticidin S	D-10, D-11
Deutomycin		*Stm. flavochramogenes var. deutoensis*	C: 38.74, H: 6.91, N: 9.43 (M.W. 3000–5000)								S	*S	L (ip)	Sugar-peptide group *Mycoplasma*	D-12
Dextromycin	Neomycin Streptothricin B	*Stm. sp.* A-C404	$C_{23}H_{46}O_{13}N_6$		S	S							M (ip)	Aminocyclitol group	1, 2
Dianemycin		*Stm. hygroscopicus*	$C_{51}H_{88}O_{16}$		S	S	S	S				*S	T (sc)	Monensin group *Mycoplasma*	D-13
Diazomycin	Duazomycin	*Stm. ambofaciens*	$C_{17}H_{23}O_8N_7$					S	S		S			See Duazomycin	
6-Diazo-5-Oxo-L-Norleucin	DON	*Stm. ambofaciens* *Stm. phaeochromogenes*	$C_6H_9O_3N_3$	D-8	S			S	S		S		MT	Azaamino acid group Mode act.: N	1, 2, 3, D-14
Dienomycin A		*Stm. sp.* (MC67-Cl)	$C_{20}H_{27}NO_2 \cdot HCl$	D-9			W						M (ip)		D-15
Dienomycin B		*Stm. sp.* (MC67-Cl)	$C_{18}H_{23}NO_2 \cdot HCl$	D-9									L (ip)		D-15

Name	Synonym	Producer	Formula	Struct. Ref. No.	G+	G-	Myco	Fungi	Virus	Protoz	Tumor	Others	Tox.	Remarks	Ref.
Dienomycin C		Stm. sp. (MC67-Cl)	$C_{16}H_{21}NO \cdot HCl$	D-11									T (ip)		D-15
Dihydronancimycin				D-9										See Nancimycin	
Dihydrostreptomycin	23572-A	Stm. humidus	$C_{21}H_{41}O_{12}N_7 \cdot 3HCl$	D-12	S	S	S						L	Aminocyclitol group	1, 2, 3
5,8-dihydroxy-2,7-Dimethoxy-1,4-naphthoquinone	Spinochrome M	Stm. sp. 12396	$C_{12}H_{10}O_6$	D-13	S			S					L (ip)		D-16
2,7-Dimethoxy-5-hydroxy-1,4-naphthoquinone	Spinochrome M	Stm. sp. 12396	$C_{12}H_{10}O_5$	D-14				S							D-16
Dihydroxyphenazine	1,6-phenazinediol	Stm. thioluteus	$C_{12}H_8O_2N_2$	D-15				S						pH Indicator	1
Dinactin		Stm. sp. ETH A-23112 / Stm. werraensis / Stm. sp. SC 3763	$C_{42}H_{68}O_{12}$		S		S	S				*S	L (ip)	Cyclopolylactone, Macrotetrolide group Kills insect (Axukibean weevil) & *Priicularia oryzae* Mode act.: O	1, 2, 3, D-17
Distacyne		Stm. distallicus NCIB-8936	$C_{22}H_{25}O_6$		W		W								1
Distamycin A		Stm. distallicus NCIB-8936	$C_{22}H_{27}O_4N_9$	D-16					S		S		MT	Netropsin	1, 2, D-18, D-19
Diumycin A, B		Stm. umbrinus ATCC 15972	C H N P O A: 46.5 7.2 4.5 2.0 39.8 B: 45.9 6.6 4.6 1.8 41.1		S		S				S			Co-product: umbrinomycin	D-20

Name	Synonym	Producer	Formula	Struct. Ref. No.	G+	G−	Myco	Fungi	Virus	Protoz	Tumor	Others	Tox.	Remarks	Ref.
Doricin		*Stm. loidensis*	$C_{43}H_{52}O_{11}N_8$	D-17										Depsipeptide (mikamycin) group	1
Duamycin		*Stm. hygroscopicus subsp. duamyceticus*	$C_{36}H_{60}O_{10}$		S			S							D-21
Diazomycin A	Diazomycin A N-Acetyl-DON	*Stm. ambofaciens*	$C_8H_{11}O_4N_3$	D-18	S			*S			S			*yeast only Azaamino acid group Mode act.: N	1, 2, D-22
Diazomycin B	Diazomycin B	*Stm. ambofaciens*	$C_{17}H_{23}O_8N_7$	D-19	S			*S			S			*yeast only Azaamino acid group Mode act.: N	1, 2, D-22
Diazomycin C	Diazomycin C	*Stm. ambofaciens*	$C_{15}H_{21}O_6N_7$		S			*S			S			*Yeast only Azaamino acid group Mode act.: N	1, 2, D-22
Duramycin		*Stm. cinnamoneus forma azacoluta*	Picrate C: 51.30, H: 5.76 N: 16.85, S: 3.18					S						Peptide group	1, 2
Durhamycin		*Stm. durhamensis*	C: 63.78, H: 10.16 O: 25.54, N: trace ($C_{15}H_{30}O_5$) Prob.					S	S				T	Polyene (pentaene) group	D-23
Echanomycin	Echinomycin, Levomycin, A, X-53III, Quinomycin X-948	*Stm. sp.* *Stm. echinatus*	$C_{50}H_{60}O_{12}N_{12}S_2$		S		W	W	S		S		T (ip)	Quinoxaline group	1
Echinomycin	Quinomycin A	*Stm. sp.* *Stm. echinatus*	$C_{50}H_{60}O_{12}N_{12}S_2$	E-1	S		W	W	S		S			Quinoxaline group Mode act.: N	1, 2, 3

Name	Synonym	Producer	Formula	Struct. Ref. No.	G+	G−	Myco	Fungi	Virus	Protoz	Tumor	Others	Tox.	Remarks	Ref.
Ehrlichin		Stm. lavendulae							S				M (ip)		1, 2
Elaiomycin	No. 1252	Stm. gelaticus Stm. sp.	$C_{13}H_{26}O_3N_2$	E-2			S						MT		1, 2
Elaiophylin		Stm. melanosporus var. melanosporofaciens	$(C_6H_{10}O_2)n$		S	S	S						ML (ip)		1, 2
Emimycin		Stm. sp. 2020-1 related to Stm. griseochromogenes	$C_4H_4O_2H_2$	E-3	S	S								Antagonist of uracil	1, 2, E-1
Encaline group		Stm. halstedii									S				1, 2
Endomycin A, B		Stm. endus Stm. sp. 9-20			S	S		S						Polyene (tetraene) group Similar to Helixins Mode act.: M	1, 2
Enduracidin		Stm. fungicidicus B5477	C: 53.2±0.5, H: 6.54±0.3 N: 14.45±0.5, C: 13.36±0.5		S		S						ML	Peptide group Mode act.: C	E-2, E-2a, E-2b
Enomycin		Stm. mauvecolor				S					S		T	Peptide group	1, 2
Enteromycin	Seligocidin	Stm. achromogenes, Stm. albireticuli	$C_6H_8O_5N_2$	E-4	S	S		S					M		1, 2
Enteromycin carboxamide		Stm. sp. AO-126	$C_6H_9O_4N_3$	E-5	S	S	W							Enteromycin group	1
Ericamycin		Stm. varius (SE-548)	$C_{30-31}H_{23}O_8N$		S		W						T (ip)	pH Indicator	1

Name	Synonym	Producer	Formula	Struct. Ref. No.	G +	G -	Myco	Fungi	Virus	Protoz	Tumor	Others	Tox.	Remarks	Ref.
Erizomycin		*Stm. griseus var. erizensis*	$C_{27}H_{32}N_4O_8$			W							T		E-3
Erygrisin		*Stm. erythrogriseus*			S								MT (ip)	pH Indicator	1, 2
Erythromycin		*Stm. erythreus*	$C_{37}H_{67}O_{13}N$	E-6	S		S		S				ML	Macrolide group Mode act.: P	1, 2, 3
Erythromycin B		*Stm. erythreus*	$C_{37}H_{67}O_{12}N$	E-7	S		S		S				ML	Macrolide group	1, 2
Erythromycin C		*Stm. erythreus*	$C_{36}H_{65}O_{13}N$	E-8	S		S		S				ML	Macrolide group	1, 2
Etabetacin		*Stm. sp.* Ep/7	C: 56.78, H: 8.886 N: 12.19, O: 24.177								S		T		E-4
Etamycin	Viridogrisein K-179, F1370A	*Stm. sp. res.*, *Stm. lavendulae*, *Stm. griseus*, *Stm. griseoviridis*	$C_{44}H_{62}O_{11}N_8$	E-9	S		S						L	Depsipeptide group	1, 2
Ethesdanine	U-20,207	*Stm. caelestis*	$C_9H_{16}N_2OS$		S	S								Convertible to desdanine	E-5
Eucaline		*Acm. sp.* 13363									*S			*Mouse lympholioma NK/L1	E-6
Eulicin		*Stm. sp.* closely related to *Stm. parvus*	$C_{24}H_{52}O_2N_8$	E-10				S					T	Peptide group	1, 2
Eumimycin		*Stm. cinnamonensis var. eumimyceticus*	$C_{20}H_{40}O_9N_2$								S		M (ip)		E-7, E-8

Name	Synonym	Producer	Formula	Struct. Ref. No.	G +	G -	Myco	Fungi	Virus	Protoz	Tumor	Others	Tox.	Remarks	Ref.
Eumycetin		*Stm. sp.* res. *Stm. purpeochromogenes*	C: 62.93, H: 7.63					S	S				T (ip)		1, 2
Eurimycin	(Eurymycin)	*Stm. sp.*			S	S	S						ML (sc)		1, 2
Eurocidin (A,B,C,D)		*Stm. albireticuli, Stm. eurocidicus, Stm. reticuli, Stm. sp.*	C: 57.99, H: 8.13 N: 1.65					S		S			MT (ip)	Polyene (pentaene) group	1, 2
Eurocidin group antibiotic		*Stm. sp.* No. 991-A2, producing actinomycin						S						Polyene (pentaene) group	1, 2
Eurotin (A)		*Stm. griseus* H-5592	C: 46.94, H: 6.29 N: 4.07					S					L	Polyene (heptaene) group	1
Evericin		*Stm. sp.*			S	S	S						T	Peptide group	1, 2
Everinomicin B		*Micromonospora carbonacea*	C: 51.02, H: 6.68, N: 1.23 Cl: 3.97, O—Me: 12.80 C—Me: 7.15, N—Me: 2.33		S	S	S						ML		1
Everinomicin D		*Micromonospora carbonacea*	C: 51.72, H: 6.35, N: 1.40 Cl: 3.89, O—Me: 13.3, C—Me: 6.93, N—Me: 1.98	E-11	S		S								1
Exfoliatin		*Stm. exfoliatus*	$C_{27}H_{40}O_{16}Cl$		S		W						L (sc)		1, 2
Factor B of E129	E129 = Ostreogrycins	*Stm. ostreogriseus* NRRL 2258, NCIB 8792	C: 63.25, H: 7.10 N: 8.05, O: 21.60		S									Streptogramin like Co-product: Factor A (PA114A) and Factor X (PA114B)	1, F-1

Name	Synonym	Producer	Formula	Struct. Ref. No.	G +	G -	Myco	Fungi	Virus	Protoz	Tumor	Others	Tox.	Remarks	Ref.
Factor S of antibiotic 899	(Antibiotic 899 = staphylomycin or virginiamycin)	Stm. virginiae ATCC 13161	$C_{43}H_{49}N_7O_{10}$		S									A component of antibiotic 899 (Staphylomycin), synergistic with Factor M, other component of Staphylomycin	1, F-2
Farcinicin	Aureothricin													See Aureothricin	1
Farmiglucin	Catenulin, Chrestomycin, Hydroxymycin, Aminosidin, Paromomycin, Zygomycin A, F.I. 1600, F.I. 5853													See Aminosidin	1
Fermicidin		Stm. sp.	$C_{14}H_{21}O_4$					*¹S	*²S	S			M	Glutarimide group *¹ Yeast only *² Influenza Mode act.: P	1, 2, 3
Ferramido chloromycin		Stm. sp. AS13	$C_{127}H_{201}N_{24}O_{70}SCl_2Fe_7$		S	W							T		F-3
Ferrimycin A-1	Pilosomycin A-9578	Stm. griseoflavus A-9578 Stm. galilaceus A18822 Stm. lavendulae	$C_{41}H_{67}O_{14}N_{10}Cl_2Fe$	F-1	S	S								Sideromycin group Inhibit catalase activity (antagonized by ferioxamine)	1, 2, F-4, F-5
Ferrimycin A-2	Pilosomycin A-9578	Stm. griseoflavus A-9578 Stm. galilaceus A-822 Stm. lavendulae	2HCl, C: 45.78 H: 6.77, N: 12.75 Fe: 5.29, Cl: 6.23		S	S								Sideromycin group	1, 2

Name	Synonym	Producer	Formula	Struct. Ref. No.	G +	G −	Myco	Fungi	Virus	Protoz	Tumor	Others	Tox.	Remarks	Ref.
Ferromycin	Solemycin	*Stm. mayaensis* res *Stm. olivochromogenes*	$C_{21}H_{54}O_{14}N_7 \cdot 2HCl$		S	S	S							Streptothricin group	1
Fervenulin	Planomycin	*Stm. fervens* NRRL 2755	$C_7H_7O_2N_5$	F-2	S	W	W			S			M (ip)	Purine group	1
Filipin	14-Deoxylagosin	*Stm. filipinensis*	$C_{35}H_{58}O_{11}$	F-3				S					T (ip)	Polyene (pentaene) group Mode act.: M	1, 3
Filipin		*Stm. filipinensis*	I: $C_{35}H_{58}O_9$ II: $C_{35}H_{58}O_{10}$ III: $C_{35}H_{58}O_{11}$ IV: $C_{35}H_{58}O_{11}$	F-4				S		S			T (ip)	Polyene (pentaene) group	F-6
Flavacid		*Stm. sp.* closely related to *Stm. flavus*	C: 61.57, H: 7.77 N: 1.06, Ash: 2.2		S			S		*S			T (ip)	Polyene (hexaene) group *Trichomonas vaginalis* Mode act.: M	1, 2, 3
Flavensomycin	829	*Stm. sp.* res. *Stm. tanashiensis*	C: 63.73, H: 8.04 N: 2.27	F-5	S	S						*S	T (ip)	*Insecticidal	1, 2
Flaveolin		*Stm. sp.* res. *Stm. flaveolus*			S	S	S	S					T	pH Indicator	1, 2
Flavofungin	SA-IX	*Stm. flavofungini* producing desertomycin	$C_{30}H_{48}O_9$	F-6		S		S					T (ip)	Polyene group	1, 2, F-7
Flavomycin	Similar to neomycin	*Stm. roseoflavus* res. *Stm. microflavus*			S	S	S	S					ML	Aminocyclitol (neomycin) group Polyene-pentaene	1, 2
Flavomycoin		*Stm. roseoflavus*	$C_{41}H_{68}O_{10} \cdot 2H_2O$		S	W		S						Polyene group	F-8
Flavucidin		*Stm. sp.* 14420	$C_{34}H_{55-59}O_9N$		*S¹				S²				T (ip)	*¹ *M. flavus* *² Influenza	1, 2

Name	Synonym	Producer	Formula	Struct. Ref. No.	G +	G −	Myco	Fungi	Virus	Protoz	Tumor	Others	Tox.	Remarks	Ref
Florimycin (USSR)	Vinactin, Vinactane Viomycin													See Viomycin	1
Fluorin		Acm. fluorescens group	$C_{36}H_{54}O_{12}$	F-7										Cyclic polylactone group	1, 2
Folimycin		Stm. ncyagawaensis	C: 61.23−61.43, H: 8.74−8.92 N: 1.33−1.46					S					T (ip)		1
Formiglucine	Aminosidin, Catenulin, Chrestomycin, Farmiglucin Hydroxymycin, Paromomycin, Zygomycin A	Stm. chrestomyceticus	$C_{23}H_{45}O_{14}N_5 \cdot \frac{5}{2} H_2SO_4 \cdot 2H_2O$		S	S	S						M	Aminocyclitol group	1, 2
Formycin		Noc. interforma	$C_{10}H_{13}O_4N_5 \cdot H_2O$	F-8		*S	S				*S²		L	Purine group *Xanthomonas oryzae *² Hela, Yoshida Mode act.: N	1, 2
Formycin B	Deaminated Formycin, Laurusin, Oyamycin	Noc. interforma	$C_{10}H_{12}O_5N_4$	F-9		*S			*S²				L	Purine group *Xanthomonas oryzae *² Influeza A	1 F 8/9
Foromacidin A	Spiramycin I	Stm. ambofaciens	$C_{45}H_{78}O_{15}N_2$		S									Macrolide group	1, 2
Foromacidin B	Spiramycin II	Stm. ambofaciens	$C_{47}H_{80}O_{16}N_2$		S										1, 2
Foromacidin C	Spiramycin III	Stm. ambofaciens	$C_{48}H_{82}O_{16}N_2$		S										1, 2
Foromacidin D	(Spiramycin)	Stm. sp. A 8703	C: 59.85, H: 8.48 N: 3.35		S										1, 2

Name	Synonym	Producer	Formula	Struct. Ref. No.	G+	G-	Myco	Fungi	Virus	Protoz	Tumor	Others	Tox.	Remarks	Ref.
Fradicin	X factor	*Stm. fradiae* producing neomycin	$C_{30}H_{34}O_4N_4$ (prob.)					S					T (ip)	Polyene (hexaene) group	1, 2
Fradicin like Antibiotics	A1404	*Stm. fradiae* A1404 (producer of dextromycin)						S							1
Fradiomycin	Neomycin B, C Streptothricin B													See Neomycin	1
Framycetin	Neomycin B, EF-185	*Stm. fradiae,* *Stm. lavendulae*	$C_{23}H_{44}O_{13}N_6$		S	S	S						MT	Aminocyclitol (neomycin) group Mode act.: P	1, 2, 3
Framycin	Neomycins (prob.)												MT		1
Frenolicin		*Stm. fradiae*	$C_{18}H_{18}O_7$	F-10	S			W							1, 2
Fungichromatin	Fungichromin (Lagosin)	*Stm. sp.*	$C_{35}H_{58}O_{12} \cdot H_2O$			S		S							1, 2
Fungichromin	Fungichromatin (Lagosin)	*Stm. cellulosae*	$C_{35}H_{58}O_{12} \cdot H_2O$					S					T	Polyene (pentaene) group Mode act.: M	1, 2, 3
Fungicidin type antibiotic	RAW, 3569 Fungicidin	*Stm. aureus*						S					M (ip)		1, 2
Fungimycin	Aminomycin, NC-1968 Perimycin													See Perimycin	1
Furanomycin		*Stm. sp.* L-803 (ATCC 15795)	$C_7H_{11}NO_3$	F-11	S	S	W		*S					Isoleucine antagonist *Anticoliphage	F-9

Name	Synonym	Producer	Formula	Struct. Ref. No.	G+	G-	Myco	Fungi	Virus	Protoz	Tumor	Others	Tox.	Remarks	Ref.
Fuscomycin		*Stm. fuscus*			S	W							M	pH Indicator	1, 2
Garlirubin		*Stm. galilaeus* JA3043 (cinerubin producing) strain			S	S			S					Anthracycline group	1
Garlirubin A		*Stm. galilaeus* JA3043 (cinerubin producing) strain		G-1	S				S					Aglycon = ϵ-pyrromycinon	G-1
Garlirubin B		*Stm. galilaeus* JA3043 (cinerubin producing) strain		G-1	S				S					Aglycon = Aklavinon	G-1
Garlirubinon B₁	Bis-anhydro-aklavinon	*Stm. galilaeus* JA3043 (cinerubin producing) strain		G-1	S				S						G-1
Garlirubinon B₂	η-Pyrromycinon	*Stm. galilaeus* JA3043 (cinerubin producing) strain		G-1	S				S						G-1
Garlirubinon C	ξ-Pyrromycinon	*Stm. galilaeus* JA3043 (cinerubin producing) strain		G-1	S				S						G-1
Garlirubinon D	7-Desoxy-aklavinon	*Stm. galilaeus* JA3043 (cinerubin producing) strain	$C_{22}H_{20}O_7$	G-1	S				S						G-1
Gancidin A		*Stm. sp.* AAK-82	$C_{43}H_{58-60}O_{14}N_6$		S						S	S	MT		1, 2
Gancidin W		*Stm. sp.* AAK-82	$C_{11}H_{17-19}O_2N_2$								S	S	T		1, 2

Name	Synonym	Producer	Formula	Struct. Ref. No.	G +	G −	Myco	Fung	Virus	Protoz	Tumor	Others	Tox.	Remarks	Ref.
Gangtokmycin		Stm. gangtokensis			S	S		S					T	Polyene pentaene group	1
Gelbecidin		Stm. sp.	C 56.98, H 6.93, O 36.09 (Diff.)		S	W		*S		*S^2				*Candida albicans, Trichophton	G-12
Geldanamycin		Stm. hygroscopicus var. geldanus var. nosa (UC-5208)	$C_{29}H_{40}N_2O_9$	G-2	W	W				S	S		L (ip)	Ansamycin group Mode act.: N	G-2, G-3
Geliomycin		Acm. sp.			S		S								1
Geminimycin A-1, A-2		Stm. sp.	A-1: (C: 72.24 H: 8.02, N: 8.49)		S								L	Peptide group Mixture (A-1:B = 1:1) is effective vs. gram+, −	1, 2
Geminimycin B		Stm. sp.					S						L		1, 2
Gentamicin complex	Gentamycin (patent name)	Mim. purpurea NRRL 2985 Mim. echinospora NRRL 2953	C_1:$C_{17-18}H_{34-36}O_7N_4$		S	S	S						M	Aminocyclitol group	1, G-4, G-5, G-6
Gentamicin A		Mim. purpurea NRRL 2953		G-3	S	S							L	Mode act.: P	1, 3, G-7, G-8
Gentamicin C_1		Mim. purpurea NRRL 2953	$C_{21}H_{43}N_5O_7$	G-3	S	S									1, G-8
Gentamicin C_2		Mim. purpurea NRRL 2953	$C_{20}H_{41}N_5O_7$	G-3	S	S									1, G-8
Gentamicin C_{1a}		Mim. purpurea NRRL 2953	$C_{19}H_{39}N_5O_7$	G-3	S	S									1, G-8

Name	Synonym	Producer	Formula	Struct. Ref. No.	G +	G -	Myco	Fungi	Virus	Protoz	Tumor	Others	Tox.	Remarks	Ref.
Geomycin		Stm. xanthophaeus	$C_{46}H_{96-98}O_{16}N_{16} \cdot 8HCl$	G-4	S	S							*	*Nephrotoxic Streptothricin group	1, 2
Gerobriecin		Stm. jujuy	$C_{35}H_{55}O_{13}N$					S					L	Polyene (heptaene) group	1
Glebomycin	Bluensomycin	Stm. hygroscopicus	$C_{21}H_{39}O_{14}N_5$	G-5	S	S	S							Aminocyclitol (streptomycin) group	1, 2
Globismycin		Stm. griseus			S			S							1
Glucomycin		Stm. flavogriseus var. 12TA	$C_{18}H_{30}N_3O_{11}$		S									Consisting of peptide bond & reduced sugar	G-10
Gluconimycin		Stm. AS9, resemb. Stm. erythreus, albosporeus, niveoruber, ruber	C 55.76, H 6.55, N 10.88, O 25.41, Fe 1.40		S	S		S					T	*Candida Ferrimycin-group	G-10/11
Glumamycin		Stm. zaomyceticus	$C_{59}H_{91}O_{20}N_{13}$	G-6	S								L (ip)	Peptide group	1, 2
Gougerotin	No. 21544 substance	Stm. gougerotii	$C_{16}H_{25}O_8N_7 \cdot H_2O$	G-7	W	W							M (ip)	Pyrimidine group Mode act.: P	1, 2, 3, G-11, G-12, G-13, G-14
Gougeroxymycin		Stm. sp. (MA428-Cl)	$C_{11}H_{19}O_5N$							*S			*2	*1 Yeast only *2 i.v. of 25 mg/kg causes no death of mice	G-15
Graminomycin		Stm. sp.									S				1
Granaticin		Stm. olivaceus	$C_{22}H_{20}O_{10}$		S					S			M	pH Indicator	1, 2

Name	Synonym	Producer	Formula	Struct. Ref. No.	G +	G −	Myco	Fungi	Virus	Protoz	Tumor	Others	Tox.	Remarks	Ref.
Granaticin B				G-8											G-16, G-17
Grasseriomycin		*Stm. lavendulae*	Reineckate C: 22.04, H: 4.51 N: 19.33, Cr: 11.22		S	S	S	S							Streptothricin group	1, 2
Grisamine		*Stm. griseoflavus*	$C_{20}H_{30}O_7N_4$ or $C_{28}H_{38}O_{10}N_6$		S		S							L		1, 2
Grisein	3510	*Stm. griseus*	$C_{40}H_{61}O_{20}N_{10}SFe$		S	W								L	Sideromycin group	1, 2
Griselimycin		*Stm. griseus* *Stm. coelicus*		G-9			S								Polypeptide group	G-17/18
Griseococcin		*Stm. griseus*			S											1, 2
Griseococcin D		*Stm. griseus* *Stm. griseus*	$C_{21}H_{36}O_{12}N_4$		S											1, 2
Griseoflavin	Novobiocin	*Stm. griseoflavus*	C: 59.10, H: 6.21 N: 4.51		S		S							L		1, 2
Griseolutein A		*Stm. griseoluteus*	$C_{17}H_{14}O_6N_2$	G-10	S	S								L		1, 2
Griseolutein B		*Stm. griseoluteus*	$C_{17}H_{16}O_6N_6$	G-11	S	S	S							L		1, 2
Griseomycin	Lomycin	*Stm. griseolus*	$C_{25}H_{45}O_8N,$ $C_{28}H_{48}O_8N$	G-12	S		S							L	Macrolide group	1, 2
Griseorhodins		*Stm. griseus* J.A. 2640	A: C: 57.40, H: 4.05 O—Me: 6.10		S	S									pH Indicator	1
Griseoviridin	F-1370B	*Stm. griseoviridus* *Stm. griseus*	$C_{22}H_{27}O_7N_3S$	G-13	S	S	S							M		1, 2
Grisonomycin		*Stm. griseus,* A 10273			S	S		S								1

Name	Synonym	Producer	Formula	Struct. Ref. No.	G+	G-	Myco	Fungi	Virus	Protoz	Tumor	Others	Tox.	Remarks	Ref.
Grisorixin		Stm. griseus	$C_{40}H_{68}O_{10}$	G-14	S			S				*		*Cytotoxic Related to Nigericin	G-18, G-19, G-20
Grizin	Grisine, Grisemin, IEM-1	Stm. griseus	N: 13.6—14.9		S	S	S	S					T	Peptide group	1, 2
Grubilin		Stm. sp. BA-27						S					T	Polyene (heptaene) group	1, 2
Halomicin		Mim. halophytica (NRRL 2998) Mim. halophytica var. nigra (NRRL 3097)			S								L (sc)	Complex (A,B,C,D)	H-1
Hamycin		Stm. pimprina	C: 59.5, H: 8.3 N: 2.2					S					T (sc)	Polyene (heptaene) group Mode act.: M	1, 2, 3
Harimycin		Stm. hariensis	C: 39.29, H: 5.43 N: 12.00, O: 39.57					S							H-2
Hedamycin	Pluramycin A Rubiflavin Iyomicin B_1 NSC-70929D (016950)	Stm. griseoruber C-1150 (ATCC 15,422)	$C_{41}H_{52}O_{11}N_2$		S	*		S			S			*Induction of lysogenic E. coli	H-3
Heliomycin	Resistomycin X-340	Stm. flavochromogenes var. heliomycini	$C_{22}H_{18}O_6$	H-1	S				S				T (ip)		1, 2, H-4, H-5
Helixins	Endomycins (Prob.)	Stm. sp. A 158			S			S							1, 2
Heptafungin		Stm. achromolavendulae												Polyene group	H-6

Name	Synonym	Producer	Formula	Struct. Ref. No.	G+	G-	Myco	Fungi	Virus	Protoz	Tumor	Others	Tox.	Remarks	Ref.
Heptamycin		*Stm. sp.*						S		S				Polyene (heptaene) group	1
Hexamycin		*Stm. sp.*		H-2				S		S				Polyene (heptaene) group	H-7
Hilamycin A		*Stm. rochei*	C: 69.22, H: 9.74 N: 0.99				S				S		T (ip)		1
Hilamycin B		*Stm. rochei*					S				S		T (ip)		1
Hikizimycin		*Stm. hikiziensis*	$C_{13}H_{29}O_{10}N_3$					*S					T	*Active to phytopathogenic fungi	H-8
Histidomycin A, B		*Noc. histidans* MA 1157			S	S									H-9
Hodydamycin		*Stm. sp.* (AS-Y-400)	$C_{40}H_{50}N_3O_{14}Cl$		S	*S							L (ip)	Peptide group *Kleb. pneum. Prot. vulgaris Haemophilus influenzae	H-10
Holomycin	Des-N-methyl-thiolutin	*Stm. griseus*	$C_7H_6O_2N_2S_2$	H-3	S	S	S	S		S				Pyrrothine group	1, 2
Holothin		*Stm. griseus*	$C_5H_4ON_2S_2 \cdot HCl$	H-4			S							Pyrrothine group	1, 2
Homomycin	Hygromycin	*Stm. noboritoensis*	$C_{23}H_{29}O_{12}N$	H-5	W		S						L		1, 2
Hondamycin		*Stm. griseochromogenes* var. *albicus*	$C_{47}H_{78}O_{13}$					S					T (ip)		H-11
Hortesin		*Stm. versipellis* NRRL 2528						S						Pyrimidin group	1

Name	Synonym	Producer	Formula	Struct. Ref. No.	G+	G-	Myco	Fungi	Virus	Protoz	Tumor	Others	Tox.	Remarks	Ref.
Humidin	23572B	*Stm. humidus*	$(C_{12}H_{20}O_4)_n$					W		S			T (ip)		1, 2
Hybrimycin A$_1$, A$_2$		*Stm. fradiae* 3535		H-6	S	S									H-12
Hybrimycin B$_1$, B$_2$		*Stm. fradiae* 3535		H-6	S	S									H-12
5-hydroxy-7-chlortetra-cycline		*Stm. rimosus* ATCC 13224		H-7										See Tetracycline	H-13
Hydroxymycin	Paromomycin, 4915		$C_{25}H_{47}O_{15}N_5$		S	S				S			M	Aminocyclitol (neomycin) group	1, 2
Hydroxystreptomycin	Reticulin, D-212, NA-232-MI	*Stm. griseocarneus* *Stm. rubrireticuli* *Stm. subrutilus*	$C_{21}H_{39}O_{13}N_7 \cdot 3HCl$	H-8			Similar to S.M.						M	Aminocyclitol (streptomycin) group Mode act.: P	1, 2, 3
Hygromycin	Homomycin, Totamycin, 1703-18B	*Stm. hygroscopicus* *Stm. noboritoensis*	$C_{23}H_{29}O_{12}N$	H-5	S	S				S		*	L	*PPLO	1, 2
Hygromycin B	Marcomycin	*Stm. hygroscopicus*	$C_{15}H_{30}O_{10}N_2$	H-10	S	S	W				S			Mode act.: P	1, 2, 3
Hygroscopin A		*Stm. hygroscopicus*	$C_{13}H_{24}O_3N_2$				S	S	S		S		T (ip)		1, 2
Hygroscopin B		*Stm. hygroscopicus*	$C_{15}H_{28}O_3N_2$						S	S	S		T (ip)		1, 2
Hygroscopin C		*Stm. hygroscopicus*													1, 2
Hygrostatin		*Stm. hygrostaticus*			S		S	S		S			T		1, 2
Iaquirina I		*Stm. iakyrus*			S		S	S						Peptide group	1
Iaquirina II		*Stm. iakyrus*			S		S							Peptide group	1
Iaquirina III		*Stm. iakyrus*			S	W	S							Peptide group	1

Name	Synonym	Producer	Formula	Struct. Ref. No.	G+	G−	Myco	Fungi	Virus	Protoz	Tumor	Others	Tox.	Remarks	Ref.
Ikutamycin		Stm. diastatochromogenes. 79	$C_{68}H_{114}N_2O_{23}$					S		S			T (ip)		I-1, I-2
Ilamycin	Rufomycin A	Stm. islandicus	$C_{54}H_{75}O_{12}N_9$	I-1			S						L	Peptide group	1, 2
Ilamycin A_2		Stm. islandicus	$C_{54}H_{75}O_{12}N_9$				S							Peptide group	1, 2
Ilamycin B	Rufomycin B	Stm. islandicus	C: 61.07, H: 7.66 N: 11.78				S							Peptide group	1, 2
Ilamycin B_1	Rufomycin B_1	Stm. islandicus	$C_{54}H_{77}O_{10}N_9$	I-2			S							Peptide group	1, 2
Ilamycin B_2	Rufomycin B_2	Stm. islandicus	$C_{54}H_{77}O_{11}N_9$	I-3			S							Peptide group	1, 2
Ilamycin C		Stm. islandicus	C: 59.79, or C: 61.18 H: 7.09, or H: 7.16 N: 11.38, or N: 12.60				S							Peptide group	1
Ilamycin C_1		Stm. islandicus	$C_{54}H_{75}O_{12}N_9$				S							Peptide group	1
Ilamycin C_2		Stm. islandicus	$C_{54}H_{75}O_{12}N_9$				S							Peptide group	1
Imoticidin		Stm. albus B_{1-2} & C_{1-2}	C: 64.71, H: 9.50 Ash: 3.16					*S						*Piricularia oryzae only	1, 2
Inactone		Stm. griseus producing Cycloheximide	$C_{15}H_{21}O_4N$	I-4				W						Glutarimide group Mode act.: P	1, 2, 3, I-3
Indolmycin	PA-155A	Stm. albus	$C_{14}H_{15}O_2N_3$	I-5	S		S								1, 2, I-4
Indomycin A		Stm. sp. Ind. 927	$C_{40}H_{52}N_2O_{10}$	I-6									T		I-5
Indomycin B		Stm. sp. Ind. 927	$C_{38}H_{48}N_2O_{10}$	I-6	*S								T	*B. subtilis only	I-5
Indomycin C		Stm. sp. Ind. 927	$C_{41}H_{52}N_2O_{11}$	I-6									T		I-5
Iodinin	1,6-Phenazinediol-5,10-dioxide	Waksmania aerata	$C_{12}H_8O_4N_2$	I-7	S						S			Phenazine group	2, I-6

Name	Synonym	Producer	Formula	Struct. Ref. No.	G +	G −	Myco	Fung	Virus	Protoz	Tumor	Others	Tox.	Remarks	Ref.
Isobutyro-pyrrothine		*Stm. pimprina* producing thiolutin, Aureothricin and Hamycin *Stm. sp.* No. 2236 producing neopentaene	$C_{10}H_{12}O_2N_2S_2$	I-8	S									Pyrrothine group	1, 2, I-3
Isocycloheximide		*Stm. griseus* (also obtained by isomerization of Cyclohexmide)	$C_{15}H_{23}O_4N$	I-9				S						Glutarimide group Mode act.: P	1, 2
Isomaltose		*Stm. sp.* res. *Stm. albus*	$C_{12}H_{22}O_{11}$		S								L		1
Isoquinocycline A	PA-371γ	*Stm. aureofaciens*	$C_{33}H_{32}O_{10}N_2$	I-10	S		S							Mode act.: N	1, 2, I-7
Isoquinocycline B	PA-371ε	*Stm. aureofaciens*	C: 60.71, H: 5.72 N: 3.71, Cl: 4.84 (−HCl)		S		S							Mode act.: N	1, 2, I-7
Isorhodomycin A		*Stm. purpurascens* producing Rhodomycins	$C_{36}H_{48}O_{13}N_2 \cdot 2HCl$	I-11	S									Anthracycline group	1, 2
Isorhodomycin B		*Stm. purpurascens* producing Rhodomycins		I-11										Anthracycline group	1, 2
Iyomycin complex (A,B₁,B₂,B₃,B₅)		*Stm. phaeoverticillatus*	C: 47.64, H: 6.15 N: 9.4	I-12	W							S	M (ip)	Peptide group	1, 2, I-8
Iyomycin B₁		*Stm. phaeoverticillatus*	C: 67.36−67.45 H: 7.07−7.20 N: 3.81−3.98		S							S	T (ip)	pH Indicator	1
Iyomycin B₄		*Stm. phaeoverticillatus*	C: 55.30, H: 5.81 N: 8.94										TM (ip)	pH Indicator	1

Name	Synonym	Producer	Formula	Struct. Ref. No.	G +	G −	Myco	Fungi	Virus	Protoz	Tumor	Others	Tox.	Remarks	Ref.
Janiemycin		*Stm. macrosporeus* ATCC 21,388	C: 47.81, 47.49 H: 5.48, 5.30 N: 13.07, 13.07		S	W	S								J-1
Jossamycin	(Leucomycin A_3)	*Stm. narbonensis var. josamyceticus*	$C_{40}H_{69}NO_{14}$		S		S						L	Macrolide group	J-2
Jossamycin S		*Stm. narbonensis var. josamyceticus*			S	S	S						L	Macrolide group	J-3
Julymycin A complex						S									2
Julymycin B complex B-0, B-1, S.V.B-III, B-IV						S									2
Julymycin B-II		*Stm. shiodaensis*	$C_{38}H_{34}O_{14}$	J-1	W										
Julymycin C		*Stm. shiodaensis*	$C_{38}H_{36}O_{15} \cdot 2.5H_2O$	J-2	S	W			S		S		ML	pH Indicator	1, 2, J-4
Juvenimicins	T-1124	*Mim. chalcea var. izumensis*	A; C 63.39 ± 1.0 H 8.83 ± 0.5 N 2.80 ± 0.5 A_3; C 62.97 ± 1.0, H 8.62 ± 0.5 N 3.00 ± 0.5 B_1; C 64.0 ± 1.0, H 9.4 ± 0.5 N 2.5 ± 0.5 B_3; C 64.7 ± 1.0, H 10.2 ± 0.5 N 3.6 ± 0.5		S S S S	W S			W				ML	Macrolide group	2, J-5 J- 6

Name	Synonym	Producer	Formula	Struct. Ref. No.	G+	G-	Myco	Fungi	Virus	Protoz	Tumor	Others	Tox.	Remarks	Ref.
Kabicidin		*Stm. gougeroti*	$C_{35}H_{60}O_{13}$					*S					T L (or)	Polyene (pentaene) group	1
Kalafungin	Kalamycin, U-19718	*Stm. tanashiensis* Kala	$C_{16}H_{12}O_6$		S			S							K-1, K-2
Kalamycin	Kalafungin													See Kalafungin	
Kanamycin (A)		*Stm. kanamyceticus*	$C_{18}H_{36}O_{11}N_4$	K-1	S	S	S						L	Aminocyclitol group Mode act.: P	1, 2, 3
Kanamycin B		*Stm. kanamyceticus*	$C_{18}H_{37}O_{10}N_5 \cdot 2H_2O$	K-2	S	S	S						L	Aminocyclitol group	1, 2
Kanamycin C		*Stm. kanamyceticus*	$C_{18}H_{36}O_{11}N_4$	K-3	S	S	S						L	Aminocyclitol group	1, 2
Kanchanomycin	BA-180265	*Stm. sp.*	C: 61.66, H: 4.64 N: 5.33, OMe: 5.67		S			S			*S			*HeLa cell	1
Kanendomycin	Kanamycin B														
Karnatakin		*Stm. karnatakensis*	C: 55.3, H: 5.8 N: 7.5		S	S							ML	See Kamamycin B	1
Kasugamycin		*Stm. kasugaensis*	$C_{14}H_{25}O_9N_3$	K-4	S	W		*W					L	*Piricularia · oryzae (at pH 5.0) Mode act.: P	1, 3
Kikumycin A		*Stm. phaeochromogenes*	$C_{24}H_{35}O_9N_{11} \cdot 2H_2SO_4$		S	S	W		*S¹		*S²		M (ip)	*1 Hela Sarcoma 180 *2 Polio, Influenza virus	1

Name	Synonym	Producer	Formula	Struct. Ref. No.	G+	G-	Myco	Fungi	Virus	Protoz	Tumor	Others	Tox.	Remarks	Ref.
Kikumycin B		*Stm. phaeochromogenes*	$C_{13}H_{22}O_4N_6 \cdot H_2SO_4$		S	S			*S¹		*S²		ML (ip)	*¹ Hela Sarcoma 180 *² Polio, Influenza virus	1
Kinamycin A,B,C,D		*Stm. murayamaensis*	A: $C_{19}H_{20}N_2O_{10}$ B: $C_{20}H_{14}N_2O_7$ C: $C_{24}H_{20}N_2O_{10}$ D: $C_{21}H_{18}N_2O_8$		S	W							T		K-3
Kobenomycin		*Stm. kobenensis*	C: 51.93, H: 6.53 N: 15.75, S: 4.96		S								T	Peptide group	K-4
Koluophthisin	Viomycin (prob.)	*Stm. floridae*	$C_{18}H_{32}O_8N_9 \cdot H_2SO_4$		S	S	S						L		1
Kokubumycin		*Stm. sp.* No. 59–42	C: 55.16, H: 4.64 N: 1.14, Ash: 2.05		S	S		S					T (ip)		1
Komamycin A		*Stm. sp.* No. 606-N2	C: 57.07, H: 7.90 N: 2.63, P: 2.32		W	W		S							K-5
Komamycin B		*Stm. sp.* No. 606-N2	C: 52.48, H: 7.47 N: 4.15, P: 2.85		W	W		S							K-5
Kujimycins		*Stm. sp.* (TPR- 885)	A: $C_{40}H_{70}O_{15}$ B: $C_{42}H_{72}O_{16}$		S		S						L	Macrolide group	K-6
Kundrymycin		*Stm. metachromogenes*	$C_{45}H_{50}O_{18}$		S						*S		T	Antitumor pigment, resem. Aquayamycin *Walker 256, L1210 (asc.), P-388 (asc.) S-180 (solid)	K-7
Kunomycin		*Stm. sp.*						*S²			*S¹			*¹ anti HeLa 0.1 mcg/ml Macromolecular *² yeast	1

Name	Synonym	Producer	Formula	Struct. Ref. No.	G +	G −	Myco	Fungi	Virus	Protoz	Tumor	Others	Tox.	Remarks	Ref.
Labilomycin		Stm. albosporeus var. labilomyceticus	$C_{23}H_{34}O_8$	L-1	S		S				S		MT		1, 2
Lactenocin		Stm. fradiae NRRL 2702, 2703	$C_{39}H_{63}NO_{14}$		S									Macrolide group Co-product of Macrocin Contain mycorose & mycaminose	2, L-1
Lagosin	A-246, fungichromin	Stm. roseoluteus NRRL 2776	$C_{35}H_{58}O_{12}$	L-2				S						Polyene (pentaene) group Mode act.: M	1, 2, 3
Lankacidin	Lankavacidin (pat. name)	Stm. violaceoniger	$C_{49}H_{66}O_{16}N_2$		W	W							L (sc)	Synergistic with Lankamycin	1, 2
Lankamycin	Lankavamycin (pat. name)	Stm. violaceoniger	$C_{42}H_{72}O_{16}$	L-3	W	W							L (sc)	Synergistic with Lankacidin Mode act.: P	1, 2, 3
Lankavacidin	Lankacidin													See Lankacidin	1
Lankavamycin	Lankamycin													See Lankavacidin	1
Largomycin		Stm. pluricolorescens	F-I C: 47.93, H: 7.46 N: 6.50, S: 0.82 F-II C: 47.83, H: 8.30 N: 13.63, S: 0.25 F-III C: 46.79, H: 6.57 N: 8.64, S: 1.08								S	*S	T (F-II) ip	Chromoprotein group F-II: Similar to pluralin, iyomycin *Mycoplasma	L-2a, L-2b, L-2c
Laspartomycin		Str. viridochromogenes var. komabensis	$C_{82}H_{137-139}N_{17}O_{29}$	L-4	S								L	Peptide group	L-3a, L-3b
Lateriomycin A		Stm. griseoruber No. 71070	C: 60.13, 60.15, 60.18 H: 5.84, 6.06, 6.02 N: 1.90, 1.92, 2.01		S		W				S		T (ip)	pH Indicator	L-4

Name	Synonym	Producer	Formula	Struct. Ref. No.	G+	G-	Myco	Fungi	Virus	Protoz	Tumor	Others	Tox.	Remarks	Ref.
Lateriomycin B		*Stm. griseoruber* No. 71070	C: 65.21, H: 5.22 O: 30.09		S		W				S		T (ip)	pH Indicator	L-5
Lateriomycin F		*Stm. griseoruber* No. 71070	C: 61.96, 63.04 H: 6.45, 6.35 N: 1.39, 1.98		S	W	S				S		T	pH Indicator	L-6, L-7
Lathumycin		*Stm. lathumensis*	$C_{51-52}H_{79-81}O_{14-15}N_9$		S									Peptide group	1
Latumcidin	Abikoviramycin, Virocidin	*Stm. reticuli* var. *latumcidicus*											T	See Abikoviromycin	1, 2
Laurusin	Formycin B	*Stm. lavendulae*		L-5										See Formycin B	1
Lavendulin		*Stm. lavendulae*	Helianthate $C_{49}H_{63}O_{18}N_{13}S_3$		S	S	S	S					T	Streptothricin group	1, 2
Lemacidin		*Stm. venezuelae* A, 9692			S	S	S	S					ML (sc)		1
Lemonomycin	LL-AP 191	*Stm. candidus*	$C_{30}H_{41}O_9N_3$		S	S	S						T		1
Lenamycin	Aquamycin, Cellocidin, Acetylene dicarboxamide	*Stm.* No. 902 res. *Stm. reticuli*	$C_4H_4O_2N_2$	L-6							S				1
Leucinamycin		*Str. cinnamomeus* MA404-C2	C: 49.42, H: 6.85 O: 21.81, N: 15.79 S: 5.38		*1 S	*3 S	*2 S	*2 S		*4 S			T (ip)	Polypeptide group *1 *B. subtilis* *B. anthracis* *2 *Piricularia oryzae* *Pallicularia filamentosa* *Cryptococcus neoformans* *3 *Xantho oryzae* *4 *Trichomonas vaginalis*	L-8

Name	Synonym	Producer	Formula	Struct. Ref. No.	G +	G -	Myco	Fungi	Virus	Protoz	Tumor	Others	Tox.	Remarks	Ref.
Leucocidin		*Stm. hygrostaticus*	$C_{21-24}H_{35-40}O_{7-8}$		S		S						T		1
Leucomycin complex		*Stm. kitasatoensis*	$C_{33-38}H_{54-66}O_{11-13}N$		S		S		S	S			L	Macrolide group Commercially available Mode act.: P	1, 2
Leucomycin A₁		*Stm. kitasatoensis*	$C_{40}H_{67}O_{14}N$	L-7	S	W	S								1, 2, L-9, L-10
Leucomycin A₂		*Stm. kitasatoensis*	$C_{65}H_{110}O_{22}N$		S	W									2, L-11
Leucomycin A₃	Jossamycin	*Stm. kitasatoensis*	$C_{42}H_{69}O_{15}N$	L-7	S	W	S								L-12a, L-12b, L-12c
Leucomycin A₄		*Stm. kitasatoensis*	$C_{41}H_{67}O_{15}N$	L-7	S	W	S								L-13
Leucomycin A₅		*Stm. kitasatoensis*	$C_{39}H_{65}O_{14}N$	L-7	S	W	S								1, L-11
Leucomycin A₆		*Stm. kitasatoensis*	$C_{40}H_{65}O_{15}N$	L-7	S	W	S								1, L-14
Leucomycin A₇		*Stm. kitasatoensis*	$C_{38}H_{63}O_{14}N$	L-7	S	W	S								1, L-15
Leucomycin A₈		*Stm. kitasatoensis*	$C_{39}H_{63}O_{15}N$	L-7	S	W	S								1, L-16
Leucomycin A₉		*Stm. kitasatoensis*	$C_{37}H_{61}O_{14}N$	L-7	S	W	S								1, L-17
Leucomycin B₁		*Stm. kitasatoensis*	$C_{35}H_{59}O_{13}N$	L-8a	S		W								1, 2
Leucomycin B₂		*Stm. kitasatoensis*	$C_{38}H_{65}O_{16}N$	L-8b	S		W								1, 2
Leucomycin B₃		*Stm. kitasatoensis*	$C_{34}H_{53}O_{13}N$	L-8c	S		W								1, 2
Leucomycin B₄		*Stm. kitasatoensis*	$C_{38}H_{59}O_{16}N$	L-8c	S		W								1, 2
Leucomycin U			$C_{37}H_{61}O_{14}N$	L-7											L-18
Leucomycin V			$C_{35}H_{59}O_{13}N$	L-7											L-18

Name	Synonym	Producer	Formula	Struct. Ref. No.	G+	G-	Myco	Fungi	Virus	Protoz	Tumor	Others	Tox.	Remarks	Ref.
Isoleucomycin A₃			$C_{42}H_{69}O_{15}N$	L-7											L-18
Leucopeptin		*Stm. hachijoensis var. takahaziensis*	C: 49.13, H: 6.74, N: 14.80, S: 5.67, O: 21.35		W	W	S	W					T	Peptide group	1, 2
Levomycetin (USSR)	Chloramphenicol	*Stm. sp.*													1
Levomycin	Echinomycin Quinomycin A	*Stm. sp.*												See Echinomycin	1, 2
Levorin	26/1	*Acm. levoris* *Stm. globisporus*						W					T	Polyene (heptaene) group	1, 2
Levorin A	Candicidin														2, L-19
Levorin B															2, L-20
Levoristatin		*Acm. levoris*						S							1
Libanomycin A,B,C	Lybanomycin (A,B,C)	*Stm. libani*												See Lybanomycins	L-35
Limocrocin		*Stm. limosus*	$C_{26}H_{26}O_6N_2$	L-9	S									pH Indicator	1
Lincolnensin	Lincomycin	*Stm. lincolnensis var. lincolnensis*												See Lincomycin	1
Lincomycin		*Stm. lincolnensis var. lincolnensis*	$C_{18}H_{34}O_6N_2S \cdot HCl \cdot \frac{1}{2}H_2O$	L-10	S	S							L (ip)	Mode act.: N	1, 2, 3, L-21, L-22, L-23, L-24, L-25
Lincomycin B	U-21,699	*Stm. lincolnensis var. lincolnensis*	$C_{17}H_{32}O_6N_2S$	L-10	S										1

Name	Synonym	Producer	Formula	Struct. Ref. No.	G +	G -	Myco	Fungi	Virus	Protoz	Tumor	Others	Tox.	Remarks	Ref.
Lincomycin C	U-11,921	Stm. lincolnensis var. lincolnensis	$C_{19}H_{36}O_6N_2S$	L-10	S										1
Lincomycin D	U-11,973E	Stm. lincolnensis var. lincolnensis	$C_{17}H_{32}N_2O_6S \cdot HCl \cdot H_2O$	L-10	S									Structurally related to Lincomycin	L-26
Lincomycin K	U-20943	Stm. lincolnensis var. lincolnensis	$C_{18}H_{34}N_2O_6S \cdot HCl \cdot H_2O$	L-10	S		S								L-27
Lincomycin S	U-25,468	Stm. lincolnensis var. lincolnensis	$C_{20}H_{38}N_2O_6S \cdot HCl \cdot H_2O$	L-10	S	S								Structurally related to Lincomycin	L-28
1-Demethyl-thio-1-hydroxylincomycin		Stm. lincolnensis var. lincolnensis	$C_{17}H_{32}N_2O_7 \cdot HCl \cdot H_2O$											Transformed from Linomycin	L-29
Lipoxamycin		Stm. virginiae var. lipoxae NRRL 3630						S							L-30
Litmocidin		Noc. cyanea (Proacm. cyaneus var. antibioticus)	$C_{20}H_{22}O_9$		S		S						MT (ip)	pH Indicator	1
Litmomycin		Stm. litmogenes			S								MT	pH Indicator	1
Lomofungin	Lomondomycin	Stm. lomondensis var. lomondensis	$C_{15}H_{10}N_2O_6$		W	W		W							L-31, L-32
Lomondomycin														See Lomofungin	L-33
Lomycin	Griseomycin	Stm. griseolus												See Griseomycin	1, 2
Longisporin		Acm. (Stm.) longisporus	$C_{36}H_{58}O_{10}$		S		S							Cyclic polylactone group	1, 2
Lucensomycin	1163-FI, Etruscomycin	Stm. lucensis	$C_{36}H_{53}O_{13}N$	L-11				S					T	Polyene (tetraene) group	1, 2, L-34

Name	Synonym	Producer	Formula	Struct. Ref. No.	G+	G-	Myco	Fungi	Virus	Protoz	Tumor	Others	Tox.	Remarks	Ref.
Luridin		*Stm. luridus*			S		S		S					Streptothricin group	1, 2
Lustericin		*Stm. sp.* 10400	$C_{40}H_{64}O_3$		S	W	S	S					M (ip)	Cyclic polylactone group	1, 2
Luteomycin	H-2053	*Stm. tanashiensis,* *Stm. flaveolus*	$C_{26}H_{33}O_{12}N$ $C_{23}H_{29}O_9N \cdot HCl$		S	W	S	S	S			S	T	pH Indicator	1, 2
Lybanomycin(e) A, B, C	Libanomycin A, B, C	*Stm. libani* F.I. 2343 (IMRU, IMI 3915, 130777), 2399 (3916, 130778), 2501 (3917, 130779), 2521 (3918, 13233)	A: C 63.12, H: 8.05 N 6.49 B: C 64.34, H: 8.42 N:6.77 C: —		S S		W W						T		L-35
Luteostatin		*Stm. sp.*			S		W				S				1
Lydimycin	U-15965	*Stm. lydius* NRRL 2433	$C_{10}H_{14}N_2O_3S$		W			W							L-36
Lymphomycin		*Stm. sp.* S-66	C: 44.98, H: 5.97 N: 11.00								S		L (ip)	Acidic protein group	L-37
Lysotoxin		*Stm. lysotoxis*	$C_{19}H_{25-27}O_6N_3 \cdot 2HCl$		S	S				S			T	Peptide group Similar to Xanthomycin	1
Macarbomycin		*Stm. phaeochromogenes*	C: 45.74, H: 6.79 N: 5.06, O: 30.89 Ash: 4.3, P: 2.9		S		S							Moenomycin group	M-1a, M-1b
Macrocin		*Stm. fradiae* M48-E2724, NRRL 2702, 2703	$C_{46}H_{79}O_{17}N$	M-1	S							*S	L (ip)	Macrolide group *Mycoplasma Co-product: Lactenocin	1, 2, M-2

Name	Synonym	Producer	Formula	Struct. Ref. No.	G +	G -	Myco	Fungi	Virus	Protoz	Tumor	Others	Tox.	Remarks	Ref.
Macromomycin		*Stm. macro-momyceticus*	$(C_{49}H_{153}O_{24}N_{13}S)_{8-10}$		S			*S[1]			*S[2]			Macromolecular peptide group *[1]*Torula utilis Cryptococcus neoformans* *[2]Sarcoma 180, Leukemia L-1210	M-3
Macrotetrolide B		*Stm. flaveolus* ETH 31442	$C_{46}H_{76}O_{12}$	M-2											M-4
Macrotetrolide C		*Stm. flaveolus* ETH 31442	$C_{45}H_{74}O_{12}$	M-3											M-4
Macrotetrolide D		*Stm. flaveolus* ETH 31442	$C_{44}H_{72}O_{12}$	M-4											M-4
Macrotetrolide G		*Stm. flaveolus* ETH 31442	$C_{43}H_{70}O_{12}$	M-5											M-4
Mannosyl glucosamine		*Stm. virginiae* var., 4243-MTt$_1$	$C_{12}H_{23}NO_{10} \cdot HCl$				S								M-5
Mannosidohy-droxystreptomycin				M-6										See Streptomycin	
Mannosido-streptomycin	Streptomycin B			S-15	S									See Streptomycin	1
Manumycin		*Stm. parvulus* ETH 25000 & 28331	$C_{30}H_{38}O_7N_2$												1
Marcomycin	Hygromycin B													See Hygromycin B	
Marinamycin		*Stm. mariensis*	C: 40.3, H: 7.1, N: 10.8								S	S	ML (ip)	Peptide group	1, 2
Marinamycin-like substance		*Acm. candidus* 10484									S	S	L		M-6

Name	Synonym	Producer	Formula	Struct. Ref. No.	G+	G-	Myco	Fung	Virus	Protoz	Tumor	Others	Tox.	Remarks	Ref.
Matamycin	Althiomycin	*Stm. matensis* *Stm. bellus*	C: 43.95, H: 4.06 N: 14.45, S: 13.57		S	W									1, 2
Matchamycin		*Stm. amagasakensis*	$C_{26}H_{13}O_6N_3Cu$		W	W									M-7
Mediocidin		*Stm. mediocidicus*			W			S					T	Polyene (hexaene) group	1, 2
Megacidin		*Stm. sp.*	$C_{24}H_{38}O_{10}$		*S									Neutral macrolide group Bandamycin A like *B. megatherium*	1
Megalomicin A		*Mim. sp.*			S								L (ip)	Macrolide group	M-8
Megalomicin B		*Mim. sp.*			S								L (ip)		M-8
Megalomicin C₁		*Mim. sp.*			S								L (ip)		M-8
Megalomicin C₂		*Mim. sp.*			S								L (ip)		M-8
Mekemycin		*Stm. mekemicus*	C: 55.83, H: 5.59		S	W	S	W					M (ip)		1
Melanomycin		*Stm. melanogenes*	–Na: C: 54.74, H: 6.99, N: 9.87, Na: 1.9 Free: C: 57.84, H: 8.11, N: 6.38		S					S	S		TM	Peptide group	1, 2
Melanosporin		*Stm. melanosporus*	$C_{56-63}H_{105-117}O_{20-22}N_3$		S		S	S					T (ip)	Azalomycin F-like	1, 2
Melrosporus	U-22956	*Stm. fervens* *var. melrosporus*	$C_5H_7NO_4$		S	S	S	S			S				M-9
Mesenterin		*Noc. mesenterica*	C: 65.82, H: 7.10, N: 8.66—8.44		S		S						M (ip)		1, 2

Name	Synonym	Producer	Formula	Struct. Ref. No.	G +	G -	Myco	Fungi	Virus	Protoz	Tumor	Others	Tox.	Remarks	Ref.
Metamycin	Paromomycin													See Paromomycin	1
Methobottromycin														A component of Bottromycin complex See Bottromycin	M-10
Methymycin	11B	*Stm. eurocidicus Stm. sp.* M2140	$C_{25}H_{43}O_7N$	M-7	S	W								Macrolide group Mode act.: P	1, 2, 3, M-11
Mezzanomycin		*Stm. senocanescens*	C: 46.04, H: 5.56, N: 2.88		S		W				S		T (ip)	pH Indicator Luteomycin like	1
Miamycin		*Stm. sp.* res. *Stm. ambofaciens* simultaneously producing spiramycin	C: 61.45, H: 8.65, N: 2.32		S								L (ip)	Macrolide group Cross resistance with carbomycin & erythromycin	1, 2
Micloretin	Chloramphenicol													See Chloramphenicol	1
Micoheptin		*Acm. sp.* 44 B/1 (*Stm. netropsis* like)						S						Polyene (heptaene) group	M-12
Microcin A		*Mim. sp.*			S								L		1, 2
Microcin B		*Mim. sp.*			S								L		1, 2
Micromonosporin	BU-271	*Mim. sp.*			S										1, 2
Miharamycins		*Stm. miharaensis* SF-489			*1 S			*2 S	*3 S					*1 *Pseudomonas* *2 *Pricularia oryzae* *3 TMV	M-13, M-14
Miharamycin A		*Stm. miharaensis* SF-489	A–HCl: $C_{22}H_{38}N_{10}O_{10} \cdot 2HCl$												M-13, M-14

Name	Synonym	Producer	Formula	Struct. Ref. No.	G +	G -	Myco	Fungi	Virus	Protoz	Tumor	Others	Tox.	Remarks	Ref.
Miharamycin B		Stm. miharaensis SF-489	B–HCl: $C_{21}H_{36}N_{10}O_{11} \cdot HCl$												M-15
Mikonomycin	Myconomycin Mykonomycin													See Myconomycin	
Micropolysporin A	Mikropolisporin A (Antibiotic No. 55)	Mip. caesia No. 55			S		S							Crude, differentiated by paper chromatograph	M-16
Micropolysporin B	Mikropolisporin B (Antibiotic No. 55)	Mip. caesia No. 55			S		S							Crude, differentiated by paper chromatograph	M-16
Mikamycin A		Stm. mitakaensis	$C_{31}H_{39}O_9N_3$		S		W						L (ip)	Synergistic with B Ostreogrycin A like Mode act.: P	1, 2, 3
Mikamycin B		Stm. mitakaensis	$C_{45}H_{54}O_{10}N_8$	M-8	S									Synergistic with A Ostreogrycin B like	1, 2, 3
Mimimycin		Stm. sp. (Stm. hygroscopicus)	$C_9H_{11}NO_7$	M8/9	S	S					S		M		M-16/17
Minomycin		Stm. minoensis	$C_{18}H_{22}O_7$		S	W	W				S		T (ip)	pH Indicator Mode act.: N(RNA), P	1, 2, 3
Minocycline	7-dimethyl-6-demethyl-6-deoxy-tetracycline													See Tetracycline	
Miramycin		Stm. mirabilis			S	S							L		1, 2
Miromycin		Stm. shirahamaensis			S			S					T (ip)		1

Name	Synonym	Producer	Formula	Struct. Ref. No.	G +	G -	Myco	Fungi	Virus	Protoz	Tumor	Others	Tox.	Remarks	Ref.
Miserin	Neomycin													See Neomycin	1
Mithramycin	NSC A-2371, PA-144	*Stm. sp.*	Free: C 55.98, H 7.39 Na–: C 54.69, H 6.93		S		S				S			Aureolic acid group Mode act.: N(RNA)	1, 3
Mitiromycin A	Mitiromycin A_2 (formerly)	*Stm. verticillatus*	$C_{16}H_{19}O_6N_3$	M-9	S	S					S			Mitosane group	1, 2, M-17
Mitiromycin B	Mitomycin A	*Stm. verticillatus*	$C_{16}H_{19}O_6N_3$	M-10	S	S					S			See Mitomycin A	1
Mitiromycin C	Mitomycin B_1 (prob.) Porfiromycin	*Stm. verticillatus*	$C_{16}H_{19}O_6N_3$	M-10	S	S					S			See Mitomycin B	1
Mitiromycin D	Porfiromycin	*Stm. verticillatus*		M-10										See Porfiromycin	1
Mitiromycin E	Mitomycin C	*Stm. verticillatus*		M-10										See Mitomycin C	1
Mitocromin (A & B)	NSC-77471	*Stm. viridochromogenes* B35251, B105621	Mix: C 55.66, H 6.97, N 2.32		S	S					S		*	Anthracyclines group A & B equilibrium in mixture *Cytotoxic to Hela cells	M-18 M-19
Mitomalcin		*Stm. malayensis*	M.W. 17,400		S						S			Protein group	M-20
Mitomycin (Umezawa)	289					S								See 289	1, 2
Mitomycin A	Mitiromycin B	*Stm. caespitosus*	$C_{16}H_{19}O_6N_3$	M-10	S	S	S		S		S		T	Mitosane group Mode act.: N(DNA)	1, 2, M-21
Mitomycin B	Mitiromycin C	*Stm. caespitosus*	$C_{16}H_{19}O_6N_3$	M-10	S	S	S		S		S		T	Mitosane group	1, 2
Mitomycin C	Fraction X, Mitiromycin E	*Stm. caespitosus*	$C_{15}H_{18}O_5N_4$	M-10	S	S	S		S		S		T	Mitosane group	1, 2

Name	Synonym	Producer	Formula	Struct. Ref. No.	G+	G−	Myco	Fungi	Virus	Protoz	Tumor	Others	Tox.	Remarks	Ref.
Mitomycin R		*Stm. caespitosus*			S	S	S				S				1
Mitomycin Y		*Stm. caespitosus*			S	S	S				S				1
Miusidin		*Stm. sumaensis*	C 35.67, H 8.45, N 14.28		S	S	S	*S						L	*Yeast only	1
Moenomycin complex A, B₁, B₂, C		*Stm. bambergiensis* ATCC 13879 *Stm. ghanaensis* ATCC 14672 *Stm. ederensis* ATCC 15304 *Stm. geysiriensis* ATCC 15303	C 48.5, H 7.3, O 37.3, N 5.1, P 1.8 (Complex) A) P 1.8 C 48.5 H 7.3 N 5.3 B₁) P 1.8 C 48.6 H 7.2 N 4.9 B₂) −NH₄ P 1.8 C 42.8 H 7.2 N 4.9 C) P 1.7 C 49.1 H 7.4 N 5.3		S	S	S								P containing Promote animal growth Related to Prasionomycin, 11837 RP, Diumycins, Macarbomycin	M-22
Moenomycin complex D		*Stm. bambergiensis*	(NH₄-salt), C 43.7 H 7.1 N 5.7 P 1.9		S	S	S								Separated from B₁	M-23, M-24
Moenomycin complex E		*Stm. bambergiensis*	(K-salt), C 46.3 H 6.7 N 4.4 P 1.8 K 2.6		S	S	S								Separated from B₁	M-23, M-24
Moenomycin complex F		*Stm. bambergiensis*	(K-salt), C 44.9 H 5.4 N 4.0 P 1.7 K 4.5		S	S	S								Separated from B₁	M-23, M-24
Moenomycin complex G		*Stm. bambergiensis*	(K-salt), C 47.2 H 6.4 N 4.6 P 2.0 K 2.3		S	S	S								Separated from B₂	M-23, M-24
Moenomycin complex H		*Stm. bambergiensis*	(K-salt), C 46.1 H 6.3 N 4.1 P 1.9 K 4.3		S	S	S								Separated from B₂	M-23, M-24
Moldcidin		*Stm. sp.* J-4	$C_{26}H_{35}O_8N_2$					S								1, 2
Moldcidin A		*Stm. griseofuscus* *Stm. sp.* No. 1068	$C_{42}H_{51}O_{19}N$					S						T	Polyene (pentaene) group	1, 2
Moldcidin B	Pentamycin	*Stm. griseofuscus*	$C_{35}H_{60}O_{13}$					S							Polyene (pentaene) group	1, 2

Name	Synonym	Producer	Formula	Struct. Ref. No.	G +	G -	Myco	Fungi	Virus	Protoz	Tumor	Others	Tox.	Remarks	Ref.
Moldin		Stm. phaeochromogenes						*S					T	*Yeast only	1, 2
Monactin		Stm. sp. ETH A23112	$C_{41}H_{66}O_{12}$	M-2~5	S		S							Cyclipolylactone group Mode act.: See Macrotetrolide	1, 2, 3
Monamycin D$_1$		Stm. jamaicensis	$C_{22}H_{36-38}O_5N_4$	M-11	S								L		1, 2
Monazomycin		Stm. mashuensis	$C_{62}H_{119}O_{20}N$		S	W		S			S		T	Takacidin like	1, 2
Monensin	Monensic acid	Str. cinnamonensis	$C_{36}H_{62}O_{11} \cdot H_2O$	M-12	S		S	S		*S	S		T (or)	Momensin group *[1] Eimeria	M-25a, M-25b, M-25c, M-25d, M-25e, M-25f
													*M^2 (or)	*[2] chick	M-26
Monicamycin		Stv. cinnamomeus var. monicae	C 58.26, H 7.85, N 2.21				/	S						Polyene (heptaene) group	1
Monilin		Stm. sakaiensis Stm. sp.	$C_{15}H_{20}O_3N_6$				S	*S					T (ip)	*Yeast only	1, 2
Monomycin	Paromomycin	Acm. circulatus var. monomucini	N 9.54, 9.4		S		S	S					M		1, 2
Morimycin	Primocarcin	Stm. diastatochromogenes var. luteus Stm. sp. 1N701	$C_8H_{12}N_2O_7$				S						M		M-27, M-28
Multhiomycin		Stm. antibioticus 8446-CC	$C_{44}H_{45}O_{11}N_{11}S_5$		S								*	Similar to Taitomycin *Not less than 10 mg/kg (iv)	M-29

Name	Synonym	Producer	Formula	Struct. Ref. No.	G +	G −	Myco	Fungi	Virus	Protoz	Tumor	Others	Tox.	Remarks	Ref.
Musarin		*Acm. sp.*	$C_{35}H_{60}O_{14}N_2$		S		S	S							1, 2
Musashimycin		*Stm. sp.* UA-97	C 34.03, H 5.76, N 17.44		S	S	S						M	Streptothricin like	1
Mutabilicin	21–31	*Acm. mutabilis var. tiostreptoni* (21–31)	$(C_{42}H_{53}S_3O_{12}N_{11})_n$		S									Thiostrepton like	M-30
Mutamycin		*Stm. sp.*									S				1
Mutomycin		*Acm. atroolivaceus var. mutomycini*	$C_7H_{11-12}O_2$		*S				W		W		T (sc)	*Inhibit respiratory deficient staphylococci	1, 2
Mycelin		*Stm. diastatochromogenes* *Stm. roseoflavus*						S							1, 2
Mycelin IMO		*Stm. sp.*	C 71.29, H 5.96, N 11.31					S					T (ip)	Polyene (hexaene) group Fradicin, Mycelin like	1, 2
Mycerin	Neomycins	*Stm. fradiae*												See Neomycins	1, 2
Mycetin		*Stm. violaceus*			S		S								1, 2
Mycetin A			C 57.95, H 6.68, N 3.24				S								1, 2
Mycifradin	Neomycin													See Neomycin	1
Mycobacidin	Actithiazic acid	*Stm. lavendulae*	$C_9H_{15}O_3NS$				S						L		1, 2
Mycodicin		*Stm. sumaensis*	C 35.67, H 8.45, N 14.28, O 43.60, S,Cl none		S	S	S	S					L		M-31

Name	Synonym	Producer	Formula	Struct. Ref. No.	G +	G -	Myco	Fungi	Virus	Protoz	Tumor	Others	Tox.	Remarks	Ref.
Mycoheptin		A 44B/1 belonging to Acm. netropsis												Polyene (heptaene) group	1
Mycolutein		Stm. sp.	$C_{22}H_{24}O_6N$					*S					T (ip)	*Yeast only	1, 2
Mycomycetin		Stm. arenae					S								1, 2
Mycomycin		Noc. acidophilus	$C_{13}H_{10}O_2$	M-13	S	S	S	*S					T	Polyene group *Yeast only	1, 2
Myconomycin	Mykonomycin Mikonomycin	Stm. albogriseolus A-21066	C 59.33, H 8.35, N O, O–Me 10.92, C–Me 11.49		S		W							Neutral macrolide group Bandamycin B Chalcomycin like	1
Mycorhodin		Stm. sp.	C 58.7, H 5.2 N 2.1		S		S						M (ip)	pH Indicator	1, 2
Mycospocidin		Stm. sp.	$(C_{20}H_{33}O_9N_2)_n$		S		S	W					T (ip)		1, 2
Mycthricin A		Stm. lavendulae			S	S		S					M	Streptothricin group	1, 2
Mycothricin B		Stm. lavendulae			S	S		S					M	Streptothricin group	1, 2
Mycoticin		Stm. ruber	$C_{18}H_{30}O_5$					S					T (ip)	Polyene group Flavofungin like	1, 2
Mycotrienin		Stm. sp. (7 morphological groups)	$C_{36}H_{50}N_2O_8$	M-14				S					T (ip)	Polyene (triene) group	M-32
Myxoviromycin	Amidinomycin	Stm. sp.	–H_2SO_4: C 36.37, H 7.05, N 19.22, S 10.90 –HCl: C 26.96, H 5.34, N 13.86, Cl 17.80						S				T (sc)	See Amidinomycin	1, 2

Name	Synonym	Producer	Formula	Struct. Ref. No.	G+	G-	Myco	Fungi	Virus	Protoz	Tumor	Others	Tox.	Remarks	Ref.
Nancimycin	Rifamycin B	*Stm. albovinaceus* ATCC 12951	$C_{39}H_{51}NO_{14}$		S				S				L (sc)		1, N-1, N-2
Naramycin A	Cycloheximide													See Cycloheximide	1, 2
Naramycin B		*Stm. naraensis* *Stm. pulveraceus*	$C_{15}H_{23}O_4N$	N-1				S					M (ip)	Glutarimide group	1, 2
Narangomycin		*Stm. lavenduligriseus* ATCC 13,306	$C_{23}H_{30}O_{10}$		S			*S		S	S			*Yeast only	1
Narbomycin		*Stm. narbonensis*	$C_{28}H_{47}O_7N$	N-2	S								L (sc)	Macrolide group	1, 2
Naritheracin	Toyocamycin	*Stm. sp.* A-392-Y4	$C_{12}H_{13}O_4N_5$					*S			S			Purine group *Yeast only	1
Natamycin	Pimaricin Tennecetin	*Stm. natalensis*						S						Polyene (tetraene) group	N-3a, N-3b, N-3c
Neamine	Neomycin A													See Neomycin A	1, 2
Nebramycin complex		*Stm. tenebrarius*		N-3	S	*S	S							Amino-glycosidic Water-sol, basic compd. *Pseudomonas	N-4
Nebularin	9-(β-D-Ribofuranosyl) purine Clitocybine	*Stm. sp.* & Fungi (Clitocybe) *Agaricus nebularis*					S				S			Related to Cordycepin	1, 2
Negamycin		*Stm. sp.* No. M890-C2 *Stm. sp.* MA91—M1 *Stm. sp.* MA104-M1	$C_9H_{20}N_4O_4 \cdot H_2O$		S	S							L	Mode act.: P	N-5a, N-5b
Neoantimycin		*Stm. orinoci*	$C_{36}H_{46}N_2O_{12}$				W								N-6

Name	Synonym	Producer	Formula	Struct. Ref. No.	G +	G -	Myco	Fung	Virus	Protoz	Tumor	Others	Tox.	Remarks	Ref.
Neoaureothin		*Stm. orinoci*	$C_{28}H_{31}NO_6$					W							N-6
Neocarzinostatin		*Stm. carzinostaticus* var. F-41	C 41.34, H 9.43, N 11.43		S						S		T (ip)	Polypeptide group	1
Neocide		*Acm. sp.* (prob.)			S						S		L	Peptide group	1
Neohumidin		*Stm. multispiralis*	C 64.03, H 8.45, C_6H_6 addn. compd. C 67.22, H 8.25		S			S					T (ip)	Humidin like	1
Neomethymycin		*Stm. sp.* M-214	$C_{25}H_{43}O_7N$	N-4	S									Macrolide group	1, 2
Neomycin A	Neamine	*Stm. fradiae*, *Stm. albogriseolus*	$C_{12}H_{26}O_6N_4$	N-5	S	S	S						L	Aminocyclitol (neomycin) group Mode act.: P.N(RNA)	1, 2, 3
Neomycin B	F-ramycetin, Streptothricin B II	*Stm. fradiae*, *Stm. albogriseolus*	$C_{23}H_{46}O_{13}N_6$	N-6	S	S	S	*W					L (sc)	Aminocyclitol (neomycin) group *Yeast only	1, 2
Neomycin C	Streptothricin B 1	*Stm. fradiae*, *Stm. albogriseolus*	$C_{23}H_{46}O_{13}N_6$	N-7	S	S	S	*W					L (sc)	Aminocyclitol (neomycin) group *Yeast only	1, 2
Neomycin D	Paromamine	*Stm. fradiae*												See Paromamine	N-7
Neomycin E	Paromomycin I	*Stm. fradiae*												See Paromomycin I	N-7
Neomycin F	Paromomycin II	*Stm. fradiae*												See Paromomycin II	N-7
Neomycin I					S		W								2
Neomycin LP$_B$		A mutant strain of Neomycin producing *Stm. sp.*												N-Acetylneomycin B & C	1

Name	Synonym	Producer	Formula	Struct. Ref. No.	G +	G -	Myco	Fungi	Virus	Protoz	Tumor	Others	Tox.	Remarks	Ref.
Neomycin LP$_C$		A mutant strain of Neomycin producing *Stm. sp.*												N-Acetylneomycin B & C	1
Neonocardin		*Noc. kuroishi*			S	S							T		1, 2
Neopentaene		*Stm. sp.* 2236	C 61.3, H 8.8, N. 0					S						Polyene (pentaene) group	1
Neopluramycin		*Stm. pluricolorescens*	C$_{40}$H$_{50}$N$_2$O$_{10}$		S						S		T	Similar to Pluramycin A	N-8
Neotalomycin		*Acm. sp.* 128		*	*S									*Staphylococcus Polypeptide group (contain 12 different amino acids MW: 1300)	N-9
Netropsin	Congocidin, Sinanomycin, IA-887, K-117, T-1384	*Stm. ambofaciens Stm. chromogenus Stm. netropsis Stm. reticuli Stm. sp.* 7618	C$_{18}$H$_{26}$O$_3$N$_{10}$ · ½H$_2$SO$_4$	N-8	S	S	S	W		W			T		1, 2
Neutramycin	LL-AE705W	*Stm. rimosus*	C$_{34}$H$_{54}$O$_{14}$		S									Neutral macrolide group	1
Niddamycin	3-Desacetyl-carbomycin B, F-3463	*Stm. djakartensis*	C$_{40}$H$_{65}$O$_{14}$N	N-9	S		S						L (sc)	Macrolide group	1
Nigericin	Polyetherin A	*Stm. sp.*	C$_{39}$H$_{69}$O$_{11}$	N-10	S	W	S	S					T (ip)	Mode act.: R	1, 2, 3, N-10
Niromycin A		*Stm. albus*	C$_{14-15}$H$_{21-23}$O$_{3-4}$N					*S	S				T	Glutarimide group *Yeast only	1, 2, 3
Niromycin B		*Stm. albus*	C$_{14}$H$_{21}$O$_4$N					*S	S					Mode act.: P Glutarimide group *Yeast only	1, 2

Name	Synonym	Producer	Formula	Struct. Ref. No.	G +	G -	Myco	Fung	Virus	Protoz	Tumor	Others	Tox.	Remarks	Ref.
Nitraminoacetic acid		Stm. noursei 8054-MC$_3$	$C_2H_4N_2O_4$	N-11	[*1] S[*1]		[*2] S[*2]						T	[*1] E. coli, Xanthomonas oryzae, Pseudomonas tabaci [*2] M. phlei	N-11
Nitropropionic acid	Bovinocidin	Stm. sp.												See Bovinocidin	
Nitrosporin		Stm. nitrosporeus	$C_{20}H_{26}O_6N_2$		S								T	Macrolide group	1, 2
Nivemycin	Neomycin													See Neomycin	1
Nocardamin		Noc. sp.	$C_9H_{14}O_3N_2$	N-12			S								1, 2
Nocardianin		Noc. sp.	$C_{65-67}H_{96-104}O_{15}N_{18}$		S										1, 2
Nocardin		Noc. coeliaca					S						W		1, 2
Nocardorubin	Rufinosporin	Mim. narashino (previously Noc. narashinoensis)			S		S						M (ip)	pH Indicator	1, 2
Noformicin	MK-61	Noc. formica	$C_8H_{15}ON_5$	N-13	S				S				T (sc)	Amidinomycin like	1, 2
Nogalamycin	NSC 70845	Stm. nogalater var. nogalater NRRL 3035	$C_{35}H_{47}O_{15}N$		S	S							T (ip)	Mode act.: N(RNA)	1, 3
Nojirimycin	R-468	Stm. sp. SF425	$C_6H_{13}O_5N$	N-14	S	S								Aminosugar group	1, N-12, N-13, N-14
Nonactin	FH-3582A, N-329A	Stm. tsusimaensis Stm. viridochromogenes Stm. werraensis	$C_{40}H_{64}O_{12}$	N-15	S		S							Cyclic polylactone Mode act.: O	1, 3, N-15

Name	Synonym	Producer	Formula	Struct. Ref. No.	G +	G -	Myco	Fungi	Virus	Protoz	Tumor	Others	Tox.	Remarks	Ref.
Notomycin A₁ & A₂	Coumermycin A₁, A₂	*Stm. rishiriensis* (40473, A9795) ATCC 14812	A1: $C_{55}H_{59}N_5O_{20}$ A2: C 57.58, H 4.63, N 5.90 (in average)	N-16 N-17	S	W	S						M (sc)		N-16
Novurseothricin A		*Stm. noursei* var. J.A. 3890b			S	S	S							Streptothricin group	1
Nourseothricin B		*Stm. noursei* var. J.A. 3890b			S	S	S								1
Novobiocin	Crystallinic acid Griseoflavin, Streptonivicin, PA-93	*Stm. griseus Stm. griseoflavus Stm. niveus Stm. spheroides Stm. sp.*	$C_{31}H_{36}O_{11}N_2$	N-18	S	W							L	Mode act.: N(RNA) (DNA)	1, 2, 3
Novomycetin	Chloramphenicol													See Chloramphenical	1
Novomycin		*Stm. sp.* related to *Stm. roseochromogenes*			S	S	S						L	Streptothricin group	1, 2
Nucleocidin	T-3018	*Stm. calvus*	$C_{11}H_{16}O_8N_6S$	N-19	S	S	S			S				Purine group Mode act.: P	1, 2, 3, N-17
Nybomycin		*Stm. sp.*	$C_{16}H_{14}O_4N_2$	N-20	S	S		S					L (ip)		1, 2
Nystatin	Fungicidin	*Stm. albulus, Stm. noursei*	$C_{46}H_{77}O_{19}N$	N-21				S		S			T (ip)	Polyene (tetraene) group Mode act.: M	1, 2, 3, N-18
Ochramycin		*Stv. orinoci*	$C_{46}H_{77}N_3O_{15}$		S		S						T (ip)		O-1
Ochromycinone		*Stm. sp.*	$C_{19}H_{14}O_4$	O-1											O-2

Name	Synonym	Producer	Formula	Struct. Ref. No.	G+	G-	Myco	Fungi	Virus	Protoz	Tumor	Others	Tox.	Remarks	Ref.
Oleandomycin	PA-105, RO$_2$-7638, 69895B	Stm. antibioticus, Stm. olivochromogenes	$C_{35}H_{61}O_{12}N$	O-2	S				S	S			ML	Macrolide group	1, 2, O-3
Oleficin		Stm. parvullus	$C_{38}H_{55}NO_{10}$		S						S				O-4
Oligomycin A		Stm. sp. (Aspergillus sp. Glomerella sp.)	$C_{24}H_{40}O_6$					S					T (ip)	Mode act.: O & R	1, 2, 3, O-5a, O-5b
Oligomycin B		Stm. sp.	$C_{22}H_{36}O_6$					S					T (ip)		1
Oligomycin C		Stm. sp.	$C_{28}H_{46}O_6$					S					T (ip)		1
Olivomycins (A,B,C,D)		Acm. olivoreticuli	$C_{61-65}H_{90-98}O_{27-29}$	O-3	S				S		S		T	Aureolic acid group Mode act.: N(RNA)	1, 2, O-6
Omegamycin	Tetracycline													See Tetracycline	1
Oncostatin B(X)		Stm. sp. INA 16/58									S				1
Oncostatin C	Actinomycin C	Stm. sp.									S			Actinomycin C complex	1, 2
Oncostatin K	Actinomycin K										S			Actinomycin I complex	2
Onomycin-I		Stm. sp. J-4	$C_{43}H_{76}O_{17}N$					S						Polyene (pentaene) group	1
Orientomycin	Cycloserine	Stm. roseochromogenes K-300	$C_3H_6O_2N_2$		S	W	S						L		1, 2
Ornamycin		Stm. erumpens Stm. ornatus			S	S		S							1

Name	Synonym	Producer	Formula	Struct. Ref. No.	G +	G −	Myco	Fungi	Virus	Protoz	Tumor	Others	Tox.	Remarks	Ref.
Orosomycin	Thioaurin													See Thioaurim	1
Orymycin		*Stm. albochromogenes*	$C_{25}H_{43}O_7$		S			S					T (ip)		1
Oryzacidin A		*Stm. sp.* F-35	$C_{22}H_{34}O_{12}N_4$					S							1, 2
Oryzamycin		*Stm. shimizuensis* S-2337	C 67.67, H 8.66, N 2.80					S					T (ip)		1
Oryzoxymycin		*Stm. venezuelae* var. *oryzoxymyceticus*	$C_{21}H_{30}N_2O_{10}$			*							M	*Xanthomonas oryzae*	O-7
Ossamycin		*Stm. hygroscopicus* var. *ossamyceticus* C8158	$C_{50}H_{87}O_{14}N$							S	S		T		1
Ostreogrycin A	Mikamycin Staphylomycin M$_1$, E-129 factor A, PA-114A Pristinamycin IIA	*Stm. ostreogriseus*	$C_{28}H_{36}O_8N_3$ or $C_{28}H_{35}O_7N_3$		S								L	Depsipeptide (Mikamycin) group Mode act.: P	1, 2, 3
Ostreogrycin B (B$_1$,B$_2$,B$_3$)	E-129 factor Z	*Stm. ostreogriseus*	$C_{45}H_{54}O_{10}N_8$	O-4	S									Depsipeptide (Mikamycin) group Synergistic with Ostreogrycin A or G	1, 2
Ostreogrycin Factor B	E-129 factor B Ostreogrycin G													See Ostreogrycin G	1
Ostreogrycin G	E-129 factor B	*Stm. ostreogriseus*	$C_{28}H_{37}O_7N_3$		S									Depsipeptide (Mikamycin) group Synergistic with Ostreogrycin B	1

Name	Synonym	Producer	Formula	Struct. Ref. No.	G +	G −	Myco	Fungi	Virus	Protoz	Tumor	Others	Tox.	Remarks	Ref.
Ostreogrycin Z	E-129 factor Z, Ostreogrycin B	*Stm. ostreogriseus*	C 62.20, H 6.08, N 12.88		S									Depsipeptide (Mikamycin) group Synergistic with E-129 factor B	1
Ostreogrycin Z₁	E-129 factor Z₁, Ostreogrycin B₁	*Stm. ostreogriseus*			S									Mikamycin group, Synergistic with E-129 factor B	1
Ostreogrycin Z₂	E-129 factor Z₂, Ostreogrycin B₂	*Stm. ostreogriseus*			S									Mikamycin group Synergistic with E-129 factor B	1
Ostreogrycin Z₃	E-129 factor Z₃, Ostreogrycin B₃	*Stm. ostreogriseus*	C 60.99, H 6.31, N 11.9		S									Mikamycin group, Synergistic with E-129 factor B	1
L-4-Oxalysine		*Stm. chartreusis* MA-2856 *Stm. erythrochromogenes* MA-2872	− HCl: $C_5H_{13}N_2O_3Cl$	O-5	W	W							M (ip)	Lysine antagonist	O-8a, O-8b
Oxamycin	D-4-amino-B-oxazolidone Cycloserine									S	S			See Cycloserine	1, 2
5-Oxo-1H-pyrrolo-2,1-C 1,4-Benzo-diazepin-2-Acrylamides		*Stm. refuineus* var. *thermotolerans* (NRRL 3143)		O-6	S	S							T	Several derivatives were patented	O-9
Oxytetracycline	Terramycin (Previously)	*Stm. rimosus, Stm. platensis Stm. armillatus Stm. vendargensis Stm. henetus Stm. gilvus Stm. capuensis Stm. sp.* M-590 cf. Tetracycline producer	$C_{22}H_{24}O_9N_2$	O-7	S	S	S				S	S		Tetracycline group See Tetracycline	1, 2

Name	Synonym	Producer	Formula	Struct. Ref. No.	G +	G -	Myco	Fungi	Virus	Protoz	Tumor	Others	Tox.	Remarks	Ref.
Oyamycin	Formycin B	*Stm. roseochromogenes* var. *oyaensis*												See Formycin B	1, O-10
Pactacin	Pactamycin													See Pactamycin	1
Pactamycin	Pactacin, NSC-52947, U-15800	*Stm. pactum* var. *pactum*	$C_{28}H_{40}O_8N_4$	P-1	S	S					S		T	Mode act.: P	1, 2, 3, P-1
Parvulines A, B, C		*Stm. parvullus*			S								L (ip)	Peptide group similar to Glumamycin, Amphomycin and Aspartocyn	P-2
Paromomycin I	Aminocidin, Catenulin, Hydroxymycin, Zygomycin A₁, A₂, D-68, 4915	*Stm. rimosus* f. *paromomycinus*	$C_{23}H_{45}O_{14}N_5$	P-2	S	S	S						M	Aminocyclitol (neomycin) group, cross resistance with Neomycin or Kanamycin Mode act.: P	1, 2, 3
Paromomycin II	Aminocidin, Catenulin, Hydroxymycin, Zygomycin A₁, A₂, D-68, 4915	*Stm. rimosus* f. *paromomycinus*	$C_{23}H_{45}O_{14}N_5$	P-2	S	S	S						M	Same as I	1, 2
Pathocidin	8-Azaguanine, B-28	*Stm. albus* var. *pathocidicus* *Stm. morookaensis*	$C_4H_4ON_6$	P-3		S		S			S		MT (ip)	Purine group	1, 2
Peliomycin	NSC-76455D	*Stm. luteogriseus*	$C_{46}H_{76}O_{14}$		*S						S			*Micrococcus lysodeikticus* only	1
Penicillin N	Cephalosporin N Synnematin B	*Stm. sp.*		P-4	S	S									P-3

Name	Synonym	Producer	Formula	Struct. Ref. No.	G+	G-	Myco	Fungi	Virus	Protoz	Tumor	Others	Tox.	Remarks	Ref.
Pentaene G8		Stm. anandii ATCC 19388						S							P-4
Pentafungin		Stm. antimycoticus	$C_{41}H_{74}O_{16}N$					S					T (ip)	Polyene (pentaene) group	1
Pentamycin		Stm. pentaticus	$C_{35}H_{60}O_{13}$					S					T (ip)	Polyene (pentaene) group Oxidative degradation gave 2-methyl-2,4,6,8,10 dodecapentaendiol	1, 2
Pepthiomycin A		Stm. roseospinus	C 58.73, H 6.23, N 11.23, O 20.27, S 4.22		S	*S[1]		*S[2]						Cross resistant with Bryamycin *[1]Xanthomonas oryzae *[2]Torula utilis	P-5
Pepthiomycin B		Stm. roseospinus	C 51.43, H 5.26, N 14.81, O 23.12, S 4.73		S	*S[1]		*S[2]						*[1]Xanthomonas oryzae *[2]Torula utilis	P-5
Peptimycin		Stm. mauvecolor 1112	$(C_{38}H_{60}O_{13}N_{12})_{1-2}$								S		TM	Peptide group	1, 2
Peresimycin		Stm. kawachiensis	$C_{12}H_{27}O_8N_2$		S		W						L		1
Perimycin	Aminomycin, Fungimycin, NC-1968	Stm. coelicolor var. aminophilus	$C_{47}H_{74-76}O_{14}N_2$					S						Polyene (heptaene) group Hydrolysis gave perosamine and p-aminophenylacetone	1, 2, 3
Perlimycin		Stm. chrysomallus ATCC 11523	$C_{40}H_{60-62}O_{12}$		S						S				1
Phaeochromin		Stm. sp.	C 42.57, H 5.05, N 14.82					*S					TM	*Yeast only	1

Name	Synonym	Producer	Formula	Struct. Ref. No.	G+	G-	Myco	Fungi	Virus	Protoz	Tumor	Others	Tox.	Remarks	Ref.
Phaeofacin		Stm. phaeofaciens						S					L		1, 2
Phagocidin		Stm. sp. res. Stm. antibioticus							*S				T	*Anti phage	1, 2
Phagolessin	A-58	Stm. griseus				S			S	S			T		1, 2
Phagomycin		Stm. sp. Stm. sp.	Reineckate: C 12.38, H 6.79, N 12.14, S 14.34, Cr 5.24			*S^1			*S^2					*1 Limited strain of salmonella *2 $T_{1,3,5,7}$ phage	1, 2
Phagostatin		Stm. sp. F-300				*W^1			*S^2					*1 Salmonella *2 E. coli B phage	1, 2
Phalamycin		Stm. noursei	$C_{36}H_{41}O_{14}N_9S$			S	S	S					T.M. (ip)		1, 2
1.6-phenazinediol	Dihydroxy-phenazine	Waksmania aerata, Stm. thioluteus (Brevibacterium crystalloiodinum Pseud. iodina)	$C_{12}H_8N_2O_2$	P-5	S	S	S	S							P-6, P-7, P-8
2-amino-3H-phenoxazin-3-one	Questiomycin A	Waksmania aerata, Stm. sp.	$C_{12}H_8N_2O_2$	P-6	S	S	S	S							P-6, P- 9
2-acetamido-3H-phenoxazin-3-one		Waksmania aerata	$C_{14}H_{10}N_2O_3$	P-6	S		S	W							P-6
1.6-phenazinediol-5,10-dioxide	Iodinin	Waksmania aerata (Microbispora)	$C_{12}H_8N_2O_4$	P-7	S		W	W						Also produced by Pseudomonas iodina and Brevibacterium crystalloiodinum	P-6, P-10
Phenomycin		Stm. fervens var. phenomyceticus 564-Cl	C 47.25, H 7.30, N 15.77, S 0.79	P-8							S		TM (ip)	Polypeptide group	P-11

Name	Synonym	Producer	Formula	Struct. Ref. No.	G+	G-	Myco	Fungi	Virus	Protoz	Tumor	Others	Tox.	Remarks	Ref.
Phleocidin		*Stm. kawachiensis*	$C_{16}H_{48-50}O_{16}$		S	S	S						L		1
Phleomycins		*Stm. verticillus*	D_1: $C_{36}H_{69}O_{19}N_{13}SCu$. E: $C_{42}H_{74}O_{21}$ $N_{16}SCu$. G: $C_{26}H_{49}$ $N_{16}SCu$. H: C_{42} $H_{72}O_8N_{16}SCu$ I: $C_{59}H_{111}O_{32}N_{21}S_2$ Cu.		S	S	S	*S			S		TM	Peptide group *Rust fungi of plant Mode act.: N	1, 2, 3
Phosphonomycin		*Stm. fradiae* MA-2898, 2915 (NRRL-3357, 3417) *Stm. viridochromogenes* MA-3867, 2903, 2916, 2917, 3270, 3271, 3272 (NRRL 3414, 3413, 3415, 3416 3427, 3720, 3721) *Stm. wedmorensis* MA3269 (NRRL 3426)	$C_3H_7O_4P$	P-9	S	S								M	Mode act.: C	P-12, P-13, P-14, P-15, P-16
Phthiomycin		*Stm. luteochromogenes*	Reineckate: C 25.35, H 4.4, N 22.88		W	W	S						L	Peptide group	1, 2
Phycomycin		*Stm. sp.* C-44, C-55	$C_{23}H_{40}O_7$				S						L (ip)	Oligomycin like	1
Phyllomycin		*Stm. umbrosus*	$C_{23}H_{30}O_9N_2$					S					T	Similar to Antimycin A	1
Phytoactin		*Stm. sp.*	C 56.48, 57.14, H 8.11, 8.34, N 12.24, 12.57		S			S						Polypeptide, group	1, 2

Name	Synonym	Producer	Formula	Struct. Ref. No.	G+	G-	Myco	Fungi	Virus	Protoz	Tumor	Others	Tox.	Remarks	Ref.
Phytobacteriomycin	696	*Acm.* sp 696 res. *Acm. lavendulae*			S	S									1, 2
Phytostreptin	Polyaminohy-grostreptin (patent name)	*Stm. sp.*	C 52.44, 53.70, H 7.66, 8.44, N 12.53, 13.48		S								*	Polypeptide group *Non-toxic to tomato and bean plant	1, 2
Picromycin	Pikromycin													See Pikromycin	2
Piericidin A		*Stm. mobaraensis*	A: $C_{25}H_{37}O_4N$	P-10				S				*	T	*Insect of plant disease	1, P-18
Piericidin B		*Stm. mobaraensis* 16-22	B: $C_{26}H_{39}O_4N$	P-10				S							1, P-18
Pikromycin	Picromycin, Amaromycin, Albomycetin IMRU 3627, Bu-277 Hydroxynar-bomycin	*Stm. felleus* *Stm. venezuelae*	$C_{28}H_{47}NO_8$	P-11	S	W							L (iv)	Macrolide group	1, 2, P-17, P-19
Pillaromycin A (B$_I$, B$_{II}$, C)		*Stm. flavovirens* No. 65786	$C_{25}H_{26}O_{10}$	P-12	S						S				1, 2
Piloquinone		*Stm. pilosus*	$C_{21}H_{20}O_5$	P-13	S										1
Pilosomycin A	A-9578	*Stm. galilaeus* NRRL 2722, *Stm. griseoflavus*	A-HCl: C 50.68, H 6.99, N 13.45, Fe 3.60, Cl 2.75		S			S					L (sc)	Sideromycin group	1
Pilosomycin B	A-9578	*Stm. galilaeus* NRRL 2722, *Stm. griseoflavus*	A-HCl: C 50.68, H 6.99, N 13.45, Fe 3.60, Cl 2.75		S								L (sc)	Sideromycin group	1
Pimaricin	Tennecetin	*Stm. natalensis*	$C_{33}H_{47}O_{13}N$	P-14				S					L (ip)	Polyene (tetraene) group	1, 2

Name	Synonym	Producer	Formula	Struct. Ref. No.	G +	G -	Myco	Fungi	Virus	Protoz	Tumor	Others	Tox.	Remarks	Ref.
Piomycin(e)		*Stm. sp.* E-1443 (ATCC 21137)	$C_{17}H_{32}N_6O_{15}$					*S					L (ip)	*Piricularia oryzae, Cochliobolus miyabeanus* and other plant pathogens	P-21
Planomycin	Fervenulin	*Stm. rubrireticuli*	$C_7H_7O_2N_5$	P-15	W	S					S		TM	Toxoflavin, Xanthothricin like	1
Pleocidin I		*Stm. sp.*			S	S	S						T (ip)	Streptothricin group	1, 2
Pleocidin II		*Stm. sp.*			S	S	S						T (ip)	Streptothricin group	1, 2
Pleocidin III		*Stm. sp.*			S	S	S						T (ip)	Streptothricin group	1, 2
Pleocidin IV		*Stm. sp.*			S	S	S						T (ip)	Streptothricin group	1, 2
Pleomycin		*Stm. pleofaciens*	$C_{14}H_{12}O_8$		S	S	S						T (ip)		1, 2
Plicacetin	Amicetin B, Antibiotic C	*Stm. plicatus*	$C_{25}H_{35}O_7N_5$	P-16	S		S						T (ip)	Pyrimidine derivative	1, 2
Plurallin		*Stm. pluricolorescens*	C 50.01, H 7.23, N 10.89, S 0.94. Ash 0.161 mg/2.460 mg		*S						S		TM (ip)	pH Indicator, Glycoprotein *Sarcina lutea*	1
Pluramycin A	MA-321 A₃ (O15350)	*Stm. pluricolorescens*	A (prism): C 66.63, H 6.30, N 3.66 A (needle): C 66.87, H 6.61, N 3.80 B: C 69.71, H 9.18, N 1.23		S						S		T (ip)	pH Indicator Mode act.: N	1, 2, 3

Name	Synonym	Producer	Formula	Struct. Ref. No.	G +	G -	Myco	Fungi	Virus	Protoz	Tumor	Others	Tox.	Remarks	Ref.
Polyetherin A	Nigericin	Stm. hygroscopicus E-749	$C_{39}H_{69}O_{11}$		S		S	*S		S			T (ip)	Polycyclic polyether group *Certain plant pathogens: Pricularia oryzae etc.	P-22, N-10
Polyketoacidomycin	PKAM	Stm. sp. AS 51	$C_{39}H_{50}O_{14}N$		S	S		S					ML	Niddamycin like	P-23, 1
Polymycin		Acm. sp. 1787-9	$C_{33-35}H_{6-7}N_{14,8-15}$		S	S	S	W			W		M	Streptothricin group	P-24, 1
Polyoxin A		Stm. cacaoi var. asoensis	$C_{23}H_{32}O_{14}N_6$	P-17				*S					L	*Pellicularia sasaki	P-25, 1
Polyoxin B		Stm. cacaoi var. asoensis	$C_{17}H_{25}O_{13}N_5$	P-17				*S					L	*P. sasaki	P-25, 1
Polyoxin C		Stm. cacaoi var. asoensis	$C_{11}H_{15}O_8N_3$	P-17				*S						*P. sasaki	P-25, 1
Polyoxin D		Stm. cacaoi var. asoensis	$C_{17}H_{23}O_{14}N_5$	P-17				S						Mode act.: C	P-25, P-26, 1
Polyoxin E		Stm. cacaoi var. asoensis	$C_{17}H_{23}O_{13}N_5$	P-17				S							1, P-25
Polyoxin F		Stm. cacaoi var. asoensis	$C_{23}H_{30}O_{15}N_6$	P-17				S							1, P-25
Polyoxin G		Stm. cacaoi var. asoensis	$C_{17}H_{25}O_{12}N_5$	P-17				S							1, P-25
Polyoxin H		Stm. cacaoi var. asoensis	$C_{23}H_{32}N_6O_{13}$	P-17				*S					L	*Phytopathogenic fungi	P-25, P-27
Polyoxin I		Stm. cacaoi var. asoensis		P-17											P-27

Name	Synonym	Producer	Formula	Struct. Ref. No.	G+	G-	Myco	Fungi	Virus	Protoz	Tumor	Others	Tox.	Remarks	Ref.
Polyoxin J		Stm. cacaoi var. asoensis	$C_{17}H_{25}N_5O_{12}$	P-17				*S						*Pricularia oryzae and other plant pathogens	P-25, P-28
Polyoxin K		Stm. cacaoi var. asoensis	$C_{22}H_{30}N_6O_{13}$	P-17				S							P-25, P-28
Polyoxin L		Stm. cacaoi var. asoensis	$C_{16}H_{23}N_5O_{12}$	P-17											P-25, P-28
Porfiromycin	Methylmitomycin C, Mitiromycin D	Stm. ardus, Stm. verticillus	$C_{16}H_{20}O_5N_4$	P-18	S	S					S			Mitosane group Mode act.: N	1, 2, 3
Poryzamycin		Stm. sp.	C 73.24, H 10.06, N 0.74		S								M		1
Prasinomycin		Stm. prasinus			S	S								Moenomycin group	P-29
Prasinomycins A			C 48.52, H 6.51, N 4.42, P 2.43												
Prasinomycins B			C 50.72, H 6.70, N 4.76, P 2.30												
Prasinomycins C			C 51.36, H 7.53, N 4.34, P 2.19												
Primocarcin		Noc. fukayae	$C_8H_{12}O_3N_2$	P-19	S	S					S		TM (ip)		1, 2, P-30
Primycin		Stm. sp.	$C_9H_{37}O_7N$	P-20	S	S	S		S				T		1, 2, P-31

Name	Synonym	Producer	Formula	Struct. Ref. No.	G+	G-	Myco	Fungi	Virus	Protoz	Tumor	Others	Tox.	Remarks	Ref.
Pristinamycins (IA, IB, IIA, IIB)	RP-7293, RP-12535 IA = mikamycin B, PA-114B, ostreogrycin B, Vernamycin Bα, IB = ostreogrycin B2, vernamycin Bβ, RP-13919 (023550) IIA = mikamycin A, ostreogrycin A, staphylomycin M-1, PA-114A, RP-12536 IIB = RP-13920	*Stm. pristinae spiralis*	IA: $C_{45}H_{54}O_{10}N_8$ IB: $C_{44}H_{52}O_{10}N_8$ IIA, IIB: $C_{30}H_{37}O_8N_3$	P-21	S								M (ip)	Depsipeptide (mikamycin) group Mode act.: P	1, 2, 3, P-32
Proactinomycin A	Picromycin	*Noc. (Proactinomyces) gardneri*	$C_{27}H_{47}O_8N$		S								M	Macrolide group	1, 2
Proactinomycin B	Griseomycin		$C_{28}H_{49}O_8N$		S								M	Macrolide group	1, 2
Proactinomycin C		*Stm. albolongus*	$C_{24}H_{41}O_6N$		S								M	Macrolide group	1, 2
Proceomycin			C 51.02, H 6.21, N 3.14, S 3.83		S								T		1
Prodigiosin		*Stm. sp.*	$C_{20}H_{25}ON_3$	P-22	S			S		S			S	Pigment (originally produced by *Serratia macescens*)	2, P-33
Prodigiosin-25C		*Stm. sp. 28-24 similar to Stm. ruber*	$C_{25}H_{35}N_3O$	P-23	S								T		P-34

Name	Synonym	Producer	Formula	Struct. Ref. No.	G+	G-	Myco	Fungi	Virus	Protoz	Tumor	Others	Tox.	Remarks	Ref.
Prodigiosin-like antibiotic		Stm. sp.	$C_{25}H_{35}ON_3$		S	W		S		S				Prodigiosin is an antibiotic from bact.	1
Protoactinorhodin		Stm. violaceoruber 199		P-24	S									Biosynthesized	P-35
Protocidin(e)		Stm. sp. 964-A	$C_{29}H_{45}O_{13}N$					S						Polyene (tetraene) group	1, 2
Protomycin		Stm. reticuli var. protomycicus	$C_{19}H_{29}O_5N$	P-25				*S		S			ML	Glutarimide group *Yeast only Mode act.: P	1, 2, 3
Prunacetin A		Stm. griseus var. purpureus No. CD270 (Stm. californicus)	C 54.70, H 7.14, N 8.95, Ash 1.18, No halogen, No S		W						S			Pigment-carbohydrate-protein complex Prosthetic group is related to Griseorhodin	P-36
Pseudostreptomycin		Stm. griseus											T	Activity, 1/10 of Streptomycin	1
Psicofuranine	Angustmycin C	Stm. hygroscopicus var. decoyicus	$C_{11}H_{15}O_5N_5$		S		S				S			Purine group	1, 2
Pulvomycin		Stm. sp.	C 65.57–65.82, H 7.33–7.57, N 1.83–1.86, Ash 0.12		S		S						M (ip)	Mycolutein like	1, 2
Puromycin	Achromycin (previously)	Stm. alboniger	$C_{22}H_{29}O_5N_7$	P-26	S	W				S	S		M	Purine group Mode act.: P	1, 2, 3
Pyrazomycin		Stm. candidus	$C_9H_{13}O_6N_3$	P-27					S					C-nucleosides group Inhibitor of Orotidylic acid decarboxilase	P-37, P-38

Name	Synonym	Producer	Formula	Struct. Ref. No.	G +	G −	Myco	Fungi	Virus	Protoz	Tumor	Others	Tox.	Remarks	Ref.
Pyridomycin		*Stm. pyridomyceticus* *Stm. sp.* 6706	$C_{27}H_{32}O_8N_4$	P-28	W		S						L		1, 2, P-39
Pyrromycin		*Stm. sp.* DOA 1205	$C_{30}H_{35}O_{11}N \cdot HCl$	P-29	S									Indicator group, Anthracycline group	1, 2
Pyrromycinones	ε-Pyrromycinone Rutilantinone	*Stm. sp.*	$\zeta: C_{22}H_{20}O_8$ $\eta: C_{22}H_{16}O_7$ $\epsilon: C_{22}H_{20}O_9$	P-30	S									Indicator group, Anthracycline group	1, 2
(Quatrimycin)					*S	S							*L (ip)	An epimer of Tetracycline *Ammonium Quatrimycin	2, Q-1
Questiomycin A	6-Amino-phenoxazone	*Stm. sp.*	$C_{12}H_8O_2N_2$	Q-1	S		S	S					L (ip)		1
Questiomycin B	O-Aminophenol	*Stm. sp.*	C_6H_7ON	Q-2	S		S	S					L (ip)		1
Quinocycline A	PA-371 a'	*Stm. aureofaciens*	−HCl: C: 60.46, H: 5.8, N: 4.13, Cl: 5.37		S		S						T (ip)		1, 2
Quinocycline B	PA-371	*Stm. aureofaciens*	−HCl: C 60.5, H 5.8, N 3.9, Cl 4.9		S		S						T (ip)		1, 2
Quinomycin A	Echinomycin, Levomycin	*Stm. sp.* 1752 *Stm. aureus*	$C_{50}H_{60}O_{12}N_{12}S_2$	Q-3	S	W	W				S		T (ip)	Quinoxaline group	1, 2, Q-2
(NX)-Quinomycin A		*Stm. sp.* 731-I	$C_{51}H_{61}O_{12}N_{11}S_2$		S	W	S						T (ip)	Biosynthetic replacement of Quinomycin A chromophores	Q-3
(QN)-Quinomycin A		*Stm. sp.* 731-I	$C_{52}H_{62}O_{12}N_{10}S_2$		S	W	S						T (ip)	Biosynthetic replacement of Quinomycin A chromophores	Q-3

Name	Synonym	Producer	Formula	Struct. Ref. No.	G +	G -	Myco	Fungi	Virus	Protoz	Tumor	Others	Tox.	Remarks	Ref.
Quinomycin B		*Stm. sp.* 732 & 1752 (related to *Stm. aureus*)	$C_{52}H_{64}O_{12}N_{12}S_2$	Q-3	S	W					S		T (ip)	Quinoxaline group; Echinomycin like	1, 2, Q-2
Quinomycin B$_0$		*Stm. sp.* 732 & 1752 (related to *Stm. aureus*)													Q-2
Quinomycin C		*Stm. sp.* 732 & 1752 (related to *Stm. aureus*)	$C_{54}H_{68}O_{12}N_2S_2$	Q-3	S	W					S		T (ip)	Quinoxaline group Echinomycin like	1, 2, Q-2
Quinomycin D		*Stm. sp.* 732 & 1752 (related to *Stm. aureus*)		Q-3											Q-2
Quinomycin E		*Stm. sp.* 732 & 1752 (related to *Stm. aureus*)		Q-3											Q-2
Quinquamycin		*Stm. lavendulae* E 20-27						S							1
Quintomycin A	Lividomycin A	*Stm. lividus* ATCC 21178	$C_{29}H_{55}N_5O_{19}$	Q-4	S	S	S						L	Oligosaccharide group	Q-4, Q-5
Quintomycin B	Lividomycin B = Lividomycin	*Stm. lividus*	$C_{29}H_{55}N_5O_{18}$		S	S	S						L	Oligosaccharide group	Q-4, Q-5
Quintomycin C	Lividomycin C Paromomycin	*Stm. lividus*	$C_{23}H_{45}N_5O_{14}$		S	S	S						M	Oligosaccharide group	Q-4, Q-5
Quintomycin D	Lividomycin D	*Stm. lividus*	$C_{23}H_{45}N_5O_{13}$		S	S	S						M	Oligosaccharide group	Q-4
Rabelomycin		*Stm. olivaceus* ATCC 21549	$C_{19}H_{14}O_6$	R-1	S								M	Anthraquinone group Similar to Tetrangomycin	R-1
Racemomycin A	229A	*Stm. racemochromogenes*	$(C_{19}H_{31}O_8N_7)_2 \cdot 3H_2SO_4 \cdot 3H_2O$	R-2	S	S	S						M	Streptothricin group	1, 2

Name	Synonym	Producer	Formula	Struct. Ref. No.	G +	G -	Myco	Fungi	Virus	Protoz	Tumor	Others	Tox.	Remarks	Ref.
Racemomycin B	229B	*Stm. racemochromogenes*	$C_{60}H_{128}O_{32}N_{20} \cdot 8HCl$	R-3	S	S	W						M	Streptothricin group	1, 2
Racemomycin C	229C	*Stm. racemochromogenes*	$C_{15}H_{32}O_8N_5$		S	S							M	Streptothricin group	1, 2
Racemomycin O		*Stm. racemochromogenes*	$C_{25}H_{44}O_{10}N_8 \cdot 3H_2O$	R-4	S	S							(L)	Streptothricin group	1, 2
Ractinomycin A		*Stm. sp.* A788-A-2 (closely related *Stm. phaeochromogenes*)	C 57.3, H 4.25, N 6.07		S			S			S		T	pH indicator	1, 2
Ractinomycin B		*Stm. sp.* A788-A-2 (closely related *Stm. phaeochromogenes*)			S			S			S		T (ip)	pH Indicator	1, 2
Raisnomycin		*Stm. kentuckensis*			S	S							T		1, 2
Ramnacin		*Stm. ramnaii*	$C_{26}H_{43}O_8$		S	*S		S						*Proteus vulgaris*	1, 2
Rancomycin I	U-22,583	*Stm. lincolnensis var. lincolnensis*	$C_{17}H_{20}N_4O_9$		S	S									R-2
Rancomycin II	U-25,873 Rancinomycin	*Stm. lincolnensis var. lincolnensis* NRRL 2936	$C_{17}H_{20}N_4O_9$		S	S									R-2
Raromycin		*Stm. albochromogenes*	C 56.95–57.53, H 7.93–8.03, N 4.19–4.46		W						S		L (ip)		1, 2
Refuin		*Thermophilic actinomycete* ATCC 14760, 14761, 14762			W						S		L	High molecular	R-3, R-4
Refusin	Anthramycin													See Anthramycin	

Name	Synonym	Producer	Formula	Struct. Ref. No.	G +	G -	Myco	Fungi	Virus	Protoz	Tumor	Others	Tox.	Remarks	Ref.
Relomycin	LL-AM684β	*Stm. hygroscopicus* NRRL 3017	$C_{45}H_{79}O_{17}N$		S									Macrolide group	1
Resistomycin	Croceomycin, Heliomycin	*Stm. arabicus,* *Stm. resistomycificus*	$C_{22}H_{16}O_4$	R-5	S		S								R-5, R-6, 1, 2
Restomycin		*Stm. naniwaensis*	$C_{12}H_{14}O_8N_2$		S	S	S						L		1
Reticulin	Hydroxy-streptomycin	*Stm. reticuli* H-365												See Hydroxy-streptomycin	1, 2
Reumycin		*Acm. sp.*									S				R-7
Rhizomycin		*Stm. novoverticillus*	$C_{19}H_{11}O_5N_4$		S		S	*S					M	*Yeast	R-8
Rhodocidin		*Stm. phoenix*			S	S	S						T		1, 2
Rhodomycetin		Red mutant of *Stm. griseus*			S		S						T		1, 2
Rhodomycin (Brockmann)		*Stm. purpurascens*	A: $C_{36}H_{48}O_{12}N_2 \cdot 2HCl$ B: $C_{28}H_{33}O_{10}N \cdot HCl$	R-6	S									Anthracycline group	1, 2
γ-Rhodomycins		*Stm. purpurascens*	I											Contg. 1. γ-Rhodomycinone 2. Rhodosamine	2
		Stm. purpurascens	II											Contg. 1. γ-Rhodomycinone 2. Rhodosamine	2
		Stm. purpurascens	III											Contg. 1. γ-Rhodosamine 2. Rhodosamine 3. 2-deoxy-L-fucose	2

Name	Synonym	Producer	Formula	Struct. Ref. No.	G+	G-	Myco	Fungi	Virus	Protoz	Tumor	Others	Tox.	Remarks	Ref.
Isorhodomycin A		*Stm. purpurascens*	IV											Contg. 1,γ-Rhodomycinone 1.Rhodosamine 1,2-deoxy-L-fucose 1.Rhodinose	2
		Stm. purpurascens												Chromophor: Isorhodomycinone	2
2-Rhodomycinone		*Stm. purpurascens*	$C_{20}H_{18}O_8$	R-7											2, R-9
α₂-Rhodomycinone		*Stm. purpurascens*	$C_{20}H_{18}O_8$	R-8											R-10
β-Rhodomycinone		*Stm. purpurascens*	$C_{20}H_{18}O_8$	R-9										Aglycone of Rhodomycin A	2, R-9
β-Isorhodomycinone		*Stm. purpurascens*	$C_{20}H_{18}O_8$	R-9											R-9
γ-Rhodomycinone		*Stm. purpurascens*	$C_{20}H_{18}O_7$	R-10										Aglycone of γ-Rhodomycins	2, R-9
δ-Rhodomycinone		*Stm. purpurascens*	$C_{22}H_{20}O_9$	R-11	S									Anthracyclinone,	1,.2
ε-Rhodomycinone		*Stm. purpurascens*	$C_{22}H_{20}O_9$	R-12										Isomeric to ε-Pyrromycinone	2
ζ-Rhodomycinone		*Stm. purpurascens*	$C_{22}H_{20}O_7$	R-13											2
ζ-Isorhodomycinone		*Stm. purpurascens*	$C_{22}H_{20}O_7$	R-14											2
Rhombomycin		*Stm. antibioticus* GB			S										1
9-β-D-Ribofuranosyl-purine	Nebularine	*Stm. yokosukaensis*	$C_{10}H_{12}O_4N_4$	R-15			S				S		ML (ip)	Purine derivative	1
Rifamycin B		*Stm. mediterranei*	$C_{39}H_{49}O_{14}N$	R-16	S		S						L	Mode act.: N	1, 2, 3

Name	Synonym	Producer	Formula	Struct. Ref. No.	G+	G-	Myco	Fungi	Virus	Protoz	Tumor	Others	Tox.	Remarks	Ref.
Rifamycin C.D.		*Stm. mediterranei*	C: C 61.52, H 6.73, N 4.21 D: C 62.17, H 6.58, N 3.53		S		S								1, 2
Rifamycin L		*Stm. mediterranei*	$C_{39}H_{49}NO_{14}$	R-17										Transformable from Rifamycin S	R-11
Rifamycin D		*Stm. sp.* 4107 A_2	$C_{39}H_{49}O_{14}N$		S	W	S						L		1, 2
Rifamycin S		*Stm. mediterranei*	$C_{37}H_{45}NO_{12}$	R-18	S	S	S						M	Acid hydrolyzed from Rifamycin O	2
Rifamycin SV		*Stm. mediterranei* (ATCC 21271)	$C_{37}H_{47}NO_{12}$	R-19	S	S	S						L	Oxidized from Rifamycin Semisynthetic antibiotic (Chemical modification of Rifamycin B) also produced by mutant of *Stm. mediterranei*	S-2, R-12
Rifamycin Y		Semi-synth. by *Stm. mediterranei*	$C_{39}H_{47}O_{15}N$	R-20	S		S						L	Similar to Rifamycin B	R-13, R-14
Rifamycin O		*Stm. sp.* 4107 A_2	$C_{39}H_{49}O_{14}N$	R-21	S	W	S						L		1, 2
Rifomycin	Rifamycin													See Rifamycin	1
Rimocidin	PA-85	*Stm. rimosus*	$C_{37}H_{59}O_{13}N$					S					T	Polyene (tetraene) group Mode act.: M	1, 2, 3
Ristocetin A	Spontin	*Noc. lurida*	$-H_2SO_4$: C 52.0−53.0 H 5.5−5.6, N 4.7−5.9 S 0.8−1.3		S		S						L	Ristocetin B, Vancomycin like Mode act.: C	1, 2, 3

Name	Synonym	Producer	Formula	Struct. Ref. No.	G +	G -	Myco	Fungi	Virus	Protoz	Tumor	Others	Tox.	Remarks	Ref.
Ristocetin B	Spontin	*Noc. lurida*	$-H_2SO_4$: C 52.9, 53.5, H 5.9, 5.5, N 5.6, 6.6, S 1.3, 1.5		S		S						L	Ristocetin A, Vancomycin like Mode act.: C	1, 2, 3
Ristomycin	Ristocetins	*Noc. (Proactinomyces) fructiferi*	C 52.2, H 5.8, N 6.8		S		S						L	Hydrolysis gave Glucose, Mannose, Arabinose, Rhamnose and Phenylglycine	1, 2
Robigocidin A		*Stm. platensis* sub sp. *robigocidicus*	$C_{36}H_{50}N_2O_8$					*S						*Yeast only	R-15
Rokugomycin	Neomycins													See Neomycins	1
Roseocitrin A	212 A	*Stm. roseocitreus*			S	S	S						T	Streptothricin group	1
Roseocitrin B	212 B	*Stm. roseocitreus*			S	S	S						T	Streptothricin group	1
Roseolic acid		*Stm. sp.*	C 36.95—37.54, H 5.08—4.79, N 11.89—11.71 P 7.89—8.13								S			pH Indicator	1, 2
Roseomycin	36	*Stm. roseochromogenes*			S	S	S						L		1, 2
Roseothricin A	H-277	*Stm. roseochromogenes*	$C_{36}H_{65}O_{16}N_{15}$	R-22	S	S							MT	Streptothricin group	1, 2
Ros(s)imycin		*Stm. chrysomallus*									S				1, 2
Rotaventin		*Stm. rubrieticuli* *Stm. griseocarneus*						S					M		1, 2
Rubidin		*Stm. sp.* A_{12}	C 51.9, H 5.56		S									pH Indicator	1, 2

Name	Synonym	Producer	Formula	Struct. Ref. No.	G +	G -	Myco	Fungi	Virus	Protoz	Tumor	Others	Tox.	Remarks	Ref.
Rubidomycin	RP-13057 Danubomy-cin	*Stm. coeruleorubidus*	$C_{27}H_{34}O_{11}N \cdot HCl$		S	W					S				1, 2, R-16, R-17
Rubiflavin		*Stm. sp.*	$C_{23}H_{29-31}O_5N$		S	S					S		T	pH Indicator	1
Rubomycin		*Acm. sp. res. Acm. coeruleorubidus*									S				1
Rubomycin B	Daunomycin?														R-18
Rubomycin C	Daunomycin?														R-18
Rubradirin	U-11092	*Stm. achromogenes var. rubradiris*	$C_{51}H_{50}O_{21}N_4$		S	S								pH Indicator	1
Rubromycin		*Stm. collinus*	C 60.30, H 4.26, N O	R-23	S	W								Pigment	1, 2
Rufinosporin	Nocardorubin													See Nocardorubin	1
Rufochromomycin	RP-5287	*Stm. rufochromogenes* NRRL 2816	$(C_{25}H_{26}O_8N_4)_n$		S	S	S						T		1
Rufomycin A		*Stm. atratus*	$C_{49}H_{67}O_{11}N_9$	R-24		S	S						L (ip)	Peptide Ilamycin like	1, 2
Rufomycin B	Ilamycin B	*Stm. atratus*	$C_{49}H_{69}O_{10}N_9$	R-25		S	S						L (ip)	Peptide	1, 2
Rutamycin	A-272	*Stm. sp.*	$C_{27}H_{44}O_4$					S					T (ip)	Oligomycin like Inhibitor of Phosphoryl transfer reaction	1, 2, 3, R-19, R-20
Ruticin		*Stm. sp.* closely related to *Stm. rutgersensis*			S	S							T		1, 2

Name	Synonym	Producer	Formula	Struct. Ref. No.	G+	G−	Myco	Fungi	Virus	Protoz	Tumor	Others	Tox.	Remarks	Ref.
Ruticulomycin A, B		Stm. rubrireticuli	A: C 58.52, H 6.71, N 1.80, O–Me 7.26, N–Me 4.42, B: C 57.09, H 6.37, N 1.84, O–Me 5.87, N–Me 2.13	R-26										Anthracycline group	1
Rutilantin	Cinerubin B	Acm. sp. S-200	$C_{22}H_{21}O_9$		S	S			S				T	Anthracycline group	1, 2
Rutilantinone	Pyrromycinone			R-27										See Pyrromycinone	2
Ryemycin A₁		Stm. ryensis	$C_{22}H_{20}O_8$		W	*S					S		T	*Corynebacterium diphtheriae	1
Ryemycin A₂		Stm. ryensis	$C_{22}H_{20}O_7$		W	*S					S			*Corynebacterium diphtheriae	1
Ryemycin B₁	Cinerubin B	Stm. sp. N044 / Stm. ryensis			S						S			Anthracycline group	1
Ryemycin B₂	Cinerubin A	Stm. sp. N044 / Stm. ryensis / Stm. sp.			S						S			Anthracycline group	1
Sacromycin	Amicetin													See Amicetin	1, 2
Sanclomycine	Tetracycline													See Tetracycline	1
Sangivamycin	BA-90912	Stm. sp.	$C_{12}H_{17}O_6N_5$ / $C_{12}H_{15}O_5N_5$	S-1							W			Purine group Chemically synthesized Mode act.: N	1, 3, S-1
Saramycetin	RO2-7758, X-5079C	Stm. saraceticus NRRL-2831	−Na: C 46−48, H 4−5 N 14−16, S 12−14 Na 1−2 (M.W. 1000−1300 titration)					S					L	Polypeptide group	1
Sarcidin		Stm. achromogenes	C: 41.89, H: 5.02 N: 21.82		S								L (ip)		1, 2

Name	Synonym	Producer	Formula	Struct. Ref. No.	G +	G -	Myco	Fungi	Virus	Protoz	Tumor	Others	Tox.	Remarks	Ref.
Sarkomycin		Stm. erythrochromogenes	$C_7H_8O_3$	S-2	W						S		L	Mode act.: N & P	1, 2, 3, S-2
Sclerothricin		Stm. sclerogranulatus	$C_{16}H_{30}O_8N_6$		W	W	W						T	Streptothricin group	S-3
Scopafungin	U-29,479	Stm. hygroscopicus						S						Also antibacterial	S-4
Scopamycin A		Stm. aureofaciens ETH 28832	$C_{44}H_{72}O_{14}$					S						Similar to Folimycin	1
Scopamycin B		Stm. aureofaciens ETH 28832	$C_{44}H_{72}O_{14}$					S						Similar to A	1
Sekazin	Secasine	Acm. sp. SK-3	$C_{41-42}H_{67-69}O_{16}N$											Macrolide group Similar to Carbomycin	1
Selenomycin		Stm. brasiliense	–Acetyl: C 55.5, H 4.86, O 38.8												S-5
Seligocidin	Enteromycin	Stm. sp.	$C_6H_7O_5N_2$	S-3				*S						*Yeast only See Enteromycin	1, 2
Senfolomycin A	LL-RA6950-BA	Stm. ochrosporus	C 49.66, H 5.28 O 35.45, N 3.81 S 4.57		S			S							1
Senfolomycin B	LL-RA6950-BB	Stm. ochrosporus	C 49.15, H 5.71 O 34.76, N 3.93 S 4.44		S			S							1
Senmimycin		Stm. senmiensis	$C_{24}H_{20}O_5$					*S						*Yeast only	1
Senomycin		Stm. senoensis	Picrate: C 38.93, H 4.23, N 18.25		S	S	S	W					L		1
Septacidin		Stm. fimbriatus	$C_{30}H_{51}O_7N_7$	S-4			S	S			S		T	Purine group	1, 2

Name	Synonym	Producer	Formula	Struct. Ref. No.	G +	G -	Myco	Fungi	Virus	Protoz	Tumor	Others	Tox.	Remarks	Ref.
Sequamycin	Spiramycin													See Spiramycin	1
Shincomycin A		*Stm. sp.* R-903 res. *Stm. flavochromogenes*	$C_{52}H_{89}O_{19}N$		S							*S	L	Macrolide group Similar to Angolamycin *Mycoplasma	1
Shincomycin B		*Stm. sp.* R-903	C 58.26, H 8.44 N 1.83		S							*S		Macrolide group *Mycoplasma	1
Showdomycin		*Stm. showdoensis* (ATCC 15105)	$C_9H_{11}O_6N$	S-5	S	S					S		M	Mode act.: N	1, 2, S-6
Sideromycin	Griseins, Ferrimycin etc.														1
Sinanomycin	Netropsin	*Stm. sp.*												See Netropsin	1, 2
Sintomyce(et)in	Chloramphenicol													See Chloramphenicol	1, 2
Siomycin (complex)		*Stm. sioyaensis*	$C_{64}H_{88}O_{22}N_{16}S_4$		S		S						M	Peptide group Similar to Bryamycin Thiostrepton	1, 2
Siomycin A		*Stm. sioyaensis*	$C_{74}H_{92}O_{19}N_{19}S_5$		S		S							Peptide group Similar to Thiostrepton Major component of Siomycin complex	S-7
Siomycin B		*Stm. sioyaensis*	C 43.83, H 4.71, N 13.11 S 8.53		S		S							Peptide group	S-7
Siomycin C		*Stm. sioyaensis*	C 47.97, H 5.11, N 11.17 S 8.00		S		S							Peptide group	S-7
Siromycin	Toyocamycin													See Toyocamycin	1

Name	Synonym	Producer	Formula	Struct. Ref. No.	G+	G−	Myco	Fungi	Virus	Protoz	Tumor	Others	Tox.	Remarks	Ref.
Sistomycosin		Stm. viridosporus						*S					M	*Yeast only Polyene (tetraene) group	1, 2
Soframycin	Neomycins	Stm. lavendulae	C 46.6 H 7.5 N 12.8 O 33.1		S	S		S?						See Neomycins	1, 2
Solemycin		Stm. soluensis	$C_{21}H_{54}O_{14}N_{7}$		S	S									1
Sparsogenin	Sparsomycin													See Sparsomycin	1
Sparsomycin	Sparsogenin, NSC-59729	Stm. sparsogenes var. sparsogenes	$C_{13}H_{19}N_{3}O_{5}S_{2}$	S-6	S	S		S			S		T (ip)	Mode act.: P	1, 2, S-8
Sparsomycin A	Tubercidin	Stm. sparsogenes var. sparsogenes NRRL 2940	$C_{11}H_{14}O_{4}N_{4}$												1
Speciomycin	190/I	Stm. sp.			S	W									1, 2
Spectinomycin A	Actinospectacin													See Actinospectacin Mode act.: P	3
Speleomycin		Acm. erythreus var. speleomycini													1
Spheromycin	Novobiocin													See Novobiocin	1
Spinamycin		Stm. albospinus	$C_{16}H_{16}O_{2}N_{2} \cdot H_{2}O$	S-7				S			S				S-9, S-10
Spinathricin		Stm. sp. A 18897													1
Spiramycins	Foromacidins Provomycin Rovamycin Selectomycin Sequamycin RP-5337	Stm. ambofaciens		S-8	S				*S					Macrolide group Mode act.: P *Rickettsia	1, 2, 3, S-11

Name	Synonym	Producer	Formula	Struct. Ref. No.	G +	G -	Myco	Fungi	Virus	Protoz	Tumor	Others	Tox.	Remarks	Ref.
Spiramycin 1	Foromacidin A	Stm. ambofaciens	$C_{43}H_{74}O_{14}N_2$	S-8-1	S				*S				L (sc)	Macrolide group *Rickettsia	1, 2, S-11
Spiramycin II	Acetylspiramycin 1 Foromacidin B	Stm. ambofaciens	$C_{45}H_{76}O_{15}N_2$	S-8-2	S				*S					Macrolide group *Rickettsia	1, 2, S-11
Spiramycin III	Foromacidin C Propionyl-Spiramycin 1	Stm. ambofaciens	$C_{46}H_{78}O_{15}N_2$	S-8-3	S				*S					Macrolide group *Rickettsia	1, 2, S-11
Sporangiomycin		Planomonospora parontospora var. antibiotica	$C_{77-80}H_{101-105}N_{20-21}O_{21}S_6$		S								L	Depsipeptide or peptolide group	S-12
Sporaviridin		Sts. viridogriseum	–HCl: C 55.41, H 8.47, N 2.26 C 52.52, H 8.36, N 1.95 Cl 4.49 –HNO₃: C 51.30, H 7.83, N 3.42 –H₂SO₄: C 49.18, H 8.36, N 2.01 S 3.62		S			*W					T	*Yeast only	1
Staphylomycin M_I	Mikamycin A, Ostreogrycin A, Pristinamycin IIA, PA-114A₁	Stm. virginiae	$C_{28}H_{36}O_8N_3$ $C_{28}H_{35}O_7N_3$	S-9	S		S						L	Depsipeptide group Mode act.: P	1, 2, 3
Staphylomycin M_{II}	PA-114A₂	Stm. virginiae					S								2
Staphylomycin S	PA-114 B-2	Stm. sp. res. to Stm. virginiae	$C_{43}H_{49}O_{10}N_7$	S-10	S		S						L	Depsipeptide group (mikamycin)	1, 2
Steffimycin	U-20,661	Stm. steffiburgensis	$C_{28}H_{30}O_{13}$		S						S		L (ip)	Mode act.: N	S-13, S-14

Name	Synonym	Producer	Formula	Struct. Ref. No.	G+	G-	Myco	Fungi	Virus	Protoz	Tumor	Others	Tox.	Remarks	Ref.
Stendomycin A		Stm. sp. res. Stm. endus or Stm. antimycoticus	$C_{95}H_{172}O_{31}N_{20}$		W		S	S					T	Polypeptide group isomyristic acid (main) and isotridecanoic acid as fatty acid constituents	1, 2, S-15
Stendomycin B		Stm. sp. or Stm. antimycoticus	$C_{85}H_{152}O_{26}N_{18}$		W		S	S						Polypeptide group	1, 2
Streptavidin	MSD-235L	Stm. avidinii Stm. lavendulae	C 51.1, H 7.4, N 16.4 S 0.2		S	S								Protein group (biotin-binding) Synergistic with MSD-235S, one part of streptavidin complex	1
Streptimidone		Stm. rimosus forma paromomycinus	$C_{16}H_{23}O_4N$	S-11			S	S		S			M	Glutarimide group Mode act.: P	1, 2, 3
Streptin		Stm. sp. res. Stm. reticulus-ruber or Stm. lavendulae			S	S	S						T	Streptothricin group	1, 2
Streptocardin		Stm. sp. Nocardia sp.			S	S	S						T		1, 2
Streptocin		Stm. griseus			W	W				S			L		1, 2
Streptogan		Stm. streptoganensis	C 40.2, H 6.29, N 2.35, Ash 1.68, O (49.48)		W	S					S			Large molecule glycosidic polypeptide	S-16
Streptogramin A, B		Stm. graminofaciens	$C_{26}H_{33}O_7N_3$		S	W	S						L (ip)	Depsipeptide (mikamycin) group Mode act.: P	1, 2, 3
Streptolins	136	Stm. griseus forma farinosus	A: $C_{25}H_{46}O_9N_{10}$	S-12	S	S	S						T	Streptothricin group	1, 2
Streptolydigin	Portamycin	Stm. lydicus	$C_{32}H_{44}O_9N_2$	S-13	S								L (ip)		1, 2

Name	Synonym	Producer	Formula	Struct. Ref. No.	G+	G-	Myco	Fungi	Virus	Protoz	Tumor	Others	Tox.	Remarks	Ref.
Streptomycin		Stm. griseus Stm. bikiniensis Stm. olivaceus Stm. mashuensis Stm. rameus Stm. galbus Stm. erythrochromogenes var. narutoensis	$C_{21}H_{39}O_{12}N_7$	S-14	S	S	S						L (sc)	Aminocyclitol group Mode act.: M, N & P (misreading)	1, 2, 3
Streptomycin B	Mannosido-streptomycin													See Mannosido-streptomycin	1, 2
Mannosidohydroxy-streptomycin		Stm. sp. 86*	$C_{27}H_{49}O_{18}N_7$		S	S	S							*Different from Stm. griseus Co-product: Toyocamycin	S-17
Mannosidostrepto-mycin	Streptomycin B	Stm. griseus	$C_{27}H_{49}O_{17}N_7 \cdot 3HCl$	S-15	S	S	S						L	Aminocyclitol group	1, 2
Streptonigrin	BA-163 Nigrin	Stm. flocculus	$C_{25}H_{22}O_8N_4$	S-16	S	W	S				S			Mode act.: N, P	1, 3
Streptonivicin	Novobiocin													See Novobiocin	1, 2
N-Demethylstrepto-mycin		Stm. griseus	$C_{20}H_{36}O_{12}N_7$		W	W	W								S-18
Dihydrostreptomycin	23572A	Stm. humidus	$C_{21}H_{41}O_{12}N_7 \cdot 3HCl$	D-13	S	S	S						*L	*Less ototoxic than Streptomycin	1
Streptorubin A		Stm. rubrireticuli var. pimprina	C 70.7, H 8.1 N 8.6								S			pH Indicator	1
Streptorubin B		Stm. roseoverticillatus var. albospora	C 70.1, H 8.3 N 9.3								S			pH Indicator	1

Name	Synonym	Producer	Formula	Struct. Ref. No.	G+	G-	Myco	Fungi	Virus	Protoz	Tumor	Others	Tox.	Remarks	Ref.
Streptothricin		Stm. griseus forma farinosus, Stm. lavendulae, Stm. sp.	$C_{19}H_{34}O_8N_8$	S-17	S	S	S						*L	Streptothricin group *7245 mg/kg (iv), but shows delayed toxicity (8 mg/kg)	1
Streptothricin A		Stm. sp. Stm. lavendulae, Stm. albus, Stm. ruber												Streptothricin group	1
Streptothricin B$_I$	Neomycin C	Stm. fradiae	$C_{23}H_{46}O_{13}N_6$		S		S							Aminocyclitol (neomycin) group	1
Streptothricin B$_{II}$	Neomycin B	Stm. fradiae	$C_{23}H_{46}O_{13}N_6$		S		S							Aminocyclitol (neomycin) group	1
Streptovaricins		Stm. spectabilis	$C_{34}H_{47-49}O_{13}N$	S-*18	S		S	S					L	pH Indicator *Related to rifamycin S	1, 2, S-19, S-20
A			$C_{42}H_{53}NO_{16}$												
B			$C_{42}H_{53}NO_{15}$												
C			$C_{40}H_{53}NO_{14}$												
D			$C_{42}H_{55}NO_{15}$												
E			$[C_{40}H_{55}NO_{13}]$												
G			$C_{40}H_{51}NO_{15}$												
Streptovitacin A	4e-hydroxycycloheximide	Stm. griseus	$C_{15}H_{23}O_5N$	S-19				S			S	S	*	Glutarimide group *Symptoms at high dose levels: listlessness diarrhea and hematuria Mode act.: P	1, 2, S-21
Streptovitacin B		Stm. griseus	$C_{15}H_{23}O_5N$	S-20				*1 S		S			*2	*1 Yeast only Glutarimide group *2 Same as A	1, 2
Streptovitacin C$_1$		Stm. griseus	$C_{15}H_{23}O_5N$											Glutarimide group	1, 2

Name	Synonym	Producer	Formula	Struct. Ref. No.	G+	G-	Myco	Fungi	Virus	Protoz	Tumor	Others	Tox.	Remarks	Ref.
Streptovitacin C$_2$		*Stm. griseus*	C$_{15}$H$_{23}$O$_5$N	S-21							*			Glutarimide group No description except activity of C$_2$ against certain animal tumor cells in vivo	1, 2
Streptovitacin D		*Stm. griseus*	C$_{15}$H$_{23}$O$_5$N								*			Glutarimide group No description except activity of D against certain animal tumor cells in vivo	1
Streptovitacin E		*Stm. griseus*	C$_{15}$H$_{23}$O$_5$N											Glutarimide group No description of activity	1, 2
Streptozotocin	NSC 37917	*Stm. achromogenes* var. 128	C$_{14}$H$_{27}$O$_{12}$N$_5$	S-22	S	S							L		1, 2; S-22
Subliomycin		*Stm. sumaensis*	C$_{20}$H$_{30}$O$_{16}$		S	S	S						L	Co-product: miusidin	1
Succinimycin			Acetate C: 45.03–46.22 H: 6.63–7.02 N: 8.11–8.90, E: 4.3–4.8											Sideromycin group	1, 2
Sugordomycins	Coumermycins	*Stm. sp.* X-7763												See Coumermycins	1
Suitamycin		*Stm. griseochromogenes* var. *suitaences*	C: 58.31–58.66 H: 9.17–9.38 N: 0.80–1.14		S	S		S					T	Co-product: Blasticidin S	S-23
Sulfactin		*Stm. sp.* (res. *Stm. roseus*)	C$_{38}$H$_{55}$O$_7$N$_{11}$S$_4$ or C$_{27}$H$_{40}$O$_5$N$_5$S$_3$		S	S		S					M (ip)		1
Sulfocidin		*Stm. sp.*			S	S	S				S		T (ip)		1, 2

Name	Synonym	Producer	Formula	Struct. Ref. No.	G+	G-	Myco	Fungi	Virus	Protoz	Tumor	Others	Tox.	Remarks	Ref.
Sulfomycin I		*Stm. viridochromogenes* var. *sulfomycin*	C: 49.95, 49.81 H: 4.50, 4.49 N: 16.86, 16.64 S: 4.80, 4.58		S		W					*	T (ip)	Peptide group Cross resistance with Bryamycin *Mycoplasma	2, S-24
Sulfomycin II		*Stm. viridochromogenes* var. *sulfomycin*	C: 50.68, 50.55 H: 5.25, 4.17 N: 16.14, 16.99 S: 5.81, 5.31		S		W						T (ip)	Same as I	2, S-24
Sulfomycin III		*Stm. viridochromogenes* var. *sulfomycin*	C: 50.42, 50.23 H: 4.29, 4.32 N: 16.71, 16.14 S: 5.14, 5.46										T (ip)	Same as I	2, S-24
Sygromycin	Marcomycin				S									Mikamycin Group	1
Synergistin A	PA-114A	*Stm. olivaceus*			S									Mode act.: P	1, 3
Synergistin B	PA-114B	*Stm. olivaceus*			S									Mikamycin Group	1
Synergistin B-3	PA-114 B-3	*Stm. olivaceus*			S									Mikamycin Group	1
Synthomycin (USSR)	Chloramphenicol														1
Syntomycin (USSR)	Chloramphenicol														1
Taitomycin		*Stm. afghaniensis* *Stm. griseosporeus*	C: 53.37, H: 4.87 N: 9.50, Ash: 2.80 C: 49.76, H: 4.47 N: 12.61, S: 12.79		S	*S							L	*Neisseria	1, 2
Takacidin		*Stm. griseoverticillatus* 722	$C_{50}H_{93}O_{16}N$										T (ip)	Similar to Monazomycin	1
Takamycin		*Stm. takataensis* *Stm. reticuli* C-11						S						Polyene (heptaene) group	1

Name	Synonym	Producer	Formula	Struct. Ref. No.	G+	G-	Myco	Fungi	Virus	Protoz	Tumor	Others	Tox.	Remarks	Ref.
Tauromycetin		*Acm. tauricus* 13170									S				T-1
Tbilimycin		*Acm. chartreusis* var. *tbilisus*						S						Polyene (heptaene) group	T-2
Teleocidin A	Teleocidin (formerly)	Var. of *Stm. mediocidicus* (*Stm. sp.* 2A-1563)	C: 73.00–73.31, H: 9.08–9.17 N: 7.70–8.50 (MW 327)									*	T	*Active against: ascaris and certain fish	1
Teleocidin B	SK-toxic substance	*Stm. kitasatoensis*	Dihydroteleocidin B $C_{23}H_{43}O_2N_3$	T-1								*	T	Same as A	1
Telomycin	C-159	*Stm. canus* *Stm. sp.*	C 57.4, H 6.74 N 13.10 (MW 1000)	T-2	S								L	Polypeptide group	1, 2
Tennecetin	Pimaricin (probably)	*Stm. chattanoogensis*	$C_{34}H_{49}O_{14}N$					*S					M (ip)	*Yeast Polyene (tetraene) group	1, 2
Tenebrimycin complex (I,I',II,III,IV,V,VI)		*Stm. tenebrarius* ATCC 17920 and 17921			S	S	S						L	Aminocyclitol	T-3
Tenebrimycin II		*Stm. tenebrarius* ATCC 17920 and 17921	$C_{16}H_{36}N_4O_9$		S	S	S							Aminocyclitol	
Tenebrimycin IV		*Stm. tenebrarius* ATCC 17920 and 17921	$C_{16}H_{36}N_5O_{10}$		S	S	S							Aminocyclitol	
Tenebrimycin V		*Stm. tenebrarius* ATCC 17920 and 17921	$C_{14}H_{32}N_4O_{10}$		S	S	S							Aminocyclitol	
Tenebrimycin VI		*Stm. tenebrarius* ATCC 17920 and 17921	$C_{16}H_{36}N_4O_9$		S	S	S							Aminocyclitol	

Name	Synonym	Producer	Formula	Struct. Ref. No.	G+	G−	Myco	Fungi	Virus	Protoz	Tumor	Others	Tox.	Remarks	Ref.
Tertiomycin A		Stm. eurocidicus (producer of eurocidin and azomycin) Stm. albireticuli	$C_{42}H_{69}O_{16}N$		S									Macrolide group	1, 2
Tertiomycin B		Stm. eurocidicus	$C_{43}H_{71}O_{17}N$		S		W						L	Macrolide group	1, 2
Tetracycline		Stm. alboflavus Stm. mediolamum Stm. fuscofaciens Stm. aureofaciens, Stm. viridifaciens, Stm. feofaciens, (Stm. psammoticus) Stm. persimilis, Stm. lusitanus, var. tetracyclini, Stm. sayamaensis Stm. sp. 88	$C_{22}H_{24}O_8N_2 \cdot 3H_2O$	T-3	S	S	S		S				ML (ip)	Mode act.: P	1, 2, 3
(Bromo)tetracycline		Mutant strain of Stm. aureofaciens	$C_{22}H_{23}O_8N_2Br$	T-4	S	S	S						M	Tetracycline group	1
(7-chloro-5a(11a)-dehydro)tetracycline	Dehydrochlor-tetracycline	Stm. aureofaciens	$C_{22}H_{21}O_8N_2Cl \cdot HCl$	T-5	S	S	S							Tetracycline group	1
(7-chloro-6-Demethyl)-tetracycline	Demethylchlor-tetracycline	Stm. aureofaciens			S	S	S							Tetracycline group	2, T-4-a
(Chlor)tetracycline	Aureomycin	Stm. aureofaciens Stm. psammoticus Stm. fuscofaciens Stm. alboflavus Stm. persimilis Stm. sayamaensis	$C_{22}H_{23}O_8N_2Cl$		S	S	S						ML	Tetracycline group	1, 2
(4-N-demethyl-4-N-ethyloxy)tetracycline		Stm. rimosus NRRL 2234			S		S							Tetracycline group	T-4-b

Name	Synonym	Producer	Formula	Struct. Ref. No.	G +	G −	Myco	Fungi	Virus	Protoz	Tumor	Others	Tox.	Remarks	Ref.
(4-N-demethyl-4-N-ethyl-6-demethyl)-tetracycline		Stm. aureofaciens, Stm. viridofaciens			S	S								Tetracycline group	T-4-c
(4-N-demethyl-4-N-ethylhalo)tetracycline		Stm. aureofaciens, Stm. viridofaciens			S	S								Tetracycline group	T-4-c
(4-N-demethyl-4-N-ethyl)-tetracycline		Stm. aureofaciens, Stm. viridofaciens			S	S								Tetracycline group	T-4-c
(Demethyl)tetracycline		Stm. aureofaciens, Stm. peruviensis	$C_{21}H_{22}O_8N_2 \cdot HCl-\frac{1}{2}H_2O$	T-6	S	S	S							Tetracycline group	1, 2
(5-Hydroxy-7-chlor)-tetracycline		Stm. rimosus													
(Oxy)tetracycline		Stm. alboflavus, Stm. albofaciens, Stm. varsoviensis, Stm. henetus, Stm. rimosus, Stm. platensis, Stm. armillatus, Stm. flavus, Stm. vendargensis, Stm. sp. M590, Stm. gilvus, Stm. capuensis	$C_{22}H_{24}O_9N_2$		S	S	S						ML	Tetracycline group	1, 2
Tetraesin	(Tetrahexin)	Stm. sp. 5391	C: 65.55, H: 8.46 N: 1.32,		S			*S					L (sc)	*Yeasts Polyene (tetraene, hexaene) group	1
Tetrahexine	(Tetraesin)	Stm. sp. ATCC 14972	C: 65.55, H: 8.46 N: 1.32, O: 24.67		S			S						Polyene (tetraene, hexaene) group	T-5
Tetramycin		Stm. noursei	$C_{34}H_{53}O_{14}N$					S					M		T-6

Name	Synonym	Producer	Formula	Struct. Ref. No.	G +	G -	Myco	Fungi	Virus	Protoz	Tumor	Others	Tox.	Remarks	Ref.
Tetranactin		Stm. flaveolus		M2~M5										Related to monactin, dinactin, trinactin	T-7
		Stm. aureus	$C_{44}H_{72}O_{12}$										L (ip)	Macrotetrolide group Active to insect	T-8
Tetrangomycin	LLAE705Y	Var. strain of Stm. rimosus	$C_{19}H_{14}O_5$	T-7	S									Anthraquinone group	1
Tetrenolin		Mip. venezuelensis	$C_{11}H_{12}O_4$		S										T-8
Tetrin A		Stm. sp. ILL155-2	$C_{35}H_{53-55}O_{13}N$	T-9				S					*	Polyene (tetraene) group *No phytotoxicity against certain plants	1, 2
Tetrin B		Stm. sp. ILL155-2	$C_{34-35}H_{53-55}O_{14}N$	T-10				S						Polyene (tetraene) group	1, 2
T(h)aimycin A		Str. michiganensis var. amylolyticus	C: 60.30, H: 7.88, N: 4.71					S *1		S *2			L (oral) T (ip)	*1 Yeasts *2 Anthelmintic & antiprotozoal (Trichomonas)	T-9
T(h)aimycin B		Above	C: 67.26, H: 7.97, N: 5.98					S *1		S *2			L (oral) T (ip)		
T(h)aimycin C		Above	C: 62.50, H: 8.11, N: 4.45							*S			L (oral) M (ip)		
Theiomycetin		Stm. sp. res. Stm. lavendulae	$C_{55}H_{59-61}O_{20}N_{15-16}$		S	S							L (ip)	Peptide group	1
Thermomycin		Stm. thermophilus			S										1, 2

Name	Synonym	Producer	Formula	Struct. Ref. No.	G+	G-	Myco	Fungi	Virus	Protoz	Tumor	Others	Tox.	Remarks	Ref.
Thermorubin		Tha. antibioticus ATCC14570 Tha. sp. res. Tha. antibioticus or Tha. vulgaris	$C_{22}H_{18}O_8$		S	W							L (ip)	pH Indicator	1, 2
Thermothiocin		Thermoactinopoly-spora coremialis	$C_{60}H_{110}O_{25}N_{15}S_3$		S								L (ip)	Peptide group	1
Thermoviridin		Tha. viridis			S	W							T		1, 2
Thermycetin	19A	Stm. sp. MA-568	$C_{11}H_{13}O_9N_4$		S	S								Enteromycin group	1
Thiactin	Bryamycin													See Bryamycin	1
Thiazolidone-antibiotic	Actithiazic acid													See Actithiazic acid	1
Thioaurin	Hyden antibiotic No. 9 (HA-9) Orosomycin B-870	Stm. sp. (related to Stm. lipmanii)	$C_{14}H_{12}O_4N'_4S_4$		S	S		S					T		1, 2
Thiogriseofulvins	(+)-1-thio-griseofulvin (+)-5'-thio-hydroxy-1-griseofulvin	Stm. cinereocrocatus NRRL 3443						S						Dehydro-1-thio-griseofulvin used as a substrate	G-10
Thiolutin		Stm. albus Stm. celluloflavus Stm. sp. (producer of aureothricin)	$C_8H_8O_2N_2S_2$	T-11	S	S		*S					T (sc) L (or)	Pyrrothine group *Yeasts	1, 2
Thiomycin		Stm. sp. closely related to Stm. phaeochromo-genes var. chloromyceticus	C: 49.61, H: 5.50 N: 8.88, S: 16.26		S	S	S						T (sc)		1, 2

Name	Synonym	Producer	Formula	Struct. Ref. No.	G+	G-	Myco	Fungi	Virus	Protoz	Tumor	Others	Tox.	Remarks	Ref.
Thiopeptin B		Stm. tateyamensis 7906	$C_{72}H_{90}O_{22}N_{18}S_6$		S		W						L (ip)	Peptide group	T-11
Thiostrepton	Bryamycin, Thiactin	Stm. azureus, Stm. sp.	$C_{72}H_{83}O_{17}N_{19}S_5$	T-12	S		S								1, 2
Threomycin		Stm. sp. L-803	$C_7H_{11}O_3N$	T-13	S	W	W		*S					Amino acid group *T_2 phage	1
Tirandamycin		Stm. tirandis	$C_{22}H_{26}NO_7 \cdot Na$		S									Mode of act.: N, O	T-12
Tolypomycin Y		Stm. tolypophorus	$C_{43}H_{54}N_2O_{14}$	T-14	S									Ansamycin group Tolypomycinone as a constituent	T-13
Totomycin	Hygromycin A	Stm. crystallinus	$C_{21}H_{29}O_{11}N$		S	W									1, 2
Toyamycin	Akitamycin	Stm. toyamaensis	$C_{41}H_{65}O_{18}N$					S						Polyene (tetraene) group	1
Toyocamycin	Siromycin E-212 first substance 9-27 9-48 Naritheracin 1037	Stm. toyocaensis Stm. sp. (res. Stm. albus)	$C_{12}H_{13}O_4N_5$	T-15			S	S			S		T	Chemically synthesized Mode act.: N	1, 2, 3, T-14
Trehalosamine		Stm. sp. res. Stm. lavendulae	$C_{12}H_{23}O_{10}N$	T-16	W	S	W	*S					T	Amino sugar group *Yeasts	1, 2
Trichomycin A		Stm. hachijoensis Stm. sp.	$C_{61}H_{86}O_{21}N_2 \cdot 2H_2O$	T-17				S		S			T	Polyene (heptaene) group Mode act.: M	1, 2, 3,

Name	Synonym	Producer	Formula	Struct. Ref. No.	G+	G-	Myco	Fungi	Virus	Protoz	Tumor	Others	Tox.	Remarks	Ref.
Trichomycin B		*Stm. hachijoensis*	C: 59.49, H: 8.09 N: 2.16											Polyene group Similar to Trichomycin A	1, 2
Trichonin		*Stm. rubrireticuli*													1
Triculamin		*Stm. triculaminicus*	$C_{85}H_{154}N_{32}O_{27}$		W -		S						L	Peptide group	T-14, T-15
Trienine		*Stm. sp.* SC3725	C: 55.00—54.49 H: 7.85—8.13 N: 1.34—0.99		S		S	S			S		T	Polyene (triene) group	T-16
Trinactin		*Stm. sp.* ETHA 23112 *Stm. aureus*	$C_{43}H_{70}O_{12}$	T-18	S								L (ip)	Macrotetrolide group Active to insect (*Callosofruchus chinensis*, Azukibean weevil) Mode act.: O	1, 2, 3
Triostin complex		*Stm. sp.* S-2-210 res. *Stm. aureus*	C: 54.03, H: 6.29, N: 13.89, S: 5.60 Plate: C: 56.91, H: 6.66, N: 13.15, S: 5.86		S		S				S			Quinoxaline group related to quinomycins	1, 2
Triostin A		Above	$C_{50}H_{62}O_{12}N_{12}S_2$	T-19										Quinoxaline group	1, 2, T-17, T-18
Triostin B₀		Above		T-19										Quinoxaline group	T-17, T-18
Triostin C		Above	$C_{54}H_{70}O_{12}N_{12}S_2$	T-19	S		S				S		L (ip)	Quinoxaline group	T-17, T-18
Triostin B		Above		T-19											T-17, T-18

Name	Synonym	Producer	Formula	Struct. Ref. No.	G+	G-	Myco	Fungi	Virus	Protoz	Tumor	Others	Tox.	Remarks	Ref.
Trypanomycin		Stm. diastatochromogenes IMET JA-10081					W			S		.	ML	Anthracycline group	T-19
Tselikomycin		Stm. coelicolor			S		S						L	Amphomycin group	1
Tsushimycin		Stm. sp. Z-237	$C_{59}H_{93}O_{20}N_{13} \cdot 6H_2O$		S										Contains fatty acid	T-20
Tuberactin	Tuberactinomycin													See Tuberactinomycin	
Tuberactinomycin		Stm. griseoverticillatus var. tuberacticus	$C_{16}H_{31}N_9 \cdot 2HCl$		W	W	S						L (im)	Viomycin-like peptide group	T-21
Tubercidin		Stm. tubercidicus	$C_{11}H_{14}O_4N_4$	T-20			S	*W					T (ip)	Purine group Mode act.: N *Yeast	1, 2, 3
Tuberin		Stm. amakusaensis	$C_{10}H_{11}O_2N$	T-21			S						L (ip, sc)		1, 2
Tubermycin A		Stm. misakiensis	$C_{17}H_{16}O_2N_2$		S		S						M		1, 2
Tubermycin B	Phenazine-2-carboxylic acid	Stm. misakiensis	$C_{13}H_8O_2N_2$	T-22	S		S						L		1, 2
Tundromycin		Stm. globisporus tundromycini			S										1
Tuoromycin		Stm. tuirus A 31554 (NRRL 3120)			S			S			S			Pigment	1
Tylosin		Stm. fradiae, Stm. hygroscopicus	$C_{45}H_{77}O_{17}N$	T-23	S		S		S				L	Macrolide group	1, 2
Umbrinomycin A & B		Stm. umbrinus ATCC 15972	A: $C_{22}H_{24}O_8$ B: $C_{22}H_{24}O_8$		S		S							Co-product of Diumycin	U-1

Name	Synonym	Producer	Formula	Struct. Ref. No.	G +	G −	Myco	Fung	Virus	Protoz	Tumor	Others	Tox.	Remarks	Ref.
Umimycetin	Chloramphenicol														1
Unamycin		*Stm. fungicidicus*	C: 52.24, H: 7.77 N: 1.74					S		*W			M	Polyene (tetraene) group *Trichomonas	1, 2
Unamycin B		*Stm. fungicidicus*	C: 46.4–46.9, H: 4.46–4.25 N: 22.25–22.79 –HCl: C: 43.81 H: 4.0–4.4 N: 21.3–21.6 Cl: 6.05				S	S					*T	Purine group Similar to Toyocamycin, E-212, Vengicide *Delayed toxicity	1, 2
Uredolysin		*Stm. griseus* 528 NRRL 2607						S							1
Ussamycin		*Stm. lavendulae* UV-9			S						S			Peptide group	1, 2
Vaccinocidin		*Acm. sp.* 3933-13	N, S, P, halogen 0%		S	W			*S					pH Indicator *Small pox	1
Valacidin		*Stm. lavendulae*	C₂₆H₂₄O₈N		S	S	S							Also active to plant pathogens	1
Validamycins (A,B)		*Stm. hygroscopicus* var. *limonensis*	A: C₂₀H₃₃₋₃₇NO₁₃₋₁₄ B: C₂₀H₃₃₋₃₇NO₁₄₋₁₅	V-0				*S						*Pellicularia sasaki	V-0
Valinomycin	Aminomycin, N-329B	*Stm. fulvissimus*, *Stm. tsusimaensis*	C₅₄H₉₀O₁₈N₆	V-1	S		S	S					T (ip)	Depsipeptide group Similar to Amidomycin Mode act.: M.P.	1, 2, 3
Vancomycin		*Stm. orientalis*	C₁₄₄H₁₈₁O₅₅N₁₅Cl₄		S		S						L	Similar to Actinoidin, Ristocetin, K-288 Mode act.: C	1, 2, 3

Name	Synonym	Producer	Formula	Struct. Ref. No.	G+	G−	Myco	Fungi	Virus	Protoz	Tumor	Others	Tox.	Remarks	Ref.
Variomycin	Luteomycin 289 (similar)	*Stm. sp.*									S	S			1
Vasocidin	5-amino-1H-V-triazoro-d-pyrimidine-7-ol	*Stm. albus var. vasocidicus*						S						Pyrimidine group	V-1
Vengicid(e)		*Stm. vendargensis*	$C_{24}H_{29}O_9N_{10}$					*S						Purine deriv. Similar to Monilin, Toyocamycin, Unamycin B *Yeast only	1, 2
Venturicidin	AA-368, Venturicidin A	Three strains of *Stm. sp.* res. *Stm. griseolus*, *Stm. halstedii* or *Stm. xanthophaeus*	$C_{43}H_{71}O_{12}N$					S					T		1, 2
Venturicidin A	Venturicidin	*Stm. aureofaciens* 342	$C_{43}H_{71}NO_{12}$												V-2
Venturicidin B		*Stm. aureofaciens* 342	$C_{42}H_{70}O_{11}$					S							V-2, V-3
Venturicidin X		*Stm. aureofaciens* 342	$C_{39}H_{64}O_{10}$					S							V-3
Vernamycin A		*Stm. loidensis* ATCC 11415	$C_{20}H_{25}O_5N_2$		S	S	S							Depsipeptide group (mikamycin) Mode act.: P	1, 3
Vernamycin B (Bα, Bβ, Bγ, B)	Bα = ostreogrycin B, Bβ = ostreogrycin B₂, Bγ = ostreogrycin B₁	*Stm. loidensis*	$C_{30}H_{40}O_8N_3$ Bα: $C_{45}H_{54}O_{10}N_8$ Bβ: $C_{34}H_{52}O_{10}N_8$ Bγ: $C_{44}H_{52}O_{10}N_8$ B: $C_{43}H_{50}O_{10}N_8$	V-2	S	W	S							Depsipeptide group (mikamycin)	1

Name	Synonym	Producer	Formula	Struct. Ref. No.	G +	G −	Myco	Fungi	Virus	Protoz	Tumor	Others	Tox.	Remarks	Ref.
Vernamycin C		Stm. loidensis ATCC 11415	C: 60.27, H: 6.34, N: 12.58, N-methyl: 5.83	V-3	S									Isolated from Vernamycin B mixture	V-4
Vertimycin		Stm. sp. JA4498	$C_8H_{14}O_4$	V-4	*S						S			*Nocardia	1
Vertimycin C		Stm. verticillatus	C: 62.4, H: 6.84, N: 8.0												1
Vinacetin		Stm. sp. related to Stm. albosporeus			S	S	S							pH Indicator	1
Violacetin		Stm. sp. closely related to Stm. purpeochromogenes	$(C_9H_{14}O_8N_5 \cdot HCl)_n$		S	S	S	*S					T	*Yeast only	1, 2
Violacin	2732/3	Acm. sp. 2732/3	$(C_5H_7O_2)_n$		S									Anthracycline group (indicator)	1
Violarins	12-12, 452-7	Acm. violaceus	$C_{22-24}H_{32-34}O_{8-9}$		S	S	S	S						Anthracycline group	1, 2
Violarin B	Mycetin A, Rhodomycin, 12-12, 452-7	Acm. violaceus No. 452/7 and No. 12-12	$(C_{16}H_{28}O_9N)_n$		S				S					Anthracycline group	1, 2
Violarin I		Acm. violaceus			S								*S	*Jaundice of silkworms	2
Violarin W	Florimycin Vinactin Vinactane	Acm. violaceus	$C_{16}H_{28}O_9N$		W				S						2
Viomycin	Florimycin Vinactin Vinactane	Stm. abikoensis Stm. floridae Stm. olivoreticuli Stm. puniceus Stm. vinaceus Stm. californicus	$C_{23}H_{36}O_8N_{12}$	V-5	W	S	S						L	Peptide group Mode act.: P	1, 2, 3 V-5

Name	Synonym	Producer	Formula	Struct. Ref. No.	G +	G –	Myco	Fungi	Virus	Protoz	Tumor	Others	Tox.	Remarks	Ref.
Viractin		Stm. griseus (actidione producer)							S				L	Prophylactically active in man	2, V-6
Virginiamycin	Staphylomycin													See Staphylomycin	1, 2
Viridofulvin		Stm. viridogriseus	C: 64.86, H: 8.98					S							V-7
Viridogrisein	Etamycin, F-1370A	Stm. griseus (griseoviridin producer)												Mode act.: P See Etamycin	1, 2, 3
Virocidin	Abikoviromycin Latumcidin	Stm. flavoreticuli												See Abikoviromycin	1, 2
Virosin	Antimycin A, No. 720-A	Stm. olivochromogenes												See Antimycin A	1, 2
Virothricin		Stm. lavendulae var. virothricinus			S	S	S		S					Streptothricin group	1
Virusin 1609		Stm. lavendulae			S	S	S		S					Streptothricin group	1, 2
Vivomycin		Stm. sp. C 2989 (closely related to Stm. griseus)							S					Active only in vivo polysaccharide	1
Vulgarin		Stm. sp. H-3206 (res. Stm. flavochromogenes)	$C_{25}H_{33}O_8N_2$		S			*S					T	*Yeast only	1, 2
Werramycin	FH 3582B	Stm. werraensis	$C{:}_{65.6}H{:}_{9.2}O{:}_{24.9}$		S		S	S							1, 2
Xanthalycin A								S						Polyene (pentaene) group	1
Xanthalycin B								S						Polyene (pentaene) group	1

Name	Synonym	Producer	Formula	Struct. Ref. No.	G +	G -	Myco	Fungi	Virus	Protoz	Tumor	Others	Tox.	Remarks	Ref.
Xanthicin		*Stm. xanthochromogenes*	$C_{13}H_{15}O_5$		W	W		W					M		1, 2
Xanthocidin		*Stm. xanthocidicus*	$C_{11}H_{16}O_5$	X-1		*S							M	*Xanthomonas oryzae*	X-1, X-2
Xanthomycin A	H-1159, 534 (prob.)	*Stm. chromogenes* K152, *Stm. pseudogriseolus* *Stm. sp.* 94 *Stm. sp.* NCIB8697	$C_{23}H_{29-31}O_7N_3 \cdot 2HCl$		S	S	S						M	pH Indicator Mode act.: N(RNA)	1, 2, 3
Xanthomycin B		*Stm. rutgersensis* *Stm.* *pseudogriseolus* *Stm. sp.* 94												pH Indicator	1, 2
Xanthomycin C		*Stm. rutgersensis* *Stm.* *pseudogriseolus* *Stm. sp.* 94												pH Indicator	1
Xanthothricin	Toxoflavin	*Stm. sp.* closely related to *Stm. albus* (*Pseudomonas cocovenenans*)	$C_7H_7O_2N_5$	X-2	W	W	W						T	Purine group	1, 2
Yazumycin		*Stm. lavendulae*	$C_{15}H_{35}N_6O_{13}S$			S							L		Y-1
Yumimycin		*Stm. cinnamonensis* var. *yumimyceticus*	$C_{20}H_{40}O_9N_2$					S					M (ip)		Y-2
Zaomycin		*Stm. zaomyceticus*			S		S						L	Similar to Amphomycin	1, 2

Name	Synonym	Producer	Formula	Struct. Ref. No.	G +	G -	Myco	Fungi	Virus	Protoz	Tumor	Others	Tox.	Remarks	Ref.
Zedalan	Trans-3-(hydroxy-imino)aceto-amid acryl-amid	Stm. achromogenes var. streptozoticus NRRL 2697 (Same as Streptozotocin producer)	$C_5H_7N_3O_3$	Z-1							S		M		Z-1
Zygomycin A$_1$	Paromomycin 45449A$_1$	Stm. pulveraceus	$C_{25}H_{47}O_{15}N_5$	Z-2	S	S	S						M	Aminocyclitol (Neomycin) group	1, 2
Zygomycin A$_2$	Paromomycin II 45449A$_2$	Stm. pulveraceus	$C_{25}H_{47}O_{15}N_5$	Z-3	S	W	S						M	Aminocyclitol (Neomycin) group	1, 2
Zygomycin B	45449B	Stm. pulveraceus			S	S	S						L	Streptomycin group	1, 2
Zygomycin D-2	45449-D	Stm. pulveraceus	$C_{15}H_{23}O_4N$				*S							Glutarimide group *Yeast only	1
A-6		Stm. sp. res. Stm. fradiae			S	S									1, 2
A-14	Decoyinin Angustmycin A	Stm. sp.												See Decoyinin	1
A-20		Stm. sp.			S	S	S	S							1, 2
A-58	Phagolessin	Stm. sp.												See Phagolessin	1
A-59		Stm. sp. A-59	C 49.57, H 5.62, N 13.95	NA-1	S		S						M	Siomycin like Peptide group	1
A-59B		Stm. sp. A-59			S										1
A-67		Stm. sp.						S							1, 2
A-94	Fungicidin	Stm. sp.												See Fungicidin	1

Name	Synonym	Producer	Formula	Struct. Ref. No.	G +	G -	Myco	Fungi	Virus	Protoz	Tumor	Others	Tox.	Remarks	Ref.
A-116		*Stm. sp.*			W	W							L (sc)	Fraction A10 & C	1, 2
A-116 SA		*Stm. endus* NRRL 2736	$C_{95-98}H_{175}O_{32}N_{20}$	NA-2	S		S	*S						Peptide group *Yeast only	1, 2
A-116 SO		*Stm. endus* NRRL 2736	C 55.40, H 8.33, N 13.80	NA-3	S		S	S						Peptide group	1, 2
A-195		*Stm. spadicis* (ATCC 19017)	C 57.82, H 6.2, N 1.71 OMe 14.36, N–Me 2.36		S		S	S							NA-1
A-204		*Stm. albus* NRRL 3384	C 61.74 H 9.37 O 28.38		S		S	S							NA-2
A-206		*Acm. indigocolor* No. 206	C 55.38, 55.27 H 4.93, 4.99 N 3.23, 3.36											Celicomycin-like	NA-3
A-216		*Stm. rubrireticuli* A-216													1
A-228 (a, b)		*Stm. sp.* Ds 41	C ca60, H ca8, N ca2, S ca4					*S						Polyene (heptaene) group *Yeast only	1, 2
A-246	Lagosin Fungichromin	*Stm. sp.*		NA4	S	S	S							See Lagosin	1, 2
A-249		*Stm. lavendulae* var. *hypotoxicus*	–HCl: C 35.88, H 7.16 N 17.42, Cl 17.13		S	S	S	S						Streptothricin group	1
A-272	Rutamycin	*Stm. sp.* res. *Stm. rutgersensis*			S									See Rutomycin	1, 2
A-280		*Stm. sp.* A-280			S	W	S		W	S					1
A-396-1		*Stv. eurocidicus* A-396-I	$C_{19}H_{35}O_{13}N_3$		S		W		W				T		NA-4

Name	Synonym	Producer	Formula	Struct. Ref. No.	G +	G −	Myco	Fungi	Virus	Protoz	Tumor	Others	Tox.	Remarks	Ref.
A-396-II	Hygromycin B	Stv. eurocidicus A-396												See Hygromycin B	NA-4
A-365		Stm. sp. related to Stm. lavendulae													1
A-415-Z3	3-Carboxy-2,4-pentadienal lactol PA-147	Stm. sp.												See 3-Carboxy-2,4-pentadienal lactol	1
A-418-Z4		Stm. sp.			S									Sideromycin group	1
A/672		Acp. brasiliensis			W								L (ip)		NA-5
A-1404		Stm. fradiae A1404 Stm. diastatochromogenes 207						S					T	Similar Fradicin mycelin 207 Substance	1
A-1459		Stm. sp. 1616			S	S									NA-6
A-1787		Stm. sp. res. Stm. griseus												Sideromycin group similar Albomycin	1, 2
A-3823		Stm. cinnamonensis ATCC 15413	$C_{36}H_{64}O_{11}$		S			S		*S				*Eimeria	NA-7
A-3823 D		Stm. sp.													NA-7
A-4788		Stm. sp. A-4788			S	S								Streptothricin group	1, 2
A-5283		Stm. sp. A-5283	$C_{31-34}H_{44-51}O_{13-14}N$					*S						Polyene (tetraene) group *Yeast only	1

Name	Synonym	Producer	Formula	Struct. Ref. No.	G+	G-	Myco	Fungi	Virus	Protoz	Tumor	Others	Tox.	Remarks	Ref.
A-7907		*Stm. fradiae* A-7907	$-H_2SO_4$: $(C_{12}H_{25-27}O_5N_4 \cdot H_2SO_4)_n$		S	S		S					L	Streptothricin group Geomycin like	1
A-8265		*Stm. sp.* A-8265			S	S								Streptothricin group	1
A-9578 Component A	Pilosomycin	*Stm. sp.*												See Pilosomycin	1
A-9578 Component B	Pilosomycin	*Stm. sp.*												See Pilosomycin	1
A-9828		*Stm. sp.* A-9828	$C_{36}H_{58}O_{10}$		S		S	S	S					Cyclic polylactone group	1
A-10598		*Stm. hygroscopicus*	$C_{32}H_{48}O_{10}N_2$						S						1
A-22765		*Stm. aureofaciens* NRRL 2858	C 49.79, H 7.57, N 10.18 Fe 2.38		S								L (ip)	Sideromycin group	1
A-28829		*Stm. antibioticus* A28829	$C_{44}H_{72}O_{15}NB$		S			S		*S			T (or)	*Plasmodium berghei*	NA-8
AA-368	Venturicidin	*Stm. sp.*												See Venturicidin	1
AB-6442		*Stm. candidus* NRRL 3083	$-HCl$: C 33.30, H 6.61 N 17.20, Cl 15.16		S	S									NA-9
AB-6443		*Stm. candidus* NRRL 3083	$-H_2SO_4$: C 35.21, H 5.52 N 17.48, S 5.54		S	S									NA-9
Abbott 29119		*Stm. erythreus*	$C_{42}H_{75}O_{14}N$		S				S					Macrolide group	1
AC-98 complex A		*Stm. hygroscopicus* NRRL-3085	C 48.62, H 6.55 O 32.56, N 10.92 Cl 1.15, S 0.87		S										NA-11
AC-98 complex B		*Stm. hygroscopicus* NRRL-3085	C 46.27, H 6.15 O 30.85, N 11.68 CL 1.08, S 0.92		S										NA-11

Name	Synonym	Producer	Formula	Struct. Ref. No.	G +	G -	Myco	Fungi	Virus	Protoz	Tumor	Others	Tox.	Remarks	Ref.
AC₂ 435		*Stm. sp.* AC₂ 435	C 43.01, H 7.68					*S					L (ip)	Polyene (tetraene) group *Yeast only	1
Ac₃		*Stm. sp.* Ac 3 res. *Stm. fradiae*													2
AC-541		*Stm. hygroscopicus* NRRL 3111			S	S									NA-12
AC₆ 569	Boseimycin	*Stm. sp.*												See Boseimycin	NA-13
ADOT	2-Acetyl-2-de-carboxamide-oxytetracycline Terramycin X	*Stm. sp.*												See 2-Acetyl-2-de-carboxamide-oxy-tetracycline	1
AE-56	Ayfactin (prob)	*Stm. sp.*												Polyene (heptaene) group	1
AF 1231		*Stm. sp.* 1231	$C_{85}H_{127}O_{33}N_4$					*S		S			T	*Yeast only	1
AF2832		*Stm. filipinensis* NRRL-3217	(Mild drying conditions) C 51.08, H 7.59 O (direct) 27.88 N 11.83		S	S									NA-14
AF2833		*Stm. filipinensis* NRRL 3217	C 50.28, H 6.77 O (direct) 26.09 N 15.32		S	S									NA-14
AM-374		*Stm. sp.*	C, 54.54; H, 6.10; O 29.01; N, 7.60; Cl, 2.47		S										NA-15
AM-684		*Stm. hygroscopicus* *Stm. griseospiralis*	$C_{45}H_{77}NO_{17}$		S							*S		*Active to *Mycoplasma gallisepticum* Converted from Tylosin	NA-16

Name	Synonym	Producer	Formula	Struct. Ref. No.	G+	G-	Myco	Fungi	Virus	Protoz	Tumor	Others	Tox.	Remarks	Ref.
AO-341		*Stm. candidus* (NRRL 3147, 3148)	C: 58.0, H: 6.4, N: 13.5, O (dif.): 22.1		S									Peptide group	NA-17
AP-191α															NA-18
AP-191β															NA-18
AP-191-γ		*Stm. candidus* NRRL 3110	C 61.54, H 7.27, N 7.22, O 23.97		S	S							T (sc)	Produced with AP-191-α, & β	NA-18
ASK-753		*Stm. sp.* AS-K-753	C 54.64, H 6.98 N 6.92, O 28.96 Fe 2.5		S	S		*S					M	Peptide group Similar to Sideromycin *Yeast only	NA-19
AV-290	LL-AV 290	*Stm. candidus* NRRL 3218	Base: C 53.11, H 6.04 O 30.04, N 6.12 Cl 3.34		S	S									NA-20
AX-18		*Stm. recifensis* (*Noc. reciferi*)	$C_{34}H_{45-47}O_9N$		S										1
AYF A, B		*Stm. aureofaciens*	$C_{25}H_{35-36}O_7N$					*S					T (ip)	Polyene (heptaene) group *Yeast only	1, 2
B 44P (A, B, C)	Streptovaricin	*Stm. sp.* NO. B44-P1												Consisting of closely related components A, B, & C. See Streptovaricin	NA-21
B-73	Albonoursin	*Acm. sp.*												See Albonoursin	1
B-74		*Stm. griseus*						*S						*Plant pathogens	1
B-637		*Stm. sp.*											T	Streptothricin group	1

Name	Synonym	Producer	Formula	Struct. Ref. No.	G+	G-	Myco	Fung	Virus	Protoz	Tumor	Others	Tox.	Remarks	Ref.
B-2847R		*Stm. tolypophorus* (ATCC 21177)	$C_{43}H_{56}N_2O_{14} \cdot H_2O$ C: 64.66 ± 1.0, H: 6.67 ± 0.5, N: 3.03 ± 0.5		S	W							M		NA-22, NA-23
B-2847Y		*Stm. tolypophorus* (ATCC 21177)	$C_{43}H_{56}N_2O_{14} \cdot H_2O$ C: 60.71 ± 1.0, H: 6.59 ± 0.5, N: 3.10 ± 0.5		S	W							M	Oxidized product of B-2847R	NA-23, NA-24
B-5477-m		*Stm. fungicidicus* like	C: 49.7 ± 0.5, H: 6.29 ± 0.3, N: 13.51 ± 0.3, Cl: 9.4 ± 0.3		S								M	Basic peptide group	NA-25
B-5794 complex														Anthracycline group	1
B-10000		*Stm. sp.* B-10000	C 58.80, H 7.99		S								L (ip)	Aldgamycin-like	NA-26
B-14437		*Stm. purpureofuscus* var. *acoagulans*	$C_{12}H_{15}O_5N_5$					*S					T	*Piricularia colletotrichum* Sangivamycin-like	NA-27, NA-28
B-15565 A		*Stm. hygroscopicus* B-15565	C 59.43 H 9.12 N 3.66		S	W	W	S						Endomycin-like	NA-29
B-15565 B		*Stm. hygroscopicus* B-15565	C 59.77 H 9.05 N 3.72		S	W	W	S						Endomycin-like	NA-29
B-15645		*Stm. griseolus* B-15645	$C_{11}H_{13}N_3O_5$					*S					T	*Certain plant pathogens (*Piricularia oryzae*)	NA-30

Name	Synonym	Producer	Formula	Struct. Ref. No.	G+	G-	Myco	Fung	Virus	Protoz	Tumor	Others	Tox.	Remarks	Ref.
B-21085		Stm. collinus var. albescens	C 64.79, 65.02, H 5.28, 5.32, O 29.79		S								L (ip)	Chrysomycin-like	NA-31
B-35251 A, B, C		Stm. viridochromogenes ATCC 21343 Stm. griseo-laqueus ATCC 21344 Stm. sp. ATCC 21345									S			Also produced Daunomycin	NA-32
B-5941		Stm. fradiae var. acinicolor B58941	$C_{37}H_{59}NO_{12}$		S		S							Macrolide group Similar to Acumycin or Cirramycin B	NA-33
BA-163	Streptonigrin	Stm. flocculus ATCC 13535-13536	C 58.3, H 4.5, N 10.9, OCH_3 18.0											See Streptonigrin	1, NA-34
BA-4721	Streptonigrin													See Streptonigrin	1
BA-6903		Stm. sp.	$C_{28}H_{30}O_{10}$		S					S					1, 2
BA-8509	Duazomycin													See Duazomycin	1
BA-90912	Sangivamycin													See Sangivamycin	1
BA-17039A		Stm. longisporus subsp. griseus	$C_{24}H_{41}O_9N_5$								S			Peptide group	1
BA-17039B		Stm. longisporus subsp. griseus	$C_{20}H_{28}O_5$		S										1
BA-180265	Kanchanomycin													See Kanchanomycin	1
BA-181314		Stm. sp.	$(C_{28}H_{26}O_{11}N_2)_n$		S						S				1, 2
BA-181314A		Stm. aspergilloides	$C_{28}H_{26}O_{11}N_2$		S			S							NA-35

Name	Synonym	Producer	Formula	Struct. Ref. No.	G +	G -	Myco	Fungi	Virus	Protoz	Tumor	Others	Tox.	Remarks	Ref.
BD-12	LL-AB664	*Stm. luteocolor*	$C_{19}H_{35}N_7O_{12} \cdot 2HCl$		S	S	S						M	Streptothricin group	NA-36, N-37
BH 890		*Stm. misionensis*	C, 58.52; H, 8.72, N, 1.78; 0, 30.97					S							NA-38
BT-3-3	Thermorubin	*Tha. sp.* BT-3-3	C 63.44; H 4.07		S	S	S								1
BU-271		*Stm. sp.*											T	Similar to Micro-monosporin	1, 2
BU-277	Pikromycin Amaromycin													See Pikromycin	1
BY-81	IMRU 3627 LL-AC541 E-749-C	*Stm. olivoreticuli*	$C_{23}H_{45}N_9O_{11} \cdot HCl$		S	S	S						M	Streptothricin group	NA-36, NA-37
BU-306		*Acm. sp.*			S	S					S		T	Actinomycin group	1, 2
C	Plicacetin Amicetin B													See Plicacetin	1
C-11		*Stm. sp.* C-11	C 59.51—60.48 H 7.70—7.93 N 2.64—3.05					*S						Polyene (heptaene) group *Yeast only	1
C-73	Actiphenol	*Stm. griseus* SC 3675 *Stm. albulus*	$C_{15}H_{17}O_4N$	NA-5				*S			S			Related to C-73X, Glutarimide group *Yeast mainly	1
C-73X		*Stm. griseus* SC 3675	$C_{15}H_{17}O_5N$	NA-6				*S			S			Related to C-73, Glutarimide *Yeast mainly	NA-39
C-159	Telomycin													See Telomycin	1
C-381	Antimycoin													See Antimycoin	1
C-637	Leucomycin													See Leucomycin	1

Name	Synonym	Producer	Formula	Struct. Ref. No.	G +	G -	Myco	Fungi	Virus	Protoz	Tumor	Others	Tox.	Remarks	Ref.
C-1292		*Stm. sp.* ATCC 13748	C 44.61, H 5.85, N 2.00 S & P								S			Polysaccharide group	1
C-2989	Vivomycin	*Stm. sp.* C 2989							S					See Vivomycin	NA-40
Ch-777		*Acm. hetropsis*						S						3 components, produced with Netropsin	NA-41
D	Bamicetin N-noramicetin													See Bamicetin	1
D-13	Amicetin	*Stm. vinaceus-drappus*	−H$_2$O: C 56.91; H 6.97; N 13.51		S	W	S							Pyrimidine group	1, 2
D-52	Celesticetin													See Celesticetin	1
D-68	Paromomycin													See Paromomycin	1
D-73 (I, II)		*Stm. albulus*	I: C$_{15}$H$_{23}$O$_4$N II: C 63.57, H 8.22, N 5.23	NA-7*										*Steroisomeric forms of Cycloheximide	1
D-212	Hydroxystreptomycin													See Hydroxy-Streptomycin	1

Name	Synonym	Producer	Formula	Struct. Ref. No.	G +	G −	Myco	Fungi	Virus	Protoz	Tumor	Others	Tox.	Remarks	Ref.
DCS	2,5-Bis(Amino-oxymethyl)-3,6-diketo-piperazine	Stm. nagasakiensis	$C_6H_{10}O_2N_2$								S		L (ip)	Amino acid group	1
DON	6-Diazo-5-Oxo-L-norleucin													Azaamino acid group	1, 2
DX27		Stm. gabonae Aubriot 272 (ATCC 15282)	$C_{32}H_{41}O_6N_2S$		S		S		S				M (ip)		NA-42
E-73	4-Acetoxycy-clohexeximide	Stm. albulus	$C_{17}H_{25}O_6N$	NA-8										See 4-Acetoxycyclo-heximide	1, 2
E-129	Ostreogrysin													See Ostreogrysin	1, 2
E-212 first substance	Toyocamycin 9-27, 9-48 Siromycin, Naritheracin														
E-212 lutea factor		Stm. sp. res. Stm. albus (Myxoviromycin-producing strain)													1
E-212 niger factor		Stm. sp. res. Stm. albus	C 49.14; H 4.34 N 23.77					*S						Similar to Cycloheximide *Candida only	1
E-300		Stm. sp.							S				M (ip)		1, 2
E-416		Stm. sp.			*S				S				T (ip)	*Limited Gram pos. bacteria	1

Name	Synonym	Producer	Formula	Struct. Ref. No.	G+	G-	Myco	Fungi	Virus	Protoz	Tumor	Others	Tox.	Remarks	Ref.
E-733A	Cycloserin, D-Amino-3-isoxazolidone, Orientomycin, Oxamycin, JN-21, K-300, PA-94, 106-7													See Cycloserin	1
E-733B	O-Carbamyl-D-Serine													See O-Carbamyl-D-Serine	1
E-749-C	LL-AC541 (BY-81)	Stm. hygroscopicus	–HCl: C 36.53, 37.20 H 6.50, 5.70, N 18.28, 19.15, Cl 12.40, 13.10											Streptothricin group	NA-43
EF-185	Framycetin Neomycin B													See Framycetin	1
EI$_5$		Stm. sp.			S								*	Streptothricin group Similar to Actinorubin *Similar to Actinorubin, less toxic than Streptothricin	1
F-10 A		Stm. luteoreticuli F-10						S						Similar to Fraction A: Mycolutein Fraction C: Aureothricin	1
F-10 B								S							1
F-10 C					S	S		S							1
F-17 C	Azacolutin													See Azacolutin	1

Name	Synonym	Producer	Formula	Struct. Ref. No.	G+	G-	Myco	Fungi	Virus	Protoz	Tumor	Others	Tox.	Remarks	Ref.
F-20		Stm. reticulus-ruber			S	S	S								1
F-43		Stm. sp.	C 54.33, H 5.95, N 14.13		S	S			S					Quinoxaline group Actinoleukin Levomycin like	1, 2
F-256		Stm. sp.			S								T (ip)		1, 2
F-1370A	Etamycin, viridogrisein	Stm. conganensis ATCC 13528	$-$K: $C_{44}H_{47}O_{15}N_8K$		S	W	S						L (sc)	Depsipeptide group	1
F-1370B	Griseoviridine	Stm. conganensis ATCC 13528	$C_{23}H_{30}O_8N_3S$ ($C_{22}H_{27}O_7N_3S$)		S	W	S						L (sc)		1
F-3463	Niddamycin													See Niddamycin	1
FH-3582A	Nonactin N-329A													See Nonactin	1, 2
FH-3582B	Werramycin													See Werramycin	1, 2
FI-1163	Lucensomycin Etruscomycin													See Lucensomycin	1
FI-1600	Aminosidin Catenulin, Chrestomycin Farmiglucin Hydroxymycin Paromomycin Zygomycin A, FI-5853													See Aminosidin	1
FI-1762	Daunomycin													See Daunomycin	1
FI-2604A	Axenomycin A	Stm. lisandrin	C 61.74, H 8.52 O 28.87					S		S			ML (or)		NA-44
FI-2604B	Axenomycin B	Stm. lisandrin	C 62.8, H 8.58 O 27.8					S		S			ML (or)		NA-44

Name	Synonym	Producer	Formula	Struct. Ref. No.	G +	G -	Myco	Fungi	Virus	Protoz	Tumor	Others	Tox.	Remarks	Ref.
FI-5853	Aminosidin, Catenulin, Chrestomycin, Farmiglucin, Hydroxymycin, Paromomycin, Zygomycin A, FI-1600													See Aminosidin	1, 2
FN 1636	L-1,4-cyclo-Hexadiene-1-Alanine	*Stm. diastatochromogenes* var. *sakaii*	$C_9H_{13}NO_2$	N-9				*S						*Microporium audouni, Ustilago Trichophyton* Glutarimide group	NA-45
G-72	Chartreusin X-465A 747, 1293													See Chartreusin	1
G-83		*Stm. coriofaciens* ATCC 14155	$(C_{21}H_{34}O_8N)_n$					S					T	Polyene (heptaene) group	1
G-253B		*Stm. reticuli* var. *shimofusaensis*	C 54.46, H 6.28, N 15.39 $(C_{15}H_{20}O_5N_4)$		S	S					S			Mitomycin C like	1, NA-46
G-253 B$_1$		*Stm. reticuli* var. *shimofusaensis*	$C_{15}H_{20}N_4O_5$		S	S	S	*S			S			*Candida albicans* Mitomycin related	NA-46
G-253 B$_2$		*Stm. reticuli* var. *shimofusaensis*	$C_{15}H_{21}N_3O_5$		S	S	S	*S			S			*Candida albicans* Mitomycin related	NA-47
G-253 C		*Stm. reticuli* var. *shimofusaensis*	C 56.03, H 5.85, N 14.15 $(C_{15}H_{19}O_5N_3)$		S	S	S	S			S		T	Mitomycin like	1
G-253 C$_1$		*Stm. reticuli* var. *shimofusaensis*	$C_{17}H_{22}N_4O_6$		S	S	S	*S			S			*Candida albicans* Mitomycin related	NA-46, NA-47
GB-229		*Stm. sp.*	—HCl: N 13.47, Cl 20.4		S	S	S	S					T	Streptothricin group	1, 2
H-146															1

Name	Synonym	Producer	Formula	Struct. Ref. No.	G +	G -	Myco	Fungi	Virus	Protoz	Tumor	Others	Tox.	Remarks	Ref.
H-277	Roseothricin													See Roseothricin	1
H-1159	Xanthomycin, 534 (prob.)													See Xanthomycin	1
H-2053	Luteomycin													See Luteomycin	1
HA-9	Thioaurin, Orosomycin, B-870													See Thioaurin	1, 2
HBF-386	Actinomycin C, Actinochrysin, Actinomycin S-67													See Actinomycin C	1
HON	Hydroxy-Oxo-Norvaline	*Stm. akiyoshiensis*	$C_5H_9O_4N$	NA-10			S						L		1, 2
Hyden Antibiotic 9	Thioaurin, Orosomycin, B-870													See Thioaurin	1, 2
I-337 A	Chloramphenicol	*Stm. sp.* I-337	C 40.95, H 3.67, N 8.61											See Chloramphenicol	NA-48
I-337 B		*Stm. sp.* I-337			S								L	See Bottromycin A$_2$ & B	NA-48
I-337 C	Fradicin	*Stm. sp.* I-337						S					T	See Fradicin	NA-48
IA-887		*Stm. netropsis*	−HCl: C 39.20, H 6.01 N 24.78, Cl 21.30		S		S	S					T	Netropsin like	1, 2
IEM-1	Grizin													See Grizin	1
IMRU-3627	Pikromycin, Amaromycin, BU-277													See Pikromycin	1

Name	Synonym	Producer	Formula	Struct. Ref. No.	G+	G-	Myco	Fungi	Virus	Protoz	Tumor	Others	Tox.	Remarks	Ref.
J$_4$-A		*Stm. sp.*	C 61.47, H 7.41, N 5.15		S			S							1
J$_4$-B		*Stm. fungicidicus*						*S							1
JA-6599		*Stm. hygroscopicus* JA-6599			S		S	S				*S		*Yeast only Macrolide group *Mycoplasma	NA-49
JN-21	Cycloserine, D-4-Amino-3-Isoxazolidone, Orientomycin, Oxamycin, E-733A, K-300, PA-94, 106-7													See Cycloserin	1
K$_{13}$		*Stm. hygroscopicus*						S					T	Similar to K$_{27}$, Hygromycin B	1
K-16		*Stm. rimosus* (CBS 569.66, 570.66)	C$_{14}$H$_{18}$O$_{10}$N$_4$							S			L (ip)		NA-50, NA-51
K-27	Hygromycin B													See Hygromycin B	NA-50, NA-51
K-117														Similar to Netropsin	1
K-125a		*Stm. sp.* K-125a			S			W					T (ip)		1
K-178		*Stm. albus* K-178	C$_{41}$H$_{84\text{-}85}$O$_{11\text{-}12}$		S		S						T (ip)		1
K-288	Vancomycin	*Stm. haranomachiensis*	C 45.25, H 5.93 N 8.04		S								L		1, 2

Name	Synonym	Producer	Formula	Struct. Ref. No.	G+	G-	Myco	Fungi	Virus	Protoz	Tumor	Others	Tox.	Remarks	Ref.
K-300	Cycloserine, D-4-Amino-3-Isoxazolidone, Orientomycin, Oxamycin, E-733A, JN-21, PA-94, 106-7													See Cycloserine	1
K-349		Stm. sp.												pH Indicator Similar to Luteomycin	1
K-358	Nigericin, X206	Stm. albus K-358	–Na: C 64.43, H 8.9, Ash 5.0		S	W							T (ip)		1
LA-5352		Stm. sp. LA-5352												Sideromycin group	1, 2
LA-5937		Stm. bobiliae var. sporificans Stm. sp. LA-5937	C 47.8, H 6.01, N 14.95		S								MT	Sideromycin group	1, 2
LA-7017		Stm. sp.	C 56.99, H 7.18		S								T	Aureolic acid group	1, 2
LL-A491		Stm. sp. Lederle A491	$C_{72\pm2}H_{144\pm8}O_{25\pm1}N$		S									Monazomycin-like	NA-52
LL-AB664		Stm. candidus	C 40.67, 38.69 H 6.08, 5.94 N 18.07, 17.63		W	S		S						Streptothricin-like	NA-53
LL-AC-541		Stm. hygroscopicus (NRRL 3111)	C, 35.68; H, 6.04; N, 18.58; O, 24.35; Cl, 12.31 (HCl-salt)		W	S		S					T	Streptothricin group but without β-Lysine as a constituent	NA-54, NA-55
LL-AE-705W	Neutramycin													See Neutramycin	1

Name	Synonym	Producer	Formula	Struct. Ref. No.	G +	G −	Myco	Fungi	Virus	Protoz	Tumor	Others	Tox.	Remarks	Ref.
LL-AE-705Y	Tetrango-mycin													See Tetrangomycin	1
LL-AF 283α		*Stm. filipinensis*	C 51.08, H 7.49 N 11.83, O 27.88		W	W	W						L		NA-56
LL-AF 283β		*Stm. filipinensis*	C 50.28, H 6.77 O 26.09 N 15.32,				W						L		NA-56
LL-AL 471-E	Aldgamycin E													See Aldgamycin E	1
LL-AM 684β	Relomycin													See Relomycin	1
LL-AO 341A		*Stm. candidus*	C 58.0, H 6.4 N 13.5, O 22.1		S										NA-57
LL-AO 341B		*Stm. candidus*		NA-11											NA-58
LL-AP 191	Lemonomycin													See Lemonomycin	1
LL-AV-290	AV-290	*Stm. candidus* NRRL 3218	C 53.11, H 6.04 O 30.04, N 6.12, Cl 3.34		S	W	W								NA-59
LL-RA 6950-BB	Senfolomycin A, B													See Senfolomycin A, B1	
M-II		*Acm. violaceus*							S						1
M5-18903		*Stm. caelestis*	C 56.32, H 7.44		S		S						T		1
M-141	Actinospec-tacin													Aureolic acid group	1
M-188		*Stm. caelestis*	$C_{38}H_{61}O_{14}N$		S									See Actinospectacin	1, 2
M-259	NSC-51954	*Stm. nigellus*	C 63.00, H 6.43, N 11.95		S	W							T (ip)		1
M-319		*Noc. sulphurea* NRRL 2822	$C_{23}H_{21}O_7N$		S		S								1

Name	Synonym	Producer	Formula	Struct. Ref. No.	G+	G-	Myco	Fungi	Virus	Protoz	Tumor	Others	Tox.	Remarks	Ref.
M-411		*Stm. impexus*	C 43.01, H 6.2, N 13.28, O 25.89, Cl 10.24, Na 1.40		S									pH Indicator	NA-60
M-741		*Stv. septatum*						W			S				1
M-770		*Acm. violaceus var. rubescens*			S	S		*S					ML (ip)	*Yeast only	1
M-4209	Carbomycin Carbomycin A Magnamycin													See Carbomycin	1
MA-1267		*Stm. griseoflavus var. (MA-1267)*	$C_9H_{17}N_3O_6$		S	W	W								NA-61
MK-61	Noformicin													See Noformicin	1
MM8		*Stm. sp.* ATCC1293						S							NA-62
MSD-92		*Stm. sp.*	$C_8H_9O_3N_5$	NA-12	S	S							T (ip)	Purine group	1
MSD-235L	Streptavidin													See Streptavidin	1
MSD-235S		*Stm. avidinii* *Stm. lavendulae*	$C_{17}H_{28}O_3N_3$			S									1
MSD-819	6-Chloro-2-quinoxaline carboxylic acid 1,4-dione	*Stm. ambofaciens var. (MA-2870)*	$C_9H_5N_2O_4Cl$	NA-13	S	S									NA-63, NA-64
N-44A-21		*Stm. sp.* N-44A-21	C 50.19−50.44 H 7.64−7.78 N 12.85−13.49		S								L (ip)		1
N-63		*Stm. sp.* similar to *Stm. albus*										*	L	*Anti-parasite	1

Name	Synonym	Producer	Formula	Struct. Ref. No.	G +	G -	Myco	Fungi	Virus	Protoz	Tumor	Others	Tox.	Remarks	Ref.
N-109 1st substance		*Stm. tanashiensis*			S									pH Indicator Similar to Luteomycin	1
N-102 2nd substance		*Stm. tanashiensis*			S									pH Indicator	1
N-329A	Nonactin, FH-3582A													See Nonactin	1, 2
N-329B	Valinomycin, Aminomycin													See Valinomycin	1, 2
NA-232-M1	Hydroxystreptomycin, Reticulin, D-212													See Hydroxystreptomycin	1
NC-1968	Perimycin, Aminomycin, Fungimycin													See Perimycin	1, 2
NK-1001		*Stm. kanamyceticus* var. (12-48)	$C_{18}H_{35}N_3O_{12} \cdot H_2O$		W S									Kanamycin-related, NK1003 produced by addn of NK1001	NA-65
NK-1003		*Stm. kanamyceticus* var. (27-6)	$C_{12}H_{25}N_3O_7$		W S	W	W							Kanamycin-related, produced by addn of NK1001	NA-66
NK-1012		*Stm. kanamyceticus* var. (21-16 & 22-46)			S	S	S							Kanamycin-related	NA-67
NK-1012-1			$C_{18}H_{36}N_4O_{11}$		S	S	S								NA-67
NK-1012-2			$C_{12}H_{25}N_3O_7$		W										NA-67

Name	Synonym	Producer	Formula	Struct. Ref. No.	G +	G −	Myco	Fungi	Virus	Protoz	Tumor	Others	Tox.	Remarks	Ref.
NK-1012-3			$C_{18}H_{36}N_4O_{11}$		S	W									NA-67
NK-1013-1			$C_{22}H_{41}N_{12}O_5$			S									NA-67
NK-1013-2			$C_{20}H_{39}N_5O_{11}$		W	W									NA-67
NP-522	Indolmycin	Stm. hygroscopicus	$C_{14}H_{15-17}O_2N_3$		S	S	S						L (sc)		1
NSC-37917	Streptozotocin													See Streptozotocin	1
NSC-51954	M-259, A-6413													See M-259	1
NSC-52947	Pactamycin, Pactacin, U-15800													See Pactamycin	1
NSC-76455D	Peliomycin													See Peliomycin	1
NSC A-649		Stm. sp. NSC A-649	−Mg: C 55.55, H 7.04, MgSO₄ 1.9		S						S		T (ip)	Aureolic acid group	1
NSC-A-2371	Mithramycin, PA-144													See Mithramycin	1
NSC-B-152222	Demetric acid													See Demetric acid	1
O-2		Stm. sp.	$C_{18}H_{26}O_4N_2$		S		S				S		M (ip)	pH Indicator	1
O-5		Stm. sp.	Reineckate: C 41.17, H 5.10, N 15.59, Ash 0.305 mg/2.850 mg		S	S	S				S		T	pH Indicator	1
O-20														Streptothricin group	1
P-9		Stm. sp.						*S					**	*Plant disease **No phytotoxic	1, 2

Name	Synonym	Producer	Formula	Struct. Ref. No.	G +	G -	Myco	Fungi	Virus	Protoz	Tumor	Others	Tox.	Remarks	Ref.
P42-13		*Act. tumemacerans*			S			S			S				NA-68
PA-85	Rimocidin														1
PA-86		*Stm. rimosus*	C 60.30, H 8.30, N 3.41					S						Polyene (tetraene) group	1, 2
PA-93	Novobiocin, Albamycin, Crystallinic acid Biotexin, Griseoflavin, Cardelmycin, Inamycin, Cathocin, Streptonivicin, Cathomycin, Vulcamycin														1, 2
PA-94	Cycloserin, D-4-Amino-3-isoxazolidone, Orientomycin, Oxamycin, E733A, JN-21 K-300, 106-7														1, 2
PA-95	Actithiazic acid, Acidomycin, Cinnamonin, Mycobacidin, Thiazolidione Antibiotic														1

Name	Synonym	Producer	Formula	Struct. Ref. No.	G+	G−	Myco	Fungi	Virus	Protoz	Tumor	Others	Tox.	Remarks	Ref.
PA-105	Oleandomycin, Amimycin, Matromycin, Romicil, RO2-7638, 69895B														1, 2
PA-108	PA-1008	*Stm. sp.*	$C_{38}H_{63}O_{14}N$		S									Macrolide group	1, 2
PA-114A	Synergistin A, Staphylomycin M$_1$	*Stm. olivaceus*	$C_{35}H_{42}O_9N_4$ or $C_{25}H_{31}O_6N_3$		S		S							Depsipeptide (Mikamycin) group	1, 2
PA-114B	Synergistin B	*Stm. olivaceus*	$C_{52}H_{63}O_{12}N_9$		S								L (sc)	Depsipeptide (Mikamycin) group	1, 2
PA-114B-3	Synergistin B-3	*Stm. olivaceus*	C 62.77, H 6.52, N 12.61		S									Depsipeptide (Mikamycin) group	1, 2
PA-126 P$_2$	Actinomycin P$_2$														1
PA-128		*Stm. sp.*	$C_{37-46}H_{61-75}O_{13-16}N$		S	S				S					1, 2
PA-132		*Stm. sp.*	$C_{16}H_{18-20}O_5$		S	S		W		S			T (sc)		1, 2
PA-133A	PA-1033A	*Stm. griseofaciens* ATCC 13180	$C_{25}H_{43}O_6N$		S									Macrolide group	1, 2
PA-133B	PA-1033B	*Stm. griseofaciens* ATCC 13180	$C_{25}H_{45}O_{10}N$		S									Macrolide group	1, 2
PA-144	Mithramycin NSC A-2371														1
PA-147	3-Carboxy-3,4-pentadienol lactol A 415-Z3														1

Name	Synonym	Producer	Formula	Struct. Ref. No.	G+	G-	Myco	Fungi	Virus	Protoz	Tumor	Others	Tox.	Remarks	Ref.
PA-148	Angolamycin	Stm. sp.	$C_{38}H_{65}O_{15}N$		S									Macrolide group	1, 2
PA-150		Stm. sp.	$C_{54}H_{82}O_{18}N_2$					S					T (sc)	Polyene (heptaene) group	1
PA-153		Stm. sp.	$C_{37}H_{61}O_{14}N$					S					L (sc)	Polyene (pentaene) group	1
PA-155A	Former name of Indolmycin														1, 2
PA-155B		Stm. albus BA-3972			S										1
PA-155X		Stm. albus BA-3972													1
PA-166		Stm. glaucus ATCC 12730	$C_{35}H_{53}O_{14}N$					S					L (sc)	Polyene (tetraene) group	1, 2
PA-180		Stm. fradiae ATCC 14443			W							S^1	$^*_2$	*¹ Inhibitory to plant, duckweed (Lemna minor) *² Non-phytotoxic	NA-69
PA-371	Quinocyclines														1
PA-616		Stm. parvisporogenes ATCC 12568	C 62.0, H 7.8, N 2.7 (Kjeldahl)					S						Polyene (heptaene) group	1
PA-1008	PA-108													See PA-108	1
PA-1033A	PA-133A,													See PA-133A	1, 2
PA-1033B	PA-133B													See PA-133B	1, 2
PA-7478		Stm. rimosus ATCC 13224	–HCl: C 54.9, H 5.5 N 2.6, Cl 7.4		S								L	Tetracycline group	1

Name	Synonym	Producer	Formula	Struct. Ref. No.	G +	G -	Myco	Fungi	Virus	Protoz	Tumor	Others	Tox.	Remarks	Ref.
R-42(A,B,S)	A: Cyclo-heximide S: Strepto-mycin													B: Tetraene antifungal	1
R-285	(Amicetin B)	*Stm. sp.* R-285	$C_{25}H_{35}O_7N_5$		S		S						L		1
R-451A		*Mim. carbonacea* NRRL 2972 *Mim. carbonacea* var. *aurantiaca* NRRL 2997			S								L	Glycosidic group	NA-70
R-451B		*Mim. carbonacea* NRRL 2972 *Mim. carbonacea* var. *aurantiaca* NRRL 2997	C 51.02, H 6.68 N 1.23, Cl 3.97		S								L	Glycosidic group	NA-70
R-451B(X)		*Mim. carbonacea*	C 51.02, H 6.68 N 1.23, O 34.72 Cl 3.97		S									Glycosidic group	NA-71
R-451C		*Mim. carbonacea* NRRL 2972 *Mim. carbonacea* var. *aurantiaca* NRRL 2997			S								L	Glycosidic group	NA-70
R-451D		*Mim. carbonacea* NRRL 2972 *Mim. carbonacea* var. *aurantiaca* NRRL 2997	C 52.02, H 6.81 N 0.72, Cl 5.04		S								L	Glycosidic group	NA-70

Name	Synonym	Producer	Formula	Struct. Ref. No.	G +	G −	Myco	Fungi	Virus	Protoz	Tumor	Others	Tox.	Remarks	Ref.
R-451D(X)		*Mim. carbonacea*	C 51.79, H 6.35, N 1.40 O 36.35, Cl 3.98		S									Glycosidic group	NA-71
R-451E		*Mim. carbonacea* NRRL 2972 *Mim. carbonacea* var. *aurantiaca* NRRL 2997	C 65.17, H 7.54, N 1.31, O 24.23		S								L	Glycosidic group	NA-70
R-468	Nojirimycin	*Stm. sp.* belonging to *Stm. roseochromogenes* series			S	S							L		1
R-491A		*Stm. sp.* R-491	C 62.72, H 8.31 N 2.61		S									Macrolide group	1
R-491B		*Stm. sp.* R-491	C 58.26, H 8.44 N 1.83		S									Macrolide group	1
R-719	Kikumycin														1
RA-6950β (A, B)		*Stm. ochrosporus* NRRL 3146	A B C 49.76 49.36 H 5.38 5.73 O 35.44 34.74 N 3.80 3.73 S 4.52 4.44		S										NA-72
RO-2-7638	Oleandomycin, Amimycin, Matromycin, Romicil, PA-105, 69895B														1
RO-2-7758	Saramycetin, X-5079C														1
RP-5278	Rufochrom-omycin														1

Name	Synonym	Producer	Formula	Struct. Ref. No.	G +	G -	Myco	Fungi	Virus	Protoz	Tumor	Others	Tox.	Remarks	Ref.
RP-5337	Spiramycin														1
RP-6237 I		*Stm. sp.* 6670 similar to *Stm. kitasatoensis*	C 60.5, H 8.5 N 1.85, O 27.8		S									Macrolide group Similar to Leucomycin complex	1
RP-6237 II		*Stm. sp.* 6670 similar to *Stm. kitasatoensis*	C 60.5, H 8.4 N 1.6, O 29.3		S									Macrolide group Similar to Leucomycin complex	1
RP-6798	Demethyl-tetracyline														1
RP-7071		*Stm. sp.*	C 58.3, H 8.0 N 1.65					S						Polyene (tetraene) group	1
RP-7080		*Stm. sp.*	C 38.3, H 4.45 N 14.7		W					S			T (sc)	Enteromycin group	1
RP-7293	Pristinamycin														1
RP-8036		*Stm. canadiensis*	C 47.9, H 7.1, O 38.3, N 4.5, P 2.2		S	*S								*Neisseria*	NA-73
RP-9671		*Stm. actuosus*	C 49.6, H 4.0, N 14.4, S 15.75–15.80		S		S						T (ip)	Polypeptide group	1
RP 9768		*Stm. livescens*	C 57.5, H 7.5, O 34.7		*S	W					S		T	*Cross resist to actinomycin	NA-74
RP-9865		*Stm. sp.* 31723 8,899			S						S			Mixture of 13,213 RP. 13,057 RP & 13,330 RP gave one hydrogenated product, 13567 RP., aglycone	NA-75
RP-9971		*Stm. gascariensis* NRRL 2955	C 59.30–59.35 H 7.55–7.85 N 1.90–1.95					S			S		L (or)		1

Name	Synonym	Producer	Formula	Struct. Ref. No.	G +	G -	Myco	Fungi	Virus	Protoz	Tumor	Others	Tox.	Remarks	Ref.
RP-10192	7-Chloro-6-demethyl-tetracycline														1
RP-11072		Stm. caelicus, 9461 Stm. griseus 20129	$(C_{29}H_{49}O_7N_5)_n$		S		S						L (or)	Peptide group	1
RP-11837		Stm. viridans NRRL 3087	C 46.9, 49.36 H 7.9, 7.08 N 3.9, 5.7 P 2.25, 2.06		S	*S								*Only *Neisseria* Related to Moenomycin, Prasinomycin	1, NA-76
RP-12535	Pristinamycin IA													See Pristinamycin	
RP-12536	Pristinamycin IIA														
RP-13057 in RP-9,865	Rubidomycin	Stm. sp. 31,723 8,899	C 55.35, H 6.45, O 30.1, N 2.55, Cl 6.05		*S									*Bac. subtilis, Kl. pneumoniae	1
RP-13213 in RP 9,865		Stm. sp. 31,723 8,899 Stm. coeruleorubidus	C 58.6, H 6.2, N 2.2		S	S					*S		*T	*100 times more active and toxic than RP-13,057	1
RP-13252		Stm. croceus	C 63, H 5.95, O 20.85, N 9.90		S								T		NA-77
RP-13330 in RP-9865		Stm. coeruleorubidus			S	S					S				1
RP-13919	Pristinamycin IB													See Pristinamycin	
RP-13920	Pristinamycin IIB													See Pristinamycin	

Name	Synonym	Producer	Formula	Struct. Ref. No.	G+	G-	Myco	Fungi	Virus	Protoz	Tumor	Others	Tox.	Remarks	Ref.
RP-16511		*Stm. hygroscopicus* var. *tepidalitus*	C 65.8, H 7.8, O 22.75, N 3.55		S								L (ip)		NA-78
RP-16978		*Stm. livescens*	C 57.2, H 7.3, O 35.5		*S	W					S		T	*Limited	NA-76
RP-17967		*Stm. roseopullatus* (DS 20.073)	C: 61.5, H: 5.25, O: 20.0, N: 12.5		S	S	S				S		T		NA-79
RP-18631		*Stm. hygroscopicus* DS9751 or *Stm. albocinerescens* DS21647	C, 59.4 H, 5.45; O, 25.1; N, 4.0; Cl, 5.1		S										NA-80
RP-19402		*Stm. peruviensis* 6227	$-$Na: C 46.25; H 6.9; O 36.7; N 4.1; P 2.4; S 0.6; Na 3.05		S	W	W						L (ip)		NA-81
S-4C-33		*Stm. sp.* 4C-33	$C_{42}H_{67}NO_6$		S								L (ip)		NA-82
S-39		*Stm. sp.* res. to *Stm. albus*						S							1
S-339		*Stm. sp.*									S		T		1
S-520		*Stm. diastaticus*	$C_{40}H_{59-60}O_{10}N_{8-9}Cl$				S						M (ip)	Peptide group	NA-83, NA-84, NA-85, NA-86
S-583-A-II	Rhodomycin B	*Stm. purpurascens*	Hydrochloride: $C_{28}H_{33}O_{10}N \cdot HCl$	NA-14	S									Anthracycline, related to Rhodomycin	NA-87
S-583-A-III	Rhodomycin A		Dihydrochloride: $C_{36}H_{48}O_{12}N_2 \cdot 2HCl$	NA-15											

Name	Synonym	Producer	Formula	Struct. Ref. No.	G+	G−	Myco	Fungi	Virus	Protoz	Tumor	Others	Tox.	Remarks	Ref.
S-583 B		*Stm. purpurascens*	$C_{59}H_{78}O_{20}N_2 \cdot 2HCl.4\frac{1}{2}H_2O$		S								T (ip)	Anthracycline group Similar to Rhodomycin A, Rhodomycin B	NA-87, NA-88
S-685		*Stm. sp.* (*Stm. lavendulae* like)	Picrate: $C_{16}H_{16}N_8O_{11}$			*S								*Xanthomonas oryzae*	NA-89
SA-IX	Flavofungin														1
SAX-10		*Stm. aureus* ATCC 3309	C 64.78; H 6.43; N 2.51		S		W							pH Indicator Luteomycin like	1
SE-352	V-factor														1
SF-666		*Stm. setonensis*	$C_7H_{14}O_6$	NA-16	W								L	Monosaccharide group	NA-90
SF-666 B		*Stm. setonensis*	$C_7H_{14}O_6$		W								L		NA-90
SF-689		*Stm. platensis*	$C_{59-60}H_{99}O_{14}N_5$		S								T		NA-91
SF-701		*Stm. griseochromogenes*	$C_{18}H_{36}N_7O_{11}$		S	S	S	*S					M	Streptothricin-like *Candida albicans, Cryptococcus neoformans*	NA-92, NA-93
SF-733		*Stm. ribosidificus*	$C_{17}H_{34}N_4O_{10}$	NA-17	S	S	S						L	Aminoglyrosidic	NA-94
SF-767-A		*Stm. microsporeus*	$C_{23}H_{44}N_4O_{15}$		S	S	S							Basic, water-soluble, oligosaccharide	NA-95
SF-767-L		*Stm. microsporcus*	$C_{23}H_{46}N_4O_{16}$		S	S	S							Basic, water-soluble, oligosaccharide	NA-96
SK-229A		*Stm. griseus var.* 229	$C_{52}H_{80}O_{25}$ or $C_{54}H_{80}O_{25}$		S						S			Aureolic acid group	1

Name	Synonym	Producer	Formula	Struct. Ref. No.	G+	G-	Myco	Fungi	Virus	Protoz	Tumor	Others	Tox.	Remarks	Ref.
SK-229B		Stm. griseus var. 229	$C_{54}H_{84}O_{25}$ or $C_{59}H_{88}O_{26}$								S			Aureolic acid group	1
SKCC-1377		Stm. sp.			S								T (ip)	pH Indicator	1, 2
SP-30		Mim. halophytica (NRRL 2998)			S								L	Complex of A,B,C,D	NA-97
SQ-15859		Stm. chrysomallus	$C_{40}H_{64}O_{12}$	NA-18	S	S					S		T (ip)	Cyclic polylactone group	1, 2
T-12		Strain T-12/3 of a thermophilic Mim. sp.			S		*S							*Limited	NA-98
T-82		Stm. sp.	C 54.39; H 7.72; N 6.37; C 40.09; H 6.32; N 9.07		S			S					L		1
T-1124	Juvenimicins	Mim. chalcea var. izumensis	C 63.66, 63.12 H 8.67, 8.99 N 3.07, 2.53		S	S							ML	Macrolide group	J-6
T-124-C	Juvenimicin-A3	Mim. chalcea Var. izumensis	C 63.22, 62.79 H 8.54, 8.70 N 3.08, 2.92		S	S								Macrolide group	J-6
T-1384	Netropsin Congocidin, IA-887 Sinanomycin, K-117														1, 2

Name	Synonym	Producer	Formula	Struct. Ref. No.	G+	G−	Myco	Fungi	Virus	Protoz	Tumor	Others	Tox.	Remarks	Ref.
T-2636 A, **B**, C, D		*Stm. rochei* var. *volubilis*			S*								ML (ip)	Macrolide group *A+B, B+C, B+D synergistic	NA-99
T-2636 A															
T-2636A	Bundlin B	*Stm. rochei* var. *volubilis*	$C_{27}H_{35}O_8N$	NA-20	S								L (ip)	Macrolide group	NA-99 NA-100
T-2636 B		*Stm. rochei* var. *volubilis*	$C_{42}H_{74}O_{17}$		W								L (ip)	Macrolide group	NA-100
T-2636 C	Bundlin A Lankacidin	*Stm. rochei* var. *volubilis*	$C_{27}H_{33}NO_7$	NA-21	W								L (ip)	Macrolide	NA-100
T-2636D															NA-100
T-2636 D		*Stm. rochei* var. *volubilis*	$C_{27}H_{37}NO_8$	NA-22	S								L (ip)	Macrolide group	NA-100
T-2636F		*Stm. rochei* var. *volubilis*	$C_{25}H_{35}NO_7$	NA-23	S										NA-100
T-2636M		*Stm. rochei* var. *voluvilis*						S							NA-100
T-3018	Nucleocidin	*Stm. sp.* T-3018	$C_{11}H_{15-16}O_6N_6S_8$		S	S					S				1
T-7545 A		*Stm. hygroscopicus* var. *limonens* IFO-12703 or IFO-12704						S							NA-102
B															

Name	Synonym	Producer	Formula	Struct. Ref. No.	G +	G -	Myco	Fungi	Virus	Protoz	Tumor	Others	Tox.	Remarks	Ref.
TA-435A		*Stm. sp.* O-80	C 61.90, 61.83, H 6.90, 6.94 S −, Halogen −		W		W				S		T (ip)	Ayamycin A_3 like	1
TA 2407		*Stm. burnaeogriseus*	$C_{18}H_{28}O_4$		S			W					L (ip)	Polylactone group	NA-103
TA 2590		*Stm. pluricolorescens*	C: 46.75, 47.69 H: 7.16, 7.41 N: 12.68, 12.68 Ash: 0.98 (MW 20,000−25,000)								S		M (ip)		NA-104
U-11092	Rubradirin														1
U-11921	Lincomycin C	*Stm. lincolnensis* var. *lincolnensis*	$C_{19}H_{36}O_6N_2S$	NA-24	S		S							Lincomycin group	1, 2
U-11973	N-Demethyl lincomycin	*Stm. lincolnensis* var. *lincolnensis*	$C_{17}H_{32}O_6N_2S$	NA-25	S		S							Lincomycin group	1, 2
U-11994	Desdamethine	*Stm. caelestis*												See Desdamethine	NA-104
U-12241	Cirolerosus														1
U-12898	Bluensomycin Glebomycin														1, 2
U-13714	Canarius														1
U-15774		*Stm. achromogenes*	$C_5H_7O_3N_3$	NA-26										Enteromycin group	1
U-15800	Pactamycin Pactacin, NSC-52947														1
U-19718	Kalafungin														See Kalafungin	NA-106
U-20207	Ethesdanine														See Ethesdanine	NA-105

Name	Synonym	Producer	Formula	Struct. Ref. No.	G+	G-	Myco	Fungi	Virus	Protoz	Tumor	Others	Tox.	Remarks	Ref.
U-20661	Steffimycin	*Stm. steffisburgensis*	$C_{28}H_{30}O_{13}$		W						*S		L (ip)	*Inhibit the growth of KB-cells	NA-107 NA-108
U-20904	Cis-β-Carboxy-acrylamidine	*Acm. sp.*	$C_4H_6O_2N_2$	NA-27				S			S		M (ip)		1
U-20943		*Stm. lincolnensis var. lincolnensis*	$C_{18}H_{34}O_6N_2S \cdot HCl.H_2O$	NA-28										Lincomycin group	1, 2
U-21699	Lincomycin B	*Stm. lincolnensis var. lincolnensis*	$C_{17}H_{32}O_6N_2S$	NA-29	S		S							Lincomycin group	1, 2
U-22662	Desdanine													See Desdanine	NA-105
U-22956		*Stv. fervens var. melrosporus*	$C_5H_7O_4N$	NA-30	S	S							M (ip)	Enteromycin group	1, NA-106
U-24544		*Stm. griseus*	$C_{27}H_{32}N_4O_8$		W	W							L (ip)		NA-109
VD 844		*Stm. sp.*	$C_7H_6N_2O_2S_2$	NA-31	S	S							T		NA-110
V-factor		*Stm. sp.* E352									*			*Bacterial leaf blight of rice plant in vivo (not in vitro)	1
W-2		*Stm. diastatochro-mogenes*						*S					T (ip)	Similar to Trichomycin *Only yeasts	1

Name	Synonym	Producer	Formula	Struct. Ref. No.	G +	G -	Myco	Fungi	Virus	Protoz	Tumor	Others	Tox.	Remarks	Ref.
W-112		Stm. sp.							S						1
W 847 (A,B,C$_1$,C$_2$)		Mim. sp. W847 (NRRL 3275)	A: $C_{44}H_{80}N_2O_{15}$ B: $C_{46}H_{82}N_2O_{16}$ C$_1$: $C_{48}H_{84}N_2O_{17}$ C$_2$: $C_{49}H_{86}N_2O_{17}$		S S S S								ML	Macrolide group Cross resistance to Erythromycin. etc.	NA-111
WC-3628		Stm. sp. WC-3628	$C_{42}H_{73}O_{16}N$		W	W	W								1
WR-3/17		Stm. albus			S		S	*S					T (ip)	*Plant pathogens	1, 2
X Factor	Fradicin														1
X activity	γ-Activity X	Stm. aureofaciens			*S									*B. subtilis	1
X-45	Actinomycin B	Acm. sp.													1, 2
X-53III	Echinomycin, Echanomycin, Levomycin, Quinomycin A, X-948														1
X-63		Stm. sp.			S			S						Polyene (heptaene) group Similar to Trichomycin, Hamycin	NA-112
X-206		Stm. sp.	$C_{46-47}H_{80-82}O_{13}$		S		S						T (sc)		1, 2
X-340	Resistomycin, Heliomycin	Stm. sp.	$C_{19}H_{16}O_5$	NA-32	S	W							L (or)		1

Name	Synonym	Producer	Formula	Struct. Ref. No.	G +	G −	Myco	Fungi	Virus	Protoz	Tumor	Others	Tox.	Remarks	Ref.
X-464	Nigericin Polyetherin A	*Stm. sp.*	$C_{25}H_{40}O_7$		S		S						T (ip)		1, 2, NA-113
X-465 A, B	Chartreusin 747, 1293														1, 2
X-537A		*Stm. sp.*	$C_{34}H_{52}O_8$	NA-33	S		S						T (ip)		1, 2, NA-114
X-948	(Echinomycin)	*Stm. echinatus*	C 55.19, H 5.94, N 15.22		S									Quinoxaline group	1, 2
X-1008		*Stm. sp.* X-1008	C 56.40, H 6.52, N 13.69		S									Quinoxaline group	1, 2
X-1285		*Stm. sp.* 1285			S	S				*S			ML (or)	*Entamoeba histolytica*	1
X-5079C	Saramycetin RO2-7758	*Stm. sp.* X-5079C		NA-34				*S					L	polypeptide	1, 2,
YL-704A, B		*Stm. platensis* var. MCRL 0388	A: $C_{43}H_{71}NO_{15}$ B: $C_{41}H_{67}NO_{15}$		S W	S W	S S						L	Macrolide group,	NA-115, NA-116
Y-107		*Stm. sp.* Y-107			S	S							MT	pH Indicator Similar to Luteomycin	1
α		*Stm. verticillatus*			S		S	*S						Mitosane group *Yeast mainly	1
β		*Stm. verticillatus*			S		S							Mitosane group	1
γ		*Stm. verticillatus*			S		S							Mitosane group	1
γ₂	Porfiromycin	*Stm. verticillatus*			S		S							Mitosane group	1
1-81-d-1S		*Stm. albus*	$C_{38}H_{63-65}O_{12}N$		S		S	S						Camphomycin like	1, 2

Name	Synonym	Producer	Formula	Struct. Ref. No.	G +	G -	Myco	Fungi	Virus	Protoz	Tumor	Others	Tox.	Remarks	Ref.
2-229	229, Racemomycins	*Stm. sp.* 2-229 res. *Stm. phaeochromogenes*												Streptothricin group	1
4		*Acm. griseoflavus*				*S	S							*Eberthella typhosa	1
6A-36		*Stm. sp.* 6A-36	$C_{18}H_{18}O_3$		S		S						L	Chartreusin like	1
9-27	Toyocamycin				S		S								1
9-28	Toyocamycin														1
10CM		*Stm. sp.* res. *Stm. albus*			S	W	S						L		1, 2
11A	(Azomycin)	*Stm. sp.* 11	C 31.68, H 2.25 N 36.87		S	W	W								1
11B	Methymycin	*Stm. eurocidicus*			S										1
12-12	Violarins, 452-7,														1

Name	Synonym	Producer	Formula	Struct. Ref. No.	G +	G -	Myco	Fungi	Virus	Protoz	Tumor	Others	Tox.	Remarks	Ref.
15-A, B, C, D, E, F		Stm. sp. 3223	A: C 41.65, H 7.23 N 8.38, O 42.72		S	S							ML (ip)		NA-117
			B: C 38.35, H 6.45 N 8.10, O 47.10		S	S									
			C: C 39.18, H 6.81 N 10.74, O 43.27		S	S									
			D: C 36.31, H 6.42 N 10.90, O 46.37		S	S									
			E: C 38.34, H 6.86 N 12.31, O 42.49		S	S									
			F: C 44.45, H 7.61 N 12.89, O 35.05		S	S									
17-41A		Acm. sp. 17-41			S									Polypeptide group	1
17-41B		Acm. sp. 17-41						S						Polyene(pentaene) group	1
18-45								S							NA-118
18-80								S							NA-118
19-21		Stm. sp. 19-21	C 50.61, H 6.03 N 2.54				S						L		1
19A		Stm. sp. MA 568 (Thermophilic)	$C_{11} H_{12-14} O_9 N_4$		S	S								Enteromycin group	1, 2
Antibiotic 21-31	Mutabicillin	Acm. mutabilis var. tiostreptoni												Thiostreptone group	NA-119
23-21		Stm. sp.												Actinomycin group	1
24														Streptothricin group	1

Name	Synonym	Producer	Formula	Struct. Ref. No.	G +	G -	Myco	Fungi	Virus	Protoz	Tumor	Others	Tox.	Remarks	Ref.
24-10		Stm. sp. 24-10	C 57.55, H 6.3		S		S								2. NA-120
26/I	Levorin						S								1, 2
30-10		Stm. sp.							S						1, 2
35NT		Stm. celluloflavus var. 35NT	C 44.39, N 7.70 N 6.62, O 42.29		S		S								NA-121
36	Roseomycin													Streptothricin group	1
39															1
43-127		Stm. lavendulae 43-127	HCl– : C 37.11, H 6.21 N 15.81, Cl 15.71, O 24.26		S	S	S						L		NA-122
58		Stm. fasciculus ATCC 12703	C 59.94, H 8.29 N 1.96				S	S						Polyene (pentaene) group	1
62-2		Stm. reticuli	C 59.81, H 8.07		S								L (ip)		NA-123
69		Acm. sp. 69	C 60.77, H 7.70 N 1.37								S				1
74 a, β, γ		Stm. sp. KS-274						*S			S			*Trichophyton	1
83	S.K. toxic substance	Acm. sp. 83	C 73.86, H 8.96, N 8.73								S				1
101 a	Streptovaricin														1
106-7	Cycloserine,														1

Name	Synonym	Producer	Formula	Struct. Ref. No.	G +	G –	Myco	Fungi	Virus	Protoz	Tumor	Others	Tox.	Remarks	Ref.
116 a	(Althiomycin) (Matamycin)														1
120														Streptothricin group	1
123		*Stm. sp.*	C 63.78, H 10.16, Ash 0.21		S			S					M (ip)	Polyene (pentaene) group	1
135/1		*Acm. sp.* 135/1												pH Indicator	1
136														Streptothricin group	1
156														Streptothricin group	1
190/1	Speciomycin													Streptothricin group	1
202C		*Stm. sp.* 202-C			W			S					*	Similar to Flavensomycin on paper chromatography *No symptoms of phytotoxic effect were found to pinto bean plants at ca. 10 mcg/ml	1
207	Mycelin IMO	*Stm. sp.*	C 70.46–71.29, H 5.96–6.14, N 11.30–11.31					S					T (ip)	Fradicin group	1
212A, B	Roseocitrin A, B														1
232		*Acm. sp.* 232 (related to *Acm. variabilis*)	C, 56.43; H 7.06		S									Olivomycin-like	NA-142

Name	Synonym	Producer	Formula	Struct. Ref. No.	G+	G-	Myco	Fungi	Virus	Protoz	Tumor	Others	Tox.	Remarks	Ref.
235A															NA-124
255(HA)															NA125
259														Streptothricin group	1
289	Mitomycin (Umezawa)	Stm. sp. 289	$C_{26}H_{33}O_{12}N.H_2SO_4$	NA-35	S						S			pH Indicator	1
289 F		Stm. phaeoverticillatus var. takatsukiensis ATCC 21395-7			S						S				NA-126
289 FO					S						S				NA-126
294		Acm. flaveolus													NA-127
323/58		Acm. sp. 323/58	C 51.17–51.43, H 7.52–7.84, N 6.11–6.32		S			S			S				1
362		Stm. sp. ATCC 13694		NA-36		W							L	Polypeptide group	1
369														Glutarimide group	1
446		Noc. mesenterica	C 60.47, H 7.99, N 2.02		S								L	Macrolide group	1
467		Mim. halophytica var. nigra NRRL-3097	C 60.73, H 7.53, N 3.70, O 27.20, Cl 0.04, S 0.80		S								L (ip)		NA-128
452-7	Violarin B 1212														1

Name	Synonym	Producer	Formula	Struct. Ref. No.	G +	G −	Myco	Fungi	Virus	Protoz	Tumor	Others	Tox.	Remarks	Ref.
460 (A, B, C) or (Fraction I, II, III)		*Mim. chalcea var. flavida* (NRRL 3222)	A B C C 49.15 37.98 42.75 H 7.65 6.82 6.93 N 3.28 – 7.27 O 39.92 – 43.05		*S	W								*Comp. C. is weak	NA-129, NA-130, NA-131
534	(Xanthomycin)	*Stm. pseudogriseolus*	−3HCl: C 48.54, H 6.04, N 7.84, Cl 13.86		S		S			S			T		1, 2
539		*Stm. sp.* 539	−HCl: C 28.42, H 5.64, N 12.21, Cl 20.48		S	S	S	W					M	Streptothricin group	1
583		*Stm. orientalis*	C 42.87, H 6.87, N 20.62 ($C_{19}H_{27}N_9O_3$)		S	S	S	W					T		NA-132
587/13		*Acm. sp.* related to *Stm. lavendulae*			S	S	S	S					*	*Nephrotoxic to animal	1, 2
593A		*Stm. griseoluteus*	$C_7H_{11}ON_2Cl$			S					S				NA-133
616	PA616														1
660-15	(Albofungin)														1
695		*Acm. sp.* 695			S										NA-134
696	Phytobacteriomycin								S						1
719		*Acm. violaceus*													1
720A	Antimycins	*Stm. sp.* related to *Stm. olivochromogenes*	$C_{28}H_{40}O_9N_2$					*S					*	*Plant pathogen (*P. oryzae*)	1, 2
720B		*Stm. sp.* related to *Stm. olivochromogenes*											(		1, 2

Name	Synonym	Producer	Formula	Struct. Ref. No.	G +	G −	Myco	Fungi	Virus	Protoz	Tumor	Others	Tox.	Remarks	Ref.
721		*Stm. sp.*			S								ML (sc)		1, 2
732-C	Triostin C														1
747	Chartreusin														1
757		*Stm. sp.* 757						S			S			Polyene (heptaene) group	1, 2
829	Flavensomycin														1
833A	(−)(Cis-1,2-epoxypropylphosphonoic acid	*Stm. viridochromogenes* NRRL-3413, 3414, 3415, 3427 *Stm. wedmorensis* ATCC-21239 NRRL-3426	$C_3H_7O_4P$	NA-37	S	S									NA-132
899	Staphylomycins														1, 2
956		*Acm. fradiae*												Similar to Neomycin	1, 2
1037	(Toyocamycin)	*Stm. sp.* 1037	C 49.33 49.47, H 4.56–4.90, N 23.75 24.14					S					T		1, NA-136
1130/12		*Acm. xantholiticus*												Polyene (pentaene) group	1
1160		*Acm. griseus*			S	S							L	Colimycin-like	NA-137
1163 FI	Lucensomycin Etruscomycin														1, 2
1212	12-12, 452-7 Violarins	*Stm. sp.* (blue-violet strain)			S										1

Name	Synonym	Producer	Formula	Struct. Ref. No.	G+	G−	Myco	Fungi	Virus	Protoz	Tumor	Others	Tox.	Remarks	Ref.
1252	Elaiomycin														1
1293	Chartreusin														1
1321		*Acm. sp.* 1321													1
1415		*Stm. sp.* 1415	C 59.44–59.57 H 7.98–8.04, N 2.04	NA-38	S								L	Polypeptide group	1, 2
1418-Al		*Stm. sp.* 1418-Al producing trans-cinnamic acid amide									S		T		1
1483A			C 35.58, H 6.36 N 17.42, O 26.90, Cl 11.13		S	S							MT		NA-138
1579		*Stm. efflvius*	C 58.36, H 8.24 N 2.04			S		S		S					1
1618		A mutant of *Stm. longispororuber*			S	*S								*Erwinia carotovora* only	NA-139
1645P₁		*Stm. sp.* Cepa 1645-IAUR	C 55.84, H 7.67 N 2.19					S					T	Polyene (heptaene) group	1, 2
1645P₂		*Stm. sp.* Cepa 1645-IAUR						S							1, 2
1683		*Stm. sp.* 1683			S		S				S			Anthracycline group Similar to Rhodomycin	1
1703-18B	Hygromycin														1, 2

Name	Synonym	Producer	Formula	Struct. Ref. No.	G +	G -	Myco	Fungi	Virus	Protoz	Tumor	Others	Tox.	Remarks	Ref.
1762		Stm. sanguineus 1762	C 62.05, H 9.15					S							1
1900C															2
1900C$_2$															2
1900S															2
1943		Stm. sp.			S	S							*	Similar to Streptomycin *Ototoxic to cats	1, 2
2230A		Stm. sp. 2230	C$_{29}$H$_{55}$N$_5$O$_{19}$		S	S	S						L		NA-140
2230 B		Stm. sp. 2230	C$_{29}$H$_{55}$N$_5$O$_{18}$		S	S	S						L		NA-140
2410	24-10														2, NA-120
2339 & 2789		Acm. levoris 2339 & 2789												Similar to 26/1, Candicidin, Trichomycin	1
2527		Cultures of thermophylic actinomycetes													NA-141
2703	Actinomycin	Acm. sp.									S			Actinomycin group	1
2732/3	Violacin														1
2814A	Netropsin														1, 2
2814 H		Stm. sp. IA 2814	C 56.89–56.91, H 8.25–8.06, N 1.88–1.96					S						Polyene (heptaene) group	1, 2

Name	Synonym	Producer	Formula	Struct. Ref. No.	G +	G -	Myco	Fungi	Virus	Protoz	Tumor	Others	Tox.	Remarks	Ref.
2814 P		*Stm. sp.* 2814	C 57.52–57.54, H 8.23–8.31, N 1.87–1.76					S						Polyene (heptaene) group	1, 2
2844-31A		*Acm. prunicolor* 2844-31			S									Pigment group	1
2911/1, 2911/2		*Acm. globisporus* var. *roseus* 2911			S	S									1
3014		*Acm. sp.* 3014 (related to *Acm. nigrificans*)	C 57.08, H 7.58		S									Aburamycin-like	NA-142
3510	Griseins	*Stm. griseus* 3510												Sinderomycin group	1
3569	Fungicidin type antibiotic														1
4279		*Acm.* 4279			S	S		S							NA-143
4418	Fraction II = Iomycin B	*Acm. griseorubiginosus* v. *spiralis*									S			Pluramycin, Iomycin group	NA-144
4695											S				1
4738A	Cyanomycin													Pyrrothine group	1
4738B	Hydroxymycin														1
4915															1, 2

Name	Synonym	Producer	Formula	Struct. Ref. No.	G+	G-	Myco	Fungi	Virus	Protoz	Tumor	Others	Tox.	Remarks	Ref.
5888 I, II, III		Act. galilaeus, 5888	I: C 66.64, H 5.28								S			Contain Aclavine and Cinerubine A	NA-145
			II: C 64.2, H 4.96								S			Contain Aclavine and Cinerubine A	NA-145
			III: C 61.88, H 4.84								S			Contain Aclavine and Cinerubine A	NA-145
5901		Stm. sp. 5901						S							1
6270		Acm. flavochromogenes	$C_{29}H_{37}O_{6-7}N_6S$		S						S			Similar to Echinomycin but not identical	1, 2
6431-36		Acm. griseus 6431-36					S								NA-146
6613	(Etamycin)	Acm. daghestanicus			S										1, 2
7193	11296	Acm. sp. 7193 (related to Acm. halstedi)	C, 55.86; H, 7.36		S									Olivomycin like	NA-142
11296	7193	Acm. sp. 11296 (related to Acm. globisporus var. flavofuseus)	C, 56.07; H, 7.19		S									Olivomycin like	NA-142
12782		Acm. sp. 12782			S	S	S							Netropsin group	NA-147
14725I		Acm. kurssanovii	C 61.47, H 6.82 N 13.08		*S									Mikamycin group *Synergistic with 14725 II	1
14725II		Acm. kurssanovii	C 62.07, H 6.85 N 7.59		*S									Mikamycin group *Synergistic with 14725I	1

Name	Synonym	Producer	Formula	Struct. Ref. No.	G +	G −	Myco	Fungi	Virus	Protoz	Tumor	Others	Tox.	Remarks	Ref.
17731	Ornamycin														1
21544	Gougerotin														1
22765														Sideromycin group	1, 2
23572A	Dihydrostreptomycin														1
23572B	Humidin														1
45449A$_1$, A$_2$, B	Zygomycin A$_1$, A$_2$, B														1, 2
45449D	Zygomycin D													Glutarimide group	1
59266		*Stm. sp.*	59266	C 55.34, H 6.05 N 17.54, S 5.40	S		W	*S					T (ip)	Echinomycin like *Limited	NA-148
69895B	Oleandomycin														1

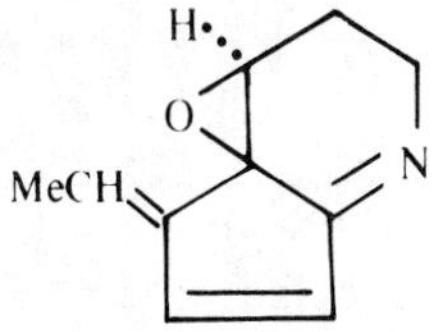

A-1

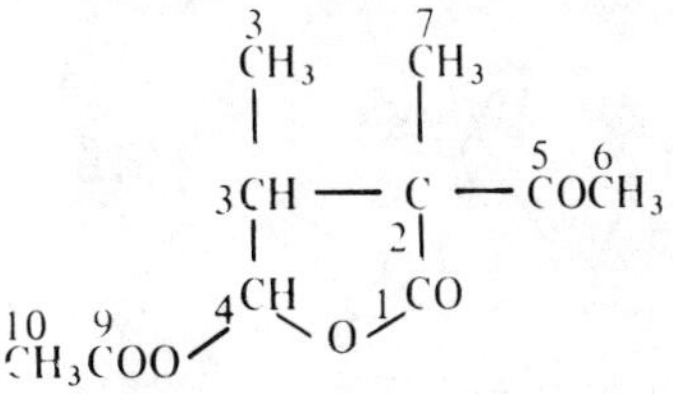

A-2

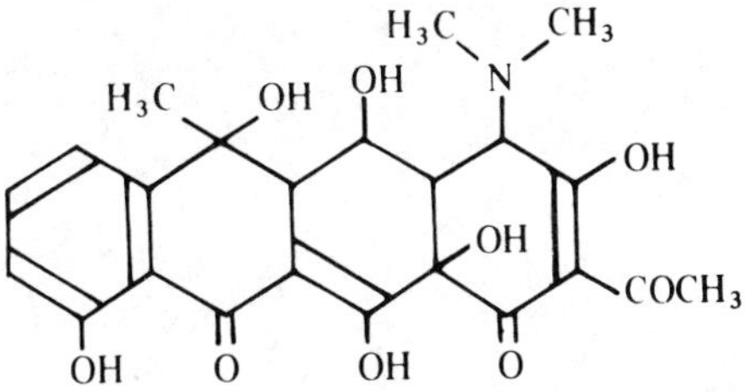

A-3

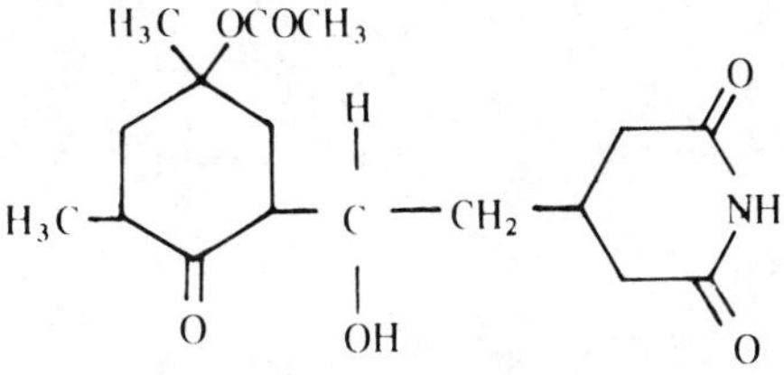

A-4

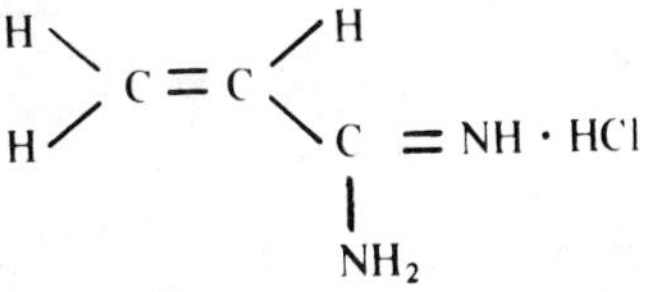

A-5

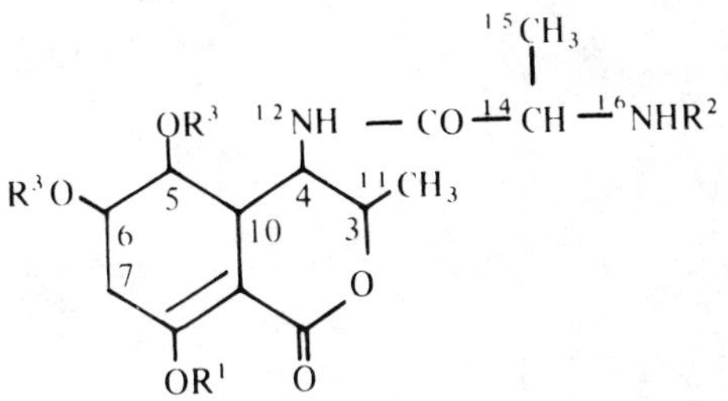

1. $R^1 = R^2 = R^3 = H$

A-6

A-7

L-Thr:	L-Threonine
D-Val:	D-Valine
L-Pro:	L-Proline
Sar:	Sarcosine
L-MeVal:	N-Methyl-L-valine
D-alleu:	D-allo-Isoleucine

HOPro:	4-Hydroxy-L-proline
COPro:	L-γ-Oxoproline (4-Ketoproline)
MeIleu:	N-Methyl isoleucine
MeAla:	N-Methyl alanine
aHOPro:	L-allo-hydroxyproline
Ileu:	Isoleucine

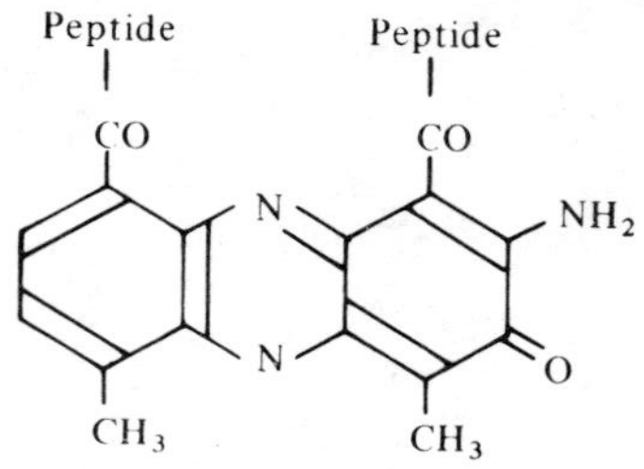

Chemical structure (phenoxazinone chromophore with two Peptide chains): labels — Peptide, CO, NH_2, O, CH_3, N, N, Peptide, CO, CH_3.

Type	Component	Synonym	Probably identical	M.P. °C	D-Thr	D-Val	L-Pro	Sar	L-MeVal	D-alleu	HOPro	COPro	MeIleu	MeAla	aHOPro	Structure of peptide
A	A_I	$I, B_I, X_{0\beta}$		(248–250) 237–238	2	2	1	2	2		1					-Thr-Val-HOPro-Sar-MeVal
																-Thr-Val-Pro-Sar-MeVal
	A_{II}	II, B_{II}			2	2		4	2							-Thr-Val-Sar-Sar-MeVal
																-Thr-Val-Sar-Sar-MeVal
	A_{III}	$III, B_{III}, X_{0\tau}$			2	2	1	3	2							-Thr-Val-Sar-Sar-MeVal
																-Thr-Val-Pro-Sar-MeVal
	A_{IV}	$IV, B_{IV}, D_{IV}, C_1, I_1, X_1$		235–236	2	2	2	2	2							-Thr-Val-Pro-Sar-MeVal
	A_V	V, B_V, X_2		245–246.5	2	2	1	2	2			1				-Thr-Val-COPro-Sar-MeVal
																-Thr-Val-Pro-Sar-MeVal
B	X-45			(250–252)												
	B_1	C_1														
	B_2	X_2														
	B_I	$I, A_I, X_{0\beta}$		237.5–238	2	2	1	2	2		1					-Thr-Val-HOPro-Sar-MeVal
																-Thr-Val-Pro-Sar-MeVal
	B_{II}	II, A_{II}			2	2		4	2							-Thr-Val-Sar-Sar-MeVal
																-Thr-Val-Sar-Sar-MeVal
	B_{III}	$III, A_{III}, X_{0\tau}$			2	2	1	3	2							-Thr-Val-Sar-Sar-MeVal
																-Thr-Val-Pro-Sar-MeVal
	B_{IV}	$IV, A_{IV}, D_{IV}, C_1, I_1, X_1$		235.5–236.5	2	2	2	2	2							-Thr-Val-Pro-Sar-MeVal
	B_V	V, A_V, X_2		246–246.5	2	2	1	2	2			1				-Thr-Val-COPro-Sar-MeVal
																-Thr-Val-Pro-Sar-MeVal

Type	Component	Synonym	Probably identical	M.P. °C	D-Thr	D-Val	L-Pro	Sar	L-MeVal	D-alleu	HOPro	COPro	Melleu	MeAla	aHOPro	Structure of peptide
C	C_0															
	$C_0\,\alpha$															
	$C_0\,\beta$		S-67, HBF386	(254)												
	C_1	IV, A_{IV}, B_{IV}, D_{IV}, I_1, X_1		241–243	2	2	2	2	2							-Thr-Val-Pro-Sar-MeVal -Thr-Val-Pro-Sar-MeVal
	$C_1\,a$															-Thr-alleu-Pro-Sar-MeVal
	C_2	VI		237–239	2	1	2	2	2	1						-Thr-Val-Pro-Sar-MeVal Isomer of C_2
	$C_2\,a$			233–235												-Thr-alleu-Pro-Sar-MeVal
	C_3	VII		232–235	2	2	2	2	2	2						-Thr-alleu-Pro-Sar-MeVal
	$C_3\,a$	D_{IV}		(241.5–243)												
	C_4															
D	D_{IV}	IV, A_{IV}, B_{IV}, C_1, I_1, X_1		235.5–236.5	2	2	2	2	2							
E	E_1				2	2	2	2	1	2			1			-Thr-alleu-Pro-Sar-Melleu -Thr-alleu-Pro-Sar-MeVal
	E_2				2	2	2	2		2			2			-Thr-alleu-Pro-Sar-Melleu -Thr-Alleu-Pro-Sar-Melleu
F	F_0				+	+	+	+	+							
	F_1				2	1	1	4	2	1						-Thr-alleu-Sar-Sar-MeVal -Thr-Val-Sar-Sar-MeVal
	F_2				2	1	1	3	2	1						-Thr-alleu-Sar-Sar-MeVal -Thr-Val-Pro-Sar-MeVal
	F_3				2			4	2	2						-Thr-alleu-Sar-Sar-MeVal
	F_4				2		1	3	2	2						-Thr-alleu-Sar-Sar-MeVal -Thr-alleu-Sar-Sar-MeVal
	F_5															-Thr-alleu-Pro-Sar-MeVal
	F_6															
	F_7															
	F_8	II														
	F_9	III, $X_{0\tau}$			+	+	+	+	+	+						

Type	Component	Synonym	Probably identical	M.P. °C	D-Thr	D-Val	L-Pro	Sar	L-MeVal	D-alleu	HOPro	COPro	MeIleu	MeAla	aHOPro	Structure of peptide
H	I_0															
I(J)	I_{0a}														+	
	I_1	IV, A_{IV}, B_{IV}, C_1, X_1		242–243	+	+		+	+							-Thr-Val-Pro-Sar-MeVal
	I_{1a}															
	I_2			240–242	+	+	+	+	+							-Thr-Val-Pro-Sar-MeVal
	I_3				2	2	2	2	2							
J	J_1	Actinoflavin	4A-2	248–250	+	+	+	+	+							
K					+	+	+	+	+							
L	L_1				+	+	+	+	+	+						
	L_2				+		+	+	+	+						
M			B or X mixture		+						+					Ileu +
P	P_1															
	P_2															
	P_3															
S	S_0			233–235												
	S_1			234–235	+	+	+	+	+		+					
	S_2			238–240												
	S_3															
U	U_1															
	U_2															
	U_3															
	U_4															
X	X_0															
	X_{0a}															
	$X_{0\beta}$	I, A_I, B_I		245–247	2	2	1	2	2		1					-Thr-Val-HOPro-Sar-MeVal -Thr-Val-Pro-Sar-MeVal

Type	Component	Synonym	Probably identical	M.P. °C	L-Thr	D-Val	L-Pro	Sar	L-MeVal	D-alleu	HOPro	COPro	MeIleu	MeAla	aHOPro	Structure of peptide
	$X_{0\tau}$	III, A_{III}, B_{III}			2	2	1	3	2							-Thr-Val-Sar-Sar-MeVal
																-Thr-Val-Pro-Sar-MeVal
	$X_{0\beta}$				2	2	1	2	2						1	-Thr-Val-aHOPro-Sar-MeVal
																-Thr-Val-Pro-Sar-MeVal
	X_1	IV, A_{IV}, B_{IV}, D_{IV}, C_1, I_1		241–242	2	2	2	2	2							-Thr-Val-Pro-Sar-MeVal
																-Thr-Val-Pro-Sar-MeVal
	X_{1a}				2	2		3	2							-Thr-Val-Sar-Sar-MeVal
												1				-Thr-Val-COPro-Sar-MeVal
	X_2			244–246	2	2	1	2	2			1				-Thr-Val-COPro-Sar-MeVal
																-Thr-Val-Pro-Sar-MeVal
Z	X_3															
	X_4															
	Z_0				+	+		+	+					+		
	Z_1				+	+		+	+					+		
	Z_2			(260–264)	+	+		+	+					+		
	Z_3			ca. 250	+	+		+	+					+		
	Z_4			256–260	+	+		+	+					+		
	Z_5				2	2	1	2	2							
I		A_I, B_I, $X_{0\beta}$		261–267	2	2	1	2	2		1					-Thr-Val-HOPro-Sar-MeVal
				237–238												-Thr-Val-Pro-Sar-MeVal
II		A_{II}, B_{II}		215–216	2	2		4	2							-Thr-Val-Sar-Sar-MeVal
																-Thr-Val-Sar-Sar-MeVal
III		A_{III}, B_{III}, $X_{0\tau}$		237–238	2	2	1	3	2							-Thr-Val-Sar-Sar-MeVal
																-Thr-Val-Pro-Sar-MeVal
IV		A_{IV}, B_{IV}, D_{IV}, C_1, I_1, X_1		235.5–236.5	2	2	2	2	2							-Thr-Val-Pro-Sar-MeVal
																-Thr-Val-Pro-Sar-MeVal
V		A_V, B_V, X_2		245–246.5	2	2	1	2	2			1				-Thr-Val-COPro-Sar-MeVal
																-Thr-Val-Pro-Sar-MeVal
VI		C_2		237–239	2	1	2	2	2	1						-Thr-alleu-Pro-Sar-MeVal
																-Thr-Val-Pro-Sar-MeVal
VII		C_3		232–235	2		2	2	2	2						-Thr-alleu-Pro-Sar-MeVal
																-Thr-alleu-Pro-Sar-MeVal

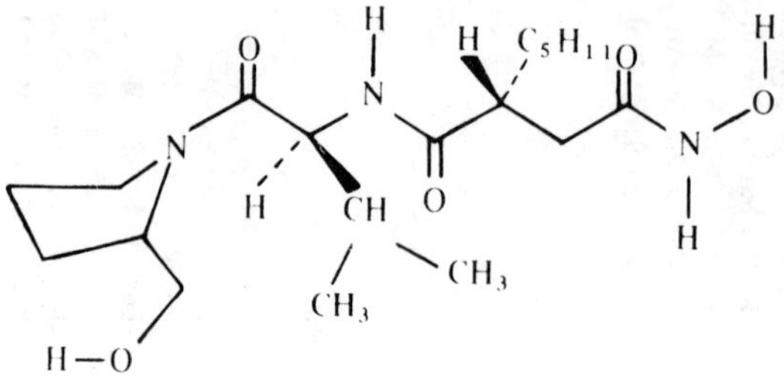

A-8

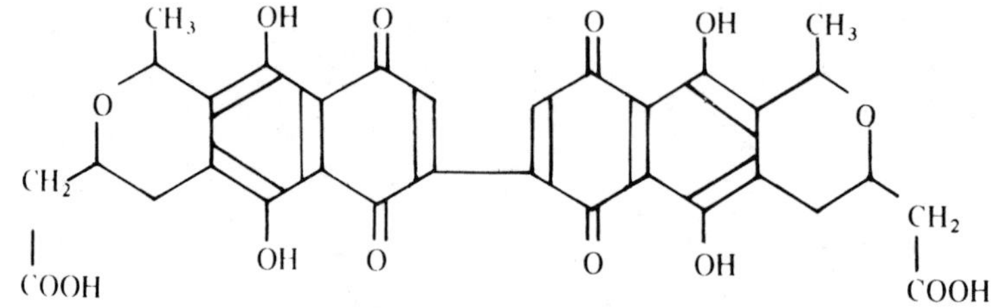

A-9

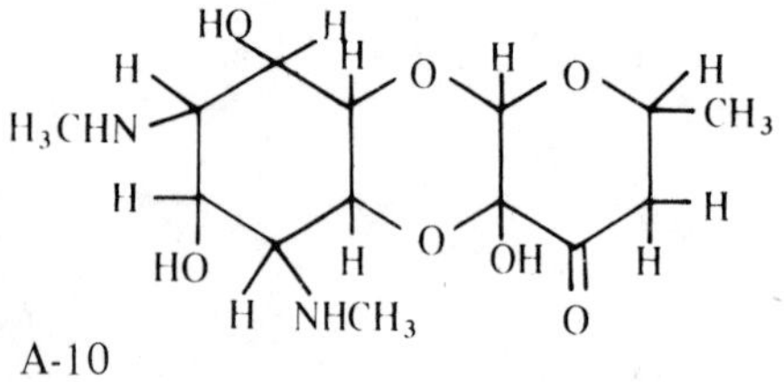

A-10

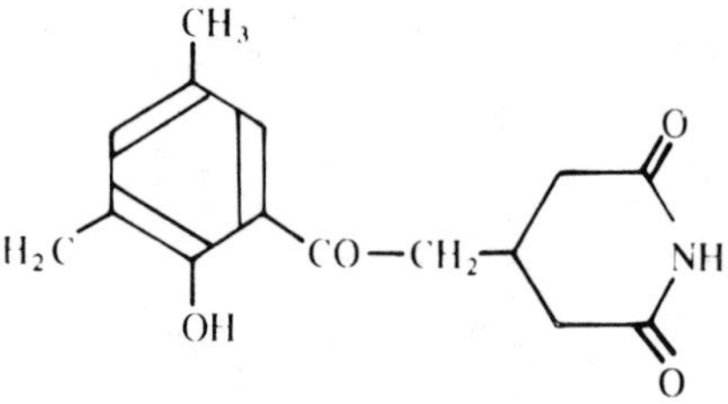

A-11

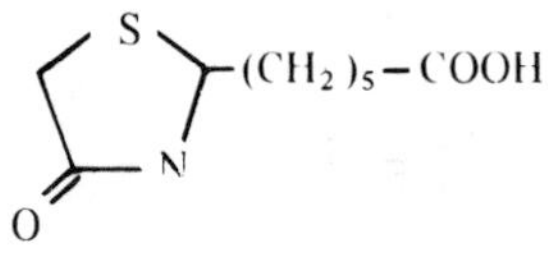

A-12

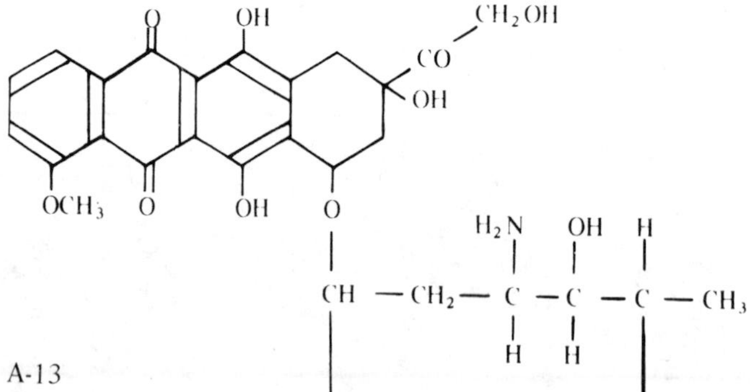

A-13

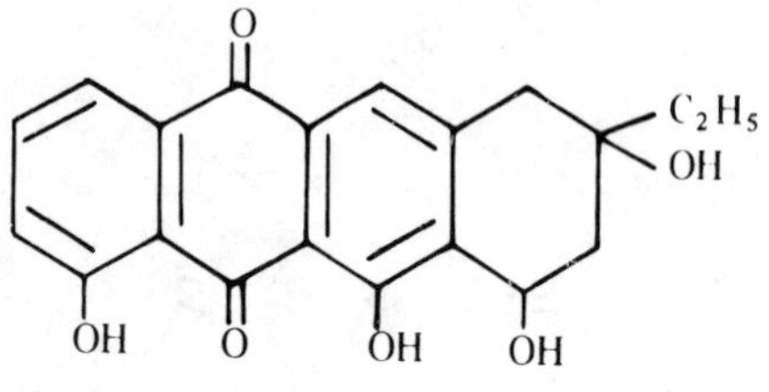

A-14   Acid hydrolysis gave aklavinone (I) and basic
sugar ($C_8H_{17}O_4N$)

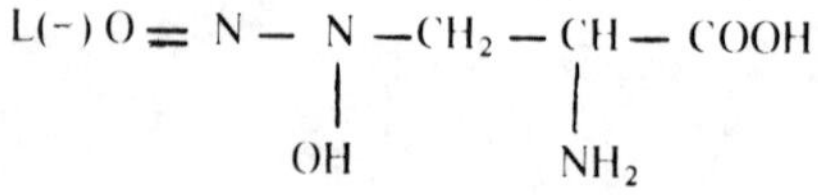

A-15

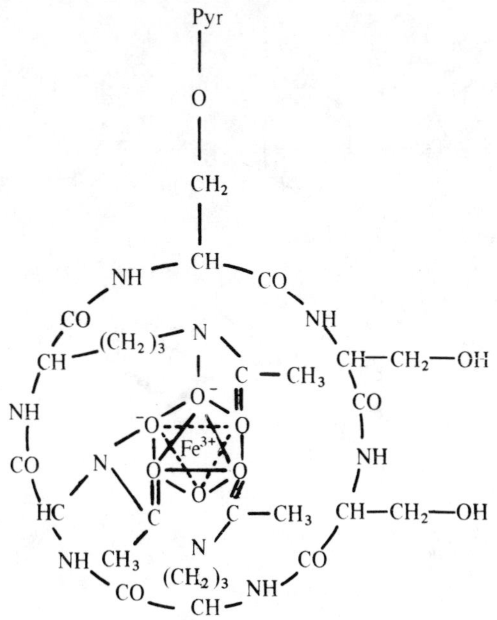

Pyr

Pry: Pyrimidine part contg. S.

A-16

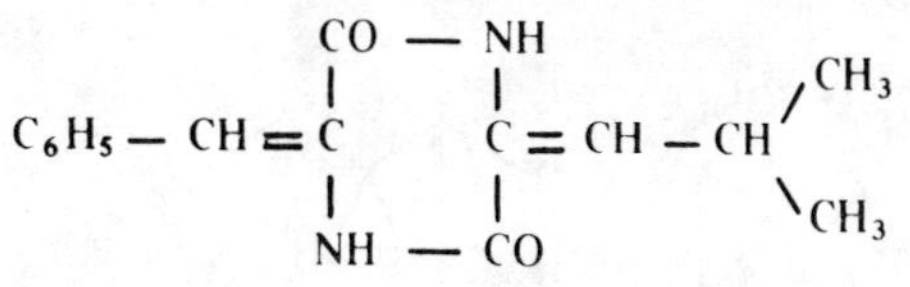

A-17

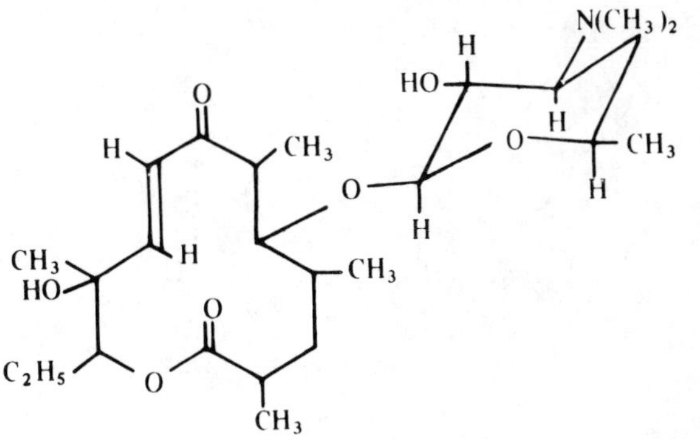

A-18

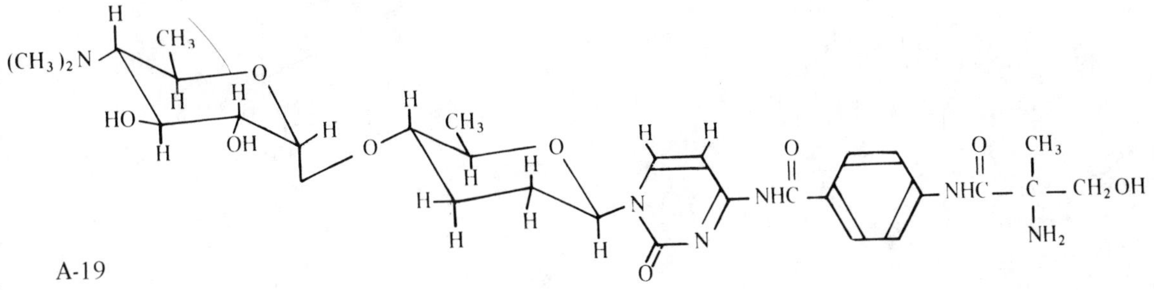

A-19

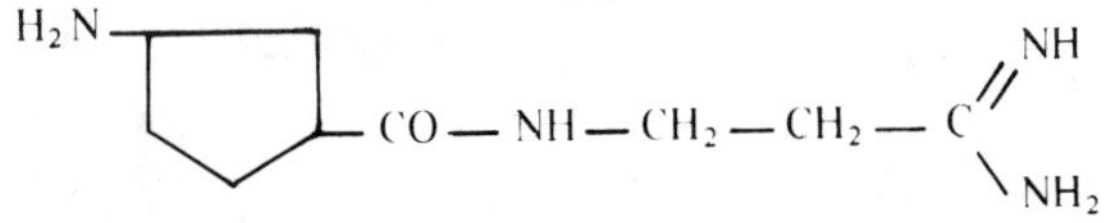

A-20

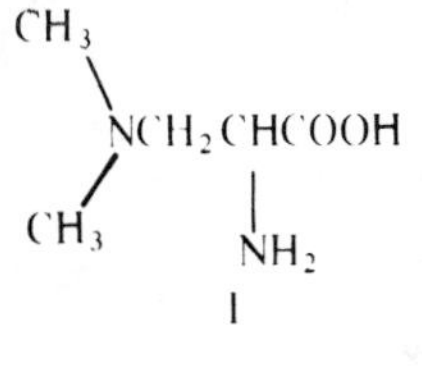

A-21

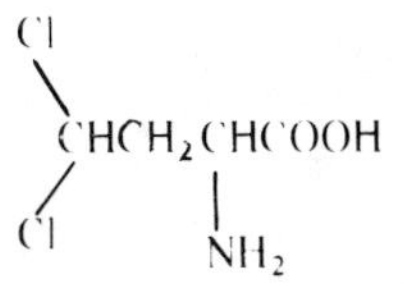

A-22

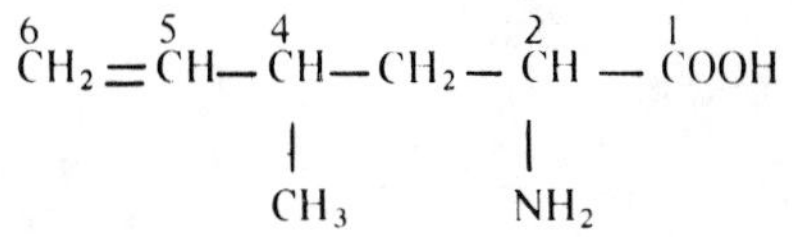

A-23

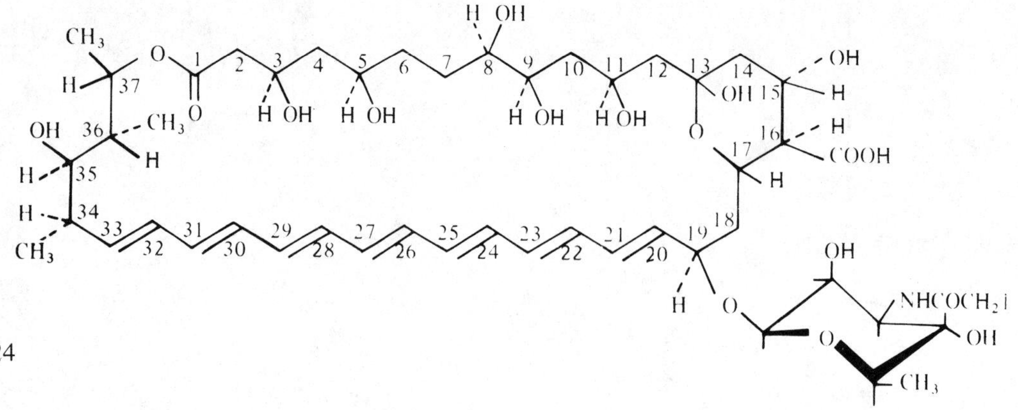

A-24

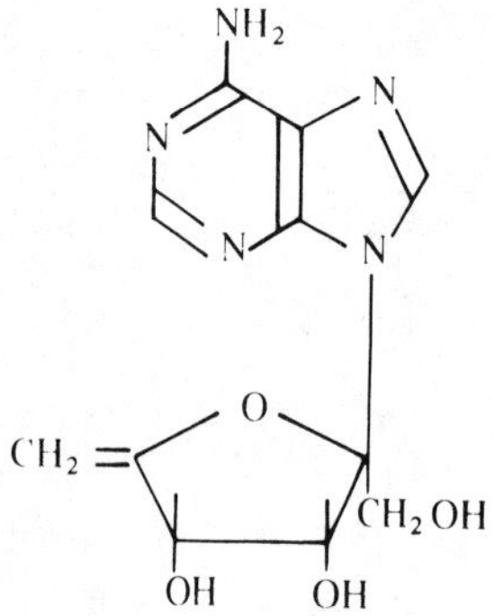

A-25

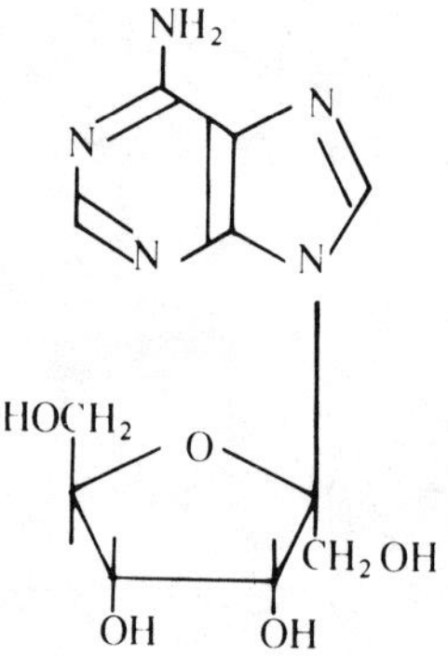

A-26

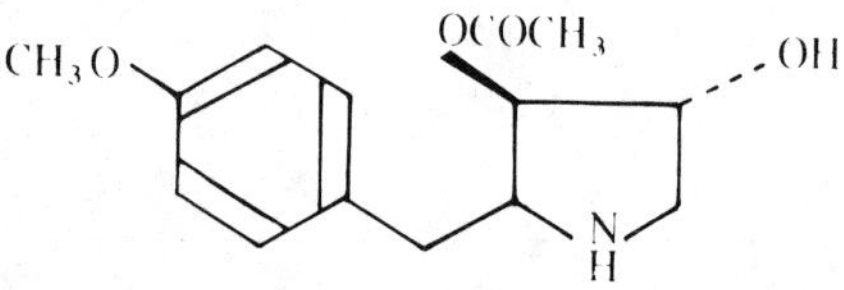

A-27

A-28

A-29

A-30

A-31

A-32

A-33

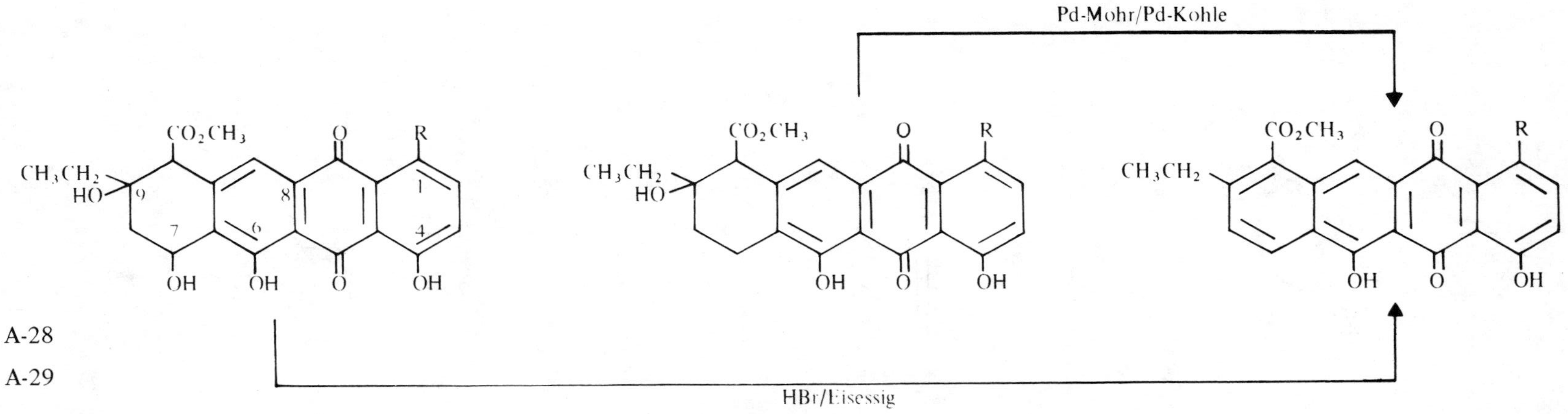
Pd-Mohr/Pd-Kohle
HBr/Eisessig
CO2CH3
CH3CH2
HO
R
O
OH
1
4
6
7
8
9

1:	Aklavinon; R = H	2:	7-Desoxy-aklavinon; R = H	3:	Bis-anhydro-aklavinon; R = H
1a:	ϵ-Pyrromycinon; R = OH	2a:	ζ-Pyrromycinon; R = OH	3a:	η-Pyrromycinon; R = OH

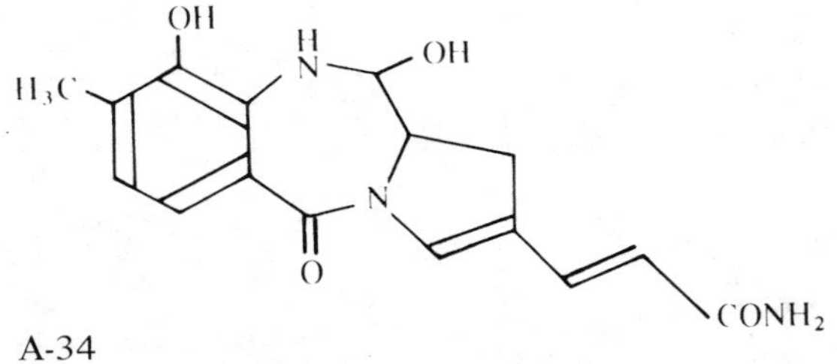
OH
H3C
H
N
OH
N
O
CONH2

A-34

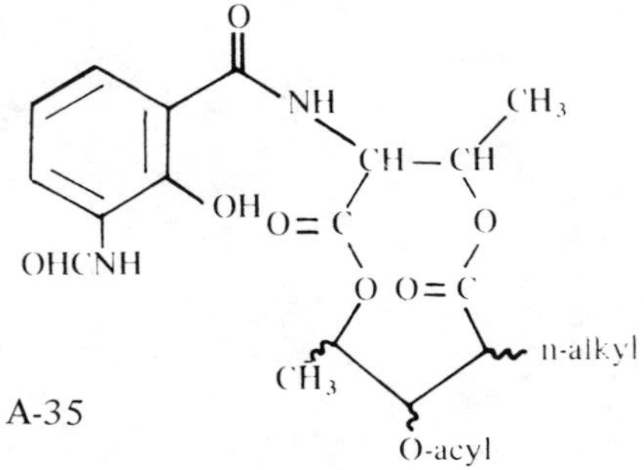
O
NH
CH3
CH—CH
OH
O=C
O
OHCNH
O O=C
n-alkyl
CH3
O-acyl

A-35

a) or b) would explain absence of peak M_0 (identical with $M_{1,2}$ or $M_{3,4}$) all four would explain mass peak of 562 observed in intact antimycin A complex.

A-36

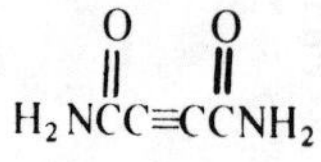

$$H_2NCC\equiv CCNH_2$$

(with two carbonyl oxygens shown above the structure)

A-37

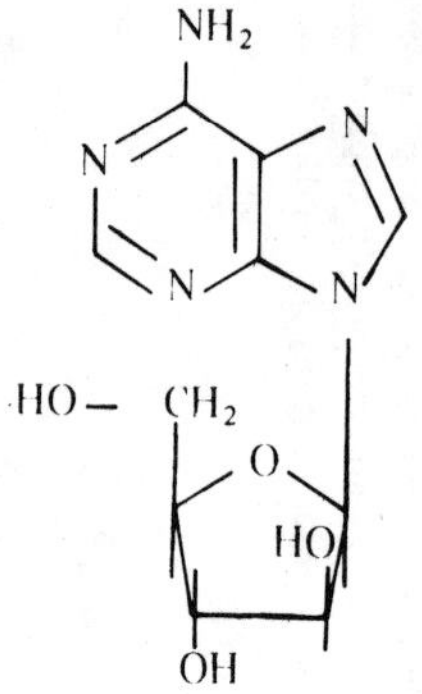

(adenosine structure with NH₂ purine base and ribose sugar)

A-38

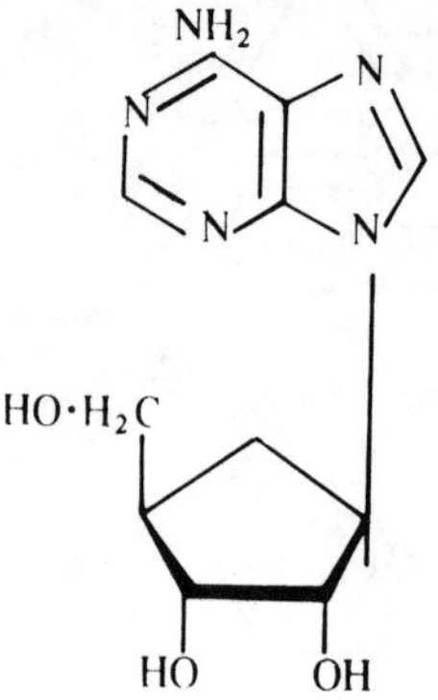

(adenine nucleoside analog structure with cyclopentane ring, HO·H₂C, HO, OH)

A-38/39

$$CH_3-C=\overset{\overset{\displaystyle H}{|}}{C}-COOH$$

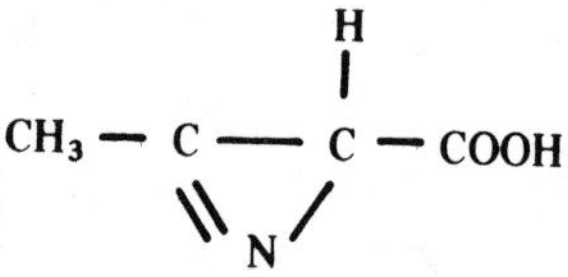

Consisting of five principal components, three of which have the following structures.

a: R₁, R₂ = D-alloisleucine
b: R₁ = D-alloisoleucine, R₂ = D-valine
c: R₁, R₂ = D-valine

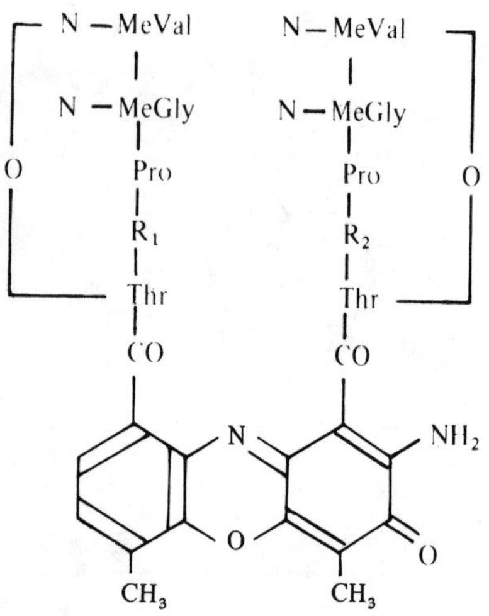

A-40

(structure with S-S ring, NHCOC₂H₅, N-CH₃, and O)

A-39

```
      N—MeVal        N—MeVal
       |              |
      N—MeGly        N—MeGly
       |              |
   O   Pro        Pro   O
       |              |
       R₁            R₂
       |              |
      Thr            Thr
       |              |
      CO             CO
```

(actinomycin phenoxazinone chromophore structure with NH₂, O, N, and two CH₃ groups)

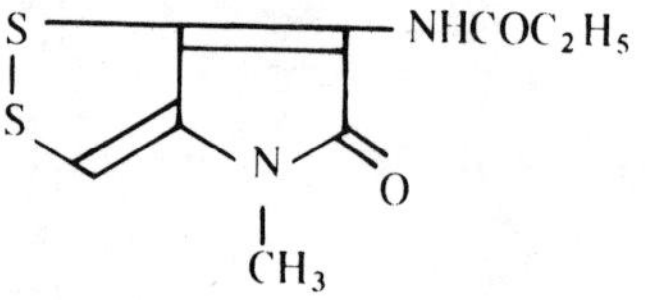

A-41

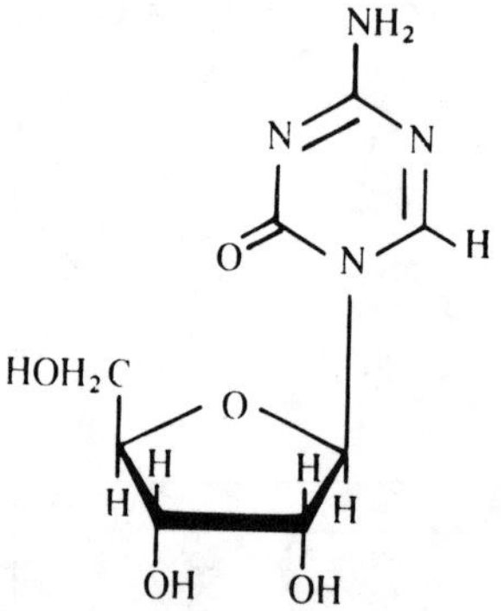

(5-azacytidine structure with NH₂ triazine base, O, and ribose with HOH₂C, OH, OH)

A-42

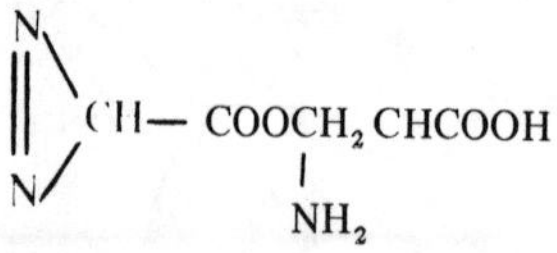

$$\underset{\underset{NH_2}{|}}{CH}-COOCH_2CHCOOH$$

(with triazole ring attached)

A-43

(imidazole ring with N, NH, and NO₂ substituent)

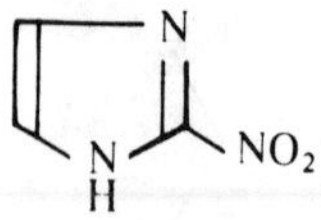

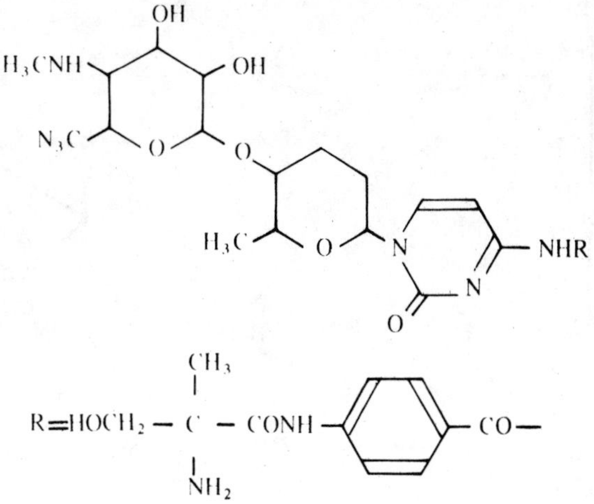

B-1

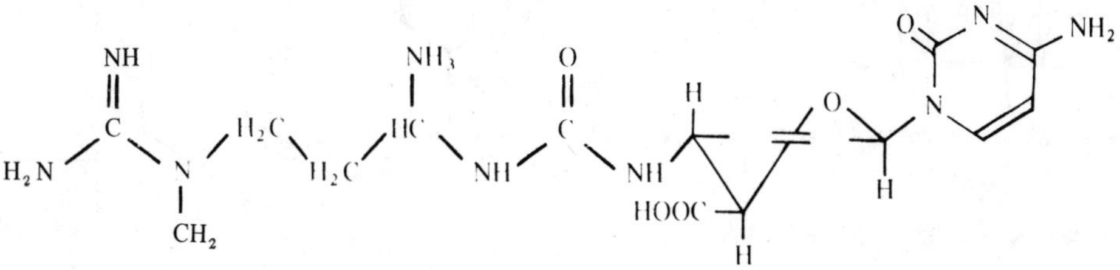

$$R=HOCH_2-\underset{\underset{NH_2}{|}}{\overset{\overset{CH_3}{|}}{C}}-CONH-\text{(benzene ring)}-CO-$$

B-2

B-3

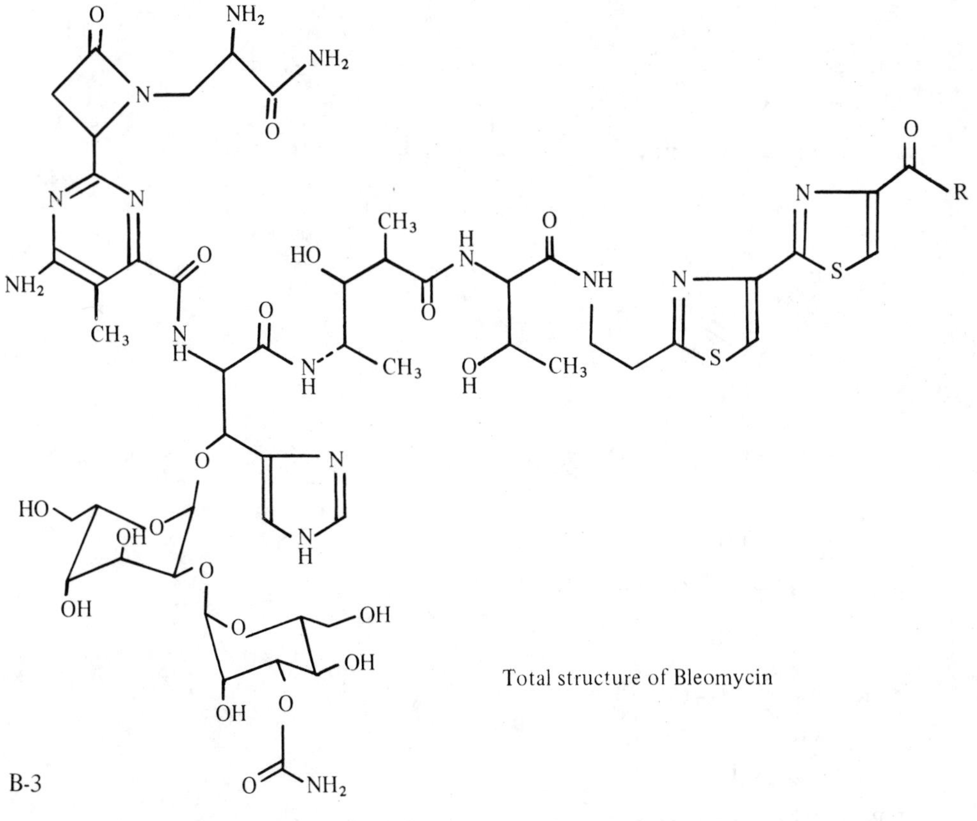

R; Amines

Total structure of Bleomycin

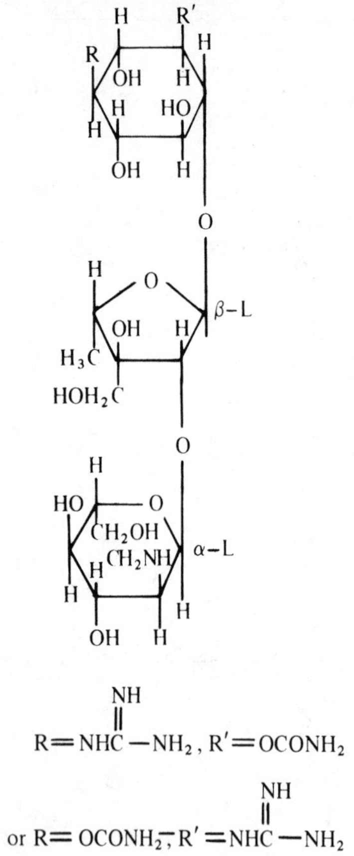

B-4/5

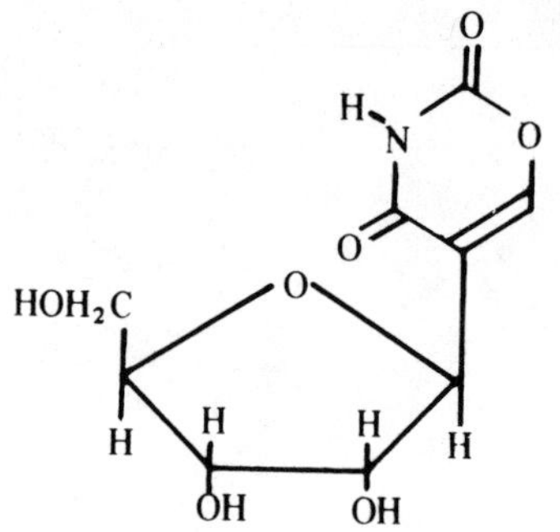

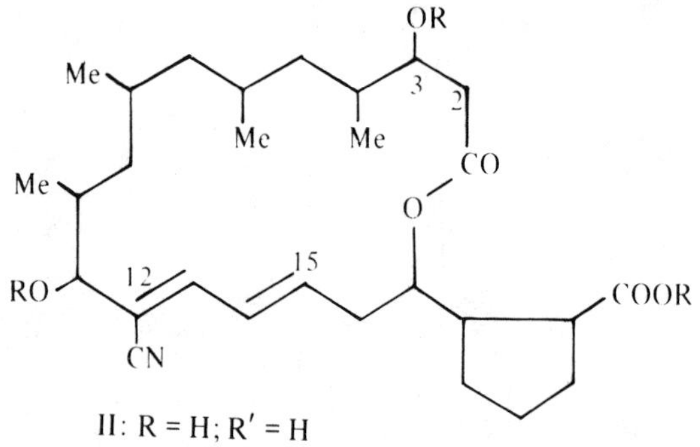

$R = NHC\!-\!NH_2$, $R' = OCONH_2$
or $R = OCONH_2$, $R' = NHC\!-\!NH_2$
(each with $\overset{NH}{\|}$ above the C)

B-4

II: R = H; R' = H

B-5

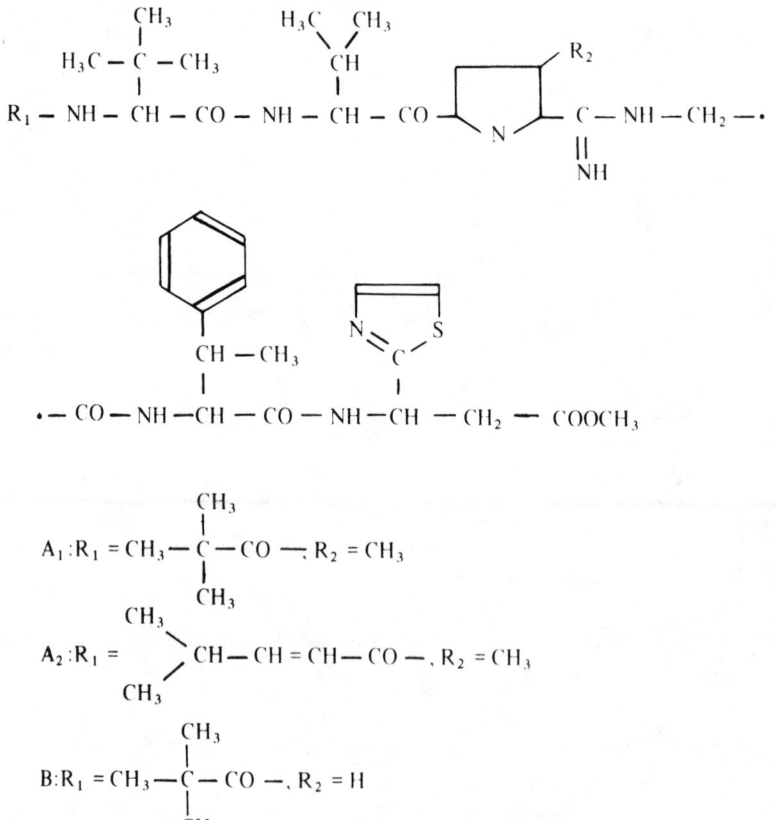

$A_1 : R_1 = CH_3\!-\!\underset{CH_3}{\overset{CH_3}{C}}\!-\!CO\!-\!,\ R_2 = CH_3$

$A_2 : R_1 = \underset{CH_3}{\overset{CH_3}{\diagup}}CH\!-\!CH = CH\!-\!CO\!-\!,\ R_2 = CH_3$

B-6, 7, 8

$B : R_1 = CH_3\!-\!\underset{CH_3}{\overset{CH_3}{C}}\!-\!CO\!-\!,\ R_2 = H$

B-9 NO₂CH₂CH₂COOH

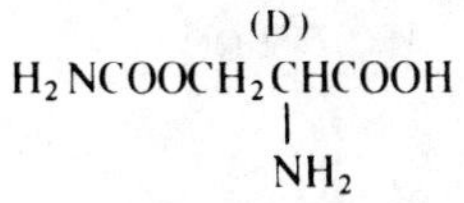

$$NO_2CH_2CH_2COOH$$

(D)

$$H_2NCOOCH_2CHCOOH$$
$$\underset{NH_2}{|}$$

C-2

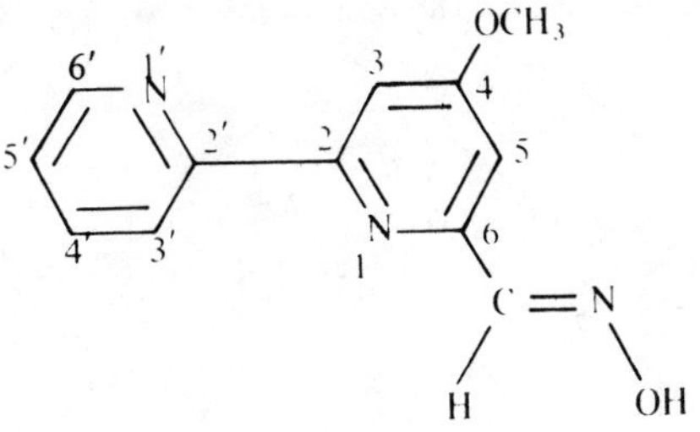

C-1

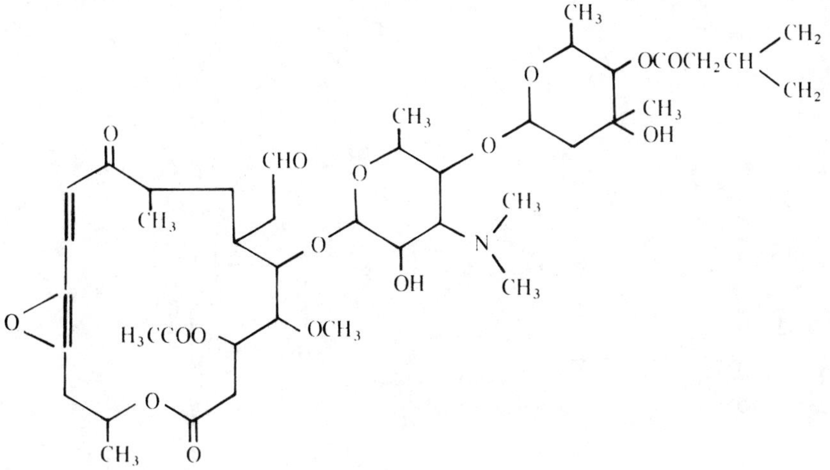

C-3

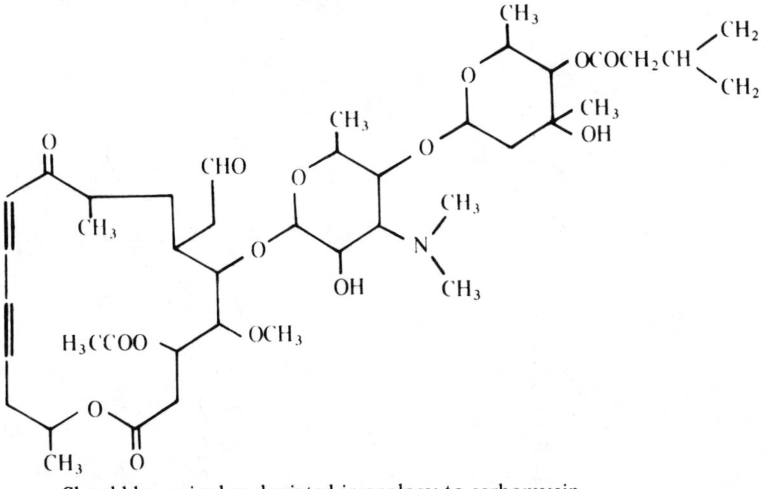

Should be revised as depicted in analogy to carbomycin.

C-4

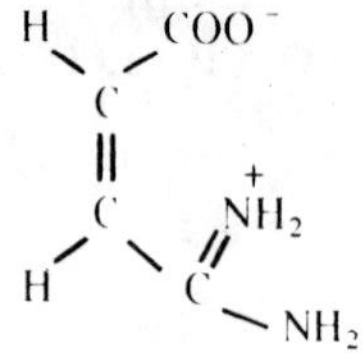

C-5

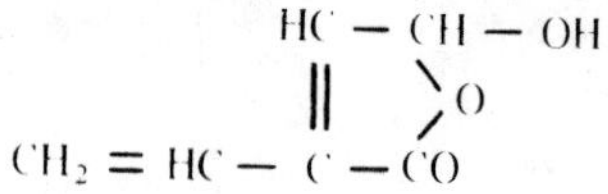

C-6

Hydrolysis product

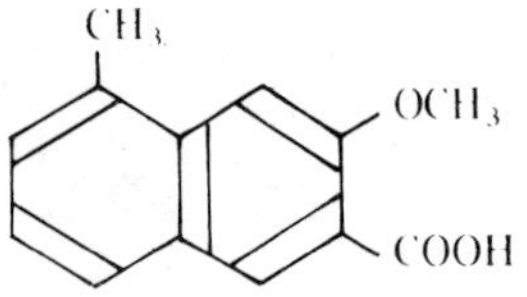

C-7

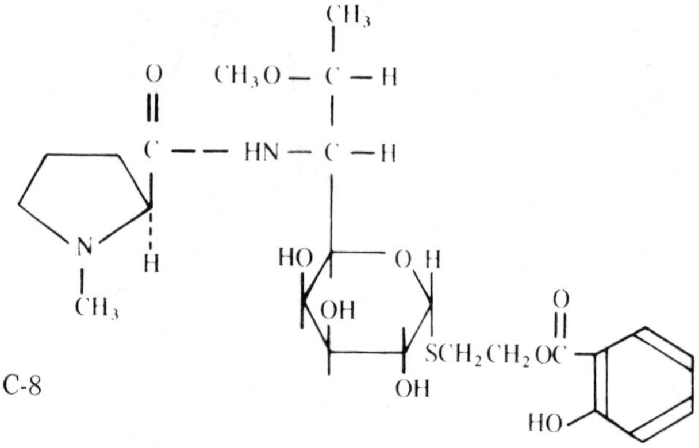

C-8

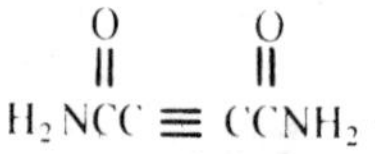

C-9

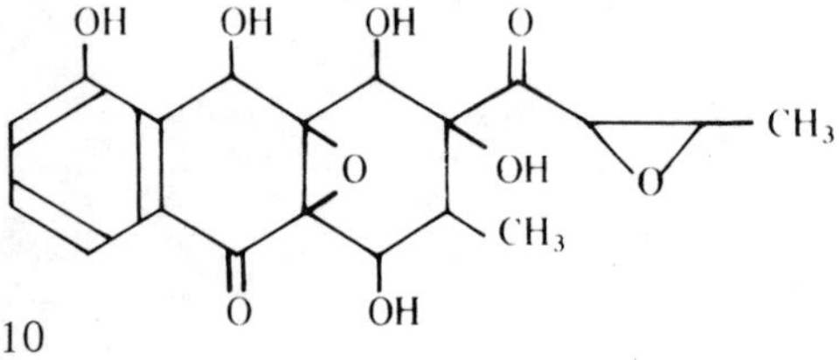

C-10

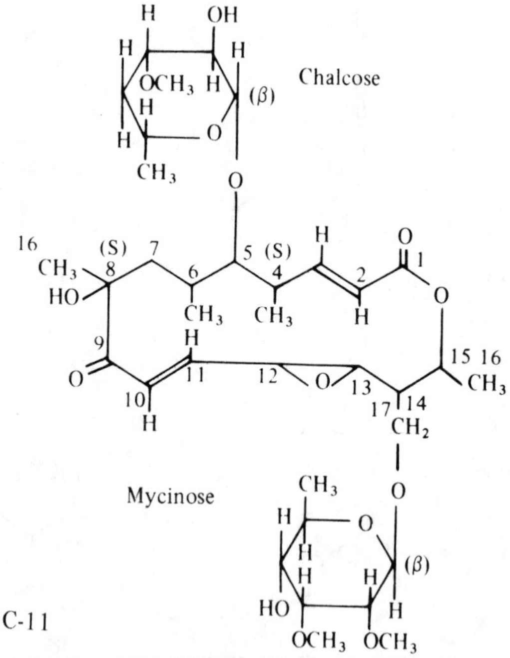

Chalcose (β)

Mycinose (β)

C-11

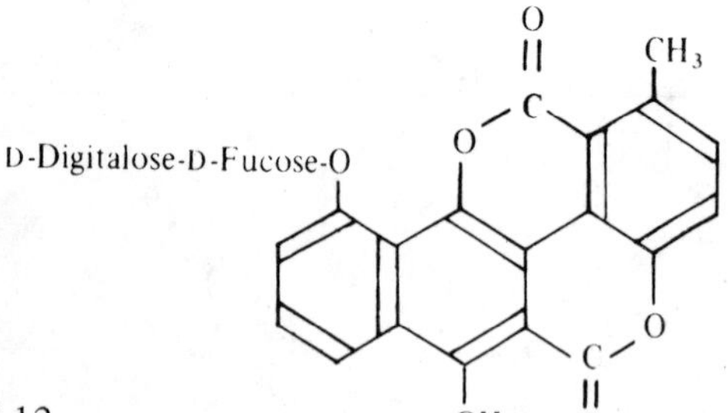

D-Digitalose-D-Fucose-O

C-12

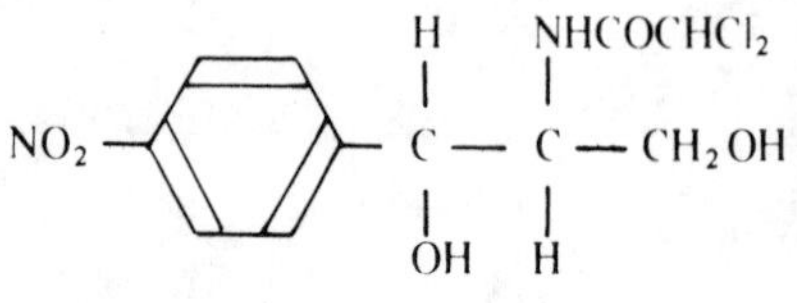

C-13

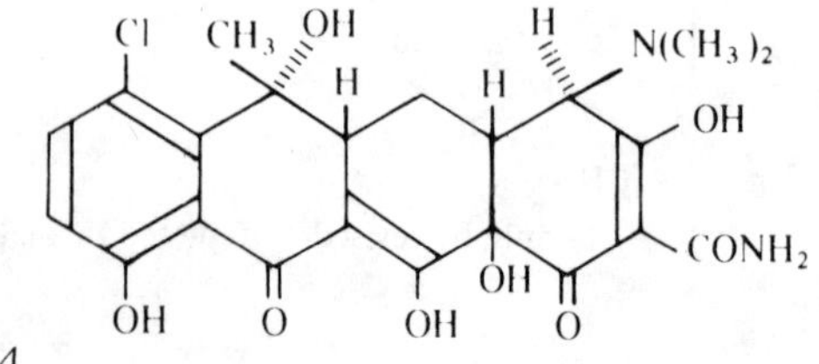

C-14

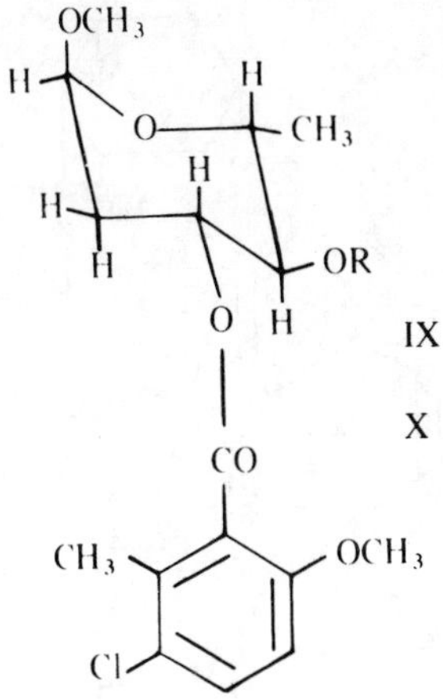

C-15

IX   R = H

X   R = CO–CH₃

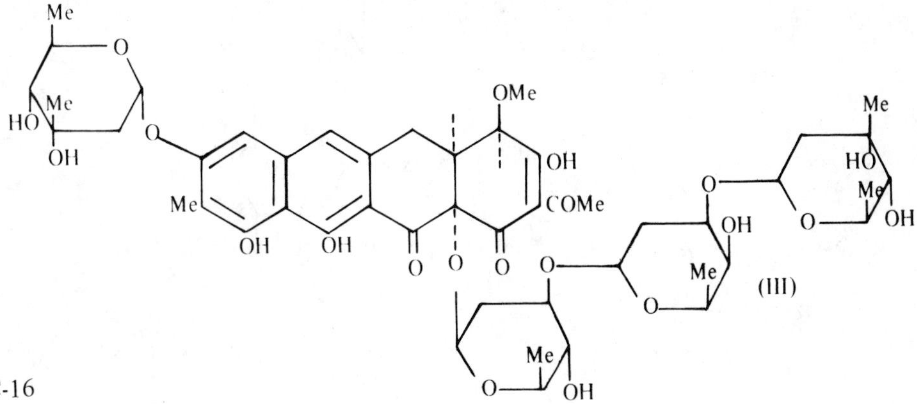

C-16

(III)

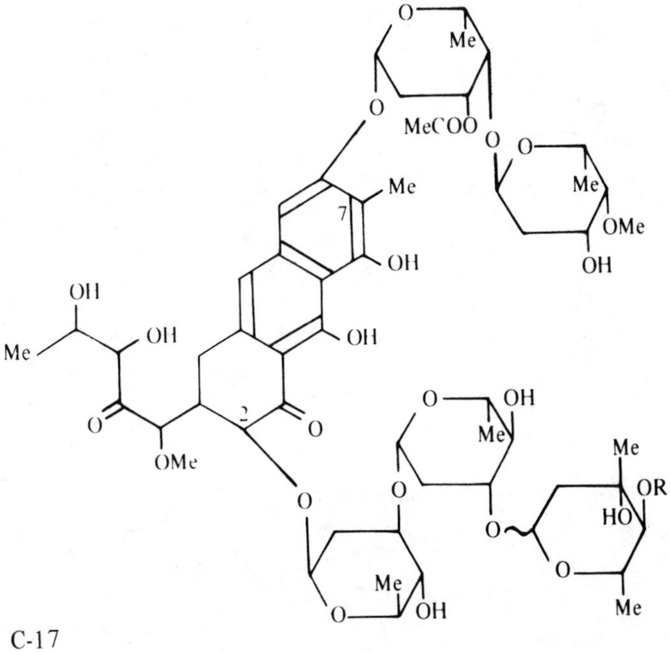

C-17

R = isobutyryl

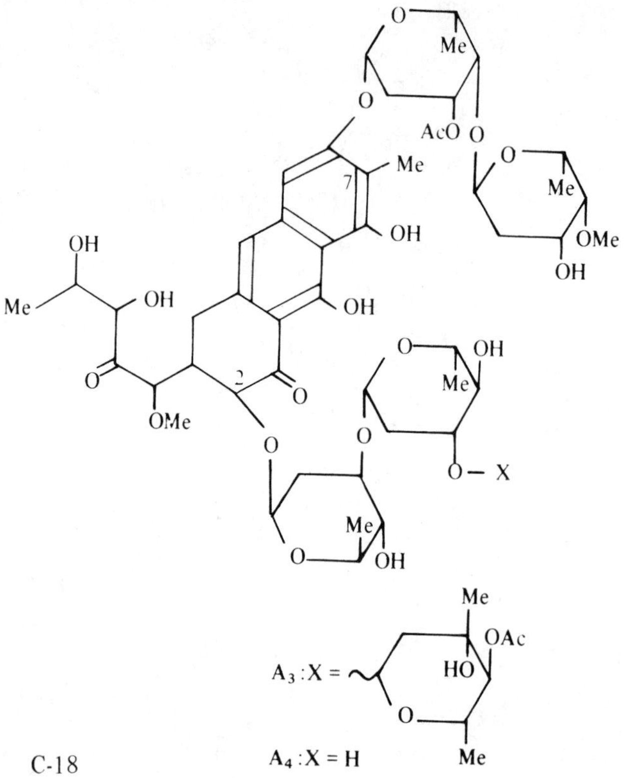

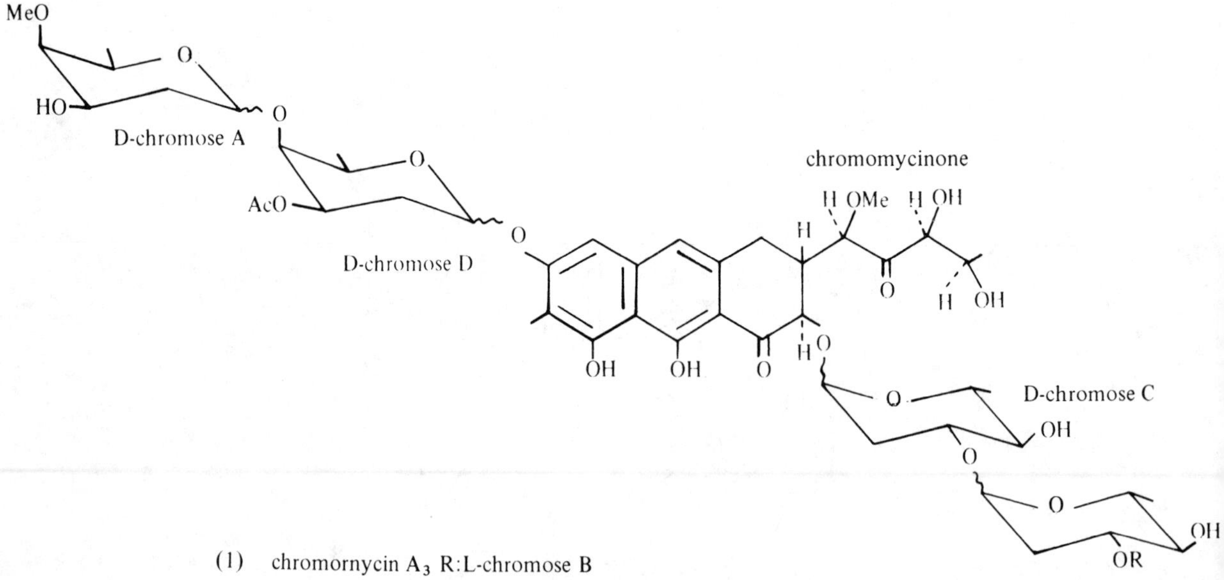

C-18

$A_3 : X = $

$A_4 : X = H$

(1)   chromornycin A_3 R:L-chromose B

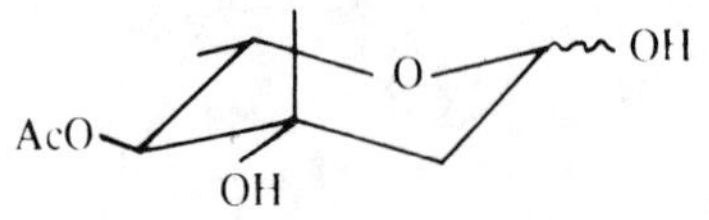

(2) chromomycin A_2 R:L-chromose B′

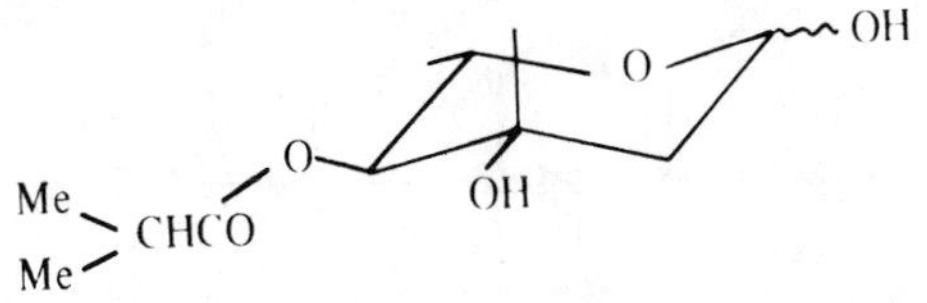

(3) chromomycin A_4 R:H

(13) monodeacetylchromomycin A_3 R:deacetylchromose B

(14) CHR–DCCA≡chromomycin A_4 (3) R:H

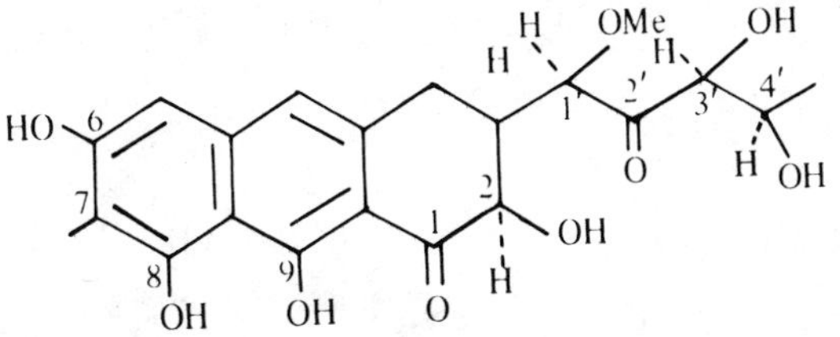

(4)

C_2–R
C_3–S   A–D–CHR–C–B
$C_{1'}$–R   (9)
$C_{3'}$–S   A–D–CHR–C–C–B
$C_{4'}$–R   (9a)

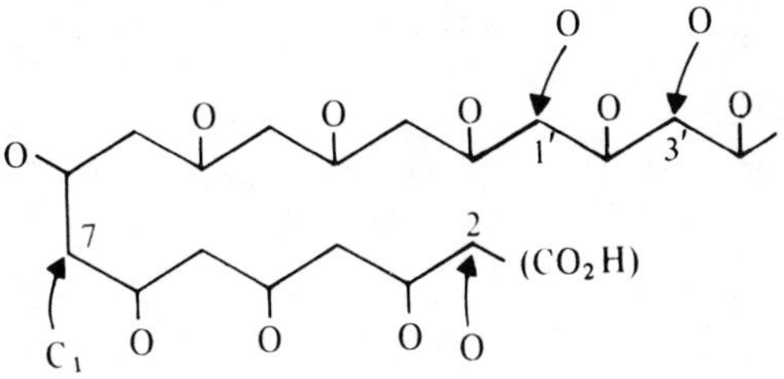

(5)

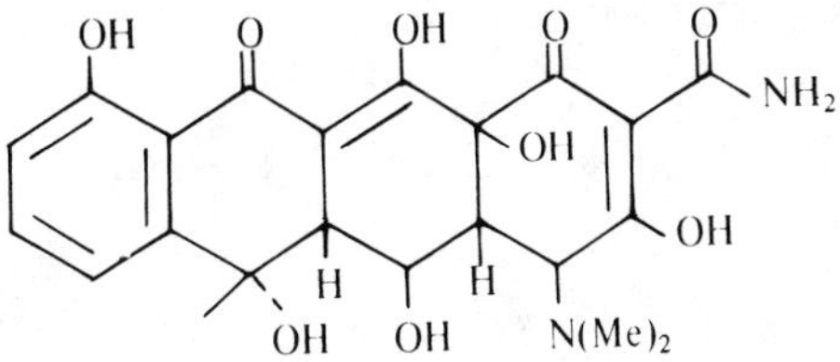

(6)

C-19, 20, 21

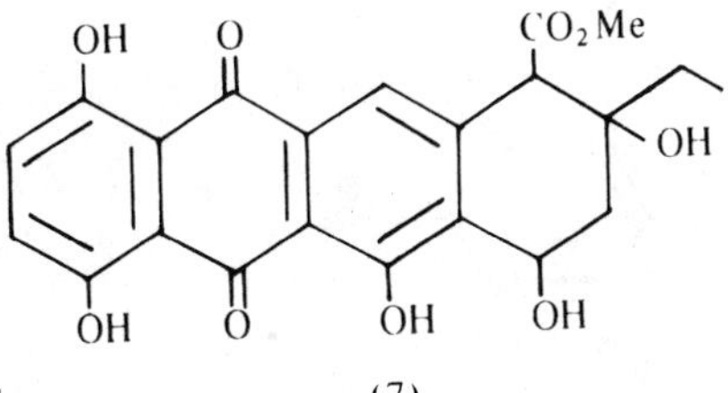

(7)

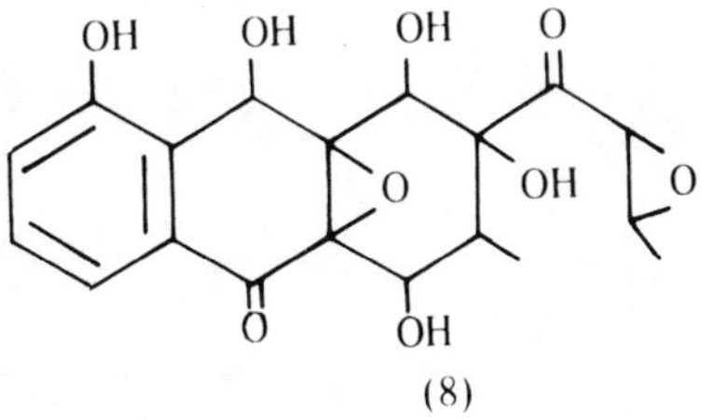

(8)

C-22

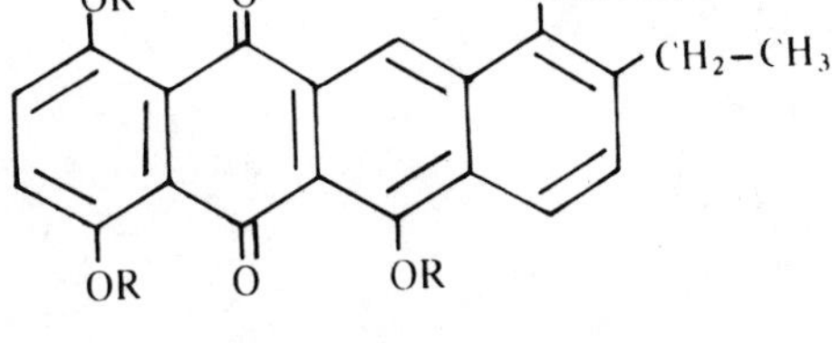

(I) R = H

C-23

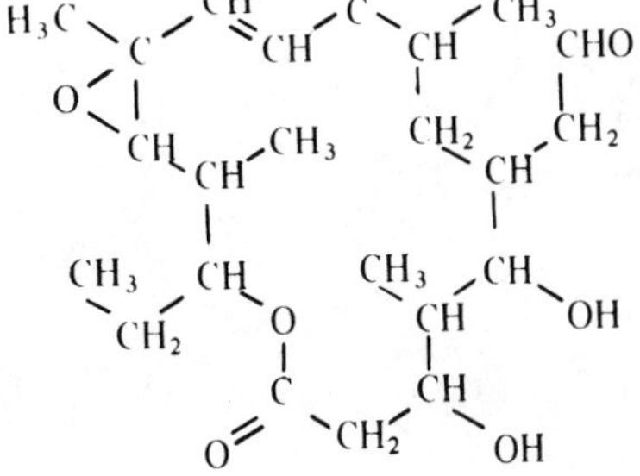

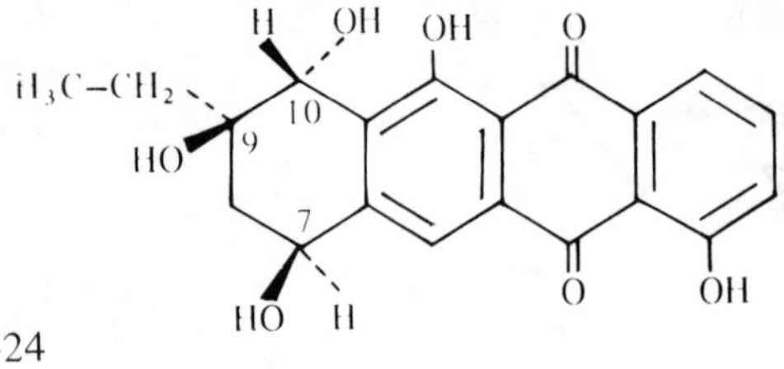

C-24

C-25

C-7; OH⟹H or C-10; OH⟹H in C-24

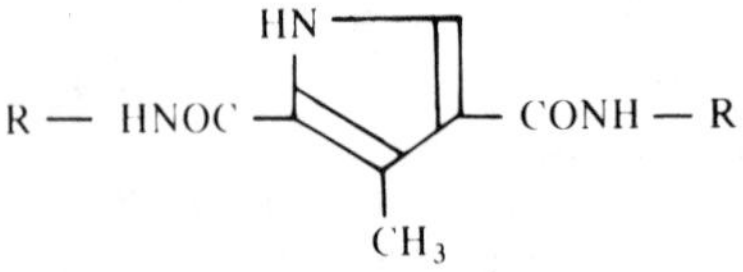

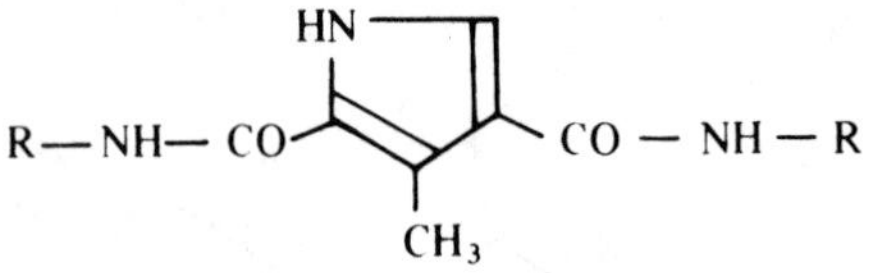

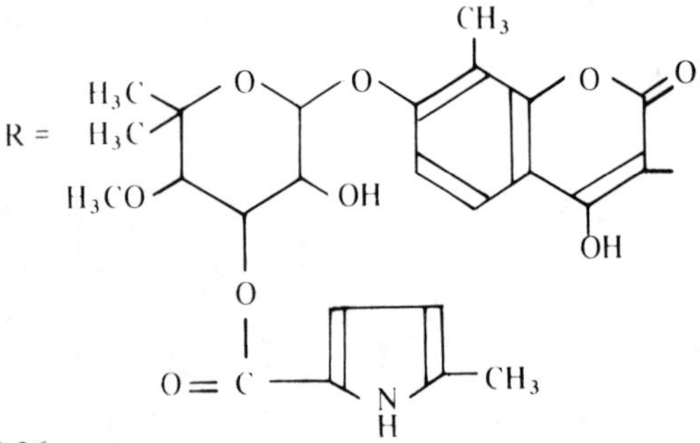

C-26

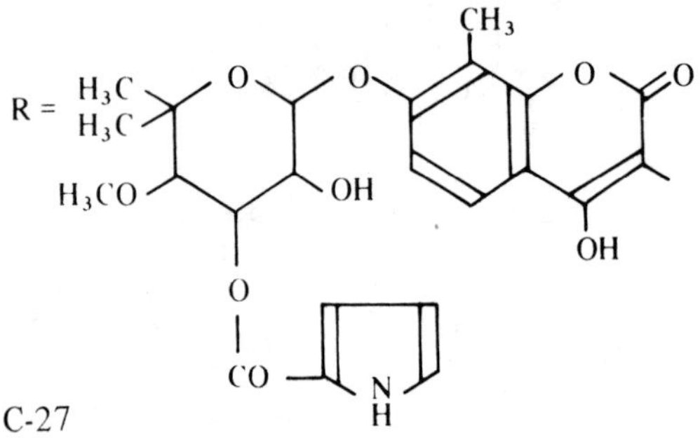

C-27

Schematically structure (*m*, *n* and *p* are the number of molecule):

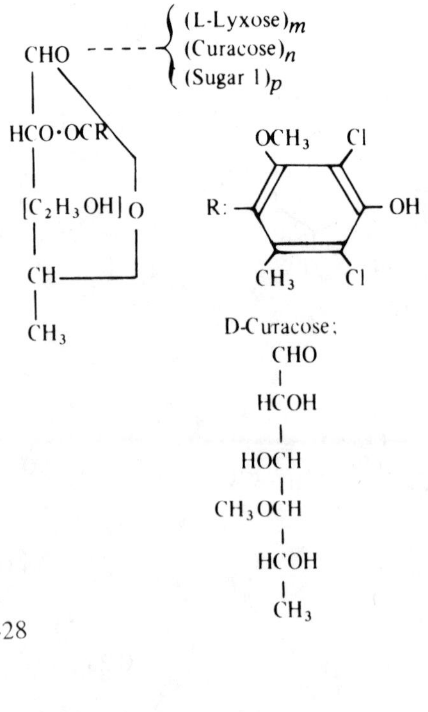

(L-Lyxose)$_m$
(Curacose)$_n$
(Sugar 1)$_p$

D-Curacose:

CHO
|
HCOH
|
HOCH
|
CH$_3$OCH
|
HCOH
|
CH$_3$

C-28

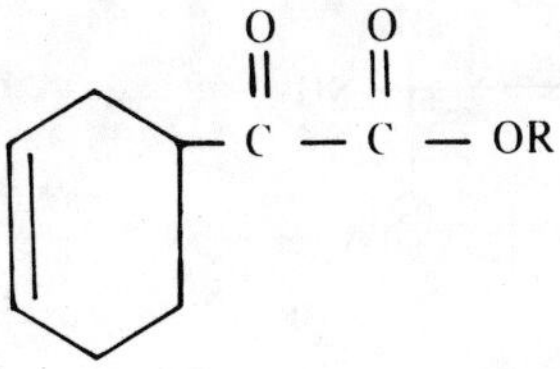

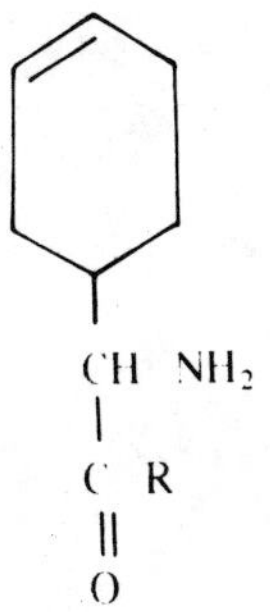

Compd. II, R
  $CH_3(CH_2)_9O-$
  C_6H_5O-
  $2-CH_3C_6H_4O-$
  $4-CH_3(CH_2)_3C_6H_4O-$
  $C_6H_5CH_2O-$
  $4-O_2N-C_6H_4O-$
Compd. III, R
  $CH_3(CH_2)_9O-$
  C_6H_5O-
  $2-CH_3C_0H_4O-$
  $4-CH_3(CH_2)_3C_6H_4O-$
  $C_6H_5CH_2O-$
  $4-O_2N-C_6H_4O-$

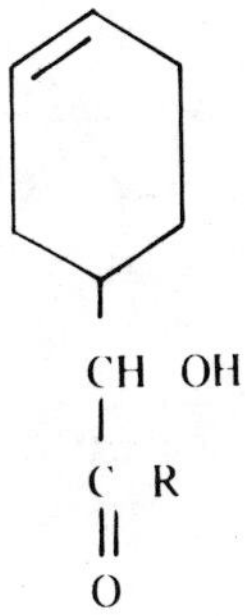

Compd. II   Compd. III

C-29

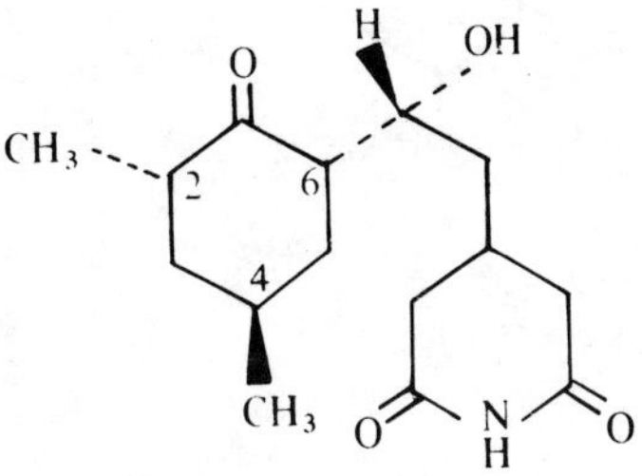

C-30

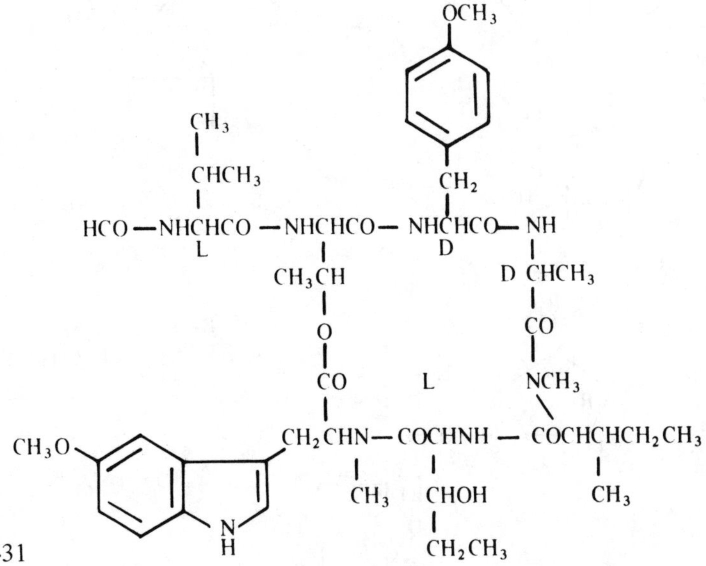

C-31

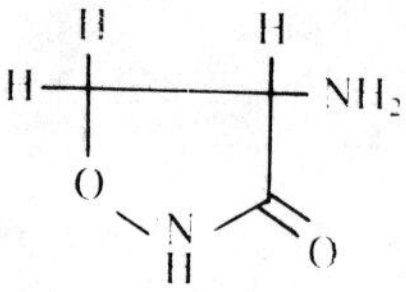

C-32

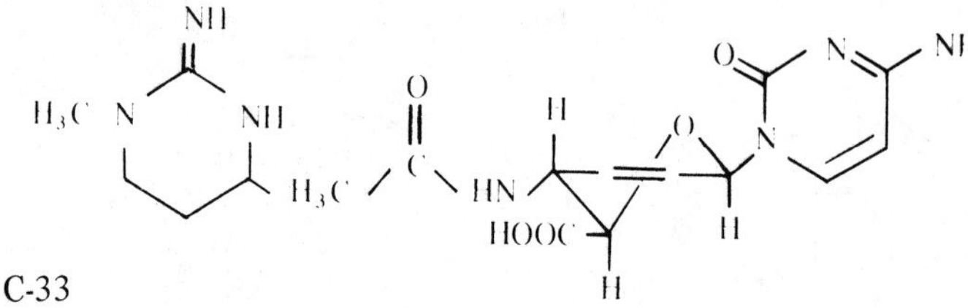

C-33

C-34   Acid hydrolysis gives cytoviridin, $(C_{10}H_{12}O_4N_4)$

Hydrolysis products.

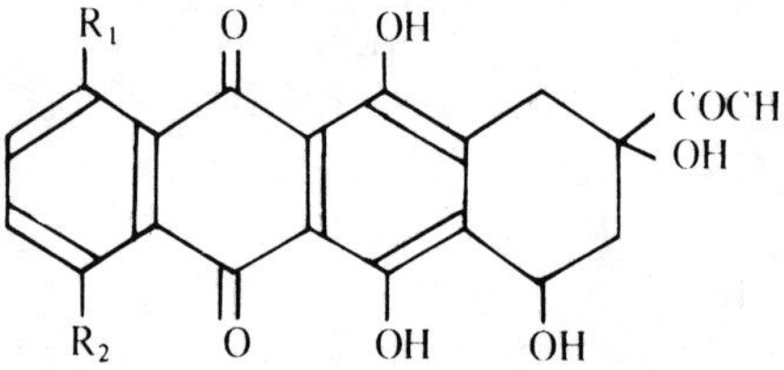

Daunomycinone
$R_1 = H, R_2 = OCH_3$
or $R_1 = OCH_3, R_2 = H$

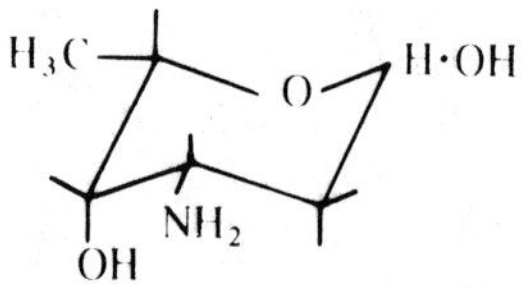

Daunosamine

D-1

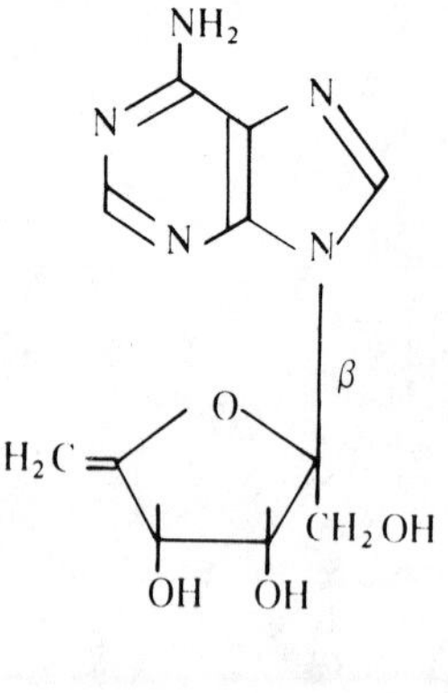

D-2

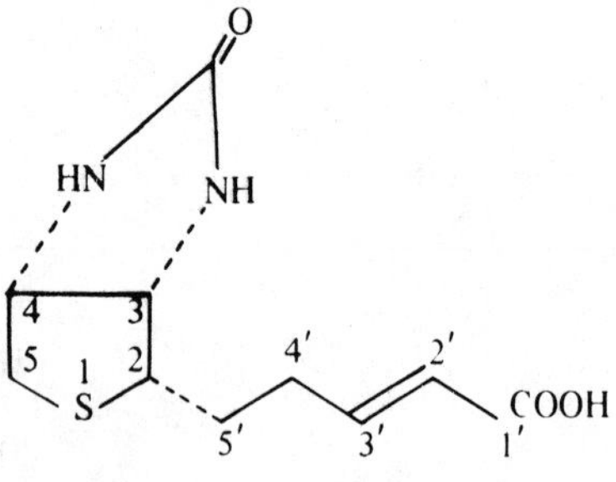

D-3

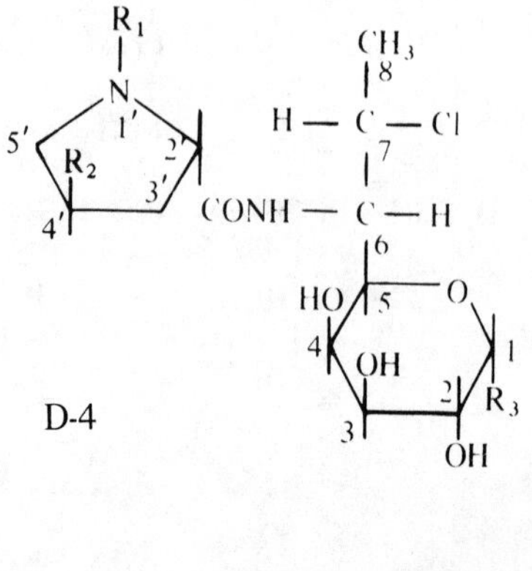

$R_1 = H$
$R_2 = CH_2CH_2CH_3$
$R_3 = SCH_3$

D-4

Proposed partial structure for demetric acid.

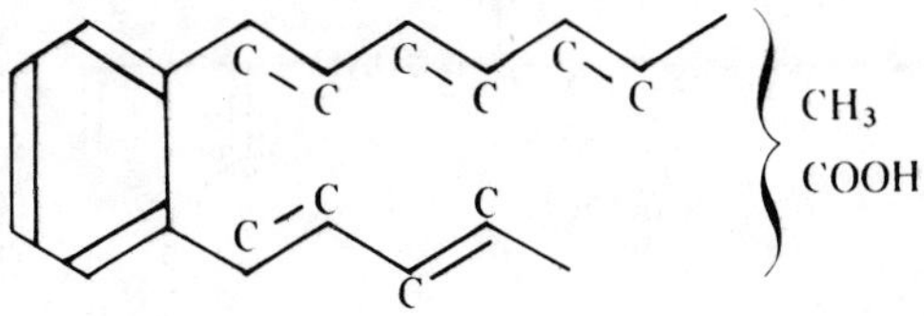

D-5

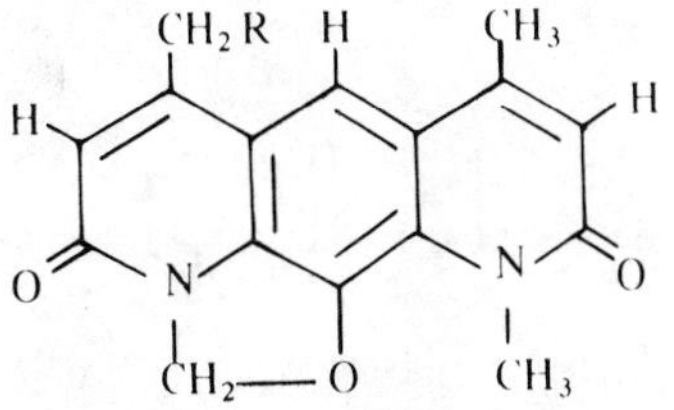

R = H, compound I
    (deoxynybomycin)
R = OH, nybomycin

D-6

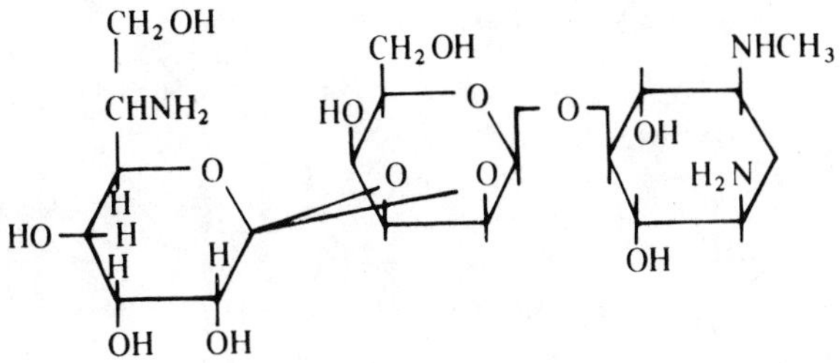

D-7

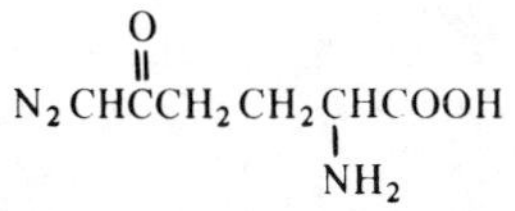

D-8

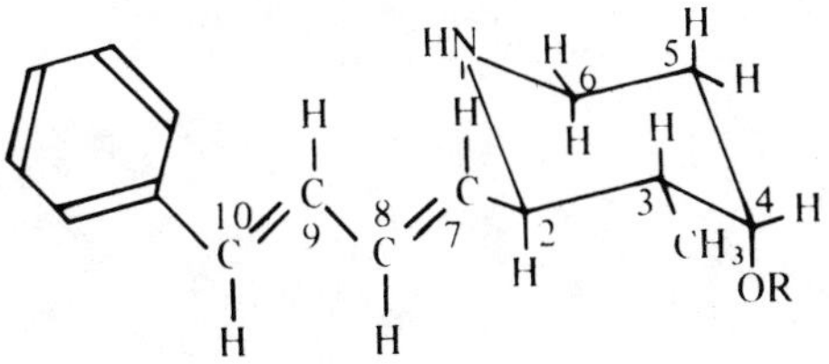

A: $R = OCCH(CH_3)_2$

B: $R = OCCH_3$

C: $R = H$

D-9, 10, 11

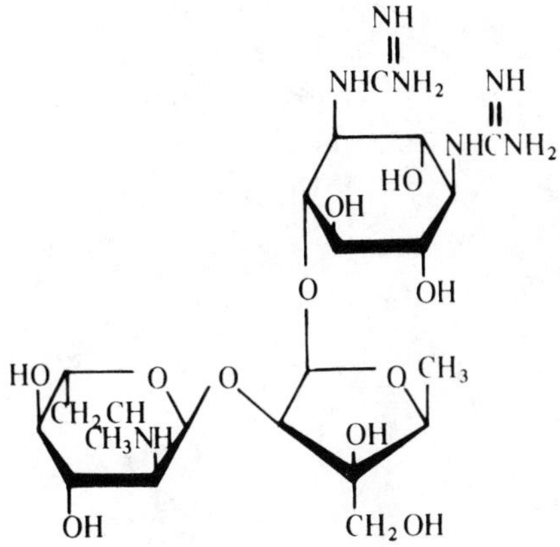

D-12

Scheme I

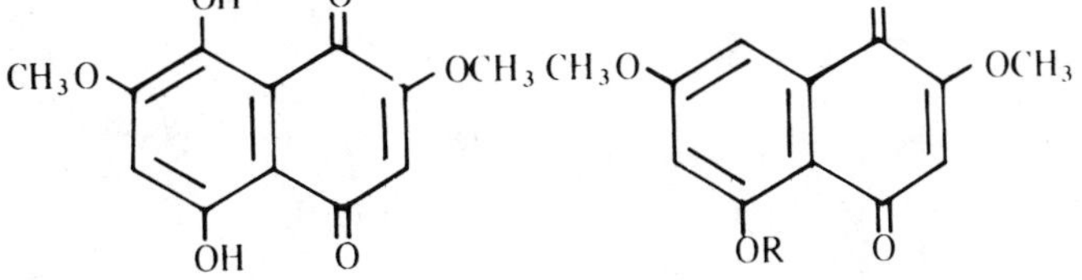

2

3, R = H
4, R = CH_3

D-13, 14

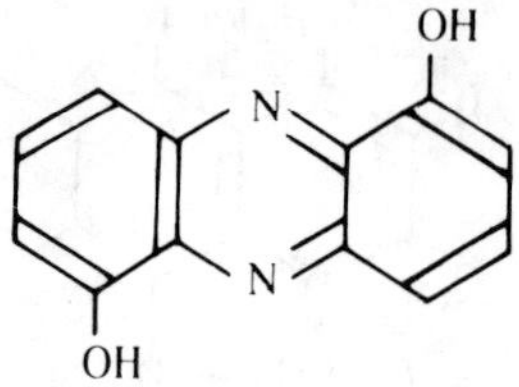

D-15

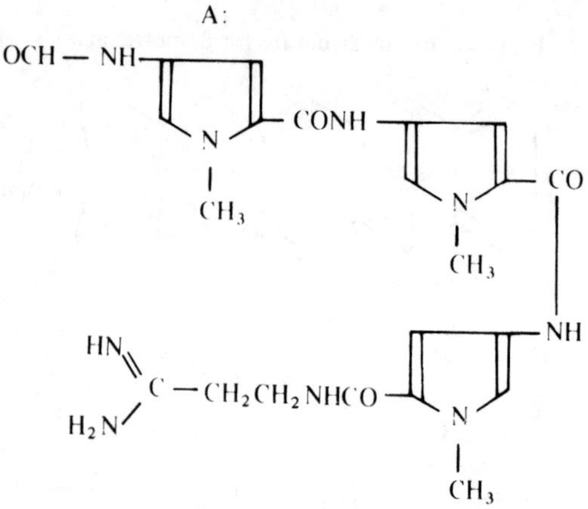

A:

D-16

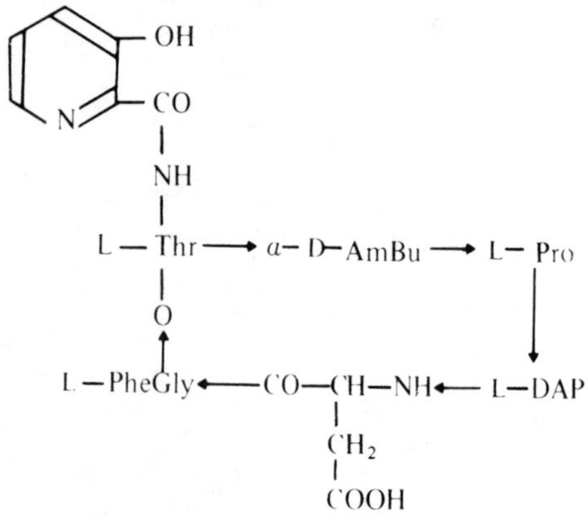

D-17

D-AmBu: D-α-aminobutyric acid

L-DAP: L-*p*-dimehtylamino-
              N-methylphenylalamine

$N_2CHCOCH_2CH_2CHCOOH$
                    $NHCOCH_3$

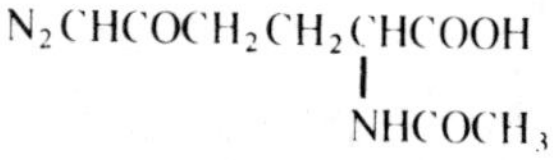

D-18

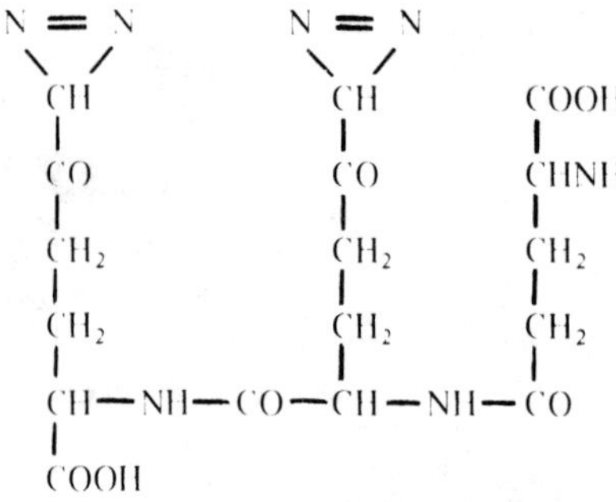

D-19

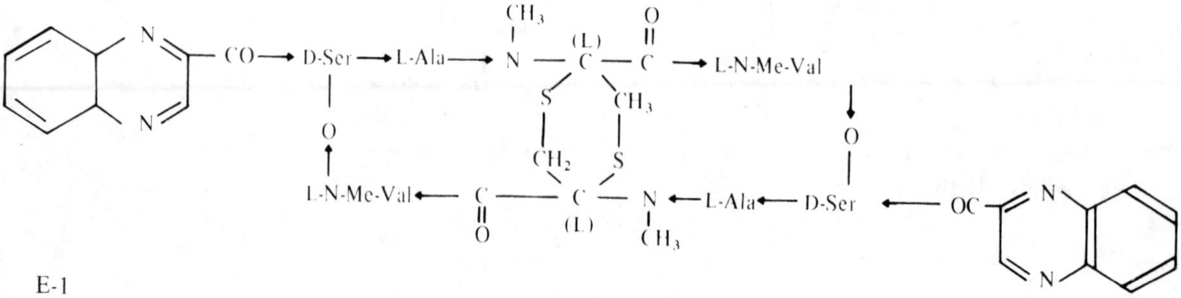

E-1

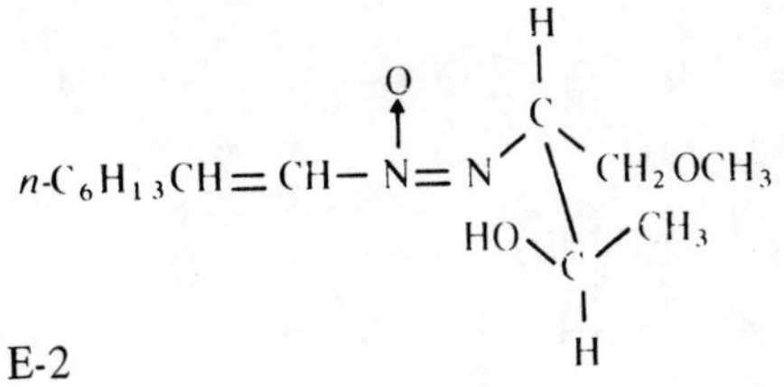

E-2

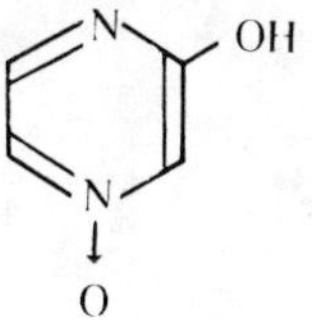

E-3

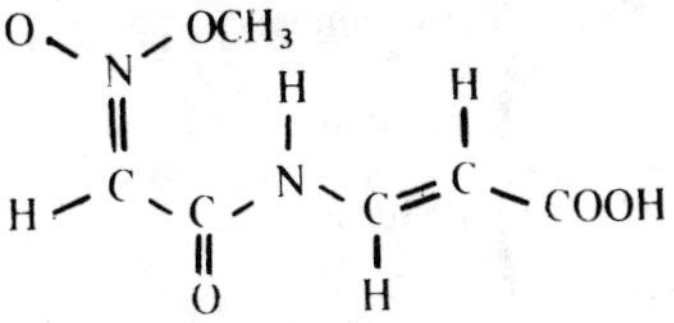

E-4

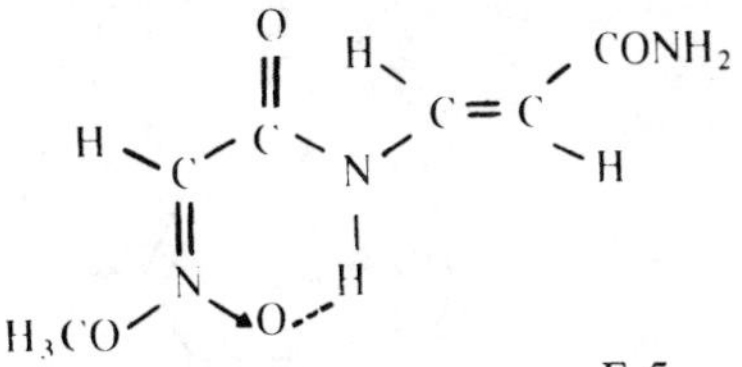

E-5

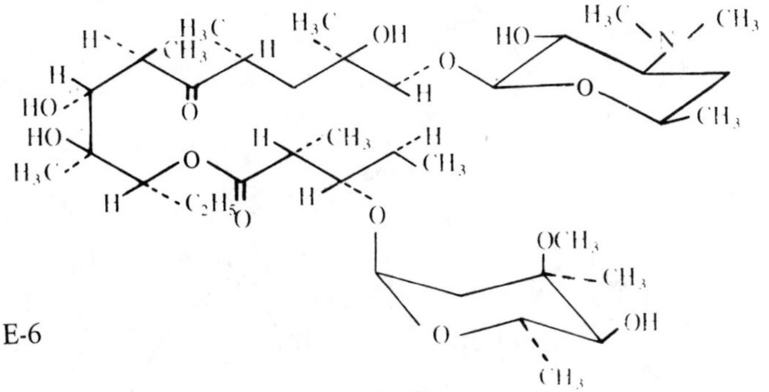

E-6

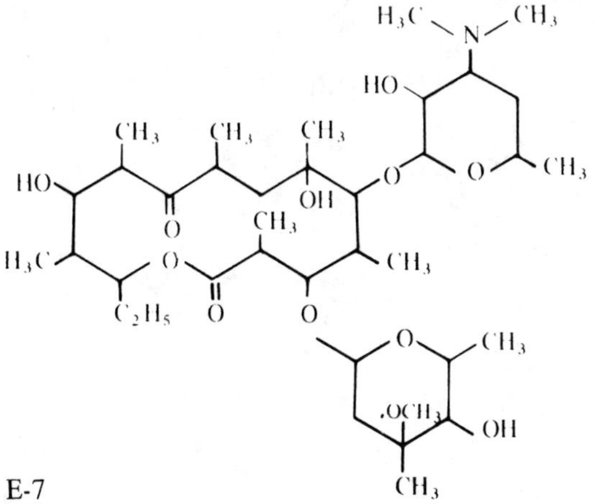

E-7

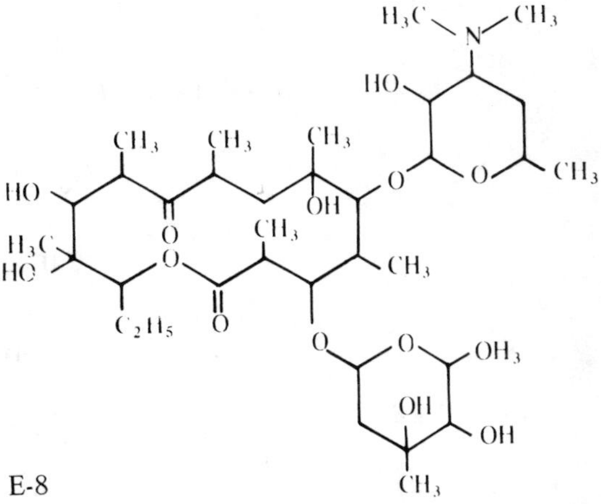

E-8

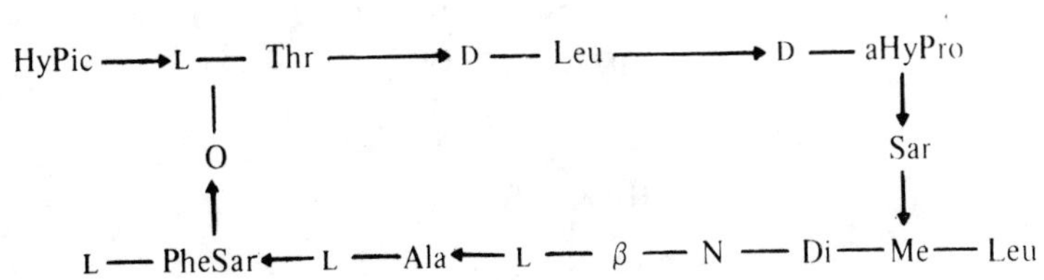

E-9

HyPic = 3-hydroxypicolinic acid
PheSar = α-phenylsarcosine
aHyPro = allo-hydroxyproline

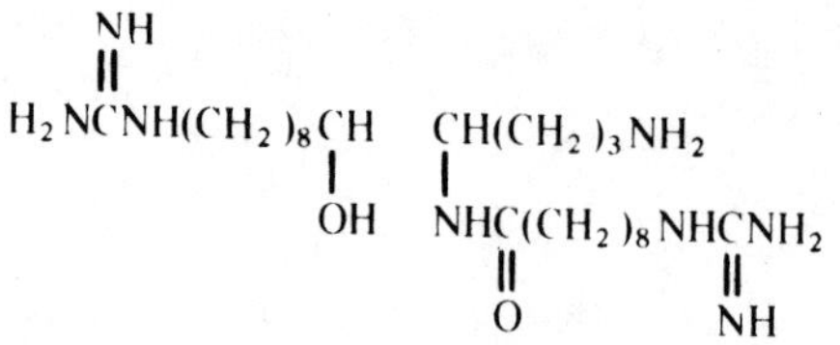

$$\text{H}_2\text{NCNH(CH}_2)_8\text{CH} \quad \text{CH(CH}_2)_3\text{NH}_2$$

NH (on first C)

E-10

$$\text{OH} \quad \text{NHC(CH}_2)_8\text{NHCNH}_2$$

$$\text{O} \quad \text{NH}$$

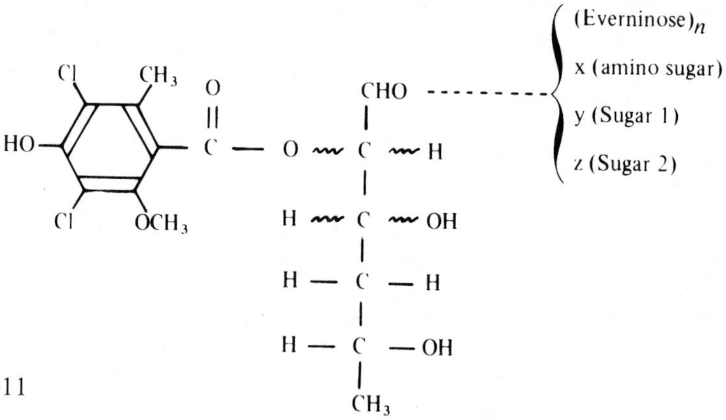

{ (Everninose)$_n$
x (amino sugar)
y (Sugar 1)
z (Sugar 2) }

E-11

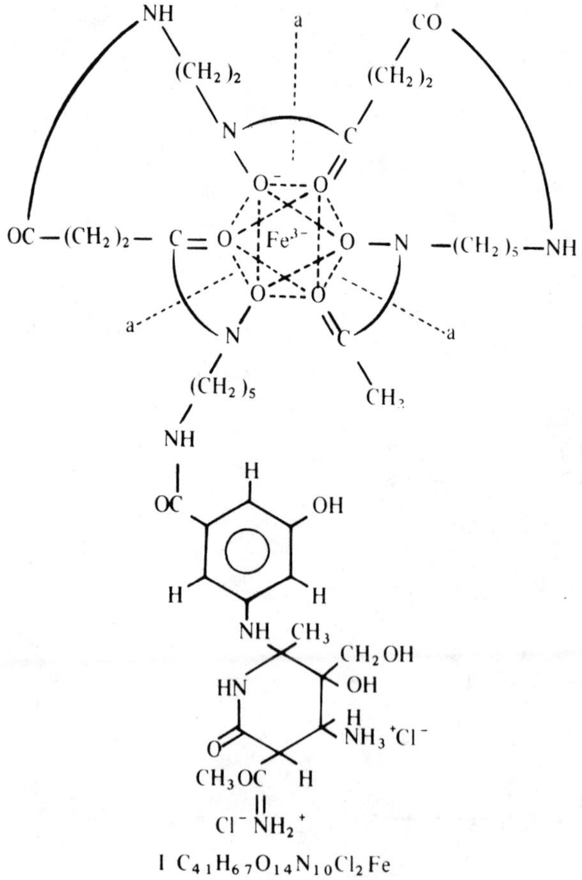

F-1

$$\text{I } C_{41}H_{67}O_{14}N_{10}Cl_2Fe$$

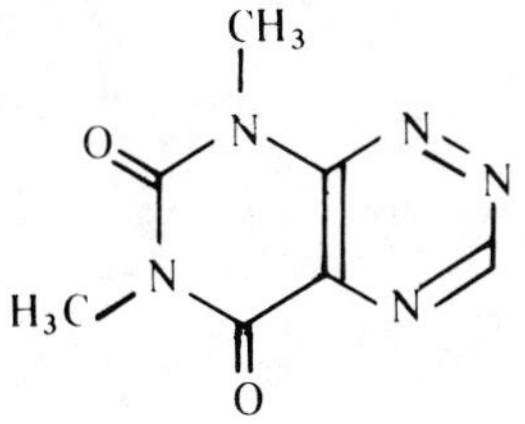

F-2

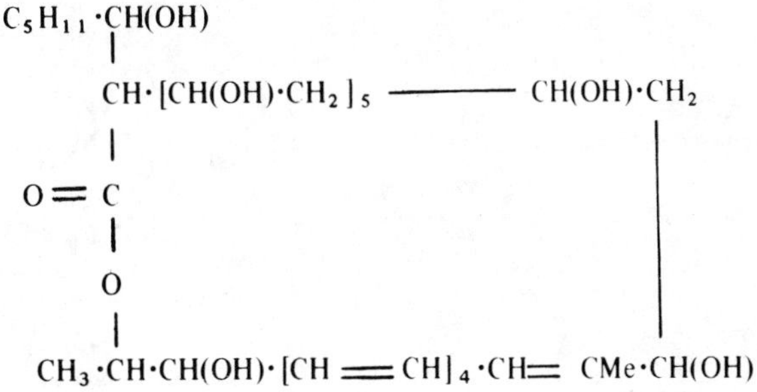

C₅H₁₁·CH(OH)

F-3

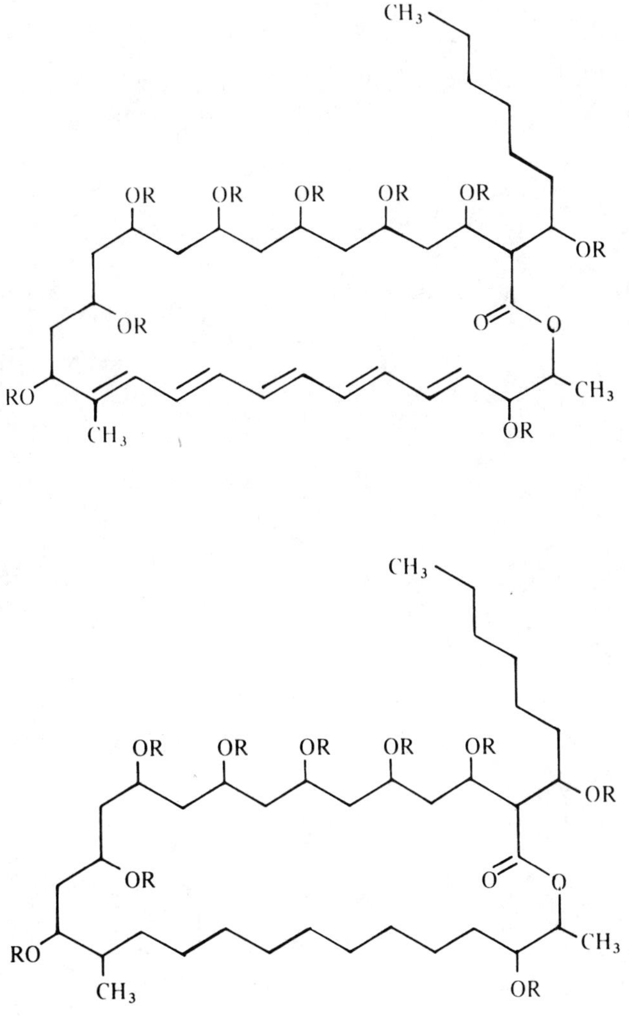

Filipin (filipin III) and its derivatives.

F-4

I: R = H   III: R = H
II: R = COCH₃   IV: R = COCH₃

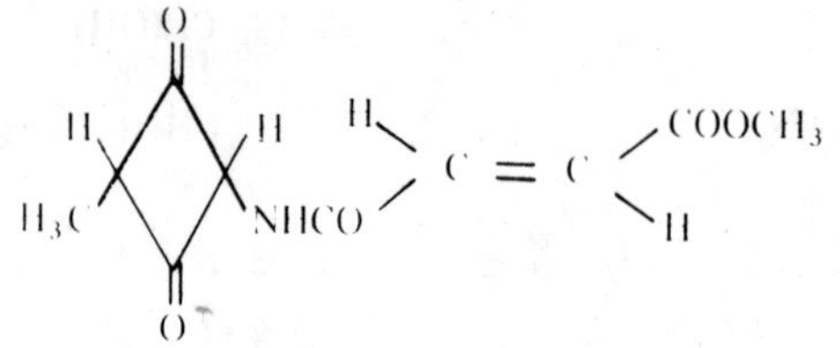

F-5　　Flavensomycinic acid (methanolysis product).

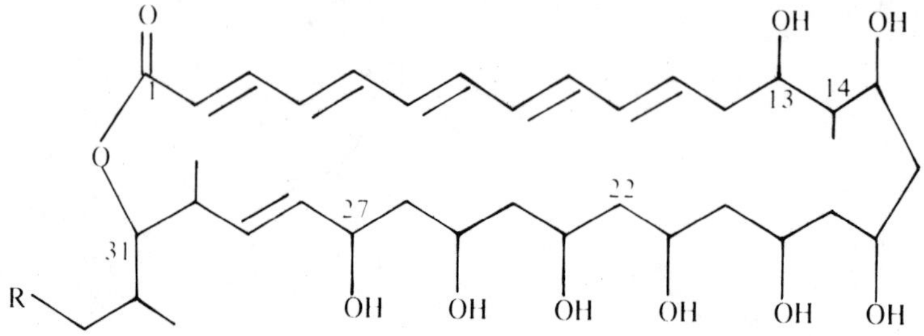

1a, R = H (1a:1b = ca 10:1)
1b, R = Me

F-6

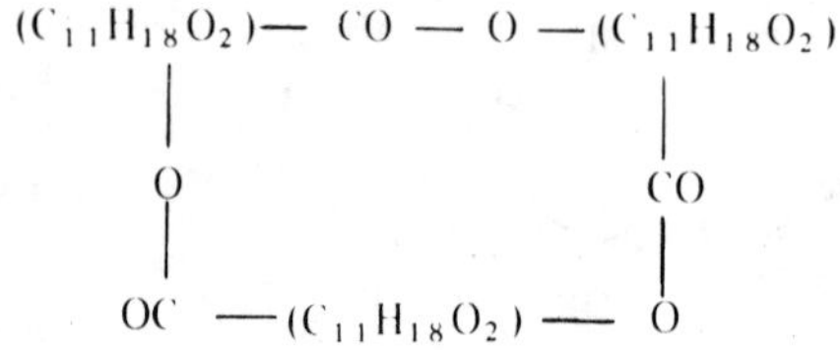

F-7

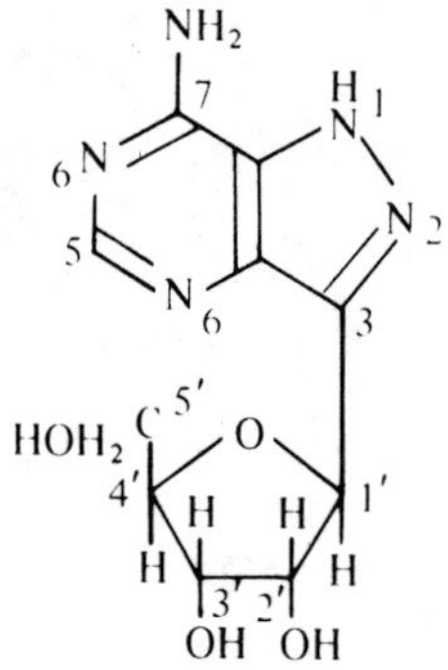

F-8

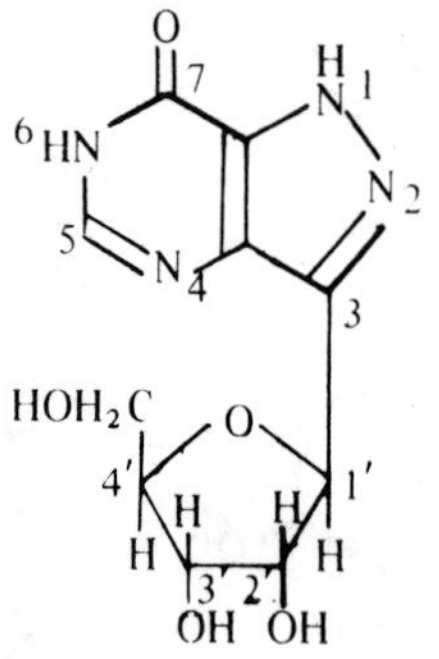

F-9

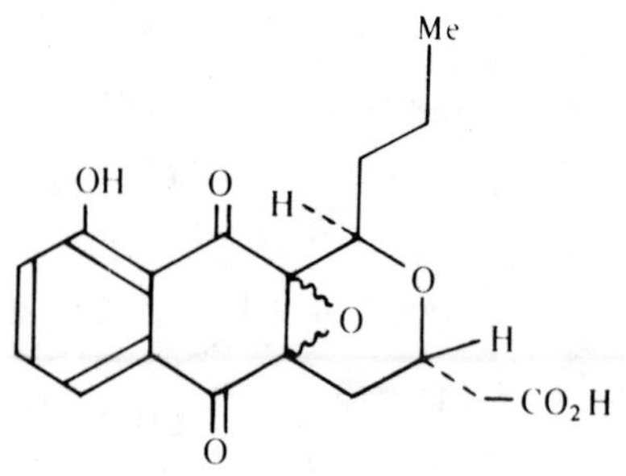

F-10

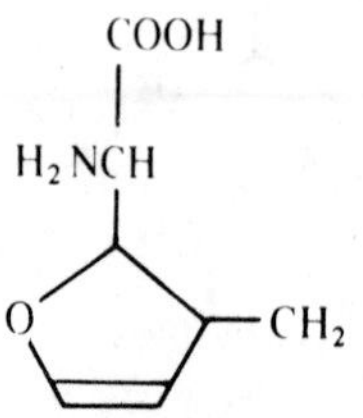

F-11　　III

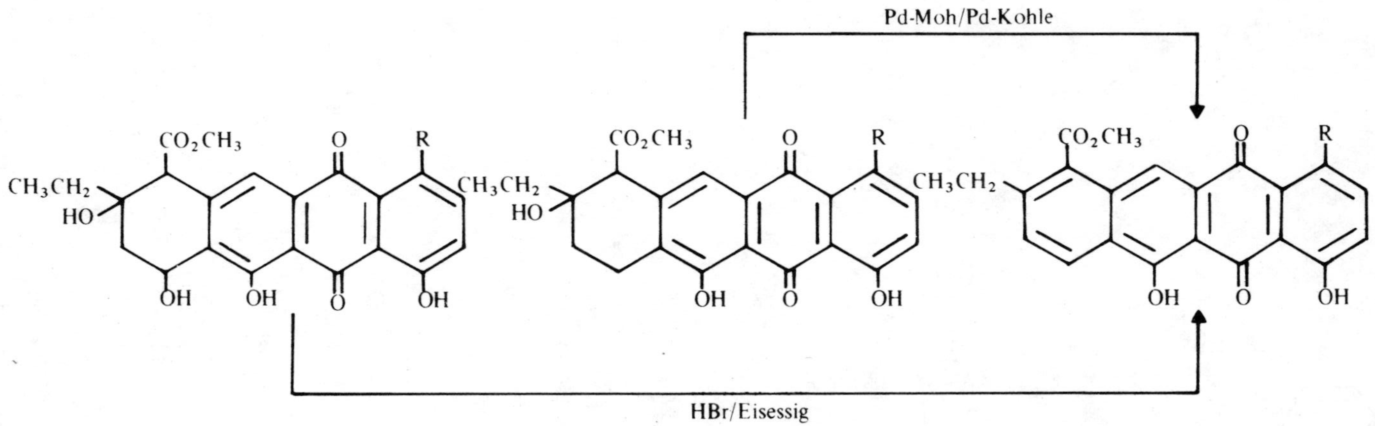

Pd-Moh/Pd-Kohle

HBr/Eisessig

G-1

1:	Aklavinon;	2:	7-Desoxy-aklavinon;	3:	Bis-anhydro-aklavinon;
	R = H		R = H		R = H
1a:	ε-Pyrromycinon;	2a:	ζ-Pyrromycinon;	3a:	η-Pyrromycinon;
	R = OH		R = OH		R = OH

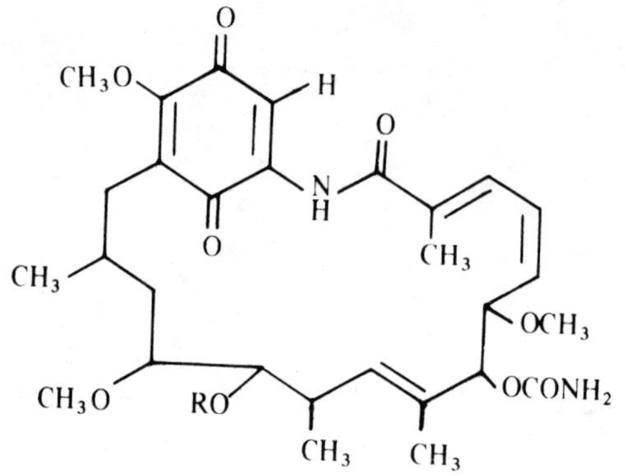

G-2   1, R = H

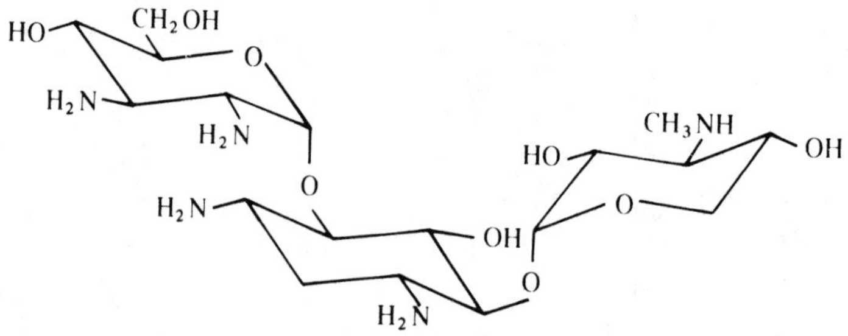

GENTAMICIN A

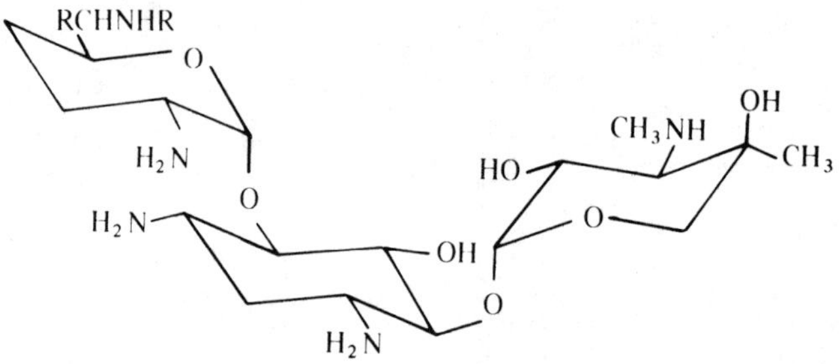

GENTAMICINS C$_1$:R, R' = CH$_3$
C$_2$:R = CH$_3$, R' = H
C$_{1a}$:R, R' = H

G-3

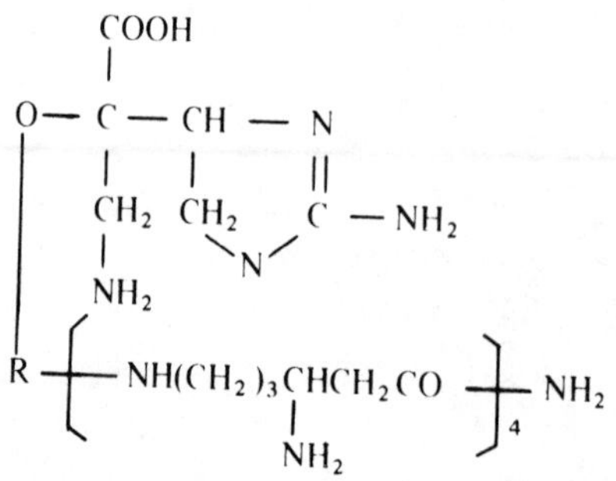

G-4   R = C$_{18}$H$_{37-39}$O$_{12}$N$_3$

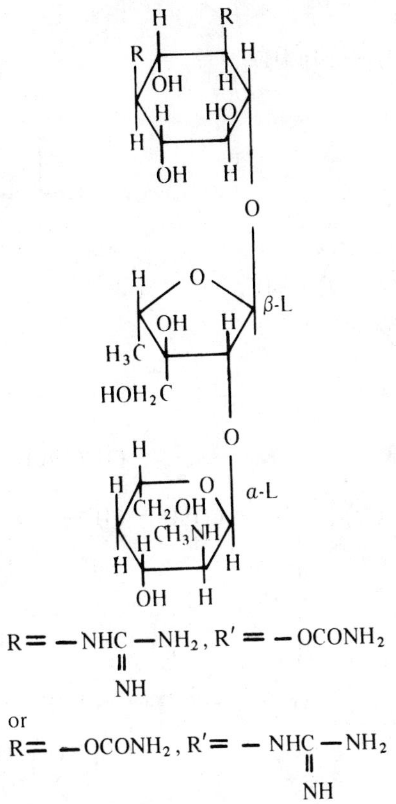

$$R = -NHC-NH_2, \quad R' = -OCONH_2$$
$$\overset{\parallel}{NH}$$

or

$$R = -OCONH_2, \quad R' = -NHC-NH_2$$
$$\overset{\parallel}{NH}$$

G-5

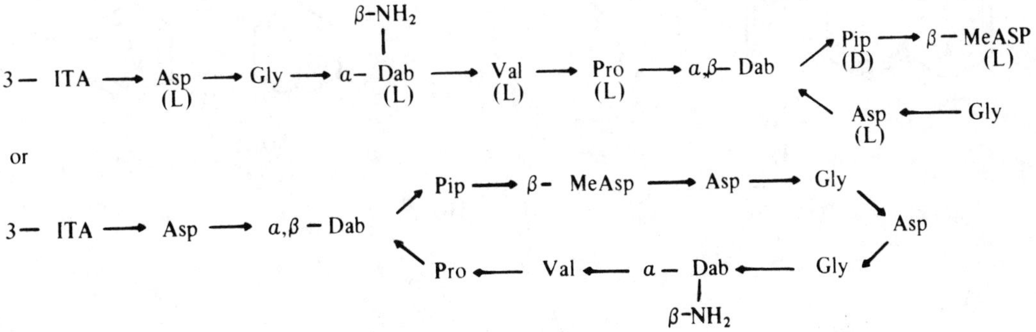

3-ITA: 3-isotridecanoic acid
Pip: pipecolic acid
Dab: α,β-diaminobutyric acid

G-6

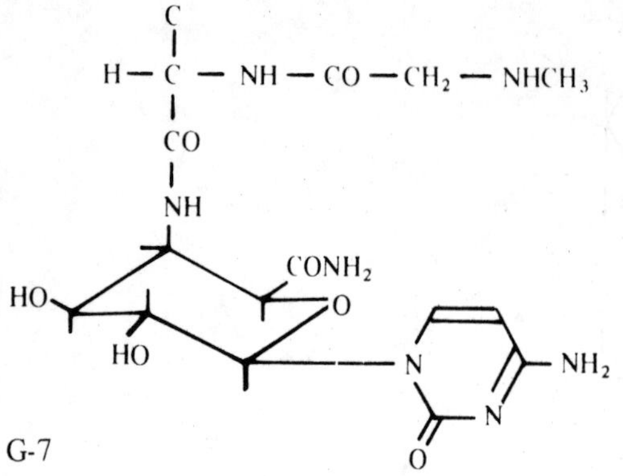

G-7

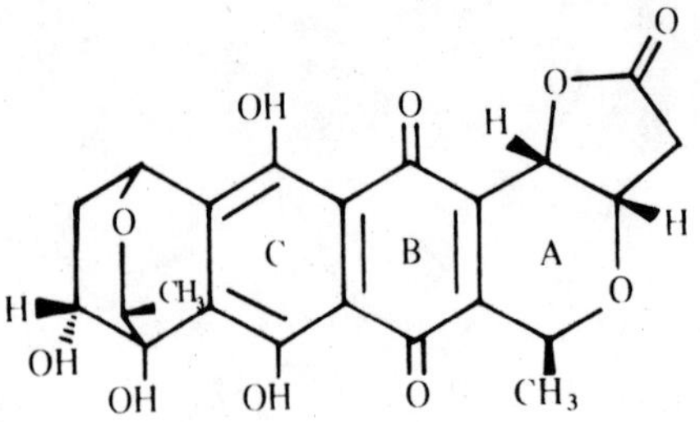

G-8

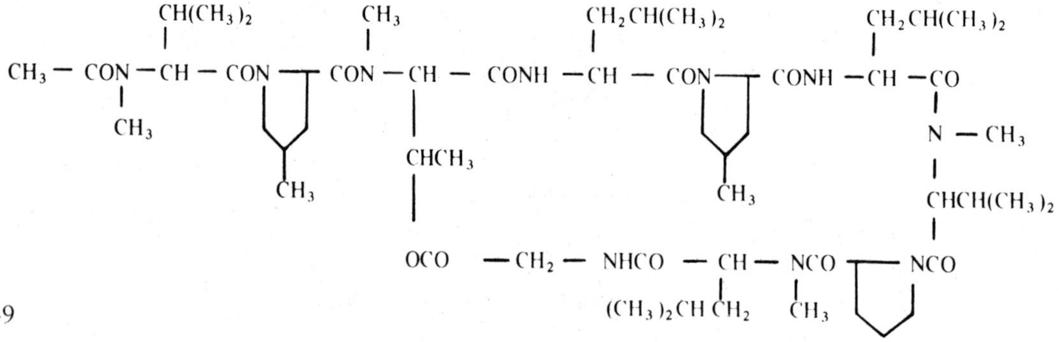

G-9

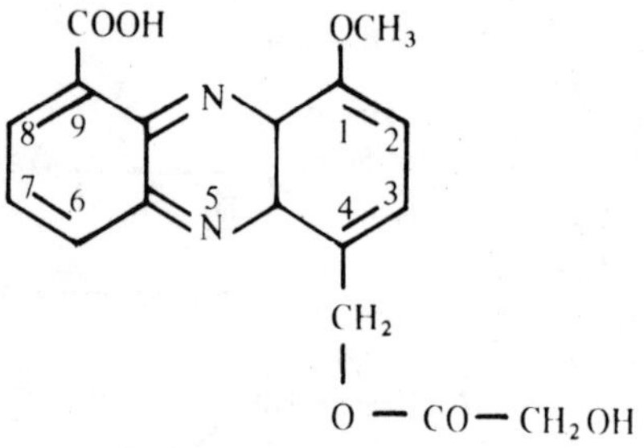

G-10

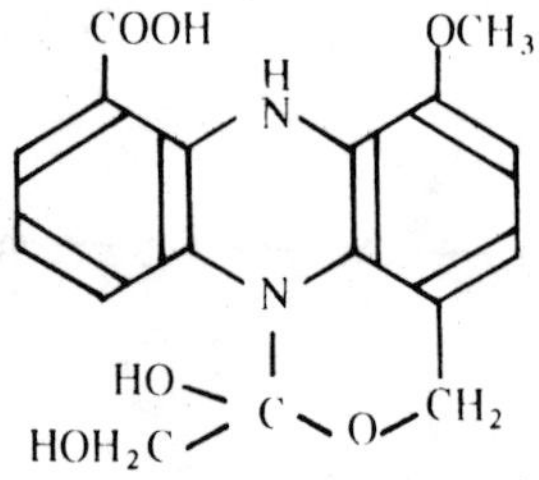

G-11

A: Partial structure

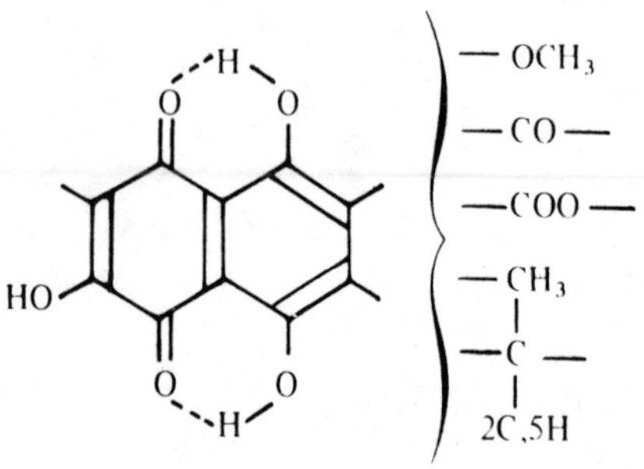

G-12

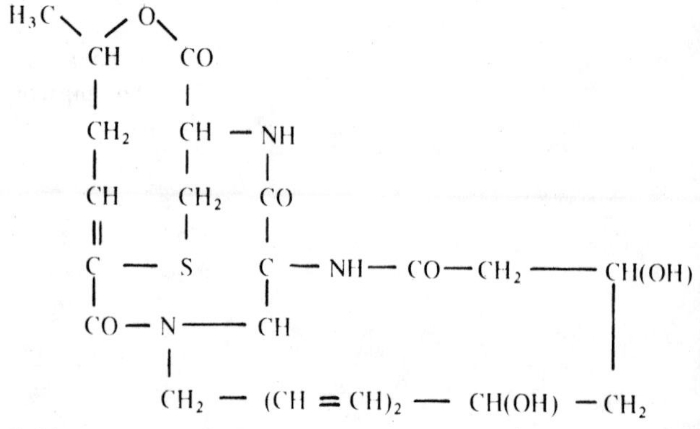

G-13

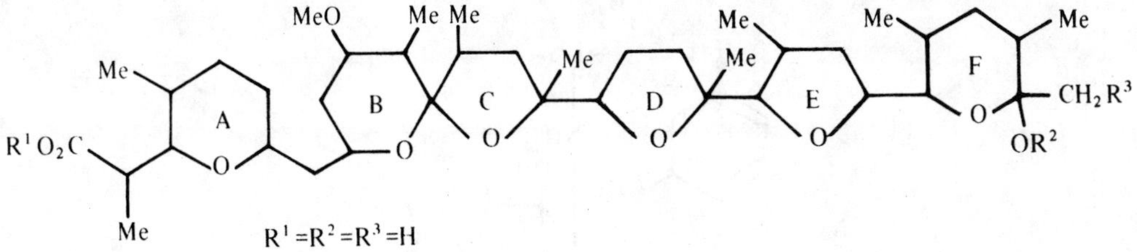

G-14

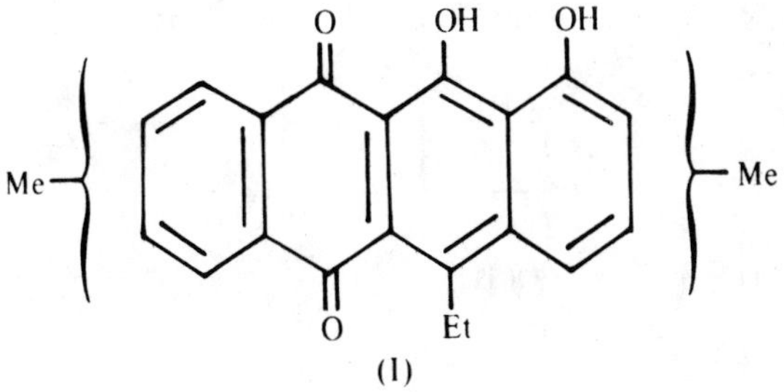

H-1

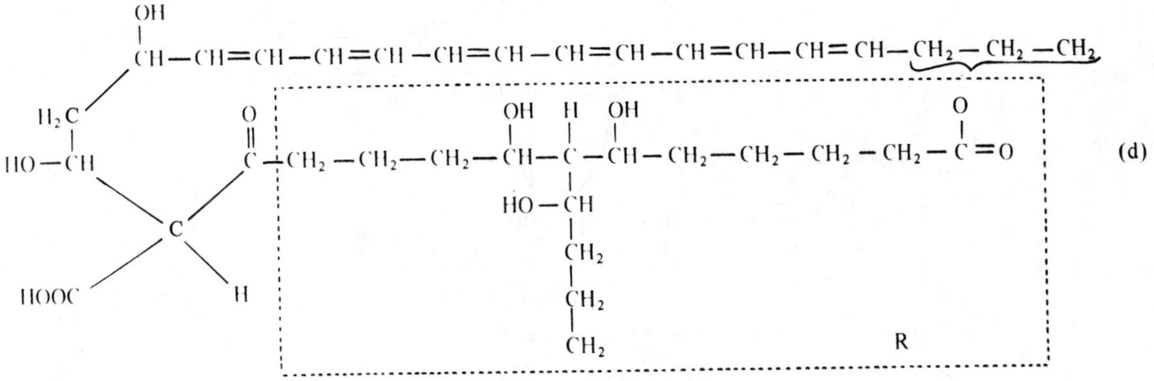

H-2

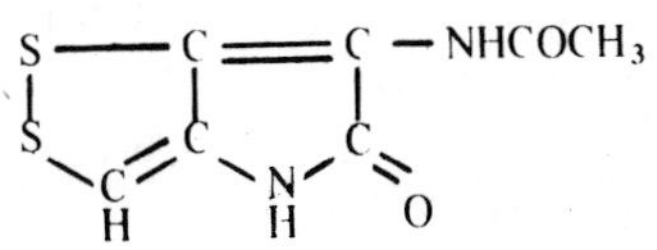

H-3

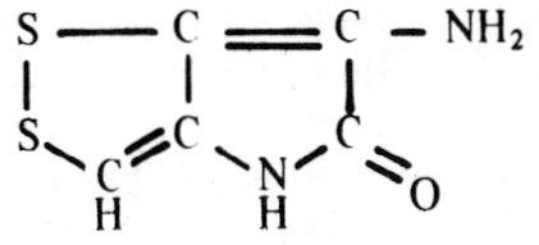

H-4

H-5

Figure 1. — Antibiotics cited.

Antibiotic	Aminocyclitol	R^1	R^2	R^3	R^4
Neomycin B	Deoxystreptamine	H	H	H	CH_2NH_2
Neomycin C	Deoxystreptamine	H	H	CH_2NH_2	H
Hybrimycin A1	Streptamine	H	OH	H	CH_2NH_2
Hybrimycin A2	Streptamine	H	OH	CH_2NH_2	H
Hybrimycin B1	Epistreptamine	OH	H	H	CH_2NH_2
Hybrimycin B2	Epistreptamine	OH	H	CH_2NH_2	H

H-6

5-Hydroxy-7-chlortetracycline

H-7

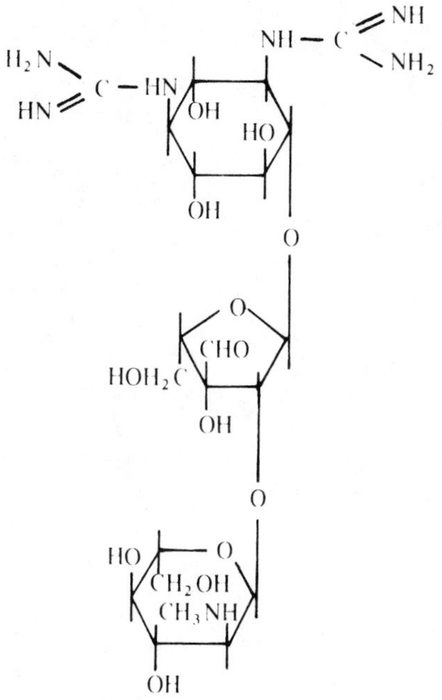

H-8

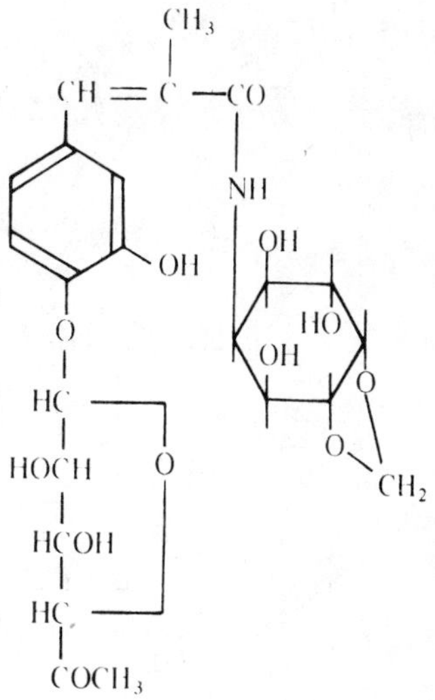

H-9

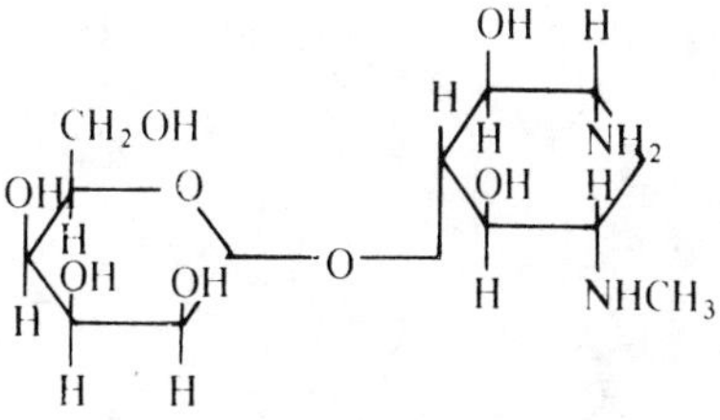

Hygromycin B₂ (mild acid hydrolysis product)

H-10

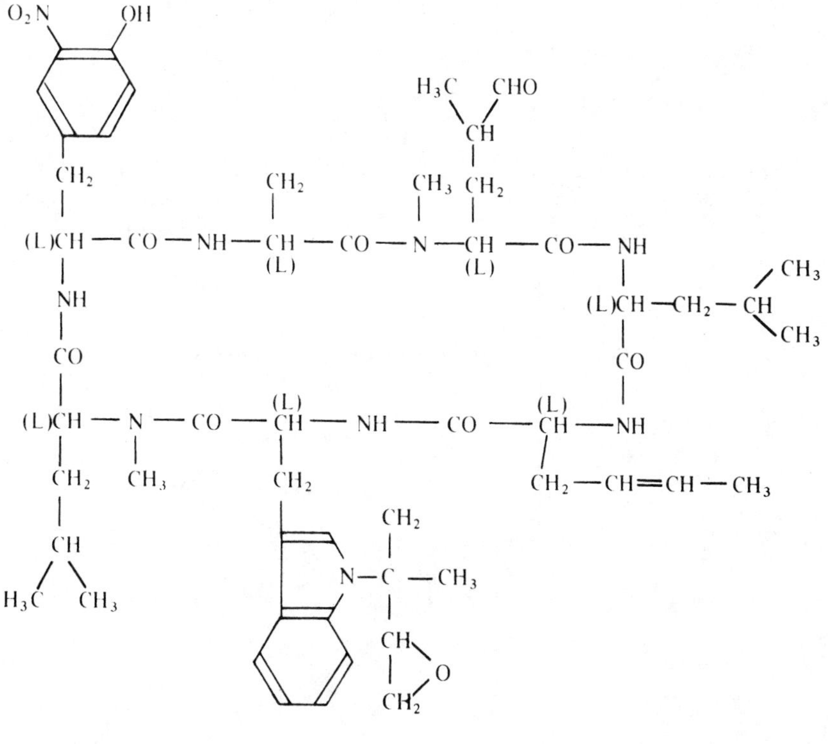

I-1

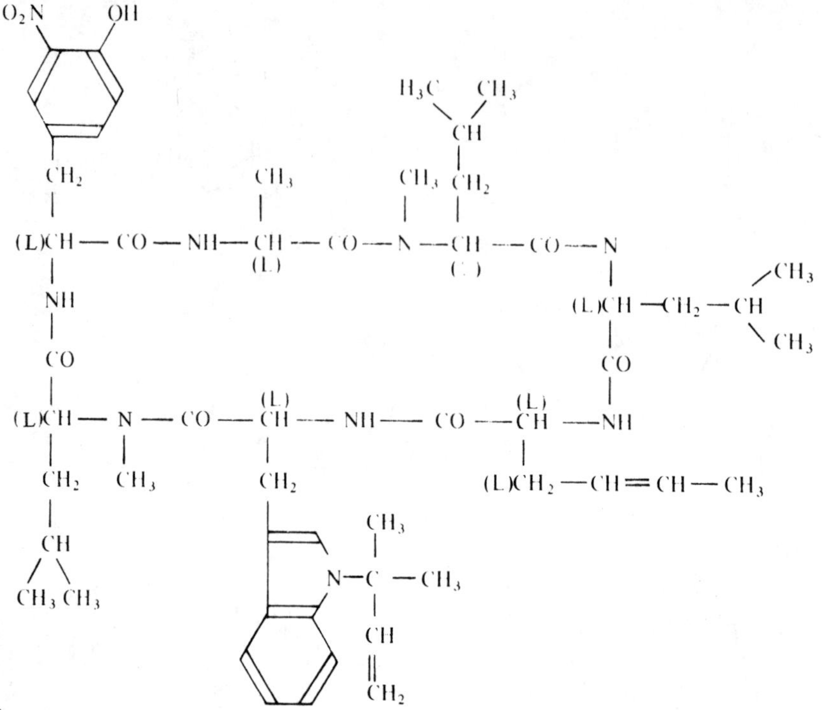

O₂N   OH

I-2

$$
\begin{array}{c}
\text{O}_2\text{N}\quad\text{OH} \\
\text{CH}_2 \\
\text{(L)CH}-\text{CO}-\text{NH}-\overset{\text{CH}_3}{\text{CH}}-\text{CO}-\text{N}-\overset{\text{CH}_3\ \text{CH}_2}{\text{CH}}-\text{CO}-\text{N} \\
(L) \qquad\qquad (\ ) \\
\text{NH} \\
\text{CO} \\
\text{(L)CH}-\text{N}-\text{CO}-\overset{(L)}{\text{CH}}-\text{NH}-\text{CO}-\overset{(L)}{\text{CH}}-\text{NH} \\
\text{CH}_2\quad\text{CH}_3 \qquad \text{CH}_2 \\
\text{CH} \\
\text{CH}_3\ \text{CH}_3
\end{array}
$$

H₃C   CH₃ / CH

(L)CH—CH₂—CH ⟨ CH₃ / CH₃

(L)CH₂—CH=CH—CH₃

CH₃
N—C—CH₃
CH
‖
CH₂

I-2

O₂N   OH

H₃C   CH₃
CH
CH₃ CH₂

(L)CH—CO—NH—CH—CO—N—CH—CO—NH
(L)   (L)

NH
CO
(L)CH—N—CO—CH—NH—CO—CH—NH
CH₃ CH₃   (L)   (L)
CH₂   CH₂—CH=CH—CH₃
CH
H₃C CH₃

(L)CH—CH₂—CH ⟨ CH₃ / CH₃
CO

CH₃
N—C—CH₂
CH
| O
CH₂

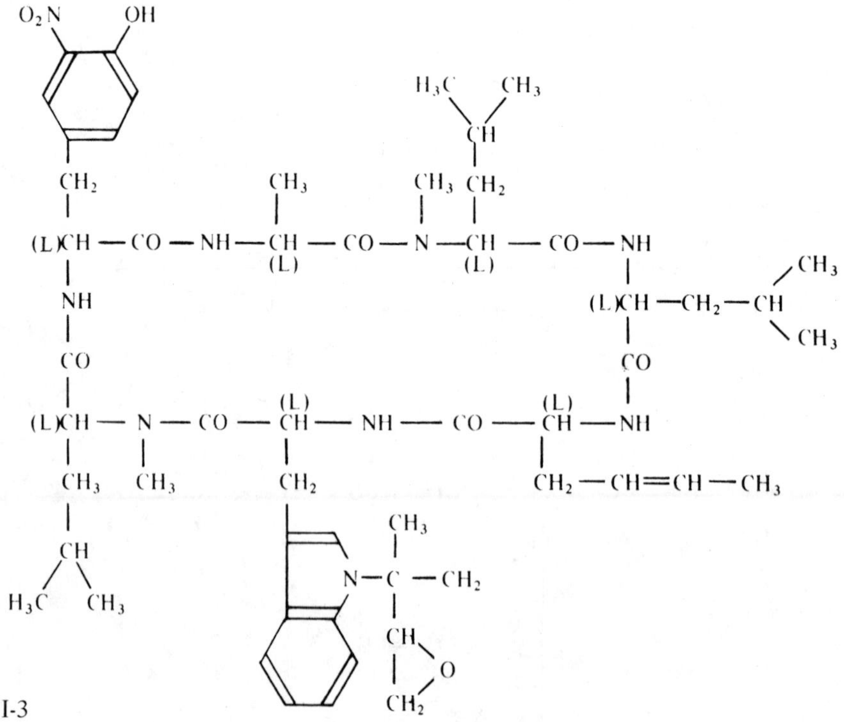

I-3

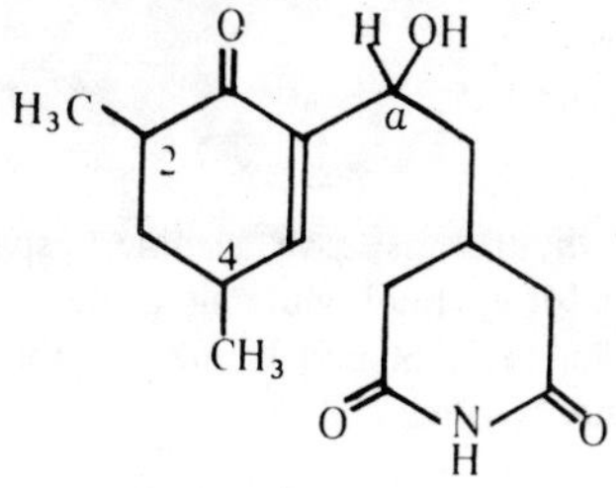

I-4   (2S: 4S; aR)

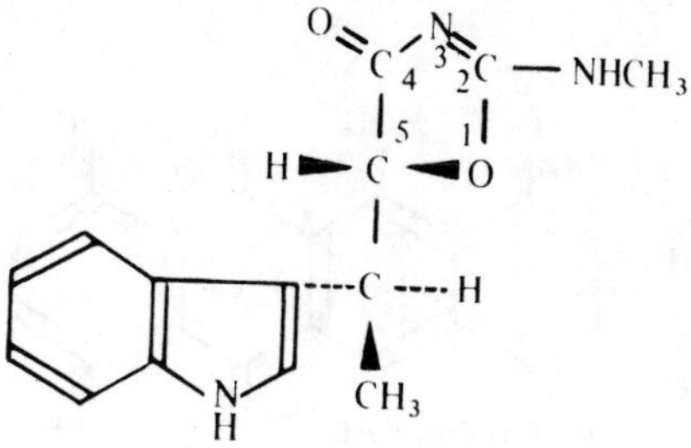

I-5

Partial a-Indomycinon

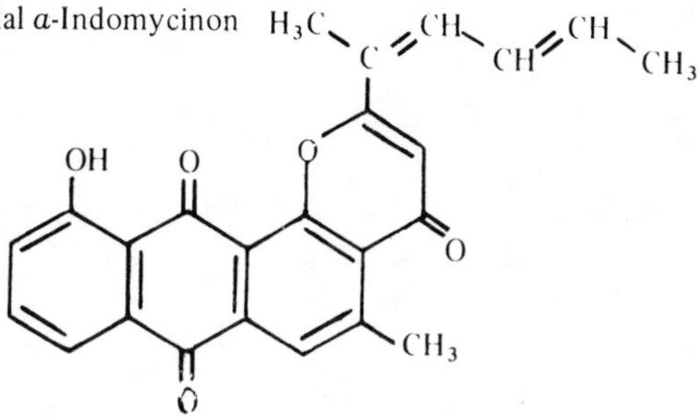

I-6   (I)

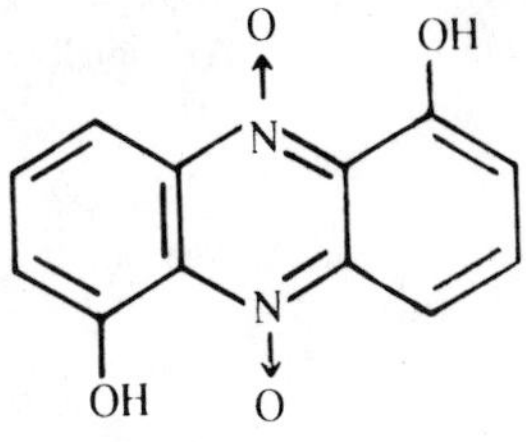

I-7

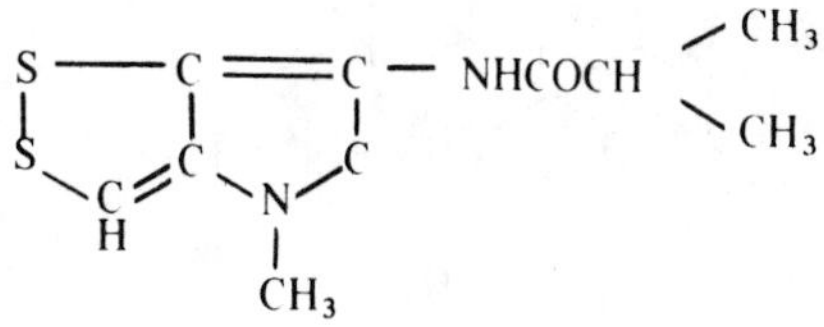

I

I-8

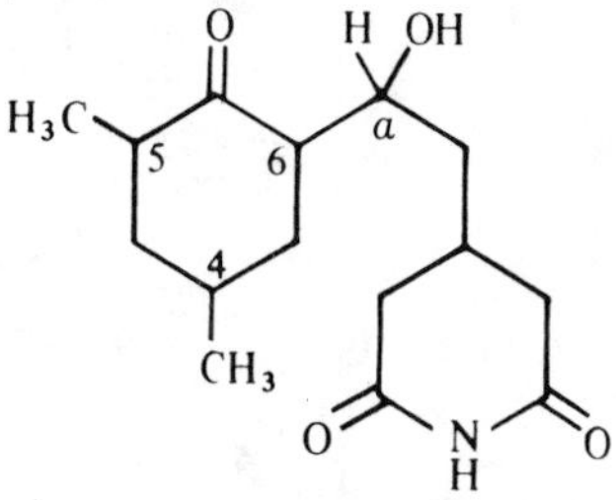

I-9   (2R: 4S: 6R: aR)

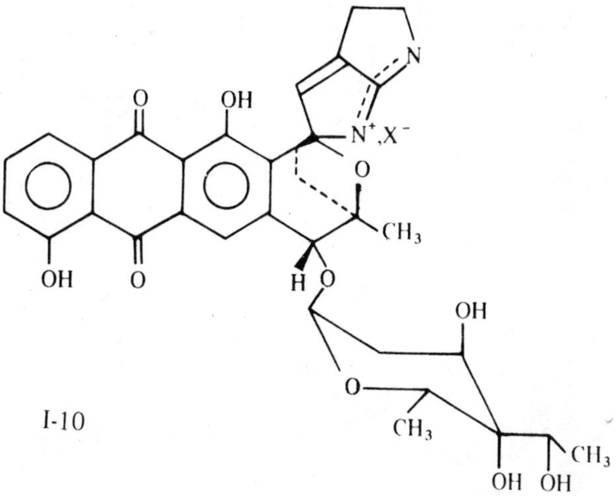

I-10

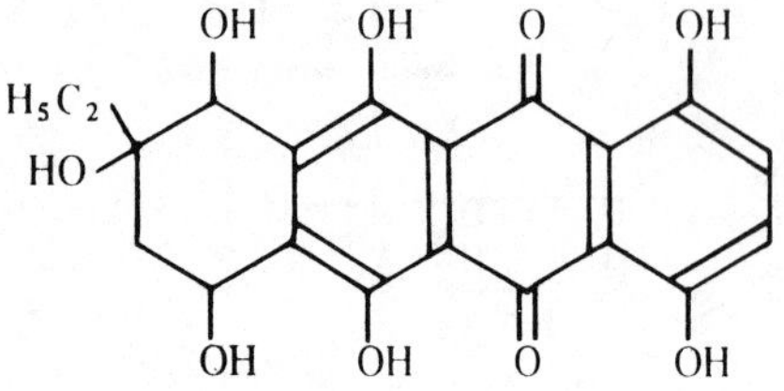

β-Isorhodomycinone

Hydrolysis products: isorhodomycinone (β-isorhodomycinone from **A**) and two moles of sugar (prob. rhodosamine).

Acid hydrolysis gave alanine, aspartic acid, cysteine (or cystine), glutamic acid, glycine, histidine, leucine, phenylalanine, proline, serine, threonine, tyrosine and valine.

I-12

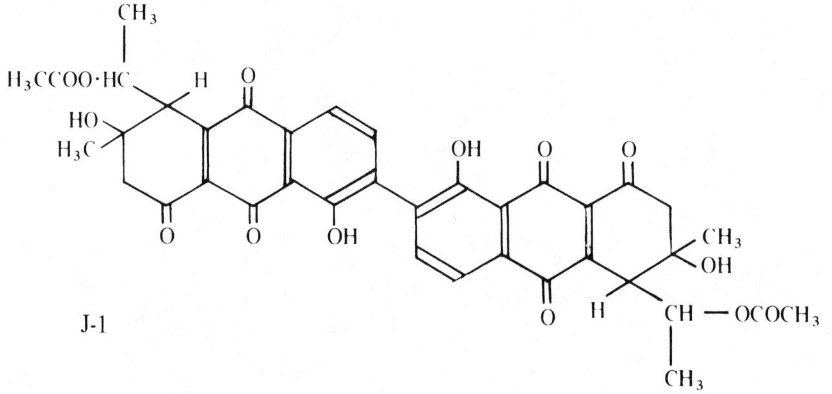

I-11

J-1

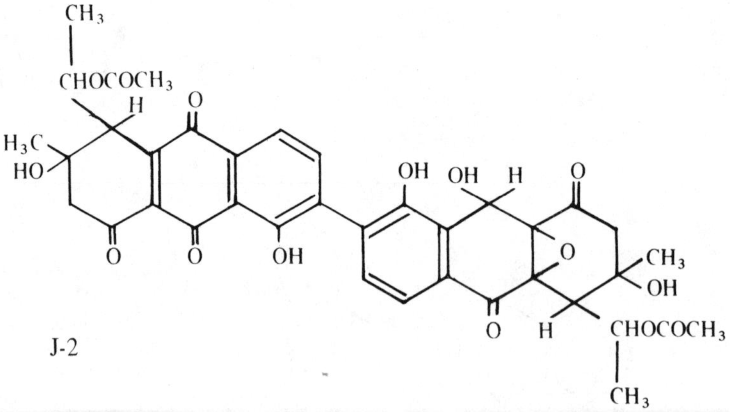

J-2

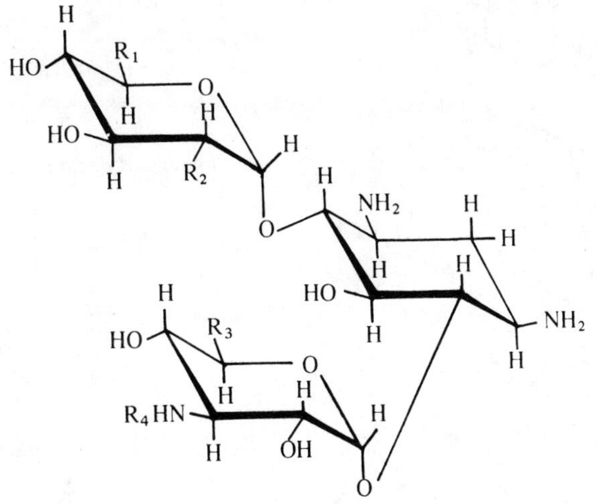

K-1, 2, 3

		R_1	R_2	R_3	R_4
1	Kanamycin A	CH_2NH_2	OH	CH_2OH	H
2	Kanamycin B	CH_2NH_2	NH_2	CH_2OH	H
3	Kanamycin C	CH_2OH	NH_2	CH_2OH	H

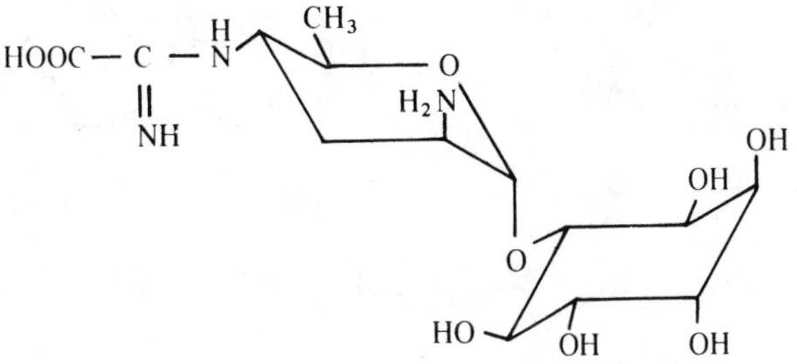

K-4

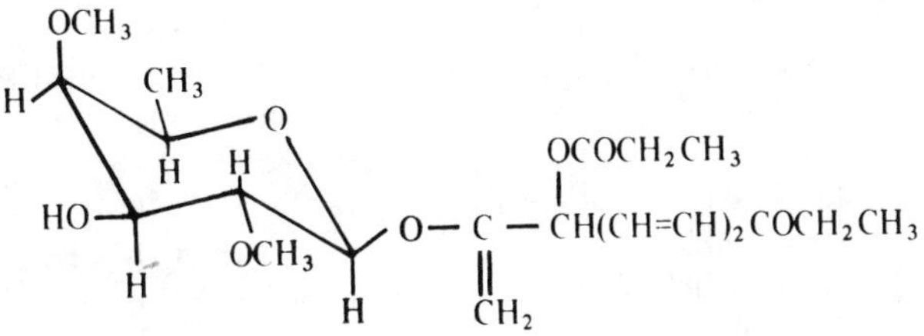

L-1

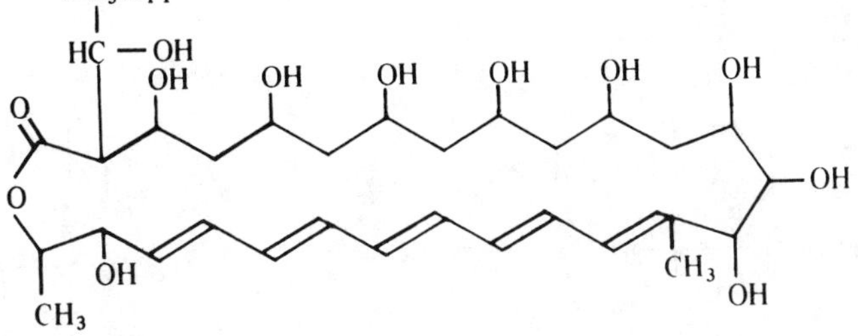

L-2

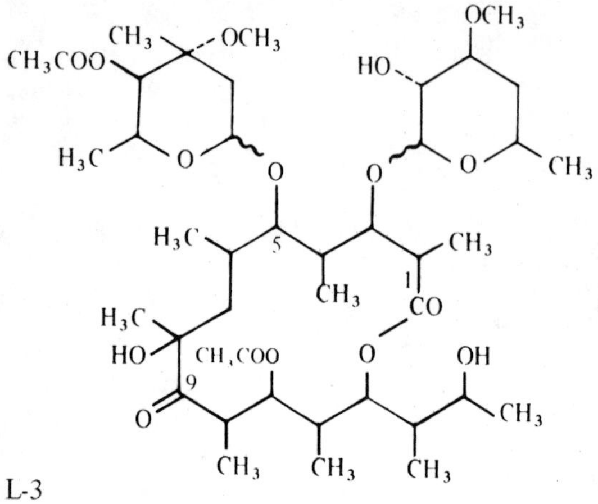

L-3

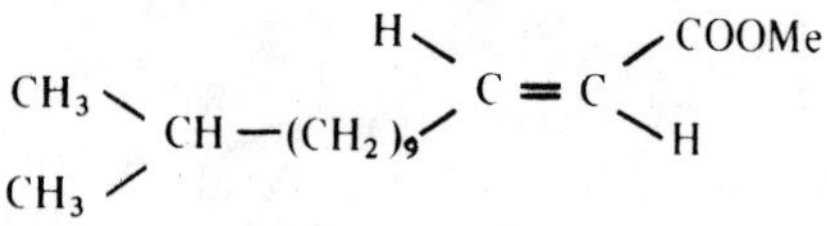

L-6 $H_2NOCC{\equiv}CCONH_2$

Major fatty acid ester of Laspartomycin

L-4

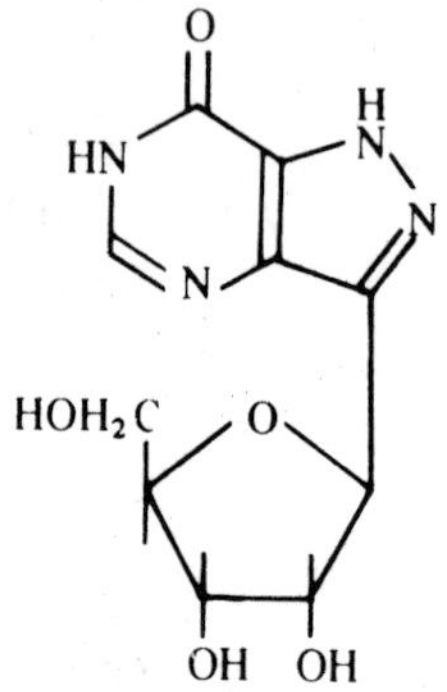

L-5

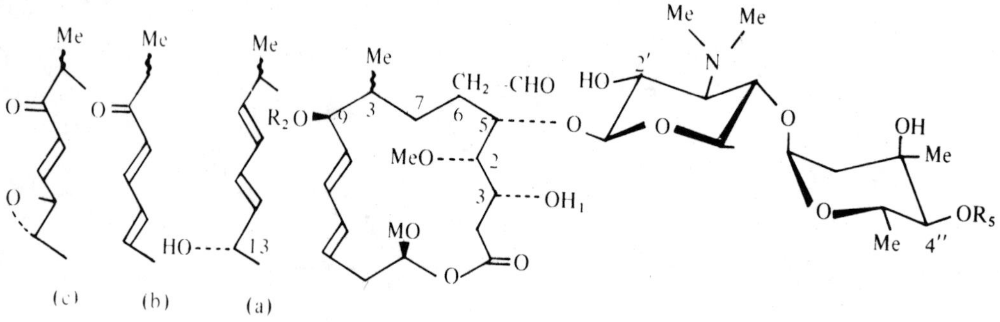

		R_1	R_2	R_3
Leucomycin A_1	(as shown)	H	H	$CO \cdot CH_2 \cdot CH(CH_3)_2$
Leucomycin A_3	(as shown)	$CO \cdot CH_3$	H	$CO \cdot CH_2 \cdot CH(CH_3)_2$
Leucomycin A_4	(as shown)	$CO \cdot CH_3$	H	$CO \cdot CH_2 \cdot CH_2 \cdot CH_3$
Leucomycin A_5	(as shown)	H	H	
Leucomycin A_6	(as shown)	$CO \cdot CH_3$	H	$CO \cdot CH_2 \cdot CH_3$
Leucomycin A_7	(as shown)	H	H	$CO \cdot CH_2 \cdot CH_3$
Leucomycin A_8	(as shown)	$CO \cdot CH_3$	H	$CO \cdot CH_3$
Leucomycin A_9	(as shown)	H	H	$CO \cdot CH_3$
Leucomycin U	(as shown)	$CO \cdot CH_3$	H	H
Leucomycin V	(as shown)	H	H	H
Iso Leucomycin A_3	(cf. offseta)	$CO \cdot CH_3$	—	$CO \cdot CH_2 \cdot CH(CH_3)_2$

L-7

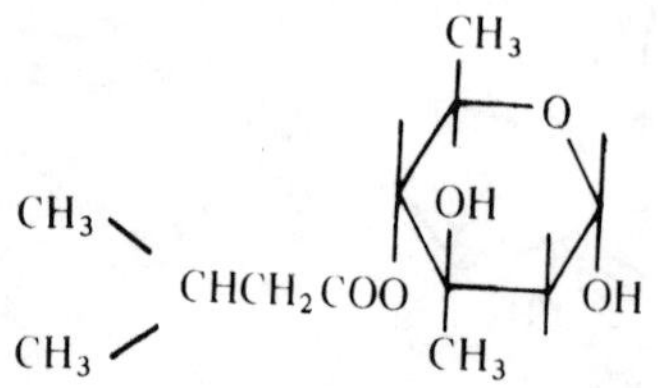

Mycarose-4-isovaleriate (mild acid hydrolysis product).

L-8a

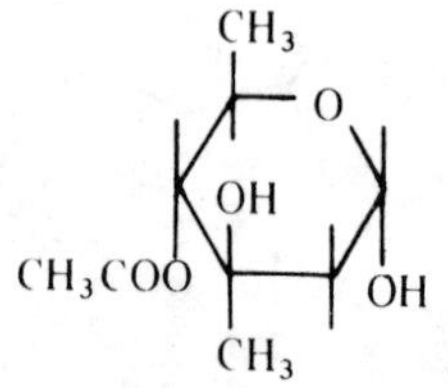

4-O-Acetylmycarose (mild acid hydrolysis product).

L-8b

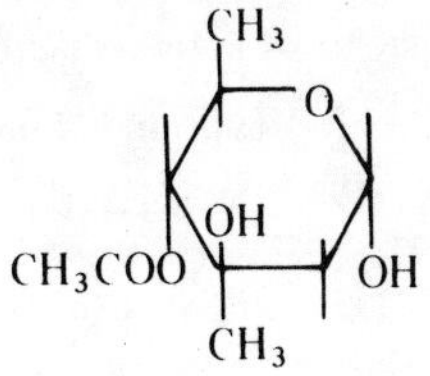

c

L-8c   4-O-Acetylmycarose (mild acid hydrolysis product).

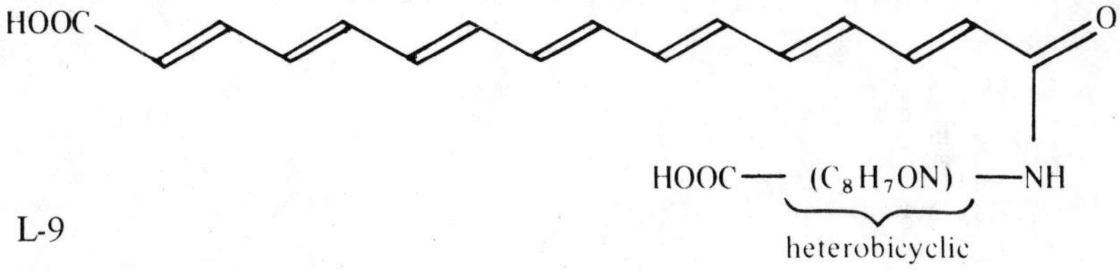

$HOOC - (C_8H_7ON) - NH$

L-9

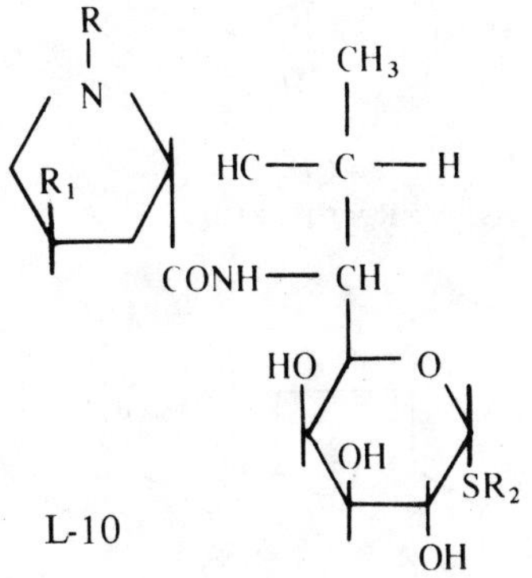

heterobicyclic

A. $R = CH_3$; $R_1 = CH_2CH_2CH_3$; $R_2 = CH_3$
B. $R = CH_3$; $R_1 = CH_2CH_3$; $R_2 = CH_3$
C. $R = CH_3$; $R_1 = CH_2CH_2CH_3$; $R_2 = CH_2CH_3$
D. $R = H$; $R_1 = CH_2CH_2CH_3$; $R_2 = CH_3$
K. $R = H$; $R_1 = CH_2CH_2CH_3$; $R_2 = CH_2CH_3$
S. $R = C_2H_5$; $R_1 = C_3H_7$; $R_2 = C_2H_5$

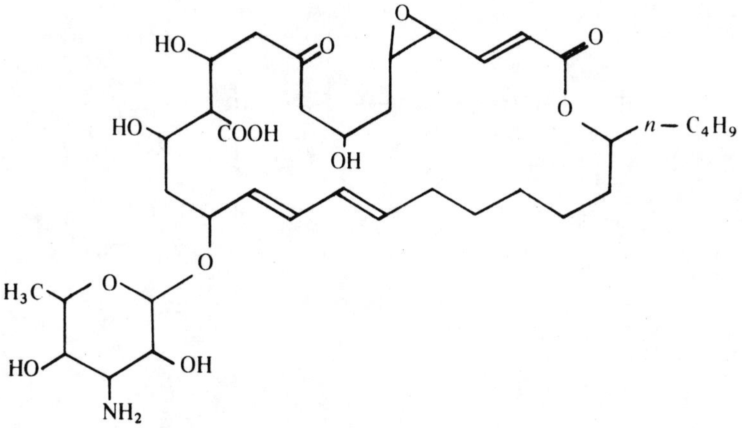

L-10

L-11

partial structure
of hydrolysis
product

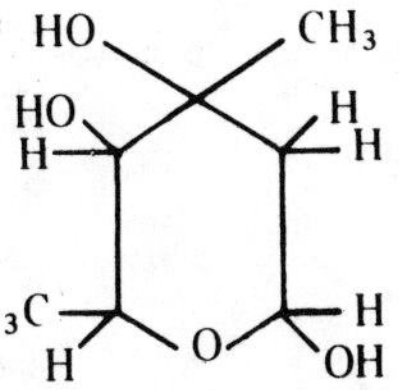

M-1

M-2—5 Components of known macrotetrolides

Compound	Type and Number of Components			
	(+)-Ns (I)	(−)-Ns (III)	(+)-Hs (II)	(−)-Hs (IV)
Nonactin	2	2	—	—
Monactin	1	2	1	—
Dinactin	—	2	2	—
Trinactin	—	1	2	—
(hypoth. Tetranactin)	—	—	2	2

Components of New Macrotetrolides

Compound	(+)-Ns	(−)-Ns	(+)-Hs	(−)-Hs	(+)-Bs	(−)-Bs
G $C_{43}H_{70}O_{12}$	—	2	1	—	1	—
D $C_{41}H_{72}O_{12}$	—	1	2	—	—	1
C $C_{45}H_{71}O_{12}$	—	1	1	—	1	1
B $C_{46}H_{70}O_{12}$	—		2		2	

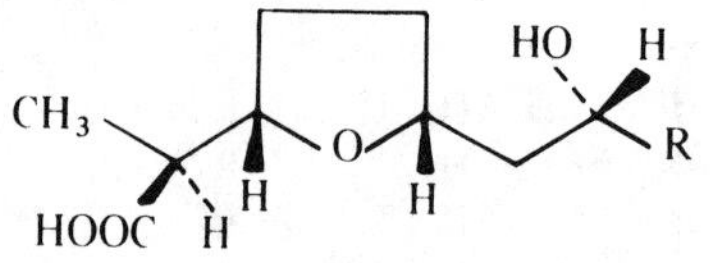

(+)Ns I: R = CH_3
(+)Hs II: R = CH_2—CH_3

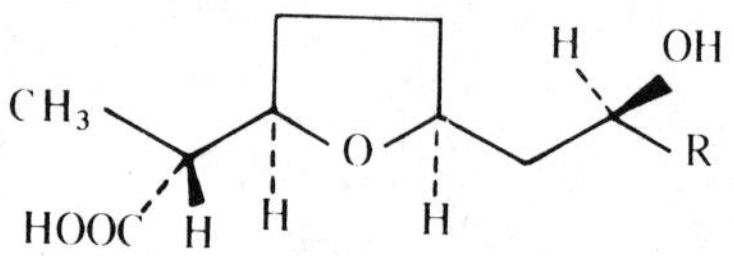

(−)Ns III: R = CH_3
(−)Hs IV: R = CH_2—CH_3

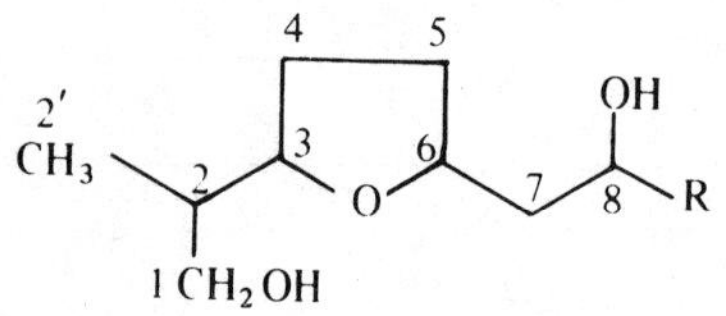

V: R = CH_3
VI: R = CH_2—CH_3
VII: R = $CH(CH_3)_2$

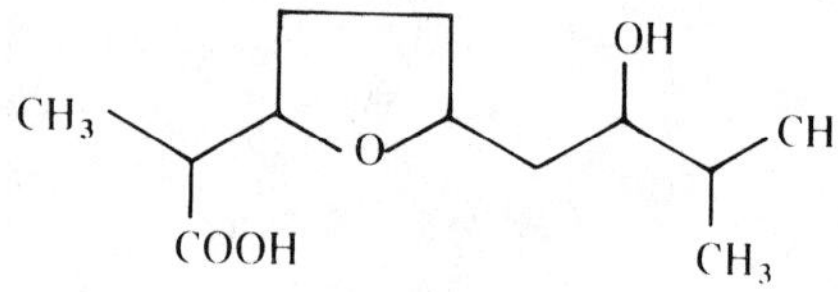

(Bs) VIII

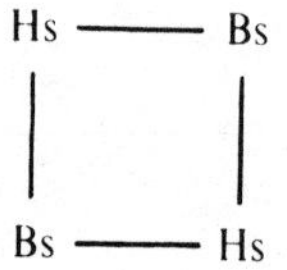

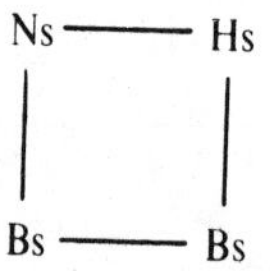

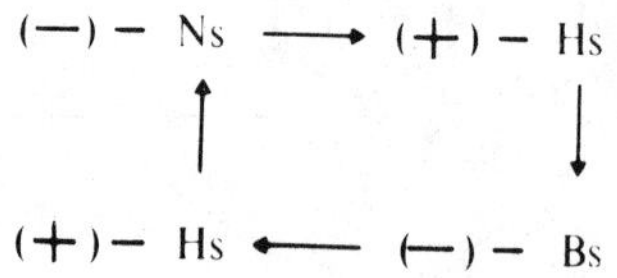

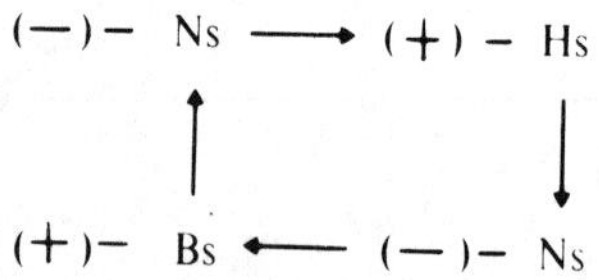

M-2, 3, 4, 5

M-6

M-7

M-8

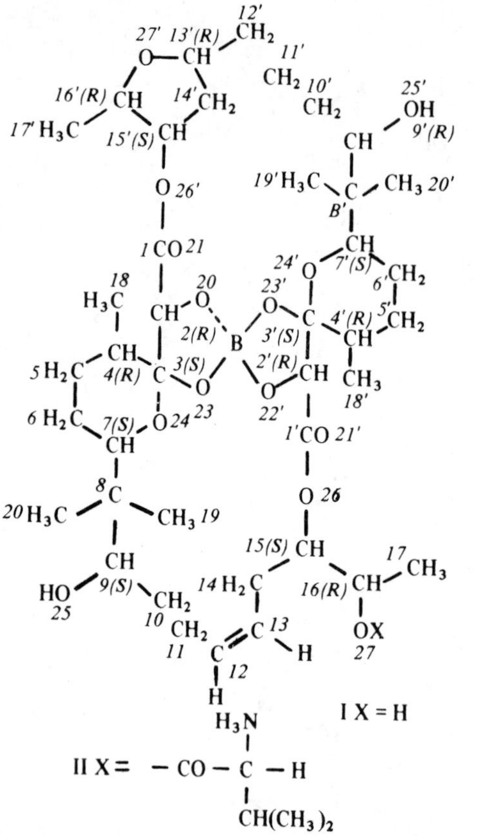

M-8/9

IV $C_{18}H_{32}O_5$

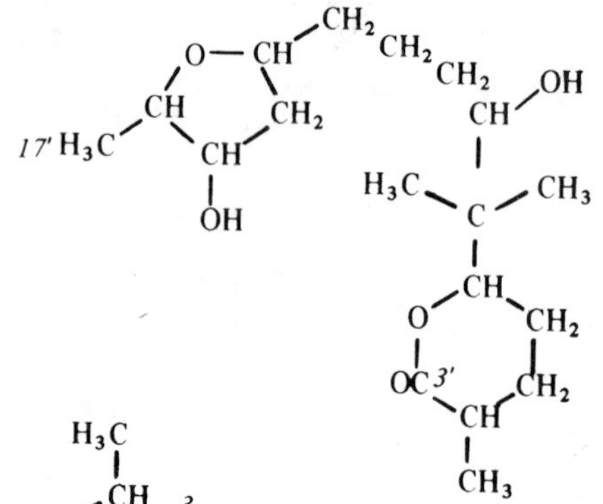

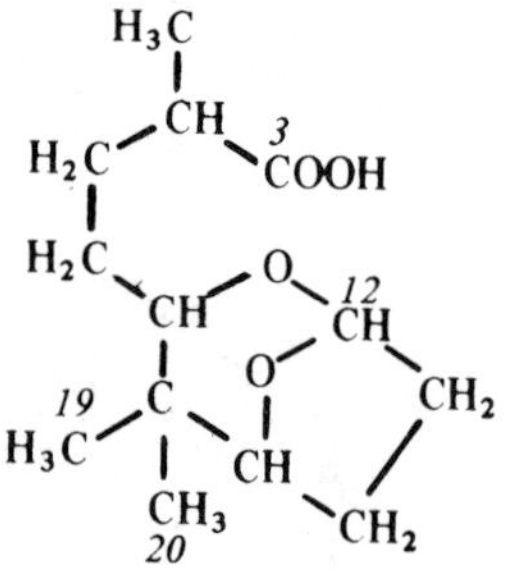

III $C_{13}H_{24}O_4$

I X = H

II X= $- CO - C - H$
                    |
              $CH(CH_3)_2$

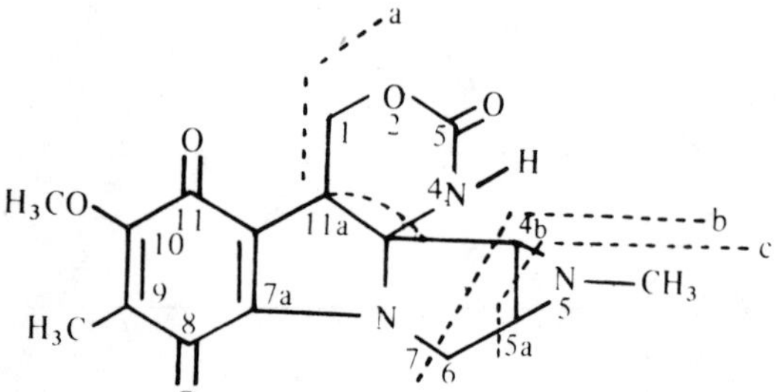

M-9

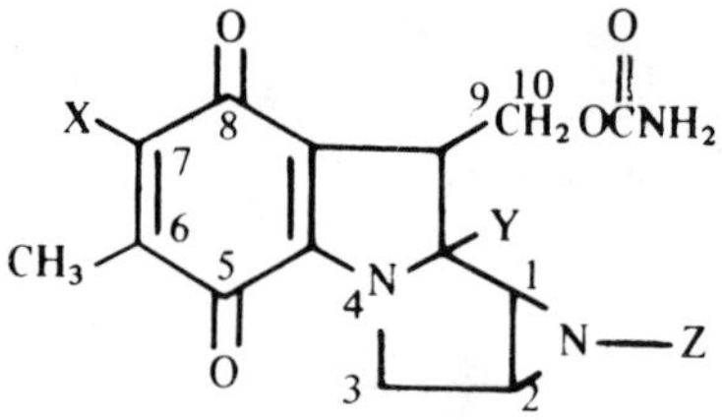

I–IV

I, mitomycin A; X = H_3CO; Y = OCH_3; Z = H
II, mitomycin B; X = H_3CO; Y = OH; Z = CH_3
III, mitomycin C; X = H_2N; Y = OCH_3; Z = H
    porfiromycin; X = H_2N; Y = OCH_3; Z = CH_3

M-10

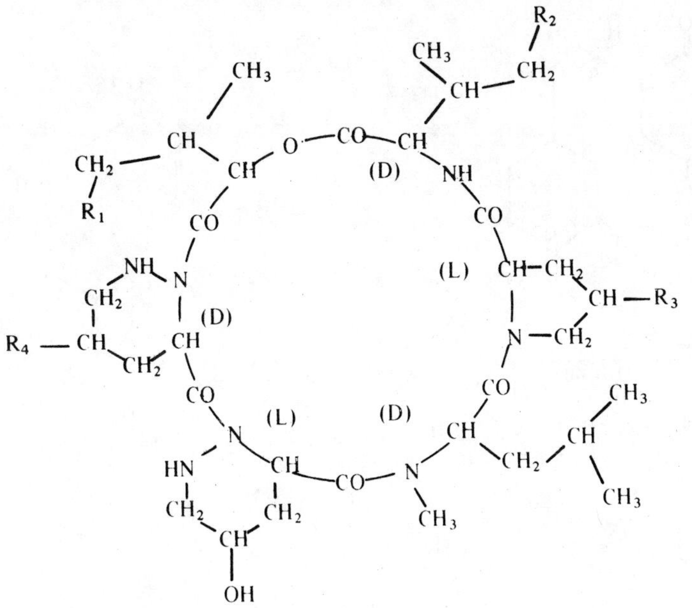

Monamycin $D_1(V)$, $R_1 = CH_3$, $R_2 = H$, $R_3 = CH_3$, $R_4 = H$.

M-11 (Substitutions $R_1 = H$ or CH_3; $R_2 = H$ or CH_3; $R_3 = H$ or CH_3; $R_4 = H$ or Cl)

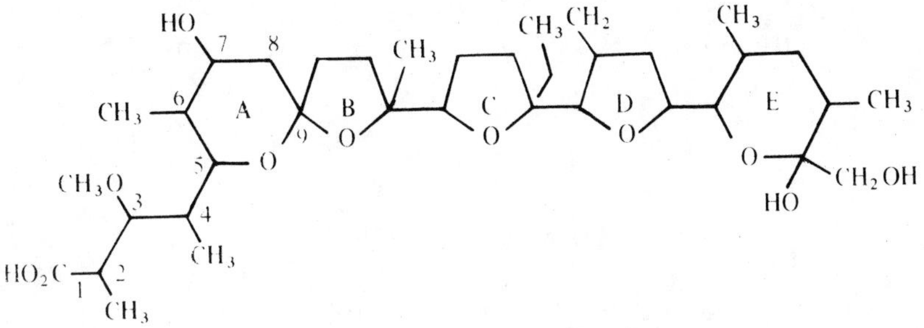

M-12

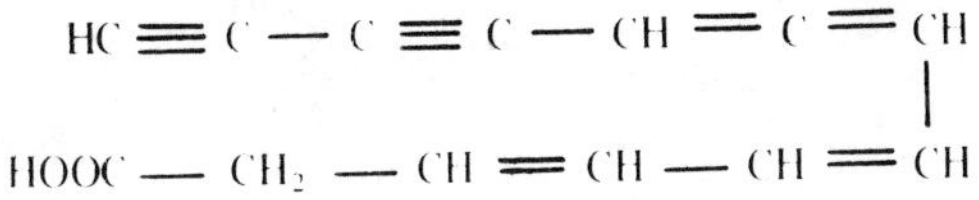

M-13

$C_{56}H_{50}N_2O_6 =$ Mycotrienin

M-14

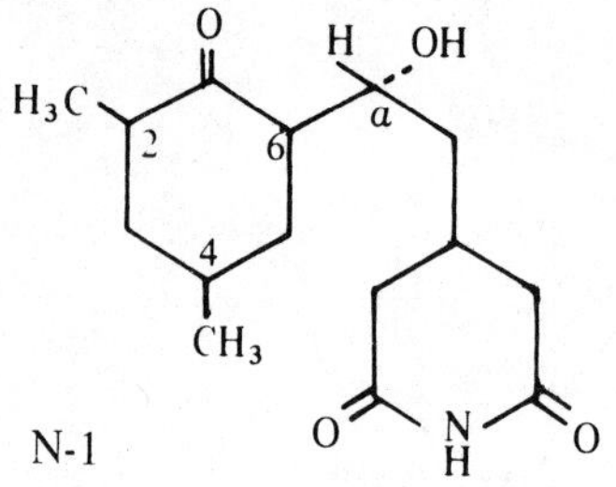

N-1

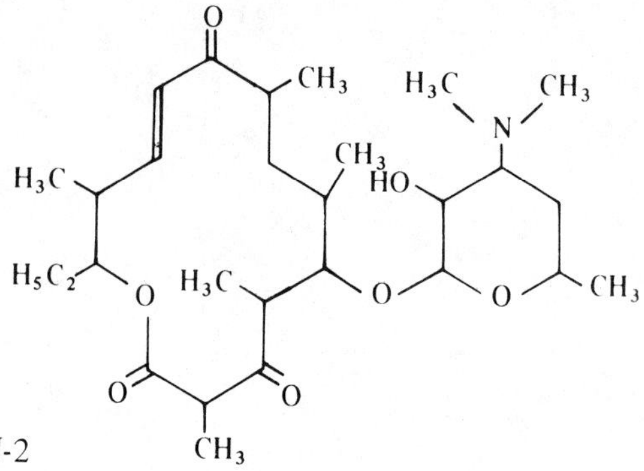

N-2

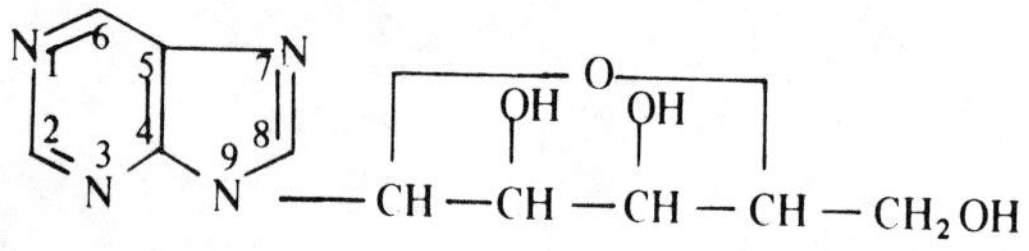

N-3

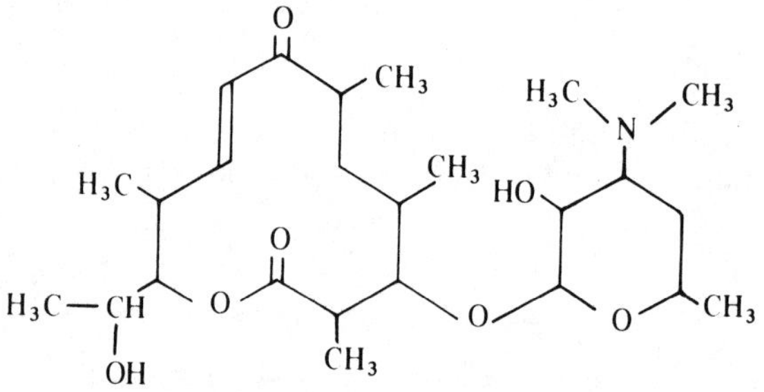

N-4

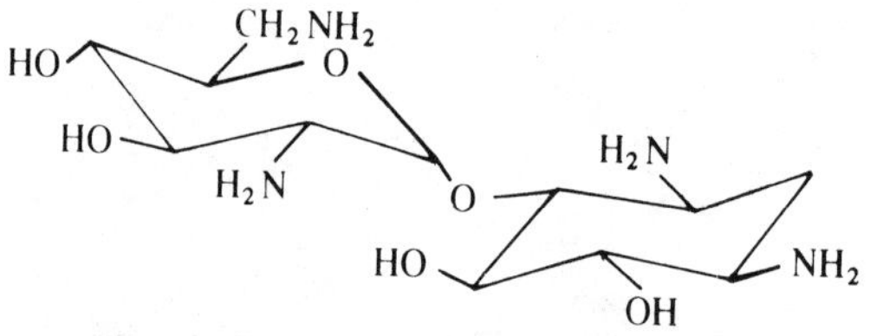

N-5

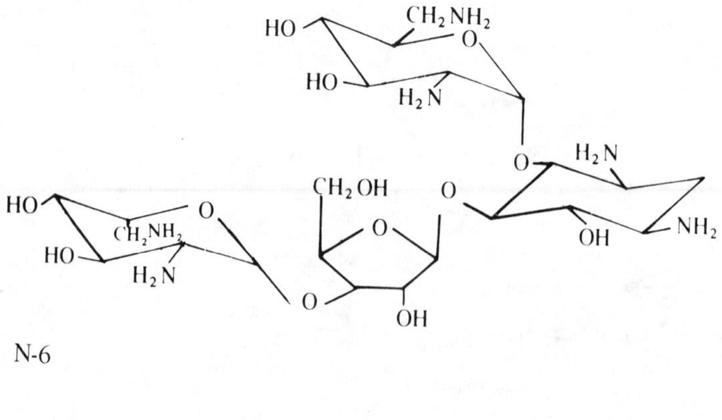

N-6

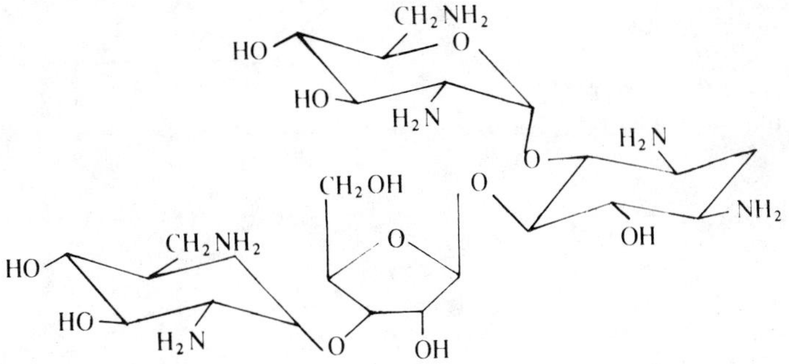

N-7

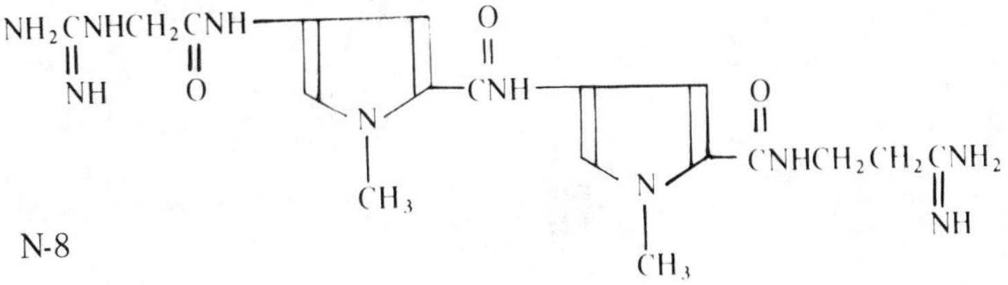

N-8

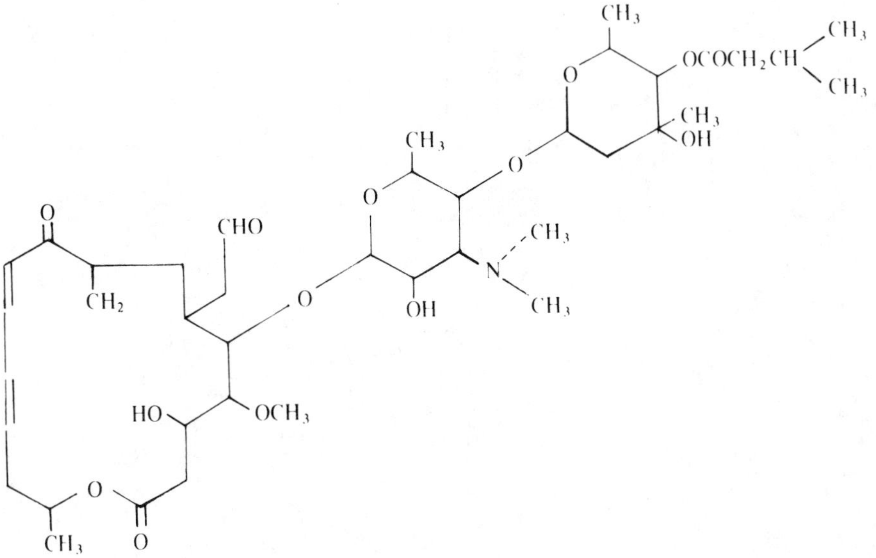

N-9   Should be revised as depicted in connection with carbomycin and carbomycin B.

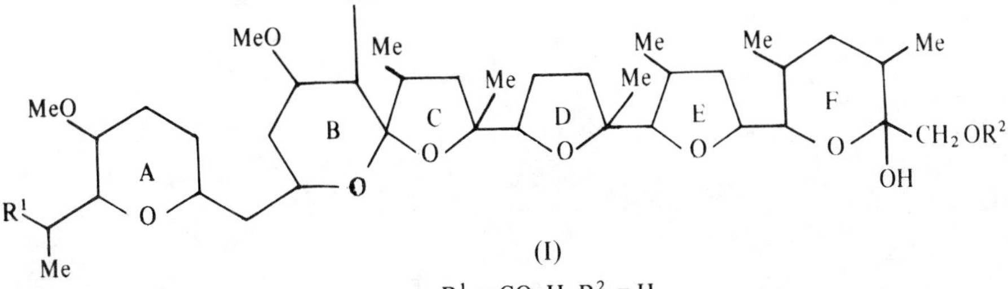

(I)

a: $R^1 = CO_2H$, $R^2 = H$
b: $R^1 = CO_2Me$, $R^2 = H$
c: $R^1 = CO_2Me$, $R^2 = Ac$
d: $R^1 = CO_2Et$, $R^2 = H$

N-10

$$HN(NO_2) - CH_2 - COOH$$
(m.p. 106°C, yield 84%)

N-11

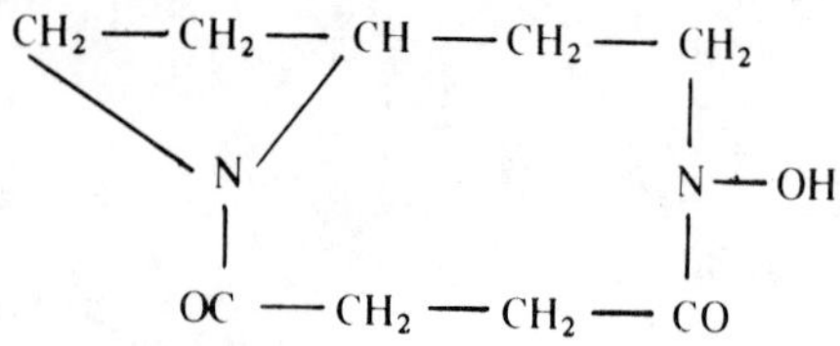

N-12

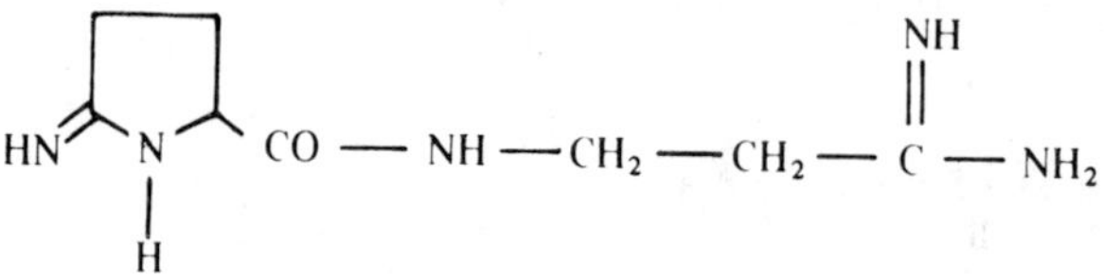

N-13

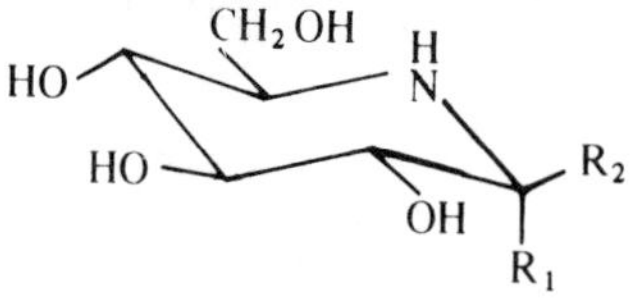

α-Anomer: $R_1 = OH$, $R_2 = H$
β-Anomer: $R_1 = H$, $R_2 = OH$

N-14

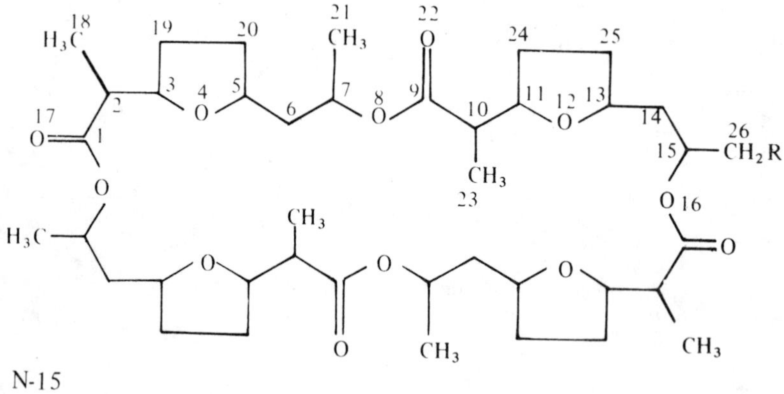

R:H   Nonactin
R:CH$_3$ Monactin

N-15

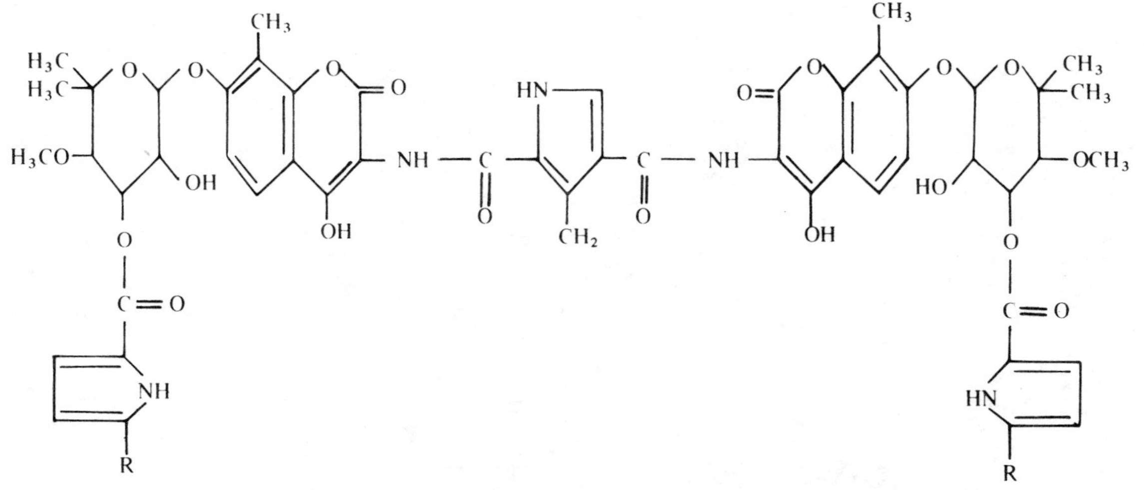

R:CH₃ A₁
R:H A₂

N-20

N-19

N-16,17

wherein both R groups represent methyl in the case of notomycin A₁ and represent hydrogen in notomycin A₂.

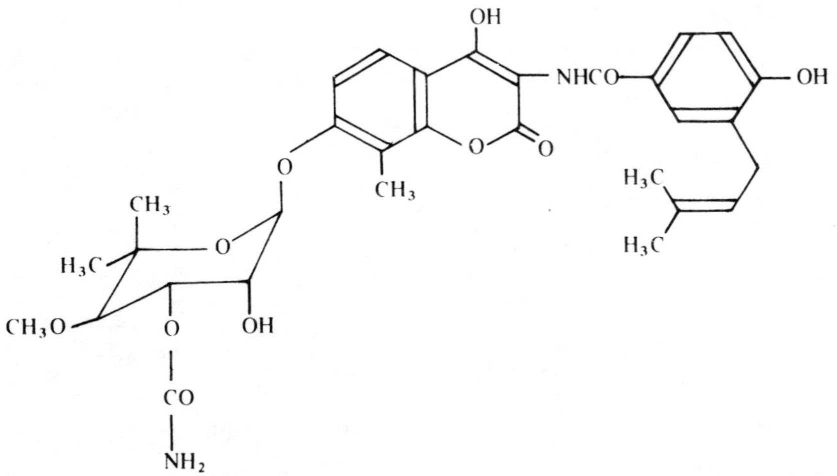

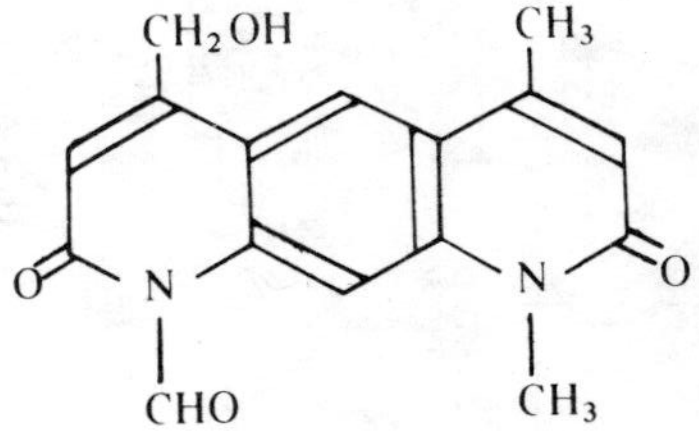

N-18

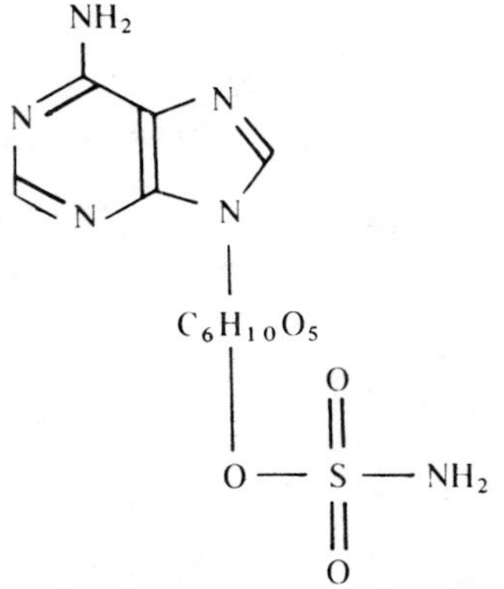

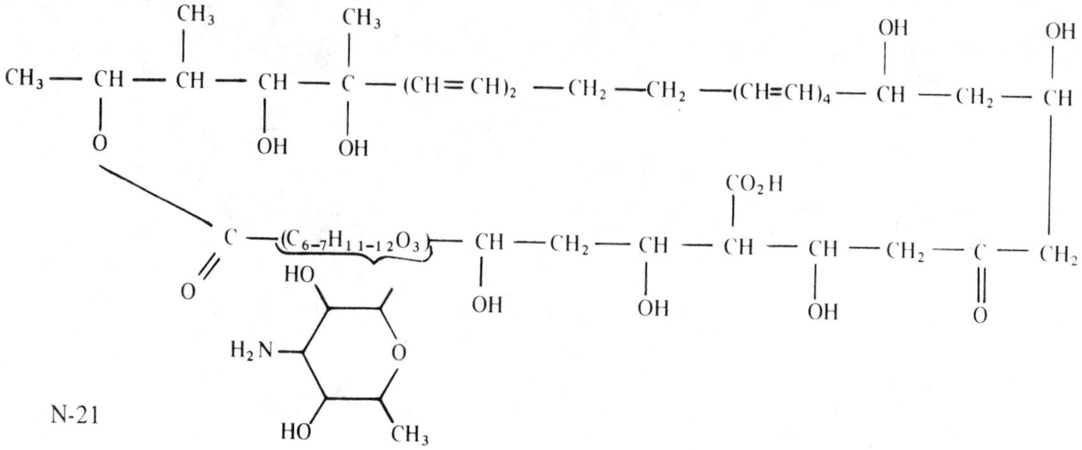

N-21

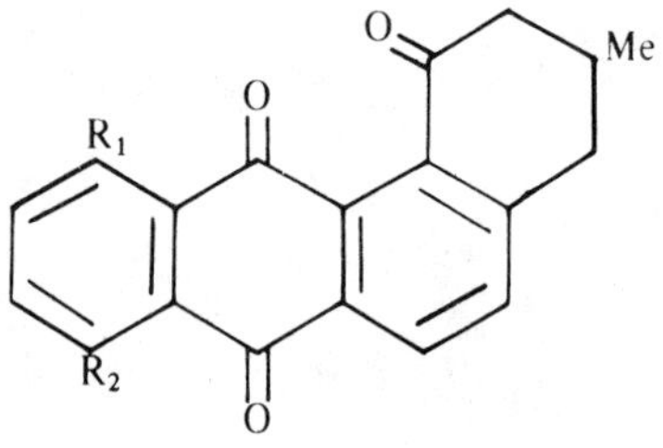

O-1

(V) $R_1 = OH, R_2 = H$
(VI) $R_1 = H, R_2 = OH$

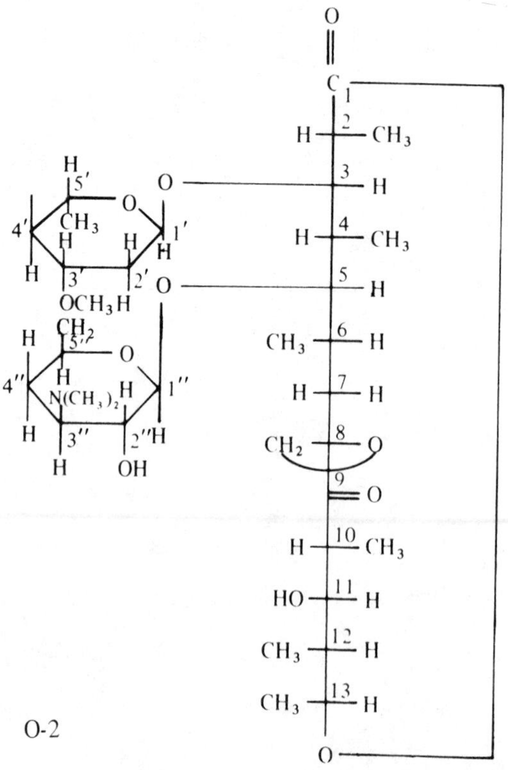

O-2

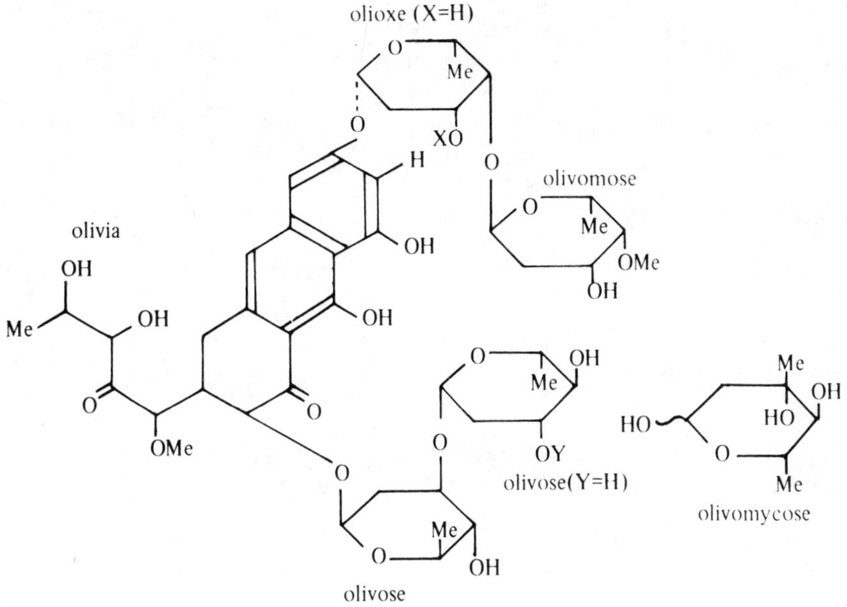

olioxe (X=H)

olivomose

olivia

olivose(Y=H)

olivomycose

olivose

A: X = Ac, Y = a-4-isobutyrylolivomycosyl
B: X = Ac, Y = a-4-acetylolivomycosyl
C: X = H, Y = a-4-isobutyrylolivomycosyl
D: X = Ac, Y = H

O-3

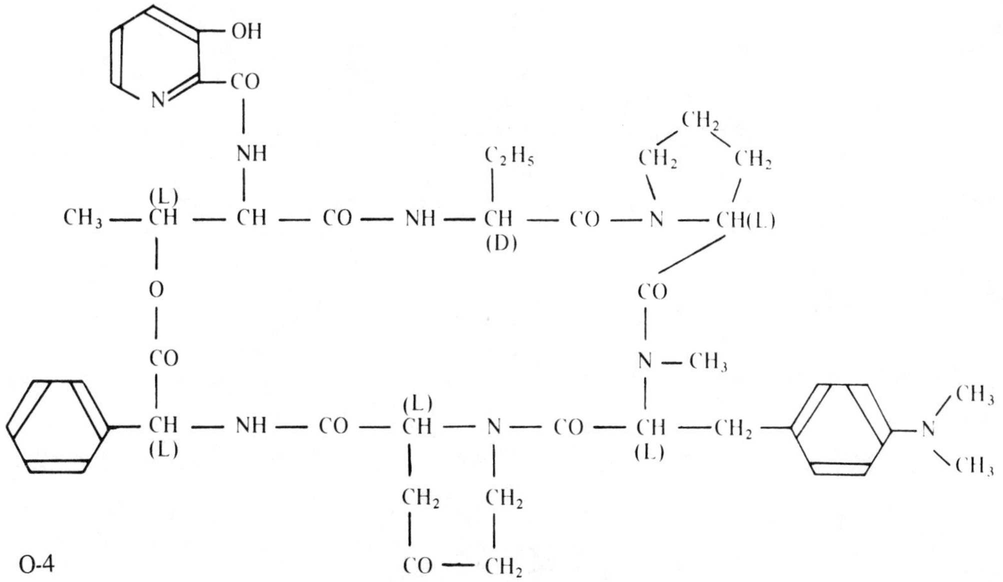

O-4

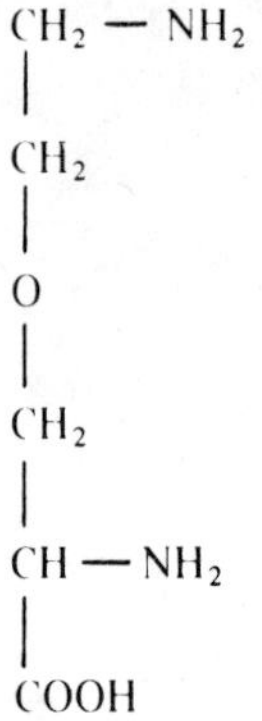

CH₂ — NH₂
|
CH₂
|
O
|
CH₂
|
CH — NH₂
|
COOH

4-Oxalysine

O-5

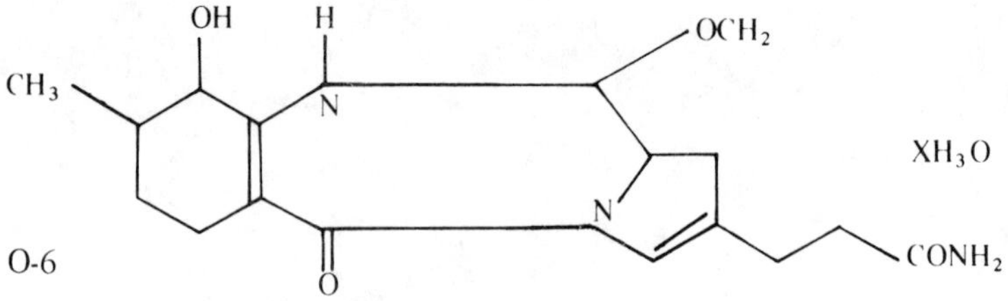

O-6

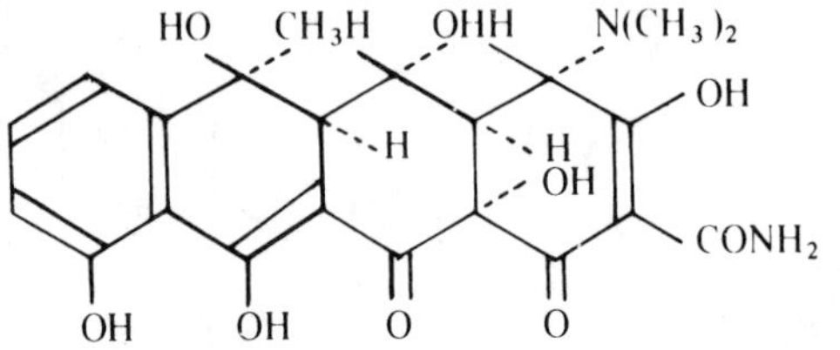

XH₃O

O-7

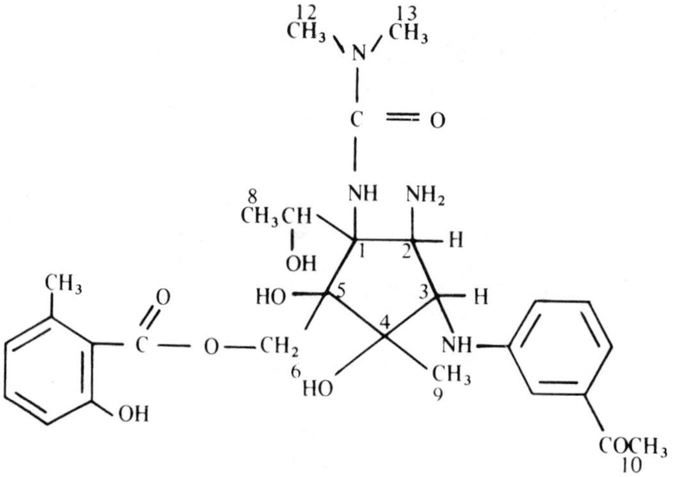

P-1

P-2

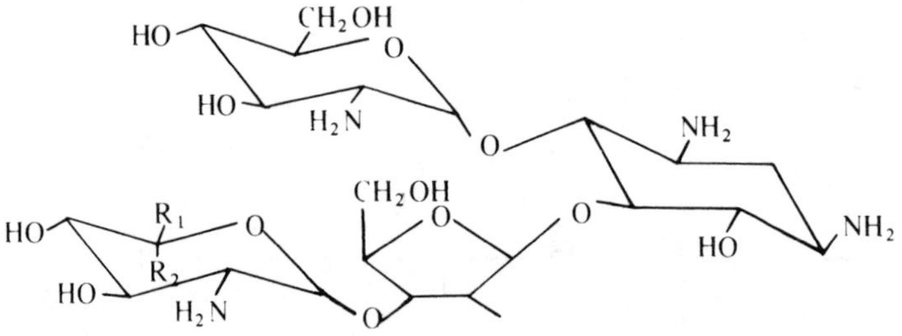

Paromomycin: $R_1 = H, R_2 = CH_2NH_2$
Paromomycin II: $R_1 = CH_2NH_2, R_2 = H$

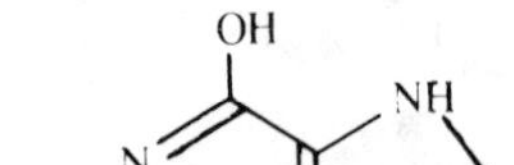

P-3

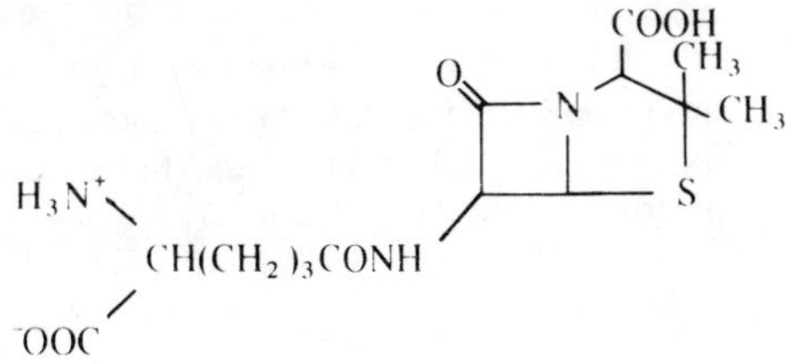

P-4

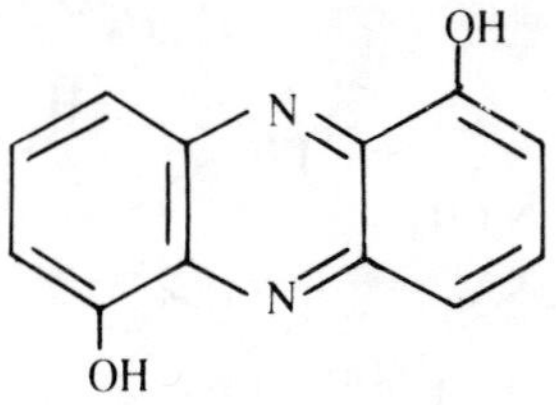

P-5

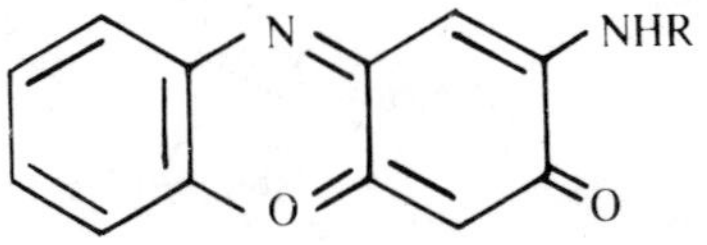

P-6

III, R = –H
IV, R = –COCH₃

$$III, R = -H$$
$$IV, R = -COCH_3$$

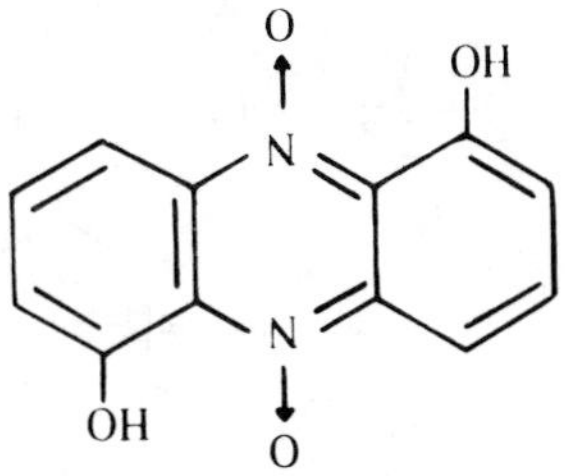

P-7

Aspartic acid	10	Alanine	18	Phenylalanine	1
Threonine	6	Valine	3	Lysine	11
Serine	8	Methionine	0	Histidine	4
Glutamic acid	4	Isoleucine	3	Arginine	4 (or 5)
Proline	2	Leucine	4		
Glycine	3 (or 4)	Tyrosine	3		

P-8

(–)-cis-1,2,-epoxy-
propylphosphoric acid

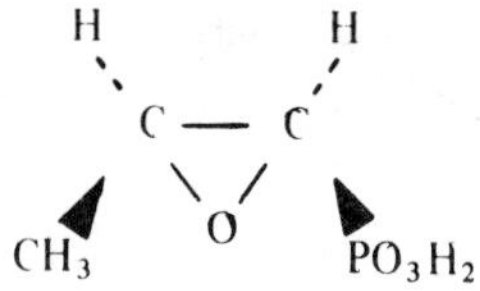

P-9

A:

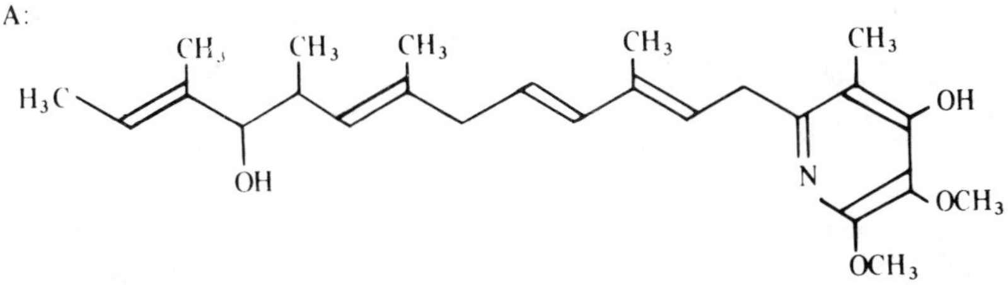

B:

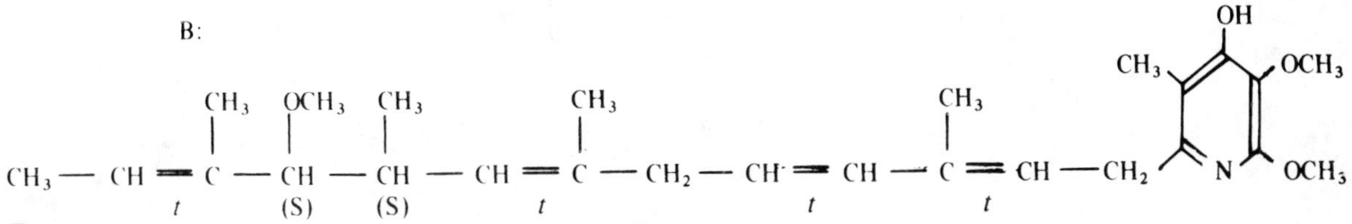

P-10

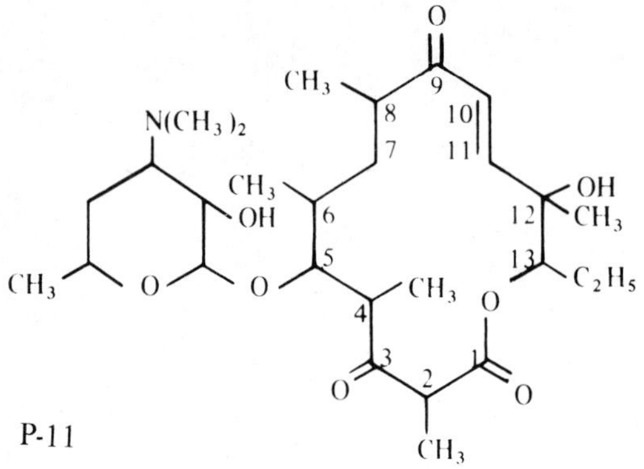

P-11

Acid hydrolysis of pillaromycin A gave pillaro-mycinone ($C_{20}H_{10\ 18}O_7$) and pillarose, 2,5-dide-oxypentose ($C_5H_{10}O_3$). Catalytic reduction of pillaromycinone gave pillaronone.

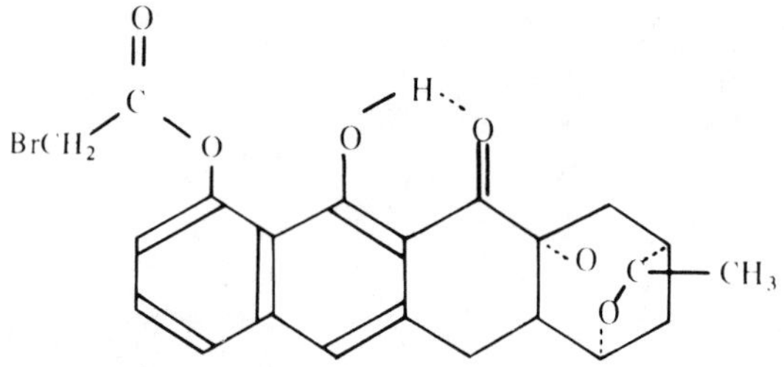

P-12   Monobromoacetate of pillaronone

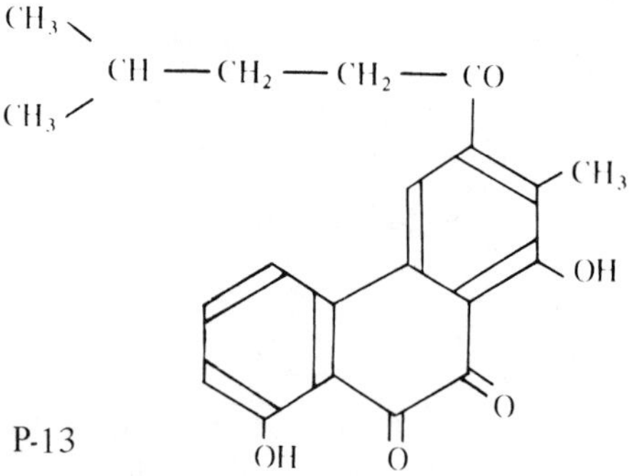

P-13

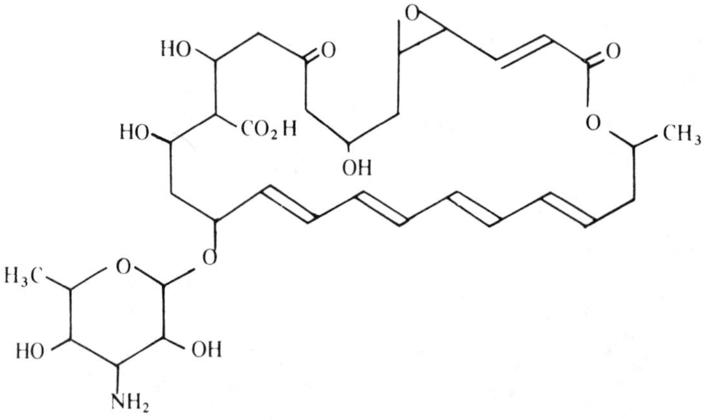

P-14

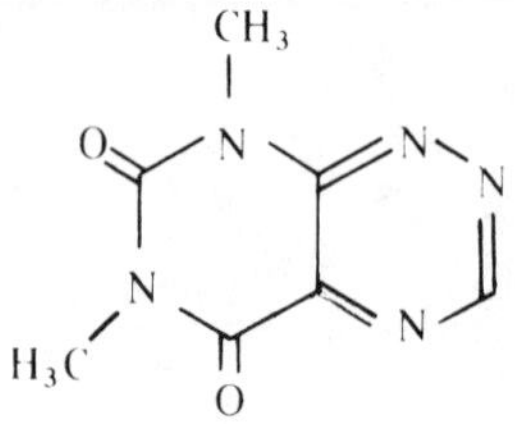

P-15

P-16 Should be revised as depicted in analogy to amicetin.

Polyoxin	R₁		R₂	R₃
1a,	A	CH₂OH	COOH	OH
b,	B	CH₂OH	HO	OH
d,	D	COOH	HO	OH
e,	E	COOH	HO	H
f,	F	COOH	COOH	OH
g,	G	CH₂OH	HO	H
b,	H	CH₃	COOH	OH
i,	J	CH₃	HO	OH
k,	K	H	COOH	OH
l,	L	H	HO	OH
1′c,	C	HO	COOH	
i,	I			

P-17

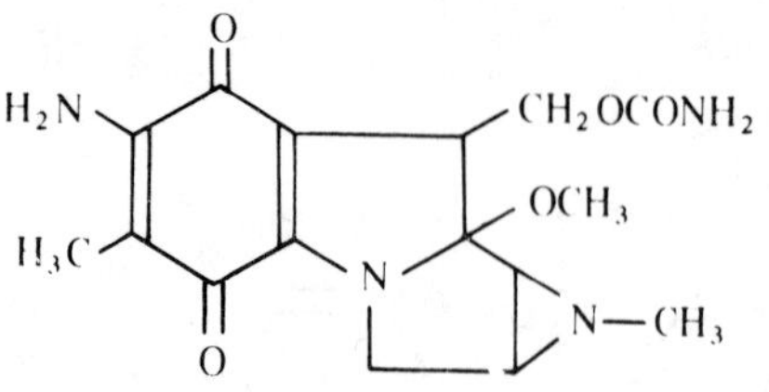

P-18

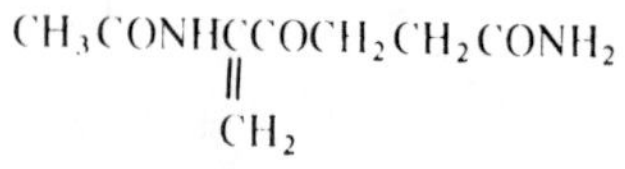

$$CH_3CONHCCOCH_2CH_2CONH_2$$
$$\|$$
$$CH_2$$

P-19

Primycin (1, R = H)

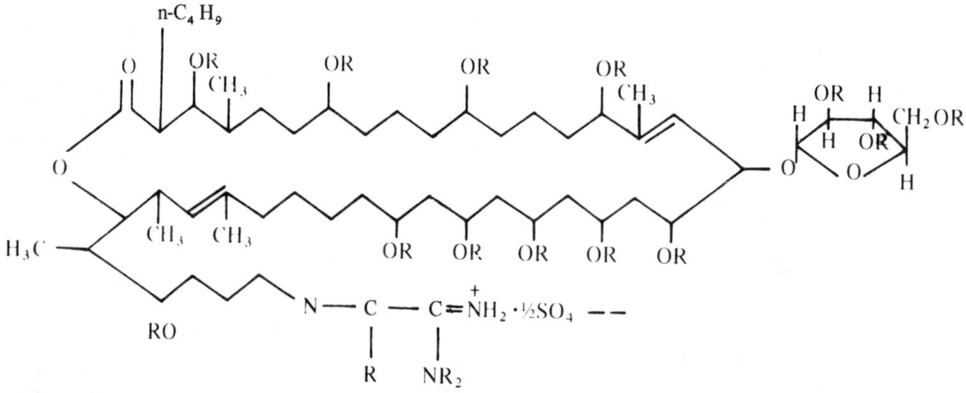

P-20

1

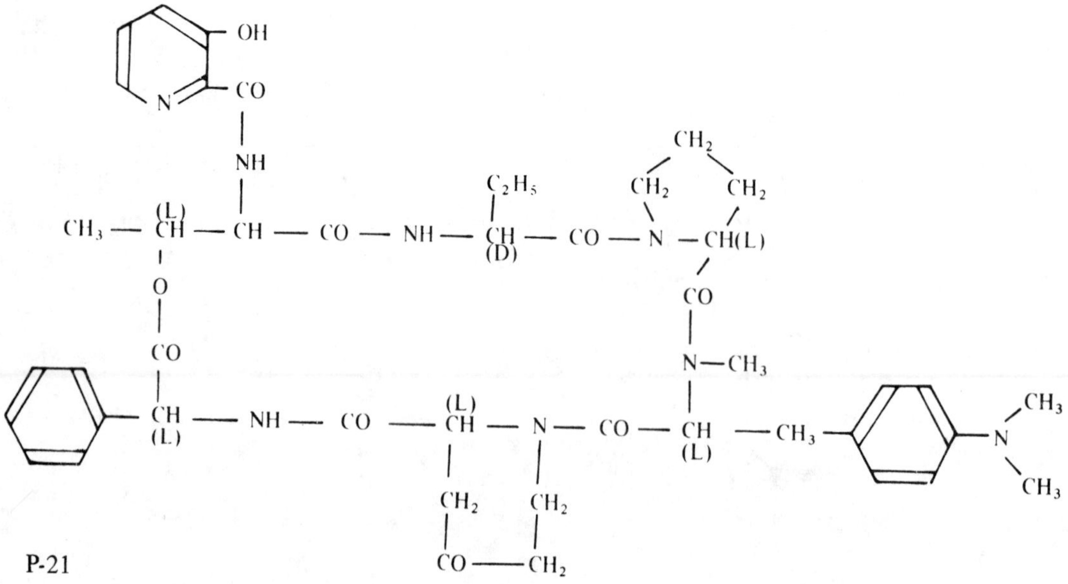

P-21

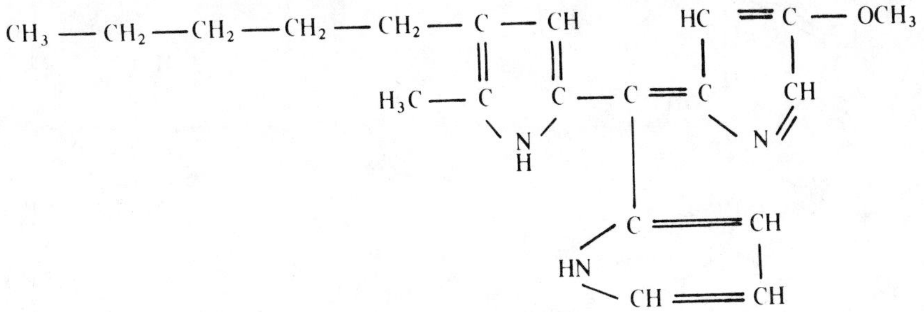

P-22 Prodigiosin

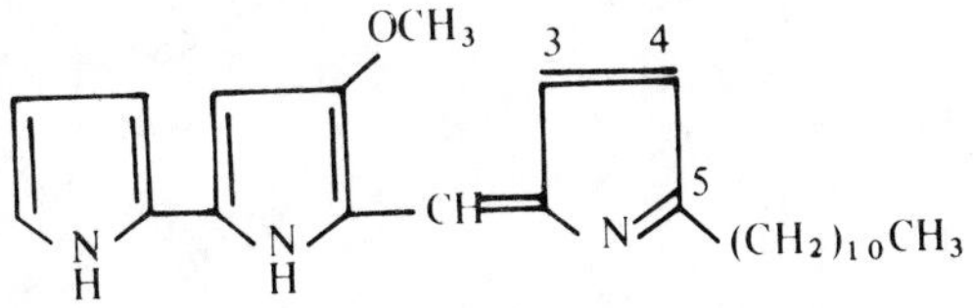

P-23 XI
Prodigiosin-25 C

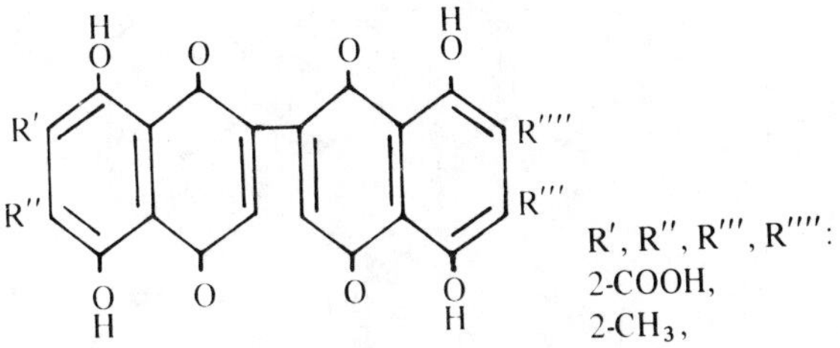

P-24 ACTINORHODIN

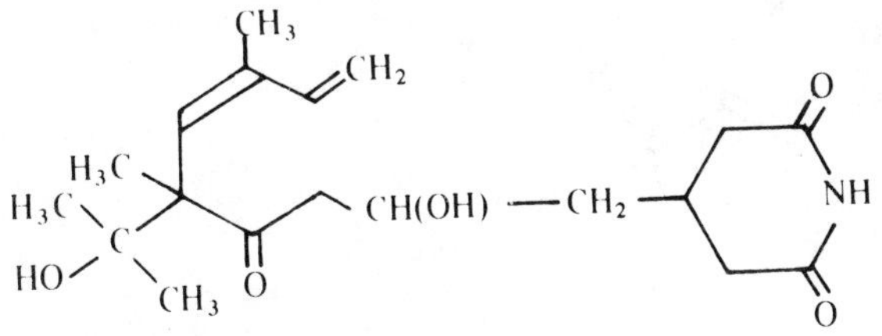

R′, R″, R‴, R‴′:
2-COOH,
2-CH₃,
C₈H₁₆O₂

P-25

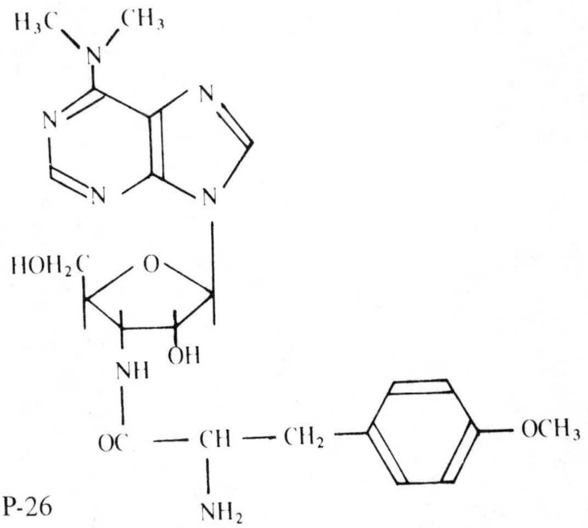

P-26

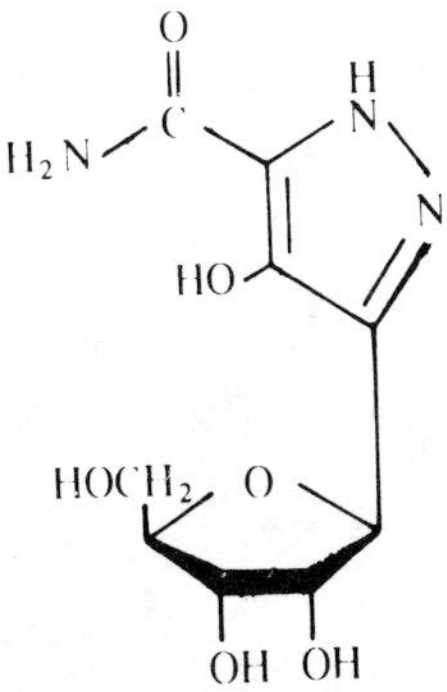

P-27

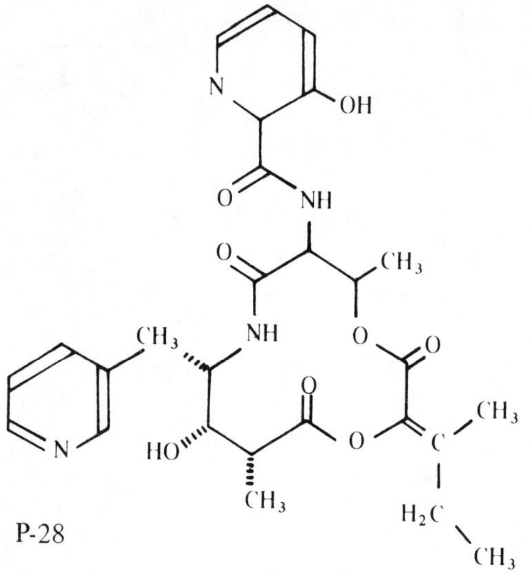

P-28

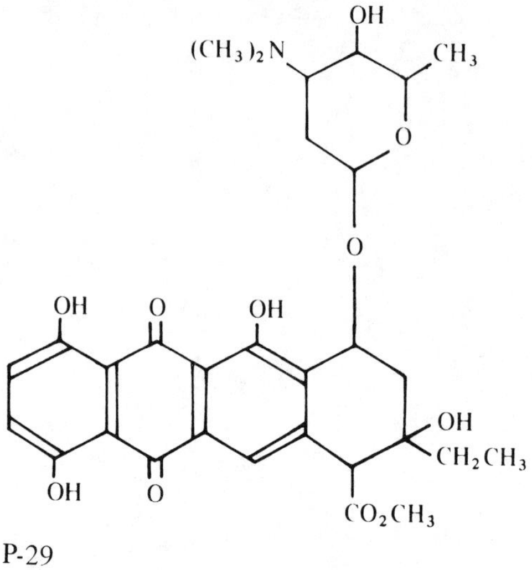

P-29

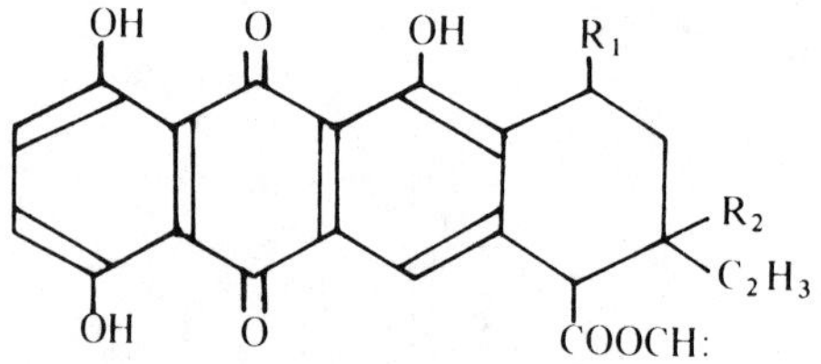

$\zeta: R_1 = H, \quad R_2 = OH$
$\epsilon: R_1 = OH, R_2 = OH$

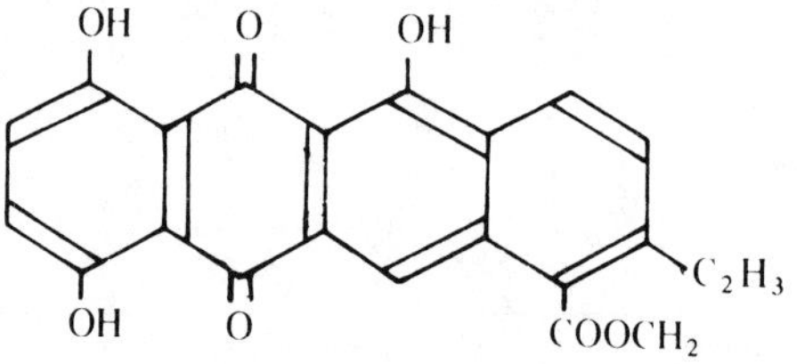

P-30

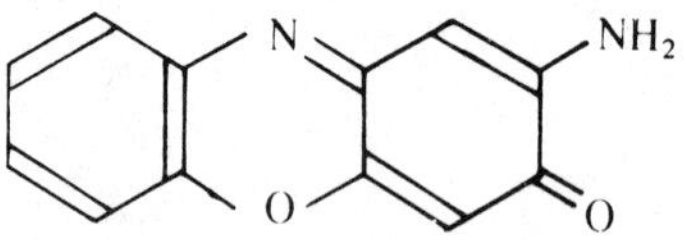

Q-1

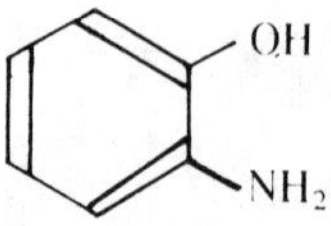

Q-2

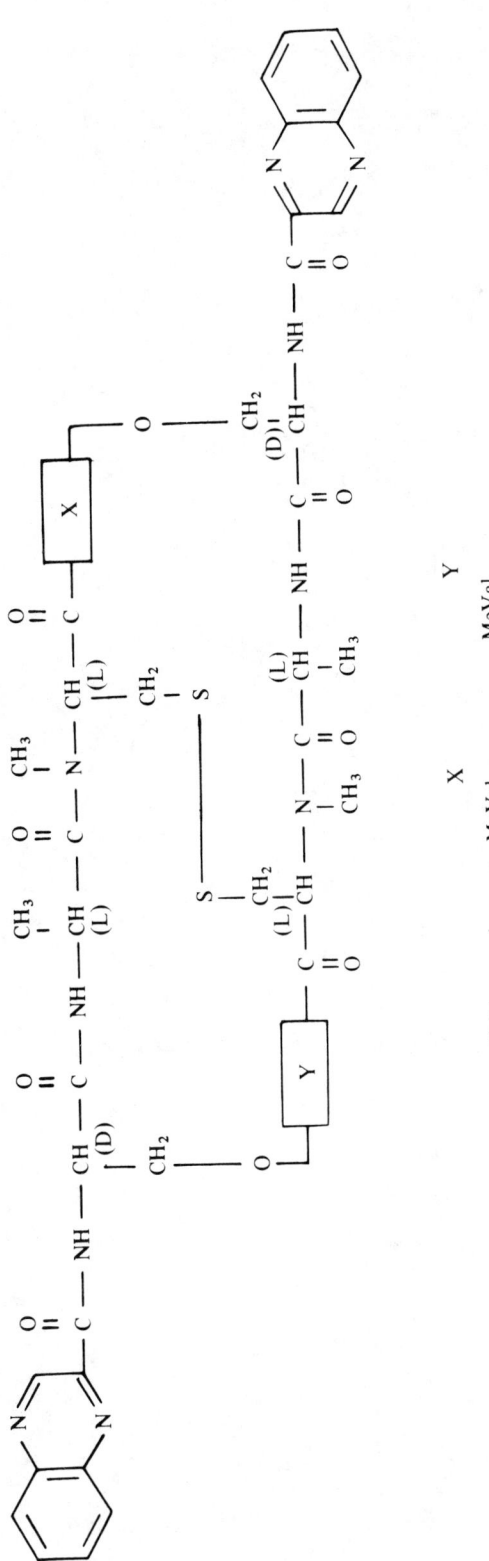

	X	Y
Echinomycin (Quinomycin A)	MeVal	MeVal
Quinomycin B₀	MeVal	DimeAlloileu
Quinomycin C	DimeAlloileu	DimeAlloileu
Quinomycin D	MeVal	McAlloileu
Quinomycin B	MeAlloileu	McAlloileu
Quinomycine E	McAlloileu	DimeAlloileu

MeVal: N—Methyl-valine
Dimelloileu: N—Dimethyl-allo-isoleucine
MeAlloileu: N—Methyl-allo-isoleucine

Q-3

Lividomycin:
$C_{29}H_{35}N_5O_{18}$

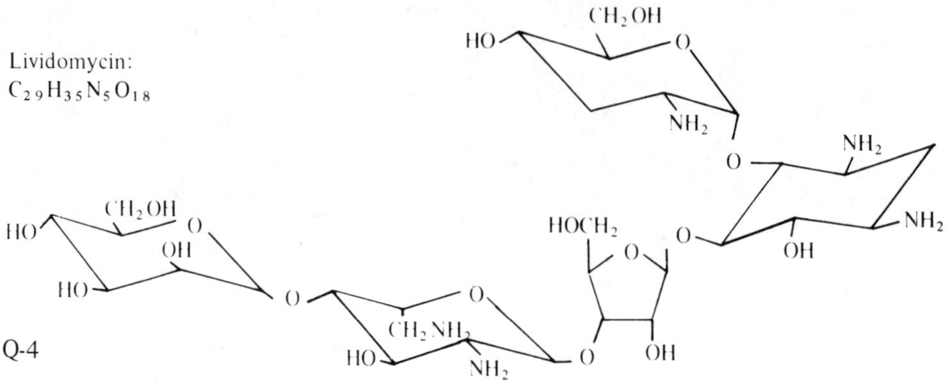

Q-4

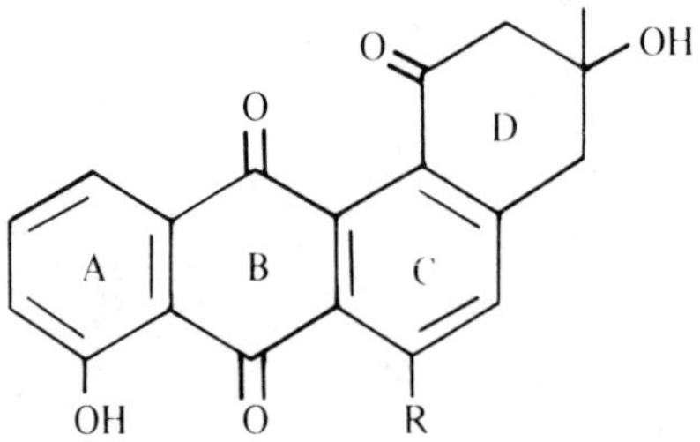

R-1

I R=OH
II R=H

or

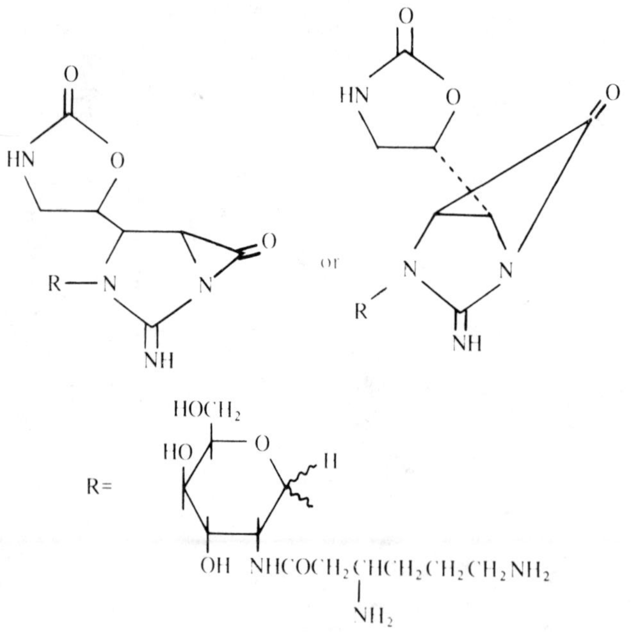

R-2

$R=$

R-3

Acid hydrolysis gave β-lysine and streptolidine identical with geamine or roseonine.

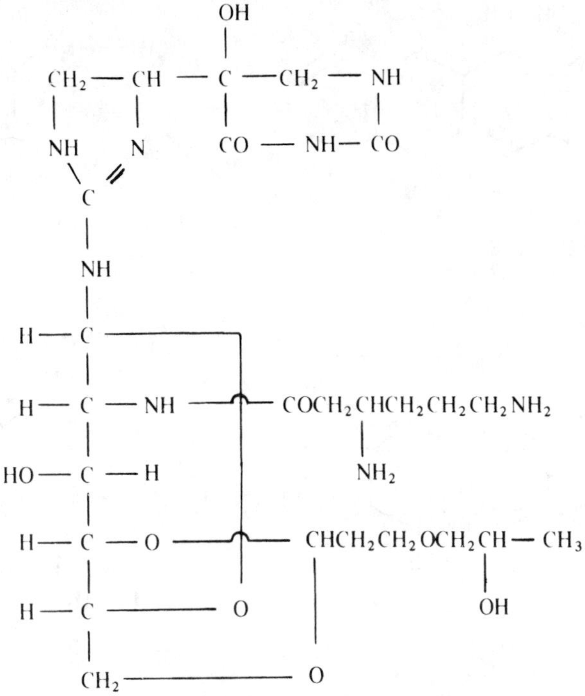

R-4

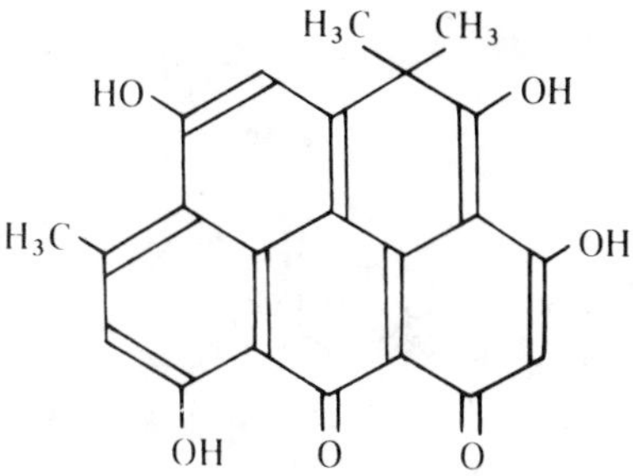

$CH_2 - CH - C - CH_2 - NH$

OH

$NH \quad N \quad CO - NH - CO$

C

NH

$H - C$

$H - C - NH \quad COCH_2CHCH_2CH_2CH_2NH_2$

NH_2

$HO - C - H$

$H - C - O \quad CHCH_2CH_2OCH_2CH - CH_3$

$H - C \quad O \quad OH$

$CH_2 \quad O$

R-5

R-7

1a: R = H
b: R = COCF$_3$

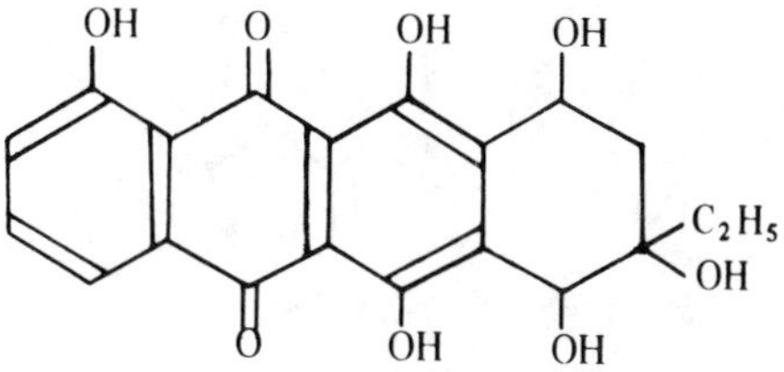

(I)

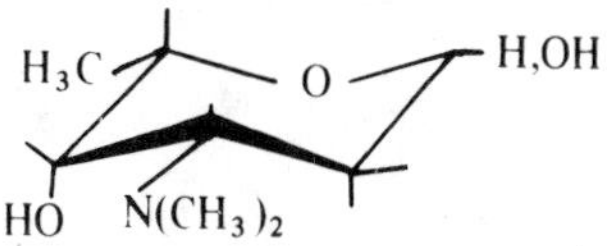

(II)

On hydrolysis A gave one mole of β-rhodo-mycinone (I) and two moles of rhodosamine (II), while B gave one mole of β-rhodomycinone and one mole of rhododosamine.

R-6

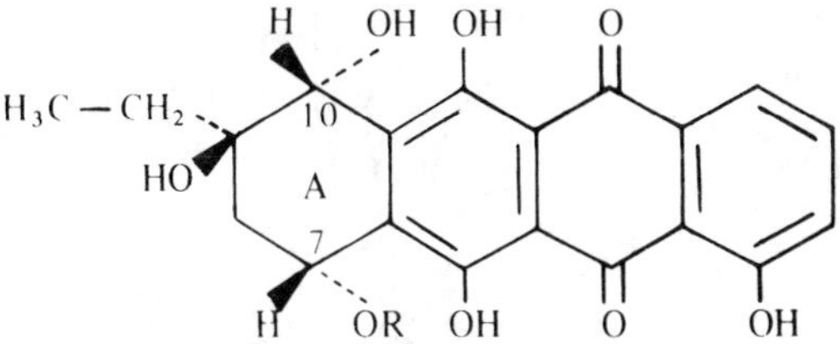

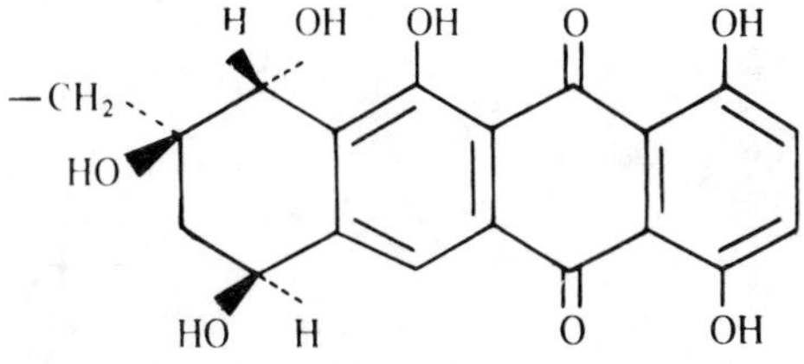

R-8

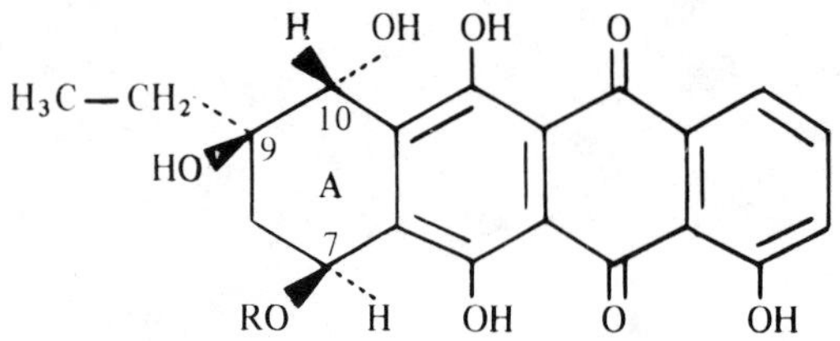

2a: R = H
 b: R = COCF$_3$
β-Rhodomycimone

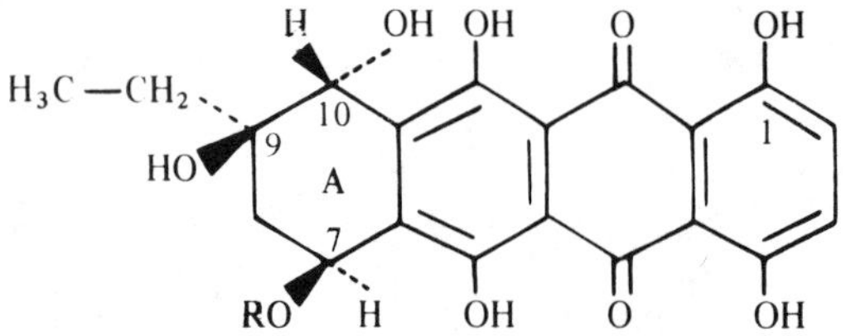

R-9

4a: R = H
 b: R = COCF$_3$
 c: R = H; umgekehrte
    Konfiguration an C-7
β-Iso-Rhodomycimone

R-13, 14

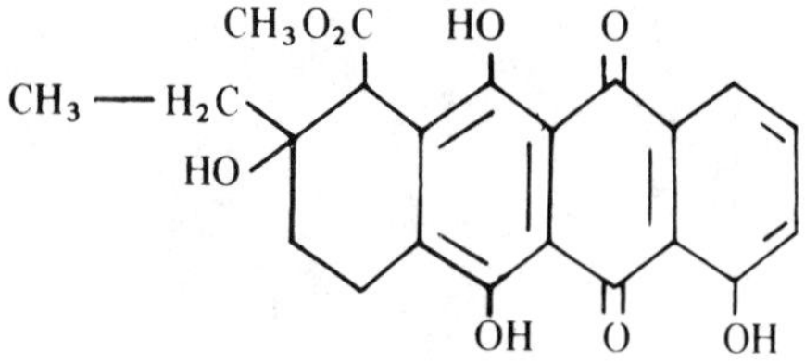

ζ-Rhodomycinone   (V)
ζ-Isorhodomycinone
OH at C-1, H at C-4   (Va)

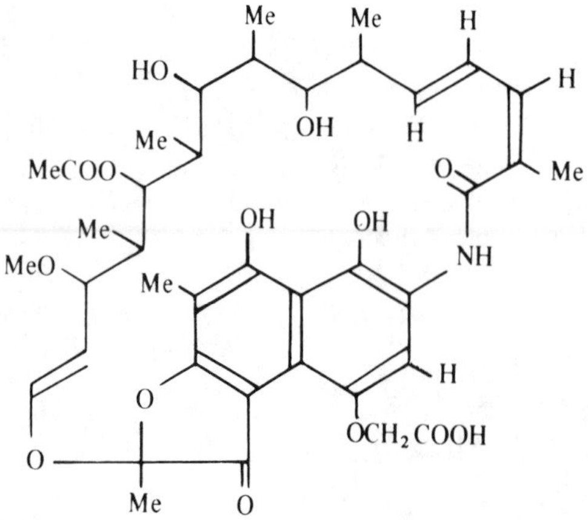

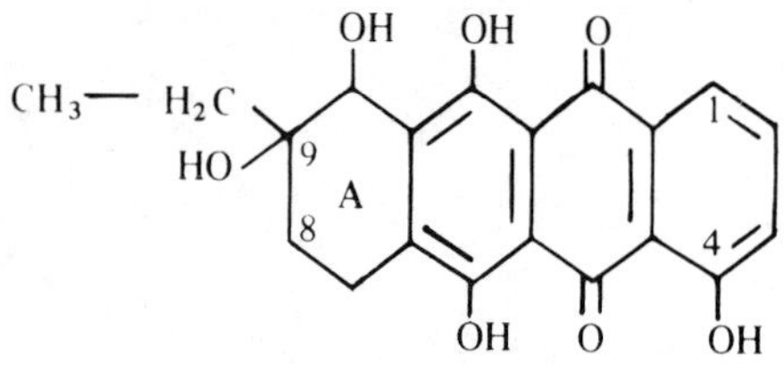

R-10   γ-Rhodomycinone   (II)

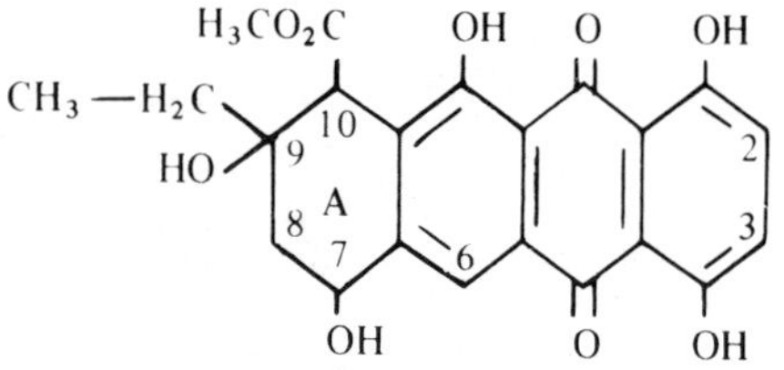

R-11   δ-Rhodomycinone   (IV)

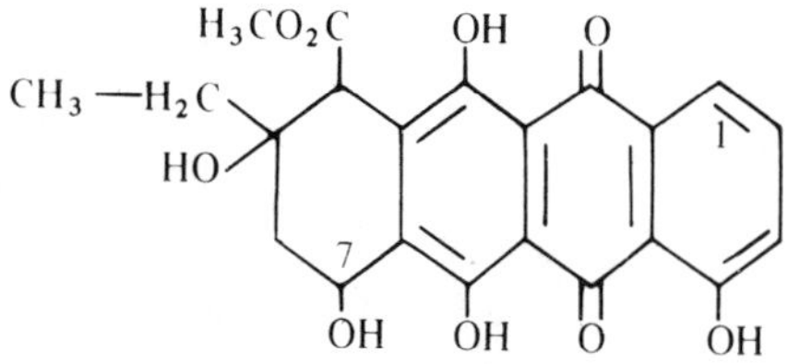

R-12   ε-Rhodomycinone

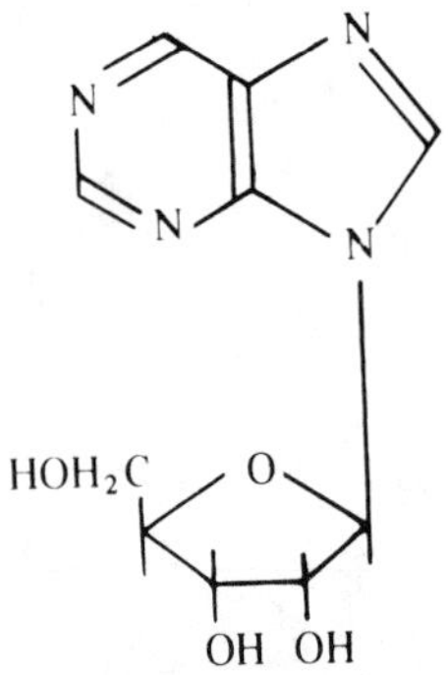

R-15

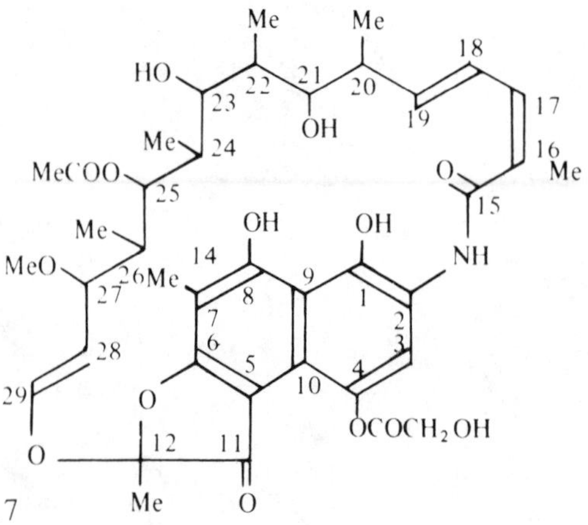

R-16

R-17

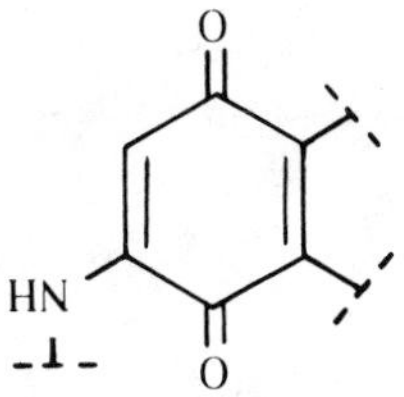

R-18

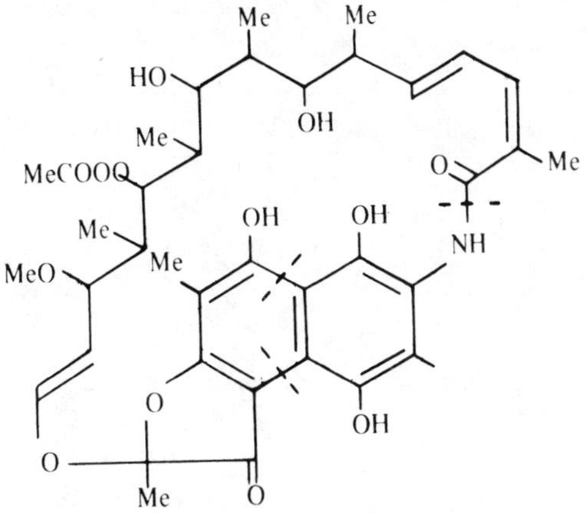

R-19

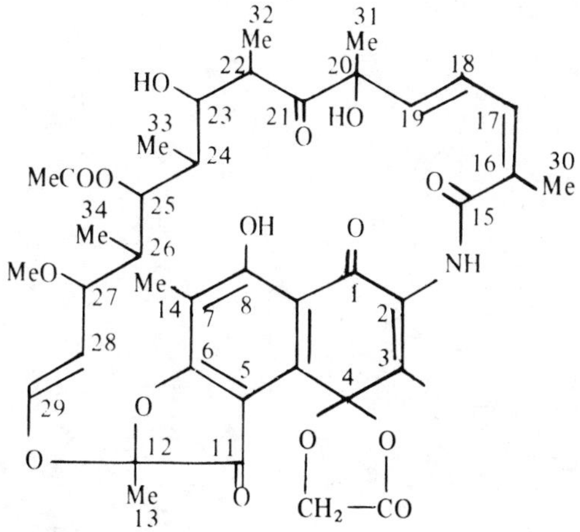

I Rifamycin YO

R-20

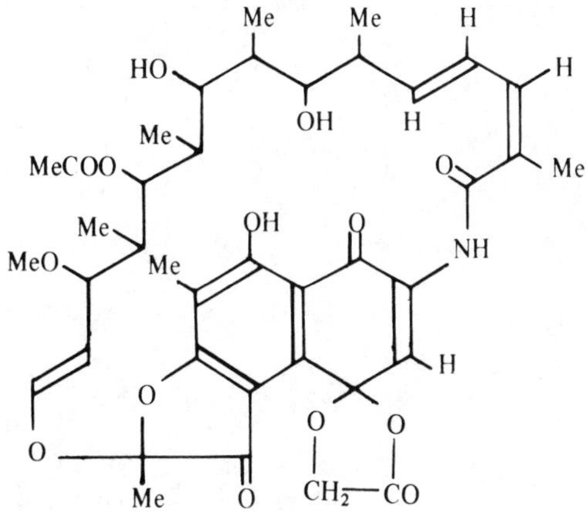

R-21

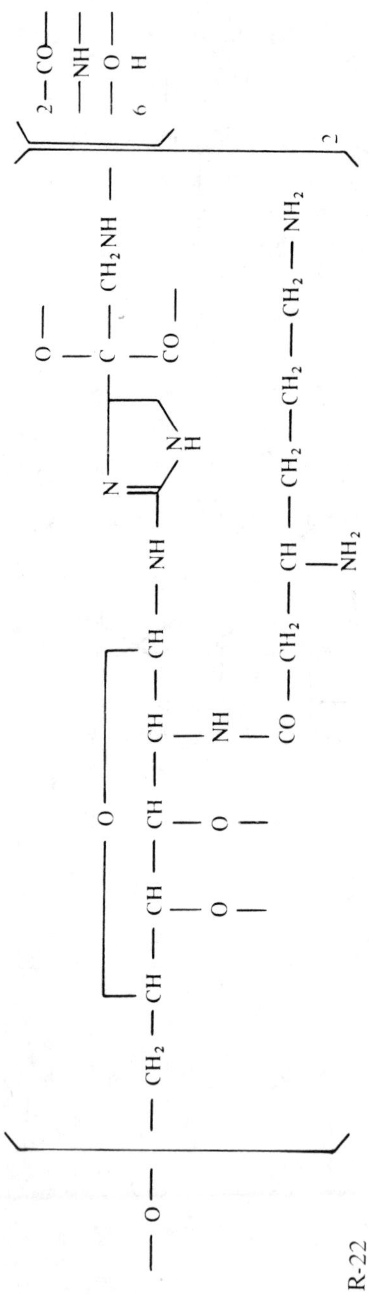

R-22

Partial structure

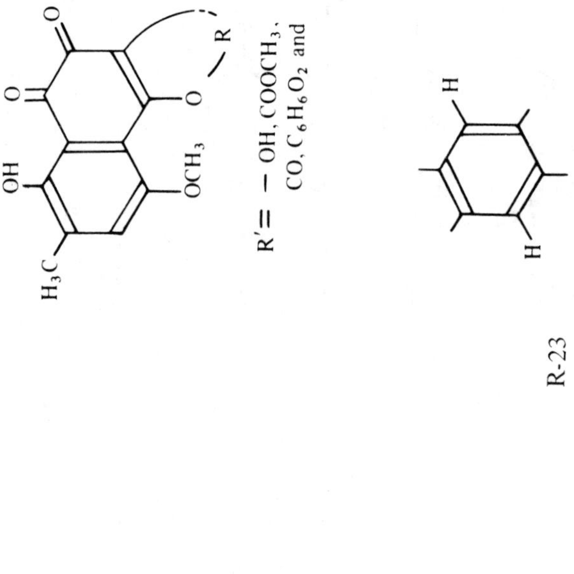

$R' = -OH, COOCH_3,$
$CO, C_6H_6O_2$ and

R-23

R-24

R-25

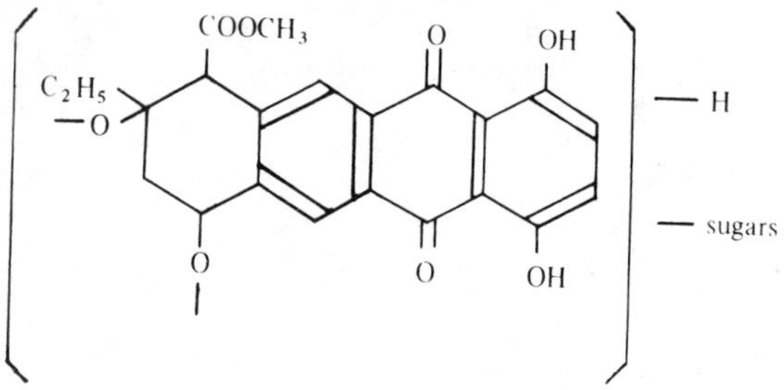

A and B differ from one another in the sugar portion.

R-26

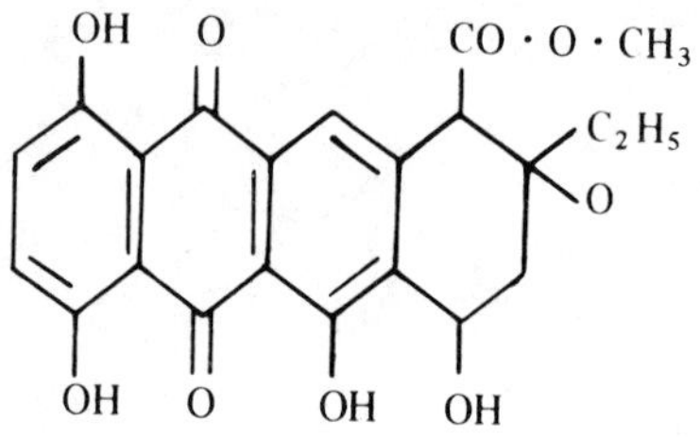

R-27

(1)

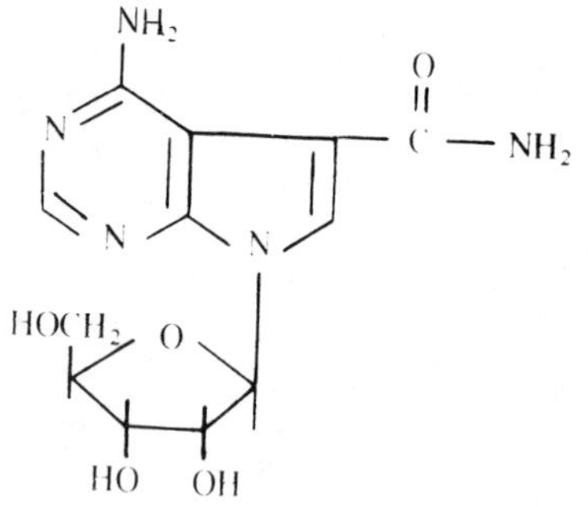

S-1

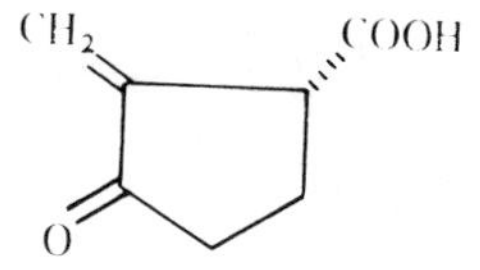

S-2

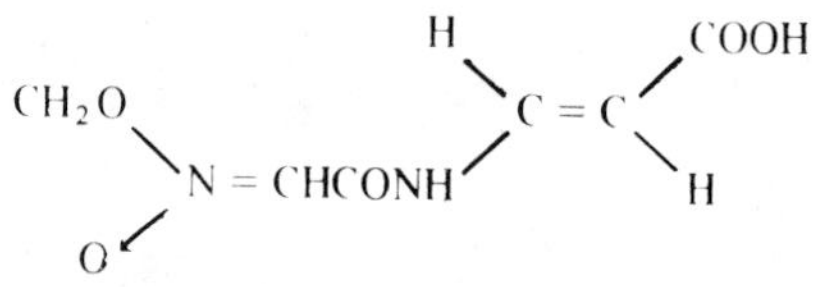

S-3

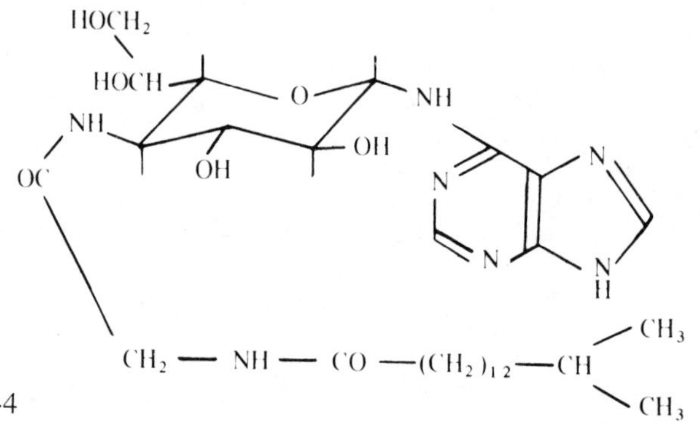

S-4

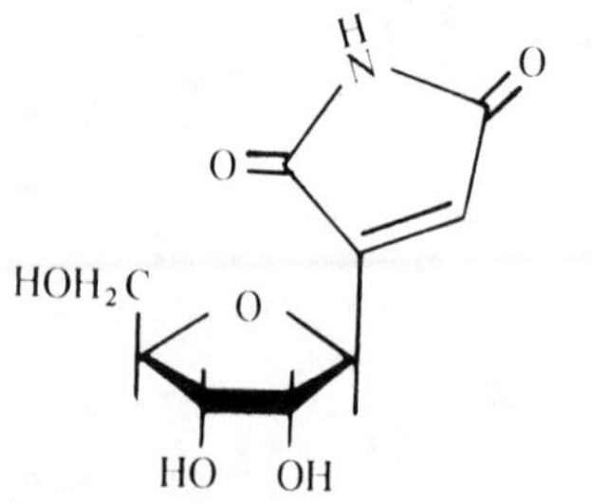

Showdomycin
3-(β-D-ribofuranosyl)maleimide

S-5

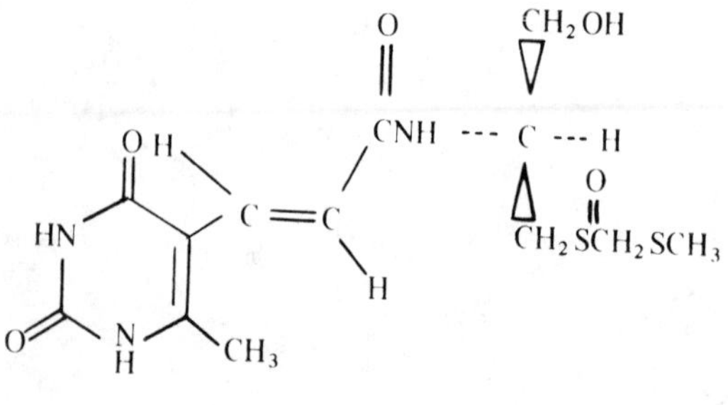

S-6

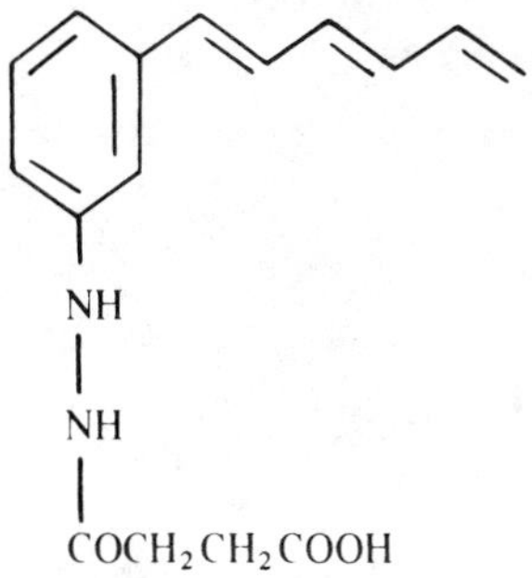

S-7

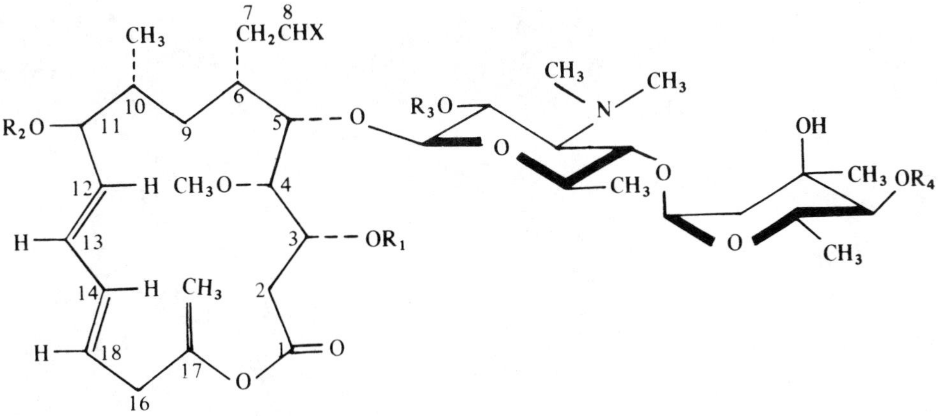

	X	R_1	R_2	R_3	R_4
S-8-1	S-8-1 = O	H	Z	H	H
S-8-2	S-8-2 = O	$COCH_3$	Z	H	H
S-8-3	S-8-3 = O	$COCH_2CH_3$	Z	H	H

M_1:

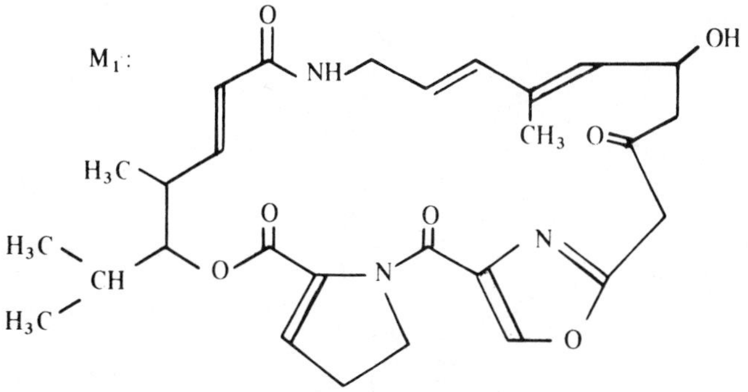

Proposed for ostreogrycin A.

S-9

S-10

S-11

S-12

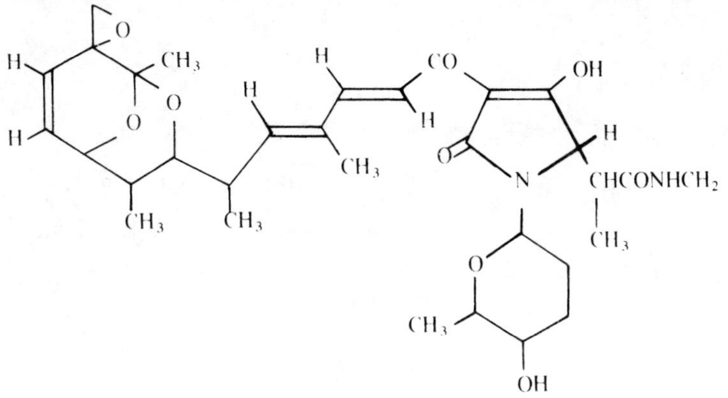

S-13

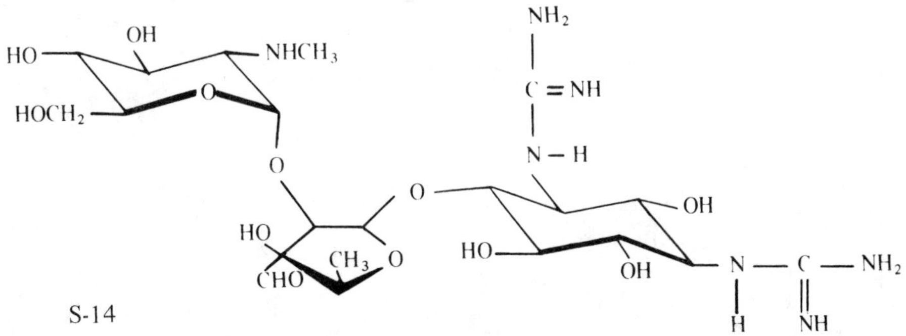

S-14

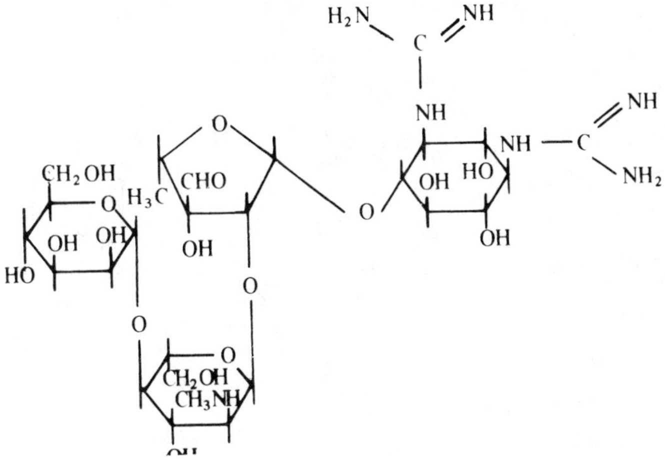

S-15

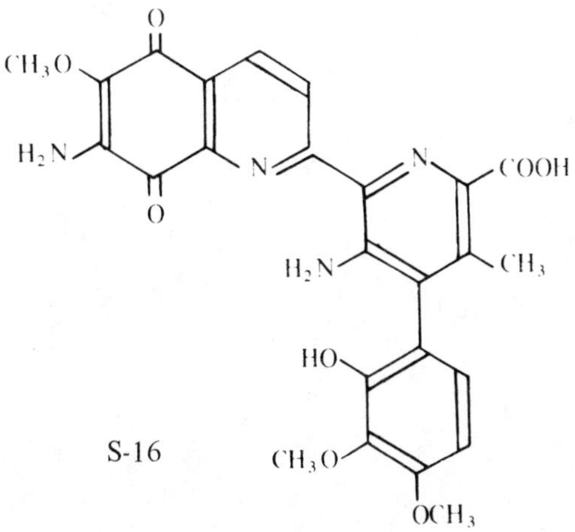

S-16

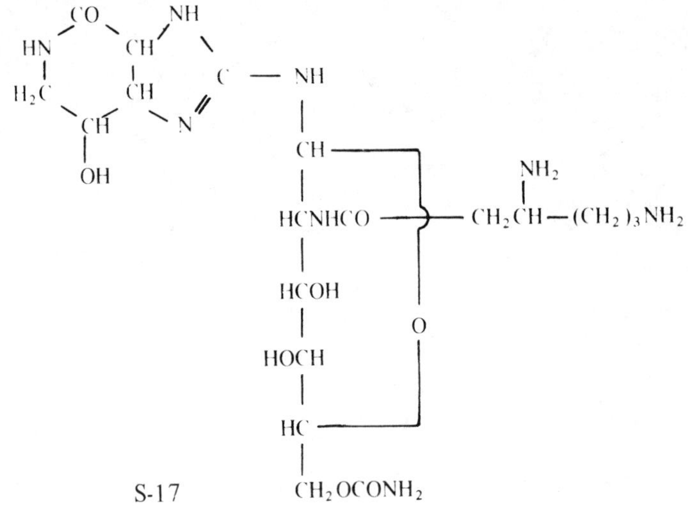

S-17

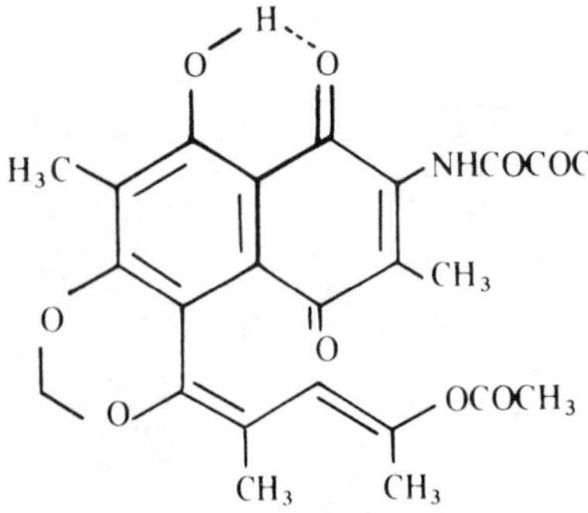

S-18

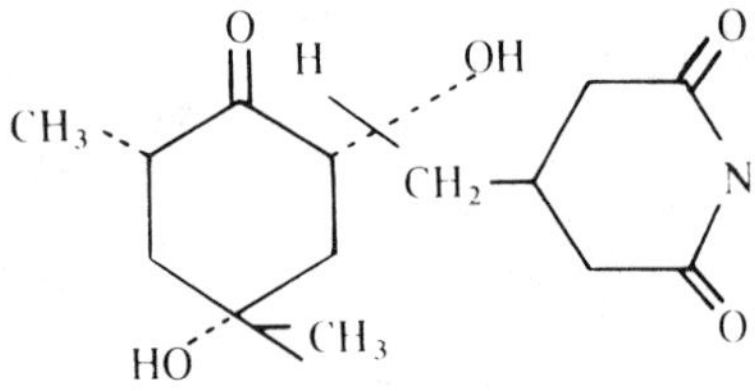

S-19

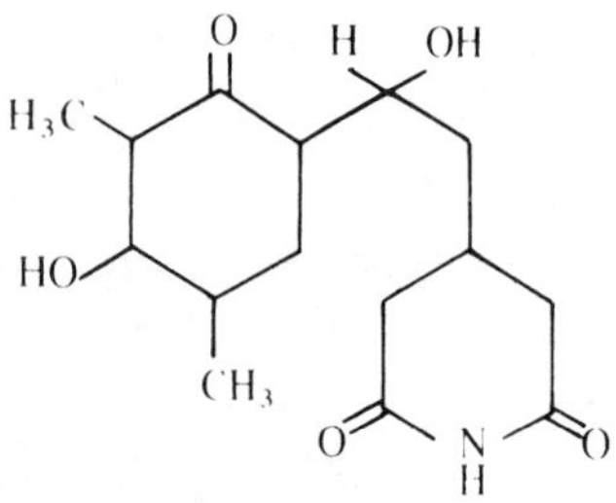

S-20

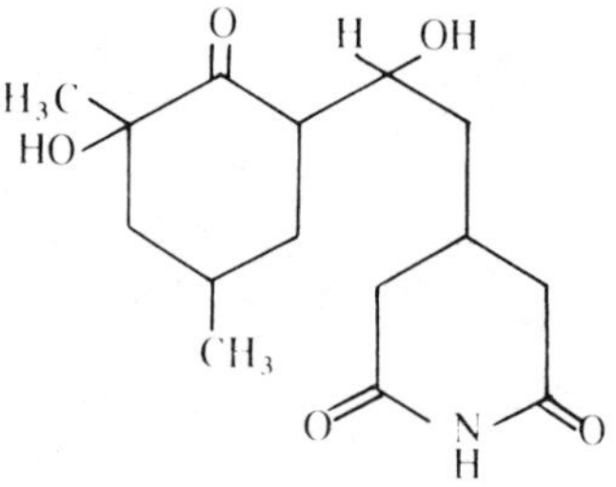

S-21

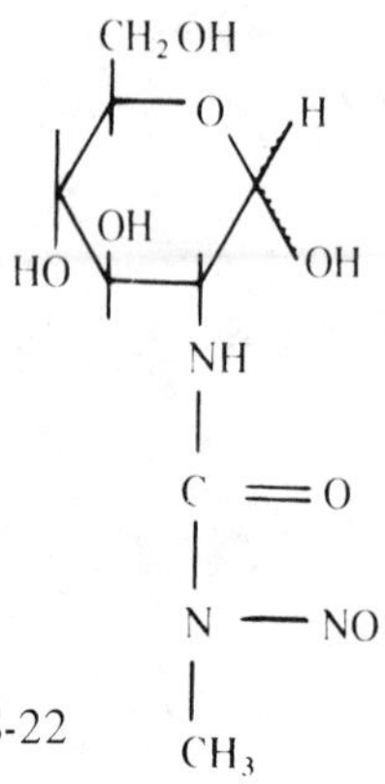

S-22

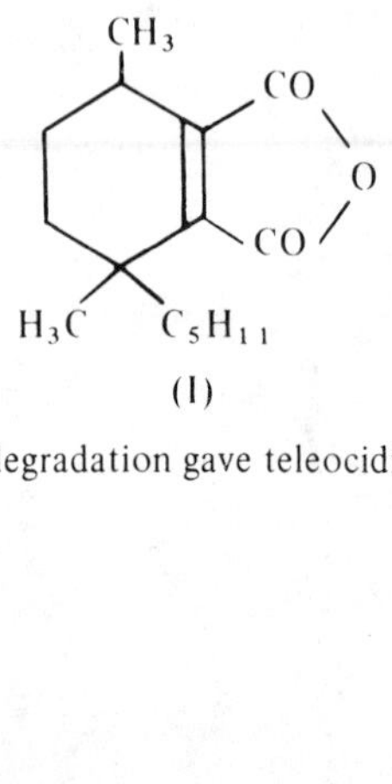

Oxidative degradation gave teleocidic anhydride (I).

T-1

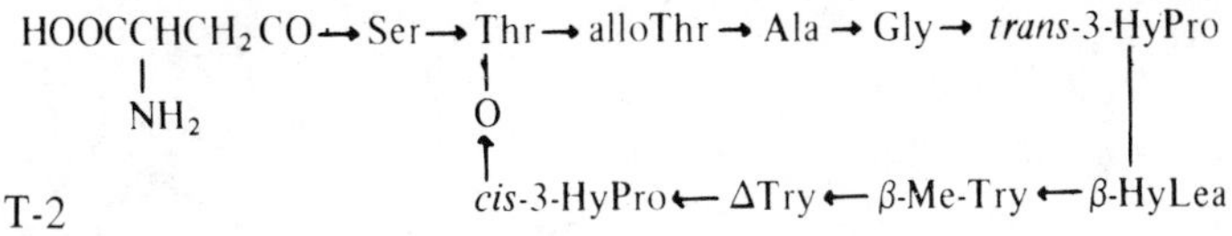

T-2

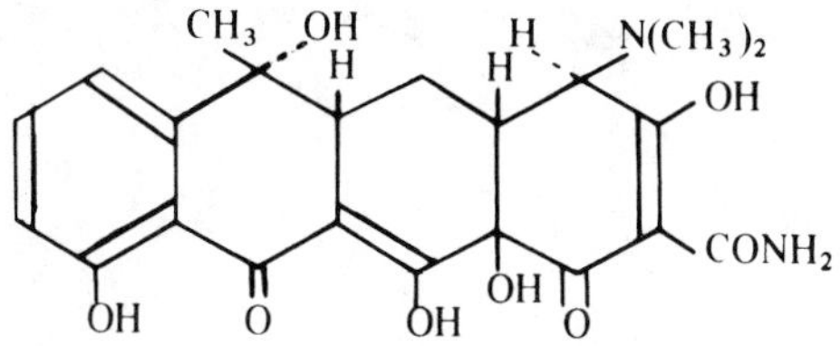

$$HOOCCHCH_2CO \rightarrow Ser \rightarrow Thr \rightarrow alloThr \rightarrow Ala \rightarrow Gly \rightarrow \textit{trans}\text{-3-HyPro}$$

with NH$_2$ below, O, and

$$cis\text{-3-HyPro} \leftarrow \Delta Try \leftarrow \beta\text{-Me-Try} \leftarrow \beta\text{-HyLea}$$

T-3

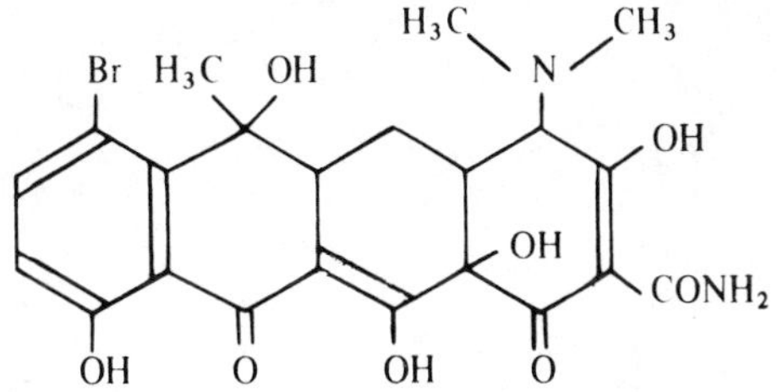

T-4

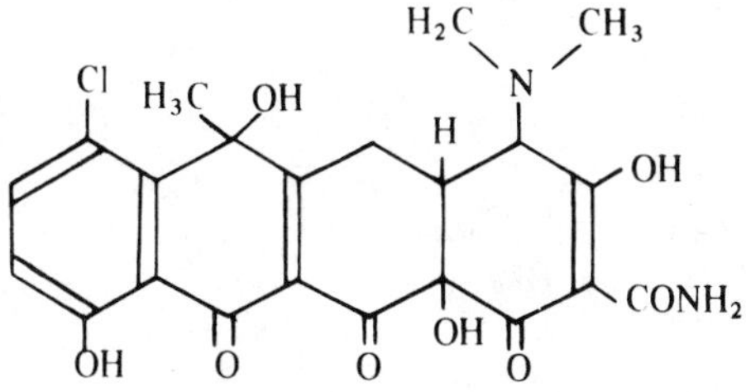

T-5

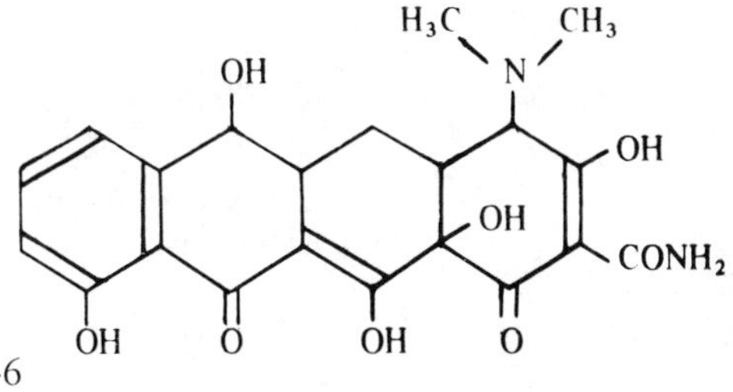

T-6

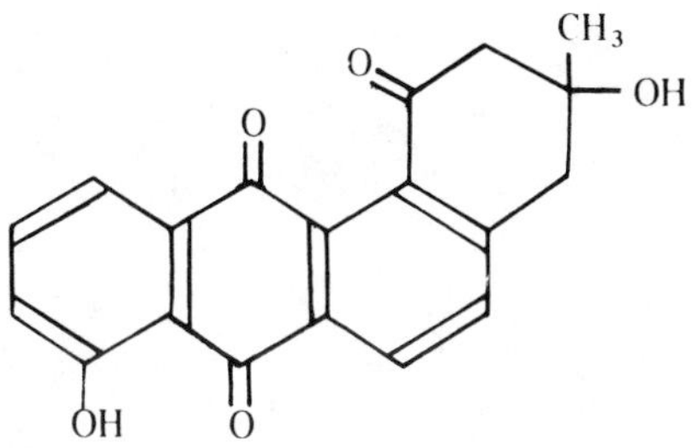

T-7

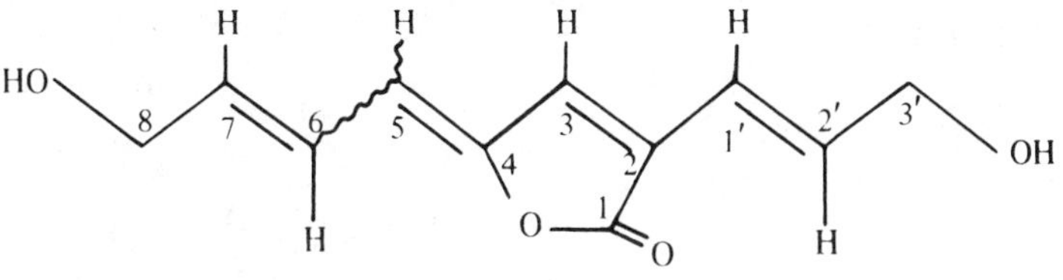

T-8

(I)

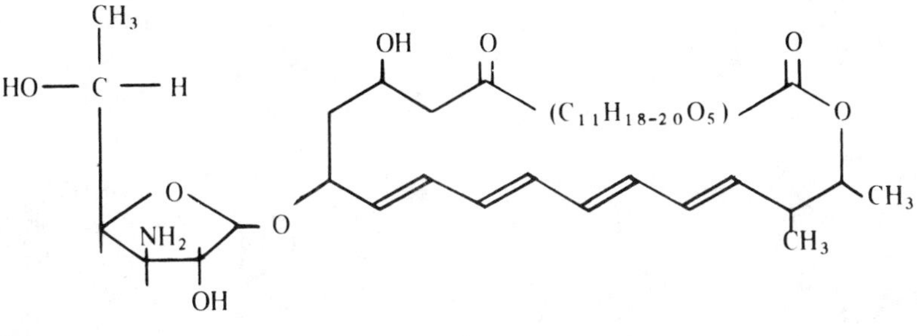

T-9

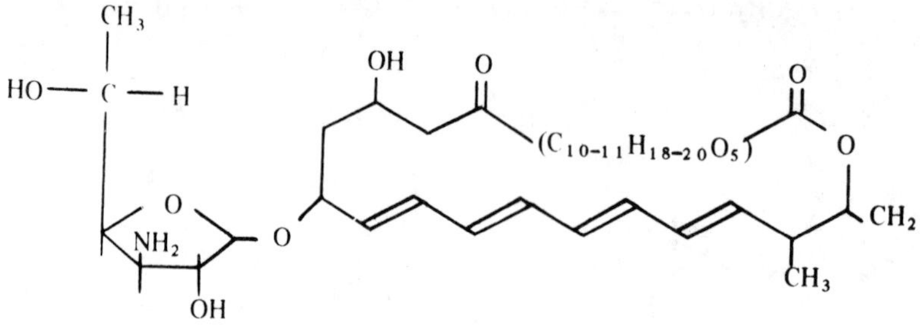

T-10

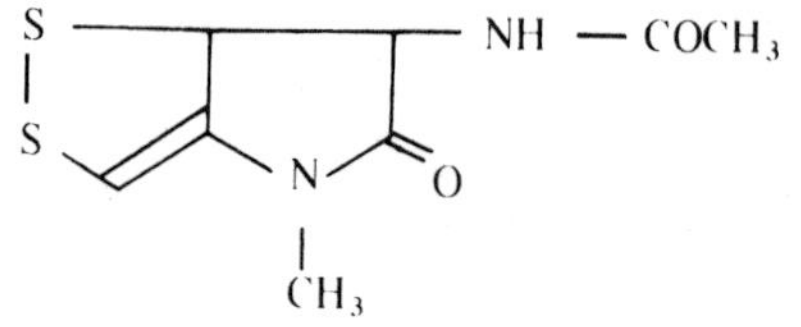

T-11

Acid hydrolysis gave L-alanine (2 moles), D-cystine, L-isoleucine, L-threonine, 8-hydroxyquinaldic acid, thiostreptoic acid (I) and thiostreptine (II).

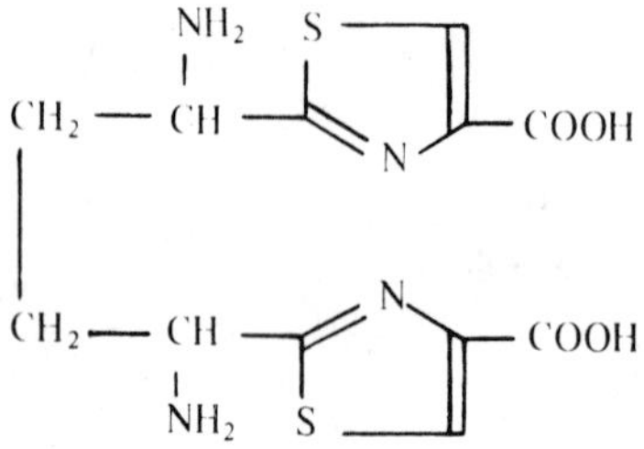

(I) 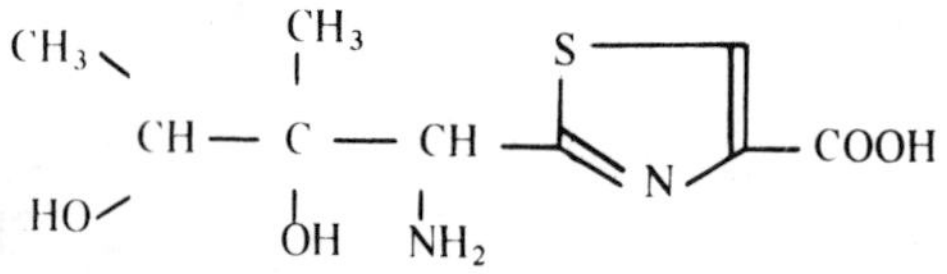   (II)

T-12

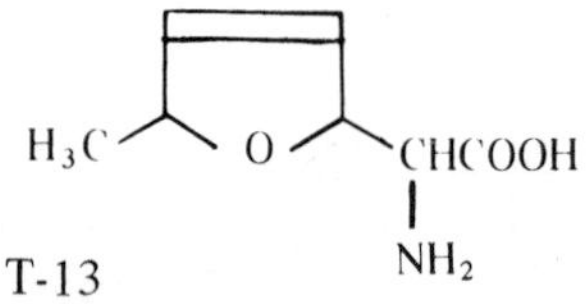

T-13

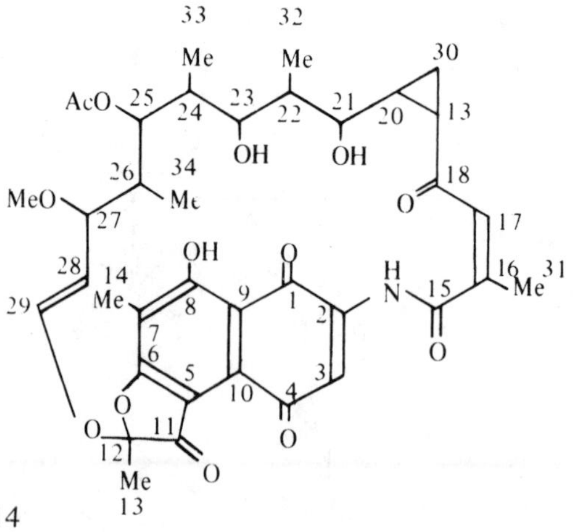

T-15

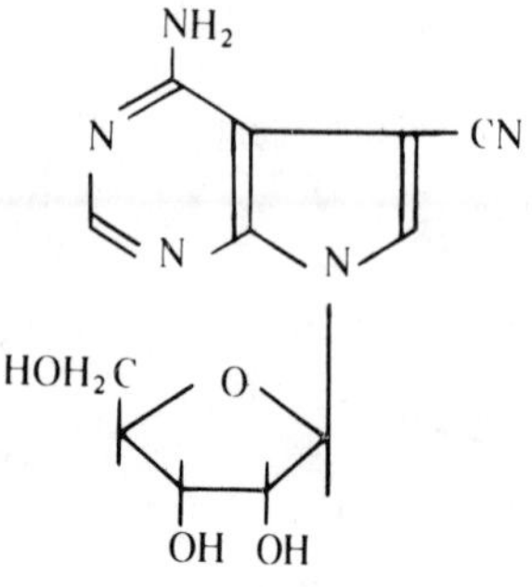

T-14

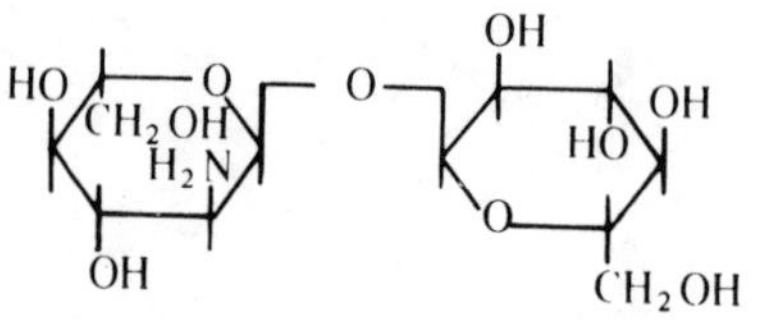

T-16

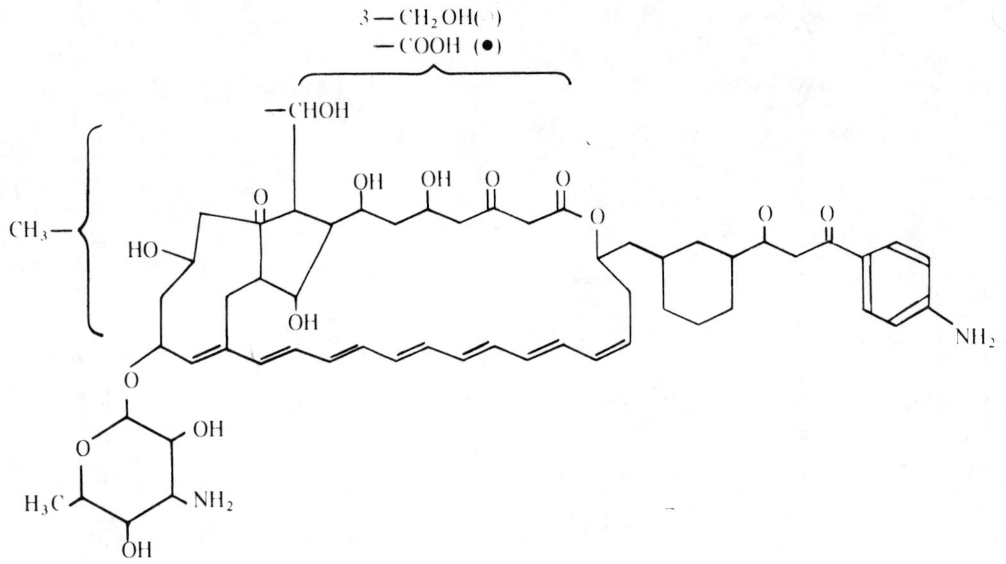

$3 - CH_2OH(\circ)$
$- COOH \; (\bullet)$

One of oxygen marked by * is epoxide, positions marked by $\circ$ have $-CH_2OH$ and one of position marked by $\bullet$ has $-COOH$.

T-17

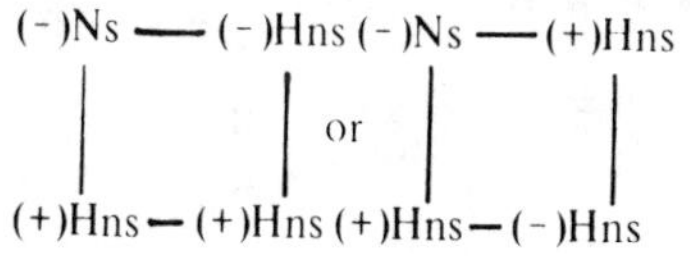

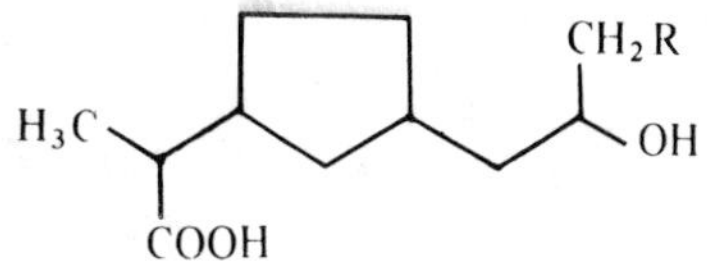

Nonactic acid (Ns): R = H
Homononactic acid (Hns): R = CH$_3$

T-18

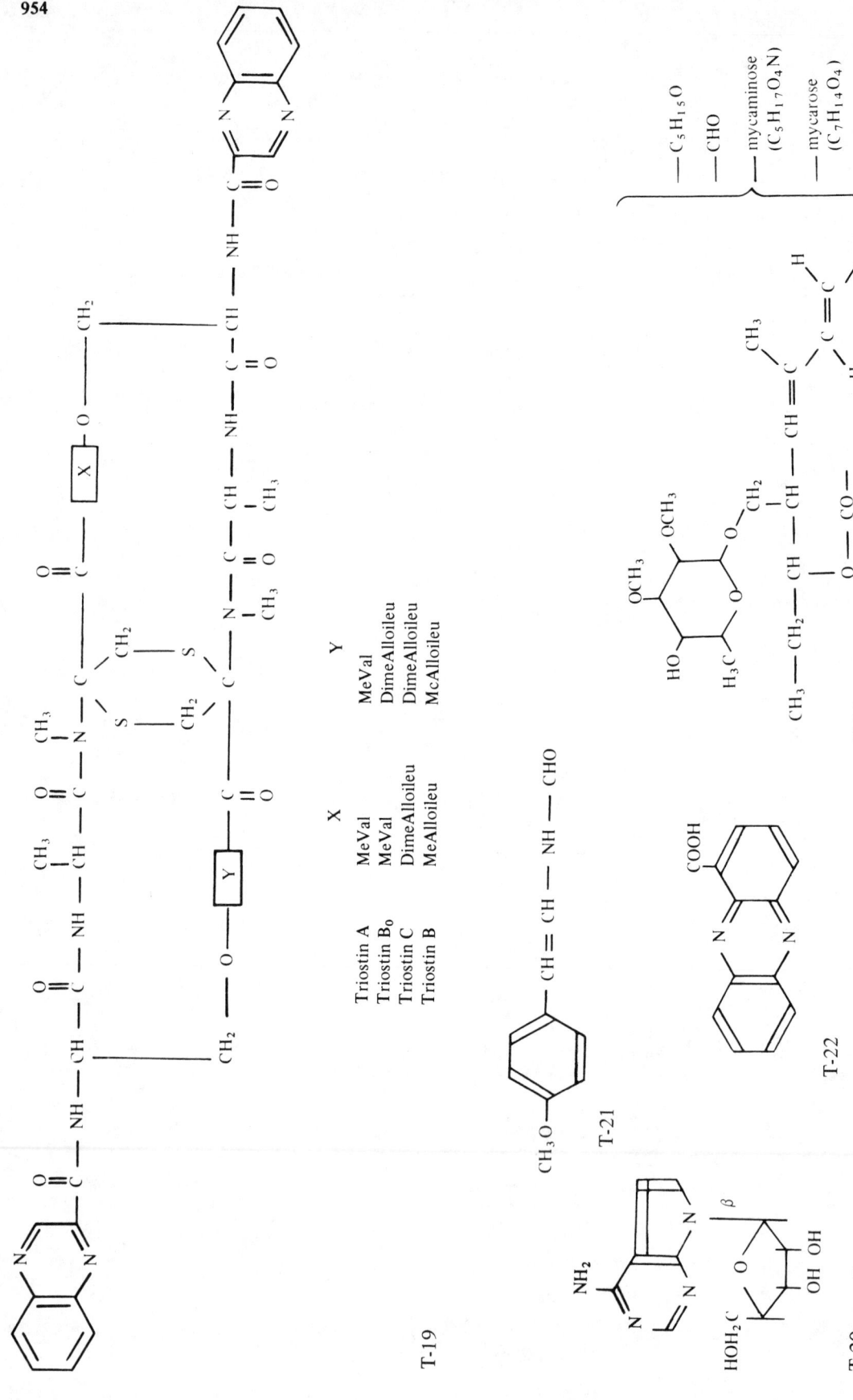

	X	Y
Triostin A	MeVal	MeVal
Triostin B₀	MeVal	DimeAlloileu
Triostin C	DimeAlloileu	DimeAlloileu
Triostin B	MeAlloileu	McAlloileu

T-19

T-20

T-21

T-22

T-23

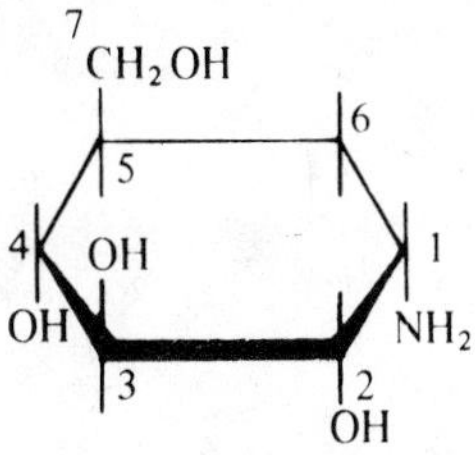

V-0

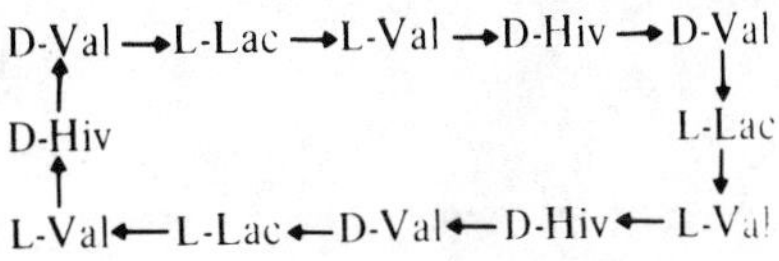

V-1

D-Val →L-Lac →L-Val →D-Hiv →D-Val
D-Hiv   L-Lac
L-Val ←L-Lac ←D-Val ←D-Hiv ←L-Val

Val: valine, Lac: lactic acid, Hiv: a-hydroxyisovaleric acid.

Vernamycin B$_a$ (ostreogrycin B): X = CH$_2$CH$_3$, R = CH$_3$
Vernamycin B$_\beta$ (osterogrycin B$_2$): X = CH$_2$CH$_3$, R = H
Vernamycin Bγ (ostreogrycin B$_1$): X = CH$_3$, R = CH$_3$
Vernamycin B$_\delta$: X = CH$_3$, R = H

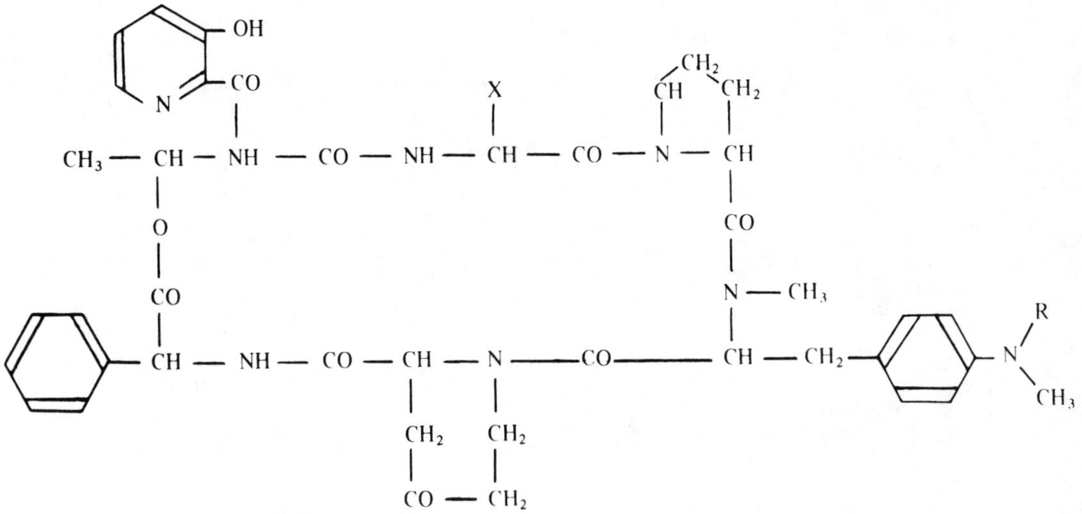

V-2

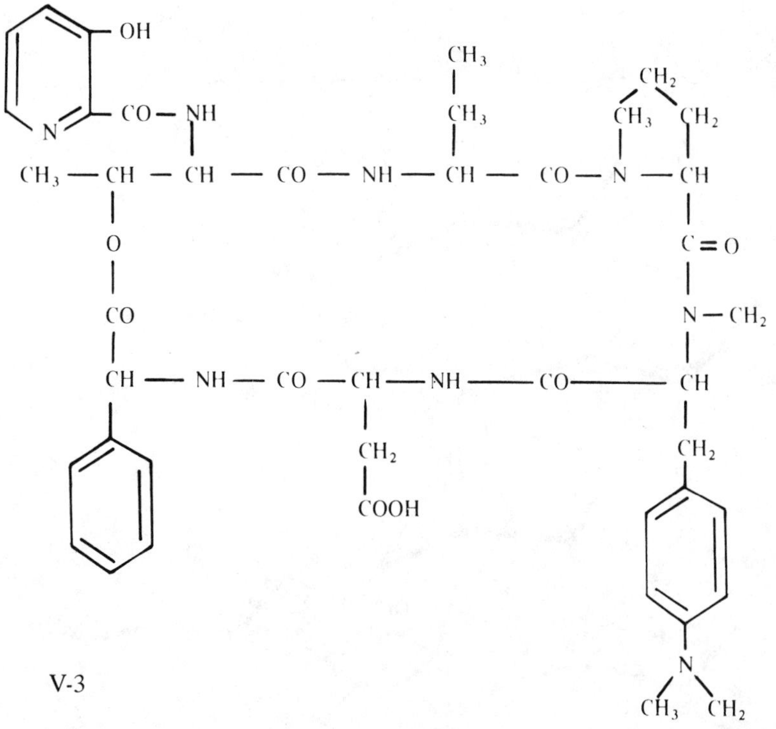

V-3

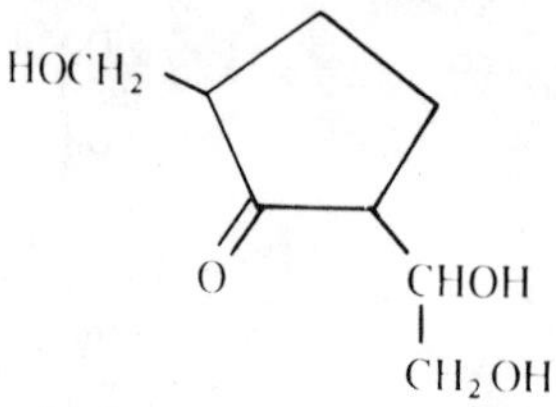

V-4

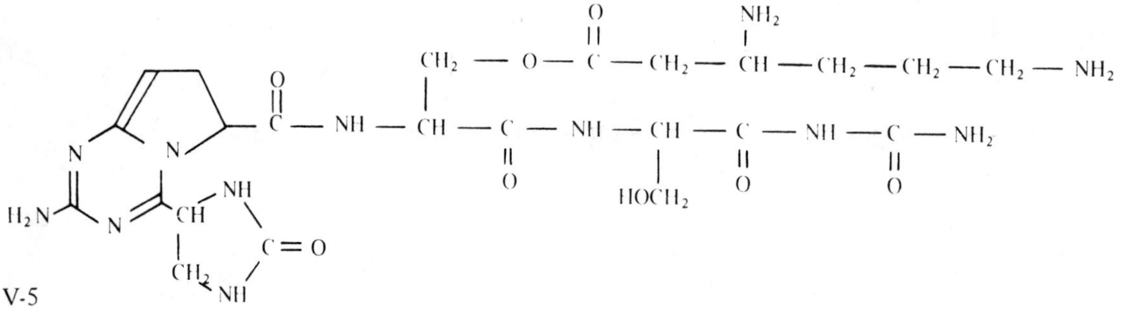

V-5

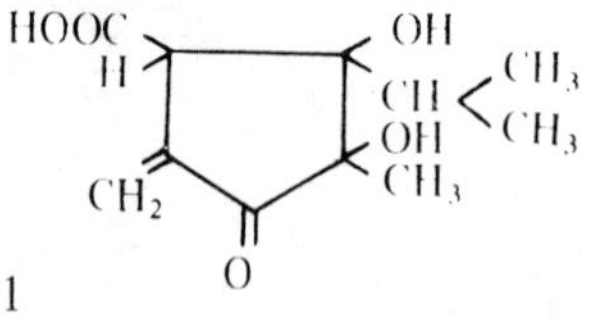

X-1

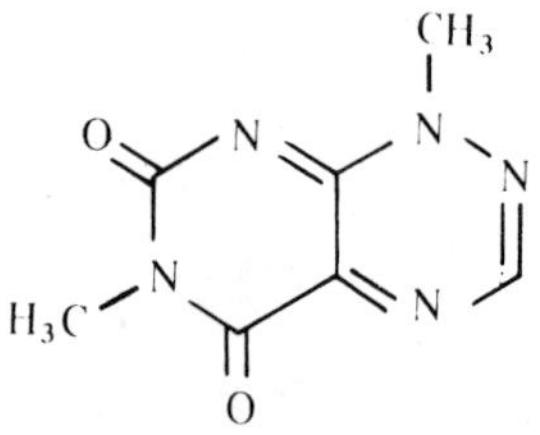

X-2

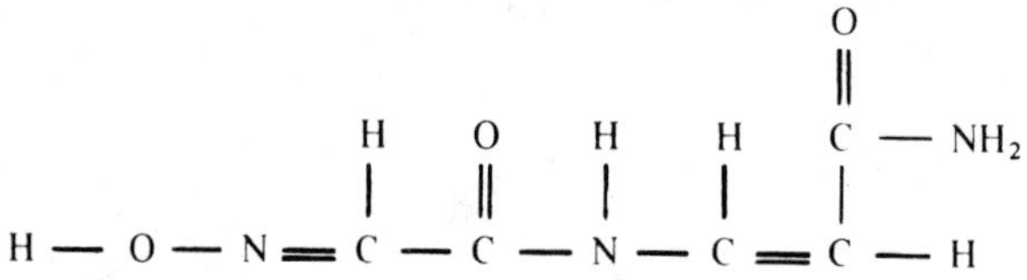

Z-1

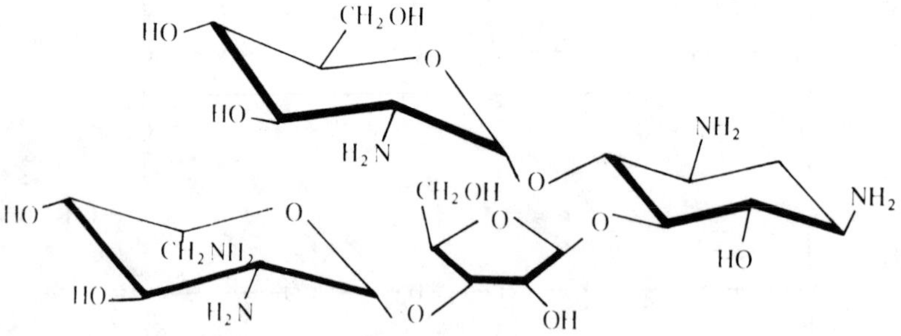

Z-2

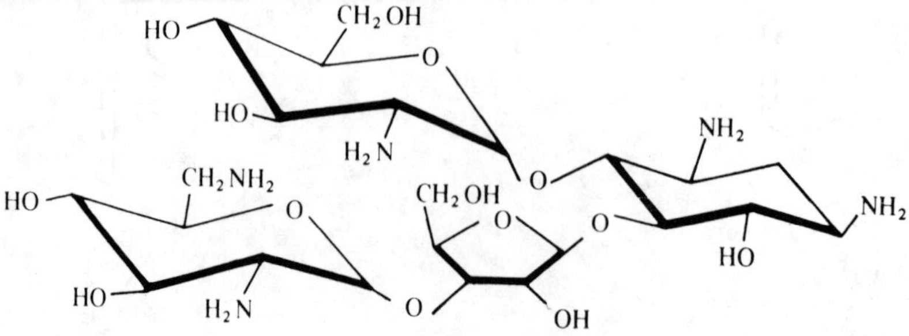

Z-3

Acid hydrolysis gave alanine, aspartic acid, cystine, glutamic acid, lysine, proline, threonine, tryptophan and valine.

NA-1

Hydrolysis gave alanine, glycine, isoleucine, proline, serine, threonine and valine.

NA-2

Hydrolysis gave alanine, glycine, isoleucine, proline, serine, threonine and valine.

NA-3

Acid hydrolysis gave geamine, gulosamine and β-lysine.

NA-4

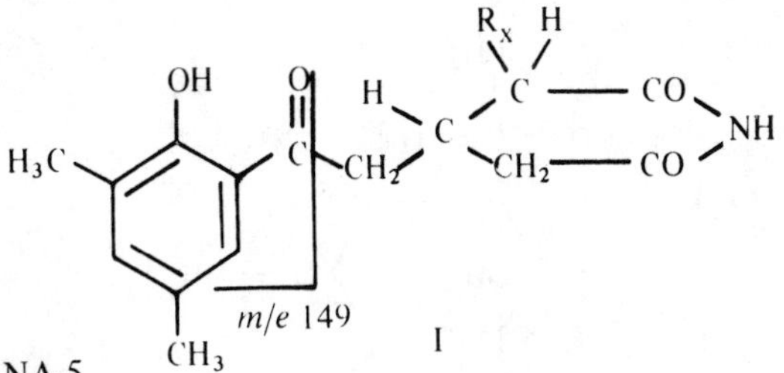
OH
H$_3$C
CH$_3$
m/e 149
I

NA-5 C–73, R_x = H

NA-6 C–73X, R_x = OH

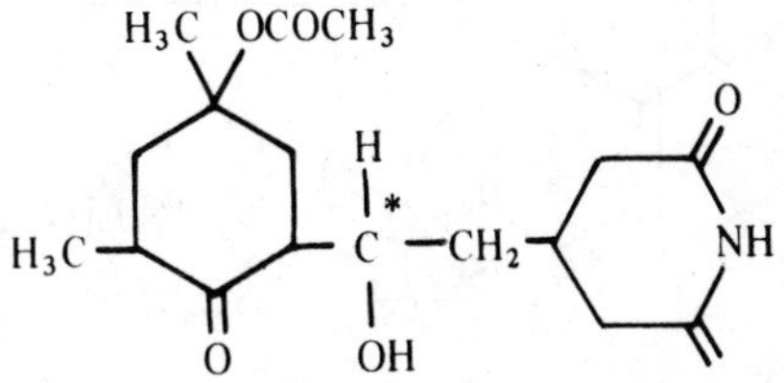
H$_3$C OCOCH$_3$
H$_3$C
OH
NH

NA-7

NA-8

* NA-7: steroisomer

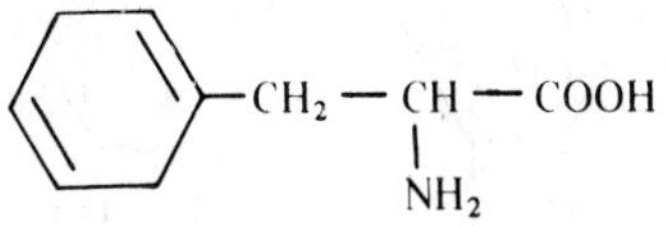

NA-9

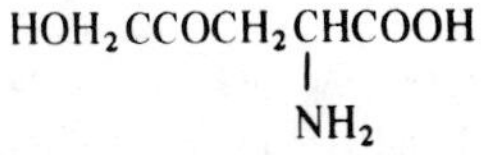
HOH$_2$CCOCH$_2$CHCOOH
NH$_2$

NA-10

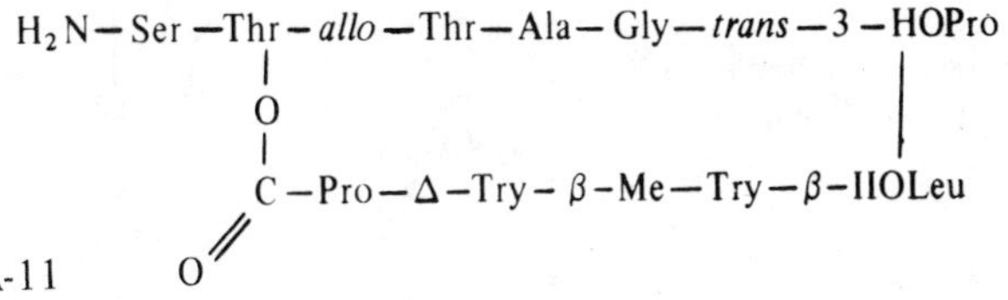
H$_2$N–Ser–Thr–allo–Thr–Ala–Gly–trans–3–HOPro
C–Pro–Δ–Try–β–Me–Try–β–HOLeu

NA-11

One of the following four structures.

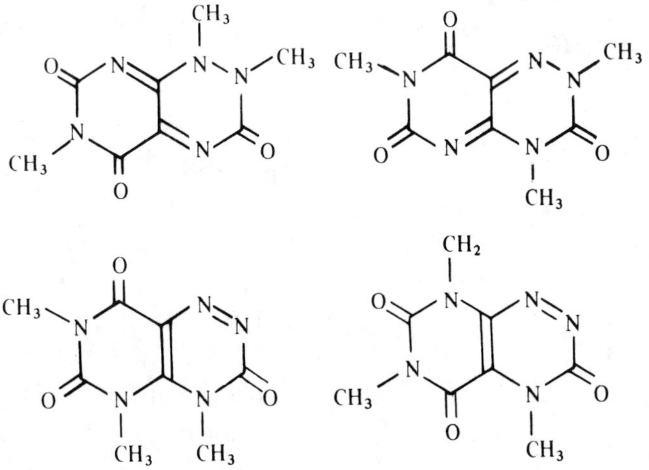

NA-12

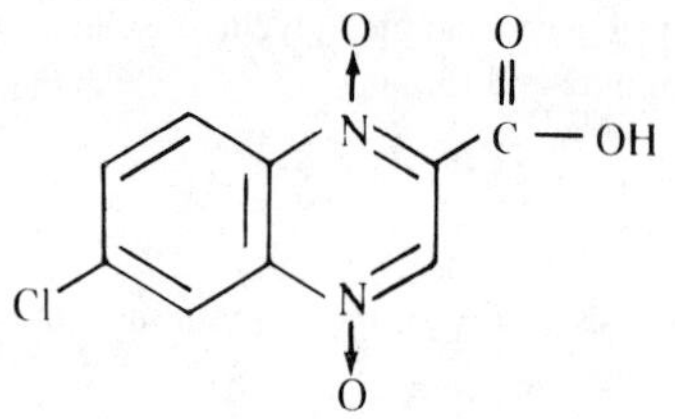

NA-13

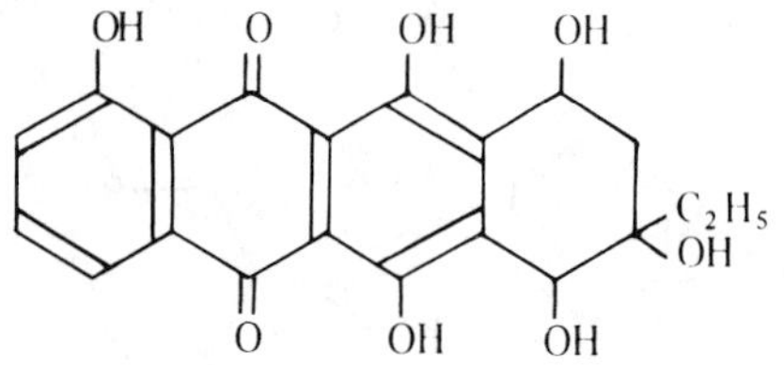

NA-14

(I)

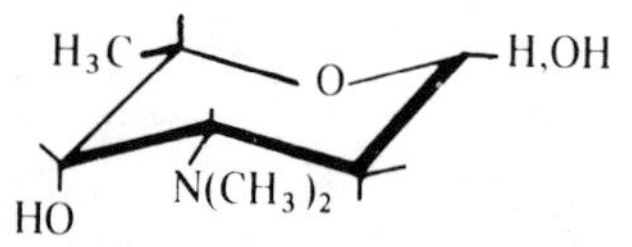

(II)

On hydrolysis A gave one mole of β-rhodomycinone (I) and two moles of rhodosamine (II), while B gave one mole of β-rhodomycinone and one mole of rhododosamine.

NA-15   NA-16

A cyclic structure composed of 4 moles of the following hydroxy acid with ester linkages.

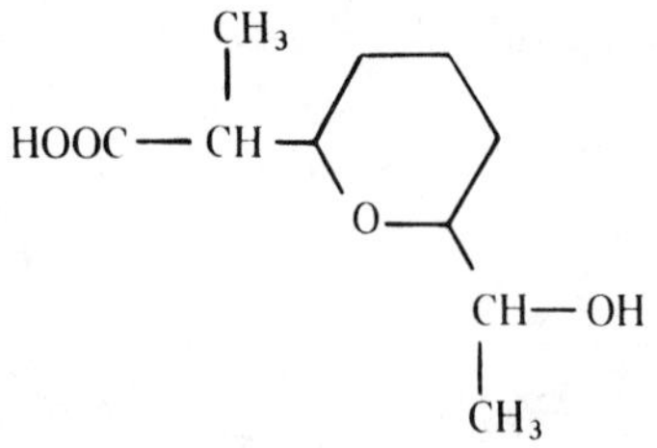

NA-19

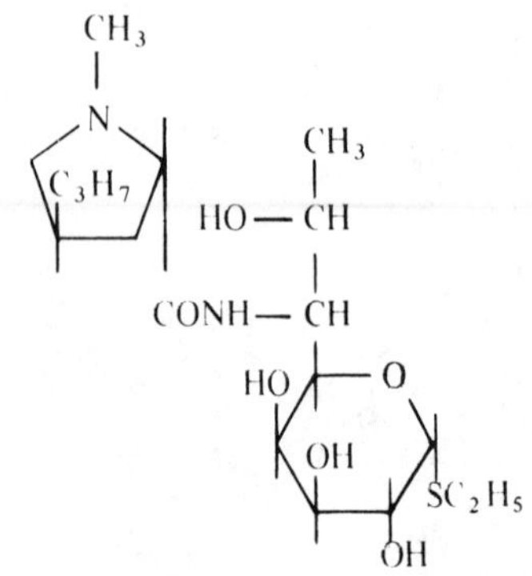

NA-24

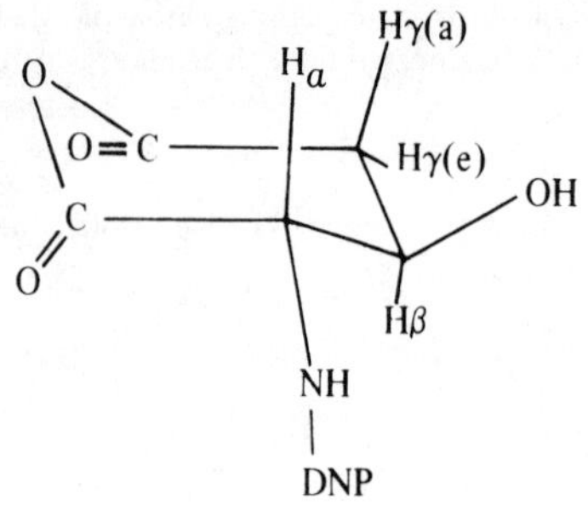

NA-17

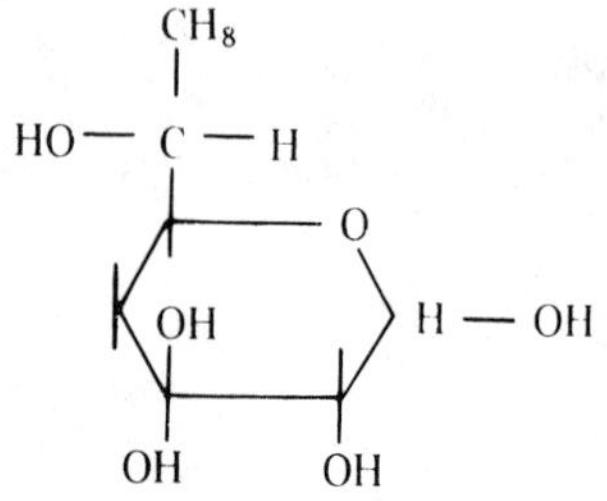

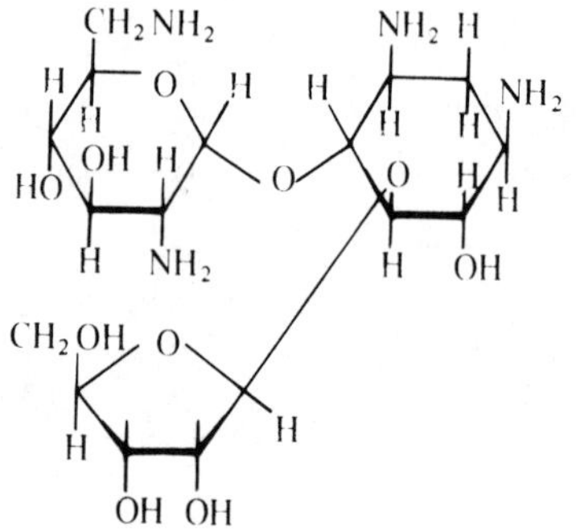

NA-18

NA-20

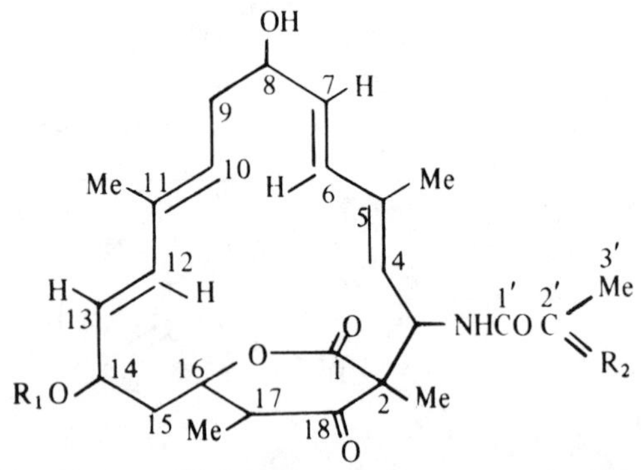

	R_1	R_2
T-2636 A (I)	$-COMe$	$= O$
T-2636 C (II)	$-H$	$= O$
T-2636 D (III)	$-COMe$	$-H, -OH$
T-2636 F (IV)	$-H$	$-H, -OH$

NA-20   NA-21   NA-22   NA-23

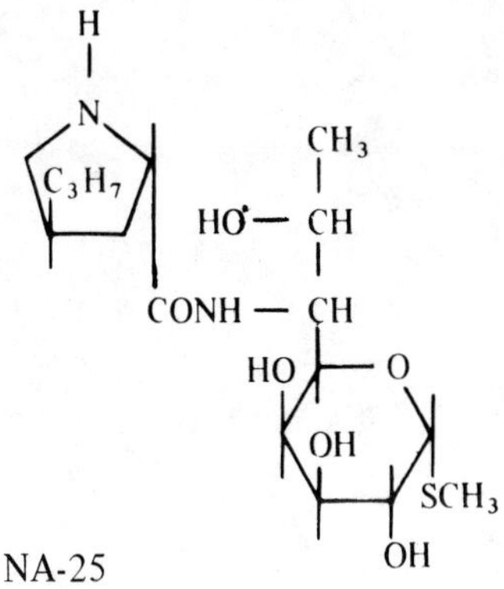

NA-25

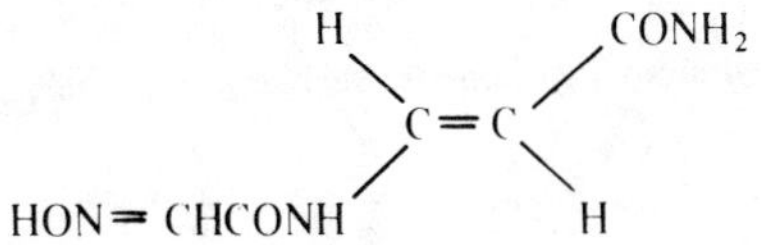

NA-26

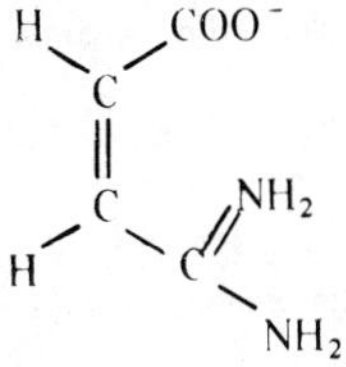

NA-27

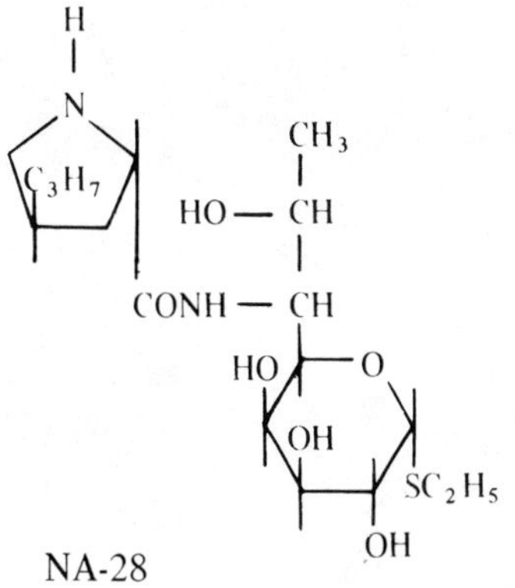

NA-28

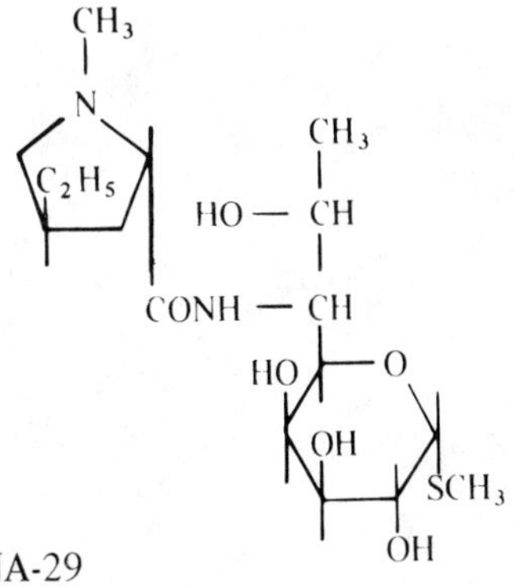

NA-29

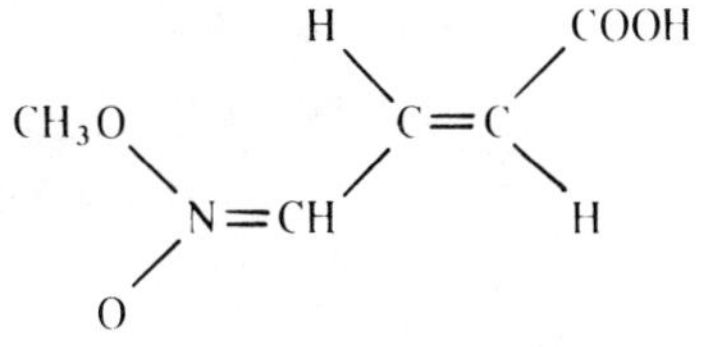

NA-30

$R_1 = CH_3$
$R_2 = CHO$

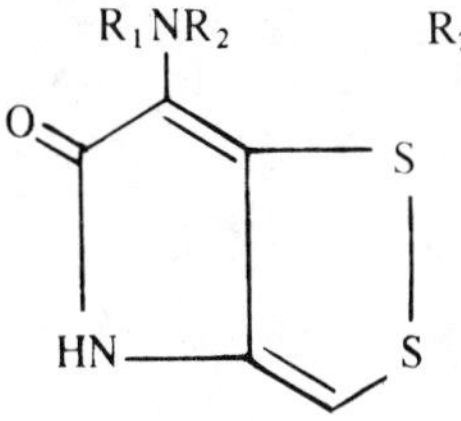

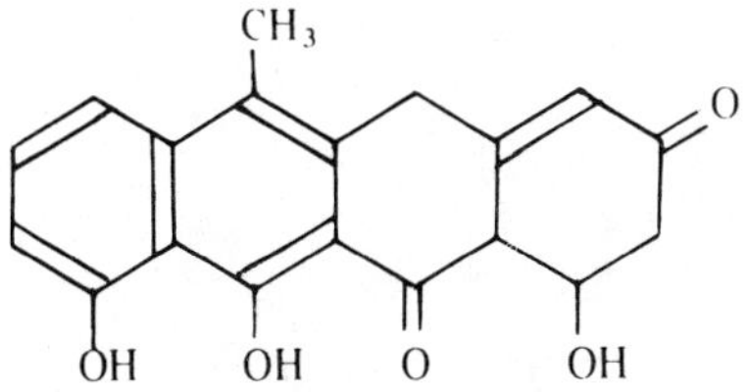

NA-31

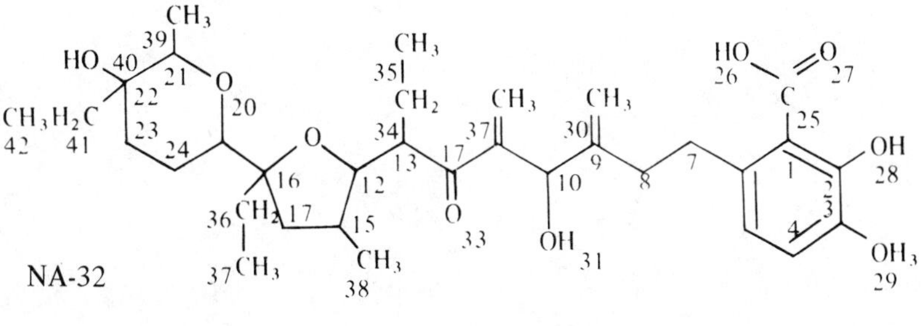

NA-32

Acid hydrolysis gave aspartic acid, cystine, glycine,

NA-33

Acid hydrolysis gave alanine, arginine, aspartic acid, glutamic acid, glycine, histidine, isoleucine, leucine, lysine, methionine, ornithine, proline threonine and tryptophan.

NA-35

Acid hydrolysis gave alanine, glycine, leucine, phenylalanine, proline, valine and γ-aminobutyric acid (prob.).

NA-36

Acid hydrolysis gave teomycic acid ($C_{17}H_{23}O_7N$).

NA-34

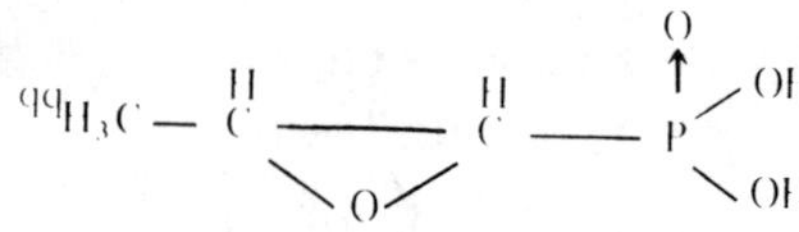

$$^{qq}H_3C — \overset{H}{C} \underset{O}{———} \overset{H}{C} — \overset{O}{\underset{}{P}} \big\langle \begin{matrix} OH \\ OH \end{matrix}$$

NA-37

REFERENCE

A-1   Aizawa, Nakamura, Shirato, Taguchi, Yamaguchi and Misato, *J. Antibiot. (Tokyo) Ser. A, 22,* 457 (1969).

A-2   Yamaguchi, Taguchi, Huang, and Misato, *J. Antibiot. (Tokyo) Ser. A., 22,* 463 (1969).

A-3   Seino, Sugawara, Shirato, and Misato, *J. Antibiot. (Tokyo) Ser. A., 23,* 204 (1970).

A-4   Gurevich, Kolosov, Korobko, and Onoprienko, *Tetrahedron Lett.,* p. 2209 (1968).

A-5   Kono, Takeuchi, Yonehara, Marumo, and Saito, *J. Antibiot. (Tokyo) Ser. A., 23,* 572 (1970).

A-6   Hashimoto, Kito, Takeuchi, Hamada, Maeda, Okami, and Umezawa, *J. Antibiot (Tokyo) Ser. A, 21,* 37 (1968).

A-7   Lancini and Sensi, *Experientia, 20,* 83 (1964).

A-8   Miller, and Hochstein, *J. Org. Chem. 27,* 2525 (1962).

A-9   Siegel and Sisler, *Nature, 200,* 675 (1963).

A-10   Siegel and Sisler, *Biochim. Biophys. Acta, 87,* 70 (1964).

A-11   Siegel and Sisler, *Biochim. Biophys. Acta, 103,* 558 (1965).

A-12   Bennet, Ward, and Brockman, *Biochim. Biophys. Acta, 103,* 478 (1965).

A-13   Munk, Sodano, Mclean, and Haskell, *J. Am. Chem. Soc., 89,* 4158 (1967).

A-14   Lancini and Sensi, *Experientia, XX/2,* 83 (1964).

A-15   Meiji Nyugyo Co., *Japanese Patent S-35-18443* (1960).

A-16   Yagishita, Utahara, Maeda, Hamada, and Umezawa, *J. Antibiot. (Tokyo) Ser. A, 21,* 444 (1968).

A-17   Struck, Thorpe, Jrand, and Shealy, *Tetrahedron Lett.,* p. 1589 (1967).

A-18   Antosz, Nelson, Herald, Jr., and Munk, *J. Am. Chem. Soc., 92,* 4933 (1970).

A-19   Pittillo, Schabel, Jr., and Quinnelly, *Antibiot. Chemother. 11,* 501 (1961).

A-20   Ward, Reich, and Goldberg, *Science, 149,* 1259 (1965).

A-21   Ghuysen, *Arch. Intern. Physiol. Biochim., 65,* 173 (1957).

A-22   Hurwitz, Furth, Malamy, and Alexander, *Proc. Natl. Acad. Sci. USA, 48,* 1222 (1962).

A-23   Goldberg and Rainowitz, *Science, 136,* 315 (1963).

A-24   Reich, Goldberg, and Rabinowitz, *Nature, 196,* 743 (1962).

A-25   Hartmann, Coy, and Kniese: *Ztschr. Physiol. Chem., 330,* 227 (1962).

A-26   Beyer, A. G.: *U.S. Patent, 3,219,544* (1965).

A-27   Furukawa, Inoue, Asano, and Kawamota, *J. Antibiot. (Tokyo) Ser. A, 21,* 568 (1968).

A-28   Khokhlov, Cherches, Reshetov, Smirnova, Sorokina, Prokoptzeva, Koloditskaya, Smirnov, Navashin, and Fomina, *J. Antibiot. (Tokyo) Ser. A, 22,* 541 (1969).

A-29   Vazquez, *Biochim. Biophys Acta, 114,* 277 (1966).

A-30   Felicetti, Colombo, and Baglioni, *Biochim. Biophys. Acta, 119,* 120 (1966).

A-31   Vazquez and Monro, *Biochim. Biophys. Acta, 142,* 155 (1967).

A-32   Hamada, Kawashima, Miyake, and Okamoto, *J. Antibiot. (Tokyo) Ser. A, 6,* 158 (1953).

A-33   Stim, Arnwine, and Foster, *J. Bacteriol, 77,* 566 (1959).

A-34   Pittillo and Foster, *J. Bacteriol., 66,* 118 (1953).

A-35   Umezawa, Oikawa, Okami, and Maeda, *J. Bacteriol., 66,* 118 (1953).

A-36   Soc. Farm. Italia: *Netherland Patent 68,04925,* Oct. (1968).

A-37   Hai-lau and Su, *Japanese Patent 45-26713* (1970): *Acta Microbiol. Sinica, 11,* 160 (1965).

A-38   Lepetit Co.: *Japanese Patent S43-5718,* Mar. (1968); *Nature, 211,* 1198 (1966), *Farm. Ed. Sci., 21* (41), 269 (1966); F.D. 20120, *British Patent 1,115,041.*

A-39   Alazopeptin and Coggin, *J. Bacteriol., 89,* 212 (1965).

A-40   Sakagami, *Japanese Patent 45-2,076.*

A-41   Furumai, Nagahama, and Okuda, *J. Antibiot. (Tokyo) Ser. A, 21,* 85 (1968).

A-42   Grunberger, Shormora, and Sorm, *Biokhimiya, 22,* 141 (1957).

A-43   Sazykin and Borisova, *Antibiotiki, 7,* 975 (1962).

A-44   Ellestad, Kunstmann, Lancaster, Mitscher, and Morton, *Tetrahedron Lett.,* 3893, (1967).

A-45   Hasegawa, Higashide, Shibata, Kameda, and Mizuno, *Annu. Rep. Takeda Res. Lab., 25,* 15 (1966).

A-46   Fujimoto, Kinoshita, Suzuki, and Umezawa, *J. Antibiot. (Tokyo) Ser. A, 23,* 271 (1970).

A-47   Ogura, Otagoshi, Sano, and Hata, *Chem. Pharm. Bull., 15,* 682 (1967).

A-48   Miller, Stapley, and Woodruff, *Antimicrob. Agents Chemother., – 1967,* 407 (1968).

A-49   Bloch and Goutsogeorgopoulos, *Biochemistry, 5,* 3345 (1966).

A-50   Broch, *J. Bacteriol., 85,* 527 (1963).

A-51   Takasawa, Utahara, Osato, Maeda, and Umezawa, *J. Antibiot. (Tokyo) Ser. A, 21,* 567 (1968).

A-52   Argoudelis, Herr, Mason, Pyke, and Zieseri: *Biochemistry 6,* 165 (1967).

A-53   Kelly, Martin, and Hanka, *Can. J. Chem., 47,* 2504 (1969).

A-54   Bardi, Boretti, and DiMarco, *Biochem. Pharmacol., 1,* 165 (1961).

A-55   Soc. Farm. Italia, *U.S. Patent 3,454,469,* July (1969).

A-56   Lichtenstein and Leaf, *J. Clin. Invest., 8,* 1328 (1965).

A-57   Vazquez, *Biochim. Biophys. Acta, 114,* 277 (1966).

A-58   Vazquez and Monro, *Biochim. Biophys. Acta, 142,* 155 (1967).

A-59 Monro and Vazquez, *J. Mol. Biol., 28,* 161 (1967).

A-60 Tanaka, Miyairi, and Umezawa, *J. Antibiot. (Tokyo) Ser. A, 13,* 265 (1960).

A-61 Tanaka, *J. Antibiot. (Tokyo) Ser. A., 16,* 163 (1963).

A-62 Slechta, *Biochem. Biophys. Res. Commun., 3,* 596 (1960).

A-63 Fukuyama and Moyd, *Biochemistry, 3,* 1488 (1964).

A-64 Kuramitsu and Moyd, *Biochim. Biophys. Acta, 85,* 504 (1964).

A-65 Vazquez and Monro, *Biochim. Biophys. Acta, 142,* 155 (1967).

A-66 Schaefer and Wheatley, *Chem. Commun.,* p. 578 (1967).

A-67 Eckardt, *Chem. Ber., 100,* 2561 (1967).

A-68 Shionogi Co., *Japanese Patent S-39-11049,* June (1964).

A-69 Kluepfel, Sehgal, and Vezina, *J. Antibiot. (Tokyo) Ser. A, 23,* 75 (1970).

A-70 Schilling, Berti, and Kluepfel: *J. Antibiot. (Tokyo) Ser. A,* 61 (1970).

A-71 Hatefi, Haavik, and Griffith, *J. Biol. Chem., 237,* 1681 (1962).

A-72 King and Takemori, *Biochim. Biophys. Acta, 64,* 194 (1962).

A-73 Ouchi, *J. Antibiot. (Tokyo) Ser. A, 3*(8), 517 (1949).

A-74 Kurosawa, *J. Antibiot. (Tokyo) Ser. A, 4*(3), 183 (1950).

A-75 CIBA Co., *Swiss Patent 417,631* (1967).

A-76 Sezaki, Hara, Ayukawa, Takeuchi, Okami, Hamada, Nagatsu, and Umezawa, *J. Antibiot. (Tokyo) Ser. A., 21,* 91 (1968).

A-77 Haseltine, Lake Co., *British Patent 1,159,290* (1969).

A-78 Kusaka, Yamamoto, Shibata, Muroi, Kishi, and Mizuno: *J. Antibiot. (Tokyo) Ser. A., 21,* 255 (1968).

A-79 Kishi, Muroi, Kusuka, Nishikawa, Kamiya, and Mizuno, *Chem. Commun.,* p. 852 (1967).

A-79/80 Stapley, Hendlin, Jackson, and Miller, *J. Antibiot. (Tokyo) Ser. A, 24,* 42 (1971); Miller, Tristram, and Wolf, *J. Antibiot. (Tokyo) Ser. A, 24,* 48 (1971).

A-80 *Med. Prom. SSSR,* (6) 36 (1967).

A-81 Beechey, Williams, Holloway, Knight, and Roberton, *Biochem. Biophys. Res. Commun., 26,* 339 (1967).

A-82 Società Farmacologica, *Netherland Patent 6908269,* Dec. (1969).

A-83 Bergy and Herr, *Antimicrob. Agents Chemother. –* 1966, 625 (1967).

A-84 Hanka, Evans, Mason, and Dietz, *Antimicrob. Agents Chemother. –* 1966, 619 (1967).

A-85 Sankyo Co., *Japanese Patent S-41-13791,* Aug. (1966).

A-86 Arai and Hamano, *J. Antibiot. (Tokyo) Ser. A., 23,* 107 (1970).

A-87 Sugawara, *J. Antibiot. (Tokyo) Ser. A, 20,* 93 (1967).

A-88 Hartman, Levenburg, and Buchanan, *J. Biol. Chem., 221,* 1057 (1956).

A-89 Tomisek, Kelly, and Skipper, *Arch. Biochem. Biophys., 64,* 437 (1956).

A-90 Levenberg, Melnick, and Buchanan, *J. Biol. Chem., 225,* 163 (1957).

A-91 Kammen and Burlbert, *Biochim. Biophys. Acta, 30,* 195 (1958).

A-92 Abrrams and Bentley, *Arch. Biochem. Biophys., 79,* 91 (1959).

A-93 Buchanan, Hartman, Herrmann, and Day, *J. Cell. Comp. Physiol., 54,* (Suppl. 1) 139 (1959).

A-94 Hurlbert and Kammen, *J. Biol. Chem., 235,* 443 (1960).

A-95 Srinivasan and Weiss, *Biochim. Biophys. Acta, 51,* 597 (1961).

A-96 Moore and Hurlbert, *Cancer Res., 1,* 257 (1961).

A-97 Kaken Chemical Co., *Japanese Patent S45-6073,* Feb. (1970).

B-1 Misato, Huang, Katagiri, Ueda, Fukatsu, and Niida, *J. Antibiot. (Tokyo) Ser. A, 20,* 254 (1967).

B-2 Kozlenko, *Russian Patent 169,753* (1965), in Farmdoc 18, 717.

B-3 Sakagami, *Japanese Patent 45-2077* (1970).

B-4 Sakagami, *Japanese Patent 43-97* (1968).

B-5 Kono, Takeuchi, and Yonehara, *J. Antibiot. (Tokyo) Ser. A, 21,* 433 (1968).

B-6 Yamaguchi, Yamamoto, and Tanaka, *J. Biochem., 57,* 667 (1965).

B-7 Yamaguchi and Tanaka, *J. Biochem., 60,* 632 (1966).

B-8 Van Tamelen, Dickie, Loomans, Dewey, and Strong, *J. Am. Chem. Soc., 83,* 1639 (1961).

B-9 Tappel, *Biochem. Pharmacol., 3,* 289 (1960).

B-10 Takita, Muraoka, Maeda, and Umezawa, *J. Antibiot. (Tokyo) Ser. A, 21,* 79 (1968).

B-11 Koyama, Nakamura, Muraoka, Takita, Maeda, Umezawa, and Iitaka, *Tetrahedron Lett.,* p. 4635 (1968).

B-12 Takita, Maeda, Umezawa, Omoto, and Umezawa, *J. Antibiot. (Tokyo) Ser. A, 22,* 237 (1969).

B-13 Suzuki, Nagai, Yamaki, Tanaka, and Umezawa, *J. Antibiot. (Tokyo) Ser. A, 22,* 446 (1969).

B-14 Nagai, Suzuki, Tanaka, and Umezawa, *J. Antibiot. (Tokyo) Ser A, 22,* 569 (1969).

B-15 Nagai, Suzuki, Tanaka, and Umezawa, *J. Antibiot. (Tokyo) Ser A, 22,* 624 (1969).

B-16 Kinoshita and Tanaka, *J. Antibiot. (Tokyo) Ser. A, 23,* 311 (1970).

B-17 Suzuki, Nagai, Akutsu, Yamaki, Tanaka, and Umezawa, *J. Antibiot. (Tokyo) Ser. A, 23,* 473 (1970).

B-18 Takita, Muraoka, Fujii, Itoh, Maeda, Umezawa, *J. Antibiot. (Tokyo) 25,* 197 (1972).

B-19 Gorini and Kataja, *Biochem. Biophys. Res. Commun., 18,* 656 (1965).

B-20 Davies, Gorini, and Davis, *Mol. Pharmacol., 1,* 93 (1965).

B-20/21 Hütter, Keller-Schierlein, Nuesch, and Zahner, *Arch. Mikrobiol., 51,* 1 (1965); Hutter, Keller-Schierlein, Knusel, Prelog, Rodgers, Suter, Vogel, Vosel, and Zähner, *Helv. Chim. Acta, 50*(6), 1553 (1967); Dunitz, Hawley, Miklos, White, Berlin, Marusic, and Prelog, *Helv. Chim. Acta, 54,* (6), 1709 (1971).

B-21 Berger, Jampolsky, and Goldberg, *Arch. Biochem. Biophys., 22,* 476 (1949).

B-22 Keller-Schierlein, *Helv. Chim. Acta, 50,* 731 (1967).

B-23 Sinha and Nandi, *Experientia, 24,* 795 (1968).

B-24 Sinha, *J. Antibiot. (Tokyo) Ser. A, 23,* 360 (1970).

B-25 Brufani, Keller-Schierlein, Löffler, Mansperger, and Zähner, *Helv. Chim. Acta, 51,* 1293 (1968).

B-26 Miller, Stapley, and Woodruff, *Antimicrob. Agents Chemother.,* p. 407 (1968).

B-27 Tanaka, Sashikata, Yamaguchi, and Umezawa, *J. Biochem., 60,* 405 (1966).

B-28 Lin and Tanaka, *J. Biochem., 63,* 1 (1968).

B-29 Nakamura, Yajima, Lin, and Umezawa, *J. Antibiot. (Tokyo) Ser. A, 20,* 1 (1967).

B-30 Sakagami, Sekine, Yamabayashi, Kitaura, Ueda, and Kosaka, *J. Antibiot. (Tokyo) Ser. A, 19,* 99 (1966).

B-31 Sakagami, Yamabayashi, and Sekine, *J. Antibiot. (Tokyo) Ser. A, 19,* 104 (1966).

B-32 Sakagami, *Japanese Patent 45-2077.*

B-33 Dudnik, *Antibiotiki, 10,* 112 (1965).

B-34 Shoji, Sakazaki, Mayama, Kawamura, and Yasuda, *J. Antibiot. (Tokyo) Ser. A, 23,* 295 (1970).

C-1 Stark, Hoehn, and Knox, *Antimicrob. Agents Chemother., 1967,* 314 (1968).

C-2 Diveker, Read, and Vining, *Can. J. Chem., 45,* 1215 (1967).

C-3 Davies, Gorini, and Davis, *Mol. Pharmacol., 1,* 93 (1965).

C-4 Lynch and Neuhaus, *J. Bact., 91,* 449 (1966).

C-5 Tanaka, *Biochem. Biophys. Res. Commun., 12,* 68 (1963).

C-6 Stanley, Owen, Bhuyan, and Kupiecki, *Antimicrob. Agents Chemother., 1965,* 808 (1966); 812 (1966).

C-7 Terawaki and Greenberg, *Nature, 209,* 481 (1966).

C-8 Gopalkrishman, Narasimhachari Joshi, and Thirumalachar, *Nature, 218,* 597 (1968).

C-9 Rendi and Ochoa, *J. Biol. Chem., 237,* 3711 (1962).

C-10 Jardetzky, *J. Biol. Chem., 238,* 2498 (1963).

C-11 Schierlein, Muntwyler, Pache, and Zahner, *Helv. Chim. Acta, 52,* 127 (1969).

C-12 Berlin, Kolosov, Vasina, and Vertseva, *Chem. Commun.,* p. 762 (1968).

C-13 Berlin, Kolosov, Vasina, and Yartseva, *Chem. Commun.,* p. 762 (1968).

C-14 Hartmann, Goller, Koschel, Kersten, and Kersten, *Biochem. Z., 341,* 126 (1964).

C-15 Ward, Reich, and Goldberg, *Science, 149,* 1259 (1965).

C-16 Kaziro and Kamiyama, *Biochem. Biophys. Res. Commun., 19,* 433 (1965).

C-17 Miyamoto, Kawamatsu, Kawashima, Shinohara, Tanaka, Tatsuoka, and Nakanishi, *Tetrahedron, 23,* 421 (1967).

C-18 Miyamoto, Morita, Kawamatsu, Kawashima, and Nakanishi, *Tetrahedron, 23,* 411 (1967).

C-19 Miyamoto, Kawamatsu, Kawashima, Shinohara, Tanaka, and Tatsuoka, *Tetrahedron, 23,* 421 (1967).

C-20 Ayukawa, Hamada, Kojiri, Takeuchi, and Hara, *J. Antibiot. (Tokyo) Ser. A, 22,* 303 (1969).

C-21 Bettolo, *Tetrahedron Lett.,* p. 471 (1968).

C-22 Hartmann, Goller, Koschel, Kersten, and Kersten, *Biochem. Z., 341,* 126 (1964).

C-23 Kersten and Kersten, *Biochem. Z., 341,* 174 (1965).

C-24 Koshiyama, Tsukiura, Fujisawa, Konishi, Hatori, Tomita, and Kawaguchi, *J. Antibiot. (Tokyo) Ser. A, 22,* 61 (1969).

C-25 Fujisawa, Matsumoto, Ohmori, Hoshiya, and Kawaguchi, *J. Antibiot. (Tokyo) Ser. A, 22,* 65 (1969).

C-26 Tsukiura, Konishi, Saka, Naito, and Kawaguchi, *J. Antibiot. (Tokyo) Ser. A, 22,* 89 (1969).

C-27 Kusakabe, Yamauchi, Nagatsu, Abe, Akasaki, and Shirato, *J. Antiobiot. (Tokyo) Ser. A, 22,* 112 (1969).

C-28 Brockmann and Niemeyer, *Chem. Ber., 101,* 1341 (1968).

C-29 Meiji Seika Co., *Japanese Patent S45-12278,* May (1970).

C-30 Upjohn Co., *U.S. Patent 3,350,269,* Oct. (1967).

C-31 Chas Pfizer & Co., *Japanese Patent S42-8919,* April (1967).

C-32 Siegel and Sisler, *Biochim. Biophys. Acta, 103,* 558 (1965).

C-33 Benett, Lward, and Brockman, *Biochim. Biophys. Acta, 103,* 478 (1965).

C-34 Godtfredsen and Vangedal, *Tetrahedron, 26,* 4931 (1970).

C-35 Strominger, Threnn, and Scott, *J. Am. Chem. Soc., 81,* 3803 (1959).

C-36 Roze and Strominger, *Mol. Pharmacol., 2,* 92 (1966).

D-1 Societe Pharm. Italy, *Japanese Patent S-44-18913,* Aug. (1969).

D-2 Rusconi and Calendi, *Biochim. Biophys. Acta, 119,* 413 (1966).

D-3 Hanka, Bergy, and Kelly, *Science, 154,* 1667 (1966).

D-4 Argoudelis, Coats, Mason, and Sebek, *J. Antibiot. (Tokyo) Ser. A, 22,* 309 (1969).

D-5 Devault, Schmitz, and Hooper, *Antimicrob. Agents Chemother.,* p. 796 (1965).

D-6 Miyazaki, Yoshida, Hidaka, Takeuchi, and Yonehara, *J. Antibiot. (Tokyo) Ser. A, 22,* 393 (1969).

D-7 Appeared in New Names, *J.A.M.A., 212,* 466 (1970).

D-8 Naganawa, Wakashiro, Yagi, Kondo, Takita, Hamada, Maeda, and Umezawa, *J. Antibiot. (Tokyo) Ser. A, 23,* 365 (1970).

D-9   Meyer and Mason, *Antimicrob. Agents Chemother.*, − 1965, 850 (1966).

D-10   Yonehara, Seto, Aizawa, Hidaka, Shimazu, and Ohtake, *J. Antibiot. (Tokyo) Ser. A, 21,* 369 (1968).

D-11   Otake, Kakinuma, and Yonehara, *J. Antibiot. (Tokyo) Ser. A, 21,* 371 (1968).

D-12   Ishida et al., Koyama et al., *Japanese Patent S45-26715,* Sept. (1970).

D-13   Hamill, Hoehn, Pittenger, Chamberlin, and Gorman, *J. Antibiot. (Tokyo) Ser. A, 22,* 161 (1969).

D-14   Coggin and Martin, *J. Bact., 89,* 1348 (1965).

D-15   Umezawa, Tsuchiya, Tatsuta, Horiuchi, and Usui, *J. Antibiot. (Tokyo) Ser. A, 23,* 20 (1970).

D-16   Gerber and Wieclawek, *J. Org. Chem., 31,* 1496 (1966).

D-17   Oishi, Sugawa, Okutomi, Suzuki, Hayashi, Sawada, and Ando, *J. Antibiot. (Tokyo) Ser. A, 23,* 105 (1970).

D-18   Arcamone, Orezzi, Barbieri, Nicolella, and Penco, *Gazz. Chim. Ital., 97,* 1097 (1967).

D-19   Penco, Redaelli, and Arcamone, *Gazz. Chim. Ital., 97,* 1110 (1967).

D-20   Squibb Co., *U.S. Patent 3,476,268,* Feb. (1970); *Belgian Patent 708601,* June (1968).

.D-21   Kaken Co. and Institute of Physical and Chemical Research, *Japanese Patent S-45-26719* (1970).

D-22   Coggin and Martin, *J. Bact., 89,* 1348 (1965).

D-23   Gordon and Lapa, *Appl. Microbiol., 14,* 754 (1966).

E-1   De Zeeuw and Tynan, *J. Antibiot. (Tokyo) Ser. A, 22,* 386 (1969).

E-2   Higashide, Hatano, Shibata, and Nakazawa, *J. Antibiot. (Tokyo) Ser. A, 21,* 126 (1968).

E-2a   Asai, Muroi, Sugita, Kawashima, Mizuno, and Miyake, *J. Antibiot. (Tokyo) Ser. A, 21,* 138 (1968).

E-2b   Tsuchiya, Kondo, Oishi, and Yamazaki, *J. Antibiot. (Tokyo) Ser. A, 21,* 147 (1968).

E-3   Herr and Reusser, *U.S. Patent 3,367,833,* Feb. (1968).

E-4   Bortell, Bonmassar, Montagnani, Marelli, and Zannini, *Chem. Abst., 67,* 8951 4Z (1967).

E-5   Meyer and Mason, *Antimicrob. Agents Chemother.,* p. 850 (1965).

E-6   Maksimova, Toropova, Kovalenkova, and Gauze, *Antibiotiki, 10,* 201 (1965).

E-7   Arai, *Japanese Patent S-40-25237,* Nov. (1965).

E-8   Arai, *Annu. Rep. Fuhai, 16,* 48 (1963).

F-1   Glaxo Group Ltd., *U.S. Patent 3,311,528,* Mar. (1967).

F-2   RIT (Recherche et Industrie Therapeutique Belgium, *U.S. Patent 3,325,359,* June (1967).

F-3   Shimi and Shoukry, *J. Antibiot. (Tokyo) Ser A, 19,* 110 (1966).

F-4   Bickel, Mertens, Prelog, Seibl, and Walser, *Tetrahedron,* Suppl. 8, Part I, p.171 (1966).

F-5   Burnham, *J. Gen Microbiol., 32,* 117 (1963).

F-6   Pandey and Rinehart, *J. Antibiot. Tokyo, 23,* No 8. 414 (1970).

F-7   Bognar, Brown, Lockley, Makleit, Toube, Weedon and Zsupan, *Tetrahedron Lett., 7,* 471 (1970).

F-8   Schlegel and Thrum, *Experientia, 24,* 11 (1968).

F-8/9   Ishizuka, Sawa, Hori, Takayama, Takeuchi, Umezama, *J. Antib. 21,* 5 (1968).

F-9   Katagiri, Tori, Kimura, Yoshida, Nagasaki, and Minato, *J. Med. Chem., 10,* 1149 (1967).

G-1   Sckardt, *Chem. Ber. 100,* 2561 (1967).

G-1/2   Squibb Co., *German Patent 1942694* (1971).

G-2   Deboer, Meulman, Wnuk, and Peterson, *J. Antibiot. (Tokyo) Ser. A, 23,* 442 (1970).

G-3   Sasaki and Rinehart, *J. Am. Chem. Soc., 92,* 7591 (1970).

G-4   Cooper and Yudis, *Chem. Commun.,* 821 (1967).

G-5   Maehr and Schaffner, *J. Am. Chem. Soc., 89,* 6787 (1967).

G-6   Milanesi and Ciferri, *Biochemistry, 5,* 3926 (1966).

G-7   Davies, Luigi, Gorini, and Davis, *Mol. Pharmacol., 1,* 93 (1965).

G-8   Rinehart, *J. Infect. Dis., 119,* 345 (1969).

G-9   Terlain and Thomas, *Compt. Rend. Ser. C., 269,* 1546 (1969).

G-10   Ono Pharmaceutical Co., *Japanese Patent S-44-21395,* Sept. (1969).

G-10/11 Shimi and Dewedar, *Arch. Mikrobiol., 54,* 246 (1966).

G-11   Clark and Gunther, *Biochim. Biophys. Acta, 76,* 636 (1963).

G-12   Casjens and Morris, *Biochim. Biophys. Acta, 108,* 677 (1965).

G-13   Clark and Chang, *J. Biol. Chim., 240,* 4734 (1965).

G-14   Sinohara and Sky-Peck, *Biochem. Biophys. Res. Commun., 18,* 98 (1965).

G-15   Wang, Kanda, and Umezawa, *J. Antibiot. (Tokyo) Ser. A, 22,* 211 (1969).

G-16   Barcza, Brufani, Keller-Schierlein, and Zähner, *Helv. Chim. Acta, 49,* 1736 (1966).

G-17   Keller-Schierlein, Brufani, and Barcza, *Helv. Chim. Acta. 51,* 1257 (1968).

G-17/18 Terlain, Bernard, Thomas, Jean, Pierre, *Bull. Soc. Chim., 6,* Z357 (1971).

G-18   Gachon, Kergomard, and Veschambre, *Chem. Commun.,* p. 1421 (1970).

G-19   Alleaume and Hickel, *Chem. Commun.,* p. 1422 (1970).

G-20   Gyogyszerkutata Intezet, *Hungarian Patent 157,600,* July (1970).

H-1   Weinstein, Luedemann, Oden, and Wagman, *Antimicrob. Agents Chemother.,* p. 435 (1967).

H-2   Microbial Chemistry Research Foundation, *Japanese Patent S-45-25414,* Dec. (1967).

H-3   Schmitz, Crook, and Bush, *Antimicrob. Agents Chemother.,* 606, (1966).

H-4   Grinev et al., *Zh. Obshch. Khim., 33,* 315 (1963); *Chem. Abstr., 58,* 13863 (1963).

H-5   Grinev et al., *Zh. Obshch. Khim., 33,* 127 (1963); *Chem. Abstr., 59,* 383 (1963).

H-6   Gyogyszerkutato Intezet, *Hungarian Patent 155,813,* Mar. (1969); *Chem. Abstr. 71,* 208 (1969).

H-7   Eisenbrandt, *Z. Chem., 7,* 311 (1967).

H-8   Sanraku Ocean Co., *Japanese Patent S-45-39038,* (1970).

H-9   Stapley, Demny, Miller, and Woodruff, *Antimicrob. Agents Chemother.,* p. 595 (1966).

H-10   Shimi, Dewedar, and Shoukry, *J. Antibiol., 23*(8), 388 (1970).

H-11   Sakagami, Ueda, Yamabayashi, Tsurumaki, and Kumon, *J. Antibiol., 22*(11), 521 (1969).

H-12   Shier, Rinehart, and Gottlieb, *Proc. Natl. Acad. Sci. USA, 63,* 198 (1969).

H-13   Martin, Mitscher, Miller, Shu, and Bohonos, *Antimicrob. Agents Chemother.,* p. 563 (1966).

I-1   Sakagami, Ueda, and Yamabayashi, *J. Antibiot. (Tokyo) Ser. A, 20,* 299 (1967).

I-2   Sakagami, *Japanese Patent S-45-16793,* June (1970).

I-3   Siegel, Sisler, and Johnson, *Biochem. Pharmacol., 15,* 1218 (1966).

I-4   Chan and Hill, *J. Org. Chem., 35,* 3519 (1970).

I-5   Brockman, *Angew. Chem., 80,* 493 (1968).

I-6   Gerber and Lechevalier, *Biochemistry, 3,* 598 (1964).

I-7   Artman, Behr, Beissner, Honikel, and Sippel, *Angew. Chem., 80,* 710 (1968); *Angew. Chem. Int. Ed., 7,* 693 (1968).

I-8   Hoshino, Umezawa, Mimura, and Hata, *J. Antibiot. (Tokyo) Ser. A, 20,* 30 (1967).

J-1   Meyers, Welsenborn, Pansy, Slusarchyk, Saltza, Rathnum, and Parker, *J. Antibiot. (Tokyo) Ser. A, 23,* 502 (1970).

J-2   Osono, Oka, Watanabe, Numazaki, Moriyama, Ishida, Suzuki, Okami, and Umezawa, *J. Antibiot. (Tokyo) Ser. A, 20,* 174, 181 (1967).

J-3   Umezawa, Osono, Moriyama, and Washizaki, *Japanese Patent 45-5032.*

J-4   Matsuura, Shiratori, Harada, and Katagiri, *J. Antibiot. (Tokyo) Ser. A, 20,* 282 (1967).

J-5   Shionogi Seiyaku Co., *Japanese Patent S43-5720,* (1969).

J-6   Takeda Chemical Co., *Japanese Patent 47-4514* (1971); *German Patent 2034245* (1971).

K-1   Bergy, *J. Antibiot. (Tokyo) Ser. A, 21,* 454 (1968).

K-2   Upjohn Co., *U.S. Patent 3,300,382,* Jan. (1967).

K-3   Ito, Matsuya, Ōmura, Otani, Nakagawa, Takeshima, Iwai, Ohtani, and Hata, *J. Antibiot. (Tokyo) Ser. A, 23,* 315 (1970).

K-4   Okamoto, Mayama, Tanaka, Tawara, Shimaoka, Kato, Nishimura, Ebata, and Ohtsuka, *J. Antibiot. (Tokyo) Ser. A, 315,* (1968).

K-5   Kowa Co., *Japanese Patent S45-8636,* Mar. (1970).

K-6   Ōmura, Namiki, Shibata, Muro, Nakayoshi, and Sawada, *J. Antibiot. (Tokyo) Ser. A, 22,* 500 (1969).

K-7   Bush, Cassidy, Crook, and German, *J. Antibiot. (Tokyo) Ser. A, 24,* 143, 149 (1971).

L-2a   Yamaguchi, Furumai, Sato, Okuda, and Ishida, *J. Antibiot. (Tokyo) Ser. A, 23,* 369 (1970).

L-2b   Yamaguchi, Kashida, Nawa, Yajima, Miyagishima, Ito, Okuda, Ishida, and Kumagai, *J. Antibiot. (Tokyo) Ser. A, 23,* 373 (1970).

L-2c   Yamaguchi, Seto, Oura, Arai, Enomoto, Ishida, and Kumagai, *J. Antibiot. (Tokyo) Ser. A, 23,* 382 (1970).

L-3a   Naganawa, Hamada, Maeda, Okami, Takeuchi, and Umezawa, *J. Antibiot. (Tokyo) Ser. A, 21,* 55 (1968).

L-3b   Naganawa, Takita, Maeda, and Umezawa, *J. Antibiot. (Tokyo) Ser. A, 23,* 423 (1970).

L-4   Takeda Chemical Industries, *Japanese Patent S41-13789,* Aug. (1966).

L-5   Higashide, Hasegawa, and Shibata, *J. Antibiot. (Tokyo) Ser. A, 22,* 409 (1969).

L-6   Takeda Chemical Industries, *Japanese Patent S-45-20558,* July (1970).

L-7   Takeda Chemical Industries, *French Patent 1,523,522,* Mar. (1968).

L-8   Mizuno, Ohkubo, Yokoyama, Hamada, Maeda, and Umezawa, *J. Antibiot. (Tokyo) Ser. A, 20,* 194 (1967).

L-9   Hata, Ōmura, Katagiri, Ogura, Naya, Abe, and Watanabe, *Chem. Pharm. Bull., 15,* 358 (1967).

L-10   Ōmura, Katagiri, and Hata, *J. Antibiot. (Tokyo) Ser. A, 21,* 199 (1968).

L-11   Kitasato Inst., *Japanese Patent S-45-27395,* Sept. (1970).

L-12a   Ōmura, Ogura, and Hata, *Tetrahedron Lett., 7,* 609 (1967).

L-12b   Ōmura, Ogura, and Hata, *Tetrahedron Lett., 14,* 1267 (1967).

L-12c   Ōmura, Hironaka, and Hata, *J. Antibiot. (Tokyo) Ser. A, 23,* 511 (1970).

L-13   Ōmura, Katagiri, and Hata, *J. Antibiot. (Tokyo) Ser. A, 20,* 234 (1967).

L-14   Hata et al. (Kitasato Institute), *Japanese Patent S-45-27395,* Sept. (1970).

L-15   Hata et al. (Kitasato Institute), *Japanese Patent S-45-27397,* Sept. (1970).

L-16   Hata et al. (Kitasato Institute), *Japanese Patent S-45-27798,* Sept. (1970).

L-17   Hata et al. (Kitasato Institute), *Japanese Patent S-45-27799,* Sept. (1970).

L-18   Ōmura, Reported in 14th Pharm. Soc. (Kanto Branch), p. 13 (1970).

L-19   Bosshardt and Bickel, *Experientia, 23,* 442 (1968).

L-20   Soc. Farm. Italia, *Belgian Patent 714243,* Oct. (1968).

L-21   Slomp and Mackellar, *J. Am. Chem. Soc., 89,* 2454 (1967).

L-22   Schroeder, Bannister, and Hoeksema, *J. Am. Chem. Soc., 89,* 2448 (1967).

L-23   Herr and Slomp, *J. Am. Chem. Soc., 89,* 2444 (1967).

L-24   Magerlein, Birkenmeyer, Herr, and Kagen, *J. Am. Chem. Soc., 89,* 245 (1967).

L-25   Jonston and Allen, *Biochem. Biophys. Res. Commun., 14,* 241 (1964).

L-26   Upjohn Co., *U.S. Patent 3,329,568,* July (1967).

L-27   Upjohn Co, *Japanese Patent S-43-8719,* Apr. (1968).

L-28   Upjohn Co, *U.S. Patent 3,395,139,* July (1968).

L-29   Argondelis and Mason, *J. Antibiot. (Tokyo) Ser. A, 22,* 289 (1969).

L-30   Upjohn Co., *German Offen 2,020,231,* Nov. (1970).

L-31   Johnson and Dietz, *Appl. Microbiol.,* May (1969).

L-32   Bergy, *J. Antibiot. (Tokyo) Ser. A, 22,* 126 (1969).

L-33   Upjohn Co., *Japanese Patent S-45-21635,* July (1970).

L-34   Gaudiano, Bravo, Quilico, Golding, and Lickards, *Gazz. Chim. Ital., 96,* 1470 (1967).

L-35   Soc. Farm. Italy, *Belgian Patent 714,243,* Oct. (1968).

L-36   Upjohn Co., *Japanese Patent S-45-16791,* June (1970).

L-37   Ishida, Suzuki, Maeda, Ozu, and Kumagai, *J. Antibiot. (Tokyo) Ser. A, 22,* 218 (1969).

M-1a   Institute Microbial Chemistry, *Japanese Patent S45-17593,* June (1970).

M-1b   Takahashi, Okanishi, Utahara, Nitta, Maeda, and Umezawa, *J. Antibiot. (Tokyo) Ser. A, 23,* 48 (1970).

M-2   Eli Lilly Co., *U.S. Patent 3,326,759,* June (1967).

M-3   Chimura, Ishizuka, Hamada, Hori, Kimura, Iwanaga, Takeuchi, and Umezawa, *J. Antibiot. (Tokyo) Ser. A, 21,* 44 (1968).

M-4   Gerlach, Hütter, Keller-Schierlein, Seibl, and Zähner, *Helv. Chim. Acta, 50,* 1782 (1967).

M-5   Uramoto, Otake, and Yonehara, *J. Antibiot. (Tokyo) Ser. A, 20,* 236 (1967).

M-6   Klitzunova, Ivanitzkaya, Kudinova, Prozorovskaya, Kudrina, and Gauze, *Antibiotiki, 11,* 967 (1966).

M-7   Shionogi Pharmaceutical Co., *Japanese Patent S45-14879,* May (1970).

M-8   Weitz, Moss, Oden, and Weinstein, *J. Antibiot. (Tokyo) Ser. A, 22,* 265 (1969).

M-9   Upjohn Co., *Japanese Patent S42-21759,* Oct. (1967).

M-10   Miller, Stapley, and Woodruff, *Antimicrob. Agents Chemother., 1967,* 407 (1968).

M-11   Manwaring, Rickards, and Smith, *Tetrahedron, 13,* 1029 (1970).

M-12   Borousky, Malyshkina, Kotenko, and Soloviev, *Antibiotiki, 10,* 776 (1965).

M-13   Tsuruoka, Yumoto, Ezaki, and Niida, *Sci. Rep. Meiji Seika, 9,* 1 (1967).

M-14   Shomura, Hamamoto, Ōhashi, Amano, Yoshida, Moriyama, and Niida, *Sci. Rep. Meiji Seika, 9,* 5 (1967).

M-15   Noguchi, Kohmoto, Yasuda, Hashimoto, and Niida, *Sci. Rep. Meiji Seika, 9,* 11 (1967).

M-16   Blinov, Bodkova, Kalakutzky, and Krassilnikov, *Antibiotiki, 11,* 587 (1966).

M-16/17   Kaken Kagaku Co., *German Patent 2043946* (1971).

M-17   Morton, Van Lear, and Fulmor, *J. Am. Chem. Soc. 92,* 2588 (1970).

M-18   Liu, Cullen, and Rao, *J. Antibiot. (Tokyo) Ser. A, 22,* 608 (1969).

M-19   *Pharmascope, 10,* 6 (1970).

M-20   McBride, Axelrod, Cullen, Marsh, Sodano, and Rao, *Proc. Soc. Exp. Biol. Med., 130,* 1188 (1969).

M-21   Tulinsky and Van den Hende, *J. Am. Chem. Soc., 89,* 2905 (1967).

M-22   Wallhausser, Nesemann, Prave, and Steigler, *Antimicrob. Agents Chemother., 1965,* 734 (1966).

M-23   Hoechst, A. G., *Japanese Patent S45-24793,* Aug. (1970).

M-24   Schacht and Huber, *J. Antibiot. (Tokyo) Ser. A, 22,* 597 (1969).

M-25a   Haney and Hoehn, *Antimicrob. Agents Chemother., 1967,* 349 (1968).

M-25b   Stark, Knox, and Westhead, *Antimicrob. Agents Chemother., 1967,* 353 (1968).

M-25c   Agtarap and Chamberlin, *Antimicrob. Agents Chemother., 1967,* 359 (1968).

M-25d   Gorman, Chamberlin, and Hamill, *Antimicrob. Agents Chemother., 1967,* 363 (1968).

M-25e   Shumard and Callender, *Antimicrob. Agents Chemother., 1967,* 369 (1968).

M-25f   Donoho and Kline, *Antimicrob. Agents Chemother., 1967,* 763 (1968).

M-26   Chamberlin and Steinrauf, *J. Am. Chem. Soc., 89,* 5737 (1967).

M-27   Kaken Chemical Co. and Kumiai Chemical Co., *Japanese Patent S45-955,* Jan. (1970).

M-28   Abe, Akasaki, Seino, and Shirato, *J. Antibiot. (Tokyo) Ser. A, 20,* 167 (1967).

M-29   Tanaka, Endo, Shimazu, Yoshida, Suzuki, Ōtake, and Yonehara, *J. Antibiot. (Tokyo) Ser. A, 23,* 231 (1970).

M-30   Frolova, Yulikova, Kuzovkov, and Oparysheva, *Antibiotiki, 11,* 887 (1966).

M-31   Ono Pharmaceutical Co., *Japanese Patent S35-5294,* May (1960).

M-32   Coronelli, Pasqualucci, Thiemann, and Tamoni, *J. Antibiot. (Tokyo) Ser. A, 20,* 329 (1967).

N-1   Schwarz, *J. Antibiot. (Tokyo) Ser. A, 20,* 238 (1967).

N-2   Olin Mathieson Co., *U.S. Patent 2,999,048,* Sept. (1961).

N-3a   Struk et al., *Antibiot. Annu.,* p. 878, (1957/58).

N-3b   Gist en spirit, *Netherland Patent 87323,* (1958).

N-3c   Patrick et al., *Tetrahedron Lett.,* p. 3551 (1966).

N-4   Stark, Hoehn, and Knox, *Antimicrob. Agents Chemother., 1967,* 314 (1968).

N-5a Hamada, Takeuchi, Kondo, Ikeda, Naganawa, Maeda, Okami, and Umezawa, *J. Antibiot. (Tokyo) Ser. A, 23,* 170 (1970).

N-5b Mizuno, Nitta, and Umezawa, *J. Antibiot. (Tokyo) Ser. A, 23,* 581 (1970).

N-6 Cassinelli, Grein, Orezzi, Pennella, and Sanfilippo, *Arch. Mikrobiol., 55,* 358 (1967).

N-7 Hessler, Jahnke, Robertson, Tsuji, Rinehart, and Shier, *J. Antibiot. (Tokyo) Ser. A, 23,* 464 (1970).

N-8 Kondo, Wakashiro, Hamada, Maeda, Takeuchi, and Umezawa, *J. Antibiot. (Tokyo) Ser. A,* 23, 354, 1970).

N-9 Belova and Stolpnik, *Antibiotiki, 11,* 21 (1966); Silaev, *Antibiotiki, 12,* 755 (1967).

N-10 Kubota and Matsutani, *J. Chem. Soc. Ser. C,* p. 695 (1970).

N-11 Miyazaki, Kono, Shimazu, Takeuchi, and Yonehara, *J. Antibiot. (Tokyo) Ser. A, 21,* 279 (1968).

N-12 Ishida, Kumagai, Niida, Hamamoto, and Shomura, *J. Antibiot. (Tokyo) Ser. A, 22,* 62 (1967).

N-13 Ishida, Kumagai, Niida, Tsuruoka, and Yumoto, *J. Antibiot. (Tokyo) Ser. A, 22,* 66 (1967).

N-14 Inouye, Tsuruoka, Ito, and Niida, *Tetrahedron, 23,* 2125 (1968).

N-15 Kilbourn, Dunitz, Pioda, and Simon, *Biology, 30,* 559 (1967).

N-16 Bristol Co., *U.S. Patent 3,403,078,* Sept. (1968).

N-17 Ohtsuka, Murao, Ubasawa, and Ikehara, *J. Am. Chem. Soc., 91,* 1535 (1969).

N-18 Borowski, Zielineski, Falkowski, Ziminski, Golik, Kotodziejczyk, Jereczek, Gdulewicz, Shenin, and Kofienko, *Tetrahedron Lett.,* p. 685 (1971).

O-1 Cassinelli, Grein, Orezzi, Pennella, and Sanfillipo, *Arch. Mikrobiol., 55,* 358 (1967).

O-2 Bowie and Johnson, *Tetrahedron Lett., 16,* 1449 (1967).

O-3 Higashide, Hasegawa, Shibata, Mizuno, Imanishi, and Miyake, *J. Antibiot. (Tokyo) Ser. A, 18,* 26 (1965).

O-4 Gyögyszerkutato Intezet, *Hung. Teljes,* p. 423, June (1970).

O-5a Lardy, Johnson, and Murray, *Biochemistry, 3,* 1961 (1964).

O-5b Beechey, Williams, Halloway, Knight, and Robertson, *Biochem. Biophys. Res. Commun., 26*(3), 339 (1967).

O-6 Bakhaeva, Berlin, Chuprunova, Kolosov, Peck, Piotrovich, Shemyakin, and Vasina, *Chem. Commun.,* 10 (1967).

O-7 Hashimoto, Kondo, Takita, Hamada, Takeuchi, Ōkami, and Umezawa, *J. Antibiot. (Tokyo) Ser. A, 21,* 653 (1968).

O-8a McCord, Ravel, Skinner, and Shive, *J. Am. Chem. Soc., 79,* 5693 (1957).

O-8b Stapley, Miller, Mata, and Hendlin, *A.A.C.–1967,* 401 (1968).

O-9 Hoffman-LaRoche Co., *U.S. Patent 3,361,742,* Jan. (1968).

O-10 Meiji Seika Co., *Japanese Patent S41-13792,* (1966).

P-1 Wiley, Jahnke, Mackellar, Kelly, and Argoudelis, *J. Org. Chem., 35,* 1420 (1970).

P-2 Gyogys-Zerkutato Intezet, *Hung. Teljes,* p. 422, June (1970).

P-3 Miller, Stapley, and Chaiet, *Bacteriol. Proc.,* p. A49 (1962).

P-4 Batra and Bajaj, *Indian J. Exp. Biol., 3,* 240 (1965).

P-5 Mizuno, Hamada, Maeda, and Umezawa, *J. Antibiot. (Tokyo) Ser. A, 21,* 429 (1968).

P-6 Gerber and Lechevalier, *Biochemistry, 3,* 598 (1964).

P-7 Akabori and Nakamura, *J. Antibiot. (Tokyo) Ser. A, 12,* 17 (1959).

P-8 Irie, Kurosawa, and Nagaoka, *Bull. Chem. Soc. Jap. 33,* 1057 (1960).

P-9 Anzai, Isono, Okuma, and Suzuki, *J. Antibiot. (Tokyo) Ser. A, 13,* 125 (1960).

P-10 Clemo and McIlwain, *J. Chem. Soc.,* p. 479 (1938).

P-11 Nakamura, Yajima, Hamada, Nishimura, Ishizuka, Takeuchi, Tanaka, and Umezawa, *J. Antibiot. (Tokyo) Ser. A, 20,* 210 (1967).

P-12 Hendlin et al., *Science, 166,* 122 (1969).

P-13 Christensen et al., *Science, 166,* 124 (1969).

P-14 Stapley et al., *Antimicrob. Agents Chemother., 1969,* 284 (1970).

P-15 Hendlin et al., *Antimicrob. Agents Chemother., 1969,* 297 (1970).

P-16 Miller et al., *Antimicrob. Agents Chemother., 1969,* 310 (1970).

P-17 Rickards and Smith, *Chem. Commun.,* 1049 (1968).

P-18 Takahashi, Suzuki, Kimura, Miyamoto, Tamura, Mitsui, and Fukami, *Agr. Biol. Chem., 32,* 1115 (1968).

P-19 Muxfeldt, Shrader, Hansen, and Brockmann, *J. Am. Chem. Soc., 90,* 4748 (1968).

P-20 Ceder and Hansson, *Tetrahedron, 23,* 3753 (1967).

P-21 Hokko Chem. Ind. (Japan), *Netherland Patent 6,713,997,* Oct. (1967).

P-22 Shoji, Kozuki, Matsutani, Kubota, Nishimura, Mayama, Motokawa, Tanaka, Shimaoka, and Otsuka, *J. Antibiot. (Tokyo) Ser. A, 21,* 402 (1968).

P-23 Shimi, Imam, and Shehata, *J. Antibiot. (Tokyo) Ser. A, 20,* 204 (1967).

P-24 Solovieva et al., *Antibiotiki, 5,* 5 (1960).

P-25 Isono, Asahi, and Suzuki, *J. Am. Chem. Soc., 91,* 7490 (1969).

P-26 Endo et al., *J. Bacteriol., 104,* 189 (1970).

P-27 Isono, Nagatsu, Kawashima, Kobinata, Sasaki, and Suzuki, *Agr. Biol. Chem., 31,* 190 (1967).

P-28 Isono, Kobinata, and Suzuki, *Agr. Biol. Chem., 32,* 792 (1968).

P-29 Weisenborn, Bouchard, Smith, Pansy, Maestrone, Miraglia, and Meyers, *Nature,* p. 1092 (1967).

P-30 Abe, Akasaki, Seino, and Shirato, *J. Antibiot. (Tokyo) Ser. A, 20,* 167 (1967).

P-31   Aberhart, Fehr, Jain, Mayo, Motl, Baczynskyi, Gracey, McLean, and Szilagyi, *J. Am. Chem. Soc., 92,* 5816 (1970).

P-32   Preud'homme, Belloc, Charpentée, and Tarridec, *C. R. Acad. Sci. (Paris), 260,* 1309 (1965).

P-33   Perry, *Nature, 191,* 77 (1961).

P-34   Harashima, Tsuchida, Tanaka, and Nagatsu, *Agr. Biol. Chem., 31,* 481 (1967).

P-35   Bradley, *Dev. Ind. Microbiol., 3,* 362 (1962).

P-36   Asahi, Kushikata, and Takamiya, *J. Antibiot. (Tokyo) Ser. A, 20,* 334 (1967).

P-37   Gerzon, Williams, Hoehn, Gorman, and DeLong, 2nd Int. Congr. Heterocyclic Chem., Abstract 30-C (1969).

P-38   Streightoff, Nelson, Cline, Gerzon, Williams, and DeLong, Abstracts, 9th Int. Conf. Antimicrob. Agents Chemother., p. 8 (1969).

P-39   Koyama and Iitaka, *Tetrahedron Lett.,* p. 3587 (1967).

Q-1   Murray, Kaplan, Granatek, and Buckwalter, *Antibiot. Chemother., 7,* 569 (1957).

Q-2   Otsuka and Shoji, *Tetrahedron, 23,* 1535 (1967).

Q-3   Yoshida, Kimura, and Katagiri, *J. Antibiot. (Tokyo) Ser. A, 21,* 465 (1968).

Q-4   Kowa Co., Ltd., *Japanese Patent S-45-26080,* Aug. (1970).

Q-5   Japan Medical Gazette, p. 5, February 20 (1971).

R-1   Liu, Parker, Slusarchyk, Greenwood, Graham, and Meyers, *J. Antibiot. (Tokyo) Ser. A, 23,* 437 (1970).

R-2   Upjohn Co., *U.S. Patent 3,476,857,* Nov. (1969).

R-3   Tendler and Korman, *Nature,* p. 4892 (1963).

R-4   Department of Health, Education and Welfare, *U.S. Patent 3,366,540,* Jan. (1968).

R-5   Rosenbrook, *J. Org. Chem., 32,* 2924 (1967).

R-6   Bailey, Falsham, Ollis, Watanabe, Dhar, Khan, and Vora, *Chem. Commun.,* 374 (1968).

R-7   Navashin, Fomina, Koroleva, Terentieva, and Stegelman, *Antibiotiki, 10,* 892 (1967).

R-8   Meiji Seika Co., *Japanese Patent S45-17155,* June (1970).

R-9   Brockmann and Niemeyer, *Chem. Ber., 100,* 3578 (1967).

R-10   Brockmann and Niemeyer, *Chem. Ber., 101,* 1341 (1968).

R-11   Lancini, Gallo, Sartori, and Sensi, *J. Antibiot. (Toyko) Ser. A, 22,* 369 (1969).

R-12   Lancini and Hengeller, *J. Antibiot. (Tokyo) Ser. A, 22,* 637 (1969).

R-13   Leitich, Prelog, and Sensi, *Experientia, 23,* 505 (1967).

R-14   Brufani, Fedeli, Giacomello, and Vaciago, *Experientia, 23,* 508 (1967).

R-15   Kaken Chemical Co. and Institute of Physical and Chemical Research, *Japanese Patent S45-17598,* June (1970).

R-16   Rhone-Poulenc, *French Patent 1,551,195* (1968).

R-17   Despois et al., *Arzneim. Forsch., 17,* 934 (1967).

R-18   Brazhnikova et al., *Antibiotiki,* p. 781 (1968).

R-19   Beechey, Williams, Holloway, Knight, and Roberton, *Biochem. Biophys. Res. Commun., 26,* 339 (1967).

R-20   Lardy, Witonsky, and Johnson, *Biochemistry, 4,* 552 (1965).

S-1   Tolman, Robins, and Townsend, *J. Am. Chem. Soc. 90,* 524 (1968).

S-2   Hill, Foley, and Gardella, *J. Org. Chem., 32,* 2330 (1967).

S-3   Kōno, Makino, Takeuchi, and Yonehara, *J. Antibiot. (Tokyo) Ser. A, 22,* 583 (1969).

S-4   New Names in *J.A.M.A., 212,* 466 (1970).

S-5   Gruppo Lepetit SpA, *Belgian Patent 751903,* Nov. (1970).

S-6   Komatsu and Tanaka, *Agr. Biol. Chem., 35,* 526 (1971).

S-7   Ebata, Miyazaki, and Otsuka, *J. Antibiot. (Tokyo) Ser. A, 22,* 364 (1969).

S-8   Wiley and MacKellar, *J. Am. Chem. Soc., 92,* 417 (1970).

S-9   Wang, Hamada, Okami, and Umezawa, *J. Antibiot. (Tokyo) Ser. A, 19,* 216 (1966).

S-10   Naganawa, Takita, Maeda, and Umezawa, *J. Antibiot. (Tokyo) Ser. A, 21,* 241 (1968).

S-11   Ōmura, Nakagawa, Otani, Hata, Ogura, and Furuhata, *J. Am. Chem. Soc., 91,* 3401 (1969).

S-12   Thiemann, Coronell, Pagani, Beretta, Tamoni, and Arioli, *J. Antibiot. (Tokyo) Ser. A, 21,* 526 (1968).

S-13   Bergy and Reusser, *Experientia, 23,* 254 (1966).

S-14   Reusser, *Biochem. Pharmacol., 18,* 287 (1968).

S-15   Bodanszky, Muramatsu, and Bodanszky, *J. Antibiot. (Tokyo) Ser. A, 20,* 384 (1967).

S-16   Bristol-Myers Co., *U.S. Patent 3,334,015,* Aug. (1967).

S-17   Arcamone, Cassinelli, D'Amico, and Orezzi, *Experientia, 24,* 444 (1968).

S-18   Heding, *Acta Chem. Scand., 22,* 1649 (1968).

S-19   Rinehart, Martin, and Coverdale, *J. Am. Chem. Soc., 88,* 3149 (1966).

S-20   Rinehart, Coverdale, and Martin, *J. Am. Chem. Soc., 88,* 3150 (1966).

S-21   Johnson, Duquette, and Hennis, *J. Org. Chem., 33,* 904 (1968).

S-22   Herr, Jalinke, and Argoudelis, *J. Am. Chem. Soc., 89,* 4808 (1967).

S-23   Takeda Chemical Industries, *Japanese Patent S-42-514,* Jan. (1967).

S-24   Egawa, Umino, Tamura, Shimizu, Kaneko, Sakurazawa, Awataguchi, and Okuda, *J. Antibiot. (Tokyo) Ser. A, 22,* 12 (1969).

T-1   Ivanitskaya, Upiter, Sveshnikova, and Gauze, *Antibiotiki, 10,* 973 (1965).

T-2 Shenin, Sokolova, and Konev, *Antibiotiki, 15,* 9 (1970).

T-3 Eli Lilly & Co., *Belgian Patent 697, 319,* Oct. (1967).

T-4a American Cyanamid Co., *Netherland Patent 6616081,* May (1967).

T-4b Olin-Mathieson Co., *Japanese Patent S-41-19599,* Nov. (1966).

T-4c E. R. Squibb & Sons Inc., *U.S. Patent 3,364,123,* Jan. (1968).

T-5 Lepetit S.p.A., *French Patent 1,395,876,* April (1965).

T-6 Thrum, Bradler, Dornberger, and Fuegner, *German (East) Patent 70,706,* Jan. (1970).

T-7 Gerlack, Hütter, Keller-Schierlein, Seibl, and Lähner, *Helv. Chim. Acta, 50,* 1782 (1967).

T-8 Gallo, Coronelli, Vigevani, Lancini, *Tetrahedron, 25,* 5677 (1969).

T-9 Cassinelli, Cotta, D'Amico, Bruna, Grein, Mazzoleni, Ricciardi, and Tintinelli, *Arch. Mikrobiol., 70,* 197 (1970).

T-10 American Cyanamid Co., *U.S. Patent 3,532,714,* Oct. (1970).

T-11 Miyairi, Miyoshi, Aoki, Kohsaka, Ikushima, Kunugita, Sakai, and Imanaka, *J. Antibiot. (Tokyo) Ser. A, 23,* 113 (1970).

T-12 Reusser, *Infect. Immun. 2,* 77 (1970).

T-13 Kishi, Asai, Muroi, Harada, Mizuta, Terao, Miki, and Mizuno, *Tetrahedron Lett.,* p. 91 (1969).

T-14 Suzuki, Asahi, Nagatsu, Kawashima, and Suzuki, *J. Antibiot. (Tokyo) Ser. A, 20,* 126 (1967).

T-15 Institute of Physical and Chemical Research, *Japanese Patent S-45-9234,* April (1970).

T-16 Aszalos, Robinson, Lemanski, and Berk, *J. Antibiot. (Tokyo) Ser. A,, 21,* 611 (1968).

T-17 Ōtsuka and Shōji, *J. Antibiot (Tokyo) Ser. A, 19,* 128 (1966).

T-18 Ōtsuka and Shōji, *Tetrahedron, 23,* 535 (1967).

T-19 VEB Jenapharm, *German Offen 2,009,116,* Oct. (1970).

T-20 Shōji, Kozaki, Okamoto, Sakazaki, and Ōtsuka, *J. Antibiot. (Tokyo) Ser. A, 21,* 439 (1968).

T-21 Nagata, Ando, Izumi, Sakakibara, Take, Hayano, and Abe, *J. Antibiot. (Tokyo) Ser. A, 21,* 681 (1968).

U-1 Squibb Co., *U.S. Patent 3,496,268,* Feb. (1970); *Belgian Patent 708,601,* June (1968).

V-0 Iwasa, Yamamoto, and Shibata, *J. Antibiot. (Tokyo) Ser. A, 23,* 595 (1970). Iwasa, Higashide, Yamamoto, and Shibata, *J. Antibiot. (Tokyo) Ser. A, 24,* 107 (1971). Iwasa Higashide, and Shibata, *J. Antibiot. (Tokyo) Ser. A, 24,* 114 (1971). Iwasa, Kameda, Asai, Horii, and Migano, *J. Antibiot. (Tokyo) Ser. A, 24,* 119 (1971). Kamiya, Wada, Horii, and Nishikawa, *J. Antibiot. (Tokyo) Ser. A, 24,* 317 (1971).

V-1 Institute of Physical and Chemical Research, *Japanese Patent 25298,* Jan. (1961).

V-2 Brufani, Keller-Schierlein, Löffler, Mansperger, and Zähner, *Helv. Chim. Acta, 51,* 1293 (1968).

V-3 Ciba Ltd., *Netherland Patent 6,809,205,* Dec. (1968).

V-4 E. R. Squibb & Sons, Inc., *U.S. Patent 3,299,047,* Jan. (1967).

V-5 Tanaka and Igusa, *J. Antibiot. (Tokyo) Ser. A, 21,* 239 (1968).

V-6 Leach, Hackman, and Byers, *Nature, 204,* 788 (1964).

V-7 Narasimbachari and Swami, *Chemotherapy, 13,* 181 (1968).

X-1 Asahi, Nagatsu, and Suzuki, *J. Antibiot. (Tokyo) Ser. A, 19,* 195 (1966).

X-2 Asahi, Nagatsu, Mizuno, and Suzuki, Abstract of Japan Agr. Chem. Meeting, p. 22 (1970).

Y-1 Akasaki, Abe, Seino, and Shirato, *J. Antibiot. (Tokyo) Ser. A, 21,* 98 (1968).

Y-2 Arai, *Japanese Patent 25237,* Nov. (1965).

Z-1 Upjohn Co., *Japanese Patent 43-5717,* Mar. 1 (1968).

NA-1 Eli Lilly Co., *South Africa Patent 67/4864* (1967); (Farmdoc 37,906).

NA-2 Eli Lilly Co., *Belgian Patent 728,382* (1968); (Farmdoc 40,135).

NA-3 Chukanova, Morozova, Denisova, and Blinov, *Antibiotiki,* p. 194 (1965).

NA-4 Shoji, Kozuki, Mayama, Kawamura, and Matsumoto, *J. Antibiot. (Tokyo) Ser. A, 23,* 291 (1970).

NA-5 Thiemann, Beretta, Coronelli, and Pagami, *J. Antibiot. (Tokyo) Ser. A, 22,* 119 (1969).

NA-6 Ankerfarm S. p. A., *Belgian Patent 674154* (1966).

NA-7 Eli Lilly Co., *Japanese Patent S45-113,* Jan. (1970).

NA-8 Ciba Co., *Japanese Patent S45-26075,* Aug. (1970).

NA-9 American Cyanamid Co., *U.S. Patent 3,495,003,* Feb. 10 (1970).

NA-11 American Cyanamid Co., *U.S. Patent 3,495,004,* Feb. 10 (1970).

NA-12 American Cyanamid Co., *U.S. Patent 3,522,349,* July (1970).

NA-13 Sinha and Nandi, *Experientia, 24*(8), 795 (1968).

NA-14 American Cyanamid Co., *U.S. Patent 3,452,136,* June 24 (1969).

NA-15 *South Africa Patent 7002415* (1970).

NA-16 American Cyanamid Co., *U.S. Patent 3,321,368,* May (1967).

NA-17 American Cyanamid Co., *U.S. Patent 3,377,244,* Apr. (1968).

NA-18 American Cyanamid Co., *U.S. Patent 3,344,025,* Sept. (1967).

NA-19 Shimi, Imam, and Haroun, *J. Antibiot. (Tokyo) Ser. A, 22,* 106 (1969).

NA-20 American Cyanamid Co., *U.S. Patent 3,338,786,* Aug. (1967).

NA-21 Yamazaki, *J. Antibiot. (Tokyo) Ser. A, 21,* 204 (1968).

NA-22 Takeda Co., *Japanese Patent, S45-25275,* Aug. (1970).

NA-23 Takeda Chemical Industries, *Netherland Patent 6,802,679,* Aug. (1968).

NA-24 Takeda Chemical Industries, *Japanese Patent S45-24962*, Aug. (1970).

NA-25 Takeda Chemical Industries, *Japanese Patent S45-114*, Jan. (1970).

NA-26 Hasegawa, Higashide, Shibata, and Mizuno, *Ann. Rep. Takeda Res. Lab., 25*, 15 (1966).

NA-27 Takeda Chemical Industries, *Japanese Patent S45-19638*, July (1970).

NA-28 Kusaka, Iwasa, Shibata, Yamana, and Kishi, *J. Takeda Res. Lab., 29*, 406 (1970).

NA-29 Iwasa, Fujii, Yamamoto, Shibata, Kameda, Horii, and Mizuno, *Takeda Annu. Rep. 27*, 74 (1968).

NA-30 Takeda Chemical Industries, *Japanese Patent S45-20559*, July (1970).

NA-31 Takeda Chemical Industries, *Japanese Patent S46-271*, Jan. (1971).

NA-32 Chas. Pfizer Co., *German Offen. 1,954,047*, June (1970).

NA-33 Kusaka, Yamamoto, and Suzuki, *Takeda Kenkyusho Ho, 29*, 239 (1970).

NA-34 Chas. Pfizer Co., *German Patent 1089928*, Sept. (1960); *Japanese Patent 6597*, May (1963).

NA-35 Chas. Pfizer Co., *U.S. Patent 3,328,248*.

NA-36 Ito, Ohashi, Sakurai, Sakurazuka, Yoshida, Awatoguchi, and Okuda, *J. Antibiot. (Tokyo) Ser. A, 21*, 307 (1968).

NA-37 Furumai, Kaneko, Matsuzawa, Sato, and Okuda, *J. Antibiot. (Tokyo) Ser. A, 21*, 283 (1968).

NA-38 American Cyanamid Co., *South Africa Patent 7002462* (1970); *Belgian Patent 750045*, Nov. (1970).

NA-39 Aszalos, Heberecht, and Cohen, *J. Med. Chem., 10*, 281 (1967).

NA-40 Dickinson, Griffiths, Mason, and Mills, *Nature, 206*, (4981), 265 (1965).

NA-41 Lysenko and Bondarenko, *Mikrobiologia, 32*, 201 (1970).

NA-42 Société Anonyme Industrie Biologique Francaise, *U.S. Patent 3,320,128* (1967).

NA-43 Shoji, Kozuki, Ebata, and Otsuka, *J. Antibiot. (Tokyo) Ser. A, 21*, 509 (1968).

NA-44 Societa Farmaceutici Italia, Nerland 41,590 (1969); (Farmdoc 41590).

NA-45 Fujisawa Pharmaceutical Co. and Nissan Chemical Co., *Japanese Patent S45-26712*, Sept. (1970).

NA-46 Nomura, Yamamoto, Umesawa, Matsumae, and Hata, *J. Antibiot. (Tokyo) Ser. A, 20*(2), 55 (1967).

NA-47 Kitasato Institute, *Japanese Patent S45-117*, Jan. (1970).

NA-48 Abe, Otani, Nakabawa, Omura, Masumae, and Hata, *Jap. J. Microbiol., 23*(8), (1968).

NA-49 Veb Jenapharm, *Nederland Patent 7003002*, Sept. (1970).

NA-50 Koniklijke Nederland, Gisten Spiritufabrik, *British Patent 1,178,783* (1970).

NA-51 Nederlandse Centrale Organisatie Voor Toegepast Natuurwetenschappelijk Onderzoek, *Netherland Patent, 105,150*, Jan. (1963).

NA-52 Mitscher, Shay, and Bohonos, *Appl. Microbiol., 15*(5), 1002, (1967).

NA-53 Sax, Monnikendam, Borders, Shu, Mitscher, Hausmann, and Patterson, *Antimicrob. Agents Chemother.*, p. 442 (1967).

NA-54 Zbinovsky, Hausmann, Wetzel, Borders, and Patterson, *Appl. Microbiol., 16*, 614 (1968).

NA-55 Borders, Hausmann, Nelzel, and Patterson, *Tetrahedron Lett.*, p. 4187 (1967).

NA-56 Martin, Mitscher, Shu, Porter, Bohonos, Devoe, and Patterson, *Antimicrob. Agents Chemother.*, p. 422 (1967).

NA-57 American Cyanamid Co., *U.S. Patent 3,377,244* (1968).

NA-58 Whaley, Patterson, Kunstmann, and Bohonos, *Antimicrob. Agents Chemother.*, p. 591 (1966).

NA-59 American Cyanamid Co., *U.S. Patent 3,338,786* (1967); (Farmdoc 28,295).

NA-60 Abbott Co., *British Patent 1,041,766* (1966).

NA-61 Meiji Seika Co., *Japanese Patent S45-6075*, Feb. (1970).

NA-62 Armstrong, Grove, Turner, and Ward, *Nature, 206*(4982), 399 (1965).

NA-63 Stapley et al., *Antimicrob. Agents Chemother.*, 250 (1969).

NA-64 Miller, Walker, Trenner, Arison, and Wolf, *Antimicrob. Agents Chemother., 1968*, 255 (1969).

NA-65 Meiji Seika Co., *Japanese Patent S45-6877*, Mar. (1970).

NA-66 Meiji Seika Co., *Japanese Patent S45-16792*, June (1970).

NA-67 Meiji Seika Co., *Japanese Patent S45-6077*, Feb. (1970).

NA-68 Tokhtamuratov, Silaev, and Khodzhibaeva, *Antibiotiki, 9*, 205 (1964).

NA-69 Chas. Pfizer Co., *U.S. Patent 3,268,418*, Aug. (1966).

NA-70 Scherico Ltd., *British Patent 1,021,402*, Mar. (1966).

NA-71 Scherico Ltd., *British Patent 1,074,235*, July (1967).

NA-72 American Cyanamid Co., *U.S. Patent 3,377,243*, Apr. (1968).

NA-73 Rhone-Poulenc Co., *Jap. Patent S45-116*, Jan. (1970).

NA-74 Rhone-Poulenc Co., *French Patent 1,455,303*, Sept. (1966).

NA-75 Rhone-Poulenc Co., *French Patent 1,527,892*, April (1968).

NA-76 Mancy et al., Abstr. 9th Int. Congr. Microbiol., Moscow, p. 165 (1966).

NA-77 Rhone-Poulenc Co., *South African Patent 67/1832* (1967); (Farmdoc 28754).

NA-78 Rhone-Poulenc Co., *South African Patent 67/1147* (1967); (Farmdoc 28551).

NA-79 Rhone-Poulenc Co., *Netherland Patent 6,906,827*, Nov. (1969).

NA-80 Societé des Usines Chimiques, *French Patent 1584984*, Jan. (1970).

NA-81 Rhone-Poulenc Co., *Netherland Patent 68,02093* (1968); (Farmdoc 33586).

NA-82 Sumitomo Chemical Industries, *Japanese Patent S45-112* (1970).

NA-83 Shoji, Kozuki, Mayama, and Shimaoka, *J. Antibiot. (Tokyo) Ser. A, 23*, 429 (1970).

NA-84   Shoji and Sakazaki, *J. Antibiot. (Tokyo) Ser. A, 23,* 432 (1970).

NA-85   Shoji and Sakazaki, *J. Antibiot. (Tokyo) Ser. A, 23,* 418 (1970).

NA-86   Shionogi Pharmaceutical Co., *Japanese Patent S45-27800,* Sept. (1970).

NA-87   Shoji, Kozuki, Nishimura, Mayama, Motokawa, Tanaka, and Otsuka, *J. Antibiot. (Tokyo) Ser. A, 21,* 643, (1968).

NA-88   Shionogi Pharmaceutical Co., *Japanese Patent S45-17600,* June (1970).

NA-89   Sumitomo Chemical Co., *Japanese Patent S45-5435,* Feb. (1970).

NA-90   Meiji Seika Co., *Japanese Patent S45-21631,* July (1970).

NA-91   Meiji Seika Co., *Japanese Patent S45-6070,* Feb. (1970).

NA-92   Tsuruoka, Shomura, Ezaki, Niwa, and Niida, *J. Antibiot. (Tokyo) Ser. A, 21,* 237 (1968).

NA-93   Meiji Seika Co., *Japanese Patent S45-6878,* March (1970).

NA-94   Akita, Tsuruoka, Ezaki, and Niida, *J. Antibiot. (Tokyo) Ser. A, 23,* 173 (1970).

NA-95   Meiji Seika Co., *Japanese Patent S45-17157,* June (1970).

NA-96   Meiji Seika Co., *Japanese Patent S45-17159,* June (1970).

NA-97   Sherico Co. (Swiss), *German Patent 1,227,194,* Oct. (1966).

NA-98   Kosmachev, *Mikrobiologiya, 31,* 66 (1962).

NA-99   Takeda Chemical Industries, *Netherland Patent 6,807,199,* May (1968).

NA-100   Harada, Higashide, Fugono, and Kishi, *Tetrahedron Lett., 27,* 2239 (1969).

NA-101   Takeda Chemical Industries, *Japanese Patent S45-6071* (1970).

NA-102   Takeda Chemical Industries, *German Offen. 1,954,110,* May (1970).

NA-103   Tanabe Pharmaceutical Co., *Japanese Patent S45-956,* Jan. 13 (1970).

NA-104   Tanabe Co., *Japanese Patent S45-17594* (1970).

NA-105   Meyer et al., *Antimicrob. Agents Chemother., 1965,* 850 (1966).

NA-106   Masson et al., *Antimicrob. Agents Chemother., 1964,* 110 (1965).

NA-107   Upjohn Co., *U.S. Patent 3,300,382,* (1967).

NA-108   Bergy and Reusser, *Experientia, 23*(4), (1967).

NA-109   Herr and Reusser, *Appl. Microbiol.,* p. 1142 (1967).

NA-110   Daehne, Godtfredsen, and Tybring, *J. Antibiot. (Tokyo) Ser. A, 22,* 233 (1969).

NA-111   Sherico Ltd., *Netherland Patent 6,807,363,* Nov. (1968).

NA-112   Kannan, Khan, Kutty, and Vora, *J. Antibiot. (Tokyo) Ser. A, 20,* 293 (1967).

NA-113   Stempel, Westley, and Benz, *J. Antibiot. (Tokyo) Ser. A, 22,* 384 (1969).

NA-114   Johnson, Herrin, Liu, and Paul, *Chem. Commun. (J. Chem. Soc. Sect. D), 2,* 72, (1970).

NA-115   Tanabe Seiyaku Co., Ltd., *South African Patent 7003035,* (1970).

NA-116   Suzuki, Takamori, Kinumaki, Sugawara, and Okuda, *Tetrahedron Lett.,* p. 435 (1971).

NA-117   Schering Co., *U.S. Patent 3,458,626,* July 29 (1969).

NA-118   Tokhtamuratov and Silaev, *Antibiotiki, 12,* 887 (1967).

NA-119   Rudaya and Bychkova, *Antibiotiki, 11,* 410 (1966).

NA-120   Bringi, Bhatt, and Thirumalachar, *Hindustan Antibiot. Bull., 2,* 120 (1960).

NA-121   Tanaka, *J. Ferm. Assoc. Jap., 44,* 384 (1966).

NA-122   Physics and Chemistry Institute, *Japanese Patent S45-9836* (1970).

NA-123   Ito, Yashumura, and Nawada, *Annual Paper of Pharmaceutical Society of Japan,* p. 225 (1966).

NA-124   Merck & Co., *Belgian Patent 653,769,* (1965).

NA-125   Dundarov, Andonov, Georgieva, and Khlebarova, *Antibiotiki, 12,* 569 (1967).

NA-126   Daiichi Seiyaku Co., Ltd., *South African Patent 6905196,* Feb. (1970).

NA-127   Brazhnikova, *Antibiotiki, 15,* 675 (1970).

NA-128   Luedemann et al., *German Patent 1,271,896,* July (1968).

NA-129   Scherico Ltd., *Netherland Patent 6,711,212,* Feb. (1968).

NA-130   Schering Co., *U.S. Patent 3,454,696,* July (1969).

NA-131   Scherico Co. (Swiss), *Netherland Patent 6,711,212,* Feb. (1968).

NA-132   Kyowa Fermentation Industries Co., *Japanese Patent S45-17596* (1970).

NA-133   Gitterman, Rickes, Wolf, Madas, Zimmerman, Stoudt, and Demmy, *J. Antibiot. (Tokyo) Ser. A, 23,* 305 (1970).

NA-134   Kominkov, Burdarov, and Gushterov, *God. Sofiiskiya Univ. Biol. Fak., 52,* 75 (1969).

NA-135   Merck Co., *Japanese Patent S45-9828,* Apr. (1970).

NA-136   Yamamoto, Fujii, Nakazawa, Miyake, Hitomi, and Imanishi, *Annu. Rep. Takeda Res. Lab., 16,* 28 (1957).

NA-137   Bogdanova, Konev, Sannikov, Soloviev, Soklov, and Tsyganov, *Antibiotiki, 10,* 195 (1965).

NA-138   Kaken Chemical Co. and Kumiai Chemical Co., *Japanese Patent S45-6072* (1970).

NA-139   Frolova and Parshina, *Akad. Nauk Kaz. USSR 8,* 133 (1965); *Chem. Abstr., 64,* 7323 (1966).

NA-140   Kowa Co., *Japanese Patent S45-5035* (1970).

NA-141   Radzhapov, Silaev, and Agre, *Antibiotiki. 11,* 909 (1966).

NA-142   Ukholina, Krugliak, Borisova, Kovsharova, and Proshliakova, *Mikrobiologiya, 34,* 147 (1965).

NA-143   Christensen, Rosenlundhansen, Alflund, and Noer, *Acta Chem. Scand., 11,* 755 (1957).

NA-144   Kudinova, Babenko, Ukholina, Maksimova, Nechaeva, Terekhova, and Rossolimo, *Antibiotiki, 13,* 201 (1968).

NA-145   Brazhikova, Kudinova, Mezentsev, Fedorova, Ukholina Kochetkova, Maksimova, Nechaeva, and Rossolimo, *Antibiotiki, 13,* 963 (1968).

NA-146   Rudaya, Solovieva, Vikhrova, Gromova, Goncharskaya, Germanova, Braginskaya, and Navashin, *Antibiotiki, 14,* 254 (1969).

NA-147   Muraveiskaya, Pevzer, Shapovalova, Philipposyan, Ukholina, Kovalenkova, Nechaeva, Sveshnikova, Lavrova, and Babenko, *Antibiotiki, 11,* 234 (1966).

NA-148   Nakazawa, Shibata, Wada, Kanzaki, Higashide, Miyake, Hitomi, and Takewaka, *Annu. Rep. Takeda Res. Inst., 19,* 53 (1960).

GENERAL REFERENCES

1. Umezawa, H., Ed., Index of Antibiotics from Actinomycetes, University of Tokyo Press, Tokyo, and University Park Press, State College, Pennsylvania, 1967.
2. Korzybski, T. G., Antibiotics, Polish Scientific Publishers, Warszawa, and Pergamon Press, Oxford, 1967.
3. Gottlieb, D. and Shaw, P. D., Antibiotics, Springer-Verlag, Berlin, 1967.

COMPOUNDS INHIBITING VIRUS MULTIPLICATION

DR. DONALD C. DELONG and B. LOUISE CRANDALL

Virology became a science because viruses cause disease. In modern biology, this fact is sometimes either forgotten or ignored. An important part of progress in virology, and in medicine in general, involves the search for chemotherapeutic agents that prevent or cure disease. The ability to provide prophylatic or therapeutic treatment of virus disease by low-molecular weight compounds is now satisfactorily demonstrated by the use of 1-methylisatin 3-thiosemicarbazone for smallpox, IDUR for herpes keratitis, and adamantyl amine for influenza. It is no longer questionable that compounds can specifically inhibit virus multiplication without altering the host. Compounds that inhibit virus multiplication are also valuable tools to those studying the biochemistry and the molecular biology of the transfer of genetic information.

The following list of compounds inhibiting virus multiplication was compiled as a guide to types of compounds that have been shown to have such an effect. Although the table was designed to be comprehensive, it is impossible to assure completeness, i.e., very few nucleoside-type antivirals and only one compound of most active series are included. The most useful information is considered to be the structure and a reference to the original work. Compounds described are from three main sources: (1) those isolated from microbial fermentations, (2) those isolated from other natural sources, and (3) those made synthetically. The systems used for measurement of activity are listed as an implication of interest in the compound. An entry is also made concerning the chemical type of the inhibited virus (DNA, RNA, or both) as an indication of the spectrum of activity of the compound. Of course, when mention is made that a compound is active, for example, against RNA viruses, one should not conclude that the compound is active against all RNA viruses. Viral inhibitors, like antibiotics, are often highly specific in their spectrum of activity. The reader should consult the references for additional information.

ABBREVIATIONS

Sources

F = fermentation product; P = natural product, usually of plant origin; S = synthetic chemical.

Type of Activity

A = animal virus inhibited in experimental animal assays; E = enzyme system (virus-induced enzymes or enzyme carried by viruses); M = human virus disease altered in experimental or natural infections; O = egg sytem used to determine effect on virus growth; T = tissue culture system for measuring activity of inhibiting viruses.

Type of Virus

DNA = deoxyribonucleic acid; RNA = ribonucleic acid.

Systematic Name (Common Name)	Structure	Source	Type of Activity	Type of Virus	Comments	Reference
A 10598	Unknown	F	T, A	RNA + DNA	Antiviral antibiotic without antibacterial activity	1
5-Acetoxy-5,5a,13,13a-tetrahydro-13-hydroxy-7H,8H,15H,16H-7a,15a-epidithiobisoxepino(3',4':4,5)pyrrolo(1,2-a:1',2'-d)pyrazine-7,15-dione (Aranotin)	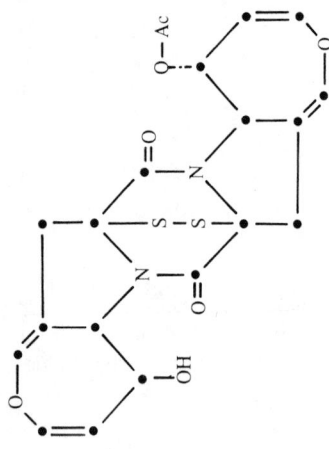	F	E, T, A	RNA	Inhibits virus-induced RNA-dependent RNA polymerase	2
3-Acetyl-5-(*sec*-butyl)-4-hydroxy-3-pyrrolin-2-one (Tenuazonic acid)		F	T	DNA + RNA	Antiviral antibiotic without antibacterial or antifungal activity	3
1-Adamantane amine (Amantadine, Symmetrel)		S	T, A, M	RNA (influenza)	Prevents penetration or uncoating of virus	4
2-Adamantane amine		S	T, A	RNA	Very similar to 1-adamantane amine	5

Systematic Name (Common Name)	Structure	Source	Type of Activity	Type of Virus	Comments	Reference
1-Adamantylguanidine	(adamantane with $-NH-C(=NH)-NH$ substituent)	S	A	RNA	Increased basicity by comparison with adamantane amine, but antiviral activity did not increase	6
1-Allyl-3,5-diethyl-6-chloro-uracil	(uracil ring with CH_3-CH_2-, CH_2-CH_3, Cl, and $CH_2-CH=CH_2$ substituents)	S	T, A, M	DNA	Claimed not to be an antimetabolite, since the allylic double bond is essential	7
N-(2'-Amidinoethyl)-3-aminocyclopentanecarboxamide (Amidinomycin, myxoviromycin)	(cyclopentane with NH_2 and $O=C-NH-(CH_2)_2-C(=NH)-NH_2$)	F	A, T	RNA	Antibiotic with both antibacterial and antiviral activity	8
2-[N-(2-Amidinoethyl)carbamoyl]-5-iminopyrrolidine (Noformicin)	(iminopyrrolidine ring with $O=C-NH-CH_2-CH_2-C(=NH)-NH_2$)	F	T	RNA + DNA	Low toxicity in cell culture	9
N''-(2-Amidinoethyl)-4-formamido-1,1',1''-trimethyl-N,4':N',4''-ter(pyrrole-2-carboxamide) (Distamycin A)	($O-HC-NH-$ tris(N-methylpyrrole-2-carboxamide) chain terminating in $-CO-NH-(CH_2)_2-C(=NH)-NH_2$)	F	T, E	DNA	Interacts directly with DNA	10

Systematic Name (Common Name)	Structure	Source	Type of Activity	Type of Virus	Comments	Reference
2-Amino-4-guanidinooxybutyric acid (L-Canavanine)		S	T	RNA	An amino acid analog of arginine	11
6-Aminonicotinamide		S	T	DNA	Antimetabolite of niacin; central-nervous-system toxicity?	12
L(−)2-Amino-3-(N-nitroso-N-hydroxyamino)propionic acid (Alanosine)		F	T, A	DNA + RNA	Active agent against oncogenic viruses	13
7-Amino-3-β-D-ribofuranosyl-1H-pyrazolo(4,3-d)pyrimidine (Formycin A)		F	T	RNA	Inhibits RNA synthesis in infected cells; also a member of the C-nucleoside class of antibiotics	14
N^1,N^1-Anhydrobis-(β-hydroxyethyl)biguanide (Flumidin)		S	A, T, M?	RNA	May be prophylactically active against respiratory viruses in man	15

Systematic Name (Common Name)	Structure	Source	Type of Activity	Type of Virus	Comments	Reference
9-β-D-Arabinofuranosyladenine (Ara A)	(structure)	S, F	T, A	DNA	Oxidation to the hypoxanthine analog allows retention of part of the activity	16
1-β-D-Arabinofuranosylcyclosine (Ara C, cytarabine)	(structure)	F	A, T, M	DNA	Also suppresses the immune response	17
L-Asparaginase	Unknown (protein)	F	A	RNA	Inhibits oncogenic viruses; also an antitumor agent	18

Systematic Name (Common Name)	Structure	Source	Type of Activity	Type of Virus	Comments	Reference
α-Azatricyclo(4.3.1.1)undecane (4-Azahomoadamantane)		S	T, A	RNA	Variation of the stable amine class without improved activity	19
1,2-Benzapyrone (Coumarin)		P	Plant systems	RNA	Also a plant growth inhibitor	20
N-(2-Benzylcarbamylethyl)-N-isonicotinylhydrazine (Nialamide)		S	A	DNA	Also a monoamine oxidase inhibitor	21
Benzyloxycarbonyl-D-phenyl-alaninyl-D-phenylalanine (SU-3963)		S	T	RNA	Effective after penetration	22
1,3-Bis(2-chloroethyl)-1-nitroso-urea		S	A	RNA	Inhibits oncogenic viruses	23

Systematic Name (Common Name)	Structure	Source	Type of Activity	Type of Virus	Comments	Reference
1,2-Bis(5-methoxy-2-benzimidazolyl)-1,2-ethanediol	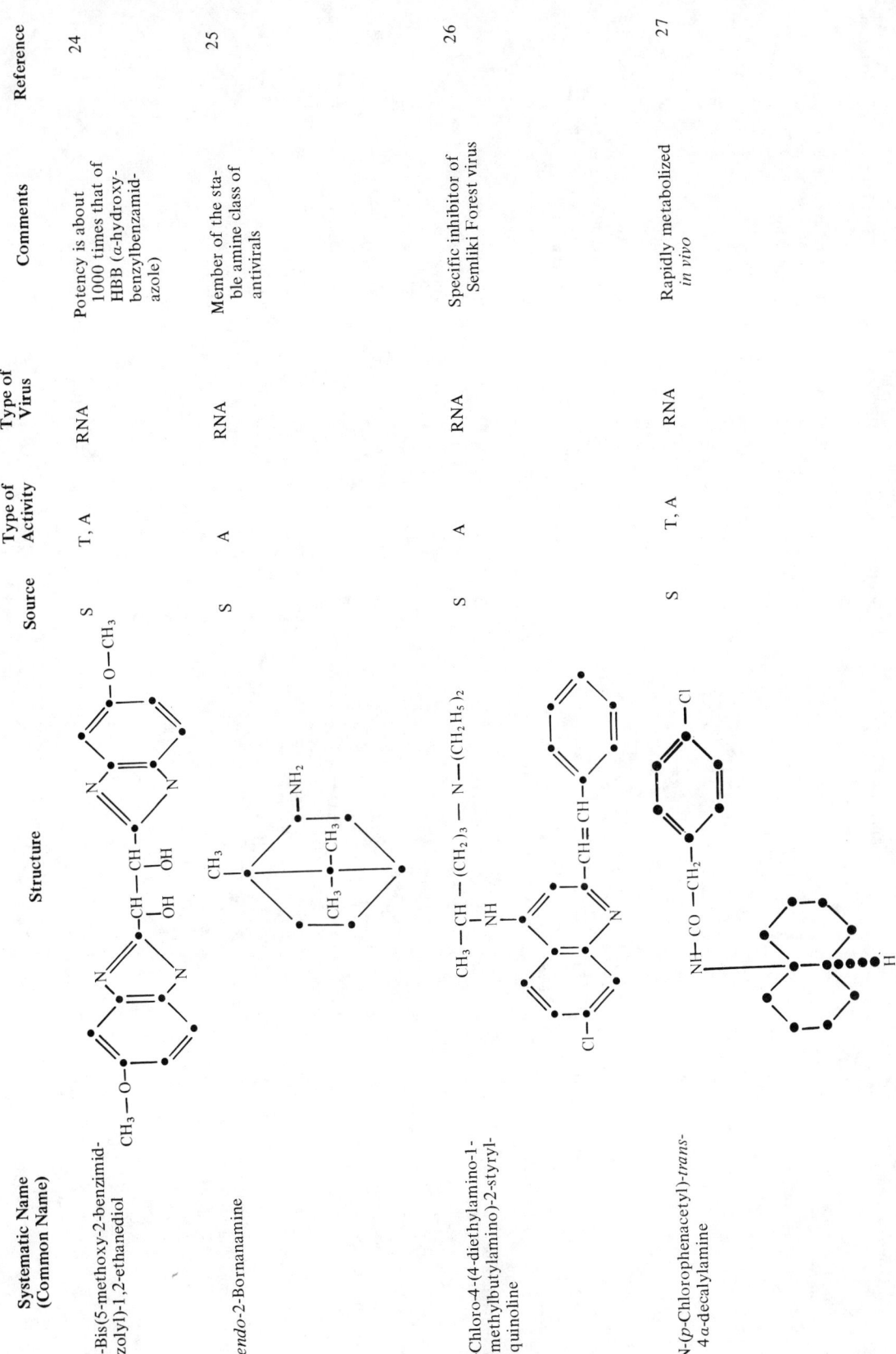	S	T, A	RNA	Potency is about 1000 times that of HBB (α-hydroxybenzylbenzamidazole)	24
dl-endo-2-Bornanamine		S	A	RNA	Member of the stable amine class of antivirals	25
7-Chloro-4-(4-diethylamino-1-methylbutylamino)-2-styrylquinoline		S	A	RNA	Specific inhibitor of Semliki Forest virus	26
N-(p-Chlorophenacetyl)-trans-4-α-decalylamine		S	T, A	RNA	Rapidly metabolized in vivo	27

Systematic Name (Common Name)	Structure	Source	Type of Activity	Type of Virus	Comments	Reference
1-(p-Chlorophenoxymethyl)-3,4-dihydroisoquinoline hydrochloride		S	A, E, T, M	RNA	Detected as an inhibitor of influenza neuraminidase	28
N-Cyclohexyl-N'-allylbenzamidine		S	T, A	RNA	Better activity against mouse influenza infections than adamantane amine	29
Cyclooctylamine		S	T, A	RNA + DNA	Inhibits the penetration or uncoating of myxoviruses	30
2,6-Dibutoxy-3-methyl-Δ^3-dihydropyran		S	T, A	RNA + DNA	Virucidal activity	31
5,6-Dichloro-1-(2'-deoxy-β-D-ribofuranosyl)benzimidazole		S	T	RNA + DNA	Inhibits RNA synthesis	32

Systematic Name (Common Name)	Structure	Source	Type of Activity	Type of Virus	Comments	Reference
5-(3',4'-Dichlorophenyl)-5-ethylhexahydropyrimidine-2,4-6-trione (DEHT)		S	T, A	RNA	Barbituric acid derivative	33
1-[3',4'-Dichlorophenyl-3-(2-pyrimidyl)] urea		S	T, A	RNA + DNA	Active by injection, but not by oral routes	34
2-Diethylaminoethyl-4-methyl-piperazine-1-carboxylate		S	A	RNA	Toxic due to peripheral vascular damage	35
3,4-Dihydro-1-isoquinolineacetamide (DIQA)		S	A	RNA + DNA	Active *in vivo* without *in vitro* activity	36
4-[2,3-Diisocyano-4-(4-methoxyphenyl)-1,3-butadienyl]-phenol (Xanthocillin X monomethyl ether)		F	T	RNA	Also an antitumor antibiotic	37

Systematic Name (Common Name)	Structure	Source	Type of Activity	Type of Virus	Comments	Reference
N,N'-Dimethyl-epidithiapiperazinedione		S	T	RNA	Synthetic compound with the nucleus of the epidithiapiperazinedione antibiotics	38
2,4-Dioxo-5'-thiazolidineacetic acid		S	T	DNA	Specific inhibitor of herpesvirus	39
3,3'-Diselenodialanine (Selenocystine)		S	T, A	RNA	Inhibits virus-induced RNA-dependent RNA polymerase	40
3-(4-Ethylbenzenesulfonamido)-propionamidine		S	A	RNA	Mouse activity comparable to that of adamantyl amine	41
5-Ethyldeoxyuridine		S	T, A	DNA	Non-mutagenic, but of equal antivaccinia activity as IDUR	42

Systematic Name (Common Name)	Structure	Source	Type of Activity	Type of Virus	Comments	Reference
Ethyl-2-methylthio-4-methyl-5-pyrimidinecarboxylate		S	T	RNA	Produces resistant and dependent strains of poliovirus	43
2-Ethylthioisonicotinamide		S	O	RNA	Exerts virucidal action	44
p-Fluorophenylalanine		S	T	RNA	Amino acid analog that inhibits viral induced enzyme synthesis	45
1-(4-Fluorophenyl)-1-phenyl-2-propynyl N-cyclohexylcarbamate (FPPC)		S	A	RNA	Inhibits oncogenic viruses and transplantable tumors	46

Systematic Name (Common Name)	Structure	Source	Type of Activity	Type of Virus	Comments	Reference
5-Formyl-5,6-dihydro-3-methoxy-carbonyl-6-methyl-4H-pyran-4-acetate hemicalcium salt (Calcium elenolate)		P	T, A	RNA + DNA	Contact virucide	47
Guanidine		S	T	RNA	Used extensively as a biochemical tool for studying virus multiplication	48
Gymnemic acid A	Unknown	P	T, A	RNA	Effective against an early event in the viral infectious cycle	49
2-(α-Hydroxybenzyl)benzimidazole (HBB)		S	T	RNA	Inhibits viral RNA synthesis	50
N^6-(2-Hydroxyethyl)adenine		S	T, A	RNA	Found to be inactive in humans	51
6-(4-Hydroxy-6-methoxy-7-methyl-3-oxo-5-phthalanyl)-4-methyl-4-hexenoic acid (Mycophenolic acid)		F	T, A	RNA + DNA	Inhibits oncogenic viruses in animals	52

Systematic Name (Common Name)	Structure	Source	Type of Activity	Type of Virus	Comments	Reference
4-Hydroxy-3-β-D-ribofuranosyl-5-pyrazolecarboxamide (Pyrazomycin)		F	T, A	RNA + DNA	Antagonizes uridine requirement for viral nucleic acid synthesis	53
5-Iodo-2'-deoxyuridine (IDUR, idoxuridine)		S	T, A	DNA (herpes)	Antimetabolite of thymidine	54
N¹-Isonicotinoyl-N²-(3 methyl-4-chlorobenzoyl)hydrazine (IMCBH)		S	T	DNA	Inhibits release of vaccinia virus in tissue culture	55

Systematic Name (Common Name)	Structure	Source	Type of Activity	Type of Virus	Comments	Reference
Julimycin B-II		F	A	RNA	Stimulates host defense other than interferon	56
α-Keto-β-ethoxybutyraldehyde (Kethoxal)		S	T, A	DNA	Contact virucide	57
1-(p-Methoxyphenoxymethyl)-3,4-dihydroisoquinoline hydrochloride		S	E, T, A	RNA	Detected as an inhibitor of influenza neuraminidase	28
3-(4-Methoxyphenyl)-2-(α-methylbenzylidenehydrazono)-4-oxo-5-thiazolidineacetic acid		S	T, M	DNA	Acts on late phases of herpesvirus multiplication	58

Systematic Name (Common Name)	Structure	Source	Type of Activity	Type of Virus	Comments	Reference
N-Methyl-1-adamantanecarbox-amide octachloro chlorination product		S	T, A	RNA	Specific therapeutic activity for neuro-tropic influenza	59
α-Methyl-1-adamantanemethyl-amine (Rimantadine)		S	T, A, M	RNA	Improved analog of 1-adamantane amine	60
N-Methyladamantane-2-spiro-3'-pyrrolidine		S	T, A	RNA	Broader spectrum of antiviral activity than other adaman-tane compounds	61
2-(α-Methylbenzylidenehydra-zono)-4-oxo-3-(p-tolyl)-5-thia-zolidineacetic acid		S	T, M	RNA + DNA	Treatment by topical ointment	62

Systematic Name (Common Name)	Structure	Source	Type of Activity	Type of Virus	Comments	Reference
α-Methylbicyclo(2.2.2)octane-4-methyl-1-methylamine		S	T, A	RNA	Stable amine with specific activity about 8 times that of adamantane amine against mouse infection with influenza A/swine	63
1,1'-(1-Methylethanediylidene-dinitrilo)diguanidine (Methyl GAG)		S	O, A, T	RNA + DNA	Also an effective anti-tumor agent	64
N-Methylisatin-β-(4,4'-dibutyl-thiosemicarbazine		S	E, T	RNA	Blocks viral RNA synthesis without accumulation of viral RNA intermediates	65
1-Methylisatin 3-thiosemicarbazone (Marboran, methisazone)		S	T, A, M	DNA	Inhibits "late" virus mRNA function	66
2-Methyl-4-[(5-methyl-5H-as-triazino[5,6-b]indol-3-yl)amino]-2-butanol		S	T, A	RNA	Modification of isatin thiosemicarbazone that led to increased spectrum	67

Systematic Name (Common Name)	Structure	Source	Type of Activity	Type of Virus	Comments	Reference
α-Methyl-2-phenoxathiinmeth-anol (MPM)	[structure]	S	T	RNA	Induces poliovirus variants that are drug-dependent	68
2-(1-Methyl-4-pyridyl)benzothi-azole	[structure]	S	T	RNA	Modification of HBB led to active compounds with different spectra	69
4'-[2-Nitro-1-(p-tolylthio)ethyl]-acetanilide	[structure]	S	T, A	DNA	Inactive in man	70
Paecilomycerol	Unknown	F	T	RNA	No antibacterial or antifungal activity	71
l-4-Phenyl-4-(2-chlorobenzyl)-5-oxohexanoic acid (Caprochlorone)	[structure]	S	O, A	RNA	Probably blocks penetration of myxoviruses; also has action on virus release	72

Systematic Name (Common Name)	Structure	Source	Type of Activity	Type of Virus	Comments	Reference
1-Phenyl-2-methyl-5-hydroxy-indole		S	T	RNA	Virucidal activity	73
4'-Phenyl-2-pyrrolidinoimino-acetophenone		S	O	DNA + RNA	Also active against *Trichophyton mentagrophytes*	74
Potassium benzylaminothiomethanesulfonate		S	T, O, A	DNA + RNA	Liberates isothiocyanate, which inhibits virus growth	75
2-β-D-Ribofuranosyl-*as*-triazine-3,5(2H,4H)-dione (Azauridine)		S	T	DNA	Used clinically for psoriasis	76

Systematic Name (Common Name)	Structure	Source	Type of Activity	Type of Virus	Comments	Reference
Rifampicin, rifampin		F	T	DNA	Also antibacterial (vaccinia)	77
Salicylhydroxamic acid (SHA)		S	T	DNA	Inhibitor of DNA synthesis	78
1,1,3,3-Tetracyanopropene sodium salt (TCNP)		S	T, A	RNA + DNA	Virucidal	79
5a,6,10,10a-Tetrahydro-6-hydroxy-3-hydroxymethyl-2-methyl-3,10a-epidithiopyrazino(1,2-a)indole-1,4(2H,3H)-dione (Gliotoxin)		F	T	RNA	Inhibitor of viral RNA synthesis	80

No.	Compound	Structure				Remarks
81	Tetramethyldipicrylamine (TMP)	(structure)	S	T	RNA	Inactivated by serum binding
82	3,3'-(Tetramethylendiimino)bis-propionaldehyde	$O=C-(CH_2)_2-NH-(CH_2)_4-NH-(CH_2)_2-C-H$, $=O$	S	T	RNA	Virucidal due to the aldehyde groups
83	2-Thiouracil	(structure)	S	P	RNA	Reacts with sulfhydryl groups of virus protein
84	2,3,5-Triiodobenzoic acid	(structure)	S	P	RNA	Also a plant growth regulator
85	dl-Tropyl tropate (Atropine, dl-hyoscyamine)	(structure)	P	A	RNA	Anticholinergic drug that decreases mucous secretions; decreased the severity of mouse influenza infection

Systematic Name (Common Name)	Structure	Source	Type of Activity	Type of Virus	Comments	Reference
1-β-D-Arabinofuranosyl-cytosine 3',5' cyclic phosphate (Cyclic ara-CMP)		S	T, A	DNA	This compound can't be deaminated and crosses the cell membrane in the active form.	86
1-β-D-Ribofuranosyl-1,2,4-triazole-3-carboxamide (Virazole)		S	T, A	RNA + DNA	Inhibits the conversion of IMP to xanthosine 5' phosphate in the infected cell	87
2,7-Bis[2-(diethylamino)-ethoxy]fluoren-9-one dihydrochloride		S	T, A, M	RNA + DNA	Low-molecular-weight inducer of interferon; good oral activity in mice, but low activity in man	88

Systematic Name (Common Name)	Structure	Source	Type of Activity	Type of Virus	Comments	Reference
p-Ureido-*p'*-amino-diphenylsulfone		S	Chick	DNA, Marek's disease virus		89
Streptovaricin complex	Factor A: W = OH; X = H, OH; Y = Ac; Z = OH Factor B: W = H; X = H, OH; Y = Ac; Z = OH Factor C: W = H; X = H, OH; Y = H; Z = OH Factor D: W = H; X = H, OH; Y = H; Z = H Factor E: W = H; X = =O; Y = H; Z = OH Factor G: W = OH; X = H, OH; Y = H; Z = OH	F	T, A	DNA + RNA	Reverse-transcriptase inhibitor	90
Inosine complex with benzoic acid, 4-acetamido-3-(dimethylamino)-2-propyl ester (Isoprinosine)		S	T, A, M(?)	RNA + DNA	Alters the binding of the viral mRNA to the host ribosome	91

Systematic Name (Common Name)	Structure	Source	Type of Activity	Type of Virus	Comments	Reference
N,N-dioctadecyl-N'-N'-bis-(2-hydroxyethyl)propane-diamine		S	A	RNA + DNA	Stimulates interferon production	92

REFERENCES

1. Delong, D. C., Boniece, W. S., Cline, J. C., Johnson, I. S., *Ann. N.Y. Acad. Sci., 130,* 440 (1965).
2. Nagarajan, R., Huckstep, L. L., Lively, D. H., DeLong, D. C., Marsh, M. M., and Neuss, N., *J. Am. Chem. Soc., 90,* 2980 (1968).
3. Miller, F. A., Rightsel, W. A., Sloan, B. J., Ehrlich, J., French, J. C., Bartz, Q. R., and Dixon, G. J., *Nature (London), 200,* 1338 (1963).
4. Davies, W. L., Grunert, R. R., Haff, R. F., McGahen, J. W., Neumayer, E. M., Paulshock, M., Watts, J. C., Wood, T. R., Herman, E. C., and Hoffman, C. E., *Science, 144,* 862 (1964).
5. Goedemans, W. T., and Peters, A., in *Abstracts of the 5th International Congress on Chemotherapy, Vienna, Austria, 1967,* p. 377. Verlag der Wiener Medizinischen Akademie, Vienna, Austria (1967).
6. Geluk, H. W., Schut, J., and Schlatmann, J. L. M. A., *J. Med. Chem., 12,* 712 (1969).
7. Gauri, K. K., and Rohde, B., *Klin. Wochenschr., 47,* 375 (1969).
8. Nakamura, S., Umezawa, H., and Ishida, N., *J. Antibiot. (Tokyo) Ser. A., 14,* 163 (1961).
9. Furusawa, E., Cutting, W., and Furst, A., *Chemotherapia, 8,* 95 (1964).
10. Fournel, J., Ganter, P., Koenig, F., de Ratuld, Y., and Werner, G. H., in *Antimicrobial Agents and Chemotherapy – 1965,* pp. 599–604, G. L. Hobby, Ed. American Society for Microbiology, Washington, D.C. (1966).
11. Ranki, M., and Kaariainen, L., *Ann. Med. Exp. Biol. Fenn., 47,* 65 (1969).
12. Lee, S. H. S., Dobson, P. R., and Van Roogen, C. E., *Chemotherapia, 11,* 163 (1966).
13. Murphy, Y. K. S., Thiemann, J. E., Coronelli, C., and Sensi, P., *Nature (London), 211,* 1198 (1966).
14. Nishimura, C., and Tsukeda, H., *Progress in Antimicrobial and Anticancer Chemotherapy,* Vol. 2, pp. 20–25. University of Tokyo Press, Tokyo, Japan (1970).
15. Sjoberg, B., *Muench. Med. Wochenschr., 10,* 485 (1960).
16. Miller, F. A., Sloan, B. J., and Silverman, C. A., in *Antimicrobial Agents and Chemotherapy – 1969,* pp. 192–195, G. L. Hobby, Ed., American Society for Microbiology, Washington, D. C. (1970).
17. Renis, H. E., and Buthala, D. A., *Ann. N. Y. Acad. Sci., 130,* 343 (1965).
18. Campbell, W. F., and Levine, A. L., *Life Sci. Part II, 8,* 1033 (1969).
19. Korsloot, J. G., Keizer, V. G., and Schlatmann, J. L., *Recl. Trav. Chim. Pays-Bas, 88,* 447 (1969).
20. Mishra, M. D., Ghosh, A., and Raychandhuri, S. P., *Acta Virol., 12,* 173 (1968).
21. Cifuentes-Bernal, H., Albornoz-Plata, A., Nuñez-Olarte, E., Triana- Aguilar, S., and Buitrazo, B., *Toxicol. Appl. Pharmacol., 12,* 508 (1968).
22. Miller, F. A., Dixon, G. J., Arnett, G., Dice, J. R., Rightsel, W. A., Schabel, F. M., and McLean, I. W., Jr., *Appl. Microbiol., 16,* 1489 (1968).
23. Sidwell, R., Dixon, G. J., Sellers, L. M., and Schabel, F. M., Jr., *Appl. Microbiol., 13,* 579 (1965).
24. Akihama, S., Okude, M., Sato, K., and Iwabuchi, S., *Nature (London), 217,* 562 (1968).
25. Mosimann, W., Borgulya, J., and Bernauer, K., *Experientia (Basel), 25,* 726 (1969).
26. Chan, J. C., and Gadebusch, H. H., *Experientia (Basel), 25,* 329 (1969).
27. Swallow, D. L., Edwards, P. N., and Finter, N. B., *Ann. N.Y. Acad. Sci., 173,* 292 (1970).
28. Tute, M. S., Brammer, K. W., Kaye, B., and Broadbent, R. W., *J. Med. Chem., 13,* 44 (1970).
29. Toyoshima, S., Seto, Y., Fujeta, H., Tonegi, H., Abe, J., Watanabe, T., and Fujimoto, K., in *Progress in Antimicrobial and Anticancer Chemotherapy,* Vol. 2, pp. 57–60. University of Tokyo Press, Tokyo, Japan (1970).
30. Flagg, W. B., Stanfield, F. J., Haff, R. F., Stewart, R. C., Stedman, R. J., Gold, J., and Ferlauto, R. J., in *Antimicrobial Agents and Chemotherapy – 1968,* pp. 194–200, G. L. Hobby, Ed. American Society for Microbiology, Washington, D.C. (1969).
31. Perskin, G. N., Bogdanova, N. S., and Mahking, S. M., in *Progress in Antimicrobial and Anticancer Chemotherapy,* Vol. 2, pp. 49–52. University of Tokyo Press, Tokyo, Japan (1970).
32. Diwan, A., Gowdy, C. N., Robbins, R. K., and Prusoff, W. H., *J. Gen. Virol., 3,* 393 (1968).
33. DeLong, D. C., Doran, W. J., Baker, L. A., and Nelson, J. D., *Ann. N.Y. Acad. Sci., 173,* 516 (1970).
34. Paget, C. J., Ashbrook, C. W., Stone, R. L., and DeLong, D. C., *J. Med. Chem., 12,* 1097 (1969).
35. Angier, R. B., Murdock, K. C., Curran, W. V., Sollenberger, P. Y., and Casey, J. P., *J. Med. Chem., 11,* 720 (1968).
36. Grunberg, E., Prince, H. N., *Ann. N.Y. Acad. Sci., 173,* 122 (1970).
37. Ando, K., Suzuki, S., Takatsuki, A., Arima, K., and Tamura, G., *J. Antibiot. (Tokyo), 21,* 582 (1968).
38. Trown, P. W., *Biochem. Biophys. Res. Commun., 33,* 402 (1968).
39. Schauer, P., Likar, M., Tisler, M., Krbavcic, A., and Pollak, A., *Pathol. Microbiol., 28,* 382 (1965).
40. Ho, P. P. K., Walters, C. P., Streightoff, F., Baker, L. A., and DeLong, D. C., in *Antimicrobial Agents and Chemotherapy – 1967,* pp. 636–641, G. L. Hobby, Ed., American Society for Microbiology, Washington, D. C. (1968).
41. Ueda, T., Nagahara, K., Takahashi, K., and Sato, S., *Chem. Pharm. Bull. (Tokyo), 17,* 2065 (1969).
42. Swierkowski, M., and Shugar, D., *J. Med. Chem., 12,* 533 (1969).
43. Yamazi, Y., Takahashi, M., and Todome, Y., *Proc. Soc. Exp. Biol. Med., 133,* 674 (1970).
44. Cretnic, S., and Koruncev, D., *Chim. Ther., 4,* 246 (1969).
45. Levintow, L., Thoren, M. M., Darnell, J. E., Jr., and Hooper, J. L., *Virology, 16,* 220 (1962).

46. DeLong, D. C., Baker, L. A., Easton, N. R., and Dillard, R. D., *Proc. Soc. Exp. Biol. Med., 127,* 845 (1968).

47. Renis, H. E., in *Antimicrobial Agents and Chemotherapy – 1969,* p. 167–172, G. L. Hobby, Ed. American Society for Microbiology, Washington, D.C. (1970).

48. Tamm, I., and Eggers, T. H. J., *Science, 142,* 24 (1963).

49. Sinsheimer, J. E., Rao, G. S., McIlhenney, H. M., Smith, R. V., Maassab, H. F., and Cochran, K. W., *Experientia (Basel), 24,* 302 (1968).

50. Eggers, H. J., and Tamm, I., *J. Exp. Med., 113, 657* (1961).

51. Underwood, G. E., Baker, C. A., and Weed, S. D., *Proc. Soc. Exp. Biol. Med., 122,* 167 (1966).

52. Williams, R. H., Lively, D. H., DeLong, D. C., Cline, J. C., Sweeney, M. J., Poore, G. A., and Larsen, S. H., *J. Antibiot. (Tokyo), 21,* 463 (1968).

53. Streightoff, F., Nelson, J. D., Cline, J. C., Gerzon, K., Williams, R. H., and DeLong, D. C., in *Abstracts of the 9th Interscience Conference on Antimicrobial Agents and Chemotherapy, Washington, D.C., 1969,* p. 8. American Society for Microbiology, Washington, D. C. (1969).

54. Prussoff, W. H., *Biochim. Biophys. Acta, 32,* 295 (1959).

55. Kato, N., Eggers, H. J., and Rolly, H., *J. Exp. Med., 129,* 795 (1969).

56. Katagiri, K., Nishiyama, S., and Sato, K., in *Progress in Antimicrobial and Anticancer Chemotherapy,* Vol. 2, pp. 11–13. University of Tokyo Press, Tokyo, Japan (1970).

57. Underwood, G. E., *Proc. Soc. Exp. Biol. Med., 129,* 235 (1968).

58. Likar, M., Schauer, P., Novak, I., in *Abstracts of the 5th International Congress on Chemotherapy, Vienna, Austria, 1967,* p. 1245. Verlag der Wiener Medizinischen Akademie, Vienna, Austria (1967).

59. McGahen, J. W., *Fed. Proc., 25,* 562 (1966).

60. Tsunoda, A., Maassab, H. F., Cochran, K. W., and Eveland, W. C., in *Antimicrobial Agents and Chemotherapy – 1965,* pp. 553–560, G. L., Hobby, Ed. American Society for Microbiology, Washington, D. C. (1966).

61. Peters, A., de Bock, C. A., Paerels, G. B., and Schlatmann, J., in *Progress in Antimicrobial and Anticancer Chemotherapy,* Vol. 2, pp. 71–74. University of Tokyo Press, Tokyo, Japan (1970).

62. Schauer, P., Likar, M., and Klemenc-Sebek, S., *Ann. N.Y. Acad. Sci., 173,* 603 (1970).

63. Whitney, J. G., Gregory, W. A., Kauer, J. C., Roland, J. R., Snyder, J. A., Benson, R. E., and Hermann, E. C., *J. Med. Chem., 13,* 254 (1970).

64. Kuchler, C., Kuchler, W., and Schulze, W., *Acta Virol., 12,* 441 (1968).

65. Pearson, G. D., and Zimmerman, E. F., *Virology, 38,* 641 (1969).

66. Thompson, R. L., Davis, J., Russell, P. B., and Hitchings, G. H., *Proc. Soc. Exp. Biol. Med., 84,* 496 (1953).

67. Gladych, J. M. Z., Hunt, J. H., Jack, D., Haff, R. F., Boyle, J. J., Stewart, R. C., and Ferlanto, R. J., *Nature (London), 221,* 286 (1969).

68. Paget, C. J., Dennis, E. M., Nelson, J., and DeLong, D. C., *J. Med. Chem., 13,* 620 (1970).

69. Vaczi, L., Hadhazy, G., Hideg, K., Kimura, T., Gergely, L., Hankovszky, O. H., and Toth, F. D., *Acta Virol., 12,* 371 (1968).

70. Underwood, G. E., *Ann. N.Y. Acad. Sci., 173,* 782 (1970).

71. Kato, A., Ando, K., Kimura, T., Tamura, K., and Arima, K., *J. Antibiot. (Tokyo), 22,* 419 (1969).

72. Liu, O. C., Carter, J. E., Malsberger, R. G., Delanctis, A. N., and Hampil, B., *J. Immunol., 78,* 222 (1957).

73. Grinev, A. N., Shrvdov, V. I., Panisheva, E. K., Bogdanova, N. S., Nikolaiva, I. S., and Pershin, G. N., *Pharm. Chem. J., 9,* 497 (1969).

74. Massarani, E., Nardi, D., Pozzi, R., and Degen, L., *J. Med. Chem., 13,* 157 (1970).

75. Rao, P. L. N., and Ramanathan, S., *Br. J. Pharmacol. Chemother., 31,* 19 (1967).

76. Rada, B., and Blaskovic, D., *Acta Virol., 10,* 1 (1966).

77. Subak-Sharpe, J. H., Timbury, M. C., and Williams, J. F., *Nature (London), 222,* 341 (1969).

78. Scheele, C. M., and Pfefferkorn, E. R., *Proc. Soc. Exp. Biol. Med., 128,* 902 (1968).

79. Grunberg, E., and Prince, H. N., in *Antimicrobial Agents and Chemotherapy – 1967,* pp. 642–645, G. L. Hobby, Ed. American Society for Microbiology, Washington, D.C. (1968).

80. Rightsel, W. A., Schneider, H. G., Sloan, B. J., Graf, P. R., Miller, F. A., Bartz, Q. R., Ehrlich, J., and Dixon, G. J., *Nature (London), 204,* 1333 (1964).

81. Rosenbaum, M. J., Sullivan, E. J., Meyer, T. S., Moore, C. E., and Fritsch, A. J., in *Antimicrobial Agents and Chemotherapy – 1968,* pp. 207–212, G. L. Hobby, Ed. American Society for Microbiology, Washington, D.C. (1969).

82. Kremzner, L. T., and Harter, D. H., *Biochem. Pharmacol., 19,* 2451 (1970).

83. Brockman, R. W., and Anderson, E. P., in *Metabolic Inhibitors* pp. 239–285, R. M. Hochster, and J. H. Quastel, Eds. Academic Press, New York (1963).

84. Weeraratne, V., *Life Sci., 4,* 1923 (1965).

85. Streightoff, F., Redman, C. E., and DeLong, D. C., in *Antimicrobial Agents and Chemotherapy – 1966,* pp. 503–508, G. L. Hobby, Ed. American Society for Microbiology, Washington, D.C. (1967).

86. Sidwell, R. W., Simon, L. N., Huffman, J. H., Allen, L. B., Long, R. A., and Robbins, R. K., *Nature New Biology, 242,* 204 (1973).

87. Sidwell, R. W., Huffman, J. H., Khare, G. P., Allen, L. B., Witkowski, J. T., and Robbins, R. K., *Science, 177,* 705 (1972).

88.   Krueger, R. F., and Mayer, G. D., *Science, 169,* 1213 (1970).
89.   Shen, T. Y., Johnston, D. B. R., Jensen, N. P., Ruyle, W. V., Friedman, J. J., Fordice, M. W., McPherson, J. F., Maag, T. A., Burg, R. W., Pellegrino, R. M., Jewell, M. E., Morris, C. A., Easterbrooks, H. L., and Skelly, B. J., in *Abstracts of the 162nd Meeting of the American Chemical Society, Washington, D.C., September, 1971,* MEDI 44.
90.   Carter, W. A., Brockman, W. W., and Borden, E. C., *Nature New Biology, 232,* 212 (1971).
91.   Gordon, P., and Brown, E. R., *Can. J. Microbiol., 18,* 1463 (1972).
92.   Hoffman, W. W., Korst, J. J., Niblack, J. F., and Cronin, T. H., *Antimicrob. Agents Chemother., 3,* 498 (1973).

PHARMACOLOGICALLY ACTIVE AGENTS FROM MICROBIAL SOURCES

DR. S. L. NEIDLEMAN

INTRODUCTION

The metabolites present in fermentations represent the efforts of a versatile and inventive biosynthetic chemist, the microorganism. The biological activities that can be detected in these metabolites depend upon and are limited only by the assay system employed. The greatest emphasis in past years has been in the area of antibiotics, but a tide of research pursuing other diverse bioactivities is currently rising. The search for pharmacologically active metabolites represents a large proportion of this surge.

To be exhaustive, this review would have to be extremely extensive; minimally, it would have to consider (a) compounds isolated primarily for their pharmacological activity, (b) antibiotics for which pharmacological activity has been demonstrated, and (c) microbial toxins.

For the purpose of this review, (a) is covered rather extensively in Table 1; (b) is represented by selected examples in Table 2; and (c) is presented in Table 3, "Fungal Toxins", and Table 4, "Bacterial Exotoxins", but again on a selective basis. It will be noted that the data included range from those obtained with pure compounds to those obtained from crude fermentation extracts.

Very inclusive references are available for each of these three general areas of interest.[75-77] Additionally, several reports devoted to general screening techniques have been listed.[69-74]

TABLE 1
PHARMACOLOGICALLY ACTIVE MICROBIAL METABOLITES

Microorganism	Substance	Pharmacological Activity	Reference
Aerobacter cloacae	Polysaccharide	Anti-inflammatory	1
Amanita muscaria and other fungi	Muscarine	Parasympathomimetic	75
Bacillus anthracis	Crude filtrates Extract	Epinephrine-like Cardioactive	2, 3 4
Bacillus licheniformis	Crude extract	ACTH-like	5
Bacillus subtilis	Crude extract	ACTH-like	5
Claviceps species	Ergot alkaloids	α-Adrenergic blockers, smooth-muscle stimulators, CNS-active	6, 75
Corticium caeruleum	Rugulovasins A & B	Hypotensive	7
Fusarium oxysporum	Fusaric acid (5-butylpicolinic acid)	Hypotensive	8
Fusarium species	Isoprenoidal derivatives of methyl resorcylaldehyde	Analgesic, hypocholesteremic	9
	Crude extracts	Emetic	10

TABLE 1 (Continued)
PHARMACOLOGICALLY ACTIVE MICROBIAL METABOLITES

Microorganism	Substance	Pharmacological Activity	Reference
Gibberella zea	Zeagenin	Blood β-protein-lowering	11
	Zearalenone	Estrogenic	12, 13
Lactobacillus leichmannii	Crude extract	ACTH-like	5
Lentinus edodes	Lentysine	Hypolypidemic	14
Lenzites trabea	Rugulovasins A and B	Hypotensive	7
Naematoloma fasciculare (FR) Karst	Naemotolin	Coronary vasodilator	15
Oospora astringenes	Oosponal, oospolactone, oosponglycol	Muscle contraction	16
Oudemansiella radicata	Oudenone	Hypotensive	17
Panaeolus species	Serotonin, 5-hydroxytryptophan	Vasoconstrictor	19
Pellicularia filamentosa	Rugulovasins A and B	Hypotensive	7
Penicillium concavorugulosum	Rugulovasins A and B	Hypotensive	7
Penicillium corylophiloides	Rugulovasins A and B	Hypotensive	7
Penicillium purpurogenum	Rugulovasins A and B	Hypotensive	7
Penicillium rugulosum	Rugulovasins A and B	Hypotensive	7
Phialocephala repens	Phialocin	Anticoagulant	18
Psilocybe mexicana and other fungi	Psilocybin, psilocin	Psychotomimetic	19, 20
Rhizoctonia leguminicola	Slaframine	Parasympathomimetic	21
Salmonella enteriditis	Crude extract	ACTH-like	22
Salmonella typhi	Crude extract	ACTH-like	22
Serratia marcescens	Crude extract	ACTH-like	5
Serratia species	Protease (TSP)	Anti-inflammatory	23
Streptomyces albireticuli	Leupeptins	Anti-plasmin, anti-inflammatory, inhibitor of trypsin, papain, blood coagulation (human), other enzymes	24, 27
Streptomyces argenteolus var. *toyonakensis*	Pepstatin	Pepsin inhibitor	33, 34

TABLE 1 (Continued)
PHARMACOLOGICALLY ACTIVE MICROBIAL METABOLITES

Microorganism	Substance	Pharmacological Activity	Reference
Streptomyces cacaoi var. *asoensis*	Neutral proteinase	Anti-inflammatory	25
Streptomyces caespitosus (also other cited pepstatin producers)	Pepstatin	Pepsin inhibitor	26
Streptomyces chartreusis	Leupeptins	As for *Streptomyces albireticuli*	24, 27
Streptomyces griseolus	Neutral proteinase	Anti-inflammatory	25
Streptomyces griseus	Protease	Insulin-like	28
Streptomyces hygroscopicus	Chymostatin	Chymotrypsin inhibitor	29
Streptomyces kinoluteus	Kinonase B_1, proteinase	Anti-inflammatory	30
Streptomyces lavendulae	Chymostatin Leupeptins	Chymotrypsin inhibitor As for *Streptomyces albireticuli*	31 24, 27
Streptomyces nigrifaciens var. FFD-101	Nigrifactin	Antihistaminic	32
Streptomyces noboritoensis	Leupeptins	As for *Streptomyces albireticuli*	24, 27
Streptomyces roseochromogenes	Leupeptins	As for *Streptomyces albireticuli*	24, 27
Streptomyces roseus	Leupeptins	As for *Streptomyces albireticuli*	24, 27
Streptomyces testaceus	Pepstatin	Pepsin inhibitor	33, 34
Streptomyces thioluteus	Leupeptins	As for *Streptomyces albireticuli*	24, 27
Streptomyces verticillatus var. *zynogenes*	Retikinonase I and II, neutral proteinases	Anti-inflammatory	35
Streptomyces EF-44-201	Four S-PI components	Pepsin inhibitors	36
Streptomyces species (several)	Enzymes	Degrade blood group A and B protein	37
Zygosporium masonii	Zygosporin A	Anti-inflammatory	38
Zygosporium mycophilum	Zygosporin A	Anti-inflammatory	38

TABLE 2
SELECTED ANTIBIOTICS WITH PHARMACOLOGICAL ACTIVITY

Microorganism	Substance	Pharmacological Activity	Reference
Bacillus brevis	Colisan	Antispasmotic	39
Bacillus colistinus	Colistin	Antispasmotic	40
Monosporium bonorden	Monorden	Tranquilizing	41
Penicillium brevi-compactum	Mycophenolic acid	Immunosuppressive	42
Penicillium griseofulvum	Griseofulvin	Anti-inflammatory	43
Penicillium patulum	Patulin	Antispasmotic	44
Pseudomonas aeruginosa	Aeruginic acid	Anti-inflammatory, hypertensive	45
Streptomyces achromogenes	Streptozotocin	Diabetogenic	46
Streptomyces alanosinicus	Alanosine	Immunosuppressive	47
Streptomyces cellulosae	Fungichromin	Cardiotonic	75
Streptomyces endus	Endomycin	Cardiotonic	75
Streptomyces eurocidicus and other species	Eurocidine	Cardiotonic	75
Streptomyces hachijoensis	Trichomycin	Cardiotonic	75
Streptomyces natalensis	Pimaricin	Cardiontonic	75
Stretpomyces nodosus	Amphotericin	Cardiotonic	75
Streptomyces noursei	Nystatin	Cardiotonic	75
Streptomyces orchidaceus and other *Streptomyces* sp.	D-cycloserine	CNS	48
Streptomyces penticus	Pentamycin	Cardiotonic	75
Streptomyces pimprina	Hamycin	Cardiotonic	75
Streptomyces pluricolorescens	Chrothiomycin	Inhibits tyrosine hydroxylase, dopamine-β-hydroxylase	49
	Pluramycin	Immunosuppressive	50
Streptomyces reticuli	Aquayamycin	Hypotensive	51, 52
Streptomyces viridoflavus	Candidin	Cardiotonic	75
Streptomyces sp.	Lagosin	Cardiotonic	75

TABLE 3
FUNGAL TOXINS

Microorganism	Substance	Toxicity	Reference
Aspergillus flavus	Tremorgen	Trembling	54
Aspergillus flavus and other *Aspergillus* sp.	Aflaxotoxins	Hepatotoxic, hepato-carcinogenic	53
Aspergillus ochraceus	Ochratoxin A	Fatty infiltration of liver	55
Cephalosporium crotocinigenium (and *Penicillium* sp.)	Crotocin	Hemorrhagic, CNS-toxic	56
Fusarium tricinctum	Scirpenes	Hemorrhagic	57
Myrothecium roridum	Roridins A to E	Inflammatory	58
Myrothecium verrucaria	Muconomycins A and B	Inflammatory	59
	Verrucarins A to J	Inflammatory	57
Penicillium citreoviride	Citreoviridin	CNS-toxic	60
Penicillium citrinum	Citrinin	Nephrotoxic	61
Penicillium cyclopium	Tremorgen	Trembling	54
Penicillium islandicum	Islanditoxin, luteoskyrin	Hepatotoxic	62
Penicillium palitans	Tremorgen	Trembling	63
Penicillium rubrum	Rubratoxins A and B	Hepatotoxic	53
Pithomyces chartarum	Sporidesmins	Hepatotoxic, skin eczema, edema	53
Sclerotinium sclerotiorum	Psoralens	Phototoxic	64
Streptomyces sp.	Teleocidin	Irritant	65
Trichothecium roseum	Trichothecin	Inflammatory	66

TABLE 4
BACTERIAL EXOTOXINS

Microorganism	Substance	Toxicity	Reference
Bacillus anthracis	Complex toxin	Edema	67, 68
Bordetella pertussis	Whooping-cough toxin	Necrotizing	67, 68
Clostridium botulinum	Six type-specific neurotoxins	Paralytic	67, 68
Clostridium navyi	α-Toxin	Necrotizing	67, 68
	β-Toxin	Necrotizing, hemolytic	
	γ-Toxin	Necrotizing, hemolytic	
	δ-Toxin	Hemolytic	
	ϵ-Toxin	Hemolytic	
	ζ-Toxin	Hemolytic	

TABLE 4 (Continued)
BACTERIAL EXOTOXINS

Microorganism	Substance	Toxicity	Reference
Clostridium oedematiens	α-Toxin	Necrotizing	67, 68
	β-Toxin	Hemolytic, necrotizing	
	γ-Toxin	Hemolytic, necrotizing	
	δ-Toxin	Hemolytic	
	ε-Toxin	Hemolytic	
	ς-Toxin	Hemolytic	
Clostridium septicum	α-Toxin	Hemolytic	67, 68
	β-Toxin	DNAase	
Clostridium sordellii	Toxin	Edema	67, 68
Clostridium tetani	Tetanospasmin	Neurotoxic	67, 68
	Tetanolysin	Hemolytic, cardiotoxic	
Corynebacterium diphtheriae	Shick toxin	Necrotizing	67, 68
Shigella dysenteriae	Neurotoxin	Hemorrhagic, paralytic	67, 68
Staphylococcus aureus	α-Toxin	Necrotizing, hemolytic	67, 68
	β-Toxin	Hemolytic	
	γ-Toxin	Hemolytic	
	δ-Toxin	Hemolytic	
	ε-Toxin	Hemolytic	
	Enterotoxin	Emetic	
	Leukocidin	Leukocidic	
	Hyaluronidase	Spreading factor	
Streptococcus pyogenes	Dick toxin	Erythrogenic	67, 68
	Streptolysin O	Hemolytic	
	Streptolysin S	Hemolytic	
	Streptokinase	Fibrinolytic	
	Streptodornase	DNAase	
	Hyaluronidase	Spreading factor	
Vibrio cholerae	Enterotoxin	Diarrheagenic, hemolytic	67, 68

REFERENCES

1. Ono, H., and Sugiura, M., *J. Pharm. Soc. Jap., 89,* 1741 (1969).
2. Williams, R. P., Hill, H. R., Hawkins, D., Chao, K.-C., Neuenshwander, J., and Lipscombe, H. S., *Fed. Proc., 26,* 1545 (1967).
3. Massingill, J. L., Jr., and Hodgkins, J. E., *Phytochemistry, 6,* 977 (1967).
4. Liu, C. T., and Williams, R. P., *Arch. Int. Pharmacodyn. Ther., 189,* 336 (1971).
5. Nelson, J. W., O'Connell, P. W., and Haines, W. J., *Science, 119,* 379 (1954).
6. Abe, M., and Yamatodani, S., *Prog. Ind. Microbiol., 5,* 205 (1960).
7. Abe, M., *Agric. Biol. Chem., 33,* 469 (1969).
8. Hidaka, H., Nagatsu, T., Takeya, K., Takeuchi, T., Suda, H., Kojiri, K., Matsusaki, M., and Umezawa, H., *J. Antibiot. (Tokyo), 22,* 228 (1969).
9. American Cyanamid Co., *U.S. Patent 3,546,073* (1970).
10. Prentice, N., and Dickson, A. D., *Biotechnol. Bioeng., 10,* 413 (1968).
11. Fujisawa Pharmaceutical Co. Ltd, *Japanese Patent 7,021,634* (1970).

12. Stob, M., Baldwin, R. S., Tuite, J., Andrews, F. N., and Gillette, K. G., *Nature, 196,* 1318 (1962).
13. Urry, W. H., Whermeister, H. L., Hodge, E. R., and Hidy, P. H., *Tetrahedron Lett.,* p. 3109 (1966).
14. Rokujo, T., *Life Sci., 9,* 379 (1970).
15. Tanabe Seiyaku KK., *Japanese Patent No. 7,016,795* (1970).
16. Ohashi, S., Yamaguchi, M., and Kobayashi, Y., *Proc. Jap. Acad., 38,* 66 (1962).
17. Ohno, M., Okamoto, M., Kawabe, N., Umezawa, H., Takeuchi, T., Iinuma, H., and Takahashi, S., *J. Am. Chem. Soc., 93,* 1285 (1971).
18. Kondo, S., *Japanese Patent 7,102,115* (1971).
19. Benedict, R. G., and Brady, L. R., in *Fermentation Advances,* pp. 63–68, D. Perlman, Ed. Academic Press, New York (1969).
20. Picker, J., and Rickards, R. W., *Aust. J. Chem., 23,* 853 (1970).
21. Aust, S. D., Broquist, H. P., and Rinehart, K. L., Jr., *Biotechnol. Bioeng., 10,* 403 (1968).
22. Chedid, L., and Boyer, F., *Acad. Sci., 246,* 2937 (1958).
23. Yamasaki, H., Tsuji, H., and Saeki, K., *Nippon Yakurigaku Zasshi, 63,* 302 (1967).
24. Aoyagi, T., Miyata, S., Nanbo, M., Kojima, F., Matsuzaki, M., Ishizuka, M., Takeuchi, T., and Umezawa, H., *J. Antibiot. (Tokyo), 22,* 558 (1969).
25. Nakamura, S., Fukuda, H., Hamada, M., and Umezawa, H., *Chem. Pharm. Bull. (Tokyo), 18,* 2577 (1970).
26. Zardon Hojin Biseibutsu Kagaku Kenkyu Kai, *Belgian Patent 751,931* (1969).
27. Kawamura, K., Kondo, S.–I., Maeda, K., and Umezawa, H., *Chem. Pharm. Bull. (Tokyo), 17,* 1902 (1969).
28. Kuo, J. F., Holmlund, C. E., Dill, I. K., and Bohonos, N., *Arch. Biochem. Biophys., 117,* 269 (1966).
29. Umezawa, H., *J. Antibiot. (Tokyo), 23,* 425 (1970).
30. Nakamura, S., Marumoto, Y., Miyata, H., Tsukada, I., Tanaka, N., Ishizuka, M., and Umezawa, H., *Chem. Pharm. Bull. (Tokyo), 17,* 2044 (1969).
31. Umezawa, H., *J. Antibiot. (Tokyo), 23,* 425 (1970).
32. Terashima, T., Kuroda, Y., and Kaneko, Y., *Agric. Biol. Chem., 34,* 753 (1970).
33. Umezawa, H., Aoyagi, T., Morishima, H., Matsuzaki, M., Hamada, M., and Takeuchi, T., *J. Antibiot. (Tokyo), 23,* 259 (1970).
34. Morishima, H., Takita, T., Aoyagi, T., Takeuchi, T., and Umezawa, H., *J. Antibiot. (Tokyo), 23,* 263 (1970).
35. Nakamura, S., Hamada, M., Ishizuka, M., and Umezawa, H., *Chem. Pharm. Bull. (Tokyo), 18,* 2112 (1970).
36. Murao, S., and Satoi, S., *Agric. Biol. Chem., 34,* 1265 (1970).
37. Oishi, K., and Aida, K., *Agric. Biol. Chem., 35,* 1101 (1971).
38. Shionogi and Co. Ltd, *French Patent 138,350* (1969).
39. Leon, S. A., and Bergmann, F., *Biotechnol. Bioeng., 10,* 429 (1968).
40. Naranjo, P., and de Naranjo, E., *Antimicrob. Agents Chemother.,* p. 245 (1965).
41. McCapra, F., Scott, A. I., Delmotte, P., Delmotte–Plaquee, J., and Bhacca, N. S., *Tetrahedron Lett.,* p. 869 (1964).
42. Mitsui, A., and Suzuki, S., *J. Antibiot. (Tokyo), 22* 358 (1969).
43. Tridon, P., Weber, M., and Toussain, P., *Ann. Med. Nancy, 5,* 65 (1966).
44. Eliasson, R., *Experientia, 14,* 460 (1958).
45. Godo Shusei, KK., *Netherlands Patent 7,013,705* (1971).
46. Dulin, W. E., and Wyse, B. M., *Proc. Soc. Exp. Biol. Med., 20,* 422 (1969).
47. Fumarola, D., *Pharmacology (Basel), 3,* 215 (1970).
48. Mayer, O., Janku, I., and Krsiak, M., *Arzneim.-Forsch., 21,* 298 (1971).
49. Ayukawa, S., Hamada, M., Kojiri, K., Takeuchi, T., Hara, T., Nagatsu, T., and Umezawa, H., *J. Antibiot. (Tokyo), 22,* 303 (1969).
50. Yamaki, H., Tanaka, N., and Umezawa, H., *J. Antibiot. (Tokyo), 22,* 315 (1969).
51. Ayukawa, S., Takeuchi, T., Sezaki, M., Hara, T., Umezawa, H., and Nagatsu, T., *J. Antibiot. (Tokyo), 21,* 350 (1968).
52. Nagatsu, T., Ayukawa, S., and Umezawa, H., *J. Antibiot. (Tokyo), 21,* 354 (1968).
53. Wilson, B. J., Wilson, C. H., and Hayes, A. W., *Nature, 220,* 77 (1968).
54. Lillehoj, E. B., Ciegler, A., and Detroy, R. W., Fungal Toxins, in *Essays in Toxicology,* Vol. 2, pp. 1–136, F. R. Blood, Ed. Academic Press, New York (1970).
55. Van der Merwe, K. J., Steyn, P. S., Fourie, L., de Scott, B., and Theron, J. J., *Nature, 205,* 1112 (1965).
56. Mirocha, C. J., Christensen, C. M., and Nelson, G. H., *Biotechnol. Bioeng., 10,* 469 (1968).
57. Bamberg, J. R., Marasas, W. F., Riggs, N. V., Smalley, E. B., and Strong, F. M., *Biotechnol. Bioeng., 10,* 445 (1968).
58. Härri, E., Loeffler, W., Sigg, H. P., Stähelin, H., Stoll, C., Tamm, C., and Weisinger, D., *Helv. Chim. Acta, 45,* 839 (1962).
59. Guarino, A. M., Mendillo, A. B., and De Feo, J. J., *Biotechnol. Bioeng., 10,* 457 (1968).
60. Sakabe, N., Goto, T., and Hirata, Y., *Tetrahedron Lett.,* p. 1825 (1964).
61. Sakai, F., *Folia Pharmacol. Jap., 51,* 431 (1955).
62. Miyake, M., and Saito, M., in *Mycotoxins in Food Stuffs,* pp. 133–146, C. N. Wogan, Ed., M.I.T. Press, Cambridge, Massachusetts (1965).
63. Ciegler, A., *Appl. Microbiol., 18,* 128 (1969).

64. Perone, V. B., Scheel, L. D., and Meitus, R. A., *J. Invest. Dermatol., 42,* 267 (1964).
65. Nakata, H., Harada, H., and Hirata, Y., *Tetrahedron Lett.,* p. 2515 (1966).
66. Freeman, G. G., Gill, J. E., and Waring, W. S., *J. Chem. Soc.,* p. 1105 (1959).
67. Davis, B. D., Dulbeco, R., Eisen, H. N., Ginsberg, H. S., and Wood, W. B., in *Microbiology,* p. 612. Harper and Row, New York (1967).
68. van Heyningen, W. E., in *Microbial Toxins,* Vol. 1, pp. 1–28, S. Kadis, T. C. Montie and S. J. Ajl, Eds. Academic Press, New York (1971).
69. Malone, M. H., *Lloydia, 30,* 250 (1967).
70. Terashima, T., Kuroda, Y., and Kaneko, Y., *Agric. Biol. Chem., 34,* 747 (1970).
71. Richou, R., Jacquet, J., Lallouette, P., and Boutikonnes, P., *C. R. Hebd. Seances Acad. Sci. Ser. D Sci. Nat. (Paris), 272,* 1819 (1971).
72. Tyler, V. E., and Stuntz, D. E., *Lloydia, 25,* 225 (1962).
73. Tyler, V. E., Jr., and Stuntz, D. E., *Lloydia, 25,* 158 (1962).
74. Spilsbury, J. F., and Wilkinson, S., *J. Chem. Soc.,* p. 2085 (1961).
75. Perlman, D., and Peruzzotti, G. P., in *Advances in Applied Microbiology,* Vol. 12, pp. 277–294, D. Perlman, Ed. Academic Press, New York (1970).
76. Kadis, S., Montie, T. C., and Ajl, S. J. (Eds.), *Microbial Toxins,* Vols. 1–8. Academic Press, New York (1970–1971).
77. Korzybski, T., Kowszyk-Gindifer, A., and Kurylowicz, W. (Eds.), *Antibiotics,* Vols. 1–2. Pergamon Press, New York (1967).

COMPOUNDS PRODUCED BY INDUSTRIAL FERMENTATION

DR. D. PERLMAN

TABLE 1
PRODUCTS OF COMMERCIAL-SCALE FERMENTATION PROCESSES IN THE U.S. (1970)

Compound	Microbial Source	Applications
Organic Acids		
Citric acid	*Aspergillus niger*	Food acidulant; pharmaceuticals
Gluconic acid	*Aspergillus niger*	Food acidulant; pharmaceuticals
Glutamic acid	*Brevibacterium* species	Food additive
Itaconic acid	*Aspergillus terreus*	Plastics
2-Ketogluconic acid	*Pseudomonas* species	Intermediate in the preparation of food anti-oxidants
Kojic acid	*Aspergillus* species	Insecticide
Lactic acid	*Lactobacillus delbruekii*	Food acidulant; pharmaceuticals
Lysine	*Micrococcus glutamicus*	Animal feed additive
Solvents		
Acetone	*Clostridium* species	Industrial chemical
n-Butanol	*Clostridium* species	Industrial chemical
Ethanol	*Saccharomyces* species	Industrial chemical
Enzymes		
Amylases	*Aspergillus* species	Desizing of cloth; pharmaceuticals
	Bacillus species	
Amyloglucosidase	*Aspergillus* species	Food products
Catalase	*Aspergillus* species	Food products
Glucose isomerase	*Aspergillus* species	Food products
Glucose oxidase	*Aspergillus* species	Food products
Invertase	*Saccharomyces* species	Food products
Lactase	*Saccharomyces* species	Food products
Lipase	*Aspergillus* species	Food products
Pectinase	*Aspergillus niger*	Food products
Protease	*Aspergillus oryzae*	Food products; washing powder additive
	Bacillus subtilis	
Rennet(-like)	*Endothia parasitica*	Cheese
Streptokinase–streptodornase	*Streptococcus hemolyticus*	Medical use (dissolving of blood clots)
Enzyme Transformations		
Glycerol to dihydroxyacetone	*Acetobacter* species	Cosmetic use
Benzaldehyde to phenylacetylcarbinol	*Saccharomyces cerevisiae*	Intermediate for ephedrine
Sorbitol to L-sorbose	*Acetobacter suboxydans*	Intermediate for ascorbic acid
Steroid dehydrogenation	*Corynebacterium* species	Intermediates for steroidal anti-inflammatory agents
	Septomyxa species	
Steroid hydroxylation	*Curvularia lunata*	Intermediates for steroidal anti-inflammatory agents
	Rhizopus nigricans	
	Streptomyces roseochromogenes	
Penicillin to 6-aminopenicillinic acid	*Bacillus* species	Intermediates for semisynthetic penicillins

TABLE 1 (Continued)
PRODUCTS OF COMMERCIAL-SCALE FERMENTATION PROCESSES IN THE U.S. (1970)

Compound	Microbial Source	Applications
Vitamins and Growth Factors		
Gibberellic acid	*Gibberella fujikuroi*	Plant growth stimulator
Vitamin B$_{12}$	*Propionibacterium* species	Human therapy; animal growth
	Pseudomonas denitrificans	factor
Antibiotics		
Actinomycin D	*Streptomyces antibioticus*	Antitumor therapy
Amphotericin B	*Streptomyces nodosus*	Antifungal therapy; animal feed supplement
Amphomycin	*Streptomyces canus*	Antibacterial therapy (Gram-positive)
Antimycin A	*Streptomyces* species	Telocide
Bacitracin	*Bacillus subtilis*	Antibacterial therapy (Gram-positive); animal feed supplement
Candicidin B	*Streptomyces griseus*	Antifungal therapy, topical
Capreomycin	*Streptomyces capreolus*	Antitubercular therapy
Cephalosporins, semisynthetic	*Cephalosporium acremonium* (to produce cephalosporin C, the starting material for the semisynthetic cephalosporins)	
Cephalothin		Antibacterial therapy (Gram-positive and Gram-negative)
Cephaloridine		Antibacterial therapy (Gram-positive and Gram-negative)
Cephaloglycine		Antibacterial therapy (Gram-positive and Gram-negative)
Cephalexin		Antibacterial therapy (Gram-positive and Gram-negative)
Cycloheximide	*Streptomyces griseus*	Antifungal agent, agricultural
Cycloserine	*Streptomyces orchidaceus*	Antitubercular therapy
Erythromycin	*Streptomyces erythreus*	Antibacterial therapy (Gram-positive, PPLO); animal feed supplement
Gentamicin	*Micromonospora purpurea*	Antibacterial therapy (Gram-negative)
Gramicidin	*Bacillus brevis*	Antibacterial therapy, topical (Gram-positive)
Hygromycin B	*Streptomyces hygroscopicus*	Animal feed supplement, antihelminthic
Kanamycin	*Streptomyces kanamyceticus*	Antibacterial therapy (Gram-negative, TB)
Lincomycin	*Streptomyces lincolnensis*	Antibacterial therapy (Gram-positive)
Clindamycin	Semisynthetic derivative from lincomycin	Antibacterial therapy (Gram-positive)
Mithramycin	*Streptomyces* species	Antitumor therapy
Neomycin	*Streptomyces fradiae*	Antibacterial therapy (Gram-positive and Gram-negative)
Nystatin	*Streptomyces noursei*	Antifungal therapy, topical
Oleandomycin	*Streptomyces antibioticus*	Antibacterial therapy (Gram-positive)
Paromomycin	*Streptomyces rimosus*	Antiprotozoal therapy
Penicillins, direct fermentation		
Benzylpenicillin (penicillin G)	*Penicillium chrysogenum*	Antibacterial therapy (Gram-positive)
Phenoxymethylpenicillin (penicillin V)	*Penicillium chrysogenum*	Antibacterial therapy (Gram-positive)
Allylmercaptomethylpenicillin (penicillin O)	*Penicillium chrysogenum*	Antibacterial therapy (Gram-positive)

TABLE 1 (Continued)
PRODUCTS OF COMMERCIAL-SCALE FERMENTATION PROCESSES IN THE U.S. (1970)

Compound	Microbial Source	Applications
Antibiotics (continued)		
Penicillins, semisynthetic		
Ampicillin		Antibacterial therapy (Gram-positive and Gram-negative)
Carbenicillin		Antibacterial therapy (Gram-positive and Gram-negative)
Cloxacillin		Antibacterial therapy (Gram-positive)
Dicloxacillin		Antibacterial therapy (Gram-positive)
Hetacillin		Antibacterial therapy (Gram-positive and Gram-negative)
Methicillin		Antibacterial therapy (Gram-positive)
Nafcillin		Antibacterial therapy (Gram-positive)
Oxacillin		Antibacterial therapy (Gram-positive)
Phenethicillin		Antibacterial therapy (Gram-positive)
Polymyxin B (also colistin)	*Bacillus polymyxa*	Antibacterial therapy (Gram-negative)
Streptomycin	*Streptomyces griseus*	Antibacterial therapy (Gram-negative, TB)
Dihydrostreptomycin	Chemical derivative of streptomycin	Antibacterial therapy (Gram-negative, TB)
Tetracyclines, direct fermentation		
Tetracycline	*Streptomyces aureofaciens*	Antibacterial therapy (Gram-positive and Gram-negative)
7-Chlortetracycline	*Streptomyces aureofaciens*	Antibacterial therapy (Gram-positive and Gram-negative); animal feed supplement
7-Chloro-6-demethyltetracycline	*Streptomyces aureofaciens*	Antibacterial therapy (Gram-positive and Gram-negative)
5-Hydroxytetracycline	*Streptomyces rimosus*	Antibacterial therapy (Gram-positive and Gram-negative); animal feed supplement
Tetracyclines, chemical modification		
Tetracycline	Modification of chlortetracycline	Antibacterial therapy (Gram-positive and Gram-negative)
6-Methylene-6-deoxy-5-hydroxy-tetracycline	Modification of 5-hydroxytetracycline	Antibacterial therapy (Gram-positive and Gram-negative)
6-Methyl-6-deoxy-5-hydroxytetracycline	Modification of 5-hydroxytetracycline	Antibacterial therapy (Gram-positive and Gram-negative)
7-Amino-6-demethyltetracycline	Modification of 7-chloro-6-demethyltetracycline	Antibacterial therapy (Gram-positive and Gram-negative)
Thiostrepton	*Streptomyces azureus*	Antibacterial therapy in veterinary practice (Gram-positive)
Tylosin	*Streptomyces fradiae*	Animal feed supplement
Tyrocidine	*Bacillus brevis*	Antibacterial therapy, topical
Vancomycin	*Streptomyces orientalis*	Antibacterial therapy (Gram-positive)
Viomycin	*Streptomyces floridae*	Antibacterial therapy (Gram-negative, TB)
Miscellaneous		
Insecticide	*Bacillus thuringiensis*	Toxic to selected groups of insects
Monensin	*Streptomyces cinnamonensis*	Coccidiostat
Zearalenone	*Fusarium* species	Estrogenic activitiy

TABLE 2
PRODUCTS PRODUCED ON A COMMERCIAL SCALE IN
THE U.S. BY BOTH FERMENTATION AND
CHEMICAL SYNTHESIS (1965—1970)

Acetone[a]	Ethanol[a]	Lactic acid[a]
n-Butanol[a]	Gluconic acid	Riboflavin[a]
	Glutamic acid[b]	

[a] These compounds can be produced more economically by chemical means.
[b] Recovered from beet sugar process waste.

TABLE 3
PRODUCTS NO LONGER PRODUCED IN THE U.S.
BY FERMENTATION PROCESSES (1971)

Acetone	Fumagillin	Riboflavin
n-Butanol	Lactic acid	Ristocetin
Ethanol	Novobiocin	Stendomycin

TABLE 4
PRODUCTS OF COMMERCIAL-SCALE FERMENTATION PROCESSES
IN COUNTRIES OTHER THAN THE U.S. (1971)

Compound	Country	Applications
Amino Acids		
L-Aspartic acid	Japan	
L-Isoleucine	Japan	
L-Phenylalanine	Japan	
L-Proline	Japan	
L-Threonine	Japan	
L-Tryptophan	Japan	
L-Tyrosine	Japan	
Nucleic Acids		
Guanylic acid	Japan	Food additive
Inosinic acid	Japan	Food additive
Enzyme Transformations		
Cholesterol to androstenedione	Netherlands, Japan	
Antibiotics		
Actinomycin C_3	Germany	Antitumor therapy
Alazomycin F	Japan	Antifungal therapy
Aminosidine	Italy	Antibacterial therapy (Gram-negative)
Blasticidin S	Japan	Antifungal agent, agricultural
Bleomycin	Japan	Antitumor therapy
Carzinophilin	Japan	Antitumor therapy

TABLE 4 (Continued)
PRODUCTS OF COMMERCIAL-SCALE FERMENTATION PROCESSES
IN COUNTRIES OTHER THAN THE U.S. (1971)

Compound	Country	Applications
Antibiotics (continued)		
Cephalosporins		
Cefazolin	Japan	Antibacterial therapy (Gram-positive and Gram-negative)
Cephradine	Philippines	Antibacterial therapy (Gram-positive and Gram-negative)
Chromomycin A_3	Japan	Antitumor therapy
Fusidic acid	Denmark	Antibacterial therapy (Gram-positive)
Gramicidin J	Japan	Antibacterial therapy (Gram-positive)
Griseofulvin	England	Antifungal therapy
Hamycin	India	Antifungal therapy
Josamycin	Japan	Antibacterial therapy (Gram-positive)
Kasugamycin	Japan	Antibacterial agent (agricultural)
Leucensomycin	Italy	Antifungal therapy
Leucomycin	Japan	Antibacterial therapy (Gram-positive)
Mikamycins	Japan	Animal feed supplements
Mitomycin C	Japan	Antitumor therapy
Meonomycin	Germany	Animal feed supplement
Nisin	England	Food preservative
Penicillins		
Azidocillin	Sweden	Antibacterial therapy (Gram-positive and Gram-negative)
Epicillin	Philippines	Antibacterial therapy (Gram-positive and Gram-negative)
Flucloxacillin	England	Antibacterial therapy (Gram-positive)
Pivampicillin	Denmark	Antibacterial therapy (Gram-positive and Gram-negative)
Pimaricin	Netherlands	Antifungal therapy, topical; cheese wrapper
Polyoxin	Japan	Antifungal agent (agricultural)
Pristinamycins	France	Antibacterial therapy (Gram-positive)
Pyrrolnitrin	Japan	Antifungal therapy, topical
Rifampin	Italy	Antibacterial therapy (Gram-positive, TB)
Rifamycin SV	Italy	Antibacterial therapy (Gram-positive, TB)
Sarkomycin	Japan	Antitumor therapy
Spiramycin	France, Japan	Antibacterial therapy (Gram-positive)
Staphylomycins	Belgium	Antibacterial therapy (Gram-positive)
Trichomycin	Japan	Antifungal therapy (topical)
Variotin	Japan	Antifungal therapy (topical)
Miscellaneous		
Dextran	Sweden	Blood plasma expander
Ergot alkaloids	Switzerland	Human therapy
2,3-Butylene glycol	Europe	Solvent

MISCELLANEOUS
INFORMATION

MEASUREMENT OF DISSOLVED OXYGEN

DR. LLOYD E. McDANIEL.

Dissolved-oxygen levels in biological materials can be measured with membrane-protected oxygen probes of the amperometric type. The original instruments consisted of bare metal electrodes, but these were not completely satisfactory because they quickly became poisoned (polarized) by medium constituents. It was the introduction of the membrane probe by Clark et al.[1] that led to the development of suitable measuring devices. Application of the "unsealed" principle of construction by Johnson and coworkers[2,3] resulted in a satisfactory heat-sterilizable model. Membrane probes are of two general types; one requires application of a potential (Clark type) and the other is "self generating" and thus requires no applied voltage (Mancy[4] or Johnson type). A commercial model of the former type is produced by Instrumentation Laboratory, Inc., 113 Hartwell Avenue, Lexington, MA 02173, and one of the latter type is made by New Brunswick Scientific Co., 1130 Somerset Street, New Brunswick, NJ 08903. Several non-sterilizable models are also commercially available. Readings from membrane-protected probes indicate oxygen tensions and thus can give values only relative to saturation. Saturation concentrations for the particular substrate involved must be known for conversion of probe readings to concentration units (e.g., ppm, mmoles/liter, etc.). Membranes may also be employed as coils of tubing through which oxygen-free gas (e.g., N_2 gas) is passed; the oxygen that diffuses through the membranes is measured by a paramagnetic or other oxygen analyzer.[5]

REFERENCES

1. Clark, L. C., Jr., Wold, R., Granger, D., and Taylor, Z., *J. Appl. Physiol., 6,* 189 (1953).
2. Johnson, M. J., Borkowski, J., and Engblom, C., *Biotechnol. Bioeng., 6,* 457 (1964).
3. Borkowski, J. D., and Johnson, M. J., *Biotechnol. Bioeng., 9,* 635 (1967).
4. Mancy, K. H., Okun, D. A., and Reilley, C. N., *J. Electroanal. Chem., 4,* 65 (1962).
5. Phillips, D. H., and Johnson, M. J., *J. Biochem. Microbiol. Technol. Eng., 3,* 261 (1961).

MICROBIOLOGICAL ASSAYS

DR. THOMAS B. PLATT and DR. DAVID M. ISAACSON

The measurement of antibiotic or vitamin concentrations by determination of their effects on growth of pure cultures of microorganisms has been in routine use for more than thirty years. The tables on the following pages outline methods of assay for commercially available antibiotics in pharmaceutical products, for certain vitamins in pharmaceutical products and foodstuffs, and for the polyene antifungal agent, amphotericin B, in blood. More detailed information and alternative methods can be found in the general references listed below:

1. Kavanagh, F. (Ed.), *Analytical Microbiology.* Academic Press, New York (1963).
2. Grove, D. C., and Randall, W. A., *Assay Methods of Antibiotics, A Laboratory Manual.* Medical Encyclopedia, Inc., New York (1955).
3. Regulations for Tests and Methods of Assay of Antibiotic Drugs, *The United States Federal Register,* Part 141 − 149. Government Printing Office, Washington, D.C.
4. W. Horwitz, (Ed.), *Official Methods of Analysis.* Association of Official Analytical Chemists, Washington, D.C. (1970).

The basic methods employed in microbiological assay are agar diffusion and turbidimetry. In the former method, as used for antibiotic assay, the properly diluted test solutions are applied to the surface of agar medium inoculated with the appropriate test culture. During incubation of the test system the antibiotic diffuses in an ever-widening circle around the point of application, resulting in a circular area in which growth has been prevented in an otherwise solidly grown lawn of the test culture. The diameter of this zone of inhibition is proportional to the concentration of the antibiotic and generally yields a linear dose − response curve when log concentration is plotted versus diameter. In the United States, test solutions are usually applied by filling stainless steel cylinders (6 mm I.D., 8 mm, O.D., and 10 mm long) that rest on the surface of a Petri plate containing two layers of agar medium. The lower (or base) layer is sterile, and the upper layer is seeded with the test organism. The basic design consists of three plates with six cylinders per plate for each concentration of standard, with the exception of the reference standard concentration, and three plates for each sample. Three of the six cylinders in each plate are filled with the reference concentration of standard; the other cylinders of each set of three plates contain the appropriate test solution. The purpose of the reference response throughout the assay plates is to correct for plate-to-plate variation. In general, a complete standard curve must be used each time an assay is performed. Potency is calculated as follows:

1. Average the reference responses in all of the standard-curve plates.
2. Average the reference and test solution responses in each set of three plates for standards and unknowns.
3. Determine the correction for each standard-curve concentration by subtracting the average response for each set of three plates from the average reference response for the entire curve.
4. Determine the corrected standard-curve responses by adding algebraically these corrections to the average response of each appropriate average standard response.
5. Construct a best-fit standard curve by plotting log concentration of standards versus the corrected responses.
6. Determine the best-fit reference response from the curve (this usually will be close to the actual average reference response of the standard-curve plates) and use it to derive a correction for each average sample response in the manner described in step 3.
7. Using the corrected sample responses, determine the potency of the samples from the standard curve.

Turbidimetric assays are carried out by adding test solutions to liquid media inoculated with the test organism and measuring the resulting growth as turbidity after incubation. Most antibiotic assays of this type employ relatively high concentrations of inoculum, short incubation periods, and non-aseptic conditions of performance. By contrast, vitamin assays use smaller numbers of cells in the inocula, overnight incubation, and, ideally, aseptic conditions to avoid possible interference by contaminants.

Three or four replicate tubes of each standard and sample concentration normally are used for turbidimetric assay, and each rack must have its own standard curve. The standard curve for each rack is plotted and used only for the sample tubes in that rack. For accurate results, it is imperative that constant incubation temperature be provided for the short-incubation assays of antibiotics.

Turbidimetric assay results are calculated by interpolating from a plot of standard concentration or log standard concentration versus percent transmission or absorbance, depending upon which of these gives the most linear dose — response line.

Methods for Microbiological Assay of Vitamins

DR. THOMAS B. PLATT

TABLE 1
MEDIA FOR MICROBIOLOGICAL ASSAY OF VITAMINS

Vitamin	Assay Organism	Maintenance	Inoculum	Assay	Reference
B_{12}	*Lactobacillus leichmannii* ATCC 7830	Difco 0319 BBL 11414	Difco 0320 BBL 11416	Difco 0457 BBL 11001	1
Biotin	*Lactobacillus plantarum* ATCC 8014	Difco 0319 BBL 11414	1% glucose plus 1% yeast extract	Difco 0419	2
Folic acid	*Streptococcus faecalis* ATCC 8043	Difco 0900	Assay medium plus folic acid, 4 ng/ml	Difco 0967	1
Nicotinic acid amide and analogs	*Lactobacillus plantarum* ATCC 8014	Difco 0319 BBL 11414	Difco 0320 BBL 11416	Difco 0322	1
Panthenol	*Acetobacter suboxydans* ATCC 621H	Difco 0321	Difco 0213 plus pantoic acid,[a] 10 ng/ml, plus Difco 0212	Difco 0994 plus Difco 0212	3
Pantothenate	*Lactobacillus plantarum* ATCC 8014	Difco 0319 BBL 11414	Difco 0320 BBL 11416	Difco 0604 BBL 11492	4
Pyridoxine	*Saccharomyces carlsbergensis* ATCC 9080	Difco 0900	Assay medium plus pyridoxine, 1 ng/ml	Difco 0951	5, 6[b]
Riboflavin	*Lactobacillus casei* ATCC 7469	Difco 0900	Assay medium plus riboflavin, 0.1 μg	Difco 0324	1, 4

[a] Dilute 10 ml of panthenol standard stock solution (see Table 3) 1:2 in 0.2N NaOH, heat for 30 minutes at 121°C, then dilute to 1 μg/ml in phosphate buffer, pH 6.0 (see Reagent 11, Table 4).
[b] Modified.

TABLE 2
MAINTENANCE AND PREPARATION OF INOCULA FOR MICROBIOLOGICAL ASSAY OF VITAMINS[a]

| Vitamin[b] | Maintenance | | Daily Inocula[d] | | | | |
	Form	Incubation Temperature,[c] °C ± 0.5	Broth Culture Incubation Time,[e] hours	Wash	Suspension[f] in	Suspend in	Other Manipulation[g]
B₁₂	Stab	37	24	3 times	Sterile assay medium	10 ml of sterile assay medium	1:10 in sterile assay medium
Biotin	Stab	37	20	Once	Sterile saline	10 ml of sterile saline	1:100 in sterile saline
Folic acid	Stab	30	24	Once	Sterile saline	10 ml of sterile saline	
Nicotinic acid amide	Stab	37	20	Once	Sterile saline	10 ml of sterile saline	1:100 in sterile saline
Panthenol	Slant	30	24	Once	Sterile saline	12 ml of sterile saline	Incubate inoculum culture on shaker to enhance aeration
Pantothenate	Stab	37	20	Once	Sterile saline	10 ml of sterile saline	1:100 in sterile saline
Pyridoxine	Slant	30	24[h]	Once[i]	Sterile assay medium	10 ml of sterile saline	
Riboflavin	Stab	37	20	Once	Sterile assay medium	10 ml of sterile saline	

a   For media, see Table 1.
b   Test cultures are given in Table 1.
c   All maintenance cultures should be transferred weekly and incubated for 24 hours.
d   The source of all inocula except pyridoxine should be loop transfer from a fresh maintenance culture to 10 ml of sterile inoculum medium.
e   Incubate at the temperature of the maintenance culture.
f   All inoculum broths are centrifuged to recover the cells; supernatant culture fluid is discarded; further resuspension–centrifugation cycles are as indicated.
g   All assay tubes are inoculated aseptically with one drop of the final suspension.
h   50 ml in a 250-ml Erlenmeyer flask; the culture may be used for one week if stored at 5°C.
i   10 ml of inoculum broth.

TABLE 3
STANDARDS FOR VITAMIN ASSAY

Vitamin	Source	Drying Conditions	Preparation of Stock Solution			Standard-Curve Concentrations[a]	
			Solvent[b]	Final Concentration	Storage Time Limit[g]	Diluent[b]	ng/ml or/Tube
B₁₂	USP	4 hours over silica gel at room temperature	7	1 ng/ml	None	2	0.04, 0.08, 0.12, 0.16, 0.2
Biotin	C[c]	None	16	100 µg/ml		2	0.2, 0.4, 0.6, 0.8, 1.2
Folic acid	USP	None	18	1 mg/ml	4 weeks	2	3.0, 4.5, 6.0, 9.0, 13.5
Nicotinic acid amide	USP	1 hour *in vacuo* at 40°C	2	100 µg/ml	None	2	20, 40, 60, 80, 100
Panthenol	HLR[d]	None	2	100 µg/ml D-isomer[e]	None	11	400, 800, 1200, 1600, 2000
Pantothenate	USP (calcium)	3 hours at 60°C	7	1 mg/ml[f]	1 week	2	20, 30, 40, 60, 80
Pyridoxine	USP	None	6	100 µg/ml	None	19	75, 150, 225, 300, 375
Riboflavin	USP	2 hours at 105°C	12	100 µg/ml	1 year	2	40, 80, 120, 160, 200

a   1.0 ml of standard-curve solutions should be pipetted into each assay tube; 10 ml of single-strength assay medium should be added to each tube; all tubes should be autoclaved at 121°C for 10 to 15 minutes and rapidly cooled.
b   See Table 4.
c   Pure D-biotin from any commercial source.
d   Crystalline DL-panthenol, Hoffman-La Roche, Nutley, N.J.
e   Panthenol must be hydrolyzed to pantoic acid; dilute the stock solution 1:2 in 0.2N NaOH, heat for 30 minutes at 121°C, cool, then dilute to 2.5 µg/ml in Reagent 11 (see Table 4).
f   Free acid.
g   When stored in the cold and protected from light; non-alcoholic solutions should be stored under toluene.

TABLE 4
REAGENTS FOR VITAMIN ASSAY

1. Metabisulfite buffer

Na_2HPO_4	12.9 g
Citric acid	11.0 g
$Na_2S_2O_5$	10.0 g

Distilled water to 1 liter
Do not sterilize by heating

2. Sterile distilled water

3. 95% Ethanol

4. Approximately $0.1N$ NH_4OH solution containing 0.2% Tween 80

5. Approximately $0.3N$ NH_4OH solution

6. Approximately $0.1N$ HCl solution

7. 25% Ethanol

8. Acetate—ethanol mixture

Sodium acetate	25 g
Ethanol, 95%	100 ml

Distilled water to 1 liter

9. Mylase — buffer mixture

Glacial acetic acid	14.3 ml
Distilled water	85 ml

Adjust pH to 4.2 with
10N NaOH and suspend

Mylase P (Wallerstein Labs)	1.33 g

10. Approximately $0.2N$ NaOH soluton

11. 1% Phosphate buffer, pH 6.0

K_2HPO_4	2 g
KH_2PO_4	8 g

Distilled water to 1 liter
Sterilize at 121°C for 30 minutes

12. Approximately $0.02N$ acetic acid solution

13. 25% (w/v) Sodium acetate in distilled water

14. Approximately $1N$ H_2SO_4

15. $0.12N$ Sodium citrate solution, pH 5.0

16. 50% Ethanol

17. Approximately $6N$ H_2SO_4 solution

18. 2% NH_4OH

19. 0.1M Phosphate buffer, pH 4.5

KH_2PO_4	13.6 g

Distilled water to 1 liter
Sterilize at 121°C for 30 minutes

20. Anion-exchange resin, hydroxide form, about 50% resin in water, IRA-400

21. 0.1M Phosphate buffer, pH 7.2

KH_2PO_4	13.6 g
NaOH	2.8 g

Distilled water to 1 liter

TABLE 5
SAMPLE PREPARATION PROCEDURES FOR VITAMIN ASSAY

		Preparation Steps			Treatment Steps[b]						
Vitamin	Type of Sample	Amount of Sample	Dilute in Reagent No.[a]	Solvent Ratio	1	2	3	4	5	6	7
B_{12}	Crude or bulk product	25 to 30 mg	1	10 ml/mg	B	D-121-10	G				
	Tablet	3 tablets	1	250 ml/tablet	B	D-121-10	G				
	Solution	1 dose	1	250 ml/dose	B	D-121-10	G				
Biotin	Bulk (synthetic) product, Natural products	25 to 30 mg	16	10 ml/mg	A						
		200 to 300 mg	17	0.05 ml/mg	B	D-121-120	E-7	F-2-100 ml			
Folic acid	Bulk product	25 to 50 mg	5	1 ml/mg	A						
	Tablets	3 tablets	3	50 ml/3 tablets	B	C-4-450	B at 60°C	G			
	Natural products	An amount estimated to contain 0.5 to 0.15 μg of folic acid activity	21	100 ml/sample	I-10	G	D-121 15	C[c]			
Nicotinic acid amide	Bulk product	20 to 30 mg	2	10 ml/mg	A						
	Natural products	1 g	14	90 ml/g	B	D-121-30	E-6.5	C-6 until precipitate formation ceases	G	E-6.8	
Panthenol	Bulk product	20 to 30 mg	2	10 ml/mg	F-10-equal volume	D-121-30 I-10	G				
	Solution	5 ml containing 0.5 to 0.1 mg/ml	2	1 ml/mg	C-20-equal volume	G		F-resin washings-50	C-20-equal volume	D-121-30	
Pantothenate	Bulk product	40 to 50 mg	7	1 ml/mg	A						
	Tablets	5 tablets	8	100 ml/tablet	B	G		G			
	Natural products	0.5 to 1 g	9	2.5 ml/100 mg	B	H-37-24	E-6.8	G			

TABLE 5 (Continued)
SAMPLE PREPARATION PROCEDURES FOR VITAMIN ASSAY

Vitamin	Type of Sample	Amount of Sample	Dilute in Reagent No.[a]	Solvent Ratio	Treatment Steps[b]					
					1	2	3	4	5	6	7
Pyridoxine	Bulk product	20 to 30 mg	6	10 ml/mg	A					
	Tablets	3 tablets	6	160 ml/tablet	B	G				
Riboflavin	Bulk product	40 to 50 mg	12	10 ml/mg	A					
	Dried yeast	0.8 to 1 g	6	90 ml/g	D-121-30	C-13-6 ml	E-4.5	F-2-100 ml	G	E-6.8

[a] See Table 4 for identification of the reagents.
[b] Dilute the solution produced after treatment to the standard concentrations with the solvent specified for the appropriate vitamin standard-curve concentrations given in Table 3. Treatment steps: A = none; B = disperse by blending; C = add reagent No. . . , . . . ml; D = heat at . . °C for . . . minutes; E = adjust pH to . . .; F = dilute in reagent No. . . to . . . ml; G = filter or centrifuge; H = incubate under toluene at . . . °C for . . . hours; I = shake gently for . . . minutes.
[c] 100 mg of Difco B459 dehydrated chicken pancreas per gram of sample.

TABLE 6
ASSAY INCUBATION CONDITIONS AND TURBIDIMETRIC
READOUT OF RESPONSES

| Vitamin | Incubation[a] | | Wavelength, nm, for Readout of % Transmission |
	Temperature, °C	Approximate Time, hours	
B_{12}	37	24—48	560
Biotin	37	16—20	540
Folic acid	35—37	16—18	570
Nicotinic acid amide	35—37	16—18	525
Panthenol	30	18—24	525
Pantothenate	37	16—18	570
Pyridoxine	30	19—21	570
Riboflavin	35—37	18—24	550

[a] Growth response also may be determined (except for panthenol and pyridoxine) by titration of the acid produced after 72 hours of incubation.[1]

TABLE 7
INTERFERENCE WITH VITAMIN ASSAYS

Vitamin	Assay Organism	Interfering Substances	Remedial Measures
B_{12}	*Lactobacillus leichmannii*	Deoxyribotides, deoxyribosides, factor III, pseudovitamin B_{12}	Differential assay for deoxyribosides and deoxyribotides[7] Removal by solvent extraction[8]
Biotin	*Lactobacillus plantarum*	Fatty acids and lipoidal substances in the presence of aspartic acid	Remove lipids by ether extraction at pH 4.5[2]
Folic acid	*Streptococcus faecalis*	10-Formylpteroic acid, pteroyl-heptaglutamic acid, leucovorin, pteroic acid, pteroyltriglutamic acid	Assay should be used only when the identity of the folic acid activity is known[9]
Nicotinic acid amide	*Lactobacillus plantarum*	Nicotinamide has activity equal to nicotinic acid[10] N-Substituted nicotinamides have up to 43% of nicotinamide activity[11]	None; considered a relatively specific method
Panthenol	*Acetobacter suboxydans*	Pantoic acid, pantothenic acid, pantoyl lactone	Remove interfering substances by treating test solutions with anion-exchange resin[3]
Pantothenate	*Lactobacillus plantarum*	Pantethine, phosphorylated pantethine and pantothenate, Co A	Differential assay of pantethine[12] Enzyme treatment and differential assays[14]
Pyridoxine	*Saccharomyces carlsbergensis*	Pyridoxamine and pyridoxal have activities equal to pyridoxine; phosphorylated forms are much less active[14]	Separate the compounds by chromatography on Dowex 50[15]

TABLE 7 (Continued)
INTERFERENCE WITH VITAMIN ASSAYS

Vitamin	Assay Organism	Interfering Substances	Remedial Measures
Riboflavin	*Lactobacillus casei*	Fatty acids and lecithin have a sparing effect on the riboflavin requirement; ethyl substituents in place of methyl at the 6 and 7 positions of 9-(1'-D-ribityl) isoalloxazine have activity almost equal to that of riboflavin; isoriboflavin, some araboflavins, and lyxoflavin are somewhat active in place of riboflavin or have a sparing effect on it[10]	An internal standard may partially correct for fatty acids and lecithin effects

REFERENCES

1. Horwitz, W. (Ed.), *Official Methods of Analysis,* 11th ed. Association of Official Analytical Chemists, Washington, D.C. (1970).
2. Skeggs, H. R., in *Analytical Microbiology,* p. 421, F. Kavanagh, Ed. Academic Press, New York (1963).
3. Weiss, M. S., Sonnenfeid, I., De Ritter, E., and Rubin, S. H., *Anal. Chem., 23,* 1687 (1951).
4. *The United States Pharmacopeia,* 18th Revision U.S. Pharmacopeial Convention, Inc., Washington, D.C. (1970).
5. Hurley, N. A., *J. Assoc. Anal. Chem., 43,* 43 (1960).
6. Parrish, W. P., Loy, H. W., Jr., and Kline, O. L., *J. Assoc. Anal. Chem., 39,* 157, (1956).
7. Skeggs, H. R., in *Analytical Microbiology,* p. 551, F. Kavanagh, Ed. Academic Press, New York (1963).
8. McLaughlin, J. M., Rogers, C. G., Middleton, E. J., and Campbell, J. A., *Can J. Biochem. Physiol., 36,* 195 (1958).
9. Eigen, E., and Shockman, G. D., in *Analytical Microbiology,* p. 431, F. Kavanagh, Ed. Academic Press, New York (1963).
10. Koser, S. A., *Vitamin Requirements of Bacteria and Yeasts.* Charles C Thomas, Springfield, Illinois (1968).
11. Cote, L., and Oleson, J. J., *J. Bacteriol., 61,* 463 (1951).
12. Bird, O. D., in *Analytical Microbiology,* p. 497, F. Kavanagh, Ed. Academic Press, New York (1963).
13. Brown, G. M., *J. Biol. Chem., 234,* 379 (1959).
14. Sauberlich, H. E., in *The Vitamins,* 2nd ed., p. 33, W. H. Sebrell, Jr. and R. B. Harris, Eds. Academic Press, New York (1968).
15. Skeggs, H. R., in *Analytical Microbiology,* p. 793, F. Kavanagh, Ed. Academic Press, New York (1963).

Methods for Bioassay of Antibiotics

DR. DAVID M. ISAACSON

TABLE 1

TEST ORGANISMS USED FOR THE BIOASSAY OF SELECTED ANTIBIOTICS[a]

Test Organism Number	Test Organism	Test Organism Number	Test Organism
1	*Bacillus cereus* var. *mycoides* (ATCC 11778)	15	*Pseudomonas pyocyanea* (ATCC 25619)
2	*Bacillus subtilis* (ATCC 6633)	16	*Pseudomonas tabaci*
3	*Bacillus subtilis* (ATCC 10707)	17	*Saccharomyces cerevisiae* (ATCC 2601)
4	*Bordetella bronchiseptica* (ATCC 4617)	18	*Saccharomyces cerevisiae* (ATCC 9763)
5	*Candida tropicalis* (ATCC 13803)	19	*Sarcina lutea* (ATCC 9341)
6	*Corynebacterium xerosis* (NTCC 9755)	20	*Sarcina lutea* resistant to dihydro-streptomycin (ATCC 9341a)
7	*Escherichia coli* (ATCC 10536)	21	*Sarcina subflava* (ATCC 7468)
8	*Klebsiella pneumoniae* (ATCC 10031)	22	*Sarcina subflava* resistant to dihydro-streptomycin (ATCC 7468/d)
9	*Micrococcus flavus* (ATCC 10240)	23	*Staphylococcus aureus* (ATCC 6538P)
10	*Micrococcus flavus* resistant to dihydro-streptomycin (ATCC 10240A)	24	*Staphylococcus aureus* resistant to dihydrostreptomycin (ATCC 6538-DR)
11	*Micrococcus flavus* resistant to neomycin (ATCC 14452)	25	*Staphylococcus aureus* resistant to novobiocin (ATCC 12692)
12	*Microsporum gypseum* (ATCC 14683)	26	*Staphylococcus epidermidis* (ATCC 12228)
13	*Paecilomyces varioti* (ATCC 22319)	27	*Streptococcus faecalis* (ATCC 10541)
14	*Pseudomonas pyocyanea* (ATCC 23389)		

[a]See Table 5 for the specific microbiological assay methods employing these test organisms.

TABLE 2

METHODS FOR PREPARATION OF SUSPENSIONS OF THE TEST ORGANISMS[a]

Method A

Maintain the test culture by weekly transfers on slants of medium 1. Suspend a 24-hour slant culture (grown at 35°C) with 3 ml of sterile solution K. Transfer the suspension onto the surface of 250 ml of medium 1 in a Roux bottle. Incubate for 24 hours at 32 to 37°C. Wash the resulting cell growth from the agar surface with 50 ml of solution K. Store the cell suspension at 5°C.

Method B

Maintain the test culture by weekly transfers on slants of medium 1. Suspend a 24-hour slant culture (grown at 35°C) with 3 ml of sterile solution K. Transfer the suspension onto the surface of 250 ml of medium 1 in a Roux bottle. Incubate for 24 hours at 32 to 37°C. Wash the resulting cell growth from the agar surface with 50 ml of sterile solution K. Transfer the suspension to a sterile centrifuge bottle. Centrifuge at 2000 x g for 10 minutes. Decant the supernatant. Resuspend the cells in 60 ml of sterile solution K. Heat the suspension for 30 minutes at 70°C. Store the spore suspension at 5°C.

TABLE 2 (Continued)
METHODS FOR PREPARATION OF SUSPENSIONS OF THE TEST ORGANISMS[a]

Method C
Maintain the test culture by weekly transfers on slants of medium 1. Suspend a 24-hour slant culture (grown at 35°C) with 3 ml of sterile solution K. Transfer the suspension onto the surface of 250 ml of medium 1 in a Roux bottle. Incubate for 24 hours at 32 to 37°C. Wash the resulting cell growth from the agar surface with 50 ml of sterile solution K. Transfer the suspension to a sterile centrifuge bottle. Heat for 30 minutes at 70°C, then centrifuge at 2000 x g for 10 minutes. Decant the supernatant. Resuspend the cells with 30 ml of sterile distilled water. Repeat centrifugation and wash two additional times. Resuspend the cells in 60 ml of sterile distilled water. Store the spore suspension at 5°C.

Method D
Same as Method A, except incubate the slants for 24 hours at 30°C and incubate the Roux bottles for 48 hours at 30°C.

Method E
Maintain the test organisms in 100-ml portions of medium 3. To prepare the test broth culture, transfer a loopful of the stock culture to 100 ml of medium 3. Incubate for 18 hours at 37°C. Store the test broth culture at 5°C.

Method F
Same as Method C, except incubate the Roux bottle for 7 days at 37°C. Suspend the spores in 100 ml of sterile distilled water.

Method G
Maintain the test culture by weekly transfers on slants of medium 16. Incubate for 24 hours at 37°C. Suspend a 24-hour slant culture with 10 ml of medium 30. Adjust the cell density to 80% light transmission at 660 nm.

Method H
Maintain the test culture by biweekly transfers on slants of medium 33. Incubate at 25°C. Suspend growth with 10 ml of sterile solution K. Transfer the slant wash to 3-liter conical flasks containing 200 ml of medium 14. Incubate for 3 to 4 weeks at 25°C. If sporulation is 80%, harvest the spores on the mycelial layer with a sterile spatula. Place the harvested spores in 50 ml of sterile solution K. Store up to two months at 5°C.

Method I
Maintain the test culture by biweekly transfers on slants containing 20 ml of medium 18. Incubate at 25°C. Wash the spores from the slants with 5 ml of solution K per slant. Store the spore suspension at 5°C.

Method J
Maintain the test culture with weekly transfers on medium 20 at 27°C. To prepare the test broth culture, make a loop transfer of the slant culture to medium 34. Incubate for 24 hours at 27°C. Prepare a 1% suspension with fresh medium 34.

Method K
Maintain the test culture on slants of medium 24. Incubate for 24 hours at 28°C. To prepare the test inoculum, transfer a loop of the slant culture to 30 ml of medium 25. Incubate for 48 hours at 28°C. Store the test broth culture at 5°C.

Method L
Maintain the test culture on slants of medium 5. Incubate for 24 hours at 37°C. To prepare the test inoculum, transfer a loop of the slant culture to medium 3. Incubate for 20 hours at 37°C without shaking. Store the test broth culture at 5°C.

Method M
Maintain the test culture on slants of medium 16. Incubate for 24 hours at 30°C. To prepare the test inoculum, transfer a loop of the slant culture to 100 ml of medium 3. Incubate for 18 hours at 30°C.

Method N
Maintain the test culture on slants of medium 36. Incubate for 7 days at 35°C. To prepare the test broth culture, wash a slant culture with 5 ml of medium 37 and transfer to 1000 ml of medium 35. Incubate, with shaking, for 16 hours at 35°C. Store the test broth culture for one week at 5°C.

[a] Refer to Table 3 for descriptions of the media cited above, and to Table 4 for descriptions of the solutions cited above.

TABLE 3

COMPOSITION OF CULTURE MEDIA USED FOR MICROBIOLOGICAL ASSAYS OF ANTIBIOTICS

Ingredient	Medium Number[a]																	
	1	2	3	4	5	6	7	8	9	10	11	12	13	14	15	16	17	18
Beef extract	1.5	1.5	1.5	1.5	1.5	—	—	1.5	—	1.5	2.4	—	—	—	—	—	—	—
Yeast extract	3.0	3.0	1.5	3.0	3.0	—	—	3.0	—	3.0	4.7	—	—	—	6.0	—	—	—
Malt extract	—	—	5.0	—	—	—	—	—	—	—	—	—	—	—	—	—	—	—
Peptic digest of animal tissue	6.0	6.0	—	6.0	6.0	—	—	6.0	10.0	6.0	9.4	10.0	10.0	10.0	10.0	—	10.0	5.0
Pancreatic digest of casein	4.0	—	—	—	—	17.0	17.0	4.0	—	4.0	—	—	—	—	—	15.0	—	—
Papaic digest of soybean	—	—	—	—	—	3.0	3.0	—	—	—	—	—	—	—	—	5.0	—	—
Infusion from horse meat	—	—	—	—	—	—	—	—	—	—	—	—	—	—	—	—	454	—
Sucrose	1.0	—	—	—	—	—	—	—	—	—	—	—	—	—	—	—	—	—
Dextrose	—	—	1.0	1.0	—	2.5	2.5	1.0	10.0	1.0	10.0	—	—	—	—	—	—	—
Sodium glutamate	—	—	—	—	—	—	—	—	—	—	—	40.0	40.0	40.0	—	—	—	20.0
Monobasic potassium phosphate	—	—	1.32	—	—	—	—	—	—	—	—	—	—	—	—	—	—	—
Dibasic potassium phosphate	—	—	3.68	—	—	2.5	2.5	—	—	—	—	—	—	—	—	—	—	—
Sodium chloride	—	—	3.5	—	—	5.0	5.0	—	—	·	—	—	—	—	—	5.0	5.0	—
Magnesium chloride hexahydrate	—	—	—	—	—	—	—	—	—	—	10.0	—	—	—	—	—	—	—
Sodium hydroxide	—	—	—	—	—	—	—	—	—	—	—	—	—	—	—	—	—	—
Chloramphenicol (activity)	—	—	—	—	—	—	—	—	—	—	—	0.05	0.05	—	—	—	—	—
Cycloheximide	—	—	—	—	—	—	—	—	—	—	—	—	0.2	—	—	—	—	—
Polysorbate 80 (ml)	—	—	—	—	—	—	10.0	—	—	20.0	—	—	—	—	—	—	—	—
Agar	15.0	15.0	—	15.0	15.0	20.0	20.0	15.0	15.0	15.0	23.5	15.0	15.0	—	15.0	15.0	20.0	17.0
pH ±0.05 (after sterilization)	6.55	6.55	7.0	7.9	5.9	7.25	7.25	7.9	5.65	7.9	6.1	5.65	5.65	5.65	7.9	7.3	7.1	6.5

TABLE 3 (Continued)
COMPOSITION OF CULTURE MEDIA USED FOR MICROBIOLOGICAL ASSAYS OF ANTIBIOTICS

Ingredient	Medium Number[a]																		
	19	20	21	22	23	24	25	26	27	28	29	30	31	32	33	34	35	36	37
Beef extract	–	–	–	1.5	3.0	–	–	–	–	3.0	1.5	3.0	3.0	–	–	10.0	1.5	1.5	1.5
Yeast extract	–	2.0	–	3.0	–	–	–	2.5	1.0	–	3.0	–	–	5.0	–	–	1.5	3.0	6.5
Malt extract	–	–	–	–	–	30.0	20.0	–	–	–	–	–	–	–	–	–	–	–	–
Peptic digest of animal tissue	–	5.0	5.0	6.0	5.0	–	–	–	1.0	5.0	6.0	5.0	5.0	–	10.0	–	5.0	6.0	5.0
Pancreatic digest of casein	17.0	–	–	4.0	–	–	–	–	–	–	–	–	–	15.0	–	–	–	–	10.0
Papaic digest of soybean	3.0	–	–	–	–	–	–	–	–	–	–	–	–	–	–	–	–	–	–
Infusion from horse meat	–	–	–	–	–	–	–	–	–	–	–	–	–	–	–	–	–	–	–
Sucrose	2.5	2.0	–	–	–	–	–	–	–	–	–	–	–	–	–	–	–	–	–
Dextrose	–	2.0	5.0	1.0	–	–	–	10.0	10.0	–	–	–	–	10.0	40.0	10.0	11.0	1.0	11.0
Sodium glutamate	–	–	–	–	–	–	–	–	–	–	–	–	–	–	–	–	1.32	–	–
Monobasic potassium phosphate	–	–	–	–	–	–	–	8.5	–	–	–	–	–	–	–	–	–	–	–
Dibasic potassium phosphate	2.5	2.0	–	–	–	–	–	–	–	–	–	–	–	–	–	–	3.68	–	–
Sodium chloride	5.0	2.0	–	–	–	–	–	–	–	–	10.0	–	–	2.5	–	–	3.5	–	–
Magnesium chloride hexahydrate	–	–	–	–	–	–	–	–	–	–	–	–	–	–	–	–	–	–	–
Sodium hydroxide	–	–	–	–	–	–	–	1.5	–	–	–	–	–	–	–	–	–	–	–
Chloramphenicol (activity)	–	–	–	–	–	–	–	–	–	–	–	–	–	–	–	–	–	–	–
Cycloheximide	–	–	–	–	–	–	–	–	–	–	–	–	–	–	–	–	–	–	–
Polysorbate 80 (ml)	–	–	–	–	–	–	–	–	–	–	–	–	–	–	–	–	–	–	–
Agar	–	12.0	10.0	15.0	15.0	15.0	–	15.0	15.0	15.0	15.0	–	15.0	24.0	15.0	–	6.0	15.0	–
pH ± 0.05 (after sterilization)	7.3	6.8	7.0	7.9	7.9	5.5	4.7	7.0	7.5	8.3	9.0	6.9	6.8	6.0	5.65	6.8	6.0	6.6	6.8

a   Refer to Tables 2 and 5 for specific applications of the media listed in this table. Each value in the body of the table represents the concentration (in grams per liter) of the designated ingredient in the specific culture medium identified by the number at the top of the column.

TABLE 4
COMPOSITION OF SOLVENTS, BUFFERS, AND DILUENTS
USED FOR MICROBIOLOGICAL ASSAYS OF ANTIBIOTICS

Solution	Name	Ingredients[a]	Amount
A	1% Phosphate buffer pH 6.0 ± 0.05	Dibasic potassium phosphate Monobasic potassium phosphate Distilled water to make	2.0 g 8.0 g 1000 ml
B	0.1*M* Phosphate buffer pH 7.9 ± 0.1	Dibasic potassium phosphate Monobasic potassium phosphate Distilled water to make	16.73 g 0.523 g 1000 ml
C	0.1*M* Phosphate buffer pH 4.5 ± 0.05	Monobasic potassium phosphate Distilled water to make	13.6 g 1000 ml
D	10% Phosphate buffer pH 6.0 ± 0.05	Dibasic potassium phosphate Monobasic potassium phosphate Distilled water to make	20.0 g 80.0 g 1000 ml
E	0.2*M* Phosphate buffer pH 10.5 ± 0.1	Dibasic potassium phosphate 10*N* Potassium hydroxide Distilled water to make	35.0 g 2.0 ml 1000 ml
F	0.1*M* Phosphate buffer pH 7.0 ± 0.1	Dibasic potassium phosphate Monobasic potassium phosphate Distilled water to make	13.6 g 4.0 g 1000 ml
G	1% Phosphate buffer with 0.1% sodium nitrate pH 6.0 ± 0.05	Dibasic potassium phosphate Monobasic potassium phosphate Sodium nitrate	2.0 g 8.0 g 1.0 g
H	0.01*N* Methanolic hydrochloric acid	1.0*N* Hydrochloric acid Methyl alcohol to make	10.0 ml 1000 ml
I	80% Isopropyl alcohol solution	Isopropyl alcohol Distilled water to make	800 ml 1000 ml
J	1% Sodium carbonate	Sodium carbonate Distilled water to make	10.0 g 1000 ml
K	0.9% Saline	Sodium chloride Distilled water to make	9.0 g 1000 ml
L	50% Methanol	Methyl alcohol Distilled water to make	500 ml 1000 ml
M	Dimethylformamide solution pH 6.7 ± 0.3	Dimethylformamide Adjust pH with 10*N* potassium hydroxide and 10*N* hydrochloric acid immediately before use	1000 ml
N	80% Dimethylsulfoxide	Dimethylsulfoxide Distilled water to make	800 ml 1000 ml
O	25% Dimethylsulfoxide	Dimethylsulfoxide Distilled water to make 1000 ml	250 ml 1000 ml

[a] Adjust phosphate buffers to correct pH with 18*N* phosphoric acid or 10*N* potassium hydroxide.

TABLE 5
SELECTED METHODS FOR THE MICROBIOLOGICAL ASSAY OF ANTIBIOTICS

Antibiotic	Test Method	Test Organism Number (Refer to Table 1)	Method of Preparation of Test Inoculum (Refer to Table 2)	Milliliters of Test Inoculum per Liter of Assay Medium	Agar-Diffusion Assays — Base Layer — Medium Number (Refer to Table 3)	Agar-Diffusion Assays — Base Layer — Milliliters of Medium[c]	Agar-Diffusion Assays — Seed Layer — Medium Number (Refer to Table 3)	Agar-Diffusion Assays — Seed Layer — Milliliters of Medium[d]	Turbidimetric Assays — Medium Number (Refer to Table 3)	Turbidimetric Assays — Milliliters of Medium per Tube
Actinomycin C	AD[a]	2	F	0.5	25	21	25	4	—	—
Amphomycin	AD	11	A	5.0	2	21	1	4	—	—
Amphotericin B	AD	18	D	2.0	—	—	11	8	—	—
Ampicillin	AD	19	A	10.0	8	21	8	4	—	—
Bacitracin	AD	21 or 22[g]	A	2.0	2	21	1	4	—	—
	AD	9 or 10[h]	A	2.0	2	21	1	4	—	—
Candicidin	TU[b]	18	A	2.0	—	—	—	—	19	9
Capreomycin	TU	8	A	1.0	—	—	—	—	3	9
Carbenicillin	AD	15	A	0.5	6	21	7	4	—	—
Carbomycin	AD	19	A	4.0	1	21	1	4	—	—

TABLE 5 (Continued)

SELECTED METHODS FOR THE MICROBIOLOGICAL ASSAY OF ANTIBIOTICS

Antibiotic	Preparation of Standard		Assay range,[e] units or µg/ml	Incubation Temperature,[f] °C	References
	Initial Concentration and Solvent Used (Refer to Table 4)	Diluent for Further Dilutions (Refer to Table 4)			
Actinomycin C	1000 µg/ml in acetone	A	2.5–40.0 (10.0)	37	1
Amphomycin	100 µg/ml in B	B	6.4–15.6 (10.0)	32–35	2
Amphotericin B	1000 µg/ml in dimethyl sulfoxide	E	0.64–1.56 (1.0)	30	2, 29
Ampicillin	100 µg/ml in distilled water	B	0.64–1.56 (1.0)	32–35	2, 29
Bacitracin	100 units/ml in A	A	0.64–1.56 (1.0)	32–35	2, 29
	100 units/ml in A	A	64–156 (100)	32–35	2, 29
Candicidin	1000 µg/ml in dimethyl sulfoxide	Distilled water	0.03–0.12 (0.06)	25	2
Capreomycin	1000 µg/ml in distilled water	Distilled water	64–156 (100)	37	2, 3, 4
Carbenicillin	1000 µg/ml in A	A	64–156 (100)	37	2
Carbomycin	1000 µg/ml in methanol	B	0.6–1.5 (1.0)	30–32	5, 6

TABLE 5 (Continued)
SELECTED METHODS FOR THE MICROBIOLOGICAL ASSAY OF ANTIBIOTICS

Antibiotic	Test Method	Test Organism			Assay Media					
		Test Organism Number (Refer to Table 1)	Method of Preparation of Test Inoculum (Refer to Table 2)	Milliliters of Test Inoculum per Liter of Assay Medium	Agar-Diffusion Assays				Turbidimetric Assays	
					Base Layer		Seed Layer			
					Medium Number (Refer to Table 3)	Milliliters of Medium[c]	Medium Number (Refer to Table 3)	Milliliters of Medium[d]	Medium Number (Refer to Table 3)	Milliliters of Medium per Tube
Cephalexin	AD	23	A	0.5	2	21	1	4	—	—
Cephaloglycin	AD	23	A	2.0	2	21	1	4	—	—
Cephaloridine	AD	23	A	3.0	2	21	1	4	—	—
Cephalothin	AD	23	A	0.5	2	21	1	4	—	—
Cephradine	AD	19	A	0.5	—	—	1	10	—	—
Cephapirin	AD	19	G	10.0	—	—	31	10	—	—
Chloramphenicol	AD	19	A	20.0	1	21	1	4	—	—
	TU	7	A	1.0	—	—	—	—	3	9
Chlortetracycline	AD	1	C	4.0	4	21	4	4	—	—
	TU	23	A	1.0	—	—	—	—	3	9

TABLE 5 (Continued)
SELECTED METHODS FOR THE MICROBIOLOGICAL ASSAY OF ANTIBIOTICS

Antibiotic	Preparation of Standard		Assay range,[e] units or µg/ml	Incubation Temperature,[f] °C	References
	Initial Concentration and Solvent Used (Refer to Table 4)	Diluent for Further Dilutions (Refer to Table 4)			
Cephalexin	1000 µg/ml in A	A	12.8–31.2 (20.0)	37	2
Cephaloglycin	100 µg/ml in distilled water	C	6.4–15.6 (10.0)	32–35	2
Cephaloridine	1000 µg/ml in A	A	0.64–1.56 (1.0)	32–35	2
Cephalothin	1000 µg/ml in A	A	0.64–1.56 (1.0)	32–35	2, 29
Cephradine	1000 µg/ml in A	A	0.05–0.50 (0.1)	30–32	32[o]
Cephapirin	1000 µg/ml in A	A	0.4–12.5 (3.2)	26	7, 30
Chloramphenicol	10,000 µg/ml in ethanol	A	32.0–78.1 (50.0)	32–35	2
	10,000 µg/ml in ethanol	A	2.0–3.12 (2.5)	37	2, 29
Chlortetracycline	1000 µg/ml in 0.01N HCl	C	0.064–0.156 (0.100)	30	2, 30
	1000 µg/ml in 0.01N HCl	C	0.038–0.094 (0.060)	37	2

TABLE 5 (Continued)
SELECTED METHODS FOR THE MICROBIOLOGICAL ASSAY OF ANTIBIOTICS

Antibiotic	Test Method	Test Organism			Assay Media					
		Test Organism Number (Refer to Table 1)	Method of Preparation of Test Inoculum (Refer to Table 2)	Milliliters of Test Inoculum per Liter of Assay Medium	Agar-Diffusion Assays				Turbidimetric Assays	
					Base Layer		Seed Layer			
					Medium Number (Refer to Table 3)	Milliliters of Medium[c]	Medium Number (Refer to Table 3)	Milliliters of Medium[d]	Medium Number (Refer to Table 3)	Milliliters of Medium per Tube
Clindamycin	AD	19	A	15.0	8	21	8	4	—	—
Cloxacillin	AD	23	A	2.0	2	21	1	4	—	—
Colistimethate, sodium	AD	4	A	0.4	6	21	7	4	—	—
Colistin	AD	4	A	0.4	6	21	7	4	—	—
Cycloserine	AD	23	A	0.4	2	10	1	4	—	—
	TU	23	A	1.0	—	—	—	—	3	9
Dactinomycin	AD	2	B	0.2	4	10	4	4	—	—
Demeclocycline	AD	1	C	4.0	5	21	5	4	—	—
	TU	23	A	1.0	—	—	—	—	3	9
Dicloxacillin	AD	23	A	2.0	2	21	1	4	—	—

TABLE 5 (Continued)
SELECTED METHODS FOR THE MICROBIOLOGICAL ASSAY OF ANTIBIOTICS

Antibiotic	Preparation of Standard		Assay range,[e] units or μg/ml	Incubation Temperature,[f] °C	References
	Initial Concentration and Solvent Used (Refer to Table 4)	Diluent for Further Dilutions (Refer to Table 4)			
Clindamycin	1000 μg/ml in distilled water	B	0.64–1.56 (1.0)	37	2
Cloxacillin	1000 μg/ml in A	A	3.2–7.81 (5.0)	32–35	2, 29
Colistimethate, sodium	10,000 μg/ml in distilled water	D	0.64–1.56 (1.0)	37	2, 29
Colistin	10,000 μg/ml in distilled water	D	0.64–1.56 (1.0)	37	2, 30
Cycloserine	1000 μg/ml in distilled water	A	32.0–78.1 (50.0)	30	2, 29
	1000 μg/ml in distilled water	Distilled water	32.0–78.1 (50.0)	37	2
Dactinomycin	10,000 μg/ml in methanol	B	0.5–2.0 (1.0)	37	2, 29
Demeclocycline	1000 μg/ml in 0.1N HCl	C	0.064–0.156 (0.10)	30	2
	1000 μg/ml in 0.1N HCl	C	0.064–0.156 (0.10)	37	2
Dicloxacillin	1000 μg/ml in A	A	3.2–7.81 (5.0)	32–35	2

TABLE 5 (Continued)
SELECTED METHODS FOR THE MICROBIOLOGICAL ASSAY OF ANTIBIOTICS

Antibiotic	Test Method	Test Organism Number (Refer to Table 1)	Method of Preparation of Test Inoculum (Refer to Table 2)	Milliliters of Test Inoculum per Liter of Assay Medium	Agar-Diffusion Assays — Base Layer Medium Number (Refer to Table 3)	Milliliters of Medium[c]	Agar-Diffusion Assays — Seed Layer Medium Number (Refer to Table 3)	Milliliters of Medium[d]	Turbidimetric Assays Medium Number (Refer to Table 3)	Milliliters of Medium per Tube
Dihydrostreptomycin	AD	2	B	0.5	4	21	4	4	–	–
	TU	8	A	1.0	–	–	–	–	3	9
Doxycycline	AD	1	C	4.0	5	21	5	4	–	–
	TU	23	A	1.0	–	–	–	–	3	9
Erythromycin	AD	19	A	10.0	8	21	8	4	–	–
Fusidate, sodium	AD	6	M	10.0	32	10	32	5	–	–
	TU	23	A	2.5	–	–	–	–	3	9
Gentamicin	AD	26	A	0.3	8	21	8	4	–	–
Gramicidin	TU	27	E	10.0	–	–	–	–	3	9
Griseofulvin	AD	12	H	10.0	12	21	13	4	–	–
Hamycin	AD	13	I	60.0	–	–	17	20	–	–

TABLE 5 (Continued)
SELECTED METHODS FOR THE MICROBIOLOGICAL ASSAY OF ANTIBIOTICS

Antibiotic	Preparation of Standard Initial Concentration and Solvent Used (Refer to Table 4)	Diluent for Further Dilutions (Refer to Table 4)	Assay range,[e] units or μg/ml	Incubation Temperature,[f] °C	References
Dihydrostreptomycin	1000 μg/ml in B	B	0.64–1.56 (1.0)	37	2, 30
	1000 μg/ml in distilled water	Distilled water	24.0–37.5 (30.0)	37	2
Doxycycline	1000 μg/ml in 0.1N HCl	B	0.064–0.156 (0.10)	30	2, 31
	1000 μg/ml in 0.1N HCl	C	0.064–0.156 (0.10)	37	2
Erythromycin	10,000 μg/ml in methanol	B	0.64–1.56 (1.0)	37	2, 29, 30
Fusidate, sodium	1000 μg/ml in A	A	0.10–3.0 (1.8)	32–35	8, 9, 10
	1000 μg/ml in J	Distilled water	80–125 (100)	37	2
Gentamicin	1000 μg/ml in B	B	0.064–0.156 (0.10)	37	2, 29
Gramicidin	1000 μg/ml in 95% ethanol	95% ethanol[j]	0.028–0.057 (0.10)	37	2
Griseofulvin	1000 μg/ml in dimethyl-formamide[i]	B	3.2–7.81 (5.0)	30[k]	2, 11, 29
Hamycin	1000 μg/ml in dimethyl sulfoxide	K[l]	0.05–0.10 (0.075)	30	12

TABLE 5 (Continued)
SELECTED METHODS FOR THE MICROBIOLOGICAL ASSAY OF ANTIBIOTICS

Antibiotic	Test Method	Test Organism			Assay Media					
		Test Organism Number (Refer to Table 1)	Method of Preparation of Test Inoculum (Refer to Table 2)	Milliliters of Test Inoculum per Liter of Assay Medium	Agar-Diffusion Assays				Turbidimetric Assays	
					Base Layer		Seed Layer			
					Medium Number (Refer to Table 3)	Milliliters of Medium^c	Medium Number (Refer to Table 3)	Milliliters of Medium^d	Medium Number (Refer to Table 3)	Milliliters of Medium per Tube
Hetacillin	AD	19	A	5.0	8	21	8	4	—	—
Hygromycin B	AD	2	C	0.5	19	10	19	5	—	—
Kanamycin	AD	23	A	0.1	8	21	8	4	—	—
Kanamycin B	AD	2	B	0.5	4	21	4	4	—	—
Kasugamycin	AD	16	J	5.0	21	10	21	10	—	—
Leucomycin	AD	19	A	10.0	—	—	11	8	—	—
	AD	2	B	5.0	22	21	22	4	—	—
Lincomycin	AD	19	A	10.0	8	21	8	4	—	—
Methacycline	AD	1	C	4.0	5	21	5	4	—	—
	TU	23	A	1.0	—	—	—	—	3	9

TABLE 5 (Continued)
SELECTED METHODS FOR THE MICROBIOLOGICAL ASSAY OF ANTIBIOTICS

Antibiotic	Initial Concentration and Solvent Used (Refer to Table 4)	Diluent for Further Dilutions (Refer to Table 4)	Assay range,[e] units or µg/ml	Incubation Temperature,[f] °C	References
Hetacillin	1000 µg/ml in B	B	0.064–0.156 (0.10)	32	2, 13
Hygromycin B	1000 units/ml in F	F	20.0–100 (50.0)	37	14
Kanamycin	1000 µg/ml in B	B	3.2–7.81 (5.0)	32–35	2, 29
Kanamycin B	1000 µg/ml in B	B	0.64–1.56 (1.0)	37	2
Kasugamycin	1000 µg/ml in F	F	100–1000 (500)	27	15
Leucomycin	10,000 µug/ml in ethanol, 1000 µg/ml in A	B	3.2–7.81 (5.0)	32–35	2
	3000 µg/ml in methanol	Distilled water	7.0–35.0 (20.0)	32–35	16
Lincomycin	1000 µg/ml in distilled water	B	1.28–3.12 (2.0)	37	2, 29
Methacycline	1000 µg/ml in H	C	0.064–0.156 (0.10)	30	2, 17
	1000 µg/ml in H	C	0.032–0.078 (0.05)	37	2

TABLE 5 (Continued)
SELECTED METHODS FOR THE MICROBIOLOGICAL ASSAY OF ANTIBIOTICS

Antibiotic	Test Method	Test Organism Number (Refer to Table 1)	Method of Preparation of Test Inoculum (Refer to Table 2)	Milliliters of Test Inoculum per Liter of Assay Medium	Base Layer Medium Number (Refer to Table 3)	Base Layer Milliliters of Medium[c]	Seed Layer Medium Number (Refer to Table 3)	Seed Layer Milliliters of Medium[d]	Turbidimetric Medium Number (Refer to Table 3)	Turbidimetric Milliliters of Medium per Tube
Methicillin	AD	23	A	2.0	2	21	1	4	—	—
Minocycline	TU	23	A	2.0	—	—	—	—	3	9
Mithramycin	AD	23	A	1.0	5	10	5	4	—	—
Mitomycin C	AD	2	B	5.0	2	21	23	4	—	—
Nafcillin	AD	23	A	2.0	2	21	1	4	—	—
Neomycin	AD	23	A	0.4	8	21	8	4	—	—
	AD	26	A	2.0	8	21	8	4	—	—
Novobiocin	AD	26	A	20.0	2	21	1	4	—	—
Nystatin	AD	17	D	10.0	—	—	11	8	—	—
	TU	5	N	30.0	—	—	—	—	35	5

TABLE 5 (Continued)
SELECTED METHODS FOR THE MICROBIOLOGICAL ASSAY OF ANTIBIOTICS

Antibiotic	Preparation of Standard — Initial Concentration and Solvent Used (Refer to Table 4)	Diluent for Further Dilutions (Refer to Table 4)	Assay range,[e] units or μg/ml	Incubation Temperature,[f] °C	References
Methicillin	1000 μg/ml in A	A	6.4–15.6 (10.0)	32–35	2, 29
Minocycline	1000 μg/ml in 0.1N HCl	C	0.032–0.078 (0.05)	37	2, 18
Mithramycin	1000 μg/ml in distilled water	A	0.5–2.0 (1.0)	32	2
Mitomycin C	200 μg/ml in distilled water[l]	F	0.5–4.0 (1.5)	32–35	19
Nafcillin	1000 μg/ml in A	A	1.28–3.12 (2.0)	32–35	2, 20, 29
Neomycin	1000 μg/ml in B	B	6.4–15.6 (10.0)	32–35	2, 30
	1000 μg/ml in B	B	0.64–1.56 (1.0)	37	2, 29
Novobiocin	10,000 μg/ml in ethanol, 1000 μg/ml in B	D	0.32–0.781 (0.5)	35	2, 30
Nystatin	1000 μg/ml in dimethyl-formamide[i]	D[l]	12.8–31.2 (20.0)	30	2, 29
	360 μg/ml in dimethyl-sulfoxide[l]	O[j,p]	0.05–0.12 (0.09)	30	33[o]

TABLE 5 (Continued)
SELECTED METHODS FOR THE MICROBIOLOGICAL ASSAY OF ANTIBIOTICS

	Test Organism				Assay Media					
					Agar-Diffusion Assays				Turbidimetric Assays	
					Base Layer		Seed Layer			
Antibiotic	Test Method	Test Organism Number (Refer to Table 1)	Method of Preparation of Test Inoculum (Refer to Table 2)	Milliliters of Test Inoculum per Liter of Assay Medium	Medium Number (Refer to Table 3)	Milliliters of Medium[c]	Medium Number (Refer to Table 3)	Milliliters of Medium[d]	Medium Number (Refer to Table 3)	Milliliters of Medium per Tube
Oleandomycin	AD	26	A	2.0	8	21	8	4	—	—
	TU	23	A	1.0	—	—	—	—	3	9
Oxacillin	AD	23	A	1.0	2	21	1	4	—	—
Oxytetracycline	AD	1	C	4.0	5	21	5	4	—	—
	TU	23	A	1.0	—	—	—	—	3	9
Paromomycin	AD	26	A	2.0	8	21	8	4	—	—
	AD	2	B	0.5	15	21	15	4	—	—
Penicillin G	AD	23	A	2.0	2	21	1	4	—	—
Phenethicillin	AD	19	A	4.0	8	21	8	4	—	—
Phenoxymethyl-penicillin	AD	23	A	2.0	2	21	1	4	—	—

TABLE 5 (Continued)
SELECTED METHODS FOR THE MICROBIOLOGICAL ASSAY OF ANTIBIOTICS

Antibiotic	Preparation of Standard Initial Concentration and Solvent Used (Refer to Table 4)	Diluent for Further Dilutions (Refer to Table 4)	Assay range,[e] units or μg/ml	Incubation Temperature,[f] °C	References
Oleandomycin	10,000 μg/ml in ethanol	B	3.2–7.81 (5.0)	37	2, 30
	1000 μg/ml in B	B	2.0–3.12 (2.5)	37	2
Oxacillin	1000 μg/ml in A	A	3.2–7.81 (5.0)	32–35	2, 29
Oxytetracycline	1000 μg/ml in 0.1N HCl	C	0.64–1.56 (1.0)	30	2
	1000 μg/ml in 0.1N HCl	C	0.16–0.39 (0.25)	37	2, 29
Paromomycin	1000 μg/ml in B	B	0.64–1.56 (1.0)	37	2
	1000 μg/ml in B	B	1.28–3.12 (2.0)	37	2
Penicillin G	1000 units/ml in A	A	0.64–1.56 (1.0)	37	2, 29
Phenethicillin	1000 units/ml in distilled water	A	0.064–0.156 (0.10)	32–35	2
Phenoxymethyl-penicillin	1000 units/ml in methanol	A	0.64–1.56 (1.0)	37	2, 21, 29

TABLE 5 (Continued)
SELECTED METHODS FOR THE MICROBIOLOGICAL ASSAY OF ANTIBIOTICS

Antibiotic	Test Method	Test Organism			Assay Media					
					Agar-Diffusion Assays				Turbidimetric Assays	
					Base Layer		Seed Layer			
		Test Organism Number (Refer to Table 1)	Method of Preparation of Test Inoculum (Refer to Table 2)	Milliliters of Test Inoculum per Liter of Assay Medium	Medium Number (Refer to Table 3)	Milliliters of Medium[c]	Medium Number (Refer to Table 3)	Milliliters of Medium[d]	Medium Number (Refer to Table 3)	Milliliters of Medium per Tube
Pimaricin	AD	18	K	10.0	26	25	26	6	—	—
Polymyxin	AD	4	A	0.4	6	21	7	4	—	—
Puromycin	AD	19	A	1.0	2	21	1	4	—	—
Rifampin	AD	2	B	1.0	2	21	2	4	—	—
Rifamycin SV	AD	19	A	6.0	2	21	1	4	—	—
Ristocetin	AD	1	C	4.0	5	21	5	4	—	—
	TU	23	A	3.0	—	—	—	—	3	9
Rolitetracycline	AD	1	C	4.0	5	21	5	4	—	—
	TU	23	A	1.0	—	—	—	—	3	9
Soframycin	TU	23	A	2.0	—	—	—	—	3	9

TABLE 5 (Continued)
SELECTED METHODS FOR THE MICROBIOLOGICAL ASSAY OF ANTIBIOTICS

Antibiotic	Preparation of Standard Initial Concentration and Solvent Used (Refer to Table 4)	Diluent for Further Dilutions (Refer to Table 4)	Assay range,[e] units or μg/ml	Incubation Temperature,[f] °C	References
Pimaricin	1000 units/ml in methanol; 1:1 in water	L[l]	15–100 (50.0)	30[m]	22
Polymyxin	20,000 units/ml in distilled water	D	6.4–15.6 (10.0)	37	2; 29, 30
Puromycin	1000 μg/ml in B	B	0.64–1.56 (1.0)	30	2
Rifampin	1000 μg/ml in methanol	A	3.2–7.81 (5.0)	30	2
Rifamycin SV	1000 μg/ml in F	F	0.05–0.6 (0.3)	32–35	23
Ristocetin	1000 μg/ml in A	G	6.4–15.6 (10.0)	30	2
	1000 μg/ml in A	A	9.0–25 (15.0)	37	2, 24
Rolitetracycline	1000 μg/ml in methanol	C	0.64–1.56 (1.0)	30	2
	1000 μg/ml in methanol	A	0.16–0.39 (0.25)	37	2
Soframycin	500 μg/ml in distilled water	Distilled water	1.92–4.65 (3.0)	37	2

TABLE 5 (Continued)
SELECTED METHODS FOR THE MICROBIOLOGICAL ASSAY OF ANTIBIOTICS

Antibiotic	Test Method	Test Organism			Assay Media					
					Agar-Diffusion Assays				Turbidimetric Assays	
					Base Layer		Seed Layer			
		Test Organism Number (Refer to Table 1)	Method of Preparation of Test Inoculum (Refer to Table 2)	Milliliters of Test Inoculum per Liter of Assay Medium	Medium Number (Refer to Table 3)	Milliliters of Medium^c	Medium Number (Refer to Table 3)	Milliliters of Medium^d	Medium Number (Refer to Table 3)	Milliliters of Medium per Tube
Spectinomycin	AD	8	A	2.0	—	—	27	8	—	—
	TU	7	A	0.2	—	—	—	—	3	9
Spiramycin	AD	2	F	10.0	28	18	28	4.5	—	—
Streptomycin	AD	2	B	0.5	4	21	4	4	—	—
	TU	8	A	1.0	—	—	—	—	3	9
Tennecitin	AD	18	D	3.0	—	—	1:	8	—	—
Tetracycline	AD	1	C	4.0	5	21	5	4	—	—
	TU	23	A	1.0	—	—	—	—	3	9
Thiostrepton	AD	23	A	1.0	—	—	11	8	—	—
	AD	23	L	1.0	29	10	29	5	—	—

TABLE 5 (Continued)
SELECTED METHODS FOR THE MICROBIOLOGICAL ASSAY OF ANTIBIOTICS

Antibiotic	Preparation of Standard		Assay range,[e] units or μg/ml	Incubation Temperature,[f] °C	References
	Initial Concentration and Solvent Used (Refer to Table 4)	Diluent for Further Dilutions (Refer to Table 4)			
Spectinomycin	1000 μg/ml in B	B	31.25–500 (150)	37[n]	25
	1000 μg/ml in B	B	12.8–31.2 (20.0)	37	2
Spiramycin	1000 units/ml in methanol	B	12.5–200 (50.0)	30	26
Streptomycin	1000 μg/ml in B	B	0.64–1.56 (1.0)	30	2, 29
	1000 μg/ml in distilled water	Distilled water	24.0–37.5 (30.0)	37	2, 29
Tennecitin	4000 μg/ml in dimethyl-sulfoxide	A	0.64–1.56 (1.0)	37	2
Tetracycline	1000 μg/ml in 0.1N HCl	C	0.64–1.56 (1.0)	32–35	2
	1000 μg/ml 0.1N HCl	C	0.160–0.390 (0.250)	37	2, 29, 30
Thiostrepton	1000 μg/ml in M	B	0.64–1.56 (1.0)	32–37	2
	1000 μg/ml in dimethyl-sulfoxide	N	1.25–5.0 (2.5)	37	27

TABLE 5 (Continued)
SELECTED METHODS FOR THE MICROBIOLOGICAL ASSAY OF ANTIBIOTICS

Antibiotic	Test Method	Test Organism			Assay Media					
		Test Organism Number (Refer to Table 1)	Method of Preparation of Test Inoculum (Refer to Table 2)	Milliliters of Test Inoculum per Liter of Assay Medium	Agar-Diffusion Assays				Turbidimetric Assays	
					Base Layer		Seed Layer			
					Medium Number (Refer to Table 3)	Milliliters of Medium[c]	Medium Number (Refer to Table 3)	Milliliters of Medium[d]	Medium Number (Refer to Table 3)	Milliliters of Medium per Tube
Troleandomycin	AD	26	A	2.0	10	21	10	4	–	–
	TU	7	A	0.5	–	–	–	–	3	9
Tylosin	AD	19	A	5.0	8	21	8	4	–	–
	AD	20	A	10.0	8	21	8	4	–	–
Tyrothricin	TU	27	E	10.0	–	–	–	–	3	9
Vancomycin	AD	1	C	2.0	5	21	5	4	–	–
Viomycin	AD	2	B	0.5	4	21	4	4	–	–
	TU	8	A	0.5	–	–	–	–	3	9

a AD = agar-diffusion assay.
b TU = turbidimetric assay.
c Volume of uninoculated agar per 90-mm Petri plate.
d Volume of inoculated agar per 90-mm Petri plate, layered onto the solidified base layer.
e The value in parentheses is the reference concentration; samples are diluted to this concentration. In the turbidimetric assays, 1 ml of each concentration is added to each tube unless otherwise noted.
f Incubation time: 18 to 24 hours for agar-diffusion assays, 3 to 6 hours for turbidimetric assays.
g Test organism 22 is used for tests of solutions containing bacitracin and dihydrostreptomycin.
h Test organism 10 is used for tests of solutions containing bacitracin and dihydrostreptomycin.

TABLE 5 (Continued)

SELECTED METHODS FOR THE MICROBIOLOGICAL ASSAY OF ANTIBIOTICS

Antibiotic	Preparation of Standard		Assay range,[e] units or μg/ml	Incubation Temperature,[f] °C	References
	Initial Concentration and Solvent Used (Refer to Table 4)	Diluent for Further Dilutions (Refer to Table 4)			
Troleandomycin	1000 μg/ml in l	E	9.6–23.4 (15.0)	37	2
	1000 μg/ml in l	A	16.0–39.0 (25.0)	37	2
Tylosin	10,000 μg/ml in methanol	B	3.2–7.81 (5.0)	32[o]	2
	10,000 μg/ml in B	B	3.2–7.81 (5.0)	32–35	2
Tyrothricin	1000 μg/ml in ethanol	Ethanol[i]	0.14–0.285 (0.20)	37	2
Vancomycin	400 μg/ml in distilled water	C	6.4–15.6 (10.0)	30	2, 29
Viomycin	1000 μg/ml in distilled water	B	32.0–78.1 (50.0)	37	2
	5000 μg/ml in distilled water	Distilled water	64.0–156 (100)	37	2, 29

i The solution is further diluted in solvent to 20 times the final concentration of each concentration of the standard curve, so that each final dilution is 1 + 19 with the indicated buffer.
j Add only 0.1 ml of solution to each tube.
k Incubation time: 48 hours.
l Solutions are prepared and stored in low-actinic glassware.
m Prediffuse for 2 hours at 5°C before incubation.
n Prediffuse for 5 hours at 5°C before incubation.
o Modified.
p Dilute to 30 μg/ml in diluent O, then dilute to the assay range with medium 35.

REFERENCES

1. *Minimum Requirements of Antibiotic Products,* p. 479. Ministry of Health and Welfare, Japanese Government, Tokyo, Japan (1961).

2. Arret, B., Johnson, D. P., and Kirshbaum, A., Outline of Details for Microbiological Assays of Antibiotics: Second Revision, *J. Pharm. Sci., 60,* 1689 (1971).

3. Stark, W. M., Higgens, C. E., Wolfe, R. N., Hoehn, M. M., and McGuire, J. M., Capreomycin, a new Antimycobacterial Agent Produced by *Streptomyces capreolus, sp. n.,* in *Antimicrobial Agents and Chemotherapy – 1962.,* pp. 596 – 606, J. C. Sylvester, Ed. American Society for Microbiology, Ann Arbor, Michigan (1963).

4. Black, H. R., Griffith, R. S., and Peabody, A. M., Absorption, Excretion and Metabolism of Capreomycin in Normal and Diseased States, *Ann. N. Y. Acad. Sci., 135,* 974 (1966).

5. Grove, D. C., and Randall, W. A., *Assay Methods of Antibiotics, A Laboratory Manual,* p. 104. Medical Encyclopedia, Inc., New York (1955).

6. Dony, J., Technique of Determination of Magnamycin, *Ann. Pharm. Fr., 12,* 307 (1954).

7. Bran, J. L., Levison, M. E., and Kaye, D., Clinical and *in-vitro* Evaluation of Cephapirin, A New Cephalosporin Antibiotic, *Antimicrob. Agents Chemother., 1,* 35 (1972).

8. Hilson, G. R. F., *In vitro* Studies of a New Antibiotic (Fucidin), *Lancet, 1,* 932 (1962).

9. Godtfresdsen, W., Roholt, K., and Tybring, L., Fucidin, A New Orally Active Antibiotic, *Lancet, 1,* 928, (1962).

10. Saggers, B. A., and Lawson, D., *In vivo* Penetration of Antibiotics into Sputum in Cystic Fibrosis, *Arch. Dis. Child., 43,* 404 (1968).

11. Knoll, E. W., Bowman, F. W., and Kirshbaum, A., Plate Assays for Griseofulvin in Pharmaceutical Preparations and Body Fluids, *J. Pharm. Sci., 52,* 586 (1968).

12. Piggott, W. R., Williams, T. W., Jr., Witorsch, P., and Emmons, C. W., Bioassay for Hamycin, in *Antimicrobial Agents and Chemotherapy – 1965,* pp. 353–357, G. L. Hobby, Ed., American Society for Microbiology, Washington, D.C. (1966).

13. Sutherland, R., and Robinson, O. P. W., Laboratory and Pharmacological Studies in Man With Hetacillin and Ampicillin, *Br. Med. J., 2,* 804 (1967).

14. Dennin, L. J., Mygromycin B, in *Analytical Microbiology,* pp. 303–307, F. Kavanagh, Ed. Academic Press, New York (1963).

15. Hamade, M., Hashimoto, T., Yokoyama, S., Miyake, M., Takeuchi, T., Okami, Y., and Umezawa, H., Antimicrobial Activity of Kasugamycin, *J. Antibiot. (Tokyo) Ser. A., 18,* 104 (1965).

16. *Minimum Requirements of Antibiotic Products,* p. 400. Ministry of Health and Welfare, Japanese Government, Tokyo, Japan (1961).

17. English, A. R., McBride, T. J., and Riggio, R., Biological Studies of 6-Methylene Oxytetracycline, A New Tetracycline, in *Antimicrobial Agents and Chemotherapy–1961,* pp. 462–473, J. C. Sylvester, Ed. American Society for Microbiology, Ann Arbor, Michigan (1962).

18. Redin, G. S., Antibacterial Activity in Mice of Minocycline, A New Tetracycline, in *Antimicrobial Agents and Chemotherapy–1961,* pp. 371–376, J. C. Sylvester, Ed. American Society for Microbiology, Ann Arbor, Michigan (1962).

19. *Minimum Requirements of Antibiotic Products,* p. 481. Ministry of Health and Welfare, Japanese Government, Tokyo, Japan (1961).

20. Anonymous, *Fed. Regist., 37 (Part 149d),* 4907, (1972).

21. Biological Assays, in *Pharmacopoeia of India,* pp. 1012–1018. Government of India Press, Nasik, India (1966).

22. Brewer, G. A., and Platt, T. B., Antibiotics in *Encyclopedia of Industrial Chemical Analysis,* Vol. 5, p. 576, F. D. Snell and E. L. Hilton, Eds. Interscience Publications, John Wiley and Sons, New York (1967).

23. Bergamini, N., and Fowst, G., Rifamycin SV, A Review, *Arneimittel-Forschung, 15,* 951, (1965).

24. Girolami, R. L., Ristocetin, in *Analytical Microbiology,* pp. 353–360, F. Kavanagh, Ed. Academic Press, New York (1963).

25. Hanka, L. J., Mason, D. J., and Sokolski, W. T., Actinospectacin, A New Antibiotic, II, Microbiological Assay, *Antibiot. Chemother., 11,* 123 (1961).

26. Brewer, G. A., and Platt, T. B., Antibiotics, in *Encyclopedia of Industrial Chemical Analysis,* Vol. 5, p. 529, F. D. Snell and E. L. Hilton, Eds. Interscience Publications, John Wiley and Sons, New York (1967).

27. Levin, J. D., and Pagano, J. F., Thiostrepton, in *Analytical Microbiology,* pp. 365–368, F. Kavanagh, Ed. Academic Press, New York (1963).

28. Horwitz, W. (Ed.), in *Official Methods of Analysis of the Association of Official Analytical Chemists,* 11th ed., pp. 261–262, 752–763. Association of Official Analytical Chemists, Washington, D.C. (1970).

29. *The Pharmacopeia of the United States of America,* 18th Revision, pp. 857-864. U. S. Pharmacopeial Convention, Inc., Bethesda, Maryland (1970).

30. Gavin, J. J., Biological Methods, in *Handbook of Analytical Chemistry,* Sect. 9, L. Meites, Ed. McGraw-Hill, New York (1963).

31. von Wittenau, M. S., and Twomey, T. M., The Disposition of Doxycycline by Man and Dog, *Chemotherapy, 16,* 217 (1971).

32. Grove, D. C., and Randall, W. A., in *Assay Methods of Antibiotics, A Laboratory Manual,* pp. 14–16. Medical Encyclopedia, Inc., New York (1961).

33. Gerke, J. R., and Madigan, M. E., Amphotericin B and Other Polyene Antifungal Antibiotics: Photometric Assay and Factors Influencing Activity, *Antibiot. Chemother., 11,* 225 (1961).

BIOENGINEERING SYMBOLS

Letter Symbols Recommended
by the Division of Microbial Chemistry and Technology,
American Chemical Society, 1964

FERMENTATION KINETICS AND CONTINUOUS FERMENTATION

Reaction Kinetics and Stoichiometry

c	Concentration (number, mass, or moles per unit volume)
f	Frequency (occurrences per unit time)
k	Specific reaction rate (reaction order must be defined)
t	Time
Y	Yield constant (reactant ratio must be defined)
$\emptyset$	Ratio; unusual occurrences per normal occurrence (dimensionless)

System Capacity Characteristics

D	Dilution rate; reciprocal time
q	Volumetric flow rate (volume per unit time)
V	Volume

Specialized Notation

Indicated by *suggestive subscripts*, convenient to the discussion in which they are used; for example:

k_M	Maximum specific reaction rate
t_G	Generation time
t_R	Retention time
$\emptyset_M$	Mutation ratio

Stagewise Notation

Indicated by *subscripts subordinate to specialized subscripts* when the latter are used; for example:

0	Initial feed
1	First stage
2	Second stage
N	N^{th} stage

Reactant Notation

Indicated by *suggestive superscripts*, convenient to the discussion in which they are used; for example:

C	Contaminant
M	Mutant
O	Oxygen
P	Product

MASS TRANSFER

Properties and Stoichiometry

c   Concentration (number, mass, or moles per unit volume)
D   Diffusivity (length squared per unit time)
H   Henry's Law constant (equilibrium concentrations must be defined)
p   Pressure; atmospheres
x   Mole fraction in liquid phase (dimensionless)
y   Mole fraction in gas phase (dimensionless)

System Geometry

a   Area per unit volume
S   Cross-sectional area
V   Volume
Z   Height
ΔZ Change in head due to air hold-up

Rate Expressions

k   Film mass transfer coefficient (mass or moles per unit time, unit area, and unit driving force)
k_R Specific respiration rate (mass or moles per unit time, and unit weight or number)
K   Over-all mass transfer coefficient (mass or moles per unit time, unit area, and unit driving force)
N   Mass or moles per unit time and unit volume
K_s Sulfite index (quantity reacted per unit time and unit volume)
P_v Agitator power input per unit volume
q   Volumetric air-flow rate (volume per unit time)
V_s Superficial-air velocity (length per unit time)
$\hat{\quad}$ Space velocity of air (volume per volume and unit time)
η Absorption efficiency (dimensionless)

Specialized Notation

Suggestive superscripts, convenient to the discussion in which they are used, are recommended for reactant designations; for example:

O   Oxygen
P   Product
T   Tissue

Suggestive subscripts are recommended for indication of phase and position; for example:

G   Gas bulk
L   Liquid bulk
W   Cell wall

Equilibrium values are indicated by an *asterisk* (*).

Data reproduced by permission of the American Chemical Society, Division of Microbial Chemistry and Technology.

Unified Symbols for Continuous Cultivation of Microorganisms

Symbol	Dimension	Meaning or Application
V, v	L^3	Volume
F, f		Flow rate
D	t^{-1}	Dilution rate
S, s	ML^{-3}	Concentration of limiting substrate
Z, z	ML^{-3}	Concentration of other substrates
X, x	ML^{-3}	Concentration of microorganisms by mass
N, n	L^{-3}	Number of microorganisms per unit volume
μ	t^{-1}	Specific growth rate
μ_{max}	t^{-1}	Maximum specific growth rate
q	$\frac{ds}{dt}\frac{1}{x}^{(x)}$	A metabolic coefficient
k		Other rate constants
$Y = \frac{dx}{ds}^{(x)}$		Yield (or economic coefficient)
$a = \frac{dp}{ds}^{(x)}$		Product yield
P, p	ML^{-3}	Concentration of product
t	t	Time
g	t	Doubling time
θ	t	Mean residence time $(1/D)$
$K_s K_z$		Saturation constants for Substrates S and Z
a		Ratio of recycle flow to total flow prior to stream division
a'		Ratio of recycle flow to net flow
b		Ratio of mass of cells recycled to mass of cells prior to stream division

Superscripts and Subscripts

$\wedge$, max	For maximal value of any variable
$1, 2, 3 \ldots \omega$	May be attached to the symbol of any variable to indicate its value in stage 1, 2, 3, etc.
o	Initial values
T	Used for point variables in a tubular system
A, B, C $\ldots$	Used to distinguish the values of variables or parameters associated with species or mutants A, B, C, etc.

Note: $^{(x)}$ = dimensionless.

Data taken from: Malek, I., Beran, K., and Hospodka, G. (Eds.), *Continuous Cultivation of Microorganisms* (1964). Reproduced by permission of Academic Press, New York.

INDICES

MICROORGANISMS AND THEIR PRODUCTS

A

O

COMPOUNDS OF MICROBIAL ORIGIN

A

Aabomycin A, 719
Abbott 29119, 838
Abikoviromycin, 719
Ablastmycin, 719
Aburamycin, 719
Aburamycin isomer, 719
AC-98 complex, 838
Acaranoic acid, 372, 462
Acarenoic acid, 462
Acetaldehyde, 3
(±)-2-Acetamido-2,5-dihydro-5-ketofuran, 55
S-(2-Acetamido ethyl) cysteine, 85
2-Acetamido-3H-phenoxazin-3-one, 333, 796
Acetic acid, 28
Acetoin, 4
Acetomycin, 55, 719
Acetone, 3, 1010
Acetopyrrothin, 719
3'-Acetoxy-2'-carboxy-4-hydroxy-6-methoxy-2-
     methoxycarbonyl-5'-methyldiphenyl ether, 178
4-Acetoxycycloheximide, 391, 719
7β-Acetoxy-22-hydroxyhopane, 496
5-Acetoxy-4-ketohexanoic acid, 38
Acetylaranotin, 348, 353
Acetylcholine, 290
Acetylcytochalasin D, 61, 505
2-Acetyl-2-decarboxamidochlorotetracycline, 265
2-Acetyl-2-decarboxamidooxytetracycline, 264, 719
2-Acetyl-2-decarboxamidotetracycline, 265, 719
4-Acetyl-6,8-dihydroxy-5-methylisocoumarin, 377
3-O-Acetyleburicoic acid, 148
Acetylenedicarboxamide, 719
Acetylglucosyldehydrotumulosic acid, 149
Acetylglucosyltumulosic acid, 149
3-Acetyl-5-isobutyltetramic acid, 59
3-Acetyl-5-isopropyltetramic acid, 59
16β-O-Acetylleucotylic acid, 496
6α-O-Acetylleucotylin, 497
Acetylportentol, 466
3-O-Acetyltumulosic acid, 148
Achromoviromycin, 719
Achromycin, 719
Acidocillin, 1011
Acidomycin, 361, 719
cis-Aconitic acid, 31
Acrylamidine, 720
Acrylic acid, 28
Actidione, 697, 720
Actiduins, 720
Actinobolin, 57, 720
Actinochrysin, 720
Actinoflavin, 720
Actinoflocin, 720
Actinogan, 720
Actinoidin, 720
Actinoleukin, 720

Actinolysin, 720
Actinomycelin, 720
Actinomycetin, 721
Actinomycins, 334–336, 721–723
Actinomycin C₃, 1010
Actinomycin D, 1008
Acintomycin J, 3
Actinones, 723
Actinonin, 509, 723
Actinorhodin, 724
(+)-Actinorhodin, 207
Actinorubin, 724
Actinospectacin, 12, 724
Actinovorin, 724
Actinoxanthin, 724
Actiphenol, 391, 724
Actithiazic acid, 48, 361, 724
Acumycin, 724
Adamantane amine, 974
1-Adamantylguanidine, 975
Adenine, 323
Adipic acid, 30
Adriamycin, 247, 724
Aerobacter clocae polysaccharide, 999
Aeroplysinin, 441
Aeruginic acid, 1002
Aeruginosins, 332
Aflatoxins, 525–528, 545, 549–551, 581, 1003
Agaricic acid, 36
Agarin, 289, 528, 545, 551
Agaritine, 275, 283, 294
Agritol, 577
Agroclavine, 401, 528, 545, 551
Agrocybin, 65, 528, 545, 552
Ahygroscopin, 724
Akimycin, 725
Akitamycin, 94, 725
Aklavin, 238, 725
Aklavinone, 237
Aklavinone glycosides, 238
L-Alanine, 291
β-Alanine, 291
Alanosine, 725, 976, 1002
Alazomycin F, 1010
Alazopeptine, 725
Albimycin, 725
Albocycline, 725
Albofungin, 726
Albomycetin, 726
δ-Albomycins, 455
ε-Albomycin, 456
Albomycin complex, 726
Albomycin-grisein antibiotics, 450, 455, 456
Albonoursin, 351
Albonurusin, 726
Alboverticillin, 726

D

G

O

Y

Z